Indrajit Ray

A Second Course in Business Statistics: Regression Analysis

FOURTH EDITION

A Second Course in Business Statistics: Regression Analysis

WILLIAM MENDENHALL
University of Florida

TERRY SINCICH
University of South Florida

DELLEN PUBLISHING COMPANY

an imprint of

MACMILLAN PUBLISHING COMPANY

New York

MAXWELL MACMILLAN CANADA

Toronto

On the cover: A detail from a mixed-media, three-dimensional screen created by San Francisco designer Robert Hutchinson. The entire six-panel screen is 24 feet long and 8 feet high. It is one of many custom-made art screens created by Mr. Hutchinson.

Macmillan Publishing Company
113 Sylvan Avenue, Englewood Cliffs, NJ 07632

Library of Congress Cataloging-in-Publication Data
Mendenhall, William.
   A second course in business statistics: regression analysis
William Mendenhall, Terry Sincich.—4th ed.
      Includes index.
      ISBN 0-02-380520-X
      1. Commercial statistics   2. Statistics.   3. Regression analysis.
   I. Sincich, Terry.   II. Title.
   HF1017.M46   1993
   519.5—dc20   92-31637
                                                        CIP

Printing: 3 4 5 6 7 8 9   Year: 5 6 7

ISBN 0-02-380520-X

Preface

This book is designed for two types of business statistics courses. The early chapters, combined with a selection of the case study chapters, are designed for use in the second half of a two-semester (or two-quarter) introductory statistics sequence for undergraduate business majors. Or, the book can be used for a course in applied regression analysis for MBA or Ph.D. students in business administration.

At first glance, these two uses for the book may seem inconsistent. How could a text be appropriate for both undergraduate and graduate students? The answer lies in the content. In contrast to a course in statistical theory, the level of mathematical knowledge required for an applied regression analysis course is minimal. Consequently, the difficulty encountered in learning the mechanics is much the same for both undergraduate and graduate students. The challenge is in the application—diagnosing practical problems, deciding on the appropriate linear model for a given situation, and knowing which inferential technique will answer a manager's practical questions. This takes *experience*, and it explains why a student can take an undergraduate course in applied regression analysis and still benefit from covering the same ground in a graduate course.

It is difficult to identify the amount of material that should be included in the second semester of a two-semester sequence in introductory business statistics. Optionally, a few lectures should be devoted to Chapter 1 (A Review of Basic Concepts) to make certain that all students possess a common background knowledge of the basic concepts covered in a first-semester (first-quarter) course. Chapter 2 (Introduction to Regression Analysis), Chapter 3 (Simple Linear Regression), Chapter 4 (Multiple Regression), Chapter 5 (Model Building), Chapter 6 (Some Regression Pitfalls), and Chapter 7 (Residual Analysis) provide the core for an applied regression analysis course. These chapters could be supplemented by the addition of Chapter 9 (Time Series Modeling and Forecasting), Chapter 10 (Principles of Experimental Design), or Chapter 11 (The Analysis of Variance for Designed Experiments).

In our opinion, the quality of an applied course is not measured by the number of topics covered or the amount of material memorized by the students. The measure is how well they can apply the techniques covered in the course to the solution of real business problems. Consequently, we advocate moving on to new topics only after the students have demonstrated ability (through testing) to apply the techniques under discussion. In-class consulting sessions, where a case study is presented and the students have the opportunity to diagnose the problem and recommend an appropriate method of analysis, are very helpful in teaching applied regression analysis. This approach is particularly useful in helping students master the difficult topic of model selection and model building (Chapters 4–7) and relating questions about the model to the real-world questions of a business manager. The case study chapters (Chapters 12–16) illustrate the type of material that might be used for this purpose.

A course in applied regression analysis for graduate students would start in the same manner as the undergraduate course, but would move more rapidly over the review material and would more than likely be supplemented by Appendix A (The Mechanics of a Multiple Regression Analysis), Appendix C (Performing Regression Analysis Using the Computer: SAS, SPSS, and Minitab), optional Chapter 8 (Special Topics in Regression), and other chapters selected by the instructor. As in the undergraduate course, we recommend the use of case studies and in-class consulting sessions to help students develop an ability to formulate appropriate statistical models and to interpret the results of their analyses.

Although the scope and coverage remain the same, the fourth edition contains many substantial changes, additions, and enhancements:

1. **More computer printouts**   Throughout the text, we have greatly increased the number of SAS, SPSS, and Minitab printouts. A printout now accompanies every statistical technique presented, allowing the instructor to emphasize interpretations of the statistical results rather than the calculations required to obtain the results.

2. **Chapter 1: Stem-and-leaf displays**   A discussion of how to construct and use stem-and-leaf displays has been added to the section on describing data graphically (Section 1.3).

3. **Chapter 2: Collecting data**   A new section on collecting data for regression (Section 2.4) presents an early emphasis on the difference between observational data and experimental data.

4. **Chapter 3: t test for correlation coefficient**   In Section 3.7, the test for zero correlation in the population is now based on the familiar Student's t statistic rather than the equivalent (but more awkward) sample correlation coefficient r.

5. **Chapter 4: PRESS statistic**   In addition to R^2, C_p, and the MSE criterion, we also present the prediction sum of squares (PRESS) criterion for selecting a "best" model using the all-possible-regressions selection procedure (Section 4.12).

6. **Chapter 4: Comprehensive example on multiple regression**   The key ideas and techniques of the chapter are applied to a practical problem on detecting collusive bidding in road construction (Section 4.13).

7. **Chapter 6: Standardized β coefficients**   The interpretation of standardized β coefficients is now included in the section on Parameter Estimability and Interpretation (Section 6.4).

8. **Chapter 8: Log-odds ratios in logistic regression**   Interpretations of the β parameters in a logistic regression model are added to Section 8.6. These are stated in terms of changes in both log-odds ratios and odds ratios.

9. **Chapter 9: ARMA models**   A new section (Section 9.11) on time series forecasting models using only past values of the dependent variable has been added. The general autoregressive moving average (ARMA) model is presented.

10. **Chapter 10: Principles of experimental design**   A new chapter presents an overview of designed experiments and the principles of noise-reducing and volume-increasing designs.

11. **Chapter 11: Regression approach to ANOVA**   Although we present both the traditional ANOVA approach and the regression approach to analyzing data from designed experiments, our emphasis is on the regression approach. For each design, we give the corresponding regression models and show how to conduct the ANOVA F tests using the models.

12. **Chapter 11: Checking ANOVA assumptions**   Guidelines on using residuals to check the standard ANOVA assumptions have been added to the chapter (Section 11.9).

13. **Case Study 12: Residential property sale price data updated**   The data set for the case study on predicting sale prices of residential properties has been updated to reflect current economic trends.

14. **Case Study 14: New case study of a designed experiment**   A new case study concerning the reluctance to transmit bad news is extracted from a recent article in the *Journal of Experimental & Social Psychology*.

15. **Appendix C: Computer commands**   The SAS, SPSS, and Minitab commands for conducting a regression analysis have been updated to reflect the new versions of the software.

16. **More exercises with real data**   Many new "real-life" applied exercises and examples have been added throughout the text. Most of these are extracted from business and economic journals.

17. **Short answers to odd-numbered exercises**   Answers to even-numbered exercises have been eliminated from the answer section in the back of the text; only short answers to odd-numbered exercises are provided. This gives the instructor the option of using even-numbered exercises for homework assignments, quizzes, exams, etc.

The main features of this text continue to be:

1. **Readability**   We have purposely tried to make this a teaching (rather than a reference) text. Concepts are explained in a logical intuitive manner using worked examples.

2. **Emphasis on model building**   The formulation of an appropriate statistical model is fundamental to any regression analysis. This topic is treated in Chapters 4–7 and is emphasized throughout the text.

3. **Emphasis on developing skill to use regression analysis**   In addition to teaching the basic concepts and methodology of regression analysis, this text stresses its use, as a tool, in solving business problems. Consequently, a major objective of the text is to develop a skill in applying regression analysis to appropriate real-life situations.

4. **Numerous real data-based examples and exercises**   The text contains many worked examples that illustrate important aspects of model construction, data analysis, and the interpretation of results. Numerous exercises based on data and problems extracted from news articles, magazines, and journals are located at the ends of key sections and at the ends of chapters.

5. **Case study chapters**   The text contains five case study chapters, each of which addresses a real-life business problem. The student can see how regression analysis was used to answer the practical questions posed by the prob-

lem, then proceed with the formulation of appropriate statistical models to the analysis and interpretation of sample data.

6. **Data sets**   The text contains three complete data sets that are associated with the case studies (Chapter 12–16). These can be used by instructors and students to practice model building and data analysis.

7. **Computer orientation**   Instructions on how to use the three popular statistical software packages, SAS, SPSS, and Minitab, are provided in Appendix C. The printouts of the respective packages are presented and discussed throughout the text.

The text is also accompanied by the following supplementary material:

1. **Student's solutions manual** (by Nancy S. Boudreau)   A student's exercise solutions manual presents the full solutions to the odd-numbered exercises contained in the text.

2. **Instructor's solutions manual** (by Mark Dummeldinger)   The instructor's exercise solutions manual presents the full solutions to the other half (the even-numbered) exercises contained in the text. For adopters, the manual is complimentary from the publisher.

3. **Appendix data sets available on diskette**   For adopters, the three data sets in Appendices E–G are available on either a $3\frac{1}{2}''$ or $5\frac{1}{4}''$ IBM PC diskette (ASCII format), complimentary from the publisher.

4. **Exercise data sets on diskette**   Also for adopters, the data for all exercises containing 15 or more observations are available on either a $3\frac{1}{2}''$ or $5\frac{1}{4}''$ IBM PC diskette (ASCII format). A list of these exercises follows this preface.

5. **ASP statistical software** (by George Blackford)   ASP is an integrated statistical package available on either a $3\frac{1}{2}''$ or $5\frac{1}{4}''$ IBM PC diskette. The software is completely menu-driven and runs on any IBM-compatible PC with at least 256K of memory and at least one floppy disk drive. ASP is an excellent alternative for students who have no previous experience in computers and want a "user-friendly" statistical software package, or to instructors who want to expose the students to computers without taking the time to teach them how to program in SAS, SPSS, or Minitab. An ASP user's guide is available from the publisher upon request.

Acknowledgments

We want to thank the many people who contributed time, advice, and other assistance to this project. We owe particular thanks to the many reviewers who provided suggestions and advice at the onset of the project and for the succeeding editions:

Mohammed Askalani, Mankato State University (Minnesota)

Ken Boehm, Pacific Telesis (California)

James Daly, California State Polytechnic Institute at San Luis Obispo

Assane Djeto, University of Nevada–Las Vegas

Robert Elrod, Georgia State University

James Ford, University of Delaware

Carol Ghomi, University of Houston

James Holstein, University of Missouri at Columbia

Steve Hora, Texas Technological University

Thomas Johnson, North Carolina State University

Ann Kittler, Ryerson College (Toronto)

James T. McClave, University of Florida

John Monahan, North Carolina State University

Kris Moore, Baylor University

Farrokh Nasri, Hofstra University

Robert Pavur, University of North Texas

P. V. Rao, University of Florida

Tom Rothrock, Info Tech, Inc.

Ray Twery, University of North Carolina at Charlotte

Joseph Van Matre, University of Alabama at Birmingham

William Weida, United States Air Force Academy

Dean Wichern, Texas A&M University

James Willis, Louisiana State University

Susan Reiland deserves special recognition for managing the production of the text. We are particularly grateful to Charles Bond (Texas Christian University), Evan Anderson (North Carolina State Data Center), Jim McClave (University of Florida), Tom Rothrock (Info Tech, Inc.), Ron Alderman (Hillsborough County Appraiser), and Mike Jacob (Florida Power Corporation), who provided data sets and/or background information used in the case studies (Chapters 12–16). And, finally, we give special thanks to Faith Sincich for a superb job of proofreading and transforming our notes into typed copy.

Exercise Data Sets Available on Disk

EXERCISE	PAGE	
1.6	11	Generic sample data ($n = 28$)
1.7	11	Foreign revenue of multinational corporations ($n = 20$)
1.8	12	Voltage readings for production runs at two locations ($n = 30$)
1.10	13	Sanitation inspection scores for cruise ships ($n = 72$)
1.11	14	Minority hirings of major colleges and universities ($n = 60$)
1.14	20	Days of maturity for money funds ($n = 50$)
1.16	21	Tar, nicotine, and carbon monoxide contents of cigarettes ($n = 25$)
1.19	22	Student loan default rates of colleges ($n = 66$)
1.31	40	Fifty random samples of 6 random digits ($n = 300$)
1.65	61	Investment/quad values for plants ($n = 27$)
1.89	76	Cash compensations of top CEOs ($n = 20$)
3.9	103	Depth and drill time for dry holes in rock ($n = 17$)
3.10	104	P/E ratios and R&D expenditures for companies ($n = 20$)
3.14	108	Unflooded area ratio and heat transfer enhancement for integral-fin tubes ($n = 24$)
3.25	115	Foreign revenue and assets for multinational firms ($n = 20$)
3.40	132	Breast height diameters and heights of white spruce trees ($n = 36$)
3.50	148	Total weight and number of bags for flour shipments ($n = 15$)
3.52	150	Factors and length of stay for coronary care patients ($n = 50$)
3.57	153	Raises and ratings of USF administrators ($n = 15$)
4.6	177	Man-hours, boiler capacity, design pressure, boiler type, and drum type for manufactured boiler drums ($n = 36$)
4.10	188	Prices, equity returns, and dividend rates for nuclear and nonnuclear stocks ($n = 28$)
4.42	230	Daily output levels for various activities at a department store ($n = 52$)

EXERCISE   PAGE

Table 4.10	237	Bid-rigging data ($n = 235$)
5.14	281	Temperature, pressure, and quality data for chemically produced products ($n = 27$)
5.16	293	Cash compensations of CEOs in four industries ($n = 32$)
5.37	320	Salesperson sales and commission data ($n = 15$)
6.10	343	(*see* Exercise 1.16)
6.11	344	Annual demand for motor fuel, weekly earnings, and average gas prices, 1965–1980 ($n = 16$)
6.12	345	Weekly labor hours, weight shipped, percent shipped, and average weight ($n = 20$)
6.18	347	Appraisals and sale prices for residential properties ($n = 15$)
6.19	348	Strike activity and related variables in manufacturing ($n = 49$)
7.12	378	Leased fee values and sizes of Hawaiian properties ($n = 20$)
7.22	396	Employee work-hours missed and annual wages ($n = 15$)
7.23	397	Weight percentile and number of cigarettes smoked per day in homes of children with cystic fibrosis ($n = 25$)
7.27	406	Purchasing value of the dollar, 1970–1988 ($n = 19$)
7.28	407	Monthly factory sales of passenger cars ($n = 24$)
7.35	411	Monthly demand for 10-speed bikes and related variables ($n = 15$)
7.36	414	Quarterly data for absentee rate of employees ($n = 20$)
7.39	416	Annual Florida tax collections, 1971–1985 ($n = 15$)
8.4	424	Annual corn production, 1950–1986 ($n = 37$)
8.5	425	Unit cost and lot size for shipments ($n = 15$)
8.8	432	Residential fires: damage and distance from station ($n = 15$)
8.12	440	Executive salary and years of experience data ($n = 50$)
8.14	444	PC ownership status and annual income for households ($n = 20$)
8.15	444	Hiring status, education, experience, and sex of applicants ($n = 28$)
9.1	477	Quarterly housing starts, 1986–1990 ($n = 20$)
9.3	478	Annual crude oil imports, 1973–1988 ($n = 16$)
9.4	478	Monthly IBM stock prices, 1988–1990 ($n = 36$)

EXERCISE   PAGE

9.6   479   Quarterly S&P 500 Average, 1983–1990 ($n = 32$)

9.7   480   Annual price of gold, 1971–1990 ($n = 20$)

9.9   486   Weekly sales of single-family homes ($n = 15$)

9.12   488   Annual price of galvanized steel, 1971–1989 ($n = 19$)

9.25   505   Quarterly GNP values, 1980–1989 ($n = 40$)

9.29   507   Annual Dow Jones Industrial Average, 1968–1987 ($n = 20$)

9.37   514   Annual total Florida tax collections, 1962–1985 ($n = 24$)

9.39   520   Monthly total retail sales in U.S., 1988–1990 ($n = 36$)

9.40   521   Monthly occupancy rates for Atlanta and Phoenix hotels/motels ($n = 24$)

9.42   522   Annual NASA space shuttle expenditures, 1973–1988 ($n = 16$)

9.43   523   Annual U.S. beer production, 1973–1989 ($n = 17$)

9.46   524   (see Exercise 9.3)

11.6   566   (see Exercise 5.16)

11.7   568   Numbers of trials until accurate decoding in each of four different training groups ($n = 20$)

11.8   568   Number of hits required to crack four different brands of golf balls ($n = 40$)

11.14   571   Total beer sales in three advertising periods ($n = 18$)

11.19   585   Prices of grocery items at each of three supermarkets ($n = 180$)

11.21   586   Preference scores of subjects rating video display combinations ($n = 70$)

11.23   588   Candidate, rater, and performance for in-tray tasks ($n = 21$)

11.25   590   Estimated market shares of beer brands using each of three methods ($n = 18$)

11.27   606   Gaskets produced, machine, and material type for production runs ($n = 18$)

11.30   607   Pricing ratio, peak period length, and satisfaction scores for electric customers ($n = 36$)

11.34   610   Failures, burn-in hours, and inspection levels for lots ($n = 81$)

11.38   618   Yield strength, charging time, alloy type, and condition of tensile specimens ($n = 24$)

EXERCISE   PAGE

11.61   642   Daily traffic density at five locations ($n = 48$)

11.62   643   Grass blades found at effluent stations for each of six months ($n = 18$)

11.63   643   Hourly wage rates of workers at each of three corporations ($n = 18$)

11.64   644   Union affiliation, incentive plan, and productivity of production workers ($n = 36$)

11.65   644   Bus travel times under four plans ($n = 17$)

11.67   645   Display type, period, and sales of diet beverage ($n = 24$)

11.68   646   Temperature, exposure time, and water removed for paper specimens ($n = 36$)

11.70   646   Salespersons, region, and annual value increase for properties ($n = 36$)

11.71   647   Delivery times for shipments by each of three carriers ($n = 15$)

11.72   647   Home appraisals in each of two communities ($n = 16$)

11.74   649   Room temperature, incoming parts rate, and productivity of workers ($n = 32$)

11.76   650   Room temperature, incoming parts rate, and productivity of workers ($n = 40$)

11.77   651   Appraised values of properties as determined by each of four appraisers ($n = 40$)

11.79   653   Pay rate, length of workday, and productivity for workers ($n = 18$)

11.80   654   Shift, operator, foil amount, machine speed and light output of flashbulbs ($n = 32$)

Contents

CHAPTER 1 **A Review of Basic Concepts (Optional)** 1

1.1 Statistics and Data 2
1.2 Populations, Samples, and Random Sampling 4
1.3 Describing Data Sets Graphically 7
1.4 Describing Data Sets Numerically 15
1.5 The Normal Probability Distribution 24
1.6 Sampling Distributions and the Central Limit Theorem 29
1.7 Estimating a Population Mean 34
1.8 Testing a Hypothesis About a Population Mean 43
1.9 Inferences About the Difference Between Two Population Means 51
1.10 Comparing Two Population Variances 63

CHAPTER 2 **Introduction to Regression Analysis** 81

2.1 Modeling a Response 82
2.2 Overview of Regression Analysis 85
2.3 Regression Applications 87
2.4 Collecting the Data for Regression 88

CHAPTER 3 **Simple Linear Regression** 93

3.1 Introduction 94
3.2 The Straight-Line Probabilistic Model 94
3.3 Fitting the Model: The Method of Least Squares 97
3.4 Model Assumptions 105
3.5 An Estimator of σ^2 106
3.6 Assessing the Utility of the Model: Making Inferences About the Slope β_1 109
3.7 The Coefficient of Correlation 115
3.8 The Coefficient of Determination 121
3.9 Using the Model for Estimation and Prediction 127
3.10 Simple Linear Regression: An Example 134
3.11 Regression Through the Origin (Optional) 141
3.12 A Summary of the Steps To Follow in a Simple Linear Regression Analysis 149

CHAPTER 4 **Multiple Regression**   161

4.1   The General Linear Model   162
4.2   Model Assumptions   164
4.3   Fitting the Model: The Method of Least Squares   165
4.4   Estimation of σ^2, the Variance of ε 167
4.5   Inferences About the β Parameters   169
4.6   The Multiple Coefficient of Determination, R^2 179
4.7   Testing the Utility of a Model: The Analysis of Variance F Test   181
4.8   Using the Model for Estimation and Prediction   191
4.9   Other Linear Models   194
4.10 Testing Portions of a Model   217
4.11 Stepwise Regression   224
4.12 Other Variable Selection Techniques (Optional)   232
4.13 Multiple Regression: An Example   236
4.14 A Summary of the Steps To Follow in a Multiple Regression Analysis   244

CHAPTER 5 **Model Building**   255

5.1   Introduction: Why Model Building Is Important   256
5.2   The Two Types of Independent Variables: Quantitative and Qualitative   257
5.3   Models with a Single Quantitative Independent Variable   259
5.4   First-Order Models with Two or More Quantitative Independent Variables   265
5.5   Second-Order Models with Two or More Quantitative Independent Variables   268
5.6   Coding Quantitative Independent Variables (Optional)   275
5.7   Models with One Qualitative Independent Variable   281
5.8   Models with Two Qualitative Independent Variables   284
5.9   Models with Three or More Qualitative Independent Variables   295
5.10 Models with Both Quantitative and Qualitative Independent Variables   299
5.11 Model Building: An Example   313

CHAPTER 6 **Some Regression Pitfalls**   321

6.1   Introduction   322
6.2   Observational Data Versus Designed Experiments   322
6.3   Deviating from the Assumptions   324
6.4   Parameter Estimability and Interpretation   325
6.5   Multicollinearity   329
6.6   Extrapolation: Predicting Outside the Experimental Region   335
6.7   Data Transformations   337

CHAPTER 7   **Residual Analysis**   351

7.1 Introduction   352
7.2 Plotting Residuals and Detecting Lack of Fit   352
7.3 Detecting Unequal Variances   366
7.4 Checking the Normality Assumption   379
7.5 Detecting Outliers and Identifying Influential Observations   384
7.6 Detecting Residual Correlation: The Durbin–Watson Test   399

CHAPTER 8   **Special Topics in Regression (Optional)**   419

8.1 Introduction   420
8.2 Piecewise Linear Regression   420
8.3 Inverse Prediction   425
8.4 Weighted Least Squares   433
8.5 Modeling Qualitative Dependent Variables   441
8.6 Logistic Regression   445
8.7 Ridge Regression   453
8.8 Robust Regression   456
8.9 Model Validation   458

CHAPTER 9   **Time Series Modeling and Forecasting**   463

9.1 What Is a Time Series?   464
9.2 Time Series Components   464
9.3 Forecasting Using Smoothing Techniques (Optional)   466
9.4 Forecasting: The Regression Approach   481
9.5 Autocorrelation and Autoregressive Error Models   489
9.6 Other Models for Autocorrelated Errors (Optional)   493
9.7 Constructing Time Series Models   495
9.8 Fitting Time Series Models with Autoregressive Errors   500
9.9 Forecasting with Time Series Autoregressive Models   508
9.10 Seasonal Time Series Models: An Example   515
9.11 Forecasting Using Lagged Values of the Dependent Variable (Optional)   518

CHAPTER 10   **Principles of Experimental Design**   527

10.1 Introduction   528
10.2 Experimental Design Terminology   528
10.3 Controlling the Information in an Experiment   530
10.4 Noise-Reducing Designs   532

10.5 Volume-Increasing Designs   539

10.6 Selecting the Sample Size   545

10.7 The Importance of Randomization   547

CHAPTER 11

The Analysis of Variance for Designed Experiments   551

11.1 Introduction   552

11.2 The Logic Behind an Analysis of Variance   553

11.3 Completely Randomized Designs   554

11.4 Randomized Block Designs   573

11.5 Two-Factor Factorial Experiments   590

11.6 More Complex Factorial Designs (Optional)   612

11.7 Follow-Up Analysis: Tukey's Multiple Comparisons of Means   621

11.8 Other Multiple Comparisons Methods (Optional)   630

11.9 Checking ANOVA Assumptions   637

CASE STUDY 12

Modeling the Sale Prices of Residential Properties in Four Neighborhoods   657

12.1 The Problem   658

12.2 The Data   658

12.3 The Models   658

12.4 Model Comparisons   661

12.5 Interpreting the Prediction Equation   665

12.6 Predicting the Sale Price of a Property   670

12.7 Conclusions   671

CASE STUDY 13

An Analysis of Bidding Competition   673

13.1 The Problem   674

13.2 The Data   674

13.3 Models for Stable and Unstable Market Conditions   674

13.4 The Regression Analyses: Testing to Detect Unstable Market Conditions   677

13.5 A Residual Analysis of the Data   682

13.6 Summary   689

CASE STUDY 14

Reluctance to Transmit Bad News: The MUM Effect   693

14.1 The Problem   694

14.2 The Design   694

14.3 Analysis of Variance Models and Results   695
14.4 Follow-Up Analysis   696
14.5 Conclusions   698

CASE STUDY 15   **An Investigation of Factors Affecting the Sale Price of Condominium Units Sold at Public Auction**   699

15.1 The Problem   700
15.2 The Data   701
15.3 The Models   702
15.4 The Regression Analyses   704
15.5 An Analysis of the Residuals from Model 3   708
15.6 What the Model 3 Regression Analysis Tells Us   712
15.7 Comparing the Mean Sale Price for Two Types of Units (Optional)   720
15.8 Conclusions   721

CASE STUDY 16   **Modeling Daily Peak Electricity Demands**   725

16.1 The Problem   726
16.2 The Data   726
16.3 The Models   727
16.4 The Regression and Autoregression Analyses   731
16.5 Forecasting Daily Peak Electricity Demand   735
16.6 Summary   737

APPENDIX A   **The Mechanics of a Multiple Regression Analysis**   739

A.1 Introduction   739
A.2 Matrices and Matrix Multiplication   740
A.3 Identity Matrices and Matrix Inversion   745
A.4 Solving Systems of Simultaneous Linear Equations   750
A.5 The Least Squares Equations and Their Solution   752
A.6 Calculating SSE and s^2 758
A.7 Standard Errors of Estimators, Test Statistics, and Confidence Intervals for β_0, β_1, ..., β_k 759
A.8 A Confidence Interval for a Linear Function of the β Parameters; A Confidence Interval for $E(y)$ 762
A.9 A Prediction Interval for Some Value of y to Be Observed in the Future   769

APPENDIX B   **A Procedure for Inverting a Matrix**   775

APPENDIX C **Performing Regression Analysis Using the Computer: SAS, SPSS, and Minitab Commands**   781

C.1 Introduction   781
C.2 Data Entry: SAS   782
C.3 Data Entry: SPSS   786
C.4 Data Entry: Minitab   788
C.5 Relative Frequency Distributions, Descriptive Statistics, Correlations, and Plots   791
C.6 Multiple Regression   793
C.7 Stepwise Regression   795
C.8 Residual Analysis and Regression Diagnostics   797
C.9 Analysis of Variance   799

APPENDIX D **Useful Statistical Tables**   803

Table 1 Normal Curve Areas   804
Table 2 Critical Values for Student's t 805
Table 3 Critical Values for the F Statistic: $F_{.10}$ 806
Table 4 Critical Values for the F Statistic: $F_{.05}$ 808
Table 5 Critical Values for the F Statistic: $F_{.025}$ 810
Table 6 Critical Values for the F Statistic: $F_{.01}$ 812
Table 7 Random Numbers   814
Table 8 Critical Values for the Durbin–Watson d Statistic ($\alpha = .05$)   817
Table 9 Critical Values for the Durbin–Watson d Statistic ($\alpha = .01$)   818
Table 10 Critical Values for the χ^2 Statistic   819
Table 11 Percentage Points of the Studentized Range, $q(p, \nu)$, Upper 5%   821
Table 12 Percentage Points of the Studentized Range, $q(p, \nu)$, Upper 1%   823

APPENDIX E **Data Set**   825

Real Estate Appraisals and Sales Data for Seven Neighborhoods in Tampa, Florida
(*see* Case Study 12)

APPENDIX F **Data Set**   833

1970–1975 Low Bid Prices for Bread Supplied to Florida Public Schools
(*see* Case Study 13)

APPENDIX G **Data Set**   839

Condominium Sales Data (*see* Case Study 15)

Answers to Odd-Numbered Exercises   843

Index   855

A Review of Basic Concepts (Optional)

CONTENTS

1.1 Statistics and Data

1.2 Populations, Samples, and Random Sampling

1.3 Describing Data Sets Graphically

1.4 Describing Data Sets Numerically

1.5 The Normal Probability Distribution

1.6 Sampling Distributions and the Central Limit Theorem

1.7 Estimating a Population Mean

1.8 Testing a Hypothesis About a Population Mean

1.9 Inferences About the Difference Between Two Population Means

1.10 Comparing Two Population Variances

OBJECTIVE

To review the basic concepts of statistics that are essential prerequisites to the study of regression analysis

Although we assume students have had a prerequisite introductory course in statistics, courses vary somewhat in content and in the manner in which they present statistical concepts. To be certain that we are starting with a common background, we will use this chapter to review some basic definitions and concepts. Coverage is optional.

1.1 Statistics and Data

According to *The Random House College Dictionary* (1988 ed.), statistics is "the science that deals with the collection, classification, analysis, and interpretation of numerical facts or data." In short, statistics is the **science of data**—a science that will enable you to be proficient data producers and efficient data users.

Definition 1.1

Statistics is the science of data. This involves collecting, classifying, summarizing, organizing, analyzing, and interpreting data.

Data are obtained by measuring some characteristic or property of the objects (usually people or things) of interest to us. These objects upon which the measurements (or observations) are made are called **experimental units** and the properties being measured are called **variables** (since, in virtually all studies of interest, the property varies from one observation to another).

Definition 1.2

An **experimental unit** is an object (person or thing) upon which we collect data.

Definition 1.3

A **variable** is a characteristic (property) that differs or varies from one observation to the next.

All data (and consequently, the variables we measure) are either **quantitative** or **qualitative** in nature. Quantitative data are data that can be measured on a numerical scale. In general, qualitative data take values that are nonnumerical; they can only be classified. The statistical tools that we use to analyze data depend on whether the data is quantitative or qualitative. Thus, it is important to be able to distinguish between the two types of data.

> **Definition 1.4**
>
> **Quantitative data** are observations measured on a numerical scale.

> **Definition 1.5**
>
> Nonnumerical data that can only be classified into one of a group of categories are said to be **qualitative data**.

EXAMPLE 1.1

The data in Table 1.1, obtained from *Business Week*'s 1990 Executive Compensation Scoreboard, contains information on the annual salaries of 10 of the highest paid corporate executives in the United States. For each executive, five variables are recorded: (1) company, (2) industry group, (3) total pay (in thousands of dollars), (4) return to shareholders (in dollars) on a $100 investment made 3 years earlier, and (5) pay-for-performance rating measured on a scale of 1 (excellent) to 5 (poor).

a. Identify the experimental units.
b. Classify the variables measured as quantitative or qualitative.

TABLE 1.1 **Data on 10 of the Highest Paid Executives in 1990**

CEO	COMPANY (1)	INDUSTRY (2)	TOTAL PAY (3)	RETURN (4)	RATING (5)
P. Fireman	Reebok	Consumer	14,606	169	5
R. Richey	Torchmark	Financial	12,666	227	5
M. Davis	Paramount	Consumer	11,635	166	5
R. Goizueta	Coca-Cola	Consumer	10,715	214	4
M. Eisner	Walt Disney	Consumer	9,589	262	5
A. Busch	Anheuser-Busch	Consumer	8,861	155	5
W. McGowan	MCI	Telecommunications	8,666	704	2
J. Moffett	Freeport McMoRan	Industrial	7,300	196	4
D. Petersen	Ford Motor	Consumer	7,147	180	4
P. Vagelos	Merck	Consumer	6,764	197	4

Source: "Pay stubs of the rich and corporate." *Business Week*, May 7, 1990.

Solution

a. Because the data are recorded for each corporate executive, the 10 executives in Table 1.1 are the experimental units.
b. The first two variables (company and industry group) are qualitative because the data they produce are nonnumerical values; they can only be classified into categories or groups. The next two variables (total pay and shareholder return) are quantitative because they are measured on a numerical scale. The fifth variable (performance rating), although coded as a number (1–5) is really qualitative in nature. The performance categories are: excellent, above average,

average, below average, and poor. For convenience, *Business Week* has chosen to assign numbers (i.e., 1 for "excellent," 2 for "above average," etc.) to the categories to obtain a performance rating.* This does result in a meaningful variable, however, since the higher the rating, the poorer the executive's performance. In contrast, assigning numbers to industry group (e.g., 1 for "automotive," 2 for "apparel," etc.) would not result in a meaningful quantitative variable.

1.2

Populations, Samples, and Random Sampling

When you examine a data set in the course of your business, you will be doing so because the data characterize some phenomenon of interest to you. In statistics, the data set that is the target of your interest is called a **population**. Notice that a statistical population does not refer to a group of people; it refers to a set of measurements. This data set, which is typically large, exists in fact or is part of an ongoing operation and hence is conceptual. Some examples of business phenomena and their corresponding populations are shown in Table 1.2.

T A B L E 1.2   **Some Typical Populations**

	PHENOMENON	EXPERIMENTAL UNITS	POPULATION	TYPE
a.	Current year, new residential construction prices	Residential properties sold this year	Set of prices of all new residential properties	Existing
b.	Starting salary of a graduating MBA this year	MBAs graduating this year	Set of starting salaries of all MBAs who graduated this year	Existing
c.	Profit per job in a construction company	Jobs	Set of profits for all jobs performed recently or to be performed in the near future	Part existing, part conceptual
d.	Quality of items produced on an assembly line	Manufactured items	Set of quality measurements for all items manufactured in the recent past and in the future	Part existing, part conceptual

> Definition 1.6
>
> A **population** is a collection (or set) of data that describe some phenomenon of interest to you.

---

*To obtain the performance rating, *Business Week* compared an executive's return-to-pay ratio with those of other executives within the same industry group and assigned a rating of 1 to those executives with the highest ratios (relative to the others in the group), a rating of 2 to those with the next highest ratios, etc.

Many populations are too large to measure (because of time and cost); others cannot be measured because they are part conceptual, such as the set of quality measurements (population **d** in Table 1.2). Thus, we are often required to select a subset of values from a population and to make inferences about the population based on information contained in a **sample**. This is one of the major objectives of modern statistics.

Definition 1.7

A **sample** is a subset of data selected from a population.

Probability theory is used to infer the nature of a population from information contained in a sample. We observe the sample data and then consider the likelihood of observing these particular measurements for populations possessing various characteristics. Generally speaking, we infer that the sample was selected from the population most likely to have produced the observed sample. For example, if you toss a coin 10 times and observe 10 heads, you should infer either that the coin that generated the sample was biased in favor of heads or that something went wrong with your sampling (coin-tossing) procedure. Since the probability of observing a particular sample depends on how the sample was selected, the sampling procedure plays an important role in statistical inference.

The most common type of sampling procedure is one that gives every different sample of fixed size in the population an equal probability (chance) of selection. Such a sample is called a **random sample**.

Definition 1.8

A **random sample** of n experimental units is one selected from the population in such a way that every different sample of size n has an equal probability (chance) of selection.

How can a random sample be generated? If the population is not too large, each observation may be recorded on a piece of paper and placed in a suitable container. After the collection of papers is thoroughly mixed, the researcher can remove n pieces of paper from the container; the elements named on these n pieces of paper are the ones to be included in the sample. Lottery officials utilize such a technique in generating the winning numbers for Florida's weekly 6/49 Lotto game. Forty-nine white Ping-Pong balls (the population), each identified from 1 to 49 in black numerals, are placed into a clear plastic drum and mixed by blowing air into the container. The Ping-Pong balls bounce at random until a total of six balls "pop" into a tube attached to the drum. The numbers on the six balls (the random sample) are the winning Lotto numbers.

This method of random sampling is fairly easy to implement if the population is relatively small. It is not feasible, however, when the population consists of a large number of observations. Since it is also very difficult to achieve a thorough mixing, the procedure only approximates random sampling. Most scientific studies, however, rely on computers (with built-in random-number generators) to automatically generate the random sample. Almost all of the commercial statistical software packages available today (e.g., SAS, SPSS, Minitab) have procedures for generating random samples.

EXERCISES

1.1 Do you want to avoid an Internal Revenue Service audit of your personal income tax return? If so, then try living in Newark, New Jersey, or in Boston. The Research Institute of America (RIA) found that only .55% of returns in those two cities were audited in 1987, in contrast to 1.45% in Anchorage, Alaska, 1.44% in San Francisco, and 1.36% in Manhattan (The *Wall Street Journal*, Mar. 22, 1989). For this RIA study, identify or describe the following:
   a. Population   b. Sample
   c. Experimental unit   d. Inference

1.2 When Nissan introduced its new Infiniti luxury cars in 1989, its television ad campaign was renowned for a novel gimmick: The automobiles were nowhere in sight. The Infiniti ads, which depicted lushly photographed trees, boulders, lightning bolts, and ocean waves (but no cars), were found by a nationwide Gallup poll of 1,000 consumers to be the best-recalled commercial on television (*Time*, Jan. 22, 1990).
   a. Describe the population of interest to the pollsters.
   b. Identify the sample.
   c. What is the inference made by the Gallup poll?

1.3 The *Journal of Performance of Constructed Facilities* (Feb. 1990) reported on the performance dimensions of water distribution networks in the Philadelphia area. For one part of the study, the following data were collected for a sample of water pipe sections:
   1. Pipe diameter (inches)
   2. Pipe material
   3. Age (year of installation)
   4. Location
   5. Pipe length (feet)
   6. Stability of surrounding soil (unstable, moderately stable, or stable)
   7. Corrosiveness of surrounding soil (corrosive or noncorrosive)
   8. Internal pressure (pounds per square inch)
   9. Percentage of pipe under land cover
   10. Breakage rate (number of times pipe had to be repaired because of breakage)
   Identify the data as quantitative or qualitative.

1.4 Marketers are keenly interested in the factors that motivate coupon usage by consumers. A study reported in the *Journal of Consumer Marketing* (Spring 1988) asked a sample of 290 shoppers to respond to the following questions:
   a. Do you collect and redeem coupons?
   b. Are you price-conscious while shopping?

c. On average, how much time per week do you spend clipping and collecting coupons? Classify the responses to the questions as quantitative or qualitative data.

1.5   Do most state lottery winners who win big payoffs quit their jobs within one year of winning? Not according to a study conducted by sociologist and professor, H. Roy Kaplan (*Journal of the Institute for Socioeconomic Studies*, Sept. 1985). Kaplan mailed questionnaires to over 2,000 lottery winners who won at least $50,000 in the past 10 years. Of the 576 who responded, only 11% had quit their jobs during the first year after striking it rich. In this study, identify

a. the population
b. the sample
c. the inference made about the population

1.3

Describing Data Sets Graphically

The word *inference* implies description. For example, to infer the nature of a company, we would describe its product, annual sales volume, number of employees, annual profit, and so forth. Similarly, to infer the nature of a data set, such as a population or a sample, we need to be able to describe the set.

A useful graphical method for describing quantitative data is provided by a **relative frequency distribution**. This type of graph shows the proportions of the total set of measurements that fall in various intervals on the scale of measurement. For example, Figure 1.1 shows the prices of new residential properties sold in a particular region in 1991. The area over a particular interval under a relative frequency distribution curve is proportional to the fraction of the total number of measurements that fall in that interval. In Figure 1.1, the fraction of the total number of residential house prices that falls between $70,000 and $80,000 is proportional to the shaded area. **If we take the total area under the distribution curve as equal to 1, then the shaded area is equal to the fraction of house prices that fall between $70,000 and $80,000.**

FIGURE 1.1

Relative frequency distribution: Prices of new residential properties

The variable measured in generating a population, denoted by the symbol y, is called a **random variable**. Observing a single value of y is equivalent to selecting a single measurement from the population. The probability that it will assume a value in an interval, say, a to b, is given by its relative frequency or **probability distribution**. The total area under a probability distribution curve is always assumed to equal 1. Hence, the probability that a measurement on y will fall in the interval between a and b is equal to the shaded area shown in Figure 1.2 on page 8.

FIGURE 1.2

Probability distribution for a random
variable

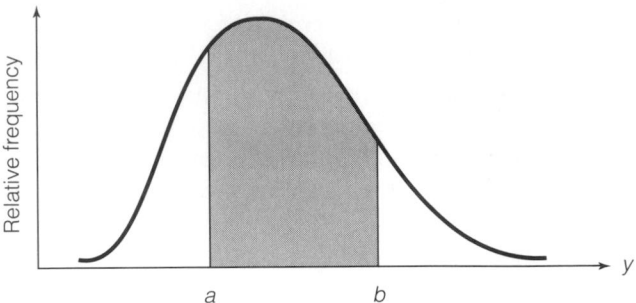

Since the theoretical probability distribution for a population is usually
unknown, we resort to obtaining a sample from the population: Our objective
is to describe the sample and use this information to make inferences about the
probability distribution of the population. **Stem-and-leaf plots** and **histograms**
are two of the most popular graphical methods for describing samples. Both
display the frequency (or relative frequency) of observations that fall into specified
intervals (or classes) of the variable's values.

For small data sets (say, 30 or fewer observations) with measurements with
only a few digits, stem-and-leaf plots can be constructed easily by hand. Histograms, on the other hand, are better suited to the description of larger data sets,
and they permit greater flexibility in the choice of classes. Both, however, can
be generated using the computer, as illustrated in the following examples.

| EXAMPLE 1.2

The data in Table 1.3 represent sale prices (in thousands of dollars) for a random
sample of 25 residential properties sold in Tampa, Florida, in 1990. (The data
are from Appendix E. We analyze the data in Appendix E more thoroughly in
Case Study 12.) A Minitab printout of a stem-and-leaf plot for the 25 sale prices
is shown in Figure 1.3. Interpret the figure.

FIGURE 1.3

Minitab stem-and-leaf display for sale
prices in Table 1.3

```
Stem-and-leaf of salepric  N  = 25
Leaf Unit = 1.0


      1      3  6
      2      4  2
      5      5  079
      9      6  3568
     (5)     7  12467
     11      8  249
      8      9  45
      6     10  169
      3     11  2
      2     12  9
      1     13
      1     14  8
```

**TABLE 1.3 Sale Prices for a
Sample of Properties from
Appendix E**

SALE PRICE (HUNDREDS OF DOLLARS)				
66	59	106	50	63
89	129	74	82	84
71	95	72	57	76
109	77	68	101	65
42	36	148	94	112

Solution

In a stem-and-leaf plot, each measurement is partitioned into a stem and a leaf.
Minitab has selected the last digit in the sale price to represent the leaf and the
preceding digits to represent the stem. For example, the value 148 (representing

a sale price of $148,000) is partitioned into a stem of 14 and a leaf of 8, as illustrated here:

STEM	LEAF
14	8

The stems are listed in order in the second column of the plot, Figure 1.3, starting with the smallest stem of 3 and ending with the largest stem of 14. The respective leaves are placed to the right in increasing order in the appropriate stem row. For example, the stem row of 9 in Figure 1.3 has two leaves, 4 and 5, representing the sale prices $94,000 and $95,000, respectively. Notice that the stem row of 7 (representing sale prices in the $70,000s) has the most leaves (5). Thus, 5 of the 25 sale prices (or 20%) have values in the $70,000s. Notice also that 20 of the 25 sale prices (80%) fall between $50,000 and $110,000. (That is, 20 of the sale prices have stems ranging from 5 to 10.)

EXAMPLE 1.3

Figure 1.4 is a SAS printout of a relative frequency histogram describing the sale prices (in $ thousands) of over 8,000 residential properties sold in Tampa, Florida, in 1990. (The data in Appendix E were extracted from this larger data set.)

FIGURE 1.4

SAS histogram for the sale prices (in $ thousands)

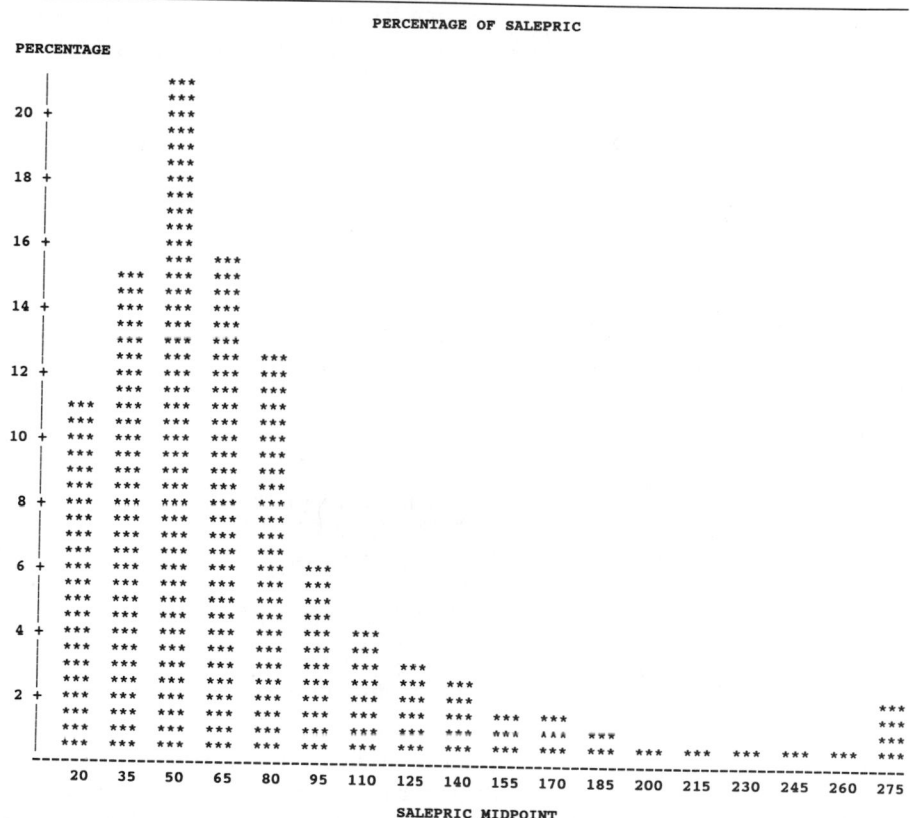

a. Interpret the graph.
b. Visually estimate the proportion of sale prices in the data set between $42,500 and $102,500.

Solution

a. In constructing a histogram, the values of sale price are divided into intervals of equal length, called **classes**. The midpoints of these classes, 20, 35, 50, etc., are shown on the horizontal axis of Figure 1.4. Thus, the classes are 12.5–27.5, 27.5–42.5, etc. The relative frequency (or percentage) of sale prices falling in each class interval is represented by the vertical bars over the class.

You can see from Figure 1.4 that the sale prices tend to pile up near $50,000; the class interval from $42,500 to $57,500 has the greatest relative frequency.

Figure 1.4 also shows a tendency for the data to tail out to the high side because of a few extremely large sale prices. Distributions of data with this feature are said to be skewed to the right, or **positively skewed**. (Similarly, distributions of data are skewed left, or **negatively skewed**, if they tend to tail out to the low side because of a few unusually small measurements.)

b. The interval $42,500 to $102,500 spans four sale price classes: 42,500–57,500; 57,500–72,500; 72,500–87,500; and 87,500–102,500. The proportion of sale prices between $42,500 and $102,500 is equal to the sum of the relative frequencies associated with these four classes. These four class relative frequencies are (approximately) .21, .155, .125, and .06, respectively. Consequently, the approximate proportion of sale prices between $42,500 and $102,500 is

$$(.21 + .155 + .125 + .06) = .55$$
$$= 55\%$$

The steps required to construct stem-and-leaf plots and histograms by hand are summarized here. The computer commands for generating these graphs are provided in Appendix C.

| Constructing a Stem-and-Leaf Display

STEP 1 Decide how the stems and leaves will be defined.

STEP 2 List the stems in order in a column, starting with the smallest stem and ending with the largest.

STEP 3 Proceed through the data set, placing the leaf for each observation in the appropriate stem row. (You may want to place the leaves of each stem in increasing order.)

> Constructing a Histogram
>
> STEP 1 Examine the data to determine the smallest and largest measurements.
>
> STEP 2 Divide the interval between the smallest and largest measurements into between 5 and 20 equal subintervals called **classes**, so that each measurement falls in one and only one subinterval. (Although the choice of the number of classes is arbitrary, you will obtain a better description of the data if you use a small number of subintervals for small data sets and a large number of subintervals for large data sets.)
>
> STEP 3 Compute the class frequencies or the class relative frequencies.
>
> STEP 4 Using a vertical axis of about three-fourths of the length of the horizontal axis, plot each relative frequency (or frequency) on the vertical axis as a rectangle or bar over the corresponding subinterval on the horizontal axis.

EXERCISES

1.6 Consider the following sample data:

5.9	5.3	1.6	7.4	8.6	1.2	2.1
4.0	7.3	8.4	8.9	6.7	4.5	6.3
7.6	9.7	3.5	1.1	4.3	3.3	8.4
1.6	8.2	6.5	1.1	5.0	9.4	6.4

a. Using the first digit as a stem, construct a stem-and-leaf display.
b. Construct a relative frequency histogram for the sample data.

1.7 Multinational corporations are firms with both domestic and foreign assets or investments. The foreign revenue (as a percentage of total revenue) generated by each of the top 20 U.S.-based multinational firms is listed in the accompanying table.

Exxon	73.2	Procter & Gamble	39.9
IBM	58.9	Philip Morris	19.6
GM	26.6	Eastman Kodak	40.9
Mobil	64.7	Digital	54.1
Ford	33.2	GE	12.4
Citicorp	52.3	United Technologies	32.9
EI duPont	39.8	Amoco	26.1
Texaco	42.3	Hewlett-Packard	53.3
ITT	43.3	Xerox	34.6
Dow Chemical	54.1	Chevron	20.5

Source: *Forbes*, July 23, 1990, pp. 362–363.

a. Construct a stem-and-leaf display for the data.
b. Interpret the stem-and-leaf display obtained in part **b**.
c. What proportion of the 20 multinational firms generated at least 50% of their revenue from foreign investments or sales?

1.8   A Harris Corporation/University of Florida study was undertaken to determine whether a manufacturing process performed at a remote location can be established locally. Test devices (pilots) were set up at both the old and new locations, and voltage readings on the process were obtained. A "good" process was considered to be one with voltage readings of at least 9.2 volts (with larger readings better than smaller readings). The table contains voltage readings for 30 production runs at each location.

OLD LOCATION			NEW LOCATION		
9.98	10.12	9.84	9.19	10.01	8.82
10.26	10.05	10.15	9.63	8.82	8.65
10.05	9.80	10.02	10.10	9.43	8.51
10.29	10.15	9.80	9.70	10.03	9.14
10.03	10.00	9.73	10.09	9.85	9.75
8.05	9.87	10.01	9.60	9.27	8.78
10.55	9.55	9.98	10.05	8.83	9.35
10.26	9.95	8.72	10.12	9.39	9.54
9.97	9.70	8.80	9.49	9.48	9.36
9.87	8.72	9.84	9.37	9.64	8.68

Source: Harris Corporation, Melbourne, Fla.

a. Construct a relative frequency histogram for the voltage readings of the old process.
b. Construct a stem-and-leaf display for the voltage readings of the old process. Which of the two graphs in parts **a** and **b** is more informative?
c. Construct a frequency histogram for the voltage readings of the new process.
d. Compare the two graphs in parts **a** and **c**. (You may want to draw the two histograms on the same graph.) Does it appear that the manufacturing process can be established locally (i.e., is the new process as good as or better than the old)?

1.9   A study was conducted to evaluate the advertisement awareness and sales effectiveness of advertising campaigns for 18 confectionary brands (*Journal of the Market Research Society*, Jan. 1986). For each brand, an ad awareness index (maximum = 100) was determined from a consumer survey, whereas a sales effectiveness index (maximum = 100) was estimated from market shares. The accompanying frequency distributions were used to summarize the data.
a. How many of the 18 brands had an ad awareness index of 40 or less?
b. How many of the 18 brands had a sales effectiveness index of 70 or more?
c. Use the information provided by the frequency distribution to construct a relative frequency distribution for the 18 ad awareness indexes. Interpret the graph.
d. Use the information provided by the frequency distribution to construct a relative frequency distribution for the 18 sales effectiveness indexes.

Distribution of 18 Campaigns by Ad Awareness Effectiveness (maximum – 100)

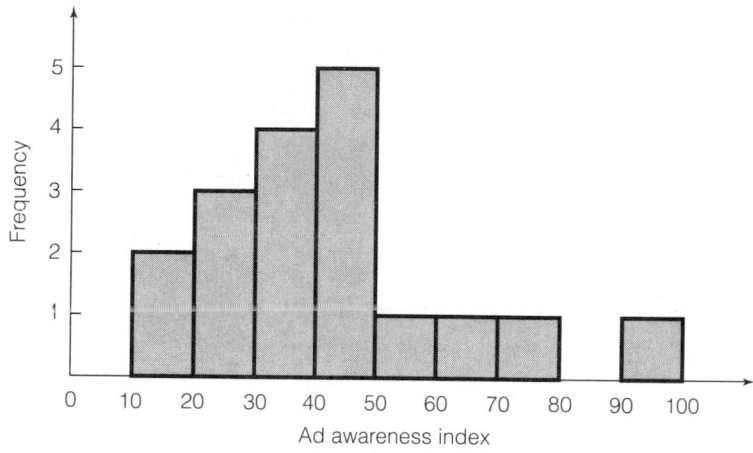

Distribution of 18 Campaigns by Sales Effectiveness (maximum = 100)

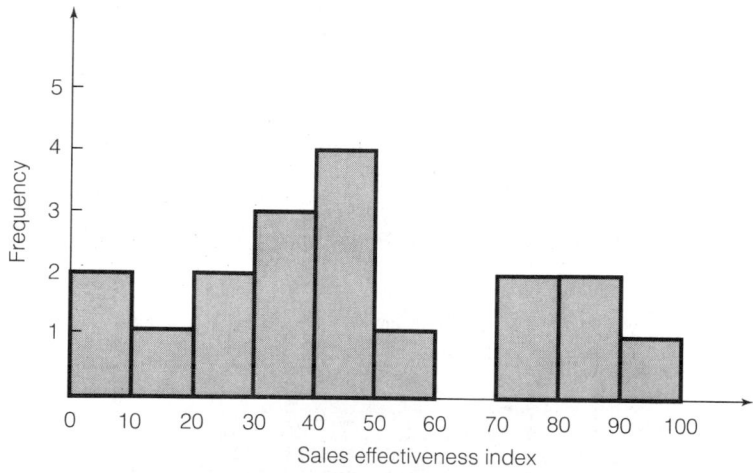

Source: Broadbent, S. and Colman, S. "Advertising effectiveness: Across brands." *Journal of the Market Research Society*, Vol. 28, No. 1, 1986, pp. 15–23.

1.10 Under a voluntary cooperative inspection program, all passenger cruise ships arriving at U.S. ports are subject to unannounced inspection. The purpose of these inspections is to achieve levels of sanitation that will minimize the potential for gastrointestinal disease outbreaks on these ships. Ships are rated on a 0- to 100-point scale depending on how well they meet the Center for Disease Control sanitation standards. In general, the lower the score, the lower the level of sanitation. The table on page 14 lists the sanitation inspection scores for 72 international cruise ships during 1990.

SHIP	SCORE	SHIP	SCORE	SHIP	SCORE
Americana	98	Jubilee	93	Sea Lion	88
Amerikanis	97	Mardi Gras	86	Seabourn Spirit	94
Azure Seas	94	Meridian	92	Seabourn Pride	80
Britanis	82	Nantucket Clipper	86	Seabreeze I	87
Caribe I	91	New Shoreham II	80	Seaward	94
Carla C	90	Nieuw Amsterdam	89	Sky Princess	87
Carnivale	89	Noordam	96	Skyward	88
Celebration	91	Nordic Empress	92	Song of America	96
Club Med 1	90	Nordic Prince	93	Song of Norway	91
Costa Riviera	89	Norway	92	Southward	93
Crown Del Mar	88	Pacific Princess	87	Sovereign of the Seas	92
Cunard Countess	87	Pride of Mississippi	57	Star Princess	91
Daphne	78	Queen Elizabeth 2	89	Starship Atlantic	93
Dawn Princess	87	Queen of Bermuda	89	Starship Majestic	91
Discovery I	91	Regent Star	88	Starship Oceanic	97
Dolphin IV	95	Regent Sun	91	Starward	93
Emerald Seas	93	Rotterdam	89	Sun Viking	88
Enchanted Isle	87	Royal Princess	89	Sunward II	87
Fair Princess	76	Royal Viking Sky	97	Tropicale	93
Fantasy	88	Royal Viking Star	99	Tropicana	91
Festivale	95	Royal Viking Sun	93	Viking Princess	86
Holiday	96	Sagafjord	95	Viking Serenade	90
Horizon	92	Scandinavian Saga	92	Westerdam	87
Island Princess	90	Scandinavian Sun	83	Yorktown Clipper	92

Source: Center of Environmental Health and Injury Control, Miami, Fla. (reported in *Tampa Tribune*, Nov. 25, 1990).

a. A Minitab stem-and-leaf display of the data is shown here. Identify the stems and leaves of the graph.

```
Stem-and-leaf of ISCORE    N  = 72
Leaf Unit = 1.0


     1     5 7
     1     6
     1     6
     1     7
     3     7 68
     7     8 0023
    31     8 6667777777788888889999999
   (30)    9 0000111111112222223333333333444
    11     9 55566677789
```

b. A score of 86 or higher at the time of inspection indicates the ship is providing an accepted standard of sanitation. Use the Minitab graph to estimate the proportion of ships that have an accepted sanitation standard.

c. Locate the inspection score of 57 (Pride of Mississippi) on the stem-and-leaf display.

1.11   Are major colleges and universities lax in hiring minorities to fill top positions in their athletic programs? A *USA Today* survey of 62 Division I schools found that only 12.5% of the jobs in each athletic department are held by minorities (Blacks, Hispanics, Native Americans, and Asians). In contrast, the 1990 census shows minorities represent 19.7% of the U.S. population (*USA Today*, Mar.

19, 1991). The results of the survey are reproduced in the table. The 62 schools were selected based on the Top 25 polls for men's and women's basketball during the 1989–90 season and the final 1990 Top 25 football poll. (Northwestern declined to respond, and Seton Hall did not supply figures.)

SCHOOL	TOTAL POSITIONS	POSITIONS HELD BY MINORITY	PERCENT MINORITY	SCHOOL	TOTAL POSITIONS	POSITIONS HELD BY MINORITY	PERCENT MINORITY
Georgetown	30	8	26.7	Louisiana Tech	26	3	11.5
Houston	43	11	25.6	Syracuse	52	6	11.5
Miami	55	14	25.5	Arkansas	44	5	11.4
Arizona	67	15	22.4	Northern Illinois	53	6	11.3
Long Beach State	53	11	20.8	Alabama	54	6	11.1
USC	65	13	20.0	Western Kentucky	18	2	11.1
Pittsburgh	54	10	18.5	UNLV	55	6	10.9
Oklahoma State	34	6	17.6	Connecticut	37	4	10.8
Oklahoma	63	11	17.5	South Carolina	49	5	10.2
Washington	63	11	17.5	East Tennessee State	30	3	10.0
Southern Mississippi	35	6	17.1	Texas-El Paso	30	3	10.0
Stanford	71	12	16.9	Rutgers	42	4	9.5
Iowa	72	12	16.7	North Carolina State	44	4	9.1
Georgia Tech	52	8	15.4	Texas[a]	67	6	9.0
Michigan State	60	9	15.0	Michigan	58	5	8.6
Illinois	54	8	14.8	Penn State	82	7	8.5
Kentucky	54	8	14.8	St. John's	60	5	8.3
Ohio State	64	9	14.1	Nebraska	64	5	7.8
Colorado	43	6	14.0	Mississippi State	52	4	7.7
LSU	59	8	13.6	Providence	13	1	7.7
Purdue	53	7	13.2	Mississippi	40	3	7.5
New Mexico State	38	5	13.2	Florida State	53	4	7.5
UCLA	84	11	13.1	Indiana	57	4	7.0
Clemson	48	6	12.5	Tennessee[a]	73	5	6.8
North Carolina	57	7	12.3	Duke	45	3	6.7
Kansas	49	6	12.2	Virginia	60	4	6.7
Utah	33	4	12.1	BYU	51	3	5.9
Louisville	34	4	11.8	Princeton	44	2	4.5
Georgia	60	7	11.7	Notre Dame	47	2	4.3
Florida	61	7	11.5	Stephen F. Austin	27	1	3.7

[a]Numbers combined from separate men's and women's athletic programs
Source: *USA Today*, Mar. 19, 1991.

a. Do the data represent a sample or a population? Explain.

b. Describe the data on percentage of minority positions in the athletic department of the 60 colleges and universities with a graphical technique. Interpret the graph.

1.4

Describing Data Sets Numerically

Numerical descriptive measures provide a second (and often more powerful) method for describing a set of data. These measures, which locate the center of the data set and its spread, actually enable you to construct an approximate mental image of the distribution of the data set.

Note: Most of the formulas used to compute numerical descriptive measures require the summation of numbers. For example, we may want to sum the observations in a data set, or we may want to square each observation and then sum the squared values. The symbol Σ (sigma) is used to denote a summation operation.

For example, suppose we denote the n sample measurements on a random variable y by the symbols $y_1, y_2, y_3, \ldots, y_n$. Then the sum of all n measurements in the sample is represented by the symbol

$$\sum_{i=1}^{n} y_i$$

This is read "summation y, y_1 to y_n" and is equal to the value

$$y_1 + y_2 + y_3 + \cdots + y_n$$

One of the most common measures of central tendency is the **mean**, or arithmetic average, of a data set. Thus, if we denote the sample measurements by the symbols $y_1, y_2, y_3, \ldots$, the sample mean is defined as follows:

Definition 1.9

The **mean** of a sample of n measurements $y_1, y_2, \ldots, y_n$ is

$$\bar{y} = \frac{\displaystyle\sum_{i=1}^{n} y_i}{n}$$

The mean of a population, or equivalently, the expected value of y, $E(y)$, is usually unknown in a practical situation (we will want to infer its value based on the sample data). Most texts use the symbol μ to denote the mean of a population. Thus, we will use the following notation:

Notation

Sample mean: $\bar{y}$
Population mean: $E(y) = \mu$

The spread or variation of a data set is measured by its **range**, its **variance**, or its **standard deviation**.

Definition 1.10

The **range** of a sample of n measurements $y_1, y_2, \ldots, y_n$ is the difference between the largest and smallest measurements in the sample.

EXAMPLE 1.4

If a sample consists of measurements 3, 1, 0, 4, 7, find the sample mean and the sample range.

Solution

The sample mean and range are

$$\bar{y} = \frac{\sum\limits_{i=1}^{n} y_i}{n} = \frac{15}{5} = 3$$

$$\text{Range} = 7 - 0 = 7$$

The variance of a set of measurements is defined to be the average of the *squares of the deviations* of the measurements about their mean. Thus, the population variance, which is usually unknown in a practical situation, would be the mean or expected value of $(y - \mu)^2$, or $E[(y - \mu)^2]$. We use the symbol σ^2 to represent the variance of a population:

$$E[(y - \mu)^2] = \sigma^2$$

The quantity usually termed the **sample variance** is defined in the box.

Definition 1.11

The **variance** of a sample of n measurements $y_1, y_2, \ldots, y_n$ is defined to be

$$s^2 = \frac{\sum\limits_{i=1}^{n} (y_i - \bar{y})^2}{n - 1} = \frac{\sum\limits_{i=1}^{n} y_i^2 - n\bar{y}^2}{n - 1}$$

Note that the sum of squares of deviations in the sample variance is divided by $(n - 1)$, rather than n. Division by n produces estimates that tend to underestimate σ^2. Division by $(n - 1)$ corrects this problem.

EXAMPLE 1.5

Refer to Example 1.4. Calculate the sample variance for the sample 3, 1, 0, 4, 7.

Solution

We first calculate

$$\sum\limits_{i=1}^{n} (y_i - \bar{y})^2 = \sum\limits_{i=1}^{n} y_i^2 - n\bar{y}^2 = 75 - 5(3)^2 = 30$$

where $\bar{y} = 3$ from Example 1.4.

Then

$$s^2 = \frac{\sum\limits_{i=1}^{n} (y_i - \bar{y})^2}{n - 1} = \frac{30}{4} = 7.5$$

The concept of a variance is important in theoretical statistics, but its square root, called a **standard deviation**, is the quantity most often used to describe data variation.

| Definition 1.12

The **standard deviation** of a set of measurements is equal to the square root of their variance. Thus, the standard deviations of a sample and a population are

Sample standard deviation: *s*

Population standard deviation: σ

The standard deviation of a set of data takes on meaning in light of a theorem (Tchebysheff's theorem) and a rule of thumb.* Basically, they give us the following guidelines:

| Guidelines for Interpreting a Standard Deviation

1. For *any* data set (population or sample), at least three-fourths of the measurements will lie within 2 standard deviations of their mean.
2. For *most* data sets of moderate size (say, 25 or more measurements) with a mound-shaped distribution, approximately 95% of the measurements will lie within 2 standard deviations of their mean.

EXAMPLE 1.6

Often, travelers who have no intention of showing up fail to cancel their hotel reservations in a timely manner. These travelers are known, in the parlance of the hospitality trade, as "no-shows." To protect against no-shows and late cancellations, hotels invariable overbook rooms. A recent study reported in the *Journal of Travel Research* examined the problems of overbooking rooms in the hotel industry. The data in Table 1.4, extracted from the study, represent daily numbers of late cancellations and no-shows for a random sample of 30 days at a large (500-room) hotel. Based on this sample, how many rooms, at minimum, should the hotel overbook each day?

Solution

To answer this question, we need to know the range of values where most of the daily numbers of no-shows fall. We must compute $\bar{y}$ and s, and examine the shape of the relative frequency distribution for the data.

---

*For a more complete discussion and a statement of Tchebysheff's theorem, see the references listed at the end of this chapter.

TABLE 1.4 Hotel No-Shows for a Sample of 30 Days

18	16	16	16	14	18	16	18	14	19
15	19	9	20	10	10	12	14	18	12
14	14	17	12	18	13	15	13	15	19

Source: Toh, R. S. "An inventory depletion overbooking model for the hotel industry." *Journal of Travel Research*, Vol. 23, No. 4, Spring 1985, p. 27. The *Journal of Travel Research* is published by the Travel and Tourism Research Association (TTRA) and the Business Research Division, University of Colorado at Boulder.

Figure 1.5 is a Minitab printout that shows a stem-and-leaf display and descriptive statistics of the sample data. Notice from the stem-and-leaf display that the distribution of daily no-shows is mound-shaped, and only slightly skewed on the low (top) side of Figure 1.5. Thus, guideline 2 in the previous box should give a good estimate of the percentage of days that fall within 2 standard deviations of the mean.

FIGURE 1.5

Minitab printout: Describing the no-show data, Example 1.6

```
Stem-and-leaf of noshows    N  = 30
Leaf Unit = 0.10

     1      9  0
     3     10  00
     3     11
     6     12  000
     8     13  00
    13     14  00000
    (3)    15  000
    14     16  0000
    10     17  0
     9     18  00000
     4     19  000
     1     20  0
```

	N	MEAN	MEDIAN	TRMEAN	STDEV	SEMEAN
noshows	30	15.133	15.000	15.231	2.945	0.538

	MIN	MAX	Q1	Q3
noshows	9.000	20.000	13.000	18.000

The mean and standard deviation of the sample data, shaded on the Minitab printout, are $\bar{y} = 15.133$ and $s = 2.945$. From guideline 2 in the box, we know that about 95% of the daily number of no-shows fall within 2 standard deviations of the mean, i.e., within the interval

$$\bar{y} \pm 2s = 15.133 \pm 2(2.945)$$
$$= 15.133 \pm 5.890$$

or between 9.243 no-shows and 21.023 no-shows. (If we count the number of measurements in this data set, we find that actually 29 out of 30, or 96.7%, fall in this interval.)

From this result, the large hotel can infer that there will be at least 9.243 (or, rounding up, 10) no-shows per day. Consequently, the hotel can overbook at least 10 rooms per day and still be highly confident that all reservations can be honored.

Numerical descriptive measures calculated from sample data are called **statistics**. Numerical descriptive measures of the population are called **parameters**. In a practical situation, we will not know the population relative frequency distribution (or equivalently, the population distribution for y). We will usually assume that it has unknown numerical descriptive measures, such as its mean μ and standard deviation σ, and by inferring (using **sample statistics**) the values of these parameters, we infer the nature of the population relative frequency distribution. Sometimes we will assume that we know the shape of the population relative frequency distribution and use this information to help us make our inferences. When we do this, we are postulating a model for the population relative frequency distribution, and we must keep in mind that the validity of the inference may depend on how well our model fits reality.

Definition 1.13

Numerical descriptive measures of a population are called **parameters**.

Definition 1.14

A **sample statistic** is a quantity calculated from the observations in a sample.

EXERCISES

1.12   Compute $\bar{y}$, s^2, and s for each of the following data sets:
  a. 1, 5, 0, 2, 5, 7, 1   **b.** 1, 2, 0, 0, 5, 4
  c. 10, 8, 12, 2   **d.** 3, 4, 10, 2

1.13   Compute $\bar{y}$, s^2, and s for each of the following data sets:
  a. 1, 1, 20, 20, 8   **b.** 2, 100, 104, 2
  c. $-1, -3, -2, 0, -3, -3$ **d.** $\frac{1}{5}, \frac{1}{5}, \frac{1}{5}, \frac{2}{5}, .2, \frac{4}{5}$

1.14   Each week, the syndicated "Money Funds Report" appears in daily newspapers across the country. The report lists average maturity and 7-day yields for over 200 taxable money funds that are available to individual investors. Shown here are the average maturity (in days) for the week ending May 7, 1991, for each of 50 money funds with assets of $100 million or more.

57	60	77	24	51	61	79	35	26	6
37	12	13	20	41	60	63	54	67	47
65	67	38	30	23	52	73	43	43	51
58	2	12	28	35	63	44	65	1	24
29	8	19	52	48	76	31	24	30	54

a. Compute $\bar{y}$, s^2, and s for the data set.

b. What percentage of the measurements would you expect to find in the interval $\bar{y} \pm 2s$?

c. Count the number of measurements that actually fall within the interval of part b, and express the interval count as a percentage of the total number of measurements. Compare this result with the answer to part b.

1.15 Refer to the *Journal of Performance of Constructed Facilities* (Feb. 1990) study of water distribution networks, Exercise 1.3. The internal pressure readings (measured in pounds per square inch, psi) for a sample of pipe sections had a mean of 7.99 psi and a standard deviation of 2.02 psi.

a. Use this information to construct an interval that captures about 95% of the pressure readings sampled.

b. Would you expect to observe an internal pressure reading of 20 psi? Explain.

1.16 The Federal Trade Commission (FTC) annually ranks varieties of domestic cigarettes according to their tar, nicotine, and carbon monoxide contents. The U.S. surgeon general considers each of these three substances hazardous to a smoker's health. Each year, the FTC tests and ranks approximately 200 brands of cigarettes. The table contains the tar, nicotine, and carbon monoxide contents (in milligrams) for a sample of 25 (filter) brands in a recent year.

BRAND	TAR	NICOTINE	CARBON MONOXIDE
Alpine	14.1	.86	13.6
Benson & Hedges	16.0	1.06	16.6
Bull Durham	29.8	2.03	23.5
Camel Lights	8.0	.67	10.2
Carlton	4.1	.40	5.4
Chesterfield	15.0	1.04	15.0
Golden Lights	8.8	.76	9.0
Kent	12.4	.95	12.3
Kool	16.6	1.12	16.3
L&M	14.9	1.02	15.4
Lark Lights	13.7	1.01	13.0
Marlboro	15.1	.90	14.4
Merit	7.8	.57	10.0
Multifilter	11.4	.78	10.2
Newport Lights	9.0	.74	9.5
Now	1.0	.13	1.5
Old Gold	17.0	1.26	18.5
Pall Mall Lights	12.8	1.08	12.6
Raleigh	15.8	.96	17.5
Salem Ultra	4.5	.42	4.9
Tareyton	14.5	1.01	15.9
True	7.3	.61	8.5
Viceroy Rich Lights	8.6	.69	10.6
Virginia Slims	15.2	1.02	13.9
Winston Lights	12.0	.82	14.9

Source: Federal Trade Commission.

a. Calculate $\bar{y}$, s^2, and s for the sample of 25 tar contents.
b. What percentage of the measurements would you expect to find in the interval $\bar{y} \pm 2s$?
c. Count the number of measurements that actually fall within the interval of part **b**, and express the interval count as a percentage of the total number of measurements. Compare this result with the answer to part **b**.
d. Repeat parts **a–c** for the sample of nicotine contents.
e. Repeat parts **a–c** for the sample of carbon monoxide contents.

1.17   Given a data set whose largest value is 900 and smallest value is 50, what would you estimate the standard deviation to be? Explain the logic behind the procedure you used to estimate the standard deviation.

1.18   It is well known that worker absenteeism is costly and leads to decreased production efficiency at most firms. A study was recently conducted to investigate employee absenteeism at a medium-size assembly and packaging plant in the United States. Workers in each of three departments—packaging, assembly, and maintenance—were monitored over a 2-year period and the number of days of unanticipated absences because of sickness was recorded each week. The accompanying table gives the mean and standard deviation of number of days absent per 100 employees for each department.

	PACKAGING DEPARTMENT	ASSEMBLY DEPARTMENT	MAINTENANCE DEPARTMENT
Mean number of days per 1,000 employees	39.24	17.38	14.56
Standard deviation	9.88	5.16	6.88

Source: Moch, M. K. and Fitzgibbons, D. E. "The relationship between absenteeism and production efficiency: An empirical assessment." *Journal of Occupational Psychology*, Vol. 58, 1985, pp. 39–47.

a. Use the information in the table to sketch your mental images of the three relative frequency distributions. Construct them on the same graph so that you can see how they appear relative to each other.
b. Estimate the proportion of weeks in which between 19.48 and 59.00 days of unanticipated absences per 100 workers occur in the packaging department.
c. In a typical week, how many days of unanticipated absences per 100 workers would you expect to occur in the assembly department?
d. Repeat part **c** for the maintenance department.

1.19   Beginning in 1991, the nation's Department of Education began taking corrective and punitive actions against colleges and universities with high student-loan default rates. Those schools with default rates above 60% face suspension from the government's massive student-loan program, whereas schools with default rates between 40% and 60% are mandated to reduce their default rates by 5% a year or face a similar penalty (*Tampa Tribune*, June 21, 1989). A list of 66 colleges and universities in Florida with their student-loan default rate is provided in the table.
    An SPSS printout giving descriptive statistics for the data set is shown on page 23.
a. Locate the mean default rate on the printout.
b. Locate the variance and standard deviation of the default rates on the printout.

COLLEGE/UNIVERSITY	DEFAULT RATE	COLLEGE/UNIVERSITY	DEFAULT RATE
Florida College of Business	76.2	Brevard CC	9.4
Ft. Lauderdale College	48.5	College of Boca Raton	9.1
Florida Career College	48.3	Florida International Univ.	8.7
United College	46.8	Santa Fe CC	8.6
Florida Memorial College	46.2	Edison CC	8.5
Bethune Cookman College	43.0	Palm Beach Junior College	8.0
Edward Waters College	38.3	Eckerd College	7.9
Florida College of Medical		University of Tampa	7.6
and Dental Careers	32.6	Lakeland College of	
International Fine Arts College	26.5	Business	7.2
Tampa College	23.9	Pensacola Junior College	6.8
Miami Technical College	23.3	University of Miami	6.7
Tallahassee CC	20.6	Florida Institute of Tech.	6.7
Charron Williams College	20.2	University of West Florida	6.3
Florida CC	19.1	Palm Beach Atlantic College	6.0
Miami-Dade CC	19.0	University of Central Florida	5.7
Broward CC	18.4	Seminole CC	5.6
Daytona Beach CC	16.9	Polk CC	5.6
Lake Sumter CC	16.7	Phillips Junior College	5.6
Florida Technical College	16.6	Nova University	5.5
Florida A&M University	15.8	Rollins College	5.5
Prospect Hall College	15.1	St. Leo College	5.5
Hillsborough CC	14.4	Gulf Coast CC	5.4
Pasco-Hernando CC	13.5	Southern College	5.3
Orlando College	13.5	Flagler College	4.7
Jones College	13.1	Florida Atlantic University	4.4
Webber College	11.8	University of South Florida	4.2
Warner Southern College	11.8	Manatee Junior College	4.1
Central Florida CC	11.8	Florida State University	4.0
Indian River CC	11.8	University of North Florida	3.9
St. Petersburg CC	11.3	Barry University	3.1
Valencia CC	10.8	University of Florida	3.1
Florida Southern College	10.3	Stetson University	2.9
Lake City CC	9.8	Jacksonville University	1.5

```
Number of Valid Observations (Listwise) =       66.00

Variable   DEFRATE

Mean           14.682          S.E. Mean         1.741
Std Dev        14.141          Variance        199.974
Kurtosis        5.427          S.E. Kurt          .582
Skewness        2.204          S.E. Skew          .295
Range          74.700          Minimum            1.50
Maximum        76.20           Sum             969.000

Valid Observations -      66    Missing Observations -        0
```

c. What proportion of measurements would you expect to find within two standard deviations of the mean?

d. Determine the proportion of measurements (default rates) that actually fall within the interval of part c. Compare this result with your answer to part c.

e. Suppose the college with the highest default rate (Florida College of Business—76.2%) was omitted from the analysis. Would you expect the mean to increase or decrease? Would you expect the standard deviation to increase or decrease?

f. Calculate the mean and standard deviation for the data set with Florida College of Business excluded. Compare these results with your answer to part **e**.

g. Answer parts **c** and **d** using the recalculated mean and standard deviation. This problem illustrates the dramatic effect a single observation can have on the analysis.

1.5

The Normal Probability Distribution

One of the most commonly used models for a theoretical population relative frequency distribution is the **normal probability distribution** shown in Figure 1.6. The normal distribution is symmetric about its mean μ, and its spread is determined by the value of its standard deviation σ. Three normal curves with different means and standard deviations are shown in Figure 1.7.

FIGURE 1.6

A normal probability distribution

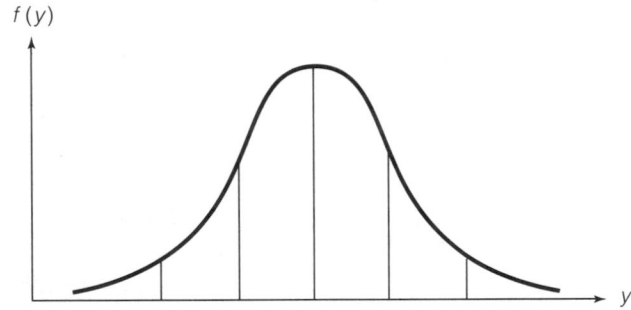

FIGURE 1.7

Several normal distributions with different means and standard deviations

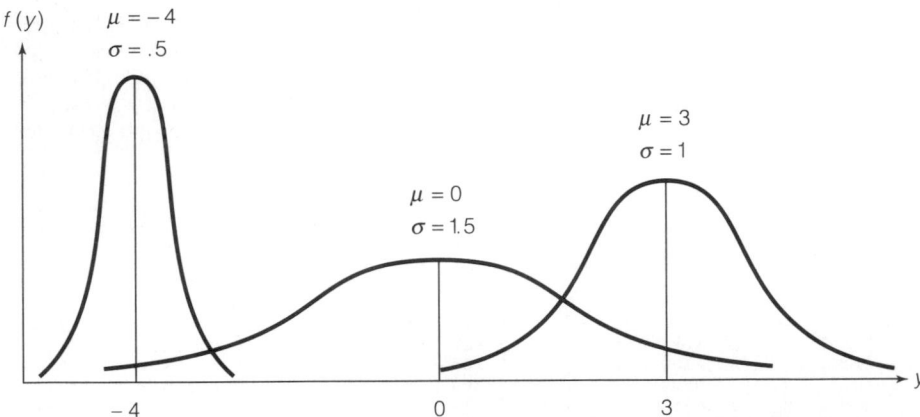

Computing the area over an interval under the normal probability distribution can be a difficult task.* Consequently, we will use the computed areas listed in

---

*Students with knowledge of calculus should note that the probability that y assumes a value in the interval $a < y < b$ is $P(a < y < b) = \int_a^b f(y)\, dy$, assuming the integral exists. The value of this definite integral can be obtained to any desired degree of accuracy by approximation procedures. For this reason, it is tabulated for the user.

Table 1 of Appendix D. A partial reproduction of this table is shown in Table 1.5. As you can see from the normal curve above the table, the entries give areas under the normal curve between the mean of the distribution and a standardized distance,

$$z = \frac{y - \mu}{\sigma}$$

to the right of the mean. Note that z is the number of standard deviations σ between μ and y. The distribution of z, which has mean $\mu = 0$ and standard deviation $\sigma = 1$, is called a **standard normal distribution**.

TABLE 1.5 Reproduction of Part of Table 1 of Appendix D

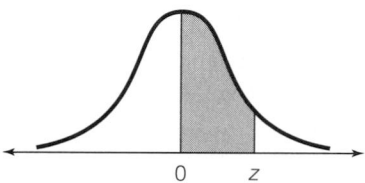

z	.00	.01	.02	.03	.04	.05	.06	.07	.08	.09
.0	.0000	.0040	.0080	.0120	.0160	.0199	.0239	.0279	.0319	.0359
.1	.0398	.0438	.0478	.0517	.0557	.0596	.0636	.0675	.0714	.0753
.2	.0793	.0832	.0871	.0910	.0948	.0987	.1026	.1064	.1103	.1141
.3	.1179	.1217	.1255	.1293	.1331	.1368	.1406	.1443	.1480	.1517
.4	.1554	.1591	.1628	.1664	.1700	.1736	.1772	.1808	.1844	.1879
.5	.1915	.1950	.1985	.2019	.2054	.2088	.2123	.2157	.2190	.2224
.6	.2257	.2291	.2324	.2357	.2389	.2422	.2454	.2486	.2517	.2549
.7	.2580	.2611	.2642	.2673	.2704	.2734	.2764	.2794	.2823	.2852
.8	.2881	.2910	.2939	.2967	.2995	.3023	.3051	.3078	.3106	.3133
.9	.3159	.3186	.3212	.3238	.3264	.3289	.3315	.3340	.3365	.3389
1.0	.3413	.3438	.3461	.3485	.3508	.3531	.3554	.3577	.3599	.3621
1.1	.3643	.3665	.3686	.3708	.3729	.3749	.3770	.3790	.3810	.3830
1.2	.3849	.3869	.3888	.3907	.3925	.3944	.3962	.3980	.3997	.4015
1.3	.4032	.4049	.4066	.4082	.4099	.4115	.4131	.4147	.4162	.4177
1.4	.4192	.4207	.4222	.4236	.4251	.4265	.4279	.4292	.4306	.4319
1.5	.4332	.4345	.4357	.4370	.4382	.4394	.4406	.4418	.4429	.4441

EXAMPLE 1.7

Suppose y is a normal random variable with $\mu = 50$ and $\sigma = 15$. Find $P(30 < y < 70)$, the probability that y will fall within the interval $30 < y < 70$.

Solution

Refer to Figure 1.8 on page 26. Note that $y = 30$ and $y = 70$ lie the same distance from the mean $\mu = 50$, with $y = 30$ below the mean and $y = 70$ above it. Then, because the normal curve is symmetric about the mean, the probability A_1 that y falls between $y = 30$ and $\mu = 50$ is equal to the probability A_2 that y falls between $\mu = 50$ and $y = 70$. The z score corresponding to $y = 70$ is

$$z = \frac{y - \mu}{\sigma} = \frac{70 - 50}{15} = 1.33$$

FIGURE 1.8

Normal probability distribution: $\mu = 50$, $\sigma = 15$

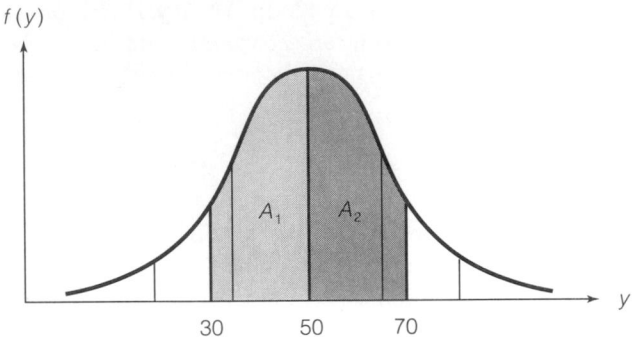

Therefore, the area between the mean $\mu = 50$ and the point $y = 70$ is given in Table 1 of Appendix D (and Table 1.5) at the intersection of the row corresponding to $z = 1.3$ and the column corresponding to .03. This area (probability) is $A_2 = .4082$. Since $A_1 = A_2$, A_1 also equals .4082, and it follows that the probability that y falls in the interval $30 < y < 70$ is $P(30 < y < 70) = 2(.4082) = .8164$. The z scores corresponding to $y = 30$ ($z = -1.33$) and $y = 70$ ($z = 1.33$) are shown in Figure 1.9.

FIGURE 1.9

A distribution of z scores (a standard normal distribution)

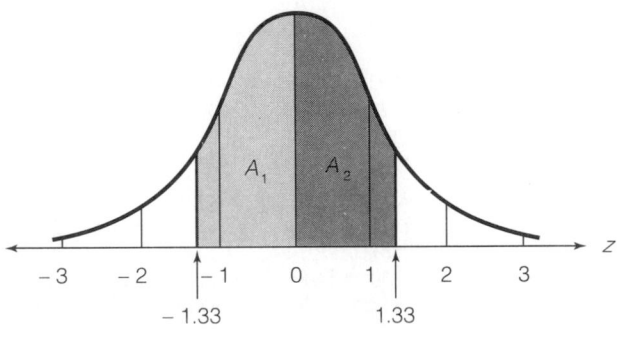

EXAMPLE 1.8

Use Table 1 of Appendix D to determine the area to the right of the z score 1.64 for the standard normal distribution. That is, find $P(z \geq 1.64)$.

Solution

The probability that a normal random variable will fall more than 1.64 standard deviations to the right of its mean is indicated in Figure 1.10. Because the normal distribution is symmetric, half of the total probability (.5) lies to the right of the mean and half to the left. Therefore, the desired probability is

$$P(z \geq 1.64) = .5 - A$$

where A is the area between $\mu = 0$ and $z = 1.64$, as shown in the figure. Referring to Table 1, we find that the area A corresponding to $z = 1.64$ is .4495. So

$$P(z \geq 1.64) = .5 - A = .5 - .4495 = .0505$$

FIGURE 1.10

Standard normal distribution:
$\mu = 0$, $\sigma = 1$

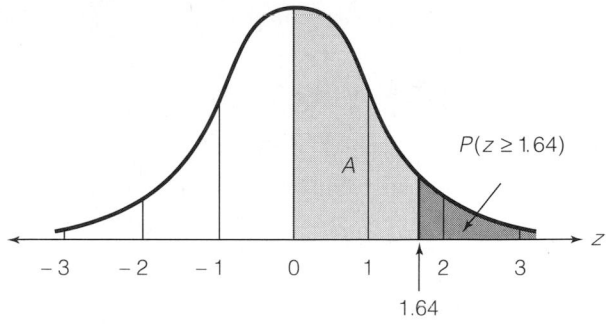

We will not be making extensive use of the table of areas under the normal curve, but you should know some of the common tabulated areas. In particular, you should note that the area between $z = -2.0$ and $z = 2.0$, which gives the probability that y falls in the interval $\mu - 2\sigma < y < \mu + 2\sigma$, is .9544 and agrees with guideline 2 of Section 1.4.

EXERCISES

1.20   Use Table 1 of Appendix D to calculate the area under the standard normal probability distribution between the following pairs of z values:
  a. $z = 0$ and $z = 2$ **b.** $z = 0$ and $z = 1.5$
  c. $z = 0$ and $z = 3$ **d.** $z = 0$ and $z = .5$
  e. $z = -1.5$ and $z = 1$ **f.** $z = -.5$ and $z = .75$
  g. $z = -2$ and $z = -1$ **h.** $z = -3$ and $z = 1.5$
  i. $z = .5$ and $z = 3.5$ **j.** $z = -.5$ and $z = 2$

1.21   Use Table 1 of Appendix D to find each of the following:
  a. $P(-1 \leq z \leq 1)$ **b.** $P(-1.96 \leq z \leq 1.96)$
  c. $P(-1.645 \leq z \leq 1.645)$ **d.** $P(-3 \leq z \leq 3)$

1.22   Find a value of z, call it z_0, such that
  a. $P(z \geq z_0) = .05$ **b.** $P(z \geq z_0) = .025$
  c. $P(z \leq z_0) = .025$ **d.** $P(z \geq z_0) = .0228$

1.23   Find a value of z, call it z_0, such that
  a. $P(z \leq z_0) = .0013$ **b.** $P(-z_0 \leq z \leq z_0) = .95$
  c. $P(-z_0 \leq z \leq z_0) = .90$ **d.** $P(-z_0 \leq z \leq z_0) = .6826$
  e. $P(-z_0 \leq z \leq 0) = .0596$

1.24   Given that the random variable y has a normal probability distribution with mean 100 and variance 64, draw a sketch (i.e., graph) of the frequency function of y. Locate μ and the interval $\mu \pm 2\sigma$ on the graph. Find the following probabilities:
  a. $P(\mu - 2\sigma \leq y \leq \mu + 2\sigma)$ **b.** $P(y \geq 108)$
  c. $P(y \leq 92)$ **d.** $P(92 \leq y \leq 116)$
  e. $P(92 \leq y \leq 96)$ **f.** $P(76 \leq y \leq 124)$

1.25   Use Table 1 of Appendix D to calculate the area under the standard normal distribution between the following pairs of z values:
    a. -1.96 and 1.96 **b.** -1.645 and 1.645 **c.** -3 and 3
    d. -3 and 2 **e.** 1 and 3 **f.** -1.5 and 3

1.26   How reliable are security analysts' forecasts of corporate earnings growth? According to David Dreman of *Forbes* magazine, "Astrology might be better" (*Forbes*, Mar. 26, 1984).* Dreman reported on a study of annual earnings estimates made by institutional brokerage analysts covering the more widely followed companies between 1977 and 1981. The study revealed that "the average annual error by the analysts was a staggering 31.3% over the 5-year period." Suppose the annual forecast error by security analysts is normally distributed with a mean of 31.3% and a standard deviation of 10%. Suppose you obtain a security analyst's forecast of annual earnings for a particular corporation.
    a. What is the probability that the forecast error will be between 20% and 25%?
    b. What is the probability that the forecast error will be greater than 50%?

1.27   Pacemakers are vital for controlling the heartbeat of cardiac patients, and over 120,000 of the devices are implanted each year. A pacemaker is made up of several biomedical components that must be of a high quality for the pacemaker to work. For manufacturers of pacemakers, it is vitally important to use parts that meet the manufacturer's specifications. One particular plastic part, called a connector module, mounts on the top of the pacemaker. Connector modules are required to have a length between .304 inch and .322 inch to work properly. Any module with length outside these limits are "out-of-spec." *Quality* (Aug. 1989) reported on one supplier of connector modules that had been shipping out-of-spec parts to the manufacturer for 12 months.
    a. The lengths of the connector modules produced by the supplier were found to follow an approximate normal distribution with mean $\mu = .3015$ inch and standard deviation $\sigma = .0016$ inch. Use this information to find the probability that the supplier produces an out-of-spec part.
    b. Once the problem was detected, the supplier's inspection crew began to employ an automated data-collection system designed to improve product quality. After two months, the process was producing connector modules with mean $\mu = .3146$ inch and standard deviation $\sigma = .0030$ inch. Find the probability that an out-of-spec part will be produced. Compare your answer to part **a**.

1.28   The U.S. Department of Agriculture (USDA) has recently patented a process that uses a bacterium for removing bitterness from citrus juices (*Chemical Engineering*, Feb. 3, 1986). In theory, almost all the bitterness could be removed by the process, but for practical purposes the USDA aims at 50% overall removal. Suppose a USDA spokesman claims that the percentage of bitterness removed from an 8-ounce glass of freshly squeezed citrus juice is normally distributed with mean 50.1% and standard deviation 10.4%. To test this claim, the bitterness removal process is applied to a randomly selected 8-ounce glass of citrus juice. Find the probability that the process removes less than 33.7%. Based on this probability, what can you infer about the spokesman's claim?

1.29   Value Line is an advisory service that provides investors with a forecast of the movement of the stock market. Each week Value Line selects one stock that it believes has the highest probability of moving upward in the coming months. In the early 1980s, Value Line's stock selection strategy was highly successful. However, one year its selections performed below par. According to an issue of Value Line's *Selection & Opinion*, which lists the performance of its stock selections to date, the mean percentage change in stock price since the date of recommendation of the 52 stocks was -19.9 and the standard deviation was 18.6. Assume that the percentage change in stock price for this period can be modeled by the normal probability distribution.

---

*Dreman, D. "Astrology might be better." *Forbes*, March 26, 1984, p. 242.

a. What is the probability that the percentage change of a stock recommended by Value Line during this period is greater than 0? (This is the probability that the stock will "gain" or move upward.)

b. What is the probability that the percentage change of a stock recommended by Value Line during this period is less than -50?

c. Determine the percentage change below which only 5% of the stocks recommended by Value Line fell during the period.

1.30   The inherent risk involved with marketing a new product is a key consideration for market researchers. Various statistical models have been developed to aid speculators and businesses in making such decisions. One area of marketing research that uses decision models is called *break-even analysis*. The underlying assumption of break-even analysis is that the demand for a product is normally distributed, with known mean and standard deviation. Of interest is the relationship between actual demand D and the break-even point BE, where BE is defined as the number of units of the product the company must sell in order to "break even" on the investment. Shih* (1981) developed two practical decision criteria using break-even analysis.

> *Decision Rule A:*   Market the new product if the chance is better than 50% that demand D will exceed the break-even point BE, i.e., if $P(D \geq BE) > .5$.
>
> *Decision Rule B:*   For a specified level of risk p, where $0 \leq p \leq 1$, market the new product if $P(D \geq BE) > p$.

Note that Decision Rule A is a special case of Decision Rule B, with specified level of risk $p = .5$. Suppose a company will utilize break-even analysis to decide whether to market a new type of ceiling fan. From past experience, the company knows that the number of ceiling fans of this type sold per year follows a normal distribution with $\mu = 4,000$ and $\sigma = 500$. Marketing researchers have also determined that the company needs to sell 3,500 units to break even for the year.

a. According to Decision Rule A, should the company market the new ceiling fans?

b. Use your knowledge of the normal distribution to show that if $P(D \geq BE) > .5$, then it must be true that $\mu \geq BE$.

c. Suppose that the minimum level of risk the company is willing to tolerate is $p = .8$. Use Decision Rule B to arrive at a decision.

d. Refer to part c. Find the value of BE such that the probability of at least breaking even is equal to the specified level of risk, $p = .8$.

1.6

Sampling Distributions and the Central Limit Theorem

Since we will use sample statistics to make inferences about population parameters, it is natural that we would want to know something about the reliability of the resulting inferences. For example, if we use a statistic to estimate the value of a population mean μ, we will want to know how close to μ our estimate is likely to fall. To answer this question, we need to know the probability distribution of the statistic.

The probability distribution for a statistic based on a random sample of n measurements could be generated in the following way. For purposes of illustration, we will suppose we are sampling from a population with $\mu = 10$ and $\sigma = 5$, the sample statistic is $\bar{y}$, and the sample size is $n = 25$. Draw a single

---

*Shih, W. "A general decision model for cost-volume-profit analysis under uncertainty: A reply." *The Accounting Review*, Vol. 56, No. 2, 1981, pp. 404–408.

random sample of 25 measurements from the population and suppose that $\bar{y} = 9.8$. Return the measurements to the population and try again. That is, draw another random sample of $n = 25$ measurements and see what you obtain for an outcome. Now, perhaps, $\bar{y} = 11.4$. Replace these measurements, draw another sample of $n = 25$ measurements, calculate $\bar{y}$, and so on. If this sampling process were repeated over and over again an infinitely large number of times, you would generate an infinitely large number of values of $\bar{y}$ that could be arranged in a relative frequency distribution. This distribution, which would appear as shown in Figure 1.11, is the probability distribution (or **sampling distribution**, as it is commonly called) of the statistic $\bar{y}$.

FIGURE 1.11

Sampling distribution for $\bar{y}$ based on a sample of $n = 25$ measurements

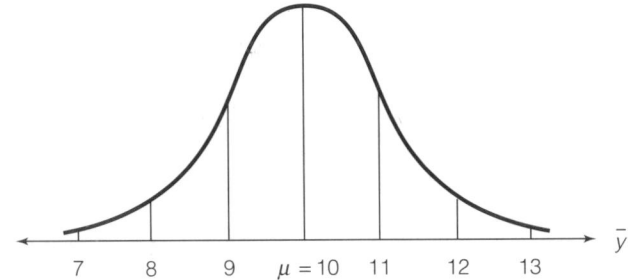

Definition 1.15

The **sampling distribution** of a sample statistic calculated from a sample of n measurements is the probability distribution of the statistic.

In actual practice, the sampling distribution of a statistic is obtained mathematically or by simulating the sampling on a computer using the procedure described previously.

If $\bar{y}$ has been calculated from a sample of $n = 25$ measurements selected from a population with mean $\mu = 10$ and standard deviation $\sigma = 5$, the sampling distribution shown in Figure 1.11 provides all the information you may wish to know about its behavior. For example, the probability that you will draw a sample of 25 measurements and obtain a value of $\bar{y}$ in the interval $9 \leq \bar{y} \leq 10$ will be the area under the sampling distribution over that interval.

Generally speaking, if we use a statistic to make an inference about a population parameter, we want its sampling distribution to center about the parameter (as is the case in Figure 1.11) and the standard deviation of the sampling distribution, called the **standard error of estimate**, to be as small as possible.

Two theorems provide information on the sampling distribution of a sample mean.

> **Theorem 1.1**
>
> If $y_1, y_2, \ldots, y_n$ represent a random sample of n measurements from a large (or infinite) population with mean μ and standard deviation σ, then, regardless of the form of the population relative frequency distribution, the mean and standard error of estimate of the sampling distribution of $\bar{y}$ will be
>
> *Mean:* $E(\bar{y}) = \mu$
>
> *Standard error of estimate:* $\sigma_{\bar{y}} = \dfrac{\sigma}{\sqrt{n}}$

> **Theorem 1.2   The Central Limit Theorem**
>
> For large sample sizes, the mean $\bar{y}$ of a sample from a population with mean μ and standard deviation σ has a sampling distribution that is approximately normal, **regardless of the probability distribution of the sampled population**. The larger the sample size, the better will be the normal approximation to the sampling distribution of $\bar{y}$.

Theorems 1.1 and 1.2 together imply that for sufficiently large samples, the sampling distribution for the sample mean $\bar{y}$ will be approximately normal with mean μ and standard error $\sigma_{\bar{y}} = \sigma/\sqrt{n}$. The parameters μ and σ are the mean and standard deviation of the sampled population.

To illustrate Theorems 1.1 and 1.2, we select 1,000 random samples of $n = 5$ measurements from a population with an **exponential relative frequency distribution** with $\mu = 1$ and $\sigma = 1$, as shown in Figure 1.12.

FIGURE 1.12

Exponential distribution with $\mu = 1$ and $\sigma = 1$

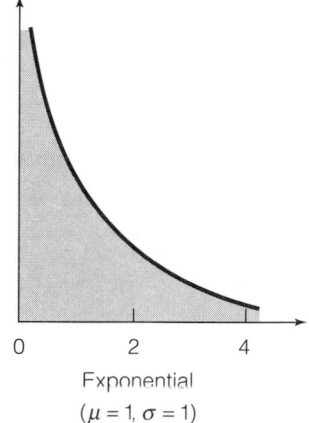

0   2   4

Exponential
($\mu = 1, \sigma = 1$)

The sample mean $\bar{y}$ was calculated for each of the 1,000 samples. The resulting relative frequency histogram (generated using SPSS) is shown in Figure 1.13a. You can see that the relative frequency distribution of 1,000 sample means (which approximates the sampling distribution of $\bar{y}$) is approximately normal, and that, in fact, the mean of the approximating normal distribution is

$$\mu = 1$$

The standard error is

$$\sigma_{\bar{y}} = \frac{\sigma}{\sqrt{n}} = \frac{1}{\sqrt{5}} = .45$$

Similarly, Figures 1.13b, c, and d show SPSS histograms for the sampling distribution of $\bar{y}$ based on samples of size $n = 15$, 25, and 50, respectively. Note that the normal approximation improves when the sample size increases from $n = 5$ to $n = 50$. Also, although the mean of the four sampling distributions shown

FIGURE 1.13

Sampling distributions of $\bar{y}$: Exponential population

a. $n = 5$

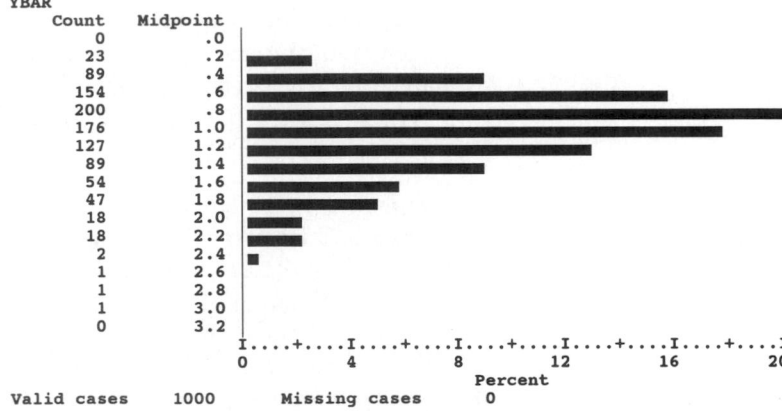

b. $n = 15$

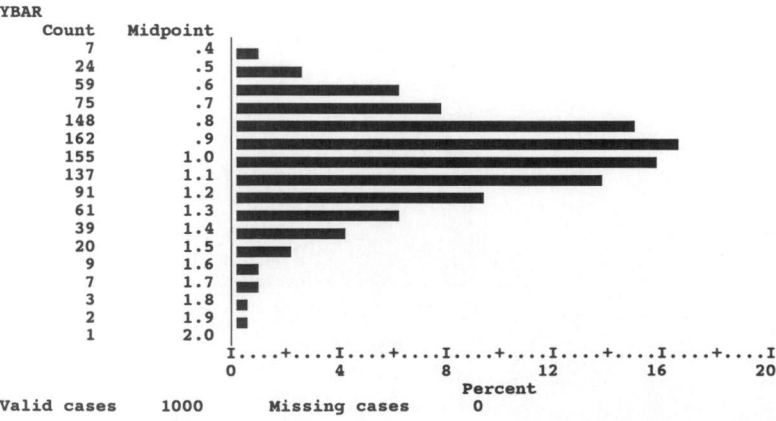

c. $n = 25$

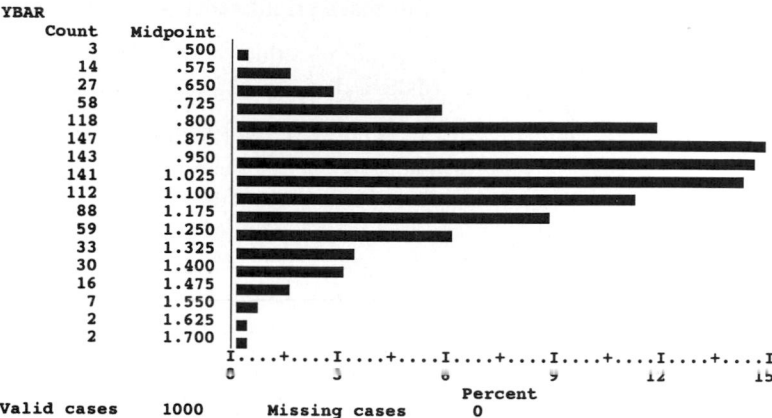

d. $n = 50$

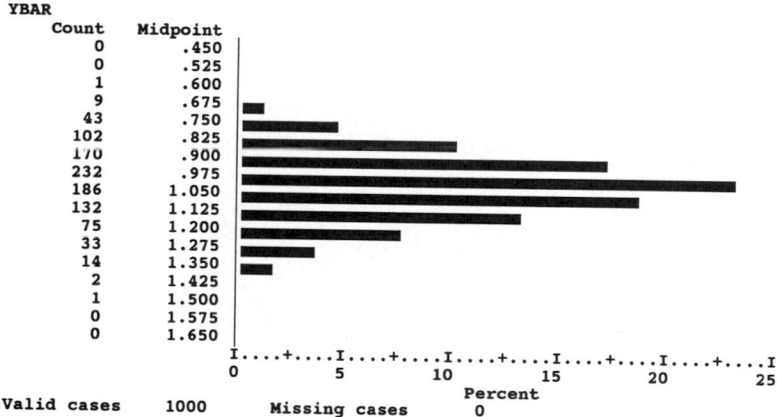

in Figure 1.13 is $\mu = 1$, the standard deviation $\sigma_{\bar{y}}$ decreases as n increases. For example, for $n = 50$,

$$\sigma_{\bar{y}} = \frac{\sigma}{\sqrt{n}} = \frac{1}{\sqrt{50}} = .14$$

compared to $\sigma_{\bar{y}} = .45$ for $n = 5$.

The central limit theorem can also be used to justify the fact that the *sum* of the sample measurements possesses a sampling distribution that is approximately normal for large sample sizes. In fact, since many statistics are obtained by summing or averaging random quantities, the central limit theorem helps to explain why many statistics have mound-shaped (or approximately normal) sampling distributions.

As we proceed, we will encounter many different sample statistics, and we will need to know their sampling distributions to evaluate the reliability of each one for making inferences. These sampling distributions will be described as the need arises.

1.7

Estimating a Population Mean

We can make an inference about a population parameter in two ways:

1. Estimate its value.
2. Make a decision about its value (i.e., test a hypothesis about its value).

In this section we will illustrate the concepts involved in estimation, using the estimation of a population mean as an example. Tests of hypotheses will be discussed in Section 1.8.

To estimate a population parameter, we choose a sample statistic that has two desirable properties: (1) a sampling distribution that centers about the parameter and (2) a small standard error. If the mean of the sampling distribution of a statistic equals the parameter we are estimating, we say that the statistic is an **unbiased estimator** of the parameter. If not, we say that it is **biased**.

In Section 1.6, we noted that the sampling distribution of the sample mean is approximately normally distributed for moderate to large sample sizes and that it possesses a mean μ and standard error $\sigma/\sqrt{n}$. Therefore, as shown in Figure 1.14, $\bar{y}$ is an unbiased estimator of the population mean μ, and the probability that $\bar{y}$ will fall within $1.96\sigma_{\bar{y}} = 1.96\sigma/\sqrt{n}$ of the true value of μ is approximately .95.*

FIGURE 1.14

Sampling distribution of $\bar{y}$

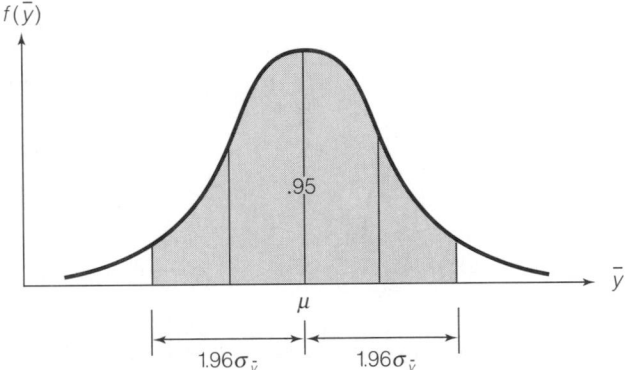

Since $\bar{y}$ will fall within $1.96\sigma_{\bar{y}}$ of μ approximately 95% of the time, it follows that the interval

$$\bar{y} - 1.96\sigma_{\bar{y}} \quad \text{to} \quad \bar{y} + 1.96\sigma_{\bar{y}}$$

will enclose μ approximately 95% of the time in repeated sampling. This interval is called a 95% **confidence interval**, and .95 is called the **confidence coefficient**. Notice that μ is fixed and that the confidence interval changes from sample to sample. The probability that a confidence interval calculated using the formula

$$\bar{y} \pm 1.96\sigma_{\bar{y}}$$

will enclose μ is approximately .95. Thus, the confidence coefficient measures the confidence that we can place in a particular confidence interval.

---

*Additionally, $\bar{y}$ has the smallest standard error among all unbiased estimators of μ. Consequently, we say that $\bar{y}$ is the **minimum variance unbiased estimator (MVUE)** for μ.

Confidence intervals can be constructed using any desired confidence coefficient. For example, if we define $z_{\alpha/2}$ to be the value of a standard normal variable that places the area $\alpha/2$ in the right-hand tail of the z distribution (see Figure 1.15), then a $100(1 - \alpha)\%$ confidence interval for μ is given in the box:

Large-Sample $100(1 - \alpha)\%$ Confidence Interval for μ

$$\bar{y} \pm z_{\alpha/2}\sigma_y$$

where $z_{\alpha/2}$ is the z value with an area $\alpha/2$ to its right (see Figure 1.15) and $\sigma_{\bar{y}} = \sigma/\sqrt{n}$. The parameter σ is the standard deviation of the sampled population and n is the sample size. If σ is unknown, its value may be approximated by the sample standard deviation s. The approximation is valid for large samples (e.g., $n \geq 30$) only.

FIGURE 1.15

Locating $z_{\alpha/2}$ on the standard normal curve

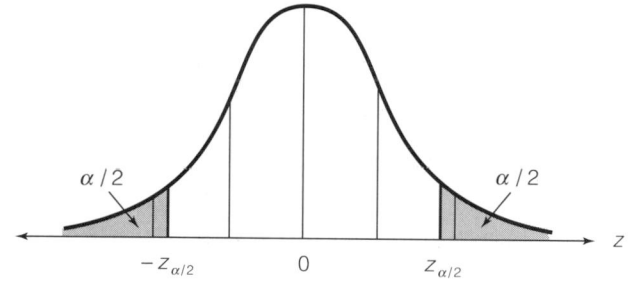

The confidence interval shown in the box is called a large-sample confidence interval because the sample size must be large enough to ensure approximate normality for the sampling distribution of $\bar{y}$. Also, and even more important, you will rarely, if ever, know the value of σ, so its value must be estimated using the sample standard deviation s. This approximation for σ will be adequate only when $n \geq 30$.

Typical confidence coefficients and corresponding values of $z_{\alpha/2}$ are shown in Table 1.6.

TABLE 1.6 Commonly Used Values of $z_{\alpha/2}$

CONFIDENCE COEFFICIENT $(1 - \alpha)$	α	$\alpha/2$	$z_{\alpha/2}$
.90	.10	.05	1.645
.95	.05	.025	1.96
.99	.01	.005	2.576

EXAMPLE 1.9

Unoccupied seats on flights cause airlines to lose revenue. Suppose a large airline wants to estimate the mean number of unoccupied seats per flight over the past year. The records of 225 flights are randomly selected from the files, and the number of unoccupied seats is noted for each of the sampled flights. The sample mean and standard deviation are

$$\bar{y} = 11.6 \text{ seats} \qquad s = 4.1 \text{ seats}$$

Estimate μ, the mean number of unoccupied seats per flight during the past year, using a 90% confidence interval.

Solution

The general form of the 90% confidence interval for a population mean is

$$\bar{y} \pm z_{\alpha/2}\sigma_{\bar{y}} = \bar{y} \pm z_{.05}\sigma_{\bar{y}}$$
$$= \bar{y} \pm 1.645\left(\frac{\sigma}{\sqrt{n}}\right)$$

For the 225 records sampled, we have

$$11.6 \pm 1.645\left(\frac{\sigma}{\sqrt{225}}\right)$$

Since we do not know the value of σ (the standard deviation of the number of unoccupied seats per flight for all flights during the year), we use our best approximation, the sample standard deviation s. (Since the sample size, $n = 225$, is large, the approximation is valid.) Then the 90% confidence interval is, approximately,

$$11.6 \pm 1.645\left(\frac{4.1}{\sqrt{225}}\right) = 11.6 \pm .45$$

or, from 11.15 to 12.05. That is, the airline can be 90% confident that the mean number of unoccupied seats per flight was between 11.15 and 12.05 during the year sampled.

The large-sample method for making inferences about a population mean μ assumes that either σ is known or the sample size is large enough ($n \geq 30$) for the sample standard deviation s to be used as a good approximation to σ. The technique for finding a $100(1 - \alpha)\%$ confidence interval for a population mean μ for small sample sizes requires that the sampled population have a normal probability distribution. The formula, which is similar to the one for a large-sample confidence interval for μ, is

$$\bar{y} \pm t_{\alpha/2}s_{\bar{y}}$$

where $s_{\bar{y}} = s/\sqrt{n}$ is the estimated standard error of $\bar{y}$. The quantity $t_{\alpha/2}$ is directly analogous to the standard normal value $z_{\alpha/2}$ used in finding a large-sample confidence interval for μ except that it is an upper-tail t value obtained from a Student's t distribution. Thus, $t_{\alpha/2}$ is an upper-tail t value such that an area $\alpha/2$ lies to its right.

Like the standardized normal (z) distribution, a Student's t distribution is symmetric about the value $t = 0$, but it is more variable than a z distribution. The variability depends on the number of **degrees of freedom, df**, which in turn depends on the number of measurements available for estimating σ^2. The smaller the number of degrees of freedom, the greater will be the spread of the t distribution. For this application of a Student's t distribution, df $= n - 1$. As the sample size increases (and df increases), the Student's t distribution looks more and more like a z distribution, and for $n \geq 30$, the two distributions will be nearly identical. A Student's t distribution based on df $= 4$ and a standard normal distribution are shown in Figure 1.16. Note the corresponding values of $z_{.025}$ and $t_{.025}$.

FIGURE 1.16

The $t_{.025}$ value in a t distribution with 4 df and the corresponding $z_{.025}$ value

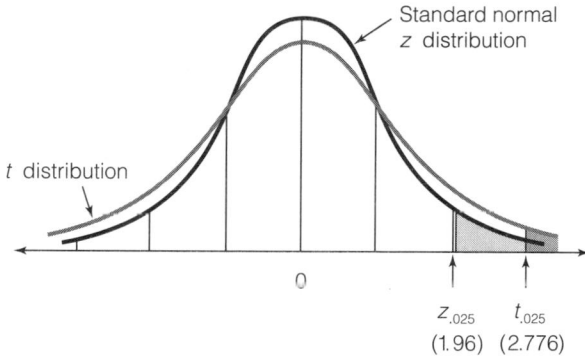

The upper-tail values of the Student's t distribution are given in Table 2 of Appendix D. An abbreviated version of the t table is presented in Table 1.7. To find the upper-tail t value based on 4 df that places .025 in the upper tail of the t distribution, we look in the row of the table corresponding to df = 4 and the column corresponding to $t_{.025}$. The t value is 2.776 and is shown in Figure 1.16.

TABLE 1.7 Reproduction of a Portion of Table 2 of Appendix D

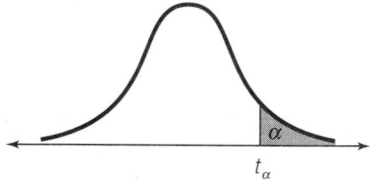

DEGREES OF FREEDOM	$t_{.100}$	$t_{.050}$	$t_{.025}$	$t_{.010}$	$t_{.005}$
1	3.078	6.314	12.706	31.821	63.657
2	1.886	2.920	4.303	6.965	9.925
3	1.638	2.353	3.182	4.541	5.841
4	1.533	2.132	2.776	3.747	4.604
5	1.476	2.015	2.571	3.365	4.032
6	1.440	1.943	2.447	3.143	3.707
7	1.415	1.895	2.365	2.998	3.499
8	1.397	1.860	2.306	2.896	3.355
9	1.383	1.833	2.262	2.821	3.250
10	1.372	1.812	2.228	2.764	3.169
11	1.363	1.796	2.201	2.718	3.106
12	1.356	1.782	2.179	2.681	3.055
13	1.350	1.771	2.160	2.650	3.012
14	1.345	1.761	2.145	2.624	2.977
15	1.341	1.753	2.131	2.602	2.947

The process of finding a small-sample confidence interval for μ is given in the next box.

Small-Sample Confidence Interval for μ

$$y \pm t_{\alpha/2}s_{\bar{y}}$$

where $s_{\bar{y}} = s/\sqrt{n}$ and $t_{\alpha/2}$ is a t value based on $(n-1)$ degrees of freedom, such that the probability that $t > t_{\alpha/2}$ is $\alpha/2$.

Assumption: The relative frequency distribution of the sampled population is approximately normal.

EXAMPLE 1.10

The Geothermal Loop Experimental Facility, located in the Salton Sea in southern California, is a U.S. Department of Energy operation for studying the feasibility of generating electricity from the hot, highly saline water of the Salton Sea. Operating experience has shown that these brines leave silica scale deposits on metallic plant piping, causing excessive plant outages. Jacobsen et al. (*Journal of Testing and Evaluation*, Vol. 9, No. 2, Mar. 1981, pp. 82–92) have found that scaling can be reduced somewhat by adding chemical solutions to the brine. In one screening experiment, each of five antiscalants was added to an aliquot of brine, and the solutions were filtered. A silica determination (parts per million of silicon dioxide) was made on each filtered sample after a holding time of 24 hours, with the following results:

229   255   280   203   229

Estimate the mean amount of silicon dioxide present in the five antiscalant solutions. Use a 95% confidence interval.

Solution

The first step in constructing the confidence interval is to compute the mean, $\bar{y}$, and standard deviation, s, of the sample of five silicon dioxide amounts. These values, $\bar{y} = 239.2$ and $s = 29.3$, are provided in the Minitab printout, Figure 1.17.

FIGURE 1.17

Minitab descriptive statistics for Example 1.10

	N	MEAN	MEDIAN	TRMEAN	STDEV	SEMEAN
ppm	5	239.2	229.0	239.2	29.3	13.1

	MIN	MAX	Q1	Q3
ppm	203.0	280.0	216.0	267.5

For a confidence coefficient of $1 - \alpha = .95$, we have $\alpha = .05$ and $\alpha/2 = .025$. Since the sample size is small ($n = 5$), our estimation technique requires the assumption that the amount of silicon dioxide present in an antiscalant solution has an approximately normal distribution (i.e., the sample of five silicon amounts is selected from a normal population).

Substituting the values for $\bar{y}$, s, and n into the formula for a small-sample confidence interval for μ, we obtain

$$\bar{y} \pm t_{\alpha/2}(s_{\bar{y}}) = \bar{y} \pm t_{.025}\left(\frac{s}{\sqrt{n}}\right)$$

$$= 239.2 \pm t_{.025}\left(\frac{29.3}{\sqrt{5}}\right)$$

where $t_{.025}$ is the value corresponding to an upper-tail area of .025 in the Student's t distribution based on $(n - 1) = 4$ degrees of freedom. From Table 2 of Appendix D, the required t value is $t_{.025} = 2.776$. Substituting this value yields

$$239.2 \pm t_{.025}\left(\frac{29.3}{\sqrt{5}}\right) = 239.2 \pm (2.776)\left(\frac{29.3}{\sqrt{5}}\right)$$

$$= 239.2 \pm 36.4$$

or, 202.8 to 275.6 ppm.

Thus, if the distribution of silicon dioxide amounts is approximately normal, then we can be 95% confident that the interval (202.8, 275.6) encloses μ, the true mean amount of silicon dioxide present in an antiscalant solution. Remember, the 95% confidence level implies that if we were to employ our interval estimator on repeated occasions, 95% of the intervals constructed would capture μ.

EXAMPLE 1.11

Suppose you want to reduce the width of the confidence interval obtained in Example 1.10. Specifically, you want to estimate the mean silicon dioxide content of an aliquot of brine correct to within 10 ppm with confidence coefficient approximately equal to .95. How many aliquots of brine would you have to include in your sample?

Solution

We will interpret the phrase, "correct to within 10 ppm . . . equal to .95" to mean that we want half the width of a 95% confidence interval for μ to equal 10 ppm. That is, we want

$$t_{.025}\left(\frac{s}{\sqrt{n}}\right) = 10$$

To solve this equation for n, we need approximate values for $t_{.025}$ and s. Since we know from Example 1.10 that the confidence interval was wider than desired for $n = 5$, it is clear that our sample size must be larger than 5. Consequently, $t_{.025}$ will be very close to 2 and this value will provide a good approximation to $t_{.025}$. A good measure of the data variation is given by the standard deviation computed in Example 1.10. We substitute $t_{.025} \approx 2$ and $s \approx 29.3$ into the equation and solve for n:

$$t_{.025}\left(\frac{s}{\sqrt{n}}\right) = 10$$

$$2\left(\frac{29.3}{\sqrt{n}}\right) = 10$$

$$\sqrt{n} = 5.86$$

$$n = 34.3 \quad \text{or approximately} \quad n = 34$$

Remember that this sample size is an approximate solution because we approximated the value of $t_{.025}$ and the value of s that might be computed from the prospective data. Nevertheless, $n = 34$ will be reasonably close to the sample size needed to estimate the mean silicon dioxide content correct to within 10 ppm.

EXERCISES

1.31   The table contains 50 random samples of random digits, $y = 0, 1, 2, 3, \ldots, 9$, where the probabilities corresponding to the values of y are given by the formula $p(y) = \frac{1}{10}$. Each sample contains $n = 6$ measurements.

SAMPLE	SAMPLE	SAMPLE	SAMPLE
8, 1, 8, 0, 6, 6	7, 6, 7, 0, 4, 3	4, 4, 5, 2, 6, 6	0, 8, 4, 7, 6, 9
7, 2, 1, 7, 2, 9	1, 0, 5, 9, 9, 6	2, 9, 3, 7, 1, 3	5, 6, 9, 4, 4, 2
7, 4, 5, 7, 7, 1	2, 4, 4, 7, 5, 6	5, 1, 9, 6, 9, 2	4, 2, 3, 7, 6, 3
8, 3, 6, 1, 8, 1	4, 6, 6, 5, 5, 6	8, 5, 1, 2, 3, 4	1, 2, 0, 6, 3, 3
0, 9, 8, 6, 2, 9	1, 5, 0, 6, 6, 5	2, 4, 5, 3, 4, 8	1, 1, 9, 0, 3, 2
0, 6, 8, 8, 3, 5	3, 3, 0, 4, 9, 6	1, 5, 6, 7, 8, 2	7, 8, 9, 2, 7, 0
7, 9, 5, 7, 7, 9	9, 3, 0, 7, 4, 1	3, 3, 8, 6, 0, 1	1, 1, 5, 0, 5, 1
7, 7, 6, 4, 4, 7	5, 3, 6, 4, 2, 0	3, 1, 4, 4, 9, 0	7, 7, 8, 7, 7, 6
1, 6, 5, 6, 4, 2	7, 1, 5, 0, 5, 8	9, 7, 7, 9, 8, 1	4, 9, 3, 7, 3, 9
9, 8, 6, 8, 6, 0	4, 4, 6, 2, 6, 2	6, 9, 2, 9, 8, 7	5, 5, 1, 1, 4, 0
3, 1, 6, 0, 0, 9	3, 1, 8, 8, 2, 1	6, 6, 8, 9, 6, 0	4, 2, 5, 7, 7, 9
0, 6, 8, 5, 2, 8	8, 9, 0, 6, 1, 7	3, 3, 4, 6, 7, 0	8, 3, 0, 6, 9, 7
8, 2, 4, 9, 4, 6	1, 3, 7, 3, 4, 3		

a. Use the 300 random digits to construct a relative frequency distribution for the data. This relative frequency distribution should approximate $p(y)$.
b. Calculate the mean of the 300 digits. This will give an accurate estimate of μ (the mean of the population) and should be very near to $E(y)$, which is 4.5.
c. Calculate s^2 for the 300 digits. This should be close to the variance of y, $\sigma^2 = 8.25$.
d. Calculate $\bar{y}$ for each of the 50 samples. Construct a relative frequency distribution for the sample means to see how close they lie to the mean of $\mu = 4.5$. Calculate the mean and standard deviation of the 50 means.

1.32 Refer to Exercise 1.31. To see the effect of sample size on the standard deviation of the sampling distribution of a statistic, combine pairs of samples (moving down the columns of the table) to obtain 25 samples of $n = 12$ measurements. Calculate the mean for each sample.

    a. Construct a relative frequency distribution for the 25 means. Compare this with the distribution prepared for Exercise 1.31 that is based on samples of $n = 6$ digits.

    b. Calculate the mean and standard deviation of the 25 means. Compare the standard deviation of this sampling distribution with the standard deviation of the sampling distribution in Exercise 1.31. What relationship would you expect to exist between the two standard deviations?

1.33 As part of a study to determine the relationship between the length of time patients wait in the physician's office and certain demand and cost factors, Sloan and Lorant* (1977) obtained data on the typical patient waiting times for 4,500 physicians in the five largest specialties—general practice, general surgery, internal medicine, obstetrics/gynecology, and pediatrics. They reported a mean waiting time of 24.7 minutes and a standard deviation of 19.3 minutes. One important aspect in the study of waiting times is the effect of policy changes on the waiting-time distribution. Suppose that 4,500 typical waiting times were collected over a period of time at a health clinic and that the clinic wants to change its operating procedure to increase the number of patients treated by a physician in a given period of time. To determine what effect the change in operating procedure will have on the mean patient waiting time, the clinic places the new procedure in operation and then collects a sample of $n = 100$ waiting times. The sample yielded the following results:

$$\bar{y} = 19.8 \text{ minutes} \qquad s = 20.6 \text{ minutes}$$

    a. Assuming that the distribution of patient waiting times has not changed (i.e., $\mu = 24.7$ and $\sigma = 19.3$), describe the sampling distribution of $\bar{y}$, the mean of the sample of 100 patient waiting times.

    b. Find $P(\bar{y} \leq 19.8)$.

    c. Based on the probability in part **b**, what can you infer about the waiting-time distribution under the new operating procedure? Explain.

1.34 By definition, an *entrepreneur* is "one who undertakes to start and conduct an enterprise or business, assuming full control and risks" (Funk and Wagnall's *Standard Dictionary*). Thus, a distinguishing characteristic of entrepreneurs is their propensity for taking risks. R. H. Brockhaus (1980)[†] used a choice dilemma questionnaire (CDQ) to measure the risk-taking propensities of successful entrepreneurs. He found that entrepreneurs had a mean CDQ score of 71 and a standard deviation of 12. (Lower scores are associated with a greater propensity for taking risks.) In a random sample of $n = 50$ entrepreneurs, let $\bar{y}$ be the sample mean CDQ score.

    a. Describe the sampling distribution of $\bar{y}$.

    b. Find $P(69 \leq \bar{y} \leq 72)$.

    c. Find $P(\bar{y} \leq 67)$.

    d. Would you expect to observe a sample mean CDQ score of 67 or lower? Explain.

1.35 Explain what is meant by the following statement: "We are 95% confident that our interval estimate contains μ."

---

*Sloan, F. A. and Lorant, J. H. "The role of patient waiting time: Evidence from physicians' practices." *Journal of Business*, Oct. 1977, Vol. 50, pp. 486–507.

[†]Brockhaus, R. H. "Risk-taking propensity of entrepreneurs." *Academy of Management Journal*, Sept. 1980, Vol. 23, No. 3, pp. 509–520.

1.36   When a university professor attempts to publish a research article in a professional journal, the manuscript goes through a rigorous review process. Usually, anywhere from three to five reviewers read and critique the article, then pass judgment on whether the article should be published. Recently, a study was undertaken to seek information on how reviewers for research journals pursue their activities (*Academy of Management Journal*, Mar. 1989). A sample of 73 reviewers for the Academy of Management's *Journal* (*AMJ*) and *Review* (*AMR*) were asked how many hours they spent per paper to complete a typical review. The sample mean and standard deviation were $\bar{y} = 5.4$ hours and $s = 3.6$ hours.

    a. Find a point estimate for μ, the true mean number of hours spent by a reviewer in conducting a complete review of a paper submitted to *AMJ* or *AMR*.

    b. Compute a 99% confidence interval for μ.

    c. Interpret the interval, part b.

1.37   Adult students are enrolling in colleges and universities in ever-increasing numbers, and many are majoring in marketing. Recently, a study was conducted to determine the attitudes of marketing faculty toward the adult students in their classes (*Journal of Marketing Education*, Summer 1987). A sample of 290 faculty, drawn at random from the American Marketing Association's membership directory, responded to a series of attitudinal statements, the first of which was, "Adult students (i.e., undergraduates 24 years or older) participate more actively in classroom discussions than do younger students." Attitudes were measured using a 5-point Likert scale (1 = strongly agree, 2 = agree, 3 = no opinion, 4 = disagree, and 5 = strongly disagree). For the participation statement, the mean attitudinal score for the sample was 1.94 and the standard deviation was .92.

    a. Estimate the true mean attitudinal score of marketing faculty with regard to classroom participation of adult students using a 98% confidence interval. Interpret the result.

    b. How could you reduce the width of the confidence interval in part a?

1.38   Because external audits can become quite expensive, many companies are creating or augmenting internal audit departments to lower auditing costs. Wanda A. Wallace,* CPA and professor at Southern Methodist University, conducted a study of the audit departments of 32 diverse companies. Her main interest was to determine the effect that internal audit departments had on external audit fees. The mean external audit fee paid by the 32 companies in the study period was $779,030, and the standard deviation was $1,083,162. Calculate a 95% confidence interval for the mean external audit fee paid by all companies in the study period.

1.39   In what ways are the distributions of the z statistic and t statistic alike? How do they differ?

1.40   Let t_0 be a particular value of t. Use Table 2 of Appendix D to find t_0 values such that the following statements are true:

    a. $P(t \geq t_0) = .025$ where $n = 10$ b. $P(t \geq t_0) = .01$ where $n = 5$

    c. $P(t \leq t_0) = .005$ where $n = 20$ d. $P(t \leq t_0) = .05$ where $n = 12$

1.41   Each year, thousands of manufacturer sales promotions are conducted by North American package goods companies, yet promotion managers are frequently dissatisfied with their results. An exploratory study was conducted by K. G. Hardy† to examine the objectives and impact of such sales promotions.

---

*Wallace, W. A. "Internal auditors can cut outside CPA costs." *Harvard Business Review*, Mar.–Apr. 1984, Vol. 62, No. 2, pp. 16–20.

†Hardy, K. G. "Key success factors for manufacturer's sales promotions in package goods." *Journal of Marketing*, Vol. 50, No. 7, July 1986, p. 16.

A sample of Canadian package goods companies provided information on examples of past sales promotions, including trade promotions. For the 21 "successful" trade promotions (where "success" is determined by the company managers) identified in the sample, the mean incremental profit was $53,000 and the standard deviation was $95,000.

a. Find a 90% confidence interval for the true mean incremental profit of "successful" trade promotions.

b. State the assumptions required for this confidence interval to be valid.

c. Give two ways in which you could reduce the width of the interval in part a. Which would you recommend?

1.42   Many North American cities have built or are considering building light rail transit (LRT) systems as an alternative to the heavy rail transit systems that use large passenger trains and subways. LRT systems are similar to the early 1900s streetcar, but are typically longer, quieter, faster, and more comfortable. In one study, the characteristics of LRT operations were examined in 10 cities that have built or are planning to build LRT systems (*Journal of the American Planning Association*, Spring 1984). One characteristic of importance to urban planners is the farebox recovery rate, computed by dividing passenger revenues by operating costs. The sample of 10 cities had a mean farebox recovery rate of .604 and a standard deviation of .163.

a. Estimate the true mean farebox recovery rate for LRT systems in North American cities, using a 95% confidence interval.

b. What assumption is required for the confidence interval procedure of part a to be valid?

1.43   During the budget preparation process at a hospital, forecasts of inpatient utilization (measured in patient days) for the coming fiscal year play a pivotal role. Planners at Sisters of St. Joseph of Peace Health and Hospital Services, Bellevue, Washington, have developed a budget early warning technique (BEWT) for forecasting future fiscal year utilizations.* The method involves collecting past data on monthly utilization and calculating, for each month, the ratio of monthly utilization to the utilization for the entire fiscal year preceding the month. Confidence intervals established on this utilization ratio enable planners to establish budget forecasts. At one of the smaller (50 beds) hospitals, the April utilization ratios for each of the past 10 years were calculated and found to have a sample mean of .0817 and a sample standard deviation of .0069. Establish a 99% confidence interval for μ, the true mean April utilization ratio. Interpret the interval.

1.8

Testing a Hypothesis About a Population Mean

The procedure involved in testing a hypothesis about a population parameter can be illustrated with the procedure for a test concerning a population mean μ.

A statistical test of a hypothesis is composed of four parts:

1. A **null hypothesis**, denoted by the symbol H_0, which is the hypothesis that we postulate is true

2. An **alternative** (or **research**) **hypothesis**, denoted by the symbol H_a, which is counter to the null hypothesis and is what we want to support

3. A **test statistic**, calculated from the sample data, that functions as a decision maker

---

*MacStravic, R. S. "An early warning technique." *Hospital & Health Services Administration*, Jan./Feb. 1986, pp. 86–98.

4. A **rejection region**—if the value of a test statistic, calculated from a particular sample, falls in the rejection region, we reject the null hypothesis and accept the alternative hypothesis.

The test statistic for testing the null hypothesis that a population mean μ equals some specific value, say μ_0, is the sample mean $\bar{y}$ or the standardized normal variable

$$z = \frac{\bar{y} - \mu_0}{\sigma_{\bar{y}}} \qquad \text{where} \quad \sigma_{\bar{y}} = \frac{\sigma}{\sqrt{n}}$$

The logic used to decide whether sample data *disagree* with this hypothesis can be seen in the sampling distribution of $\bar{y}$ shown in Figure 1.18. If the population mean μ is equal to μ_0 (i.e., if the null hypothesis is true), then the mean $\bar{y}$ calculated from a sample should fall, with high probability, within $2\sigma_{\bar{y}}$ of μ_0. If $\bar{y}$ falls too far away from μ_0, or if the standardized distance

$$z = \frac{\bar{y} - \mu_0}{\sigma_{\bar{y}}}$$

is too large, we conclude that the data disagree with our hypothesis, and we reject the null hypothesis.

FIGURE 1.18

The sampling distribution of $\bar{y}$ for $\mu = \mu_0$

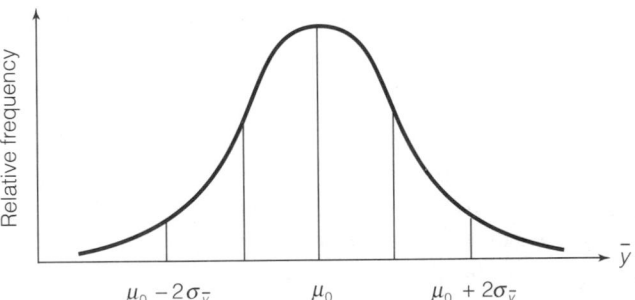

If we want to detect the alternative hypothesis that $\mu > \mu_0$, we locate the boundary of the rejection region in the upper tail of the z distribution, as shown in Figure 1.19a, at the point z_α. Similarly, to detect $\mu < \mu_0$, we place the rejection region in the lower tail of the z distribution, as shown in Figure 1.19b. These are called **one-tailed statistical tests**. To detect either $\mu > \mu_0$ or $\mu < \mu_0$—that is, $\mu \neq \mu_0$—we split α equally between the two tails of the z distribution and reject the null hypothesis if $z > z_{\alpha/2}$ or $z < -z_{\alpha/2}$, as shown in Figure 1.19c. This is called a **two-tailed statistical test**.

The z test, summarized in the next box, is called a *large-sample test* because we will rarely know σ and hence will need a sample size that is large enough so that the sample standard deviation s will provide a good approximation to σ. Normally, we recommend that the sample size be $n \geq 30$.

FIGURE 1.19 Location of the rejection region for various alternative hypotheses

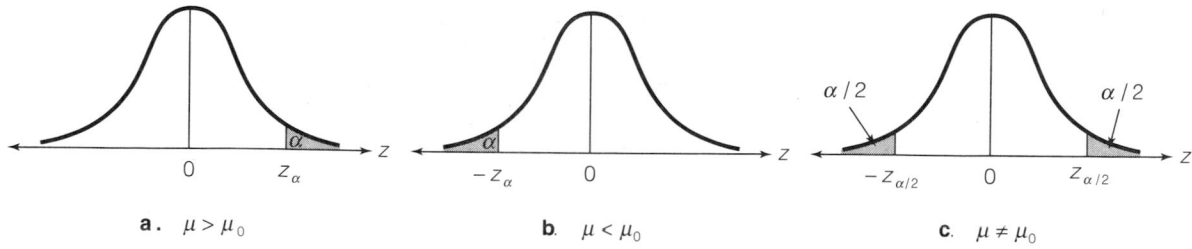

a. $\mu > \mu_0$ b. $\mu < \mu_0$ c. $\mu \neq \mu_0$

Large-Sample ($n > 30$) Test of Hypothesis About μ

ONE-TAILED TEST

H_0: $\mu = \mu_0$

H_a: $\mu < \mu_0$
  (or H_a: $\mu > \mu_0$)

Test statistic: $z = \dfrac{\bar{y} - \mu_0}{\sigma_{\bar{y}}}$

Rejection region: $z < -z_\alpha$
  (or $z > z_\alpha$)

where z_α is chosen so that
$P(z > z_\alpha) = \alpha$

TWO-TAILED TEST

H_0: $\mu = \mu_0$

H_a: $\mu \neq \mu_0$

Test statistic: $z = \dfrac{\bar{y} - \mu_0}{\sigma_{\bar{y}}}$

Rejection region: $z < -z_{\alpha/2}$
  or $z > z_{\alpha/2}$

where $z_{\alpha/2}$ is chosen so that
$P(z > z_{\alpha/2}) = \alpha/2$

We illustrate with an example.

EXAMPLE 1.12

Building specifications in a certain city require that the average breaking strength of residential sewer pipe be more than 2,400 pounds per foot of length (i.e., per lineal foot). A sampling of the strengths of 70 sections of pipe produced by a manufacturer yielded a sample mean and standard deviation of

$$\bar{y} = 2,430 \qquad s = 190$$

Do these statistics, calculated from the sample data, present sufficient evidence to indicate that the manufacturer's pipe meets the city's specifications? Use $\alpha = .05$.

Solution

Since we wish to determine whether $\mu > 2,400$, the elements of the test are

H_0: $\mu = 2,400$ H_a: $\mu > 2,400$

Test statistic: $z = \dfrac{\bar{y} - 2,400}{\sigma_{\bar{y}}} = \dfrac{\bar{y} - 2,400}{\sigma/\sqrt{n}} \approx \dfrac{\bar{y} - 2,400}{s/\sqrt{n}}$

Rejection region: $z > 1.645$ for $\alpha = .05$

Substituting the sample statistics into the test statistic, we have

$$z \approx \frac{\bar{y} - 2{,}400}{s/\sqrt{n}} = \frac{2{,}430 - 2{,}400}{190/\sqrt{70}} = 1.32$$

The sample mean of 2,430, although greater than 2,400, is only $1.32\sigma_{\bar{y}}$ above that value. Therefore, the sample does not provide sufficient evidence to conclude that the sewer pipe meets the city's strength specifications.

The reliability of a statistical test is measured by the probability of making an incorrect decision. For example, the probability of rejecting the null hypothesis and accepting the alternative hypothesis when the null hypothesis is true (called a **Type I error**) is α, the tail probability used in locating the rejection region. A second type of error could be made if we accepted the null hypothesis when, in fact, the alternative hypothesis is true (a **Type II error**). Thus, you never "accept" the null hypothesis unless you know the probability of making a Type II error. Since this probability (denoted by the symbol β) is often unknown, it is a common practice to defer judgment if a test statistic falls in the nonrejection region.

A small-sample test of the null hypothesis $\mu = \mu_0$ using a Student's t statistic is based on the assumption that the sample was randomly selected from a population with a normal relative frequency distribution. The test is conducted in exactly the same manner as the large-sample z test except that we use

$$t = \frac{\bar{y} - \mu_0}{s_{\bar{y}}} = \frac{\bar{y} - \mu_0}{s/\sqrt{n}}$$

as the test statistic and we locate the rejection region in the tail(s) of a Student's t distribution with df $= n - 1$. We summarize the technique for conducting a small-sample test of hypothesis about a population mean in the box.

Small-Sample Test of Hypothesis About μ

ONE-TAILED TEST

$H_0: \quad \mu = \mu_0$

$H_a: \quad \mu < \mu_0$
  (or $\mu > \mu_0$)

Test statistic: $\quad t = \dfrac{\bar{y} - \mu_0}{s/\sqrt{n}}$

Rejection region: $\quad t < -t_\alpha$
  (or $t > t_\alpha$)

where t_α is based on $(n - 1)$ df

TWO-TAILED TEST

$H_0: \quad \mu = \mu_0$

$H_a: \quad \mu \neq \mu_0$

Test statistic: $\quad t = \dfrac{\bar{y} - \mu_0}{s/\sqrt{n}}$

Rejection region: $\quad t < -t_{\alpha/2}$
  or $t > t_{\alpha/2}$

where $t_{\alpha/2}$ is based on $(n - 1)$ df

Assumption: The population from which the sample is drawn is approximately normal.

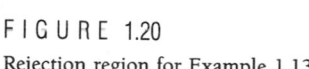EXAMPLE 1.13

Scientists have labeled benzene, a chemical solvent commonly used to synthesize plastics, as a possible cancer-causing agent. Studies have shown that people who work with benzene more than 5 years have 20 times the incidence of leukemia than the general population. As a result, the federal government has lowered the maximum allowable level of benzene in the workplace from 10 parts per million (ppm) to 1 ppm (reported in *Florida Times-Union*, Apr. 2, 1984). Suppose a steel manufacturing plant, which exposes its workers to benzene daily, is under investigation by the National Institute for Occupational Safety and Health (NIOSH). Twenty air samples, collected over a period of 1 month and examined for benzene content, yielded the following summary statistics:

$$\bar{y} = 2.1 \text{ ppm} \qquad s = 1.7 \text{ ppm}$$

Is the steel manufacturing plant in violation of the new government standards? Test the hypothesis that the mean level of benzene at the steel manufacturing plant is greater than 1 ppm, using $\alpha = .05$.

Solution

NIOSH wants to establish the research hypothesis that the mean level of benzene, μ, at the steel manufacturing plant exceeds 1 ppm. The elements of this small sample one-tailed test are

H_0: $\mu = 1$

H_a: $\mu > 1$

Test statistic: $t = \dfrac{\bar{y} - \mu_0}{s/\sqrt{n}}$

Assumption: The relative frequency distribution of the population of benzene levels for all air samples at the steel manufacturing plant is approximately normal.

Rejection region: For $\alpha = .05$ and df $= (n - 1) = 19$, reject H_0 if

$$t > t_{.05} = 1.729 \quad \text{(see Figure 1.20)}$$

FIGURE 1.20

Rejection region for Example 1.13

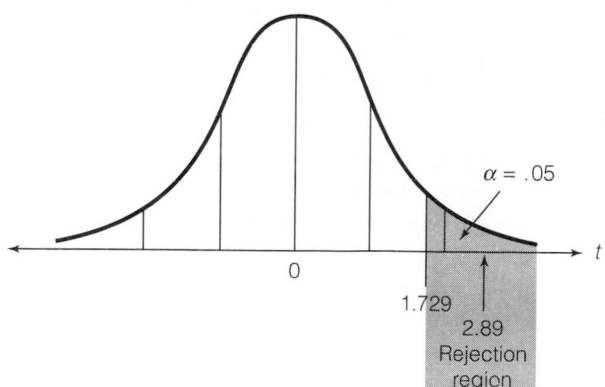

$\alpha = .05$

1.729

2.89
Rejection
region

We now calculate the test statistic:

$$t = \frac{\bar{y} - 1}{s/\sqrt{n}} = \frac{2.1 - 1}{1.7/\sqrt{20}} = 2.89$$

Since the calculated t falls in the rejection region, NIOSH conculdes that $\mu > 1$ ppm and the plant is in violation of the new government standards. The reliability associated with this inference is $\alpha = .05$. This implies that if the testing procedure was applied repeatedly to random samples of data collected at the plant, NIOSH would falsely reject H_0 for only 5% of the tests. Consequently, the OHSC is highly confident (95% confident) that the plant is violating the new standards.

To conclude, we mention a second method that is useful in reporting the results of a statistical test. Some data analyzers indicate the degree to which the test statistic contradicts the null hypothesis (and hence supports the alternative hypothesis). This quantity, called the **observed significance level**, or **p-value**, of the test, is the probability of observing a value of the test statistic at least as contradictory to the null hypothesis as the observed value of the test statistic, *assuming the null hypothesis is true.* For example, suppose you conducted a large-sample z test to detect the alternative hypothesis $\mu > 100$, and the computed value of the test statistic was 2.12. The level of significance for this one-tailed test is the probability of observing a z value larger than 2.12 if $\mu = 100$, or

$$P(z > 2.12) = .0170$$

If the test were two-tailed (i.e., the alternative hypothesis were $\mu \neq 100$), then values more contradictory to the null hypothesis $\mu = 100$ would be z values greater than 2.12 or less than -2.12. The level of significance for this test would be $2(.0170) = .0340$.

Decisions about H_0 and H_a can be made by comparing the p-value of the test to your desired value of α, as shown in the box.

| Reporting Test Results as p-Values: How to Decide Whether to Reject H_0

1. Choose the maximum value of α that you are willing to tolerate.
2. If the observed significance level (p-value) of the test is less than the maximum value of α, then reject the null hypothesis.

Most statistical software packages automatically compute the exact p-value of a test. For example, the SAS printout for the t test of Example 1.13 is shown in Figure 1.21. The p-value for a two-tailed test is shaded under the heading **Prob>** **|T|** in Figure 1.21. The p-value for a one-tailed test is equal to the reported value divided by 2. Thus, the p-value for the desired test, H_0: $\mu = 1$ vs. H_a: $\mu > 1$ is

$$p = \frac{.0088}{2} = .0044$$

This implies that we will reject H_0 for any α level that exceeds p-value $= .0044$.

FIGURE 1.21

SAS printout for t test of Example 1.13

```
Analysis Variable : BENZLEV

N Obs                 T  Prob>|T|
--------------------------------------
    20         2.8937350   0.0088
--------------------------------------
```

EXERCISES

1.44 What is the difference between a research hypothesis and a null hypothesis?

1.45 Define each of the following:
   a. Type I error   b. Type II error   c. α d. β

1.46 In hypothesis testing, who or what determines the size of the rejection region?

1.47 If you test a hypothesis and reject the null hypothesis in favor of your research hypothesis, does your test prove that the research hypothesis is correct? Explain.

1.48 When do you risk making a Type I error? A Type II error?

1.49 For each of the following rejection regions, sketch the sampling distribution for z and indicate the location of the rejection region:
   a. $z > 1.96$ b. $z > 1.645$ c. $z > 2.576$ d. $z < -1.29$
   e. $z < -1.645$ or $z > 1.645$ f. $z < -2.576$ or $z > 2.576$

1.50 If the rejection region is defined as in Exercise 1.49, what is the probability that a Type I error will be made in each case?

1.51 Farm and power equipment dealers are typically dependent on a primary supplier organization for many of their business needs. These suppliers often demand control over many of the dealers' decisions. To determine the degree to which dealers are dependent on suppliers, a national survey of 226 farm and power equipment dealers was conducted. The study revealed the following summary statistics on the total number of suppliers engaged by the dealers:

$$\bar{y} = 3.12 \qquad s = 191$$

(*Academy of Management Journal*, Mar. 1989). Use this information to test the hypothesis that the true mean number of suppliers engaged by farm and power equipment dealers exceeds 2. Compute the p-value of the test and interpret the result.

1.52 The "major bottom indicator" of the New York Stock Exchange (NYSE) was coined by Gerald Appel* to describe a stock market recovery signal. Appel noticed that when there is a large negative difference (-4.0 points or more) between the weekly close of the NYSE index and the 10-week moving average, the market is almost always deeply oversold and ready for an upward turn. Appel recommends buying stock 3 weeks after the date of the largest negative divergence. *Barron's* reported the percentage gains for stock held 26 weeks after the "buy" signal for a sample of 8 dates. These values are reproduced in the table on page 50. If the major bottom indicator is, in fact, a trustworthy indicator of upward

---

*Appel, G. "Have we hit bottom? A trusty indicator signals a possible turn in the market." *Barron's*, Feb. 27, 1984, p. 24.

turns in the market in the long run, then the true mean percentage gain for the 26-week holding period will be greater than 0%. Test this hypothesis and state any assumptions necessary for the procedure to be valid. Let $\alpha = .05$.

DATE OF BUY SIGNAL	PERCENTAGE GAIN IN NYSE INDEX 26 WEEKS AFTER BUY
June 12, 1970	21.1
December 21, 1973	−7.7
October 4, 1974	30.6
November 17, 1978	7.1
November 16, 1979	3.6
April 11, 1980	32.7
October 16, 1981	−2.7
July 9, 1982	33.8

1.53   A machine is set to produce nails with a mean length of 1 inch. Nails that are too long or too short do not meet the customer's specifications and must be rejected. To avoid producing too many rejects, the nails produced by the machine are sampled from time to time and tested as a check to see whether the machine is still operating properly, i.e., producing nails with a mean length of 1 inch. Suppose 50 nails have been sampled, and $\bar{y} = 1.02$ inches and $s = .04$ inch. Does the sample evidence indicate that the machine is producing nails with a mean not equal to 1 inch, i.e., is the production process out of control? Let $\alpha = .01$.

1.54   Most supermarket chains give each store manager a detailed plan showing exactly where each product belongs on each shelf. Since more products are picked up from shelves at eye level than from any others, the most profitable and fastest-moving items are placed at eye level to make them easy for shoppers to reach. Traditionally, the eye-level shelf has been slightly under 5 feet from the floor— just the right height for the average female shopper, 5-feet 4-inches tall. But nowadays, more men are shopping than ever before. Since the average male shopper is 5-feet 10-inches tall, how will this affect the eye-level shelf? To investigate this, a random sample of 50 supermarkets from across the country were selected and the height of the eye-level shelf (i.e., the shelf with the most popular products) was recorded for each. The results were

$$\bar{y} = 62 \text{ inches} \qquad s = 5 \text{ inches}$$

a. Is there evidence that the average height of the eye-level shelf at supermarkets is now higher than 60 inches from the floor (i.e., higher than the traditional height of 5 feet)?

b. Calculate the level of significance (p-value) of the test.

1.55   The effect of machine breakdowns on the performance of a manufacturing system was investigated using computer simulation (*Industrial Engineering*, Aug. 1990). The simulation study focused on a single machine tool system with several characteristics, including a mean interarrival time of 1.25 minutes, a constant processing time of 1 minute, and a machine that breaks down 10% of the time. After $n = 5$ independent simulation runs of length 160 hours, the mean throughput per 40-hour week was $\bar{y} = 1,908.8$ parts. For a system with no breakdowns, the mean throughput for a 40-hour week will be equal to 1,920 parts. Assuming the standard deviation of the five sample runs was $s = 18$ parts per 40-hour week, test the hypothesis that the true mean throughput per 40-hour week for the system is less than 1,920 parts. Test using $\alpha = .05$.

1.56 How do the makers of Kleenex know exactly how many tissues to put in a box? According to the *Wall Street Journal* (Sept. 21, 1984), the marketing experts at Kimberly Clark Corporation have "little doubt that the company should put 60 tissues in each pack." The researchers determined that 60 is "the average number of times people blow their nose during a cold" by asking hundreds of customers to count and record their Kleenex use. Suppose a random sample of 250 Kleenex customers yielded the following summary statistics on the number of times they blew their noses when they had a cold:

$$\bar{y} = 57 \qquad s = 26$$

Is this sufficient evidence to dispute the researchers' claim? Test at $\alpha = .05$.

1.57 An experiment was conducted in England to compare methods of estimating the diet metabolizable energy content of commercial cat foods (*Feline Practice*, Feb. 1986). Method A assumes a digestibility coefficient for crude protein of .91 (called the Atwater factor for crude protein). To determine the validity of this method, the researchers monitored the diets of 28 adult domestic short-haired cats. The cats were fed a diet of commercial canned cat food over a 3-week period. At the end of the trial, the apparent digestibility coefficient was determined for each cat, with these results:

$$\bar{y} = .81 \qquad s = .042$$

Test the hypothesis that the mean digestibility coefficient for crude protein in cats is less than .91, the value assumed by method A. Use $\alpha = .01$.

1.9

Inferences About the Difference Between Two Population Means

The reasoning employed in constructing a confidence interval and performing a statistical test for comparing two population means is identical to that discussed in Sections 1.7 and 1.8. The procedures are based on the assumption that we have selected *independent* random samples from the two populations. The parameters and sample sizes for the two populations, the sample means, and the sample variances are shown in Table 1.8. The objective of the sampling is to make an inference about the difference ($\mu_1 - \mu_2$) between the two population means.

TABLE 1.8

	POPULATION	
	1	2
Sample size	n_1	n_2
Population mean	μ_1	μ_2
Population variance	σ_1^2	σ_2^2
Sample mean	$\bar{y}_1$	$\bar{y}_2$
Sample variance	s_1^2	s_2^2

Because the sampling distribution of the difference between the sample means ($\bar{y}_1 - \bar{y}_2$) is approximately normal for large samples, the large-sample techniques are based on the standardized normal z statistic. Since the variances of the populations, σ_1^2 and σ_2^2, will rarely be known, we will estimate their values using s_1^2 and s_2^2.

To employ these large-sample techniques, we recommend that both samples sizes be large (i.e., each at least 30). The large-sample confidence interval and test are summarized in the boxes.

Large-Sample Confidence Interval for $(\mu_1 - \mu_2)$

$$(\bar{y}_1 - \bar{y}_2) \pm z_{\alpha/2}\sigma_{(\bar{y}_1-\bar{y}_2)}{}^* = (\bar{y}_1 - \bar{y}_2) \pm z_{\alpha/2}\sqrt{\frac{\sigma_1^2}{n_1} + \frac{\sigma_2^2}{n_2}}$$

Assumptions: The two samples are randomly and independently selected from the two populations. The sample sizes, n_1 and n_2, are large enough so that $\bar{y}_1$ and $\bar{y}_2$ each have approximately normal sampling distributions and so that s_1^2 and s_2^2 provide good approximations to σ_1^2 and σ_2^2. This will be true if $n_1 \geq 30$ and $n_2 \geq 30$.

Large-Sample Test of Hypothesis About $(\mu_1 - \mu_2)$

ONE-TAILED TEST   TWO-TAILED TEST

H_0: $(\mu_1 - \mu_2) = D_0$ H_0: $(\mu_1 - \mu_2) = D_0$

H_a: $(\mu_1 - \mu_2) < D_0$ H_a: $(\mu_1 - \mu_2) \neq D_0$

[or H_a: $(\mu_1 - \mu_2) > D_0$]

where D_0 = Hypothesized difference between the means (this is often 0)

Test statistic: $z = \dfrac{(\bar{y}_1 - \bar{y}_2) - D_0}{\sigma_{(\bar{y}_1-\bar{y}_2)}}$ *Test statistic:* $z = \dfrac{(\bar{y}_1 - \bar{y}_2) - D_0}{\sigma_{(\bar{y}_1-\bar{y}_2)}}$

where $\sigma_{(\bar{y}_1-\bar{y}_2)} = \sqrt{\dfrac{\sigma_1^2}{n_1} + \dfrac{\sigma_2^2}{n_2}}$

Rejection region: $z < -z_\alpha$ *Rejection region:* $z < -z_{\alpha/2}$

[or $z > z_\alpha$ when   or $z > z_{\alpha/2}$

H_a: $(\mu_1 - \mu_2) > D_0$]

Assumptions: Same as for the large-sample confidence interval above.

EXAMPLE 1.14

The management of a restaurant wants to determine whether a new advertising campaign has increased its mean daily income (gross). The daily income for 50 randomly selected business days prior to the campaign's beginning was recorded. After conducting the advertising campaign and allowing a 20-day period for the advertising to take effect, the restaurant management recorded the income for

---

*The symbol $\sigma_{(\bar{y}_1-\bar{y}_2)}$ is used to denote the standard error of the distribution of $(\bar{y}_1 - \bar{y}_2)$.

another 30 randomly selected business days. These two samples will allow management to make an inference about the effect of the advertising campaign on the restaurant's daily income. A summary of the results of the two samples is shown in the table. Do these samples provide sufficient evidence for the management to conclude that the mean income has been increased by the advertising campaign? Test using $\alpha = .05$.

BEFORE CAMPAIGN	AFTER CAMPAIGN
$n_1 = 50$	$n_2 = 30$
$\bar{y}_1 = \$1,255$	$\bar{y}_2 = \$1,330$
$s_1 = \$215$	$s_2 = \$238$

Solution

We can best answer this question by performing a test of hypothesis. Defining μ_1 as the mean daily income before the campaign and μ_2 as the mean daily income after the campaign, we will attempt to support the research (alternative) hypothesis that $\mu_1 < \mu_2$ [i.e., that $(\mu_1 - \mu_2) < 0$]. Thus, we will test the null hypothesis $(\mu_1 - \mu_2) = 0$, rejecting this hypothesis if $(\bar{y}_1 - \bar{y}_2)$ equals a large negative value. The elements of the test are as follows:

H_0: $(\mu_1 - \mu_2) - 0$ (i.e., $D_0 = 0$)

H_a: $(\mu_1 - \mu_2) < 0$ (i.e., $\mu_1 < \mu_2$)

Test statistic: $z = \dfrac{(\bar{y}_1 - \bar{y}_2) - D_0}{\sigma_{(\bar{y}_1 - \bar{y}_2)}} = \dfrac{(\bar{y}_1 - \bar{y}_2) - 0}{\sigma_{(\bar{y}_1 - \bar{y}_2)}}$

Rejection region: $z < -z_\alpha = -1.645$ (see Figure 1.22)

FIGURE 1.22

Rejection region for advertising campaign of Example 1.14

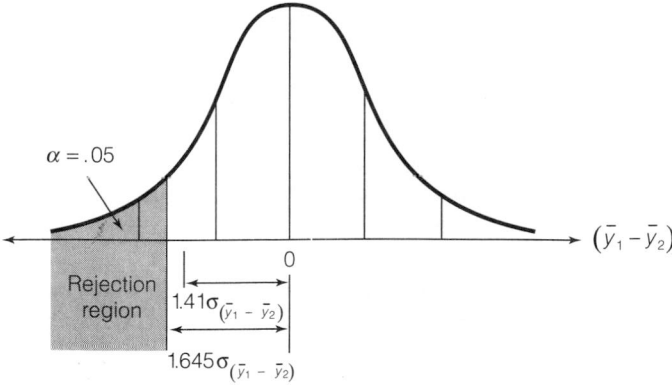

We now calculate

$$z = \frac{(\bar{y}_1 - \bar{y}_2)}{\sigma_{(\bar{y}_1 - \bar{y}_2)}} = \frac{(1,255 - 1,330)}{\sqrt{\dfrac{\sigma_1^2}{n_1} + \dfrac{\sigma_2^2}{n_2}}}$$

$$\approx \frac{-75}{\sqrt{\dfrac{s_1^2}{n_1} + \dfrac{s_2^2}{n_2}}} = \frac{-75}{\sqrt{\dfrac{(215)^2}{50} + \dfrac{(238)^2}{30}}} = \frac{-75}{53.03} = -1.41$$

As you can see in Figure 1.22, the calculated z value does not fall in the rejection region. The samples do not provide sufficient evidence, at $\alpha = .05$, for the restaurant management to conclude that the advertising campaign has increased the mean daily income.

EXAMPLE 1.15

Find a 95% confidence interval for the difference in mean daily incomes before and after the advertising campaign of Example 1.14, and discuss the implications of the confidence interval.

Solution

The 95% confidence interval for $(\mu_1 - \mu_2)$ is

$$(\bar{y}_1 - \bar{y}_2) \pm z_{\alpha/2} \sqrt{\frac{\sigma_1^2}{n_1} + \frac{\sigma_2^2}{n_2}}$$

Once again, we will substitute s_1^2 and s_2^2 for σ_1^2 and σ_2^2, because these quantities will provide good approximations to σ_1^2 and σ_2^2 for samples as large as $n_1 = 50$ and $n_2 = 30$. Then, the 95% confidence interval for $(\mu_1 - \mu_2)$ is

$$(1{,}255 - 1{,}330) \pm 1.96 \sqrt{\frac{(215)^2}{50} + \frac{(238)^2}{30}} = -75 \pm 103.95$$

Thus, we estimate the difference in mean daily income to fall in the interval $-\$178.95$ to $\$28.95$. In other words, we estimate that μ_2, the mean daily income *after* the advertising campaign, could be larger than μ_1, the mean daily income *before* the campaign, by as much as $\$178.95$ per day or it could be less than μ_1 by as much as $\$28.95$ per day.

Now what should the restaurant management do? You can see that the sample sizes collected in the experiment were not large enough to detect a difference in $(\mu_1 - \mu_2)$. To be able to detect a difference (if a difference exists), management will have to repeat the experiment and increase the sample sizes. This will reduce the width of the confidence interval for $(\mu_1 - \mu_2)$. The restaurant management's best estimate of $(\mu_1 - \mu_2)$ is the point estimate $(\bar{y}_1 - \bar{y}_2) = -\$75$. Thus, management must decide whether the cost of conducting the advertising campaign is outweighed by a possible gain in mean daily income estimated at $\$75$ (but which might be as large as $\$178.95$ or as low as $-\$28.95$). Based on this analysis, management will decide whether to continue the experiment or reject the new advertising program as a poor investment.

The small-sample statistical techniques are based on the assumptions that both populations have normal probability distributions and that the variation within the two populations is of the same magnitude, i.e., $\sigma_1^2 = \sigma_2^2$. When these assumptions are approximately satisfied, we can employ a Student's t statistic to find a confidence interval and test a hypothesis concerning $(\mu_1 - \mu_2)$. The techniques are summarized in the accompanying boxes.

Small-Sample Confidence Interval for $(\mu_1 - \mu_2)$ (Independent Samples)

$$(\bar{y}_1 - \bar{y}_2) \pm t_{\alpha/2} \sqrt{s_p^2 \left(\frac{1}{n_1} + \frac{1}{n_2} \right)}$$

where

$$s_p^2 = \frac{(n_1 - 1)s_1^2 + (n_2 - 1)s_2^2}{n_1 + n_2 - 2}$$

is a "pooled" estimate of the common population variance and $t_{\alpha/2}$ is based on $(n_1 + n_2 - 2)$ df.

Assumptions: 1. Both sampled populations have relative frequency distributions that are approximately normal.
2. The population variances are equal.
3. The samples are randomly and independently selected from the populations.

Small-Sample Test of Hypothesis About $(\mu_1 - \mu_2)$
(Independent Samples)

ONE-TAILED TEST

H_0: $(\mu_1 - \mu_2) = D_0$
H_a: $(\mu_1 - \mu_2) < D_0$
    [or H_a: $(\mu_1 - \mu_2) > D_0$]

Test statistic: $t = \dfrac{(\bar{y}_1 - \bar{y}_2) - D_0}{\sqrt{s_p^2 \left(\dfrac{1}{n_1} + \dfrac{1}{n_2} \right)}}$

Rejection region: $t < -t_\alpha$
    [or $t > t_\alpha$ when
    H_a: $(\mu_1 - \mu_2) > D_0$]

TWO-TAILED TEST

H_0: $(\mu_1 - \mu_2) = D_0$
H_a: $(\mu_1 - \mu_2) \neq D_0$

Test statistic: $t = \dfrac{(\bar{y}_1 - \bar{y}_2) - D_0}{\sqrt{s_p^2 \left(\dfrac{1}{n_1} + \dfrac{1}{n_2} \right)}}$

Rejection region: $t < -t_{\alpha/2}$
    or $t > t_{\alpha/2}$

where t_α is based on $(n_1 + n_2 - 2)$ df

Assumptions: Same as for the small-sample confidence interval for $(\mu_1 - \mu_2)$ in the previous box

---

EXAMPLE 1.16

A key aspect of organizational buying is negotiation. S. W. Clopton investigated several issues pertaining to buyer–seller negotiations. One aspect of the analysis involved a comparison of two types of bargaining strategies—competitive bargaining and coordinative bargaining. A *competitive strategy* is characterized by inflexible behavior aimed at forcing concessions, whereas a *coordinative strategy* uses problem-solving in negotiations with a high degree of trust and cooperation. A sample of organizational buyers were recruited to participate in a particular negotiation experiment. In one negotiation setting where the maximum profit was fixed, eight buyers used the competitive bargaining strategy and eight buyers used the coordinative bargaining strategy. The individual savings for the two groups of buyers are provided in Table 1.9. In theory, the mean buyer savings for the competitive strategy will be less than the corresponding mean for the coordinative strategy. Test the theory using $\alpha = .025$.

TABLE 1.9   **Data for Example 1.16**

COMPETITIVE BARGAINING	COORDINATIVE BARGAINING
$1,857	$1,544
1,700	2,640
1,829	1,645
2,644	2,275
1,566	2,137
663	2,327
1,712	2,152
1,679	2,130

Source: Clopton, S. W. "Seller and buying firm factors affecting industrial buyers' negotiation behavior and outcomes." *Journal of Marketing Research*, Feb. 1984, pp. 39–53, published by the American Marketing Association.

Solution

We want to test the following hypothesis:

H_0: $(\mu_1 - \mu_2) = 0$ (i.e., no difference in mean buyer savings)

H_a: $(\mu_1 - \mu_2) < 0$ (i.e., the mean buyer savings for competitive strategy is less than the mean for coordinative strategy)

where μ_1 and μ_2 are the true mean savings of buyers using the competitive and coordinative bargaining strategies, respectively. Since the samples selected for the study are small ($n_1 = n_2 = 8$), the following assumptions are required:

1. The populations of buyer savings under the competitive and coordinative strategies both have approximately normal distributions.
2. The variances of the populations of buyer savings for the two bargaining strategies are equal.
3. The samples were independently and randomly selected.

If these three assumptions are valid, the test statistic will have a t distribution with $(n_1 + n_2 - 2) = (8 + 8 - 2) = 14$ degrees of freedom. With a significance level of $\alpha = .025$, the rejection region is given by

$t < -t_{.025} = -2.145$ (see Figure 1.23)

FIGURE 1.23

Rejection region for Example 1.16

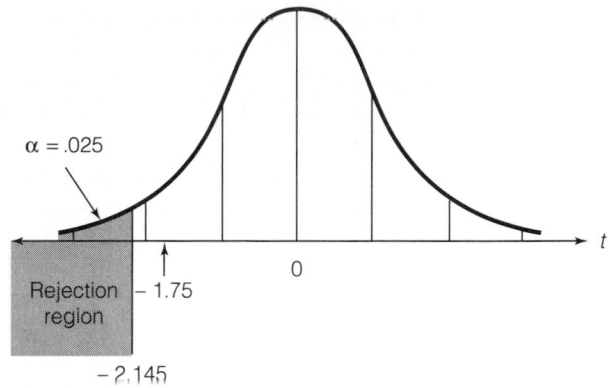

To compute the test statistic, we need to find $\bar{y}_1, \bar{y}_2, s_1,$ and s_2. These summary statistics are given in the Minitab printout, Figure 1.24, as $\bar{y}_1 = 1,706, \bar{y}_2 = 2,106, s_1 = 538,$ and $s_2 = 357$.

FIGURE 1.24

Minitab printout for Example 1.16

```
TWOSAMPLE T FOR Compete VS Coordin

            N       MEAN      STDEV    SE MEAN
Compete   8        1706        538        190
Coordin   8        2106        357        126

95 PCT CI FOR MU Compete - MU Coordin: (-890, 90)

TTEST MU Compete = MU Coordin (VS NE): T= -1.75   P=0.10   DF=   14

POOLED STDEV =           457
```

Since we have assumed that the two populations have equal variances (i.e., that $\sigma_1^2 = \sigma_2^2 = \sigma^2$), we first compute an estimate of this common variance. Our pooled estimate is given by

$$s_p^2 = \frac{(n_1 - 1)s_1^2 + (n_2 - 1)s_2^2}{n_1 + n_2 - 2} = \frac{(8 - 1)(538)^2 + (8 - 1)(357)^2}{8 + 8 - 2}$$

$$= 208{,}446.5$$

Using this pooled sample variance in the computation of the test statistic, we obtain

$$t = \frac{(\bar{y}_1 - \bar{y}_2) - D_0}{\sqrt{s_p^2\left(\dfrac{1}{n_1} + \dfrac{1}{n_2}\right)}} = \frac{(1{,}706 - 2{,}106) - 0}{\sqrt{208{,}446.5\left(\dfrac{1}{8} + \dfrac{1}{8}\right)}}$$

$$= -1.75$$

Since the computed value of t does not fall within the rejection region, we do not reject the null hypothesis (at $\alpha = .025$). There is insufficient evidence that the mean savings of bargainers using the competitive strategy is less than the corresponding mean for bargainers using the coordinative strategy.

The analysis of the data in Table 1.9 could be performed using the t test procedure available in a statistical software package. The Minitab analysis is shown at the bottom of Figure 1.24. The p-value of the two-tailed test, shaded on the printout, is .10; hence, the p-value of the one-tailed test is

$$p = \frac{.10}{2} = .05$$

Since this one-tailed p-value exceeds $\alpha = .025$, we cannot reject H_0.

EXAMPLE 1.17

Refer to Example 1.16. Suppose you wanted to estimate the difference in mean buyer savings $(\mu_1 - \mu_2)$ correct to within \$200 with 90% confidence. How many buyers of each strategy must be sampled to obtain this estimate?

Solution

This problem is similar to Example 1.11. We want to estimate $(\mu_1 - \mu_2)$ with a confidence interval that has a half-width less than or equal to \$200. That is, we want

$$t_{.05}\sqrt{s_p^2\left(\frac{1}{n_1} + \frac{1}{n_2}\right)} = 200$$

We will use equal numbers for each strategy, i.e., we will let $n_1 = n_2$. Since the number of degrees of freedom for $t_{.05}$ depends on n (and hence is unknown) and since the value of s_p^2 depends on the data we collect, we will approximate both these values. You can see from the tabulated values of $t_{.05}$ (Table 2 of Appendix D) that $t_{.05}$ approaches $z_{.05} = 1.645$ as n gets close to 30. Consequently, $t_{.05} \approx 1.64$. The best approximation to s_p^2 would be the value computed from the samples in Example 1.16, $s_p^2 = 208{,}446.5$. Substituting these values into the equation and solving for $n_1 = n_2 = n$, we obtain

$$t_{.05}\sqrt{s_p^2\left(\frac{1}{n_1} + \frac{1}{n_2}\right)} = 200$$

$$1.64\sqrt{208{,}446.5\left(\frac{1}{n} + \frac{1}{n}\right)} = 200$$

$$\sqrt{n} = 5.29$$

$$n = 28.03$$

Therefore, we will need to sample approximately 28 buyers of each strategy to estimate the difference in mean buyer savings correct to within \$200 with confidence coefficient .90. This is an approximate solution for n_1 and n_2, but it will be fairly close to the exact solution.

The two-sample t statistic is a powerful tool for comparing population means when the necessary assumptions are satisfied. It has also been shown to be useful when the sampled populations are only approximately normally distributed. And, when the sample sizes are equal, the assumption of equal population variances can be relaxed. That is, when $n_1 = n_2$, σ_1^2 and σ_2^2 can be quite different and the test statistic will still have (approximately) a Student's t distribution.

EXERCISES

1.58 Describe the sampling distribution of $(\bar{y}_1 - \bar{y}_2)$.

1.59 The *American Journal of Small Businesses* (Winter 1988) reported on a survey designed to compare female managers at large firms with those at small firms (fewer than 100 employees). Previous studies tend to indicate that female managers in large and small companies are quite similar. In this study, independent random samples of 86 female managers at small firms and 91 female managers at large firms were compared on several job-related variables. The following question was asked: "How many times have you been promoted in the last three years?" The responses for the two groups of female managers are summarized in the table.

SMALL FIRMS	LARGE FIRMS
$n_1 = 86$	$n_2 = 91$
$y_1 = 1.0$	$y_2 = .9$
$s_1 = 1.1$	$s_2 = 1.1$

Source: Anderson, R. L. and Anderson, K. P. "A comparison of women in small and large companies." *American Journal of Small Businesses*, Vol. 12, No. 3, Winter 1988, p. 28 (Table 2).

a. Compute a point estimate for the difference between the mean number of promotions awarded to female managers at small firms and at large firms.
b. Compare the mean number of promotions awarded to the two groups of female managers with a 90% confidence interval.
c. Interpret the interval, part b.
d. How could the researchers reduce the width of the interval, part b?

1.60 Are Japanese managers and their workers more motivated than their American counterparts? Recent claims in the literature tout the superiority of the Japanese-inspired "Theory Z" management for business productivity. The Theory Z philosophy emphasizes trusting, intimate, and subtle relationships between management and workers; directness and confrontation are avoided. To investigate this phenomenon, researchers surveyed middle-aged Japanese and American business managers. The Japanese sample consisted of 100 managers selected at a 2-day management seminar in Tokyo and Osaka, whereas the American sample was made up of 211 managers employed in the Bell System. Each manager was administered the Sarnoff Survey of Attitudes Toward Life (SSATL), which measures motivation for upward mobility in three areas: advancement, forward striving, and money. The SSATL scores are summarized in the table (higher scores indicate a greater motivation for upward mobility).

	AMERICAN MANAGERS	JAPANESE MANAGERS
Sample size	211	100
Mean SSATL score	65.75	79.83
Standard deviation	11.07	6.41

Source: Howard, A., Shudo, K., and Umeshima, M. "Motivation and values among Japanese and American managers." *Personnel Psychology*, Winter 1983, Vol. 36, No. 4, pp. 883–898.

a. Do the data provide sufficient evidence to indicate that Japanese managers, on average, are more motivated for upward mobility than American managers? Test using $\alpha = .05$.

b. Find a 95% confidence interval for the difference in mean SSATL scores for American and Japanese managers. Interpret the interval.

1.61   According to many experts, the single most important goal of any corporation should be maximization of shareholder wealth (measured as stock-price maximization). But do the stocks of companies that adopt such a long-term financial strategy perform better than other stocks? To answer this question, Pantalone and Welch gathered stock price and corporate strategy information on a sample of 111 companies listed in the *Standard & Poor's 500*. Of these, 23 were identified as shareholder-wealth maximizers and 88 as having alternative corporate goals based on a survey questionnaire sent to chief executives. The accompanying table gives summary information on the return on equity (measured in percent) earned by the two groups of companies over a 5-year period (1977–1981). Is there sufficient evidence to indicate that companies that aim to maximize shareholder wealth have a higher mean return on equity than companies with alternative corporate goals? Test using $\alpha = .05$.

	MAXIMIZE SHAREHOLDER WEALTH	ALTERNATIVE CORPORATE GOALS
Number of companies	23	88
Mean return on equity	16.55	14.80
Standard deviation	7.61	5.17

Source: Pantalone, C. and Welch, J. B. "The usefulness of public information about corporate goals." *Quarterly Journal of Business & Economics*, Vol. 25, No. 4, Autumn 1986, pp. 29–39 (Table II).

1.62   The term *Machiavellian* was derived from the 16th-century Florentine writer Niccolo Machiavelli, who described ways of manipulating others to accomplish one's objective. Critics often accuse marketers of being manipulative and unethical, or Machiavellian in nature. Researchers at Texas Tech University explored the question of whether "marketers are more Machiavellian than others." The Machiavellian scores (measured by the Mach IV Scale) for a sample of marketing professionals were recorded and compared to the Machiavellian scores for other groups of people, including a sample of college students in an earlier study. The results are summarized in the accompanying table (higher scores are associated with Machiavellian attitudes).

a. Construct a 99% confidence interval for the difference in mean Machiavellian scores between marketing professionals and college students.

	MARKETING PROFESSIONALS	COLLEGE STUDENTS
Sample size	1,076	1,782
Mean score	85.7	90.7
Standard deviation	13.2	14.3

Source: Hunt, S. D. and Chonko, L. B. "Marketing and Machiavellianism." *Journal of Marketing,* Summer 1984, Vol. 48, No. 3, pp. 30–42. Reprinted by permission of the American Marketing Association.

    b. Based on the interval constructed in part **a**, comment on the statement that "marketers are more Machiavellian than college students."

1.63 Does competition between separate research and development (R&D) teams in the U.S. Department of Defense, working independently on the same project, improve performance? To answer this question, performance ratings were assigned to each of 58 multisource (competitive) and 63 sole-source R&D contracts (*IEEE Transactions on Engineering Management,* Feb. 1990). With respect to quality of reports and products, the competitive contracts had a mean performance rating of 7.62, whereas the sole-source contracts had a mean of 6.95.

    a. Set up the null and alternative hypothesis for determining whether the mean quality performance rating of competitive R&D contracts exceeds the mean for sole-source contracts.

    b. Find the rejection region for the test using $\alpha = .05$.

    c. The *p*-value for the test was reported to be between .02 and .03. What is the appropriate conclusion?

1.64 To use the *t* statistic to test for differences between the means of two populations, what assumptions must be made about the two sampled populations? What assumptions must be made about the two samples?

1.65 Executives of an industrial plant want to determine which of two types of fuel—gas or electric—will produce more useful energy at the lower cost. One measure of economical energy production, called the plant investment per delivered quad, is calculated by taking the amount of money (in dollars) invested in the particular utility by the plant, and dividing by the delivered amount of energy (in quadrillion British thermal units). The smaller this ratio, the less an industrial plant pays for its delivered energy. Random samples of 11 plants using electrical utilities and 16 plants using gas utilities were taken, and the plant investment/quad ratio was calculated for each. A Minitab printout of the analysis of the data is given here; the original data are shown in the table at the top of page 62. Do these data provide sufficient evidence at the $\alpha = .05$ level of significance to indicate a difference in the average investment/quad between the plants using gas and those using electrical utilities? What assumptions are required for the procedure you used to be valid?

```
TWOSAMPLE T FOR electric VS gas

             N      MEAN     STDEV    SE MEAN
electric    11      52.4     62.4        19
gas         16      37.7     49.0        12

95 PCT CI FOR MU electric - MU gas: (-30, 59)

TTEST MU electric = MU gas (VS NE): T= 0.68   P=0.50   DF=  25

POOLED STDEV =        54.8
```

ELECTRIC				GAS			
204.15	.57	62.76	89.72	.78	16.66	74.94	.01
.35	85.46	.78	.65	.54	23.59	88.79	.64
44.38	9.28	78.60		.82	91.84	7.20	66.64
				.74	64.67	165.60	.36

1.66   Suppose you are the personnel manager for a company and you suspect a difference in the mean
length of work time lost because of sickness for two types of employees: those who work at night
versus those who work during the day. In particular, you suspect that the mean time lost for the
night shift exceeds the mean time lost for the day shift. To check your theory, you randomly sample
the records for 10 employees for each shift category and record the number of days lost because of
sickness within the past year. The data are shown in the accompanying table.

NIGHT SHIFT, 1		DAY SHIFT, 2	
21	2	13	18
10	19	5	17
14	6	16	3
33	4	0	24
7	12	7	1
$\bar{y}_1 = 12.8$		$\bar{y}_2 = 10.4$	
$\sum_{i=1}^{n} y_i^2 = 2{,}436$		$\sum_{i=1}^{n} y_i^2 = 1{,}698$	

a. Calculate s_1^2 and s_2^2.

b. Show that the pooled estimate of the common population standard deviation, σ, is 8.86. Look at
the range of the observations within each of the two samples. Does it appear that the estimate,
8.86, is a reasonable value for σ?

c. If μ_1 and μ_2 represent the mean number of days per year lost due to sickness for the night and
day shifts, respectively, test the null hypothesis $H_0: \mu_1 = \mu_2$ against the alternative $H_a: \mu_1 > \mu_2$.
Use $\alpha = .05$. Do the data provide sufficient evidence to indicate that $\mu_1 > \mu_2$?

d. What assumptions must be satisfied so that the t test from part c is valid?

1.67   An experiment was conducted to determine whether individuals could be taught how to make
decisions rationally and if such training would improve the quality of their career decisions. A
questionnaire was administered to a sample of 147 community college students in California enrolled
in career guidance classes in order to assess the decision-making style of each. Of these, 69 were
classified as rational decision makers, a style characterized by a systematic and logical approach to
decision making. Several weeks later, these 69 subjects were randomly divided into two groups. The
experimental group (34 students) received instruction in rational decision making, whereas the control
group (35 students) did not. At the end of the instruction period, all subjects completed a multiple-
choice test designed to assess the extent to which an individual knows how to apply rational principles
in job-decision situations. The results are summarized in the accompanying table. (Higher scores
indicate higher adherence to the rational style of decision making.)

	EXPERIMENTAL GROUP	CONTROL GROUP
Sample size	34	35
Mean	50.26	47.34
Standard deviation	6.67	11.52

Source: Krumboltz, J. D., Kinnier, R. T., Rude, S. S., Sherba, D. S., and Hamel, D. A. "Teaching a rational approach to career decision making: Who benefits most?" *Journal of Vocational Behavior*, Vol. 29, No. 8, August 1986, pp. 1–6 (Table 1).

a. Do the samples provide sufficient evidence to conclude that students who receive training in rational decision making score higher, on average, on the multiple-choice exam than students who receive no training? Test using $\alpha = .05$.

b. Construct a 90% confidence interval for the difference between the mean test scores of the two groups of students. Interpret the interval. Is the result consistent with your answer in part **a**?

1.68 According to a popular model of managerial behavior, the current state of automation in a manufacturing firm influences managers' perceptions of problems of automation. To investigate this proposition, researchers at Concordia University (Montreal) surveyed managers at firms with a high level of automation and at firms with a low level of automation (*IEEE Transactions on Engineering Management*, Aug. 1990). Each manager was asked to give his or her perception of the problems of automation at the firm. Responses were measured on a 5-point scale (1 = No problem, ... , 5 = Major problem). Summary statistics for the two groups of managers, provided in the table, were used to test the hypothesis of no difference in the mean perceptions of automation problems between managers of highly automated and less automated manufacturing firms. Conduct the test for the researchers, assuming that the perception variances for the two groups of managers are equal. Use $\alpha = .01$.

	SAMPLE SIZE	MEAN	STANDARD DEVIATION
Low level	17	3.274	.762
High level	8	3.280	.721

Source: Farhoomand, A. F., Kira, D. and Williams, J. "Managers' perceptions towards automation in manufacturing." *IEEE Transactions on Engineering Management*, Vol. 37, No. 3, Aug. 1990, p. 230.

1.10
Comparing Two Population Variances

Suppose you want to use the two-sample t statistic to compare the mean productivity of two paper mills. However, you are concerned that the assumption of equal variances of the productivity for the two plants may be unrealistic. It would be helpful to have a statistical procedure to check the validity of this assumption.

The common statistical procedure for comparing population variances σ_1^2 and σ_2^2 is to make an inference about the ratio, σ_1^2/σ_2^2, using the ratio of the sample variances, s_1^2/s_2^2. Thus, we will attempt to support the research hypothesis that the ratio σ_1^2/σ_2^2 differs from 1 (i.e., the variances are unequal) by testing the null hypothesis that the ratio equals 1 (i.e., the variances are equal).

$$H_0: \quad \frac{\sigma_1^2}{\sigma_2^2} = 1 \quad (\sigma_1^2 = \sigma_2^2) \qquad H_a: \quad \frac{\sigma_1^2}{\sigma_2^2} \neq 1 \quad (\sigma_1^2 \neq \sigma_2^2)$$

We will use the test statistic

$$F = \frac{s_1^2}{s_2^2}$$

To establish a rejection region for the test statistic, we need to know how s_1^2/s_2^2 is distributed in repeated sampling. That is, we need to know the sampling distribution of s_1^2/s_2^2. As you will subsequently see, the sampling distribution of s_1^2/s_2^2 depends on two of the assumptions already required for the t test, namely:

1. The two sampled populations are normally distributed.
2. The samples are randomly and independently selected from their respective populations.

When these assumptions are satisfied and when the null hypothesis is true (i.e., $\sigma_1^2 = \sigma_2^2$), the sampling distribution of s_1^2/s_2^2 is an **F distribution** with $(n_1 - 1)$ df and $(n_2 - 1)$ df, respectively. The shape of the F distribution depends on the degrees of freedom associated with s_1^2 and s_2^2, i.e., $(n_1 - 1)$ and $(n_2 - 1)$. An F distribution with 7 and 9 df is shown in Figure 1.25. As you can see, the distribution is skewed to the right.

FIGURE 1.25

An F distribution with 7 and 9 df

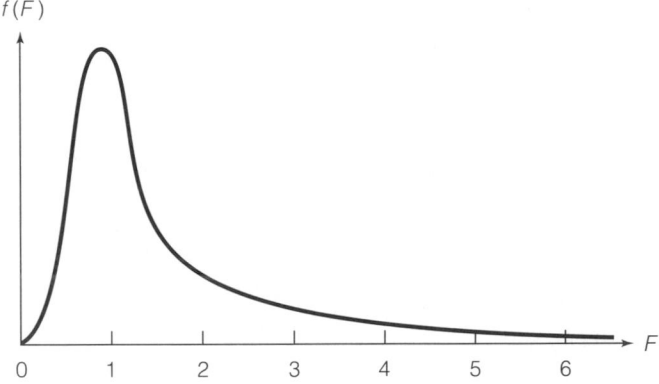

$f(F)$

0   1   2   3   4   5   6 F

When the population variances are unequal, we expect the ratio of the sample variances, $F = s_1^2/s_2^2$, to be either very large or very small. Therefore, we will need to find F values corresponding to the tail areas of the F distribution to establish the rejection region for our test of hypothesis. The upper-tail F values can be found in Tables 3, 4, 5, and 6 of Appendix D. Table 4 is partially reproduced in Table 1.10. It gives F values that correspond to $\alpha = .05$ upper-tail areas for different degrees of freedom. The columns of the tables correspond to various degrees of freedom for the numerator sample variance s_1^2, whereas the rows correspond to the degrees of freedom for the denominator sample variance s_2^2. Thus, if the numerator degrees of freedom is 7 and the denominator degrees of freedom is 9, we look in the seventh column and ninth row to find $F_{.05} = 3.29$.

As shown in Figure 1.26, $\alpha = .05$ is the tail area to the right of 3.29 in the F distribution with 7 and 9 df. That is, if $\sigma_1^2 = \sigma_2^2$, the probability that the F statistic will exceed 3.29 is $\alpha = .05$.

T A B L E 1.10 Reproduction of Part of Table 4 of Appendix D: $\alpha = .05$

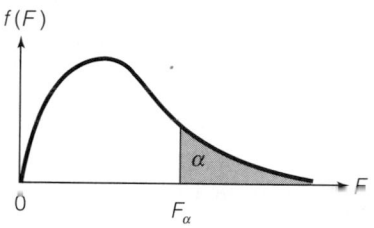

ν_1 ν_2	NUMERATOR DEGREES OF FREEDOM								
	1	2	3	4	5	6	7	8	9
1	161.4	199.5	215.7	224.6	230.2	234.0	236.8	238.9	240.5
2	18.51	19.00	19.16	19.25	19.30	19.33	19.35	19.37	19.38
3	10.13	9.55	9.28	9.12	9.01	8.94	8.89	8.85	8.81
4	7.71	6.94	6.59	6.39	6.26	6.16	6.09	6.04	6.00
5	6.61	5.79	5.41	5.19	5.05	4.95	4.88	4.82	4.77
6	5.99	5.14	4.76	4.53	4.39	4.28	4.21	4.15	4.10
7	5.59	4.74	4.35	4.12	3.97	3.87	3.79	3.73	3.68
8	5.32	4.46	4.07	3.84	3.69	3.58	3.50	3.44	3.39
9	5.12	4.26	3.86	3.63	3.48	3.37	3.29	3.23	3.18
10	4.96	4.10	3.71	3.48	3.33	3.22	3.14	3.07	3.02
11	4.84	3.98	3.59	3.36	3.20	3.09	3.01	2.95	2.90
12	4.75	3.89	3.49	3.25	3.11	3.00	2.91	2.85	2.80
13	4.67	3.81	3.41	3.18	3.03	2.92	2.83	2.77	2.71
14	4.60	3.74	3.34	3.11	2.96	2.85	2.76	2.70	2.65

DENOMINATOR DEGREES OF FREEDOM

F I G U R E 1.26

An F distribution for 7 and 9 df: $\alpha = .05$

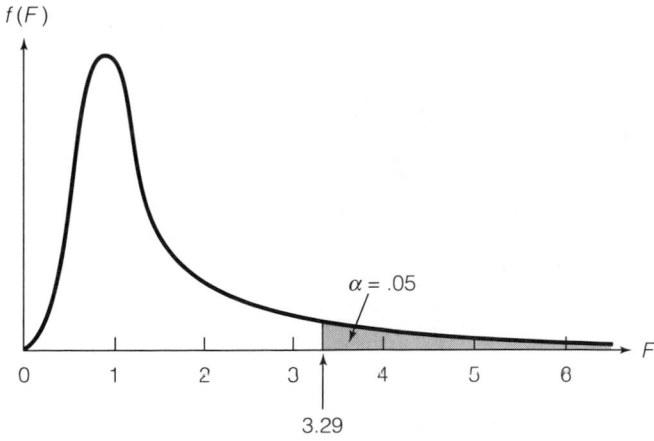

Suppose we want to compare the variability in production for two paper mills and we have obtained the following results:

SAMPLE 1	SAMPLE 2
$n_1 = 13$ days	$n_2 = 18$ days
$\bar{y}_1 = 26.3$ production units	$\bar{y}_2 = 19.7$ production units
$s_1 = 8.2$ production units	$s_2 = 4.7$ production units

To form the rejection region for a two-tailed F test we want to make certain that the upper tail is used, because only the upper-tail values of F are shown in Tables 3, 4, 5, and 6. To accomplish this, **we will always place the larger sample variance in the numerator of the F test**. This doubles the tabulated value for α, since we double the probability that the F ratio will fall in the upper tail by always placing the larger sample variance in the numerator. In effect, we make the test two-tailed by putting the larger variance in the numerator rather than establishing rejection regions in both tails.

Thus, for our production example, we have a numerator s_1^2 with df $= n_1 - 1 = 12$ and a denominator s_2^2 with df $= n_2 - 1 = 17$. Therefore, the test statistic will be

$$F = \frac{\text{Larger sample variance}}{\text{Smaller sample variance}} = \frac{s_1^2}{s_2^2}$$

and we will reject $H_0: \sigma_1^2 = \sigma_2^2$ for $\alpha = .10$ if the calculated value of F exceeds the tabulated value:

$$F_{.05} = 2.38 \quad \text{(see Figure 1.27)}$$

FIGURE 1.27

Rejection region for production example F distribution

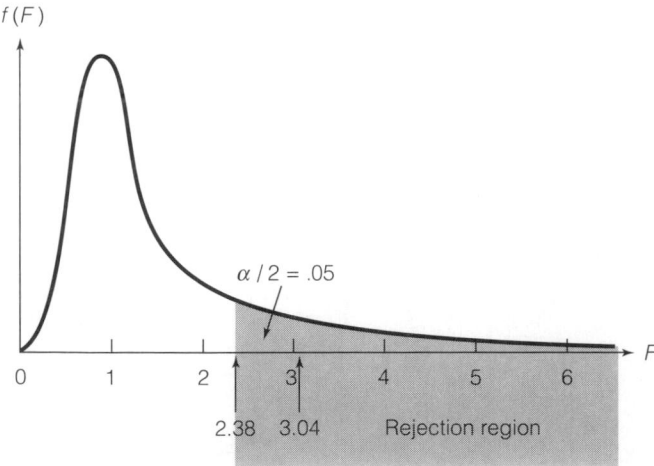

Now, what do the data tell us? We calculate

$$F = \frac{s_1^2}{s_2^2} = \frac{(8.2)^2}{(4.7)^2} = 3.04$$

and compare it to the rejection region shown in Figure 1.27. Since the calculated F value, 3.04, falls in the rejection region, the data provide sufficient evidence to indicate that the population variances differ. Consequently, we would be reluctant to use the two-sample t statistic to compare the population means, since the assumption of equal population variances is apparently untrue.

What would you have concluded if the value of F calculated from the samples had not fallen in the rejection region? Would you conclude that the null hypothesis of equal variances is true? No, because then you risk the possibility of a Type II error (accepting H_0 when H_a is true) without knowing the probability of this error (the probability of accepting H_0: $\sigma_1^2 = \sigma_2^2$ when it is false). Since we will not consider the calculation of β for specific alternatives in this text, when the F statistic does not fall in the rejection region, we simply conclude that **insufficient sample evidence exists to refute the null hypothesis that $\sigma_1^2 = \sigma_2^2$.**

The F test for equal population variances is summarized in the next box.

| F Test for Equal Population Variances |

ONE-TAILED TEST

H_0: $\dfrac{\sigma_1^2}{\sigma_2^2} = 1$ (or $\sigma_1^2 = \sigma_2^2$)

H_a: $\dfrac{\sigma_1^2}{\sigma_2^2} > 1$ (or $\sigma_1^2 > \sigma_2^2$)

Test statistic: $F = \dfrac{s_1^2}{s_2^2}$

Rejection region: $F > F_\alpha$
where the degrees of freedom for F are:

    $n_1 - 1 =$ df numerator sample variance

    $n_2 - 1 =$ df denominator sample variance

TWO-TAILED TEST

H_0: $\dfrac{\sigma_1^2}{\sigma_2^2} = 1$ (or $\sigma_1^2 = \sigma_2^2$)

H_a: $\dfrac{\sigma_1^2}{\sigma_2^2} \neq 1$ (or $\sigma_1^2 \neq \sigma_2^2$)

Test statistic:

$$F = \frac{\text{Larger sample variance}}{\text{Smaller sample variance}} = \frac{s_1^2}{s_2^2}$$

where the samples are numbered so that $s_1^2 > s_2^2$

Rejection region: $F > F_{\alpha/2}$
where the degrees of freedom for F are:

    $n_1 - 1 =$ df larger (numerator) sample variance

    $n_2 - 1 =$ df smaller (denominator) sample variance

Assumptions: 1. Both sampled populations are normally distributed.
             2. The samples are random and independent.

EXAMPLE 1.18

Refer to the *Journal of Marketing Research* study described in Example 1.16. The researcher's objective is to compare the mean buyer savings for two bargaining strategies, competitive and coordinative. Since both samples of buyers are small ($n_1 = 8$ and $n_2 = 8$), the t test for $(\mu_1 - \mu_2)$ requires that both sampled populations of buyer savings be normal with equal variances. To check the latter assumption, the researcher wants to determine (at $\alpha = .10$) whether the variation in the savings of buyers using the competitive strategy differs from the variation in the savings of buyers using the coordinative strategy.

Solution

The elements of the test are as follows:

$$H_0: \quad \frac{\sigma_1^2}{\sigma_2^2} = 1 \quad (\text{i.e., } \sigma_1^2 = \sigma_2^2)$$

$$H_a: \quad \frac{\sigma_1^2}{\sigma_2^2} \neq 1 \quad (\text{i.e., } \sigma_1^2 \neq \sigma_2^2)$$

Test statistic: $\quad F = \dfrac{\text{Larger sample variance}}{\text{Smaller sample variance}} = \dfrac{s_1^2}{s_2^2}$

To find the rejection region, we proceed as follows: The numerator degrees of freedom are df $= n_2 - 1 = 8 - 1 = 7$ and the denominator degrees of freedom are $n_1 - 1 = 8 - 1 = 7$. From Table 4 of Appendix D, we find

$$F_{\alpha/2} = F_{.05} \approx 3.79$$

Summary statistics for the data are shown in the Minitab printout, Figure 1.28. We now calculate

$$F = \frac{s_1^2}{s_2^2} = \frac{(538)^2}{(357)^2} = 2.27$$

This F value is not in the rejection region. Therefore, there is insufficient evidence at the $\alpha = .10$ level to refute the assumption of equal population variances.

FIGURE 1.28

Minitab printout for Example 1.18

	N	MEAN	MEDIAN	TRMEAN	STDEV	SEMEAN
Compete	8	1706	1706	1706	538	190
Coordin	8	2106	2145	2106	357	126

	MIN	MAX	Q1	Q3
Compete	663	2644	1594	1850
Coordin	1544	2640	1766	2314

The following example shows that the F statistic is sometimes used to compare population variances in their own right, rather than just to check the validity of an assumption.

EXAMPLE 1.19

Suppose an investor wants to compare the risks associated with two different stocks, where the risk of a given stock is measured by the variation in daily price changes. Suppose we obtain random samples of 25 daily price changes for stock 1 and 25 for stock 2. The sample results are summarized in the table.

STOCK 1	STOCK 2
$n_1 = 25$	$n_2 = 25$
$\bar{y}_1 = .250$	$\bar{y}_2 = .125$
$s_1 = .76$	$s_2 = .46$

Compare the risks associated with the two stocks by testing the null hypothesis that the variances of the price changes for the stocks are equal. Use $\alpha = .10$.

Solution

$$H_0: \frac{\sigma_1^2}{\sigma_2^2} = 1 \qquad H_a: \frac{\sigma_1^2}{\sigma_2^2} \neq 1$$

Test statistic: $F = \dfrac{\text{Larger sample variance}}{\text{Smaller sample variance}} = \dfrac{s_1^2}{s_2^2}$

Rejection region: $F > F_{.05}$

where F is based on $n_1 - 1 = 24$ df and $n_2 - 1 = 24$ df

Assumptions:
1. The changes in daily stocks have relative frequency distributions that are approximately normal.
2. The stock samples are randomly and independently selected from a set of daily stock reports.

We calculate

$$F = \frac{s_1^2}{s_2^2} = \frac{(.76)^2}{(.46)^2} = 2.73$$

The calculated F exceeds the rejection value of 1.98. Therefore, we conclude that the variances of daily price changes differ for the two stocks. In fact, it appears that the risk, as measured by the variance of daily price changes, is greater for stock 1 than for stock 2. How much reliability can we place in this inference? Only one time in 10 (since $\alpha = .10$), on the average, would this statistical test lead us to conclude erroneously that σ_1^2 and σ_2^2 were different when in fact they were equal.

The previous examples demonstrate how to conduct a two-tailed F test when the alternative hypothesis is $H_a: \sigma_1^2 \neq \sigma_2^2$. One-tailed tests for determining whether one population variance is larger than another population variance (i.e., $H_a: \sigma_1^2 > \sigma_2^2$) are conducted similarly. However, the α value no longer needs to be doubled since the area of rejection lies only in the upper (or lower) tail area of the F distribution. The procedure for conducting an upper-tailed F test is outlined in the previous box. Whenever you conduct a one-tailed F test, be sure to write H_a in the form of an upper-tailed test. This can be accomplished by numbering the populations so that the variance hypothesized to be larger in H_a is associated with population 1 and the hypothesized smaller variance is associated with population 2.

EXERCISES

1.69 Under what conditions does the sampling distribution of s_1^2/s_2^2 have an F distribution?

1.70 Use Tables 3, 4, 5, and 6 of Appendix D to find F_α for α, numerator df, and denominator df equal to:

a. .05, 8, 7   **b.** .01, 15, 20   **c.** .025, 12, 5
d. .01, 5, 25   **e.** .10, 5, 10   **f.** .05, 20, 9

1.71 Refer to the English study on estimating the metabolizable energy (ME) content of commercial cat foods in Exercise 1.57. In addition to the sample of 28 cats that were fed canned food, a second (independent) sample of 29 cats were fed a commercial brand of dry cat food. The ME content for each cat was measured (in kilocalories per gram) using method A, with the results shown in the table.

	CANNED FOOD	DRY FOOD
Sample size	28	29
Mean ME content	.96	3.70
Standard deviation	.26	.48

Source: Kendall, P. T., Burger, I. N., and Smith, P. M. "Methods of estimation of the metabolizable energy content of cat foods." *Feline Practice*, Vol. 15, No. 2, February 1986, pp. 38–44.

Conduct a test to determine whether the variation in ME content of cats fed canned food differs from the variation in ME content of cats fed dry food. Use $\alpha = .10$.

1.72 U.S. corporations, especially automakers, have felt the effects of increased competition from foreign corporations. According to the *Wall Street Journal* (Feb. 9, 1984), many U.S. corporations are countering foreign competition by merging with other large firms to "lead to more efficient, more effective, and more competitive combinations." One of the major concerns of these types of mergers, however, is the effect on the variability of returns to shareholders after stock acquisition takes place. Consider a U.S. tool company that has recently merged with another U.S. corporation to combat foreign competition. Samples of monthly returns, before and after the merger, yielded the summary statistics provided in the table. Is there evidence of a difference in the variation of the monthly returns before and after the merger? Test using $\alpha = .02$.

	BEFORE MERGER	AFTER MERGER
Months sampled	10	6
Mean return	8.5%	7.1%
Standard deviation	4.3%	4.8%

1.73 A study was conducted to compare the variation in the price of wholesale residual petroleum sold in rural (low-density) and urban (high-density) counties. In particular, the variable of interest was the natural logarithm of the ratio of the county price to state price, i.e., log (county price/state price). Based on independent random samples of 10 rural counties and 23 urban counties, the following descriptive statistics (shown in the table) were obtained. Is there evidence of a difference between the variance in the log-price ratios of rural and urban counties?

	n	$\bar{y}$	s
Rural	10	.239	.310
Urban	23	.117	.199

Source: Saavedra, P., et al., "Geographical stratification of petroleum retailers and resellers." Paper presented at Joint Statistical Meetings, Anaheim, Calif., Aug. 1990.

1.74 The quality control department of a paper company measures the brightness (a measure of reflectance) of finished paper on a periodic basis throughout the day. Two instruments that are available to measure the paper specimens are subject to error, but they can be adjusted so that the mean readings for a control paper specimen are the same for both instruments. Suppose you are concerned about the precision of the two instruments, namely, that the variation in readings from instrument 2 exceeds that for instrument 1. To check this theory, five measurements of a single paper sample are made on both instruments. The data are shown in the table. Do the data provide sufficient evidence to indicate that instrument 2 is less precise than instrument 1? Test using $\alpha = .05$.

INSTRUMENT 1	INSTRUMENT 2
29	26
28	34
30	30
28	32
30	28

Summary

The preceding sections summarize many of the basic concepts and methods presented in an introductory statistics course. We presented the concepts of a **population, random sampling**, and the ultimate objective of most statistical investigations, **making an inference about a population based on information contained in a sample**. Because inference implies description, we first considered methods for describing a set of data—two graphical methods (**relative frequency histogram** and **stem-and-leaf plot**) and **numerical descriptive methods** that provide measures of centrality and variability for a data set.

We noted that quantities computed from sample data—**statistics**—are used to estimate population numerical descriptive measures—**parameters**—and to make decisions about their values. To evaluate the properties of these statistics, we need to know the probabilities that they will assume specific sets of values in repeated sampling, i.e., we need to know their **probability sampling distributions**. If we know the sampling distribution for a statistic, we can make probabilistic statements that measure the **reliability** of the statistic when it is used as an estimator or as the basis of a decision.

Finally, we summarized the basic concepts involved in **interval estimation** and tests of hypotheses. In particular, we presented **large- and small-sample confidence intervals** and **statistical tests for making inferences about a single population mean** and for **comparing two population means or variances based on independent random sampling**.

To aid in the formulation of confidence intervals and test statistics, we present two summary tables. Table 1.11 contains a list of parameters and their corresponding estimators and standard errors. Once you have identified the parameter of interest in Table 1.11, use Table 1.12 to formulate confidence intervals and test statistics.

TABLE 1.11   Some Population Parameters and Corresponding Estimators and Standard Errors

PARAMETER (θ)	ESTIMATOR ($\hat{\theta}$)	STANDARD ERROR ($\sigma_{\hat{\theta}}$)	ESTIMATE OF STANDARD ERROR ($s_{\hat{\theta}}$)
μ Mean (average)	$\bar{y}$	$\dfrac{\sigma}{\sqrt{n}}$	$\dfrac{s}{\sqrt{n}}$
$\mu_1 - \mu_2$ Difference between means (averages)	$\bar{y}_1 - \bar{y}_2$	$\sqrt{\dfrac{\sigma_1^2}{n_1} + \dfrac{\sigma_2^2}{n_2}}$	$\sqrt{\dfrac{s_1^2}{n_1} + \dfrac{s_2^2}{n_2}}$, $n_1 \geq 30$, $n_2 \geq 30$ $\sqrt{s_p^2\left(\dfrac{1}{n_1} + \dfrac{1}{n_2}\right)}$, either $n_1 < 30$ or $n_2 < 30$ where $s_p^2 = \dfrac{(n_1 - 1)s_1^2 + (n_2 - 1)s_2^2}{n_1 + n_2 - 2}$
$\dfrac{\sigma_1^2}{\sigma_2^2}$ Ratio of variances	$\dfrac{s_1^2}{s_2^2}$	(not necessary)	(not necessary)

TABLE 1.12   Formulation of Confidence Intervals for a Population Parameter θ and Test Statistics for H_0: $\theta = \theta_0$, where $\theta = \mu$ or $(\mu_1 - \mu_2)$

SAMPLE SIZE	CONFIDENCE INTERVAL	TEST STATISTIC
Large	$\hat{\theta} \pm z_{\alpha/2}s_{\hat{\theta}}$	$z = \dfrac{\hat{\theta} - \theta_0}{s_{\hat{\theta}}}$
Small	$\hat{\theta} \pm t_{\alpha/2}s_{\hat{\theta}}$	$t = \dfrac{\hat{\theta} - \theta_0}{s_{\hat{\theta}}}$

Note: The test statistic for testing H_0: $\sigma_1^2/\sigma_2^2 = 1$ is $F = s_1^2/s_2^2$ (see the box on page 67).

SUPPLEMENTARY EXERCISES

1.75   For each of the following data sets, compute $\bar{y}$, s, and s^2.
a. 11, 2, 2, 1, 9   b. 22, 9, 21, 15
c. 1, 0, 1, 10, 11, 11, 0   d. 4, 4, 4, 4

1.76   Tchebysheff's theorem states that at least $1 - (1/K^2)$ of a set of measurements will lie within K standard deviations of the mean of the data set. Use Tchebysheff's theorem to find the fraction of a set of measurements that will lie within:
a. 2 standard deviations of the mean ($K = 2$)
b. 3 standard deviations of the mean
c. 1.5 standard deviations of the mean

1.77 How strong an effect do characteristics such as brand name or store name have on a buyer's perception of the quality of a product? Numerous studies have been conducted to investigate this phenomenon, but the results seem to vary depending on the method used to analyze the data, type of product, price, etc. An article in the *Journal of Marketing Research* (Aug. 1989) summarized the results of 15 recent studies that investigated the effect of brand name on product quality and 17 recent studies that examined the effect of store name on quality. In all studies (the experimental units), an effect-size index was computed. The index ranges from 0 to 1; values closer to 0 indicate weaker effects and values near 1 indicate stronger effects. Stem-and-leaf displays of effect-size index for the two groups of studies are illustrated here. Compare and contrast the two figures. Which variable, brand name or store name, seems to have the stronger effect on perceived quality? Explain.

BRAND NAME (15 STUDIES)		STORE NAME (17 STUDIES)	
STEM	LEAF	STEM	LEAF
.6	0	.6	
.5	7	.5	
.4		.4	3 4
.3	4	.3	
.2	5 5	.2	
.1	0 1 1 2 4	.1	2
.0	3 3 5 5 7	.0	0 0 0 1 1 2 2 3 3 4 6 7 8 8

Source: Rao, A. R. and Monroe, K. B. "The effect of price, brand name, and store name on buyers' perceptions of product quality: An integrative review." *Journal of Marketing Research*, Vol. 26, Aug. 1989, p. 354 (Table 2).

1.78 Use Table 1 of Appendix D to find each of the following:
   a. $P(z \geq 2)$ b. $P(z \leq -2)$ c. $P(z \geq -1.96)$
   d. $P(z \geq 0)$ e. $P(z \leq -.5)$ f. $P(z \leq -1.96)$

1.79 Suppose the random variable y has mean $\mu = 30$ and standard deviation $\sigma = 5$. How many standard deviations away from the mean of y are each of the following y values?
   a. $y = 10$ b. $y = 32.5$ c. $y = 30$ d. $y = 60$

1.80 The variation in the rates of return on a bond is often used to measure the level of risk associated with buying the bond. The greater the variation, the higher the level of risk. Writing in his monthly "Your Finances" column for the *American Bar Association Journal*, Tom Potter presented the results of a simulation to compare the performance of zero-coupon bonds (those for which the interest coupons have been removed and which, therefore, pay no interest) to bonds with coupon payments attached. The rates of return on zero-coupon bonds are determined solely by the changes in bond prices, whereas the return on coupon-attached bonds depends also on the market interest rate. The simulation study yielded means and standard deviations of the rates of return for both zero-coupon and coupon-attached bonds at a fixed interest rate. The results for a market interest rate of 14% are shown in the table.

	ZERO-COUPON BONDS	COUPON-ATTACHED BONDS
Mean	12.18	12.48
Standard deviation	20.80	14.56

Source: Potter, T. "Your finances." *American Bar Association Journal*, Aug. 1984, Vol. 70, No. 8, pp. 160–161.

a. How many standard deviations away from the mean is a zero-coupon bond with a rate of return of −20%?

b. How many standard deviations away from the mean is a coupon-attached bond with a rate of return of −20%?

c. From your knowledge of the standard deviation, if a bond has a rate of return of −20%, is it more likely to be a zero-coupon bond or a coupon-attached bond? Explain.

1.81   Find a value of z, call it z_0, such that
a. $P(z \geq z_0) = .5$ b. $P(z \leq z_0) = .5199$
c. $P(z \geq z_0) = .3300$ d. $P(z_0 \leq z \leq .59) = .5845$

1.82   The random variable y has a normal distribution with $\mu = 80$ and $\sigma = 10$. Find the following probabilities:
a. $P(y \leq 75)$ b. $P(y \geq 90)$ c. $P(60 \leq y \leq 70)$
d. $P(y \geq 75)$ e. $P(y = 75)$ f. $P(y \leq 105)$

1.83   A firm believes the internal rate of return for its proposed investment can best be described by a normal distribution with mean 15% and standard deviation 3%. What is the probability that the internal rate of return for the investment will be:
a. Greater than 20% or less than 10%?
b. At least 6%?
c. More than 16.5%?

1.84   The *Tampa Bay Business Journal* recently surveyed the 25 largest certified public accountant (CPA) firms in the Tampa, Florida, area. The data in the accompanying table give the number of CPAs employed by each firm.

110	60	102	86	106
63	24	29	16	16
20	28	25	25	20
18	14	8	6	16
12	11	10	11	6

Source: *Tampa Bay Business Journal*, Mar. 8–14, 1991.

a. Compute $\bar{y}$, s^2, and s for the data set.
b. What percentage of the measurements would you expect to find in the interval $\bar{y} \pm 2s$?
c. Count the number of measurements that actually fall within the interval of part **b**, and express the interval count as a percentage of the total number of measurements. Compare this result with the answer to part **b**.

1.85   Foresters "cruising" British Columbia's boreal forest have determined that the diameter at breast height of white spruce trees in a particular community is approximately normal, with mean 17 meters and standard deviation 6 meters.*
a. Find the probability that the breast height diameter of a randomly selected white spruce in the forest community is less than 12 meters.
b. Suppose you observe a white spruce with a breast height diameter of 12 meters. Is this an unusual event? Explain.

*Scholz, H. "Fish Creek Community Forest: Exploratory statistical analysis of selected data." Working Paper, Northern Lights College, British Columbia, Canada.

c. Find the probability that the breast height diameter of a randomly selected white spruce in the forest community will exceed 37 meters.

d. Suppose you observe a tree in the forest community with a breast height diameter of 38 meters. Is this tree likely to be a white spruce? Explain.

1.86 When long-term resource allocation decisions (e.g., building a domed stadium or constructing a major highway system) meet with unexpected major setbacks, the decision maker must decide whether to abandon the project before irretrievable costs are incurred or to continue in the face of probable losses. Surprisingly, researchers have found that decision makers all too often persist or even allocate more resources in these situations. To explain this pattern of resource allocation behavior, a study was conducted at the University of Arizona.* Twenty undergraduate business school students were asked to make resource allocation decisions for fictional Sunburst Investments, which had undertaken construction of an office park and tennis club. For one portion of the study, each student was asked to rate the options of finishing and not finishing the project on a 7-point scale (1 = very negative, 7 = very positive). The differences between the ratings of the two options (finishing option rating minus not finishing option rating) were calculated, and the mean and standard deviation of the 20 sample differences obtained were

$$\bar{y} = -2.05 \qquad s = 1.85$$

a. Conduct a test to determine whether subjects rate the finishing option more negative, on average, than the option of not finishing the project (i.e., test the hypothesis that the mean difference in ratings is less than 0). Use $\alpha = .10$.

b. What assumptions are required for the test in part a to be valid?

1.87 Multinational corporations are firms with both domestic and foreign assets/investments. The foreign revenue (as a percentage of total revenue) generated by each of the top 20 U.S.-based multinational firms is listed in the accompanying table.

Exxon	73.2	Procter & Gamble	39.9
IBM	58.9	Philip Morris	19.6
GM	26.6	Eastman Kodak	40.9
Mobil	64.7	Digital	54.1
Ford	33.2	GE	12.4
Citicorp	52.3	United Technologies	32.9
EI duPont	39.8	Amoco	26.1
Texaco	42.3	Hewlett-Packard	53.3
ITT	43.3	Xerox	34.6
Dow Chemical	54.1	Chevron	20.5

Source: Forbes, July 23, 1990, pp. 362–363.

a. Construct a 90% confidence interval for the mean foreign revenue percentage of all U.S.-based multinational firms.

b. Interpret the interval, part a.

c. What assumption is required for the interval estimation procedure to be valid?

---

*Northcraft, G. B. and Neale, M. A. "Opportunity costs and the framing of resource allocation decisions." *Organizational Behavior and Human Decision Processes*, Vol. 37, No. 6, June 1986, pp. 348–356.

1.88   The National Institute for Occupational Safety and Health (NIOSH) recently completed a study to evaluate the level of exposure of workers to the chemical dioxin, 2, 3, 7, 8-TCDD. The distribution of TCDD levels in parts per trillion (ppt) of production workers at a Newark, New Jersey, chemical plant had a mean of 293 ppt and a standard deviation 847 ppt (*Chemosphere*, Vol. 20, 1990). A graph of the distribution is shown here. In a random sample of $n = 50$ workers selected at the New Jersey plant, let $\bar{y}$ represent the sample mean TCDD level.

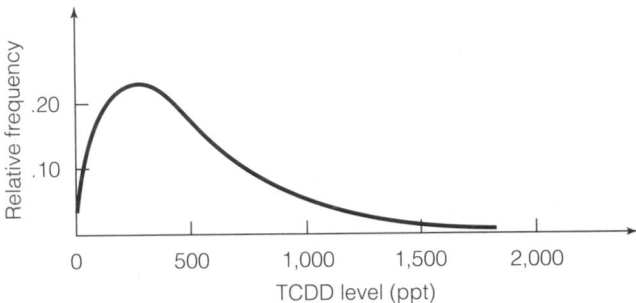

a. Find the mean and standard deviation of the sampling distribution of $\bar{y}$.
b. Draw a sketch of the sampling distribution of $\bar{y}$. Locate the mean on the graph.
c. Find the probability that $\bar{y}$ exceeds 550 ppt.

1.89   What are the top corporate executives being paid? To answer this question, *Business Week* magazine conducts a survey of corporate executives each year. *Business Week*'s 1990 survey of executives at 356 companies revealed only a 3.4% increase in salaries and bonuses from 1988 to 1989, the smallest

CORPORATE EXECUTIVE (COMPANY)	TOTAL 1989 CASH COMPENSATION (IN THOUSANDS)
Craig O. McCaw (McCaw Cellular)	$53,944
Steven J. Ross (Time Warner)	34,200
Donald A. Pels (Lin Broadcasting)	22,791
Jim P. Manzi (Lotus Development)	16,363
Paul Fireman (Reebok International)	14,606
Ronald K. Richey (Torchmark)	12,666
Martin S. Davis (Paramount)	11,635
Roberto C. Goizueta (Coca-Cola)	10,715
Michael D. Eisner (Walt Disney)	9,589
August A. Busch III (Anheuser-Busch)	8,861
William G. McGowan (MCI)	8,666
James R. Moffett (Freeport McMoRan)	7,300
Donald E. Petersen (Ford Motor)	7,147
P. Roy Vagelos (Merck)	6,764
W. Michael Blumenthal (Unisys)	6,511
S. Parker Gilbert (Morgan Stanley)	5,510
Harry A. Merlo (Louisiana-Pacific)	5,314
Reuben Mark (Colgate-Palmolive)	5,004
Robert J. Pfeiffer (Alexander & Baldwin)	4,943
William P. Stiritz (Ralston Purina)	4,854

Source: "Pay stubs of the rich and corporate." *Business Week*, May 7, 1990, p. 57.

increase since 1970. The top 20 corporate executives and their 1989 total cash compensations (salary plus bonus plus long-term compensation) are shown in the table. Assume that these represent a sample of the highest-paid corporate executives in the United States.

a. Calculate the mean and standard deviation of the sample of 1989 total cash compensations.
b. Describe the sampling distribution of the sample mean $\bar{y}$.
c. What is the probability that $\bar{y}$ will fall within 2 standard deviations of μ, the true mean compensation paid to the top corporate executives in 1989?

1.90 Let t_0 be a particular value of t. Use Table 2 of Appendix D to find the values such that the following statements are true:
a. $P(t \le t_0) = .10$ when $n = 23$
b. $P(t \ge t_0) = .005$ when $n = 3$
c. $P(t \le -t_0 \text{ or } t \ge t_0) = .05$ when $n = 7$
d. $P(t \le -t_0 \text{ or } t \ge t_0) = .01$ when $n = 24$

1.91 If rejection of the null hypothesis for a particular test would cause your firm to go out of business, would you want α to be small or large? Explain.

1.92 Due to recent advances in educational telecommunications, many colleges and universities are utilizing instruction by interactive television for "distance" education. For example, each semester Ball State University televises six graduate business courses to students at remote off-campus sites (*Journal of Education for Business*, Jan./Feb. 1991). To compare the performance of the off-campus MBA students at Ball State (who take the televised classes) to the on-campus MBA students (who have a "live" professor), a test devised by the American Assembly of Collegiate Schools of Business (AACSB) was administered to a sample of both groups of students. (The test included seven exams covering accounting, business strategy, finance, human resources, marketing, management information systems, and production and operations management.) The AACSB test scores (50 points maximum) are summarized in the table. Based on these results, the researchers report that "there was no significant difference between the two groups of students."

	MEAN	STANDARD DEVIATION
On-campus students	41.93	2.86
Off-campus TV students	44.56	1.42

Source: Arndt, T. L. and LaFollette, W. R. "Interactive television and the nontraditional student." *Journal of Education for Business*, Jan./Feb. 1991, p. 184.

a. Note that the sample sizes were not given in the journal article. Assuming 50 students are sampled from each group, perform the desired analysis. Do you agree with the researchers findings?
b. Repeat part a, but assume 15 students are sampled from each group.

1.93 To conduct the hypothesis test required in Exercise 1.92b, you had to assume that $\sigma_1^2 = \sigma_2^2$. Perform an F test with $\alpha = .10$ to check that assumption. Explain the significance of your result.

1.94 A company is interested in estimating the mean number of days of sick leave, μ, taken by all its employees. The firm's statistician selects at random 100 personnel files and notes the number of sick days taken by each employee. The following sample statistics are computed:

$$\bar{y} = 12.2 \text{ days} \qquad s = 10 \text{ days}$$

a. Estimate μ using a 90% confidence interval.

b. How many personnel files would the statistician have to select in order to estimate μ to within 2 days with 99% confidence?

c. Do the data support the research hypothesis that μ, the mean number of sick days taken by the employees, is greater than 10.9 days? Test using $\alpha = .05$.

1.95 C. S. Patterson (Concordia University, Montreal) conducted an investigation of the financing policies and practices of large regulated public utilities (*Financial Management*, Summer 1984). One goal of Patterson's research was to determine whether public utilities operate at a debt level that maximizes shareholder wealth. A sample of 47 publicly held electric utilities identified in *Moody's Manual* as having revenues of $300 million or more took part in the study. The actual 1982 debt ratios (defined as long-term debt divided by total capital) of the companies were recorded, with the following results:

$$\bar{y} = .485 \qquad s = .029$$

Prior to giving their actual 1982 debt ratios, the companies estimated the mean debt ratio at which they should operate to maximize shareholder wealth as .459. Is there sufficient evidence to indicate that the actual mean 1982 debt ratio of public utilities differs from the optimum value of .459? Test using $\alpha = .10$.

1.96 Studies have shown that in a nonbusiness (e.g., academic) setting, those who tend to have job mobility are predominantly better performers. To examine the performance turnover relationship in a business setting, G. F. Dreher (University of Kansas) examined the personnel records of a large national oil company. Dreher's sample consisted of 174 employees who were classified as "stayers" (those who stayed with the company from 1964 through 1979) and 355 former employees who were classified as "leavers" (those who left the company at varying points during the 15-year period). The company's annual performance appraisals corresponding to the initial years of service were used to form an initial performance rating for each employee. Summary statistics on initial performance for the two groups of employees are provided in the table. Is there evidence of a difference between the mean initial performance ratings of "stayers" and "leavers"? Test using $\alpha = .01$.

STAYERS	LEAVERS
$n_1 = 174$	$n_2 = 355$
$\bar{y}_1 = 3.51$	$\bar{y}_2 = 3.24$
$s_1 = .51$	$s_2 = .52$

Source: Dreher, G. H. "The role of performance in the turnover process." *Academy of Management Journal*, Mar. 1982, Vol. 25, No. 1, pp. 137–147.

1.97 Refer to Exercise 1.96. Dreher also measured the rates of career advancement (number of promotions per year) for each employee. The data are summarized in the table. Suppose we want to compare the variation in rates of career advancement for the two groups. Do the data provide sufficient evidence to indicate that the variation in rates of career advancement differs for "leavers" and "stayers"? Test using $\alpha = .02$.

STAYERS	LEAVERS
$n_1 = 174$	$n_2 = 355$
$\bar{y}_1 = .43$	$\bar{y}_2 = .31$
$s_1 = .20$	$s_2 = .31$

References

McClave, J. T. and Benson, P. G. *A First Course in Business Statistics,* 5th ed. San Francisco: Dellen, 1992.

Mendenhall, W. *Introduction to Probability and Statistics*, 8th ed. Boston: Duxbury, 1989.

Sincich, T. *Business Statistics by Example*, 4th ed. San Francisco: Dellen, 1992.

Introduction to Regression Analysis

2

CONTENTS

2.1 Modeling a Response

2.2 Overview of Regression Analysis

2.3 Regression Applications

2.4 Collecting the Data for Regression

OBJECTIVE

To explain the concept of a statistical model; to describe applications of regression

Many applications of inferential statistics are much more complex than the methods presented in Chapter 1. Often, you will want to use sample data to investigate the relationships among a group of variables, ultimately to create a model for some variable (e.g., profit, productivity, price, etc.) that can be used to predict its value in the future. The process of finding a mathematical model (an equation) that best fits the data is part of a statistical technique known as **regression analysis**.

2.1
Modeling a Response

Suppose a tax assessor wants to adjust the appraised values of all residential properties in Tampa, Florida. One way to do this is to select a random sample of residential sales during the past year, note the percentage increase y in each over the current appraised value, and then use these percentage increases to estimate the true mean percentage increase of all residential properties in Tampa. The assessor could then adjust the appraised value of each residential property in the city by this estimated percentage.

Adjusting the appraised value of every residential property in Tampa by the mean percentage increase in property value is tantamount to using the mean percentage increase as a **model** for the true percentage increase in the value of each residential property.

In regression, the variable y to be modeled is called the **dependent (or response) variable** and its true mean (or **expected value**) is denoted $E(y)$. In this example,

y = percentage increase in residential property value

$E(y)$ = mean percentage increase of all residential properties

---

Definition 2.1
The variable to be predicted (or modeled), y, is called the **dependent (or response) variable**.

The assessor knows that the actual value of y for a particular property will depend on location, square footage of heated space, lot size, and many other factors. Consequently, the real percentage increases in value for all residential properties may have the distribution shown in Figure 2.1. Thus, the assessor is modeling the percentage increase in value y for a particular property by stating that y is equal to the mean increase $E(y)$ plus or minus some random amount, which is unknown to the assessor; that is,

$y = E(y) +$ Random error

Since the assessor does not know the value of the random error for a particular property, a reasonable strategy would be to increase the assessed value by the estimate of the mean amount $E(y)$.

FIGURE 2.1

Distribution of percentage increase in residential property values

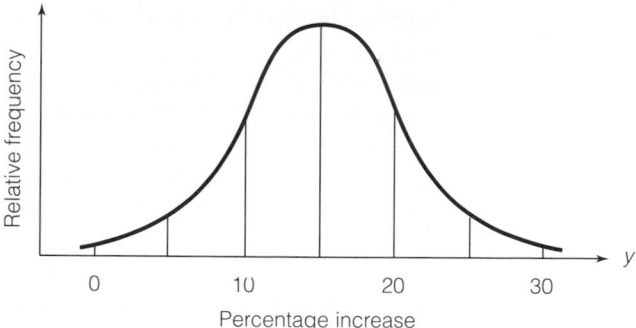

This model is called a **probabilistic model** for y. The adjective *probabilistic* comes from the fact that, when certain assumptions about the model are satisfied, we can make a probability statement about the magnitude of the deviation between y and $E(y)$. For example, if y is normally distributed with mean 15% and standard deviation 5% (as shown in Figure 2.1), then the probability that y will fall within 2 standard deviations (i.e., 10%) of its mean is .95. The probabilistic model shown in the box is the foundation of all models considered in this text.

| General Form of Probabilistic Model in Regression

$$y = E(y) + \varepsilon$$

where

y = Dependent variable

$E(y)$ = Mean (or expected) value of y

ε = Unexplainable, or random, error

In practice we will need to use sample data to estimate the parameters of the probabilistic model—namely, the mean $E(y)$ and the random error ε. In Chapter 3 we will learn a standard assumption in regression: the mean error is 0. Based on this assumption, our best estimate of ε is 0. Thus, we need only estimate $E(y)$.

The simplest method of estimating $E(y)$ is to use the technique of Section 1.7. For example, the tax assessor could select a random sample of residential sales during the past year and note the percentage increase, y, in each sale over the current assessed value. The sample mean y could be used as an estimate of the true mean percentage increase $E(y)$. Denoting the predicted value of y as $\bar{y}$, the prediction equation for the simple model is $\hat{y} - \bar{y}$. Therefore, with this simple model, the sample mean percentage increase $\bar{y}$ is used to predict the percentage increase y for any property.

Suppose the assessor estimates the mean increase in residential property values to be $\bar{y} = 10\%$ and, for the sake of argument, let us assume this is a very accurate estimate of $E(y)$. Further, suppose you own a house in the city and you are advised by an experienced appraiser that an increase of 10% in the assessed value of your property (and subsequent property taxes) is too high. How can you justify the disagreement between the appraiser and the tax assessor? Both possess models that predict property values, but they are led to different conclusions.

If you place yourself in the position of the appraiser, you will be able to answer this question. The appraiser has actually seen the house and is, therefore, using a more sophisticated model than is the assessor. In particular, the appraiser is taking into consideration a number of variables, called **independent variables**,* that are highly related to the value of a residential property (such as location, square footage, lot size, and so forth), and is using these variables to obtain a more accurate model for the percentage increase in residential property values. This model might be based purely on experience and exist only in the appraiser's mind. Or, the appraiser might have obtained a mathematical model (an equation) relating percentage price increase to square footage, location, and so forth, by sampling individual residential sales and measuring the percentage increase, square footage, location, and so forth, of each. The process of finding the mathematical model that relates y to a set of independent variables and best fits the data is part of the process known as **regression analysis**.

Definition 2.2

The variables used to predict (or model) y are called **independent variables** and are denoted by the symbols x_1, x_2, x_3, etc.

For example, suppose the appraiser decided to relate percentage price increase y to a single independent variable x, defined as the square footage of heated space in the residence. The appraiser might select a random sample of residential sales, record y and x for each sale, and then plot them on a graph as shown in Figure 2.2. Finding the equation of the smooth curve that best fits the data points is part of a regression analysis. Once obtained, this equation (a graph of which is superimposed on the data points in Figure 2.2) provides a model for estimating the mean percentage increase for houses with any specific amount of heated floor space. The appraiser can use the model to predict the percentage price increase for an individual residential property that has a known amount of floor space. As you can see from Figure 2.2, the model would also predict with some error (most of the points do not lie exactly on the curve), but the error of prediction will be much less than the error obtained using the model represented in Figure 2.1. As shown in Figure 2.1, a good estimate of the percentage increase for a

---

*The word *independent* should not be interpreted in a probabilistic sense. The phrase *independent variable* is used in regression analysis to refer to a predictor variable for the response y.

FIGURE 2.2

Relating percentage price increase in
residential property to heated floor space

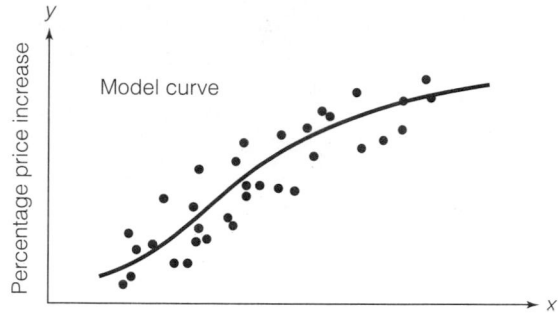

Square footage of heated floor space

residential property would be a value near the center of the distribution, say, the mean. Since this prediction does not take square footage of floor space into account, the error of prediction will be larger than the error of prediction for the model of Figure 2.2. Consequently, we would state that the model utilizing information provided by the independent variable, square footage, is superior to the model represented in Figure 2.1.

2.2

Overview of Regression Analysis

Regression analysis is a branch of statistical methodology concerned with relating a response y to a set of independent, or predictor, variables $x_1, x_2, \ldots, x_k$. The goal is to build a good model—a prediction equation relating y to the independent variables—that will enable us to predict y for given values of $x_1, x_2, \ldots, x_k$, and to do so with a small error of prediction. When using the model to predict y for a particular set of values of $x_1, x_2, \ldots, x_k$, we will want a measure of the reliability of our prediction. That is, we will want to know how large the error of prediction might be. All these elements are parts of a regression analysis, and the resulting prediction equation is often called a **regression model**.

For example, a property appraiser might like to relate percentage price increase y to the two quantitative independent variables x_1, square footage of heated space, and x_2, lot size. This model could be represented by a **response surface** (see Figure 2.3 on page 86) that traces the mean percentage price increase $E(y)$ for various combinations of x_1 and x_2. To predict the percentage price increase y for a given residential property with $x_1 = 2,000$ square feet of heated space and lot size $x_2 = .7$ acre, you would locate the point $x_1 = 2,000, x_2 = .7$ on the x_1, x_2-plane (see Figure 2.3). The height of the surface above that point gives the mean percentage increase in price $E(y)$; this is a reasonable value to use to predict the percentage price increase for a property with $x_1 = 2,000$ and $x_2 = .7$.

The response surface is a convenient method for modeling a response y that is a function of two quantitative independent variables, x_1 and x_2. The mathematical equivalent of the response surface shown in Figure 2.3 might be given by the **deterministic model**

$$E(y) = \beta_0 + \beta_1 x_1 + \beta_2 x_2 + \beta_3 x_1 x_2 + \beta_4 x_1^2 + \beta_5 x_2^2$$

FIGURE 2.3

Mean percentage price increase as a function of heated square footage, x_1, and lot size, x_2

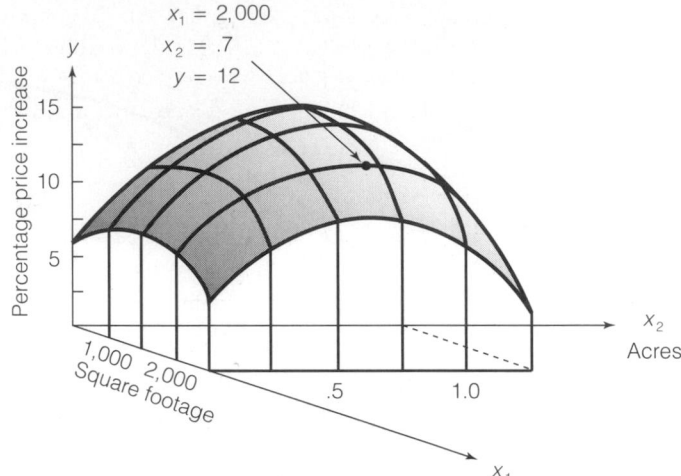

where $E(y)$ is the mean percentage price increase for a set of values x_1 and x_2, and $\beta_0, \beta_1, \ldots, \beta_5$ are constants (or weights) with values that would have to be estimated from the sample data. Note that the model for $E(y)$ is deterministic because, if the constants $\beta_0, \beta_1, \ldots, \beta_5$ are known, the values of x_1 and x_2 determine exactly the value of $E(y)$.

Replacing $E(y)$ with $\beta_0 + \beta_1 x_1 + \beta_2 x_2 + \beta_3 x_1 x_2 + \beta_4 x_1^2 + \beta_5 x_2^2$ in the probabilistic model for y, we obtain the full equation for y:

$$y = \beta_0 + \beta_1 x_1 + \beta_2 x_2 + \beta_3 x_1 x_2 + \beta_4 x_1^2 + \beta_5 x_2^2 + \varepsilon$$

Now the property appraiser would obtain a sample of residential properties and record square footage, x_1, and lot size, x_2, in addition to percentage increase y in assessed value (see Section 2.4). Subjecting the sample data to a regression analysis will yield estimates of the model parameters and enable the appraiser to predict percentage increase y for a particular property. The prediction equation takes the form

$$\hat{y} = \hat{\beta}_0 + \hat{\beta}_1 x_1 + \hat{\beta}_2 x_2 + \hat{\beta}_3 x_1 x_2 + \hat{\beta}_4 x_1^2 + \hat{\beta}_5 x_2^2$$

where $\hat{y}$ is the predicted value of y, and $\hat{\beta}_0, \hat{\beta}_1, \ldots, \hat{\beta}_5$ are estimates of the model parameters.

In practice, the appraiser would construct a deterministic model for $E(y)$ that takes into account other quantitative variables, as well as qualitative independent variables, such as location and type of construction. In the following chapters we will show how to construct a model relating a response to both quantitative and qualitative independent variables, and we will fit the model to a set of sample data using a regression analysis.

The preceding description of regression analysis is oversimplified, but it provides a preliminary view of the methodology that is the subject of this text. In addition to predicting y for specific values of $x_1, x_2, \ldots, x_k$, a regression model can also be used to estimate the mean value of y for given values of $x_1, x_2, \ldots,$

x_k and to answer other questions concerning the relationship between y and one or more of the independent variables. The practical values attached to these inferences will be illustrated by examples in the following chapters.

We conclude this section with a summary of the major steps involved in a regression analysis.

Regression Modeling: Six-Step Procedure

1. Hypothesize a model for $E(y)$.
2. Collect the sample data.
3. Use the sample data to estimate unknown parameters in the model.
4. Specify the probability distribution of the random error term and estimate any unknown parameters of this distribution.
5. Statistically check the usefulness of the model.
6. When satisfied that the model is useful, use it for prediction, estimation, and so on.

2.3
Regression Applications

Regression analysis of data is a very powerful statistical tool. It provides a technique for building a statistical predictor of a response and enables you to place a bound (an approximate upper limit) on your error of prediction. For example, suppose you manage a construction company and you would like to predict the profit y per construction job as a function of a set of independent variables x_1, $x_2, \ldots, x_k$. If you could find the right combination of independent variables and could postulate a reasonable mathematical equation to relate y to these variables, you could possibly deduce which of the independent variables were causally related to profit per job and then control these variables to achieve a higher profit. In addition, you could use the forecasts in corporate planning. The following examples illustrate a few of the many useful applications of regression analysis in business and economics.

EXAMPLE 2.1

Real estate. As suggested in Section 2.2, regression models for predicting property values are used by tax assessors, appraisers, and real estate investment managers to forecast (or predict) the value of a parcel of real estate.

EXAMPLE 2.2

Management. To reward their executives appropriately, many large corporations receive advice from consulting firms regarding the amount of compensation that each executive should receive. To provide this advice, the consulting firm collects information on the compensation y received by a large number of corporate executives. For each of these executives, the firm records the values of many independent variables, some of which are the following:

1. Experience (years)
2. College education (years)
3. Number of employees supervised
4. Corporate assets (dollars)
5. Age of the executive
6. Whether the executive is on the company's board of directors (1 if yes; 0 if no)
7. Whether the executive has international responsibility (1 if yes; 0 if no)

The consulting firm then uses a regression analysis to build a good prediction equation for y, an executive's annual compensation, as a function of the independent variables listed here. If it is successful, the company can sell its services to both participating and nonparticipating corporations, providing them with reasonable compensation projections for their executives.

EXAMPLE 2.3

Finance.   Some commodity traders have successfully used regression analysis to forecast the price y of a commodity. For example, we might expect the supply of wheat, and hence its price y, to be directly related to (1) the amount of wheat planted in the United States, the Soviet Union, Europe, and so forth; (2) the weather conditions in various locations; and (3) the projected sizes of the crops of competing grains. Using the recorded prices of wheat at a specific time over a period of a year as well as corresponding information on the three independent variables listed, the commodity trader could use a regression analysis to build a model relating wheat price y to the independent variables. The model could be used to predict the price of wheat at the time of year used in the data collection, based on the present values of the independent variables. If the general economic conditions remained stable and representative of the conditions prevalent over the period during which the sample data were observed, the regression analysis would provide a measure of the prediction error that would apply to predictions of future wheat prices.

2.4

Collecting the Data for Regression

Recall from Section 2.2 that the initial step in regression analysis is to hypothesize a deterministic model for the mean response, $E(y)$, as a function of one or more independent variables. Once a model for $E(y)$ has been hypothesized, the next step is to collect the sample data that will be used to estimate the unknown model parameters (β's). This entails collecting observations on both the response y and the independent variables, $x_1, x_2, \ldots, x_k$, for each experimental unit in the sample. Thus, a sample to be analyzed by regression includes observations on several variables $(y, x_1, x_2, \ldots, x_k)$, not just a single variable.

The data for regression can be of two types: **observational** or **experimental**. Observational data are obtained if no attempt is made to control the values of the independent variables (x's). For example, suppose you want to relate an executive's compensation y to the set of predictors listed in Example 2.2. One way to obtain the data for regression is to select a random sample of $n = 100$ executives and record the value of y and the values of each of the predictor variables. The data for the first five executives in the sample is displayed in Table 2.1. Note that in this example, the x values, such as experience, college education, number of employees supervised, etc., for each executive are not specified in advance of observing salary y; that is, the x values were uncontrolled. Therefore, the sample data is observational.

Definition 2.3

If the values of the independent variables (x's) in regression are uncontrolled (i.e., not set in advance before the value of y is observed) but are measured without error, the data are **observational**.

TABLE 2.1 **Observational Data**

	EXECUTIVE				
	1	2	3	4	5
Annual compensation, y ($)	85,420	61,333	107,500	59,225	98,400
Experience, x_1 (years)	8	2	7	3	11
College education, x_2 (years)	4	8	6	7	2
No. of employees supervised, x_3	13	6	24	9	4
Corporate assets, x_4 (millions)	1.60	.25	3.14	.10	2.22
Age, x_5 (years)	42	30	53	36	51
Board of directors, x_6	0	0	1	0	1
International responsibility, x_7	1	0	1	0	0

How large a sample should be selected when regression is applied to observational data? In Section 1.7 we learned that when estimating a population mean, the sample size n will depend on (1) the (estimated) population standard deviation, (2) the confidence level, and (3) the desired half-width of the confidence interval used to estimate the mean. Because regression involves estimation of the mean response, $E(y)$, the sample size will depend on these three factors. The problem, however, is not as straightforward as in Section 1.7, since $E(y)$ is modeled as a function of a set of independent variables, and the additional parameters in the model (i.e., the β's) must also be estimated. In regression the sample size should be large enough so that the β's are both estimable and testable. This will not occur unless n is at least as large as the number of β parameters included in the model for $E(y)$. To ensure a sufficiently large sample, a good rule of thumb is to select n greater than or equal to 10 times the number of β parameters in the model.

For example, suppose the consulting firm wants to use the following model for annual compensation, y, of a corporate executive:

$$E(y) = \beta_0 + \beta_1 x_1 + \beta_2 x_2 + \cdots + \beta_7 x_7$$

where $x_1, x_2, \ldots, x_7$ are defined in Example 2.2. Excluding β_0, there are seven β parameters in the model; thus, the firm should include at least $10 \times 7 = 70$ corporate executives in its sample.

The second type of data in regression, *experimental data*, are generated by designed experiments where the values of the independent variables are set in advance (i.e., controlled) before the value of y is observed. For example, if a production supervisor wants to investigate the effect of two quantitative independent variables, say, temperature x_1 and pressure x_2, on the purity of batches of a chemical, the supervisor might decide to employ three values of temperature (100°C, 125°C, and 150°C) and three values of pressure (50, 60, and 70 pounds per square inch) and to produce and measure the impurity y in one batch of chemical for each of the $3 \times 3 = 9$ temperature–pressure combinations (see Table 2.2). For this experiment, the settings of the independent variables are controlled, in contrast to the uncontrolled nature of observational data in the real estate sales example.

Definition 2.4

If the values of the independent variables (x's) in regression are controlled using a designed experiment (i.e., set in advance before the value of y is observed), the data are **experimental**.

TABLE 2.2 **Experimental Data**

TEMPERATURE, x_1	PRESSURE, x_2	IMPURITY, y
100	50	2.7
	60	2.4
	70	2.9
125	50	2.6
	60	3.1
	70	3.0
150	50	1.5
	60	1.9
	70	2.2

In business studies, it is usually not possible to control the values of the x's; consequently, most data collected for regression applications are observational. (Consider the regression analysis in Example 2.2. Clearly, it is impossible or impractical to control the values of the independent variables.) Therefore, you may want to know why we distinguish between the two types of data. We will learn (Chapter 6) that inferences made from regression studies based on observational data have more limitations than those based on experimental data. In particular, we will learn that establishing a cause-and-effect relationship between variables is much more difficult with observational data than with experimental data.

The majority of the examples and exercises in Chapters 3–9 are based on observational data. In Chapters 10–11 we describe regression analyses based on data collected from a designed experiment.

Summary

Better business management requires a better understanding of the phenomena that affect the variables that measure a business's health—namely, profit, sales, inventory, various measures of product demand, and others. To achieve this understanding, we seek the assistance of mathematical models that relate the mean value of a response (such as profit) to various variables such as the advertising budget, size of inventory, and so forth. Since we know that even a perfect mathematical description of this relationship will still predict the response with error, we include a random component in the model to account for the many other variables that have been purposely or inadvertently excluded.

The mathematical relationship that we have described forms a model for the relative frequency distribution of the population of response measurements that would be generated when the process is in a specific state. For example, a model might represent a relative frequency distribution of the population of monthly profits that a business might generate, now and in the immediate future, when the business is operating with a \$1,000,000 inventory and a \$500,000 annual advertising budget. Estimating the unknown parameters for this population, i.e., the unknown parameters in the model, and using the model to make predictions with known reliability, is the objective of a **regression analysis**.

The sample data for regression can be either **observational** (in which the values of the x's are uncontrolled) or **experimental** (in which the x's are set in advance of observing y). For practical reasons most regression applications in business are based on observational data.

Simple Linear Regression

CONTENTS

3.1   Introduction

3.2   The Straight-Line Probabilistic Model

3.3   Fitting the Model: The Method of Least Squares

3.4   Model Assumptions

3.5   An Estimator of σ^2

3.6   Assessing the Utility of the Model: Making Inferences About the Slope β_1

3.7   The Coefficient of Correlation

3.8   The Coefficient of Determination

3.9   Using the Model for Estimation and Prediction

3.10 Simple Linear Regression: An Example

3.11 Regression Through the Origin (Optional)

3.12 A Summary of the Steps to Follow in a Simple Linear Regression Analysis

OBJECTIVE

To present the basic concepts of regression analysis based on a simple linear relation between a response y and a single predictor variable x

3.1
Introduction

As noted in Chapter 2, much business research is devoted to the topic of **modeling**, i.e., trying to describe how variables are related. For example, an econometrician might be interested in modeling the relationship between the level of consumption expenditure and the level of disposable personal income. An advertising agency might want to know the relationship between a firm's sales revenue and the amount spent on advertising. And an investment firm may be interested in relating the performance of the stock market to the current discount rate of the Federal Reserve Board.

The simplest graphical model for relating a response variable y to a single independent variable x is a straight line. In this chapter we will discuss **simple linear (straight-line) models** and will show how to fit them to a set of data points using the **method of least squares**. We will then show how to judge whether a relationship exists between y and x, and how to use the model either to estimate $E(y)$, the mean value of y, or to predict a future value of y for a given value of x. The totality of these methods is called a **simple linear regression analysis**.

Most models for business response variables are much more complicated than implied by a straight-line relationship. Nevertheless, the methods of this chapter are very useful, and they set the stage for the formulation and fitting of more complex models in succeeding chapters. Thus, this chapter will provide an intuitive justification for the techniques employed in a regression analysis, and it will identify most of the types of inferences that we will want to make using a **multiple regression analysis** later in this book.

3.2
The Straight-Line Probabilistic Model

An important consideration in merchandising a product is the amount of money spent on advertising. Suppose you want to model the monthly sales revenue y of an appliance store as a function of the monthly advertising expenditure x. The first question to be answered is this: Do you think an exact (deterministic) relationship exists between these two variables? That is, can the exact value of sales revenue be predicted if the advertising expenditure is specified? We think you will agree that this is not possible for several reasons. Sales depend on many variables other than advertising expenditure—for example, time of year, state of the general economy, inventory, and price structure. However, even if many variables are included in the model (the topic of Chapter 4), it is still unlikely that we can predict the monthly sales *exactly*. There will almost certainly be some variation in sales due strictly to **random phenomena** that cannot be modeled or explained.

Consequently, we need to propose a probabilistic model for sales revenue that accounts for this random variation:

$$y = E(y) + \varepsilon$$

The random error component, ε, represents all unexplained variations in sales caused by important but omitted variables or by unexplainable random phenonmena.

As you will subsequently see, the random error ε will play an important role in testing hypotheses or finding confidence intervals for the deterministic portion of the model and will enable us to estimate the magnitude of the error of prediction when the model is used to predict some value of y to be observed in the future.

We begin with the simplest of probabilistic models—a **first-order linear model*** that graphs as a straight line. The elements of the straight-line model are summarized in the box.

A First-Order (Straight-Line) Model

$$y = \beta_0 + \beta_1 x + \varepsilon$$

where

$y =$ **Dependent** variable (variable to be modeled—sometimes called the **response** variable)

$x =$ **Independent** variable (variable used as a **predictor** of y)

$E(y) = \beta_0 + \beta_1 x =$ Deterministic component

$\varepsilon =$ (epsilon) = Random error component

$\beta_0 =$ (beta zero) = y-intercept of the line, i.e., point at which the line intercepts or cuts through the y-axis (see Figure 3.1)

$\beta_1 =$ (beta one) = Slope of the line, i.e., amount in increase (or decrease) in the mean of y for every 1-unit increase in x (see Figure 3.1)

FIGURE 3.1

The straight-line model

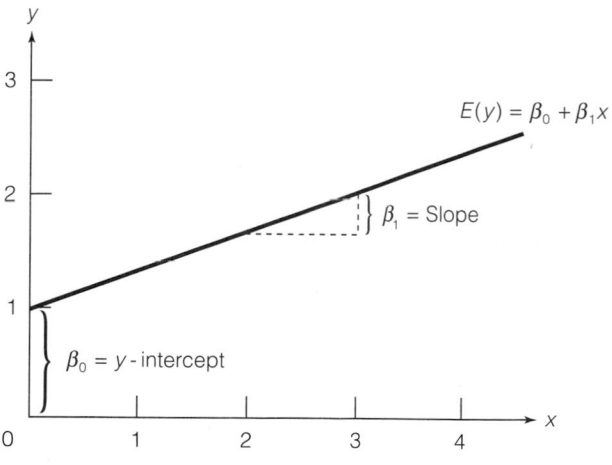

---

*A general definition of the expression *first-order* is given in Section 7.3.

In Section 3.4 we make the standard assumption that the average of the random errors is zero, i.e., $E(\varepsilon) = 0$. Then the deterministic component of the straight-line probabilistic model represents the line of means $E(y) = \beta_0 + \beta_1 x$. Note that we use Greek symbols β_0 and β_1 to represent the y-intercept and slope of the line. They are population parameters with numerical values that will be known only if we have access to the entire population of (x, y) measurements.

Recall from Section 2.2 that it is helpful to think of regression modeling as a six-step procedure:

STEP 1 Hypothesize the deterministic component of the probabilistic model.

STEP 2 Collect the sample data.

STEP 3 Use the sample data to estimate unknown parameters in the model.

STEP 4 Specify the probability distribution of the random error term, and estimate any unknown parameters of this distribution.

STEP 5 Statistically check the usefulness of the model.

STEP 6 When satisfied that the model is useful, use it for prediction, estimation, and so on.

In this chapter we will skip step 1 and deal only with the straight-line model. Chapters 4 and 7 will discuss how to build more complex models.

EXERCISES

3.1   In each case graph the line that passes through the points.
    a. (0, 2) and (2, 6)   **b.** (0, 4) and (2, 6)
    c. (0, − 2) and (−1, −6)   **d.** (0, −4) and (3, −7)

3.2   The equation for a straight line (deterministic) is

$$y = \beta_0 + \beta_1 x$$

If the line passes through the point (0, 1), then $x = 0$, $y = 1$ must satisfy the equation. That is,

$$1 = \beta_0 + \beta_1(0)$$

Similarly, if the line passes through the point (2, 3), then $x = 2$, $y = 3$ must satisfy the equation:

$$3 = \beta_0 + \beta_1(2)$$

Use these two equations to solve for β_0 and β_1, and find the equation of the line that passes through the points (0, 1) and (2, 3).

3.3   Find the equations of the lines passing through the four sets of points given in Exercise 3.1.

3.4   Plot the following lines:
    a. $y = 3 + 2x$ **b.** $y = 1 + x$ **c.** $y = -2 + 3x$
    d. $y = 5x$ **e.** $y = 4 - 2x$

3.5   Give the slope and y-intercept for each of the lines defined in Exercise 3.4.

3.3

Fitting the Model: The Method of Least Squares

Suppose an appliance store conducts a 5-month experiment to determine the effect of advertising on sales revenue. The results are shown in Table 3.1. (The number of measurements is small and the measurements themselves are unrealistically simple to avoid arithmetic confusion in this initial example.) The straight-line model is hypothesized to relate sales revenue y to advertising expenditure x. That is,

$$y = \beta_0 + \beta_1 x + \varepsilon$$

The question is this: How can we best use the information in the sample of five observations in Table 3.1 to estimate the unknown y-intercept β_0 and slope β_1?

To gain some information on the approximate values of these parameters, it is helpful to plot the sample data. Such a plot, called a scattergram, locates each of the five data points on a graph, as in Figure 3.2. Note that the scattergram suggests a general tendency for y to increase as x increases. If you place a ruler on the scattergram, you will see that a line may be drawn through three of the five points, as shown in Figure 3.3 on page 98. To obtain the equation of this visually fitted line, notice that the line intersects the y-axis at $y = -1$, so the y-

TABLE 3.1

MONTH	ADVERTISING EXPENDITURE x, hundreds of dollars	SALES REVENUE y, thousands of dollars
1	1	1
2	2	1
3	3	2
4	4	2
5	5	4

FIGURE 3.2

Scattergram for data in Table 3.1

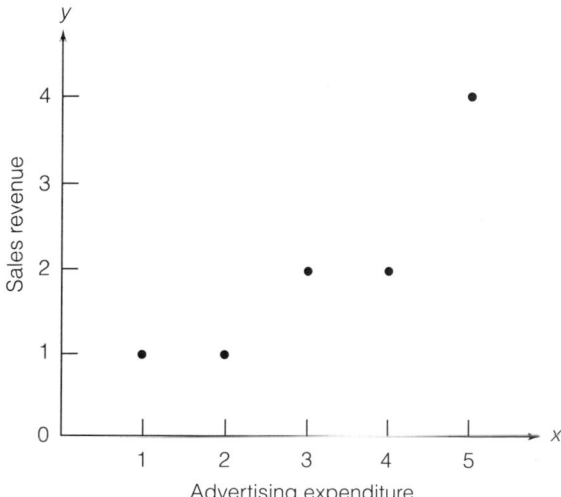

Visual straight-line fit to data in Table 3.1

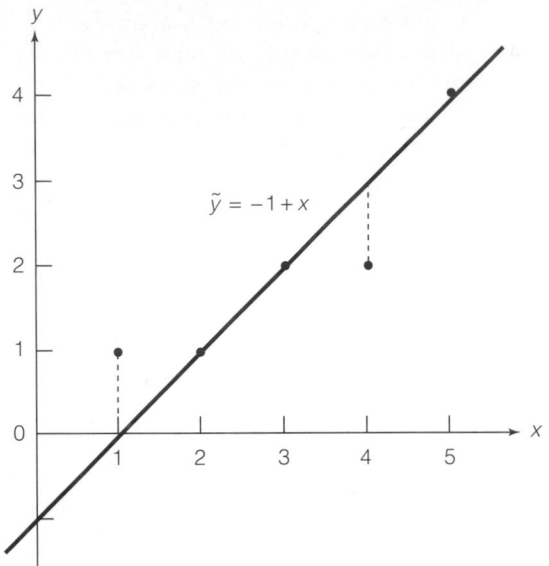

$$\tilde{y} = -1 + x$$

intercept is -1. Also, y increases exactly 1 unit for every 1-unit increase in x, indicating that the slope is $+1$. Therefore, the equation is

$$\tilde{y} = -1 + 1(x) = -1 + x$$

where $\tilde{y}$ is used to denote the predictor of y based on the visually fitted model.

One way to decide quantitatively how well a straight line fits a set of data is to determine the extent to which the data points deviate from the line. For example, to evaluate the visually fitted model in Figure 3.3, we calculate the magnitude of the **deviations**, i.e., the differences between the observed and the predicted values of y. These deviations, or **errors**, are the vertical distances between observed and predicted values of y (see Figure 3.3). The observed and predicted values of y, their differences, and their squared differences are shown in Table 3.2. Note that the **sum of the errors** equals 0 and the **sum of squares of the errors (SSE)**, which gives greater emphasis to large deviations of the points from the line, is equal to 2.

T A B L E 3.2 **Comparing Observed and Predicted Values for the Visual Model**

x	y	$\tilde{y} = -1 + x$	$(y - \tilde{y})$	$(y - \tilde{y})^2$
1	1	0	$(1 - 0) =$ 1	1
2	1	1	$(1 - 1) =$ 0	0
3	2	2	$(2 - 2) =$ 0	0
4	2	3	$(2 - 3) = -1$	1
5	4	4	$(4 - 4) =$ 0	0
			Sum of errors = 0	Sum of squared errors (SSE) = 2

You can see by shifting the ruler around the graph that is possible to find many lines for which the sum of the errors is equal to 0, but it can be shown that there is one (and only one) line for which the *SSE is a minimum*. This line is called the **least squares line**, **regression line**, or **least squares prediction equation**.

To find the least squares line for a set of data, assume that we have a sample of n data points that can be identified by corresponding values of x and y, say, (x_1, y_1), (x_2, y_2), ..., (x_n, y_n). For example, the $n = 5$ data points shown in Table 3.2 are $(1, 1)$, $(2, 1)$, $(3, 2)$, $(4, 2)$, and $(5, 4)$. The straight-line model for the response y in terms of x is

$$y = \beta_0 + \beta_1 x + \varepsilon$$

The line of means is

$$E(y) = \beta_0 + \beta_1 x$$

and the fitted line, which we hope to find, is represented as

$$\hat{y} = \hat{\beta}_0 + \hat{\beta}_1 x$$

The "hats" can be read as "estimator of." Thus, $\hat{y}$ is an estimator of the mean value of y, $E(y)$, and a predictor of some future value of y; and $\hat{\beta}_0$ and $\hat{\beta}_1$ are estimators of β_0 and β_1, respectively.

For a given data point, say, the point (x_i, y_i), the observed value of y is y_i and the predicted value of y would be obtained by substituting x_i into the prediction equation:

$$\hat{y}_i = \hat{\beta}_0 + \hat{\beta}_1 x_i$$

The deviation of the ith value of y from its predicted value, called the **ith residual**, is

$$(y_i - \hat{y}_i) = [y_i - (\hat{\beta}_0 + \hat{\beta}_1 x_i)]$$

Then the sum of squares of the deviations of the y values about their predicted values (i.e., the **sum of squares of residuals**) for all of the n data points is

$$SSE = \sum_{i=1}^{n} [y_i - (\hat{\beta}_0 + \hat{\beta}_1 x_i)]^2$$

The quantities $\hat{\beta}_0$ and $\hat{\beta}_1$ that make the SSE a minimum are called the **least squares estimates** of the population parameters β_0 and β_1, and the prediction equation $\hat{y} = \hat{\beta}_0 + \hat{\beta}_1 x$ is called the **least squares line**.

Definition 3.1

The **least squares line** is one that has a smaller SSE than any other straight-line model.

The values of $\hat{\beta}_0$ and $\hat{\beta}_1$ that minimize the SSE are given by the formulas in the box.*

Formulas for the Least Squares Estimates

Slope: $\hat{\beta}_1 = \dfrac{SS_{xy}}{SS_{xx}}$

y-intercept: $\hat{\beta}_0 = \bar{y} - \hat{\beta}_1\bar{x}$

where

$$SS_{xy} = \sum_{i=1}^{n} \frac{(x_i - \bar{x})(y_i - \bar{y})}{n} = \sum_{i=1}^{n} x_iy_i - n\bar{x}\bar{y}$$

$$SS_{xx} = \sum_{i=1}^{n} \frac{(x_i - \bar{x})^2}{n} = \sum_{i=1}^{n} x_i^2 - n(\bar{x})^2$$

n = Sample size

Preliminary computations for finding the least squares line for the advertising–sales example are contained in Table 3.3. We can now calculate[†]

$$SS_{xy} = \sum x_iy_i - n\bar{x}\bar{y} = 37 - 5\,(3)(2) = 37 - 30 = 7$$
$$SS_{xx} = \sum x_i^2 - n(\bar{x})^2 = 55 - 5\,(3)^2 = 55 - 45 = 10$$

TABLE 3.3 **Preliminary Computations for the Advertising–Sales Example**

x_i	y_i	x_i^2	x_iy_i
1	1	1	1
2	1	4	2
3	2	9	6
4	2	16	8
5	4	25	20
Totals: $\Sigma x_i = 15$	$\Sigma y_i = 10$	$\Sigma x_i^2 = 55$	$\Sigma x_iy_i = 37$
Means: $\bar{x} = 3$	$\bar{y} = 2$		

---

*Students who are familiar with calculus should note that the values of β_0 and β_1 that minimize SSE = $\Sigma\,(y_i - \hat{y}_i)^2$ are obtained by setting the two partial derivatives $\partial SSE/\partial\beta_0$ and $\partial SSE/\partial\beta_1$ equal to 0. The solutions to these two equations yield the formulas shown in the box. (The complete derivation is provided in Appendix A.) Furthermore, we denote the *sample* solutions to the equations by $\hat{\beta}_0$ and $\hat{\beta}_1$, whereas the "^" (hat) denotes that these are sample estimates of the true population intercept β_0 and slope β_1.

[†]Since summations will be used extensively from this point on, we will omit the limits on Σ when the summation includes all the measurements in the sample, i.e., when the summation is $\Sigma_{i=1}^n$, we will write Σ.

Then, the slope of the least squares line is

$$\hat{\beta}_1 = \frac{SS_{xy}}{SS_{xx}} = \frac{7}{10} = .7$$

and the y-intercept is

$$\hat{\beta}_0 = \bar{y} - \hat{\beta}_1 \bar{x}$$
$$= 2 - (.7)(3) = 2 - 2.1 = -.1$$

The least squares line is then

$$\hat{y} = \hat{\beta}_0 + \hat{\beta}_1 x = -.1 + .7x$$

The graph of this line is shown in Figure 3.4.

FIGURE 3.4

The line $\hat{y} = -.1 + .7x$ fit to the data

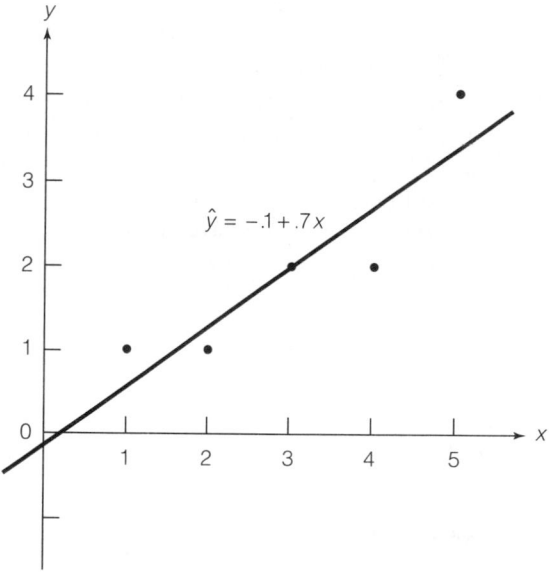

Our interpretation of the least squares slope, $\hat{\beta}_1 = .7$, is that the mean of sales revenue y will increase .7 unit for every 1-unit increase in advertising expenditure x. Since y is measured in units of $1,000 and x in units of $100, our interpretation is that mean monthly sales revenue increases $700 for every $100 increase in monthly advertising expenditure. (We will attach a measure of reliability to this inference in Section 3.6.)

The least squares intercept, $\hat{\beta}_0 = -.1$, is our estimate of mean sales revenue y when advertising expenditure is set at $x = $0. Since sales revenue can never be negative, why does such a nonsensical result occur? The reason is that we are attempting to use the least squares model to predict y for a value of x ($x = 0$) that is outside the range of the sample data and impractical. (We have more to say about predicting outside the range of the sample data—called **extrapo-lation**—in Section 3.9.) Consequently, $\hat{\beta}_0$ will not always have a practical interpretation. Only when $x = 0$ is within the range of the x values in the sample and is a practical value will $\hat{\beta}_0$ have a meaningful interpretation.

TABLE 3.4 **Comparing Observed and Predicted Values for the Least Squares Model**

x	y	$\hat{y} = -.1 + .7x$	$(y - \hat{y})$	$(y - \hat{y})^2$
1	1	.6	$(1 - .6) = \quad .4$	.16
2	1	1.3	$(1 - 1.3) = -.3$	.09
3	2	2.0	$(2 - 2.0) = \quad 0$	.00
4	2	2.7	$(2 - 2.7) = -.7$	.49
5	4	3.4	$(4 - 3.4) = \quad .6$	.36
			Sum of errors $= \quad 0$	SSE $= 1.10$

The observed and predicted values of y, the deviations of the y values about their predicted values, and the squares of these deviations are shown in Table 3.4. Note that the sum of squares of the deviations, SSE, is 1.10, and (as we would expect) this is less than the SSE $= 2.0$ obtained in Table 3.2 for the visually fitted line.

To summarize, we have defined the best-fitting straight line to be the one that satisfies the least squares criterion; that is, the sum of the squared errors will be smaller than for any other straight-line model. This line is called the **least squares line**, and its equation is called the **least squares prediction equation**.

EXERCISES

3.6 Use the method of least squares to fit a straight line to these six data points:

x	1	2	3	4	5	6
y	1	2	2	3	5	5

a. What are the least squares estimates of β_0 and β_1?
b. Plot the data points and graph the least squares line on the scattergram.

3.7 Use the method of least squares to fit a straight line to these five data points:

x	-2	-1	0	1	2
y	4	3	3	1	-1

a. What are the least squares estimates of β_0 and β_1?
b. Plot the data points and graph the least squares line on the scattergram.

3.8 Each year, *Fortune* ranks the top American cities according to their ability to provide high-quality, low-cost labor for companies that are relocating. One important measure used to form the rankings is the labor market stress index (y), which indicates the availability of workers in the city. (The higher the index, the tighter the labor market.) A second important variable is the unemployment rate (x). The values of these two variables for each of the top 10 cities in 1990 are listed in the table.

RANK	CITY	LABOR MARKET STRESS INDEX, y	UNEMPLOYMENT RATE, x
1	Salt Lake City	107	4.5%
2	Minneapolis–St. Paul	107	3.8%
3	Atlanta	100	5.1%
4	Sacramento	100	4.9%
5	Austin (Texas)	80	5.4%
6	Columbus (Ohio)	100	4.8%
7	Dallas/Fort Worth	100	5.5%
8	Phoenix	93	4.3%
9	Jacksonville (Florida)	87	5.7%
10	Oklahoma City	80	4.6%

Source: *Fortune*, Oct. 22, 1990, pp. 58–63.

a. Construct a scattergram for the data.
b. Find the least squares prediction equation.
c. Graph the least squares line on the scattergram.
d. Interpret the values of $\hat{\beta}_0$ and $\hat{\beta}_1$.

3.9 Two processes for hydraulic drilling of rock are dry drilling and wet drilling. In a dry hole, compressed air is forced down the drill rods to flush the cuttings and drive the hammer; in a wet hole, water is forced down. An experiment was conducted to determine whether the time y it takes to dry drill a distance of 5 feet in rock increases with depth x. The results for one portion of the experiment are shown in the accompanying table.

DEPTH AT WHICH DRILLING BEGINS (FEET), x	TIME TO DRILL 5 FEET (MINUTES), y
0	4.90
25	7.41
50	6.19
75	5.57
100	5.17
125	6.89
150	7.05
175	7.11
200	6.19
225	8.28
250	4.84
275	8.29
300	8.91
325	8.54
350	11.79
375	12.12
395	11.02

Source: Penner, R. and Watts, D. G. "Mining information." *The American Statistician*, Vol. 45, No. 1, Feb. 1991, p. 6 (Table 1).

a. Construct a scattergram for the data.
b. Find the least squares prediction equation.
c. Graph the least squares line on the scattergram.
d. Interpret the values of $\hat{\beta}_0$ and $\hat{\beta}_1$.

3.10   For a company to maintain a competitive edge in the marketplace, spending on research and development (R&D) is essential. To determine the optimum level for R&D spending and its effect on a company's value, a simple linear regression analysis was performed. Data collected for the largest R&D spenders (based on 1981–1982 averages) were used to fit the straight-line model

$$y = \beta_0 + \beta_1 x + \varepsilon$$

where

y = Price/earnings (P/E) ratio
x = R&D expenditures/sales (R/S) ratio

The data for 20 of the companies used in the study are provided in the table.

COMPANY	P/E RATIO y	R/S RATIO x	COMPANY	P/E RATIO y	R/S RATIO x
1	5.6	.003	11	8.4	.058
2	7.2	.004	12	11.1	.058
3	8.1	.009	13	11.1	.067
4	9.9	.021	14	13.2	.080
5	6.0	.023	15	13.4	.080
6	8.2	.030	16	11.5	.083
7	6.3	.035	17	9.8	.091
8	10.0	.037	18	16.1	.092
9	8.5	.044	19	7.0	.064
10	13.2	.051	20	5.9	.028

Source: Wallin, C. C. and Gilman, J. J. "Determining the optimum level for R&D spending." *Research Management*, Vol. 14, No. 5, Sept./Oct. 1986, pp. 19–24 (adapted from Figure 1, p. 20).

a. Construct a scattergram for the data.
b. Find the least squares prediction equation.
c. Plot the least squares line on your scattergram.
d. Use the least squares line to predict the P/E ratio for a company with an R/S ratio of .070. [*Note:* We will find a measure of reliability for this prediction in Section 3.9.]

3.11   Civil engineers often use the straight-line equation $E(y) = \beta_0 + \beta_1 x$ to model the relationship between the mean shear strength $E(y)$ of masonry joints and precompression stress x. To test this theory, a series of stress tests was performed on solid bricks arranged in triplets and joined with mortar (*Proceedings of the Institute of Civil Engineers*, Mar. 1990). The precompression stress was varied for each triplet and the ultimate shear load just before failure (called the shear strength) was recorded. The stress results for seven triplets (measured in N/mm^2) is shown in the accompanying table.
a. Plot the seven data points in a scattergram. Does the relationship between shear strength and precompression stress appear to be linear?
b. Use the method of least squares to estimate the parameters of the linear model.
c. Interpret the values of $\hat{\beta}_0$ and $\hat{\beta}_1$.

TRIPLET TEST	1	2	3	4	5	6	7
Shear Strength, y	1.00	2.18	2.24	2.41	2.59	2.82	3.06
Precompression Stress, x	0	.60	1.20	1.33	1.43	1.75	1.75

Source: Riddington, J. R. and Ghazali, M. Z. "Hypothesis for shear failure in masonry joints." *Proceedings of the Institute of Civil Engineers, Part 2,* Mar. 1990, Vol. 89, p. 96 (Fig. 7).

3.4

Model Assumptions

In the advertising–sales example presented in Section 3.3, we assumed that the probabilistic model relating the firm's sales revenue y to advertising dollars x is

$$y = \beta_0 + \beta_1 x + \varepsilon$$

Recall that the least squares estimate of the deterministic component of the model $\beta_0 + \beta_1 x$ is

$$\hat{y} = \hat{\beta}_0 + \hat{\beta}_1 x = -.1 + .7x$$

Now we turn our attention to the random component ε of the probabilistic model and its relation to the errors of estimating β_0 and β_1. In particular, we will see how the probability distribution of ε determines how well the model describes the true relationship between the dependent variable y and the independent variable x.

We will make four basic assumptions about the general form of the probability distribution of ε:

ASSUMPTION 1 The mean of the probability distribution of ε is 0. That is, the average of the errors over an infinitely long series of experiments is 0 for each setting of the independent variable x. This assumption implies that the mean value of y, $E(y)$, for a given value of x is $E(y) = \beta_0 + \beta_1 x$.

ASSUMPTION 2 The variance of the probability distribution of ε is constant for all settings of the independent variable x. For our straight-line model, this assumption means that the variance of ε is equal to a constant, say, σ^2, for all values of x.

ASSUMPTION 3 The probability distribution of ε is normal.

ASSUMPTION 4 The errors associated with any two different observations are independent. That is, the error associated with one value of y has no effect on the errors associated with other y values.

The implications of the first three assumptions can be seen in Figure 3.5 (page 106), which shows distributions of errors for three particular values of x, namely, x_1, x_2, and x_3. Note that the relative frequency distributions of the errors are normal, with a mean of 0, and a constant variance σ^2 (all the distributions shown have the same amount of spread or variability). A point that lies on the straight line shown in Figure 3.5 represents the mean value of y for a given value of x. We will denote this mean value as $E(y)$. Then, the line of means is given by the equation

$$E(y) = \beta_0 + \beta_1 x$$

FIGURE 3.5

The probability distribution of ε

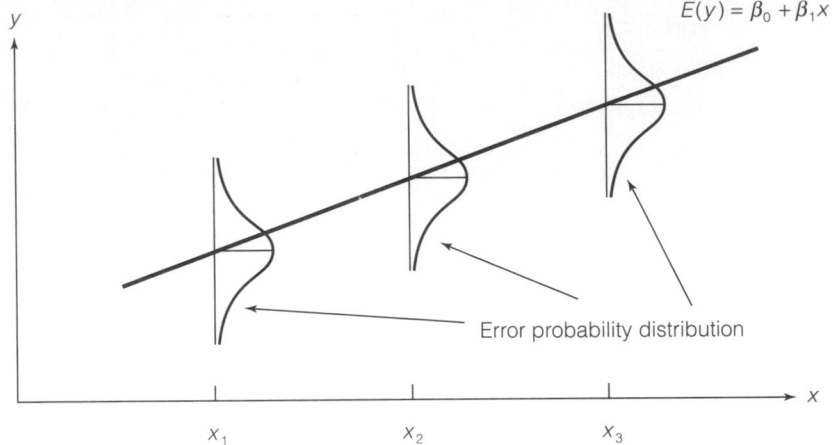

Various techniques exist for checking the validity of these assumptions, and there are remedies to be applied when the assumptions appear to be invalid. We discuss these techniques in detail in Chapter 6. In actual practice, the assumptions need not hold exactly for least squares estimators and test statistics (to be described subsequently) to possess the measures of reliability that we would expect from a regression analysis. The assumptions will be satisfied adequately for many applications encountered in business.

3.5

An Estimator of σ^2

It seems reasonable to assume that the greater the variability of the random error ε (which is measured by its variance σ^2), the greater will be the errors in the estimation of the model parameters β_0 and β_1, and in the error of prediction when $\hat{y}$ is used to predict y for some value of x. Consequently, you should not be surprised, as we proceed through this chapter, to find that σ^2 appears in the formulas for all confidence intervals and test statistics that will be used.

In most practical situations, σ^2 will be unknown, and we must use the data to estimate its value. The best (proof omitted) estimate of σ^2 is s^2, which is obtained by dividing the sum of squares of residuals,

$$\text{SSE} = \sum (y_i - \hat{y}_i)^2$$

by the number of degrees of freedom associated with this quantity. We use 2 df to estimate the y-intercept and slope in the straight-line model, leaving $(n - 2)$ df for the error variance estimation (see the formulas in the box).

In the advertising–sales example, we previously calculated SSE $= 1.10$ for the least squares line $\hat{y} = -.1 + .7x$. Recalling that there were $n = 5$ data points, we have $n - 2 = 5 - 2 = 3$ df for estimating σ^2. Thus,

$$s^2 = \frac{\text{SSE}}{n - 2} = \frac{1.10}{3} = .367$$

is the estimated variance, and

Estimation of σ^2

$$s^2 = \frac{\text{SSE}}{\text{Degrees of freedom for error}} = \frac{\text{SSE}}{n-2}$$

where

$$\text{SSE} = \sum (y_i - \hat{y}_i)^2$$

$$= \text{SS}_{yy} - \hat{\beta}_1 \text{SS}_{xy} \quad \text{(calculation formula)}$$

$$\text{SS}_{yy} = \sum (y_i - \bar{y})^2 = \sum y_i^2 - n(\bar{y})^2$$

Warning. When performing these calculations, you may be tempted to round the calculated values of SS_{yy}, $\hat{\beta}_1$, and SS_{xy}. Be certain to carry at least six significant figures for each of these quantities to avoid substantial errors in the calculation of the SSE.

$$s = \sqrt{.367} = .61$$

is the estimated standard deviation of ε.

You may be able to obtain an intuitive feeling for s by recalling the interpretation given to a standard deviation in Chapter 1 and remembering that the least squares line estimates the mean value of y for a given value of x. Since s measures the spread of the distribution of y values about the least squares line, we should not be surprised to find that most of the observations lie within $2s$ or $2(.61) = 1.22$ of the least squares line. For this simple example (only five data points), all five data points fall within $2s$ of the least squares line. In Section 3.9, we will use s to evaluate the error of prediction when the least squares line is used to predict a value of y to be observed for a given value of x.

Interpretation of s, the Estimated Standard Deviation of ε

We expect most of the observed y values to lie within $2s$ of their respective least squares predicted values, $\hat{y}$.

EXERCISES

3.12 Suppose you fit a least squares line to nine data points and calculate SSE = .219.
   a. Find s^2, the estimator of σ^2, the variance of the random error term ε.
   b. Calculate s and interpret the result.

3.13 Calculate SSE, s^2, and s for the least squares line plotted in the following exercises. Interpret the value of s.
   a. Exercise 3.6   **b.** Exercise 3.7   **c.** Exercise 3.8
   d. Exercise 3.9   **e.** Exercise 3.10   **f.** Exercise 3.11

3.14   A study was conducted to model the thermal performance of integral-fin tubes used in the refrigeration and process industries (*Journal of Heat Transfer*, Aug. 1990). Twenty-four specially manufactured integral-fin tubes with rectangular-shaped fins made of copper were used in the experiment. Vapor was released downward into each tube and the vapor-side heat transfer coefficient (based upon the outside surface area of the tube) was measured. The dependent variable for the study is the heat transfer enhancement ratio, y, defined as the ratio of the vapor-side coefficient of the fin tube to the vapor-side coefficient of a smooth tube evaluated at the same temperature. Theoretically, heat transfer will be related to the area at the top of the tube that is "unflooded" by condensation of the vapor. The data in the table are the unflooded area ratio (x), and heat transfer enhancement (y) values recorded for the 24 integral-fin tubes.

UNFLOODED AREA RATIO, x	HEAT TRANSFER ENHANCEMENT, y	UNFLOODED AREA RATIO, x	HEAT TRANSFER ENHANCEMENT, y
1.93	4.4	2.00	5.2
1.95	5.3	1.77	4.7
1.78	4.5	1.62	4.2
1.64	4.5	2.77	6.0
1.54	3.7	2.47	5.8
1.32	2.8	2.24	5.2
2.12	6.1	1.32	3.5
1.88	4.9	1.26	3.2
1.70	4.9	1.21	2.9
1.58	4.1	2.26	5.3
2.47	7.0	2.04	5.1
2.37	6.7	1.88	4.6

Source: Marto, P. J., et al. "An experimental study of R-113 film condensation on horizontal integral-fin tubes." *Journal of Heat Transfer*, Vol. 112, Aug. 1990, p. 763 (Table 2).

a. Fit a least squares line to the data.
b. Plot the data and graph the least squares line as a check on your calculations.
c. Calculate SSE and s^2.
d. Calculate s and interpret its value.

3.15   The Consumer Attitude Survey, performed by the Bureau of Economic and Business Research, University of Florida, is conducted using random-digit telephone dialings to Florida households. The reliability of such a telephone survey depends on the refusal rate—that is, the percentage of dialed households that refuse to take part in the study. One factor thought to be related to refusal rate is personal income. The accompanying table gives the refusal rate y and personal income per capita x for 12 randomly selected Florida counties from a recent survey.

a. Construct a scattergram for the data.
b. Find the least squares prediction equation.
c. Graph the least squares line on the scattergram.
d. Use the least squares prediction equation to predict the refusal rate for a Florida county with a per capita income of $8,000. [*Note:*   We will find a measure of the reliability of this prediction in Section 3.9.]
e. Calculate SSE and s^2.
f. Calculate s and interpret its value.

COUNTY	REFUSAL RATE y	PER CAPITA INCOME x
1	.296	$ 7,737
2	.498	12,330
3	.386	12,058
4	.327	9,927
5	.500	6,904
6	.333	9,463
7	.429	11,466
8	.422	10,000
9	.441	10,052
10	.191	8,636
11	.526	7,445
12	.405	9,059

Source: Bureau of Economic and Business Research, University of Florida.

3.16 A company keeps extensive records on its new salespeople on the premise that sales should increase with experience. A random sample of seven new salespeople produced the data on experience and sales shown in the table.

MONTHS ON JOB x	MONTHLY SALES y, thousands of dollars
2	2.4
4	7.0
8	11.3
12	15.0
1	.8
5	3.7
9	12.0

a. Fit a least squares line to the data.
b. Plot the data and graph the least squares line.
c. Predict the sales that a new salesperson would be expected to generate after 6 months on the job. After 9 months.
d. Calculate SSE and s^2.
e. Calculate s and interpret its value.

3.6

Assessing the Utility of the Model: Making Inferences About the Slope β_1

Refer to the advertising–sales data of Table 3.1 and suppose that the appliance store's sales revenue is *completely unrelated* to the advertising expenditure. What could be said about the values of β_0 and β_1 in the hypothesized probabilistic model

$$y = \beta_0 + \beta_1 x + \varepsilon$$

if x contributes no information for the prediction of y? The implication is that the mean of y, i.e., the deterministic part of the model $E(y) = \beta_0 + \beta_1 x$, does not change as x changes. Regardless of the value of x, you always predict the same value of y. In the straight-line model, this means that the true slope, β_1, is equal to 0. Therefore, to test the null hypothesis that x contributes no information for the prediction of y against the alternative hypothesis that these variables are linearly related with a slope differing from 0, we test

$$H_0: \quad \beta_1 = 0 \qquad H_a: \quad \beta_1 \neq 0$$

If the data support the alternative hypothesis, we will conclude that x does contribute information for the prediction of y using the straight-line model [although the true relationship between $E(y)$ and x could be more complex than a straight line]. Thus, to some extent, this is a test of the utility of the hypothesized model.

The appropriate test statistic is found by considering the sampling distribution of $\hat{\beta}_1$, the least squares estimator of the slope β_1.

Sampling Distribution of $\hat{\beta}_1$

If we make the four assumptions about ε (see Section 3.4), then the sampling distribution of $\hat{\beta}_1$, the least squares estimator of the slope, will be a normal distribution with mean β_1 (the true slope) and standard deviation

$$\sigma_{\hat{\beta}_1} = \frac{\sigma}{\sqrt{SS_{xx}}} \qquad \text{(See Figure 3.6.)}$$

FIGURE 3.6

Sampling distribution of $\hat{\beta}_1$

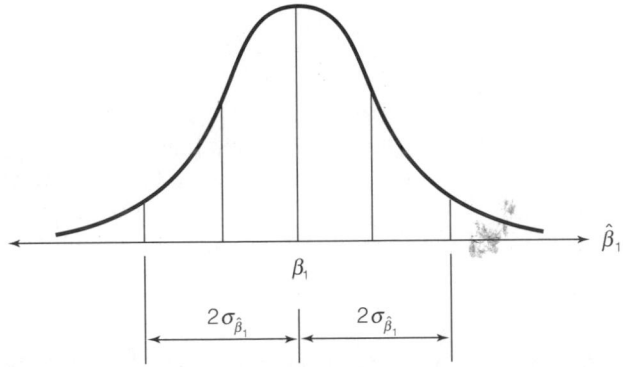

Since σ will usually be unknown, the appropriate test statistic will generally be a Student's t statistic formed as follows:

$$t = \frac{\hat{\beta}_1 - \text{Hypothesized value of } \beta_1}{s_{\hat{\beta}_1}} \qquad \text{where} \quad s_{\hat{\beta}_1} = \frac{s}{\sqrt{SS_{xx}}}$$

$$= \frac{\hat{\beta}_1 - 0}{s/\sqrt{SS_{xx}}}$$

Note that we have substituted the estimator s for σ, and then formed $s_{\hat{\beta}_1}$ by dividing s by $\sqrt{SS_{xx}}$. The number of degrees of freedom associated with this t statistic is the same as the number of degrees of freedom associated with s. Recall that this will be $(n - 2)$ df when the hypothesized model is a straight line (see Section 3.5).

The test of the utility of the model is summarized in the next box.

A Test of Model Utility: Simple Linear Regression

ONE-TAILED TEST

H_0: $\beta_1 = 0$

H_a: $\beta_1 < 0$
    (or H_a: $\beta_1 > 0$)

TWO-TAILED TEST

H_0: $\beta_1 = 0$

H_a: $\beta_1 \neq 0$

Test statistic: $t = \dfrac{\hat{\beta}_1}{s_{\hat{\beta}_1}} = \dfrac{\hat{\beta}_1}{s/\sqrt{SS_{xx}}}$

Rejection region: $t < -t_\alpha$
    (or $t > t_\alpha$)

Rejection region: $t < -t_{\alpha/2}$ or
    $t > t_{\alpha/2}$

where t_α is based on $(n - 2)$ df

where $t_{\alpha/2}$ is based on $(n - 2)$ df

Assumptions: The four assumptions about ε listed in Section 3.4.

For the advertising–sales example, we will choose $\alpha = .05$ and, since $n = 5$, df $= (n - 2) = 5 - 2 = 3$. Then the rejection region for the two-tailed test is

$$t < -t_{.025} = -3.182 \quad \text{or} \quad t > t_{.025} = 3.182$$

We previously calculated $\hat{\beta}_1 = .7$, $s = .61$, and $SS_{xx} = 10$. Thus,

$$t = \frac{\hat{\beta}_1}{s/\sqrt{SS_{xx}}} = \frac{.7}{.61/\sqrt{10}} = \frac{.7}{.19} = 3.7$$

Since this calculated t value falls in the upper-tail rejection region (see Figure 3.7 on page 112), we reject the null hypothesis and conclude that the slope β_1 is not 0. The sample evidence indicates that x contributes information for the prediction of y using a linear model for the relationship between sales revenue and advertising.

What conclusion can be drawn if the calculated t value does not fall in the rejection region? We know from previous discussions of the philosophy of hypothesis testing that such a t value does *not* lead us to accept the null hypothesis. That is, we do not conclude that $\beta_1 = 0$. Additional data might indicate that β_1 differs from 0, or a more complex relationship may exist between x and y, requiring the fitting of a model other than the straight-line model. We will discuss several such models in Chapter 4.

Another way to make inferences about the slope β_1 is to estimate it using a confidence interval. This interval is formed as shown in the next box.

FIGURE 3.7

Rejection region and calculated t value for testing whether the slope $\beta_1 = 0$

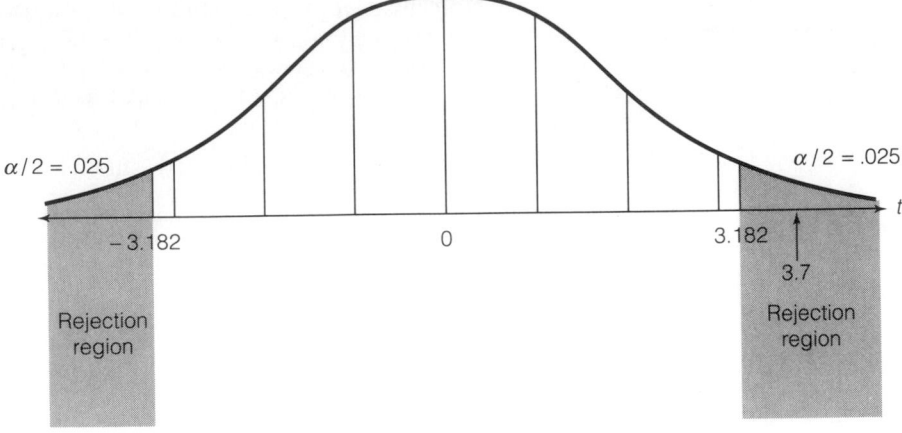

$\alpha/2 = .025$ $\alpha/2 = .025$

-3.182 0   3.182

3.7

Rejection region   Rejection region

A $100(1 - \alpha)\%$ Confidence Interval for the Simple Linear Regression Slope β_1

$$\hat{\beta}_1 \pm t_{\alpha/2} s_{\hat{\beta}_1} \quad \text{where} \quad s_{\hat{\beta}_1} = \frac{s}{\sqrt{SS_{xx}}}$$

and $t_{\alpha/2}$ is based on $(n - 2)$ df

For the advertising–sales example, a 95% confidence interval for the slope β_1 is

$$\hat{\beta}_1 \pm t_{.025} s_{\hat{\beta}_1} = .7 \pm 3.182 \left(\frac{s}{\sqrt{SS_{xx}}} \right)$$

$$= .7 \pm 3.182 \left(\frac{.61}{\sqrt{10}} \right) = .7 \pm .61$$

Thus, we estimate that the interval from .09 to 1.31 includes the slope parameter β_1.

Remembering that y is recorded in units of $1,000 and x in units of $100, we can say, with 95% confidence, that the mean monthly sales revenue will increase between $90 and $1,310 for every $100 increase in monthly advertising expenditure.

Since all the values in this interval are positive, it appears that β_1 is positive and that the mean of y, $E(y)$, increases as x increases. However, the rather large width of the confidence interval reflects the small number of data points (and, consequently, a lack of information) in the experiment. We would expect a narrower interval if the sample size were increased.

EXERCISES

3.17 Do the data provide sufficient evidence to indicate that β_1 differs from 0 for the least squares analyses in the following exercises? Use $\alpha = .05$.
   a. Exercise 3.6   b. Exercise 3.7   c. Exercise 3.8
   d. Exercise 3.9   e. Exercise 3.10   f. Exercise 3.11

3.18 Do the data in Exercise 3.14 provide sufficient evidence to indicate that heat transfer enhancement, y, increases as unflooded area ratio, x, increases (i.e., that $\beta_1 > 0$)? Test using $\alpha = .10$.

3.19 Do the data in Exercise 3.15 provide sufficient evidence to indicate that refusal rate y is linearly related to per capita income x? Test using $\alpha = .01$.

3.20 Do the data in Exercise 3.16 support the theory that sales increase as experience of a salesperson increases? Test using $\alpha = .05$.

3.21 Refer to the shareholder-wealth maximization study in Exercise 1.61. The researchers conducted an analysis of return on equity to compare companies that aim at maximizing shareholder wealth to those with alternative corporate goals (*Quarterly Journal of Business & Economics*, Autumn 1986). Monthly holding-period returns were averaged for the stocks of shareholder wealth-maximizer companies and for companies holding alternative goals, and the difference between the averages (called monthly portfolio return differences) were calculated for each of $n = 72$ months. The return differences, y, were then regressed on monthly *Standard & Poor's 500* Composite Stock Index, x, using the straight-line model

$$y = \beta_0 + \beta_1 x + \varepsilon$$

The regression results are summarized here:

$$\hat{y} = .00008 - .01748x \qquad SSE = .322 \qquad SS_{xx} = 2.914$$

   a. Conduct a test to determine whether the monthly portfolio return differences are linearly related to the monthly *Standard & Poor's 500* Stock Index. Use $\alpha = .10$.
   b. Construct a 90% confidence interval for the slope of the straight-line model.
   c. Interpret the confidence interval in part b, and explain what it tells you about the relationship between monthly portfolio return differences and the monthly *Standard & Poor's 500* Stock Index.

3.22 Some economists fear that the current unemployment compensation system in the United States distorts the number of layoffs in a downturn of the business cycle. The hypothesis is that the unemployment compensation subsidy causes firms to lay off more people than they would if they knew those laid off would receive no outside subsidy. The table at the top of page 114 gives the unemployment compensation subsidy rate x (as a percentage of total revenues) and the layoff rate y (number of workers per 1,000) for 11 industries.
   a. Fit the model $E(y) = \beta_0 + \beta_1 x$.
   b. Do the data provide sufficient evidence to indicate that the unemployment compensation subsidy rate x contributes information for the prediction of the layoff rate y? Test using $\alpha = .05$.
   c. Find a 95% confidence interval for the mean increase in layoff rate y for each percentage-point increase in subsidy rate x. [*Hint:* Find a 95% confidence interval for β_1.]

INDUSTRY	SUBSIDY RATE x	LAYOFF RATE y
Apparel	57%	12.54
Chemicals	32	1.78
Construction	31	7.10
Electrical machinery	29	8.38
Fabricated metals	27	11.72
Food	36	5.10
Machinery	32	4.44
Misc. manufacturing	61	9.82
Primary metals	23	7.34
Retail	27	1.98
Wholesale trade	33	1.86

Source: Tropel, R. H. "On layoffs and unemployment insurance." *American Economic Review*, 1983, Vol. 83, pp. 541–559.

3.23 Buyers are often influenced by bulk advertising of a particular product. For example, suppose you have a product that sells for 25¢. If it is advertised at 2/50¢, 3/75¢, or 4/$1, some people may think they are getting a bargain. To test this theory, a store manager advertised an item for equal periods of time at five different bulk rates and observed the volume sold, as listed in the table. Do the data provide sufficient evidence to indicate that sales increase as the number in the bulk increases?

ADVERTISED NUMBER IN BULK SALE x	VOLUME SOLD y
1	27
2	36
3	34
4	63
5	52

3.24 Starting salary is considered to be an important indicator of success in the job market by many graduating MBAs. The factors that influence starting salary are many and varied, but one variable intrinsically interesting to MBA students is Graduate Management Aptitude Test (GMAT) score. The starting salaries and GMAT scores for a sample of 10 MBAs who graduated from the University of Florida's MBA program in 1987 are given in the table.

STARTING SALARY y	GMAT x	STARTING SALARY y	GMAT x
$40,000	510	$28,000	520
33,000	510	34,000	560
40,000	550	30,000	530
35,000	600	32,500	530
28,000	600	31,000	590

Source: Graduate College of Business Administration, University of Florida.

a. Find the least squares prediction equation for the model

$$y = \beta_0 + \beta_1 x + \varepsilon$$

Plot the data on a scattergram and graph the least squares line.

b. Find SSE and s^2 for the data.

c. Is there sufficient evidence to indicate that GMAT score x contributes information for the prediction of starting salary y? Test using $\alpha = .10$.

d. Construct a 95% confidence interval for the true slope β_1. Interpret the interval.

3.25 In Exercise 1.7 we presented data on the foreign revenue of 20 multinational firms (i.e., firms with both domestic and foreign investments). Is there a positive linear relationship between the foreign revenue (measured as a percentage of total revenue) and the foreign assets (measured as a percentage of total assets) of multinational firms? Use the data in the accompanying table to conduct the analysis. Test using $\alpha = .10$.

FIRM	FOREIGN REVENUE (%)	FOREIGN ASSETS (%)
Exxon	73.2	55.8
IBM	58.9	48.6
GM	26.6	25.2
Mobil	64.7	51.1
Ford	33.2	26.9
Citicorp	52.3	39.4
EI duPont	39.8	29.5
Texaco	42.3	26.6
ITT	43.3	23.6
Dow Chemical	54.1	44.9
Procter & Gamble	39.9	32.2
Philip Morris	19.6	14.8
Eastman Kodak	40.9	28.0
Digital	54.1	44.2
GE	12.4	8.8
United Technologies	32.9	26.7
Amoco	26.1	32.7
Hewlett-Packard	53.3	38.7
Xerox	34.6	25.4
Chevron	20.5	22.6

Source: *Forbes*, July 23, 1990, pp. 362–363.

3.7

The Coefficient of Correlation

The claim is often made that the crime rate and the unemployment rate are "highly correlated." Another popular belief is that the gross national product (GNP) and the rate of inflation are "correlated." Some people even believe that the Dow Jones Industrial Average and the lengths of fashionable skirts are "correlated." Thus, the term *correlation* implies a relationship or "association," between two variables.

The **Pearson product moment correlation coefficient** r, defined in the next box, provides a quantitative measure of the strength of the linear relationship between x and y, just as does the least squares slope $\hat{\beta}_1$. However, unlike the slope, the correlation coefficient r is *scaleless*. The value of r is always between -1 and $+1$, regardless of the units of measurement used for the variables x and y.

> **Definition 3.2**
>
> The **Pearson product moment coefficient of correlation** r is a measure of the strength of the *linear* relationship between two variables x and y. It is computed (for a sample of n measurements on x and y) as follows:
>
> $$r = \frac{SS_{xy}}{\sqrt{SS_{xx}SS_{yy}}}$$

Note that r is computed using the same quantities used in fitting the least squares line. Since both r and $\hat{\beta}_1$ provide information about the utility of the model, it is not surprising that there is a similarity in their computational formulas. In particular, note that SS_{xy} appears in the numerators of both expressions and, since both denominators are always positive, r and $\hat{\beta}_1$ will always be of the same sign (either both positive or both negative). A value of r near or equal to 0 implies little or no linear relationship between y and x. In contrast, the closer r is to 1 or -1, the stronger the linear relationship between y and x. And, if $r = 1$ or $r = -1$, all the points fall exactly on the least squares line. Positive values of r imply that y increases as x increases; negative values imply that y decreases as x increases. Each of these situations is portrayed in Figure 3.8.

EXAMPLE 3.1

A firm wants to know the correlation between the size of its sales force and its yearly sales revenue. The records for the past 10 years are examined, and the results listed in Table 3.5 are obtained. Calculate the coefficient of correlation r for the data.

T A B L E 3.5 **Data and Preliminary Calculations, Example 3.1**

YEAR	NUMBER OF SALESPEOPLE x	ANNUAL SALES REVENUE y, hundred thousand dollars
1983	15	1.35
1984	18	1.63
1985	24	2.33
1986	22	2.41
1987	25	2.63
1988	29	2.93
1989	30	3.41
1990	32	3.26
1991	35	3.63
1992	38	4.15

$$\sum x = 268 \qquad \sum y = 27.73$$
$$\bar{x} = 26.8 \qquad \bar{y} = 2.773$$
$$\sum x^2 = 7{,}668 \qquad \sum y^2 = 83.8733$$
$$\sum xy = 800.62$$

FIGURE 3.8

Values of *r* and their implications

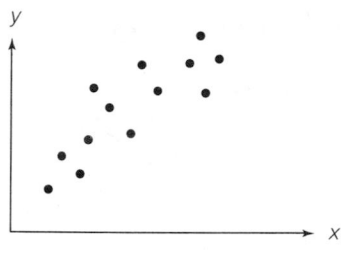

a. Positive *r*: *y* increases as *x* increases

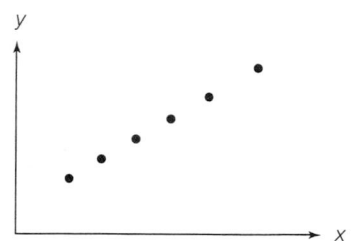

b. *r* = 1: a perfect positive relationship between *y* and *x*

c. Negative *r*: *y* decreases as *x* increases

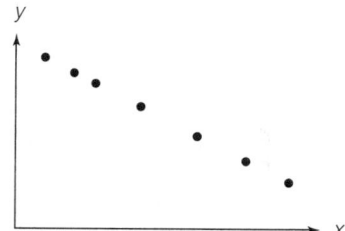

d. *r* = −1: a perfect negative relationship between *y* and *x*

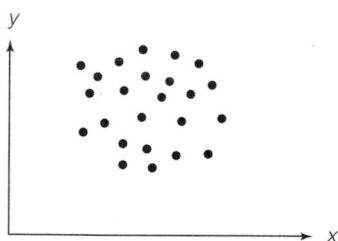

e. *r* near zero: little or no linear relationship between *y* and *x*

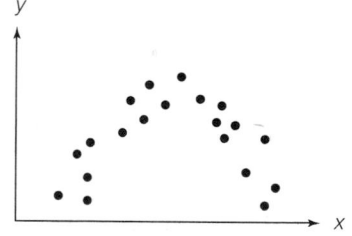

f. *r* near zero: little or no linear relationship between *y* and *x*

Solution

We need to calculate SS_{xy}, SS_{xx}, and SS_{yy} as follows:

$$SS_{xy} = \sum x_i y_i - n\bar{x}\bar{y} = 800.62 - 10(26.8)(2.773) = 57.456$$

$$SS_{xx} = \sum x_i^2 - n\bar{x}^2 = 7{,}668 - 10(26.8)^2 = 485.6$$

$$SS_{yy} = \sum y_i^2 - n\bar{y}^2 = 83.8733 - 10(2.773)^2 = 6.97801$$

Then, the coefficient of correlation is

$$r = \frac{SS_{xy}}{\sqrt{SS_{xx}SS_{yy}}} = \frac{57.456}{\sqrt{(485.6)(6.97801)}} = \frac{57.456}{58.211} = .99$$

Thus, the size of the sales force and sales revenue are very highly correlated—at least over the past 10 years. The implication is that a strong positive linear relationship exists between these variables (see Figure 3.9). We must be careful, however, not to jump to any unwarranted conclusions. For instance, the firm may be tempted to conclude that the best thing it can do to increase sales is to hire a large number of new salespeople. The implication of such a conclusion is that there is a *causal* relationship between the two variables. However, **high correlation does not imply causality**. The fact is, many things have probably contributed both to the increase in the size of the sales force and to the increase in sales revenue. The firm's expertise has undoubtedly grown, the economy has inflated (so that 1992 dollars are not worth as much as 1983 dollars), and perhaps the scope of products and services sold by the firm has widened. We must be careful not to infer a causal relationship on the basis of high sample correlation. The only safe conclusion when a high correlation is observed in the sample data is that a linear trend may exist between x and y.

FIGURE 3.9
Scattergram for Example 3.1

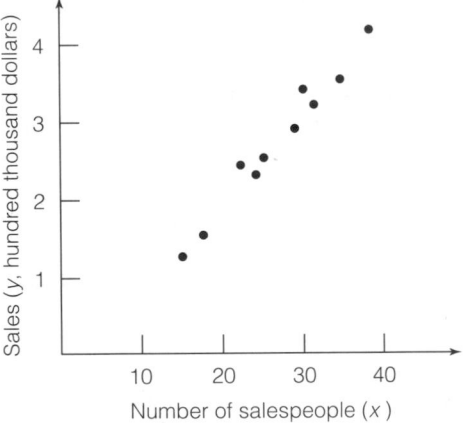

Warning
High correlation does *not* imply causality. If a large positive or negative value of the sample correlation coefficient r is observed, it is incorrect to conclude that a change in x causes a change in y. The only valid conclusion is that a linear trend *may* exist between x and y.

Keep in mind that the correlation coefficient r measures the correlation between x values and y values in the sample, and that a similar linear coefficient of correlation exists for the population from which the data points were selected. The **population correlation coefficient** is denoted by the symbol ρ (rho). As you might expect, ρ is estimated by the corresponding sample statistic, r. Or, rather than estimating ρ, we might want to test

$$H_0: \quad \rho = 0 \quad \text{against} \quad H_a: \quad \rho \neq 0$$

That is, we might want to test the hypothesis that x contributes no information for the prediction of y, using the straight-line model against the alternative that the two variables are at least linearly related. However, we have already performed this identical test in Section 3.6 when we tested $H_0: \beta_1 = 0$ against $H_a: \beta_1 \neq 0$.

It can be shown (proof omitted) that $r = \hat{\beta}_1 \sqrt{SS_{xx}/SS_{yy}}$. Thus, $\hat{\beta}_1 = 0$ implies $r = 0$, and vice versa. Consequently, the null hypothesis $H_0: \rho = 0$ is equivalent to the hypothesis $H_0: \beta_1 = 0$. When we tested the null hypothesis $H_0: \beta_1 = 0$ in connection with the advertising–sales example, the data led to a rejection of the null hypothesis for $\alpha = .05$. This implies that the null hypothesis of a zero linear correlation between the two variables (advertising and sales) can also be rejected at $\alpha = .05$. The only real difference between the least squares slope $\hat{\beta}_1$ and the coefficient of correlation r is the measurement scale. Therefore, the information they provide about the utility of the least squares model is to some extent redundant. Furthermore, the slope $\hat{\beta}_1$ gives us additional information on the amount of increase (or decrease) in y for every 1-unit increase in x. For this reason, we recommend using the slope to make inferences about the existence of a positive or negative linear relationship between two variables.

For those who prefer to test for a linear relationship between two variables using the coefficient of correlation r, we outline the procedure in the following box.

Test of Hypothesis for Linear Correlation

ONE-TAILED TEST   TWO-TAILED TEST

$H_0: \quad \rho = 0$ $H_0: \quad \rho = 0$

$H_a: \quad \rho > 0$ $H_a: \quad \rho \neq 0$
$\quad \quad$ (or $\rho < 0$)

$$\text{Test statistic:} \quad t = \frac{r\sqrt{n - 2}}{\sqrt{1 - r^2}}$$

Rejection region: $\quad t > t_\alpha$ Rejection region: $\quad |t| > t_{\alpha/2}$
$\quad \quad \quad \quad$ (or $t < -t_\alpha$)

where the distribution of t depends on $(n - 2)$ df.

Assumption: The sample of (x, y) values is randomly selected from a normal population.

The next example illustrates how the correlation coefficient r may be a misleading measure of the strength of the association between x and y in situations where the true relationship is nonlinear.

EXAMPLE 3.2

Underinflated or overinflated tires can increase tire wear and decrease gas mileage. A manufacturer of a new tire tested the tire for wear at different pressures with the results shown in Table 3.6. Calculate the coefficient of correlation r for the data. Interpret the result.

TABLE 3.6 **Data for Example 3.2**

PRESSURE, x (pounds per sq. inch)	MILEAGE, y (thousands)
30	29.5
30	30.2
31	32.1
31	34.5
32	36.3
32	35.0
33	38.2
33	37.6
34	37.7
34	36.1
35	33.6
35	34.2
36	26.8
36	27.4

Solution

Rather than perform the calculations by hand, we use a computer to find the value of r. A SAS printout of the correlation analysis is shown in Figure 3.10. The value of r, shaded on the printout, is $r = -.114$. This relatively small value for r describes a weak linear relationship between pressure (x) and mileage (y). The manufacturer, however, would be remiss in concluding that tire pressure has little or no impact on wear of the tire. On the contrary, the relationship between pressure and wear is fairly strong, as the scattergram in Figure 3.11 illustrates. Note that the relationship is not linear, but curvilinear; the underinflated tires (low pressure values) and overinflated tires (high pressure values) both lead to low mileages.

FIGURE 3.10

SAS printout of correlation analysis of data in Table 3.6

Pearson Correlation Coefficients / Prob > |R| under Ho: Rho=0 / N = 14

	X	Y
X	1.00000 0.0	-0.11371 0.6987
Y	-0.11371 0.6987	1.00000 0.0

FIGURE 3.11
Scattergram of data in Table 3.6

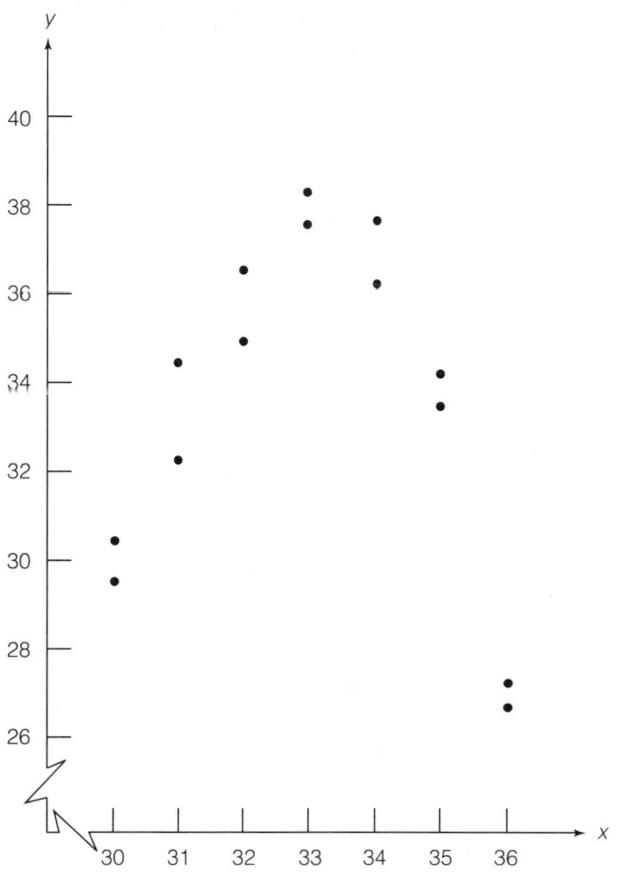

3.8

The Coefficient of Determination

Another way to measure the contribution of x in predicting y is to consider how much the errors of prediction of y were reduced by using the information provided by x.

To illustrate, suppose a sample of data produces the scattergram shown in Figure 3.12a on page 122. If we assume that x contributes no information for the prediction of y, the best prediction for a value of y is the sample mean $\bar{y}$, which graphs as the horizontal line shown in Figure 3.12b. The vertical line segments in Figure 3.12b are the deviations of the points about the mean $\bar{y}$. Note that the sum of squares of deviations for the model $\hat{y} = \bar{y}$ is

$$SS_{yy} = \sum (y_i - \bar{y})^2$$

Now suppose you fit a least squares line to the same set of data and locate the deviations of the points about the line as shown in Figure 3.12c. Compare

FIGURE 3.12

A comparison of the sum of squares of deviations for two models

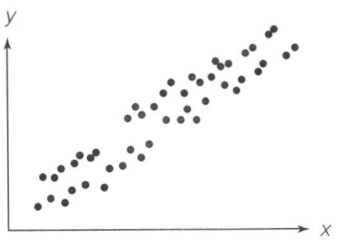

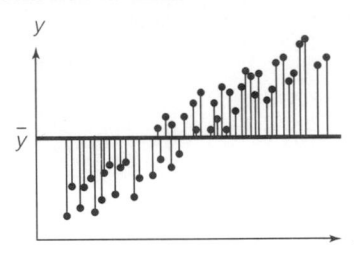

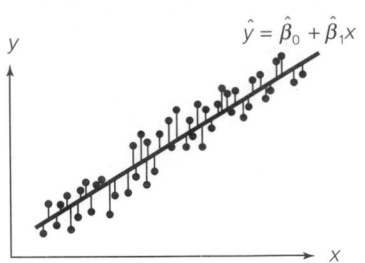

a. Scattergram of data

b. Assumption: x contributes no information for predicting y:
$$\hat{y} = \bar{y}$$

c. Assumption: x contributes information for predicting y;
$$\hat{y} = \hat{\beta}_0 + \hat{\beta}_1 x$$

the deviations about the prediction lines in Figures 3.12b and 3.12c. You can see that:

1. If x contributes little or no information for the prediction of y, the sums of squares of deviations for the two lines,
$$SS_{yy} = \sum (y_i - \bar{y})^2 \quad \text{and} \quad SSE = \sum (y_i - \hat{y}_i)^2$$
will be nearly equal.

2. If x does contribute information for the prediction of y, then SSE will be smaller than SS_{yy}. In fact, if all the points fall on the least squares line, then SSE = 0.

A convenient way of measuring how well the least squares equation $\hat{y} = \hat{\beta}_0 + \hat{\beta}_1 x$ performs as a predictor of y is to compute the reduction in the sum of squares of deviations that can be attributed to x, expressed as a proportion of SS_{yy}. This quantity, called the **coefficient of determination**, is

$$\frac{SS_{yy} - SSE}{SS_{yy}}$$

In simple linear regression, it can be shown that this quantity is equal to the square of the simple linear coefficient of correlation r.

Definition 3.3

The **coefficient of determination** is

$$r^2 = \frac{SS_{yy} - SSE}{SS_{yy}} = 1 - \frac{SSE}{SS_{yy}}$$

It represents the proportion of the sum of squares of deviations of the y values about their mean that can be attributed to a linear relationship between y and x. (In simple linear regression, it may also be computed as the square of the coefficient of correlation r.)

Note that r^2 is always between 0 and 1, because r is between -1 and $+1$. Thus, an r^2 of .60 means that the sum of squares of deviations of the y values about their predicted values has been reduced 60% by the use of $\hat{y}$, instead of $\bar{y}$, to predict y.

EXAMPLE 3.3

Calculate the coefficient of determination for the advertising–sales example. The data are repeated in Table 3.7.

TABLE 3.7

ADVERTISING EXPENDITURE x, hundreds of dollars	SALES REVENUE y, thousands of dollars
1	1
2	1
3	2
4	2
5	4

Solution

We first calculate

$$SS_{yy} = \sum y_i^2 - n\bar{y}^2 = 26 - 5(2)^2$$
$$= 26 - 20 = 6$$

From previous calculations,

$$SSE = \sum (y_i - \hat{y}_i)^2 = 1.10$$

Then, the coefficient of determination is given by

$$r^2 = \frac{SS_{yy} - SSE}{SS_{yy}} = \frac{6.0 - 1.1}{6.0} = \frac{4.9}{6.0}$$
$$= .82$$

By using the advertising expenditure x to predict y with the least squares line

$$\hat{y} = -.1 + .7x$$

the total sum of squares of deviations of the five sample y values about their predicted values has been reduced 82%. That is, 82% of the sample variation in monthly sales revenue can be "explained" by the least squares line.

In situations where a straight-line regression model is found to be a statistically adequate predictor of y, the value of r^2 can help guide the regression analyst in the search for better, more useful models. For example, Crandall and Cedercreutz (1976) use a simple linear model to relate cost of mechanical work (heating, ventilating, and plumbing) in construction to floor area. Based on the data associated with 26 factory and warehouse buildings, the least squares prediction equation given in Figure 3.13 (page 124) was found. It was concluded that floor area and mechanical cost are linearly related, since the t statistic (for testing H_0:

FIGURE 3.13

Simple linear model relating cost to floor area

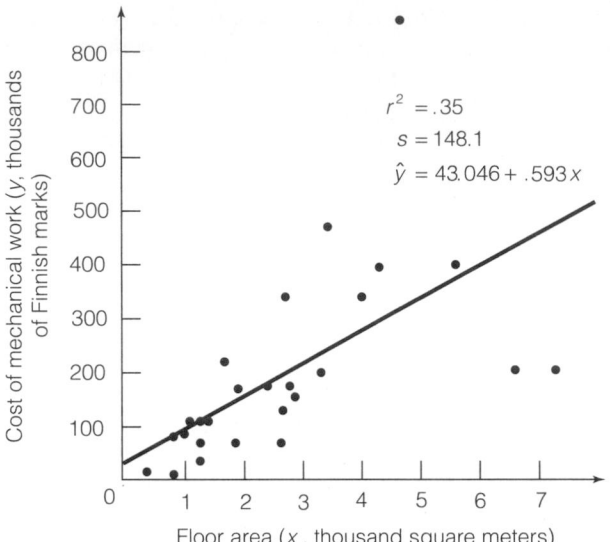

$\beta_1 = 0$) was found to equal 3.61, which is significant with an α as small as .002.* Thus, floor area should be useful when predicting the mechanical cost of a factory or warehouse. However, the value of the coefficient of determination r^2 was found to be .35. This tells us that only 35% of the variation among mechanical costs is accounted for by the differences in floor areas. This relatively small r^2 value led Crandall and Cedercreutz to include other independent variables (e.g., volume, amount of glass) in the model to account for a significant portion of the remaining 65% of the variation in mechanical cost not explained by floor area. In the next chapter we discuss this important aspect of relating a response to more than one independent variable.

EXERCISES

3.26   Find the correlation coefficient and the coefficient of determination for the sample data of each of the following exercises. Interpret your results.

    a. Exercise 3.8   **b.** Exercise 3.9   **c.** Exercise 3.10   **d.** Exercise 3.11

3.27   An investigation sought to determine whether there were certain collective behaviors, affective reactions, or performance outcomes associated with the maturity level of small groups (*Small Group Behavior*, May 1988). Fifty-eight undergraduate students enrolled in management information systems or communications courses at a medium-size university participated in the experiment. A 10-item questionnaire was used to measure the maturity level, y, of the students on a scale of 0–100, with more mature students receiving higher scores. One of several other variables measured was the number, x, of meetings held with their groups outside of regular class sessions. The correlation coefficient relating y to x was found to be $r = .46$. Is this sufficient evidence to indicate a positive correlation between group maturity and outside-of-class meetings? Test using $\alpha = .01$.

---

*Crandall, J. S. and Cedercreutz, M. "Preliminary cost estimates for mechanical work." *Building Systems Design*, Oct.–Nov. 1976, Vol. 73, pp. 35–51.

3.28 Refer to the *Research Management* study of 20 large R&D spenders, Exercise 3.10. The data on price/earnings ratio (y) and R&D expenditures/sales ratio (x) for the 20 firms are reproduced here.

COMPANY	P/E RATIO y	R/S RATIO x	COMPANY	P/E RATIO y	R/S RATIO x
1	5.6	.003	11	8.4	.058
2	7.2	.004	12	11.1	.058
3	8.1	.009	13	11.1	.067
4	9.9	.021	14	13.2	.080
5	6.0	.023	15	13.4	.080
6	8.2	.030	16	11.5	.083
7	6.3	.035	17	9.8	.091
8	10.0	.037	18	16.1	.092
9	8.5	.044	19	7.0	.064
10	13.2	.051	20	5.9	.028

Source: Wallin, C. C. and Gilman, J. J. "Determining the optimum level for R&D spending." *Research Management*, Vol. 14, No. 5, Sept./Oct. 1986, pp. 19–24 (adapted from Figure 1, p. 20).

a. Find the correlation coefficient r and interpret its value.

b. Do the data provide sufficient evidence to indicate that x and y are linearly correlated? Test using $\alpha = .05$.

3.29 In 1984, federal government outlays for elementary, secondary, and vocational education were cut, yet Scholastic Aptitude Test (SAT) scores increased. Has such a relationship existed in the past? According to *Fortune* (Oct. 29, 1984), "for the decade ending in 1984 . . . federal spending on education is strongly and negatively correlated with both verbal and math SAT scores." The correlation coefficient between verbal scores and federal spending on education for the past $n = 10$ years is $r = -.92$, whereas the correlation between math scores and federal spending is $r = -.71$.

a. Do the data support *Fortune*'s claim that verbal SAT scores and federal spending on education are "strongly and negatively correlated"? Test using $\alpha = .05$.

b. Do the data support *Fortune*'s claim that math SAT scores and federal spending on education are "strongly and negatively correlated"? Test using $\alpha = .05$.

c. Calculate the coefficient of determination for a straight-line model relating verbal SAT scores to federal spending. Interpret this value.

d. Calculate the coefficient of determination for a straight-line model relating math SAT scores to federal spending. Interpret this value.

3.30 To examine potential gender differences in the industrial sales force, a sample of 244 males and a sample of 153 females were administered a questionnaire (*Journal of Personal Selling & Sales Management*, Summer 1990). All respondents were either sales managers or sales people at one of 16 industrial firms located in the southeastern United States. Two variables of interest to the researchers were level of organizational commitment (y) and total months experience in sales (x). For the 244 males in the study, the coefficient of correlation between x and y was $r_{males} = -.35$. For the 153 females in the study, the correlation coefficient was $r_{females} = -.06$.

a. Interpret the value of r_{males}.

b. Interpret the value of $r_{females}$.

c. For each gender, test the hypothesis of no linear correlation between organizational commitment (y) and experience in years (x).

3.31   Best and Kahle (1985) investigated several factors thought to influence market share for capital-equipment businesses.* One variable considered was product quality, measured as the difference between the percentage of sales derived from products superior to competition and the percentage of sales inferior to competition. Based on data collected for 333 capital-equipment businesses, the correlation between market share and product quality was found to be $r = .373$.

a. Is there sufficient evidence to indicate that product quality and market share for capital equipment businesses are positively correlated? Test using $\alpha = .01$.

b. Calculate r^2. Interpret its value.

3.32   The accompanying table shows a portion of the experimental data obtained in a study of the radial tension strength of concrete pipe. The concrete pipe used for the experiment had an inside diameter of 84 inches and a wall thickness of approximately 8.75 inches. In addition, it was reinforced with cold drawn wire. The variable y is the load (in pounds per foot) until the first crack in a pipe specimen was observed. The variable x is the age of the specimen (in days) at the time of the test.

y	x	y	x
11,450	20	10,540	25
10,420	20	9,470	31
11,142	20	9,190	31
10,840	25	9,540	31
11,170	25		

Source: Heger, F. J. and McGrath, T. J. "Radial tension strength of pipe and other curved flexural members." *Journal of the American Concrete Institute*, Vol. 80, No. 1, 1983, pp. 33–39.

a. Construct a scattergram for the data. After examining the scattergram, do you think that x and y are correlated? If correlation is present, is it positive or negative?

b. Find the correlation coefficient r and interpret its value.

c. Do the data provide sufficient evidence to indicate that x and y are linearly correlated? Test using $\alpha = .05$.

d. Find the coefficient of determination r^2 and interpret its value.

3.33   A major portion of the effort expended in developing commercial computer software is associated with program testing. A study was undertaken to assess the potential usefulness of various product- and process-related variables in identifying error-prone software (*IEEE Transactions on Software Engineering*, Apr. 1985). A straight-line model relating the number y of module defects to the number x of unique operands in the module was fit to the data collected for a sample of software modules. The coefficient of determination for this analysis was $r^2 = .74$.

a. Interpret the value of r^2.

b. Based on this value, would you infer that the straight-line model is a useful predictor of number y of module defects? Explain.

---

*Best, R. J. and Kahle, L. R. "Managing a capital equipment business's market share." *Industrial Marketing Management*, Vol. 14, 1985, pp. 159–164.

3.9

Using the Model for Estimation and Prediction

If we are satisfied that a useful model has been found to describe the relationship between sales revenue and advertising, we are ready to accomplish the original objectives for building the model: using it to estimate or to predict sales on the basis of advertising dollars spent.

The most common uses of a probabilistic model can be divided into two categories. **The first is the use of the model for estimating the mean value of y, $E(y)$, for a specific value of x.** For our example, we may want to estimate the mean sales revenue for *all* months during which $400 ($x = 4$) is spent on advertising. **The second use of the model entails predicting a particular y value for a given x.** That is, if we decide to spend $400 next month, we want to predict the firm's sales revenue for that month.

In the case of estimating a mean value of y, we are attempting to estimate the mean result of a very large number of experiments at the given x value. In the second case, we are trying to predict the outcome of a single experiment at the given x value. In which of these model uses do you expect to have more success, i.e., which value—the mean or individual value of y—can we estimate (or predict) with more accuracy?

Before answering this question, we first consider the problem of choosing an estimator (or predictor) of the mean (or individual) y value. We will use the least squares model

$$\hat{y} = \hat{\beta}_0 + \hat{\beta}_1 x$$

both to estimate the mean value of y and to predict a particular value of y for a given value of x. For our example, we found

$$\hat{y} = -.1 + .7x$$

so that the estimated mean value of sales revenue for all months when $x = 4$ (advertising = $400) is

$$\hat{y} = -.1 + .7(4) = 2.7$$

or $2,700 (the units of y are thousands of dollars). The identical value is used to predict the y value when $x = 4$. That is, both the estimated mean value and the predicted value of y equal $\hat{y} = 2.7$ when $x = 4$, as shown in Figure 3.14.

FIGURE 3.14

Estimated mean value and predicted individual value of sales revenue y for $x = 4$

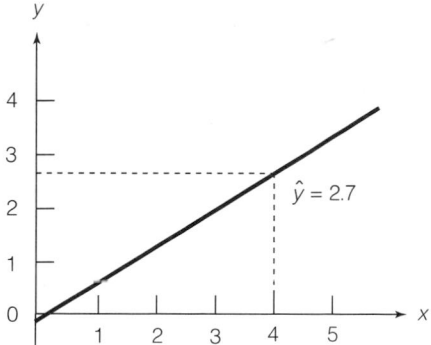

The difference in these two model uses lies in the relative accuracy of the estimate and the prediction. These accuracies are best measured by the repeated sampling errors of the least squares line when it is used as an estimator and as a predictor, respectively. These errors are given in the box.

Sampling Errors for the Estimator of the Mean of y and the Predictor of an Individual y for $x = x_p$

1. The standard deviation of the sampling distribution of the estimator $\hat{y}$ of the mean value of y at a particular value of x, say, x_p, is

$$\sigma_{\hat{y}} = \sigma\sqrt{\frac{1}{n} + \frac{(x_p - \bar{x})^2}{SS_{xx}}}$$

where σ is the standard deviation of the random error ε.

2. The standard deviation of the prediction error for the predictor $\hat{y}$ of an individual y value for $x = x_p$ is

$$\sigma_{(y-\hat{y})} = \sigma\sqrt{1 + \frac{1}{n} + \frac{(x_p - \bar{x})^2}{SS_{xx}}}$$

where σ is the standard deviation of the random error ε.

The true value of σ will rarely be known. Thus, we estimate σ by s and calculate the estimation and prediction intervals as shown in the next two boxes.

A $100(1 - \alpha)$% Confidence Interval for the Mean Value of y for $x = x_p$

$$\hat{y} \pm t_{\alpha/2}(\text{Estimated standard deviation of } \hat{y})$$

or

$$\hat{y} \pm t_{\alpha/2}s\sqrt{\frac{1}{n} + \frac{(x_p - \bar{x})^2}{SS_{xx}}}$$

where $t_{\alpha/2}$ is based on $(n - 2)$df

A $100(1 - \alpha)$% Prediction Interval for an Individual y for $x = x_p$

$$\hat{y} \pm t_{\alpha/2}[\text{Estimated standard deviation of } (y - \hat{y})]$$

or

$$\hat{y} \pm t_{\alpha/2}s\sqrt{1 + \frac{1}{n} + \frac{(x_p - \bar{x})^2}{SS_{xx}}}$$

where $t_{\alpha/2}$ is based on $(n - 2)$ df

EXAMPLE 3.4

Find a 95% confidence interval for mean monthly sales when the appliance store spends $400 on advertising.

Solution

For a $400 advertising expenditure, $x_p = 4$ and, since $n = 5$, df $= n - 2 = 3$. Then the confidence interval for the mean value of y is

$$\hat{y} \pm t_{\alpha/2}s \sqrt{\frac{1}{n} + \frac{(x_p - \bar{x})^2}{SS_{xx}}}$$

or

$$\hat{y} \pm t_{.025}s \sqrt{\frac{1}{5} + \frac{(4 - \bar{x})^2}{SS_{xx}}}$$

Recall that $\hat{y} = 2.7$, $s = .61$, $\bar{x} = 3$, and $SS_{xx} = 10$. From Table 2 of Appendix D, $t_{.025} = 3.182$. Thus, we have

$$2.7 \pm (3.182)(.61) \sqrt{\frac{1}{5} + \frac{(4 - 3)^2}{10}} = 2.7 \pm (3.182)(.61)(.55)$$

$$= 2.7 \pm 1.1$$

We estimate that the interval from $1,600 to $3,800 encloses the mean sales revenue when the store spends $400 a month on advertising. Note that we used a small amount of data for purposes of illustration in fitting the least squares line and that the width of the interval could be decreased by using a larger number of data points.

EXAMPLE 3.5

Predict the monthly sales for next month if a $400 expenditure is to be made on advertising. Use a 95% prediction interval.

Solution

To predict the sales for a particular month for which $x_p = 4$, we calculate the 95% prediction interval as

$$\hat{y} \pm t_{\alpha/2}s \sqrt{1 + \frac{1}{n} + \frac{(x_p - \bar{x})^2}{SS_{xx}}} = 2.7 \pm (3.182)(.61) \sqrt{1 + \frac{1}{5} + \frac{(4 - 3)^2}{10}}$$

$$= 2.7 \pm (3.182)(.61)(1.14) = 2.7 \pm 2.2$$

Therefore, we predict that the sales next month will fall in the interval from $500 to $4,900. As in the case of the confidence interval for the mean value of y, the prediction interval for y is quite large. This is because we have chosen a simple example (only five data points) to fit the least squares line. The width of the prediction interval could be reduced by using a larger number of data points.

A comparison of the confidence limits for the mean value of y and the prediction interval for some future value of y for various advertising expenditure

(x) values is illustrated in Figure 3.15. It is important to note that the prediction interval for an individual value of y will always be wider than the confidence interval for a mean value of y. You can see this by examining the formulas for the two intervals, and you can see it in Figure 3.15.

FIGURE 3.15

Comparison of widths of 95% confidence and prediction intervals

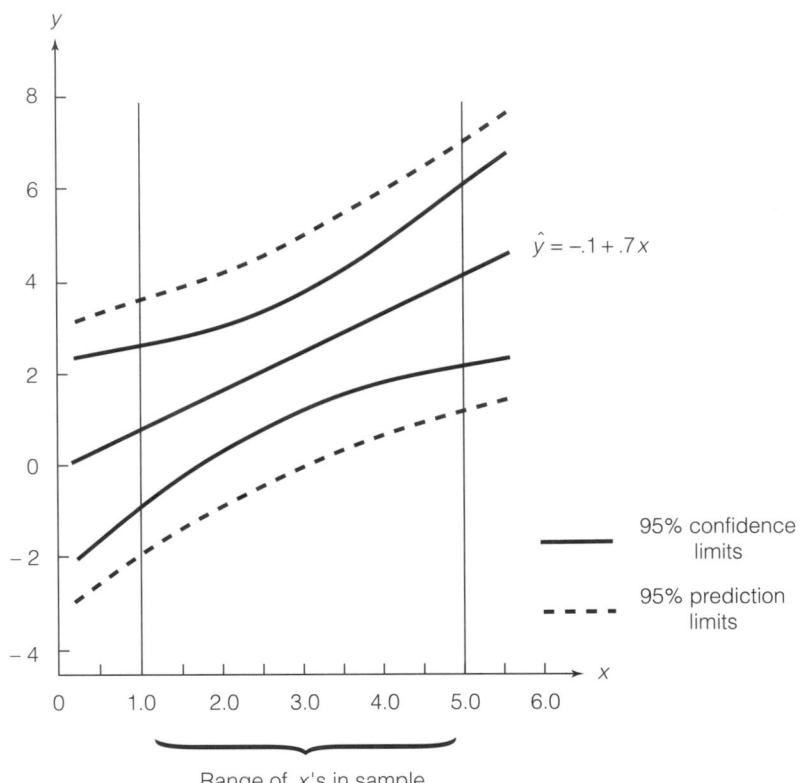

The error in estimating the mean value of y, $E(y)$, for a given value of x, say, x_p, is the distance between the least squares line and the true line of means, $E(y) = \beta_0 + \beta_1 x$. This error, $[\hat{y} - E(y)]$, is shown in Figure 3.16. In contrast, the error $(y_p - \hat{y})$ in predicting some future value of y is the sum of two errors—the error of estimating the mean of y, $E(y)$, shown in Figure 3.16, plus the random error that is a component of the value of y to be predicted (see Figure 3.17). Consequently, the error of predicting a particular value of y will always be larger than the error of estimating the mean value of y for a particular value of x. Note from their formulas that both the error of estimation and the error of prediction take their smallest values when $x_p = \bar{x}$. The farther x lies from $\bar{x}$, the larger will be the errors of estimation and prediction (see Figure 3.15). You can see why this is true by noting the deviations for different values of x between the line of means $E(y) = \beta_0 + \beta_1 x$ and the predicted line $\hat{y} = \hat{\beta}_0 + \hat{\beta}_1 x$ shown

FIGURE 3.16

Error of estimating the mean value of y for a given value of x

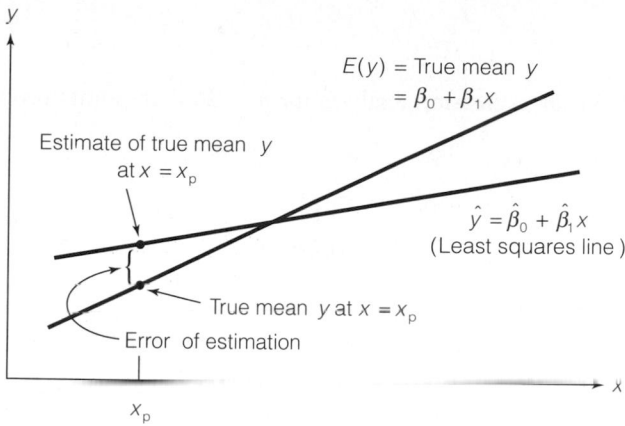

FIGURE 3.17

Error of predicting a future value of y for a given value of x

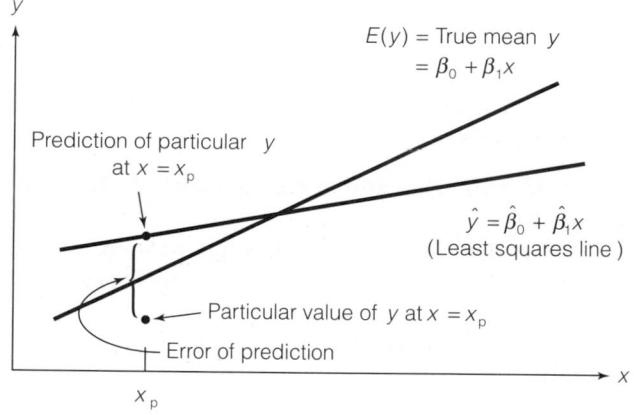

in Figure 3.17. The deviation is larger at the extremities of the interval where the largest and smallest values of x in the data set occur. In fact, when x is selected far enough away from $\bar{x}$ so that it falls outside the range of the sample data, it is dangerous to make any inferences about $E(y)$, or y, as the warning box explains.

| Warning

Using the least squares prediction equation to estimate the mean value of y or to predict a particular value of y for values of x that fall *outside the range* of the values of x contained in your sample data may lead to errors of estimation or prediction that are much larger than expected. Although the least squares model may provide a very good fit to the data over the range of x values contained in the sample, it could give a poor representation of the true model for values of x outside this region.

EXERCISES

3.34   A simple linear regression analysis for $n = 20$ data points produced the following results:

$$\hat{y} = 2.1 + 3.4x \qquad SS_{xx} = 4.77$$

$$\bar{x} = 2.5 \qquad SS_{yy} = 59.21$$

$$\bar{y} = 10.6 \qquad SS_{xy} = 16.22$$

    a. Find SSE and s^2.

    b. Find a 95% confidence interval for $E(y)$ when $x = 2.5$. Interpret this interval.

    c. Find a 95% confidence interval for $E(y)$ when $x = 2.0$. Interpret this interval.

    d. Find a 95% confidence interval for $E(y)$ when $x = 3.0$. Interpret this interval.

    e. Examine the widths of the confidence intervals obtained in parts **b**, **c**, and **d**. What happens to the width of the confidence interval for $E(y)$ as the value of x moves away from the value of $\bar{x}$?

    f. Find a 95% prediction interval for a value of y to be observed in the future when $x = 3.0$. Interpret its value.

3.35   Refer to Exercise 3.9. Find a 95% prediction interval for drill time y when drilling begins at a depth of 300 feet. Interpret your result.

3.36   Refer to Exercise 3.10. The ratio of research and development expenditures to sales (R/S) ratio for a company is .070.

    a. Find a 90% prediction interval for the company's price/earnings (P/E) ratio.

    b. Find a 90% confidence interval for the average (mean) P/E ratio of all firms with an R/S ratio of .070.

    c. Compare and comment on the sizes of the intervals in parts **a** and **b**.

    d. Could you reduce the size of either or both intervals by increasing your sample size? Explain.

3.37   Explain why for a particular x value, the prediction interval for an individual y value will always be wider than the confidence interval for a mean value of y.

3.38   Explain why the confidence interval for the mean value of y for a particular x value, say, x_p, gets wider the farther x_p is from $\bar{x}$. What are the implications of this phenomenon for estimation and prediction?

3.39   Refer to Exercise 3.15. Find a 95% prediction interval for the refusal rate of a Florida county with a per capita income of $8,000.

3.40   In forestry, the diameter of a tree at breast height (which is fairly easy to measure) is used to predict the height of the tree (a difficult measurement to obtain). Silviculturists working in British Columbia's boreal forest conducted a series of spacing trials to predict the heights of several species of trees. The data in the accompanying table are the breast height diameters (in centimeters) and heights (in meters) for a sample of 36 white spruce trees.

    a. Construct a scattergram for the data.

    b. Assuming the relationship between the variables is best described by a straight line, use the method of least squares to estimate the y-intercept and slope of the line.

    c. Plot the least squares line on your scattergram.

BREAST HEIGHT DIAMETER	HEIGHT	BREAST HEIGHT DIAMETER	HEIGHT
x, cm	y, m	x, cm	y, m
18.9	20.0	16.6	18.8
15.5	16.8	15.5	16.9
19.4	20.2	13.7	16.3
20.0	20.0	27.5	21.4
29.8	20.2	20.3	19.2
19.8	18.0	22.9	19.8
20.3	17.8	14.1	18.5
20.0	19.2	10.1	12.1
22.0	22.3	5.8	8.0
23.6	18.9	20.7	17.4
14.8	13.3	17.8	18.4
22.7	20.6	11.4	17.3
18.5	19.0	14.4	16.6
21.5	19.2	13.4	12.9
14.8	16.1	17.8	17.5
17.7	19.9	20.7	19.4
21.0	20.4	13.3	15.5
15.9	17.6	22.9	19.2

Source: Scholz, H., Northern Lights College, British Columbia.

d. Do the data provide sufficient evidence to indicate that the breast height diameter x contributes information for the prediction of tree height y? Test using $\alpha = .05$.

e. Use your least squares line to find a 90% confidence interval for the average height of white spruce trees with a breast height diameter of 20 cm. Interpret the interval.

3.41 Refer to the study of the radial tension strength of concrete pipe in Exercise 3.32. The data are reproduced in the table. The variable y is the load (in pounds per foot) until the first crack in a pipe was observed, and the variable x is the age of the specimen (in days) at the time of the test.

y	x	y	x
11,450	20	10,540	25
10,420	20	9,470	31
11,142	20	9,190	31
10,840	25	9,540	31
11,170	25		

Source: Heger, F. J. and McGrath, T. J. "Radial tension strength of pipe and other curved flexural members." *Journal of the American Concrete Institute,* Vol. 80, No. 1, 1983, pp. 33–39.

a. Find the least squares prediction equation relating load y and age x.

b. Test the hypothesis $H_0: \beta_1 = 0$ (at $\alpha = .05$), and show that the result agrees with your answer to Exercise 3.32, part c.

c. Find a 95% prediction interval for the crack load of a 35-day-old concrete specimen.

d. Why might the prediction interval in part c be less reliable than expected? Explain.

3.10

Simple Linear Regression: An Example

In the previous sections, we have presented the basic elements necessary to fit and use a straight-line regression model. In this section, we will assemble these elements by applying them in an example where we use the computer to perform the calculations.

Suppose a fire insurance company wants to relate the amount of fire damage in major residential fires to the distance between the residence and the nearest fire station. The study is to be conducted in a large suburb of a major city; a sample of 15 recent fires in this suburb is selected.

STEP 1 First, we hypothesize a model to relate fire damage y to the distance x from the nearest fire station. We will hypothesize a straight-line probabilistic model:

$$y = \beta_0 + \beta_1 x + \varepsilon$$

STEP 2 Second, we collect the (x, y) values for each of the $n = 15$ experimental units (residential fires) in the sample. The amount of damage y and the distance x between the fire and the nearest fire station are recorded for each fire in Table 3.8.

TABLE 3.8 **Fire Damage Data**

DISTANCE FROM FIRE STATION x, miles	FIRE DAMAGE y, thousands of dollars
3.4	26.2
1.8	17.8
4.6	31.3
2.3	23.1
3.1	27.5
5.5	36.0
.7	14.1
3.0	22.3
2.6	19.6
4.3	31.3
2.1	24.0
1.1	17.3
6.1	43.2
4.8	36.4
3.8	26.1

STEP 3 Next, we enter the data into a computer and use a statistical software package to estimate the unknown parameters in the deterministic component of the hypothesized model. The SAS printout for the simple linear regression analysis is shown in Figure 3.18.

The least squares estimates of β_0 and β_1 are found (shaded) under the column labeled **Parameter Estimate** (in the middle portion of the printout) in the rows labeled **INTERCEP** and **X**, respectively. Note that the estimate of the slope is

$$\hat{\beta}_1 = 4.919331$$

FIGURE 3.18

SAS printout for the fire damage linear regression

Dependent Variable: Y

Analysis of Variance

Source	DF	Sum of Squares	Mean Square	F Value	Prob>F
Model	1	841.76636	841.76636	156.886	0.0001
Error	13	69.75098	5.36546		
C Total	14	911.51733			

Root MSE	2.31635	R-square	0.9235	
Dep Mean	26.41333	Adj R-sq	0.9176	
C.V.	8.76961			

Parameter Estimates

Variable	DF	Parameter Estimate	Standard Error	T for H0: Parameter=0	Prob > |T|
INTERCEP	1	10.277929	1.42027781	7.237	0.0001
X	1	4.919331	0.39274775	12.525	0.0001

Obs	X	Dep Var Y	Predict Value	Std Err Predict	Lower95% Predict	Upper95% Predict	Residual
1	3.4	26.2000	27.0037	0.600	21.8344	32.1729	-0.8037
2	1.8	17.8000	19.1327	0.834	13.8141	24.4514	-1.3327
3	4.6	31.3000	32.9068	0.791	27.6186	38.1951	-1.6068
4	2.3	23.1000	21.5924	0.711	16.3577	26.8271	1.5076
5	3.1	27.5000	25.5279	0.602	20.3573	30.6984	1.9721
6	5.5	36.0000	37.3342	1.057	31.8334	42.8351	-1.3342
7	0.7	14.1000	13.7215	1.177	8.1087	19.3342	0.3785
8	3	22.3000	25.0359	0.608	19.8622	30.2097	-2.7359
9	2.6	19.6000	23.0682	0.655	17.8678	28.2686	-3.4682
10	4.3	31.3000	31.4311	0.720	26.1908	36.6713	-0.1311
11	2.1	24.0000	20.6085	0.757	15.3442	25.8729	3.3915
12	1.1	17.3000	15.6892	1.044	10.1999	21.1785	1.6108
13	6.1	43.2000	40.2858	1.259	34.5906	45.9811	2.9142
14	4.8	36.4000	33.8907	0.845	28.5640	39.2175	2.5093
15	3.8	26.1000	28.9714	0.632	23.7843	34.1585	-2.8714
16	3.5	.	27.4956	0.604	22.3239	32.6672	.

Sum of Residuals	-4.44089E-14
Sum of Squared Residuals	69.7510
Predicted Resid SS (Press)	93.2117

and the estimate of the y-intercept is

$$\hat{\beta}_0 = 10.277929$$

Thus, the least squares equation is

$$\hat{y} = 10.278 + 4.919x$$

This prediction equation is graphed in Figure 3.19 (page 136), along with a plot of the data points.

STEP 4 Now, we specify the probability distribution of the random error component ε. The assumptions about the distribution will be identical to those listed in Section 3.4. Although we know that these assumptions are not completely satisfied (they rarely are for any practical problem), we are willing to assume they are approximately satisfied for this example. The estimate of the variance

FIGURE 3.19

Least squares model for the fire damage data

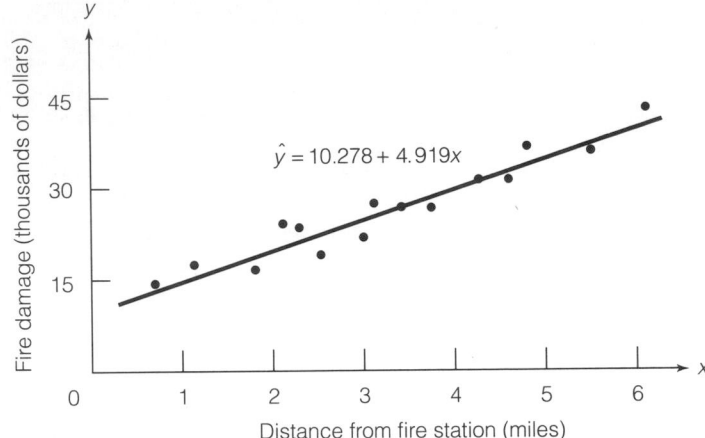

σ^2 of ε is given (shaded) in the top portion of the printout in the column labeled **Mean Square** and the row labeled **Error**. This value is

$$s^2 = \text{MSE} = 5.36546$$

The estimated standard deviation of ε, given next to the heading **ROOT MSE**, is

$$s = \sqrt{5.36546} = 2.31635$$

The value of s implies that most of the observed fire damage (y) values will fall within approximately $2s = 4.64$ thousand dollars of their respective predicted values.

STEP 5

a. *Test of model utility*   We can now check the utility of the hypothesized model, that is, whether x really contributes information for the prediction of y using the straight-line model. First, test the null hypothesis that the slope β_1 is 0, i.e., that there is no linear relationship between fire damage and the distance from the nearest fire station, against the alternative that x and y are positively linearly related. We test:

$$H_0: \quad \beta_1 = 0$$
$$H_a: \quad \beta_1 > 0$$

The value of the test statistic is shaded in the middle portion of the printout under the column labeled **T for H0: Parameter=0** and the row corresponding to **X**. The value, a t statistic, is

$$t = 12.525$$

The p-value of the test is reported on the printout under the column heading **Prob > |T|** in the X row. This value (shaded) is a two-tailed p-value. The

p-value for a one-tailed test of H_0: $\beta_1 = 0$ is found by dividing the value reported in the SAS printout in half. Thus, the p-value for our test is

$$p = \frac{.0001}{2} = .00005$$

This small p-value leaves little doubt that distance between the fire and the fire station contributes information for the prediction of fire damage and that fire damage increases as the distance increases.

b. *Confidence interval for slope*   We gain additional information about the relationship by forming a confidence interval for the slope β_1. A 95% confidence interval is $\hat{\beta}_1 \pm (t_{.025})s_{\hat{\beta}_1}$, where the value of $\hat{\beta}_1$ and its standard error, $s_{\hat{\beta}_1}$, are shown (shaded) on the printout. The value of $t_{.025}$, based on $n - 2 = 13$ df, is 2.160. Therefore, the 95% confidence interval is

$$\hat{\beta}_1 \pm (t_{.025})s_{\hat{\beta}_1} = 4.919 \pm (2.160)(.3927)$$

$$= 4.919 \pm .849 = (4.070, 5.768)$$

We estimate that the interval from \$4,070 to \$5,768 encloses the mean increase (β_1) in fire damage per additional mile distance from the fire station.

c. *Numerical descriptive measures of model adequacy*   The coefficient of determination is found next to the heading **R-Square** in the middle portion of the printout. This value (shaded) is

$$r^2 = .9235$$

which implies that about 92% of the sample variation in fire damage (y) is explained by the distance x between the fire and the fire station.

The coefficient of correlation r, which measures the strength of the linear relationship between y and x, is not shown on the SAS printout and must be calculated. Using the facts that $r = \sqrt{r^2}$ in simple linear regression and that r and $\hat{\beta}_1$ have the same sign, we find

$$r = +\sqrt{r^2} = \sqrt{.9235} = .96$$

The high correlation confirms our conclusion that β_1 differs from 0; it appears that fire damage and distance from the fire station are linearly correlated. All signs point to a strong linear relationship between x and y.

STEP 6   We are now prepared to use the least squares model. Suppose the insurance company wants to predict the fire damage if a major residential fire were to occur 3.5 miles from the nearest fire station, i.e., $x_p = 3.5$. The predicted value is shown (shaded) at the bottom of the SAS printout in the row corresponding to **X** = 3.5 and the column headed **Predict Value**. This value is

$$\hat{y} = 27.4956$$

Lower and upper prediction limits for this value are given under the columns labeled **Lower 95% Predict** and **Upper 95% Predict**, respectively. These values (shaded) are 22.3239 and 32.6672. Thus, the model yields a 95% prediction interval of \$22,324 to \$32,667 for fire damage in a major residential fire 3.5 miles from the nearest fire station.

Caution: We would not use this prediction model to make predictions for homes less than .7 mile or more than 6.1 miles from the nearest fire station. A look at the data in Table 3.8 reveals that all the x values fall between .7 and 6.1. Recall from Section 3.9 that it is dangerous to use the model to make predictions outside the region in which the sample data fall. A straight line might not provide a good model for the relationship between the mean value of y and the value of x when stretched over a wider range of x values.

In Chapter 4 we will discuss the interpretation of those portions of the SAS printout not mentioned here. However, the important elements of a simple linear regression analysis have been located, and you should be able to use this discussion as a guide to interpreting simple linear regression computer printouts. Details on how to perform a simple linear regression analysis on the computer using any one of the three program packages, SAS, SPSS, or Minitab, are provided in Appendix C.

EXERCISES

3.42   A portion of the Minitab printout for the model $E(y) = \beta_0 + \beta_1 x$ fit to $n = 10$ data points is shown here:

```
The regression equation is
y = 55.29 - .7716 x

Predictor        Coef       Stdev     t-ratio        p
Constant      55.2942      7.4529        7.42    0.001
x              -.7716       .0978       -7.89    0.001

s = 1.4321      R-sq = 88.6%       R-sq(adj) = 87.2%

Analysis of Variance

SOURCE         DF          SS          MS        F        p
Regression      1     127.6935     127.6935    62.26   0.0001
Error           8      16.4065       2.0508
Total           9     144.1000
```

a. Give the least squares prediction equation.
b. Locate the values of SSE, s^2, s, and r^2 on the printout.
c. What is the value of the test statistic for testing model adequacy?
d. Locate and interpret the p-value of the test for model adequacy.
e. Give the lower and upper endpoints for a 95% prediction interval for y when $x = 58$. Interpret the interval.

3.43 The Federal Trade Commission (FTC) annually ranks American cigarette brands according to carbon monoxide and nicotine content (measured in milligrams). To determine the relationship between carbon monoxide content (y) and nicotine content (x), the FTC fit the straight-line model $y = \beta_0 + \beta_1 x + \varepsilon$ to data collected for a sample of 10 brands of cigarettes. The SAS printout is reproduced here.

```
                               ANALYSIS OF VARIANCE

                          SUM OF            MEAN
           SOURCE    DF   SQUARES          SQUARE      F VALUE      PROB>F

           MODEL      1   283.93484       283.93484    325.699      0.0001
           ERROR      8   6.97415579        0.87176947
           C TOTAL    9   290.90900

                ROOT MSE    0.933686       R-SQUARE     0.9760
                DEP MEAN    9.79           ADJ R-SQ     0.9730
                C.V.        9.53714

                            PARAMETER ESTIMATES

                       PARAMETER      STANDARD     T FOR H0:
      VARIABLE   DF      ESTIMATE       ERROR      PARAMETER=0   PROB > |T|

      INTERCEP   1     -0.53490343     0.64380408    -0.831        0.4302
      X          1     15.52617057     0.86031171    18.047        0.0001
```

OBS	ID	ACTUAL	PREDICT VALUE	STD ERR PREDICT	LOWER95% PREDICT	UPPER95% PREDICT	RESIDUAL
1	0.86	13.6000	12.8176	0.3396	10.5265	15.1087	0.7824
2	0.17	1.2000	2.1045	0.5182	-0.3579	4.5670	-0.9045
3	0.72	9.8000	10.6439	0.2990	8.3831	12.9048	-0.8439
4	0.1	1.2000	1.0177	0.5687	-1.5034	3.5388	0.1823
5	0.4	5.4000	5.6756	0.3730	3.3570	7.9942	-0.2756
6	0.93	12.7000	13.9044	0.3730	11.5858	16.2230	-1.2044
7	0.67	10.5000	9.8676	0.2953	7.6094	12.1258	0.6324
8	1.02	15.4000	15.3018	0.4248	12.9363	17.6673	0.0982
9	0.57	10.0000	8.3150	0.3064	6.0490	10.5811	1.6850
10	1.21	18.1000	18.2518	0.5541	15.7481	20.7555	-0.1518

```
SUM OF RESIDUALS              -6.66134E-16
SUM OF SQUARED RESIDUALS       6.974156
PREDICTED RESID SS (PRESS)     9.781261
```

a. Give the least squares prediction equation.
b. Locate the values of SSE, s^2, s, and r^2 on the printout.
c. What is the value of the test statistic for testing model adequacy?
d. Locate and interpret the p-value of the test for model adequacy.
e. Give the lower and upper endpoints for a 95% prediction interval for y when $x = .4$ milligram. Interpret the interval.

3.44 In retailing ready-to-wear fashion, price concession—in the form of a markdown—is the traditional vehicle for selling slow-moving merchandise. To investigate the effect of a markdown on the rate of sale of slow-moving items, P. G. Carlson (Emory University) studied the markdown policy at Rich's department store in Atlanta. Nine styles of budget junior dresses were selected for analysis. For each style, the initial rate of sale, x, and the postmarkdown rate of sale, y, were recorded. (The initial rate of sale is the average rate of sale from the day the style was put on the floor to the time of the first markdown, while the postmarkdown rate of sale is the average rate of sale from the day of the first markdown to the time of either the second markdown or the end of the fashion life.) The nine data points were subjected to a simple linear regression analysis. The SPSS printout appears here.

```
_____

            * * * *   M U L T I P L E   R E G R E S S I O N   * * * *

Equation Number 1    Dependent Variable..    Y


Variable(s) Entered on Step Number
    1..    X


Multiple R            .8229
R Square              .6772
Adjusted R Square     .6311
Standard Error        .1044

Analysis of Variance
                      DF      Sum of Squares      Mean Square
Regression            1              .1603             .1603
Residual              7              .0764             .0109

F =      14.7064          Signif F =   .0100

----------------- Variables in the Equation ------------------

Variable             B         SE B        Beta        T   Sig T

X                2.2985       .5999       .8229      3.831  .0200
(Constant)      -0.0292       .0902                 -0.324  .5000


End Block Number   1    All requested variables entered.
_____
```

Source: Carlson, P. G. "Fashion retailing: The sensitivity of rate of sale to markdown." *Journal of Retailing,* Spring 1983, Vol. 59, No. 1, pp. 67–76.

a. Write the least squares prediction equation.

b. Locate SSE on the printout and interpret its value.

c. Locate s on the printout and interpret its value.

d. Is there sufficient evidence to indicate that postmarkdown rate of sale y and initial rate of sale x are linearly related? Test at $\alpha = .05$.

3.45   The SAS printout for the straight-line model relating sale price y of a residential property in neighborhood D and total appraised value of x of the property (both measured in thousands of dollars) for a sample of 20 properties selected from Appendix E is provided here.

a. Give the least squares prediction equation.

b. Locate the values of SSE, s^2, s, and r^2 on the printout.

c. What is the value of the test statistic for testing model adequacy?

d. Locate and interpret the p-value of the test for model adequacy.

e. Give the lower and upper endpoints for a 95% prediction interval for y when $x = 40$. Interpret the interval.

ANALYSIS OF VARIANCE

SOURCE	DF	SUM OF SQUARES	MEAN SQUARE	F VALUE	PROB>F
MODEL	1	58.31306210	58.31306210	3.729	0.0694
ERROR	18	281.48694	15.63816322		
C TOTAL	19	339.80000			

ROOT MSE	3.954512	R-SQUARE	0.1716	
DEP MEAN	41.1	ADJ R-SQ	0.1256	
C.V.	9.621683			

PARAMETER ESTIMATES

VARIABLE	DF	PARAMETER ESTIMATE	STANDARD ERROR	T FOR H0: PARAMETER=0	PROB > |T|
INTERCEP	1	30.11691649	5.75599468	5.232	0.0001
X	1	0.31605996	0.16367390	1.931	0.0694

OBS	ID	ACTUAL	PREDICT VALUE	STD ERR PREDICT	LOWER95% PREDICT	UPPER95% PREDICT	RESIDUAL
1	31	44.0000	39.9148	1.0764	31.3044	48.5251	4.0852
2	31	45.0000	39.9148	1.0764	31.3044	48.5251	5.0852
3	32	44.0000	40.2308	0.9922	31.6652	48.7964	3.7692
4	33	40.0000	40.5469	0.9295	32.0124	49.0814	-0.5469
5	35	36.0000	41.1790	0.8852	32.6653	49.6927	-5.1790
6	40	45.0000	42.7593	1.2330	34.0568	51.4619	2.2407
7	29	40.0000	39.2827	1.2914	30.5428	48.0225	0.7173
8	40	40.0000	42.7593	1.2330	34.0568	51.4619	-2.7593
9	41	39.0000	43.0754	1.3522	34.2951	51.8557	-4.0754
10	36	45.0000	41.4951	0.9076	32.9710	50.0192	3.5049
11	27	43.0000	38.6505	1.5463	29.7299	47.5711	4.3495
12	31	33.0000	39.9148	1.0764	31.3044	48.5251	-6.9148
13	39	40.0000	42.4433	1.1251	33.8055	51.0810	-2.4433
14	43	42.0000	43.7075	1.6141	34.7340	52.6810	-1.7075
15	28	36.0000	38.9666	1.4151	30.1426	47.7906	-2.9666
16	36	37.0000	41.4951	0.9076	32.9710	50.0192	-4.4951
17	33	41.0000	40.5469	0.9295	32.0124	49.0814	0.4531
18	32	42.0000	40.2308	0.9922	31.6652	48.7964	1.7692
19	30	38.0000	39.5987	1.1774	30.9302	48.2672	-1.5987
20	48	52.0000	45.2878	2.3420	35.6320	54.9436	6.7122

SUM OF RESIDUALS		8.88178E-14
SUM OF SQUARED RESIDUALS		281.4869
PREDICTED RESID SS (PRESS)		388.2499

3.11

Regression Through the Origin (Optional)

In practice, occasionally it is known in advance that the true line of means $E(y)$ passes through the point $(x = 0, y = 0)$, called the **origin**. For example, a chain of convenience stores may be interested in modeling sales y of a new diet soft drink as a linear function of amount x of the new product in stock for a sample of stores. Or, a graphic designer may be interested in the linear relationship between the number x of company logos produced in a month and the total monthly logo production cost y for the past 12 months. In both cases, it is known that the regression line must pass through the origin. The convenience store chain knows that if one of its stores chooses not to stock the new diet soft drink, it will have zero sales of the new product. Likewise, if the graphic designer fails to produce any company logos during a month, total production costs will be 0.

Formulas for Regression Through the Origin

$$y = \beta_1 x + \varepsilon$$

Least squares slope:

$$\hat{\beta}_1 = \frac{\sum x_i y_i}{\sum x_i^2}$$

Estimate of σ^2:

$$s^2 = \frac{\text{SSE}}{n-1}, \quad \text{where SSE} = \sum y_i^2 - \hat{\beta}_1 \sum x_i y_i$$

Estimate of $\sigma_{\hat{\beta}_1}$:

$$s_{\hat{\beta}_1} = s \Big/ \sqrt{\sum x_i^2}$$

Estimate of $\sigma_{\hat{y}}$ for estimating $E(y)$ when $x = x_p$:

$$s_{\hat{y}} = s \left(\frac{x_p}{\sqrt{\sum x_i^2}} \right)$$

Estimate of $\sigma_{(y-\hat{y})}$ for predicting y when $x = x_p$:

$$s_{(y-\hat{y})} = s \sqrt{1 + \frac{x_p^2}{\sum x_i^2}}$$

For situations in which we know that the regression line passes through the origin, the y-intercept is $\beta_0 = 0$ and the probabilistic straight-line model takes the form

$$y = \beta_1 x + \varepsilon$$

When the regression line passes through the origin the formula for the least squares estimate of the slope β_1 differs from the formula given in Section 3.3. Several other formulas required to perform the regression analysis also are different. These new computing formulas are provided in the accompanying box.

Note that the denominator of s^2 is $n - 1$, not $n - 2$ as in the previous sections. This is because we need to estimate only a single parameter β_1 rather than both β_0 and β_1. Consequently, we have one additional degree of freedom for estimating σ^2, the variance of ε. Tests and confidence intervals for β_1 are carried out exactly as outlined in the previous sections, except that the t distribution is based on $(n - 1)$ df. The test statistic and confidence intervals are given in the next box.

| Tests and Confidence Intervals for Regression Through the Origin

Test statistic for H_0: $\beta_1 = 0$:

$$t = \frac{\hat{\beta}_1 - 0}{s_{\hat{\beta}_1}} = \frac{\hat{\beta}_1}{s \big/ \sqrt{\sum x_i^2}}$$

$100(1 - \alpha)\%$ *confidence interval for* β_1:

$$\hat{\beta}_1 \pm t_{\alpha/2} s_{\hat{\beta}_1} = \hat{\beta}_1 \pm t_{\alpha/2} s \big/ \sqrt{\sum x_i^2}$$

$100(1 - \alpha)\%$ *confidence interval for* $E(y)$:

$$\hat{y} \pm t_{\alpha/2} s_{\hat{y}} = \hat{y} \pm t_{\alpha/2} s \left(\frac{x_p}{\sqrt{\sum x_i^2}} \right)$$

$100(1 - \alpha)\%$ *prediction interval for* y:

$$\hat{y} \pm t_{\alpha/2} s_{(y-\hat{y})} = \hat{y} \pm t_{\alpha/2} s \sqrt{1 + \frac{x_p^2}{\sum x_i^2}}$$

where the distribution of t is based on $(n - 1)$ df

We illustrate the procedure for conducting simple linear regression through the origin in the following example.

EXAMPLE 3.6

Suppose we want to fit a straight-line model relating the number of items produced by a particular manufacturing process, x, and the total variable cost, y, involved in the production. For this process, total variable cost will be 0 when total output is 0. Table 3.9 gives the total output and total variable cost for a sample of $n = 7$ production days. Use the data to fit a straight-line regression model through the origin and calculate SSE.

TABLE 3.9 **Sample data for Example 3.6**

TOTAL OUTPUT x	TOTAL VARIABLE COST y, dollars
10	10
15	12
20	20
20	21
25	22
30	20
30	19

Solution

The model we want to fit is $y = \beta_1 x + \varepsilon$. Preliminary calculations for estimating β_1 and calculating SSE are contained in Table 3.10.

TABLE 3.10   Preliminary Calculations for Example 3.6

x_i	y_i	x_i^2	$x_i y_i$	y_i^2
10	10	100	100	100
15	12	225	180	144
20	20	400	400	400
20	21	400	420	441
25	22	625	550	484
30	20	900	600	400
30	19	900	570	361
Totals		$\Sigma\, x_i^2 = 3{,}550$	$\Sigma\, x_i y_i = 2{,}820$	$\Sigma\, y_i^2 = 2{,}330$

The estimate of the slope is

$$\hat{\beta}_1 = \frac{\Sigma\, x_i y_i}{\Sigma\, x_i^2} = \frac{2{,}820}{3{,}550} = .7943662$$

and the least squares line is

$$\hat{y} = .794x$$

The value of SSE for the line is

$$\text{SSE} = \Sigma\, y_i^2 - \hat{\beta}_1 \Sigma\, x_i y_i$$
$$= 2{,}330 - (.7943662)(2{,}820) = 89.8873$$

The graph of the least squares line with the observations is shown in Figure 3.20.

FIGURE 3.20

The line $\hat{y} = .794x$ fit to the data in Table 3.9

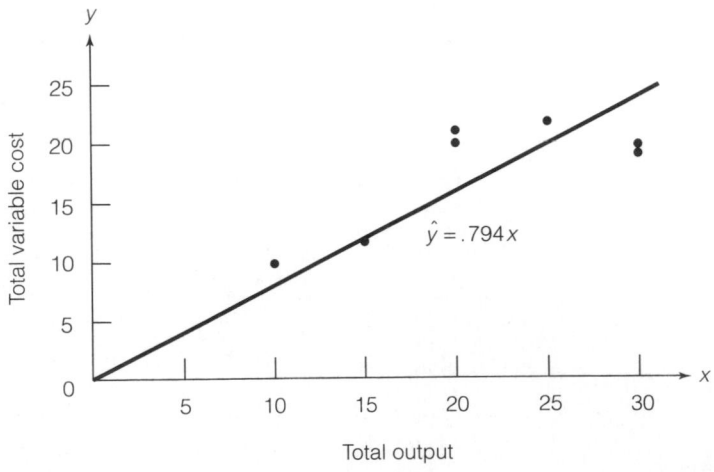

Total output

EXAMPLE 3.7

Refer to Example 3.6. Conduct the appropriate test for model adequacy. If the model is deemed adequate, predict the total variable cost y for a day in which $x = 23$ items are produced with a 95% prediction interval.

Solution

The appropriate test for model adequacy is

$$H_0: \quad \beta_1 = 0 \qquad H_a: \quad \beta_1 > 0$$

(We choose to do an upper-tailed test since it is reasonable to assume that if a linear relationship exists between total output x and total variable cost y, it is a positive one.)

To calculate the test statistic, we first compute s, where

$$s = \sqrt{\frac{SSE}{n-1}} = \sqrt{\frac{89.8873}{6}} = 3.87056$$

Then the test statistic is

$$t = \frac{\hat{\beta}_1}{s / \sqrt{\sum x_i^2}} = \frac{.7943662}{3.87056 / \sqrt{3{,}550}} = 12.23$$

For $\alpha = .05$, we will reject the null hypothesis if $t > t_{.05}$, where the distribution of t is based on $(n - 1) = 6$ df. From Table 2 of Appendix D, $t_{.05} = 1.943$. Thus, we will reject H_0 if

$$t > 1.943$$

Since the calculated value of t, 12.23, exceeds the critical value, there is sufficient evidence (at $\alpha = .05$) to conclude that the model is adequate for predicting total variable cost y.

To calculate a 95% prediction interval for total variable cost y when 23 units are produced, we first substitute $x = 23$ into the least squares prediction equation $\hat{y} = .794x$:

$$\hat{y} = .794(23) = 18.26$$

From Table 2 of Appendix D, $t_{.025} = 2.447$ (based on 6 df). Then, our 95% prediction interval is

$$\hat{y} \pm t_{.025}s \sqrt{1 + \frac{x_p^2}{\sum x_i^2}}$$

$$= 18.26 \pm 2.447(3.87056) \sqrt{1 + \frac{(23)^2}{3{,}550}}$$

$$= 18.26 \pm 2.447(3.87056)(1.07192)$$

$$= 18.26 \pm 10.15 \quad \text{or} \quad (8.11, 28.41)$$

We predict with 95% confidence that the total variable cost will range between $8.11 and $28.41 on a day in which 23 units are produced.

Warning: There are several situations where it is dangerous to fit the model $E(y) = \beta_1 x$. If you are not certain that the regression line passes through the origin, it is a safe practice to fit the more general model $E(y) = \beta_0 + \beta_1 x$. If the line of means does, in fact, pass through the origin, the estimate of β_0 will differ from the true value $\beta_0 = 0$ by only a small amount. For all practical purposes, the least squares prediction equations will be the same.

On the other hand, you may know that the regression passes through the origin (see Example 3.6), but are uncertain about whether the true relationship between y and x is linear or curvilinear. In fact, most theoretical relationships are *curvilinear*. Yet, we often fit a linear model to the data in such situations because we believe that a straight line will make a good approximation to the mean response $E(y)$ over the region of interest. The problem is that this straight line is not likely to pass through the origin (see Figure 3.21). By forcing the regression line through the origin, we may not obtain a very good approximation to $E(y)$. For these reasons, regression through the origin should be used with extreme caution.

FIGURE 3.21

Using a straight line to approximate a curvilinear relationship when the true relationship passes through the origin

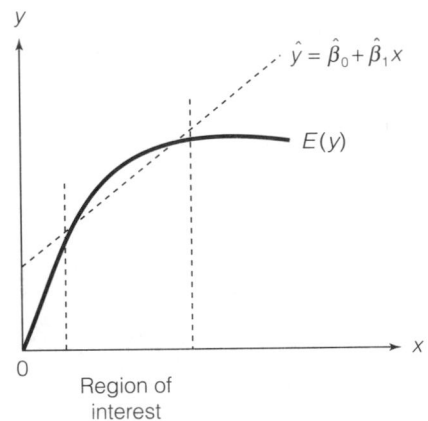

EXERCISES

3.46   Consider the eight data points shown in the table.

x	-4	-2	0	2	4	6	8	10
y	-12	-7	0	6	14	21	24	31

a. Fit a straight-line model through the origin, i.e., fit $E(y) = \beta_1 x$.
b. Calculate SSE, s^2, and s.
c. Do the data provide sufficient evidence to indicate that x and y are positively linearly related?
d. Construct a 95% confidence interval for β_1.
e. Construct a 95% confidence interval for $E(y)$ when $x = 7$.
f. Construct a 95% prediction interval for y when $x = 7$.

3.47 Consider the five data points shown in the table.

x	0	1	2	3	4
y	0	-8	-20	-30	-35

a. Fit a straight-line model through the origin, i.e., fit $E(y) = \beta_1 x$.
b. Calculate SSE, s^2, and s.
c. Do the data provide sufficient evidence to indicate that x and y are negatively linearly related?
d. Construct a 95% confidence interval for β_1.
e. Construct a 95% confidence interval for $E(y)$ when $x = 1$.
f. Construct a 95% prediction interval for y when $x = 1$.

3.48 Consider the 10 data points shown in the table.

x	30	50	70	90	100	120	140	160	180	200
y	4	10	15	21	21	22	29	34	39	41

a. Fit a straight-line model through the origin, i.e., fit $E(y) = \beta_1 x$.
b. Calculate SSE, s^2, and s.
c. Do the data provide sufficient evidence to indicate that x and y are positively linearly related?
d. Construct a 95% confidence interval for β_1.
e. Construct a 95% confidence interval for $E(y)$ when $x = 125$.
f. Construct a 95% prediction interval for y when $x = 125$.

3.49 A pharmaceutical company has developed a new drug designed to reduce a smoker's reliance on tobacco. Since certain dosages of the drug may reduce one's pulse rate to dangerously low levels, the product-testing division of the pharmaceutical company wants to model the relationship between decrease in pulse rate, y (beats/minute), and dosage, x (cubic centimeters). Different dosages of the drug were administered to eight randomly selected patients, and 30 minutes later the decrease in each patient's pulse rate was recorded. The results are given in the accompanying table. Initially, the company considered the model $y = \beta_1 x + \varepsilon$, since in theory, a patient who receives a dosage of $x = 0$ should show no decrease in pulse rate ($y = 0$).

PATIENT	DOSAGE x, cubic centimeters	DECREASE IN PULSE RATE y, beats/minute
1	2.0	12
2	4.0	20
3	1.5	6
4	1.0	3
5	3.0	16
6	3.5	20
7	2.5	13
8	3.0	18

a. Fit a straight-line model that passes through the origin.

b. Is there evidence of a linear relationship between drug dosage and decrease in pulse rate? Test at $\alpha = .10$.

c. Find a 99% prediction interval for the decrease in pulse rate corresponding to a dosage of 3.5 cubic centimeters.

3.50 Consider the relationship between the total weight of a shipment of 50-pound bags of flour, y, and the number of bags in the shipment, x. Since a shipment containing $x = 0$ bags (i.e., no shipment at all) has a total weight of $y = 0$, a straight-line model of the relationship between x and y should pass through the point $x = 0$, $y = 0$. Hence, the appropriate model might be

$$y = \beta_1 x + \varepsilon$$

From the records of past flour shipments, 15 shipments were randomly chosen and the data in the table recorded.

WEIGHT OF SHIPMENT	NUMBER OF 50-POUND BAGS IN SHIPMENT
5,050	100
10,249	205
20,000	450
7,420	150
24,685	500
10,206	200
7,325	150
4,958	100
7,162	150
24,000	500
4,900	100
14,501	300
28,000	600
17,002	400
16,100	400

a. Find the least squares line for the given data under the assumption that $\beta_0 = 0$. Plot the least squares line on a scattergram of the data.

b. Find the least squares line for the given data using the model

$$y = \beta_0 + \beta_1 x + \varepsilon$$

(i.e., do not restrict β_0 to equal 0). Plot this line on the same scatterplot you constructed in part **a**.

c. Refer to part **b**. Why might $\hat{\beta}_0$ be different from 0 even though the true value of β_0 is known to be 0?

d. The estimated standard error of $\hat{\beta}_0$ is equal to

$$s\sqrt{\frac{1}{n} + \frac{\bar{x}^2}{SS_{xx}}}$$

Use the t statistic,

$$t = \frac{\hat{\beta}_0 - 0}{s\sqrt{\dfrac{1}{n} + \dfrac{\bar{x}^2}{SS_{xx}}}}$$

to test the null hypothesis H_0: $\beta_0 = 0$ against the alternative H_a: $\beta_0 \neq 0$. Use $\alpha = .10$. Should you include β_0 in your model?

3.51   To satisfy the Public Service Commission's energy conservation requirements, an electric utility company must develop a reliable model for projecting the number of residential electric customers in its service area. The first step is to study the effect of changing population on the number of electric customers. The information shown in the accompanying table was obtained for the service area from 1983 to 1992.

Since a service area with 0 population obviously would have 0 residential electric customers, one could argue that regression through the origin is appropriate.

YEAR	POPULATION IN SERVICE AREA x, hundreds	RESIDENTIAL ELECTRIC CUSTOMERS IN SERVICE AREA y
1983	262	14,041
1984	319	16,953
1985	361	18,984
1986	381	19,870
1987	405	20,953
1988	439	22,538
1989	472	23,985
1990	508	25,641
1991	547	27,365
1992	592	29,967

a. Fit the model $y = \beta_1 x + \varepsilon$ to the data.
b. Is there evidence that x contributes information for the prediction of y? Test using $\alpha = .01$.
c. Now fit the more general model $y = \beta_0 + \beta_1 x + \varepsilon$ to the data. Is there evidence (at $\alpha = .01$) that x contributes information for the prediction of y?
d. Which model would you recommend?

3.12

A Summary of the Steps to Follow in a Simple Linear Regression Analysis

We have introduced an extremely useful tool in this chapter—**the method of least squares** for fitting a prediction equation to a set of data. This procedure, along with associated statistical tests and estimations, is called a **regression analysis**. In six steps we showed how to use sample data to build a model relating a dependent variable y to a single independent variable x.

> ### Steps to Follow in a Simple Linear Regression Analysis
>
> 1. The first step is to hypothesize a **probabilistic model**. In this chapter, we confined our attention to the **first-order (straight-line) model**
>
> $$y = \beta_0 + \beta_1 x + \varepsilon$$
>
> 2. The second step is to collect the (x, y) pairs for each experimental unit in the sample.
>
> 3. The third step is to use the method of least squares to estimate the unknown parameters in the **deterministic component**, $\beta_0 + \beta_1 x$. The least squares estimates yield a model $\hat{y} = \hat{\beta}_0 + \hat{\beta}_1 x$ with a **sum of squared errors (SSE)** that is smaller than the SSE for any other straight-line model.
>
> 4. The fourth step is to specify the probability distribution of the **random error component ε**.
>
> 5. The fifth step is to assess the utility of the hypothesized model. Included here are making inferences about the **slope β_1**, calculating the **coefficient of correlation r**, and calculating the **coefficient of determination r^2**.
>
> 6. Finally, if we are satisfied with the model, we are prepared to use it. We used the model to **estimate the mean y value**, $E(y)$, for a given x value and to **predict an individual y value** for a specific value of x.

The concepts introduced in this chapter will be developed more fully in Chapter 4.

SUPPLEMENTARY EXERCISES

3.52 A medical item used to administer to a hospital patient is called a *factor*. For example, factors can be intravenous tubing, intravenous fluid, needles, shave kits, bedpans, diapers, dressings, medications, and even code carts. The coronary care unit at Bayonet Point Hospital (St. Petersburg, Florida) investigated the relationship between the number of factors per patient, x, and the patient's length of stay (in days), y. The data for a random sample of 50 coronary care patients are given in the table, followed by a SAS printout of the simple linear regression analysis.

a. Construct a scattergram of the data.

b. Find the least squares line for the data and plot it on your scattergram.

c. Define β_1 in the context of this problem.

d. Test the hypothesis that the number of factors per patient (x) contributes no information for the prediction of the patient's length of stay (y) when a linear model is used (use $\alpha = .05$). Draw the appropriate conclusions.

e. Find a 90% confidence interval for β_1. Interpret your results.

NUMBER OF FACTORS x	LENGTH OF STAY y (days)	NUMBER OF FACTORS x	LENGTH OF STAY y (days)
231	9	354	11
323	7	142	7
113	8	286	9
208	5	341	10
162	4	201	5
117	4	158	11
159	6	243	6
169	9	156	6
55	6	184	7
77	3	115	4
103	4	202	6
147	6	206	5
230	6	360	6
78	3	84	3
525	9	331	9
121	7	302	7
248	5	60	2
233	8	110	2
260	4	131	5
224	7	364	4
472	12	180	7
220	8	134	6
383	6	401	15
301	9	155	4
262	7	338	8

Source: Bayonet Point Hospital, Coronary Care Unit.

Dependent Variable: Y

Analysis of Variance

Source	DF	Sum of Squares	Mean Square	F Value	Prob>F
Model	1	126.58393	126.58393	28.683	0.0001
Error	48	211.83607	4.41325		
C Total	49	338.42000			

Root MSE	2.10077	R-square	0.3740
Dep Mean	6.54000	Adj R-sq	0.3610
C.V.	32.12193		

Parameter Estimates

Variable	DF	Parameter Estimate	Standard Error	T for H0: Parameter=0	Prob > |T|
INTERCEP	1	3.306032	0.67297426	4.913	0.0001
X	1	0.014755	0.00275502	5.356	0.0001

Obs	X	Dep Var Y	Predict Value	Std Err Predict	Lower95% Predict	Upper95% Predict	Residual
1	200	.	6.2570	0.302	1.9898	10.5242	.

Sum of Residuals	9.769963E-15
Sum of Squared Residuals	211.8361
Predicted Resid SS (Press)	234.7934

    f. Find the coefficient of correlation for the data. Interpret your results.

    g. Find the coefficient of determination for the linear model you constructed in part **b**. Interpret your result.

    h. Find a 95% prediction interval for the length of stay of a coronary care patient who is administered a total of $x = 200$ factors.

    i. Explain why the prediction interval obtained in part **h** is so wide. How could you reduce the width of the interval?

3.53   Refer to the experiment on estimating the diet metabolizable energy (ME) content of commercial cat foods in Exercise 1.57. In one phase of the study, simple linear regression analysis was used to relate the estimated ME content, x (in kilocalories per gram), to the actual ME content y for cats fed a diet of canned and dry foods (*Feline Practice*, Feb. 1986). The results for the two groups of cats are reported in the table.

DIET	$\hat{\beta}_0$	$\hat{\beta}_1$	$s_{\hat{\beta}_0}$	$s_{\hat{\beta}_1}$	r	s
Canned foods ($n = 28$)	.02	.96	.04	.04	.97	.05
Dry foods ($n = 29$)	.47	.84	.26	.09	.88	.15

    a. Is there sufficient evidence to indicate that estimated ME content x is a useful predictor of actual ME content y for cats fed a diet of canned foods? Test using $\alpha = .05$.

    b. Is there sufficient evidence to indicate that estimated ME content x is a useful predictor of actual ME content y for cats fed a diet of dry foods? Test using $\alpha = .05$.

    c. Calculate the coefficient of determination r^2 for both regressions. Interpret these values.

    d. Interpret the values of s for both regressions. What is the practical implication of this result?

3.54   Consider the data on air and rail passengers provided in the accompanying table.

YEAR	PASSENGERS CARRIED BY SCHEDULED AIR CARRIERS IN THE UNITED STATES y, millions	PASSENGERS CARRIED BY RAILROADS IN THE UNITED STATES x, millions
1950	19	488
1955	42	433
1960	62	327
1965	103	306
1970	169	289
1975	205	275
1980	297	350
1985	380	82

Source: 1986 Statistical Abstract of the United States, U.S. Department of Commerce, Bureau of the Census.

    a. Find the correlation coefficient and coefficient of determination for the data and interpret your result.

    b. Do the data provide sufficient evidence to indicate correlation between x and y?

3.55 In the business world, *Machiavellian* is a term often used to describe one who employs aggressive, manipulative, exploiting, and devious moves to achieve personal and corporate objectives. Hunt and Chonko (*Journal of Marketing*)* investigated Machiavellian tactics in marketing. One question concerned the relationship between age and Machiavellianism. Do young marketers tend to be more Machiavellian than older marketers? A sample of 1,076 members of the American Marketing Association were administered a questionnaire that measured tendency toward Machiavellianism. (The higher the score, the greater the tendency toward Machiavellianism.) The correlation coefficient between age and Machiavellian score was found to be $r = -.20$.

a. Is there evidence of a negative linear relationship between age of marketers and Machiavellian score? Test using $\alpha = .01$.

b. Calculate the coefficient of determination r^2 for the linear relationship between age of marketers and Machiavellian score. Interpret this value.

3.56 Many successful businesses have adopted high market share as a strategic objective. However, Michael Hergert, writing in *Business Economics*,[†] warns that "generalizations about the attractiveness of seeking high market share may prove misleading to strategic planners." Hergert investigated the relationship between 1980 market share and profitability (measured as return on assets) for a sample of approximately 5,400 businesses registered with the Securities and Exchange Commission. A simple linear regression analysis relating return on assets y and market share x yielded the following results:

$$\hat{y} = .103 + .128x$$
$$r^2 = .001 \qquad t = 2.51$$

a. Is there evidence to indicate that return on assets y and market share x are positively linearly related? Test using $\alpha = .05$.

b. Interpret the value of r^2.

c. Based on your answers to parts **a** and **b**, would you recommend a corporate strategy based on the premise that "a bigger market share yields a higher profit"?

3.57 At major colleges and universities, administrators (e.g., deans, chairpersons, provosts, vice presidents, and presidents) are among the highest-paid state employees. Is there a relationship between the raises administrators receive and their performance on the job? This was the question of interest to a group of faculty union members at the University of South Florida called the United Faculty of Florida (UFF). The UFF compared the April 1990 ratings of 15 University of South Florida administrators (as determined by faculty in a survey) to their subsequent raises in August 1990. The data for the analysis is listed in the table at the top of page 154. [*Note:* Ratings are measured on a 5-point scale, where 1 = very poor and 5 = very good.] According to the UFF, the "relationship is inverse, i.e., the lower the rating by the faculty, the greater the raise. Apparently, bad administrators are more valuable than good administrators."[‡] (With tongue in cheek, the UFF refers to this phenomenon as "the SOB effect.") The UFF based its conclusions on a simple linear regression analysis of the data in the table, where y = administrator's raise and x = average rating of administrator.

---

*Hunt, S. D. and Chonko, L. B. "Marketing and Machiavellianism." *Journal of Marketing*, Summer 1984, Vol. 48, No. 3, pp. 30–42.

†Hergert, M. "Market share and profitability: Is bigger really better?" *Business Economics*, Oct. 1984, pp. 45–48.

‡*UFF Faculty Forum*, University of South Florida Chapter, Vol. 3, No. 5, May 1991.

ADMINISTRATOR	RAISE[a]	AVERAGE RATING (5-PT SCALE)[b]
1	$18,000	2.76
2	16,700	1.52
3	15,787	4.40
4	10,608	3.10
5	10,268	3.83
6	9,795	2.84
7	9,513	2.10
8	8,459	2.38
9	6,099	3.59
10	4,557	4.11
11	3,751	3.14
12	3,718	3.64
13	3,652	3.36
14	3,227	2.92
15	2,808	3.00

Sources: [a]Faculty and A&P Salary Report, University of South Florida, Resource Analysis and Planning, 1990.
  [b]Administrative Compensation Survey, *Chronicle of Higher Education*, Jan. 1991.

a. Initially, the UFF conducted the analysis using all 15 data points in the table. Fit a straight-line model to the data. Is there evidence to support the UFF's claim of an inverse relationship between raise and rating?

b. A second simple linear regression was performed using only 14 of the data points in the table. The data for administrator #3 was eliminated because he was promoted to dean in the middle of the 1989–1990 academic year. (No other reason was given for removing this data point from the analysis.) Perform the simple linear regression analysis using the remaining 14 data points in the table. Is there evidence to support the UFF's claim of an inverse relationship between raise and rating?

c. Based on the results of the regression, part **b**, the UFF computed estimated raises for selected faculty ratings of administrators. These are shown in the following table. What problems do you perceive with using this table to estimate administrators' raises at the University of South Florida?

RATINGS		RAISE
Very Poor	1.00	$15,939
	1.50	13,960
Poor	2.00	11,980
	2.50	10,001
Average	3.00	8,021
	3.50	6,042
Good	4.00	4,062
	4.50	2,083
Very Good	5.00	103

d. The ratings of administrators listed in this table were determined by surveying the faculty at the University of South Florida. All faculty are mailed the survey each year, but the response rate is typically low (approximately 10%–20%). The danger with such a survey is that only disgruntled faculty, who are more apt to give a low rating to an administrator, will respond. Many of these faculty also believe that they are underpaid and that the administrators are overpaid. Comment on how such a survey could bias the results shown here.

e. Based on your answers to the previous questions, would you support the UFF's claim?

3.58 Most investment firms provide estimates of systematic risks of securities, called *betas*. A stock's beta measures the relationship between the stock's rate of return and the average rate of return for the market as a whole. The term *beta* derives its name from the beta coefficient for the slope in simple linear regression, where the dependent variable is the stock's rate of return y and the independent variable is the market rate of return x. Stocks with beta values (i.e., slopes) greater than 1 are considered "aggressive" securities since their rates of return are expected to move (upward or downward) faster than the market as a whole. In contrast, stocks with beta values less than 1 are called "defensive" securities since their rates of return move much slower than the market. A stock with a beta value near 1 is called a "neutral" security for its rate of return mirrors the market. The data in the table are monthly rates of return (in percent) for a particular stock and the market as a whole for seven randomly selected months.

MONTH	STOCK RATE OF RETURN y	MARKET RATE OF RETURN x
1	12.0	7.2
2	−1.3	.0
3	2.5	2.1
4	18.6	11.9
5	9.0	5.3
6	−3.8	−1.2
7	−10.0	−4.7

a. Fit the straight-line model $E(y) = \beta_0 + \beta_1 x$.
b. Find r^2 and s and interpret their values.
c. Is there sufficient evidence to indicate that market rate of return x is a useful predictor of stock rate of return y? Test using $\alpha = .01$.
d. Construct a 99% confidence interval for β_1, the mean increase in stock rate of return for every 1% increase in market rate of return.
e. Based on your answer to part d, would you classify this stock as aggressive, defensive, or neutral?

3.59 Refer to Exercise 3.58. Does a stock's beta value depend on the length of the horizon over which the rates of return are calculated? Since some brokerage firms base their beta values on monthly data and others on annual data, the question is an important one for investors. Haim Levy (University of Florida) investigated the relationship between length of horizon (in months) and average beta value for each of the three types of stocks. Varying the length of horizon from 1 to 30 months, Levy calculated rates of return for 144 stocks over the years 1946–1975. The stocks were divided into 38 aggressive, 38 defensive, and 68 neutral stocks based on their beta values. The table on page 156 gives the average beta value for different horizons for each of the stock types.
a. Find the least squares simple linear regression equation relating length of horizon x to average beta value y for (i) aggressive stocks, (ii) defensive stocks, and (iii) neutral stocks.
b. For each type of stock, test the hypothesis that length of horizon x is a useful linear predictor of average beta value y. Test using $\alpha = .05$.

LENGTH OF HORIZON	BETA VALUES		
	Aggressive Stocks	Defensive Stocks	Neutral Stocks
1	1.37	.50	.98
3	1.42	.44	.95
6	1.53	.41	.94
9	1.69	.39	1.00
12	1.83	.40	.98
15	1.67	.38	1.00
18	1.78	.39	1.02
24	1.86	.35	1.14
30	1.83	.33	1.22

Source: Levy, H. "Measuring risk and performance over alternative investment horizons." *Financial Analysts Journal*, Mar.–Apr. 1984, pp. 61–68.

c. For each type of stock, construct a 95% confidence interval for the slope of the line. Which stocks have beta values that increase linearly as length of horizon increases? Which stocks have beta values that decrease linearly as length of horizon increases?

3.60 Paper-and-pencil honesty tests of employees are common among retail stores, financial institutions, and warehouse operations where employees have access to cash and merchandise. These tests are less costly than polygraphs and can be used in states where preemployment polygraph examinations are illegal. P. R. Sackett and M. M. Harris (University of Illinois) reviewed a number of studies that examined the validity of paper-and-pencil honesty tests (*Personnel Psychology*, Summer 1984).* In one study, $n = 80$ applicants for retail management positions were given both an honesty test and a polygraph examination. The correlation coefficient between the scores of the two tests was $r = .48$. Another, independent study of $n = 17$ warehouse employees showed a correlation coefficient of $r = -.41$ between honesty test scores and dollar amount of money and merchandise stolen.

a. Is there sufficient evidence to indicate that paper-and-pencil honesty test scores and polygraph examination scores of retail management applicants are positively correlated? Test using $\alpha = .05$.

b. Calculate and interpret the coefficient of determination r^2 for the linear relationship between honesty test scores and polygraph examination scores.

c. Is there sufficient evidence to indicate that paper-and-pencil honesty test scores of warehouse employees are correlated with dollar amount of theft? Test using $\alpha = .05$.

d. Calculate and interpret the coefficient of determination r^2 for the linear relationship between honesty test scores and dollar amount of theft.

3.61 "In the analysis of urban transportation systems it is important to be able to estimate expected travel time between locations." Cook and Russell collected data in the city of Tulsa on the urban travel times and distances between locations for two types of vehicles, large hoist compactor trucks and passenger cars. A simple linear regression analysis was conducted for both sets of data (y = urban travel time in minutes, x = distance between locations in miles) with the results given in the accompanying table.

a. Is there sufficient evidence to indicate that distance between locations is linearly related to urban travel time for passenger cars? Test at $\alpha = .05$.

b. Is there sufficient evidence to indicate that distance between locations is linearly related to urban travel time for trucks? Test at $\alpha = .01$.

*Sackett, P. R. and Harris, M. M. "Honesty testing for personnel selection: A review and critique." *Personnel Psychology*, Summer 1984, Vol. 37, No. 2, pp. 221–243.

PASSENGER CARS	TRUCKS
$\hat{y} = 2.50 + 1.93x$	$\hat{y} = 1.85 + 3.86x$
$r^2 = .676$; p-value $< .05$	$r^2 = .758$; p-value $< .01$

Source: Cook, T. M. and Russell, R. A. "Estimating urban travel times: A comparative study." *Transportation Research*, June 1980, 14A, pp. 173–175.

c. Interpret the values of r^2 for the two prediction equations.

d. Estimate the mean urban travel time for all passenger cars traveling a distance of 3 miles on Tulsa's highways.

e. Predict the urban travel time for a particular truck traveling a distance of 5 miles on Tulsa's highways

f. Explain how we could attach a measure of reliability to the inferences derived in parts **d** and **e**.

3.62 To what extent is the age of members of a board of directors related to a firm's financial performance? To explore this question, three Pennsylvania State University professors collected data on the composition of the boards of directors and financial performance of 399 firms listed in the *Fortune 500* (*Quarterly Journal of Business and Economics*, Autumn 1984). The correlation coefficient between the average age of corporate boards of directors and the ratio of operating earnings to assets (a measure of financial performance) was found to be $r = -.1153$.

a. Interpret the value of r.

b. Is there evidence to indicate that average age of corporate boards of directors is linearly related to ratio of operating earnings to assets? Test using $\alpha = .01$.

3.63 How much influence does the board of directors have on the length of tenure of the chief executive of a large corporation? Donald L. Helmich* examined the relationship between the size of the board (i.e., the number of members on the board) and the tenure lengths of newly appointed corporate presidents for a sample of 54 petrochemical companies. From each corporation the chief executive office tenures during the period 1947–1977 and the number of directors serving on the board during each year of the 30-year period were obtained. Using this information, Helmich measured the following variables for each of the 54 companies:

y = Average succession rate at the corporation (a measure of the rate of replacement of the chief executive)

x = Variation coefficient for corporation board size (a measure of the variability of board size adjusted for annual and long-term linear trends due to organizational growth or decline)

Helmich subjected the 54 (x, y) data points to a simple linear regression analysis. The slope of the least squares line was found to be $\hat{\beta}_1 = .254$, whereas a test of the null hypothesis $H_0: \beta_1 = 0$ resulted in a p-value less than .05. Based on this result, Helmich concluded that "board size variation had a significant positive influence upon the rate of succession in the office of chief executive over the 30-year interval of study for the total sample of 54 companies. That is, the higher the variation in board size, the more rapid the change in leadership in the chief executive office." Do you agree with this assessment?

---

*Helmich, D. L. "Board size variation and rates of succession in the corporate presidency." *Journal of Business Research*, Mar. 1980, Vol. 8, pp. 51–63. Reprinted by permission. Copyright 1980 by Elsevier North Holland, Inc.

3.64 Refer to Exercise 3.63. Knowledge of the (apparent) positive linear relationship between variation in board size and rate of succession becomes important during the process of succession planning by directors in the corporate structure. However, what may be even more important is knowing whether this positive relationship is exhibited by profitable as well as unprofitable companies. To examine this phenomenon, Helmich categorized each of the sampled companies as unsuccessful or successful, based on profit performance. The linear regression was then performed independently on each of the two groups ($n = 27$ observations in each group) of data. The results are summarized in the table.

UNSUCCESSFUL COMPANIES	SUCCESSFUL COMPANIES
$\hat{\beta}_1 = .603$	$\hat{\beta}_1 = -.230$
p-value $< .05$	p-value $< .05$

a. Is there evidence of a linear relationship between rate of succession and variation in board size for unsuccessful companies? If so, does the relationship appear to be positive, as suggested by the previous results?

b. Answer part a for the group of successful companies.

3.65 Use Table 7 in Appendix D to select a random sample of $n = 20$ residential properties from the 2,440 properties in Appendix E. Record the sale price y and the total appraised property value (land plus improvements) x for each property.

a. Fit a simple linear regression model to the data.

b. Do the data provide sufficient evidence to indicate that the total appraised value contributes information for the prediction of sale price?

c. Calculate the coefficient of determination and interpret its value.

d. Use the prediction equation to find 90% prediction interval for the sale price of a property that has a total appraised value of $70,000.

ON YOUR OWN...

The gross national product (GNP) is one of the nation's best-known economic indicators. Many economists have developed models to forecast future values of the GNP. There are surely a large number of variables that should be included if an accurate prediction is to be made. For the moment, however, consider the simple case of choosing one important variable to include in a simple straight-line model for GNP.

First, list three independent variables, x_1, x_2, and x_3, that you think might be (individually) strongly related to the GNP. Next, obtain 10 yearly values (preferably the last 10) of the three independent variables and the GNP.*

a. Use the least squares formulas given in this chapter to fit three straight-line models—one for each independent variable—for predicting the GNP.

*The assumption that the random errors are independent is debatable for time series data. For purposes of illustration, we will assume they are approximately independent. The problem of dependent errors is discussed in Chapters 6 and 9.

b. Interpret the sign of the estimated slope coefficient $\hat{\beta}_1$ in each case, and test the utility of the model by testing H_0: $\beta_1 = 0$ against H_a: $\beta_1 \neq 0$.

c. Calculate the coefficient of determination r^2 for each model. Which of the independent variables predicts the GNP best over the 10 sample years when a straight-line model is used? Is this variable necessarily best in general (i.e., for all years)? Explain.

References

Chou, Ya-lun. *Statistical Analysis with Business and Economic Applications*, 2nd ed. New York: Holt, Rinehart, and Winston, 1975. Chapter 17.

Draper, N. and Smith, H. *Applied Regression Analysis*, 3rd ed. New York: Wiley, 1987. Chapter 1.

Miller, R. B. and Wichern, D. W. *Intermediate Business Statistics: Analysis of Variance, Regression, and Time Series*. New York: Holt, Rinehart, and Winston, 1977. Chapter 5.

Neter, J., Wasserman, W., and Kutner, M. H. *Applied Linear Statistical Models*, 3rd ed. Homewood, Ill.: Richard D. Irwin, 1990. Chapters 2–3.

Multiple Regression

CONTENTS

4.1 The General Linear Model

4.2 Model Assumptions

4.3 Fitting the Model: The Method of Least Squares

4.4 Estimation of σ^2, the Variance of ε

4.5 Inferences About the β Parameters

4.6 The Multiple Coefficient of Determination, R^2

4.7 Testing the Utility of a Model: The Analysis of Variance F Test

4.8 Using the Model for Estimation and Prediction

4.9 Other Linear Models

4.10 Testing Portions of a Model

4.11 Stepwise Regression

4.12 Other Variable Selection Techniques (Optional)

4.13 Multiple Regression: An Example

4.14 A Summary of the Steps to Follow in a Multiple Regression Analysis

OBJECTIVE

To extend the methods of Chapter 3; to develop a procedure for predicting a response y based on the values of two or more independent variables; to illustrate the types of practical inferences that can be drawn from this type of analysis

4.1

The General Linear Model

Most practical applications of regression analysis utilize models that are more complex than the first-order (straight-line) model. For example, a realistic probabilistic model for monthly sales revenue would include more than just the advertising expenditure discussed in Chapter 3 in order to provide a good predictive model for sales. Factors such as season, inventory on hand, sales force, and productivity are a few of the many variables that might influence sales. Thus, we would want to incorporate these and other potentially important independent variables into the model if we need to make accurate predictions.

Probabilistic models that include terms involving x^2, x^3 (or higher-order terms), or more than one independent variable are called **multiple regression models**, or **linear statistical models**. The general form of these models is shown in the box.

The General Linear Model

$$y = \beta_0 + \beta_1 x_1 + \beta_2 x_2 + \cdots + \beta_k x_k + \varepsilon$$

where

y is the dependent variable

$x_1, x_2, \ldots, x_k$ are the independent variables

$E(y) = \beta_0 + \beta_1 x_1 + \beta_2 x_2 + \cdots + \beta_k x_k$ is the deterministic portion of the model

β_i determines the contribution of the independent variable x_i

Note: Remember that the symbols $x_1, x_2, \ldots, x_k$ may represent higher-order terms. For example, x_1 might represent the current interest rate, x_2 might represent x_1^2, and so forth.

The dependent variable y is now written as a function of k independent variables, $x_1, x_2, \ldots, x_k$. The random error term is added to make the model probabilistic rather than deterministic. The value of the coefficient β_i determines the contribution of the independent variable x_i, given that the other $(k - 1)$ independent variables are held constant, and β_0 is the y-intercept. The coefficients $\beta_0, \beta_1, \ldots, \beta_k$ will usually be unknown, since they represent population parameters.

At first glance it might appear that the regression model shown here would not allow for anything other than straight-line relationships between y and the independent variables, but this is not true. Actually, $x_1, x_2, \ldots, x_k$ can be functions of variables as long as the functions do not contain unknown parameters. For example, the dollar sales y in new housing in a region could be a function of the independent variables

$x_1 =$ Mortgage interest rate

$x_2 =$ (Mortgage interest rate)$^2 = x_1^2$

$x_3 =$ Unemployment rate in the region

and so on. You could even insert a cyclical term (if it would be useful) of the form $x_4 = \sin t$, where t is a time variable. The multiple regression model is quite versatile and can be made to model many different types of response variables.

The steps we followed in developing a straight-line model are applicable to the multiple regression model.

STEP 1 Collect the sample data, i.e., the values of y, x_1, x_2, . . . , x_k, for each experimental unit in the sample.

STEP 2 Hypothesize the form of the model. Choose which independent variables to include in the model.

STEP 3 Estimate the unknown parameters β_0, β_1, . . . , β_k.

STEP 4 Specify the probability distribution of the random error component ε and estimate its variance σ^2.

STEP 5 Check the utility of the model.

STEP 6 Finally, if the model is deemed adequate, use the fitted model to estimate the mean value of y or to predict a particular value of y for given values of the independent variables.

Hypothesizing the form of the model (Step 2) is the subject of Chapter 5. In this chapter we will assume that the form of the model is known, and we will discuss steps 3–6 for a given model.

CASE STUDY 4.1

Towers, Perrin, Forster & Crosby (TPF&C), an international management consulting firm, has developed a unique and interesting application of multiple regression analysis. Many firms are interested in evaluating their management salary structure, and TPF&C uses multiple regression models to accomplish this salary evaluation. The Compensation Management Service, as TPF&C calls it, measures both the internal and external consistency of a company's pay policies to determine whether they reflect the management's intent.

The dependent variable y used to measure executive compensation is annual salary. The independent variables used to explain salary structure include the executive's age, education, rank, and bonus eligibility; number of employees under the executive's direct supervision; as well as variables that describe the company for which the executive works, such as annual sales, profit, and total assets.

The initial step in developing models for executive compensation is to obtain a sample of executives from various client firms, which TPF&C calls the Compensation Data Bank. The data for these executives are used to estimate the model coefficients (the β parameters), and these estimates are then substituted into the linear model to form a prediction equation. To predict a particular executive's compensation, TPF&C substitutes into the prediction equation the values of the independent variables that pertain to the executive (the executive's age, rank, and so on). This application of multiple regression analysis will be developed more fully in Section 4.9.

4.2

Model Assumptions

We noted in Section 4.1 that the multiple regression model is of the form

$$y = \beta_0 + \beta_1 x_1 + \beta_2 x_2 + \cdots + \beta_k x_k + \varepsilon$$

where y is the response variable that you want to predict; $\beta_0, \beta_1, \ldots, \beta_k$ are parameters with unknown values; $x_1, x_2, \ldots, x_k$ are independent information-contributing variables that are measured without error; and ε is a random error component. Since $\beta_0, \beta_1, \ldots, \beta_k$ and $x_1, x_2, \ldots, x_k$ are nonrandom, the quantity

$$\beta_0 + \beta_1 x_1 + \beta_2 x_2 + \cdots + \beta_k x_k$$

represents the deterministic portion of the model. Therefore, y is composed of two components—one fixed and one random—and, consequently, y is a random variable.

$$y = \overbrace{\beta_0 + \beta_1 x_1 + \beta_2 x_2 + \cdots + \beta_k x_k}^{\substack{\text{Deterministic} \\ \text{portion of model}}} + \overbrace{\varepsilon}^{\substack{\text{Random} \\ \text{error}}}$$

We will assume (as in Chapter 3) that the random error can be positive or negative and that for any setting of the x values, $x_1, x_2, \ldots, x_k$, ε has a normal probability distribution with mean equal to 0 and variance equal to σ^2. Further, we assume that the random errors associated with any (and every) pair of y values are probabilistically independent. That is, the error ε associated with any one y value is independent of the error associated with any other y value. These asssumptions are summarized in the accompanying box.

Assumptions About the Random Error ε

1. For any given set of values of $x_1, x_2, \ldots, x_k$, ε has a normal probability distribution with mean equal to 0 [i.e., $E(\varepsilon) = 0$] and variance equal to σ^2 [i.e., $\text{Var}(\varepsilon) = \sigma^2$].
2. The random errors are independent (in a probabilistic sense).

The assumptions that we have described for a multiple regression model imply that the mean value $E(y)$ for a given set of values of $x_1, x_2, \ldots, x_k$ is equal to

$$E(y) = \beta_0 + \beta_1 x_1 + \beta_2 x_2 + \cdots + \beta_k x_k$$

Models of this type are called **linear statistical models** because $E(y)$ is a *linear function* of the unknown parameters $\beta_0, \beta_1, \ldots, \beta_k$.

All the estimation and statistical test procedures described in this chapter depend on the data satisfying the assumptions described in this section. Since we will rarely, if ever, know for certain whether the assumptions are actually satisfied in practice, we will want to know how well a regression analysis works, and how much faith we can place in our inferences when certain assumptions are not satisfied. We will have more to say on this topic in Chapters 6 and 7. First, we need to discuss the methods of a regression analysis more thoroughly and show how they are used in a practical situation.

4.3

Fitting the Model: The Method of Least Squares

The method of fitting multiple regression models is identical to that of fitting the first-order (straight-line) model—namely, the method of least squares. That is, we choose the estimated model

$$\hat{y} = \hat{\beta}_0 + \hat{\beta}_1 x_1 + \cdots + \hat{\beta}_k x_k$$

that minimizes

$$\text{SSE} = \sum (y_i - \hat{y}_i)^2$$

As in the case of the straight-line model, the sample estimates $\hat{\beta}_0, \hat{\beta}_1, \ldots, \hat{\beta}_k$ will be obtained as solutions to a set of simultaneous linear equations.*

The primary difference between fitting the simple and multiple regression models is computational difficulty. The $(k + 1)$ simultaneous linear equations that must be solved to find the $(k + 1)$ estimated coefficients $\hat{\beta}_0, \hat{\beta}_1, \ldots, \hat{\beta}_k$ are often difficult (sometimes physically impossible) to solve with a pocket or desk calculator. Consequently, we resort to the use of computers. As with the straight-line model, many statistical software packages have been developed to fit a multiple regression model by the method of least squares. We will present output from several of the more popular computer packages (SAS, SPSS, and Minitab) in examples and exercises. Since these printouts are similar to most other statistical software packages, you should have little trouble interpreting regression output from other packages as you encounter them in the future.

To illustrate, suppose we theorize that monthly electrical usage y in all-electric homes is related to the size x of the home by the model

$$y = \beta_0 + \beta_1 x + \beta_2 x^2 + \varepsilon$$

To estimate the unknown parameters β_0, β_1, and β_2, values of y and x were collected for each of 10 homes during a particular month. The data are shown in Table 4.1.

Notice that we include a term involving x^2 in the model because we expect curvature in the graph of the response model relating x to y. The term involving x^2 is called a **second-order**, or **quadratic**, term. Figure 4.1 (page 166) illustrates that the electrical usage appears to increase in a curvilinear manner with the size

TABLE 4.1

SIZE OF HOME x, square feet	MONTHLY USAGE y, kilowatt-hours
1,290	1,182
1,350	1,172
1,470	1,264
1,600	1,493
1,710	1,571
1,840	1,711
1,980	1,804
2,230	1,840
2,400	1,956
2,930	1,954

---

*Students who are familiar with calculus should note that $\hat{\beta}_0, \hat{\beta}_1, \ldots, \hat{\beta}_k$ are the solutions to the set of equations $\partial \text{SSE}/\partial\beta_0 = 0$, $\partial \text{SSE}/\partial\beta_1 = 0, \ldots, \partial \text{SSE}/\partial\beta_k = 0$. The solution, given in matrix notation, is presented in Appendix A.

FIGURE 4.1

Scattergram of the home size–electrical usage data

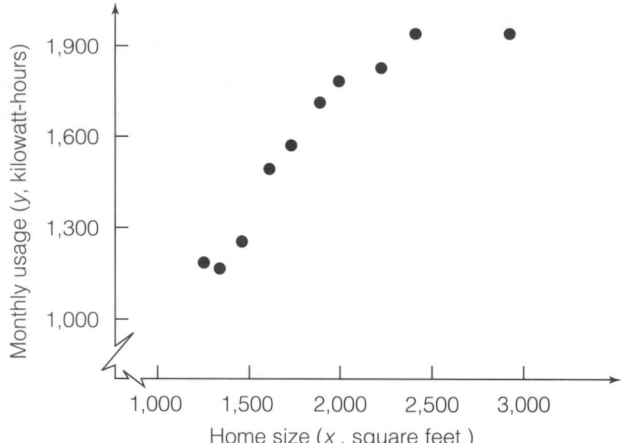

of the home. This provides some support for the inclusion of the second-order term x^2 in the model.

Part of the SAS printout for the analysis of the data in Table 4.1 is reproduced in Figure 4.2. The least squares estimates of the β parameters (shaded on the printout) are $\hat{\beta}_0 = -1,216.1$, $\hat{\beta}_1 = 2.3989$, and $\hat{\beta}_2 = -.00045$. Therefore, the equation that minimizes the SSE for the data is

$$\hat{y} = -1,216.1 + 2.3989x - .00045x^2$$

The minimum value of SSE, 15,332.6, is also shaded on the printout. [*Note:* Much detail on the printout has not yet been discussed. Throughout this chapter we will continue to shade the aspects of the printout under discussion.]

Note that the graph of the multiple regression model (Figure 4.3, a response curve) provides a good fit to the data of Table 4.1. Furthermore, the small value of $\hat{\beta}_2$ does *not* imply that the curvature is insignificant, since the numerical value of $\hat{\beta}_2$ is dependent on the scale of the measurements. We will test the contribution of the second-order coefficient β_2 in Section 4.5.

FIGURE 4.2

Portion of the SAS printout for the home size–electrical usage data

```
Dependent Variable: Y

                    Analysis of Variance

                        Sum of          Mean
Source        DF       Squares         Square      F Value    Prob>F

Model          2  831069.54637  415534.77319      189.710    0.0001
Error          7   15332.55363    2190.36480
C Total        9  846402.10000

      Root MSE        46.80133    R-square        0.9819
      Dep Mean      1594.70000    Adj R-sq        0.9767
      C.V.             2.93480

                    Parameter Estimates

                   Parameter      Standard     T for H0:
Variable   DF       Estimate         Error    Parameter=0    Prob > |T|

INTERCEP    1   -1216.143887  242.80636850       -5.009        0.0016
X           1       2.398930    0.24583560        9.758        0.0001
XX          1      -0.000450    0.00005908       -7.618        0.0001
```

FIGURE 4.3

Least squares model for the home size–electrical usage data

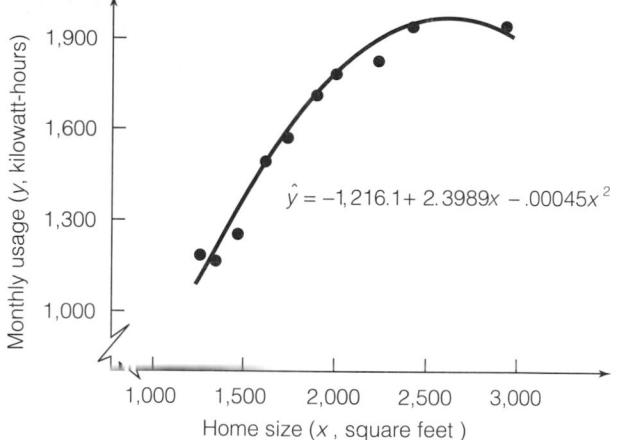

$$\hat{y} = -1,216.1 + 2.3989x - .00045x^2$$

The ultimate goal of this multiple regression analysis is to use the fitted model to predict electrical usage y for a home of a specific size (area) x. And, of course, we will want to give a prediction interval for y so that we will know how much faith we can place in the prediction. That is, if the prediction model is used to predict electrical usage y for a given size of home x, what will be the error of prediction? To answer this question, we need to estimate σ^2, the variance of ε.

4.4
Estimation of σ^2, the Variance of ε

You will recall that σ^2 is the variance of the random error ε. If $\sigma^2 = 0$, all the random errors will equal 0 and the prediction equation $\hat{y}$ will be identical to $E(y)$, i.e., $E(y)$ will be estimated without error. In contrast, a large value of σ^2 implies large (absolute) values of ε and larger deviations between the prediction equation $\hat{y}$ and the mean value $E(y)$. Consequently, the larger the value of σ^2, the greater will be the error in estimating the model parameters $\beta_0, \beta_1, \ldots, \beta_k$ and the error in predicting a value of y for a specific set of values of $x_1, x_2, \ldots, x_k$. Thus, σ^2 plays a major role in making inferences about $\beta_0, \beta_1, \ldots, \beta_k$, in estimating $E(y)$, and in predicting y for specific values of $x_1, x_2, \ldots, x_k$.

Since the variance σ^2 of the random error ε will rarely be known, we must use the results of the regression analysis to estimate its value. You will recall that σ^2 is the variance of the probability distribution of the random error ε for a given set of values for $x_1, x_2, \ldots, x_k$, and hence that it is the mean value of the squares of the deviations of the y values (for given values of $x_1, x_2, \ldots, x_k$) about the mean value $E(y)$.* Since the predicted value $\hat{y}$ estimates $E(y)$ for each of the data points, it seems natural to use

$$\text{SSE} = \sum (y_i - \hat{y}_i)^2$$

to construct an estimator of σ^2.

---

*Remember, we stated in Section 4.2 that $y = E(y) + \varepsilon$. Therefore, ε is equal to the deviation $y - E(y)$. Also, by definition, the variance of a random variable is the expected value of the square of the deviation of the random variable from its mean. According to our model, $E(\varepsilon) = 0$. Therefore, $\sigma^2 = E(\varepsilon^2)$.

For example, in the second-order model describing electrical usage as a function of home size, we found that SSE = 15,332.6. We now want to use this quantity to estimate the variance of ε. Recall that the estimator for the straight-line model was $s^2 = \text{SSE}/(n - 2)$, and note that the denominator is $(n -$ number of estimated β parameters), which is $(n - 2)$ in the first-order (straight-line) model. Since we must estimate one more parameter, β_2, for the second-order model $y = \beta_0 + \beta_1 x + \beta_2 x^2 + \varepsilon$, the estimator of σ^2 is

$$s^2 = \frac{\text{SSE}}{n - 3}$$

That is, the denominator becomes $(n - 3)$ because there are now three β parameters in the model. The numerical estimate for this example is

$$s^2 = \frac{\text{SSE}}{10 - 3} = \frac{15,332.6}{7} = 2,190.36$$

In many computer printouts and textbooks, s^2 is called the **mean square for error (MSE)**. This estimate of σ^2 is shown in the column titled MEAN SQUARE in the SAS printout in Figure 4.2.

For the general multiple regression model

$$y = \beta_0 + \beta_1 x_1 + \beta_2 x_2 + \cdots + \beta_k x_k + \varepsilon$$

we must estimate the $(k + 1)$ parameters $\beta_0, \beta_1, \beta_2, \ldots, \beta_k$. Thus, the estimator of σ^2 is SSE divided by the quantity $(n -$ Number of estimated β parameters).

Estimator of σ^2 for Multiple Regression Model with k Independent Variables

$$s^2 = \text{MSE} = \frac{\text{SSE}}{n - \text{Number of estimated } \beta \text{ parameters}}$$

$$= \frac{\text{SSE}}{n - (k + 1)}$$

We will use MSE, the estimator of σ^2, both to check the utility of the model (Sections 4.5 and 4.7) and to provide a measure of the reliability of predictions and estimates when the model is used for those purposes (Section 4.8). Thus, you can see that the estimation of σ^2 plays an important part in the development of a regression model.

Finally, the interpretation of $s = \sqrt{\text{MSE}}$ in multiple regression is essentially the same as that for simple linear regression. Since s estimates σ, the standard deviation of the errors of prediction, we expect most of the y values to lie within $2s$ of their least squares predicted value $\hat{y}$. On the SAS printout, Figure 4.2, $s = \sqrt{\text{MSE}} = 46.8$ is shaded next to **Root MSE**. Consequently, we expect the second-order model to predict electrical usage (y) to within $2s = 2(46.8) = 93.6$ kilowatt-hours of its true value.

4.5

Inferences About the β Parameters

Sometimes the individual β parameters in a model have particular practical significance and we want to estimate their values or test hypotheses about them. For example, if electrical usage y is related to home size x by the straight-line relationship

$$y = \beta_0 + \beta_1 x + \varepsilon$$

then β_1 has a very practical interpretation. That is, you saw in Chapter 3 that β_1 is the mean increase in kilowatt-hours of electrical usage y for a square foot increase in home size x.

As proposed in the preceding sections, suppose the electrical usage y is related to home size x by the quadratic model

$$y = \beta_0 + \beta_1 x + \beta_2 x^2 + \varepsilon$$

Then the mean value of y for a given value of x is

$$E(y) = \beta_0 + \beta_1 x + \beta_2 x^2$$

What is the practical interpretation of β_2? As noted earlier, the parameter β_2 measures the curvature of the response curve shown in Figure 4.3. If $\beta_2 > 0$, the slope of the curve will increase as x increases [see Figure 4.4a]. If $\beta_2 < 0$, the slope of the curve will decrease as x increases, as shown in Figure 4.4b.

FIGURE 4.4

The interpretation of β_2 for a second-order model

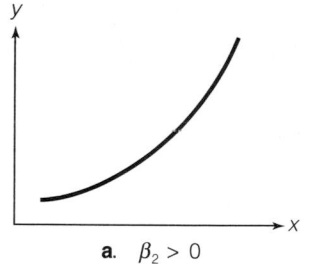

a. $\beta_2 > 0$

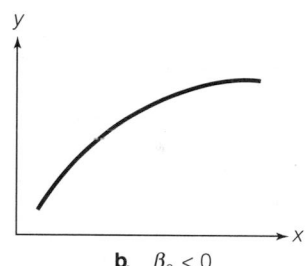

b. $\beta_2 < 0$

Intuitively, we would expect the electrical usage y to rise almost proportionally to home size x. Then, eventually, as the size of the home increases, the increase in electrical usage for a 1-unit increase in home size might begin to decrease. Thus, a forecaster of electrical usage would want to determine whether this type of curvature actually was present in the response curve, or, equivalently, the forecaster would want to test the null hypothesis

H_0: $\beta_2 = 0$ (No curvature in the response curve)

against the alternative hypothesis

H_a: $\beta_2 < 0$ (Downward curvature in the response curve)

A test of this hypothesis can be performed using a Student's t test.

The t test utilizes a test statistic that is analogous to that used to make inferences about the slope of the straight-line model (Section 3.6). The t statistic is formed by dividing the sample estimate $\hat{\beta}_2$ of the population coefficient β_2 by the estimated standard deviation of the sampling distribution of $\hat{\beta}_2$:

$$\textit{Test statistic:} \quad t = \frac{\hat{\beta}_2}{s_{\hat{\beta}_2}}$$

We use the symbol $s_{\hat{\beta}_2}$ to represent the estimated standard deviation of $\hat{\beta}_2$. The formula for computing $s_{\hat{\beta}_2}$ (presented in Appendix A) is very complex, but its computation is performed automatically as part of most standard multiple regression computer analyses. Thus, most computer packages list the estimated standard deviation $s_{\hat{\beta}_i}$ for each estimated model coefficient $\hat{\beta}_i$. In addition, they usually give the calculated t values for testing $H_0: \beta_i = 0$ for each coefficient in the model.

The rejection region for the test is found in exactly the same way as the rejection regions for the t tests in Chapters 1 and 3. That is, we consult Table 2 of Appendix D to obtain an upper-tail value of t. This is a value t_α such that $P(t > t_\alpha) = \alpha$. We can then use this value to construct rejection regions for either one- or two-tailed tests. To illustrate, in the electrical usage example, the error degrees of freedom is $(n - 3) = 7$, the denominator of the estimate of σ^2. Then the rejection region (shown in Figure 4.5) for a one-tailed test with $\alpha = .05$ is

$$\textit{Rejection region:} \quad t < -t_\alpha; \quad \alpha = .05, \quad df = 7$$
$$t < -1.895$$

FIGURE 4.5

Rejection region for test of $H_0: \beta_2 = 0$

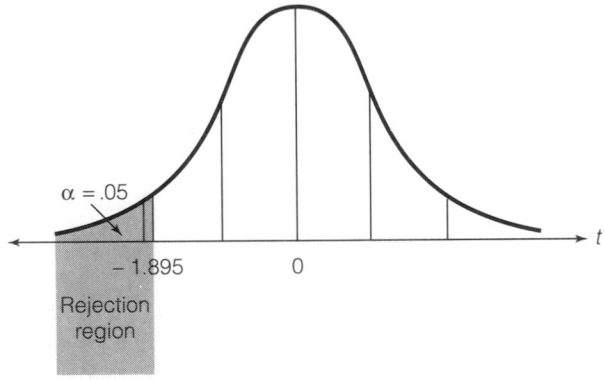

In Figure 4.6 we again show a portion of the SAS printout for the electrical usage example. The estimated standard deviations for the model coefficients appear under the column labeled **Standard Error**. The t statistics for testing the null hypotheses that the coefficients β_0, β_1, and β_2 individually equal 0 appear under the column headed T **for H0: Parameter=0**. The t value corresponding to the test of the null hypothesis $H_0: \beta_2 = 0$, shaded on the printout, is the last one in the column, i.e., $t = -7.62$. Since this value falls in the rejection region, i.e., it is less than -1.895, we conclude that the second-order term $\beta_2 x^2$ makes an important contribution to the prediction model of electrical usage.

The SAS printout shown in Figure 4.6 also lists the two-tailed observed significance levels (i.e., p-values) for each t value. These values appear under the column headed **Prob > |T|** . The observed significance level corresponding to

FIGURE 4.6

SAS printout for electrical usage example

Dependent Variable: Y

Analysis of Variance

Source	DF	Sum of Squares	Mean Square	F Value	Prob>F
Model	2	831069.54637	415534.77319	189.710	0.0001
Error	7	15332.55363	2190.36480		
C Total	9	846402.10000			

Root MSE	46.80133	R-square	0.9819	
Dep Mean	1594.70000	Adj R-sq	0.9767	
C.V.	2.93480			

Parameter Estimates

Variable	DF	Parameter Estimate	Standard Error	T for H0: Parameter=0	Prob > |T|
INTERCEP	1	-1216.143887	242.80636850	-5.009	0.0016
X	1	2.398930	0.24583560	9.758	0.0001
XX	1	-0.000450	0.00005908	-7.618	0.0001

the quadratic term, .0001, is also shaded; this implies that we would reject H_0: $\beta_2 = 0$ in favor of H_a: $\beta_2 \neq 0$ at any α level larger than .0001. Since our alternative was one-sided, H_a: $\beta_2 < 0$, the observed significance level is half that given in the printout, i.e., p-value $= \frac{1}{2}(.0001) = .00005$. Thus, there is very strong evidence that the mean electrical usage increases more slowly per square foot for large houses than for small houses.

We can also form a 95% confidence interval for the parameter β_2 as follows:

$$\hat{\beta}_2 \pm t_{\alpha/2} s_{\hat{\beta}_2} = -.000450 \pm (2.365)(.0000591)$$

or $(-.000590, -.000310)$. Note that the t value 2.365 corresponds to $\alpha/2 = .025$ and $(n - 3) = 7$ df. This interval constitutes a 95% confidence interval for β_2, the rate of change in curvature in mean electrical usage as home size is increased. Note that all values in the interval are negative, reconfirming the conclusion of our test.

Testing a hypothesis about a single β parameter that appears in any multiple regression model is accomplished in exactly the same manner as described for the second-order electrical usage model. The form of the t test is shown in the box at the top of page 172.

EXAMPLE 4.1

A collector of antique grandfather clocks believes that the price received for the clocks at an antique auction increases with the age of the clocks and with the number of bidders. Thus, the following model is hypothesized:

$$y = \beta_0 + \beta_1 x_1 + \beta_2 x_2 + \varepsilon$$

where

    y = Auction price

    x_1 = Age of clock (years)

    x_2 = Number of bidders

Test of an Individual Parameter Coefficient in the Multiple Regression Model

$$y = \beta_0 + \beta_1 x_1 + \beta_2 x_2 + \cdots + \beta_k x_k + \varepsilon$$

ONE-TAILED TEST   TWO-TAILED TEST

H_0: $\beta_i = 0$ H_0: $\beta_i = 0$
H_a: $\beta_i > 0$ H_a: $\beta_i \neq 0$
     (or $\beta_i < 0$)

Test statistic:* $t = \dfrac{\hat{\beta}_i}{s_{\hat{\beta}_i}}$ Test statistic:* $t = \dfrac{\hat{\beta}_i}{s_{\hat{\beta}_i}}$

Rejection region: $t > t_\alpha$ Rejection region: $t > t_{\alpha/2}$
          (or $t < -t_\alpha$)   or $t < -t_{\alpha/2}$

where

   n = Number of observations

   k = Number of independent variables in the model

and $t_{\alpha/2}$ is based on $[n - (k + 1)]$ df

Assumptions: See Section 4.2 for the assumptions about the probability distribution of the random error component ε.

A sample of 32 auction prices of grandfather clocks, along with their age and the number of bidders, is given in Table 4.2. The model is fit to the data, and a portion of the Minitab printout is shown in Figure 4.7. Test the hypothesis that the auction price increases as the number of bidders increases (and age is held constant), i.e., $\beta_2 > 0$. Use $\alpha = .05$.

FIGURE 4.7

Portion of the Minitab printout for Example 4.1

```
The regression equation is
Y = - 1339 + 12.7 X1 + 86.0 X2

Predictor       Coef       Stdev    t-ratio        p
Constant      -1339.0       173.8      -7.70    0.000
X1            12.7406      0.9047      14.08    0.000
X2             85.953       8.729       9.85    0.000

s = 133.5      R-sq = 89.2%      R-sq(adj) = 88.5%

Analysis of Variance

SOURCE         DF          SS          MS         F        p
Regression      2     4283063     2141532    120.19    0.000
Error          29      516727       17818
Total          31     4799789
```

*To test the null hypothesis that a parameter β_i equals some value other than 0, say, H_0: $\beta_i = \beta_{i0}$, use the test statistic $t = (\hat{\beta}_i - \beta_{i0})/s_{\hat{\beta}_i}$. All other aspects of the test will be described in the box.

TABLE 4.2 **Auction Price Data**

AGE x_1	NUMBER OF BIDDERS x_2	AUCTION PRICE y	AGE x_1	NUMBER OF BIDDERS x_2	AUCTION PRICE y
127	13	1,235	170	14	2,131
115	12	1,080	182	8	1,550
127	7	845	162	11	1,884
150	9	1,522	184	10	2,041
156	6	1,047	143	6	854
182	11	1,979	159	9	1,483
156	12	1,822	108	14	1,055
132	10	1,253	175	8	1,545
137	9	1,297	108	6	729
113	9	946	179	9	1,792
137	15	1,713	111	15	1,175
117	11	1,024	187	8	1,593
137	8	1,147	111	7	785
153	6	1,092	115	7	744
117	13	1,152	194	5	1,356
126	10	1,336	168	7	1,262

Solution

The hypothesis of interest concerns the parameter β_2. Specifically,

$$H_0: \quad \beta_2 = 0 \qquad H_a: \quad \beta_2 > 0$$

Test statistic: $\quad t = \dfrac{\hat{\beta}_2}{s_{\hat{\beta}_2}}$

Rejection region: For $\alpha = .05$, $t > t_{.05}$
where df $\doteq n - (k + 1) = 32 - 3 = 29$
or $t > 1.699$ (see Figure 4.8)

FIGURE 4.8
Rejection region for $H_0: \beta_2 = 0$

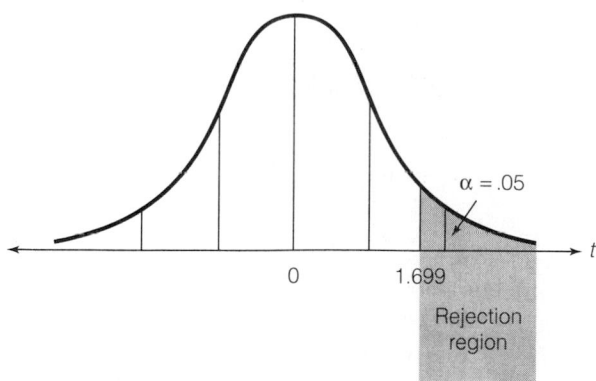

The calculated t value, $t = 9.85$, is indicated in Figure 4.7. This value exceeds 1.699 and, therefore, falls in the rejection region. Thus, the collector can conclude that the mean auction price of the clocks increases as the number of bidders increases, when age is held constant. Note that the observed significance level for the test (shaded on Figure 4.7) is p-value ≈ 0. Therefore, any reasonable value of α will lead us to reject H_0.

Note that the values $\hat{\beta}_1 = 12.7406$ and $\hat{\beta}_2 = 85.953$ (shaded in Figure 4.7) are easily interpreted. We estimate that the mean auction price increases \$12.74 per year of age of the clock when the number of bidders is held constant, and the mean price increases by \$85.95 per additional bidder, for clocks of a fixed age.

Be careful not to try to interpret the estimated intercept $\hat{\beta}_0 = -1,339.0$ in the same way as we interpreted $\hat{\beta}_1$ and $\hat{\beta}_2$. You might think that this implies a negative price for clocks 0 years of age with 0 bidders. However, these zeros are meaningless numbers in this example, since the ages range from 108 to 194 and the number of bidders ranges from 5 to 15. Keep in mind that we are modeling y within the range of values observed for the predictor variables and that interpretations of the models for values of the independent variables outside their sampled ranges can be very misleading.

Some statistical software packages use an F test to test hypotheses concerning the individual β parameters. If you conduct a two-tailed t test and reject the hypothesis if $t > t_{\alpha/2}$ or $t < -t_{\alpha/2}$, the corresponding F test will imply rejection if the computed value of F (which is equal to the square of the computed t statistic) is larger than F_α, because the square of the Student's t with ν (Greek nu) degrees of freedom is equal to an F statistic with 1 df in the numerator and ν df in the denominator. Thus, $t_{\alpha/2}^2 = F_\alpha$, where t is based on ν df and F possesses 1 numerator and ν denominator degrees of freedom, respectively. As an example, when we tested the curvature parameter β_2 in the second-order model relating electrical usage to home size, the computed t value was -7.62 (see Figure 4.6). The equivalent F statistic yields

$$F = t^2 = (-7.62)^2 = 58.06$$

Suppose we wanted to conduct a two-tailed statistical test, i.e., $H_0: \beta_2 = 0$ and $H_a: \beta_2 \neq 0$. The upper-tail rejection region for a two-tailed test with $\alpha = .05$ is

$$F > F_{.05} \quad \text{where } F_{.05} \text{ is based on } \nu_1 = 1 \text{ df and } \nu_2 = 7 \text{ df}$$

or

$$F > 5.59$$

Note that the F value, 5.59, is equal to the square of 2.365, the value of t that corresponds to $t_{.025}$ with 7 df. In other words, you can conduct a two-tailed test of the null hypothesis $H_0: \beta_i = 0$, using either a two-tailed t test or a one-tailed F test. If you want to conduct a one-tailed test to detect $H_a: \beta_i > 0$ (or $H_a: \beta_i < 0$), the F test will not suffice. You will have to conduct the test using a t statistic.

EXERCISES

4.1 How is the number of degrees of freedom available for estimating σ^2, the variance of ε, related to the number of independent variables in a regression model?

4.2 An employer believes that factory workers who are with the company longer tend to invest more in a company investment program per year than workers with less time with the company. The following model is believed to be adequate in modeling the relationship between annual amount invested y and years working for the company x:

$$y = \beta_0 + \beta_1 x + \beta_2 x^2 + \varepsilon$$

The employer checks the records for a sample of 50 factory employees for a previous year, and fits the above model to get $\hat{\beta}_2 = .0015$ and $s_{\hat{\beta}_2} = .000712$. The basic shape of a second-order model depends on whether $\beta_2 < 0$ or $\beta_2 > 0$. Test to determine whether the employer can conclude that $\beta_2 > 0$. Use $\alpha = .05$.

4.3 Real estate appraisers rely heavily on multiple regression analysis in their evaluation of property. Typically, the sale price y of a home is modeled as a function of several home-related variables (e.g., home size, home condition, location, and so forth). For example, an article in *The Real Estate Appraiser and Analyst* (Spring 1986) considered the following regression model:

$$y = \beta_0 + \beta_1 x_1 + \beta_2 x_2 + \varepsilon$$

where

$x_1 =$ Home size (in square feet)

$x_2 =$ Home condition rating (1 = poor, ... , 10 = excellent)

Data collected for $n = 10$ recent home sales were used in the analysis. The data are reproduced in the table, and an SPSS printout is provided at the top of page 176.

SALE PRICE y, $ thousands	HOME SIZE x_1, hundreds of sq. ft.	CONDITION RATING x_2, 1 to 10
60.0	23	5
32.7	11	2
57.7	20	9
45.5	17	3
47.0	15	8
55.3	21	4
64.5	24	7
42.6	13	6
54.5	19	7
57.5	25	2

Source: Andrews, R. L. and Ferguson, J. T. "Integrating judgment with a regression appraisal." *The Real Estate Appraiser and Analyst*, Vol. 52, No. 2, Spring 1986 (Table I).

a. Plot sale price y against home size x_1. Do you detect a linear relationship between the two variables? Is it positive or negative?

```
Equation Number 1      Dependent Variable..    Y

Multiple R              .99505
R Square                .99012
Adjusted R Square       .98730
Standard Error         1.08055

Analysis of Variance
                     DF       Sum of Squares        Mean Square
Regression            2          819.32795          409.66398
Residual              7            8.17305            1.16758

F =     350.86649      Signif F =   .0000

------------------ Variables in the Equation ------------------

Variable            B          SE B        Beta         T   Sig T

X2            1.278141     .144400     .332794      8.851  .0000
X1            1.870935     .076174     .923464     24.561  .0000
(Constant)    9.782271    1.630481                  6.000  .0005
```

---

b. Plot sale price y against home condition x_2. Do you detect a linear relationship between the two variables? Is it positive or negative?

c. Is there evidence to indicate that sale price and home size are linearly related? Test using $\alpha = .01$.

d. Calculate a 99% confidence interval for β_2. Interpret the result.

4.4 Refer to the *Feline Practice* (Feb. 1986) study on estimating the diet metabolizable energy (ME) content of commercial cat foods in Exercises 1.57 and 3.53. Three factors thought to influence ME content (y) of dry cat food are the crude protein (x_1), acid ether abstract (x_2), and nitrogen-free extract (x_3) content of the food. Data collected for 28 cats fed a diet of dry food were used to fit the multiple regression model

$$y = \beta_0 + \beta_1 x_1 + \beta_2 x_2 + \beta_3 x_3 + \varepsilon$$

with the following results:

$$\hat{y} = 2.44 + .45x_1 + 3.43x_2 + .10x_3$$

$$s_{\hat{\beta}_1} = .38 \qquad s_{\hat{\beta}_2} = .73 \qquad s_{\hat{\beta}_3} = .36$$

a. Is there sufficient evidence to indicate that crude protein x_1 is positively related to ME content y? Test using $\alpha = .05$.

b. Find a 95% confidence interval for β_3. Interpret your result.

c. The F statistic for testing H_0: $\beta_2 = 0$ was found to be $F = 22.08$. Is there sufficient evidence to indicate that ME content y is linearly related to acid ether abstract x_2? Test using $\alpha = .05$.

d. Refer to part c. Calculate the t statistic for testing H_0: $\beta_2 = 0$ and show that $F = t^2$.

4.5 In the mid 1800s, the U.S. census inquired about the real property and personal wealth of individual households. Using census information from 1860 and 1870, J. R. Kearl and C. L. Pope (Brigham Young University) examined the mobility of Utah households as measured by their wealth holdings (*The Review of Economics and Statistics*, May 1984). Holding occupation, time of entry into the economy, nativity, sex, place of residence, and internal migration constant, Kearl and Pope fit the quadratic model $E(y) = \beta_0 + \beta_1 x + \beta_2 x^2$, where y is personal wealth (in dollars) of a Utah household and x is age (in years) of the head of household. The results of the regression are summarized as follows:

$$\hat{y} = 52.39 + 74.21x - .71x^2$$

$$s_{\hat{\beta}_1} = 5.38 \qquad s_{\hat{\beta}_2} = 4.73 \qquad n > 20{,}000$$

a. Graph the least squares prediction equation.

b. Is there evidence of a quadratic relationship in the wealth–age relationship for Utah households during 1860–1870? Test using $\alpha = .10$.

4.6 In a production facility, an accurate estimate of man-hours needed to complete a task is crucial to management in making such decisions as the proper number of workers to hire, an accurate deadline to quote a client, or cost-analysis decisions regarding budgets. A manufacturer of boiler drums wants to use regression to predict the number of man-hours needed to erect the drums in future projects. To accomplish this, data for 35 boilers were collected. In addition to man-hours (y), the variables measured were boiler capacity (x_1 = pounds per hour or lb/hr), boiler design pressure (x_2 = pounds per square inch or psi), boiler type (x_3 = 1 if industry field erected, 0 if utility field erected), and drum type (x_4 = 1 if steam, 0 if mud). The data are provided in the table on page 178. A Minitab printout for the model $E(y) = \beta_0 + \beta_1 x_1 + \beta_2 x_2 + \beta_3 x_3 + \beta_4 x_4$ follows.

```
The regression equation is
Y = - 3783 + 0.00875 X1 + 1.93 X2 + 3444 X3 + 2093 X4

Predictor          Coef         Stdev     t-ratio         p
Constant          -3783          1205       -3.14     0.004
X1            0.0087490     0.0009035        9.68     0.000
X2               1.9265        0.6489        2.97     0.006
X3               3444.3         911.7        3.78     0.001
X4               2093.4         305.6        6.85     0.000

s = 894.6        R-sq = 90.3%      R-sq(adj) = 89.0%

Analysis of Variance

SOURCE         DF           SS          MS         F         p
Regression      4    230854848    57713712     72.11     0.000
Error          31     24809760      800315
Total          35    255664608

SOURCE         DF       SEQ SS
X1              1    175007136
X2              1       490357
X3              1     17813090
X4              1     37544264

Unusual Observations
Obs.       X1           Y       Fit Stdev.Fit   Residual   St.Resid
 19   1089490       14791     12022       523       2769      3.81R

R denotes an obs. with a large st. resid.

        Fit  Stdev.Fit         95% C.I.         95% P.I.
       1936        239   (  1449,   2424) (    47,   3825)
```

a. Test the hypothesis that boiler capacity (x_1) is positively linearly related to man-hours (y). Use $\alpha = .05$.

MAN-HOURS (y)	BOILER CAPACITY (x_1)	DESIGN PRESSURE (x_2)	BOILER TYPE	DRUM TYPE
3,137	120,000	375	Industrial	Steam
3,590	65,000	750	Industrial	Steam
4,526	150,000	500	Industrial	Steam
10,825	1,073,877	2,170	Utility	Steam
4,023	150,000	325	Industrial	Steam
7,606	610,000	1,500	Utility	Steam
3,748	88,200	399	Industrial	Steam
2,972	88,200	399	Industrial	Steam
3,163	88,200	399	Industrial	Steam
4,065	90,000	1,140	Industrial	Steam
2,048	30,000	325	Industrial	Steam
6,500	441,000	410	Industrial	Steam
5,651	441,000	410	Industrial	Steam
6,565	441,000	410	Industrial	Steam
6,387	441,000	410	Industrial	Steam
6,454	627,000	1,525	Utility	Steam
6,928	610,000	1,500	Utility	Steam
4,268	150,000	500	Industrial	Steam
14,791	1,089,490	2,170	Utility	Steam
2,680	125,000	750	Industrial	Steam
2,974	120,000	375	Industrial	Mud
1,965	65,000	750	Industrial	Mud
2,566	150,000	500	Industrial	Mud
1,515	150,000	250	Industrial	Mud
2,000	150,000	500	Industrial	Mud
2,735	150,000	325	Industrial	Mud
3,698	610,000	1,500	Utility	Mud
2,635	90,000	1,140	Industrial	Mud
1,206	30,000	325	Industrial	Mud
3,775	441,000	410	Industrial	Mud
3,120	441,000	410	Industrial	Mud
4,206	441,000	410	Industrial	Mud
4,006	441,000	410	Industrial	Mud
3,728	627,000	1,525	Utility	Mud
3,211	610,000	1,500	Utility	Mud
1,200	30,000	325	Industrial	Mud

Source: Kelly Uscategui, graduate student, University of South Florida, 1988.

b. Test the hypothesis that boiler pressure (x_2) is positively linearly related to man-hours (y). Use $\alpha = .05$.

c. Construct a 95% confidence interval for β_3.

d. In Chapter 5 we will learn that β_3 represents the difference between the mean number of man-hours required for industrial and utility field erected boilers. Use this information to interpret the confidence interval of part **c**.

e. Construct a 95% confidence interval for β_4 and interpret the result. [*Hint:* $\beta_4 = \mu_{Steam} - \mu_{Mud}$, where μ_i represents the mean number of man-hours required for drum type i.]

4.7 *Zoning* is defined as the distribution of vacant land to residential and nonresidential uses via policy set by local governments. Although the negative effects of zoning have been studied (e.g., distorting urban property markets, creating barriers to residential mobility, and impeding economic and social integration), little empirical evidence exists that identifies the factors that encourage restrictive zoning practices. A recent study, reported in the *Journal of Urban Economics* (Vol. 21, 1987), developed a series of multiple regression models that hypothesize several determinants of zoning. One of the models studied took the form

$$E(y) = \beta_0 + \beta_1 x_1 + \beta_2 x_1^2 + \beta_3 x_2$$

where

y = Percentage of vacant land zoned for residential use

x_1 = Proportion of existing land in nonresidential use

x_2 = Proportion of total tax base derived from nonresidential property

The model was fit to data collected for $n = 185$ municipal communities in northeastern New Jersey, with the following results:

INDEPENDENT VARIABLE	PARAMETER ESTIMATE	STANDARD ERROR OF ESTIMATE	t VALUE	p-VALUE
Intercept	92.26	3.07	30.05	$p < .01$
x_1	−96.35	46.59	−2.07	$p < .05$
x_1^2	166.80	120.88	1.38	$p > .10$
x_2	−75.51	13.35	−5.66	$p < .01$

Adjusted $R^2 = .25$ $F = 21.86$ $(p < .01)$

Source: Rolleston, B. S. "Determinants of restrictive suburban zoning: An empirical analysis." *Journal of Urban Economics*, Vol. 21, 1987, p. 15, Table 4.

a. Construct a 95% confidence interval for β_3. Interpret the result.
b. Test the hypothesis that a curvilinear relationship exists between percentage (y) of land zoned for residential use and proportion (x_1) of existing land in nonresidential use.

4.6

The Multiple Coefficient of Determination, R^2

Recall from Chapter 3 that the coefficient of determination, r^2, is a measure of how well a straight-line model fits a data set. To measure how well a general linear model (for example, a second-order model) fits a set of data, we compute the multiple regression equivalent of r^2, called the **multiple coefficient of determination** and denoted by the symbol R^2.

Definition 4.1

The **multiple coefficient of determination, R^2,** is defined as

$$R^2 = 1 - \frac{SSE}{SS_{yy}} \qquad 0 \le R^2 \le 1$$

where $SSE = \Sigma (y_i - \hat{y}_i)^2$, $SS_{yy} = \Sigma (y_i - y)^2$, and $\hat{y}_i$ is the predicted value of y_i for the multiple regression model.

Just as for the simple linear model, R^2 is a sample statistic that represents the fraction of the sample variation of the y values (measured by SS_{yy}) that is attributable to the regression model. Thus, $R^2 = 0$ implies a complete lack of fit of the model to the data, and $R^2 = 1$ implies a perfect fit, with the model passing through every data point. In general, the closer the value of R^2 is to 1, the better the model fits the data.

To illustrate, the value $R^2 = .9819$ for the electrical usage example is shaded in the SAS printout, Figure 4.9. This very high value of R^2 implies that 98.2% of the sample variation is attributable to, or explained by, the independent variable (home size) x. Thus, R^2 is a sample statistic that tells how well the model fits the data, and thereby represents a measure of the utility of the entire model.

FIGURE 4.9

Portion of the SAS printout for electrical usage example

Dependent Variable: Y

Analysis of Variance

Source	DF	Sum of Squares	Mean Square	F Value	Prob>F
Model	2	831069.54637	415534.77319	189.710	0.0001
Error	7	15332.55363	2190.36480		
C Total	9	846402.10000			

Root MSE	46.80133	R-square	0.9819	
Dep Mean	1594.70000	Adj R-sq	0.9767	
C.V.	2.93480			

Parameter Estimates

Variable	DF	Parameter Estimate	Standard Error	T for H0: Parameter=0	Prob > |T|
INTERCEP	1	-1216.143887	242.80636850	-5.009	0.0016
X	1	2.398930	0.24583560	9.758	0.0001
XX	1	-0.000450	0.00005908	-7.618	0.0001

A large value of R^2 computed from the *sample* data does not necessarily mean that the model provides a good fit to all of the data points in the *population*. For example, a first-order linear model that contains three parameters will provide a perfect fit to a sample of three data points and R^2 will equal 1. Likewise, you will always obtain a perfect fit ($R^2 = 1$) to a set of n data points if the model contains exactly n parameters. Consequently, if you want to use the value of R^2 as a measure of how useful the model will be for predicting y, it should be based

> **Warning**
>
> In a multiple regression analysis, use the value of R^2 as a measure of how useful a linear model will be for predicting y only if the sample contains substantially more data points than the number of β parameters in the model.

on a sample that contains substantially more data points than the number of parameters in the model.

As an alternative to using R^2 as a measure of model adequacy, the **adjusted multiple coefficient of determination**, denoted R_a^2, is often reported. The formula for R_a^2 is shown in the box.

| The Adjusted Multiple Coefficient of Determination

The **adjusted multiple coefficient of determination** is given by

$$R_a^2 = 1 - \frac{n-1}{n-(k+1)}\left(\frac{SSE}{SS_{yy}}\right)$$

$$= 1 - \frac{n-1}{n-(k+1)}(1 - R^2)$$

Unlike R^2, R_a^2 takes into account ("adjusts" for) both the sample size n and the number of β parameters in the model. R_a^2 will always be smaller than R^2, and, more importantly, cannot be "forced" to 1 by simply adding more and more independent variables to the model. Consequently, analysts prefer the more conservative R_a^2 when choosing a measure of model adequacy.

To illustrate, the value of R_a^2 is shown on the SAS printout (Figure 4.9) directly underneath the value of R^2. Note that $R_a^2 = .9767$, a value only slightly smaller than R^2. Our interpretation is that after adjusting for sample size and the number of parameters in the model, approximately 98% of the sample variation in sale price can be "explained" by the first-order model.

Despite their utility, R^2 and R_a^2 are only sample statistics. Consequently, it is dangerous to judge the usefulness of the model based solely on these values. We discuss a more formal method of checking the predictive ability of a general linear model—a statistical test of hypothesis—in the following section.

4.7 Testing the Utility of a Model: The Analysis of Variance F Test

The objective of Step 5 in a multiple regression analysis is to conduct a test of the utility of a general linear model—that is, a test to determine whether the model is adequate for predicting y. Conducting t tests on each β parameter in a model (Section 4.5) is generally not a good way to determine whether a model is contributing information for the prediction of y. If we were to conduct a series of t tests to determine whether the independent variables are contributing to the predictive relationship, we would be very likely to make one or more errors in deciding which terms to retain in the model and which to exclude.

For example, suppose you fit a model with 10 independent variables, $x_1, x_2, \ldots, x_{10}$, and decide to conduct t tests on all 10 individual β's in the model, each at $\alpha = .05$. Even if all the β parameters (except β_0) in the model are equal to 0, you will incorrectly reject the null hypothesis at least once and conclude that some β parameter is nonzero approximately 40% of the time.* In other words, the overall Type I error is about .40, not .05!

Thus, in multiple regression models for which a large number of independent variables are being considered, conducting a series of t tests may cause the experimenter to include a large number of insignificant variables and exclude some useful ones. If we want to test the utility of a multiple regression model, we will need a global test (one that encompasses all the β parameters).

In particular, for the second-order model $E(y) = \beta_0 + \beta_1 x + \beta_2 x^2$ fit to the electrical usage data, the test

$$H_0: \quad \beta_1 = \beta_2 = 0$$

$$H_a: \quad \text{At least one of the parameters } \beta_1 \text{ and } \beta_2 \text{ is nonzero}$$

would formally test the global utility of the model. The test statistic used to test this null hypothesis is

$$\text{Test statistic:} \quad F = \frac{\text{Mean square for model}}{\text{Mean square for error}}$$

$$= \frac{\text{SS(model)}/k}{\text{SSE}/[n - (k + 1)]}$$

where n is the number of data points, k is the number of parameters in the model (not including β_0), and SS(Model) = SS(Total) $-$ SSE. When H_0 is true, this F test statistic will have an F probability distribution with k df in the numerator and $[n - (k + 1)]$ df in the denominator. The upper-tail values of the F distribution are given in Tables 3, 4, 5, and 6 of Appendix D.

It can be shown (proof omitted) that an equivalent form of this test statistic is

$$F = \frac{R^2/k}{(1 - R^2)/[n - (k + 1)]}$$

Therefore, the F test statistic becomes large as the coefficient of determination R^2 becomes large. To determine how large F must be before we can conclude at a given value of α that the model is useful for predicting y, we set up the rejection region as follows:

$$\text{Rejection region:} \quad F > F_\alpha \quad \text{where} \quad \nu_1 = k \text{ df and } \nu_2 = n - (k + 1) \text{ df}$$

For the electrical usage example, $n = 10$, $k = 2$, $n - (k + 1) = 7$, and $\alpha = .05$.

---

*The proof of this result proceeds as follows:

$$P(\text{reject } H_0 \text{ at least once} \mid \beta_1 = \beta_2 = \cdots = \beta_{10} = 0)$$
$$= 1 - P(\text{Reject } H_0 \text{ no times} \mid \beta_1 = \beta_2 = \cdots = \beta_{10} = 0)$$
$$\leq 1 - [P(\text{Accept } H_0: \beta_1 = 0 \mid \beta_1 = 0) \cdot P(\text{Accept } H_0: \beta_2 = 0 \mid \beta_2 = 0) \cdots$$
$$\cdot P(\text{Accept } H_0: \beta_{10} = 0 \mid \beta_{10} = 0)]$$
$$= 1 - [(1 - \alpha)^{10}] = 1 - (.95)^{10} = .401$$

Consequently, we will reject $H_0: \beta_1 = \beta_2 = 0$ if

$$F > F_{.05} \quad \text{where} \quad \nu_1 = 2 \text{ and } \nu_2 = 7$$

or

$$F > 4.74 \quad \text{(see Figure 4.10)}$$

FIGURE 4.10

Rejection region for the *F* statistic with $\nu_1 = 2$, $\nu_2 = 7$, and $\alpha = .05$

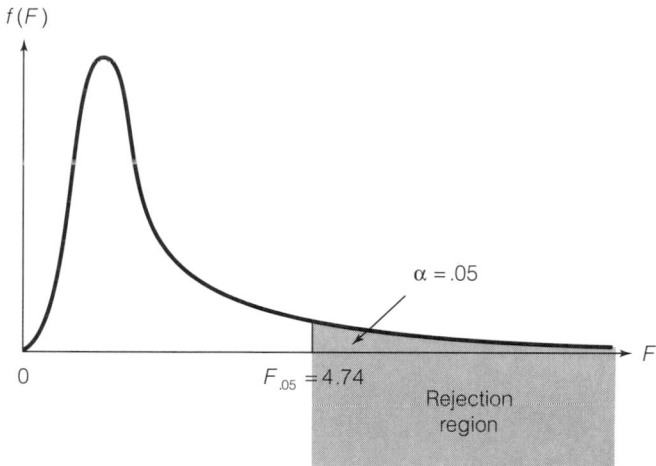

From the SAS printout shown in Figure 4.11, we find that the computed *F* (shaded in the upper right-hand corner of the printout) is 189.71. Since this value greatly exceeds the tabulated value of 4.74, we conclude that at least one of the model coefficients β_1 and β_2 is nonzero. Therefore, this global *F* test indicates that the second-order model $y = \beta_0 + \beta_1 x + \beta_2 x^2 + \varepsilon$ is useful for predicting electrical usage.

FIGURE 4.11

Portion of the SAS printout for electrical usage example

Dependent Variable: Y

Analysis of Variance

Source	DF	Sum of Squares	Mean Square	F Value	Prob>F
Model	2	831069.54637	415534.77319	189.710	0.0001
Error	7	15332.55363	2190.36480		
C Total	9	846402.10000			

Root MSE	46.80133	R-square	0.9819	
Dep Mean	1594.70000	Adj R-sq	0.9767	
C.V.	2.93480			

Parameter Estimates

Variable	DF	Parameter Estimate	Standard Error	T for H0: Parameter=0	Prob > |T|
INTERCEP	1	-1216.143887	242.80636850	-5.009	0.0016
X	1	2.398930	0.24583560	9.758	0.0001
XX	1	-0.000450	0.00005908	-7.618	0.0001

Test of the Overall Utility of a Multiple Regression Model:
The Analysis of Variance F Test

H_0: $\beta_1 = \beta_2 = \cdots = \beta_k = 0$

H_a: At least one of the parameters, $\beta_1, \beta_2, \ldots, \beta_k$, differs from 0.

Rejection region: $F > F_\alpha$

where the distribution of F depends on k numerator df and $n - (k + 1)$ denominator df

Test statistic:

$$F = \frac{\text{Mean square for model}}{\text{Mean square for error}} = \frac{\text{SS(model)}/k}{\text{SSE}/[n - (k + 1)]}$$

$$= \frac{R^2/k}{(1 - R^2)/[n - (k + 1)]}$$

where

$n =$ Number of observations

$k =$ Number of parameters in the model (excluding β_0)

$R^2 =$ Multiple coefficient of determination

Values of F_α for $\alpha = .10, .05, .025,$ and $.01$ are given in Tables 3, 4, 5, and 6 of Appendix D.

We could arrive at the same decision by checking the observed significance level (*p*-value) of the F test, given as PROB $> F$ in the SAS printout. This value (shaded in Figure 4.11) indicates that we will reject H_0 for any α greater than $p = .0001$.

EXAMPLE 4.2

Refer to Example 4.1, in which an antique collector modeled the auction price y of grandfather clocks as a function of the age of the clock, x_1, and the number of bidders, x_2. The hypothesized model was

$$y = \beta_0 + \beta_1 x_1 + \beta_2 x_2 + \varepsilon$$

A sample of 32 observations was obtained, with the results summarized in the Minitab printout repeated in Figure 4.12. Discuss the coefficient of determination R^2 for this example and then conduct the global F test of model utility using $\alpha = .05$.

Solution

The R^2 value (shaded in Figure 4.12) is .89. This implies that 89% of the variation of the y values (the auction prices) about their mean can be explained by the least squares model. We now test

H_0: $\beta_1 = \beta_2 = 0$ [*Note:* $k = 2$]

H_a: At least one of the two model coefficients is nonzero.

$$\text{Test statistic:} \quad F = \frac{\text{Mean square for model}}{\text{Mean square for error}} = \frac{\text{SS(Model)}/k}{\text{SSE}/[n - (k + 1)]}$$

$$\text{Rejection region:} \quad F > F_\alpha \quad \text{where} \quad \nu_1 = k \text{ and } \nu_2 = n - (k + 1)$$

FIGURE 4.12

Portion of the Minitab printout for Example 4.2

```
The regression equation is
Y = - 1339 + 12.7 X1 + 86.0 X2

Predictor        Coef       Stdev     t-ratio         p
Constant      -1339.0       173.8       -7.70     0.000
X1            12.7406      0.9047       14.08     0.000
X2             85.953       8.729        9.85     0.000

s = 133.5      R-sq = 89.2%      R-sq(adj) = 88.5%

Analysis of Variance

SOURCE        DF          SS          MS          F          p
Regression     2     4283063     2141532     120.19     0.000
Error         29      516727       17818
Total         31     4799789
```

For this example, $n = 32$, $k = 2$, and $n - (k + 1) = 32 - 3 = 29$. Then, for $\alpha = .05$, we will reject $H_0: \beta_1 = \beta_2 = 0$ if $F > F_{.05}$, i.e., if $F > 3.33$ (obtained from Table 4 of Appendix D). The computed value of the F test statistic is 120.19 (see Figure 4.12). Since this value of F falls in the rejection region ($F = 120.19$ greatly exceeds $F_{.05} = 3.33$ and $\alpha = .05$ greatly exceeds $p - .0001$), the data provide strong evidence that at least one of the model coefficients is nonzero. The model appears to be useful for predicting auction prices.

Can we be sure that the best prediction model has been found if the global F test indicates that a model is useful? Unfortunately, we cannot. There is no way of knowing whether the addition of other independent variables will further improve the utility of the model, as the following example indicates.

EXAMPLE 4.3

Refer to Examples 4.1 and 4.2. Suppose the collector, having observed many auctions, believes that the *rate of increase* of the auction price with age will be driven upward by a large number of bidders. Thus, instead of a relationship like that shown in Figure 4.13a (page 186), in which the rate of increase in price with age is the same for any number of bidders, the collector believes the relationship is like that shown in Figure 4.13b. Note that as the number of bidders increases from 5 to 15, the slope of the price versus age line increases. When the slope of the relationship between y and one independent variable (x_1) depends on the value of a second independent variable (x_2), as is the case here, we say that x_1 and x_2 **interact.*** A model that accounts for this type of interaction is written

$$y = \beta_0 + \beta_1 x_1 + \beta_2 x_2 + \beta_3 x_1 x_2 + \varepsilon$$

Note that the increase in the mean price $E(y)$ for each 1-year increase in age x_1 is no longer given by the constant β_1, but is now $\beta_1 + \beta_3 x_2$. That is, the amount $E(y)$ increases for each 1-unit increase in x_1 is *dependent on the number of bidders* x_2. Thus, the two variables x_1 and x_2 interact to affect y.

*A more complete discussion of interaction is given in Section 4.9 and optional Chapter 5.

FIGURE 4.13

Examples of no interaction and interaction models

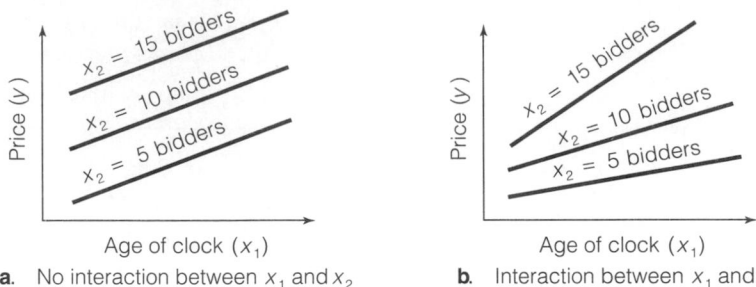

a. No interaction between x_1 and x_2 **b.** Interaction between x_1 and x_2

The 32 data points listed in Table 4.2 were used to fit the first-order model with interaction. A portion of the Minitab printout is shown in Figure 4.14.

Test the hypothesis that the price–age slope increases as the number of bidders increases, i.e., that age and number of bidders, x_2, interact positively.

FIGURE 4.14

Portion of the Minitab printout for the model with interaction, Example 4.3

```
The regression equation is
Y = 320 + 0.88 X1 - 93.3 X2 + 1.30 X1X2

Predictor        Coef       Stdev     t-ratio         p
Constant        320.5       295.1        1.09     0.287
X1              0.878       2.032        0.43     0.669
X2             -93.26       29.89       -3.12     0.004
X1X2           1.2978      0.2123        6.11     0.000

s = 88.91        R-sq = 95.4%        R-sq(adj) = 94.9%

Analysis of Variance

SOURCE        DF           SS          MS          F         p
Regression     3      4578428     1526142     193.04     0.000
Error         28       221362        7906
Total         31      4799789
```

Solution

The model is

$$y = \beta_0 + \beta_1 x_1 + \beta_2 x_2 + \beta_3 x_1 x_2 + \varepsilon$$

and the hypothesis of interest to the collector concerns the parameter β_3. Specifically,

H_0: $\beta_3 = 0$ H_a: $\beta_3 > 0$

Test statistic: $t = \dfrac{\hat{\beta}_3}{s_{\hat{\beta}_3}}$

Rejection region: For $\alpha = .05$, $t > t_{.05}$
where df $= n - (k + 1)$

In this example, $n = 32$, $k = 3$, df $= n - (k + 1) = 32 - 4 = 28$, and thus, $t_{.05} = 1.701$.

The t value corresponding to $\hat{\beta}_3$ is shaded in Figure 4.14. The value, $t = 6.11$, exceeds 1.701 and, therefore, falls in the rejection region. Thus, the collector can conclude that the rate of change of the mean price of the clocks with age increases as the number of bidders increases, i.e., x_1 and x_2 interact positively.

(The same conclusion can be reached by noting that the p-value of the test is approximately 0.) Thus, it appears that the interaction term should be included in the model.

One note of caution: Although the coefficient of x_2 is negative ($\hat{\beta}_2 = -93.26$) in Example 4.3, this does not imply that auction price decreases as the number of bidders increases. Since interaction is present, the rate of change (slope) of mean auction price with the number of bidders *depends on x_1*, the age of the clock. Thus, for example, the estimated rate of change of y with x_2 for a 150-year-old clock is

$$\text{Estimated } x_2 \text{ slope: } \quad \hat{\beta}_2 + \hat{\beta}_3 x_1 = -93.26 + 1.30(150)$$
$$= 101.74$$

In other words, we estimate that the auction price of a 150-year-old clock will increase by \$101.74 for every additional bidder. Although this rate of increase will vary as x_1 is changed, it will remain positive for the range of values of x_1 included in the sample. Use extreme care in interpreting the signs and sizes of coefficients in a multiple regression model.

After we have determined that the overall model is useful for predicting y using the F test, we may elect to conduct one or more t tests on the individual β parameters (see Section 4.5). However, the test (or tests) to be conducted should be decided a priori, i.e., prior to fitting the model. Also, we should limit the number of t tests conducted to avoid the potential problem of making too many Type I errors. Generally, the regression analyst will conduct t tests on only the "most important" β's. These are usually the β's associated with higher-order terms (x_1^2, $x_1 x_2$, etc.). We provide insight in identifying the most important β's in a linear model in Chapter 5.

Recommendation for Checking the Utility of a Multiple Regression Model

1. First, conduct a test of overall model adequacy using the F test, i.e., test

   $$H_0: \quad \beta_1 = \beta_2 = \cdots = \beta_k = 0$$

   If the model is deemed adequate (i.e., if you reject H_0), then proceed to step 2. Otherwise, you should hypothesize and fit another model. The new model may include more independent variables or higher-order terms (see Chapter 5).

2. Conduct t tests on those β parameters in which you are particularly interested (i.e., the "most important" β's). These usually involve only the β's associated with higher-order terms (x_1^2, $x_1 x_2$, etc.). However, it is a safe practice to limit the number of β's that are tested. Conducting a series of t tests leads to a high overall Type I error rate α.

> **Warning**
>
> Rejecting $H_0: \beta_1 = \beta_2 = \cdots = \beta_k = 0$ in a test of overall model adequacy does not necessarily imply that the model is best for predicting y. Another model may prove even more useful in terms of providing more reliable estimates and predictions.

EXERCISES

4.8 Refer to the *Journal of Urban Economics* study, Exercise 4.7. Recall that the model relating y to percentage of vacant land zoned for residential to nonresidential land use, x_1, and nonresidential property tax, x_2, resulted in an adjusted R^2 of .25 and an analysis of variance F statistic of 21.86, with p-value $< .01$.

a. Interpret the adjusted R^2 value.

b. Is the overall model statistically useful for predicting y?

4.9 Because the coefficient of determination R^2 always increases when a new independent variable is added to the model, it is tempting to include many variables in a model to force R^2 to be near 1. However, doing so reduces the degrees of freedom available for estimating σ^2, which adversely affects our ability to make reliable inferences. As an example, suppose you want to use 18 economic indicators to predict next year's gross national product (GNP). You fit the model

$$y = \beta_0 + \beta_1 x_1 + \beta_2 x_2 + \cdots + \beta_{17} x_{17} + \beta_{18} x_{18} + \varepsilon$$

where $y = $ GNP and $x_1, x_2, \ldots, x_{18}$ are indicators. Only 20 years of data $(n = 20)$ are used to fit the model, and you obtain $R^2 = .95$.

a. Test to see whether this impressive-looking R^2 is large enough for you to infer that this model is useful, i.e., that at least one term in the model is important for predicting GNP. Use $\alpha = .05$.

b. Calculate R_a^2 and interpret its value.

4.10 Stock market analysts are continually searching for reliable predictors of stock price. Consider the problem of modeling the price per share, y, of electric utility stocks. Two variables thought to influence stock price are return on average equity, x_1, and annual rate of dividend, x_2. The stock prices, returns on equity, and dividend rates on a randomly selected day for a sample of 12 nuclear and 16 nonnuclear electric utility stocks are shown in the table. The interaction model

$$E(y) = \beta_0 + \beta_1 x_1 + \beta_2 x_2 + \beta_3 x_1 x_2$$

was fit to the data on each type of stock (nuclear and nonnuclear). The SAS printouts are provided on page 189.

a. Write the least squares prediction equations for the two types of electric utility stock.

b. Is the model useful for predicting price of nuclear stocks? Nonnuclear stocks? Test using $\alpha = .05$.

c. Is there evidence of interaction between return on equity and dividend rate for the nuclear stock model? The nonnuclear stock model? Perform each test using $\alpha = .05$.

NUCLEAR STOCKS			NONNUCLEAR STOCKS		
y	x_1	x_2	y	x_1	x_2
21	15.1	2.36	25	15.2	2.60
31	15.0	3.00	20	13.9	2.14
26	11.2	3.00	15	15.8	1.52
11	12.1	1.96	34	12.8	3.12
24	16.3	3.00	20	6.9	2.48
8	11.9	1.40	33	14.6	3.08
18	14.9	1.80	28	15.4	2.92
23	11.8	2.56	30	17.3	2.76
13	13.4	2.06	23	13.7	2.36
14	16.2	1.94	24	12.7	2.36
35	17.1	2.96	25	15.3	2.56
13	13.3	2.20	26	15.2	2.80
			26	12.0	2.72
			20	15.3	1.92
			20	13.7	1.92
			13	13.3	1.60

Source: United Business Investment Report.

SAS printout for Exercise 4.10:
Nuclear stocks

TYPE=NUC

ANALYSIS OF VARIANCE

SOURCE	DF	SUM OF SQUARES	MEAN SQUARE	F VALUE	PROB>F
MODEL	3	640.93485	213.64495	13.217	0.0018
ERROR	8	129.31515	16.16439437		
C TOTAL	11	770.25000			

ROOT MSE	4.020497	R-SQUARE	0.8321	
DEP MEAN	19.75	ADJ R-SQ	0.7692	
C.V.	20.35695			

PARAMETER ESTIMATES

VARIABLE	DF	PARAMETER ESTIMATE	STANDARD ERROR	T FOR H0: PARAMETER=0	PROB > |T|
INTERCEP	1	-17.55634046	40.05327713	-0.438	0.6727
X1	1	0.51898803	2.93598713	0.177	0.8641
X2	1	10.88942280	15.57141350	0.699	0.5042
X1X2	1	0.13221521	1.12491646	0.118	0.9093

SAS printout for Exercise 4.10:
Nonnuclear stocks

TYPE=NO

ANALYSIS OF VARIANCE

SOURCE	DF	SUM OF SQUARES	MEAN SQUARE	F VALUE	PROB>F
MODEL	3	478.30855	159.43618	60.851	0.0001
ERROR	12	31.44145069	2.62012089		
C TOTAL	15	509.75000			

ROOT MSE	1.618679	R-SQUARE	0.9383	
DEP MEAN	23.875	ADJ R-SQ	0.9229	
C.V.	6.779806			

PARAMETER ESTIMATES

VARIABLE	DF	PARAMETER ESTIMATE	STANDARD ERROR	T FOR H0: PARAMETER=0	PROB > |T|
INTERCEP	1	-44.68177311	25.23972659	-1.770	0.1021
X1	1	2.87957851	1.74113100	1.654	0.1241
X2	1	25.06218058	10.02876655	2.499	0.0280
X1X2	1	-0.95900631	0.69103996	-1.388	0.1904

4.11 Marketers are keenly interested in the factors that motivate coupon usage by consumers. Three dominant motivational factors are thought to be (1) price reduction, (2) time and effort required to collect coupons, and (3) self-satisfaction. Using questionnaire data collected for a sample of $n = 290$ shoppers, a trio of marketing researchers examined the relationship between coupon usage and these factors (*The Journal of Consumer Marketing*, Spring 1988). The multiple regression model took the form

$$E(y) = \beta_0 + \beta_1 x_1 + \beta_2 x_2 + \beta_3 x_3$$

where

y = Coupon redemption rate

x_1 = Price-consciousness score

x_2 = Time-value score

x_3 = Satisfaction/pride score

The results are summarized as follows (*t* values for testing β's in parentheses):

$\hat{\beta}_1 = .09784\ (1.444)$ $R^2 = .11671$
$\hat{\beta}_2 = -.13134\ (-1.695)$ $F = 9.6893$
$\hat{\beta}_3 = .20019\ (2.571)$

a. Conduct an overall test of model accuracy. Use $\alpha = .10$.
b. In theory, coupon users are more price-conscious than nonusers. Test the theory using $\alpha = .10$.
c. Interpret the negative β estimate for time-value score (x_2).

4.12 In Exercise 3.56 we gave the results of M. Hergert's simple linear regression analysis relating 1980 return on assets y to market share x for a sample of 5,400 businesses (*Business Economics*, Oct. 1984). The main objective of the analysis was to investigate the conventional wisdom in business strategy that "a better market share yields a higher profit." The extremely small value of R^2 leads Hergert to suggest that "corporate strategy should be based on this premise only with great caution." In addition to the straight-line model, Hergert fit the quadratic model $E(y) = \beta_0 + \beta_1 x + \beta_2 x^2$. The results of the multiple regression are shown here:

$$\hat{y} = .093 + .441x - .409x^2 \qquad s_{\hat{\beta}_2} = .171 \qquad R^2 = .002$$

a. Graph the least squares prediction equation. (Let market share x range from 0% to 60%.)
b. Intepret the value of R^2.
c. Is there sufficient evidence to indicate that market share x is a useful predictor of return on assets y? Test using $\alpha = .05$.
d. As a result of the multiple regression analysis, Hergert concludes that "profits rise with size (of market share) up to some intermediate level and taper off thereafter." Do you agree with this statement? [*Hint:* Test for downward curvature using $\alpha = .05$.]

4.13 A study was conducted at Union Carbide to identify the optimal catalyst preparation conditions in the conversion of monoethanolamine (MEA) to ethylenediamine (EDA), a substance used commercially in soaps.* For each of 10 selected catalysts, the following experimental variables were measured:

---

*Hansen, J. L. and Best, D. C. "How to pick a winner." Paper presented at Joint Statistical Meetings, American Statistical Association and Biometric Society, Aug. 1986, Chicago, Illinois.

y = Rate of conversion of MEA to EDA

x_1 = Atom ratio of metal used in the experiment

x_2 = Reduction temperature

$x_3 = \begin{cases} 1 & \text{if high acidity support used} \\ 0 & \text{if low acidity support used} \end{cases}$

The data for the $n = 10$ experiments were used to fit the model $E(y) = \beta_0 + \beta_1 x_1 + \beta_2 x_2 + \beta_3 x_3$. The results are summarized here:

$$\hat{y} = 40.2 - .808x_1 - 6.38x_2 - 4.45x_3 \qquad R^2 = .899$$

$$s_{\hat{\beta}_1} = .231 \qquad s_{\hat{\beta}_2} = 1.93 \qquad s_{\hat{\beta}_3} = .99$$

a. Is there sufficient evidence to indicate that the model is useful for predicting rate of conversion y? Test using $\alpha = .01$.
b. Conduct a test to determine whether atom ratio x_1 is a useful predictor of rate of conversion y. Use $\alpha = .05$.
c. Construct a 95% confidence interval for β_2. Interpret the interval.

4.8

Using the Model for Estimation and Prediction

In Section 3.9 we discussed the use of the least squares line for estimating the mean value of y, $E(y)$, for some value of x, say, $x = x_p$. We also showed how to use the same fitted model to predict, when $x = x_p$, some value of y to be observed in the future. Recall that the least squares line yielded the same value for both the estimate of $E(y)$ and the prediction of some future value of y. That is, both are the result obtained by substituting x_p into the prediction equation $\hat{y} = \hat{\beta}_0 + \hat{\beta}_1 x$ and calculating $\hat{y}$. There the equivalence ends. The confidence interval for the mean $E(y)$ was narrower than the prediction interval for y, because of the additional uncertainty attributable to the random error ε when predicting some future value of y.

These same concepts carry over to the multiple regression model. For example, suppose we want to estimate the mean electrical usage for a given home size, say, $x_p = 1,500$ square feet. Assuming the quadratic model represents the true relationship between electrical usage and home size, we want to estimate

$$E(y) = \beta_0 + \beta_1 x_p + \beta_2 x_p^2$$
$$= \beta_0 + \beta_1(1,500) + \beta_2(1,500)^2$$

Substituting into the least squares prediction equation yields the estimate of $E(y)$:

$$\hat{y} = \hat{\beta}_0 + \hat{\beta}_1(1,500) + \hat{\beta}_2(1,500)^2$$
$$= -1,216.144 - 2.3989(1,500) - .00045(1,500)^2$$
$$= 1,369.7$$

To form a confidence interval for the mean, we need to know the standard deviation of the sampling distribution for the estimator $\hat{y}$. For multiple regression models, the form of this standard deviation is rather complex. However, some regression packages allow us to obtain the confidence intervals for mean values of y at any given setting of the independent variables. A portion of the SAS output for the electrical usage example is shown in Figure 4.15. The mean values and corresponding 95% confidence intervals are shown in the columns labeled **Predict Value**, **Lower95% Mean**, and **Upper95% Mean**. In the last row (shaded) in Figure 4.15, we observe that $\hat{y} = 1,369.7$ when $x_p = 1,500$, which agrees with our earlier calculation. The corresponding 95% confidence interval for the true mean of y is shown to be 1,325.0 to 1,414.3 (see Figure 4.16).

FIGURE 4.15

SAS printout for estimated mean values and corresponding intervals

Obs	X	Dep Var Y	Predict Value	Std Err Predict	Lower95% Mean	Upper95% Mean	Residual
1	1290	1182.0	1129.6	30.072	1058.5	1200.7	52.4359
2	1350	1172.0	1202.2	25.851	1141.1	1263.3	-30.2136
3	1470	1264.0	1337.8	19.832	1290.9	1384.7	-73.7916
4	1600	1493.0	1470.0	17.392	1428.9	1511.2	22.9586
5	1710	1571.0	1570.1	17.976	1527.6	1612.6	0.9359
6	1840	1711.0	1674.2	19.919	1627.1	1721.3	36.7685
7	1980	1804.0	1769.4	21.887	1717.6	1821.2	34.5998
8	2230	1840.0	1895.5	23.348	1840.3	1950.7	-55.4654
9	2400	1956.0	1949.1	23.611	1893.2	2004.9	6.9431
10	2930	1954.0	1949.2	44.734	1843.4	2055.0	4.8287
11	1500	.	1369.7	18.892	1325.0	1414.3	.

FIGURE 4.16

Confidence interval for mean electrical usage when $x_p = 1,500$

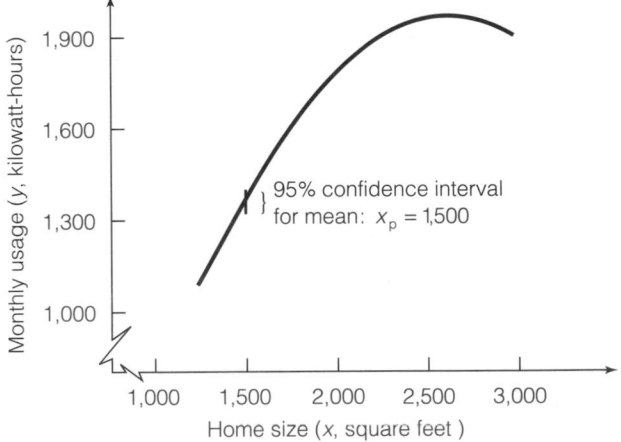

If we were interested in predicting electrical usage for a particular 1,500-square-foot home, $\hat{y} = 1,369.7$ would be used as the predicted value. However, the prediction interval for a particular value of y will be wider than the confidence interval for the mean value. This is reflected by the printout shown in Figure 4.17, which gives the predicted value of y and corresponding 95% prediction intervals. Note that the predicted value for $x_p = 1,500$ is 1,369.7 (shaded in the last row), and the prediction interval extends from 1,250.3 to 1,489.0. This interval is shown graphically in Figure 4.18.

FIGURE 4.17

SAS printout for predicted values and
corresponding prediction intervals

Obs	X	Dep Var Y	Predict Value	Std Err Predict	Lower95% Predict	Upper95% Predict	Residual
1	1290	1182.0	1129.6	30.072	998.0	1261.1	52.4359
2	1350	1172.0	1202.2	25.851	1075.8	1328.6	-30.2136
3	1470	1264.0	1337.8	19.832	1217.6	1458.0	-73.7916
4	1600	1493.0	1470.0	17.392	1352.0	1588.1	22.9586
5	1710	1571.0	1570.1	17.976	1451.5	1688.6	0.9359
6	1840	1711.0	1674.2	19.919	1554.0	1794.5	36.7685
7	1980	1804.0	1769.4	21.887	1647.2	1891.6	34.5998
8	2230	1840.0	1895.5	23.348	1771.8	2019.1	-55.4654
9	2400	1956.0	1949.1	23.611	1825.1	2073.0	6.9431
10	2930	1954.0	1949.2	44.734	1796.1	2102.3	4.8287
11	1500	.	1369.7	18.892	1250.3	1489.0	.

FIGURE 4.18

Prediction interval for electrical usage
when $x_p = 1,500$

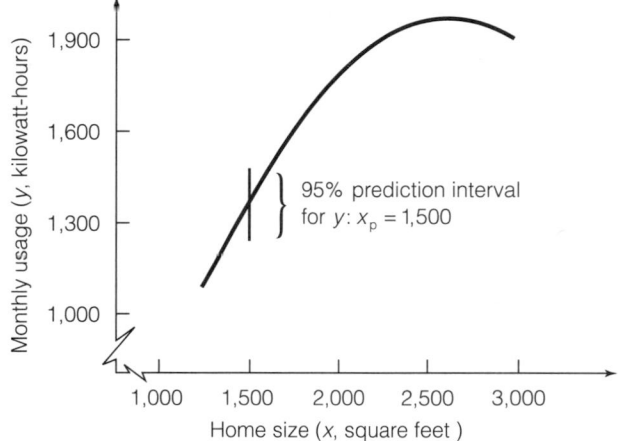

Unfortunately, not all statistical software packages have the capability to produce confidence intervals for means and prediction intervals for particular y
values. This is a rather serious oversight, since the estimation of mean values
and the prediction of particular values represent the culmination of our model building efforts: using the model to make inferences about the dependent variable y.

EXERCISES

4.14   Refer to the sale price model in Exercise 4.3. The SAS printout with 95% confidence intervals for
$E(y)$ is shown on page 194.
  a. Locate the lower and upper limits for a 95% confidence interval for $E(y)$ when $x_1 = 17$ and x_2
     $= 3$ (observation #4).
  b. Interpret the interval in part a.
  c. Would you recommend using the model to estimate the mean sale price for homes with 3,000
     square feet? Explain.

SAS printout for Exercise 4.14

ANALYSIS OF VARIANCE

SOURCE	DF	SUM OF SQUARES	MEAN SQUARE	F VALUE	PROB>F
MODEL	2	819.32795	409.66398	350.866	0.0001
ERROR	7	8.17304573	1.16757796		
C TOTAL	9	827.50100			

ROOT MSE	1.080545	R-SQUARE	0.9901	
DEP MEAN	51.73	ADJ R-SQ	0.9873	
C.V.	2.088817			

PARAMETER ESTIMATES

VARIABLE	DF	PARAMETER ESTIMATE	STANDARD ERROR	T FOR H0: PARAMETER=0	PROB > |T|
INTERCEP	1	9.78227061	1.63048067	6.000	0.0005
X1	1	1.87093528	0.07617357	24.561	0.0001
X2	1	1.27814078	0.14440032	8.851	0.0001

OBS	ID	ACTUAL	PREDICT VALUE	STD ERR PREDICT	LOWER95% MEAN	UPPER95% MEAN	RESIDUAL
1	23	60.0000	59.2045	0.4714	58.0899	60.3191	0.7955
2	11	32.7000	32.9188	0.8200	30.9799	34.8578	-0.2188
3	20	57.7000	58.7042	0.6375	57.1969	60.2116	-1.0042
4	17	45.5000	45.4226	0.4919	44.2595	46.5857	0.0774
5	15	47.0000	48.0714	0.6019	46.6481	49.4948	-1.0714
6	21	55.3000	54.1845	0.4276	53.1735	55.1955	1.1155
7	24	64.5000	63.6317	0.5705	62.2826	64.9808	0.8683
8	13	42.6000	41.7733	0.5710	40.4231	43.1235	0.8267
9	19	54.5000	54.2770	0.4206	53.2824	55.2717	0.2230
10	25	57.5000	59.1119	0.7657	57.3013	60.9226	-1.6119

SUM OF RESIDUALS	6.75016E-14
SUM OF SQUARED RESIDUALS	8.173046
PREDICTED RESID SS (PRESS)	21.09523

4.15 Refer to the stock price model in Exercise 4.10. The SAS printout with 95% prediction intervals for price per share y of nuclear stocks is given at the top of page 195.

   a. Locate the lower and upper limits for a 95% prediction interval for y when $x_1 = 13.3$ and $x_2 = 2.20$ (observation #12).

   b. Interpret the interval in part a.

   c. Would you recommend using the model to predict price per share of a nuclear stock with a dividend rate of 1.10? Explain.

4.9
Other Linear Models

In the preceding sections, we have demonstrated the methods of multiple regression analysis by fitting several different models, including a quadratic model and an interaction model. In this section we formally introduce other types of general linear models that are useful for relating a response variable y to a set of data.*

---

*A more complete discussion of general linear models and their role in model building is provided in Chapter 5.

SAS printout for Exercise 4.15

ANALYSIS OF VARIANCE

SOURCE	DF	SUM OF SQUARES	MEAN SQUARE	F VALUE	PROB>F
MODEL	3	640.93485	213.64495	13.217	0.0018
ERROR	8	129.31515	16.16439437		
C TOTAL	11	770.25000			

ROOT MSE	4.020497	R-SQUARE	0.8321	
DEP MEAN	19.75	ADJ R-SQ	0.7692	
C.V.	20.35695			

PARAMETER ESTIMATES

VARIABLE	DF	PARAMETER ESTIMATE	STANDARD ERROR	T FOR H0: PARAMETER=0	PROB > |T|
INTERCEP	1	-17.55634046	40.05327713	-0.438	0.6727
X1	1	0.51898803	2.93598713	0.177	0.8641
X2	1	10.88942280	15.57141358	0.699	0.5042
X1X2	1	0.13221521	1.12491646	0.118	0.9093

OBS	ID	ACTUAL	PREDICT VALUE	STD ERR PREDICT	LOWER95% PREDICT	UPPER95% PREDICT	RESIDUAL
1	15.1	21.0000	20.6910	1.4471	10.8374	30.5447	0.3090
2	15	31.0000	28.8464	1.8979	18.5940	39.0989	2.1536
3	11.2	26.0000	25.3670	3.2339	13.4687	37.2654	0.6330
4	12.1	11.0000	13.2023	1.9170	2.9301	23.4745	-2.2023
5	16.3	24.0000	30.0368	2.3630	19.2826	40.7909	-6.0368
6	11.9	8.0000	6.0675	3.2551	-5.8616	17.9966	1.9325
7	14.9	18.0000	13.3236	2.2922	2.6512	23.9959	4.6764
8	11.8	23.0000	20.4386	1.9934	10.0902	30.7870	2.5614
9	13.4	13.0000	15.4800	1.3474	5.7018	25.2581	-2.4800
10	16.2	14.0000	16.1320	2.9501	4.6325	27.6315	-2.1320
11	17.1	35.0000	30.2433	2.7022	19.0724	41.4141	4.7567
12	13.3	13.0000	17.1715	1.2640	7.4528	26.8903	-4.1715

SUM OF RESIDUALS   -8.19345E-14
SUM OF SQUARED RESIDUALS   129.3152
PREDICTED RESID SS (PRESS)   321.047

Models with Quantitative x's

We begin with a discussion of models using **quantitative** (numerical) independent variables. Suppose that the mean value $E(y)$ of a response y is related to two quantitative variables, x_1 and x_2, by the model

$$E(y) = 1 + 2x_1 - x_2$$

Note that when $x_2 = 0$, the relationship between $E(y)$ and x_1 is given by

$$E(y) = 1 + 2x_1 - (0) = 1 + 2x_1$$

A graph of this relationship (a straight line) is shown in Figure 4.19 on page 196. Similar graphs of the relationship between $E(y)$ and x_1 for $x_2 = 1$,

$$E(y) = 1 + 2x_1 - (1) = 2x_1$$

and for $x_2 = 2$,

$$E(y) = 1 + 2x_1 - (2) = -1 + 2x_1$$

are also shown in Figure 4.19.

FIGURE 4.19
Graphs of $E(y) = 1 + 2x_1 - x_2$ for $x_2 = 0, 1, 2$

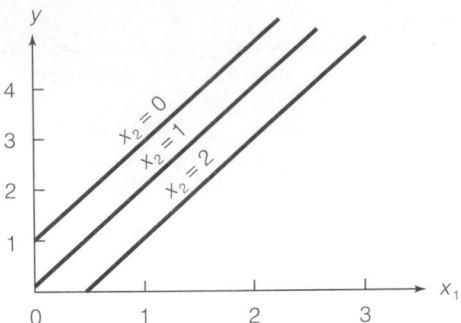

The model $E(y) = 1 + 2x_1 - x_2$ is an example of a **first-order linear model** in two quantitative independent variables, x_1 and x_2. A first-order linear model in five quantitative independent variables is shown in the box.

A First-Order Linear Model Relating $E(y)$ to $x_1, x_2, \ldots, x_5$

$$E(y) = \beta_0 + \beta_1 x_1 + \beta_2 x_2 + \cdots + \beta_5 x_5$$

Figure 4.19 exhibits a characteristic of all first-order models: If you graph $E(y)$ versus any one variable—say, x_1—for fixed values of the other variables, the response curve will always be a *straight line*. If you repeat the process for other values of the fixed independent variables, you will obtain a set of *parallel* straight lines. This indicates that the effect on $E(y)$ of a change in x_1 is independent of the other variables in the model. When this situation occurs (as it always does for a first-order model), we say that the independent variables in the model **do not interact**.

Now, suppose that the mean value $E(y)$ of a response y is related to two quantitative variables, x_1 and x_2, by the model

$$E(y) = 1 + 2x_1 - x_2 + x_1 x_2$$

This model contains the second-order cross product term, $x_1 x_2$, in addition to all the terms of the first-order model. Figure 4.20 shows the graphs of the relationship between $E(y)$ and x_1 for $x_2 = 0, 1,$ and 2. The straight-line equations relating $E(y)$ to x_1 are:

For $x_2 = 0$: $E(y) = 1 + 2x_1 - (0) + x_1(0) = 1 + 2x_1$

For $x_2 = 1$: $E(y) = 1 + 2x_1 - (1) + x_1(1) = 3x_1$

For $x_2 = 2$: $E(y) = 1 + 2x_1 - (2) + x_1(2) = -1 + 4x_1$

The effect of adding a term involving the cross product $x_1 x_2$ can be seen in Figure 4.20. In contrast to Figure 4.19, the lines relating $E(y)$ to x_1 are no longer parallel. The effect on $E(y)$ of a change in x_1 is now dependent on the value of

FIGURE 4.20

Graphs of $E(y) = 1 + 2x_1 - x_2 + x_1x_2$
for $x_2 = 0, 1, 2$

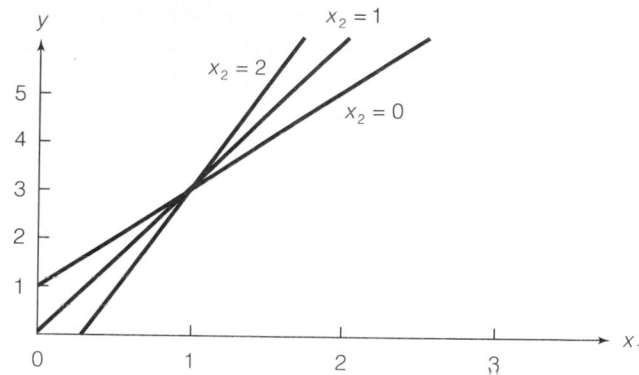

x_2. When this situation occurs, we say that x_1 and x_2 **interact**. An **interaction model** is a model that contains first-order terms in the independent variables as well as two-way cross product terms such as x_1x_2, x_1x_3, ... An interaction model with three quantitative independent variables is shown in the box.

An Interaction Model Relating $E(y)$ to Three Quantitative
Independent Variables

$$E(y) = \beta_0 + \beta_1 x_1 + \beta_2 x_2 + \beta_3 x_3 + \beta_4 x_1 x_2 + \beta_5 x_1 x_3 + \beta_6 x_2 x_3$$

Finally, consider relating the mean value $E(y)$ of a response y to two quantitative independent variables, x_1 and x_2, by the model

$$E(y) = 1 + 2x_1 - x_2 + x_1x_2 + x_1^2 + 3x_2^2$$

This model contains all of the terms contained in the interaction model plus the second-order terms x_1^2 and x_2^2. Figure 4.21 (page 198) shows a computer-generated graph of the relationship between $E(y)$ and x_1 for $x_2 = 0, 1$, and 2.

The response curves in Figure 4.21 rise (or fall) in the same manner as the lines shown in Figure 4.20. However, the graphs are curvilinear and the spacing between the curves has changed. These changes were produced by adding the second-order terms (those involving x_1^2 and x_2^2) to the model.

The model $E(y) = 1 + 2x_1 - x_2 - x_1x_2 + x_1^2 + 3x_2^2$ is an example of a **second-order model** in two quantitative independent variables. A second-order model contains all of the terms in a first-order model and, in addition, the second-order terms involving cross products (interaction terms) and squares of the independent variables. A second-order model with three quantitative independent variables is shown in the box. (Note that an interaction model is a special case of a second-order model, where the β coefficients of x_1^2, x_2^2, ..., are all equal to 0.)

FIGURE 4.21　Computer graph of $E(y) = 1 + 2x_1 - x_2 + x_1x_2 + x_1^2 + 3x_2^2$ for $x_2 = 0, 1, 2$

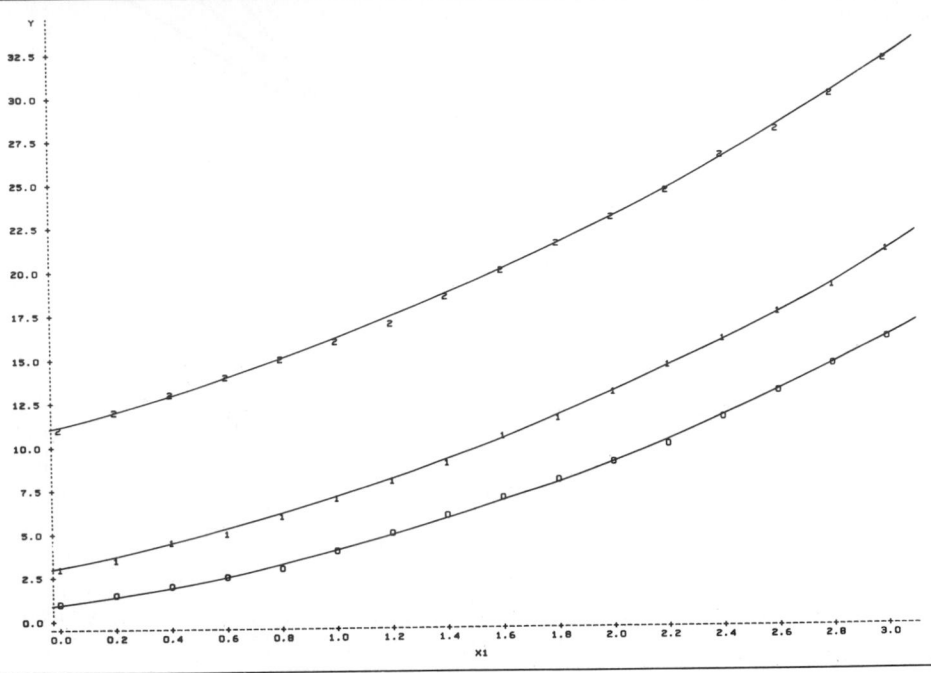

A Second-Order Model with Three Quantitative Independent Variables

$$E(y) = \beta_0 + \beta_1x_1 + \beta_2x_2 + \beta_3x_3 + \beta_4x_1x_2 + \beta_5x_1x_3$$
$$+ \beta_6x_2x_3 + \beta_7x_1^2 + \beta_8x_2^2 + \beta_9x_3^2$$

How can you choose an appropriate linear model to fit a set of data? Since most relationships in the real world are curvilinear (at least to some extent), a good first choice would be a second-order linear model. If you are fairly certain that the relationships between $E(y)$ and the individual independent variables are approximately first-order and that the independent variables do not interact, you could select a first-order model for the data. If you have prior information that suggests there is moderate or very little curvature over the region in which the independent variables are measured, you could use the interaction model described earlier. However, keep in mind that for all multiple regression models, the number of data points must exceed the number of parameters in the model. Thus, you may be forced to use a first-order model rather than a second-order model simply because you do not have sufficient data to estimate all of the parameters in the second-order model.

A practical example of choosing and fitting a linear model with two quantitative independent variables follows.

EXAMPLE 4.4

Although a regional express delivery service bases the charge for shipping a package on the package weight and distance shipped, its profit per package depends on the package size (volume of space that it occupies) and the size and nature of the load on the delivery truck. The company recently conducted a study to investigate the relationship between the cost, y, of shipment (in dollars) and the variables that control the shipping charge—package weight, x_1 (in pounds), and distance shipped, x_2 (in miles). Twenty packages were randomly selected from among the large number received for shipment and a detailed analysis of the cost of shipment was made for each package, with the results shown in Table 4.3.

TABLE 4.3 Cost of Shipment Data for Example 4.4

PACKAGE	x_1	x_2	y	PACKAGE	x_1	x_2	y
1	5.9	47	2.60	11	5.1	240	11.00
2	3.2	145	3.90	12	2.4	209	5.00
3	4.4	202	8.00	13	.3	160	2.00
4	6.6	160	9.20	14	6.2	115	6.00
5	.75	280	4.40	15	2.7	45	1.10
6	.7	80	1.50	16	3.5	250	8.00
7	6.5	240	14.50	17	4.1	95	3.30
8	4.5	53	1.90	18	8.1	160	12.10
9	.60	100	1.00	19	7.0	260	15.50
10	7.5	190	14.00	20	1.1	90	1.70

a. Give an appropriate linear model for the data.
b. Fit the model to the data and give the prediction equation.
c. Find the value of SSE and specify its number of degrees of freedom.
d. Find the value of R_a^2 and interpret it.
e. Is the model useful for the prediction of shipping cost y? Find the value of the F statistic on the printout and give the observed significance level (p-value) for the test.
f. Find a 95% prediction interval for the cost of shipping a 5-pound package a distance of 100 miles.

Solution

a. Since we have no reason to expect that the relationship between y and x_1 and x_2 would be first-order, we will allow for curvature in the response surface and fit the second-order model

$$y = \beta_0 + \beta_1 x_1 + \beta_2 x_2 + \beta_3 x_1 x_2 + \beta_4 x_1^2 + \beta_5 x_2^2 + \varepsilon$$

The mean value of the random error term ε is assumed to equal 0. Therefore, the mean value of y is

$$E(y) = \beta_0 + \beta_1 x_1 + \beta_2 x_2 + \beta_3 x_1 x_2 + \beta_4 x_1^2 + \beta_5 x_2^2$$

b. The SAS printout for fitting the model to the $n - 20$ data points is shown in Figure 4.22 on page 200. You can see from the printout that the parameter estimates (shaded in Figure 4.22) are:

$$\hat{\beta}_0 = .827016 \quad \hat{\beta}_1 = -.609137 \quad \hat{\beta}_2 = .004021$$
$$\hat{\beta}_3 = .007327 \quad \hat{\beta}_4 = .089751 \quad \hat{\beta}_5 = .00001507$$

Therefore, the prediction equation that relates the predicted shipping cost, $\hat{y}$, to weight of package, x_1, and distance shipped, x_2, is

$$\hat{y} = .827016 - .609137x_1 + .004021x_2 + .007327x_1x_2 + .089751x_1^2$$
$$+ .00001507x_2^2$$

FIGURE 4.22

SAS printout for the multiple regression analysis of Example 4.4

Analysis of Variance

Source	DF	Sum of Squares	Mean Square	F Value	Prob>F
Model	5	449.34076	89.86815	458.388	0.0001
Error	14	2.74474	0.19605		
C Total	19	452.08550			

Root MSE	0.44278	R-square	0.9939	
Dep Mean	6.33500	Adj R-sq	0.9918	
C.V.	6.98940			

Parameter Estimates

Variable	DF	Parameter Estimate	Standard Error	T for H0: Parameter=0	Prob > |T|
INTERCEP	1	0.827016	0.70228935	1.178	0.2586
X1	1	-0.609137	0.17990408	-3.386	0.0044
X2	1	0.004021	0.00799842	0.503	0.6230
X1X2	1	0.007327	0.00063743	11.495	0.0001
X1SQ	1	0.089751	0.02020542	4.442	0.0006
X2SQ	1	0.000015070	0.00002243	0.672	0.5127

Obs	X1	X2	Dep Var Y	Predict Value	Std Err Predict	Lower95% Predict	Upper95% Predict	Residual
1	5.9	47	2.6000	2.6114	0.299	1.4655	3.7573	-0.0114
2	3.2	145	3.9000	4.0964	0.211	3.0443	5.1486	-0.1964
3	4.4	202	8.0000	7.8238	0.199	6.7823	8.8654	0.1762
4	6.6	160	9.2000	9.4828	0.182	8.4562	10.5093	-0.2828
5	0.75	280	4.4000	4.2666	0.399	2.9880	5.5452	0.1334
6	0.7	80	1.5000	1.2730	0.245	0.1874	2.3587	0.2270
7	6.5	240	14.5000	13.9229	0.213	12.8693	14.9764	0.5771
8	4.5	53	1.9000	1.9063	0.228	0.8378	2.9748	-0.00629
9	0.6	100	1.0000	1.4862	0.226	0.4202	2.5523	-0.4862
10	7.5	190	14.0000	13.0560	0.216	11.9998	14.1122	0.9440
11	5.1	240	11.0000	10.8562	0.196	9.8175	11.8949	0.1438
12	2.4	209	5.0000	5.0559	0.200	4.0140	6.0979	-0.0559
13	0.3	160	2.0000	2.0332	0.260	0.9316	3.1347	-0.0332
14	6.2	115	6.0000	6.3863	0.195	5.3482	7.4243	-0.3863
15	2.7	45	1.1000	0.9383	0.257	-0.1600	2.0366	0.1617
16	3.5	250	8.0000	8.1527	0.212	7.1001	9.2054	-0.1527
17	4.1	95	3.3000	3.2101	0.180	2.1847	4.2356	0.0899
18	8.1	160	12.1000	12.3066	0.301	11.1577	13.4555	-0.2066
19	7	260	15.5000	16.3603	0.311	15.1999	17.5207	-0.8603
20	1.1	90	1.7000	1.4749	0.199	0.4335	2.5163	0.2251
21	5	100	.	4.2414	0.184	3.2133	5.2695	.

Sum of Residuals	-1.04361E-14	
Sum of Squared Residuals	2.7447	
Predicted Resid SS (Press)	6.8568	

c. The **Sum of Squares** for **Error**, shaded on the printout, is

$$SSE = 2.74474$$

based on 14 degrees of freedom (**DF** for **Error** in the printout). Recall that SSE has degrees of freedom equal to $n -$ (Number of estimated β parameters).

d. The value of R_a^2 shown in Figure 4.22 is **Adj R-sq** = .9918. This means that after adjusting for sample size and the number of model parameters, about 99% of the total variation, $SS(\text{Total}) = SS_{yy} = \Sigma(y - \bar{y})^2$, is explained by the model; the remainder is explained by random error.

e. The test statistic for testing whether the model is useful for predicting shipping cost is

$$F = \frac{\text{Mean square for model}}{\text{Mean square for error}} = \frac{SS(\text{Model})/k}{SSE/[n - (k + 1)]}$$

where $n = 20$ is the number of data points and $k = 5$ is the number of parameters (excluding β_0) contained in the model.

This value of F has been computed for us as $F = 458.388$ and is shaded on the printout (Figure 4.22) in the row corresponding to **MODEL**. The observed significance level (p-value) for the test is shown to the right of the F value on the printout as **PROB > F** = .0001. This means that if the model contributed no information for the prediction of y, the probability of observing a value of the F statistic as large as 458.39 would be only .0001. Thus, we would reject the null hypothesis that the model contributes no information for the prediction of y for all values of α larger than .0001.

f. The predicted value of y for $x_1 = 5.0$ pounds and $x_2 = 100$ miles is

$$\hat{y} = .827016 - .609137(5.0) + .004021(100) + .007327(5.0)(100)$$
$$+ .089751(5.0)^2 + .00001507(100)^2$$

This quantity has been computed and is shown (shaded) on the printout as $\hat{y} = 4.2414$. The corresponding 95% prediction interval (shaded) is given as 3.2133 to 5.2695. Therefore, if we were to select a 5-pound package and ship it 100 miles, we would expect the actual cost to fall between $3.21 and $5.27.

Models with Qualitative *x*'s

Linear models can also be written to include **qualitative** (or **categorical**) independent variables. Qualitative variables, unlike quantitative variables, cannot be measured on a numerical scale. Therefore, we need to code the values of the qualitative variable (called **levels**) as numbers before we can fit the model. These coded qualitative variables are called **dummy variables** since the numbers assigned to the various levels are arbitrarily selected.

For example, consider a salary discrimination case where there exists a claim of sex discrimination—specifically, the claim that male executives at a large company receive higher average salaries than female executives with the same credentials and qualifications.

To test this claim, we might propose a multiple regression model for executive salaries using the sex of an executive as one of the independent variables. The dummy variable used to describe sex may be coded as follows:

$$x_3 = \begin{cases} 1 & \text{if male} \\ 0 & \text{if female} \end{cases}$$

The advantage of using a 0–1 coding scheme is that the β coefficients associated with the dummy variables are easily interpreted. To illustrate, consider the following model for executive salary y:

$$E(y) = \beta_0 + \beta_1 x$$

where

$$x = \begin{cases} 1 & \text{if male} \\ 0 & \text{if female} \end{cases}$$

This model allows us to compare the mean executive salary $E(y)$ for males with the corresponding mean for females:

Males $(x = 1)$: $E(y) = \beta_0 + \beta_1(1) = \beta_0 + \beta_1$

Females $(x = 0)$: $E(y) = \beta_0 + \beta_1(0) = \beta_0$

First note that β_0 represents the mean salary for females (say, μ_F). When using a 0–1 coding convention, β_0 will always represent the mean response associated with the level of the qualitative variable assigned the value 0 (called the **base level**). The difference between the mean salary for males and the mean salary for females, $\mu_M - \mu_F$, is represented by β_1—that is,

$$\mu_M - \mu_F = (\beta_0 + \beta_1) - (\beta_0) = \beta_1$$

Therefore, with the 0–1 coding convention, β_1 will always represent the difference between the mean response for the level assigned the value 1 and the mean for the base level. Thus, for the executive salary model we have

$$\beta_0 = \mu_F$$
$$\beta_1 = \mu_M - \mu_F$$

If β_1 exceeds 0, then $\mu_M > \mu_F$ and evidence of sex discrimination at the company exists.

The model relating a mean response $E(y)$ to a qualitative independent variable at two levels is shown in the first box on page 203.

For models that involve qualitative independent variables at more than two levels, additional dummy variables must be created. In general, the number of dummy variables used to describe a qualitative variable will be one less than the number of levels of the qualitative variable. The second box on page 203 presents a model that includes a qualitative independent variable at three levels.

A Model Relating $E(y)$ to a Qualitative Independent Variable with Two Levels

$$E(y) = \beta_0 + \beta_1 x$$

where

$$x = \begin{cases} 1 & \text{if level A} \\ 0 & \text{if level B} \end{cases}$$

Interpretation of β's:

$\beta_0 = \mu_B$ (Mean for base level)

$\beta_1 = \mu_A - \mu_B$

A Model Relating $E(y)$ to a Qualitative Independent Variable with Three Levels

$$E(y) = \beta_0 + \beta_1 x_1 + \beta_2 x_2$$

where

$$x_1 = \begin{cases} 1 & \text{if level A} \\ 0 & \text{if not} \end{cases} \qquad x_2 = \begin{cases} 1 & \text{if level B} \\ 0 & \text{if not} \end{cases} \qquad \text{Base level} = \text{Level C}$$

Interpretation of β's:

$\beta_0 = \mu_C$ (Mean for base level)

$\beta_1 = \mu_A - \mu_C$

$\beta_2 = \mu_B - \mu_C$

EXAMPLE 4.5

Refer to the problem of modeling the shipment cost, y, of a regional express delivery service, described in Example 4.4. Suppose we want to model $E(y)$ as a function of cargo type, where cargo type has three levels—fragile, semifragile, and durable. Costs for 15 packages of approximately the same weight and same distance shipped, but of different cargo types, are listed in Table 4.4 (page 204).

a. Write a linear model relating $E(y)$ to cargo type.

b. Interpret the β coefficients in the model.

c. A Minitab printout for the model, part **a**, is shown in Figure 4.23. Conduct the F test for overall model utility using $\alpha - .05$. Explain the practical significance of the result.

TABLE 4.4   Data for Example 4.4

PACKAGE	COST, y	CARGO TYPE	x_1	x_2
1	$17.20	Fragile	1	0
2	11.10	Fragile	1	0
3	12.00	Fragile	1	0
4	10.90	Fragile	1	0
5	13.80	Fragile	1	0
6	6.50	Semifragile	0	1
7	10.00	Semifragile	0	1
8	11.50	Semifragile	0	1
9	7.00	Semifragile	0	1
10	8.50	Semifragile	0	1
11	2.10	Durable	0	0
12	1.30	Durable	0	0
13	3.40	Durable	0	0
14	7.50	Durable	0	0
15	2.00	Durable	0	0

Solution

a. Since the qualitative variable of interest, cargo type, has three levels, we need to create $(3 - 1) = 2$ dummy variables. First, select (arbitrarily) one of the levels to be the base level—say, durable cargo. Then each of the remaining levels is assigned the value 1 in one of the two dummy variables as follows:

$$x_1 = \begin{cases} 1 & \text{if fragile} \\ 0 & \text{if not} \end{cases} \qquad x_2 = \begin{cases} 1 & \text{if semifragile} \\ 0 & \text{if not} \end{cases}$$

(Note that for the base level, durable cargo, $x_1 = x_2 = 0$.) The values of x_1 and x_2 for each package are given in Table 4.4. Then, the appropriate model is

$$E(y) = \beta_0 + \beta_1 x_1 + \beta_2 x_2$$

FIGURE 4.23

Minitab printout for dummy variable regression, Example 4.5

```
The regression equation is
Y = 3.26 + 9.74 X1 + 5.44 X2

Predictor      Coef      Stdev    t-ratio        P
Constant      3.260      1.075       3.03    0.010
X1            9.740      1.521       6.41    0.000
X2            5.440      1.521       3.58    0.004

s = 2.404      R-sq = 77.4%      R-sq(adj) = 73.7%

Analysis of Variance

SOURCE        DF         SS         MS        F        P
Regression     2     238.25     119.13    20.61    0.000
Error         12      69.37       5.78
Total         14     307.62
```

b. To interpret the β's, first write the mean shipment cost $E(y)$ for each of the three cargo types as a function of the β's:

Fragile ($x_1 = 1$, $x_2 = 0$):

$$E(y) = \beta_0 + \beta_1(1) + \beta_2(0) = \beta_0 + \beta_1 = \mu_F$$

Semifragile ($x_1 = 0$, $x_2 = 1$):

$$E(y) = \beta_0 + \beta_1(0) + \beta_2(1) = \beta_0 + \beta_2 = \mu_S$$

Durable ($x_1 = 0$, $x_2 = 0$):

$$E(y) = \beta_0 + \beta_1(0) + \beta_2(0) = \beta_0 = \mu_D$$

Then we have

$$\beta_0 = \mu_D \quad \text{(Mean of the base level)}$$
$$\beta_1 = \mu_F - \mu_D$$
$$\beta_2 = \mu_S - \mu_D$$

Note that the β's associated with the non-base levels of cargo type (fragile and semifragile) represent differences between a pair of means. As always, β_0 represents a single mean—the mean response for the base level (durable).

c. The F test for overall model utility tests the null hypothesis

$$H_0 = \beta_1 = \beta_2 = 0$$

Note that $\beta_1 = 0$ implies that $\mu_F = \mu_D$ and $\beta_2 = 0$ implies that $\mu_S = \mu_D$. Therefore, $\beta_1 = \beta_2 = 0$ implies that $\mu_F = \mu_S = \mu_D$. Thus, a test for model utility is equivalent to a test for equality of means, i.e.,

$$H_0: \quad \mu_F = \mu_S = \mu_D$$

From the Minitab printout, Figure 4.23, $F = 20.61$. Since the p-value of the test (.000) is less than $\alpha = .05$, the null hypothesis is rejected. Thus, there is evidence of a difference between any two of the three mean shipment costs, i.e., cargo type is a useful predictor of shipment cost y.

Multiplicative Models

In all the models presented so far, the random error component has been assumed to be *additive*. An additive error is one for which the response is equal to the mean $E(y)$ plus random error,

$$y = E(y) + \varepsilon$$

Another useful type of model for business and economic data is the **multiplicative model**. In this model, the response is written as a *product* of its mean and the random error component, i.e.,

$$y = [E(y)] \cdot \varepsilon$$

Researchers have found multiplicative models to be useful when the change in the response y, for every 1-unit change in an independent variable x, is better represented by a percentage increase (or decrease) rather than a constant amount increase (or decrease).* For example, economists often want to predict a percentage change in the price of a commodity or a percentage increase in the salary of a worker. Consequently, a multiplicative model is employed rather than an additive model.

A multiplicative model in two independent variables can be specified as:

$$y = (e^{\beta_0})(e^{\beta_1 x_1})(e^{\beta_2 x_2})(e^{\varepsilon})$$

where β_0, β_1, and β_2 are population parameters that must be estimated from the sample data and e^x is a notation for the antilogarithm of x. Note, however, that the multiplicative model is not a general linear model as defined in Section 4.1. To use the method of least squares to fit the model to the data, we must transform the model into the form of a linear model. Taking the natural logarithm of both sides of the equation, we obtain

$$\log(y) = \beta_0 + \beta_1 x_1 + \beta_2 x_2 + \varepsilon$$

which is now in the form of a general linear (additive) model.

When the dependent variable is $\log(y)$, rather than y, the β parameters and other key regression quantities have slightly different interpretations, as the next example illustrates.

EXAMPLE 4.6

Let us return to the executive compensation example introduced in Case Study 4.1. Recall that the management consultant firm of Towers, Perrin, Forster & Crosby (TPF&C) uses a multiple regression model to project executive salaries. Suppose the list of independent variables given in Table 4.5 is to be used to build a model for the salary, y, of a corporate executive. TPF&C have found that executive compensation models that use the logarithm of salary as the dependent variable provide better predictive models than those using the salary as the dependent variable. This is probably because salaries tend to be incremented in *percentages* rather than dollar values. Thus, the multiplicative model we propose (in its linear form) is:

$$\log(y) = \beta_0 + \beta_1 x_1 + \beta_2 x_2 + \beta_3 x_3 + \beta_4 x_4 + \beta_5 x_5 + \beta_6 x_1^2 + \beta_7 x_3 x_4 + \varepsilon$$

Note that we have included a second-order term, x_1^2, to account for a possible curvilinear relationship between $\log(\text{salary})$ and years of experience, x_1. Also, the interaction term $x_3 x_4$ is included to account for the fact that the relationship between the number of employees supervised, x_4, and corporate salary may depend on sex, x_3. For example, as the number of supervised employees increases, a male's salary (with all other factors being equal) might rise more rapidly than a female's. (If this is found to be true, the firm will take steps to remove the apparent discrimination against female executives.)

---

*Multiplicative models are also found to be useful when the standard regression assumption of equal variances is violated. We discuss this application of multiplicative models in Chapter 7.

TABLE 4.5 List of Independent Variables for Executive Compensation Example

INDEPENDENT VARIABLE	DESCRIPTION
x_1	Years of experience
x_2	Years of education
x_3	1 if male; 0 if female
x_4	Number of employees supervised
x_5	Corporate assets (millions of dollars)
x_6	x_1^2
x_7	$x_3 x_4$

TABLE 4.6 Values of Independent Variables for a Particular Executive

$x_1 = 12$ years of experience
$x_2 = 16$ years of education
$x_3 = 0$ (female)
$x_4 = 400$ employees supervised
$x_5 = \$160.1$ million (the firm's asset value)
$x_1^2 = 144$
$x_3 x_4 = 0$

A sample of 100 executives is selected, and the variables y and $x_1, x_2, \ldots, x_5$ are recorded. The sample is then used as input for a SAS regression routine; the output is shown in Figure 4.24.

a. Find the least squares prediction equation and interpret the estimate of β_2.
b. Locate the estimate of s and interpret its value.
c. Locate R_a^2 and interpret its value.
d. Conduct a test of overall model utility.
e. Test for evidence of sex discrimination at the firm.
f. Use the model to predict the salary of an executive with the characteristics shown in Table 4.6.

FIGURE 4.24

Portion of the SAS printout for executive compensation example

Analysis of Variance

Source	DF	Sum of Squares	Mean Square	F Value	Prob>F
Model	7	27.06425	3.86632	1823.73	0.0001
Error	92	0.19551	0.00212		
C Total	99	27.25976			

Root MSE	0.0461	R-square	0.9928	
Dep Mean	12.570	Adj R-sq	0.9923	
C.V.	0.3660			

Parameter Estimates

Variable	DF	Parameter Estimate	Standard Error	T for H0: Parameter=0	Prob > |T|
INTERCEP	1	8.87878	0.04612	192.49	0.0001
X1	1	0.04460	0.00166	26.83	0.0001
X2	1	0.03326	0.00270	12.31	0.0001
X3	1	0.11892	0.01724	6.89	0.0001
X4	1	0.00033	0.00001	19.97	0.0001
X5	1	0.00201	0.00002	73.25	0.0001
X1X1	1	-0.00071	0.00004	-15.11	0.0001
X3X4	1	0.00031	0.00002	16.16	0.0001

Obs	X1	X2	X3	X4	X5	Dep Var Y	Predict Value	Std Err Predict	Lower95% Predict	Upper95% Predict	Residual
100	1	12	16	0	160.1	.	10.2977	.0483	10.2030	10.3924	.

Solution

a. The least squares model is

$$\widehat{\log(y)} = 8.88 + .045x_1 + .033x_2 + .119x_3 + .00033x_4 + .002x_5$$
$$- .00071x_6 + .00031x_7$$

Because we are using the logarithm of salary as the dependent variable, the β estimates have different interpretations than previously discussed. In general, a parameter β in a multiplicative (log) model represents the percentage increase (or decrease) in the dependent variable for a 1-unit increase in the corresponding independent variable. The percentage change is calculated by taking the antilogarithm of the β estimate and subtracting 1, i.e., $e^{\beta} - 1$.* For example, the percentage change in executive compensation associated with a 1-unit (i.e., 1-year) increase in years of education x_2 is $(e^{\hat{\beta}_2} - 1) = (e^{.033} - 1) = .034$. Thus, when all other independent variables are held constant, we estimate executive salary to increase 3.4% for each additional year of education.

b. The estimate of the variance σ^2 is given in the SAS printout as

$$s^2 = \text{MSE} = \frac{\text{SSE}}{n - (k + 1)} = \frac{\text{SSE}}{10 - (7 + 1)} = .00212$$

and the estimate of the standard deviation σ, also given on the SAS printout as **Root MSE**, is $s = \sqrt{s^2} = .046$. Our interpretation is that most of the observed $\log(y)$ values (logarithms of salaries) lie within $2s = 2(.046) = .092$ of their least squares predicted values, $\log(\hat{y})$. A more practical interpretation (in terms of salaries) is obtained, however, if we take the antilog of this value and subtract 1, similar to the manipulation in part a. That is, we expect most of the observed executive salaries to lie within $e^{2s} - 1 = e^{.092} - 1 = .096$, or 9.6% of their respective least squares predicted values.

c. The adjusted R^2 value given on the SAS printout is $R_a^2 = .9923$. This implies that, after taking into account sample size and the number of independent variables, over 99% of the variation in the logarithm of salaries for these 100 sampled executives is accounted for by the model.

d. The test for overall model utility is conducted as follows:

H_0: $\beta_1 = \beta_2 = \cdots = \beta_7 = 0$

H_a: At least one of the model coefficients is nonzero.

Test statistic: $F = \dfrac{\text{Mean square for model}}{\text{MSE}} = 1,823.73$ (see Figure 4.24)

Rejection region: For $\alpha = .05, F > F_{.05}$

where $\nu_1 = k = 7$ and $\nu_2 = n - (k + 1) = 92$

---

*The result is derived by expressing the percentage change in salary y, as $(y_1 - y_0)/y_0$, where y_1 = the value of y when, say, $x = 1$, and y_0 = the value of y when $x = 0$. Now let $y^* = \log(y)$ and assume the log model is $y^* = \beta_0 + \beta_1 x$. Then

$$y = e^{y^*} = e^{\beta_0} e^{\beta_1 x} = \begin{cases} e^{\beta_0} & \text{when } x = 0 \\ e^{\beta_0} e^{\beta_1} & \text{when } x = 1 \end{cases}$$

Substituting, we have

$$\frac{y_1 - y_0}{y_0} = \frac{e^{\beta_0} e^{\beta_1} - e^{\beta_0}}{e^{\beta_0}} = e^{\beta_1} - 1$$

where from Table 4 of Appendix D, $F_{.05} \approx 2.1$. Since $F = 1{,}823.73$ exceeds the tabulated value of F, we conclude that the model does contribute information for predicting executive salaries. It appears that at least one of the β parameters in the model differs from 0. Note that the observed significance level of the F test, $p = .0001$, confirms this result.

e. If the firm is (knowingly or unknowingly) discriminating against female executives, then the mean salary for females (denoted μ_F) will be less than the mean salary for males (denoted μ_M) with the same qualifications (e.g., years of experience, years of education, etc.) From our earlier discussion of dummy variables, this difference will be represented by β_3, the β coefficient multiplied by x_3. Since $x_3 = 1$ if male, 0 if female, then $\beta_3 = (\mu_M - \mu_F)$ for fixed values of x_1, x_2 and x_5, and $x_4 = 0$. Consequently, a test of

$$H_0: \quad \beta_3 = 0 \quad \text{versus} \quad H_a: \quad \beta_3 > 0$$

is one way to test the discrimination hypothesis.* The p-value for this one-tailed test is one-half the p-value shown on the SAS printout, i.e., $.0001/2 = .00005$. With such a small p-value, there is strong evidence to reject H_0 and claim that some form of sex discrimination exists at the firm.

f. The least squares model can be used to obtain a predicted value for the logarithm of salary. Substituting the values of the x's shown in Table 4.6, we obtain

$$\widehat{\log(y)} = \hat{\beta}_0 + \hat{\beta}_1(12) + \hat{\beta}_2(16) + \hat{\beta}_3(0) + \hat{\beta}_4(400) + \hat{\beta}_5(160.1) +$$
$$\hat{\beta}_6(144) + \hat{\beta}_7(0)$$

This predicted value is given at the bottom of the SAS printout, Figure 4.24, $\widehat{\log(y)} = 10.298$. The 95% prediction interval, from 10.203 to 10.392, is also given. To predict the salary of an executive with these characteristics, we take the antilog of these values. That is, the predicted salary is $e^{10.298} = \$29{,}700$ (rounded to the nearest hundred) and the 95% prediction interval is from $e^{10.203}$ to $e^{10.392}$ (or from \$27,000 to \$32,600). Thus, an executive with the characteristics in Table 4.6 should be paid between \$27,000 and \$32,600 to be consistent with the sample data.

Note: The linear models described in this section form the basis for building models with quantitative independent variables and models with qualitative independent variables. More complex models, such as those with interactions between qualitative variables and those with both quantitative and qualitative variables (including interactions) may be required in practice, however. Chapter 5 presents a detailed discussion of model building with both quantitative and qualitative variables.

---

*A test for discrimination could also include testing the interaction term, $\beta_7 x_3 x_1$. If, as number of employees supervised (x_4) increases, the rate of increase in salary for males exceeds the rate for females, then $\beta_7 > 0$. Thus, rejecting $H_0: \beta_7 = 0$ in favor of $H_a: \beta_7 > 0$ would also suggest discrimination against female executives.

EXERCISES

4.16   Write a first-order linear model relating the mean value of y, $E(y)$, to two quantitative independent variables.

4.17   Write a first-order linear model relating the mean value of y, $E(y)$, to four quantitative independent variables.

4.18   Write a second-order linear model relating the mean value of y, $E(y)$, to two quantitative independent variables.

4.19   Write a second-order linear model relating the mean value of y, $E(y)$, to three quantitative independent variables.

4.20   Write a model relating $E(y)$ to a qualitative independent variable with two levels, A and B. Interpret the β parameters.

4.21   Write a model relating $E(y)$ to a qualitative independent variable with four levels, A, B, C, and D. Interpret the β parameters.

4.22   Consider the first-order equation

$$y = 1 + 2x_1 + x_2$$

a. Graph the relationship between y and x_1 for $x_2 = 0$, 1, and 2.
b. Are the graphed curves in part a first-order or second-order?
c. How do the graphed curves in part a relate to each other?
d. If a linear model is first-order in two independent variables, what type of geometric relationship will you obtain when $E(y)$ is graphed as a function of one of the independent variables for various values of the other independent variable?

4.23   Consider the first-order equation

$$y = 1 + 2x_1 + x_2 - 3x_3$$

a. Graph the relationship between y and x_1 for $x_2 = 1$ and $x_3 = 3$.
b. Repeat part a for $x_2 = -1$ and $x_3 = 1$.
c. If a linear model is first-order in three independent variables, what type of geometric relationship will you obtain when $E(y)$ is graphed as a function of one of the independent variables for various values of the other independent variables?

4.24   Consider the second-order model

$$y = 1 + x_1 - x_2 + 2x_1^2 + x_2^2$$

a. Graph the relationship between y and x_1 for $x_2 = 0$, 1, and 2.
b. Are the graphed curves in part a first-order or second-order?
c. How do the graphed curves in part a relate to each other?
d. Do the independent variables x_1 and x_2 interact? Explain.

4.25   Consider the second-order model

$$y = 1 + x_1 - x_2 + x_1x_2 + 2x_1^2 + x_2^2$$

a. Graph the relationship between y and x_1 for $x_2 = 0$, 1, and 2.
b. Are the graphed curves in part a first-order or second-order?

c. How do the graphed curves in part **a** relate to each other?

d. Do the independent variables x_1 and x_2 interact? Explain.

e. Note that the model used in this exercise is identical to the noninteraction model of Exercise 4.24, except that it contains the term involving x_1x_2. What does the term x_1x_2 introduce into the model?

4.26   One of the most promising methods for extracting crude oil employs a carbon dioxide (CO_2) flooding technique. CO_2, when flooded into oil pockets, enhances oil recovery by displacing the crude oil. In a microscopic investigation of the CO_2 flooding process, flow tubes were dipped into sample oil pockets containing a known amount of oil. The oil pockets were flooded with CO_2 and the percentage of oil displaced was recorded. The experiment was conducted at three different flow pressures and three different dipping angles. The displacement test data are recorded in the table.

PRESSURE x_1, pounds per square inch	DIPPING ANGLE x_2, degrees	OIL RECOVERY y, percentage
1,000	0	60.58
1,000	15	72.72
1,000	30	79.99
1,500	0	66.83
1,500	15	80.78
1,500	30	89.78
2,000	0	69.18
2,000	15	80.31
2,000	30	91.99

Source: Wang, G. C. "Microscopic investigation of CO_2 flooding process." *Journal of Petroleum Technology*, Vol. 34, No. 8, Aug. 1982, pp. 1789–1797. Copyright © 1982, Society of Petroleum Engineers, American Institute of Mining. First published in the *JPT* Aug. 1982.

ANALYSIS OF VARIANCE

SOURCE	DF	SUM OF SQUARES	MEAN SQUARE	F VALUE	PROB>F
MODEL	3	843.19083	281.06361	44.670	0.0005
ERROR	5	31.45996667	6.29199333		
C TOTAL	8	874.65080			

ROOT MSE	2.508385	R-SQUARE	0.9640	
DEP MEAN	76.90667	ADJ R-SQ	0.9425	
C.V.	3.261596			

PARAMETER ESTIMATES

VARIABLE	DF	PARAMETER ESTIMATE	STANDARD ERROR	T FOR H0: PARAMETER=0	PROB > |T|
INTERCEP	1	54.50000000	5.03415841	10.826	0.0001
X1	1	0.007696667	0.003238311	2.377	0.0634
X2	1	0.55411111	0.25996282	2.132	0.0862
X1X2	1	0.000113333	0.000167226	0.678	0.5280

a. Write the complete second-order model relating percentage oil recovery y to pressure x_1 and dipping angle x_2.

b. Plot the sample data on a scattergram, with percentage oil recovery y on the vertical axis and pressure x_1 on the horizontal axis. Connect the points corresponding to the same value of dipping angle x_2. Based on the scattergram, do you believe a complete second-order model is appropriate?

c. The SAS printout for the interaction model

$$y = \beta_0 + \beta_1 x_1 + \beta_2 x_2 + \beta_3 x_1 x_2 + \varepsilon$$

is provided on page 211. Give the prediction equation for this model.

d. Construct a plot similar to the scattergram of part b, but use the predicted values from the interaction model on the vertical axis. Compare the two plots. Do you believe the interaction model will provide an adequate fit?

e. Check model adequacy using a statistical test with $\alpha = .05$.

f. Is there evidence of interaction between pressure x_1 and dipping angle x_2? Test using $\alpha = .05$.

4.27   Refer to Exercise 3.15 and the Consumer Attitude (Telephone) Survey conducted by the University of Florida's Bureau of Economic and Business Research (BEBR). Suppose we want to model the refusal rate, y (i.e., the percentage of dialed households in a county that refuse to take part in the survey), as a function of the county's personal income per capita, x_1, and percentage of residents with a college education, x_2.

ANALYSIS OF VARIANCE

SOURCE	DF	SUM OF SQUARES	MEAN SQUARE	F VALUE	PROB>F
MODEL	5	0.06174787	0.01234957	1.803	0.2465
ERROR	6	0.04109780	0.006849633		
C TOTAL	11	0.10284567			

ROOT MSE	0.08276251	R-SQUARE	0.6004	
DEP MEAN	0.3961667	ADJ R-SQ	0.2674	
C.V.	20.89083			

PARAMETER ESTIMATES

VARIABLE	DF	PARAMETER ESTIMATE	STANDARD ERROR	T FOR H0: PARAMETER=0	PROB > |T|
INTERCEP	1	1.98524969	0.83441322	2.379	0.0548
X1	1	-0.000431242	0.000255467	-1.688	0.1424
X2	1	0.03724751	0.08995885	0.414	0.6932
X1X2	1	-0.000002983	0.0000069142	-0.431	0.6812
X1SQ	1	2.72363E-08	2.20347E-08	1.236	0.2626
X2SQ	1	-0.000294703	0.000580117	-0.508	0.6296

OBS	ID	ACTUAL	PREDICT VALUE	STD ERR PREDICT	LOWER95% PREDICT	UPPER95% PREDICT	RESIDUAL
1	7737	0.2960	0.2630	0.0814	-0.0211	0.5470	0.0330
2	12330	0.4980	0.5042	0.0748	0.2313	0.7771	-.006186
3	12058	0.3860	0.4166	0.0712	0.1495	0.6838	-0.0306
4	9927	0.3270	0.3410	0.0400	0.1161	0.5659	-0.0140
5	6904	0.5000	0.5234	0.0690	0.2598	0.7871	-0.0234
6	9463	0.3330	0.4112	0.0757	0.1366	0.6857	-0.0782
7	11466	0.4290	0.4034	0.0368	0.1818	0.6249	0.0256
8	10000	0.4220	0.3564	0.0361	0.1354	0.5774	0.0656
9	10052	0.4410	0.3700	0.0378	0.1474	0.5927	0.0710
10	8636	0.1910	0.3234	0.0498	0.0871	0.5597	-0.1324
11	7445	0.5260	0.4634	0.0598	0.2135	0.7133	0.0626
12	9059	0.4050	0.3780	0.0397	0.1534	0.6026	0.0270

SUM OF RESIDUALS	-1.94289E-16
SUM OF SQUARED RESIDUALS	0.0410978
PREDICTED RESID SS (PRESS)	1.363713

a. Write a first-order linear model for refusal rate y.

b. Write an interaction model for refusal rate y.

c. Write a second-order model for refusal rate y.

d. The second-order model of part **c** was fit to data on 12 Florida counties (data supplied by BEBR). The resulting SAS printout is provided. Find R^2 on the printout and interpret its value.

e. Is there evidence that the model is useful for predicting refusal rate y? Test using $\alpha = .05$.

f. The SAS printout also gives a 95% prediction interval for the refusal rate for each of the 12 counties. County 8 had a per capita income of $x_1 = \$10,000$ (ID) and $x_2 = 29.74\%$ of residents with a college education. Interpret the prediction interval for this county. How do you explain the large width of the interval?

4.28   In recent years, many companies have converted to the metric system of measurement. To quantify some of the characteristics of companies that have converted to metric production, B. D. Phillips, H. A. G. Lakhani, and S. L. George analyzed data on 350 small manufacturers collected for a U.S. Metric Board study (*Technological Forecasting and Social Change*, Apr. 1984). One of the research objectives was to investigate the relationship between the percentage y of metric work performed by a company, age x_1 of the company (in years), and cost x_2 of metric conversion, where

$$x_2 = \begin{cases} 1 & \text{if cost over } \$10,000 \\ 0 & \text{if not} \end{cases}$$

A first-order linear model was fit to the $n = 350$ data points with the following results:

$$\hat{y} = 70.9770 - .2167x_1 - 13.2768x_2$$
$$R^2 = .0576 \qquad t \text{ (for } H_0\text{: } \beta_1 = 0) = -1.56 \qquad t \text{ (for } H_0\text{: } \beta_2 = 0) = -2.71$$

a. Sketch the least squares relationship between $\hat{y}$ and x_1 for the two levels of metric conversion cost.

b. Is there evidence that the model is useful for predicting percentage y of metric work performed? Test using $\alpha = .01$.

c. Is there evidence that percentage y of metric work performed decreases as age x_1 of the company increases, for companies with the same cost x_2 of conversion? Test using $\alpha = .05$.

d. Write an interaction model for percentage y of metric work performed.

e. How will interaction between age x_1 and cost x_2, if determined to be significant, affect the graphs constructed in part **a**?

4.29   Researchers recently conducted an analysis of bus travel demand in Albuquerque, New Mexico, a city selected because of its unique multicentered "Sun Tran" public transit system. One aspect of the study involved the development of a multiple regression model for predicting y, the home-origin trip rate (that is, the number of home-origin trips per 1,000 residents) of a Sun Tran subzone urban area. The following five independent variables, all designed to measure transit level of service (travel time), were entered into the model:

x_1 = Composite in-vehicle travel time to reach major destination (minutes)

x_2 = Composite transit wait time (minutes)

x_3 = Composite number of transfers required to reach major destination

x_4 = Number of transit routes serving the Sun Tran zone

$$x_5 = \begin{cases} 1 & \text{if Sun Tran zone at end of major regional transportation corridor} \\ 0 & \text{if not} \end{cases}$$

Data collected from a survey of the city's bus passengers in each of 298 Sun Tran planning analysis zones were used to fit the first-order model

$$E(y) = \beta_0 + \beta_1 x_1 + \beta_2 x_2 + \beta_3 x_3 + \beta_4 x_4 + \beta_5 x_5$$

with the results shown in the accompanying Minitab printout.

```
The regression equation is
Y = 22.02 - .181 X1 - .250 X2 - 4.69 X3 + 3.67 X4 + 22.52 X5

Predictor        Coef       Stdev      t-ratio        p
Constant        22.019        *           *           *
X1              -0.181       0.039      -4.64       0.000
X2              -0.250       0.121      -2.07       0.038
X3              -4.691       1.702      -2.76       0.006
X4               3.674       0.403       9.12       0.000
X5              22.520       3.596       6.26       0.000

s = 8.657       R-sq = 59.9%       R-sq(adj) = 59.3%

Analysis of Variance

SOURCE        DF          SS           MS          F         p
Regression     5        32774       6554.8       87.45     0.000
Error        292        21886        74.95
Total        297        54660
```

Source: Adapted from Nelson, D. and O'Neil, K. "Analyzing demand for grid system transit." *Transportation Quarterly,* Vol. 37, No. 1, Jan. 1983, pp. 41–56.

a. Write the least squares prediction equation.
b. Compute and interpret the value of R^2.
c. Compute and interpret the value of s.
d. Is the model useful for predicting home-origin trip rate y? Test using $\alpha = .05$.
e. Is there evidence that home-origin trip rate y decreases as in-vehicle travel time x_1 increases and the remaining independent variables are held constant? Test using $\alpha = .05$.
f. Construct a 95% confidence interval for β_4. Interpret the interval.
g. Construct a 95% confidence interval for β_5. Interpret the inverval.

4.30 In theory, most academics advocate group decision making as a way to solve conflicts among a manager's subordinates. Many managers reject this proposition in practice, however, believing that conflict in groups is counterproductive. A study was conducted in Australia to examine this contradiction between accepted normative theory and current practice (*Organizational Behavior and Human Decision Processes*, Vol. 39, 1987). For one part of the study, multiple regression analysis was used to test "the proposition that the effective use of group discussion methods to resolve conflict depends on the manager's ability and willingness to encourage subordinates to confront conflict." A sample of 89 upper-level managers were asked to complete a questionnaire that measured (on a 7-point Likert scale):

y = Average performance of manager's subordinates (i.e., subordinate performance)

x_1 = Manager's preferred level of subordinate participation in decision making when conflict is present (i.e., group decision method)

x_2 = Average of subordinates' perceptions of manager's inclination to legitimize conflict (i.e., conflict legitimization)

The interaction model $E(y) = \beta_0 + \beta_1 x_1 + \beta_2 x_2 + \beta_3 x_1 x_2$ was fit to the 89 data points, with the following results (t values in parentheses):

$$\hat{y} = 7.09 - .44x_1 - .01x_2 + .06x_1 x_2 \qquad R^2 = .22$$
$$(-1.86) \quad (-.01) \quad (1.85)$$

a. Conduct a test to determine whether the model is adequate for predicting subordinate performance y. Use $\alpha = .10$.

b. Use the least squares prediction equation to graph the estimated relationships between subordinate performance (y) and group decision method (x_1) for low-conflict legitimization ($x_2 = 1$) and high-conflict legitimization ($x_2 = 7$). Interpret the graphs.

c. Conduct a test to determine whether the relationship between subordinate performance (y) and manager's use of a group decision method (x_1) depends on a manager's legitimization of conflict (x_2). Use $\alpha = .10$.

d. Based on the result of part c, would you recommend that the researchers conduct t tests on β_1 and β_2? Explain.

4.31 As a result of the U.S. surgeon general's warnings about the health hazards of smoking, Congress banned television and radio advertising of cigarettes in January 1971. The banning of prosmoking messages, however, also led to the virtual elimination of antismoking messages. In theory, if these antismoking commercials are more effective than prosmoking commercials, the net effect of the Congressional ban will be to increase the consumption of cigarettes and, therefore, benefit the tobacco industry. To test this hypothesis, researchers at the University of Houston built a cigarette demand model based on data collected for 46 states over the 18-year period from 1963 to 1980 (*The Review of Economics and Statistics*, Feb. 1986). For each state-year, the following independent variables were recorded:

x_1 = Natural log of price of a carton of cigarettes

x_2 = Natural log of minimum price of a carton of cigarettes in any neighboring state (This variable was included to measure the effect of "bootlegging" cigarettes in nearby states with lower tax rates.)

x_3 = Natural log of real disposable income per capita

x_4 = Per capita index of expenditures for cigarette advertising on television and radio (This value is 0 for the years 1971–1980, when the ban was in effect.)

The dependent variable of interest is y, the natural log of per capita consumption of cigarettes by persons of smoking age (14 years and older). The multiple regression model

$$E(y) = \beta_0 + \beta_1 x_1 + \beta_2 x_2 + \beta_3 x_3 + \beta_4 x_4$$

was fit to the $n = 828$ observations (48 states $\times$ 18 years) with the following results:

$$R^2 = .95 \qquad s = .047$$

a. Test the hypothesis that the model is useful for predicting y. Use $\alpha = .05$.

b. Interpret the value of s.

c. Give the null and alternative hypotheses appropriate for testing whether a decrease in per capita cigarette advertising expenditures is accompanied by an increase in per capita consumption of cigarettes over the period from 1963–1980.

d. The value of $\hat{\beta}_4$ was determined to be .033. Interpret this value.

e. Does the value $\hat{\beta}_4 = .033$ support the alternative hypothesis in part c? Explain.

4.32 A study reported in *Human Factors* (Apr. 1990) investigated the effects of recognizer accuracy and vocabulary size on the performance of a computerized speech recognition device. Accuracy (x_1) of the device, measured as the percentage of correctly recognized spoken utterances, was set at three levels: 90%, 95%, and 99%. Vocabulary size (x_2), measured as the percentage of words needed for the task, was also set at three levels: 75%, 87.5%, and 100%. The dependent variable of primary interest was task completion time (y, in minutes), measured from when a user of the recognition device spoke the first input until the recognizer displayed the last spoken word of the task. Data collected for $n = 162$ trials were used to fit a complete second-order model for task completion time (y), as a function of the quantitative independent variables accuracy (x_1) and vocabulary (x_2). The coefficient of determination for the model was $R^2 = .75$.

a. Write the complete second-order model for $E(y)$.

b. Interpret the value of R^2.

c. Conduct a test of overall model adequacy. Use $\alpha = .05$.

4.33 Real estate appraisers are often called upon to estimate the value of residential rental properties. The task is more difficult than home appraisals because the amenities (e.g., swimming pools, tennis courts) and services (e.g., maid service) provided by the rental unit must be taken into account. Regression analysis was employed to determine the extent to which certain physical, location, amenity and service, and external factors influence the dependent variable, apartment rent (*The Real Estate*

VARIABLE	β ESTIMATE	t VALUE
Intercept	183.50	5.81
Physical Characteristics		
Number of bedrooms	79.52	9.92
Size of complex (total units)	.16	4.67
Number of units in complex of same size	−.15	−1.65
Age (years)	−1.79	−3.40
Amenities/Services		
Covered parking (yes/no)	40.94	4.07
Modern kitchen (yes/no)	13.27	1.28
Maid service (yes/no)	21.22	1.74
Miscellaneous amenities (yes/no)	19.08	2.02
Utilities paid (yes/no)	36.92	1.66
Pool (yes/no)	13.76	1.40
No students (yes/no)	−8.08	−.42
No children (yes/no)	−9.61	−.88
No pets (yes/no)	−6.47	−.78
External Factors		
Traffic congestion (yes/no)	−23.66	−2.22
Proximity to employment (yes/no)	4.18	1.18
Number of bus lines	−4.84	−1.59
Distance to bus stop (feet)	−18.75	−1.38
Location		
Area 3	29.72	1.39
Area 5	22.80	1.83

Notes: All yes/no variables are coded as 0–1 dummy variables, with 1 = yes and 0 = no. Location is coded using 2 dummy variables (area 3 and area 5); the base level is area 1.

Source: Sirmans, G. F., Sirmans, C. F., and Benjamin, J. D. "Examining the variability of apartment rent." *The Real Estate Appraiser & Analyst*, Summer 1990, p. 46 (Table II).

Appraiser & Analyst, Summer 1990). A regression model was fit to data collected on a sample of 188 apartment units in three areas in Lafayette, Louisiana. The independent variables in the model, their β estimates, and associated t values are given in the table on page 216. Also, the coefficient of determination for the model is $R^2 = .71$.

a. Interpret the β estimates in the model.
b. Based on the reported t values, which variables appear to be the best predictors of rental price?
c. Refer to part b. Why is it dangerous to determine the best predictors of y based on the series of t tests?
d. Interpret the value of R^2.
e. Is the overall model useful for predicting apartment rental price? Test using $\alpha = .05$.

4.10

Testing Portions of a Model

In Example 4.4, we fit a second-order model for the mean shipment cost $E(y)$ of an express delivery package as a function of two quantitative variables—package weight, x_1, and distance shipped, x_2. The second-order model was

$$E(y) = \beta_0 + \beta_1 x_1 + \beta_2 x_2 + \beta_3 x_1 x_2 + \beta_4 x_1^2 + \beta_5 x_2^2$$

If we assume that the relationship between shipment cost y, package weight x_1, and distance shipped x_2 is first-order, then the following interaction model is appropriate:

$$E(y) = \beta_0 + \beta_1 x_1 + \beta_2 x_2 + \beta_3 x_1 x_2$$

Now, suppose we wanted to test whether the curvature terms contribute information for the prediction of y. This can be done by testing the hypothesis that the parameters for the curvature terms, x_1^2 and x_2^2, equal 0:

H_0: $\beta_4 = \beta_5 = 0$

H_a: At least one of the two parameters, β_4 and β_5, is nonzero.

In Section 3.6 we presented the t test for a single coefficient, and in Section 4.7 we gave the F test for all the β parameters (except β_0) in the model. We now need a test for a subset of the β parameters in the model. The test procedure is intuitive. First, we use the method of least squares to fit the interaction model, and calculate the corresponding sum of squares for error, SSE_1 (the sum of squares of the deviations between observed and predicted y values). Next, we fit the second-order model, and calculate its sum of squares for error, SSE_2. Then, we compare SSE_1 to SSE_2 by calculating the difference $SSE_1 - SSE_2$. If the curvature terms contribute to the model, then SSE_2 should be much smaller than SSE_1, and the difference $SSE_1 - SSE_2$ will be large. The larger the difference, the greater the weight of evidence that the variables package weight and distance shipped affect the mean shipment cost in a curvilinear manner.

The sum of squares for error will always decrease when new terms are added to the model since the total sum of squares, $SS_{yy} = \Sigma(y - \bar{y})^2$, remains the same. The question is whether this decrease is large enough to conclude that it is due to more than just an increase in the number of model terms and to chance. To test the null hypothesis that the curvature coefficients β_4 and β_5 simultaneously equal 0, we use an F statistic calculated as follows:

$$F = \frac{\text{Drop in SSE/Number of } \beta \text{ parameters being tested}}{s^2 \text{ for larger model}}$$

$$= \frac{(\text{SSE}_1 - \text{SSE}_2)/2}{\text{SSE}_2/[n - (5 + 1)]}$$

When the assumptions listed in Sections 3.4 and 4.2 about the error term ε are satisfied and the β parameters for curvature are all 0 (H_0 is true), this F statistic has an F distribution with $\nu_1 = 2$ and $\nu_2 = n - 6$ df. Note that ν_1 is the number of β parameters being tested and ν_2 is the number of degrees of freedom associated with s^2 in the larger, second-order model.

If the quadratic terms do contribute to the model (H_a is true), we expect the F statistic to be large. Thus, we use a one-tailed test and reject H_0 when F exceeds some critical value, F_α, as shown in Figure 4.25.

FIGURE 4.25

Rejection region for the F test
$H_0: \beta_4 = \beta_5 = 0$

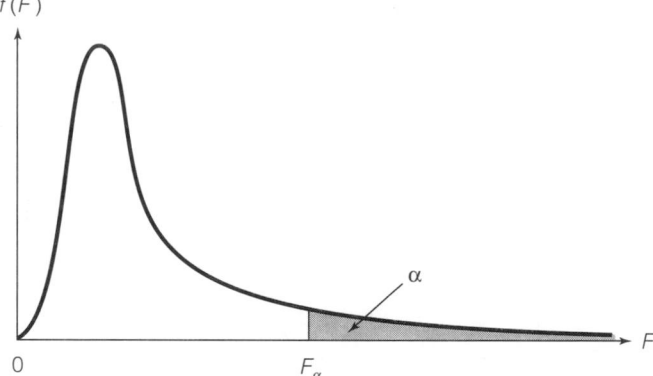

The F test can be used to determine whether any set of terms should be included in a model by testing the null hypothesis that a particular subset of β parameters simultaneously equal 0. For example, we may want to test to determine whether a set of interaction terms for quantitative variables or a set of main effect terms for a qualitative variable should be included in a model. To do this, we fit a pair of models. The first model, called the **complete model**, contains the terms we want to test and any additional terms. The second model, called the **reduced model**, is identical to the complete model except that the terms being tested are dropped. The F test appropriate for testing the null hypothesis that all of a subset of β parameters are equal to 0 is summarized in the next box.

EXAMPLE 4.7

In Example 4.4, we fit the second-order model for a set of $n = 20$ data points relating shipment cost to package weight and distance shipped. The SAS printout for this model, called the complete model, is shown in Figure 4.22. Figure 4.26

F Test for Testing the Null Hypothesis: Subset of β Parameters Equal Zero

Reduced model: $E(y) = \beta_0 + \beta_1 x_1 + \cdots + \beta_g x_g$

Complete model: $E(y) = \beta_0 + \beta_1 x_1 + \cdots + \beta_g x_g + \beta_{g+1} x_{g+1} + \cdots + \beta_k x_k$

H_0: $\beta_{g+1} = \beta_{g+2} = \cdots = \beta_k = 0$

H_a: At least one of the β parameters being tested is nonzero.

Test statistic:
$$F = \frac{(SSE_R - SSE_C)/(k - g)}{SSE_C/[n - (k + 1)]}$$
$$= \frac{(SSE_R - SSE_C)/\# \ \beta\text{'s tested}}{MSE_C}$$

where

SSE_R = Sum of squared errors for the reduced model

SSE_C = Sum of squared errors for the complete model

MSE_C = Mean square error for the complete model

$k - g$ = Number of β parameters specified in H_0 (i.e., number of β's tested)

$k + 1$ = Number of β parameters in the complete model (including β_0)

n = Total sample size

Rejection region: $F > F_\alpha$

and

$\nu_1 = k - g$ = Degrees of freedom for the numerator

$\nu_2 = n - (k + 1)$ = Degrees of freedom for the denominator

---

on page 220 shows the SAS printout for the straight-line interaction model (the reduced model) fit to the same $n = 20$ data points. Referring to the printouts, we find the following:

Straight-line interaction (reduced) model:

$SSE_R = 6.63330508$ (see Figure 4.25)

Second-order (complete) model:

$SSE_C = 2.74473841$ (see Figure 4.22)

Test the hypothesis that the quadratic terms do not contribute information for the prediction of y.

FIGURE 4.26   SAS printout for straight-line interaction model of Example 4.7

```
                        ANALYSIS OF VARIANCE

                       SUM OF        MEAN
         SOURCE    DF   SQUARES       SQUARE      F VALUE     PROB>F

         MODEL      3  445.45219    148.48406    358.154      0.0001
         ERROR     16    6.63330508   0.41458157
         C TOTAL   19  452.08550

               ROOT MSE    0.6438801    R-SQUARE    0.9853
               DEP MEAN    6.335        ADJ R-SQ    0.9826
               C.V.       10.16385


                        PARAMETER ESTIMATES

                       PARAMETER     STANDARD     T FOR HO:
         VARIABLE   DF  ESTIMATE      ERROR        PARAMETER=0   PROB > |T|

         INTERCEP   1  -0.14050074   0.64810001   -0.217        0.8311
         X1         1   0.01908803   0.15821160    0.121        0.9055
         X2         1   0.77208456   0.39056785    1.977        0.0656
         X1X2       1   0.77957444   0.08976644    8.684        0.0001
```

Solution

The test statistic is

$$F = \frac{(\text{SSE}_R - \text{SSE}_C)/2}{\text{SSE}_C/(20 - 6)}$$

$$= \frac{(6.63330508 - 2.74473841)/2}{2.74473841/14} = \frac{1.944283}{.196053} = 9.92$$

The critical value of F for $\alpha = .05$, $\nu_1 = 2$, and $\nu_2 = 14$ is found in Table 4 (Appendix D) to be

$$F_{.05} = 3.74$$

Since the calculated $F = 9.92$ exceeds 3.74, we are confident in concluding that the quadratic terms contribute to the prediction of y, shipment cost per package. The curvature terms should be retained in the model.

EXERCISES

4.34 Consider the second-order model relating $E(y)$ to three quantitative independent variables, x_1, x_2, and x_3:

$$E(y) = \beta_0 + \beta_1 x_1 + \beta_2 x_2 + \beta_3 x_3 + \beta_4 x_1 x_2 + \beta_5 x_1 x_3 + \beta_6 x_2 x_3 + \beta_7 x_1^2 + \beta_8 x_2^2 + \beta_9 x_3^2$$

a. Specify the parameters involved in a test of the hypothesis that no curvature exists in the response surface.

b. State the hypothesis of part a in terms of the model parameters.

c. What hypothesis would you test to determine whether x_3 is useful for the prediction of $E(y)$?

4.35 Refer to the *Journal of Personal Selling & Sales Management* (Summer 1990) study of gender differences in the industrial sales force, Exercise 3.30. Recall that a sample of 244 male sales managers and a sample of 153 female sales managers participated in the survey. One objective of the research was to assess how supervisory behavior affects intrinsic job satisfaction. Initially, the researchers fit the following reduced model to the data on each gender group:

$$E(y) = \beta_0 + \beta_1 x_1 + \beta_2 x_2 + \beta_3 x_3 + \beta_4 x_4$$

where

y = Intrinsic job satisfaction (measured on a scale of 0 to 40)

x_1 = Age (years)

x_2 = Education level (years)

x_3 = Firm experience (months)

x_4 = Sales experience (months)

To determine the effects of supervisory behavior, four variables (all measured on a scale of 0 to 50) were added to the model: x_5 = contingent reward behavior, x_6 = noncontingent reward behavior, x_7 = contingent punishment behavior, and x_8 = noncontingent punishment behavior. Thus, the complete model is

$$E(Y) = \beta_0 + \beta_1 x_1 + \beta_2 x_2 + \beta_3 x_3 + \beta_4 x_4 + \beta_5 x_5 + \beta_6 x_6 + \beta_7 x_7 + \beta_8 x_8$$

a. For each gender, specify the null hypothesis and rejection region (α = .05) for testing whether any of the four supervisory behavior variables affect intrinsic job satisfaction.

b. The R^2 values for the four models (reduced and complete model for both samples) are given in the accompanying table. Interpret the results. For each gender, does it appear that the supervisory behavior variables have an impact on intrinsic job satisfaction? Explain.

MODEL	R^2	
	Males	Females
Reduced	.218	.268
Complete	.408	.496

Source: Schul, et al. "Assessing gender differences in relationships between supervisory behaviors and job related outcomes in industrial sales force." *Journal of Personal Selling & Sales Management*, Vol X, Summer 1990, p. 9 (Table 4).

c. The F statistics for comparing the two models are: F_{males} = 13.00 and $F_{females}$ = 9.05. Conduct the tests, part **a**, and interpret the results.

4.36 Refer to Exercise 4.35. One way to test for gender differences in the industrial sales force is to incorporate a dummy variable for gender into the model for intrinsic job satisfaction, y, and then fit the model to the data for the combined sample of males and females.

a. Write a model for y as a function of the independent variables, x_1–x_8, and the gender dummy variable. Include interactions between gender and each of the other independent variables in the model.

b. Based on the model, part **a**, what is the null hypothesis for testing whether gender has an effect on job satisfaction?

c. Explain how to conduct the test, part **b**.

4.37   Research was conducted at Temple University to examine the effect of several factors on managerial performance (*Journal of Vocational Behavior*, Oct. 1986). A sample of 100 management personnel from several divisions within a government agency took part in the study. Each manager completed a questionnaire designed to measure the following variables:

y = Performance rating (1 = unacceptable, 5 = outstanding)

$$x_1 = \begin{cases} 1 & \text{if male} \\ 0 & \text{if female} \end{cases}$$

x_2 = Job tenure (years)

x_3 = Manager–subordinate work relationship rating
    (1 = unsatisfactory, 5 = excellent)

x_4 = Effort level (average number of hours per week invested in job)

$$x_5 = \begin{cases} 1 & \text{if middle/upper-level manager} \\ 0 & \text{if lower-level manager} \end{cases}$$

x_6 = Subordinate-related managerial behavior score (low scores indicate little or
    no effort spent on counseling, evaluating, and training subordinates)

The data collected on the 100 managers were then used to fit several regression models of managerial performance.

a. Initially, the model

$$E(y) = \beta_0 + \beta_1 x_1 + \beta_2 x_2 + \beta_3 x_3 + \beta_4 x_4$$

was considered to account for the influence of sex, job tenure, manager–subordinate work relationship and effort level on performance rating. For this model, SSE = 352 and R^2 = .11. Calculate the F statistic for testing model adequacy. Is the model useful for predicting performance rating y? Use α = .05.

b. Terms for managerial level and subordinate-related behavior (i.e., $\beta_5 x_5 + \beta_6 x_6$) were added to the model in part **a**, resulting in SSE = 341 and R^2 = .14. Do these terms contribute additional information for the prediction of performance rating y? Test using α = .05.

c. A third model was also considered:

$$E(y) = \beta_0 + \beta_1 x_1 + \beta_2 x_2 + \beta_3 x_3 + \beta_4 x_4 + \beta_5 x_5 + \beta_6 x_6 + \beta_7 x_5 x_6$$

This model resulted in SSE = 321 and R^2 = .19. Test the hypothesis that the interaction between managerial level x_5 and subordinate-related behavior x_6 is not important, i.e., test H_0: $\beta_7 = 0$. Use α = .05.

d. Interpret the result of part **c** in terms of the problem.

4.38   Since 1978, when the U.S. airline industry was deregulated, researchers have questioned whether the deregulation has ensured a truly competitive environment. If so, the profitability of any major airline would be related only to overall industry conditions (e.g., disposable income and market share) but not to any unchanging feature of that airline. This profitability hypothesis was tested in

Transportation Journal (Winter 1990) using multiple regression. Data for $n = 234$ carrier-years were used to fit the model

$$E(y) = \beta_0 + \beta_1 x_1 + \beta_2 x_2 + \beta_3 x_3 + \cdots + \beta_{30} x_{30}$$

where

$$y = \text{Profit rate}$$

$$x_1 = \text{Real personal disposable income}$$

$$x_2 = \text{Industry market share}$$

$$x_3 - x_{30} = \text{Dummy variables (coded 0–1) for the 29 air carriers investigated in the study}$$

The results of the regression are summarized in the table. Interpret the results. Is the profitability hypothesis supported?

VARIABLE	β-ESTIMATE	t VALUE	p-VALUE
Intercept	1.2642	.09	.9266
x_1	−.0022	−.99	.8392
x_2	4.8405	3.57	.0003
$x_3 - x_{30}$	(not given)	—	—

$R^2 = .3402$
$F(\text{model}) = 3.49, \quad p\text{-value} = .0001$
$F(\text{carrier dummies}) = 3.59, \quad p\text{-value} = .0001$

Source: Leigh, L. E. "Contestability in deregulated airline markets: Some empirical tests." *Transportation Journal*, Winter 1990, p. 55 (Table 4).

4.39 A chain of drug stores wants to model mean profit per week, $E(y)$, as a function of three advertising factors: type of design, choice of newspaper, and percentage discount offered on sale items. A first proposal is the model

$$E(y) = \beta_0 + \beta_1 x_1 + \beta_2 x_1^2 + \beta_3 x_2 + \beta_4 x_3 + \beta_5 x_2 x_3 + \beta_6 x_1 x_2 + \beta_7 x_1 x_3$$
$$+ \beta_8 x_1 x_2 x_3 + \beta_9 x_1^2 x_2 + \beta_{10} x_1^2 x_3 + \beta_{11} x_1^2 x_2 x_3$$

where

$$x_1 = \text{Percentage discount}$$

$$x_2 = \begin{cases} 1 & \text{if design } D_1 \\ 0 & \text{if design } D_2 \end{cases}$$

$$x_3 = \begin{cases} 1 & \text{if newspaper } N_1 \\ 0 & \text{if newspaper } N_2 \end{cases}$$

a. Specify the parameters that would be involved in a test of the hypothesis, "The design of the advertising and the choice of newspaper have no effect on the mean value of the weekly profits."

b. Refer to part a. State the hypothesis that you would make regarding the parameter values.

c. Give the parameters that would be involved in a test of the hypothesis, "The relationship between mean weekly profit and percent discount for each of the combinations of newspaper and design is first-order (i.e., a straight line)."

4.11

Stepwise Regression

Refer to the problem of predicting executive salaries, Example 4.6. Perhaps the biggest problem in building a model to describe executive salaries is choosing the important independent variables to be included in the model. The list of potentially important independent variables is extremely long, and we need some objective method of screening out those that are not important.

The problem of deciding which of a large set of independent variables to include in a model is common, for instance, trying to determine which variables influence the profit of a firm, affect product quality, or are related to the state of the economy.

A systematic approach to building a model with a large number of independent variables is difficult because the interpretation of multivariable interactions and higher-order polynomials (squared terms, cubic terms, and so forth) is tedious. Therefore, we turn to a screening procedure known as **stepwise regression**.

The most commonly used stepwise regression procedure, available in most popular statistical software packages, works as follows. The user first identifies the response, y, and the set of potentially important independent variables, x_1, $x_2, \ldots, x_k$, where k will generally be large. (Note that this set of variables could represent both first- and higher-order terms, as well as any interaction terms that might be important information contributors.) The response and independent variables are then entered into the computer, and the stepwise procedure begins.

STEP 1 The computer fits all possible one-variable models of the form

$$E(y) = \beta_0 + \beta_1 x_1$$

to the data. For each model, the test of the null hypothesis

$$H_0: \quad \beta_1 = 0$$

against the alternative hypothesis

$$H_a: \quad \beta_1 \neq 0$$

is conducted using the t (or the equivalent F) test for a single β parameter. The independent variable that produces the largest (absolute) t value is declared the best one-variable predictor of y.* Call this independent variable x_1.

STEP 2 The stepwise program now begins to search through the remaining $(k - 1)$ independent variables for the best two-variable model of the form

$$E(y) = \beta_0 + \beta_1 x_1 + \beta_2 x_i$$

This is done by fitting all two-variable models containing x_1 and each of the other $(k - 1)$ options for the second variable x_i. The t values for the test H_0: $\beta_2 = 0$ are computed for each of the $(k - 1)$ models (corresponding to the remaining independent variables x_i, $i = 2, 3, \ldots, k$), and the variable having the largest t is retained. Call this variable x_2.

---

*Note that the variable with the largest t value will also be the one with the largest Pearson product moment correlation, r (Section 3.7), with y.

At this point, some software packages diverge in methodology. The better packages now go back and check the t value of $\hat{\beta}_1$ *after $\hat{\beta}_2 x_2$ has been added to the model*. If the t value has become nonsignificant at some specified α level (say, $\alpha = .10$), the variable x_1 is removed and a search is made for the independent variable with a β parameter that will yield the most significant t value in the presence of $\hat{\beta}_2 x_2$. Other packages do not recheck $\hat{\beta}_1$, but proceed directly to step 3.

The best-fitting model may yield a different value for $\hat{\beta}_1$ than that obtained in step 1, because $\hat{\beta}_1$ and $\hat{\beta}_2$ may be correlated. Thus, both the value of $\hat{\beta}_1$ and, therefore, its significance will usually change from step 1 to step 2. For this reason, the software packages that recheck the t values at each step are preferable.

STEP 3 The stepwise procedure now checks for a third independent variable to include in the model with x_1 and x_2. That is, we seek the best model of the form

$$E(y) = \beta_0 + \beta_1 x_1 + \beta_2 x_2 + \beta_3 x_i$$

To do this, we fit all the $(k - 2)$ models using x_1, x_2, and each of the $(k - 2)$ remaining variables, x_i, as a possible x_3. The criterion is again to include the independent variable with the largest t value. Call this best third variable x_3.

The better programs now recheck the t values corresponding to the x_1 and x_2 coefficients, replacing the variables that have t values that have become nonsignificant. This procedure is continued until no further independent variables can be found that yield significant t values (at the specified α level) in the presence of the variables already in the model.

The result of the stepwise procedure is a model containing only the main effects with t values that are significant at the specified α level. Thus, in most practical situations, only several of the large number of independent variables will remain. However, it is very important not to jump to the conclusion that all the independent variables important for predicting y have been identified or that the unimportant independent variables have been eliminated. Remember, the stepwise procedure is using only *sample estimates* of the true model coefficients (β's) to select the important variables. An extremely large number of single β parameter t tests have been conducted, and the probability is very high that one or more errors have been made in including or excluding variables. That is, we have very probably included some unimportant independent variables in the model (Type I errors) and eliminated some important ones (Type II errors).

There is a second reason why we might not have arrived at a good model. When we choose the variables to be included in the stepwise regression, we may often omit high-order terms (to keep the number of variables manageable). Consequently, we may have initially omitted several important terms from the model. Thus, we should recognize stepwise regression for what it is—an objective screening procedure.

Now, we will consider interactions and quadratic terms (for quantitative variables) among variables screened by the stepwise procedure. It would be best to develop this response surface model with a second set of data independent of that used for the screening, so the results of the stepwise procedure can be partially verified with new data. However, this is not always possible, because in many business modeling situations only a small amount of data is available.

Remember, do not be deceived by the impressive-looking t values that result from the stepwise procedure—it has retained only the independent variables with the largest t values. Also, be certain to consider second-order terms in systematically developing the prediction model. The first-order model given by the stepwise procedure may be greatly improved by the addition of interaction and quadratic terms.

| Warning

Be cautious when using the results of stepwise regression to make inferences about the relationship between $E(y)$ and the independent variables in the resulting first-order model. First, an extremely large number of t tests have been conducted, leading to a high probability of making either one or more Type I or Type II errors. Second, the stepwise model does not include any higher-order or interaction terms. Stepwise regression should be used only when necessary, i.e., when you want to determine which of a large number of potentially important independent variables should be used in the model-building process.

EXAMPLE 4.8

In Example 4.6 we fit a multiple regression model for executive salaries as a function of experience, education, sex, etc. A preliminary step in the construction of this model was the determination of the most important independent variables. Ten independent variables (seven quantitative and three qualitative) were considered, as shown in Table 4.7. It would be very difficult to construct a second-order model with 10 independent variables. Therefore, use the sample of 100 executives from Example 4.6 to decide which of the 10 variables should be included in the construction of the final model for executive salaries.

TABLE 4.7   Independent Variables in the Executive Salary Example

INDEPENDENT VARIABLE	DESCRIPTION
x_1	Experience (years)—quantitative
x_2	Education (years)—quantitative
x_3	Sex (1 if male, 0 if female)—qualitative
x_4	Number of employees supervised—quantitative
x_5	Corporate assets (millions of dollars)—quantitative
x_6	Board member (1 if yes, 0 if no)—qualitative
x_7	Age (years)—quantitative
x_8	Company profits (past 12 months, millions of dollars)—quantitative
x_9	International responsibility (1 if yes, 0 if no)—qualitative
x_{10}	Company's total sales (past 12 months, millions of dollars)—quantitative

Solution

We will use stepwise regression with the first-order terms of the seven quantitative independent variables and the main effects of the three qualitative independent variables to identify the most important variables. The dependent variable y is the natural logarithm of the executive salaries. The SAS stepwise regression printout is shown in Figure 4.27. Note that the first variable included in the model is x_4, number of employees supervised by the executive. At the second step, x_5, corporate assets, enters the model. At the sixth step, x_6, a dummy variable for the qualitative variable board member or not, is brought into the model. However, because the significance (.2295) of the F statistic (SAS uses the $F = t^2$ statistic in the stepwise procedure rather than the t statistic) for x_6 is above the preassigned $\alpha = .10$, x_6 is then removed from the model. Thus, at step 7 the procedure indicates that the five-variable model including x_1, x_2, x_3, x_4, and x_5 is best. That is, none of the other independent variables can meet the $\alpha = .10$ criterion for admission to the model.

FIGURE 4.27

SAS stepwise regression printout for Example 4.8

```
STEP 1
   Variable X4 Entered        R-Square = 0.42071677       C(P) = 1274.7576

                     DF     Sum of Squares    Mean Square       F     Prob > F

   Regression         1        11.46854285    11.46854285    71.17     0.0001
   Error             98        15.79113802     0.16113696
   Total             99        27.25977087

                            B Value           Std Error         F     Prob > F

   Intercept              10.20077500
   X4                      0.00057284        0.00006790       71.17     0.0001
--------------------------------------------------------------------------------

STEP 2
   Variable X5 Entered        R-Square = 0.78299675       C(P) = 419.4947

                     DF     Sum of Squares    Mean Square       F     Prob > F

   Regression         2        21.34431198    10.67215599   175.00     0.0001
   Error             97         5.91545889     0.06098411
   Total             99        27.25977087

                            B Value           Std Error         F     Prob > F

   Intercept               9.87702903
   X4                      0.00058353        0.00004178      195.06     0.0001
   X5                      0.00183730        0.00014438      161.94     0.0001
--------------------------------------------------------------------------------

STEP 3
   Variable X1 Entered        R-Square = 0.89667614       C(P) = 152.4952

                     DF     Sum of Squares    Mean Square       F     Prob > F

   Regression         3        24.44318616     8.14772872   277.71     0.0001
   Error             96         2.81658471     0.02933942
   Total             99        27.25977087

                            B Value           Std Error         F     Prob > F

   Intercept               9.66449288
   X4                      0.00055251        0.00002914      359.59     0.0001
   X5                      0.00191195        0.00010041      362.60     0.0001
   X1                      0.01870784        0.00182032      105.62     0.0001
--------------------------------------------------------------------------------
```

(continued)

FIGURE 4.27
Continued

STEP 4

Variable X3 Entered   R-Square = 0.94815717   C(P) = 32.6757

	DF	Sum of Squares	Mean Square	F	Prob > F
Regression	4	25.84654710	8.46163678	434.37	0.0001
Error	95	1.41322377	0.01487604		
Total	99	27.25977087			
		B Value	Std Error	F	Prob > F
Intercept		9.40077349			
X4		0.00055288	0.00002075	710.15	0.0001
X5		0.00190876	0.00007150	712.74	0.0001
X1		0.02074868	0.00131310	249.68	0.0001
X3		0.30011726	0.03089939	94.34	0.0001

STEP 5

Variable X2 Entered   R-Square = 0.96039323   C(P) = 5.7215

	DF	Sum of Squares	Mean Square	F	Prob > F
Regression	5	26.18009940	5.23601988	455.87	0.0001
Error	94	1.07967147	0.01148587		
Total	99	27.25977087			
		B Value	Std Error	F	Prob > F
Intercept		8.85387930			
X4		0.00056061	0.00001829	939.84	0.0001
X5		0.00193684	0.00006304	943.98	0.0001
X1		0.02141724	0.00116047	340.61	0.0001
X3		0.31927842	0.02738298	135.95	0.0001
X2		0.03315807	0.00615303	29.04	0.0001

STEP 6

Variable X6 Entered   R-Square = 0.96100666   C(P) = 6.2699

	DF	Sum of Squares	Mean Square	F	Prob > F
Regression	6	26.19682148	4.36613691	382.00	0.0001
Error	93	1.06294939	0.01142956		
Total	99	27.25977087			
		B Value	Std Error	F	Prob > F
Intercept		8.87509152			
X4		0.00055820	0.00001835	925.32	0.0001
X5		0.00193764	0.00006289	949.31	0.0001
X1		0.02133460	0.00115963	338.48	0.0001
X3		0.31093801	0.02817264	121.81	0.0001
X2		0.03272195	0.00614851	28.32	0.0001
X6		0.03866226	0.03196369	1.46	0.2295

STEP 7

Variable X6 Removed   R-Square = 0.96039323   C(P) = 5.7215

	DF	Sum of Squares	Mean Square	F	Prob > F
Regression	5	26.18009940	5.23601988	455.87	0.0001
Error	94	1.07967147	0.01148587		
Total	99	27.25977087			
		B Value	Std Error	F	Prob > F
Intercept		8.85387930			
X4		0.00056061	0.00001829	939.84	0.0001
X5		0.00193684	0.00006304	943.98	0.0001
X1		0.02141724	0.00116047	340.61	0.0001
X3		0.31927842	0.02738298	135.95	0.0001
X2		0.03315807	0.00615303	29.04	0.0001

Thus, in our final modeling effort (Example 4.6) we concentrated on these five independent variables, and determined that several second-order terms were important in the prediction of executive salaries.

There are several other stepwise regression techniques designed to select the most important independent variables. One of these, called **forward selection**, is nearly identical to the stepwise procedure previously outlined. The only difference is that the forward selection technique provides no option for rechecking the t values corresponding to the x's that have entered the model in an earlier step. Thus, stepwise regression is preferred to forward selection in practice.

Another technique, called **backward elimination**, initially fits a model containing terms for all potential independent variables. That is, for k independent variables, the model $E(y) = \beta_0 + \beta_1 x_1 + \beta_2 x_2 + \cdots + \beta_k x_k$ is fit in step 1. The variable with the smallest t (or F) statistic for testing $H_0: \beta_i = 0$ is identified and dropped from the model if the t value is less than some specified critical value. The model with the remaining $(k - 1)$ independent variables is fit in step 2, and again, the variable associated with the smallest nonsignificant t value is dropped. This process is repeated until no further nonsignificant independent variables can be found. The real disadvantage of using the backward elimination technique is that you need a sufficiently large number of data points to fit the initial model in step 1.

EXERCISES

4.40 There are six independent variables, x_1, x_2, x_3, x_4, x_5, and x_6, that might be useful in predicting a response y. A total of $n = 50$ observations are available, and it is decided to employ stepwise regression to help in selecting the independent variables that appear to be useful. The computer fits all possible one-variable models of the form

$$E(y) = \beta_0 + \beta_1 x_i$$

where x_i is the ith independent variable, $i = 1, 2, \ldots, 6$. The information in the accompanying table is provided from the computer printout.

INDEPENDENT VARIABLE	$\hat{\beta}_1$	$s_{\hat{\beta}_1}$
x_1	1.6	.42
x_2	−.9	.01
x_3	3.4	1.14
x_4	2.5	2.06
x_5	−4.4	.73
x_6	.3	.35

a. Which independent variable is declared the best one-variable predictor of y? Explain.

b. Would this variable be included in the model at this stage? Explain.

c. Describe the next phase that a stepwise procedure would execute.

4.41   J. O. Wise and H. J. Dover use stepwise regression to identify a number of important factors (variables) that can be used to predict rural property values. They obtained their results by analyzing a sample of 105 cases from seven counties in Georgia. Part of their findings are duplicated in the table. The variable names are listed in the order in which the stepwise regression procedure identified their importance, and the t values found at each step are given for each variable. Note that both qualitative and quantitative variables have been included. Since each qualitative variable is at two levels, only one main effect term (i.e., dummy variable) could be included in the model for each factor.

Stepwise Regression Analysis of Price per Acre

VARIABLE NAME	t VALUE
Residential land (yes–no)	10.466
Seedlings and saplings (number)	6.692
Percentage ponds (percent)	4.141
Distance to state park (miles)	3.985
Branches or springs (yes–no)	3.855
Site index (ratio)	3.160
Size (acres)	1.142
Farmland (yes–no)	2.288

Source: Wise, J. O. and Dover, H. J. "An evaluation of a statistical method of appraising rural property." *Appraisal Journal*, Vol. 42, Jan. 1974, pp. 103–113.

a. Which of the eight variables listed in the table would you use to model rural property value, y?

b. Based on your answer to part a, propose a complete model for $E(y)$.

4.42   In any production process in which one or more workers are engaged in a variety of tasks, the total time spent in production varies as a function of the size of the work pool and the level of output of the various activities. For example, in a large metropolitan department store, the number of hours worked (y) per day by the clerical staff may depend on the following variables:

x_1 = Number of pieces of mail processed (open, sort, etc.)

x_2 = Number of money orders and gift certificates sold

x_3 = Number of window payments (customer charge accounts) transacted

x_4 = Number of change order transactions processed

x_5 = Number of checks cashed

x_6 = Number of pieces of miscellaneous mail processed on an "as available" basis

x_7 = Number of bus tickets sold

The accompanying table of observations gives the output counts for these activities on each of 52 working days.

OBS.	DAY OF WEEK	y	x_1	x_2	x_3	x_4	x_5	x_6	x_7
1	M	128.5	7781	100	886	235	644	56	737
2	T	113.6	7004	110	962	388	589	57	1029
3	W	146.6	7267	61	1342	398	1081	59	830
4	Th	124.3	2129	102	1153	457	891	57	1468
5	F	100.4	4878	45	803	577	537	49	335
6	S	119.2	3999	144	1127	345	563	64	918
7	M	109.5	11777	123	627	326	402	60	335
8	T	128.5	5764	78	748	161	495	57	962
9	W	131.2	7392	172	876	219	823	62	665
10	Th	112.2	8100	126	685	287	555	86	577
11	F	95.4	4736	115	436	235	456	38	214
12	S	124.6	4337	110	899	127	573	73	484
13	M	103.7	3079	96	570	180	428	59	456
14	T	103.6	7273	51	826	118	463	53	907
15	W	133.2	4091	116	1060	206	961	67	951
16	Th	111.4	3390	70	957	284	745	77	1446
17	F	97.7	6319	58	559	220	539	41	440
18	S	132.1	7447	83	1050	174	553	63	1133
19	M	135.9	7100	80	568	124	428	55	456
20	T	131.3	8035	115	709	174	498	78	968
21	W	150.4	5579	83	568	223	683	79	660
22	Th	124.9	4338	78	900	115	556	84	555
23	F	97.0	6895	18	442	118	479	41	203
24	S	114.1	3629	133	644	155	505	57	781
25	M	88.3	5149	92	389	124	405	59	236
26	T	117.6	5241	110	612	222	477	55	616
27	W	128.2	2917	69	1057	378	970	80	1210
28	Th	138.8	4390	70	974	195	1027	81	1452
29	F	109.5	4957	24	783	358	893	51	616
30	S	118.9	7099	130	1419	374	609	62	957
31	M	122.2	7337	128	1137	238	461	51	968
32	T	142.8	8301	115	946	191	771	74	719
33	W	133.9	4889	86	750	214	513	69	489
34	Th	100.2	6308	81	461	132	430	49	341
35	F	116.8	6908	145	864	164	549	57	902
36	S	97.3	5345	116	604	127	360	48	126
37	M	98.0	6994	59	714	107	473	53	726
38	T	136.5	6781	78	917	171	805	74	1100
39	W	111.7	3142	106	809	335	702	70	1721
40	Th	98.6	5738	27	546	126	455	52	502
41	F	116.2	4931	174	891	129	481	71	737
42	S	108.9	6501	69	643	129	334	47	473
43	M	120.6	5678	94	828	107	384	52	1083
44	T	131.8	4619	100	777	164	834	67	841
45	W	112.4	1832	124	626	158	571	71	627
46	Th	92.5	5445	52	432	121	458	42	313
47	F	120.0	4123	84	432	153	544	42	654
48	S	112.2	5884	89	1061	100	391	31	280
49	M	113.0	5505	45	562	84	444	36	814
50	T	138.7	2882	94	601	139	799	44	907
51	W	122.1	2395	89	637	201	747	30	1666
52	Th	86.6	6847	14	810	230	547	40	614

Source: Adapted from *Work Measurement*, by G. L. Smith, Grid Publishing Co., Columbus, Ohio, 1978 (Table 3-1).

a. Conduct a stepwise regression analysis of the data using an available statistical software package.

b. Interpret the β estimates in the resulting stepwise model.

c. What are the dangers associated with drawing inferences from the stepwise model?

4.12

Other Variable Selection Techniques (Optional)

In Section 4.11, we presented stepwise regression as an objective screening procedure for selecting the most important predictors of y. Other, more subjective, variable selection techniques have been developed in the literature for the purpose of identifying important independent variables. The most popular of these procedures are those that consider all possible regression models given the set of potentially important predictors. (Such a procedure is commonly known as an **all-possible-regressions selection procedure**.) The techniques differ with respect to the criteria for selecting the "best" subset of variables. In this section we describe four criteria widely used in practice, then give an example illustrating the four techniques.

R^2 Criterion

Consider the set of potentially important variables, $x_1, x_2, x_3, \ldots, x_k$. We learned in Section 4.6 that the multiple coefficient of determination

$$R^2 = 1 - \frac{SSE}{SS(Total)}$$

will increase when independent variables are added to the model. Therefore, the model that includes all k independent variables

$$E(y) = \beta_0 + \beta_1 x_1 + \beta_2 x_2 + \cdots + \beta_k x_k$$

will yield the largest R^2. Yet, we have seen examples (in Section 4.11) where adding terms to the model does not yield a significantly better prediction equation. The objective of the R^2 criterion is to find a subset model (i.e., a model containing a subset of the k independent variables) so that adding more variables to the model will yield only small increases in R^2. In practice, the best model found by the R^2 criterion will rarely be the model with the largest R^2. Generally, you are looking for a simple model that is as good as, or nearly as good as, the model with all k independent variables. Unlike stepwise regression, however, the decision about when to stop adding variables to the model is a subjective one.

Adjusted R^2 or MSE Criterion

One drawback to using the R^2 criterion, you will recall, is that the value of R^2 does not account for the number of β parameters in the model. If enough variables are added to the model so that the sample size n equals the total number of β's in the model, you will force R^2 to equal 1. Alternatively, we can use the adjusted R^2. It is easy to show that R_a^2 is related to MSE as follows:

$$R_a^2 = 1 - (n - 1)\left[\frac{MSE}{SS(Total)}\right]$$

Note that R_a^2 increases only if MSE decreases [since SS(Total) remains constant for all models]. Thus, an equivalent procedure is to search for the model with the minimum, or near minimum, MSE.

C_p Criterion

A recently developed procedure, which is gaining increasing acceptance among regression analysts, is based on a quantity called the **total mean square error** (**TMSE**) for the fitted regression model:

$$\text{TMSE} = E\left\{ \sum_{i=1}^{n} [\hat{y}_i - E(y_i)]^2 \right\} = \sum_{i=1}^{n} [E(\hat{y}_i) - E(y_i)]^2 + \sum_{i=1}^{n} \text{Var}(\hat{y}_i)$$

where $E(\hat{y}_i)$ is the mean response for the subset (fitted) regression model and $E(y_i)$ is the mean response for the true model. The objective is to compare the TMSE for the subset regression model with σ^2, the variance of the random error for the true model, using the ratio

$$\Gamma = \frac{\text{TMSE}}{\sigma^2}$$

Small values of Γ imply that the subset regression model has a small total mean square error relative to σ^2. Unfortunately, both TMSE and σ^2 are unknown, and we must rely on sample estimates of these quantities. It can be shown (proof omitted) that an estimator of the ratio Γ is given by

$$C_p = \frac{\text{SSE}_p}{\text{MSE}_k} + 2(p + 1) - n$$

where n is the sample size, p is the number of independent variables in the subset model, k is the total number of potential independent variables, SSE_p is the SSE for the subset model, and MSE_k is the MSE for the model containing all k independent variables. In their latest releases, the statistical software packages discussed in this text have routines that calculate the C_p statistic. In fact, the C_p value is now automatically printed at each step in the SAS stepwise regression printout (see Figure 4.27).

The C_p criterion selects as the best model the subset model with (1) a small value of C_p (i.e., a small total mean square error) and (2) a value of C_p near $p + 1$, a property that indicates that slight or no bias exists in the subset regression model.*

Thus, the C_p criterion focuses on minimizing total mean square error and the regression bias. If you are mainly concerned with minimizing total mean square error, you will want to choose the model with the smallest C_p value, as long as the bias is not large. On the other hand, you may prefer a model that yields a C_p value slightly larger than the minimum but which has slight (or no) bias.

PRESS Criterion

A fourth criterion used to select the best subset regression model is related to $\text{SSE} = \sum_{i=1}^{n} (y_i - \hat{y}_i)^2$. The PRESS (prediction sum of squares) statistic for a model

---

*A model is said to be *unbiased* if $E(\hat{y}) = E(y)$. We state (without proof) that for an unbiased regression model, $E(C_p) \approx p + 1$. In general, subset models will be biased since $k - p$ independent variables are omitted from the fitted model. However, when C_p is near $p + 1$, the bias is small and can essentially be ignored.

is calculated as follows:

$$\text{PRESS} = \sum_{i=1}^{n} [y_i - \hat{y}_{(i)}]^2$$

The symbol $\hat{y}_{(i)}$ denotes the predicted value for the ith observation obtained when the regression model is fit with the data point for the ith observation omitted (or deleted) from the sample.* Thus, the candidate model is fit to the sample data n times, each time omitting one of the data points and obtaining the predicted value of y for that data point. Since small differences $y_i - \hat{y}_{(i)}$ indicate that the model is predicting well, we desire a model with a small PRESS.

Computing the PRESS statistic may seem like a tiresome chore, since repeated regression runs (a total of n runs) must be made for each candidate model. However, most statistical software packages have options for computing PRESS automatically.†

Plots aid in the selection of the best subset regression model using the all-possible-regressions procedure. The criterion measure, either R^2, MSE, C_p, or PRESS, is plotted on the vertical axis against p, the number of independent variables in the subset model, on the horizontal axis. We illustrate all three variable selection techniques in Example 4.9.

| EXAMPLE 4.9

In Example 4.8 we applied stepwise regression to identify the most important variables for predicting executive salary from the list of 10 variables given in Table 4.7. Since it is impractical to fit all possible regression models involving subsets of these independent variables (for $k = 10$, there exist 1,023 possible subset models), for this example we fit only the first-order models shown in Table 4.8. These models were selected based, in part, on the results of the stepwise regression. The values of R^2, MSE, C_p, and PRESS are calculated for each model and appear in the appropriate column in Table 4.8. Plot each of these four quantities against p, the number of predictors in the subset model. Interpret the plots.

Solution

Plots of R^2, MSE, C_p, and PRESS against p are shown in Figures 4.28a–d, respectively. In Figure 4.28a, we see that the R^2 values tend to increase in very small amounts for models with more than $p = 4$ predictors. Similarly, Figure 4.28b shows that the MSE decreases by negligible amounts for models with more than $p = 4$ predictors. Thus, both the R^2 and MSE criteria suggest that the model containing the four predictors x_1, x_3, x_4, and x_5 is a good candidate for the best subset regression model.

---

*The quantity $y_i - \hat{y}_{(i)}$ is called the "deleted" residual for the ith observation. We discuss deleted residuals in more detail in Chapter 7.

†PRESS can also be calculated using the results from a regression run on all n data points. The formula is

$$\text{PRESS} = \sum_{i=1}^{n} \left(\frac{y_i - \hat{y}_i}{1 - h_{ii}}\right)^2$$

where h_{ii} is a function of the independent variables in the model. In Chapter 7, we show how h_{ii} (called **leverage**) can be used to detect influential observations.

TABLE 4.8 **Subset Models for the Executive Salary Example**

NUMBER OF PREDICTORS, p	VARIABLES IN THE MODEL	R^2	MSE	C_p	PRESS
1	x_4	.421	.1611	1,339.6	17.62
2	x_4, x_5	.783	.0610	443.8	7.27
3	x_1, x_4, x_5	.897	.0293	164.1	3.45
4	x_1, x_3, x_4, x_5	.948	.0149	38.5	2.63
5	x_1, x_2, x_3, x_4, x_5	.960	.0115	10.2	2.24
6	$x_1, x_2, x_3, x_4, x_5, x_8$	.962	.0111	8.2	2.95
7	$x_1, x_2, x_4, x_5, x_8, x_9, x_{10}$	.963	.0110	7.7	3.01
8	$x_1, x_2, x_3, x_4, x_5, x_8, x_9, x_{10}$	.963	.0111	9.7	3.58
9	$x_1, x_2, x_3, x_4, x_5, x_6, x_8, x_9, x_{10}$	.964	.0109	9.2	3.55
10	$x_1, x_2, x_3, x_4, x_5, x_6, x_7, x_8, x_9, x_{10}$	.964	.0110	11.2	3.71

FIGURE 4.28

Plots of R^2, MSE, C_p, and PRESS for subset regression models of Example 4.9

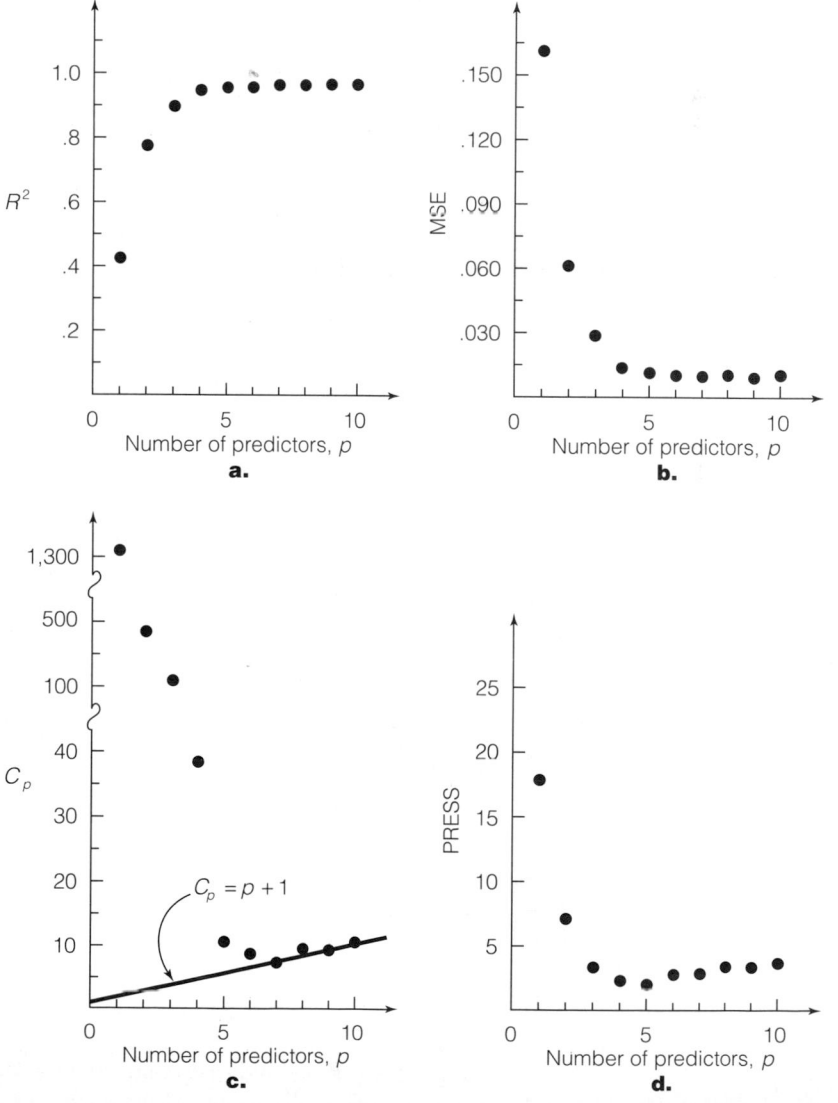

Figure 4.28c shows the plotted C_p values and the line $C_p = p + 1$. Notice that the subset models with $p \geq 5$ independent variables all have relatively small C_p values and vary tightly about the line $C_p = p + 1$. This implies that these models have a small total mean square error and a negligible bias. The model corresponding to $p = 4$, although certainly outperforming the models $p \leq 3$, appears to fall short of the larger models according to the C_p criterion. The C_p value for this model is $C_p = 38.5$ (compared to C_p values ranging from 7.7 to 11.2 for models with $p \geq 5$), and the model has a slight bias. From Figure 4.28d you can see that the PRESS is smallest for the five-variable model with x_1, x_2, x_3, x_4, and x_5 (PRESS = 2.24).

According to all four criteria, the variables x_1, x_3, x_4, and x_5 should be included in the group of the most important predictors. The decision of whether to include another variable (e.g., x_2) in this group is left to the data analyst.

In summary, variable selection procedures based on the R^2, MSE, C_p, or PRESS criterion will assist you in identifying the most important independent variables for predicting y. Keep in mind, however, that these techniques lack the objectivity of a stepwise regression procedure. Furthermore, you should be wary of concluding that the best model for predicting y has been found, since, in practice, interactions and higher-order terms are typically omitted from the list of potential important predictors.

EXERCISE

4.43   Refer to the data on units of production and time worked for a department store clerical staff in Exercise 4.42. For this exercise, consider only the independent variables x_1, x_2, x_3, and x_4 in an all-possible-regressions select procedure.

a. How many models for $E(y)$ are possible, if the model includes (i) one variable, (ii) two variables, (iii) three variables, (iv) four variables?

b. For each case in part a, use a statistical software package to find the maximum R^2, minimum MSE, minimum C_p, and minimum PRESS.

c. Plot each of the quantities R^2, MSE, C_p, and PRESS in part b against p, the number of predictors in the subset model.

d. Based on the plots in part c, which variables would you select for predicting total hours worked, y?

4.13

Multiple Regression: An Example

The basic elements of multiple regression analysis have been presented in Sections 4.1–4.12. Now we assemble these elements by applying them to a practical problem.

In the United States, commercial contractors bid for the right to construct state highways and roads. A state government agency, usually the Department of Transportation (DOT), notifies various contractors of the state's intent to build a highway. Sealed bids are submitted by the contractors, and the contractor with the lowest bid (building cost) is awarded the road construction contract. The bidding process works extremely well in competitive markets, but has the potential to increase construction costs if the markets are noncompetitive or if collusive practices are present. The latter occurred in the 1970s and 1980s in Florida. Numerous contractors either admitted or were found guilty of price-fixing, i.e., setting the cost of construction above the fair, or competitive, cost through bid-rigging or other means.

In this section, we apply multiple regression to a data set obtained from the office of the Florida Attorney General. Our objective is to build and test the adequacy of a model designed to predict the cost y of a road construction contract awarded using the sealed-bid system in Florida.

STEP 1 Based on the opinions of several experts in road construction and bid-rigging, a list of potential predictors of contract cost y is obtained. This list is shown in Table 4.9. Data collected on these eight potential predictors and contract cost for a sample of $n = 235$ contracts are shown in Table 4.10.

T A B L E 4.9 **Description of Several Potential Predictors of Contract Cost**

VARIABLE	DESCRIPTION
DOTEST (x_1)	DOT engineer's estimate of construction cost
B2B1RAT (x_2)	Ratio of second lowest bid to lowest bid
B3B1RAT (x_3)	Ratio of third lowest bid to lowest bid
BHB1RAT (x_4)	Ratio of highest bid to lowest bid
STATUS (x_5)	1 if fixed contract, 0 if competitive contract
DISTRICT (x_6)	1 if contract awarded in South Florida District, 0 if not
BTPRATIO (x_7)	Ratio of number of bidders to number of planholders
DAYSEST (x_8)	Engineer's estimate of number of work days required

T A B L E 4.10 **Data for Bid-Rigging Example**

OBS	COST	DOTEST	B2B1RAT	B3B1RAT	BHB1RAT	STATUS	DISTRICT	BTPRATIO	DAYSEST
1	1379.43	1386.29	1.01397	1.03303	1.06121	1	0	0.33333	250
2	134.03	85.71	1.00995	1.01092	1.01092	1	1	0.75000	45
3	202.33	248.89	1.12084	1.22498	1.30546	0	0	0.50000	120
4	397.12	467.49	1.00588	1.11035	1.26733	0	0	0.50000	180
5	158.54	117.72	1.01053	1.10247	1.10247	1	0	0.37500	80
6	1128.11	1008.91	1.06208	1.09137	1.09137	1	0	0.60000	200
7	400.33	472.98	1.10275	1.13560	1.13560	1	1	0.60000	70
8	581.64	785.39	1.09346	1.16794	1.33349	0	0	0.50000	200
9	353.96	370.02	1.05063	1.28312	1.47836	0	1	0.57143	75
10	138.71	174.25	1.07047	1.19279	1.27559	0	0	0.83333	70
11	383.66	410.95	1.07508	1.13970	1.13970	1	1	0.42857	60
12	3910.94	3405.94	1.02768	1.04733	1.07683	1	1	0.45455	350
13	362.92	385.96	1.01691	1.04658	1.04658	0	1	0.37500	100
14	196.50	235.41	1.16398	1.19491	1.62532	0	0	0.70000	120
15	637.99	627.41	1.07043	1.16355	1.58125	0	0	0.50000	140
16	152.06	175.40	1.07504	1.24451	1.24451	1	1	0.50000	75
17	375.00	432.33	1.05025	1.20642	1.30949	0	0	0.57143	120
18	2284.56	1499.04	1.01600	1.20033	1.20033	1	0	0.60000	270
19	551.45	497.74	1.06668	1.10932	1.10932	1	1	0.60000	100
20	239.67	194.65	1.02302	1.21276	1.21276	1	1	0.60000	65
21	207.87	167.99	1.05143	1.08977	1.15240	1	1	0.66667	60
22	640.48	767.80	1.06059	1.08447	1.27066	0	0	0.40000	90
23	230.54	260.30	1.11029	1.12570	1.12570	1	1	0.42857	125
24	299.87	247.04	1.08411	1.10180	1.10180	1	1	0.60000	80
25	2368.84	2456.77	1.17209	1.18020	1.48550	0	0	0.30769	320

(continued)

TABLE 4.10
Continued

OBS	COST	DOTEST	B2B1RAT	B3B1RAT	BHB1RAT	STATUS	DISTRICT	BTPRATIO	DAYSEST
26	496.49	879.40	1.00453	1.17145	1.38498	0	0	0.58333	140
27	1564.87	1303.40	1.00374	1.04983	1.04983	1	0	0.33333	200
28	7387.03	6107.93	1.01878	1.05413	1.05718	0	1	0.66667	340
29	195.68	199.09	1.04290	1.27466	1.27466	0	1	0.60000	50
30	830.47	715.46	1.01755	1.02450	1.08833	1	0	0.57143	135
31	179.06	208.72	1.02474	1.03067	1.60580	0	1	0.62500	90
32	150.35	199.09	1.00893	1.06483	1.55218	0	0	0.63636	100
33	240.06	429.24	1.10055	1.16394	1.72898	0	0	0.71429	120
34	586.81	709.85	1.01241	1.08838	1.38652	0	0	0.60000	120
35	537.17	676.41	1.03962	1.06037	1.25739	0	0	0.72727	225
36	392.69	490.55	1.07153	1.12770	1.64803	0	1	0.50000	180
37	216.47	406.47	1.05636	1.13153	1.57543	0	1	0.82353	130
38	1559.37	1925.31	1.07850	1.08130	1.60339	0	1	0.47619	250
39	88.31	143.07	1.09813	1.26329	1.50073	0	0	0.71429	55
40	268.45	308.09	1.14764	1.17190	1.32518	0	1	0.66667	75
41	189.02	269.55	1.00993	1.14649	1.95409	0	0	0.85714	115
42	192.81	227.70	1.04378	1.15395	1.45160	0	1	0.83333	60
43	256.22	436.79	1.00000	1.09973	1.89401	0	0	0.88235	125
44	113.61	132.39	1.01195	1.01307	1.60670	0	0	0.69231	70
45	124.99	121.61	1.00000	1.11019	1.26999	0	0	0.57143	40
46	116.57	114.21	1.00000	1.21566	1.91806	0	0	0.57143	35
47	143.13	172.71	1.08608	1.13900	1.38149	0	0	0.60000	50
48	36.19	64.44	1.25303	1.31504	1.49919	0	1	0.62500	60
49	2518.39	3124.39	1.02196	1.04278	1.31920	0	0	0.57143	255
50	1353.51	1617.53	1.12729	1.22658	1.40600	0	1	0.41176	200
51	332.82	376.37	1.00024	1.05449	1.53705	0	0	0.61538	140
52	202.50	300.32	1.18810	1.23180	1.92381	0	0	0.41667	100
53	6043.31	7074.99	1.02471	1.02528	1.22949	0	0	0.57895	350
54	2280.81	2823.87	1.08084	1.09406	1.39083	0	0	0.53333	330
55	99.92	118.99	1.16879	1.29648	1.58897	1	0	0.55556	70
56	1461.59	1774.72	1.01678	1.16689	1.54772	0	1	0.57143	220
57	1217.57	1341.08	1.01074	1.01149	1.39080	0	0	0.91667	350
58	258.44	306.44	1.00453	1.01727	1.25857	0	0	0.70000	75
59	115.42	117.94	1.06379	1.09767	1.44878	0	1	0.85714	65
60	463.93	540.01	1.07317	1.10858	1.35774	0	1	0.62500	120
61	728.86	763.02	1.02446	1.03732	1.44687	0	0	0.73333	200
62	3929.92	3941.57	1.00506	1.06041	1.23232	0	0	0.66667	400
63	181.69	194.82	1.08679	1.16042	1.37565	0	1	0.62500	60
64	479.47	487.17	1.00000	1.04831	1.35510	0	0	0.53333	140
65	93.48	92.36	1.00468	1.02595	1.19858	0	0	0.83333	60
66	2301.07	2505.60	1.06746	1.09047	1.38057	0	0	0.77778	400
67	136.06	181.09	1.13520	1.23219	1.23219	0	1	0.33333	100
68	144.06	252.92	1.30227	1.30894	1.71178	0	1	0.66667	90
69	65.17	84.75	1.20310	1.36565	1.96397	0	0	0.53846	60
70	161.33	164.03	1.07334	1.09596	1.47418	0	1	0.80000	100
71	1138.54	1254.62	1.00892	1.03438	1.37670	0	0	0.68421	200
72	84.79	97.63	1.05202	1.10115	1.36417	0	0	0.60000	50
73	749.13	859.34	1.00000	1.04832	1.49429	0	0	0.92857	180
74	43.67	41.09	1.00048	1.21624	1.33987	0	1	0.50000	315
75	2920.71	2812.50	1.05215	1.11637	1.22283	0	0	0.70000	35
76	32.63	40.05	1.06642	1.25626	2.08540	0	0	0.64286	250
77	1115.12	1148.53	1.00244	1.08541	1.45950	0	0	0.57143	50
78	50.66	59.86	1.29181	1.63089	2.39326	0	0	0.46154	450
79	2229.34	2434.60	1.00386	1.09355	1.21600	0	0	0.50000	360
80	2159.85	2698.59	1.00454	1.05874	1.32056	0	1	0.57143	60
81	45.91	61.89	1.14934	1.51780	2.10760	0	0	0.66667	30
82	127.31	137.52	1.07749	1.11993	1.38302	0	0	0.60000	50
83	147.81	143.34	1.00567	1.16406	1.16406	1	0	0.57143	150
84	470.71	447.75	1.00000	1.01164	1.37056	0	0	0.53846	120
85	188.67	202.46	1.00000	1.09388	1.69629	0	0	0.52381	300
86	4765.17	5035.94	1.00785	1.04760	1.27208	0	0	0.80000	30
87	168.53	163.67	1.05127	1.08435	1.30180	0	0	0.80000	30
88	95.86	95.63	1.06483	1.09471	1.40106	0	0	0.70000	30
89	106.92	107.89	1.03253	1.07409	1.28517	0	0	0.58333	180
90	698.48	646.43	1.01875	1.10615	1.64604	0	0	0.63636	175
91	796.00	969.69	1.14418	1.16051	1.28223	0	0	0.38462	210
92	689.73	801.61	1.06850	1.14537	1.27266	0	1	0.64286	190
93	831.84	906.84	1.00519	1.03516	1.22892	0	0	0.62500	400
94	2150.15	2161.37	1.04553	1.06808	1.29205	0	0	0.75000	100
95	169.75	187.32	1.11882	1.19117	1.19117	0	0	0.87500	130
96	923.58	887.37	1.00000	1.00012	1.29230	0	1	0.50000	350
97	2527.47	2616.81	1.01785	1.14500	1.40809	0	1	0.45455	220
98	726.58	778.27	1.04401	1.08922	1.21345	0	1	0.50000	230
99	1187.10	1573.45	1.05288	1.10879	1.35808	0	0	0.83333	40
100	138.02	149.28	1.04753	1.27350	1.30600	0	0	0.66667	40
101	147.56	162.23	1.02199	1.24037	1.26737	0	0	0.58333	75
102	94.24	116.09	1.13259	1.15708	1.54419	0	0	0.75000	225
103	580.52	675.41	1.00626	1.08359	1.37753	0	0	0.63636	150
104	445.52	435.03	1.13385	1.21180	1.41720	0	0	0.71429	70
105	110.46	120.07	1.08633	1.13347	1.96465	0	0	0.75000	65
106	45.17	68.28	1.12017	1.13175	1.13175	0	0	0.72727	425
107	800.60	1031.45	1.20374	1.23253	1.45298	0	0	0.83333	140
108	495.47	541.81	1.00323	1.13667	1.24085	0	0	0.58333	400
109	1370.06	1377.92	1.04485	1.10819	1.32836	0	0	0.66667	185
110	607.51	809.09	1.01111	1.05622	1.16072	0	0	0.75000	155
111	152.72	161.24	1.02685	1.10447	1.39143	0	0	0.66667	185
112	728.20	916.97	1.09383	1.09383	1.17993	0	0	0.75000	135
113	181.59	146.46	1.01158	1.43074	1.43074	0	0	0.50000	175
114	462.92	504.79	1.06679	1.07253	1.20283	0	0	0.50000	175

TABLE 4.10
Continued

OBS	COST	DOTEST	B2B1RAT	B3B1RAT	BHB1RAT	STATUS	DISTRICT	BTPRATIO	DAYSEST
115	169.38	144.51	1.00382	1.02281	1.02281	0	1	0.50000	145
116	2473.26	2618.24	1.01709	1.04063	1.05926	0	1	0.33333	630
117	2346.77	2447.81	1.08238	1.08388	1.34398	0	0	0.80000	455
118	170.42	196.75	1.00000	1.03571	1.06790	0	0	0.75000	75
119	77.85	109.91	1.12864	1.16302	1.16302	0	0	0.60000	80
120	4770.88	7511.33	1.04645	1.11915	1.42217	0	1	0.52000	600
121	303.34	341.22	1.09503	1.20479	1.40115	0	0	0.77778	210
122	395.98	419.00	1.21029	1.25946	1.25946	0	0	0.42857	185
123	150.55	174.48	1.05048	1.73452	1.73452	0	0	0.50000	110
124	1404.08	1573.83	1.02724	1.04925	1.24881	0	0	0.58333	550
125	1691.66	1627.08	1.03387	1.11274	1.11898	0	0	0.62500	600
126	5196.22	6365.13	1.02538	1.06196	1.30446	0	1	0.42857	805
127	3815.88	4960.82	1.01846	1.04353	1.07254	0	1	0.18182	450
128	122.62	95.38	1.04459	1.05195	1.07457	0	0	0.40000	120
129	1571.15	1759.21	1.01214	1.12645	1.27038	0	1	0.62500	385
130	4385.47	5556.82	1.27799	1.16978	1.72492	0	1	0.31818	600
131	4497.56	5186.30	1.00691	1.33380	1.51646	0	1	0.26316	750
132	23.67	32.54	1.05197	1.18120	2.22753	0	0	0.77778	60
133	1048.86	1040.53	1.05138	1.07009	1.27610	0	0	0.66667	300
134	239.51	315.11	1.05008	1.07829	1.17735	0	1	0.80000	100
135	260.64	351.80	1.02397	1.04081	1.17632	0	1	0.80000	120
136	138.89	123.31	1.00729	1.07116	1.07116	0	1	0.50000	150
137	128.38	151.89	1.00059	1.12683	1.12683	0	1	0.60000	75
138	284.98	315.97	1.04283	1.10622	1.10622	0	1	0.50000	110
139	77.32	101.04	1.00690	1.30480	1.30480	0	0	0.50000	90
140	5411.51	5086.25	1.02548	1.05641	1.29599	0	1	0.38889	815
141	3864.60	3991.03	1.02963	1.03230	1.48893	0	1	0.41176	800
142	2976.03	2832.43	1.06799	1.17414	1.56334	0	1	0.37500	440
143	257.51	205.86	1.01018	1.04526	1.05595	1	1	0.80000	130
144	36.44	42.04	1.02332	1.09791	1.09791	0	1	0.75000	45
145	182.52	223.13	1.00000	1.26493	1.26493	0	0	0.40000	90
146	1367.07	1433.62	1.07465	1.08446	1.35155	0	0	0.80000	475
147	76.81	75.25	1.01507	1.05527	1.05527	1	1	0.75000	90
148	1747.88	1493.45	1.01231	1.19420	1.31394	0	0	0.54545	445
149	5734.42	5427.31	1.02020	1.02571	1.32330	0	1	0.88889	700
150	5884.70	6097.41	1.03037	1.08206	1.26745	0	1	0.30009	900
151	346.52	331.10	1.12303	1.12316	1.47592	0	0	0.62500	190
152	646.40	818.54	1.29112	1.38469	1.59691	0	1	0.45455	400
153	760.84	718.24	1.02365	1.05158	1.05158	1	0	0.50000	185
154	169.77	175.95	1.02389	1.02754	1.05433	1	0	0.44444	125
155	138.79	151.62	1.00853	1.08094	1.08094	0	1	0.75000	60
156	346.76	394.21	1.10806	1.11556	1.11556	0	0	0.37500	150
157	1082.17	1085.87	1.07610	1.08797	1.34776	0	1	0.66667	400
158	253.68	270.93	1.07228	1.34120	1.34120	0	0	0.60000	200
159	433.15	545.26	1.00422	1.18435	1.24967	0	0	0.71429	230
160	10270.45	10467.40	1.03748	1.04612	1.28400	0	1	0.40000	720
161	1398.03	1414.73	1.09989	1.12889	1.22787	0	0	0.71429	460
162	2140.88	2152.01	1.00491	1.02566	1.02566	0	1	0.42857	400
163	6584.11	5949.35	1.02135	1.03414	1.37884	0	1	0.50000	675
164	666.77	641.52	1.14124	1.24950	1.31278	0	0	0.55556	240
165	108.09	99.56	1.08943	1.21201	1.21201	0	0	0.50000	65
166	106.10	95.29	1.00661	1.07263	1.33137	0	0	0.85714	125
167	549.64	413.08	1.02910	1.12878	1.12878	1	1	0.60000	120
168	1272.04	949.46	1.01918	1.02475	1.10157	1	0	0.55556	450
169	122.82	132.26	1.06736	1.30510	1.30510	0	0	0.30000	130
170	359.45	333.45	1.03525	1.04301	1.29319	0	1	0.50000	190
171	1731.47	1672.53	1.19267	1.21635	1.21635	0	1	0.33333	400
172	31.72	28.30	1.06263	1.22120	1.22120	0	0	0.50000	60
173	3299.96	2805.89	1.00757	1.05891	1.08968	1	1	0.44444	525
174	480.56	480.31	1.13607	1.34698	1.36586	0	0	0.66667	125
175	673.09	655.81	1.00000	1.08450	1.10744	0	0	0.83333	100
176	116.99	99.94	1.04192	1.07735	1.07735	1	0	0.75000	100
177	1157.39	891.21	1.01295	1.09076	1.09076	1	0	0.75000	450
178	166.80	131.99	1.07707	1.08335	1.08335	1	0	0.42857	120
179	668.53	596.89	1.03133	1.05542	1.05542	1	1	0.75000	120
180	7622.16	7871.19	1.06781	1.08947	1.18429	0	1	0.37500	700
181	201.32	182.94	1.04814	1.07143	1.07143	0	0	0.42857	90
182	1270.08	1306.33	1.02258	1.07352	1.07352	1	1	0.75000	195
183	1055.14	1148.65	1.03627	1.10087	1.33284	0	1	0.45455	400
184	5212.23	5090.86	1.01774	1.01786	1.25398	0	1	0.56250	500
185	5654.86	5447.59	1.04491	1.06438	1.06438	1	1	0.37500	500
186	856.46	938.14	1.06181	1.15976	1.23706	0	1	0.50000	375
187	88.98	66.06	1.15552	1.18257	1.19208	1	1	0.50000	90
188	200.00	168.99	1.10151	1.12477	1.12477	1	1	0.75000	90
189	234.04	179.74	1.03977	1.04869	1.07225	0	0	0.80000	170
190	116.56	125.85	1.05611	1.07894	1.07894	0	1	0.75000	80
191	82.11	93.04	1.00000	1.15889	1.39368	0	0	0.75000	100
192	207.81	214.25	1.07698	1.09489	1.09489	0	0	0.50000	155
193	463.28	474.89	1.00903	1.04904	1.18654	0	0	0.88889	215
194	7385.55	8460.87	1.04472	1.05852	1.05852	0	0	0.23077	505
195	91.66	100.31	1.02867	1.12586	1.18879	0	1	0.57143	90
196	546.16	622.92	1.00235	1.07635	1.07635	0	0	0.37500	165
197	740.30	810.26	1.00000	1.03483	1.14590	0	0	0.44444	175
198	888.44	883.30	1.01844	1.03710	1.18272	1	0	0.83333	250
199	656.75	750.82	1.03327	1.06556	1.11905	1	1	0.44444	180
200	1884.39	1550.49	1.01914	1.08680	1.08680	1	1	0.42857	350
201	4448.13	4197.79	1.01046	1.02215	1.02215	1	0	0.50000	660
202	258.20	181.95	1.00732	1.02541	1.04932	1	1	0.80000	130
203	1949.63	1880.83	1.05165	1.10803	1.17919	1	1	0.44444	330

TABLE 4.10
Continued

OBS	COST	DOTEST	B2B1RAT	B3B1RAT	BHB1RAT	STATUS	DISTRICT	BTPRATIO	DAYSEST
204	235.28	230.75	1.01587	1.08762	1.12906	1	0	0.83333	90
205	35.18	39.21	1.03338	1.39979	1.39979	0	1	0.60000	45
206	244.76	221.88	1.00543	1.02878	1.17723	1	0	0.80000	90
207	648.92	563.88	1.02119	1.02659	1.02659	1	0	0.60000	140
208	391.47	358.53	1.02829	1.09437	1.09437	1	0	0.60000	100
209	267.78	249.91	1.03914	1.06844	1.11820	1	0	0.36364	255
210	2130.04	2019.87	1.10956	1.16759	1.19220	0	1	0.71429	450
211	301.23	303.19	1.07610	1.10834	1.10834	1	0	0.60000	110
212	1077.90	878.72	1.04175	1.06434	1.06768	1	0	0.80000	190
213	927.38	902.03	1.11036	1.16285	1.16285	0	1	0.25000	400
214	241.70	243.97	1.04946	1.16941	1.16941	1	1	0.75000	45
215	65.79	82.36	1.18645	1.20456	1.22890	0	1	0.44444	60
216	1208.44	1230.33	1.00000	1.30919	1.49820	0	0	0.30769	295
217	9453.35	9479.73	1.02255	1.03217	1.14419	0	1	0.31579	500
218	7098.11	8296.80	1.00855	1.03726	1.19543	0	1	0.50000	510
219	912.06	1137.65	1.00000	1.28672	1.28672	0	0	0.33333	220
220	259.99	319.59	1.00717	1.06833	1.06854	1	0	0.80000	90
221	8992.25	10743.60	1.03058	1.05344	1.36599	0	1	0.45455	650
222	339.88	428.82	1.20245	1.23939	1.23939	0	1	0.37500	165
223	833.66	859.74	1.05191	1.06098	6.04598	1	0	0.33333	450
224	4833.82	6225.04	1.00000	1.06601	1.37437	0	0	0.53333	520
225	271.94	223.89	1.01232	1.03402	1.10971	1	1	0.66667	110
226	2966.28	4433.47	1.07730	1.30852	1.51367	0	0	0.25000	720
227	577.37	701.07	1.00000	1.25713	1.25713	0	0	0.25000	150
228	10480.32	10276.29	1.02502	1.03832	1.30423	0	1	0.68750	570
229	462.39	444.19	1.04262	1.04489	1.05778	0	1	0.80000	120
230	2558.19	2741.05	1.14482	1.16483	1.19685	0	1	0.44444	365
231	2814.91	2816.73	1.02002	1.11954	1.26368	0	0	0.54545	540
232	119.81	122.16	1.00000	1.06686	1.29526	0	0	0.66667	90
233	3184.86	3373.04	1.00000	1.02879	1.35838	0	1	0.58333	240
234	473.20	548.01	1.11100	1.12516	1.12516	0	0	0.37500	130
235	400.48	496.68	1.06915	1.08216	1.18507	0	1	0.50000	90

STEP 2 Since the number of potential predictors is large, we employ stepwise regression to aid in the selection of the independent variables to include in the model. The SAS stepwise regression procedure applied to the sample data resulted in the printout shown in Figure 4.29. You can see that only two of the seven variables—DOTEST (x_1) and STATUS (x_5)—were selected by the stepwise routine. Our modeling effort will focus on these two independent variables.

In Chapter 5 we will learn that a good initial choice is the complete second-order model. For one quantitative and one qualitative variable, the model has the following form:

$$E(y) = \beta_0 + \beta_1 x_1 + \beta_2 x_1^2 + \beta_3 x_5 + \beta_4 x_1 x_5 + \beta_5 x_1^2 x_5$$

where x_1 = DOTEST and x_5 = STATUS.

STEP 3 The SAS printout for the complete second-order model is shown in Figure 4.30. The β estimates, shaded on the printout, yield the following least squares prediction equation:

$$\hat{y} = -2.975 + .9155x_1 + .00000072x_1^2 - 36.725x_5 + .324x_1x_5 - .0000358x_1^2x_5$$

STEP 4 Before we can make inferences about model adequacy, we should be sure that the standard regression assumptions about the random error ε are satisfied. For given values of x_1 and x_5, the random error ε has a normal distribution with mean 0, constant variance σ^2, and are independent. We learn how to check these assumptions in Chapter 7. For now, we are satisfied with estimating σ^2 and interpreting its value.

The value of MSE, shaded on Figure 4.30, is MSE = s^2 = 87,998.9. The value of **Root MSE** (also shaded) is s = 296.65. Our interpretation is that the complete second-order model can predict contract costs to within $2s$ = 593.3 thousand dollars of its true value.

FIGURE 4.29

SAS stepwise regression printout for contract cost

Stepwise Procedure for Dependent Variable COST

Step 1   Variable DOTEST Entered   R-square = 0.97424702   C(p) = 15.18496446

	DF	Sum of Squares	Mean Square	F	Prob>F
Regression	1	864035547.29525	864035547.29525	8814.50	0.0001
Error	233	22839676.680072	98024.36343379		
Total	234	886875223.97532			

Variable	Parameter Estimate	Standard Error	Type II Sum of Squares	F	Prob>F
INTERCEP	20.90684416	24.36729323	72159.98186400	0.74	0.3918
DOTEST	0.92628789	0.00986614	864035547.29525	8814.50	0.0001

Step 2   Variable STATUS Entered   R-square = 0.97545236   C(p) = 5.66262181

	DF	Sum of Squares	Mean Square	F	Prob>F
Regression	2	865104526.39042	432552263.19521	4609.50	0.0001
Error	232	21770697.584902	93839.21372803		
Total	234	886875223.97532			

Variable	Parameter Estimate	Standard Error	Type II Sum of Squares	F	Prob>F
INTERCEP	-20.53871363	26.81797336	55040.06386988	0.59	0.4445
DOTEST	0.93077968	0.00974453	856162794.59683	9123.72	0.0001
STATUS	166.35513274	49.28829319	1068979.0951699	11.39	0.0009

All variables in the model are significant at the 0.1500 level.
No other variable met the 0.0500 significance level for entry into the model.

Summary of Stepwise Procedure for Dependent Variable COST

Step	Variable Entered Removed	Number In	Partial R**2	Model R**2	C(p)	F	Prob>F
1	DOTEST	1	0.9742	0.9742	15.1850	8814.4979	0.0001
2	STATUS	2	0.0012	0.9755	5.6626	11.3916	0.0009

FIGURE 4.30

SAS printout for complete second-order model

Dependent Variable: COST

Analysis of Variance

Source	DF	Sum of Squares	Mean Square	F Value	Prob>F
Model	5	866723465.17	173344693.03	1969.850	0.0001
Error	229	20151758.803	87998.94674		
C Total	234	886875223.98			

Root MSE	296.64616	R-square	0.9773	
Dep Mean	1268.70217	Adj R-sq	0.9768	
C.V.	23.38186			

Parameter Estimates

Variable	DF	Parameter Estimate	Standard Error	T for H0: Parameter=0	Prob > |T|
INTERCEP	1	-2.975454	30.89143173	-0.096	0.9234
DOTEST	1	0.915530	0.02917084	31.385	0.0001
DOTEST2	1	0.000000719	0.00000340	0.211	0.8330
STATUS	1	-36.724712	74.77308250	-0.491	0.6238
STA_DOT	1	0.324213	0.11917429	2.720	0.0070
STA_DOT2	1	-0.000035759	0.00002478	-1.443	0.1504

STEP 5 To check the adequacy of the complete second-order model, we conduct the analysis of variance F test. The elements of the test are as follows:

H_0: $\beta_1 = \beta_2 = \beta_3 = \beta_4 = \beta_5 = 0$

H_a: At least one $\beta \neq 0$

Test statistic: $F = 1,969.85$ (shaded in Figure 4.30)

p-value: $p = .0001$ (shaded in Figure 4.30)

Conclusion: The extremely small p-value indicates that the model is statistically adequate for predicting contract cost, y.

Are all the terms in the model statistically significant predictors? For example, is it necessary to include the curvilinear terms, $\beta_2 x_1^2$ and $\beta_5 x_1^2 x_5$, in the model? If not, the model can be simplified by dropping these curvature terms. The hypothesis we want to test is

H_0: $\beta_2 = \beta_5 = 0$

H_a: At least one of the curvature β's is nonzero.

To test this subset of β's, we need to fit a second (reduced) model. The reduced model takes the form

$$E(y) = \beta_0 + \beta_1 x_1 + \beta_3 x_5 + \beta_4 x_1 x_5$$

The SAS printout for the reduced model is shown in Figure 4.31. The SSE for the reduced model, $\text{SSE}_R = 20,334,954$ (shaded in Figure 4.31), is compared to the SSE for the complete model, $\text{SSE}_C = 20,151,759$ (shaded in Figure 4.30), using the test statistic computed here:

Test statistic: $F = \dfrac{(\text{SSE}_R - \text{SSE}_C)/\#\ \beta\text{'s tested}}{\text{MSE}_C}$

$$= \frac{(20,334,954 - 20,151,759)/2}{87,999}$$

$$= 1.04$$

Rejection region: Using $\alpha = .01$, $F_{.01} \approx 4.61$ (based on 2 numerator and 229 denominator degrees of freedom)

Conclusion: Since $F = 1.04$ falls below the critical value of 4.61, we fail to reject H_0. That is, there is insufficient evidence (at $\alpha = .01$) to indicate that the curvature terms are useful predictors of construction cost, y.

The results of the partial F test lead us to select the reduced model as the better predictor of cost. The least squares prediction equation for the reduced model is

$$\hat{y} = -6.429 + .921 x_1 + 28.671 x_5 + .163 x_1 x_5$$

Note that we cannot simplify the model any further. The t test for the interaction term $\beta_3 x_1 x_5$ is highly significant (p-value $= .0001$, shaded on Figure 4.31). Thus,

FIGURE 4.31

SAS printout for reduced model

Dependent Variable: COST

Analysis of Variance

Source	DF	Sum of Squares	Mean Square	F Value	Prob>F
Model	3	866540269.49	288846756.50	3281.227	0.0001
Error	231	20334954.484	88030.10599		
C Total	234	886875223.98			

Root MSE	296.69868	R-square	0.9771	
Dep Mean	1268.70217	Adj R-sq	0.9768	
C.V.	23.38600			

Parameter Estimates

Variable	DF	Parameter Estimate	Standard Error	T for H0: Parameter=0	Prob > |T|
INTERCEP	1	-6.428954	26.20854879	-0.245	0.8064
DOTEST	1	0.921336	0.00972347	94.754	0.0001
STATUS	1	28.670505	58.66231493	0.489	0.6255
STA_DOT	1	0.163282	0.04043122	4.039	0.0001

our best model for construction cost proposes interaction between the DOT estimate (x_1) and status (x_5) of the contract, but only a linear relationship between cost and DOT estimate.

A plot of the least squares lines for the reduced model is shown in Figure 4.32. You can see that the model proposes two straight lines (one for fixed contracts and one for competitive contracts) with different slopes. The estimated slopes of the y–x_1 lines are computed and interpreted as follows:

Competitive contracts ($x_5 = 0$): Estimated slope = $\hat{\beta}_1$ = .921
For every \$1,000 increase in DOT estimate, we estimate contract cost to increase \$921.

Fixed contracts ($x_5 = 1$): Estimated slope = $\hat{\beta}_1 + \hat{\beta}_4$ = .921 + .163 = 1.084
For every \$1,000 increase in DOT estimate, we estimate contract cost to increase \$1,084.

FIGURE 4.32

Plot of the least squares lines for the reduced model

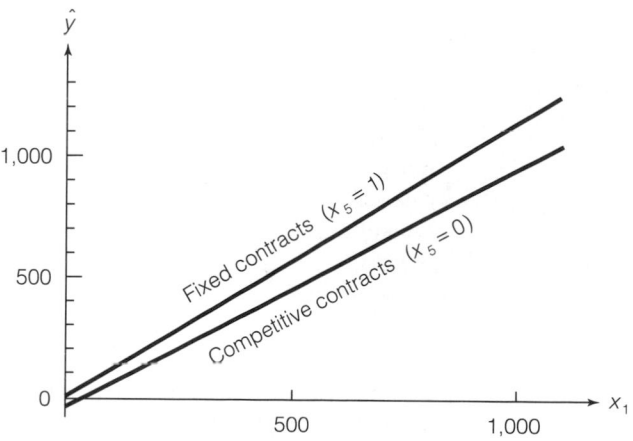

Before deciding to use the interaction model for estimation and/or prediction (step 6), we should check R^2 and s for the model. $R^2 = .9771$ (shaded on Figure 4.31) indicates that nearly 98% of the variation in the sample of construction costs can be "explained" by the model. The value of s (also shaded) implies that we can predict construction cost to within about $2s = 2(296.7) = 593.40$ thousand dollars of its true value using the model. Although the R^2 value is high, the large $2s$ value suggests that the predictive ability of the model might be improved by additional independent variables.

STEP 6 A portion of the SAS printout for the interaction (reduced) model not shown earlier is presented in Figure 4.33. The printout gives predicted values and 95% prediction intervals for the first 10 contracts in the sample. The shaded portion gives the 95% prediction interval for contract cost when the DOT estimate is \$1,386,290 ($x_1 = 1,386.29$) and the contract is fixed ($x_2 = 1$). For a contract with these characteristics, we predict the cost to fall between 933.7 thousand dollars and 2,118.0 thousand dollars, with 95% confidence.

FIGURE 4.33

SAS printout of 95% prediction intervals for reduced model

Obs	DOTEST	STATUS	Dep Var COST	Predict Value	Std Err Predict	Lower95% Predict	Upper95% Predict	Residual
1	1386.293	1	1379.4	1525.8	47.792	933.7	2118.0	−146.4
2	85.712	1	134.0	115.2	50.533	−477.8	708.2	18.8263
3	248.892	0	202.3	222.9	24.949	−363.8	809.5	−20.5498
4	467.488	0	397.1	424.3	23.989	−162.2	1010.8	−27.1655
5	117.719	1	158.5	149.9	49.844	−442.9	742.7	8.6181
6	1008.908	1	1128.1	1116.5	42.729	525.9	1707.1	11.5835
7	472.9837	1	400.3	535.2	43.916	−55.7095	1126.2	−134.9
8	785.3859	0	581.6	717.2	22.876	130.9	1303.5	−135.5
9	370.022	0	354.0	334.5	24.399	−252.1	921.0	19.4725
10	174.255	0	138.7	154.1	25.309	−432.6	740.8	−15.4045

4.14

A Summary of the Steps to Follow in a Multiple Regression Analysis

We have discussed some of the methodology of **multiple regression analysis**, a technique for modeling a dependent variable y as a function of several independent variables $x_1, x_2, \ldots, x_k$. The steps we follow in constructing and using multiple regression models, which are much the same as those for the simple straight-line models, are listed here:

1. **The set of independent variables to be included in the model is identified.** If the number of independent variables is large, you may want to use a variable selection technique such as **stepwise regression** to screen out those that do not seem important for the prediction of y.

2. **The form of the probabilistic model is hypothesized.** The model may include **second-order** terms for quantitative variables, **interaction** terms, and **dummy variables** for qualitative variables. Remember that a model with no interaction terms implies that each of the independent variables affects the response y independently of the other independent variables. **Quadratic** (or **second-order**) terms add curvature to the response curve when $E(y)$ is plotted as a function of the quantitative independent variable. Dummy variables allow the mean response to differ for the different levels of the qualitative variable.

3. **The model coefficients are estimated using the method of least squares.**

4. The probability distribution of ε is specified and σ^2 is estimated.
5. The utility of the model is checked using the analysis of variance F test and the multiple coefficient of determination R^2. The F test for testing a partial set of β parameters and t tests on individual β parameters aid in deciding the final form of the model.
6. If the model is deemed useful, it may be used to make estimates of $E(y)$ and to predict future values of y.

Subsequent chapters extend the methods of this chapter to special applications and problems encountered during a regression analysis. One of the most common problems faced by regression analysts is the problem of multicollinearity, i.e., intercorrelations among the independent variables. Methods for detecting and overcoming multicollinearity, as well as other problems, are discussed in Chapter 6. Another aspect of regression analysis is the analysis of the residuals, i.e., the deviations between the observed and the predicted values of y. An analysis of residuals (Chapter 7) may indicate that the data do not comply with the assumptions of Section 4.2 and may suggest appropriate procedures for modifying the data analysis.

SUPPLEMENTARY EXERCISES

4.44   After a regression model is fit to a set of data, a confidence interval for the mean value of y at a given setting of the independent variables will *always* be narrower than the corresponding prediction interval for a particular value of y at the same setting of the independent variables. Why?

4.45   Most homes in the United States are sold through a real estate broker, who receives a commission on each sale. As a future homebuyer, you may want to know the extent to which brokerage commissions are incorporated into the selling price of the home. Stated more simply, is the price paid by a homebuyer higher when property is sold through a real estate broker than when sold directly by an owner? Evidence collected by researchers at Louisiana State University suggests that homebuyers may, in fact, save in costs by purchasing owner-offered properties (*The Real Estate Appraiser and Analyst*, Winter 1985). Data on the sample of 111 transactions were used to fit the model

$$E(y) = \beta_0 + \beta_1 x_1 + \beta_2 x_2 + \beta_3 x_3 + \beta_4 x_4 + \beta_5 x_5 + \beta_6 x_6$$

where

y = Sale price (in dollars)

x_1 = Age of home (years)

x_2 = Living area (square feet)

x_3 = Area (square feet) of other improvements (e.g., garage, porch, etc.)

x_4 = Selling date

$x_5 = \begin{cases} 1 & \text{if home was conventionally financed} \\ 0 & \text{if not} \end{cases}$

$x_6 = \begin{cases} \text{Brokerage commission (in dollars)} & \text{if home listed with real estate broker} \\ 0 & \text{if home not listed with real estate broker} \end{cases}$

The regression results are summarized here. (The standard errors of the β estimates are given in parentheses.)

$$\hat{y} = 10.67 - 725x_1 + 27.50x_2 + 10.15x_3 + 470.12x_4 - 1,450x_5 + .43x_6 \qquad R^2 = .86$$
$$\phantom{\hat{y} = 10.67 -} (120.8) \quad (1.73) \qquad (2.67) \qquad (940.24) \quad (1,208.3) \quad (.25)$$

a. Is the model adequate for predicting the sale price of a home? Test using $\alpha = .10$.
b. Calculate a 90% confidence interval for β_1. Interpret your result.
c. Interpret the estimate of β_5.
d. Test the hypothesis that a brokerage commission increases the mean selling price of a home. Test using $\alpha = .10$.

4.46 In hopes of increasing the company's share of the fine food market, researchers for a meat-processing firm that prepares meats for exclusive restaurants are working to improve the quality of their hickory-smoked hams. One of their studies concerns the effect of time spent in the smokehouse on the flavor of the ham. Hams that were in the smokehouse for varying amounts of time were each subjected to a taste test by a panel of 10 food experts. The following model was thought to be appropriate by the researchers:

$$y = \beta_0 + \beta_1 t + \beta_2 t^2 + \varepsilon$$

where

y = Mean of the taste scores for the 10 experts

t = Time in the smokehouse (hours)

Assume the least squares model estimated using a sample of 20 hams is

$$\hat{y} = 20.3 + 5.2t - .0025t^2$$

and that $s_{\hat{\beta}_2} = .0011$. The coefficient of determination is $R^2 = .79$.
a. Is there evidence to indicate that the overall model is useful? Test at $\alpha = .05$.
b. Is there evidence to indicate that the second-order term is important in this model? Test at $\alpha = .05$.

4.47 A utility company in a major city gave the average utility bills (listed in the table) for a standard-size home during the year.

MONTH	AVERAGE MONTHLY TEMPERATURE x, °F	AVERAGE UTILITY BILL y, dollars
January	38	99
February	45	91
March	49	78
April	57	61
May	69	55
June	78	63
July	84	80
August	89	95
September	79	65
October	64	56
November	54	74
December	41	93

a. Plot the points on a scattergram.
b. Use the methods of Chapter 3 to fit the model

$$y = \beta_0 + \beta_1 x + \varepsilon$$

What would you conclude about the utility of this model?
c. Hypothesize another model that might better describe the relationship between the average utility bill and average temperature. If you have access to a computer package, fit the model and test its utility.

4.48   A company that relies on door-to-door sales wants to determine the relationship, if any, between the proportion of customers who buy its product, y, and two independent variables: price, x_1, and years of experience of the salesperson, x_2. Twenty salespeople employed by the company are randomly assigned to sell the products, five salespeople to each of four prices, ranging from $1.98 to $5.98. Each salesperson makes a sales presentation to 30 prospects, and the percentage of sales is recorded. The 20 observations (five salespeople for each of four prices) are used to fit the model

$$y = \beta_0 + \beta_1 x_1 + \beta_2 x_2 + \varepsilon$$

The least squares prediction equation is

$$\hat{y} = -.30 - .010x_1 + .10x_2$$

with $s_{\hat{\beta}_1} = .0030$, $s_{\hat{\beta}_2} = .025$, and $R^2 = .86$.
a. Interpret the values of $\hat{\beta}_1$ and $\hat{\beta}_2$.
b. Test the null hypothesis $H_0: \beta_1 = \beta_2 = 0$ that the model is not useful for predicting y. Use $\alpha = .05$.
c. Do the data support the research hypothesis that as the price of the product is increased the mean proportion of buyers will decrease, when x_2 is held fixed?
d. Is there evidence that as the experience of the salesperson increases the mean proportion of buyers increases, when x_1 is fixed?
e. Suppose it is claimed that the least squares model cannot be correct, since the value of $\hat{\beta}_0 = -.30$, and a negative proportion of buyers is clearly impossible. How do you refute this argument?

4.49   To increase the motivation and productivity of workers, an electronics manufacturer decides to experiment with a new pay incentive structure at one of two plants. The experimental plan will be tried at plant A for 6 months, while workers at plant B will remain on the original pay plan. To evaluate the effectiveness of the new plan, the average assembly time for part of an electronic system was measured for employees at both plants at the beginning and end of the 6-month period. Suppose the following model was proposed:

$$y = \beta_0 + \beta_1 x_1 + \beta_2 x_2 + \varepsilon$$

where

y = Assembly time (hours) at end of 6-month period

x_1 = Assembly time (hours) at beginning of 6-month period

$x_2 = \begin{cases} 1 & \text{if plant A} \\ 0 & \text{if plant B} \end{cases}$ (dummy variable)

A sample of $n = 42$ observations yielded

$$\hat{y} = .11 + .98x_1 - .53x_2$$

where

$$s_{\hat{\beta}_1} = .231$$
$$s_{\hat{\beta}_2} = .48$$

Test to see whether, after allowing for the effect of initial assembly time, plant A had a lower mean assembly time than plant B. Use $\alpha = .01$.

4.50   The Environmental Protection Agency wants to model the gas mileage ratings, y, of automobiles as a function of their engine size, x. A second-order model,

$$E(y) = \beta_0 + \beta_1 x + \beta_2 x^2$$

is proposed. A sample of 50 engines of varying size is selected and the miles per gallon rating of each is determined. The least squares prediction equation is

$$\hat{y} = 51.3 - 10.1x + .15x^2$$

The size, x, of the engine is measured in hundreds of cubic inches. Also, $s_{\hat{\beta}_2} = .0037$ and $R^2 = .93$.

a. Sketch this model over the interval $x = 1$ to $x = 4$.

b. Is there evidence that the second-order term in the model is contributing to the prediction of the miles per gallon rating, y? Use $\alpha = .05$.

c. Use the model to estimate the mean miles per gallon rating for all cars with 350-cubic-inch engines ($x = 3.5$).

d. Suppose a 95% confidence interval for the quantity estimated in part c is calculated to be (17.2, 18.4). Interpret this interval.

e. Suppose you purchase an automobile with a 350-cubic-inch engine and determine that the miles per gallon rating is 14.7. Is the fact that this value lies outside the confidence interval given in part d surprising? Explain.

4.51   How satisfied are you with your auto repair service? G. J. Biehal (University of Houston) examined the factors that influence the satisfaction level for auto repair services in California. One objective of the study was to show that the extent of the information search conducted by the customer prior to selecting an auto repair service affects the customer's satisfaction with the service. A sample of 208 households in Palo Alto and Menlo Park, California, were questioned about their most recent auto repair experiences. All respondents reported at least one auto repair expense that exceeded $100 during the previous year. The dependent variable, level of dissatisfaction y, was measured on a 7-point Likert scale (0 = very satisfied to 7 = very dissatisfied). The following linear model was proposed:

$$E(y) = \beta_0 + \beta_1 x_1 + \beta_2 x_2 + \beta_3 x_3 + \beta_4 x_4 + \beta_5 x_5 + \beta_6 x_6$$

where

Level of external information search: x_1 (0 = none to 7 = very high)

Repair cost: x_2 (dollars)

Problem corrected: $x_3 = \begin{cases} 1 & \text{if more than one visit required to correct the problem} \\ 0 & \text{if the problem was corrected on the first visit} \end{cases}$

Prior experience with service:

$x_4 = \begin{cases} 1 & \text{if 1–2 times} \\ 0 & \text{if not} \end{cases}$ $x_5 = \begin{cases} 1 & \text{if 3–5 times} \\ 0 & \text{if not} \end{cases}$ $x_6 = \begin{cases} 1 & \text{if more than 5 times} \\ 0 & \text{if not} \end{cases}$

The results of the multiple regression analysis are summarized in the table.

PARAMETER	ESTIMATE	p-VALUE
β_0	2.14	—
β_1	−.15	$p < .05$
β_2	.03	$p > .05$
β_3	2.54	$p < .01$
β_4	−.34	$p > .05$
β_5	−.26	$p > .05$
β_6	−.72	$p < .05$
$R^2 = .43$		

Source: Biehal, G. J. "Consumers' prior experiences and perceptions in auto repair choice." *Journal of Marketing,* Summer 1983, Vol. 47, No. 3, pp. 82–91. Reprinted by permission of the American Marketing Association.

a. Write the least squares prediction equation.
b. Is the model useful for predicting the dissatisfaction level for auto repair services? Test using $\alpha = .05$.
c. Is there evidence to indicate that the dissatisfaction level for auto repair service declines as the amount of external information search increases? Test using $\alpha = .05$.
d. Is there evidence to indicate that repair cost is a useful predictor of the dissatisfaction level for auto repair service? Test using $\alpha = .05$.
e. What hypothesis would you test to determine whether prior experience with service has no effect on the level of dissatisfaction?

4.52 To determine whether extra personnel are needed for the day, the owners of a water adventure park would like to find a model that would allow them to predict the day's attendance each morning before opening based on the day of the week and weather conditions. The model is of the form

$$E(y) = \beta_0 + \beta_1 x_1 + \beta_2 x_2 + \beta_3 x_3$$

where

$y = $ Daily admissions

$x_1 = \begin{cases} 1 & \text{if weekend} \\ 0 & \text{otherwise} \end{cases}$ (dummy variable)

$x_2 = \begin{cases} 1 & \text{if sunny} \\ 0 & \text{if overcast} \end{cases}$ (dummy variable)

$x_3 = $ Predicted daily high temperature (°F)

After collecting 30 days of data, the following least squares model is obtained:

$$\hat{y} = -105 + 25x_1 + 100x_2 + 10x_3$$

with $s_{\hat{\beta}_1} = 10$, $s_{\hat{\beta}_2} = 30$, and $s_{\hat{\beta}_3} = 4$. Also, $R^2 = .65$.
a. Interpret the model coefficients.
b. Is there sufficient evidence to conclude that this model is useful in the prediction of daily attendance? Use $\alpha = .05$.

c. Is there sufficient evidence to conclude that mean attendance increases on weekends? Use $\alpha = .10$.

d. Use the model to predict the attendance on a sunny weekday with a predicted high temperature of 95°F.

e. Suppose the 90% prediction interval for part **d** is (645, 1,245). Interpret this interval.

4.53   Refer to Exercise 4.52. The owners of the water adventure park are advised that the prediction model could probably be improved if interaction terms were added. In particular, it is thought that the *rate* of increase in mean attendance with increases in predicted high temperature will be greater on weekends than on weekdays. The following model is therefore proposed:

$$E(y) = \beta_0 + \beta_1 x_1 + \beta_2 x_2 + \beta_3 x_3 + \beta_4 x_1 x_3$$

The same 30 days of data used in Exercise 4.52 are again used to obtain the least squares model

$$\hat{y} = 250 - 700x_1 + 100x_2 + 5x_3 + 15x_1 x_3$$

with $s_{\hat{\beta}_4} = 3.0$ and $R^2 = .96$.

a. Graph the predicted day's attendance, $\hat{y}$, against the day's predicted high temperature, x_3, for a sunny weekday and for a sunny weekend day. Graph both on the same paper for x_3 between 70°F and 100°F. Note the increase in slope for the weekend day.

b. Do the data indicate that the interaction term is a useful addition to the model? Use $\alpha = .05$.

c. Use this model to predict the attendance on a sunny weekday with a predicted high temperature of 95°F.

d. Suppose the 90% prediction interval for part **c** is (800, 850). Compare this with the prediction interval for the model without interaction in Exercise 4.52, part **e**. Do the relative widths of the confidence intervals support or refute your conclusion about the utility of the interaction term (part **b**)?

e. The owners, noting that the coefficient $\hat{\beta}_1 = -700$, conclude the model is ridiculous, because it seems to imply that the mean attendance will be 700 less on weekends than on weekdays. Refute their argument.

4.54   Refer to Exercise 4.53. Suppose the second-order model

$$E(y) = \beta_0 + \beta_1 x_1 + \beta_2 x_2 + \beta_3 x_3 + \beta_4 x_1 x_3 + \beta_5 x_3^2 + \beta_6 x_1 x_3^2$$

is fit to the $n = 30$ observations on daily admissions.

a. What hypothesis would you test to determine whether the quadratic terms for predicted daily high temperature x_3 are important?

b. Use the SSEs for the interaction model (Exercise 4.53) and the second-order model given here to test the hypothesis of part **a**. Use $\alpha = .05$.

Interaction model: $SSE_1 = 585,000$
Second-order model: $SSE_2 = 530,000$

4.55   R. N. Horn (James Madison University) and W. J. McGuire (Eastern Kentucky University) used multiple regression analysis to model academic year salaries for secondary-school teachers in Philadelphia during 1981–1982. Information on a sample of 4,316 secondary-school teachers was obtained from records compiled by the Pennsylvania Department of Education. The total sample was divided by race and sex to form four subsamples: black females, black males, white females, and white males. The data for each subsample were used to fit a multiple regression model of the form

$$E(y) = \beta_0 + \beta_1 x_1 + \beta_2 x_2 + \beta_3 x_3 + \cdots + \beta_{30} x_{30}$$

Years of experience, educational level, age, marital status, and major teaching field were among the 30 independent variables in the model. The results for each subsample are summarized in the table.

	BLACK MALES	BLACK FEMALES	WHITE MALES	WHITE FEMALES
n	526	862	1,848	1,080
R^2	.8074	.7674	.7724	.7879

Source: Horn, R. N. and McGuire, W. J. "Determinants of secondary school teacher salaries in a large urban school district." *Southern Economic Journal*, Oct. 1984, pp. 481–493.

a. Is there sufficient evidence to indicate that the model is useful for predicting academic year salaries for black male secondary-school teachers in Philadelphia? Test using $\alpha = .05$.

b. Is there sufficient evidence to indicate that the model is useful for predicting academic year salaries for black female secondary-school teachers in Philadelphia? Test using $\alpha = .05$.

c. Is there sufficient evidence to indicate that the model is useful for predicting academic year salaries for white male secondary-school teachers in Philadelphia? Test using $\alpha = .05$.

d. Is there sufficient evidence to indicate that the model is useful for predicting academic year salaries for white female secondary-school teachers in Philadelphia? Test using $\alpha = .05$.

4.56 The following model was proposed for testing salary discrimination against women in a state university system:

$$E(y) = \beta_0 + \beta_1 x_1 + \beta_2 x_2 + \beta_3 x_1 x_2 + \beta_4 x_2^2$$

where

y = Annual salary (in thousands of dollars)

$x_1 = \begin{cases} 1 & \text{if female} \\ 0 & \text{if male} \end{cases}$

x_2 = Experience (years)

A portion of the computer printout that results from fitting this model to a sample of 200 faculty members in the university system is given here.

Printout for Exercise 4.56: Complete model

```
SOURCE    DF    SUM OF SQUARES    MEAN SQUARE
MODEL      4          2351.70         587.92
ERROR    195           783.90           4.02
TOTAL    199          3135.60      R-SQUARE
                                     0.7500
```

The reduced model $E(y) = \beta_0 + \beta_2 x_2 + \beta_4 x_2^2$ is fit to the same data, and the resulting computer printout is partially reproduced at the top of page 252.

Printout for Exercise 4.56: Reduced model

SOURCE	DF	SUM OF SQUARES	MEAN SQUARE
MODEL	2	2340.37	1170.185
ERROR	197	795.23	4.04
TOTAL	199	3135.60	R-SQUARE
			0.7464

Do the data provide sufficient evidence to support the claim that the mean salary of faculty members is dependent on sex? Use $\alpha = .05$.

4.57 Recent increases in gasoline prices have increased interest in modes of transportation other than the automobile. A metropolitan bus company wants to know if changes in numbers of bus riders are related to changes in gasoline prices. By using information contained in the company files and gasoline price information obtained from fuel distributors, the company planned to fit the following model:

$$y = \beta_0 + \beta_1 x_1 + \beta_2 x_2 + \beta_3 x_1 x_2 + \varepsilon$$

where

x_1 = Average wholesale price for regular gas in a given month

$x_2 = \begin{cases} 1 & \text{if the bus travels a city route only} \\ 0 & \text{if the bus travels a suburb–city route} \end{cases}$

y = Total number of riders in a bus over the month

a. For this model, how would you test to determine whether the relationship between gasoline price and the mean number of riders is different for the two different types of bus routes?

b. Suppose 12 months of data are kept, and the least squares model is

$$\hat{y} = 500 + 50x_1 + 5x_2 - 10x_1 x_2$$

Graph the predicted relationship between number of riders and gas price for city buses and for suburb–city buses. Compare the slopes.

c. If $s_{\hat{\beta}_3} = 3.0$, do the data indicate that gas price affects the number of riders differently for city and suburb–city buses? Use $\alpha = .05$.

4.58 A naval base is considering modifying or adding to its fleet of 48 standard aircraft. The final decision regarding the type and number of aircraft to be added depends on a comparison of cost versus effectiveness of the modified fleet. Consequently, the naval base would like to model the projected percentage increase y in fleet effectiveness by the end of the decade as a function of the cost x of modifying the fleet. A first proposal is the quadratic model

$$E(y) = \beta_0 + \beta_1 x + \beta_2 x^2$$

The data provided in the table were collected on 10 naval bases of similar size that recently expanded their fleets. The data were used to fit the model. The SAS printout of the multiple regression analysis is also reproduced.

a. Interpret the value of R_a^2 on the printout.

b. Find the value of s and interpret it.

c. Perform a test of overall model adequacy. Use $\alpha = .05$.

d. Is there sufficient evidence to conclude that the percentage improvement y increases more quickly for more costly fleet modifications than for less costly fleet modifications? Test with $\alpha = .05$.

PERCENTAGE IMPROVEMENT AT END OF DECADE y	COST OF MODIFYING FLEET x, millions of dollars
18	125
32	160
9	80
37	162
6	110
3	90
30	140
10	85
25	150
2	50

ANALYSIS OF VARIANCE

SOURCE	DF	SUM OF SQUARES	MEAN SQUARE	F VALUE	PROB>F
MODEL	2	1368.77501	684.38750	33.079	0.0003
ERROR	7	144.82499	20.68928481		
C TOTAL	9	1513.60000			

ROOT MSE	4.548548	R-SQUARE	0.9043	
DEP MEAN	17.2	ADJ R-SQ	0.8770	
C.V.	26.44504			

PARAMETER ESTIMATES

VARIABLE	DF	PARAMETER ESTIMATE	STANDARD ERROR	T FOR H0: PARAMETER=0	PROB > |T|
INTERCEP	1	10.65903604	14.55009061	0.733	0.4876
X	1	-0.28160568	0.28087588	-1.003	0.3494
XX	1	0.002671936	0.001253832	2.131	0.0706

e. Now consider the model

$$E(y) = \beta_0 + \beta_1 x_1 + \beta_2 x_1^2 + \beta_3 x_2 + \beta_4 x_1 x_2$$

where

x_1 = Cost of modifying the fleet

$x_2 = \begin{cases} 1 & \text{if American base} \\ 0 & \text{if foreign base} \end{cases}$

The model is fit to the $n = 10$ data points and results in SSE = 97.645. Is there sufficient evidence to indicate that type of base (American or foreign) is a useful predictor of percentage improvement y? Test using $\alpha = .05$.

ON YOUR OWN...

[*Note:* The use of a computer is required for this study.]

This is a continuation of the "On Your Own" presented in Chapter 3, in which you selected three independent variables as predictors of the gross national product and obtained 10 years of data for each. Now fit the multiple regression model (use an available computer package, if possible)

$$y = \beta_0 + \beta_1 x_1 + \beta_2 x_2 + \beta_3 x_3 + \varepsilon$$

where

y = Gross national product

x_1 = First variable you chose

x_2 = Second variable you chose

x_3 = Third variable you chose

a. Compare the coefficients $\hat{\beta}_1$, $\hat{\beta}_2$, and $\hat{\beta}_3$ to their corresponding slope coefficients in the Chapter 3 "On Your Own," where you fit three separate straight-line models. How do you account for the differences?

b. Calculate the coefficient of determination R^2, and conduct the F test of the null hypothesis H_0: $\beta_1 = \beta_2 = \beta_3 = 0$. What is your conclusion?

c. Now, increase your list of three variables to include approximately 10 that you feel would be useful in predicting the GNP. Obtain data for as many years as possible for the new list of variables and the GNP. With the aid of a computer analysis package, use a stepwise regression program to choose the important variables among those you have listed. To test your intuition, list the variables in the order you think they will be selected before you conduct the analysis. How does your list compare with the stepwise regression results?

d. After the group of 10 variables has been narrowed to a smaller group of variables by the stepwise analysis, try to improve the model by including interactions and quadratic terms. Be sure to consider the meaning of each interaction or quadratic term before adding it to the model—a quick sketch can be very helpful. See if you can systematically construct a useful model for predicting the GNP. You might want to hold out the last several years of data to test the predictive ability of your model after it is constructed. (As noted in Section 4.10, using the same data to construct and to evaluate predictive ability can lead to invalid statistical tests and a false sense of security.) If the independent variables you chose are themselves highly correlated, you may encounter some results that are difficult to explain. For example, the coefficients, $\hat{\beta}_1$, $\hat{\beta}_2$, and $\hat{\beta}_3$ in the first-order model may assume signs that run counter to what you expected. Or you may get a highly significant F value in part b, but the individual t statistics for x_1, x_2, and x_3 may all be nonsignificant. This phenomenon—a high correlation between the independent variables in a regression model—is known as **multicollinearity**. We discuss remedial measures for multicollinearity in Chapter 6.

References

Draper, N. and Smith, H. *Applied Regression Analysis*, 3rd ed. New York: Wiley, 1989.

Montgomery, D. C. and Peck, E. A. *Introduction to Linear Regression Analysis*. New York: Wiley, 1982.

Neter, J., Wasserman, W., and Kutner, M. H. *Applied Linear Statistical Models*, 3rd ed. Homewood, Ill.: Richard D. Irwin, 1990.

Weisberg, S. *Applied Linear Regression*. New York: Wiley, 1980.

Model Building

CONTENTS

5.1 Introduction: Why Model Building Is Important

5.2 The Two Types of Independent Variables: Quantitative and Qualitative

5.3 Models with a Single Quantitative Independent Variable

5.4 First-Order Models with Two or More Quantitative Independent Variables

5.5 Second-Order Models with Two or More Quantitative Independent Variables

5.6 Coding Quantitative Independent Variables (Optional)

5.7 Models with One Qualitative Independent Variable

5.8 Models with Two Qualitative Independent Variables

5.9 Models with Three or More Qualitative Independent Variables

5.10 Models with Both Quantitative and Qualitative Independent Variables

5.11 Model Building: An Example

OBJECTIVE

To show you why the choice of the deterministic portion of a linear model is crucial to the acquisition of a good prediction equation; to present some basic concepts and procedures for constructing good linear models

5.1

Introduction: Why Model Building Is Important

We have emphasized in both Chapters 3 and 4 that one of the first steps in the construction of a regression model is to hypothesize the form of the deterministic portion of the probabilistic model. This **model-building**, or model-construction, stage is the key to the success (or failure) of the regression analysis. If the hypothesized model does not reflect, at least approximately, the true nature of the relationship between the mean response $E(y)$ and the independent variables $x_1, x_2, \ldots, x_k$, the modeling effort will usually be unrewarded.

By *model building*, we mean writing a model that will provide a good fit to a set of data and that will give good estimates of the mean value of y and good predictions of future values of y for given values of the independent variables. To illustrate, several years ago, a nationally recognized educational research group issued a report concerning the variables related to academic achievement for a certain type of college student. The researchers selected a random sample of students and recorded a measure of academic achievement, y, at the end of the senior year along with data on an extensive list of independent variables, $x_1, x_2, \ldots, x_k$, that they thought were related to y. Among these independent variables were the student's IQ, scores on mathematics and verbal achievement examinations, rank in class, etc. They fit the model

$$E(y) = \beta_0 + \beta_1 x_1 + \beta_2 x_2 + \cdots + \beta_k x_k$$

to the data, analyzed the results, and reached the conclusion that none of the independent variables are "significantly related" to y. The **goodness of fit** of the model, measured by the coefficient of determination R^2, was not particularly good, and t tests on individual parameters did not lead to rejection of the null hypotheses that these parameters equaled 0.

How could the researchers have reached the conclusion that there is no significant relationship, when it is evident, just as a matter of experience, that some of the independent variables studied are related to academic achievement? For example, achievement on a college mathematics placement test should be related to achievement in college mathematics. Certainly, many other variables will affect achievement—motivation, environmental conditions, and so forth—but generally speaking, there will be a positive correlation between entrance achievement test scores and college academic achievement. So, what went wrong with the educational researchers' study?

Although you can never discard the possibility of computing error as a reason for erroneous answers, most likely the difficulties in the results of the educational study were caused by the use of an improperly constructed model. For example, the model

$$E(y) = \beta_0 + \beta_1 x_1 + \beta_2 x_2 + \cdots + \beta_k x_k$$

assumes that the independent variables $x_1, x_2, \ldots, x_k$ affect mean achievement $E(y)$ independently of each other.* Thus, if you hold all the other independent variables constant and vary only x_1, $E(y)$ will increase by the amount β_1 for

---

*Keep in mind that we are discussing the deterministic portion of the model and that the word *independent* is used in a mathematical rather than a probabilistic sense.

every unit increase in x_1. A 1-unit change in any of the other independent variables will increase $E(y)$ by the value of the corresponding β parameter for that variable.

Do the assumptions implied by the model agree with your knowledge about academic achievement? First, is it reasonable to assume that the effect of time spent on study is independent of native intellectual ability? We think not. No matter how much effort some students invest in a particular subject, their rate of achievement is low. For others, it may be high. Therefore, assuming that these two variables—effort and native intellectual ability—affect $E(y)$ independently of each other is likely to be an erroneous assumption. Second, suppose that x_5 is the amount of time a student devotes to study. Is it reasonable to expect that a 1-unit increase in x_5 will always produce the same change β_5 in $E(y)$? The changes in $E(y)$ for a 1-unit increase in x_5 might depend on the value of x_5 (for example, the law of diminishing returns). Consequently, it is quite likely that the assumption of a constant rate of change in $E(y)$ for 1-unit increases in the independent variables will not be satisfied.

Clearly, the model

$$E(y) = \beta_0 + \beta_1 x_1 + \beta_2 x_2 + \cdots + \beta_k x_k$$

was a poor choice in view of the researchers' prior knowledge of some of the variables involved. Terms have to be added to the model to account for interrelationships among the independent variables and for curvature in the response function. Failure to include needed terms causes inflated values of SSE, nonsignificance in statistical tests, and, often, erroneous practical conclusions.

In this chapter we discuss the most difficult part of a multiple regression analysis—the formulation of a good model for $E(y)$. Although many of the models presented in this chapter have already been introduced in Section 4.9, we assume the reader has little or no background in model building. This chapter serves as a basic reference guide to model building for teachers, students, and practitioners of multiple regression analysis.

5.2

The Two Types of Independent Variables: Quantitative and Qualitative

The independent variables that appear in a linear model can be one of two types—either **quantitative** or **qualitative**.

Definition 5.1

A **quantitative** independent variable is one that assumes numerical values corresponding to the points on a line. An independent variable that is not quantitative is called **qualitative**.

The gross national product, prime interest rate, number of defects in a product, and kilowatt-hours of electricity used per day are all examples of quantitative independent variables. On the other hand, suppose three different styles of packaging, A, B, and C, are used by a manufacturer. This independent variable, style of packaging, is qualitative, since it is not measured on a numerical scale. Certainly, style of packaging is an independent variable that may affect sales of a product, and we would want to include it in a model describing the product's sales, y.

Definition 5.2

The different intensity settings of an independent variable are called its **levels**.

For a quantitative variable, the levels correspond to the numerical values it assumes. For example, if the number of defects in a product ranges from 0 to 3, the independent variable assumes four levels: 0, 1, 2, and 3.

The levels of a qualitative variable are not numerical. They can be defined only by describing them. For example, the independent variable style of packaging was observed at three levels: A, B, and C.

EXAMPLE 5.1

In Chapter 4 we considered the problem of predicting executive salary as a function of several independent variables. Consider the following four independent variables that may affect executive salaries:

a. Number of years of experience
b. Sex of the employee
c. Firm's net asset value
d. Rank of the employee

For each of these independent variables, give its type and describe the nature of the levels you would expect to observe.

Solution

a. The independent variable for the number of years of experience is quantitative, since its values are numerical. We would expect to observe levels ranging from 0 to 40 (approximately) years.
b. The independent variable for sex is qualitative, since its levels can only be described by the nonnumerical labels "female" and "male."
c. The independent variable for the firm's net asset value is quantitative, with a very large number of possible levels corresponding to the range of dollar values representing various firms' net asset values.
d. Suppose the independent variable for the rank of the employee is observed at three levels: supervisor, assistant vice president, and vice president. Since we cannot assign a realistic measure of relative importance to each position, rank is a qualitative independent variable.

Quantitative and qualitative independent variables are treated differently in regression modeling. In the next section, we will see how quantitative variables are entered into a regression model.

EXERCISES

5.1   Companies keep personnel files that contain important information on each employee's background. The data in these files could be used to predict employee performance ratings. Identify the independent variables listed below as qualitative or quantitative.
  a. Age
  b. Years of experience with the company
  c. Highest educational degree
  d. Job classification
  e. Marital status
  f. Religious preference
  g. Salary
  h. Sex

5.2   Which of the assumptions about ε (Section 4.2) prohibit the use of a qualitative variable as a dependent variable? (We present a technique for modeling a qualitative dependent variable in Chapter 8.)

5.3   The *Journal of Human Stress* (Summer 1987) reported on a study of "psychological response of firefighters to chemical fire." The researchers used multiple regression to predict emotional distress as a function of the following independent variables. Identify each independent variable as quantitative or qualitative. For qualitative variables, suggest several levels that might be observed. For quantitative variables, give a range of values (levels) for which the variable might be observed.
  a. Number of preincident psychological symptoms
  b. Years experience
  c. Cigarette smoking behavior
  d. Level of social support
  e. Marital status
  f. Age
  g. Ethnic status
  h. Exposure to a chemical fire
  i. Educational level
  j. Distance lived from site of incident
  k. Gender

5.3
Models with a Single Quantitative Independent Variable

To write a prediction equation that provides a good model for a response (one that will eventually yield good predictions), we have to know how the response might vary as the levels of an independent variable change. Then we have to know how to write a mathematical equation to model it. To illustrate (with a simple example), suppose we want to model corporate profit, y, as a function of the single independent variable x, the amount of capital invested. It may be that corporate profit, y, increases in a straight line as the amount of capital invested, x, varies from \$1 to 2 million, as shown in Figure 5.1a (page 260). If this were the entire range of x values for which you wanted to predict y, the model

$$E(y) = \beta_0 + \beta_1 x$$

would be appropriate.

FIGURE 5.1

Modeling corporate profit, y, as a function of capital invested, x

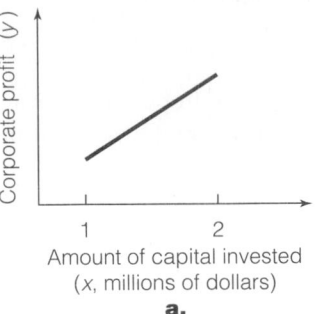

a.

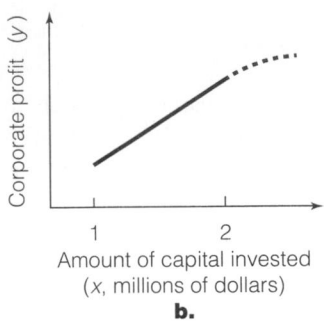

b.

Now, suppose you want to expand the range of values of x to $x = \$3$ or $x = \$4$ million of capital investment. Will the straight-line model

$$E(y) = \beta_0 + \beta_1 x$$

be satisfactory? Perhaps, but making this assumption could be risky. As the amount of capital invested, x, is increased, sooner or later the point of diminishing returns will be reached. That is, the increase in profit for a unit increase in capital invested will decrease, as shown by the dashed line in Figure 5.1b. To produce this type of curvature, you need to know the relationship between models and graphs, and how types of terms will change the shape of the curve.

A response that is a function of a single quantitative independent variable can often be modeled by the first few terms of a polynomial algebraic function. The equation relating the mean value of y to a polynomial of order p in one independent variable x is shown in the box.

Formula for a pth-Order Polynomial with One Independent Variable

$$E(y) = \beta_0 + \beta_1 x + \beta_2 x^2 + \beta_3 x^3 + \cdots + \beta_p x^p$$

where p is an integer and $\beta_0, \beta_1, \ldots, \beta_p$ are unknown parameters that must be estimated.

As we mentioned in Chapters 3 and 4, a **first-order polynomial** in x (i.e., $p = 1$),

$$E(y) = \beta_0 + \beta_1 x$$

graphs as a straight line. A **second-order polynomial** model ($p = 2$), called a **quadratic**, is given by the following equation.

A Second-Order (Quadratic) Model with One Independent Variable

$$E(y) = \beta_0 + \beta_1 x + \beta_2 x^2$$

where β_0, β_1, and β_2 are unknown parameters that must be estimated.

Graphs of two quadratic models are shown in Figure 5.2. The quadratic model is the equation of a **parabola** that opens either upward, as in Figure 5.2a, or downward, as in Figure 5.2b. (If the coefficient of x^2 is positive, it opens upward; if it is negative, it opens downward.) The parabola may be shifted upward or downward, left or right. The least squares procedure only uses the portion of the parabola that is needed to model the data. For example, if you fit a parabola to the data points shown in Figure 5.3, the portion shown as a solid curve passes through the data points. The outline of the unused portion of the parabola is indicated by a dashed curve.

FIGURE 5.2

Graphs for two second-order polynomial models

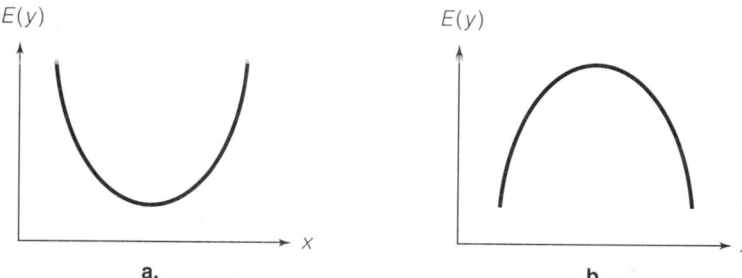

a.

b.

Figure 5.3 illustrates an important limitation on the use of prediction equations—the model is valid only over the range of x values that were used to fit the model. For example, the response might rise, as shown in the figure, until it reaches a plateau. The second-order model might fit the data very well over the range of x values shown in Figure 5.3, but would provide a very poor fit if data were collected in the region where the parabola turns downward.

FIGURE 5.3

Example of the use of a quadratic model

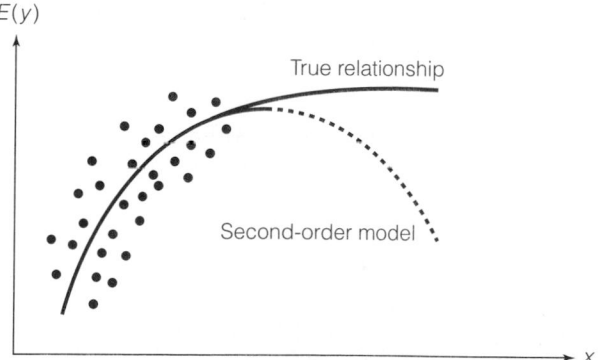

How do you decide the order of the polynomial you should use to model a response if you have no prior information on the relationship between $E(y)$ and x? If you have data, construct a scattergram of the data points and see if you can deduce the nature of a good approximating function. A pth-order polynomial, when graphed, will exhibit $(p - 1)$ peaks, or reversals in direction. Note that the graphs of the second-order model shown in Figure 5.2 each have $(p - 1)$ = 1 peak. Likewise, a third-order model will have $(p - 1) = 2$ peaks, as shown in Figure 5.4 on page 262.

FIGURE 5.4

Graph of a third-order polynomial model

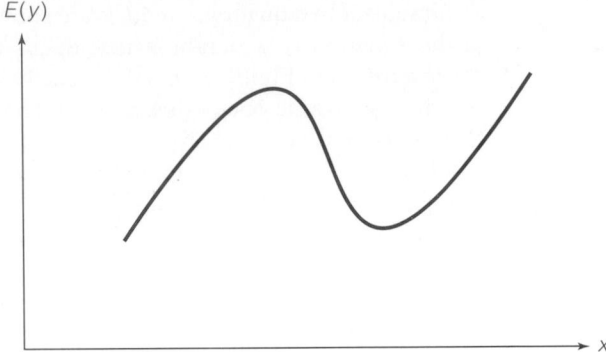

The graphs of most responses as a function of an independent variable x are, in general, curvilinear. Nevertheless, if the rate of curvature of the response curve is very small over the range of x that is of interest to you, a straight line might provide an excellent fit to the response data and function as a very useful prediction equation. If the curvature is not (or may not be) slight, you should try a second-order model. Third- or higher-order models would be used only where you expect more than one reversal in the direction of the curve. These situations are rare, except where the response is a function of time. Models for forecasting over time are presented in Chapter 9.

EXAMPLE 5.2

To operate efficiently, power companies must be able to predict the peak power load at their various stations. The peak power load is the maximum amount of power that must be generated each day to meet demand.

Suppose a power company located in the southern part of the United States decides to model daily peak power load, y, as a function of the daily high temperature, x, and the model is to be constructed for the summer months when demand is greatest. Although we would expect the peak power load to increase as the high temperature increases, the *rate* of increase in $E(y)$ might also increase as x increases. That is, a 1-unit increase in high temperature from $100°$ to $101°F$ might result in a larger increase in power demand than would a 1-unit increase from $80°$ to $81°F$. Therefore, we postulate the quadratic model

$$E(y) = \beta_0 + \beta_1 x + \beta_2 x^2$$

and we expect β_2 to be positive.

A random sample of 25 summer days is selected, and the data are shown in Table 5.1. Fit a second-order model using the data, and test the hypothesis that the power load increases at an increasing rate with temperature, i.e., that $\beta_2 > 0$.

Solution

The SAS printout shown in Figure 5.5 gives the least squares fit of the second-order model using the data in Table 5.1. The prediction equation is

$$\hat{y} = 385.048 - 8.293x + .05982x^2$$

T A B L E 5.1   **Power Load Data**

TEMPERATURE °F	PEAK LOAD megawatts	TEMPERATURE °F	PEAK LOAD megawatts	TEMPERATURE °F	PEAK LOAD megawatts
94	136.0	106	178.2	76	100.9
96	131.7	67	101.6	68	96.3
95	140.7	71	92.5	92	135.1
108	189.3	100	151.9	100	143.6
67	96.5	79	106.2	85	111.4
88	116.4	97	153.2	89	116.5
89	118.5	98	150.1	74	103.9
84	113.4	87	114.7	86	105.1
90	132.0				

F I G U R E 5.5

Portion of the SAS printout for the second-order model of Example 5.2

```
                        ANALYSIS OF VARIANCE

                        SUM OF          MEAN
   SOURCE      DF       SQUARES        SQUARE      F VALUE      PROB>F

   MODEL        2     15011.77200    7505.88600    259.687      0.0001
   ERROR       22       635.87840   28.90356374
   C TOTAL     24     15647.65040

           ROOT MSE      5.376203      R-SQUARE     0.9594
           DEP MEAN      125.428       ADJ R-SQ     0.9557
           C.V.          4.286287

                       PARAMETER ESTIMATES

                    PARAMETER      STANDARD      T FOR H0:
   VARIABLE   DF     ESTIMATE        ERROR     PARAMETER=0    PROB > |T|

   INTERCEP   1      385.04809    55.17243578      6.979        0.0001
   X          1       -8.29252680  1.29904502     -6.384        0.0001
   XX         1        0.05982337  0.007548554      7.925        0.0001
```

A plot of this equation and the observed values is given in Figure 5.6. Note that this curve passes through the data points and seems to produce (by visual examination) a set of deviations that are relatively small.

We now test to determine whether the sample value, $\hat{\beta}_2 = .05982$, is large enough to conclude in general that the power load increases at an increasing rate with temperature:

$$H_0: \quad \beta_2 = 0 \qquad H_a: \quad \beta_2 > 0$$

$$\text{Test statistic:} \quad t = \frac{\hat{\beta}_2}{s_{\hat{\beta}_2}}$$

For $\alpha = .05$, $n = 25$, $k = 2$, and df $= n - (k + 1) = 22$, we reject H_0 if

$$t > t_{.05}$$

where $t_{.05} = 1.717$ (from Table 2 of Appendix D). From Figure 5.5, the calculated value of t is 7.925. Since this value exceeds $t_{.05} = 1.717$, we reject H_0 and conclude that the mean power load increases at an increasing rate with temperature. Note that the observed significance level of the test, $p = .0001/2 = .00005$ supports this conclusion.

FIGURE 5.6

Plot of the observations and the
second-order least squares fit

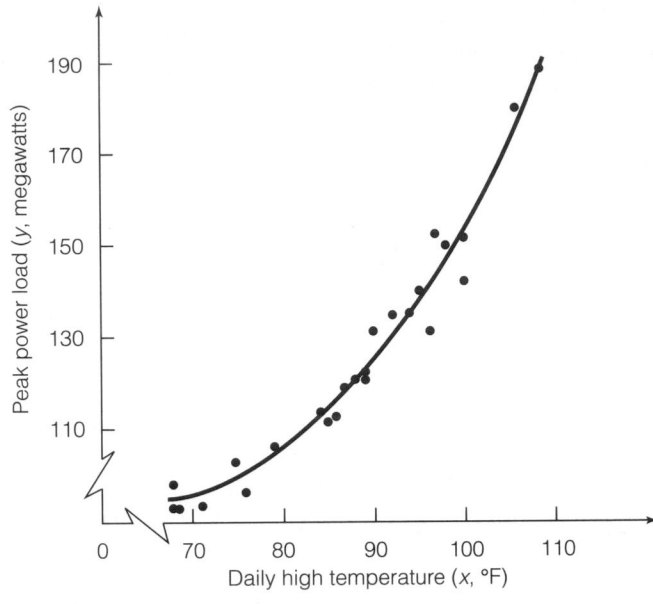

<h1>EXERCISES</h1>

5.4   In the pharmaceutical industry, a new chemical entity (NCE) is defined as a new chemical or biological
compound tested in humans for therapeutic purposes for the first time. A study published in *Managerial & Decision Economics* (Sept. 3, 1988) reported that expenditures on research and development

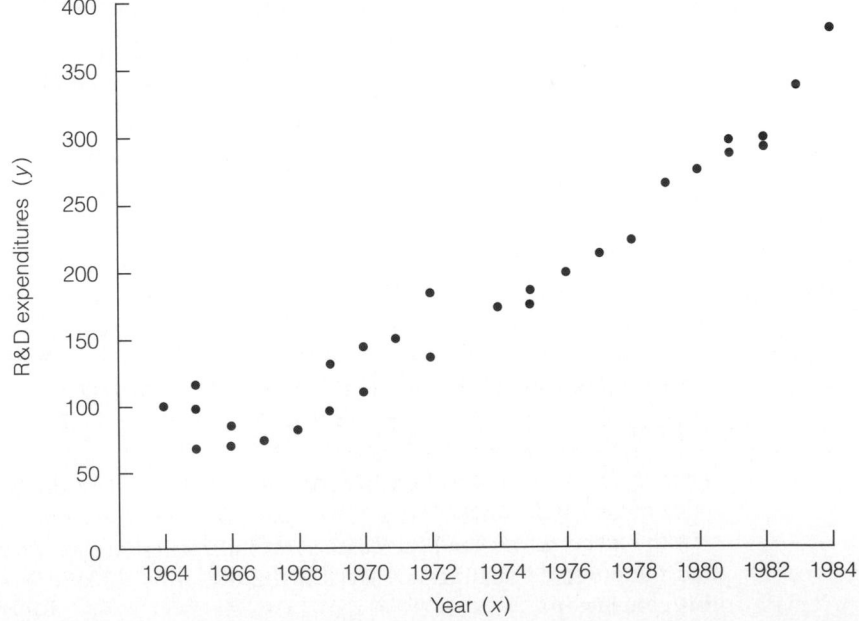

(R&D) of NCEs in the United Kingdom has increased dramatically over the 20 years 1964–1984. A plot of R&D expenditures (y) versus year (x) is shown here.

a Propose a model for $E(y)$ that would seem to fit the data well.

b. What are expected signs of the β's in the model, part **a**?

5.5 A company is considering having the employees on its assembly line work 4 days for 10 hours each instead of 5 days for 8 hours. Management is concerned that the effect of fatigue due to longer afternoons of work might increase assembly times to an unsatisfactory level. An experiment with the 4-day week is planned in which time studies will be conducted on some of the workers during the afternoons. It is believed that an adequate model of the relationship between assembly time, y, and time since lunch, x, should allow for the average assembly time to decrease for a while after lunch (as workers get back in the groove) before it starts to increase as the workers become tired. Write a model to relate $E(y)$ and x that would reflect the management's belief, and sketch the hypothesized shape of the model.

5.6 Underinflated or overinflated tires can increase the tire wear and decrease gas mileage. A new tire was tested for wear at different pressures with the results shown in the table.

PRESSURE x, pounds per square inch	MILEAGE y, thousands
30	29
31	32
32	36
33	38
34	37
35	33
36	26

a. Plot the data on a scattergram.

b. If you were given the information for $x = 30, 31, 32, 33$ only, what kind of model would you suggest? For $x = 33, 34, 35, 36$? For all the data?

5.4
First-Order Models with Two or More Quantitative Independent Variables

Like models for a single independent variable, models with two or more independent variables are classified as first-order, second-order, and so forth, but it is difficult (most often impossible) to graph the response because the plot is in a multidimensional space. For example, with one quantitative independent variable, x, the response y traces a curve. But for two quantitative independent variables, x_1 and x_2, the plot of y traces a surface over the x_1,x_2-plane (see Figure 5.7 on page 266). For three or more quantitative independent variables, the response traces a surface in a four- or higher-dimensional space. For these, we can construct two-dimensional contour curves for one independent variable or three-dimensional plots of response surfaces for two independent variables for fixed levels of the remaining independent variables, but this is the best we can do in providing a graphic description of a response.

A **first-order model** in k quantitative variables is a first-order polynomial in k independent variables. For $k = 1$, the graph is a straight line. For $k = 2$, the response surface is a plane (usually tilted) over the x_1,x_2-plane.

FIGURE 5.7

A response surface for two quantitative
independent variables

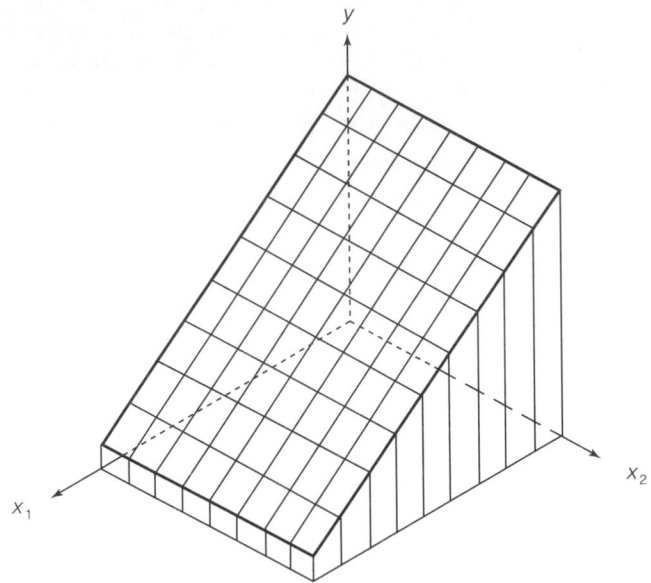

First-Order Model in k Quantitative Independent Variables

$$E(y) = \beta_0 + \beta_1 x_1 + \beta_2 x_2 + \cdots + \beta_k x_k$$

where $\beta_0, \beta_1, \ldots, \beta_k$ are unknown parameters that must be estimated.

INTERPRETATION OF MODEL PARAMETERS

β_0: y-intercept of $(k + 1)$-dimensional surface; the value of $E(y)$
when $x_1 = x_2 = \cdots = x_k = 0$

β_1:   Change in $E(y)$ for a 1-unit increase in x_1, when $x_2, x_3, \ldots, x_k$
are held fixed

β_2:   Change in $E(y)$ for a 1-unit increase in x_2, when $x_1, x_3, \ldots, x_k$
are held fixed

$\vdots$

β_k: Change in $E(y)$ for a 1-unit increase in x_k, when $x_1, x_2, \ldots,$
x_{k-1} are held fixed

If we use a first-order polynomial to model a response, we are assuming that
there is no curvature in the response surface and that the variables affect the
response independently of each other. For example, suppose the true relationship
between the mean response and the independent variables x_1 and x_2 is given by
the equation

$$E(y) = 1 + 2x_1 + x_2$$

The response surface (a plane) corresponding to this equation is shown in Figure 5.8. The graphs of this expression for $x_2 = 1$, 2, and 3 (called **contour lines**), are shown in Figure 5.9. You can see that when you substitute $x_2 = 1$ into the model, you obtain

$$E(y) = 1 + 2x_1 + x_2$$
$$= 1 + 2x_1 + 1$$
$$= 2 + 2x_1$$

FIGURE 5.8

The plane $E(y) = 1 + 2x_1 + x_2$

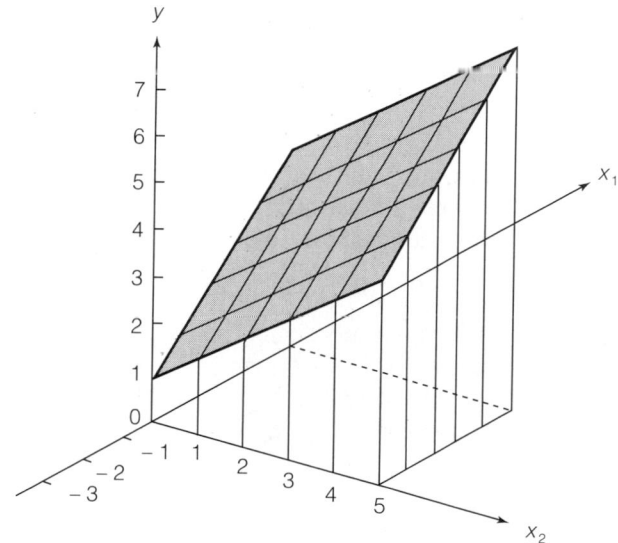

FIGURE 5.9

Contour lines of $E(y)$ for $x_2 = 1, 2, 3$ (first-order model)

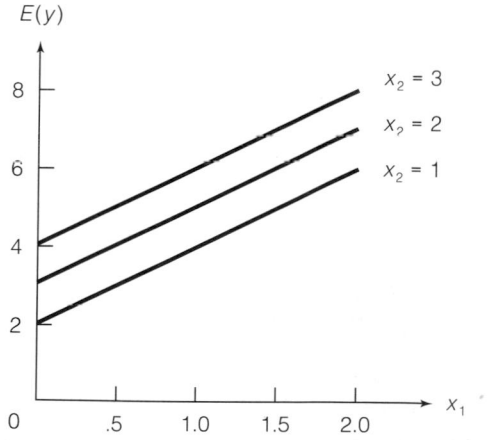

For $x_2 = 2$,

$$E(y) = 1 + 2x_1 + 2$$
$$= 3 + 2x_1$$

And for $x_3 = 3$, $E(y) = 4 + 2x_1$. In other words, regardless of the value of x_2, $E(y)$ graphs as a straight line. Changing x_2 changes only the y-intercept (the constant in the equation). Consequently, assuming that a first-order model will adequately model a response is equivalent to assuming that a 1-unit change in one independent variable will have the same effect on the mean value of y regardless of the levels of the other independent variables. That is, *the contour lines will be parallel.* In Chapter 4, we stated that independent variables that have this property *do not interact.*

Except in cases where the ranges of levels for all independent variables are very small, the implication of no curvature in the response surface and the independence of variable effects on the response restrict the applicability of first-order models.

5.5
Second-Order Models with Two or More Quantitative Independent Variables

Second-order models with two or more independent variables permit curvature in the response surface. One important type of second-order term accounts for **interaction** between two variables.* To see the effect of interaction on the model, consider the two-variable model

$$E(y) = \beta_0 + \beta_1 x_1 + \beta_2 x_2 + \beta_3 x_1 x_2$$

This interaction model traces a ruled surface (twisted plane) in a three-dimensional space (see Figure 5.10). The second-order term $\beta_3 x_1 x_2$ is called the **interaction term**, and it permits the contour lines to be *nonparallel.* For example, suppose the true equation of the response surface is

$$E(y) = 1 + 2x_1 + x_2 - x_1 x_2$$

We graph the contour lines of this response for $x_2 = 1$, 2, and 3 in Figure 5.11. Note that when we substitute $x_2 = 1$ into the model, we get

$$\begin{aligned} E(y) &= 1 + 2x_1 + x_2 - x_1 x_2 \\ &= 1 + 2x_1 + 1 - x_1(1) \\ &= 2 + x_1 \end{aligned}$$

For $x_2 = 2$,

$$\begin{aligned} E(y) &= 1 + 2x_1 + 2 - x_1(2) \\ &= 3 \end{aligned}$$

Similarly, when $x_2 = 3$, $E(y) = 4 - x_1$. Thus, when interaction is present in the model, both the *y-intercept and the slope* change as x_2 changes. Consequently, *the contour lines are not parallel. The presence of an interaction term implies that the effect of a 1-unit change in one independent variable will depend on the level of the other independent variable.* In our example (Figure 5.11), a 1-unit change in x_1 produces a 1-unit change in $E(y)$ when $x_2 = 1$, but a 1-unit change in x_1 produces *no* change in $E(y)$ when $x_2 = 2$ (i.e., the slope is 0).

---

*The order of a term involving two or more *quantitative* independent variables is equal to the sum of their exponents. Thus, $\beta_3 x_1 x_2$ is a second-order term, as is $\beta_4 x_1^2$. A term of the form $\beta_i x_1 x_2 x_3$ is a third-order term.

FIGURE 5.10

Computer-generated graph for an interaction model (second-order)

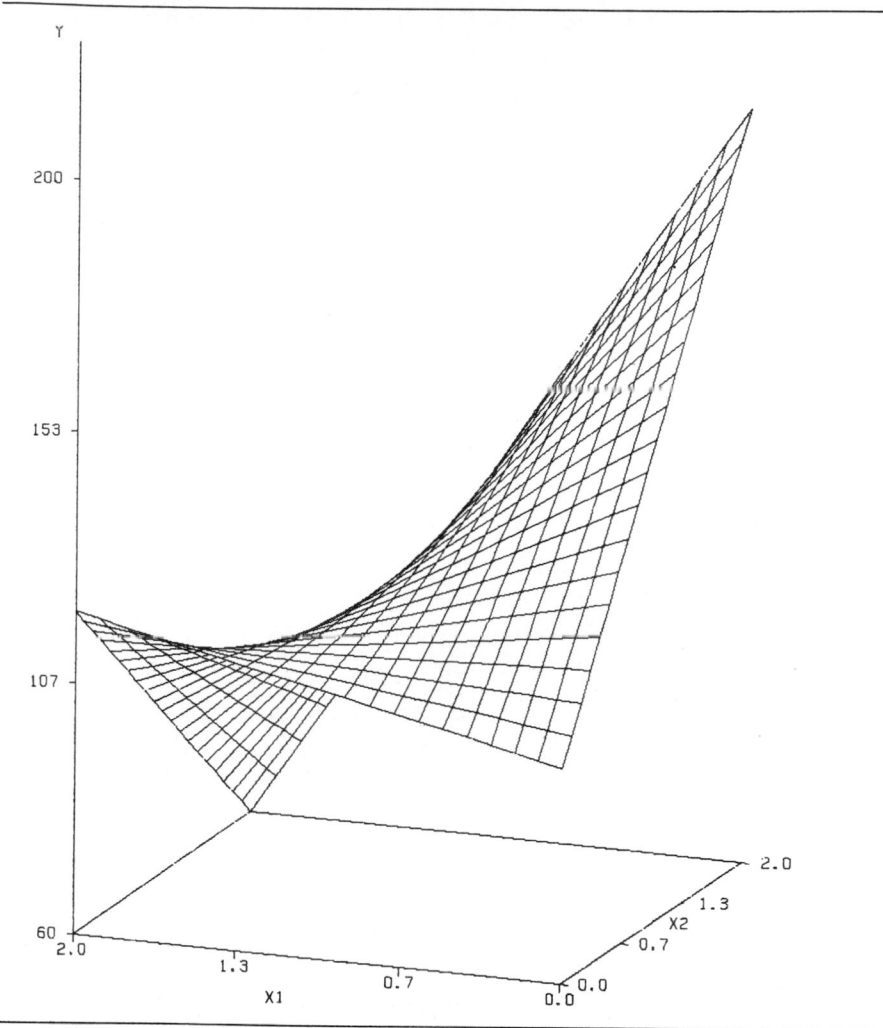

FIGURE 5.11

Contour lines of $E(y)$ for $x_2 = 1, 2, 3$ (first-order plus interaction)

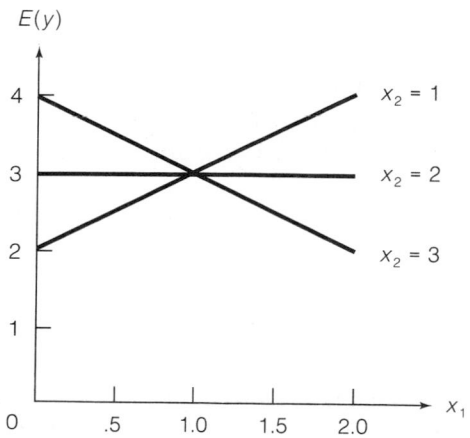

An Interaction Model (Second-Order) with Two Quantitative Independent Variables

$$E(y) = \beta_0 + \beta_1 x_1 + \beta_2 x_2 + \beta_3 x_1 x_2$$

where β_0, β_1, β_2, and β_3 are unknown parameters that must be estimated.

INTERPRETATION OF MODEL PARAMETERS

β_0: y-intercept of three-dimensional surface (see Figure 5.10); the value of $E(y)$ when $x_1 = x_2 = 0$

β_1 and β_2:   Changing β_1 and β_2 causes the surface to shift along the x_1- and x_2-axes

β_3:   Controls the rate of twist in the ruled surface (see Figure 5.10)

We can introduce even more flexibility into a model by the addition of quadratic terms. *The complete second-order model includes the constant β_0, all linear (first-order) terms, all two-variable interactions, and all quadratic terms.* This complete second-order model for two quantitative independent variables is shown in the box.

Complete Second-Order Model with Two Quantitative Independent Variables

$$E(y) = \beta_0 + \beta_1 x_1 + \beta_2 x_2 + \beta_3 x_1 x_2 + \beta_4 x_1^2 + \beta_5 x_2^2$$

where β_0, β_1, . . . , β_5 are unknown parameters that must be estimated.

INTERPRETATION OF MODEL PARAMETERS

β_0: y-intercept of three-dimensional surface (see Figure 5.12); the value of $E(y)$ when $x_1 = x_2 = 0$

β_1 and β_2:   Changing β_1 and β_2 causes the surface to shift along the x_1- and x_2-axes

β_3:   The value of β_3 controls the rotation of the surface

β_4 and β_5:   Signs and values of these parameters control the type of surface and the rates of curvature

β_4 and β_5 positive: A paraboloid that opens upward [Figure 5.12a]

β_4 and β_5 negative: A paraboloid that opens downward [Figure 5.12b]

β_4 and β_5 differ in sign: A saddle-shaped surface [Figure 5.12c]

The quadratic terms $\beta_4 x_1^2$ and $\beta_5 x_2^2$ in the second-order model imply that the response surface for $E(y)$ will possess curvature (see Figure 5.12). The interaction

FIGURE 5.12

Graphs of three second-order surfaces

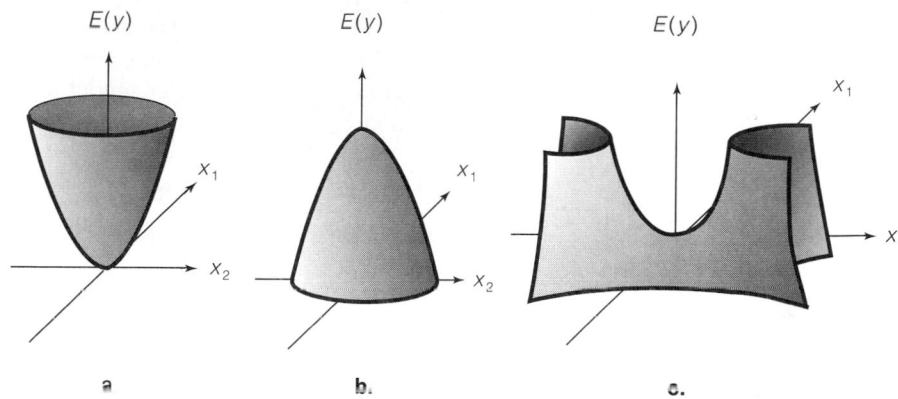

a.

b.

c.

term $\beta_3 x_1 x_2$ allows the contours depicting $E(y)$ as a function of x_1 to have different shapes for various values of x_2. For example, suppose the complete second-order model relating $E(y)$ to x_1 and x_2 is

$$E(y) = 1 + 2x_1 + x_2 - 10x_1x_2 + x_1^2 - 2x_2^2$$

Then the contours of $E(y)$ for $x_2 = -1$, 0, and 1 are shown in Figure 5.13. When we substitute $x_2 = 1$ into the model, we get

$$\begin{aligned} E(y) &= 1 + 2x_1 + x_2 - 10x_1x_2 + x_1^2 - 2x_2^2 \\ &= 1 + 2x_1 - 1 - 10x_1(-1) + x_1^2 - 2(-1)^2 \\ &= -2 + 12x_1 + x_1^2 \end{aligned}$$

For $x_2 = 0$,

$$\begin{aligned} E(y) &= 1 + 2x_1 + (0) - 10x_1(0) + x_1^2 - 2(0)^2 \\ &= 1 + 2x_1 + x_1^2 \end{aligned}$$

Similarly, for $x_2 = 1$,

$$E(y) = -8x_1 + x_1^2$$

FIGURE 5.13

Contours of $E(y)$ for $x_2 = -1, 0, 1$ (complete second-order model)

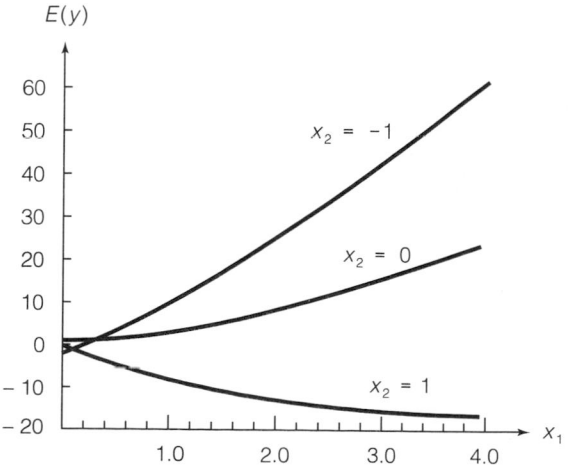

Note how the shapes of the three contour curves in Figure 5.13 differ, indicating that the β parameter associated with the x_1x_2 (interaction) term differs from 0.

The complete second-order model for three independent variables is shown in the next box.

Complete Second-Order Model with Three Quantitative Independent Variables

$$E(y) = \beta_0 + \beta_1x_1 + \beta_2x_2 + \beta_3x_3 + \beta_4x_1x_2 + \beta_5x_1x_3 + \beta_6x_2x_3$$
$$+ \beta_7x_1^2 + \beta_8x_2^2 + \beta_9x_3^2$$

where $\beta_0, \beta_1, \beta_2, \ldots, \beta_9$ are unknown parameters that must be estimated.

This second-order model in three independent variables demonstrates how you would write a second-order model for any number of independent variables. Always include the constant β_0 and then all first-order terms corresponding to $x_1, x_2, \ldots$. Then add the interaction terms for all pairs of independent variables $x_1x_2, x_1x_3, x_2x_3, \ldots$. Finally, include the second-order terms $x_1^2, x_2^2, \ldots$.

For any number, say, p, of quantitative independent variables, the response traces a surface in a $(p + 1)$-dimensional space, which is impossible to visualize. In spite of this handicap, the prediction equation can still tell us much about the phenomenon being studied.

EXAMPLE 5.3

Many companies manufacture products that are at least partially produced using chemicals (for example, steel, paint, gasoline). In many instances, the quality of the finished product is a function of the temperature and pressure at which the chemical reactions take place.

Suppose you wanted to model the quality, y, of a product as a function of the temperature, x_1, and the pressure, x_2, at which it is produced. Four inspectors independently assign a quality score between 0 and 100 to each product, and then the quality, y, is calculated by averaging the four scores. An experiment is conducted by varying temperature between 80° and 100°F and pressure between 50 and 60 pounds per square inch (psi). The resulting data ($n = 27$) are given in Table 5.2. Fit a complete second-order model to the data and sketch the response surface.

Solution

The complete second-order model is

$$E(y) = \beta_0 + \beta_1x_1 + \beta_2x_2 + \beta_3x_1x_2 + \beta_4x_1^2 + \beta_5x_2^2$$

The data in Table 5.2 were used to fit this model. A portion of the SAS output is shown in Figure 5.14.

The least squares model is

$$\hat{y} = -5{,}127.90 + 31.10x_1 + 139.75x_2 - .146x_1x_2 - .133x_1^2 - 1.14x_2^2$$

TABLE 5.2 **Temperature, Pressure, and Quality of the Finished Product**

x_1, °F	x_2, psi	y	x_1, °F	x_2, psi	y	x_1, °F	x_2, psi	y
80	50	50.8	90	50	63.4	100	50	46.6
80	50	50.7	90	50	61.6	100	50	49.1
80	50	49.4	90	50	63.4	100	50	46.4
80	55	93.7	90	55	93.8	100	55	69.8
80	55	90.9	90	55	92.1	100	55	72.5
80	55	90.9	90	55	97.4	100	55	73.2
80	60	74.5	90	60	70.9	100	60	38.7
80	60	73.0	90	60	68.8	100	60	42.5
80	60	71.2	90	60	71.3	100	60	41.4

FIGURE 5.14   SAS printout for Example 5.3

```
                        ANALYSIS OF VARIANCE

                         SUM OF           MEAN
    SOURCE      DF      SQUARES          SQUARE      F VALUE      PROB>F

    MODEL        5    8402.26454     1680.45291      596.324      0.0001
    ERROR       21      59.17842620     2.81802030
    C TOTAL     26    8461.44296

            ROOT MSE        1.678696      R-SQUARE        0.9930
            DEP MEAN       66.96296      ADJ R-SQ        0.9913
            C.V.            2.506902

                        PARAMETER ESTIMATES

                     PARAMETER       STANDARD     T FOR H0:
    VARIABLE    DF    ESTIMATE         ERROR      PARAMETER=0    PROB > |T|

    INTERCEP     1   -5127.89907     110.29602      -46.492      0.0001
    X1           1      31.09638889    1.34441322    23.130      0.0001
    X2           1     139.74722       3.14005412    44.505      0.0001
    X1X2         1      -0.14550000    0.009691956  -15.012      0.0001
    X1X1         1      -0.13338889    0.006853248  -19.464      0.0001
    X2X2         1      -1.14422222    0.02741299   -41.740      0.0001
```

A three-dimensional graph of this prediction model is shown in Figure 5.15 on page 274. Note that the mean quality seems to be greatest for temperatures of about 85°–90°F and for pressures of about 55–57 pounds per square inch.* Further experimentation in these ranges might lead to a more precise determination of the optimal temperature–pressure combination.

A look at the coefficient of determination, $R^2 = .993$, the F value for testing the entire model, $F = 596.32$, and the p-value for the test, $p = .0001$ (shaded in Figure 5.14), leaves little doubt that the complete second-order model is useful for explaining mean quality as a function of temperature and pressure. This, of course, will not always be the case. The additional complexity of second-order models is worthwhile only if a better model results. To determine whether the quadratic terms are important, we would test $H_0: \beta_4 = \beta_5 = 0$ using the partial F test outlined in Section 4.10.

---

*Students with knowledge of calculus should note that we can determine the exact temperature and pressure that maximize quality in the least squares model by solving $\partial \hat{y}/\partial x_1 = 0$ and $\partial \hat{y}/\partial x_2 = 0$ for x_1 and x_2. These estimated optimal values are $x_1 = 86.25$°F and $x_2 = 55.58$ pounds per square inch. Remember, however, that these are only sample estimates of the coordinates for the optimal value.

FIGURE 5.15

Plot of second-order least squares model
for Example 5.3

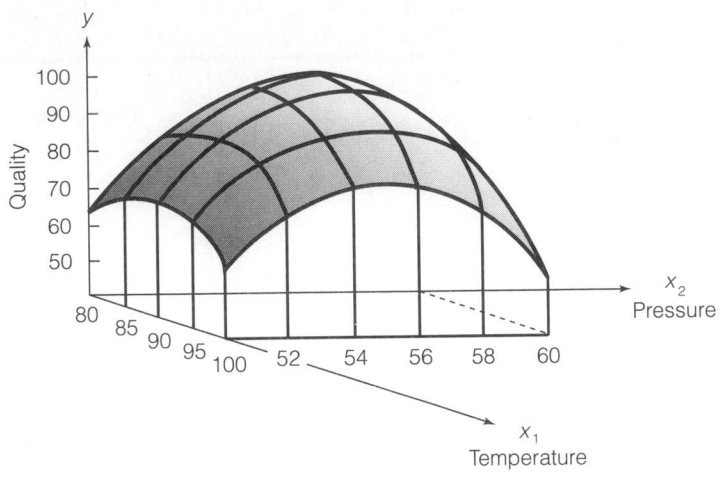

EXERCISES

5.7   Some corporations, instead of owning a fleet of cars, rent cars from a rental agency. A corporation
        may do this because it is sometimes more economical to rent new cars for a year than to buy new
        cars each year. A major rental agency wants to develop a model that will allow it to estimate the
        average annual cost to the prospective customer of renting cars, y, as a function of two independent
        variables:

   x_1 = Number of cars rented

   x_2 = Average number of miles driven per car during year (in thousands)

   a. Identify the independent variables as quantitative or qualitative.
   b. Write the first-order model for $E(y)$.
   c. Add an interaction term between x_1 and x_2 to the first-order model. Suppose interaction exists.
       Graph $E(y)$, the mean cost, versus x_2, the average mileage driven (as you would expect it to
       appear) for several values of x_1.
   d. Write the complete second-order model for $E(y)$.

5.8   Refer to Exercise 5.7. Suppose the model from part c is fit, with the following result:

   $$\hat{y} = 1 + .05x_1 + x_2 + .05x_1x_2$$

   (The units of $\hat{y}$ are thousands of dollars.) Graph the estimated cost $\hat{y}$ as a function of the average
   number of miles driven, x_2, over the range $x_2 = 10$ to $x_2 = 50$ (10–50 thousand miles) for $x_1 = 1$,
   5, and 10. Do these functions agree (approximately) with the graphs you drew for Exercise 5.7, part c?

5.9   Refer to Exercise 5.7. Suppose an additional independent variable is considered:

   x_3 = Average yearly gas price

   a. Write the first-order model plus interaction for $E(y)$ as a function of x_1, x_2, and x_3.
   b. Write the complete second-order model for $E(y)$ as a function of x_1, x_2, and x_3.

5.10 *Multinational* is the term given to an industry with foreign investors. A study of 216 manufacturing industries in Mexico found that multinational presence in a firm has a positive influence on market concentration (*World Development*, Vol. 14, 1986). The result was revealed in a multiple regression analysis on the dependent variable *y*, market concentration index, using the following quantitative independent variables:

x_1 = Market size

x_2 = Market rate of growth

x_3 = Gross production in largest plants (expressed as a percentage of total gross production)

x_4 = Capital intensity (ratio of total assets to total number of employees)

x_5 = Advertising intensity (ratio of advertising to value added)

x_6 = Foreign share (i.e., gross output produced by foreign subsidiaries)

a. Write a first-order model for $E(y)$ as a function of x_1–x_6.
b. Interpret β_6 in the model in part **a**.
c. Based on the results of the study, is β_6 positive or negative?
d. Write a second-order model for $E(y)$ that proposes interaction between the independent variables but with no curvature.
e. Using the model in part **d**, how would you test the hypothesis that effect of a multinational presence on market concentration is independent of the other independent variables in the model?

5.11 Researchers at the Upjohn Company utilized multiple regression analysis in the development of a sustained-release tablet.* One of the objectives of the research was to develop a model relating the dissolution *y* of a tablet (i.e., the percentage of the tablet dissolved over a specified period of time) to the following independent variables:

x_1 = Excipient level (i.e., amount of nondrug ingredient in the tablet)

x_2 = Process variable (e.g., machine setting under which tablet is processed)

a. Write the complete second-order model for $E(y)$.
b. Write a model that hypothesizes straight-line relationships between $E(y)$, x_1, and x_2. Assume that x_1 and x_2 do not interact.
c. Repeat part **b**, but add interaction to the model.
d. For the model in part **c**, what is the slope of the $E(y),x_1$ line for fixed x_2?
e. For the model in part **c**, what is the slope of the $E(y),x_2$ line for fixed x_1?

5.6

Coding Quantitative Independent Variables (Optional)

In fitting higher-order polynomial regression models (e.g., second- or third-order models), it is often a good practice to code the quantitative independent variables. For example, suppose one of the independent variables in a regression analysis is advertising expenditure, *A*, and *A* is observed at three levels: $500, $1,000, and $1,500. We can code (or transform) the advertising expenditure measurements using the formula

---

*Klassen, R. A. "The application of response surface methods to a tablet formulation problem." Paper presented at Joint Statistical Meetings, American Statistical Association and Biometric Society, Aug. 1986, Chicago, Ill.

$$x = \frac{A - 1,000}{500}$$

Then the coded levels $x = -1$, 0, and 1 correspond to the original levels $500, $1,000, and $1,500.

In a general sense, *coding* means transforming a set of independent variables (qualitative or quantitative) into a new set of independent variables. For example, if we observe two independent variables,

A = Advertising expenditure

I = Prime interest rate

then we can transform the two independent variables, A and I, into two new coded variables, x_1 and x_2, where x_1 and x_2 are related to A and I by two functional equations:

$$x_1 = f_1(A, I) \qquad x_2 = f_2(A, I)$$

The functions f_1 and f_2, which are frequently expressed as equations, establish a one-to-one correspondence between combinations of levels of A and I with combinations of the coded values of x_1 and x_2.

Since qualitative independent variables are not numerical, it is necessary to code their values to fit the regression model. However, you might ask why we would bother to code the quantitative independent variables. There are two related reasons for coding quantitative variables. At first glance, it would appear that a computer would be oblivious to the values assumed by the independent variables in a regression analysis, but this is not the case. To calculate the estimates of the model parameters using the method of least squares, the computer must invert a matrix of numbers, called the **coefficient** (or **information**) **matrix** (see Appendix A). Considerable rounding error may occur during the inversion process if the numbers in the coefficient matrix vary greatly in absolute value. This can produce sizable errors in the computed values of the least squares estimates, $\hat{\beta}_0, \hat{\beta}_1, \hat{\beta}_2,$ . . . . Coding makes it computationally easier for the computer to invert the matrix, thus leading to more accurate estimates.

A second reason for coding quantitative variables pertains to a problem we will discuss in detail in Chapter 6: the problem of independent variables (x's) being intercorrelated (called **multicollinearity**). When polynomial regression models (e.g., second-order models) are fit, the problem of multicollinearity is unavoidable, especially when higher-order terms are fit. For example, consider the quadratic model

$$E(y) = \beta_0 + \beta_1 x + \beta_2 x^2$$

If the range of the values of x is narrow, then the two variables, $x_1 = x$ and $x_2 = x^2$, will generally be highly correlated. As we will point out in Chapter 6, the likelihood of rounding errors in the regression coefficients is increased in the presence of these highly correlated independent variables.

The best way to cope with the rounding error problem is to:

1. Code the quantitative variable so that the new coded origin is in the center of the coded values. For example, by coding advertising expenditure, A, as

$$x = \frac{A - 1{,}000}{500}$$

we obtain coded values $-1, 0, 1$. This places the coded origin, 0, in the middle of the range of coded values (-1 to 1).

2. Code the quantitative variable so that the range of the coded values is approximately the same for all coded variables. You need not hold exactly to this requirement. The range of values for one independent variable could be double or triple the range of another without causing any difficulty, but it would not be desirable to have a sizable disparity in the ranges, say, a ratio of 100 to 1.

When the data are observational (the values assumed by the independent variables are uncontrolled), the coding procedure described in the next box satisfies, reasonably well, these two requirements. The coded variable u is similar to the standardized normal z statistic of Section 1.5. Thus, the u value is the deviation (the distance) between an x value and the mean of the x values, $\bar{x}$, expressed in units of s_x.* Since we know that most (approximately 95%) measurements in a set will lie within 2 standard deviations of their mean, it follows that most of the coded u values will lie in the interval -2 to $+2$.

| Coding Procedure for Observational Data

Let

$x =$ Uncoded quantitative independent variable
$u =$ Coded quantitative independent variable

Then if x takes values $x_1, x_2, \ldots, x_n$ for the n data points in the regression analysis, let

$$u_i = \frac{x_i - \bar{x}}{s_x}$$

where s_x is the standard deviation of the x values, i.e.,

$$s_x = \sqrt{\frac{\sum\limits_{i=1}^{n} (x_i - \bar{x})^2}{n-1}}$$

---

*The divisor of the deviation, $x - \bar{x}$, need not equal s_x exactly. Any number approximately equal to s_x would suffice.

If you apply this coding to each quantitative variable, the range of values for each will be approximately -2 to $+2$. The variation in the absolute values of the elements of the coefficient matrix will be moderate, and rounding errors generated in finding the inverse of the matrix will be reduced. Additionally, the correlation between x and x^2 will be reduced.

EXAMPLE 5.4

Table 5.3 gives observational data on the index of building construction costs per month as a function of the index of the cost of construction materials (other components of construction costs would be labor, the cost of money, and so forth). Suppose we are interested in relating monthly construction cost y to monthly index of construction materials x using a quadratic model.

TABLE 5.3 **Index of Building Construction Costs**

MONTH	CONSTRUCTION COST[a] y	INDEX OF ALL CONSTRUCTION MATERIALS[b] x
January	193.2	180.0
February	193.1	181.7
March	193.6	184.1
April	195.1	185.3
May	195.6	185.7
June	198.1	185.9
July	200.9	187.7
August	202.7	189.6

[a]*Source*: United States Department of Commerce, Bureau of the Census.
[b]*Source*: United States Department of Labor, Bureau of Labor Statistics.

a. Give the equation relating the coded variable u to the index of construction materials x using the coding system for observational data.
b. Calculate the coded values, u, for the eight x values.
c. Find the sum of the $n = 8$ values for u.

Solution

a. We first find $\bar{x}$ and s_x:

$$\bar{x} = \frac{\sum_{i=1}^{n} x_i}{n} = \frac{1,480.0}{8} = 185.0$$

$$\sum_{i=1}^{n} (x_i - \bar{x})^2 = \sum_{i=1}^{n} x_i^2 - \frac{\left(\sum_{i=1}^{n} x_i\right)^2}{n} = 273,866.54 - \frac{(1,480.0)^2}{8} = 66.54$$

$$s_x = \sqrt{\frac{\sum_{i=1}^{n} (x_i - \bar{x})^2}{n - 1}} = \sqrt{\frac{66.54}{7}} = 3.08$$

Then the equation relating u and x is

$$u = \frac{x - 185.0}{3.08}$$

b. When $x = 180.0$,

$$u = \frac{x - 185.0}{3.08} = \frac{180.0 - 185.0}{3.08} = -1.62$$

Similarly, when $x = 181.7$,

$$u = \frac{x - 185.0}{3.08} = \frac{181.7 - 185.0}{3.08} = -1.07$$

Table 5.4 gives the coded values for all $n = 8$ observations. [*Note:* You can see that all the $n = 8$ values for u lie in the interval from -2 to $+2$.]

TABLE 5.4
Coded Values of x,
Example 5.4

INDEX	CODED VALUES
x	u
180.0	−1.62
181.7	−1.07
184.1	−.29
185.3	.10
185.7	.23
185.9	.29
187.7	.88
189.6	1.49

c. If you ignore rounding error, the sum of the $n = 8$ values for u will equal 0. This is because the sum of the deviations of a set of measurements about their mean is always equal to 0.

To illustrate the advantage of coding, consider fitting the second-order model

$$E(y) = \beta_0 + \beta_1 x + \beta_2 x^2$$

to the data of Example 5.4. It can be shown that the coefficient of correlation between the two variables, x and x^2, is $r = .999$. However, the coefficient of correlation between the corresponding coded values, u and u^2, is only $r = -.203$. Thus, we can avoid potential rounding error caused by highly correlated x values by fitting, instead, the model

$$E(y) = \beta_0^* + \beta_1^* u + \beta_2^* u^2$$

Other methods of coding have been developed to reduce rounding errors and multicollinearity. One of the more complex coding systems involves fitting **orthogonal polynomials**. An orthogonal system of coding guarantees that the coded independent variables will be uncorrelated. For a discussion of orthogonal polynomials, consult the references given at the end of this chapter.

EXERCISES

5.12 As part of the first-year evaluation for new salespeople, a large food-processing firm projects the second-year sales for each salesperson based on his or her sales for the first year. Data for $n = 8$ salespeople are shown in the table.

FIRST-YEAR SALES x, thousands of dollars	SECOND-YEAR SALES y, thousands of dollars
75.2	99.3
91.7	125.7
100.3	136.1
64.2	108.6
81.8	102.0
110.2	153.7
77.3	108.8
80.1	105.4

a. Give the equation relating the coded variable u to the first-year sales, x, using the coding system for observational data.

b. Calculate the coded values, u.

c. Calculate the coefficient of correlation r between the variables x and x^2.

d. Calculate the coefficient of correlation r between the variables u and u^2. Compare this value to the value computed in part **c**.

e. If you have access to a statistical computer package, fit the model

$$E(y) = \beta_0 + \beta_1 u + \beta_2 u^2$$

5.13 Suppose you want to use the coding system for observational data to fit a second-order model to the tire pressure–automobile mileage data of Exercise 5.6, which are repeated in the table.

PRESSURE x, pounds per square inch	MILEAGE y, thousands
30	29
31	32
32	36
33	38
34	37
35	33
36	26

a. Give the equation relating the coded variable u to pressure, x, using the coding system for observational data.

b. Calculate the coded values, u.

c. Calculate the coefficient of correlation r between the variables x and x^2.

d. Calculate the coefficient of correlation r between the variables u and u^2. Compare this value to the value computed in part **c**.

e. If you have access to a statistical computer package, fit the model

$$E(y) = \beta_0 + \beta_1 u + \beta_2 u^2$$

5.14 Use the coding system for observational data to fit a complete second-order model to the data of Example 5.3, which are repeated in the table.

x_1, °F	x_2, pounds per square inch	y	x_1, °F	x_2, pounds per square inch	y	x_1, °F	x_2, pounds per square inch	y
80	50	50.8	90	50	63.4	100	50	46.6
80	50	50.7	90	50	61.6	100	50	49.1
80	50	49.4	90	50	63.4	100	50	46.4
80	55	93.7	90	55	93.8	100	55	69.8
80	55	90.9	90	55	92.1	100	55	72.5
80	55	90.9	90	55	97.4	100	55	73.2
80	60	74.5	90	60	70.9	100	60	38.7
80	60	73.0	90	60	68.8	100	60	42.5
80	60	71.2	90	60	71.3	100	60	41.4

a. Give the coded values u_1 and u_2 for x_1 and x_2, respectively.

b. Compare the coefficient of correlation between x_1 and x_1^2 with the coefficient of correlation between u_1 and u_1^2.

c. Compare the coefficient of correlation between x_2 and x_2^2 with the coefficient of correlation between u_2 and u_2^2.

d. Give the prediction equation.

5.7

Models with One Qualitative Independent Variable

Suppose we want to write a model for the mean profit, $E(y)$, per sales dollar of a construction company as a function of the sales engineer who estimates and bids on a job (for purposes of explanation, we will ignore other independent variables that might affect the response). Further, suppose there are three sales engineers, Jones, Smith, and Adams. Then sales engineer is a single qualitative variable with three levels corresponding to Jones, Smith, and Adams. Recall that with a qualitative independent variable, we cannot attach a quantitative meaning to a given level. All we can do is describe it.

To simplify our notation, let μ_A be the mean profit per sales dollar for Jones, and let μ_B and μ_C be the corresponding mean profits for Smith and Adams, respectively. Our objective is to write a single prediction equation that will give the mean value of y for the three sales engineers. This can be done as follows:

$$E(y) = \beta_0 + \beta_1 x_1 + \beta_2 x_2$$

where

$$x_1 = \begin{cases} 1 & \text{if Smith is the sales engineer} \\ 0 & \text{if Smith is not the sales engineer} \end{cases}$$

$$x_2 = \begin{cases} 1 & \text{if Adams is the sales engineer} \\ 0 & \text{if Adams is not the sales engineer} \end{cases}$$

The values of x_1 and x_2 for each of the three sales engineers are shown in Table 5.5 on page 282.

The variables x_1 and x_2 are not meaningful independent variables as in the case of the models containing quantitative independent variables. Instead, they

TABLE 5.5 Mean Response for the Model with Three Levels of Sales Engineers

ENGINEER	x_1	x_2	MEAN RESPONSE, $E(y)$
Jones	0	0	$\beta_0 = \mu_A$
Smith	1	0	$\beta_0 + \beta_1 = \mu_B$
Adams	0	1	$\beta_0 + \beta_2 = \mu_C$

are **dummy** (or **indicator**) **variables** that make the model function. To see how they work, let $x_1 = 0$ and $x_2 = 0$. This condition will apply when we are seeking the mean response for Jones (neither Smith nor Adams will be the sales engineer; hence, it must be Jones). Then the mean value of y when Jones is the sales engineer is

$$\mu_A = E(y) = \beta_0 + \beta_1(0) + \beta_2(0) = \beta_0 \quad \text{(see Table 5.5)}$$

This tells us that the mean profit per sales dollar for Jones is β_0. Equivalently, it means that $\beta_0 = \mu_A$.

Now suppose we want to represent the mean response, $E(y)$, when Smith is the sales engineer. Checking the dummy variable definitions, we see that we should let $x_1 = 1$ and $x_2 = 0$:

$$\mu_B = E(y) = \beta_0 + \beta_1 x_1 + \beta_2 x_2 = \beta_0 + \beta_1(1) + \beta_2(0) = \beta_0 + \beta_1$$

or, since $\beta_0 = \mu_A$,

$$\mu_B = \mu_A + \beta_1$$

Then it follows that the interpretation of β_1 is

$$\beta_1 = \mu_B - \mu_A$$

which is the difference in the mean profit per sales dollar for Jones and Smith.

Finally, if we want the mean value of y when Adams is the sales engineer, we let $x_1 = 0$ and $x_2 = 1$:

$$\mu_C = E(y) = \beta_0 + \beta_1(0) + \beta_2(1) = \beta_0 + \beta_2$$

or, since $\beta_0 = \mu_A$,

$$\mu_C = \mu_A + \beta_2$$

Then it follows that the interpretation of β_2 is

$$\beta_2 = \mu_C - \mu_A$$

which represents the difference in the mean profit per sales dollar between Jones and Adams.

Note that we were able to describe *three levels* of the qualitative variable with only *two dummy variables*. This is because the mean of the base level (Jones, in this case) is accounted for by the intercept β_0.

Now, carefully examine the model for a single qualitative independent variable with three levels, because we will use exactly the same pattern for any number of levels. Also, the interpretation of the parameters will always be the same.

One level is selected as the base level (we used Jones as level A). Then, for the 1–0 system of coding* for the dummy variables,

$$\mu_A = \beta_0$$

The coding for all dummy variables is as follows: To represent the mean value of y for a particular level, let that dummy variable equal 1; otherwise, the dummy variable is set equal to 0. Using this system of coding, we have

$$\mu_B = \beta_0 + \beta_1 \qquad \mu_C = \beta_0 + \beta_2$$

Because $\mu_A = \beta_0$, any other model parameter will represent the difference in means for that level and the base level:

$$\beta_1 = \mu_B - \mu_A \qquad \beta_2 = \mu_C - \mu_A$$

The general procedure is given in the accompanying box.

Procedure for Writing a Model with One Qualitative Independent Variable at k Levels

$$E(y) = \beta_0 + \beta_1 x_1 + \beta_2 x_2 + \cdots + \beta_{k-1} x_{k-1}$$

where x_i is the dummy variable for level i and

$$x_i = \begin{cases} 1 & \text{if } E(y) \text{ is the mean for level } i \\ 0 & \text{otherwise} \end{cases}$$

The number of dummy variables for a single qualitative variable is always 1 less than the number of levels for the variable.

Then, for this system of coding,

$$\mu_A = \beta_0$$
$$\mu_B = \beta_0 + \beta_1$$
$$\mu_C = \beta_0 + \beta_2$$
$$\mu_D = \beta_0 + \beta_3$$
$$\vdots$$

Also, note that

$$\beta_1 = \mu_B - \mu_A$$
$$\beta_2 = \mu_C - \mu_A$$
$$\beta_3 = \mu_D - \mu_A$$
$$\vdots$$

---

*We do not have to use a 1–0 system of coding for the dummy variables. Any two-value system will work, but the interpretation given to the model parameters will depend on the code. Using the 1–0 system makes the model parameters easy to interpret.

Thus, we can use least squares to fit predictive models in which the response y is a function of quantitative or qualitative variables. The interpretation of the parameters will differ for the two types of variables, but the objective is the same no matter what type of variable is used—to obtain a good prediction model for y.

5.8

Models with Two Qualitative Independent Variables

We will demonstrate how to write a model with two qualitative independent variables and then, in Section 5.9, will explain how to use this technique to write models with any number of qualitative independent variables.

Let us return to the example used in Section 5.7, where we wrote a model for the mean profit per sales dollar, $E(y)$, as a function of one qualitative independent variable, sales engineer. Now suppose the mean profit is also a function of the state in which the construction job is located, because of tax differences, different labor conditions, and so forth. Assume that the company operates in two states. Therefore, this second qualitative independent variable, state, will be observed at two levels. To simplify our notation, we will change the symbols for the three levels of sales engineer from A, B, C, to E_1, E_2, E_3, and we will let S_1 and S_2 represent the two states in which the company operates. The six population means of profit per sales dollar measurements (measurements of y) are symbolically represented by the six cells in the two-way table shown in Table 5.6. Each μ subscript corresponds to one sales engineer–state combination.

TABLE 5.6 **Table Showing the Six Combinations of Sales Engineer and State**

| | | STATE | |
		S_1	S_2
SALES ENGINEER	E_1	μ_{11}	μ_{12}
	E_2	μ_{21}	μ_{22}
	E_3	μ_{31}	μ_{32}

First we will write a model in its simplest form—where the two qualitative variables affect the response independently of each other. To write the model for a mean profit, $E(y)$, we start with a constant β_0 and then add *two* dummy variables for the three levels of sales engineer in the manner explained in Section 5.7. These terms, which are called the **main effect** terms for sales engineer, E, account for the effect of E on $E(y)$ when sales engineer, E, and state, S, affect $E(y)$ independently. Then,

$$E(y) = \beta_0 + \overbrace{\beta_1 x_1 + \beta_2 x_2}^{\substack{\text{Main effect} \\ \text{terms for } E}}$$

where

$$x_1 = \begin{cases} 1 & \text{if } E_2 \text{ was the sales engineer} \\ 0 & \text{if not} \end{cases}$$

$$x_2 = \begin{cases} 1 & \text{if } E_3 \text{ was the sales engineer} \\ 0 & \text{if not} \end{cases}$$

Now let level S_1 be the base level of the state variable. Since there are two levels of this variable, we will need only one dummy variable to include the state in the model:

$$F(y) = \beta_0 + \underbrace{\beta_1 x_1 + \beta_2 x_2}_{\substack{\text{Main effect} \\ \text{terms for } E}} + \underbrace{\beta_3 x_3}_{\substack{\text{Main effect} \\ \text{term for } S}}$$

where the dummy variables x_1 and x_2 are defined as above, and

$$x_3 = \begin{cases} 1 & \text{if } S_2 \text{ was the state in which the job was located} \\ 0 & \text{if } S_1 \text{ (base level) was the state in which the job was located} \end{cases}$$

If you check the model, you will see that by assigning specific values to x_1, x_2, and x_3, you create a model for the mean value of y corresponding to one of the cells of Table 5.6. We will illustrate with two examples.

EXAMPLE 5.5

Give the values of x_1, x_2, and x_3 and the model for the mean profit per sales dollar, $E(y)$, when E_1 is the sales engineer and S_1 is the state in which the job is located.

Solution

Checking the coding system, you will see that E_1 and S_1 occur when $x_1 = x_2 = x_3 = 0$. Then,

$$E(y) = \beta_0 + \beta_1 x_1 + \beta_2 x_2 + \beta_3 x_3 = \beta_0 + \beta_1(0) + \beta_2(0) + \beta_3(0)$$
$$= \beta_0$$

Therefore, the mean value of y at levels E_1 and S_1, which we represent as μ_{11}, is

$$\mu_{11} = \beta_0$$

EXAMPLE 5.6

Give the values of x_1, x_2, and x_3 and the model for the mean profit per sales dollar, $E(y)$, when E_3 is the sales engineer and S_2 is the state.

Solution

Checking the coding system, you will see that for levels E_3 and S_2,

$$x_1 = 0 \qquad x_2 = 1 \qquad x_3 = 1$$

Then, the mean profit per sales dollar when E_3 is the sales engineer and S_2 is the state, represented by the symbol μ_{32} (see Table 5.6), is

$$\mu_{32} = E(y) - \beta_0 + \beta_1 x_1 + \beta_2 x_2 + \beta_3 x_3 = \beta_0 + \beta_1(0) + \beta_2(1) + \beta_3(1)$$
$$= \beta_0 + \beta_2 + \beta_3$$

Note that in the model described above, we assumed the qualitative independent variables for sales engineer and state affect the mean response, $E(y)$, independently of each other. This type of model is called a **main effects model**. Changing the level of one qualitative variable will have the same effect on $E(y)$ for any level of the second qualitative variable. In other words, the effect of one qualitative variable on $E(y)$ is independent (in a mathematical sense) of the level of the second qualitative variable.

Main Effects Model with Two Qualitative Independent Variables, One at Three Levels (E_1, E_2, E_3) and the Other at Two Levels (S_1, S_2)

$$E(y) = \beta_0 + \overbrace{\beta_1 x_1 + \beta_2 x_2}^{\substack{\text{Main effect} \\ \text{terms for } E}} + \overbrace{\beta_3 x_3}^{\substack{\text{Main effect} \\ \text{term for } S}}$$

where

$$x_1 = \begin{cases} 1 & \text{if } E_2 \\ 0 & \text{if not} \end{cases} \qquad x_2 = \begin{cases} 1 & \text{if } E_3 \\ 0 & \text{if not} \end{cases} \qquad (E_1 \text{ is base level})$$

$$x_3 = \begin{cases} 1 & \text{if } S_2 \\ 0 & \text{if } S_1 \quad \text{(base level)} \end{cases}$$

INTERPRETATION OF MODEL PARAMETERS

β_0: μ_{11} (Mean of the combination of base levels)

β_1: $\mu_{2j} - \mu_{1j}$, for any level S_j ($j = 1, 2$)

β_2: $\mu_{3j} - \mu_{1j}$, for any level S_j ($j = 1, 2$)

β_3: $\mu_{i2} - \mu_{i1}$, for any level E_i ($i = 1, 2, 3$)

When two independent variables affect the mean response independently of each other, you may obtain the pattern shown in Figure 5.16. Note that the difference in mean profit between any two sales engineers (levels of E) is the same, *regardless* of the state in which the job is located. That is, the main effects model assumes the relative effect of sales engineer on profit is the same in both states.

If E and S do not affect $E(y)$ independently of each other, then the response function might appear as shown in Figure 5.17. Note the difference between the mean response functions for Figures 5.16 and 5.17. When E and S affect the mean response in a dependent manner (Figure 5.17), the response functions differ for each state. This means that you cannot study the effect of one variable on $E(y)$ without considering the level of the other. When this situation occurs, we say that the qualitative independent variables **interact**. In this example, interaction might be expected if one sales engineer tends to develop a knack for bidding on jobs in state S_1, while another becomes adept at bidding in state S_2.

FIGURE 5.16

Main effects model: Mean response as a function of E and S when E and S affect $E(y)$ independently

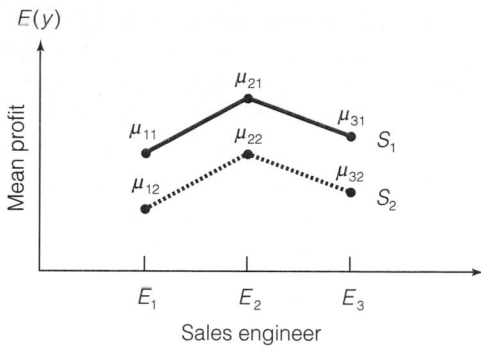

FIGURE 5.17

Interaction model: Mean response as a function of E and S when E and S interact to affect $E(y)$

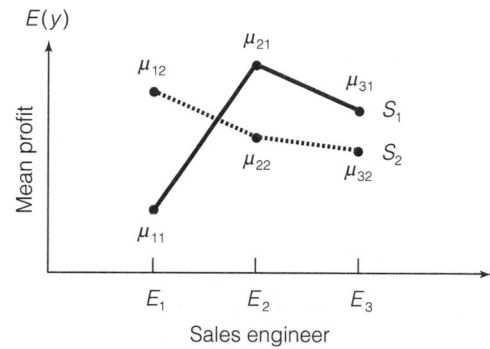

Interaction Model with Two Qualitative Independent Variables, One at Three Levels (E_1, E_2, E_3) and the Other at Two Levels (S_1, S_2)

$$E(y) = \beta_0 + \overbrace{\beta_1 x_1 + \beta_2 x_2}^{\substack{\text{Main effect} \\ \text{terms for } E}} + \overbrace{\beta_3 x_3}^{\substack{\text{Main effect} \\ \text{term for } S}} + \overbrace{\beta_4 x_1 x_3 + \beta_5 x_2 x_3}^{\substack{\text{Interaction} \\ \text{terms}}}$$

where the dummy variables x_1, x_2, and x_3 are defined in the same way as for the main effects model.

INTERPRETATION OF MODEL PARAMETERS

β_0: μ_{11} (Mean of the combination of base levels)

β_1: $\mu_{21} - \mu_{11}$ (i.e., for base level S_1 only)

β_2: $\mu_{31} - \mu_{11}$ (i.e., for base level S_1 only)

β_3: $\mu_{12} - \mu_{11}$ (i.e., for base level E_1 only)

β_4: $(\mu_{22} - \mu_{12}) - (\mu_{21} - \mu_{11})$

β_5: $(\mu_{32} - \mu_{12}) - (\mu_{31} - \mu_{11})$

When qualitative independent variables interact, the model for $E(y)$ must be constructed so that it is able (if necessary) to give a different mean value, $E(y)$, for every cell in Table 5.6. We do this by adding **interaction terms** to the main effects model. These terms will involve all possible two-way cross products between each of the two dummy variables for E, x_1, and x_2, and the one dummy variable for S, x_3. The number of interaction terms (for two independent variables) will equal the number of main effect terms for the one variable times the number of main effect terms for the other.

Note that when E and S interact, the model contains six parameters, the two main effect terms for E, one main effect term for S, $(2)(1) = 2$ interaction terms, and β_0. This will make it possible, by assigning the various combinations of values to the dummy variables x_1, x_2, and x_3, to give six different values for $E(y)$ that will correspond to the means of the six cells of Table 5.6.

EXAMPLE 5.7

In Example 5.5 we gave the mean response when E_1 was the sales engineer and S_1 was the state in which the job was located, where we assumed that E and S affected $E(y)$ independently (no interaction). Now give the value of $E(y)$ for the model where E and S interact to affect $E(y)$.

Solution

When E and S interact,

$$E(y) = \beta_0 + \beta_1 x_1 + \beta_2 x_2 + \beta_3 x_3 + \beta_4 x_1 x_3 + \beta_5 x_2 x_3$$

For levels E_1 and S_1, we have agreed (according to our system of coding) to let $x_1 = x_2 = x_3 = 0$. Substituting into the equation for $E(y)$, we have

$$E(y) = \beta_0$$

(the same as for the main effects model).

EXAMPLE 5.8

In Example 5.6 we gave the mean response for sales engineer E_3 and state S_2 when E and S affected $E(y)$ independently. Now assume that E and S interact and write a model for $E(y)$ when E_3 is the sales engineer and S_2 is the state in which the job is located.

Solution

When E and S interact,

$$E(y) = \beta_0 + \beta_1 x_1 + \beta_2 x_2 + \beta_3 x_3 + \beta_4 x_1 x_3 + \beta_5 x_2 x_3$$

To model $E(y)$ for E_3 and S_2, we set $x_1 = 0$, $x_2 = 1$, and $x_3 = 1$:

$$E(y) = \beta_0 + \beta_1(0) + \beta_2(1) + \beta_3(1) + \beta_4(0)(1) + \beta_5(1)(1)$$
$$= \beta_0 + \beta_2 + \beta_3 + \beta_5$$

This is the model for the value of μ_{32} in Table 5.6. Note the difference in $E(y)$ for the model assuming independence between E and S versus this one, which assumes interaction between E and S. The difference is β_5.

EXAMPLE 5.9

The profit per sales dollar, y, for the six combinations of sales engineer and state is shown in Table 5.7. Note that the number of construction jobs per combination varies from one for levels (E_1, S_2) to three for levels (E_1, S_1). A total of twelve jobs are sampled.

TABLE 5.7 **Profit Data for Combinations of Sales Engineer and State**

		STATE	
		S_1	S_2
SALES ENGINEER	E_1	$\$.065$ .073 .068	$\$.036$
	E_2	.078 .082	.050 .043
	E_3	.048 .046	.061 .062

a. Assume the interaction between E and S is negligible. Fit the model for $E(y)$ with interaction terms omitted.
b. Fit the complete model for $E(y)$ allowing for the fact that interactions might occur.
c. Estimate the mean profit for jobs bid by sales engineer E_3 in state S_2 using the prediction equation $\hat{y}$ for each of the two models in parts **a** and **b**. Then calculate the sample mean for this cell of Table 5.7. Explain the discrepancy between the sample mean for levels (E_3, S_2) and the estimate(s) obtained from one or both of the two prediction equations.

Solution

a. A portion of the SAS printout for the main effects model

$$E(y) = \beta_0 + \overbrace{\beta_1 x_1 + \beta_2 x_2}^{\substack{E \\ \text{main effect}}} + \overbrace{\beta_3 x_3}^{\substack{S \\ \text{main effect}}}$$

is given in Figure 5.18 on page 290. The least squares prediction equation is

$$\hat{y} = .0645 + .0067x_1 - .00230x_2 - .0158x_3$$

b. The complete model SAS printout is given in Figure 5.19 on page 291. Recall that the complete model is

$$E(y) = \beta_0 + \beta_1 x_1 + \beta_2 x_2 + \beta_3 x_3 + \beta_4 x_1 x_3 + \beta_5 x_2 x_3$$

The least squares prediction equation is

$$\hat{y} = .0687 + .0113x_1 - .0217x_2 - .0327x_3 - .0008x_1 x_3 + .0472x_2 x_3$$

FIGURE 5.18   SAS printout for main effects model of Example 5.9

DEP VARIABLE: Y

ANALYSIS OF VARIANCE

SOURCE	DF	SUM OF SQUARES	MEAN SQUARE	F VALUE	PROB>F
MODEL	3	0.000858258	0.000286086	1.513	0.2838
ERROR	8	0.001512409	0.000189051		
C TOTAL	11	0.002370667			

ROOT MSE	0.01374959	R-SQUARE	0.3620	
DEP MEAN	0.05933333	ADJ R-SQ	0.1228	
C.V.	23.17346			

PARAMETER ESTIMATES

VARIABLE	DF	PARAMETER ESTIMATE	STANDARD ERROR	T FOR H0: PARAMETER=0	PROB > |T|
INTERCEP	1	0.06445455	0.007180488	8.976	0.0001
X1	1	0.006704545	0.009940935	0.674	0.5190
X2	1	-0.002295455	0.009940935	-0.231	0.8232
X3	1	-0.01581818	0.008291313	-1.908	0.0928

OBS	ID	ACTUAL	PREDICT VALUE	STD ERR PREDICT	LOWER95% MEAN	UPPER95% MEAN	RESIDUAL
1	E1S1	0.0650	0.0645	.0071805	0.0479	0.0810	5.5E-04
2	E1S1	0.0730	0.0645	.0071805	0.0479	0.0810	.0085455
3	E1S1	0.0680	0.0645	.0071805	0.0479	0.0810	.0035455
4	E2S1	0.0780	0.0712	0.008028	0.0526	0.0897	.0068409
5	E2S1	0.0820	0.0712	0.008028	0.0526	0.0897	0.0108
6	E3S1	0.0480	0.0622	0.008028	0.0436	0.0807	-0.0142
7	E3S1	0.0460	0.0622	0.008028	0.0436	0.0807	-0.0162
8	E1S2	0.0360	0.0486	0.00927	0.0273	0.0700	-0.0126
9	E2S2	0.0500	0.0553	0.008028	0.0368	0.0739	-.005341
10	E2S2	0.0430	0.0553	0.008028	0.0368	0.0739	-0.0123
11	E3S2	0.0610	0.0463	0.008028	0.0278	0.0649	0.0147
12	E3S2	0.0620	0.0463	0.008028	0.0278	0.0649	0.0157

SUM OF RESIDUALS   2.29851E-16
SUM OF SQUARED RESIDUALS   0.001512409
PREDICTED RESID SS (PRESS)   0.003615375

c. To obtain the estimated mean response for cell (E_3, S_2), we let $x_1 = 0$, $x_2 = 1$, and $x_3 = 1$. Then, for the main effects model, we find

$$\hat{y} = .0645 + .0067(0) - .0023(1) - .0158(1) = .0464$$

The 95% confidence interval for the true mean profit per sales dollar (shown in Figure 5.18) is (.0278, .0649).

For the complete model, we find

$$\hat{y} = .0687 + .0113(0) - .0217(1) - .0327(1) - .0008(0)(1) + .0472(1)(1)$$
$$= .0615$$

The 95% confidence interval for true mean profit (Figure 5.19) is (.0557, .0673).

The mean for the cell (E_3, S_2) in Table 5.7 is

$$\bar{y}_{32} = \frac{.061 + .062}{2} = .0615$$

which is precisely what is estimated by the complete (interaction) model. However, the main effects model yields a different estimate, .0464. The reason for the

FIGURE 5.19   SAS printout for interaction model of Example 5.9

DEP VARIABLE: Y

ANALYSIS OF VARIANCE

SOURCE	DF	SUM OF SQUARES	MEAN SQUARE	F VALUE	PROB>F
MODEL	5	0.002303000	0.000460600	40.841	0.0001
ERROR	6	0.000067667	0.000011278		
C TOTAL	11	0.002370667			

ROOT MSE	0.00335824	R-SQUARE	0.9715	
DEP MEAN	0.05933333	ADJ R-SQ	0.9477	
C.V.	5.659956			

PARAMETER ESTIMATES

VARIABLE	DF	PARAMETER ESTIMATE	STANDARD ERROR	T FOR H0: PARAMETER=0	PROB > |T|
INTERCEP	1	0.06866667	0.001938881	35.416	0.0001
X1	1	0.01133333	0.003065640	3.697	0.0101
X2	1	-0.02166667	0.003065640	-7.068	0.0004
X3	1	-0.03266667	0.003877762	-8.424	0.0002
X1X3	1	-0.000833333	0.005129797	-0.162	0.8763
X2X3	1	0.04716667	0.005129797	9.195	0.0001

OBS	ID	ACTUAL	PREDICT VALUE	STD ERR PREDICT	LOWER95% MEAN	UPPER95% MEAN	RESIDUAL
1	E1S1	0.0650	0.0687	.0019389	0.0639	0.0734	-.003667
2	E1S1	0.0730	0.0687	.0019389	0.0639	0.0734	.0043333
3	E1S1	0.0680	0.0687	.0019389	0.0639	0.0734	-6.7E-04
4	E2S1	0.0780	0.0800	.0023746	0.0742	0.0858	-0.002
5	E2S1	0.0820	0.0800	.0023746	0.0742	0.0858	0.002
6	E3S1	0.0480	0.0470	.0023746	0.0412	0.0528	0.001
7	E3S1	0.0460	0.0470	.0023746	0.0412	0.0528	-1.0E-03
8	E1S2	0.0360	0.0360	.0033582	0.0278	0.0442	-8.7E-18
9	E2S2	0.0500	0.0465	.0023746	0.0407	0.0523	0.0035
10	E2S2	0.0430	0.0465	.0023746	0.0407	0.0523	-0.0035
11	E3S2	0.0610	0.0615	.0023746	0.0557	0.0673	-5.0E-04
12	E3S2	0.0620	0.0615	.0023746	0.0557	0.0673	5.0E-04

SUM OF RESIDUALS   3.76435E-16
SUM OF SQUARED RESIDUALS   .00006766667
PREDICTED RESID SS (PRESS) 0.0003151194

discrepancy is that the main effects model assumes the two qualitative independent variables affect $E(y)$ independently of each other. That is, the change in $E(y)$ produced by a change in levels of one variable is the same regardless of the level of the other variable. In contrast, the complete model contains six parameters $(\beta_0, \beta_1, \ldots, \beta_5)$ to describe the six cell populations, so that each population cell mean will be estimated by its sample mean. Thus, the complete model estimate for any cell mean is equal to the observed (sample) mean for that cell.

Example 5.9 demonstrates an important point. If we were to ignore the least squares analysis and calculate the six sample means of Table 5.7 directly, we would obtain exactly the same estimates of $E(y)$ as would be obtained by a least squares analysis for the case where the interaction between E and S is assumed to exist. We would not obtain the same estimates if the model assumes interaction does not exist.

Also, the estimates of means raise important questions. Do the data provide sufficient evidence to indicate that E and S interact? For our example, does the contribution to mean profit per sales dollar for a job in one state depend on which sales engineer estimated and bid the job? The plot of all six sample means, shown in Figure 5.20, seems to indicate interaction, since engineers E_1 and E_2 appear to operate more effectively in state S_1, while the mean profit of E_3 is higher in state S_2. Can these sample facts be reliably generalized to conclusions about the populations?

FIGURE 5.20

Graph of sample means for profit example

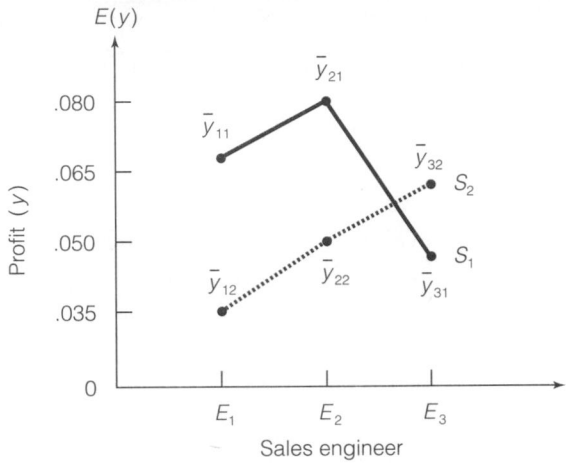

To answer this question, we will want to perform a test for interaction between the two qualitative independent variables, sales engineer and state. Since allowance for interaction between sales engineer and state in the complete model was provided by the addition of the terms $\beta_4 x_1 x_3$ and $\beta_5 x_2 x_3$, it follows that the null hypothesis that the independent variables sales engineer and state do not interact is equivalent to the hypothesis that the terms $\beta_4 x_1 x_3$ and $\beta_5 x_2 x_3$ are not needed in the model for $E(y)$—or equivalently, that $\beta_4 = \beta_5 = 0$. Conversely, the alternative hypothesis that sales engineer and state do interact is equivalent to stating that at least one of the two parameters, β_4 or β_5, differs from 0.

The appropriate procedure for testing a portion of the model parameters, an F test, was discussed in Section 4.10. The F test is carried out as follows:

H_0: $\beta_4 = \beta_5 = 0$

H_a: At least one of β_4 and β_5 differs from 0.

Test statistic: $F = \dfrac{(SSE_R - SSE_C)/g}{SSE_C/[n - (k + 1)]}$

where

$$SSE_R = \text{SSE for reduced model (main effects model)}$$
$$SSE_C = \text{SSE for complete model (interaction model)}$$
$$g = \text{number of } \beta\text{'s tested}$$
$$= \text{numerator df for the } F \text{ statistic}$$
$$n - (k + 1) = \text{df for error for complete model}$$
$$= \text{denominator df for the } F \text{ statistic}$$

For this example we have

$$\text{SSE}_R = .00151241 \quad \text{(see Figure 5.18)}$$
$$\text{SSE}_C = .00006767 \quad \text{(see Figure 5.19)}$$
$$g = 2 \quad \text{and} \quad n - (k + 1) = 6$$

Then

$$F = \frac{(.00151241 - .00006767)/2}{.00006767/6}$$

$$= \frac{.00072237}{.00001128} = 64.04$$

The critical value of F for $\alpha = .05$, $\nu_1 = 2$, and $\nu_2 = 6$ is (from Table 4 of Appendix D) $F_{.05} = 5.14$. Therefore,

Rejection region: $F > 5.14$

Since the calculated $F = 64.04$ exceeds 5.14, we are confident (at $\alpha = .05$) in concluding that the interaction terms contribute to the prediction of y, profit per sales. Equivalently, there is sufficient evidence to conclude that E and S do interact.

EXERCISES

5.15 According to the U.S. Department of Labor, the percentage of married women participating in the work force reached 56% in 1987. How does the employment status of wives affect their husbands' well-being? A study reported in the *Academy of Management* (Mar. 1989) addressed this issue by fitting the model $E(y) = \beta_0 + \beta_1 x$ to data collected for $n = 413$ professional accountants, where y = husband's job satisfaction (measured on a 5-point scale) and x is a dummy variable for employment status of wife (1 = employed, 0 = unemployed).

a. The estimate of β_1 was negative and statistically significant at $\alpha = .01$. Interpret these results.

b. The value of the coefficient of determination was $R^2 = .02$. Interpret this result.

5.16 Each year *Business Week* reports the total cash compensations (salary plus bonus) for the top corporate executives in the United States. The data in the table on page 294 (in thousands of dollars) were extracted from *Business Week's* 1990 Executive Compensation Scoreboard. To compare the mean 1990 cash compensation, $E(y)$, of executives in the four groups, the following model was fit to the data:

$$E(y) = \beta_0 + \beta_1 x_1 + \beta_2 x_2 + \beta_3 x_3$$

where

$$x_1 = \begin{cases} 1 & \text{if consumer products} \\ 0 & \text{if not} \end{cases} \qquad x_2 = \begin{cases} 1 & \text{if utilities} \\ 0 & \text{if not} \end{cases}$$

$$x_3 = \begin{cases} 1 & \text{if industrial–high tech} \\ 0 & \text{if not} \end{cases}$$

Base level = Financial services

The Minitab printout follows the data.

CONSUMER PRODUCTS	UTILITIES	INDUSTRIAL–HIGH TECH	FINANCIAL SERVICES
1,567	1,862	2,925	3,125
3,313	1,390	3,409	4,143
2,058	1,115	1,767	4,013
25,216	1,105	4,097	6,583
4,634	1,272	3,196	3,169
5,214	2,849	4,042	5,217
20,795	1,732	2,601	3,447
9,162	1,474	8,286	4,469

Source: "Executive compensation scoreboard." *Business Week*, May 7, 1990, pp. 65–108.

Minitab printout for Exercise 5.16

```
The regression equation is
Y = 4271 + 4724 X1 - 2671 X2 - 480 X3

Predictor       Coef        Stdev      t-ratio        p
Constant        4271        1651         2.59      0.015
X1              4724        2334         2.02      0.053
X2             -2671        2334        -1.14      0.262
X3              -480        2334        -0.21      0.838

s = 4669        R-sq = 27.6%     R-sq(adj) = 19.8%

Analysis of Variance

SOURCE         DF          SS           MS         F         p
Regression      3    232505648     77501880      3.56     0.027
Error          28    610272512     21795446
Total          31    842778176

SOURCE         DF       SEQ SS
X1              1    200071984
X2              1     31510622
X3              1       923041

Unusual Observations
Obs.       X1           Y      Fit Stdev.Fit   Residual   St.Resid
   4     1.00       25216     8995      1651      16221      3.71R
   7     1.00       20795     8995      1651      11800      2.70R

R denotes an obs. with a large st. resid.
```

a. Is there sufficient evidence to indicate that the model is useful for predicting cash compensation?
b. What does the result from part **a** imply about the mean cash compensation for the four groups of executives?
c. Find a 99% confidence interval for the difference between the mean 1990 cash compensation of executives in the consumer products and financial services industries.

5.17   The performance of an industry is often measured by the level of excess (or unutilized) capacity within the firm. Researchers examined the relationship between excess capacity *y* and several market variables in 273 U.S. manufacturing industries (*Quarterly Journal of Business and Economics*, Summer 1986). Two qualitative independent variables considered in the study were

   Market concentration (low, moderate, and high)
   Industry type (producer or consumer)

a. Write the main effects model for $E(y)$ as a function of the two qualitative variables.
b. Interpret the β coefficients in the main effects model.
c. Write the model for $E(y)$ that includes interaction between market concentration and industry type.
d. Interpret the β coefficients in the interaction model.
e. How would you test the hypothesis that the difference between the mean excess capacity levels of producer and consumer industry types is the same across all three market concentrations?

5.18   Suppose an oil company wants to model the mean monthly gasoline sales, $E(y)$, of its affiliated stations as a function of type of gasoline purchased—regular, premium, or lead-free—and of type of service—self-service or full-service.

a. How many dummy variables will be needed to describe each of the qualitative variables, type of gasoline and type of service?
b. Write the main effects model relating $E(y)$ to type of gasoline and type of service. Be sure to code the dummy variables.
c. Write a model for $E(y)$ that includes interaction between type of gasoline and type of service.
d. Do you think interaction would be important in this model? A plot of your intuitive estimates of the mean will help you decide.

5.19   Suppose the interaction model of Exercise 5.18 part c, is used, and the following least squares model is obtained (units of sales, y, are millions of dollars):

$$\hat{y} = 4 - 2x_1 - x_2 - x_3 + 2x_1x_3 + 3x_2x_3$$

where

$$x_1 = \begin{cases} 1 & \text{if premium} \\ 0 & \text{otherwise} \end{cases} \qquad x_2 = \begin{cases} 1 & \text{if lead-free} \\ 0 & \text{otherwise} \end{cases} \qquad x_3 = \begin{cases} 1 & \text{if full-service} \\ 0 & \text{if self-service} \end{cases}$$

a. What is the estimate of mean sales for lead-free gasoline at the full-service pumps?
b. What is the estimated difference between mean sales of regular gasoline at full-service and self-service pumps?
c. To see the effects of interaction, compare the difference between the regular self-service and full-service mean sales to the difference between the premium self-service and full-service mean sales.

5.9

Models with Three or More Qualitative Independent Variables

Models with three or more qualitative independent variables are constructed in the same manner as for two qualitative independent variables, except that you must add three-way interaction terms if you have three qualitative independent variables, three-way and four-way interaction terms for four independent variables, and so on. In this section we will explain what we mean by three-way, four-way, etc., interactions and will demonstrate the procedure for writing the model for any number, say, k, of qualitative independent variables. The pattern employed in writing the model is shown in the box on page 296.

Recall that a two-way interaction term was formed by multiplying the dummy variable associated with one of the main effect terms of one (call it the first) independent variable by the dummy variable from a main effect term of another

> ### Pattern of the Model Relating $E(y)$ to k Qualitative Independent Variables
>
> $E(y) = \beta_0 +$ Main effect terms for all independent variables
>
> $+$ All two-way interaction terms between pairs of independent variables
>
> $+$ All three-way interaction terms between different groups of three independent variables
>
> $+$
>
> $\vdots$
>
> $+$ All k-way interaction terms for the k independent variables

(the second) independent variable. Three-way interaction terms are formed in a similar way, by forming the product of three dummy variables, one from a main effect term from each of the three independent variables. Similarly, four-way interaction terms are formed by taking the product of four dummy variables, one from a main effect term from each of four independent variables. We will illustrate with three examples.

EXAMPLE 5.10

Suppose you have three qualitative independent variables, the first at three levels, A_1, A_2, and A_3, the second at three levels, B_1, B_2, and B_3, and the third at two levels, C_1 and C_2. Write a model for $E(y)$ that includes all main effect and interaction terms for the independent variables.

Solution

First write a model containing the main effect terms for the three variables:

$$E(y) = \beta_0 + \underbrace{\beta_1 x_1 + \beta_2 x_2}_{\substack{\text{Main effect} \\ \text{terms for } A}} + \underbrace{\beta_3 x_3 + \beta_4 x_4}_{\substack{\text{Main effect} \\ \text{terms for } B}} + \underbrace{\beta_5 x_5}_{\substack{\text{Main effect} \\ \text{term for } C}}$$

where

$$x_1 = \begin{cases} 1 & \text{if level } A_2 \\ 0 & \text{if not} \end{cases} \qquad x_3 = \begin{cases} 1 & \text{if level } B_2 \\ 0 & \text{if not} \end{cases} \qquad x_5 = \begin{cases} 1 & \text{if level } C_2 \\ 0 & \text{if not} \end{cases}$$

$$x_2 = \begin{cases} 1 & \text{if level } A_3 \\ 0 & \text{if not} \end{cases} \qquad x_4 = \begin{cases} 1 & \text{if level } B_3 \\ 0 & \text{if not} \end{cases}$$

The next step is to add two-way interaction terms. These will be of three types—those for the interaction between A and B, between A and C, and between B and C. Thus,

$$E(y) = \beta_0 + \underbrace{\beta_1 x_1 + \beta_2 x_2}_{\substack{\text{Main effect} \\ A}} + \underbrace{\beta_3 x_3 + \beta_4 x_4}_{\substack{\text{Main effect} \\ B}} + \underbrace{\beta_5 x_5}_{\substack{\text{Main effect} \\ C}}$$

$$\overbrace{}^{AB \text{ interaction terms}}$$
$$+ \; \beta_6 x_1 x_3 + \beta_7 x_1 x_4 + \beta_8 x_2 x_3 + \beta_9 x_2 x_4$$

$$\overbrace{\phantom{\beta_{10} x_1 x_5 + \beta_{11} x_2 x_5}}^{AC \text{ interaction terms}}$$
$$+ \; \beta_{10} x_1 x_5 + \beta_{11} x_2 x_5$$

$$\overbrace{\phantom{\beta_{12} x_3 x_5 + \beta_{13} x_4 x_5}}^{BC \text{ interaction terms}}$$
$$+ \; \beta_{12} x_3 x_5 + \beta_{13} x_4 x_5$$

Finally, since there are three independent variables, we must include terms for the interaction of A, B, and C. These terms are formed as the products of dummy variables, one from each of the A, B, and C main effect terms. The complete model for $E(y)$ is

$$E(y) = \beta_0 + \overbrace{\beta_1 x_1 + \beta_2 x_2}^{\substack{\text{Main effect}\\A} } + \overbrace{\beta_3 x_3 + \beta_4 x_4}^{\substack{\text{Main effect}\\B}} + \overbrace{\beta_5 x_5}^{\substack{\text{Main effect}\\C}}$$

$$\overbrace{}^{AB \text{ interaction terms}}$$
$$+ \; \beta_6 x_1 x_3 + \beta_7 x_1 x_4 + \beta_8 x_2 x_3 + \beta_9 x_2 x_4$$

$$\overbrace{\phantom{\beta_{10} x_1 x_5 + \beta_{11} x_2 x_5}}^{AC \text{ interaction terms}}$$
$$+ \; \beta_{10} x_1 x_5 + \beta_{11} x_2 x_5$$

$$\overbrace{\phantom{\beta_{12} x_3 x_5 + \beta_{13} x_4 x_5}}^{BC \text{ interaction terms}}$$
$$+ \; \beta_{12} x_3 x_5 + \beta_{13} x_4 x_5$$

$$\overbrace{\phantom{\beta_{14} x_1 x_3 x_5 + \beta_{15} x_1 x_4 x_5 + \beta_{16} x_2 x_3 x_5 + \beta_{17} x_2 x_4 x_5}}^{\text{Three-way } ABC \text{ interaction terms}}$$
$$+ \; \beta_{14} x_1 x_3 x_5 + \beta_{15} x_1 x_4 x_5 + \beta_{16} x_2 x_3 x_5 + \beta_{17} x_2 x_4 x_5$$

Note that the complete model in Example 5.10 contains 18 parameters, one for each of the $3 \times 3 \times 2$ combinations of levels for A, B, and C. There are 18 linearly independent linear combinations of these parameters, *one corresponding to each of the means of the $3 \times 3 \times 2$ combinations of levels of A, B, and C*. We will illustrate with an example.

EXAMPLE 5.11

Refer to Example 5.10 and give the expression for the mean value of y for observations taken at the second level of A, the first level of B, and the second level of C, i.e., at (A_2, B_1, C_2).

Solution

Check the coding for the dummy variables (given in Example 5.10) and you will see that they assume the following values:

For level A_2: $x_1 = 1$, $x_2 = 0$
For level B_1: $x_3 = 0$, $x_4 = 0$
For level C_2: $x_5 = 1$

Substituting these values into the expression for $E(y)$, we obtain

$$E(y) = \beta_0 + \beta_1(1) + \beta_2(0) + \beta_3(0) + \beta_4(0) + \beta_5(1)$$
$$+ \beta_6(1)(0) + \beta_7(1)(0) + \beta_8(0)(0) + \beta_9(0)(0)$$
$$+ \beta_{10}(1)(1) + \beta_{11}(0)(1) + \beta_{12}(0)(1) + \beta_{13}(0)(1)$$
$$+ \beta_{14}(1)(0)(1) + \beta_{15}(1)(0)(1) + \beta_{16}(0)(0)(1) + \beta_{17}(0)(0)(1)$$
$$= \beta_0 + \beta_1 + \beta_5 + \beta_{10}$$

Thus, the mean value of y observed at levels A_2, B_1, and C_2 is $\beta_0 + \beta_1 + \beta_5 + \beta_{10}$. You could find the mean values of y for the other 17 combinations of levels of A, B, and C by substituting the appropriate values of the dummy variables into the expression for $E(y)$ in the same manner. Each of the 18 means is a unique linear combination of the 18 β parameters in the model.

EXAMPLE 5.12

Suppose you want to test the hypothesis that the three qualitative independent variables discussed in Example 5.10 do not interact, i.e., the hypothesis that the effect of any one of the variables on $E(y)$ is independent of the level settings of the other two variables. Formulate the appropriate test of hypothesis about the model parameters.

Solution

No interaction among the three qualitative independent variables implies that the main effects model,

$$E(y) = \beta_0 + \overbrace{\beta_1 x_1 + \beta_2 x_2}^{\substack{\text{Main effect}\\ A}} + \overbrace{\beta_3 x_3 + \beta_4 x_4}^{\substack{\text{Main effect}\\ B}} + \overbrace{\beta_5 x_5}^{\substack{\text{Main effect}\\ C}}$$

is appropriate for modeling $E(y)$ or, equivalently, that all interaction terms should be excluded from the model. This situation will occur if

$$\beta_6 = \beta_7 = \cdots = \beta_{17} = 0$$

Consequently, we will test the null hypothesis

$$H_0: \quad \beta_6 = \beta_7 = \cdots = \beta_{17} = 0$$

against the alternative hypothesis that at least one of these β parameters differs from 0, or equivalently, that some interaction among the independent variables exists. This statistical test was described in Section 4.10.

Of what value is this section? If you are modeling a response, say, profit of a corporation, and you believe that several qualitative independent variables affect the response, then you must know how to enter these variables into your model. You must understand the implication of the interaction (or lack of it) among a subset of independent variables and how to write the appropriate terms in the model to account for it. Failure to write a good model for your response will usually lead to inflated values of the SSE and s^2 (with a consequent loss of information), and it also can lead to biased estimates of $E(y)$ and biased predictions of y.

5.10

Models with Both Quantitative and Qualitative Independent Variables

Perhaps the most interesting data analysis problems are those that involve both quantitative and qualitative independent variables. For example, suppose mean profit per sales dollar of the construction company is a function of one qualitative independent variable, sales engineer, at levels E_1, E_2, and E_3, and one quantitative independent variable, job size in millions of dollars. The company might be expected to make more profit per sales dollar on small jobs than on very large jobs due to increased competition for the larger jobs. We will proceed to build a model in stages, showing graphically the interpretation that we would give to the model at each stage. This will help you see the contribution of various terms in the model.

At first we assume that the qualitative independent variable has no effect on the response (i.e., the mean contribution to the response is the same for all three sales engineers), but the mean profit per sales dollar, $E(y)$, is related to job size. In this case, one response curve, which might appear as shown in Figure 5.21, would be sufficient to characterize $E(y)$ for all three sales engineers. The following second-order model would likely provide a good approximation to $E(y)$:

$$E(y) = \beta_0 + \beta_1 x_1 + \beta_2 x_1^2$$

where x_1 is job size in millions of dollars. This model has some distinct disadvantages. If differences in mean profit exist for the three sales engineers, they cannot be detected (because the model does not contain any parameters representing differences among sales engineers). Also, the differences would inflate the SSE associated with the fitted model and consequently would increase errors of estimation and prediction.

FIGURE 5.21

Model for $E(y)$ as a function of job size

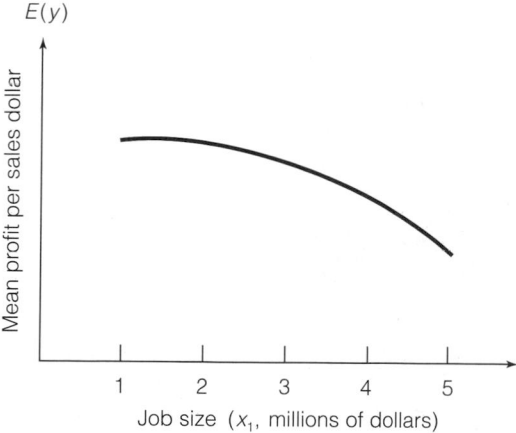

The next stage in developing a model for $E(y)$ is to assume the qualitative independent variable, sales engineer, does affect mean profit, but the effect on $E(y)$ is independent of job size. In other words, the assumption is that the two independent variables do not interact. This model is obtained by adding main effect terms for sales engineer to the second-order model we used in the first stage. Therefore, using the methods of Sections 5.7 and 5.8, we choose E_1 as the base level and add two terms to the model corresponding to levels E_2 and E_3:

$$E(y) = \beta_0 + \beta_1 x_1 + \beta_2 x_1^2 + \beta_3 x_2 + \beta_4 x_3$$

where

$$x_1 = \text{Job size} \qquad x_2 = \begin{cases} 1 & \text{if } E_2 \\ 0 & \text{if not} \end{cases} \qquad x_3 = \begin{cases} 1 & \text{if } E_3 \\ 0 & \text{if not} \end{cases}$$

What effect do these terms have on the graph for the response curve(s)? Suppose we want to model $E(y)$ for level E_1. Then we let $x_2 = 0$ and $x_3 = 0$. Substituting into the model equation, we have

$$\begin{aligned} E(y) &= \beta_0 + \beta_1 x_1 + \beta_2 x_1^2 + \beta_3(0) + \beta_4(0) \\ &= \beta_0 + \beta_1 x_1 + \beta_2 x_1^2 \end{aligned}$$

which would graph as a second-order curve similar to the one shown in Figure 5.21.

Now suppose that one of the other two sales engineers bids a job, for example, E_2. Then $x_2 = 1$, $x_3 = 0$, and

$$\begin{aligned} E(y) &= \beta_0 + \beta_1 x_1 + \beta_2 x_1^2 + \beta_3(1) + \beta_4(0) \\ &= (\beta_0 + \beta_3) + \beta_1 x_1 + \beta_2 x_1^2 \end{aligned}$$

This is the equation of exactly the same parabola that we obtained for sales engineer E_1 except that the y-intercept has changed from β_0 to $(\beta_0 + \beta_3)$. Similarly, the response curve for E_3 is

$$E(y) = (\beta_0 + \beta_4) + \beta_1 x_1 + \beta_2 x_1^2$$

Therefore, the three response curves for levels E_1, E_2, and E_3 (shown in Figure 5.22) are identical except that they are shifted vertically upward or downward in relation to each other. The curves depict the situation when the two independent variables do not interact; that is, the effect of job size on mean profit is

FIGURE 5.22

Model for $E(y)$ as a function of sales engineer and job size (no interaction)

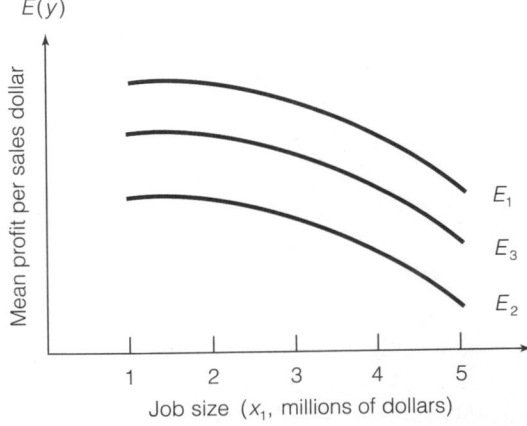

the same regardless of the sales engineer, and the effect of sales engineer on mean profit is the same for all job sizes (the relative distances between the curves is constant).

This noninteractive second-stage model has drawbacks similar to those of the simple first-stage model. It is highly unlikely that the response curves for the three sales engineers would be identical except for differing y-intercepts. Because the model does not contain parameters that measure interaction between job size and sales engineer, we cannot test to see if a relationship exists. Also, if interaction does exist, it will cause the SSE for the fitted model to be inflated and will consequently increase the errors of estimating model parameters $E(y)$.

This leads us to the final stage of the model-building process—adding interaction terms to allow the three response curves to differ in shape:

$$E(y) = \beta_0 + \overbrace{\beta_1 x_1 + \beta_2 x_1^2}^{\substack{\text{Main effect} \\ \text{terms for} \\ \text{job size}}} + \overbrace{\beta_3 x_2 + \beta_4 x_3}^{\substack{\text{Main effect} \\ \text{terms for} \\ \text{sales engineer}}}$$

$$+ \overbrace{\beta_5 x_1 x_2 + \beta_6 x_1 x_3 + \beta_7 x_1^2 x_2 + \beta_8 x_1^2 x_3}^{\text{Interaction terms}}$$

where

$$x_1 = \text{Job size} \qquad x_2 = \begin{cases} 1 & \text{if } E_2 \\ 0 & \text{if not} \end{cases} \qquad x_3 = \begin{cases} 1 & \text{if } E_3 \\ 0 & \text{if not} \end{cases}$$

Notice that this model graphs as three different second-order curves.* If E_1 is the sales engineer, we substitute $x_2 = x_3 = 0$ into the formula for $E(y)$, and all but the first three terms equal 0. The result is

$$E(y) = \beta_0 + \beta_1 x_1 + \beta_2 x_1^2$$

If E_2 is the sales engineer, $x_2 = 1$, $x_3 = 0$, and

$$E(y) = \beta_0 + \beta_1 x_1 + \beta_2 x_1^2 + \beta_3(1) + \beta_4(0) + \beta_5 x_1(1) + \beta_6 x_1(0) + \beta_7 x_1^2(1) + \beta_8 x_1^2(0)$$
$$= (\beta_0 + \beta_3) + (\beta_1 + \beta_5)x_1 + (\beta_2 + \beta_7)x_1^2$$

The y-intercept, the coefficient of x_1, and the coefficient of x_1^2 differ from the corresponding coefficients in $E(y)$ at level E_1. Finally, when E_3 is the sales engineer, $x_2 = 0$, $x_3 = 1$, and the result is

$$E(y) = (\beta_0 + \beta_4) + (\beta_1 + \beta_6)x_1 + (\beta_2 + \beta_8)x_1^2$$

A graph of the model for $E(y)$ might appear as shown in Figure 5.23 (page 302). Compare this figure with Figure 5.21, where we assumed the response curves were identical for all three sales engineers, and with Figure 5.22, where we assumed no interaction between the independent variables. Note in Figure 5.23 that the second-order curves may be completely different.

---

*Note that the model remains a second-order model for the quantitative independent variable x_1. The terms involving $x_1^2 x_2$ and $x_1^2 x_3$ appear to be third-order terms, but they are not because x_2 and x_3 are dummy variables.

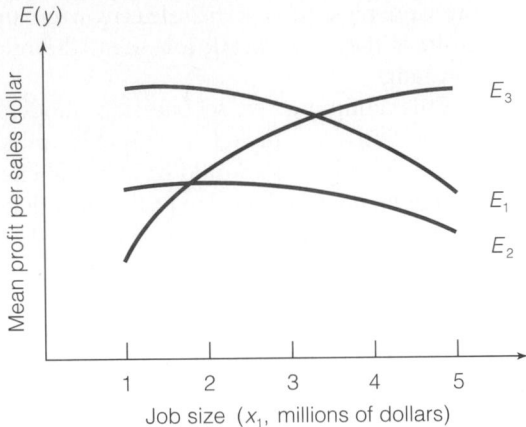

Now that you know how to write a model for two independent variables—
one qualitative and one quantitative—we ask a question. Why do it? Why not
write a separate second-order model for each level of sales engineer where $E(y)$
is a function of only job size? *One reason we wrote the single model representing
all three response curves is so that we can test to determine whether the curves are
different.* For example, we might want to know whether the effect of sales engineer
depends on job size. Thus, one sales engineer might be especially good on small
jobs, but poor on large jobs. The reverse might be true for one of the other two
sales engineers. The hypothesis that the independent variables, sales engineer
and job size, affect the response independently of one another (a case of no
interaction) is equivalent to testing the hypothesis that $\beta_5 = \beta_6 = \beta_7 = \beta_8 =
0$ [i.e., that the model in Figure 5.22 adequately characterizes $E(y)$] using the F
test discussed in Section 4.10. *A second reason for writing a single model is that
we obtain a pooled estimate of σ^2, the variance of the random error component ε.*
If the variance of ε is truly the same for each sales engineer, the pooled estimate
is superior to calculating three estimates by fitting a separate model for each sales
engineer.

   In conclusion, suppose you want to write a model relating $E(y)$ to several
quantitative and qualitative independent variables. Proceed in exactly the same
manner as for two independent variables, one qualitative and one quantitative.
First, write the model (using the methods of Sections 5.4 and 5.5) that you want
to use to describe the quantitative independent variables. Then introduce the
main effect and interaction terms for the qualitative independent variables. This
gives a model that represents a set of identically shaped response surfaces, one
corresponding to each combination of levels of the qualitative independent variables. If you could imagine surfaces in a multidimensional space, their appearance
would be analogous to the response curves of Figure 5.22. To complete the
model, add terms for the interaction between the quantitative and qualitative
variables. This is done by interacting *each* qualitative variable term with every
quantitative variable term. We will demonstrate with an example.

EXAMPLE 5.13

A chain of drug stores wished to investigate the effects of three factors on its weekly profit. The factors were:

1. Newspaper selected for the chain's weekly advertisement (two levels)
2. Design of the advertisement (two levels)
3. Percent discount of special sale items (five levels)

The different combinations of newspaper, design, and percent discount were employed in a random sequence, one combination per week, during a period of normal sales activity (holiday periods were excluded). Write a model for the profit per week, y.

Solution

The response y is affected by two qualitative factors (newspaper and design), each at two levels, and one quantitative factor (percent discount), with five levels. Each of the two advertising designs, D_1 and D_2, could be used with each of the two newspapers, N_1 and N_2, giving $2 \times 2 = 4$ possible advertising media—call them (D_1, N_1), (D_1, N_2), (D_2, N_1), (D_2, N_2). For each of these combinations, you would obtain a curve that graphs profit as a function of the quantitative factor x_1, percent discount (see Figure 5.24). The stages in writing the model for the response y shown in Figure 5.24 are given below.

FIGURE 5.24

A graphical portrayal of three factors—two qualitative and one quantitative—on profit y

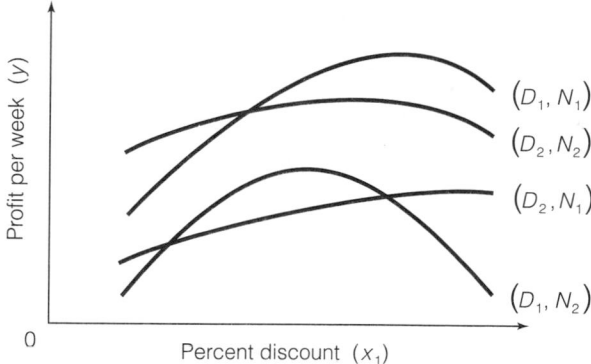

STAGE 1 *Write a model relating y to the quantitative factor(s)*   It is likely that increasing the value of the single quantitative factor x_1, percent discount, will increase sales. This should initially increase profit, but, eventually, when the discount is sufficiently large, the profit should decrease, thus producing curvature in the profit curves of Figure 5.24. Consequently, we will model the mean profit per week, $E(y)$, with the second-order model

$$E(y) = \beta_0 + \beta_1 x_1 + \beta_2 x_1^2$$

This is the model we would use if we were certain that the profit curves were identical for all design–newspaper combinations (D_i, N_j). The model would appear as shown in Figure 5.25a on page 304.

STAGE 2 *Add the terms, both main effect and interaction, for the qualitative factors*
Thus,

$$E(y) = \beta_0 + \overbrace{\beta_1 x_1 + \beta_2 x_1^2}^{\text{Terms for quantitative factor}}$$

$$+ \overbrace{\beta_3 x_2}^{\substack{\text{Main effect} \\ D}} + \overbrace{\beta_4 x_3}^{\substack{\text{Main effect} \\ N}} + \overbrace{\beta_5 x_2 x_3}^{DN \text{ interaction}}$$

$$x_2 = \begin{cases} 1 & \text{if } D_2 \text{ is employed} \\ 0 & \text{if not} \end{cases} \qquad x_3 = \begin{cases} 1 & \text{if } N_2 \text{ is employed} \\ 0 & \text{if not} \end{cases}$$

This model implies that the profit curves are identically shaped for each of the (D_i, N_j) combinations but that they possess different y-intercepts, as shown in Figure 5.25b.

FIGURE 5.25

Profit curves for stages 1 and 2

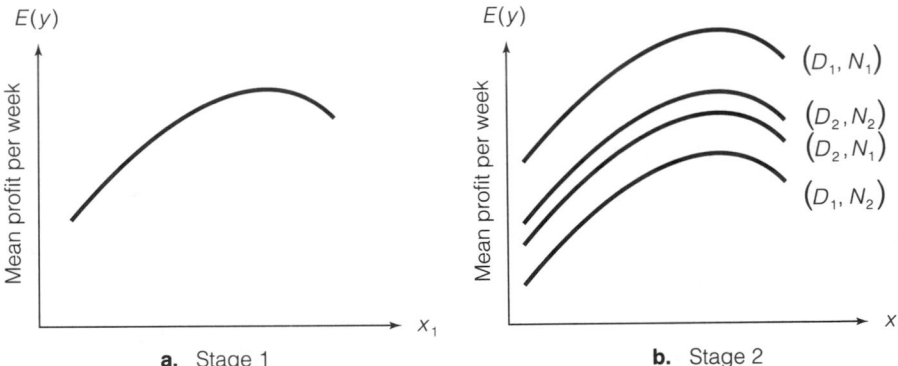

a. Stage 1

b. Stage 2

STAGE 3 *Add terms to allow for interaction between the quantitative and qualitative factors* This is done by interacting every pair of terms—one quantitative and one qualitative. Thus, the complete model, which graphs as four different-shaped second-order curves (see Figure 5.24), is

$$E(y) = \overbrace{\beta_0 + \beta_1 x_1 + \beta_2 x_1^2}^{\text{First-stage terms}}$$

$$+ \overbrace{\beta_3 x_2 + \beta_4 x_3 + \beta_5 x_2 x_3}^{\text{Second-stage terms}}$$

$$+ \overbrace{\beta_6 x_1 x_2 + \beta_7 x_1 x_3 + \beta_8 x_1 x_2 x_3 + \beta_9 x_1^2 x_2 + \beta_{10} x_1^2 x_3 + \beta_{11} x_1^2 x_2 x_3}^{\text{Third-stage terms}}$$

EXAMPLE 5.14

Use the model of Example 5.13 to find the equation relating $E(y)$ to x_1 for design D_1 and newspaper N_2.

Solution

Checking the coding for the model, we see (noted at the second stage) that when a profit observation y is made during a week when design D_1 and newspaper N_2 were used, we set $x_2 = 0$ and $x_3 = 1$. Substituting these values into the complete model, we obtain

$$E(y) = \beta_0 + \beta_1 x_1 + \beta_2 x_1^2$$
$$+ \beta_3 x_2 + \beta_4 x_3 + \beta_5 x_2 x_3$$
$$+ \beta_6 x_1 x_2 + \beta_7 x_1 x_3 + \beta_8 x_1 x_2 x_3 + \beta_9 x_1^2 x_2 + \beta_{10} x_1^2 x_3 + \beta_{11} x_1^2 x_2 x_3$$
$$= \beta_0 + \beta_1 x_1 + \beta_2 x_1^2$$
$$+ \beta_3(0) + \beta_4(1) + \beta_5(0)(1)$$
$$+ \beta_6 x_1(0) + \beta_7 x_1(1) + \beta_8 x_1(0)(1)$$
$$+ \beta_9 x_1^2(0) + \beta_{10} x_1^2(1) + \beta_{11} x_1^2(0)(1)$$
$$= (\beta_0 + \beta_4) + (\beta_1 + \beta_7)x_1 + (\beta_2 + \beta_{10})x_1^2$$

Note that this equation graphs as a portion of a parabola with y-intercept equal to $(\beta_0 + \beta_4)$. The coefficient of x_1 is $(\beta_1 + \beta_7)$, and the curvature coefficient (the coefficient of x_1^2) is $(\beta_2 + \beta_{10})$.

EXAMPLE 5.15

Suppose you have two qualitative independent variables, A and B, and A is at two levels and B is at three levels. You also have two quantitative independent variables, C and D, each at three levels. Further, suppose you plan to fit a second-order response surface as a function of the quantitative independent variables C and D, and that you want your model for $E(y)$ to allow for different shapes of the second-order surfaces for the six (2×3) combinations of levels corresponding to the qualitative independent variables A and B. Write a model for $E(y)$.

Solution

STAGE 1 *Write the second-order model corresponding to the two quantitative independent variables* If we let

$x_1 = $ Level for independent variable C

$x_2 = $ Level for independent variable D

then

$$E(y) = \beta_0 + \beta_1 x_1 + \beta_2 x_2 + \beta_3 x_1 x_2 + \beta_4 x_1^2 + \beta_5 x_2^2$$

This is the model you would use if you believed that the six response surfaces, corresponding to the six combinations of levels of A and B, were identical.

STAGE 2 *Add the main effect and interaction terms for the qualitative independent variables* These are

Main effect term for A		Main effect terms for B		AB interaction terms
$\beta_6 x_6$	$+$	$\beta_7 x_7 + \beta_8 x_8$	$+$	$\beta_9 x_6 x_7 + \beta_{10} x_6 x_8$

$$x_6 = \begin{cases} 1 & \text{if at level } A_2 \\ 0 & \text{if not} \end{cases} \qquad x_7 = \begin{cases} 1 & \text{if at level } B_2 \\ 0 & \text{if not} \end{cases} \qquad x_8 = \begin{cases} 1 & \text{if at level } B_3 \\ 0 & \text{if not} \end{cases}$$

The addition of these terms to the model produces six identically shaped second-order surfaces, one corresponding to each of the six combinations of levels of A and B. They differ only in their y-intercepts.

STAGE 3 *Add terms that allow for interaction between the quantitative and qualitative independent variables* This is done by interacting each of the five qualitative independent variable terms (both main effect and interaction) with each term (except β_0) of the quantitative first-stage model. Thus,

$$E(y) = \beta_0 + \beta_1 x_1 + \beta_2 x_2 + \beta_3 x_1 x_2 + \beta_4 x_1^2 + \beta_5 x_2^2 \qquad \left.\right\} \begin{array}{l}\text{First-stage}\\\text{model}\end{array}$$

$$\underbrace{+\ \beta_6 x_6}_{\substack{\text{Main effect}\\A}} + \underbrace{\beta_7 x_7 + \beta_8 x_8}_{\substack{\text{Main effect}\\B}} + \underbrace{\beta_9 x_6 x_7 + \beta_{10} x_6 x_8}_{AB\ \text{interaction}} \qquad \left.\right\} \begin{array}{l}\text{Portion}\\\text{added to}\\\text{form second-}\\\text{stage model}\end{array}$$

$$+\ \beta_{11} x_6 x_1 + \beta_{12} x_6 x_2 + \beta_{13} x_6 x_1 x_2 + \beta_{14} x_6 x_1^2 + \beta_{15} x_6 x_2^2 \qquad \left.\right\} \begin{array}{l}\text{Interacting}\\x_6 \text{ with the}\\\text{quantitative terms}\end{array}$$

$$+\ \beta_{16} x_7 x_1 + \beta_{17} x_7 x_2 + \beta_{18} x_7 x_1 x_2 + \beta_{19} x_7 x_1^2 + \beta_{20} x_7 x_2^2 \qquad \left.\right\} \begin{array}{l}\text{Interacting}\\x_7 \text{ with the}\\\text{quantitative terms}\end{array}$$

$$+\ \cdots \qquad\qquad\qquad\qquad\qquad\qquad\qquad\qquad\qquad\qquad\quad \cdots$$

$$+\ \beta_{31} x_6 x_8 x_1 + \beta_{32} x_6 x_8 x_2 + \beta_{33} x_6 x_8 x_1 x_2 + \beta_{34} x_6 x_8 x_1^2 + \beta_{35} x_6 x_8 x_2^2 \left.\right\} \begin{array}{l}\text{Interacting}\\x_6 x_8 \text{ with the}\\\text{quantitative terms}\end{array}$$

Note that the complete model contains 36 terms, one for β_0, five needed to complete the second-order model in the two quantitative variables, five for the two qualitative variables, and $5 \times 5 = 25$ terms for the interactions between the quantitative and qualitative variables.

To see how the model gives different second-order surfaces—one for each combination of the levels of variables A and B—consider the next example.

| EXAMPLE 5.16

Refer to Example 5.15. Find the response surface that portrays $E(y)$ as a function of the two quantitative independent variables C and D for the (A_1, B_2) combination of levels of the qualitative independent variables.

Solution

Checking the coding, we see that when y is observed at the first level of A (level A_1) and the second level of B (level B_2), the dummy variables take the following values: $x_6 = 0$, $x_7 = 1$, $x_8 = 0$. Substituting these values into the formula for the complete model (and deleting the terms that equal 0), we obtain

$$E(y) = \beta_0 + \beta_1 x_1 + \beta_2 x_2 + \beta_3 x_1 x_2 + \beta_4 x_1^2 + \beta_5 x_2^2 + \beta_7 + \beta_{16} x_1 + \beta_{17} x_2 + \beta_{18} x_1 x_2 + \beta_{19} x_1^2 + \beta_{20} x_2^2$$
$$= (\beta_0 + \beta_7) + (\beta_1 + \beta_{16}) x_1 + (\beta_2 + \beta_{17}) x_2 + (\beta_3 + \beta_{18}) x_1 x_2 + (\beta_4 + \beta_{19}) x_1^2 + (\beta_5 + \beta_{20}) x_2^2$$

Note that this is the equation of a second-order model for $E(y)$. It graphs the response surface for $E(y)$ when the qualitative independent variables A and B are at levels A_1 and B_2.

EXERCISES

5.20   Eli Lilly and Company has developed three methods (G, R_1, and R_2) for estimating the shelf life of its drug products based on potency.* One way to compare the three methods is to build a regression model for the dependent variable, estimated shelf life y (as a percentage of true shelf life), with potency of the drug (x_1) as a quantitative predictor and method as a qualitative predictor.
   a. Write a first-order, main effects model for $E(y)$ as a function of potency (x_1) and method.
   b. Interpret the β coefficients of the model, part a.
   c. Write a first-order model for $E(y)$ that will allow the slopes to differ for the three methods.
   d. Refer to part c. For each method, write the slope of the $y–x_1$ line in terms of the β's.

5.21   The liquefaction of coal is a major contributor of synthetic fuels. An experiment was conducted to evaluate the performances of a diesel engine run on synthetic (coal-derived) and petroleum-derived fuel oil (*Journal of Energy Resources Technology*, Mar. 1990). The petroleum-derived fuel used was a number 2 diesel fuel (DF-2) obtained from Phillips Chemical Company. Two synthetic fuels were used: a blended fuel (50% coal-derived and 50% DF-2) and a blended fuel with advanced timing. The brake power (kilowatts) and fuel type were varied in test runs, and engine performance was measured. The following table gives the experimental results for the performance measure, mass burning rate per degree of crank angle.

BRAKE POWER, x_1	FUEL TYPE	MASS BURNING RATE, y
4	DF-2	13.2
4	Blended	17.5
4	Advanced Timing	17.5
6	DF-2	26.1
6	Blended	32.7
6	Advanced Timing	43.5
8	DF-2	25.9
8	Blended	46.3
8	Advanced Timing	45.6
10	DF-2	30.7
10	Blended	50.8
10	Advanced Timing	68.9
12	DF-2	32.3
12	Blended	57.1

Source: Litzinger, T. A. and Buzza, T. G. "Performance and emissions of a diesel engine using a coal-derived fuel." *Journal of Energy Resources Technology*, Vol. 112, Mar. 1990, p. 32, Table 3.

*Murphy, J. R. and Weisman, D. "Using random slopes for estimating shelf life." Paper presented at Joint Statistical Meetings, Anaheim, Calif., Aug. 1990.

a. Initially, the researchers fit the first-order, main effects model

$$E(y) = \beta_0 + \beta_1 x_1 + \beta_2 x_2 + \beta_3 x_3$$

where

y = Mass burning rate

x_1 = Brake power (kW)

$$x_2 = \begin{cases} 1 & \text{if DF-2 fuel} \\ 0 & \text{if not} \end{cases}$$

$$x_3 = \begin{cases} 1 & \text{if blended fuel} \\ 0 & \text{if not} \end{cases}$$

Interpret the results shown in the accompanying Minitab printout.

Minitab printout for Exercise 5.21 (first-order main effects model)

```
The regression equation is
Y = 13.3 + 4.36 X1 - 22.6 X2 - 7.36 X3

Predictor        Coef        Stdev      t-ratio         p
Constant       13.320        6.931         1.92     0.084
X1              4.3650       0.8057         5.42     0.000
X2            -22.600        5.464        -4.14     0.002
X3             -7.360        5.464        -1.35     0.208

s = 8.057        R-sq = 81.2%      R-sq(adj) = 75.6%

Analysis of Variance

SOURCE          DF           SS          MS         F         p
Regression       3       2807.90      935.97     14.42     0.001
Error           10        649.09       64.91
Total           13       3456.99

SOURCE          DF       SEQ SS
X1               1      1603.93
X2               1      1086.22
X3               1       117.76

Unusual Observations
Obs.       X1           Y       Fit Stdev.Fit  Residual   St.Resid
  3       4.0       17.50     30.78      4.70    -13.28      -2.03R

R denotes an obs. with a large st. resid.
```

b. The interaction model

$$E(y) = \beta_0 + \beta_1 x_1 + \beta_2 x_2 + \beta_3 x_3 + \beta_4 x_1 x_2 + \beta_5 x_1 x_3$$

was fit using Minitab, with the results shown in the accompanying printout. Conduct a test to determine whether brake power and fuel type interact. Test using $\alpha = .01$.

c. Refer to the model, part b. Give the estimates of the slope of the y–x_1 line for each of the three fuel types.

Minitab printout for Exercise 5.21 (interaction model)

```
The regression equation is
Y = - 10.8 + 7.82 X1 + 19.4 X2 + 12.8 X3 - 5.68 X1X2 - 2.95 X1X3

Predictor         Coef        Stdev      t-ratio          p
Constant       -10.830        8.277       -1.31       0.227
X1               7.815        1.126        6.94       0.000
X2              19.35        10.69         1.81       0.108
X3              12.79        10.69         1.20       0.266
X1X2            -5.675        1.380       -4.11       0.003
X1X3            -2.950        1.380       -2.14       0.065

s = 5.037       R-sq = 94.1%     R-sq(adj) = 90.5%

Analysis of Variance

SOURCE         DF          SS           MS         F          p
Regression      5       3253.98       650.80     25.65     0.000
Error           8        203.01        25.38
Total          13       3456.99

SOURCE         DF        SEQ SS
X1              1       1603.93
X2              1       1086.22
X3              1        117.76
X1X2            1        330.04
X1X3            1        116.03
```

5.22 Refer to the *Journal of Human Stress* study of firefighters, Exercise 5.3. It is thought that the following complete second-order model will be adequate to describe the relationship between emotional distress and years of experience for two groups of firefighters—those exposed to a chemical fire and those unexposed*

$$E(y) = \beta_0 + \beta_1 x_1 + \beta_2 x_1^2 + \beta_3 x_2 + \beta_4 x_1 x_2 + \beta_5 x_1^2 x_2$$

where

y = Emotional distress

x_1 = Experience (years)

$x_2 = \begin{cases} 1 & \text{if exposed to chemical fire} \\ 0 & \text{if not} \end{cases}$

a. What hypothesis would you test to determine whether the *rate* of increase of emotional distress with experience is different for the two groups of firefighters?

b. What hypothesis would you test to determine whether there are differences in mean emotional distress levels that are attributable to exposure group?

c. A portion of the SAS printout that results from fitting the second-order model to a sample of 200 firefighters is shown at the top of page 310.

*In practice, we would include other variables in the model. We include only two here to simplify the exercise.

Analysis of Variance

Source	DF	Sum of Squares	Mean Square	F Value	Prob>F
Model	5	2351.70	470.34	116.42	0.0001
Error	194	783.90	4.04		
C Total	199	3135.60			

Root MSE	2.0102	R-square	0.7500
Dep Mean	24.221	Adj R-sq	0.7436
C.V.	8.299		

The reduced model $E(y) = \beta_0 + \beta_1 x_1 + \beta_2 x_1^2$ is fit to the same data, and the resulting computer printout is reproduced here. Is there sufficient evidence to support the claim that the mean emotional distress levels differ for the two groups of firefighters? Use $\alpha = .05$.

Analysis of Variance

Source	DF	Sum of Squares	Mean Square	F Value	Prob>F
Model	2	2340.37	1170.185	289.87	0.0001
Error	197	795.23	4.037		
C Total	199	3135.60			

Root MSE	2.0092	R-square	0.7464
Dep Mean	24.221	Adj R-sq	0.7438
C.V.	8.295		

5.23   Research conducted at Ohio State University focused on the factors that influence the allocation of black and white men in labor market positions (*American Sociological Review*, June 1986). Data collected for each of 837 labor market positions were used to build a regression model for y, defined as the natural logarithm of the ratio of the proportion of blacks employed in a labor market position to the corresponding proportion of whites employed (called the *black–white log odds ratio*). Positive values of y indicate that blacks have a greater likelihood of employment than whites. Several independent variables were considered, including the following:

x_1 = Market power (a quantitative measure of the size and visibility of firms in the labor market)

x_2 = Percentage of workers in the labor market who are union members

$x_3 = \begin{cases} 1 & \text{if labor market position includes craft occupations} \\ 0 & \text{if not} \end{cases}$

a. Write the first-order main effects model for $E(y)$ as a function of x_1, x_2, and x_3.
b. One theory hypothesized by the researchers is that the mean log odds ratio $E(y)$ is smaller for craft occupations than for noncraft occupations. (That is, the likelihood of black employment is less for craft occupations.) Explain how to test this hypothesis using the model in part **a**.
c. Write the complete second-order model for $E(y)$ as a function of x_1, x_2, and x_3.
d. Using the model in part **c**, explain how to test the hypothesis that level of market power x_1 has no effect on black–white log odds ratio y.
e. Consider the model

$$E(y) = \beta_0 + \beta_1 x_1 + \beta_2 x_2 + \beta_3 x_3 + \beta_4 x_1 x_3 + \beta_5 x_2 x_3$$

Holding the percentage of union members x_2 fixed, sketch the proposed contour lines for the relationship between log odds ratio y and market power x_1.

5.24   Researchers for a dog food company have developed a new puppy food that they hope will compete with the major brands. One premarketing test involved the comparison of the new food with that of two competitors in terms of weight gain. Fifteen 8-week-old German shepherd puppies, each from a different litter, were divided into three groups of five puppies each. Each group was fed one of the three brands of food.

   a. Set up a model that assumes the initial weight, x_1, is linearly related to final weight, y, but does not allow for differences among the three brands; i.e., assume the response curve is the same for the three brands of dog food. Sketch the response curve as it might appear.

   b. Set up a model that assumes the effect of initial weight is linearly related to final weight, and allows the intercept of the line to differ for the three brands. In other words, assume the initial weight and brand both affect final weight, but in an independent fashion. Sketch typical response curves.

   c. Now write the main effects plus interaction model. For this model we assume the initial weight is linearly related to final weight, but both the slope and the intercept of the line depend upon the brand. Sketch typical response curves.

5.25   An equal rights group has charged that women are being discriminated against in terms of the salary structure in a state university system. It is thought that a complete second-order model will be adequate to describe the relationship between salary and years of experience for both groups. A sample is to be taken from the records for faculty members (all of equal status) within the system and the following model is to be fit.*

$$E(y) = \beta_0 + \beta_1 x_1 + \beta_2 x_2 + \beta_3 x_1 x_2 + \beta_4 x_2^2 + \beta_5 x_1 x_2^2$$

where

     y = Annual salary (in thousands of dollars)

$$x_1 = \begin{cases} 1 & \text{if female} \\ 0 & \text{if male} \end{cases}$$

     x_2 = Experience (years)

   a. What hypothesis would you test to determine whether the *rate* of increase of mean salary with experience depends on gender?

   b. What hypothesis would you test to determine whether there are differences in the mean salaries of males and females?

5.26   During the 1970s, several industries, including the airline, trucking, natural gas, and cable television industries, were deregulated as a result of pressure from special interest groups and legislators. In the case of the airline industry, federal regulations on fares, schedules, and routes were thought to protect the airlines from price competition and generally prohibit new entry into the market. Thus, in theory, deregulation was expected to benefit consumers while having a negative impact on the airlines and their shareholders.

   To test this theory, economists analyzed the impact of the airline Deregulation Act of 1978 on the security returns in the airline industry (*Quarterly Journal of Business and Economics*, Autumn 1984). Specifically, the researchers examined the daily rates of returns of a sample of airline common stocks both prior to and following deregulation.

---

*In practice, we would include other variables in the model. We include only two here to simplify the exercise.

Thirty-two airlines engaged in air transportation for at least 1 year and listed on either the New York or the American Stock Exchange were selected for analysis. For each airline, daily stock returns were recorded for each of 120 days prior to deregulation and 120 days after deregulation. Data for the total of $n = 240$ observations (days) were then used to fit the model

$$E(y) = \beta_0 + \beta_1 x_1 + \beta_2 x_2 + \beta_3 x_1 x_2$$

where

y = Daily rate of return on the airline stock

x_1 = Average daily rate of return on the market

$$x_2 = \begin{cases} 1 & \text{if after deregulation} \\ 0 & \text{if prior to deregulation} \end{cases}$$

Thus, 32 regression analyses were conducted, one for each airline in the sample.

STOCK	$\hat{\beta}_0$	$\hat{\beta}_1$	$\hat{\beta}_2$	$\hat{\beta}_3$
AMR Corporation	.0005	2.8852*	−.0031	−.4106
Airborne Freight	.0024	.9748*	−.0012	−.6367*
Braniff International	.0007	1.4487*	−.0022	.5717
Canadian Pacific	.0011	.7687*	.0003	.3906
Continental Airlines	−.0021	2.1702*	.0004	−.7175*
Delta Airlines	−.0002	1.8557*	−.0005	−.6667*
Eastern Airlines	.0037	1.6353*	−.0069*	.9261*
Frontier	.0006	2.2188*	−.0033	−1.0355*
Greyhound	.0000	.3830*	.0020	.2163
Royal Dutch Airlines	.0004	1.3281	−.0024	−.5674
Lockheed Corporation	.0042	1.9337*	−.0053*	.6399
Northwest Airlines	−.0009	2.6056*	−.0005	−.1698
Ozark Airlines	.0032	1.1591*	−.0017	−.2921
PSA Airlines	.0033	.2555*	−.0036	1.4888*
Pan American	.0012	1.2136*	−.0025	−.0146
Piedmont	.0012	1.2138*	−.0025	−.0146
Puralotor, Inc.	.0009	.6829*	.0002	−.3787
Republic Airlines	.0023	.9524*	−.0031	1.3359*
Southwest Airlines	.0030	.8600*	.0016	−.2357
Tiger International	.0003	2.4979*	−.0009	−.7462*
Trans World Corp.	.0007	2.6399*	.0000	.1699
United Airlines	.0023	1.8137*	−.0039	.1444
US Air, Inc.	.0041	.8628	−.0029	1.4039*
W.A.F., Inc.	.0024	1.9271*	.0035	1.7067*
Western Airlines	.0012	1.5750*	−.0014	−.3987
World Airways	−.0014	3.2765*	.0015	−1.4218*

*Asterisk identifies β coefficients significant at $\alpha = .10$.
Source: Davidson, W. N., Chandy, P. R., and Walker, M. "The stock market effects of airline deregulation." *Quarterly Journal of Business and Economics*, Autumn 1984, Vol. 23, No. 4, pp. 31–45.

a. Write the equation of the line relating daily stock return y to average daily market return x_1 prior to deregulation (i.e., when $x_2 = 0$). Identify the y-intercept and slope of the line. [The slope is used to measure the stock's systematic risk and is often called the **β risk index** or **β value** for the stock. When the β value is greater than 1, the stock is classified as an *aggressive* or *risky* security since its daily rate of return is expected to move (upward or downward) faster than the market rate of return. In contrast, when the β value is less than 1, the stock is classified as a *defensive* or *stable* security since its daily rate of return moves slower than the market. A stock with a β value near 1 is called a *neutral* security, for its daily rate of return mirrors the market.]

b. Write the equation of the line relating daily stock return y to average daily market return x_1 after deregulation (i.e., when $x_2 = 1$). Identify the y-intercept and slope of the line.

c. What hypothesis would you test to determine whether the model is adequate for predicting y?

d. What hypothesis would you test to determine whether deregulation had an effect on the measure of systematic risk (i.e., β value) associated with the airline stock? [*Hint:* Use your answers to parts **a** and **b**.]

e. Of the 32 airline stocks analyzed, 26 had significant regressions (i.e., we could reject H_0: $\beta_1 = \beta_2 = \beta_3 = 0$) at $\alpha = .10$. The least squares prediction equations for these stocks are given in the table. Identify those stocks that have significant (at $\alpha = .10$) interaction between average market rate of return x_1 and deregulation x_2. For each of these stocks, how has deregulation affected the stock's β value?

5.11

Model Building: An Example

We illustrate the modeling techniques outlined in this chapter with an example based, in part, on an actual trucking deregulation study. Consider the problem of modeling the price charged for motor transport service (such as trucking) in a particular state. In the early 1980s, several states removed regulatory constraints on the rate charged for intrastate trucking services. (Florida was the first state to embark on a deregulation policy, on July 1, 1980.) One of the goals of the regression analysis is to assess the impact of state deregulation on the supply price y charged per ton-mile.

Suppose that after a careful variable-screening process (for example, stepwise regression), the following independent variables were selected as the best predictors of supply price:

1. Distance shipped (hundreds of miles)
2. Weight of product shipped (thousands of pounds)
3. Deregulation in effect (yes or no)
4. Size of market destination (large or small)

Distance shipped and weight of product are quantitative variables since they each assume numerical values (miles and pounds, respectively) corresponding to the points on a line. Deregulation and market size are qualitative, or categorical, variables that we must describe with dummy (or coded) variables. The variable assignments are given as follows:

$x_1 = $ Distance shipped

$x_2 = $ Weight of product

$x_3 = \begin{cases} 1 & \text{if deregulation in effect} \\ 0 & \text{if not} \end{cases}$

$x_4 = \begin{cases} 1 & \text{if large market} \\ 0 & \text{if small market} \end{cases}$

Note that in defining the dummy variables, we have arbitrarily chosen "no" and "small" to be the base levels of deregulation and market size, respectively.

We will begin the model-building process by specifying the three models shown below. Notice that the first model specified is the complete second-order model. Recall from Section 5.10 that the complete second-order model contains quadratic (curvature) terms for quantitative variables and interactions among the quantitative and qualitative terms. For this example, the complete second-order model (model 1) traces a parabolic surface for mean price $E(y)$ as a function of distance (x_1) and weight (x_2), and the response surfaces differ for the $2 \times 2 = 4$ combinations of the levels of deregulation and market size. Generally, the complete second-order model is a good place to start the model-building process since most real-world relationships are curvilinear. (Keep in mind, however, that you must have a sufficient number of data points to find estimates of all the parameters in the model.)

MODEL 1 $E(y) = \beta_0 + \beta_1 x_1 + \beta_2 x_2 + \beta_3 x_1 x_2 + \beta_4 x_1^2 + \beta_5 x_2^2\}$ Stage 1 terms

$\qquad + \beta_6 x_3 + \beta_7 x_4 + \beta_8 x_3 x_4$ $\}$ Stage 2 terms

$\qquad + \beta_9 x_1 x_3 + \beta_{10} x_1 x_4 + \beta_{11} x_1 x_3 x_4$

$\qquad + \beta_{12} x_2 x_3 + \beta_{13} x_2 x_4 + \beta_{14} x_2 x_3 x_4$

$\qquad + \beta_{15} x_1 x_2 x_3 + \beta_{16} x_1 x_2 x_4 + \beta_{17} x_1 x_2 x_3 x_4$ Stage 3 terms

$\qquad + \beta_{18} x_1^2 x_3 + \beta_{19} x_1^2 x_4 + \beta_{20} x_1^2 x_3 x_4$

$\qquad + \beta_{21} x_2^2 x_3 + \beta_{22} x_2^2 x_4 + \beta_{23} x_2^2 x_3 x_4$

MODEL 2 $E(y) = \beta_0 + \beta_1 x_1 + \beta_2 x_2 + \beta_3 x_1 x_2$ $\}$ Stage 1 terms

$\qquad + \beta_4 x_3 + \beta_5 x_4 + \beta_6 x_3 x_4$ $\}$ Stage 2 terms

$\qquad + \beta_7 x_1 x_3 + \beta_8 x_1 x_4 + \beta_9 x_1 x_3 x_4$

$\qquad + \beta_{10} x_2 x_3 + \beta_{11} x_2 x_4 + \beta_{12} x_2 x_3 x_4$ Stage 3 terms

$\qquad + \beta_{13} x_1 x_2 x_3 + \beta_{14} x_1 x_2 x_4 + \beta_{15} x_1 x_2 x_3 x_4$

Model 2 contains all the terms of model 1 (the complete second-order model), except that the quadratic terms (terms involving x_1^2 and x_2^2) are dropped. This model also proposes four different response surfaces for the combinations of levels of deregulation and market size, but the surfaces are twisted planes (see Figure 5.10) rather than paraboloids. A direct comparison of models 1 and 2 will allow us to test for the importance of the curvature terms.

MODEL 3 $E(y) = \beta_0 + \beta_1 x_1 + \beta_2 x_2 + \beta_3 x_1 x_2 + \beta_4 x_3 + \beta_5 x_4 + \beta_6 x_3 x_4$

Model 3 is a reduction of model 2, with the quantitative–qualitative interaction terms omitted. This model assumes that the four response surfaces corresponding to the four deregulation–market size combinations, which are twisted planes, differ only with respect to the y-intercept in the space.

Data collected for $n = 132$ shipments were used to fit the models. The results are summarized in Table 5.8.

To determine whether the quadratic terms for distance (x_1) and weight (x_2), i.e., those involving x_1^2 and x_2^2, contribute information for predicting the mean

TABLE 5.8 **Trucking Deregulation Regression Results**

MODEL	SSE	R^2	df(ERROR)
1	203,570	.83	108
2	227,520	.81	116
3	395,165	.67	125

supply price $E(y)$, we compare models 1 and 2 using the partial F test given in Section 4.10. If the quadratic terms are unimportant, then the β coefficients involving the x_1^2 and x_2^2 terms in model 1 (i.e., β_4, β_5, β_{18}, β_{19}, ..., β_{23}) will all equal 0. Thus, we test

H_0: $\beta_4 = \beta_5 = \beta_{18} = \beta_{19} = \beta_{20} = \beta_{21} = \beta_{22} = \beta_{23} = 0$

H_a: At least one of the quadratic β's differs from 0.

where the complete model is model 1 and the reduced model is model 2. The test statistic is

$$F = \frac{(SSE_2 - SSE_1)/\text{Number of } \beta \text{ parameters being tested}}{SSE_1/df(\text{Error}) \text{ for model 1}}$$

$$= \frac{(227{,}520 - 203{,}570)/8}{203{,}570/108}$$

$$= \frac{2{,}993.75}{1{,}884.91} = 1.59$$

For $\alpha = .05$, $\nu_1 =$ number of β's tested (8), and $\nu_2 = df(\text{Error})$ for model 1 (108), the critical value (from Table 4 of Appendix D) is approximately $F_{.05} = 2.02$.

Since the calculated value $F = 1.59$ does not exceed this critical value, we have insufficient evidence (at $\alpha = .05$) to conclude that the curvature terms are important predictors of supply price y. We could collect more data and retest to determine whether x_1^2 and x_2^2 contribute information for the prediction of supply price. Lacking this additional information, we will simplify our model by dropping the terms involving x_1^2 and x_2^2.*

Can we simplify the model even further by dropping the terms involving quantitative–qualitative interactions? That is, do the response surfaces for the four combinations of deregulation and market size differ only with respect to the y-intercept? If so, then the β coefficients involving the quantitative–qualitative interaction terms in model 2 (β_7, β_8, ..., β_{15}) will all equal 0.

We compare model 2 (now called the *complete* model) and model 3 (the *reduced* model) with another partial F test:

H_0: $\beta_7 = \beta_8 = \cdots = \beta_{15} = 0$

H_a: At least one of the quantitative–qualitative interaction β parameters differs from 0.

---

*There is always a danger in dropping terms from the model. Essentially, we are accepting H_0: $\beta_4 = \beta_5 = \beta_{18} = \cdots = \beta_{23} = 0$ when $P(\text{Type II error}) = P(\text{accepting } H_0 \text{ when } H_0 \text{ is false}) = \beta$ is unknown. In practice, however, many researchers are willing to risk making a Type II error rather than use a more complex model for $E(y)$ when simpler models that are nearly as good as predictors (and easier to apply and interpret) are available. Note that we employed a relatively large amount of data ($n = 132$) in fitting our models and that R^2 for model 2 is only 2% less than R^2 for model 1. If the quadratic terms are, in fact, important (i.e., we have made a Type II error), there is little lost in terms of explained variability in using model 2.

$$\text{Test statistic:} \quad F = \frac{(SSE_3 - SSE_2)/\text{Number of } \beta \text{ parameters being tested}}{SSE_2/df(\text{Error})}$$

$$= \frac{(395,165 - 227,520)/9}{227,520/116}$$

$$= \frac{18,627.22}{1,961.38} = 9.50$$

Rejection region: $F > F_{.05} = 1.96$ (from Table 4 of Appendix D) where $\nu_1 =$ number of β's tested (9) and $\nu_2 = df(\text{Error})$ for model 2 (116).

Since the calculated value $F = 9.50$ exceeds the critical value $F_{.05} = 1.96$, there is sufficient evidence (at $\alpha = .05$) to indicate that the quantitative–qualitative interaction terms are important. We should retain these terms in the model.

The results of the previous tests suggest that of the three models, model 2 is the best for modeling the mean supply price $E(y)$. Further testing may lead to a simpler model, however. For example, we might want to test the importance of the quantitative–quantitative interaction (x_1x_2) terms. If a partial F test reveals no evidence of interaction between x_1 and x_2, we may wish to drop terms involving x_1x_2 from the model.

A note of caution: Just as with t tests on individual β parameters, you should avoid conducting too many partial F tests. Regardless of the type of test (t test or F test), the more tests that are performed, the higher the overall Type I error rate will be. In practice, you should limit the number of models that you propose for $E(y)$ so that the overall Type I error rate α for conducting partial F tests remains reasonably small.*

Summary

Although this chapter on **model building** covered many topics, only experience can make you competent in this fascinating area of statistics. Successful model building requires a delicate blend of knowledge of the process being modeled, geometry, and formal statistical testing.

The first step is to **identify the response variable y and the set of independent variables**. Each independent variable is then classified as either **quantitative** or **qualitative**, and **dummy variables** are defined to represent the qualitative independent variables.

When the number of independent variables is manageable, the model-builder is ready to begin a systematic effort. At least **second-order models**, those containing **two-way interactions and quadratic terms** in the quantitative variables, should be considered. Remember that a model with no interaction terms implies

---

*A technique, suggested by Bonferroni, is often applied to maintain control of the overall Type I error rate α. If p tests are to be performed, then conduct each individual test at significance level α/p. This will guarantee an overall Type I error rate less than or equal to α. For example, conducting each of $p = 5$ tests at the $.05/5 = .01$ level of significance guarantees an overall $\alpha \leq .05$.

that each of the independent variables affects the response independently of the other independent variables. **Quadratic terms add curvature** to the contour curves when $E(y)$ is plotted as a function of the independent variable. With higher-order models, you may decide to **code** the quantitative independent variables to reduce both rounding error and the built-in multicollinearity problem.

Many problems can arise in regression modeling, and the intermediate steps are often tedious and frustrating. However, the end result of a careful and determined modeling effort is very rewarding—you will have a better understanding of the process and will have a predictive model for the dependent variable y.

SUPPLEMENTARY EXERCISES

5.27   An appliance store is interested in modeling the total sales of television sets sold per month as a function of the following independent variables: (1) warranty period, (2) color or black-and-white, (3) solid-state or tube, (4) picture-tube size, (5) brand (the store carries three brands).
   a. Determine whether each of the independent variables is quantitative or qualitative. If it is quantitative, give the approximate range of levels you might expect to observe. If it is qualitative, give the number of levels and define dummy variables to describe each variable.
   b. Write a main effects model to relate $E(y)$, the mean monthly sales for televisions sold, as a function of the five independent variables.
   c. It is believed that the mean sales will be greater for larger picture tubes than for smaller picture tubes, but the rate of increase will depend on the brand. Why does the main effects model in part **b** not incorporate this belief? How can it be incorporated into the model?

5.28   To make a product more appealing to the consumer, an automobile manufacturer is experimenting with a new type of paint that is supposed to help the car maintain its new-car look. The durability of this paint depends on the length of time the car body is in the oven after it has been painted. In the initial experiment, three groups of 10 car bodies each were baked for three different lengths of time—12, 24, and 36 hours—at the standard temperature setting. Then, the paint finish of each of the 30 cars was analyzed to determine a durability rating, y.
   a. Write a quadratic model relating the mean durability, $E(y)$, to the length of baking.
   b. Could a cubic model be fit to the data? Explain.
   c. Suppose the Research and Development Department develops three new types of paint to be tested. Thus, 90 cars are to be tested—30 for each type of paint. Write the complete second-order model for $E(y)$ as a function of the type of paint and bake time.

5.29   The manager of a supermarket wants to model the total weekly sales of beer, y, as a function of brand. (This model will enable the manager to plan the store's inventory.) The market carries three brands, B_1, B_2, and B_3.
   a. What type of independent variable is brand of beer?
   b. Write the model relating mean weekly beer sales, $E(y)$, to brand of beer. Be sure to explain any dummy variables you use.
   c. Interpret the parameters (β's) of your model in part **b**.
   d. In terms of the model parameters, what is the mean weekly sales for brand B_3?

5.30   Refer to Exercise 5.29. Suppose the manager uses brand B_1 as the base level and obtains the model

$$\hat{y} = 450 + 60x_1 - 30x_2$$

where

$$x_1 = \begin{cases} 1 & \text{if brand } B_2 \\ 0 & \text{otherwise} \end{cases} \qquad x_2 = \begin{cases} 1 & \text{if brand } B_3 \\ 0 & \text{otherwise} \end{cases}$$

a. What is the difference between the estimated mean* weekly sales for brands B_2 and B_1?

b. What is the estimated mean weekly sales B_2?

5.31   Economic research has established evidence of a positive correlation between earnings and educational attainment (*Economic Inquiry*, Jan. 1984). However, it is unclear whether higher wage rates for better educated workers reflect an individual's added value or merely the employer's use of higher education as a screening device in the recruiting process. One version of this "sheepskin screening" hypothesis supported by many economists is that wages will rise faster with extra years of education when the extra years culminate in a certificate (e.g., master's or Ph.D. degree, CPA certificate, or actuarial degree).

a. Write a first-order, main effects model for mean wage rate $E(y)$ of an employer as a function of employee's years of education and whether or not the employee is certified.

b. Write a first-order model for $E(y)$ that corresponds to the "sheepskin screening" hypothesis.

c. Write the complete second-order model for $E(y)$ as a function of the two independent variables.

5.32   Use the coding system for observational data to fit a second-order model to the data on demand y and price p given in the following table. Show that the inherent multicollinearity problem with fitting a polynomial model is reduced when the coded values of p are used.

DEMAND y, pounds	1,120	999	932	884	807	760	701	688
PRICE p, dollars	3.00	3.10	3.20	3.30	3.40	3.50	3.60	3.70

5.33   One factor that must be considered in developing a shipping system that is beneficial to both the customer and the seller is time of delivery. A manufacturer of farm equipment can ship its products by either rail or truck. Quadratic models are thought to be adequate in relating time of delivery to distance to be shipped for both modes of transportation. Consequently, it has been suggested that the following model be fit to begin the model-building process

$$E(y) = \beta_0 + \beta_1 x_1 + \beta_2 x_1^2 + \beta_3 x_2 + \beta_4 x_1 x_2 + \beta_5 x_1^2 x_2$$

where

$y = $ Time of delivery

$x_1 = $ Distance to be shipped

---

*We would generally form confidence intervals for the true means to assess the reliability of these estimates. Our objective in these exercises is to develop the ability to use the models to obtain the estimates. The corresponding confidence intervals can be obtained using the methods of Chapter 4.

$$x_2 = \begin{cases} 1 & \text{if rail} \\ 0 & \text{if truck} \end{cases}$$

a. Sketch the proposed relationships between delivery time y and distance x_1 for both modes of transportation.

b. What hypothesis would you test to determine whether the data indicate that the quadratic distance terms are useful in the model, i.e., whether curvature is present in the relationship between mean delivery time and distance?

c. What hypothesis would you test to determine whether there is a difference in mean delivery time by rail and by truck?

5.34 Refer to Exercise 5.33. Suppose the complete second-order model is fit to a total of 50 observations on time of delivery. The sum of squared errors is SSE = 226.12. Then, the reduced model

$$E(y) = \beta_0 + \beta_1 x_1 + \beta_2 x_1^2$$

is fit to the same data, and SSE = 279.34. Test to determine whether the data indicate that the mean delivery time differs for rail and truck deliveries.

5.35 A company wants to model the total weekly sales, y, of its product as a function of the variables packaging and location. Two types of packaging, P_1 and P_2, are used in each of four locations, L_1, L_2, L_3, and L_4.

a. Write a main effects model to relate $E(y)$ to packaging and location. What implicit assumption are we making about the interrelationships between sales, packaging, and location when we use this model?

b. Now write a model for $E(y)$ that includes interaction between packaging and location. How many parameters are in this model (remember to include β_0)? Compare this number to the number of packaging–location combinations being modeled.

c. Suppose the main effects and interaction models are fit for 40 observations on weekly sales. The values of SSE are:

SSE for main effects model = 422.36

SSE for interaction model = 346.65

Determine whether the data indicate that the interaction between location and packaging is important in estimating mean weekly sales. Use $\alpha = .05$. What implications does your conclusion have for the company's marketing strategy?

5.36 Many companies must accurately estimate their costs before a job is begun in order to acquire a contract and make a profit. For example, a heating and plumbing contractor may base cost estimates for new homes on the total area of the house, the number of baths in the plans, and whether central air conditioning is to be installed.

a. Write a main effects model relating the mean cost of material and labor, $E(y)$, to the area, number of baths, and central air conditioning variables.

b. Write a complete second-order model for the mean cost as a function of the same three variables.

c. How would you test the hypothesis that the second-order and interaction terms are useful for predicting mean cost?

d. The contractor samples 25 recent jobs and fits both the complete second-order model (part b) and the reduced main effects model (part a), so that a test can be conducted to determine whether the additional complexity of the second-order model is necessary. The resulting values of SSE and R^2 are shown in the table. Is there sufficient evidence to conclude that the second-order terms are important for predicting the mean cost?

	SSE	R^2
Main effects	8.548	.950
Second-order	6.133	.964

5.37 An experiment was conducted to investigate the effect of financial incentive on the sales effort of a company's sales representatives. Five salespeople were selected from each of three divisions, A, B, and C, of the company. One person from each division was offered (confidentially) a commission equal to 1% of sales. Similarly, for each division, one person was selected to receive 1.5%, one 2.0%, one 2.5%, and one 3.0% of sales. The sales, y (in hundreds of thousands of dollars), for the $n = 15$ sales representatives are shown in the table.

		COMMISSION (%)				
		1	1.5	2.0	2.5	3.0
	A	2.0	3.0	4.0	3.8	3.5
DIVISION	B	1.6	2.8	3.0	3.2	2.8
	C	1.0	2.0	2.5	2.7	2.5

a. Assume that the response curves for each division are second-order and that they may differ from division to division. Write an appropriate model for the mean sales, $E(y)$.
b. Fit the model from part **a** to the data using coded values for the quantitative factor, percent commission, and dummy variables for the qualitative factor, division.
c. Do the data provide sufficient evidence to indicate that the three sales curves differ?

References

Draper, N. and Smith, H. *Applied Regression Analysis*, 3rd ed. New York: Wiley, 1989.

Graybill, F. A. *Theory and Application of the Linear Model*. North Scituate, Mass.: Duxbury, 1976.

Mendenhall, W. *Introduction to Linear Models and the Design and Analysis of Experiments*. Belmont, Calif.: Wadsworth, 1968.

Neter, J., Wasserman, W., and Kutner, M. H. *Applied Linear Statistical Models*, 3rd ed. Homewood, Ill.: Richard D. Irwin, 1988.

Some Regression Pitfalls

CONTENTS

6.1 Introduction

6.2 Observational Data Versus Designed Experiments

6.3 Deviating from the Assumptions

6.4 Parameter Estimability and Interpretation

6.5 Multicollinearity

6.6 Extrapolation: Predicting Outside the Experimental Region

6.7 Data Transformations

OBJECTIVE

To identify several potential problems that may be encountered when constructing a model for a response y; to help you recognize when these problems exist so that you can avoid some of the pitfalls of multiple regression analysis

6.1
Introduction

Multiple regression analysis is recognized by practitioners as a powerful tool for modeling a response y. Because it is so widely used, it is also one of the most abused statistical techniques. The ease with which a multiple regression analysis can be run on the computer has opened the doors to many data analysts who have only a limited knowledge of multiple regression and statistics. In practice, building a model for some response y is rarely a simple, straightforward process. There are a number of pitfalls that trap the unwary analyst. In this chapter we discuss several problems that you should be aware of when constructing a multiple regression model.

6.2
Observational Data Versus Designed Experiments

One problem encountered in using a regression analysis for business or economic data is caused by the type of data that the analyst is often forced to collect. Recall, from Section 2.4, that the data for regression can be either *observational* (where the values of the independent variables are uncontrolled) or *experimental* (where the x's are controlled via a designed experiment). Most data collected in business studies are observational. Whether data are observational is important for the following reasons. First, as you will subsequently learn in Chapter 10, the quantity of information in an experiment is controlled not only by the *amount of data* but also by the *values of the predictor variables* $x_1, x_2, \ldots, x_k$. Consequently, if you can design the experiment (sometimes this is physically impossible), you may be able to increase greatly the amount of information in the data at no additional cost.

Second, the use of observational data creates a problem involving randomization. When an experiment has been designed and we have decided on the various settings of the independent variables to be used, the experimental units are then randomly assigned in such a way that each combination of the independent variables has an equal chance of receiving experimental units with unusually high (or low) readings. (We will illustrate this method of randomization in Chapter 11.) This procedure tends to average out any variation within the experimental units. The result is that if the difference between two sample means is statistically significant, then you can infer (with probability of Type I error equal to α) that the population means differ. But more important, you can infer that this difference was due to the settings of the predictor variables, which is what you did to make the two populations different. Thus, you can infer a cause-and-effect relationship.

If the data are observational, a statistically significant relationship between a response y and a predictor variable x does not imply a cause-and-effect relationship. It simply means that x contributes information for the prediction of y, and nothing more. This point is aptly illustrated in the following example.

EXAMPLE 6.1

The *Orlando Sentinel Star* (February 28, 1979) reported on a socioeconomic research project in an article entitled "Couples who marry with child on way end up far poorer." The article states that "White suburban couples beginning marriage with the bride already pregnant face lower income and living standards and 22 percent fewer assets by age 40 than couples with no premarital pregnancy." This conclusion was based on interviews with approximately 1,000 randomly selected white suburban married women over the period 1962–1977. The two variables measured in the study, financial reward (income) at age 40, y, and premarital pregnancy, x (where $x = 1$ if premarital pregnancy and $x = 0$ if no premarital pregnancy), were found to be negatively correlated. Note that although the quotation does not use the word *cause*, it certainly implies that the researchers have concluded that lower income can be expected to follow premarital pregnancy.

a. Are the data for the research project observational or experimental?
b. Identify any weaknesses in the study.

Solution

a. The women in the study (the experimental units) were randomly selected and no attempt was made to control the value of x, premarital pregnancy; hence, the data are observational.
b. The pitfalls provided by the researcher's observational data are apparent. The response y (income) is related to a single variable, the presence or absence of premarital pregnancy, x. Since the women questioned in the experiment were *not* randomly assigned to the two groups—an obviously impossible task—a real possibility exists that lower prospective economic achievers tended to fall in the premarital pregnancy group and higher prospective economic achievers tended to fall in the nonpremarital pregnancy group. In other words, perhaps the survey is showing that women from lower socioeconomic groups are more likely to be subject to both premarital pregnancy *and* lower long-term economic rewards. Whether this explanation or the researcher's explanation of the relationship between long-term economic gain and premarital pregnancy is valid is impossible to decide based on observational data. This demonstrates the primary weakness of observation experiments.

The point of the previous example is twofold. If you can control the values of the independent variables in an experiment, it pays to do so. If you cannot control them, you can still learn much from a regression analysis about the relationship between a response y and a set of predictors. In particular, a prediction equation that provides a good fit to your data will almost always be useful. But, **you must be careful about deducing cause-and-effect relationships between the response and the predictors in an observational experiment**.

> **Warning**
>
> With observational data, a statistically significant relationship between a response y and a predictor variable x *does not* imply a cause-and-effect relationship.

Learning about the design of experiments is useful even if most of your applications of regression analysis involve observational data. Learning how to design an experiment and control the information in the data will improve your ability to assess the quality of observational data. We introduce experimental design in Chapter 10 and present methods for analyzing the data in a designed experiment in Chapter 11.

6.3
Deviating from the Assumptions

When we apply a regression analysis to a set of data, we never know for certain that the assumptions about the random error term ε are satisfied. How far can we deviate from the assumptions and still expect a multiple regression analysis to yield results that will have the reliability stated in Chapter 4? How can we detect departures (if they exist) from the assumptions, and what can we do about them? We will provide some partial answers to these questions in this section and direct you to further discussion in succeeding chapters.

Remember (from Section 4.2) that for a given set of values of $x_1, x_2, \ldots, x_k$,

$$y = \beta_0 + \beta_1 x_1 + \beta_2 x_2 + \cdots + \beta_k x_k + \varepsilon$$

where ε is a random error. The first assumption that we made was that the mean value of the random error for *any* given set of values of $x_1, x_2, \ldots, x_k$ is $E(\varepsilon) = 0$.

One consequence of this assumption is that the mean $E(y)$ for a specific set of values of $x_1, x_2, \ldots, x_k$ is

$$E(y) = \beta_0 + \beta_1 x_1 + \beta_2 x_2 + \cdots + \beta_k x_k$$

That is,

$$\underbrace{y = E(y)}_{\substack{\text{Mean value of } y \\ \text{for specific values} \\ \text{of } x_1, x_2, \ldots, x_k}} + \underbrace{\varepsilon}_{\substack{\text{Random} \\ \text{error}}}$$

The second consequence of this assumption (which we will state without proof) is that the least squares estimators of the model parameters, $\beta_0, \beta_1, \beta_2, \ldots, \beta_k$, will be unbiased regardless of the remaining assumptions that we attribute to the random errors and their probability distributions.

The properties of the sampling distributions of the parameter estimators $\hat{\beta}_0, \hat{\beta}_1, \ldots, \hat{\beta}_k$ will depend on the remaining assumptions that we specify concerning the probability distributions of the random errors. You will recall that we assumed that for any given set of values of $x_1, x_2, \ldots, x_k$, ε has a normal probability distribution with mean equal to 0 and variance equal to σ^2. Also, we assumed that the random errors are independent (in a probabilistic sense).

It is unlikely that the assumptions stated above are satisfied exactly for many practical situations. If departures from the assumptions are not too great, experience has shown that a least squares regression analysis produces estimates—predictions and statistical test results—that have, for all practical purposes, the properties specified in Chapter 4. On the other hand, if the assumptions are flagrantly violated, any inferences derived from the regression analysis are suspect.

If the observations are likely to be correlated (as in the case of **time series data**—that is, data collected over time), we must check for correlation between the random errors (a topic to be discussed in Chapter 7) and may have to modify our methodology if correlation exists. The solution to this problem is to construct a time series model; this will be the subject of Chapter 9. If the variance of the random error ε changes from one setting of the independent x variables to another, we can sometimes transform the data so that the standard least squares methodology will be appropriate. Some techniques for detecting nonhomogeneous variances of the random errors (a condition called **heteroscedasticity**) and some methods for treating this type of data are discussed in Chapter 7. The normality assumption is the least restrictive of the assumptions when regression analysis is applied in practice. However, nonnormality can result in predicted values that deviate greatly from the observed values. A careful analysis of these extreme values, or **outliers**, is an important component of regression analysis. In Chapter 7 we give some methods for detecting outliers and determining their influence on the prediction equation.

Frequently, the data $(y, x_1, x_2, \ldots, x_k)$ are observational, i.e., we just observe an experimental unit and record values for $y, x_1, x_2, \ldots, x_k$. Do these data violate the assumption that $x_1, x_2, \ldots, x_k$ are fixed? For this particular case, if we can assume that $x_1, x_2, \ldots, x_k$ are *measured without error*, the mean value $E(y)$ can be viewed as a conditional mean. That is, it gives the mean value of y, *given* that the x variables assume a specific set of values. With this modification in our thinking, the least squares regression analysis is applicable to observational data. Keep in mind, however, that inferences about $E(y)$ have the reliability stated in Chapter 4 only for the given set of x values.

To conclude, remember that when you perform a regression analysis, the reliability you can place in your inferences depends on having satisfied the assumptions prescribed in Section 4.2. We will never know for certain that the random errors satisfy these assumptions, but we will examine the residuals [the deviations $(y_i - \hat{y}_i)$ between the observed and the corresponding predicted values of y] in Chapter 7 to see if we can discover patterns that suggest correlation, heteroscedasticity, nonnormality, or an improper choice for the deterministic portion of the model. An examination of the residuals will also have another beneficial effect: The magnitudes of the residuals will give you an idea of how well the model is predicting. This should convince you that although the assumptions may not always be satisfied exactly, a multiple regression analysis is a powerful statistical tool.

6.4
Parameter Estimability and Interpretation

Suppose we want to fit the first-order model

$$E(y) = \beta_0 + \beta_1 x$$

to relate a firm's monthly profit y to advertising expenditure x. Now, suppose we have 3 months of data, and the firm spent $1,000 on advertising during each month. The data are shown in Figure 6.1. You can see the problem: The parameters of the straight-line model cannot be estimated when all the data are concentrated at a single x value. Recall that it takes two points (x values) to fit a straight line. Thus, the parameters are not estimable when only one x value is observed.

FIGURE 6.1

Profit and advertising expenditure data: 3 months

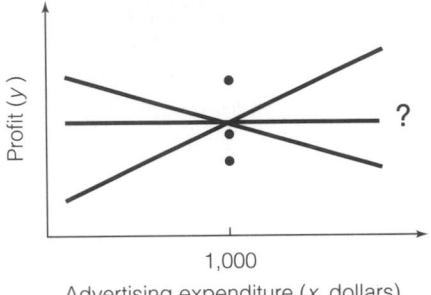

A similar problem would occur if we attempted to fit the second-order model

$$E(y) = \beta_0 + \beta_1 x + \beta_2 x^2$$

to a set of data for which only one *or two* different x values were observed (see Figure 6.2). At least three different x values must be observed before a second-order model can be fit to a set of data (that is, before all three parameters are estimable). In general, the number of levels of a quantitative independent variable x must be at least one more than the order of the polynomial in x that you want to fit. If two values of x are too close together, you may not be able to estimate a parameter because of rounding error encountered in fitting the model. Remember, also, that the sample size n must be sufficiently large to allow degrees of freedom for estimating σ^2.

FIGURE 6.2

Only two different x values observed—the second-order model is not estimable

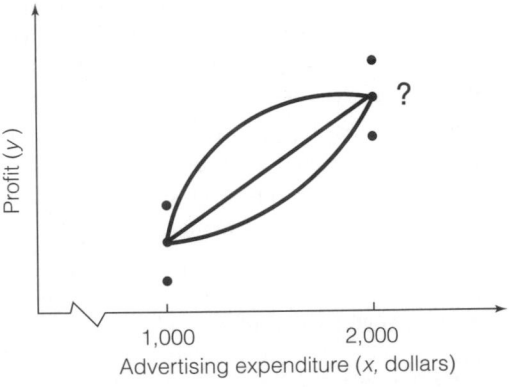

Since most business variables are not controlled by the researcher, the independent variables will almost always be observed at a sufficient number of levels to permit estimation of the model parameters. However, when the computer program you use suddenly refuses to fit a model, the problem is probably inestimable parameters.

Requirements for Fitting a pth-Order Polynomial Regression Model

$$E(y) = \beta_0 + \beta_1 x + \beta_2 x^2 + \cdots + \beta_p x^p$$

1. The number of levels of x must be greater than or equal to $(p + 1)$.
2. The sample size n must be greater than $(p + 1)$ to allow sufficient degrees of freedom for estimating σ^2.

Given that the parameters of the model are estimable, it is important to interpret the parameter estimates correctly. A typical misconception is that $\hat{\beta}_i$ always measures the effect of x_i on $E(y)$, *independent* of the other x variables in the model. This may be true for some models, but it is not true in general. We will see in Section 6.5 that when the independent variables are correlated, the values of the estimated β coefficients are often misleading. Even if the independent variables are uncorrelated, the presence of interaction changes the meaning of the parameters. For example, the underlying assumption of the first-order model

$$E(y) = \beta_0 + \beta_1 x_1 + \beta_2 x_2$$

is, in fact, that x_1 and x_2 affect the mean response $E(y)$ independently. Recall from Sections 4.5 and 5.4 that the slope parameter β_1 measures the rate of change of y for a 1-unit increase in x_1, for any given value of x_2. However, if the relationship between $E(y)$ and x_1 depends on x_2 (i.e., if x_1 and x_2 interact), then the interaction model

$$E(y) = \beta_0 + \beta_1 x_1 + \beta_2 x_2 + \beta_3 x_1 x_2$$

is more appropriate. For the interaction model, we showed that the effect of x_1 on $E(y)$, i.e., the slope, is not measured by a single β parameter, but by $\beta_1 + \beta_3 x_2$.

Generally, the interpretation of an individual β parameter becomes increasingly more difficult as the model becomes more complex. As we learned in Chapter 5, the individual β's of higher-order models usually have no practical interpretation.

Another misconception about the parameter estimates is that the magnitude of $\hat{\beta}_i$ determines the importance of x_i, that is, the larger (in absolute value) the $\hat{\beta}_i$, the more important the independent variable x_i is as a predictor of y. We learned in Chapter 4, however, that the standard error of the estimate $s_{\hat{\beta}_i}$ is critical in making inferences about the true parameter value. To reliably assess the importance of an individual term in the model we conduct a test of H_0: $\beta_i = 0$ or construct a confidence interval for β_i using formulas that reflect the magnitude of $s_{\hat{\beta}_i}$.

In addition to the parameter estimates, $\hat{\beta}_i$, some statistical software packages report the **standardized regression coefficients**,

$$\hat{\beta}_i^* = \hat{\beta}_i \left(\frac{s_{x_i}}{s_y} \right)$$

where s_{x_i} and s_y are the standard deviations of the x_i and y values, respectively, in the sample. Unlike $\hat{\beta}_i$, $\hat{\beta}_i^*$ is scaleless. These standardized regression coefficients make it more feasible to compare parameter estimates since the units are the same. However, the problems with interpreting standardized regression coefficients are much the same as those mentioned previously. Therefore, you should be wary of using a standardized regression coefficient as the sole determinant of an x variable's importance. The next example illustrates this point.

EXAMPLE 6.2

Refer to the problem of modeling the auction price y of antique grandfather clocks, Examples 4.1 and 4.2. In Example 4.1 we fit the model

$$E(y) = \beta_0 + \beta_1 x_1 + \beta_2 x_2$$

where x_1 = age of the clock and x_2 = number of bidders. The SPSS printout for the regression analysis is shown in Figure 6.3. Locate the standardized β coefficients on the printout and interpret them.

FIGURE 6.3
SPSS printout for Example 6.2

```
Equation Number 1      Dependent Variable..   Y


Multiple R            .94464
R Square              .89234
Adjusted R Square     .88492
Standard Error     133.48467

Analysis of Variance
                     DF      Sum of Squares      Mean Square
Regression            2      4283062.96010    2141531.48005
Residual             29       516726.53990      17818.15655

F =      120.18816        Signif F =   .0000


----------------- Variables in the Equation ------------------

Variable            B            SE B          Beta          T    Sig T

X2           85.952984     8.728523       .620287       9.847   .0000
X1           12.740574      .904740       .887029      14.082   .0000
(Constant) -1338.951340  173.809471                    -7.704   .0000
```

Solution

The standardized β coefficients are given on the SPSS printout in the column labeled **Beta**. These values, shaded in Figure 6.3, are

$$\hat{\beta}_1^* = .887 \quad \text{and} \quad \hat{\beta}_2^* = .620$$

Compare these values to the unstandardized β coefficients (in the **B** column):

$$\hat{\beta}_1 = 12.74 \quad \text{and} \quad \hat{\beta}_2 = 85.95$$

Based on the fact that $\hat{\beta}_2$ is nearly seven times larger than $\hat{\beta}_1$, we might be tempted to say that number of bidders (x_2) is a more important predictor of auction price than age of the clock (x_1). Once we standardize the β's (i.e., take the units of measurement and variation into account), we see that the opposite may, in fact, be true since $\hat{\beta}_1^*$ exceeds $\hat{\beta}_2^*$. Of course, from Example 4.2 we know that the two independent variables, x_1 and x_2, interact to affect y. Consequently, both age and number of bidders are important for predicting auction price and we should resist inferring that one of the variables is more important than the other.

6.5
Multicollinearity

Often, two or more of the independent variables used in the model for $E(y)$ will contribute redundant information. That is, the independent variables will be correlated with each other. For example, suppose we want to construct a model to predict the gasoline mileage rating, y, of a truck as a function of its load, x_1, and the horsepower, x_2, of its engine. In general, you would expect heavier loads to require greater horsepower and to result in lower mileage ratings. Thus, although both x_1 and x_2 contribute information for the prediction of mileage rating, some of the information is overlapping, because x_1 and x_2 are correlated. When the independent variables are correlated, we say that **multicollinearity** exists. In practice, it is not uncommon to observe correlations among the independent variables. However, a few problems arise when serious multicollinearity is present in the regression analysis.

Definition 6.1

Multicollinearity exists when two or more of the independent variables used in regression are correlated.

First, high correlations among the independent variables increase the likelihood of rounding errors in the calculations of the β estimates, standard errors, and so forth.* Second, the regression results may be confusing and misleading.

To illustrate, if the gasoline mileage rating model

$$E(y) = \beta_0 + \beta_1 x_1 + \beta_2 x_2$$

were fit to a set of data, we might find that the t values for both $\hat{\beta}_1$ and $\hat{\beta}_2$ (the least squares estimates) are nonsignificant. However, the F test for H_0: $\beta_1 = \beta_2 = 0$ would probably be highly significant. The tests may seem to be contradictory, but really they are not. The t tests indicate that the contribution of one variable, say, $x_1 =$ load, is not significant after the effect of $x_2 =$ horsepower has been discounted (because x_2 is also in the model). The significant F test, on the other hand, tells us that at least one of the two variables is making a contribution to the prediction of y (i.e., β_1, β_2, or both differ from 0). In fact, both are probably contributing, but the contribution of one overlaps with that of the other.

---

*The result is due to the fact that, in the presence of severe multicollinearity, the computer has difficulty inverting the information matrix ($X'X$). See Appendix A for a discussion of the ($X'X$) matrix and the mechanics of a regression analysis.

Multicollinearity can also have an effect on the signs of the parameter estimates. More specifically, a value of $\hat{\beta}_i$ may have the opposite sign from what is expected. For example, we expect the signs of both of the parameter estimates for the gasoline mileage rating model to be negative, yet the regression analysis for the model might yield the estimates $\hat{\beta}_1 = .2$ and $\hat{\beta}_2 = -.7$. The positive value of $\hat{\beta}_1$ seems to contradict our expectation that heavy loads will result in lower mileage ratings. We mentioned in the previous section, however, that it is dangerous to interpret a β coefficient when the independent variables are correlated. Because the variables contribute redundant information, the effect of load x_1 on mileage rating is measured only partially by $\hat{\beta}_1$. Also, we warned in Section 6.2 that we cannot establish a cause-and-effect relationship between y and the predictor variables based on observational data. By attempting to interpret the value $\hat{\beta}_1$, we are really trying to establish a cause-and-effect relationship between y and x_1 (by suggesting that a heavy load x_1 will *cause* a lower mileage rating y).

How can you avoid the problems of multicollinearity in regression analysis? One way is to conduct a designed experiment so that the levels of the x variables are uncorrelated (see Section 6.2). Unfortunately, time and cost constraints may prevent you from collecting data in this manner. For these and other reasons, most data collected in business studies are observational. Since observational data frequently consist of correlated independent variables, you will need to recognize when multicollinearity is present and, if necessary, make modifications in the analysis.

Several methods are available for detecting multicollinearity in regression. A simple technique is to calculate the coefficient of correlation r between each pair of independent variables in the model and use the procedure outlined in Section 3.7 to test for evidence of positive or negative correlation. If one or more of the r values is statistically different from 0, the variables in question are correlated and a severe multicollinearity problem may exist.* Other indications of the presence of multicollinearity include those mentioned in the beginning of this section—namely, nonsignificant t tests for the individual β parameters when the F test for overall model adequacy is significant, and estimates with opposite signs from what is expected.

A more formal method for detecting multicollinearity involves the calculation of **variance inflation factors** for the individual β parameters. One reason why the t tests on the individual β parameters are nonsignificant is because the standard errors of the estimates, $s_{\hat{\beta}_i}$, are inflated in the presence of multicollinearity. When the dependent and independent variables are appropriately trans-

---

*Remember that r measures only the pairwise correlation between x values. Three variables, x_1, x_2, and x_3, may be highly correlated as a group, but may not exhibit large pairwise correlations. Thus, multicollinearity may be present even when all pairwise correlations are not significantly different from 0.

†The transformed variables are obtained as

$$y_i^* = (y_i - \bar{y})/s_y \qquad x_{1i}^* = (x_{1i} - \bar{x}_1)/s_1 \qquad x_{2i}^* = (x_{2i} - \bar{x}_2)/s_2$$

and so on, where $\bar{y}$, $\bar{x}_1$, $\bar{x}_2$, . . . , and s_y, s_1, s_2, . . . , are the sample means and standard deviations, respectively, of the original variables.

formed,[†] it can be shown that

$$s_{\hat{\beta}_i}^2 = s^2 \left(\frac{1}{1 - R_i^2} \right)$$

where s^2 is the estimate of σ^2, the variance of ε, and R_i^2 is the multiple coefficient of determination for the model that regresses the independent variable x_i on the remaining independent variables $x_1, x_2, \ldots, x_{i-1}, x_{i+1}, \ldots, x_k$. The quantity $1/(1 - R_i^2)$ is called the *variance inflation factor* for the parameter β_i, denoted $(\text{VIF})_i$. Note that $(\text{VIF})_i$ will be large when R_i^2 is large—that is, when the independent variable x_i is strongly related to the other independent variables.

Various authors maintain that, in practice, a severe multicollinearity problem exists if the largest of the variance inflation factors for the β's is greater than 10 or, equivalently, if the largest multiple coefficient of determination R_i^2, is greater than .90.[*] Several of the statistical software packages discussed in this text have options for calculating variance inflation factors.[†]

The methods for detecting multicollinearity are summarized in the accompanying box. We illustrate the use of these statistics in Example 6.3.

| Detecting Multicollinearity in the Regression Model

$$E(y) = \beta_0 + \beta_1 x_1 + \beta_2 x_2 + \cdots + \beta_k x_k$$

The following are indicators of multicollinearity:

1. Significant correlations between pairs of independent variables in the model

2. Nonsignificant t tests for all (or nearly all) the individual β parameters when the F test for overall model adequacy H_0: $\beta_1 = \beta_2 = \cdots = \beta_k = 0$ is significant

3. Opposite signs (from what is expected) in the estimated parameters

4. A variance inflation factor (VIF) for a β parameter greater than 10, where

$$(\text{VIF})_i = \frac{1}{1 - R_i^2}, \quad i = 1, 2, \ldots, k$$

and R_i^2 is the multiple coefficient of determination for the model

$$E(x_i) = \alpha_0 + \alpha_1 x_1 + \alpha_2 x_2 + \cdots + \alpha_{i-1} x_{i-1} + \alpha_{i+1} x_{i+1} + \cdots + \alpha_k x_k$$

---

[*]See, for example, Montgomery and Peck (1982) or Neter, Wasserman, and Kutner (1990).
[†]Some software packages calculate an equivalent statistic, called the **tolerance**. The tolerance for a β coefficient is the reciprocal of the variance inflation factor, i.e.,

$$(\text{TOL})_i = \frac{1}{(\text{VIF})_i} = 1 - R_i^2$$

For $R_i^2 > .90$ (the extreme multicollinearity case), $(\text{TOL})_i < .10$. These computer packages allow the user to set tolerance limits, so that any independent variable with a value of $(\text{TOL})_i$ below the tolerance limit will not be allowed to enter into the model.

EXAMPLE 6.3

The Federal Trade Commission (FTC) annually ranks varieties of domestic cigarettes according to their tar, nicotine, and carbon monoxide contents. The U.S. surgeon general considers each of these three substances hazardous to a smoker's health. Past studies have shown that increases in the tar and nicotine contents of a cigarette are accompanied by an increase in the carbon monoxide emitted from the cigarette smoke. Table 6.1 lists tar, nicotine, and carbon monoxide contents (in milligrams) and weight (in grams) for a sample of 25 (filter) brands tested in a recent year. Suppose we want to model carbon monoxide content, y, as a function of tar content, x_1, nicotine content, x_2, and weight, x_3, using the model

$$E(y) = \beta_0 + \beta_1 x_1 + \beta_2 x_2 + \beta_3 x_3$$

The model is fit to the 25 data points in Table 6.1. A portion of the resulting SAS printout is shown in Figure 6.4. Examine the printout. Do you detect any signs of multicollinearity?

TABLE 6.1 **FTC Cigarette Data for Example 6.3**

BRAND	TAR x_1, milligrams	NICOTINE x_2, milligrams	WEIGHT x_3, grams	CARBON MONOXIDE y, milligrams
Alpine	14.1	.86	.9853	13.6
Benson & Hedges	16.0	1.06	1.0938	16.6
Bull Durham	29.8	2.03	1.1650	23.5
Camel Lights	8.0	.67	.9280	10.2
Carlton	4.1	.40	.9462	5.4
Chesterfield	15.0	1.04	.8885	15.0
Golden Lights	8.8	.76	1.0267	9.0
Kent	12.4	.95	.9225	12.3
Kool	16.6	1.12	.9372	16.3
L&M	14.9	1.02	.8858	15.4
Lark Lights	13.7	1.01	.9643	13.0
Marlboro	15.1	.90	.9316	14.4
Merit	7.8	.57	.9705	10.0
Multifilter	11.4	.78	1.1240	10.2
Newport Lights	9.0	.74	.8517	9.5
Now	1.0	.13	.7851	1.5
Old Gold	17.0	1.26	.9186	18.5
Pall Mall Light	12.8	1.08	1.0395	12.6
Raleigh	15.8	.96	.9573	17.5
Salem Ultra	4.5	.42	.9106	4.9
Tareyton	14.5	1.01	1.0070	15.9
True	7.3	.61	.9806	8.5
Viceroy Rich Lights	8.6	.69	.9693	10.6
Virginia Slims	15.2	1.02	.9496	13.9
Winston Lights	12.0	.82	1.1184	14.9

Source: Federal Trade Commission.

Solution

First, notice that a test of

$$H_0: \quad \beta_1 = \beta_2 = \beta_3 = 0$$

is highly significant. The F value (shaded on the printout) is very large ($F = 78.984$) and the observed significance level of the test (also shaded) is small ($p = .0001$). Therefore, we can reject H_0 for any α greater than .0001 and conclude that at least one of the parameters, β_1, β_2, and β_3, is nonzero. The t tests for two of the three individual β's, however, are nonsignificant. (The p-values for these tests are shaded on the printout.) Unless tar is the only one of the three variables useful for predicting carbon monoxide content, these results are the first indication of a potential multicollinearity problem.

A second clue to the presence of multicollinearity is the negative value for $\hat{\beta}_2$ and $\hat{\beta}_3$ (shaded on the printout),

$$\hat{\beta}_2 = -2.63 \qquad \hat{\beta}_3 = -.13$$

From past studies, the FTC expects carbon monoxide content y to increase when either nicotine content x_2 or weight x_3 increases—that is, the FTC expects positive relationships between y and x_2, and y and x_3, not negative ones.

A more formal procedure for detecting multicollinearity is to examine the variance inflation factors. Figure 6.4 shows the variance inflation factors (shaded) for each of the three parameters under the column labeled **VARIANCE INFLATION**. Note that the variance inflation factors for both the tar and nicotine parameters are greater than 10. The variance inflation factor for the tar parameter, $(VIF)_1 = 21.63$, implies that a model relating tar content x_1 to the remaining two independent variables, nicotine content x_2 and weight x_3, resulted in a coefficient of determination

$$R_1^2 = 1 - \frac{1}{(VIF)_1}$$

$$= 1 - \frac{1}{21.63} = .954$$

FIGURE 6.4 Portion of the SAS printout for Example 6.3

DEP VARIABLE: CO

ANALYSIS OF VARIANCE

SOURCE	DF	SUM OF SQUARES	MEAN SQUARE	F VALUE	PROB>F
MODEL	3	495.25781	165.08594	78.984	0.0001
ERROR	21	43.89258562	2.09012312		
C TOTAL	24	539.15040			

ROOT MSE	1.445726	R-SQUARE	0.9186	
DEP MEAN	12.528	ADJ R-SQ	0.9070	
C.V.	11.53996			

PARAMETER ESTIMATES

VARIABLE	DF	PARAMETER ESTIMATE	STANDARD ERROR	T FOR H0: PARAMETER=0	PROB > :T:	TOLERANCE	VARIANCE INFLATION
INTERCEP	1	3.20219002	3.46175473	0.925	0.3655		0
TAR	1	0.96257386	0.24224436	3.974	0.0007	0.04623058	21.63070592
NICOTINE	1	-2.63166111	3.90055745	-0.675	0.5072	0.04566227	21.89991722
WEIGHT	1	-0.13048185	3.88534182	-0.034	0.9735	0.74970451	1.33385886

TABLE 6.2 **Correlation Coefficients for the Three Pairs of Independent Variables in Example 6.3**

PAIR	r
x_1, x_2	.977
x_1, x_3	.491
x_2, x_3	.500

All signs indicate that a serious multicollinearity problem exists. To confirm our suspicions, we calculated the coefficient of correlation r for each of the three pairs of independent variables in the model. These values are given in Table 6.2. You can see that tar content x_1 and nicotine content x_2 appear to be highly correlated ($r = .977$), while weight x_3 appears to be moderately correlated with both tar content ($r = .491$) and nicotine content ($r = .500$). In fact, all three sample correlations have test statistics that exceed the critical t value, for a two-tailed test of H_0: $\rho = 0$ conducted at $\alpha = .05$ with $n - 2 = 23$ df (see Section 3.7).

Once you have detected that a multicollinearity problem exists, there are several alternative measures available for solving the problem. The appropriate measure to take depends on the severity of the multicollinearity and the ultimate goal of the regression analysis.

Some researchers, when confronted with highly correlated independent variables, choose to include only one of the correlated variables in the final model. One way of deciding which variable to include is by using **stepwise regression**, a topic discussed in Chapter 4. Generally, only one (or a small number) of a set of multicollinear independent variables will be included in the regression model by the stepwise regression procedure since this procedure tests the parameter associated with each variable in the presence of all the variables already in the model. For example, in fitting the gasoline mileage rating model introduced earlier, if at one step the variable representing truck load is included as a significant variable in the prediction of the mileage rating, the variable representing horsepower will probably never be added in a future step. Thus, if a set of independent variables is thought to be multicollinear, some screening by stepwise regression may be helpful.

If you are interested only in using the model for estimation and prediction, you may decide not to drop any of the independent variables from the model. In the presence of multicollinearity, we have seen that it is dangerous to interpret the individual β's for the purpose of establishing cause and effect. However, confidence intervals for $E(y)$ and prediction intervals for y generally remain unaffected *as long as the values of the independent variables used to predict y follow the same pattern of multicollinearity exhibited in the sample data.* That is, you must take strict care to ensure that the values of the x variables fall within the experimental region. (We will discuss this problem in further detail in Section 6.6.) Alternatively, if your goal is to establish a cause-and-effect relationship between y and the independent variables, you will need to conduct a designed experiment to break up the pattern of multicollinearity.

When fitting a polynomial regression model [for example, the second-order model $E(y) = \beta_0 + \beta_1 x + \beta_2 x^2$], the independent variables $x_1 = x$ and $x_2 = x^2$ will often be correlated. If the correlation is high, the computer solution may result in extreme rounding errors. For this model, the solution is not to drop one of the independent variables but to transform the x variable in such a way that the correlation between the coded x and x^2 values is substantially reduced.

| Solutions to Some Problems Created by Multicollinearity

1. Drop one or more of the correlated independent variables from the final model. A screening procedure such as stepwise regression is helpful in determining which variables to drop.

2. If you decide to keep all the independent variables in the model:
   a. Avoid making inferences about the individual β parameters (such as establishing a cause-and-effect relationship between y and the predictor variables).
   b. Restrict inferences about $E(y)$ and future y values to values of the independent variables that fall within the experimental region (see Section 6.6).

3. If your ultimate objective is to establish a cause-and-effect relationship between y and the predictor variables, use a designed experiment (see Chapters 10 and 11).

4. To reduce rounding errors in polynomial regression models, code the independent variables so that first-, second-, and higher-order terms for a particular x variable are not highly correlated (see Section 5.6).

5. To reduce rounding errors and stabilize the regression coefficients, use ridge regression to estimate the β parameters (see Section 8.7).

Coding the independent quantitative variables as described in optional Section 5.6 is a useful technique for reducing the multicollinearity inherent with polynomial regression models.

Another, more complex, procedure for reducing the rounding errors caused by multicollinearity involves a modification of the least squares method, called **ridge regression**. In ridge regression, the estimates of the β coefficients are biased [that is, $E(\hat{\beta}_i) \neq \beta_i$] but have significantly smaller standard errors than the unbiased β estimates yielded by the least squares method. Thus, the β estimates for the ridge regression are more stable than the corresponding least squares estimates. Ridge regression is a topic discussed in optional Chapter 8.

6.6
Extrapolation: Predicting Outside the Experimental Region

By the late 1960s many research economists had developed highly technical models to relate the state of the economy to various economic indices and other independent variables. Many of these models were multiple regression models, where, for example, the dependent variable y might be next year's growth in GNP and the independent variables might include this year's rate of inflation, this year's Consumer Price Index, and so forth. In other words, the model might be constructed to predict next year's economy using this year's knowledge.

Unfortunately, these models were almost unanimously unsuccessful in predicting the recession in the early 1970s. What went wrong? Well, one of the problems was that regression models were used to predict y for values of the independent variables that were outside the region in which the model was developed. For example, the inflation rate in the late 1960s, when the models were developed, ranged from 6 to 8%. When the double-digit inflation of the early 1970s became a reality, some researchers attempted to use the same models to predict the growth in GNP 1 year hence. As you can see in Figure 6.5, the model may be very accurate for predicting y when x is in the range of experimentation, but the use of the model outside that range is a dangerous practice. A $100(1 - \alpha)\%$ prediction interval for GNP when the inflation rate is, say, 10%, will be less reliable than the stated confidence coefficient $(1 - \alpha)$. How much less is unknown.

FIGURE 6.5

Using a regression model outside the experimental region

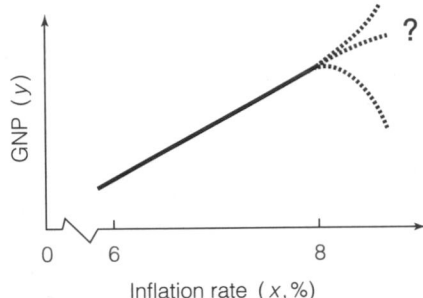

For a single independent variable x, the experimental region is simply the range of the values of x in the sample. Establishing the experimental region for a multiple regression model that includes a number of independent variables may be more difficult. For example, consider a model for GNP (y) using inflation rate (x_1) and prime interest rate (x_2) as predictor variables. Suppose a sample of size $n = 5$ was observed, and the values of x_1 and x_2 corresponding to the five values for GNP were (6, 10), (6.25, 12), (7.25, 10.25), (7.5, 13), and (8, 11.5). Notice that x_1 ranges from 6% to 8% and x_2 ranges from 10% to 13% in the sample data. You may think that the experimental region is defined by the ranges of the individual variables, i.e., $6 \leq x_1 \leq 8$ and $10 \leq x_2 \leq 13$. However, the levels of x_1 and x_2 *jointly* define the region. Figure 6.6 shows the experimental region for our hypothetical data. You can see that an observation with levels $x_1 = 8$ and $x_2 = 10$ clearly falls outside the experimental region, yet is within the ranges of the individual x values. Using the model to predict GNP for this observation may lead to unreliable results.

FIGURE 6.6

Experimental region for modeling GNP (y) as a function of inflation rate (x_1) and prime interest rate (x_2)

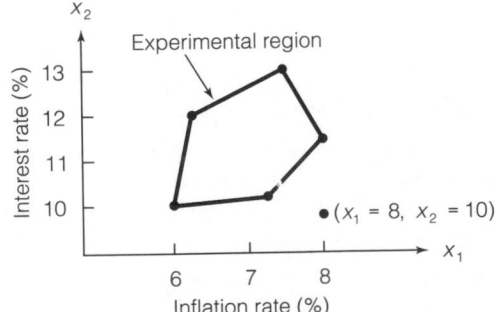

6.7

Data Transformations

The word *transform* means to change the form of some object or thing. Consequently, the phrase *data transformation* means that we have done, or plan to do, something to change the form of the data. For example, if one of the independent variables in a model is the price p of a commodity, we might choose to introduce this variable into the model as $x = 1/p$, $x = \sqrt{p}$, or $x = e^{-p}$. Thus, if we were to let $x = \sqrt{p}$, we would compute the square root of each price value, and these square roots would be the values of x that would be used in the regression analysis.

Data transformations are performed on the y values to make them more nearly satisfy the assumptions of Section 4.2 and, sometimes, to make the deterministic portion of the model a better approximation to the mean value of the transformed response. Transformations of the values of the independent variables are performed solely for the latter reason—that is, to achieve a model that provides a better approximation to $E(y)$. The purpose of this section is to discuss transformations on the independent variables. In particular, we want you to see why these transformations are sometimes useful and to see how you might use them. (Transformations on the y values for the purpose of satisfying the assumptions will be discussed in Chapter 7.)

Suppose you want to fit a model relating the demand y for a product to its price p. Also, suppose the product is a nonessential item, and you expect the mean demand to decrease as price p increases and then to decrease more slowly as p gets larger (see Figure 6.7). What function of p will provide a good approximation to $E(y)$?

FIGURE 6.7

Hypothetical relation between demand y and price p

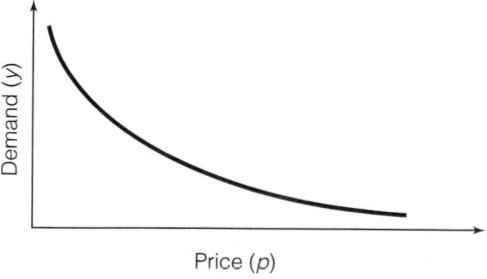

To answer this question, you need to know the graphs of some elementary mathematical functions—there is a one-to-one relationship between mathematical functions and graphs. If we want to model a relationship similar to the one indicated in Figure 6.7, we need to be able to select a mathematical function that will possess a graph similar to the curve shown.

Portions of some curves corresponding to mathematical functions that decrease as p increases are shown in Figure 6.8 on page 338. Of the four functions shown, the curves in Figures 6.8c and 6.8d will probably provide the best approximations to $E(y)$. This is because both provide graphs that show $E(y)$ decreasing and approaching (but never reaching) 0 as p increases. This suggests that the independent variable, price, should be transformed using either $x = 1/p$ or $x = e^{-p}$. Then you might try fitting the model

$$E(y) = \beta_0 + \beta_1 x$$

using the transformed data.

FIGURE 6.8   Graphs of some mathematical functions relating $E(y)$ to p

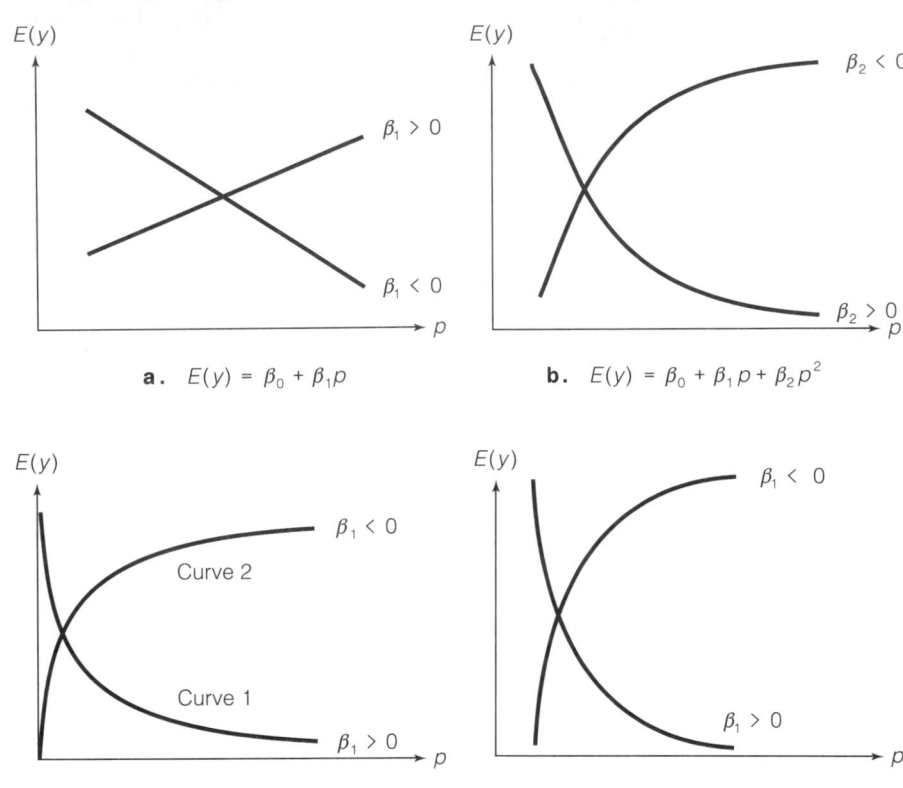

a. $E(y) = \beta_0 + \beta_1 p$

b. $E(y) = \beta_0 + \beta_1 p + \beta_2 p^2$

c. $E(y) = \beta_0 + \beta_1 \left(\frac{1}{p}\right)$

d. $E(y) = \beta_0 + \beta_1 e^{-p}$

The functions shown in Figure 6.8 produce curves that either rise or fall depending on the sign of the parameter β_1 in parts **a**, **c**, and **d**, and on β_2 and the portion of the curve used in part **b**. When you choose a model for a regression analysis, you do not have to specify the sign of the parameter(s). The least squares procedure will choose as estimates of the parameters those that minimize the sum of squares of the residuals. Consequently, if you were to fit the model shown in Figure 6.8c to a set of y values that increase in value as p increases, your least squares estimate of β_1 would be negative, and a graph of y would produce a curve similar to curve 2 in Figure 6.8c. If the y values decrease as p increases, your estimate of β_1 will be positive and the curve will be similar to curve 1 in Figure 6.8c. All the curves in Figure 6.8 shift upward or downward depending on the value of β_0.

| EXAMPLE 6.4

A supermarket chain conducted an experiment to investigate the effect of price p on the weekly demand (in pounds) for a house brand of coffee. Eight super-market stores that had nearly equal past records of demand for the product were used in the experiment. Eight prices were randomly assigned to the stores and were advertised using the same procedures. The number of pounds of coffee sold

DEMAND y, pounds	PRICE p, dollars
1,120	3.00
999	3.10
932	3.20
884	3.30
807	3.40
760	3.50
701	3.60
688	3.70

Solution

y	x = 1/p
1,120	.333
999	.323
932	.313
884	.303
807	.294
760	.286
701	.278
688	.270

FIGURE 6.9

Minitab printout for Example 6.4

during the following week was recorded for each of the stores and is shown in the table.

a. Fit the model

$$E(y) = \beta_0 + \beta_1 x$$

to the data, letting $x = 1/p$.

b. Do the data provide sufficient evidence to indicate that the model contributes information for the prediction of demand?

c. Find a 90% confidence interval for the mean demand when the price is set at \$3.20 per pound. Interpret this interval.

a. The first step is to calculate $x = 1/p$ for each data point. These values are given in the table. The Minitab computer printout* (Figure 6.9) gives

$$\hat{\beta}_0 = -1,180 \qquad \hat{\beta}_1 = 6,808$$

and

$$\hat{y} = -1,180 + 6,808x$$

$$= -1,180 + 6,808\left(\frac{1}{p}\right)$$

(You can verify that the formulas of Section 3.3 give the same answers.) A graph of this prediction equation is shown in Figure 6.10 on page 340.

```
The regression equation is
Y = - 1180 + 6808 X

Predictor      Coef       Stdev     t-ratio        p
Constant    -1180.5       107.7      -10.96    0.000
X            6808.1       358.4       19.00    0.000

s = 20.90      R-sq = 98.4%     R-sq(adj) = 98.1%

Analysis of Variance

SOURCE        DF         SS          MS        F        p
Regression     1     157718      157718   360.94    0.000
Error          6       2622         437
Total          7     160340
```

b. To determine whether x contributes information for the prediction of y, we test $H_0: \beta_1 = 0$ against the alternative hypothesis $H_a: \beta_1 \neq 0$. The test statistic is

$$t = \frac{\hat{\beta}_1 - 0}{s_{\hat{\beta}_1}} = \frac{\hat{\beta}_1}{s/\sqrt{SS_{xx}}}$$

*The Minitab program uses full decimal accuracy for $x = 1/p$. Hence, the results shown in Figure 6.9 differ from results that would be calculated using the three-decimal values for $x = 1/p$ shown in the table.

FIGURE 6.10

Graph of the demand–price curve for Example 6.4

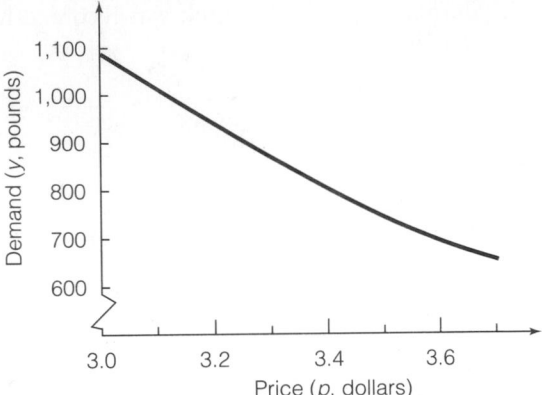

The computed value of this t statistic, given in Figure 6.9 (you can verify the computations), is $t = 19.00$. Since we wish to detect either $\beta_1 > 0$ or $\beta_1 < 0$, we will use a two-tailed test and will reject H_0 when t exceeds $t_{\alpha/2}$ based on $(n - 2) = 6$ df. Using $\alpha = .05$, $t_{.025} = 2.447$. Since the computed value of t exceeds $t_{.025}$, we reject H_0: $\beta_1 = 0$ and conclude that $x = 1/p$ contributes information for the prediction of demand y.

c. Note on the printout that $s = \sqrt{MSE} = \sqrt{437} = 20.90$, and from Table 2 of Appendix D, we have $t_{\alpha/2} = t_{.05} = 1.943$. Then the 90% confidence interval for mean demand $E(y)$ when the price per pound is $p = \$3.20$ is

$$\hat{y} \pm t_{\alpha/2}\, s \sqrt{\frac{1}{n} + \frac{(x_p - \bar{x})^2}{\sum (x_i - \bar{x})^2}}$$

where

$$\hat{y} = \hat{\beta}_0 + \hat{\beta}_1 x = -1{,}180 + 6{,}808\left(\frac{1}{p}\right)$$

$$= -1{,}180 + 6{,}808\left(\frac{1}{3.20}\right)$$

$$= 947$$

$$\bar{x} = \frac{\sum x_i}{n} = \frac{2.400}{8} = .300$$

$$\sum (x_i - \bar{x})^2 = \sum x_i^2 - \frac{\left(\sum x_i\right)^2}{n} = .723412 - \frac{(2.4)^2}{8}$$

$$= .003412$$

Substituting all these values into the formula for the confidence interval, we estimate the mean demand to be enclosed by the interval:

$$\hat{y} \pm t_{\alpha/2}s \sqrt{\frac{1}{n} + \frac{(x_p - \bar{x})^2}{\sum (x_i - \bar{x})^2}}$$

$$947 \pm (1.943)(20.90) \sqrt{\frac{1}{8} + \frac{(.313 - .300)^2}{.003412}}$$

or

$$930 \quad \text{to} \quad 964$$

with confidence coefficient .90. Thus, we are reasonably certain that this confidence interval encloses the mean demand $E(y)$.

This discussion is intended to emphasize the importance of data transformation and to explain its role in model building. Remember that the symbols, x_1, x_2, ..., x_k that appear in the linear models of this text can be transformations on the independent variables you have observed. These transformations, coupled with the methods of Chapter 5, allow you to use a great variety of mathematical functions to model the mean $E(y)$ for data.

Summary

There are several problems that you should be aware of when constructing a model for a response y. In this chapter we have identified a few of the most important of these problem areas:

1. *Establishing cause and effect*   When the data used in the regression analysis are observational (i.e., uncontrolled), it is dangerous to deduce a cause-and-effect relationship between y and the independent (predictor) variables. Only when the experiment has been designed properly can you be certain that any changes in y are due solely to the different settings of the predictor variables.

2. *Departures from the assumptions*   In a practical setting, it is unlikely that the standard least squares assumptions about the error term are satisfied exactly. When departures from the assumptions are slight, the model remains a powerful predictor of the response y. On the other hand, the model performs poorly when the assumptions of equal variances and uncorrelated errors are violated.

3. *Parameter estimability and interpretation*   To estimate all the parameters in a pth-order model, the number of levels of x must be greater than or equal to $(p + 1)$. For any model, the sample size n must be sufficiently large to allow degrees of freedom for estimating σ^2 so that a test of model adequacy can be performed. Be wary of interpreting the individual β parameters in the presence of interaction or highly correlated independent variables.

4. *Multicollinearity*   When highly correlated independent variables are present in a regression model, the results may be confusing; the t tests on the individual β's may be nonsignificant even though the F test for overall model adequacy is significant, and the β's may have signs opposite from what is expected. Also, there may be extreme rounding errors in the computation of the β estimates. Variance inflation factors aid in determining whether a serious multicollinearity problem exists. The solution to the problem depends on the severity of the multicollinearity and the ultimate goal of the regression analyst.

5. *Extrapolation*   Predicting y when the x values are outside the range of experimentation is a dangerous practice. The level of reliability associated with any

inference derived from the model will be less than the stated level of confidence $(1 - \alpha)$ since the adequacy of the model outside the experimental region is unknown.

6. *Data transformations*   To achieve a model that provides a better approximation to $E(y)$, you may need to transform the values of the independent variables or the value of y. The type of transformation you should make depends on the theoretical relationships between $E(y)$ and the independent variables.

EXERCISES

6.1   Discuss the consequences of fitting multiple regression models when the assumptions of Section 4.2 are violated.

6.2   Why is it dangerous to predict y for values of independent variables that fall outside the experimental region?

6.3   Discuss the problems that result when multicollinearity is present in a regression analysis.

6.4   How can you detect multicollinearity?

6.5   What remedial measures are available when multicollinearity is detected?

6.6   Refer to Example 6.4. Can you think of any other transformations on price that might provide a good fit to the data? Try them and answer the questions of Example 6.4 again.

6.7   The management of a manufacturing firm is considering the possibility of setting up its own market research department rather than continuing to use the services of a market research firm. Management wants to know what salary should be paid to a market researcher, based on years of experience. An independent consultant has proposed the quadratic model

$$E(y) = \beta_0 + \beta_1 x + \beta_2 x^2$$

where

y = Annual salary (thousands of dollars)

x = Years of experience

To fit the model, the consultant randomly sampled three market researchers at other firms and recorded the information given in the accompanying table. Give your opinion regarding the adequacy of the proposed model.

	y	x
Researcher 1	40	2
Researcher 2	25	1
Researcher 3	42	3

6.8   A firm that sells a special skin cream exclusively through drug stores currently operates in 15 marketing districts. As part of an expansion feasibility study, the company wants to model district sales (y) as a function of target population (x_1), per capita income (x_2), and number of drug stores (x_3) in the district. Data collected for each of the 15 districts were used to fit the first-order model

$$E(y) = \beta_0 + \beta_1 x_1 + \beta_2 x_2 + \beta_3 x_3$$

A summary of the regression results follows:

$$\hat{y} = -3,000 + 3.2x_1 - .4x_2 - 1.1x_3 \qquad R^2 = .93$$

$$s_{\hat{\beta}_1} = 2.4 \qquad s_{\hat{\beta}_2} = .6 \qquad s_{\hat{\beta}_3} = .8$$

$$r_{12} = .92 \qquad r_{13} = .87 \qquad r_{23} = .81$$

Based on these results, the company concludes that none of the three independent variables, x_1, x_2, and x_3, is a useful predictor of district sales, y. Do you agree with this statement? Explain.

6.9　A particular meat-processing plant slaughters steers and cuts and wraps the beef for its customers. Suppose a complaint has been filed with the Food and Drug Administration (FDA) against the processing plant. The complaint alleges that the consumer does not get all the beef from the steer he purchases. In particular, one consumer purchased a 300-pound steer but received only 150 pounds of cut and wrapped beef. To settle the complaint, the FDA collected data on the live weights and dressed weights of nine steers processed by a reputable meat-processing plant (not the firm in question). The results are listed in the table.

LIVE WEIGHT x, pounds	DRESSED WEIGHT y, pounds
420	280
380	250
480	310
340	210
450	290
460	280
430	270
370	240
390	250

a. Fit the model $E(y) = \beta_0 + \beta_1 x$ to the data.
b. Construct a 95% prediction interval for the dressed weight y of a 300-pound steer.
c. Would you recommend that the FDA use the interval obtained in part b to determine whether the dressed weight of 150 pounds is a reasonable amount to receive from a 300-pound steer? Explain.

6.10　Refer to the FTC cigarette data of Example 6.3. The data of Table 6.1 are reproduced at the top of page 344 for convenience.
a. Fit the model $E(y) = \beta_0 + \beta_1 x_1$ to the data. Is there evidence that tar content x_1 is useful for predicting carbon monoxide content y?
b. Fit the model $E(y) = \beta_0 + \beta_2 x_2$ to the data. Is there evidence that nicotine content x_2 is useful for predicting carbon monoxide content y?
c. Fit the model $E(y) = \beta_0 + \beta_3 x_3$ to the data. Is there evidence that weight x_3 is useful for predicting carbon monoxide content y?
d. Compare the signs of $\hat{\beta}_1$, $\hat{\beta}_2$, and $\hat{\beta}_3$ in the models of parts a, b, and c, respectively, to the signs of the β's in the multiple regression model fit in Example 6.3. The fact that the β's change dramatically when the independent variables are removed from the model is another indication of a serious multicollinearity problem.

BRAND	TAR x_1, milligrams	NICOTINE x_2, milligrams	WEIGHT x_3, grams	CARBON MONOXIDE y, milligrams
Alpine	14.1	.86	.9853	13.6
Benson & Hedges	16.0	1.06	1.0938	16.6
Bull Durham	29.8	2.03	1.1650	23.5
Camel Lights	8.0	.67	.9280	10.2
Carlton	4.1	.40	.9462	5.4
Chesterfield	15.0	1.04	.8885	15.0
Golden Lights	8.8	.76	1.0267	9.0
Kent	12.4	.95	.9225	12.3
Kool	16.6	1.12	.9372	16.3
L&M	14.9	1.02	.8858	15.4
Lark Lights	13.7	1.01	.9643	13.0
Marlboro	15.1	.90	.9316	14.4
Merit	7.8	.57	.9705	10.0
Multifilter	11.4	.78	1.1240	10.2
Newport Lights	9.0	.74	.8517	9.5
Now	1.0	.13	.7851	1.5
Old Gold	17.0	1.26	.9186	18.5
Pall Mall Light	12.8	1.08	1.0395	12.6
Raleigh	15.8	.96	.9573	17.5
Salem Ultra	4.5	.42	.9106	4.9
Tareyton	14.5	1.01	1.0070	15.9
True	7.3	.61	.9806	8.5
Viceroy Rich Lights	8.6	.69	.9693	10.6
Virginia Slims	15.2	1.02	.9496	13.9
Winston Lights	12.0	.82	1.1184	14.9

Source: Federal Trade Commission.

6.11   An economist wants to model annual per capita demand, y, for passenger car motor fuel in the United States as a function of the two quantitative independent variables, average real weekly earnings (x_1)

YEAR	PER CAPITA CONSUMPTION OF MOTOR FUEL y, gallons	AVERAGE REAL WEEKLY EARNINGS x_1, 1967 dollars	AVERAGE PRICE OF GASOLINE x_2, dollars
1965	258.88	101.01	.32
1966	271.16	101.67	.33
1967	277.35	101.84	.34
1968	291.58	103.39	.34
1969	308.09	104.38	.36
1970	320.82	103.04	.36
1971	335.63	104.96	.36
1972	350.17	109.26	.37
1973	368.10	109.23	.40
1974	346.89	104.78	.53
1975	354.17	101.45	.57
1976	361.47	102.90	.59
1977	366.49	104.13	.62
1978	376.46	104.30	.63
1979	356.29	101.02	.86
1980	323.67	95.18	1.19

Source: Per capita consumption and average price of gasoline from *Statistical Abstract of the United States*. Average real weekly earnings from *Employment and Earnings*, U.S. Department of Labor, Bureau of Labor Statistics, Oct. 1983, p. 109.

and average price of regular gasoline (x_2). Data on these three variables for the years 1965–1980 are shown in the table. Suppose the economist fits the model $E(y) = \beta_0 + \beta_1 x_1 + \beta_2 x_2$ to the data for the years 1965–1979. Would you recommend that the economist use the least squares prediction equation to predict per capita consumption of motor fuel in 1980? Explain.

6.12 A firm wants to use multiple regression in a cost analysis of its shipping department. Since most of the costs incurred by shipping result from direct labor, the firm will model weekly hours of labor (y) as a function of total weight shipped (x_1), percentage of units shipped by truck (x_2), and average weight per shipment (x_3). Data collected for a 20-week period are shown in the table.

WEEK	HOURS OF LABOR	THOUSANDS OF POUNDS SHIPPED	PERCENTAGE OF UNITS SHIPPED BY TRUCK	AVERAGE NUMBER OF POUNDS PER SHIPMENT
	y	x_1	x_2	x_3
1	100	5.1	90	20
2	85	3.8	99	22
3	108	5.3	58	19
4	116	7.5	16	15
5	92	4.5	54	20
6	63	3.3	42	26
7	79	5.3	12	25
8	101	5.9	32	21
9	88	4.0	56	24
10	71	4.2	64	29
11	122	6.8	78	10
12	85	3.9	90	30
13	50	3.8	74	28
14	114	7.5	89	14
15	104	4.5	90	21
16	111	6.0	40	20
17	110	8.1	55	16
18	100	2.9	64	19
19	82	4.0	35	23
20	85	4.8	58	25

The SAS computer printout for the model $E(y) = \beta_0 + \beta_1 x_1 + \beta_2 x_2 + \beta_3 x_3$ is shown here. The firm is concerned about the problems that occur in regression analysis when multicollinearity is present. Examine the SAS printout. Do you detect any signs of multicollinearity?

```
DEP VARIABLE: LABOR                        ANALYSIS OF VARIANCE

                             SUM OF          MEAN
               SOURCE   DF   SQUARES        SQUARE      F VALUE    PROB>F

               MODEL    3    5158.31383    1719.43794    17.866    0.0001
               ERROR   16    1539.88617    96.24288576
               C TOTAL 19    6698.20000

               ROOT MSE      9.810346     R-SQUARE    0.7701
               DEP MEAN     93.3          ADJ R-SQ    0.7270
               C.V.        10.51484

                                PARAMETER ESTIMATES

                     PARAMETER    STANDARD    T FOR HO:                              VARIANCE
    VARIABLE  DF     ESTIMATE      ERROR    PARAMETER=0   PROB > :T:    TOLERANCE    INFLATION

    INTERCEP   1     131.92425   25.69321439    5.135       0.0001          .            0
    WEIGHT     1       2.72608977  2.27500488    1.198       0.2483     0.44435417   2.25045709
    TRUCK      1       0.04721841  0.09334856    0.506       0.6199     0.91496179   1.09294182
    AVGSHIP    1      -2.58744391  0.64281819   -4.025       0.0010     0.46162372   2.16626650
```

6.13   How many levels of x are required to fit the model $E(y) = \beta_0 + \beta_1 x + \beta_2 x^2$? How large a sample size is required to have sufficient degrees of freedom for estimating σ^2?

6.14   How many levels of x_1 and x_2 are required to fit the model $E(y) = \beta_0 + \beta_1 x_1 + \beta_2 x_2 + \beta_3 x_1 x_2$? How large a sample size is required to have sufficient degrees of freedom for estimating σ^2?

6.15   How many levels of x_1 and x_2 are required to fit the model $E(y) = \beta_0 + \beta_1 x_1 + \beta_2 x_2 + \beta_3 x_1 x_2 + \beta_4 x_1^2 + \beta_5 x_2^2$? How large a sample is required to have sufficient degrees of freedom for estimating σ^2?

6.16   A physiologist wanted to investigate the relationship between the physical characteristics of preadolescent boys and their maximal oxygen uptake (measured in milliliters of oxygen per kilogram of body weight). The data shown in the table were collected on a random sample of 10 preadolescent boys. As a first step in the data analysis, the researcher fit the regression model

$$y = \beta_0 + \beta_1 x_1 + \beta_2 x_2 + \beta_3 x_3 + \beta_4 x_4 + \varepsilon$$

to the data. The output for a SAS regression analysis is given.

MAXIMAL OXYGEN UPTAKE y	AGE x_1, years	HEIGHT x_2, centimeters	WEIGHT x_3, kilograms	CHEST DEPTH x_4, centimeters
1.54	8.4	132.0	29.1	14.4
1.74	8.7	135.5	29.7	14.5
1.32	8.9	127.7	28.4	14.0
1.50	9.9	131.1	28.8	14.2
1.46	9.0	130.0	25.9	13.6
1.35	7.7	127.6	27.6	13.9
1.53	7.3	129.9	29.0	14.0
1.71	9.9	138.1	33.6	14.6
1.27	9.3	126.6	27.7	13.9
1.50	8.1	131.8	30.8	14.5

```
Dependent Variable: Y

                        Analysis of Variance

                        Sum of          Mean
Source         DF       Squares        Square      F Value     Prob>F

Model          4        0.20604        0.05151      37.204      0.0007
Error          5        0.00692        0.00138
C Total        9        0.21296

        Root MSE      0.03721     R-square      0.9675
        Dep Mean      1.49200     Adj R-sq      0.9415
        C.V.          2.49391

                        Parameter Estimates

              Parameter      Standard     T for H0:
Variable  DF   Estimate        Error    Parameter=0    Prob > |T|

INTERCEP   1   -4.774739     0.86281773    -5.534        0.0026
X1         1   -0.035214     0.01538630    -2.289        0.0708
X2         1    0.051637     0.00621522     8.308        0.0004
X3         1   -0.023417     0.01342835    -1.744        0.1416
X4         1    0.034489     0.08523877     0.405        0.7025
```

a. Is the model adequate for predicting maximal oxygen uptake?

b. It seems reasonable to assume that the greater a child's chest depth, the greater should be the maximal oxygen uptake. But note that $\hat{\beta}_4$, the estimated coefficient of chest depth, x_4, is negative. Give an explanation for this result.

c. It would seem that the weight of a child should be positively correlated to lung volume and hence to maximal oxygen uptake. Can you explain the small t value associated with $\hat{\beta}_3$?

d. Calculate the coefficient of correlation r for each pair of independent variables. Does this information confirm your suspicions in parts b and c?

6.17 Consider the data shown in the table.

x	54	42	28	38	25	70	48	41	20	52	65
y	6	16	33	18	41	3	10	14	45	9	5

a. Plot the points on a scattergram. What type of relationship appears to exist between x and y?

b. For each observation calculate log x and log y. Plot the log-transformed data points on a scattergram. What type of relationship appears to exist between log x and log y?

c. The scattergram from part b suggests that the transformed model

$$\log y = \beta_0 + \beta_1 \log x + \varepsilon$$

may be appropriate. Fit the transformed model to the data. Is the model adequate? Test using $\alpha = .05$.

d. Use the transformed model to predict the value of y when $x = 30$. [*Hint:* Use the inverse transformation $y = e^{\log y}$.]

6.18 Hamilton (1987) illustrated the multicollinearity problem with an example using the data shown in the accompanying table. The values of x_1, x_2, and y in the table represent appraised land value, appraised improvements value, and sale price, respectively, of a randomly selected residential property. (All measurements are in thousands of dollars.)

x_1	x_2	y	x_1	x_2	y
22.3	96.6	123.7	30.4	77.1	128.6
25.7	89.4	126.6	32.6	51.1	108.4
38.7	44.0	120.0	33.9	50.5	112.0
31.0	66.4	119.3	23.5	85.1	115.6
33.9	49.1	110.6	27.6	65.9	108.3
28.3	85.2	130.3	39.0	49.0	126.3
30.2	80.4	131.3	31.6	69.6	124.6
21.4	90.5	114.4			

Source: Hamilton, D. "Sometimes $R^2 > r_{yx_1}^2 + r_{yx_2}^2$: Correlated variables are not always redundant." *The American Statistician*, Vol. 41, No. 2, May 1987, pp. 129–132.

a. Calculate the coefficient of correlation between y and x_1. Is there evidence of a linear relationship between sale price and appraised land value?

b. Calculate the coefficient of correlation between y and x_2. Is there evidence of a linear relationship between sale price and appraised improvements?

c. Based on the results in parts a and b, do you think the model $E(y) = \beta_0 + \beta_1 x_1 + \beta_2 x_2$ will be useful for predicting sale price?

d. Use a statistical computer software package to fit the model in part c, and conduct a test of model adequacy. In particular, note the value of R^2. Does the result agree with your answer to part **c**?

e. Calculate the coefficient of correlation between x_1 and x_2. What does the result imply?

f. Many researchers avoid the problems of multicollinearity by always omitting all but one of the "redundant" variables from the model. Would you recommend this strategy for this example? Explain. (Hamilton notes that, in this case, such a strategy "can amount to throwing out the baby with the bathwater.")

6.19   Neil A. Palomba used multiple regression to relate the strike activity in a state (the percentage y of total working hours lost due to strikes) to three independent variables:

x_1 = Percentage of union members in nonagricultural establishments

x_2 = Percentage of all nonagricultural employment that is manufacturing

x_3 = Hourly earnings of workers on manufacturing payrolls

The data for the analysis are reproduced in the table.

STATE	y	x_1	x_2	x_3	STATE	y	x_1	x_2	x_3
Alabama	.14	18.0	30.5	2.17	Nebraska	.05	19.3	16.6	2.36
Alaska	.11	32.2	8.9	3.54	Nevada	.36	32.8	4.6	3.16
Arizona	.09	20.08	15.3	2.72	New Hampshire	.03	20.9	40.9	2.00
Arkansas	.10	26.2	29.2	1.78	New Jersey	.27	37.7	37.1	2.67
California	.16	33.8	24.9	2.96	New Mexico	.09	13.4	6.8	2.29
Colorado	.04	21.6	15.8	2.74	New York	.11	39.4	28.2	2.60
Connecticut	.08	24.6	42.5	2.62	North Carolina	.01	6.7	41.6	1.75
Delaware	.41	21.5	36.1	2.65	North Dakota	.03	14.8	5.8	2.28
Florida	.20	13.1	15.5	2.11	Ohio	.38	35.7	39.0	2.91
Georgia	.13	12.7	31.8	1.92	Oklahoma	.01	13.7	15.5	2.35
Hawaii	.02	24.2	12.1	2.14	Oregon	.12	34.8	26.5	2.85
Idaho	.11	19.2	18.8	2.50	Pennsylvania	.14	38.4	37.8	2.55
Illinois	.18	37.9	33.5	2.76	Rhode Island	.09	29.6	38.2	2.11
Indiana	.16	34.1	40.8	2.81	South Carolina	.01	7.9	42.7	1.80
Iowa	.16	20.8	25.4	2.71	South Dakota	.16	9.5	8.8	2.34
Kansas	.11	18.8	20.6	2.65	Tennessee	.23	17.6	34.6	2.03
Kentucky	.17	25.7	26.6	2.43	Texas	.06	13.3	19.4	2.42
Louisiana	.10	17.1	17.8	2.49	Utah	.66	19.7	17.6	2.77
Maine	.15	20.3	36.6	2.00	Vermont	.26	19.3	30.9	2.08
Massachusetts	.07	29.1	33.0	2.37	Virginia	.04	15.5	26.5	2.04
Michigan	.83	38.9	40.7	3.11	Washington	.16	43.1	25.6	2.98
Minnesota	.02	33.0	24.0	2.64	West Virginia	.45	42.0	27.4	2.67
Mississippi	.14	11.6	30.5	1.76	Wisconsin	.21	31.5	37.0	2.66
Missouri	.14	39.8	28.5	2.53	Wyoming	.01	19.2	7.7	2.82
Montana	.28	36.2	12.2	2.71	Maryland	—	—	—	—

Source: Palomba, N. A. "Strike activity and union membership: An empirical approach." *University of Washington Business Review,* Winter 1969.

a. Palomba initially considered the straight-line model

$$y = \beta_0 + \beta_1 x_1 + \varepsilon$$

Fit the model to the data and give your opinion on the adequacy of the model. In particular, interpret the value of $\hat{\beta}_1$.

b. Now consider the first-order model

$$y = \beta_0 + \beta_1 x_1 + \beta_2 x_2 + \beta_3 x_3 + \varepsilon$$

Fit the model to the data using an available multiple regression package, and give your opinion on the adequacy of the model.

c. Refer to part **b**. Conduct a test to determine whether the percentage of union members x_1 has a positive effect on the level of strike activity (when the other independent variables, x_2 and x_3 are held constant). Does this result contradict your result in part **a**? Explain.

d. Calculate the variance inflation factor (VIF) for β_1. [*Hint:* You will need to fit the model $x_1 = \alpha_0 + \alpha_1 x_2 + \alpha_2 x_3 + \varepsilon$.] Interpret your result.

6.20 To model the relationship between y, a dependent variable, and x, an independent variable, a researcher has taken one measurement on y at each of five different x values. Drawing on his mathematical expertise, the researcher realizes that he can fit the fourth-order polynomial model

$$E(y) = \beta_0 + \beta_1 x + \beta_2 x^2 + \beta_3 x^3 + \beta_4 x^4$$

and it will pass exactly through all five points, yielding SSE = 0. The researcher, delighted with the "excellent" fit of the model, eagerly sets out to use it to make inferences. What problems will he encounter in attempting to make inferences?

References

Draper, N. and Smith, H. *Applied Regression Analysis*, 2nd ed. New York: Wiley, 1981.

Montgomery, D. C. and Peck, E. A. *Introduction to Linear Regression Analysis*. New York: Wiley, 1982.

Mosteller, F. and Tukey, J. W. *Data Analysis and Regression: A Second Course in Statistics*. Reading, Mass.: Addison-Wesley, 1977.

Neter, J., Wasserman, W., and Kutner, M. H. *Applied Linear Statistical Models*. 3rd ed. Homewood, Ill.: Richard D. Irwin, 1990.

Residual Analysis

CONTENTS

7.1 Introduction

7.2 Plotting Residuals and Detecting Lack of Fit

7.3 Detecting Unequal Variances

7.4 Checking the Normality Assumption

7.5 Detecting Outliers and Identifying Influential Observations

7.6 Detecting Residual Correlation: The Durbin–Watson Test

OBJECTIVE

To show how residuals can be used to detect departures from the model assumptions and to suggest some procedures for coping with these problems

7.1
Introduction

We have repeatedly stated that the validity of many of the inferences associated with a regression analysis depends on the error term, ε, satisfying certain assumptions. Thus, when we test a hypothesis about a regression coefficient or a set of regression coefficients, or when we form a prediction interval for a future value of y, we must assume (1) that ε is normally distributed with a mean of 0, (2) constant variance σ^2, and (3) that all pairs of error terms are uncorrelated.* The objective of this chapter is to provide you with both graphical tools and statistical tests that will aid in checking the validity of these assumptions. In addition, these tools will help you evaluate the utility of the model and, in some cases, may suggest modifications to the model that will allow us to better describe the mean response.

In Section 7.2 we will show how to plot the residuals to reveal model inadequacies. The use of these plots and a simple test to detect unequal variances at different levels of the independent variable(s) is presented in Section 7.3. A graphical analysis of residuals for checking the normality assumption is the topic of Section 7.4. In Section 7.5 residual plots are used to detect outliers, i.e., observations that are unusually large or small relative to the others; procedures for measuring the influence these outliers may have on the fitted regression model are also presented. Finally, we discuss the use of residuals to test for time series correlation of the error term in Section 7.6.

7.2
Plotting Residuals and Detecting Lack of Fit

The error term in a multiple regression model is, in general, not observable. To see this, consider the model

$$y = \beta_0 + \beta_1 x_1 + \cdots + \beta_k x_k + \varepsilon$$

and solve for the error term

$$\varepsilon = y - (\beta_0 + \beta_1 x_1 + \cdots + \beta_k x_k)$$

Although you will observe values of the dependent variable and the independent variables $x_1, x_2, \ldots, x_k$, you will not know the true values of the regression coefficients $\beta_0, \beta_1, \ldots, \beta_k$. Therefore, the exact value of ε cannot be calculated.

After the data have been used to obtain least squares estimates $\hat{\beta}_0, \hat{\beta}_1, \ldots, \hat{\beta}_k$ of the regression coefficients, we can estimate the value of ε associated with each y value using the corresponding **regression residual**, i.e., the deviation between the observed and the predicted value of y:

$$\hat{\varepsilon}_i = y_i - \hat{y}_i$$

To accomplish this, we must substitute the values of $x_1, x_2, \ldots, x_k$ into the prediction equation for each data point to obtain $\hat{y}$, and then this value must be subtracted from the observed value of y. Remember that you encountered the regression residual in Chapters 3 and 4. In particular, the least squares estimates

---

*We assumed (Section 4.2) that the random errors associated with the linear model were independent. If two random variables are independent, it follows (proof omitted) that they will be uncorrelated. The reverse is generally untrue, except for normally distributed random variables. If two normally distributed random variables are uncorrelated, it can be shown that they are also independent.

of $\beta_0, \beta_1, \beta_2, \ldots, \beta_k$ are those that minimize the sum of squares of the residuals,

$$\sum_{i=1}^{n} \hat{\varepsilon}_i^2 = \sum_{i=1}^{n} (y_i - \hat{y}_i)^2$$

Definition 7.1

The **regression residual** is the observed value of the dependent variable minus the predicted value, or

$$\hat{\varepsilon} = y - \hat{y} = y - (\hat{\beta}_0 + \hat{\beta}_1 x_1 + \cdots + \hat{\beta}_k x_k)$$

EXAMPLE 7.1

The data in Table 7.1 show the monthly usage of electricity and the size of the home for a sample of 10 homes (the same data were presented earlier in Table 4.1). Calculate the regression residuals for both the straight-line (first-order) model and the quadratic (second-order) model.

TABLE 7.1 **Data for Monthly Electrical Usage Example**

SIZE OF HOME x, square feet	MONTHLY USAGE y, kilowatt-hours
1,290	1,182
1,350	1,172
1,470	1,264
1,600	1,493
1,710	1,571
1,840	1,711
1,980	1,804
2,230	1,840
2,400	1,956
2,930	1,954

Solution

The SAS printout for the regression analysis of the first-order model,

$$y = \beta_0 + \beta_1 x + \varepsilon$$

is shown in Figure 7.1 on page 354. The least squares model is

$$\hat{y} = 578.928 + .540304x$$

Thus, the residual for the first observation, $x = 1{,}290$ and $y = 1{,}182$, is obtained by first calculating the predicted value

$$\hat{y} = 578.928 + .540304(1{,}290) = 1{,}275.92$$

and then subtracting from the observed value:

$$\hat{\varepsilon} = y - \hat{y} = 1{,}182 - 1{,}275.92 = -93.92$$

FIGURE 7.1   SAS printout for first-order model: Example 7.1

ANALYSIS OF VARIANCE

SOURCE	DF	SUM OF SQUARES	MEAN SQUARE	F VALUE	PROB>F
MODEL	1	703957.18	703957.18	39.536	0.0002
ERROR	8	142444.92	17805.61457		
C TOTAL	9	846402.10			

ROOT MSE	133.4377	R-SQUARE	0.8317	
DEP MEAN	1594.7	ADJ R-SQ	0.8107	
C.V.	8.367573			

PARAMETER ESTIMATES

VARIABLE	DF	PARAMETER ESTIMATE	STANDARD ERROR	T FOR H0: PARAMETER=0	PROB > |T|
INTERCEP	1	578.92775	166.96806	3.467	0.0085
X	1	0.54030439	0.08592981	6.288	0.0002

Similar calculations for the other nine observations produce the residuals shown in Table 7.2.

TABLE 7.2   **Regression Residuals for First-Order Model: Example 7.1**

x	y	$\hat{y}$	$\hat{\varepsilon} = y - \hat{y}$
1,290	1,182	1,275.92	-93.92
1,350	1,172	1,308.34	-136.34
1,470	1,264	1,373.18	-109.18
1,600	1,493	1,443.41	49.59
1,710	1,571	1,502.85	68.15
1,840	1,711	1,573.09	137.91
1,980	1,804	1,648.73	155.27
2,230	1,840	1,783.81	56.19
2,400	1,956	1,875.66	80.34
2,930	1,954	2,162.02	-208.02

The SAS printout for the second-order model

$$y = \beta_0 + \beta_1 x + \beta_2 x^2 + \varepsilon$$

is shown in Figure 7.2. The least squares model is

$$\hat{y} = -1{,}216.14 + 2.39893x - .00045004x^2$$

For the first observation, $x = 1{,}290$ and $y = 1{,}182$, the predicted electrical usage is

$$\hat{y} = -1{,}216.14 + 2.39893(1{,}290) - .00045004(1{,}290)^2$$
$$= 1{,}129.56*$$

*The residuals in Tables 7.2 and 7.3 have been generated using a computer program. Therefore, the results reported here will differ slightly from hand-calculated residuals because of rounding error.

FIGURE 7.2 SAS regression printout for second-order model: Example 7.1

ANALYSIS OF VARIANCE

SOURCE	DF	SUM OF SQUARES	MEAN SQUARE	F VALUE	PROB>F
MODEL	2	831069.55	415534.77	189.710	0.0001
ERROR	7	15332.55363	2190.36480		
C TOTAL	9	846402.10			

ROOT MSE	46.80133	R-SQUARE	0.9819	
DEP MEAN	1594.7	ADJ R-SQ	0.9767	
C.V.	2.934805			

PARAMETER ESTIMATES

VARIABLE	DF	PARAMETER ESTIMATE	STANDARD ERROR	T FOR H0: PARAMETER=0	PROB > |T|
INTERCEP	1	-1216.14389	242.80637	-5.009	0.0016
X	1	2.39893018	0.24583560	9.758	0.0001
XX	1	-0.000450040	0.000059077	-7.618	0.0001

and the regression residual is

$$\hat{\varepsilon} = y - \hat{y} = 1{,}182 - 1{,}129.56 = 52.44$$

All the regression residuals for the second-order model are given in Table 7.3.

TABLE 7.3 Regression Residuals for Second-Order Model: Example 7.1

x	y	$\hat{y}$	$\hat{\varepsilon} = y - \hat{y}$
1,290	1,182	1,129.56	52.44
1,350	1,172	1,202.21	−30.21
1,470	1,264	1,337.79	−73.79
1,600	1,493	1,470.04	22.96
1,710	1,571	1,570.06	.94
1,840	1,711	1,674.23	36.77
1,980	1,804	1,769.40	34.60
2,230	1,840	1,895.47	−55.47
2,400	1,956	1,949.06	6.94
2,930	1,954	1,949.17	4.83

Graphical displays of regression residuals are useful aids to their interpretation. For example, the regression residual can be plotted on the vertical axis against one of the independent variables on the horizontal axis, or against the predicted value $\hat{y}$ (which is a linear function of the independent variables). If the assumptions concerning the error term ε are satisfied, we would expect to see residual plots that have no trends, no dramatic increases or decreases in variability, and only a few residuals (about 5%) more than two estimated standard deviations ($2s$) of ε above or below 0. It is a property of the least squares prediction equation that the mean of the regression residuals will always be 0. That is, the least squares prediction equation not only minimizes the SSE, the sum of squared errors (residuals), but also produces residuals that have mean 0.

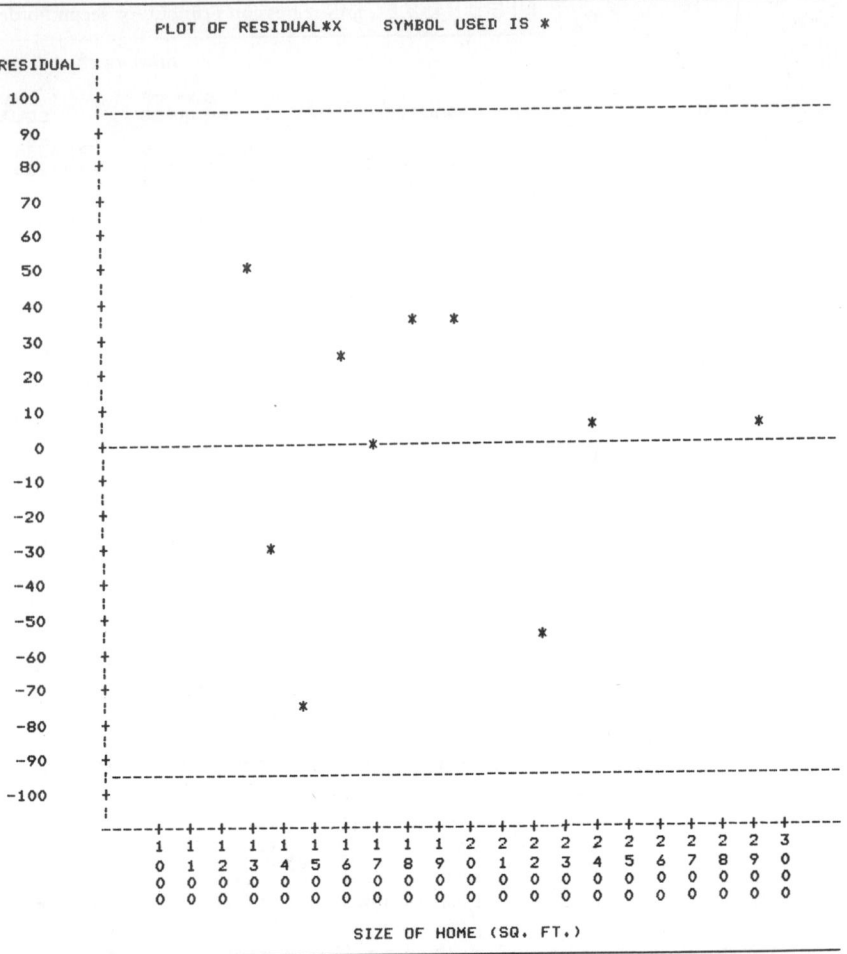

To illustrate, the residuals $\hat{\varepsilon}$ obtained by fitting the second-order model to the electrical usage data (Table 7.3) are plotted against the size of the home, x, in Figure 7.3. The middle, upper, and lower horizontal lines in Figure 7.3 locate the mean (0), $+2s$, and $-2s$, respectively, for the residuals. We detect no distinctive patterns or trends in this plot. All the residuals lie within $2s$ of the mean (0), and the variability around the mean is consistent for small and large homes.

EXAMPLE 7.2

Example 7.1 gives the residuals obtained from fitting a first-order model to the electrical usage data. Plot the residuals obtained in this regression analysis against the home size, x (x is recorded on the horizontal axis). Does the plot suggest model inadequacy or departure from the usual assumptions made about the error term ε?

Solution

The residuals for the first-order model were given in Table 7.2. The plot of these residuals versus home size is shown in Figure 7.4. The distinctive aspect of this

FIGURE 7.4

SAS plot of regression residuals for the first-order model: Electrical usage example

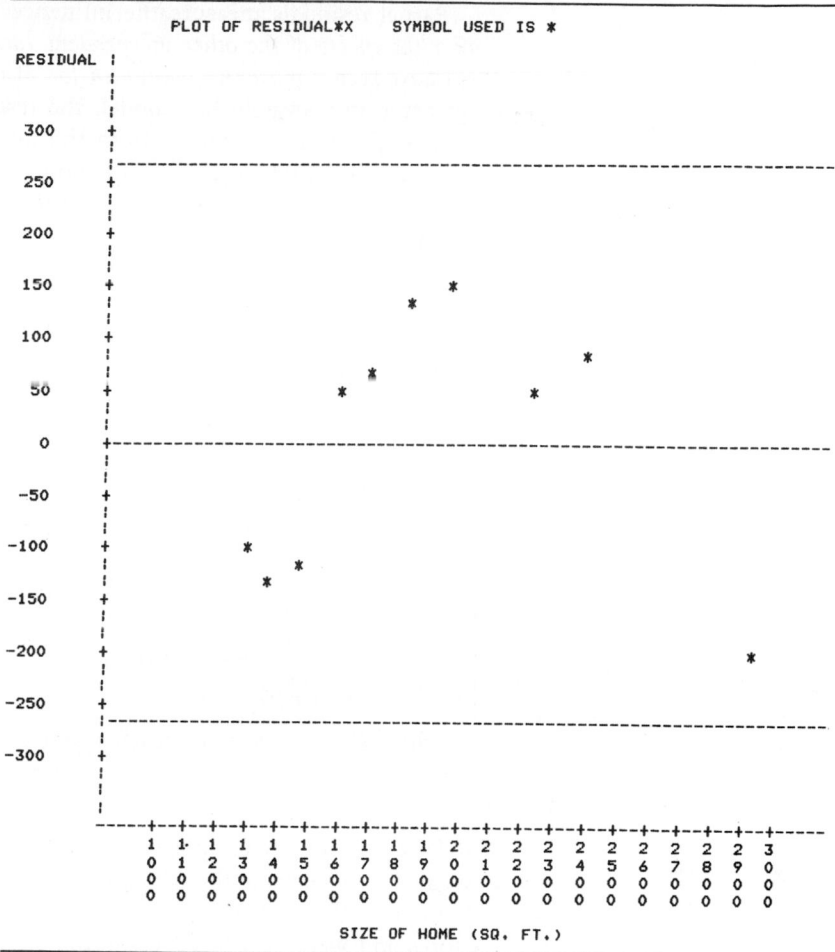

plot is the parabolic distribution of the residuals about their mean, i.e., all residuals tend to be positive for the homes of intermediate size and negative for the homes of either small or large size. This parabolic appearance of the trend in the residuals suggests that a second-order term may improve the model. As we know, the addition of a second-order term does improve the model, and the residual plot for the second-order model no longer shows an observable pattern (see Figure 7.3).

An alternative method of detecting lack of fit in models with more than one independent variable is to construct a partial residual plot. The **partial residuals** for the jth independent variable, x_j, in the model are calculated as follows:

$$\hat{\varepsilon}^* = y - (\hat{\beta}_0 + \hat{\beta}_1 x_1 + \hat{\beta}_2 x_2 + \cdots + \hat{\beta}_{j-1} x_{j-1} + \hat{\beta}_{j+1} x_{j+1} + \cdots + \hat{\beta}_k x_k)$$
$$= \hat{\varepsilon} + \hat{\beta}_j x_j$$

where $\hat{\varepsilon}$ is the usual regression residual.

Partial residuals measure the influence of x_j on the independent variable y *after the effects of the other independent variables* $(x_1, x_2, \ldots, x_{j-1}, x_{j+1}, \ldots, x_k)$ *have been removed or accounted for.* If the partial residuals $\hat{\varepsilon}^*$ are regressed against x_j in a straight-line model, the resulting least squares slope is equal to $\hat{\beta}_j$—the β estimate obtained from the full model. Therefore, when the partial residuals are plotted against x_j, the points are scattered around a line with slope equal to $\hat{\beta}_j$. Unusual deviations or patterns around this line indicate lack of fit for the variable x_j.

A plot of the partial residuals versus x_j often reveals more information about the relationship between y and x_j than the usual residual plot. In particular, a partial residual plot usually indicates more precisely how to modify the model,[†] as the next example illustrates.

Definition 7.2

The set of **partial regression residuals** for the jth independent variable x_j are calculated as follows:

$$\hat{\varepsilon}^* = y - (\hat{\beta}_0 + \hat{\beta}_1 x_1 + \hat{\beta}_2 x_2 + \cdots$$
$$+ \hat{\beta}_{j-1} x_{j-1} + \hat{\beta}_{j+1} x_{j+1} + \cdots + \hat{\beta}_k x_k)$$
$$= \hat{\varepsilon} + \hat{\beta}_j x_j$$

where $\hat{\varepsilon} = y - \hat{y}$ is the usual regression residual (see Definition 7.1).

EXAMPLE 7.3

Refer to the supermarket chain experiment in Example 6.4. Recall that the chain wanted to investigate the effect of price x_1 on the weekly demand y for a house brand of coffee at eight of its stores. Eight prices were randomly assigned to the stores and were advertised using the same procedures. A few weeks later, the chain conducted the same experiment using no advertisements. The data for the entire study are shown in Table 7.4.

Consider the model

$$E(y) = \beta_0 + \beta_1 p + \beta_2 x_2$$

where

$$x_2 = \begin{cases} 1 & \text{if advertisement used} \\ 0 & \text{if not} \end{cases}$$

a. Fit the model to the data. Is the model adequate for predicting weekly demand y?

b. Calculate and plot the residuals versus p. Do you detect any trends?

[†]Partial residual plots display the correct functional form of the predictor variables across the relevant range of interest, except in cases where severe multicollinearity exists. See Mansfield and Conerly (1987) for an excellent discussion of the use of residual and partial residual plots.

TABLE 7.4 **Data for Example 7.3**

WEEKLY DEMAND y, pounds	PRICE p, dollars per pound	ADVERTISEMENT x_2
1,120	3.00	1
999	3.10	1
932	3.20	1
884	3.30	1
807	3.40	1
760	3.50	1
701	3.60	1
688	3.70	1
1,037	3.00	0
962	3.10	0
904	3.20	0
827	3.30	0
775	3.40	0
715	3.50	0
666	3.60	0
607	3.70	0

c. Calculate the partial residuals for the independent variable p, and construct the corresponding partial residual plot. What does the plot reveal?

d. Fit the model $E(y) = \beta_0 + \beta_1 x_1 + \beta_2 x_2$, where $x_1 = 1/p$. Has the predictive ability of the model improved?

Solution

a. The SAS printout for the regression analysis is shown in Figure 7.5 (page 360). The F value for testing model adequacy, i.e., H_0: $\beta_1 = \beta_2 = 0$, is given on the printout (shaded) as $F = 368.298$ with a corresponding p-value (also shaded) of $p = .0001$. Thus, there is sufficient evidence (at any α of .0001 or greater) that the model contributes information for the prediction of weekly demand, y. Also, the coefficient of determination is $R^2 = .9827$, meaning that the model explains approximately 98% of the sample variation in weekly demand.

Recall from Example 6.4, however, that we fit a model with the transformed independent variable $x_1 = 1/p$. That is, we expect the relationship between weekly demand y and price p to be decreasing in a curvilinear fashion and approaching (but never reaching) 0 as p increases. (See curve 1 in Figure 6.8.) If such a relationship exists, the model (with untransformed price), although statistically useful for predicting demand y, will be inadequate in a practical setting.

b. The regression residuals for the model in part **a** are shown in the bottom portion of Figure 7.5. Figure 7.6 (page 361) is a computer-generated (SAS) plot of these residuals against price p. Notice that the plot reveals a parabolic trend, implying a lack of fit. Thus, the residual plot supports our hypothesis that the weekly demand–price relationship is curvilinear, not linear. However, the appropriate transformation on price (i.e., $1/p$) is not evident from the plot. In fact, the nature of the curvature in Figure 7.6 may lead you to conclude

F I G U R E 7.5   SAS printout for Example 7.3

ANALYSIS OF VARIANCE

SOURCE	DF	SUM OF SQUARES	MEAN SQUARE	F VALUE	PROB>F
MODEL	2	320515.30	160257.65	368.298	0.0001
ERROR	13	5656.70238	435.13095		
C TOTAL	15	326172.00			

ROOT MSE	20.85979	R-SQUARE	0.9827	
DEP MEAN	836.5	ADJ R-SQ	0.9800	
C.V.	2.493699			

PARAMETER ESTIMATES

VARIABLE	DF	PARAMETER ESTIMATE	STANDARD ERROR	T FOR H0: PARAMETER=0	PROB > |T|
INTERCEP	1	2848.74405	76.60151888	37.189	0.0001
P	1	-608.09524	22.75989979	-26.718	0.0001
X2	1	49.75000000	10.42989636	4.770	0.0004

OBS	ID	ACTUAL	PREDICT VALUE	RESIDUAL
1	3	1120.0	1074.2	45.7917
2	3.1	999.0	1013.4	-14.3988
3	3.2	932.0	952.6	-20.5893
4	3.3	884.0	891.8	-7.7798
5	3.4	807.0	831.0	-23.9702
6	3.5	760.0	770.2	-10.1607
7	3.6	701.0	709.4	-8.3512
8	3.7	688.0	648.5	39.4583
9	3	1037.0	1024.5	12.5417
10	3.1	962.0	963.6	-1.6488
11	3.2	904.0	902.8	1.1607
12	3.3	827.0	842.0	-15.0298
13	3.4	775.0	781.2	-6.2202
14	3.5	715.0	720.4	-5.4107
15	3.6	666.0	659.6	6.3988
16	3.7	607.0	598.8	8.2083

SUM OF RESIDUALS   8.41283E-12
SUM OF SQUARED RESIDUALS   5656.702

that the addition of the quadratic term, $\beta_3 p^2$, to the model will solve the problem. In general, a residual plot will detect curvature if it exists, but may not reveal the appropriate transformation.

c. The partial residuals (denoted $\hat{\varepsilon}^*$) for the independent variable p are calculated using the formula given in Definition 7.2:

$$\hat{\varepsilon}^* = \hat{\varepsilon} + \hat{\beta}_1 p$$

where $\hat{\beta}_1 = -608.1$ (from Figure 7.5). For example, the partial residual for the first observation is calculated by substituting $\hat{\varepsilon} = 45.79$ and $p = 3.00$ into the equation:

$$\hat{\varepsilon}^* = \hat{\varepsilon} - 608.1p$$
$$= 45.79 - 608.1(3) = -1,778.5$$

The complete set of partial residuals for p is shown in Table 7.5, accompanied by a computer-generated (SAS) partial residual plot in Figure 7.7 (page 362).

FIGURE 7.6   SAS printout of residuals against price for Example 7.3

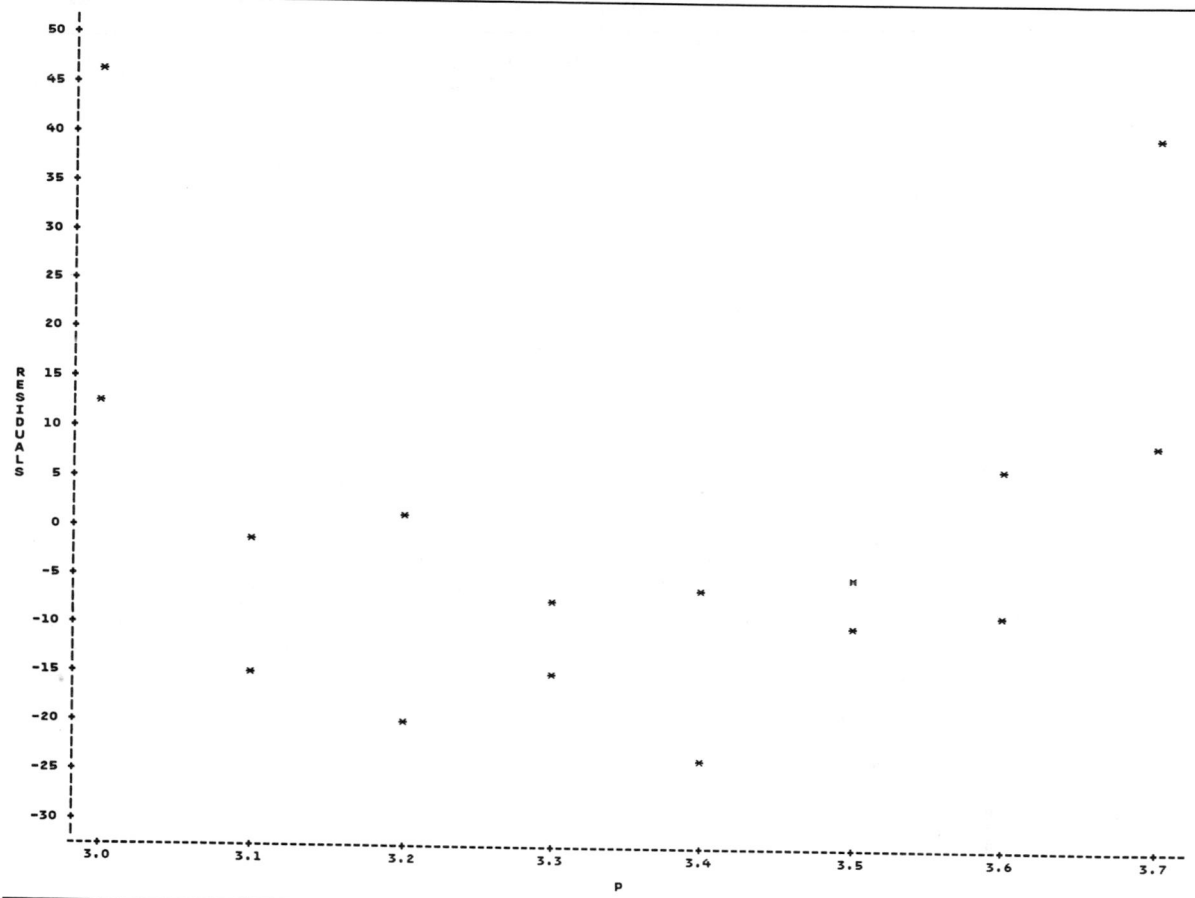

TABLE 7.5   Partial Residuals for
Price p in First-Order Model for Example 7.3

PRICE p	RESIDUAL $\hat{\varepsilon}$	PARTIAL RESIDUAL $\hat{\varepsilon}^*$	PRICE p	RESIDUAL $\hat{\varepsilon}$	PARTIAL RESIDUAL $\hat{\varepsilon}^*$
3.0	45.792	−1,778.5	3.0	12.542	−1,811.7
3.1	−14.399	−1,899.5	3.1	−1.649	−1,886.7
3.2	−20.589	−1,966.5	3.2	1.161	−1,944.7
3.3	−7.780	−2,014.5	3.3	−15.030	−2,021.7
3.4	−23.970	−2,091.5	3.4	−6.220	−2,073.7
3.5	−10.161	−2,138.5	3.5	−5.411	−2,133.7
3.6	−8.351	−2,197.5	3.6	6.399	2,182.7
3.7	39.458	−2,210.5	3.7	8.208	−2,241.7

FIGURE 7.7   Printout of partial residuals against price for Example 7.3

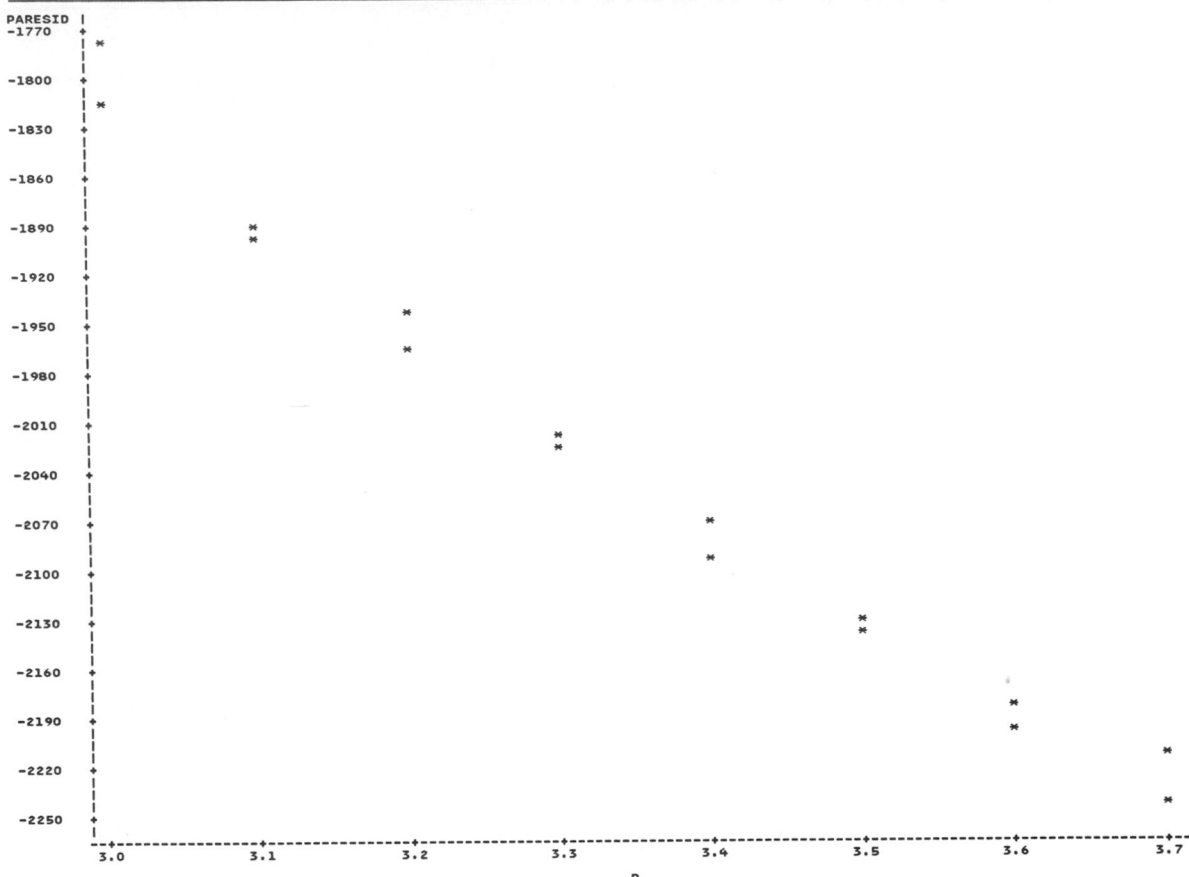

You can see that the partial residual plot also reveals a curvilinear trend but, in addition, displays the correct functional form of weekly demand–price relationship. Notice that the curve is decreasing and approaching (but never reaching) 0 as p increases. This suggests that the appropriate transformation on price is either $1/p$ or e^{-p} (see Figure 6.8).

   d. Using the transformation $x_1 = 1/p$, we refit the model to the data, and the resulting SAS printout is shown in Figure 7.8. The small p-value ($p = .0001$) for testing $H_0: \beta_1 = \beta_2 = 0$ indicates that the model is adequate for predicting y. Although the coefficient of determination increased only slightly (from $R^2 = .98$ to $R^2 = .99$), the model standard deviation (**ROOT MSE**) decreased significantly (from $s = 20.9$ to $s = 14.6$). Thus, whereas the model with untransformed price can predict weekly demand for coffee to within $2s = 2(20.9) = 41.8$ pounds, the transformed model can predict demand to within $2(14.6) = 29.2$ pounds.

FIGURE 7.8   SAS printout for Example 7.3

ANALYSIS OF VARIANCE

SOURCE	DF	SUM OF SQUARES	MEAN SQUARE	F VALUE	PROB>F
MODEL	2	323412.38	161706.19	761.765	0.0001
ERROR	13	2759.61908	212.27839		
C TOTAL	15	326172.00			

ROOT MSE	14.56978	R-SQUARE	0.9915	
DEP MEAN	836.5	ADJ R-SQ	0.9902	
C.V.	1.741755			

PARAMETER ESTIMATES

VARIABLE	DF	PARAMETER ESTIMATE	STANDARD ERROR	T FOR H0: PARAMETER=0	PROB > |T|
INTERCEP	1	-1223.99520	57.21806000	22.399	0.0001
X1	1	6787.31169	176.61334	38.430	0.0001
X2	1	49.75000000	7.28488832	6.829	0.0001

Residual (or partial residual) plots are useful for indicating potential model improvements, but they are no substitute for formal statistical tests of model terms to determine their importance. Thus, a true test of whether the second-order term contributes to the electrical usage model (Example 7.1) is the t test of the null hypothesis $H_0: \beta_2 = 0$. The appropriate test statistic, shown in the printout of Figure 7.2, indicates that the second-order term does contribute information for the prediction of electrical usage y. We have confidence in this statistical inference because we know the probability α of committing a Type I error (concluding a term is important when, in fact, it is not). In contrast, decisions based on residual plots are subjective, and their reliability cannot be measured. Therefore, we suggest that such plots only be used as indicators of *potential* problems. The final judgment on model adequacy should be based on appropriate statistical tests.*

EXERCISES

7.1   A first-order model is fit to the data shown in the table with the following result:

$$\hat{y} = 2.588 + .541x$$

x	−2	−2	−1	−1	0	0	1	1	2	2	3	3
y	1.1	1.3	2.0	2.1	2.7	2.8	3.4	3.6	4.0	3.9	3.8	3.6

---

*A more general procedure for determining whether the straight-line model adequately fits the data tests the null hypothesis $H_0: E(y) = \beta_0 + \beta_1 x$ against the alternative $H_a: E(y) \neq \beta_0 + \beta_1 x$. You can see that this test, called a test for *lack of fit*, does not restrict the alternative hypothesis to second-order models. Lack-of-fit tests are appropriate when the x values are replicated, i.e., when the sample data include two or more observations for several different levels of x. When the data are observational, however, replication rarely occurs. (Note that none of the values of x are repeated in Table 7.1.) For details on how to conduct tests for lack of fit, consult the references given at the end of this chapter.

a. Calculate the residuals for the model.

b. Plot the residuals versus x. Do you detect any trends? If so, what does the pattern suggest about the model?

7.2 A first-order model is fit to the data shown in the table with the following result:

$$\hat{y} = -3.179 + 2.491x$$

x	2	4	7	10	12	15	18	20	21	25
y	5	10	12	22	25	27	39	50	47	65

a. Calculate the residuals for the model.

b. Plot the residuals versus x. Do you detect any trends? If so, what does the pattern suggest about the model?

7.3 Refer to Example 3.2. Recall that a manufacturer of a new tire tested the tire for wear at different pressures with the results shown in the table.

PRESSURE x, pounds per sq. inch	MILEAGE y, thousands
30	29.5
30	30.2
31	32.1
31	34.5
32	36.3
32	35.0
33	38.2
33	37.6
34	37.7
34	36.1
35	33.6
35	34.2
36	26.8
36	27.4

a. Fit the straight-line model $y = \beta_0 + \beta_1 x + \varepsilon$ to the data.

b. Calculate the residuals for the model.

c. Plot the residuals versus x. Do you detect any trends? If so, what does the pattern suggest about the model?

d. Fit the quadratic model $y = \beta_0 + \beta_1 x + \beta_2 x^2 + \varepsilon$ to the data using an available statistical software package. Has the addition of the quadratic term improved model adequacy?

7.4 A recent EPA gas mileage guide gives the engine size and estimated city miles per gallon ratings for the 11 gasoline-fueled subcompact and compact cars shown in the table. (The engine sizes are in total cubic inches of cylinder volume.) To predict gas mileage from the engine size of subcompact and compact cars, the first-order model

$$y = \beta_0 + \beta_1 x + \varepsilon$$

is fit to the data. The resulting least squares model is $\hat{y} = 37.677 - .0724x$.

CAR	CYLINDER VOLUME	MILES PER GALLON
	x	y
VW Golf	97	37
Chevy Cavalier	173	19
Plymouth Horizon	97	31
Pontiac Firebird	151	23
Corvette	350	17
Honda Accord	119	27
Dodge Omni	97	31
Renault Alliance	85	35
Olds Firenza	173	19
Nissan Sentra	97	31
Ford Escort	114	32

Sources. 1986 Gas Mileage Guide, EPA Fuel Economy Estimates. U.S. Department of Energy. *Wards Automotive Yearbook*, 1986.

a. Calculate the regression residuals for this model.
b. Verify that the sum of the residuals is 0.
c. Plot these residuals against cylinder volume, x.
d. Do you detect any distinctive patterns or trends in this plot?
e. What does your answer to part c suggest about model adequacy or the usual assumptions made about the error term ε?

7.5 The real estate data for Exercise 4.3 are reproduced here. Recall that the property appraisers fit the model

$$E(y) = \beta_0 + \beta_1 x_1 + \beta_2 x_2$$

where

y = Sale price (in $ thousands)

x_1 = Home size (in square feet)

x_2 = Condition rating (1 to 10)

The resulting least squares prediction equation is

$$\hat{y} = 9.782 + 1.871x_1 + 1.278x_2$$

SALE PRICE	HOME SIZE	CONDITION RATING
y, $ thousands	x_1, hundreds of sq. ft.	x_2, 1 to 10
60.0	23	5
32.7	11	2
57.7	20	9
45.5	17	3
47.0	15	8
55.3	21	4
64.5	24	7
42.6	13	6
54.5	19	7
57.5	25	2

Source: Andrews, R. L. and Ferguson, J. T. "Integrating judgment with a regression appraisal." *The Real Estate Appraiser and Analyst*, Vol. 52, No. 2, Spring 1986 (Table I).

a. Calculate the residuals for the model.
b. Plot the residuals versus x_1. Do you detect any trends? If so, what does the pattern suggest about the model?
c. Plot the residuals versus x_2. Do you detect any trends? If so, what does the pattern suggest about the model?
d. Calculate and plot the partial residuals for x_1. Interpret the result.
e. Calculate and plot the partial residuals for x_2. Interpret the result.

7.6   A certain type of rare gem serves as a status symbol for many of its owners. In theory, as the price of the gem increases, the demand will decrease at low prices, level off at moderate prices, and increase at high prices because of the status the owners believe they gain by obtaining the gem. Although a quadratic model would seem to match the theory, the model proposed to explain the demand for the gem by its price is the first-order model

$$y = \beta_0 + \beta_1 x + \varepsilon$$

where y is the demand (in thousands) and x is the retail price per carat (dollars). This model was fit to the 12 data points given in the table. The SPSS printout of the analysis is shown here.

x	100	700	450	150	500	800	70	50	300	350	750	700
y	130	150	60	120	50	200	150	160	50	40	180	130

```
Multiple R              .23064
R Square                .05319
Adjusted R Square      -.04149
Standard Error         56.20658

Analysis of Variance
                        DF        Sum of Squares       Mean Square
Regression              1            1774.86785        1774.86785
Residual               10           31591.79882        3159.17988

F =         .56181       Signif F =   .4708


------------------ Variables in the Equation ------------------

Variable              B         SE B        Beta        T     Sig T

X                  .04516      .06025      .23064     .750    .4708
(Constant)       99.81690    29.55565                3.377    .0070
```

a. Use the least squares prediction equation to calculate the regression residuals.
b. Plot the residuals against retail price per carat, x.
c. Can you detect any trends in the residual plot? What does this imply?

7.3
Detecting Unequal Variances

Recall that one of the assumptions necessary for the validity of regression inferences is that the error term ε have constant variance σ^2 for all levels of the independent variable(s). Variances that satisfy this property are called **homoscedastic**. Unequal variances for different settings of the independent variable(s) are said to be **heteroscedastic**. Various statistical tests for heteroscedasticity have been developed. However, plots of the residuals will frequently reveal the presence of heteroscedasticity. In this section we will show how residual plots can be used to detect departures from the assumption of equal variances, and then give a simple test for heteroscedasticity. In addition, we will suggest some modifications to the model that may remedy the situation.

When business and economic data fail to be homoscedastic, the reason is often that the variance of the response y is a function of its mean $E(y)$. Some examples follow:

1. If the response y is a count that has a Poisson distribution, the variance will be equal to the mean $E(y)$. Poisson data are usually counts per unit volume, area, time, etc. For example, the number of sick days per month for an employee would very likely be a Poisson random variable. If the variance of a response is proportional to $E(y)$, the regression residuals produce a pattern about $\hat{y}$, the least squares estimate of $E(y)$, like that shown in Figure 7.9.

FIGURE 7.9

A plot of residuals for Poisson data

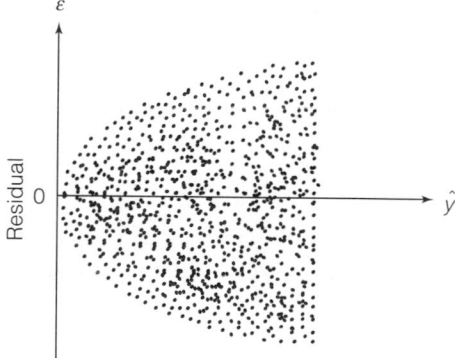

2. Many responses are proportions (or percentages) generated by **binomial experiments**. For example, the proportion of a random sample of 100 economists who forecast a recession within the next year is an example of a binomial response. Binomial proportions have variances that are functions of both the true proportion (the mean) and the sample size. In fact, if the observed proportion $y_i = \hat{p}_i$ is generated by a binomial distribution with sample size n_i and true probability p_i, the variance of y_i is

$$\text{Var}(y_i) = \frac{p_i(1 - p_i)}{n_i} = \frac{E(y_i)[1 - E(y_i)]}{n_i}$$

Residuals for binomial data produce a pattern about $\hat{y}$ like that shown in Figure 7.10 (page 368).

3. The random error component has been assumed to be **additive** in all the models we have constructed. An additive error is one for which the response is equal to the mean $E(y)$ *plus* random error,

$$y = E(y) + \varepsilon$$

FIGURE 7.10

A plot of residuals for binomial data
(proportions or percentages)

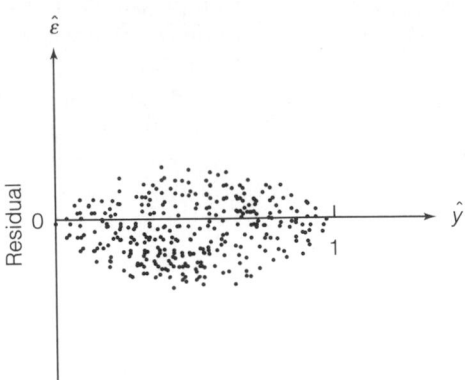

Another useful type of model for business and economic data is the **multiplicative** model. In this model, the response is written as the *product* of its mean and the random error component, i.e.,

$$y = [E(y)]\varepsilon$$

The variance of this response will grow proportionally to the *square* of the mean, i.e.,

$$\text{Var}(y) = [E(y)]^2\sigma^2$$

where σ^2 is the variance of ε. Data subject to multiplicative errors produce a pattern of residuals about $\hat{y}$ like that shown in Figure 7.11.

FIGURE 7.11

A plot of residuals for data subject to
multiplicative errors

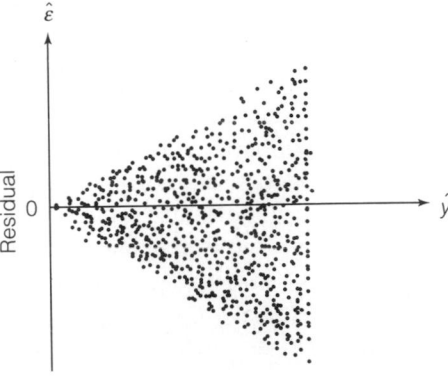

When the variance of y is a function of its mean, we can often satisfy the least squares assumption of homoscedasticity by transforming the response to some new response that has a constant variance. These are called **variance-stabilizing transformations**. For example, if the response y is a count that follows a Poisson distribution, the square root transform $\sqrt{y}$ can be shown to have approximately constant variance.[†] Consequently, if the response is a Poisson random variable, we would let

$$y^* = \sqrt{y}$$

and fit the model

$$y^* = \beta_0 + \beta_1 x_1 + \cdots + \beta_k x_k + \varepsilon$$

This model will satisfy approximately the least squares assumption of homoscedasticity.

Similar transformations that are appropriate for percentages and proportions (binomial data) or for data subject to multiplicative errors are shown in Table 7.6. The transformed responses will satisfy (at least approximately) the assumption of homoscedasticity.

T A B L E 7.6 **Stabilizing Transformations for Heteroscedastic Responses**

TYPE OF RESPONSE	VARIANCE	STABILIZING TRANSFORMATION
Poisson	$E(y)$	$\sqrt{y}$
Binomial proportion	$\dfrac{E(y)[1 - E(y)]}{n}$	$\sin^{-1}\sqrt{y}$
Multiplicative	$[E(y)]^2\sigma^2$	$\log y^a$

[a]Unless otherwise noted, natural logarithms (to the base e) will be used.

The data in Table 7.7 are the executive salaries, y, and years of experience, x, for a sample of 50 executives from a major industry. If we fit the second-order model $E(y) = \beta_0 + \beta_1 x + \beta_2 x^2$ to the data, we obtain the Minitab computer printout for the regression analysis shown in Figure 7.12 (page 370) and the prediction equation

$$\hat{y} = 20{,}242 + 522x + 53x^2$$

The computer printout suggests that the second-order model provides an adequate fit to the data. The R^2 value, .816, indicates that the model explains almost 82% of the total variation of the y values about $\bar{y}$. The global F value, $F = 103.98$, is highly significant ($p \approx 0$), indicating that the model contributes information for the prediction of y. However, an examination of the salary residuals plotted against the estimated mean salary, $\hat{y}$, as shown in Figure 7.13, reveals

---

[†]The square root transformation for Poisson responses is derived by finding the integral of $1/\sqrt{E(y)}$. In general, it can be shown (proof omitted) that the appropriate transformation for any response y is

$$y^* = \int \frac{1}{\sqrt{V(y)}}\, dy$$

where $V(y)$ is an expression for the variance of y.

TABLE 7.7   Salary and Experience Data for
50 Executives from a Major Industry

YEARS OF EXPERIENCE x	SALARY y	YEARS OF EXPERIENCE x	SALARY y	YEARS OF EXPERIENCE x	SALARY y
7	$26,075	21	$43,628	28	$99,139
28	79,370	4	16,105	23	52,624
23	65,726	24	65,644	17	50,594
18	41,983	20	63,022	25	53,272
19	62,309	20	47,780	26	65,343
15	41,154	15	38,853	19	46,216
24	53,610	25	66,537	16	54,288
13	33,697	25	67,447	3	20,844
2	22,444	28	64,785	12	32,586
8	32,562	26	61,581	23	71,235
20	43,076	27	70,678	20	36,530
21	56,000	20	51,301	19	52,745
18	58,667	18	39,346	27	67,282
7	22,210	1	24,833	25	80,931
2	20,521	26	65,929	12	32,303
18	49,727	20	41,721	11	38,371
11	33,233	26	82,641		

FIGURE 7.12

Minitab regression analysis printout for second-order model: Executive salaries

```
The regression equation is
Y = 20242 + 522 X + 53.0 XX

Predictor        Coef        Stdev      t-ratio        p
Constant        20242        4423         4.58      0.000
X               522.4        616.7         0.85      0.401
XX              53.00        19.57         2.71      0.009

s = 8123         R-sq = 81.6%     R-sq(adj) = 80.8%

Analysis of Variance

SOURCE       DF          SS            MS          F          p
Regression    2  13722605568   6861302784     103.98      0.000
Error        47   3101243136     65983896
Total        49  16823848960

SOURCE       DF      SEQ SS
X             1  13238774784
XX            1    483829792

Unusual Observations
Obs.        X          Y       Fit Stdev.Fit  Residual   St.Resid
   9       2.0      22444     21499      3453       945      0.13 X
  15       2.0      20521     21499      3453      -978     -0.13 X
  31       1.0      24833     20817      3911      4016      0.56 X
  35      28.0      99139     76421      2663     22718      2.96R

R denotes an obs. with a large st. resid.
X denotes an obs. whose X value gives it large influence.
```

FIGURE 7.13

Executive salary versus experience
example: Minitab residual plot for second-
order model with dependent variable,
salary

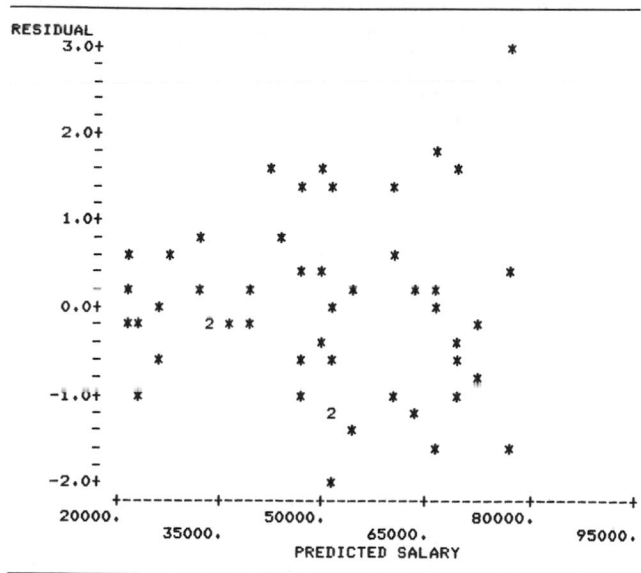

a potential problem. Note the "cone" shape of the residual variability; the size
of the residuals* increases as the estimated mean salary increases. This residual
plot indicates that a multiplicative model may be appropriate. We will explore
this possibility further in Example 7.4.

EXAMPLE 7.4

Consider the salary and experience data in Table 7.7. Use the logarithmic transformation on the dependent variable, and relate log y to years of experience, x,
using the second-order model

$$\widehat{\log y} = \beta_0 + \beta_1 x + \beta_2 x^2 + \varepsilon$$

Evaluate the adequacy of the model.

Solution

The Minitab computer printout in Figure 7.14 (page 372) gives the regression
analysis for the $n = 50$ measurements. The prediction equation used in computing
the residuals is

$$\widehat{\log y} = 9.8429 + .0497x + .000009x^2$$

---

*The vertical axis of the Minitab residual plot in Figure 7.13 represents the standardized residuals,
$\hat{\varepsilon}^*$. Minitab computes the standardized residuals as follows:

$$\hat{\varepsilon}_i^* = \frac{\hat{\varepsilon}_i}{s\sqrt{1 - h_i}}$$

where s is the standard deviation of the model and h_i is the "leverage" value for observation i. We
discuss leverage in Section 7.5.

FIGURE 7.14

Minitab regression analysis for
logarithmic transform of salary data:
Second-order model

```
The regression equation is
LOGY = 9.84 + 0.0497 X +0.000009 XX

Predictor          Coef        Stdev     t-ratio         p
Constant        9.84289      0.08479      116.08     0.000
X               0.04969      0.01182        4.20     0.000
XX            0.0000093    0.0003753        0.02     0.980

s = 0.1557      R-sq = 86.4%      R-sq(adj) = 85.8%

Analysis of Variance

SOURCE        DF          SS           MS         F         p
Regression     2      7.2119       3.6059    148.66     0.000
Error         47      1.1400       0.0243
Total         49      8.3519
```

The residual plot in Figure 7.15 indicates that the logarithmic transformation
has significantly reduced the heteroscedasticity.* Note that the cone shape is
gone; there is no apparent tendency of the residual variance to increase as mean
salary increases. We therefore are confident that inferences using the logarithmic
model are more reliable than those using the untransformed model.

FIGURE 7.15

Executive salary versus experience
example: Minitab residual plot for second-
order model with dependent variable,
log(salary)

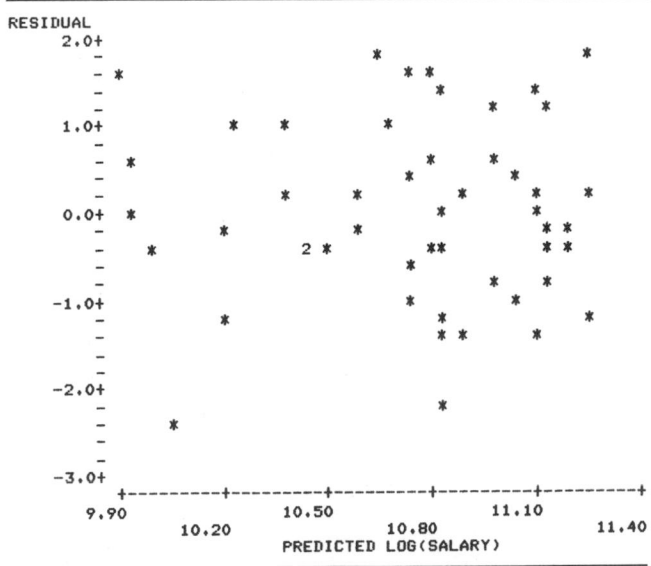

To evaluate model adequacy, we first note that $R^2 = .864$ and that about 86%
of the variation in log(salary) is accounted for by the model. The global F value
($F = 148.39$) and its associated p-value ($p \approx 0$) indicate that the model significantly improves upon the sample mean as a predictor of log(salary).

---

*A printout of the residuals is omitted.

Although the estimate $\hat{\beta}_2$ of β_2 is very small, we should check to determine whether the data provide sufficient evidence to indicate that the second-order term contributes information for the prediction of log(salary). The test of

$$H_0: \quad \beta_2 = 0 \qquad H_a: \quad \beta_2 \neq 0$$

is conducted using the t statistic shown in Figure 7.14, $t = .02$. The absolute value of this computed t value is extremely small and, clearly, does not exceed the tabulated t value for any reasonable value of α ($t_{.05} \approx 1.645$). Consequently, there is insufficient evidence to indicate that the second-order term contributes to the prediction of log(salary). There is no indication that the second-order model is an improvement over the straight-line model,

$$\log y = \beta_0 + \beta_1 x + \varepsilon$$

for predicting log(salary).

The Minitab computer printout for the first-order model (Figure 7.16) shows that the prediction equation for the first-order model is

$$\widehat{\log y} = 9.8413 + .04998x$$

The value of R^2, .863, is approximately the same as the value of R^2 obtained for the second-order model. The F statistic, computed from the mean squares in Figure 7.16, $F = 303.0$, indicates that the model contributes significantly to the prediction of log y.

FIGURE 7.16

Minitab regression analysis for logarithmic transform of salary data: First-order model

```
The regression equation is
LOGY = 9.84 + 0.0500 X

Predictor        Coef       Stdev      t-ratio         p
Constant      9.84133     0.05636       174.63     0.000
X            0.049978    0.002868        17.43     0.000

s = 0.1541      R-sq = 86.3%      R-sq(adj) = 86.1%

Analysis of Variance

SOURCE        DF          SS          MS         F         p
Regression     1      7.2118      7.2118    303.65     0.000
Error         48      1.1400      0.0238
Total         49      8.3519

Unusual Observations
Obs.       X        LOGY        Fit  Stdev.Fit   Residual    St.Resid
 19      4.0      9.6869    10.0412     0.0460    -0.3544      -2.41R
 31      1.0     10.1199     9.8913     0.0537     0.2286       1.58 X
 45     20.0     10.5059    10.8409     0.0225    -0.3350      -2.20R

R denotes an obs. with a large st. resid.
X denotes an obs. whose X value gives it large influence.
```

When the transformed model of Example 7.4 is used to predict the value of log y, the predicted value of y is the antilog, $\hat{y} = e^{\widehat{\log y}}$. The endpoints of the prediction interval are similarly transformed back to the original scale, and the interval will retain its meaning. In repeated use, the intervals will contain the observed y value $100(1 - \alpha)\%$ of the time.

Unfortunately, you cannot take antilogs to find the confidence interval for the mean value $E(y)$. The reason for this is that the mean value of log y is not equal to the logarithm of the mean of y. In fact, the antilog of the logarithmic mean of a random variable y is called its **geometric mean**. Thus, the antilogs of the endpoints of the confidence interval for the mean of the transformed response will give a confidence interval for the geometric mean. Similar care must be exercised with other types of transformations. In general, prediction intervals can be transformed back to the original scale without losing their meaning, but confidence intervals for the mean of a transformed response cannot.

The preceding examples illustrate that, in practice, residual plots can be a powerful technique for detecting heteroscedasticity. Furthermore, the pattern of the residuals often suggests the appropriate variance-stabilizing transformation to use. Keep in mind, however, that no measure of reliability can be attached to inferences derived from a graphical technique. For this reason, you may want to rely on a statistical test.

Various tests for heteroscedasticity in regression have been developed. One of the simpler techniques utilizes the F test (discussed in Chapter 1) for comparing population variances. The procedure requires that you divide the sample data in half and fit the regression model to each half. If the regression model fit to one-half the observations yields a significantly smaller or larger MSE than the model fit to the other half, there is evidence to indicate that the assumption of equal variances for all levels of the x variables in the model is being violated. (Recall that MSE, or mean square for error, estimates σ^2, the variance of the random error term.) Where you divide the data depends on where you suspect the differences in variances to be. We illustrate this procedure with an example.

EXAMPLE 7.5

Refer to Example 7.4 executive salary (y) versus years of experience (x) example. The residual plot for the quadratic model

$$E(y) = \beta_0 + \beta_1 x + \beta_2 x^2$$

indicates that the assumption of equal variances may be violated (see Figure 7.13). Conduct a statistical test of hypothesis to determine whether heteroscedasticity exists. Use $\alpha = .05$.

Solution

The residual plot shown in Figure 7.13 reveals that the residuals associated with larger values of predicted salary tend to be more variable than the residuals associated with smaller values of predicted salary. Therefore, we will divide the sample observations based on the values of $\hat{y}$, or, equivalently, the value of x (since, for the fitted model, $\hat{y}$ increases as x increases). An examination of the data in Table 7.7 reveals that approximately one-half of the 50 observed values of years of experience, x, fall below $x = 20$. Thus, we will divide the data into two subsamples as follows:

$$\begin{array}{ll}
\text{SUBSAMPLE 1} & \text{SUBSAMPLE 2} \\
x < 20 & x \geq 20 \\
n_1 = 24 & n_2 = 26
\end{array}$$

Figures 7.17a and 7.17b give the SAS printouts for the quadratic model fit to subsample 1 and subsample 2, respectively. The value of MSE is shaded in each printout.

FIGURE 7.17

a. SAS regression analysis for second-order model: Subsample 1 (years of experience < 20)

ANALYSIS OF VARIANCE

SOURCE	DF	SUM OF SQUARES	MEAN SQUARE	F VALUE	PROB>F
MODEL	2	3231228653	1615614327	51.163	0.0001
ERROR	21	663135395	31577875.97		
C TOTAL	23	3894364049			

ROOT MSE	5619.42	R-SQUARE	0.8297	
DEP MEAN	37152.75	ADJ R-SQ	0.8135	
C.V.	15.12518			

PARAMETER ESTIMATES

VARIABLE	DF	PARAMETER ESTIMATE	STANDARD ERROR	T FOR H0: PARAMETER=0	PROB > |T|
INTERCEP	1	20372.31906	3817.93090	5.336	0.0001
X	1	263.14290	861.88016	0.305	0.7631
XX	1	76.77081657	40.01457868	1.919	0.0687

b. SAS regression analysis for second-order model: Subsample 2 (years of experience ≥ 20)

ANALYSIS OF VARIANCE

SOURCE	DF	SUM OF SQUARES	MEAN SQUARE	F VALUE	PROB>F
MODEL	2	2930533276	1465266638	15.471	0.0001
ERROR	23	2178285323	94708057.53		
C TOTAL	25	5108818599			

ROOT MSE	9731.806	R-SQUARE	0.5736	
DEP MEAN	62185.85	ADJ R-SQ	0.5365	
C.V.	15.64955			

PARAMETER ESTIMATES

VARIABLE	DF	PARAMETER ESTIMATE	STANDARD ERROR	T FOR H0: PARAMETER=0	PROB > |T|
INTERCEP	1	-19228.62348	168795.40	-0.114	0.9103
X	1	2996.18859	14415.13670	0.208	0.8372
XX	1	17.03645465	304.21753	0.056	0.9558

The null and alternative hypotheses to be tested are:

$$H_0: \quad \frac{\sigma_1^2}{\sigma_2^2} = 1 \quad \text{(Assumption of equal variances satisfied)}$$

$$H_a: \quad \frac{\sigma_1^2}{\sigma_2^2} \neq 1 \quad \text{(Assumption of equal variances violated)}$$

where

σ_1^2 = Variance of the random error term, ε, for subpopulation 1 (i.e., $x < 20$)

σ_2^2 = Variance of the random error term, ε, for subpopulation 2 (i.e., $x \geq 20$)

The test statistic for a two-tailed test is given by:

$$F = \frac{\text{Larger } s^2}{\text{Smaller } s^2} = \frac{\text{Larger MSE}}{\text{Smaller MSE}} \quad \text{(see Section 1.10)}$$

where the distribution of F is based on ν_1 = df(error) associated with the larger MSE and ν_2 = df(error) associated with the smaller MSE. Recall that for a quadratic model, df(error) = $n - 3$.

From the printouts shown in Figures 7.17a and 7.17b, we have

$$\text{MSE}_1 = 31{,}577{,}876 \quad \text{and} \quad \text{MSE}_2 = 94{,}708{,}058$$

Therefore, the test statistic is

$$F = \frac{\text{MSE}_2}{\text{MSE}_1} = \frac{94{,}708{,}058}{31{,}577{,}876} = 3.00$$

Since the MSE for subsample 2 is placed in the numerator of the test statistic, this F value is based on $n_2 - 3 = 26 - 3 = 23$ numerator df and $n_1 - 3 = 24 - 3 = 21$ denominator df. For a two-tailed test at $\alpha = .05$, the critical value for $\nu_1 = 23$ and $\nu_2 = 21$ (found in Table 5 of Appendix D) is approximately $F_{.025} = 2.37$.

Since the observed value, $F = 3.00$, exceeds the critical value, there is sufficient evidence (at $\alpha = .05$) to indicate that the error variances differ.* Thus, this test supports the conclusions reached by using the residual plots in the preceding examples.

The test for heteroscedasticity outlined in Example 7.5 is easy to apply when only a single independent variable appears in the model. For a multiple regression model that contains several different independent variables, the choice of the levels of the x variables for dividing the data is more difficult, if not impossible. If you require a statistical test for heteroscedasticity in a multiple regression model, you may need to resort to other, more complex, tests.[†] Consult the references at the end of this chapter for details on how to conduct these tests.

---

*Most statistical tests require that the observations in the sample be independent. For this F test, the observations are the residuals. Even if the standard least squares assumption of independent errors is satisfied, the regression residuals will be correlated. Fortunately, when n is large compared to the number of β parameters in the model, the correlation among the residuals is reduced and, in most cases, can be ignored.

[†]For example, consider fitting the absolute values of the residuals as a function of the independent variables in the model, i.e., fit the regression model $E\{|\hat{\varepsilon}|\} = \beta_0 + \beta_1 x_1 + \beta_2 x_2 + \cdots + \beta_k x_k$. A nonsignificant global F implies that the assumption of homoscedasticity is satisfied. A significant F, however, indicates that changing the values of the x's will lead to a larger (or smaller) residual variance.

EXERCISES

7.7 Refer to Exercise 7.1. Plot the residuals for the first-order model versus $\hat{y}$. Do you detect any trends? If so, what does the pattern suggest about the model?

7.8 Refer to Exercise 7.2. Plot the residuals for the first-order model versus $\hat{y}$. Do you detect any trends? If so, what does the pattern suggest about the model?

7.9 Breakdowns of machines that produce steel cans are very costly. The more breakdowns, the fewer cans produced, and the smaller the company's profits. To help anticipate profit loss, the owners of a can company would like to find a model that will predict the number of breakdowns on the assembly line. The model proposed by the company's statisticians is the following:

$$y = \beta_0 + \beta_1 x_1 + \beta_2 x_2 + \beta_3 x_3 + \beta_4 x_4 + \varepsilon$$

where y is the number of breakdowns per 8-hour shift,

$$x_1 = \begin{cases} 1 & \text{if afternoon shift} \\ 0 & \text{otherwise} \end{cases} \qquad x_2 = \begin{cases} 1 & \text{if midnight shift} \\ 0 & \text{otherwise} \end{cases}$$

x_3 is the temperature of the plant (°F), and x_4 is the number of inexperienced personnel working on the assembly line. After the model is fit using the least squares procedure, the residuals are plotted against $\hat{y}$, as shown in the accompanying figure.

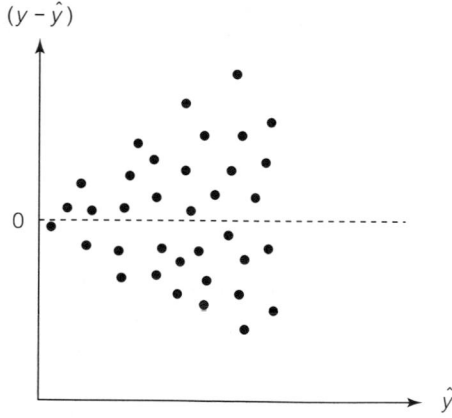

$(y - \hat{y})$

0

$\hat{y}$

a. Do you detect a pattern in the residual plot? What does this suggest about the least squares assumptions?

b. Given the nature of the response variable y and the pattern detected in part **a**, what model adjustments would you recommend?

7.10 Refer to Exercise 7.9. The regression analysis for the transformed model

$$y^* = \sqrt{y} = \beta_0 + \beta_1 x_1 + \beta_2 x_2 + \beta_3 x_3 + \beta_4 x_4 + \varepsilon$$

produces the prediction equation

$$\hat{y}^* = 1.3 + .008 x_1 - .13 x_2 + .0025 x_3 + .26 x_4$$

a. Use the equation to predict the number of breakdowns during the midnight shift if the temperature of the plant at that time is 87°F and if there is only one inexperienced worker on the assembly line.

b. A 95% prediction interval for y^* when $x_1 = 0$, $x_2 = 0$, $x_3 = 90°F$, and $x_4 = 2$ is (1.965, 2.125). For those same values of the independent variables, find a 95% prediction interval for y, the number of breakdowns per 8-hour shift.

c. A 95% confidence interval for $E(y^*)$ when $x_1 = 0$, $x_2 = 0$, $x_3 = 90°F$, and $x_4 = 2$ is (1.987, 2.107). Using only the information given in this problem, is it possible to find a 95% confidence interval for $E(y)$? Explain.

7.11 The manager of a retail appliance store wants to model the proportion of appliance owners who decide to purchase a service contract for a specific major appliance. Since the manager believes that the proportion y decreases with age x of the appliance (in years), he will fit the first-order model

$$E(y) = \beta_0 + \beta_1 x$$

A sample of 50 purchasers of new appliances are contacted about the possibility of purchasing a service contract. Fifty owners of 1-year-old machines, and 50 owners each of 2-, 3-, and 4-year-old machines are also contacted. One year later, another survey is conducted in a similar manner. The proportion y of owners deciding to purchase the service policy is shown in the table.

AGE OF APPLIANCE x, years	0	0	1	1	2	2	3	3	4	4
PROPORTION BUYING SERVICE CONTRACT, y	.94	.96	.7	.76	.6	.4	.24	.3	.12	.1

a. Fit the first-order model to the data.

b. Calculate the residuals and construct a residual plot versus $\hat{y}$.

c. What does the plot from part b suggest about the variance of y?

d. Explain how you could stabilize the variances.

e. Refit the model using the appropriate variance-stabilizing transformation. Plot the residuals for the transformed model and compare to the plot obtained in part b. Does the assumption of homoscedasticity appear to be satisfied?

7.12 In Hawaii, condemnation proceedings are underway to enable private citizens to own the property that their homes are built on. Prior to 1980, only estates were permitted to own land and homeowners leased the land from the estate (a law that dates back to the feudal period in Hawaii). To comply with the new law, a large Hawaiian estate wants to use regression analysis to estimate the fair market value of its land. A first proposal is the quadratic model

$$E(y) = \beta_0 + \beta_1 x + \beta_2 x^2$$

where

y = Leased fee value (i.e., sale price of property)

x = Size of property in square feet

Data collected for 20 property sales in a particular neighborhood, given in the accompanying table, were used to fit the model. The least squares prediction equation is

$$\hat{y} = -44.0947 + 11.5339x - .06378x^2$$

PROPERTY	LEASED FEE VALUE y, thousands of dollars	SIZE x, thousands	PROPERTY	LEASED FEE VALUE y, thousands of dollars	SIZE x, thousands
1	70.7	13.5	11	148.0	14.5
2	52.7	9.6	12	85.0	10.2
3	87.6	17.6	13	171.2	18.7
4	43.2	7.9	14	97.5	13.2
5	103.8	11.5	15	158.1	16.3
6	45.1	8.2	16	74.2	12.3
7	86.8	15.2	17	47.0	7.7
8	73.3	12.0	18	54.7	9.9
9	144.3	13.8	19	68.0	11.2
10	61.3	10.0	20	75.2	12.4

a. Calculate the predicted values and corresponding residuals for the model.

b. Plot the residuals versus $\hat{y}$. Do you detect any trends? If so, what does the pattern suggest about the model?

c. Conduct a test for heteroscedasticity. [*Hint:* Divide the data into two subsamples, $x \leq 12$ and $x > 12$, and fit the model to both subsamples.]

d. Based on your results from parts b and c, how should the estate proceed?

7.4

Checking the Normality Assumption

Recall from Section 4.2 that all the inferential procedures associated with a regression analysis are based on the assumptions that, for any setting of the independent variables, the random error ε is normally distributed with mean 0 and variance σ^2, and all pairs of errors are independent. Of these assumptions, the normality assumption is the least restrictive when we apply regression analysis in practice. That is, moderate departures from the assumption of normality have very little effect on error rates associated with the statistical tests and on the confidence coefficients associated with the confidence intervals.

Although tests are available to check the normality assumption (see, for example, Stephens, 1974), we discuss only graphical techniques in this section. The simplest way to determine whether the data violate the assumption of normality is to construct a frequency or relative frequency distribution for the residuals using the computer. If this distribution is not badly skewed, you can feel reasonably confident that the measures of reliability associated with your inferences are as stated in Chapter 4. This visual check is not foolproof because we are lumping the residuals together for all settings of the independent variables. It is conceivable (but not likely) that the distribution of residuals might be skewed to the left for some values of the independent variables and skewed to the right for others. Combining these residuals into a single relative frequency distribution could produce a distribution that is relatively symmetric. But, as noted above, we think that this situation is unlikely and that this graphical check is very useful.

To illustrate, a computer-generated frequency distribution for the $n = 50$ residuals of Example 7.4* is shown in Figure 7.18 (page 380). You can see that this distribution is mound-shaped and reasonably symmetric. Consequently, it is unlikely that the normality assumption would be violated using these data.

---

*A printout of the residuals is omitted.

FIGURE 7.18   A relative frequency distribution for the $n = 50$ residuals of Example 7.4

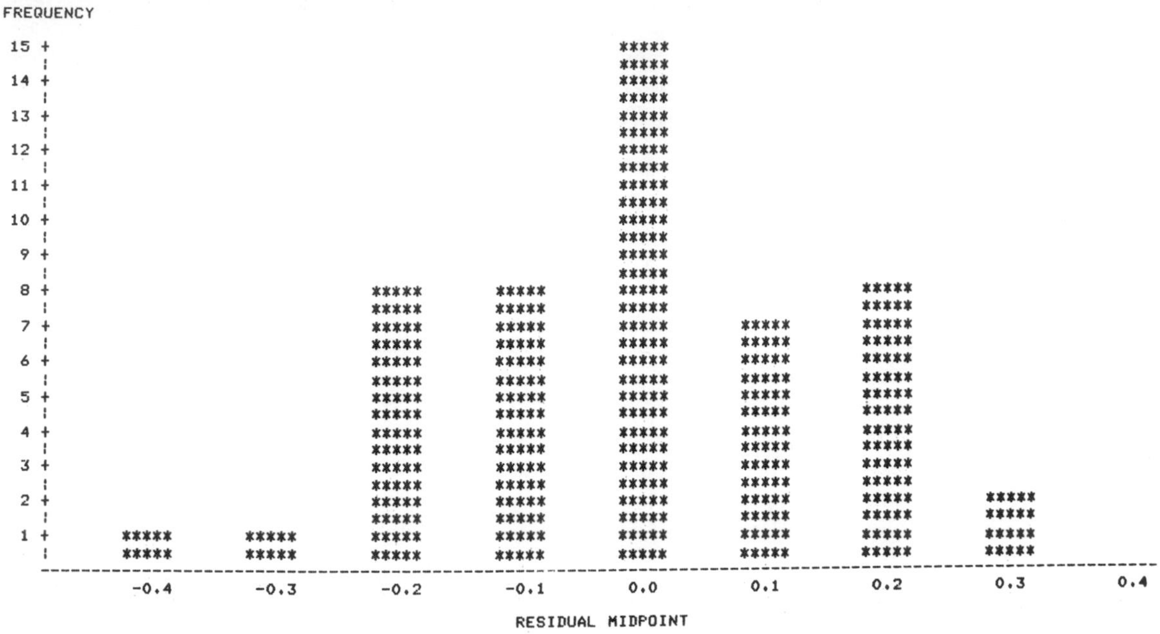

If the number of observations is small or if you do not have access to a computer, you may want to construct a stem-and-leaf display for the residuals (see Section 1.3). Figure 7.19 shows a stem-and-leaf display for the 50 residuals of Example 7.4. For our display, we chose the *stem* portion of a residual as the first digit to the right of the decimal point. The remaining portion of the residual, to the right of the stem, is the *leaf*. (For example, a residual value of .13 has a stem of .1 and a leaf of 3.) If you turn the stem-and-leaf display (Figure 7.19) on its side, it will look very much like the relative frequency distribution (Figure 7.18), mound-shaped and reasonably symmetric.

FIGURE 7.19

Stem-and-leaf display for the $n = 50$ residuals of Example 7.4

STEM	LEAF
−.3	3, 5
−.2	0, 1, 1
−.1	0, 1, 2, 5, 6, 6, 7, 8
−.0	0, 1, 2, 2, 3, 5, 5, 5, 5, 5, 6, 7, 7, 8
.0	1, 2, 3, 3, 4, 4, 5, 7, 8, 8
.1	0, 4, 5, 6, 8, 8
.2	1, 1, 3, 4, 5, 6, 6

A third graphical technique for checking the assumption of normality is to construct a **normal probability plot**. In a normal probability plot, the residuals are graphed against the expected values of the residuals under the assumption of normality. When the errors are, in fact, normally distributed, a residual value will approximately equal its expected value. Thus, a linear trend on the normal probability plot suggests that the normality assumption is nearly satisfied, whereas a nonlinear trend indicates that the assumption is most likely violated.

Most computer packages have procedures for constructing normal probability plots. Figure 7.20 (on page 382) shows the SAS normal probability plot for the residuals of Example 7.4.* Notice that the points (represented by the plotting symbol "*") fall reasonably close to a straight line (plotting symbol "+"), indicating that the normality assumption is most likely satisfied. If you do not have access to a computer package that contains a normal probability plot option, you can calculate the expected values of the residuals (under normality) using the procedure outlined in the next box.

Constructing a Normal Probability Plot for Regression Residuals

1. List the residuals in ascending order, where $\hat{\varepsilon}_i$ represents the ith ordered residual.

2. For each residual, calculate the corresponding tail area (of the standard normal distribution),

$$A = \frac{i - .375}{n + .25}$$

where n is the sample size.

3. Calculate the estimated value of $\hat{\varepsilon}_i$ under normality using the following formula:

$$E(\hat{\varepsilon}_i) \approx \sqrt{MSE}\,[Z(A)]$$

where

MSE = Mean square error for the fitted model
Z(A) = Value of the standard normal distribution (z value) that cuts off an area of A in the lower tail of the distribution.

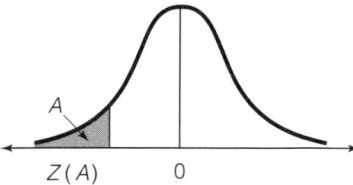

4. Plot the residuals $\hat{\varepsilon}_i$ against the estimated expected residuals, $i = 1, 2, \ldots, n$.

---

*The horizontal axis of the SAS normal probability plot (Figure 7.20) gives the expected values of the residuals in terms of number of standard deviations (positive or negative) from the mean 0.

FIGURE 7.20

Normal probability plot for the $n = 50$ residuals of Example 7.4

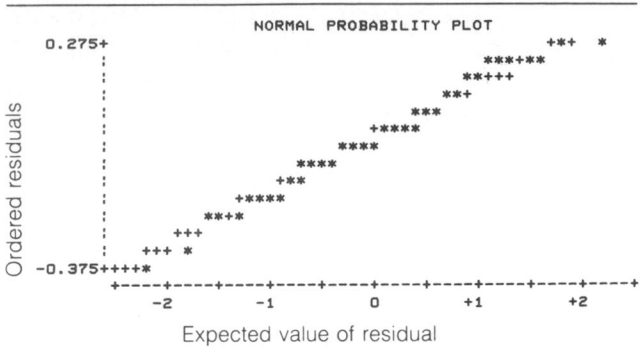

Nonnormality of the distribution of the random error ε is often accompanied by heteroscedasticity. Both these situations can frequently be rectified by applying the variance-stabilizing transformations of Section 7.3. For example, if the relative frequency distribution (or stem-and-leaf display) of the residuals is highly skewed to the right (as it would be for Poisson data), the square-root transformation on y will stabilize (approximately) the variance and, at the same time, will reduce skewness in the distribution of the residuals. Thus, for any given setting of the independent variables, the square-root transformation will reduce the larger values of y to a greater extent than the smaller ones. This has the effect of reducing or eliminating the positive skewness.

For situations in which the errors are homoscedastic but nonnormal, normalizing transformations are available. This family of transformations on the dependent variable includes $\sqrt{y}$ and $\log(y)$ (Section 7.3), as well as such simple transformations as y^2, $1/\sqrt{y}$, and $1/y$. Box and Cox (1964) have developed a procedure for selecting the appropriate transformation to use. Consult the references to learn details of the Box–Cox approach. Keep in mind, however, that regression is extremely *robust* with respect to nonnormal errors. That is, the inferences derived from the regression analysis tend to remain valid even when the assumption of normal errors is violated. Consequently, you may want to search for a normalizing transformation only when the distribution of the regression residuals is highly skewed.

EXERCISES

7.13   Refer to Exercise 3.10. Use one of the graphical techniques described in this section to check the normality assumption.

7.14   Refer to Exercise 4.42. Use one of the graphical techniques described in this section to check the normality assumption.

7.15   Refer to Exercise 7.12. Use one of the graphical techniques described in this section to check the normality assumption.

7.16 L. De Cola conducted an extensive investigation of the geopolitical and socioeconomic processes that shape the urban size distributions of the world's nations. One of the goals of the study was to determine the factors that influence population size in each nation's largest city. Using data collected for a sample of 126 countries, De Cola fit the following log model:

$$E(y) = \beta_0 + \beta_1 x_1 + \beta_2 x_2 + \beta_3 x_3 + \beta_4 x_4 + \beta_5 x_5 + \beta_6 x_6 + \beta_7 x_7 + \beta_8 x_8 + \beta_9 x_9 + \beta_{10} x_{10}$$

where

y = Log of population (in thousands) of largest city in country

x_1 = Log of area (in thousands of square kilometers) of country

x_2 = Log of radius (in hundred kilometers) of country

x_3 = Log of national population (in thousands)

x_4 = Percentage annual change in national population (1960–1970)

x_5 = Log of energy consumption per capita (in kilograms of coal equivalent)

x_6 = Percentage of nation's population in urban areas

x_7 = Log of population (in thousands) of second largest city in country

$x_8 = \begin{cases} 1 & \text{if seaport city} \\ 0 & \text{if not} \end{cases}$

$x_9 = \begin{cases} 1 & \text{if capital city} \\ 0 & \text{if not} \end{cases}$

$x_{10} = \begin{cases} 1 & \text{if city data are for metropolitan area} \\ 0 & \text{if not} \end{cases}$

[*Note:* All logarithms are to the base 10.]

The regression resulted in

$R^2 = .879$ and MSE = .036

a. Conduct a test for model adequacy. Use $\alpha = .05$.

b. The residuals for five cities selected from the total sample are given in the table. For each of these cities, calculate the estimated expected residuals under the assumption of normality.

CITY	RESIDUAL	RANK
Bangkok	.510	126
Paris	.228	110
London	.033	78
Warsaw	−.132	32
Lagos	−.392	2

Source: De Cola, L. "Statistical determinants of the population of a nation's largest city." *Economic Development and Cultural Change*, Vol. 3, No. 1, Oct. 1984, pp. 71–98.

c. A computer-generated (SAS) stem-and-leaf plot of all the residuals is shown at the top of page 384. Does it appear that the assumption of normal errors is satisfied?

SAS stem-and-leaf plot of residuals for
Exercise 7.16

```
STEM LEAF                          #
   5 1                             1
   4 5                             1
   4
   3 79                            2
   3 1224                          4
   2 5899                          4
   2 000012233344                 12
   1 5789                          4
   1 00001113444                  11
   0 5566888                       7
   0 11111122223334               14
  -0 4433333322111000             16
  -0 9999988665                   10
  -1 44433332221110               14
  -1 88865                         5
  -2 444443100                     9
  -2 77665                         5
  -3 3000                          4
  -3 97                            2
  -4
  -4 6                             1
     ----+----+----+----+
  MULTIPLY STEM.LEAF BY 10**-01
```

7.5

Detecting Outliers and Identifying Influential Observations

Although we expect almost all the regression residuals to fall within 3 standard deviations of their mean (0), sometimes one or several residuals fall outside this interval. Observations with residuals that are extremely large or small (say, more than 3 standard deviations from 0) are called **outliers**.

Outliers are usually attributable to one of several causes. The measurement associated with the outlier may be invalid. For example, the experimental procedure used to generate the measurement may have malfunctioned, the experimenter may have misrecorded the measurement, or the data might have been coded incorrectly for entry into the computer. Careful checks of the experimental and coding procedures should reveal this type of problem if it exists, so that we can eliminate erroneous observations from a data set.

T A B L E 7.8   **Data for Fast-Food Sales**

CITY	TRAFFIC FLOW thousands of cars	WEEKLY SALES y, thousands of dollars	CITY	TRAFFIC FLOW thousands of cars	WEEKLY SALES y, thousands of dollars
1	59.3	6.3	3	75.8	8.2
1	60.3	6.6	3	48.3	5.0
1	82.1	7.6	3	41.4	3.9
1	32.3	3.0	3	52.5	5.4
1	98.0	9.5	3	41.0	4.1
1	54.1	5.9	3	29.6	3.1
1	54.4	6.1	3	49.5	5.4
1	51.3	5.0	4	73.1	8.4
1	36.7	3.6	4	81.3	9.5
2	23.6	2.8	4	72.4	8.7
2	57.6	6.7	4	88.4	10.6
2	44.6	5.2	4	23.2	3.3

For example, Table 7.8 presents the sales, y, in thousands of dollars per week, for fast-food outlets in each of four cities. The objective is to model sales, y, as a function of traffic flow, adjusting for city-to-city variations that might be due to size or other market conditions. We expect a first-order (linear) relationship to exist between mean sales, $E(y)$, and traffic flow. Further, we believe that the level of mean sales will differ from city to city, but that the change in mean sales per unit increase in traffic flow will remain the same for all cities, i.e., that the factors traffic flow and cities do not interact. The model is therefore

$$E(y) = \beta_0 + \beta_1 x_1 + \beta_2 x_2 + \beta_3 x_3 + \beta_4 x_4$$

where

$$x_1 = \begin{cases} 1 & \text{if city 1} \\ 0 & \text{other} \end{cases} \qquad x_2 = \begin{cases} 1 & \text{if city 2} \\ 0 & \text{other} \end{cases}$$

$$x_3 = \begin{cases} 1 & \text{if city 3} \\ 0 & \text{other} \end{cases} \qquad x_4 = \text{Traffic flow}$$

The SAS printout for the regression analysis is shown in Figure 7.21 (page 386). The regression analysis indicates that the first-order model in traffic flow is inadequate for explaining mean sales. The coefficient of determination, R^2, is .259, indicating that only 25.9% of the total sum of squares of deviations of the sales y about their mean $\bar{y}$ is accounted for by the model. The F value, 1.67, which tests the adequacy of the model, does not indicate that the model is useful for predicting sales. The observed significance level is only .1996.

Plots of the residuals against traffic flow and city are shown in Figures 7.22 and 7.23 (pages 387 and 388), respectively. The dashed horizontal lines locate the mean (0), $+2s$, and $-2s$ for the residuals. As you can see, the plots of the residuals are very revealing. Both the plot of residuals against traffic flow in Figure 7.22 and the plot of residuals against city in Figure 7.23 indicate the presence of an outlier. One observation in city 3, with traffic flow of 75.8, is approximately 4 standard deviations from 0. (Note in Figure 7.21 that the standard deviation is 14.86, and the outlier residual is 56.46.) A further check of the observation associated with this residual reveals that the sales value entered into the computer, 82.0, does not agree with the corresponding value of sales, 8.2, that appears in Table 7.8. The decimal point was evidently dropped when the data were entered into the computer.

If the correct y value, 8.2, is substituted for the 82.0, we obtain the regression analysis shown in Figure 7.24 (page 389). Plots of the residuals against traffic flow and city are shown in Figures 7.25 and 7.26 (pages 390 and 391), respectively. The corrected computer printout indicates the dramatic effect that a single outlier can have on a regression analysis. The value of R^2 is now .979, and the F value that tests the adequacy of the model, 222.17, verifies the strong predictive capability of the model. Further analysis reveals that significant differences exist in the mean sales among cities, and that the estimated mean weekly sales increase by \$104 for every 1,000-car increase in traffic flow ($\hat{\beta}_4 = .104$). The 95% confidence interval for β_4 is

$$\hat{\beta}_4 \pm t_{.025} s_{\hat{\beta}_4} = .104 \pm (2.093)(.004094) = .104 \pm .009$$

FIGURE 7.21
SAS regression printout for model of fast-food sales

Analysis of Variance

Source	DF	Sum of Squares	Mean Square	F Value	Prob>F
Model	4	1469.76287	367.44072	1.665	0.1996
Error	19	4194.22671	220.74877		
C Total	23	5663.98958			

Root MSE	14.85762	R-square	0.2595	
Dep Mean	9.07083	Adj R-sq	0.1036	
C.V.	163.79550			

Parameter Estimates

Variable	DF	Parameter Estimate	Standard Error	T for H0: Parameter=0	Prob > |T|
INTERCEP	1	-16.459248	13.16399794	-1.250	0.2264
X1	1	1.106092	8.42256884	0.131	0.8969
X2	1	6.142771	11.67996860	0.526	0.6050
X3	1	14.489623	9.28839086	1.560	0.1353
X4	1	0.362873	0.16790819	2.161	0.0437

Obs	CITY	X4	Dep Var Y	Predict Value	Residual
1	1	59.3	6.3000	6.1652	0.1348
2	1	60.3	6.6000	6.5281	0.0719
3	1	82.1	7.6000	14.4387	-6.8387
4	1	32.3	3.0000	-3.6324	6.6324
5	1	98	9.5000	20.2084	-10.7084
6	1	54.1	5.9000	4.2783	1.6217
7	1	54.4	6.1000	4.3871	1.7129
8	1	51.3	5.0000	3.2622	1.7378
9	1	36.7	3.6000	-2.0357	5.6357
10	2	23.6	2.8000	-1.7527	4.5527
11	2	57.6	6.7000	10.5850	-3.8850
12	2	44.6	5.2000	5.8677	-0.6677
13	3	75.8	82.0000	25.5362	56.4638
14	3	48.3	5.0000	15.5571	-10.5571
15	3	41.4	3.9000	13.0533	-9.1533
16	3	52.5	5.4000	17.0812	-11.6812
17	3	41	4.1000	12.9082	-8.8082
18	3	29.6	3.1000	8.7714	-5.6714
19	3	49.5	5.4000	15.9926	-10.5926
20	4	73.1	8.4000	10.0668	-1.6668
21	4	81.3	9.5000	13.0423	-3.5423
22	4	72.4	8.7000	9.8128	-1.1128
23	4	88.4	10.6000	15.6187	-5.0187
24	4	23.2	3.3000	-8.0406	11.3406

Sum of Residuals	1.261213E-13
Sum of Squared Residuals	4194.2267
Predicted Resid SS (Press)	7303.4839

Thus, a 95% confidence interval for the mean increase in sales per 1,000-car increase in traffic flow is $95 to $113.

Outliers cannot always be explained by computer or recording errors. Extremely large or small residuals may be attributable to skewness (nonnormality) of the probability distribution of the random error, chance, or unassignable causes. Although some analysts advocate elimination of outliers, regardless of whether cause can be assigned, others encourage the correction of only those

FIGURE 7.22

SAS plot of residuals versus traffic flow

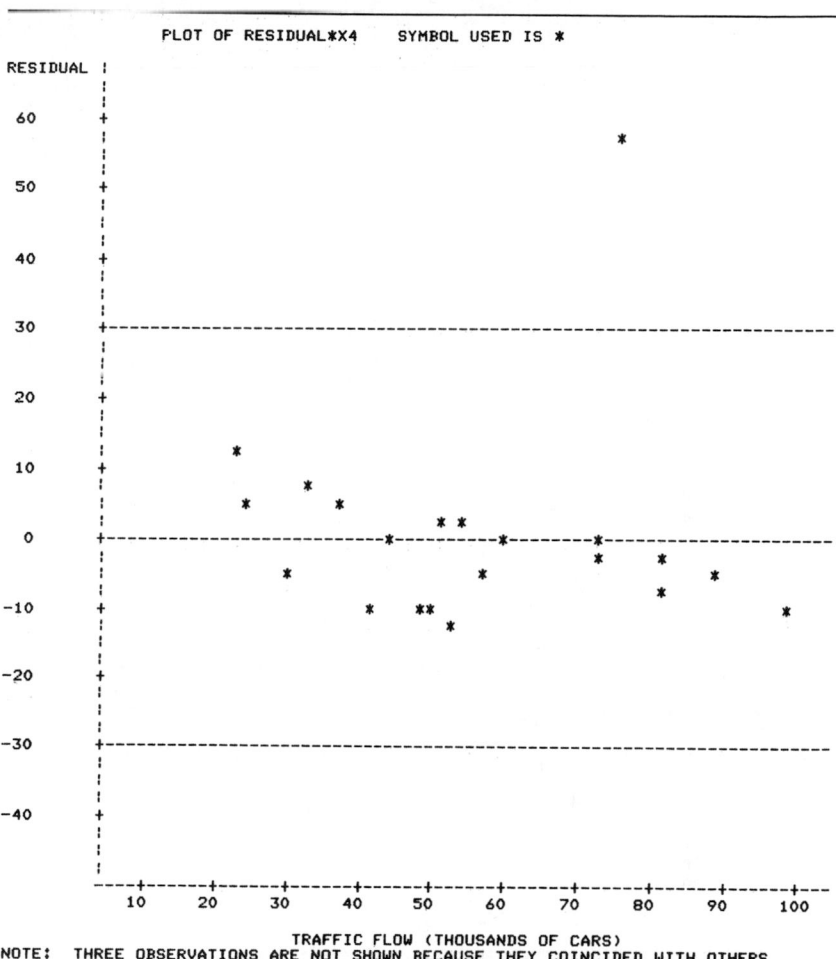

outliers that can be traced to specific causes. The best philosophy is probably a compromise between these extremes. For example, before deciding the fate of an outlier you may want to determine how much influence it has on the regression analysis. When an accurate outlier (i.e., an outlier that is not due to recording or measurement error) is found to have a dramatic effect on the regression analysis, it may be the model and not the outlier that is suspect. Omission of important independent variables or higher-order terms could be the reason why the model is not predicting well for the outlying observation. Several sophisticated numerical techniques are available for identifying outlying influential observations. We conclude this section with a brief discussion of some of these methods and an example.

FIGURE 7.23

SAS plot of residuals versus city

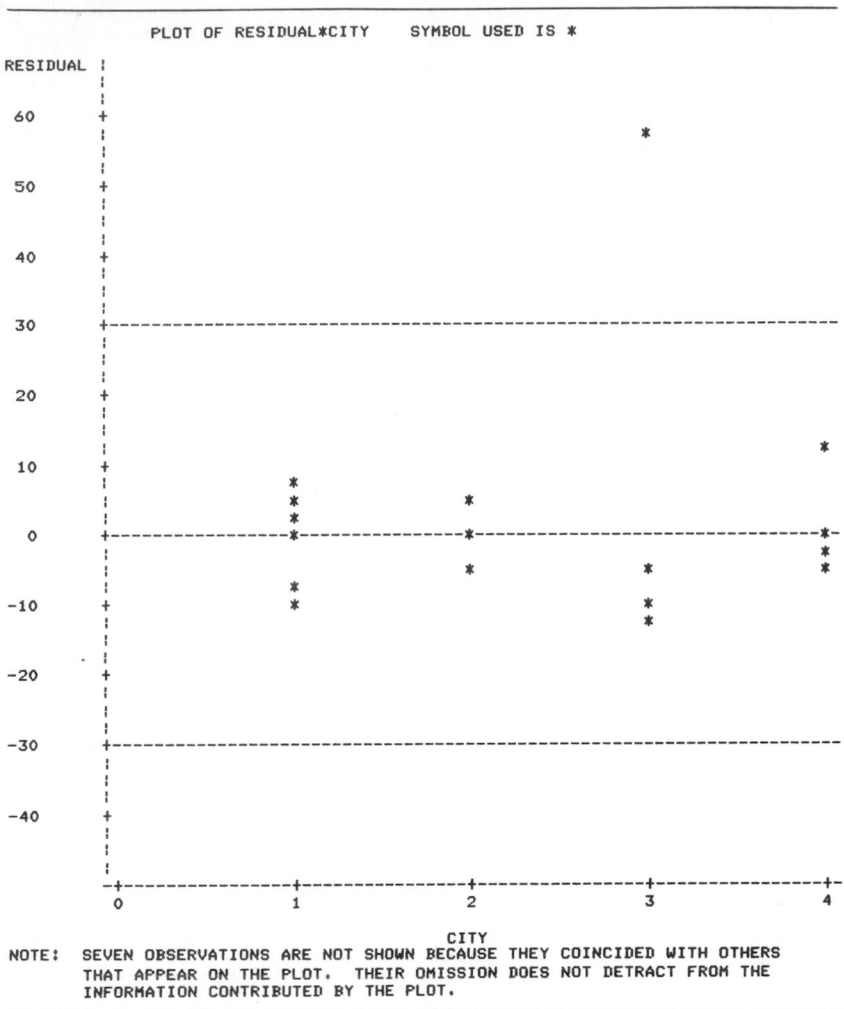

Leverage

This procedure is based on a result (proof omitted) in regression analysis that states that the predicted value for the *i*th observation, $\hat{y}_i$, can be written as a linear combination of the *n* observed values $y_1, y_2, \ldots, y_n$:

$$\hat{y}_i = h_1 y_1 + h_2 y_2 + \cdots + h_i y_i + \cdots + h_n y_n, \quad i = 1, 2, \ldots, n$$

where the weights $h_1, h_2, \ldots, h_n$ of the observed values are functions of the independent variables. In particular, the coefficient h_i measures the influence of the observed value y_i on its own predicted value $\hat{y}_i$. This value, h_i, is called the **leverage** of the *i*th observation (with respect to the values of the independent variables). Thus, leverage values can be used to identify influential observations—the larger the leverage value, the more influence the observed *y* value has on its predicted value.

FIGURE 7.24

Corrected SAS regression printout for model of fast-food sales

Analysis of Variance

Source	DF	Sum of Squares	Mean Square	F Value	Prob>F
Model	4	116.65552	29.16388	222.173	0.0001
Error	19	2.49407	0.13127		
C Total	23	119.14958			

Root MSE	0.36231	R-square	0.9791	
Dep Mean	5.99583	Adj R-sq	0.9747	
C.V.	6.04265			

Parameter Estimates

Variable	DF	Parameter Estimate	Standard Error	T for H0: Parameter=0	Prob > |T|
INTERCEP	1	1.083388	0.32100795	3.375	0.0030
X1	1	-1.215762	0.20538681	-5.919	0.0001
X2	1	-0.530757	0.28481946	-1.863	0.0779
X3	1	-1.076525	0.22650014	-4.753	0.0001
X4	1	0.103673	0.00409449	25.320	0.0001

Obs	CITY	X4	Dep Var Y	Predict Value	Residual
1	1	59.3	6.3000	6.0155	0.2845
2	1	60.3	6.6000	6.1191	0.4809
3	1	82.1	7.6000	8.3792	-0.7792
4	1	32.3	3.0000	3.2163	-0.2163
5	1	98	9.5000	10.0276	-0.5276
6	1	54.1	5.9000	5.4764	0.4236
7	1	54.4	6.1000	5.5075	0.5925
8	1	51.3	5.0000	5.1861	-0.1861
9	1	36.7	3.6000	3.6724	-0.0724
10	2	23.6	2.8000	2.9993	-0.1993
11	2	57.6	6.7000	6.5242	0.1758
12	2	44.6	5.2000	5.1765	0.0235
13	3	75.8	8.2000	7.8653	0.3347
14	3	48.3	5.0000	5.0143	-0.0143
15	3	41.4	3.9000	4.2989	-0.3989
16	3	52.5	5.4000	5.4497	-0.0497
17	3	41	4.1000	4.2575	-0.1575
18	3	29.6	3.1000	3.0756	0.0244
19	3	49.5	5.4000	5.1387	0.2613
20	4	73.1	8.4000	8.6619	-0.2619
21	4	81.3	9.5000	9.5120	-0.0120
22	4	72.4	8.7000	8.5893	0.1107
23	4	88.4	10.6000	10.2481	0.3519
24	4	23.2	3.3000	3.4886	-0.1886

Sum of Residuals	1.332268E-14
Sum of Squared Residuals	2.4941
Predicted Resid SS (Press)	3.8772

Leverage values are extremely difficult to calculate without the aid of a computer.* Fortunately, most of the statistical software packages discussed in this text have options that give the leverage associated with each observation. The leverage value for an observation is usually compared with the average leverage value of all n observations, $\bar{h}$, where

---

*In matrix notation, the leverage values are the diagonals of the H matrix (called the "hat" matrix), where $H = X(X'X)^{-1}X'$. See Appendix A for details on matrix multiplication and definition of the X matrix in regression.

Corrected SAS plot of residuals versus traffic flow

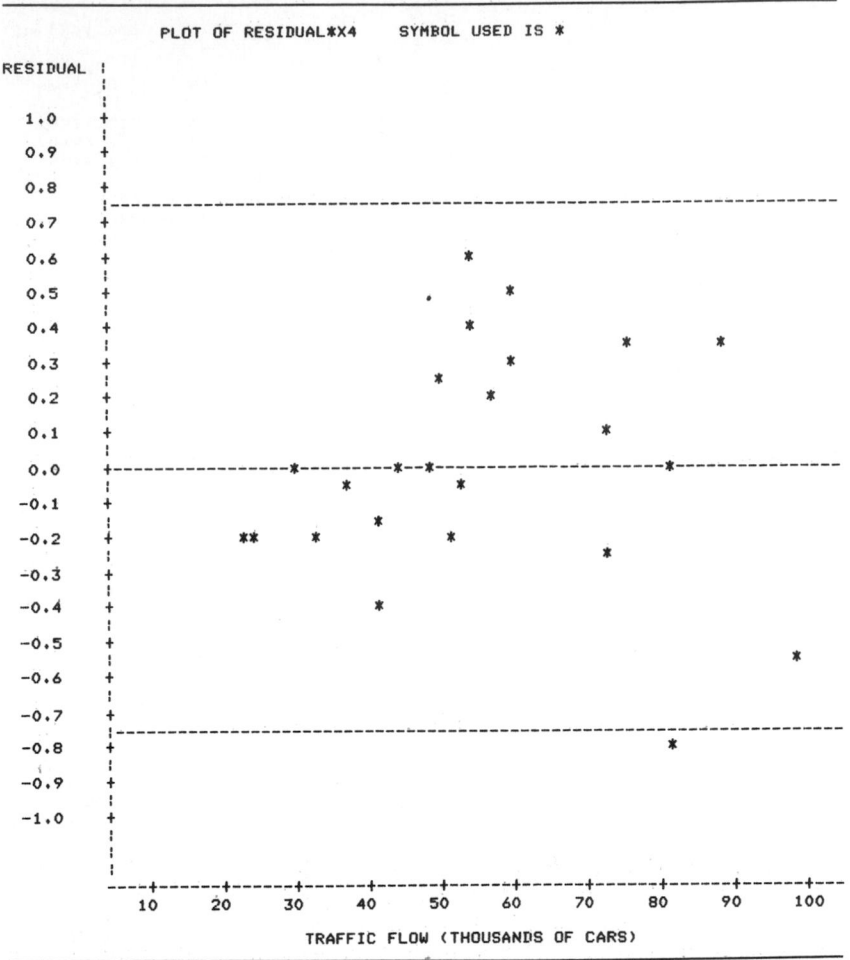

$$\bar{h} = \frac{k + 1}{n} = \frac{\text{Number of } \beta \text{ parameters in the model, including } \beta_0^*}{n}$$

A good rule of thumb identifies an observation y_i as influential if its leverage value h_i is more than twice as large as $\bar{h}$, that is, if

$$h_i > \frac{2(k + 1)}{n}$$

The Jackknife

Another technique for identifying influential observations requires that you delete the observations one at a time, each time refitting the regression model based on only the remaining $n - 1$ observations. This method is based on a statistical procedure, called the **jackknife**,* that is gaining increasing acceptance among

---

*The proof of this result is beyond the scope of this text. Consult the references given at the end of this chapter. [See Neter, Wasserman, and Kutner (1990).]

FIGURE 7.26

Corrected SAS plot of residuals versus
city

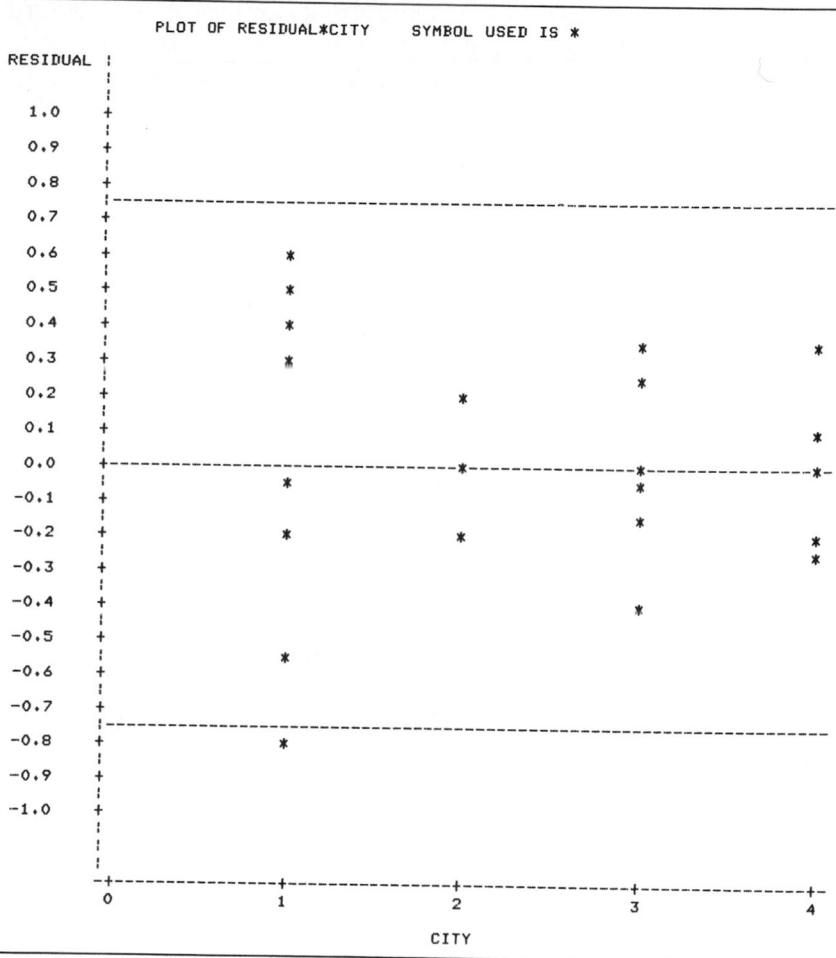

practitioners. The basic principle of the jackknife when applied to regression is
to compare the regression results using all n observations to the results with the
ith observation deleted to ascertain how much influence a particular observation
has on the analysis. Using the jackknife, several alternative influence measures
can be calculated.

The **deleted residual**, $d_i = y_i - \hat{y}_{(i)}$, measures the difference between the
observed value y_i and the predicted value $\hat{y}_{(i)}$ based on the model with the ith
observation deleted. [The notation (i) is generally used to indicate that the
observed value y_i was deleted from the regression analysis.] An observation with
an unusually large (in absolute value) deleted residual is considered to have large
influence on the fitted model.

---

*The procedure derives its name from the Boy Scout jackknife, which serves as a handy tool in a
variety of situations when specialized techniques may not be applicable. [See Belsley, Kuh, and
Welsch (1980).]

A measure closely related to the deleted residual is the difference between the predicted value based on the model fit to all n observations and the predicted value obtained when y_i is deleted, i.e., $\hat{y}_i - \hat{y}_{(i)}$. When the difference $\hat{y}_i - \hat{y}_{(i)}$ is large relative to the predicted value $\hat{y}_i$, the observation y_i is said to influence the regression fit.

A third way to identify an influential observation using the jackknife is to calculate, for each β parameter in the model, the difference between the parameter estimate based on all n observations and the estimate based on only $n - 1$ observations (with the observation in question deleted). Consider, for example, the straight-line model $E(y) = \beta_0 + \beta_1 x$. The differences $\hat{\beta}_0 - \hat{\beta}_0^{(i)}$ and $\hat{\beta}_1 - \hat{\beta}_1^{(i)}$ measure how influential the ith observation y_i is on the parameter estimates. [Using the (i) notation defined earlier, $\hat{\beta}^{(i)}$ represents the estimate of the β coefficient when the ith observation is omitted from the analysis.] If the parameter estimates change drastically, i.e., if the absolute differences are large, y_i is deemed an influential observation.

Each of the statistical software packages discussed in this text has a jackknife routine that produces one or more of the measures described above.

Cook's Distance

A measure of the overall influence an outlying observation has on the estimated β coefficients was proposed by Cook (1979). Cook's distance, D_i, is calculated for the ith observation as follows:

$$D_i = \frac{(y_i - \hat{y}_i)^2}{(k + 1)\mathrm{MSE}}\left[\frac{h_i}{(1 - h_i)^2}\right]$$

Note that D_i depends on both the residual $(y_i - \hat{y}_i)$ and the leverage h_i for the ith observation. A large value of D_i indicates that the observed y_i value has strong influence on the estimated β coefficients (since the residual, the leverage, or both will be large). Values of D_i can be compared to the values of the F distribution with $\nu_1 = k + 1$ and $\nu_2 = n - (k + 1)$ degrees of freedom. Usually, an observation with a value of D_i that falls at or above the 50th percentile of the F distribution is considered to be an influential observation. Like the other numerical measures of influence, options for calculating Cook's distance are available in most statistical software packages.

EXAMPLE 7.6

We now return to the fast-food sales example in which we detected an outlier using residual plots. Recall that the outlier was due to an error in coding the weekly sales value for observation 13 (denoted y_{13}). The SAS regression analysis is rerun with options for producing influence diagnostics. (An **influence diagnostic** is a number that measures how much influence an observation has on the regression analysis.) The resulting SAS printout is shown in Figure 7.27. Locate and interpret the measures of influence for y_{13} on the printout.

Solution

The influence diagnostics are shown in the last portion of the SAS printout in Figure 7.27. Leverage values for each observation are given under the column heading **Hat Diag H**. The leverage value for y_{13} (shaded on the printout) is $h_{13} = .2394$, whereas the average leverage for all $n = 24$ observations is

$$h = \frac{k+1}{n} = \frac{5}{24} = .2083$$

Since the leverage value .2394 does not exceed $2\bar{h} = .4166$, we would not identify y_{13} as an influential observation. At first, this result may seem confusing since we already know the dramatic effect the incorrectly coded value of y_{13} had on the regression analysis. Remember, however, that the leverage values, h_1, h_2, ..., h_{24}, are functions of the independent variables only. Since we know the values of x_1, x_2, x_3, and x_4 were coded correctly, the relatively small leverage value of .2394 simply indicates that observation 13 is not an outlier with respect to the values of the independent variables.

FIGURE 7.27   SAS regression analysis with influence diagnostics

Analysis of Variance

Source	DF	Sum of Squares	Mean Square	F Value	Prob>F
Model	4	1469.76287	367.44072	1.665	0.1996
Error	19	4194.22671	220.74877		
C Total	23	5663.98958			

Root MSE	14.85762	R-square	0.2595	
Dep Mean	9.07083	Adj R-sq	0.1036	
C.V.	163.79550			

Parameter Estimates

Variable	DF	Parameter Estimate	Standard Error	T for H0: Parameter=0	Prob > |T|
INTERCEP	1	-16.459248	13.16399794	-1.250	0.2264
X1	1	1.106092	8.42256884	0.131	0.8969
X2	1	6.142771	11.67996860	0.526	0.6050
X3	1	14.489623	9.28839086	1.560	0.1353
X4	1	0.362873	0.16790819	2.161	0.0437

Obs	Dep Var Y	Predict Value	Std Err Predict	Residual	Std Err Residual	Student Residual	-2-1-0 1 2	Cook's D
1	6.3000	6.1652	4.953	0.1348	14.008	0.010	| |	0.000
2	6.6000	6.5281	4.960	0.0719	14.005	0.005	| |	0.000
3	7.6000	14.4387	6.319	-6.8387	13.447	-0.509	*|	0.011
4	3.0000	-3.6324	6.649	6.6324	13.287	0.499	|*	0.012
5	9.5000	20.2084	8.248	-10.7084	12.358	-0.866	*|	0.067
6	5.9000	4.2783	5.013	1.6217	13.986	0.116	| |	0.000
7	6.1000	4.3871	5.005	1.7129	13.989	0.122	| |	0.000
8	5.0000	3.2622	5.107	1.7378	13.952	0.125	| |	0.000
9	3.6000	-2.0357	6.181	5.6357	13.511	0.417	|*	0.007
10	2.8000	-1.7527	9.114	4.5527	11.734	0.388	| |	0.018
11	6.7000	10.5850	8.972	-3.8850	11.843	-0.328	| |	0.012
12	5.2000	5.8677	8.590	-0.6677	12.123	-0.055	| |	0.000
13	82.0000	25.5362	7.270	56.4638	12.957	4.358	|******	1.196
14	5.0000	15.5571	5.616	-10.5571	13.755	-0.767	*|	0.020
15	3.9000	13.0533	5.734	-9.1533	13.707	-0.668	*|	0.016
16	5.4000	17.0812	5.660	-11.6812	13.737	-0.850	*|	0.025
17	4.1000	12.9082	5.748	-8.8082	13.701	-0.643	*|	0.015
18	3.1000	8.7714	6.434	-5.6714	13.392	-0.423	| |	0.008
19	5.4000	15.9926	5.619	-10.5926	13.754	-0.770	*|	0.020
20	8.4000	10.0668	6.707	-1.6668	13.258	-0.126	| |	0.001
21	9.5000	13.0423	7.027	-3.5423	13.091	-0.271	| |	0.004
22	8.7000	9.8128	6.692	-1.1128	13.265	-0.084	| |	0.000
23	10.6000	15.6187	7.500	-5.0187	12.826	-0.391	| |	0.010
24	3.3000	-8.0406	9.996	11.3406	10.992	1.032	|**	0.176

(continued)

FIGURE 7.27   Continued

Obs	Rstudent	Hat Diag H	Cov Ratio	Dffits	INTERCEP Dfbetas	X1 Dfbetas	X2 Dfbetas	X3 Dfbetas	X4 Dfbetas
1	0.0094	0.1112	1.4742	0.0033	-0.0001	0.0020	0.0000	0.0000	0.0001
2	0.0050	0.1114	1.4747	0.0018	-0.0001	0.0011	0.0000	0.0000	0.0001
3	-0.4984	0.1809	1.4939	-0.2342	0.1256	-0.1339	-0.0539	-0.0510	-0.1455
4	0.4891	0.2003	1.5339	0.2447	0.1410	0.0780	-0.0604	-0.0572	-0.1633
5	-0.8606	0.3081	1.5482	-0.5743	0.3965	-0.2848	-0.1700	-0.1609	-0.4592
6	0.1129	0.1138	1.4735	0.0405	0.0054	0.0224	-0.0023	-0.0022	-0.0063
7	0.1192	0.1135	1.4724	0.0427	0.0053	0.0237	-0.0023	-0.0022	-0.0062
8	0.1213	0.1181	1.4799	0.0444	0.0094	0.0234	-0.0040	-0.0038	-0.0108
9	0.4079	0.1731	1.5134	0.1866	0.0964	0.0680	-0.0413	-0.0391	-0.1116
10	0.3791	0.3763	2.0190	0.2945	0.0859	-0.0178	0.1667	-0.0348	-0.0995
11	-0.3202	0.3647	2.0048	-0.2426	0.0614	-0.0127	-0.1967	-0.0249	-0.0711
12	-0.0536	0.3342	1.9667	-0.0380	0.0017	-0.0004	-0.0286	-0.0007	-0.0020
13	179.3101	0.2394	0.0000	100.6096	-55.1618	11.4109	23.6508	69.3701	63.8990
14	-0.7589	0.1429	1.3061	-0.3098	-0.0000	0.0000	0.0000	-0.1873	0.0000
15	-0.6578	0.1489	1.3673	-0.2752	-0.0480	0.0099	0.0206	-0.1435	0.0556
16	-0.8439	0.1451	1.2625	-0.3477	0.0374	-0.0077	-0.0160	-0.2237	-0.0433
17	-0.6327	0.1497	1.3806	-0.2654	-0.0489	0.0101	0.0209	-0.1370	0.0566
18	-0.4141	0.1875	1.5381	-0.1990	-0.0838	0.0173	0.0359	-0.0710	0.0971
19	-0.7616	0.1430	1.3049	-0.3111	0.0096	-0.0020	-0.0041	-0.1919	-0.0112
20	-0.1224	0.2038	1.6389	-0.0619	-0.0237	0.0469	0.0318	0.0409	-0.0084
21	-0.2639	0.2237	1.6557	-0.1417	-0.0278	0.0974	0.0591	0.0797	-0.0461
22	-0.0817	0.2028	1.6408	-0.0412	-0.0164	0.0314	0.0215	0.0276	-0.0049
23	-0.3824	0.2548	1.6888	-0.2236	-0.0104	0.1378	0.0743	0.1054	-0.1037
24	1.0336	0.4527	1.7946	0.9400	0.9216	-0.6183	-0.6154	-0.6930	-0.7023

Sum of Residuals   1.261213E-13
Sum of Squared Residuals   4194.2267
Predicted Resid SS (Press)   7303.4839

A better overall measure of the influence of y_{13} on the fitted regression model is Cook's distance, D_{13}. Recall that Cook's distance is a function of both leverage and the magnitude of the residual. This value, $D_{13} = 1.189$ (shaded) is given in the column labeled **Cook's D** located on the right side of the printout. You can see that the value is extremely large relative to the other values of D_i in the printout. [In fact, $D_{13} = 1.189$ falls in the 65th percentile of the F distribution with $\nu_1 = k + 1 = 5$ and $\nu_2 = n - (k + 1) = 24 - 5 = 19$ degrees of freedom.] This implies that the observed value y_{13} has substantial influence on the estimates of the model parameters.

A statistic related to the deleted residual of the jackknife procedure is the **Studentized deleted residual** given under the column heading **Rstudent**. The Studentized deleted residual, denoted d_i^*, is calculated by dividing the deleted residual d_i by its standard error s_{d_i}:

$$d_i^* = \frac{d_i}{s_{d_i}}$$

The Studentized deleted residual for y_{13} (shaded on the printout) is $d_{13}^* = 53.8929$. This extremely large value[†] is another indication that y_{13} is an influential observation.

The **Dffits** column gives the difference between the predicted value when all 24 observations are used and when the ith observation is deleted. The difference,

---

[†]Under the assumptions of Section 4.2, the Studentized deleted residual d_i^* has a sampling distribution that is approximated by a Student's t distribution with $(n - 1) - (k + 1)$ df.

$\hat{y}_i - \hat{y}_{(i)}$, is divided by its standard error so that the differences can be compared more easily. For observation 13, this scaled difference (shaded on the printout) is $\hat{y}_{13} - \hat{y}_{(13)} = 30.2389$, an extremely large value relative to the other differences in predicted values. Similarly, the changes in the parameter estimates when observation 13 is deleted are given in the **Dfbetas** columns (shaded) immediately to the right of **Dffits** on the printout. (Each difference is also divided by the appropriate standard error.) The large magnitude of these differences provides further evidence that y_{13} is very influential on the regression analysis.

Several techniques designed to dampen the influence an outlying observation has on the regression analysis are available. One method produces estimates of the β's that minimize the sum of the absolute deviations, $\Sigma_{i=1}^{n} |y_i - \hat{y}_i|$.* Because the deviations $(y_i - \hat{y}_i)$ are not squared, this method places less emphasis on outliers than the method of least squares. Regardless of whether you choose to eliminate an outlier or dampen its influence, careful study of residual plots and influence diagnostics are essential for outlier detection.

EXERCISES

7.17 Refer to the data and model of Exercise 7.1. The MSE for the model is .1267. Plot the residuals versus $\hat{y}$. Identify any outliers on the plot.

7.18 Refer to the data and model of Exercise 7.2. The MSE for the model is 17.2557. Plot the residuals versus $\hat{y}$. Identify any outliers on the plot.

7.19 Refer to the data and model of Exercise 7.4. The MSE for the model is 19.45. Plot the residuals versus $\hat{y}$. Identify any outliers on the plot.

7.20 Refer to the grandfather clock example (Example 4.1) in Chapter 4. The least squares model used to predict auction price, y, from age of the clock, x_1, and number of bidders, x_2, was determined to be

$$\hat{y} = -1,336.722 + 12.7362x_1 + 85.8151x_2$$

a. Use this equation to calculate the residuals of each of the prices given in Table 4.2.
b. Calculate the mean and the variance of the residuals. The mean should equal 0 and the variance should be close to the value of MSE given in the portion of the SAS printout shown in Figure 4.7.
c. Determine the proportion of the residuals that fall outside two estimated standard deviations ($2s$) of 0.
d. If you have access to a multiple regression computer package, rerun the analysis and request influence diagnostics. Interpret the measures of influence given on the printout.

7.21 Refer to the study of the population of the world's largest cities in Exercise 7.16. A multiple regression model for log of the population (in thousands) of each country's largest city was fit to data collected

---

*The method of absolute deviations requires linear programming techniques that are beyond the scope of this text. Consult the references given at the end of the chapter for details on how to apply this method.

SAS residual plot for Exercise 7.21

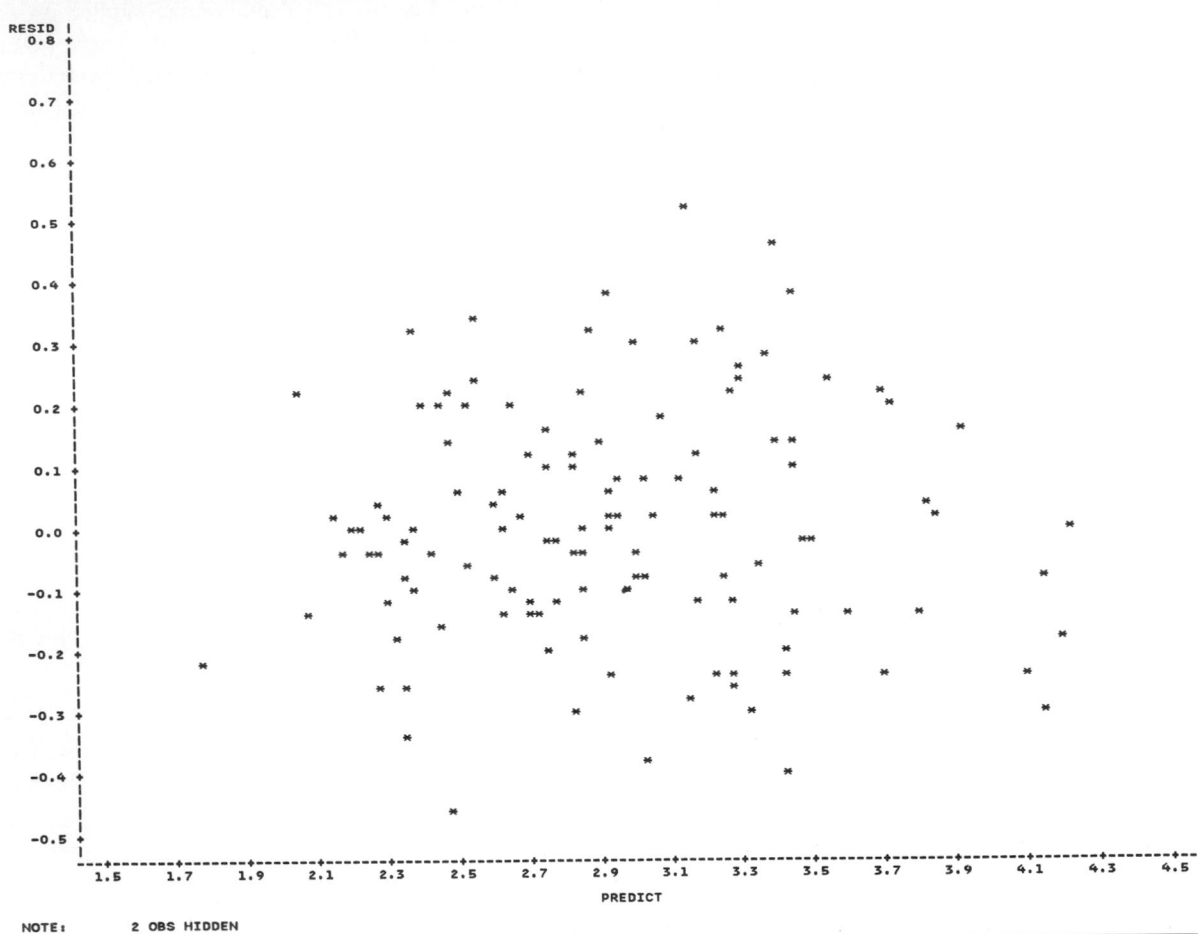

NOTE: 2 OBS HIDDEN

on 126 nations and resulted in MSE = .036. A computer-generated (SAS) plot of the regression residuals versus $\hat{y}$ is shown here. Identify any outliers on the plot.

7.22   A large manufacturing firm wants to determine whether a relationship exists between the number of work-hours an employee misses per year and the employee's annual wages. A sample of 15 employees produced the data in the accompanying table. A first-order model was fit to the data with the following results:

$$\hat{y} = 222.64 - 9.60x \qquad r^2 = .073$$

a. Interpret the value of r^2.
b. Calculate and plot the regression residuals. What do you notice?
c. After searching through its employees' files, the firm has found that employee #13 had been fired but that his name had not been removed from the active employee payroll. This explains the large

EMPLOYEE	WORK-HOURS MISSED y	ANNUAL WAGES x, thousands of dollars
1	49	12.8
2	36	14.5
3	127	8.3
4	91	10.2
5	72	10.0
6	34	11.5
7	155	8.8
8	11	17.2
9	191	7.8
10	6	15.8
11	63	10.8
12	79	9.7
13	543	12.1
14	57	21.2
15	82	10.9

accumulation of work-hours missed (543) by that employee. In view of this fact, what is your recommendation concerning this outlier?

d. Use an available multiple regression program to measure how influential the observation for employee #13 is on the regression analysis.

e. Refit the model to the data, excluding the outlier, and find the least squares line. Calculate r^2 and comment on model adequacy.

7.23 *The New England Journal of Medicine* (Sept. 20, 1990) reported on a study of passive exposure to environmental tobacco smoke in children with cystic fibrosis. The researchers investigated the correlation between a child's weight percentile, y, and the number of cigarettes smoked per day in the child's home, x. The accompanying table lists the data for the 25 boys in the sample.

WEIGHT PERCENTILE, y	NO. OF CIGARETTES SMOKED PER DAY, x	WEIGHT PERCENTILE, y	NO. OF CIGARETTES SMOKED PER DAY, x
6	0	43	0
6	15	49	0
2	40	50	0
8	23	49	22
11	20	46	30
17	7	54	0
24	3	58	0
25	0	62	0
17	25	66	0
25	20	66	23
25	15	83	0
31	23	87	44
35	10		

Source: Rubin, B. K. "Exposure of children with cystic fibrosis to environmental tobacco smoke." *The New England Journal of Medicine*, Sept. 20, 1990, Vol. 323, No. 12, p. 785 (data extracted from Figure 3).

Dependent Variable: Y

Analysis of Variance

Source	DF	Sum of Squares	Mean Square	F Value	Prob>F
Model	1	304.88209	304.88209	0.500	0.4864
Error	23	14011.11791	609.17904		
C Total	24	14316.00000			

Root MSE	24.68155	R-square	0.0213
Dep Mean	37.80000	Adj R-sq	-0.0213
C.V.	65.29511		

Parameter Estimates

Variable	DF	Parameter Estimate	Standard Error	T for H0: Parameter=0	Prob > |T|
INTERCEP	1	41.152655	6.84296599	6.014	0.0001
X	1	-0.261926	0.37024180	-0.707	0.4864

Obs	Dep Var Y	Predict Value	Std Err Predict	Residual	Std Err Residual	Student Residual	-2-1-0 1 2	Cook's D
1	6.0000	41.1527	6.843	-35.1527	23.714	-1.482	**	0.091
2	43.0000	41.1527	6.843	1.8473	23.714	0.078		0.000
3	6.0000	37.2238	5.003	-31.2238	24.169	-1.292	**	0.036
4	49.0000	41.1527	6.843	7.8473	23.714	0.331		0.005
5	2.0000	30.6756	11.215	-28.6756	21.986	-1.304	**	0.221
6	50.0000	41.1527	6.843	8.8473	23.714	0.373		0.006
7	8.0000	35.1284	6.215	-27.1284	23.886	-1.136	**	0.044
8	49.0000	35.3903	5.997	13.6097	23.942	0.568	*	0.010
9	11.0000	35.9141	5.610	-24.9141	24.036	-1.037	**	0.029
10	46.0000	33.2949	8.057	12.7051	23.329	0.545	*	0.018
11	17.0000	39.3192	5.383	-22.3192	24.087	-0.927	*	0.021
12	54.0000	41.1527	6.843	12.8473	23.714	0.542	*	0.012
13	24.0000	40.3669	6.126	-16.3669	23.909	-0.685	*	0.015
14	58.0000	41.1527	6.843	16.8473	23.714	0.710	*	0.021
15	25.0000	41.1527	6.843	-16.1527	23.714	-0.681	*	0.019
16	62.0000	41.1527	6.843	20.8473	23.714	0.879	*	0.032
17	17.0000	34.6045	6.691	-17.6045	23.757	-0.741	*	0.022
18	66.0000	41.1527	6.843	24.8473	23.714	1.048	**	0.046
19	25.0000	35.9141	5.610	-10.9141	24.036	-0.454		0.006
20	66.0000	35.1284	6.215	30.8716	23.886	1.292	**	0.057
21	25.0000	37.2238	5.003	-12.2238	24.169	-0.506	*	0.005
22	83.0000	41.1527	6.843	41.8473	23.714	1.765	***	0.130
23	31.0000	35.1284	6.215	-4.1284	23.886	-0.173		0.001
24	87.0000	29.6279	12.562	57.3721	21.246	2.700	*****	1.275
25	35.0000	38.5334	5.044	-3.5334	24.161	-0.146		0.000

Obs	Rstudent	Hat Diag H	Cov Ratio	Dffits	INTERCEP Dfbetas	X Dfbetas
1	-1.5244	0.0769	0.9686	-0.4399	-0.4399	0.3046
2	0.0762	0.0769	1.1834	0.0220	0.0220	-0.0152
3	-1.3120	0.0411	0.9804	-0.2716	-0.1627	-0.0442
4	0.3244	0.0769	1.1727	0.0936	0.0936	-0.0648
5	-1.3255	0.2065	1.1812	-0.6762	0.2058	-0.6072
6	0.3660	0.0769	1.1697	0.1056	0.1056	-0.0731
7	-1.1433	0.0634	1.0398	-0.2975	-0.0453	-0.1808
8	0.5599	0.0590	1.1292	0.1403	0.0281	0.0797
9	-1.0383	0.0517	1.0474	-0.2424	-0.0741	-0.1152
10	0.5361	0.1066	1.1920	0.1852	-0.0195	0.1463
11	-0.9236	0.0476	1.0635	-0.2064	-0.1936	0.0823
12	0.5333	0.0769	1.1540	0.1539	0.1539	-0.1066
13	-0.6764	0.0616	1.1178	-0.1733	-0.1718	0.1027
14	0.7026	0.0769	1.1326	0.2027	0.2027	-0.1404
15	-0.6730	0.0769	1.1367	-0.1942	-0.1942	0.1345
16	0.8746	0.0769	1.1058	0.2524	0.2524	-0.1748
17	-0.7335	0.0735	1.1240	-0.2066	-0.0134	-0.1395
18	1.0501	0.0769	1.0737	0.3030	0.3030	-0.2099
19	-0.4461	0.0517	1.1319	-0.1041	-0.0318	-0.0495
20	1.3126	0.0634	1.0036	0.3415	0.0520	0.2075
21	-0.4974	0.0411	1.1146	-0.1030	-0.0617	-0.0168
22	1.8561	0.0769	0.8851	0.5356	0.5356	-0.3709
23	-0.1691	0.0634	1.1639	-0.0440	-0.0067	-0.0267
24	3.1959	0.2590	0.6880	1.8896	-0.6678	1.7376
25	-0.1431	0.0418	1.1385	-0.0299	-0.0253	0.0061

a. A SAS regression printout (with residuals) for the straight-line model relating y to x is shown here. Examine the residuals. Do you detect any outliers?

b. Influence diagnostics are also given on the SAS printout. Interpret these results.

7.6

**Detecting Residual
Correlation: The Durbin–
Watson Test**

Many types of business data are observed at regular time intervals. The Consumer Price Index (CPI) is computed and published monthly, the profits of most major corporations are published quarterly, and the *Fortune* 500 list of largest corporations is published annually. Data like these, which are observed over time, are called **time series**. We will often want to construct regression models where the data for the dependent and independent variables are time series.

Regression models of time series may pose a special problem. Because business time series tend to follow economic trends and seasonal cycles, the value of a time series at time t is often indicative of its value at time $(t + 1)$. That is, the value of a time series at time t is **correlated** with its value at time $(t + 1)$. If such a series is used as the dependent variable in a regression analysis, the result is that the random errors are correlated, and this violates one of the assumptions basic to the least squares inferential procedures. Consequently, we cannot apply the standard least squares inference-making tools and have confidence in their validity. Modifications of the methods, which allow for correlated residuals in time series regression models, will be presented in Chapter 9. In this section we present a method of testing for the presence of residual correlation.

Consider the time series data in Table 7.9 which gives sales data for the 35-year history of a company. The computer printout shown in Figure 7.28 (page 400) gives the regression analysis for the first-order linear model

$$y = \beta_0 + \beta_1 t + \varepsilon$$

You will note that this model seems to fit the data very well, since $R^2 = .98$ and the F value (1,615.72) that tests the adequacy of the model is significant. The hypothesis that the coefficient β_1 is positive is accepted at almost any α level ($t = 40.2$ with 33 df).

T A B L E 7.9   **A Firm's Annual Sales Revenue
(thousands of dollars)**

YEAR	SALES	YEAR	SALES	YEAR	SALES
t	y	t	y	t	y
1	4.8	13	48.4	25	100.3
2	4.0	14	61.6	26	111.7
3	5.5	15	65.6	27	108.2
4	15.6	16	71.4	28	115.5
5	23.1	17	83.4	29	119.2
6	23.3	18	93.6	30	125.2
7	31.4	19	94.2	31	136.3
8	46.0	20	85.4	32	146.8
9	46.1	21	86.2	33	146.1
10	41.9	22	89.9	34	151.4
11	45.5	23	89.2	35	150.9
12	53.5	24	99.1		

The residuals $\hat{\varepsilon} - y - (\hat{\beta}_0 + \hat{\beta}_1 t)$ are plotted in Figure 7.29 (page 401). Note that there is a distinct tendency for the residuals to have long positive and negative runs. That is, if the residual for year t is positive, there is a tendency for the

FIGURE 7.28

SAS printout for regression analysis of annual sales model

Analysis of Variance

Source	DF	Sum of Squares	Mean Square	F Value	Prob>F
Model	1	65875.20817	65875.20817	1615.724	0.0001
Error	33	1345.45355	40.77132		
C Total	34	67220.66171			

Root MSE	6.38524	R-square	0.9800	
Dep Mean	77.72286	Adj R-sq	0.9794	
C.V.	8.21540			

Parameter Estimates

Variable	DF	Parameter Estimate	Standard Error	T for H0: Parameter=0	Prob > |T|
INTERCEP	1	0.401513	2.20570829	0.182	0.8567
T	1	4.295630	0.10686692	40.196	0.0001

Durbin-Watson D	0.821
(For Number of Obs.)	35
1st Order Autocorrelation	0.590

Obs	T	Dep Var Y	Predict Value	Residual
1	1	4.8000	4.6971	0.1029
2	2	4.0000	8.9928	-4.9928
3	3	5.5000	13.2884	-7.7884
4	4	15.6000	17.5840	-1.9840
5	5	23.1000	21.8797	1.2203
6	6	23.3000	26.1753	-2.8753
7	7	31.4000	30.4709	0.9291
8	8	46.0000	34.7666	11.2334
9	9	46.1000	39.0622	7.0378
10	10	41.9000	43.3578	-1.4578
11	11	45.5000	47.6534	-2.1534
12	12	53.5000	51.9491	1.5509
13	13	48.4000	56.2447	-7.8447
14	14	61.6000	60.5403	1.0597
15	15	65.6000	64.8360	0.7640
16	16	71.4000	69.1316	2.2684
17	17	83.4000	73.4272	9.9728
18	18	93.6000	77.7229	15.8771
19	19	94.2000	82.0185	12.1815
20	20	85.4000	86.3141	-0.9141
21	21	86.2000	90.6097	-4.4097
22	22	89.9000	94.9	-5.0054
23	23	89.2000	99.2	-10.0010
24	24	99.1	103.5	-4.3966
25	25	100.3	107.8	-7.4923
26	26	111.7	112.1	-0.3879
27	27	108.2	116.4	-8.1835
28	28	115.5	120.7	-5.1792
29	29	119.2	125.0	-5.7748
30	30	125.2	129.3	-4.0704
31	31	136.3	133.6	2.7339
32	32	146.8	137.9	8.9383
33	33	146.1	142.2	3.9427
34	34	151.4	146.5	4.9471
35	35	150.9	150.7	0.1514

Sum of Residuals	-8.10019E-13
Sum of Squared Residuals	1345.4535
Predicted Resid SS (Press)	1484.2108

residual for year $(t + 1)$ to be positive. These cycles are indicative of possible positive correlation between residuals. For most economic time series models, we want to test the null hypothesis

H_0:   No residual correlation

FIGURE 7.29

SAS plot of residuals for the sales data:
Least squares model

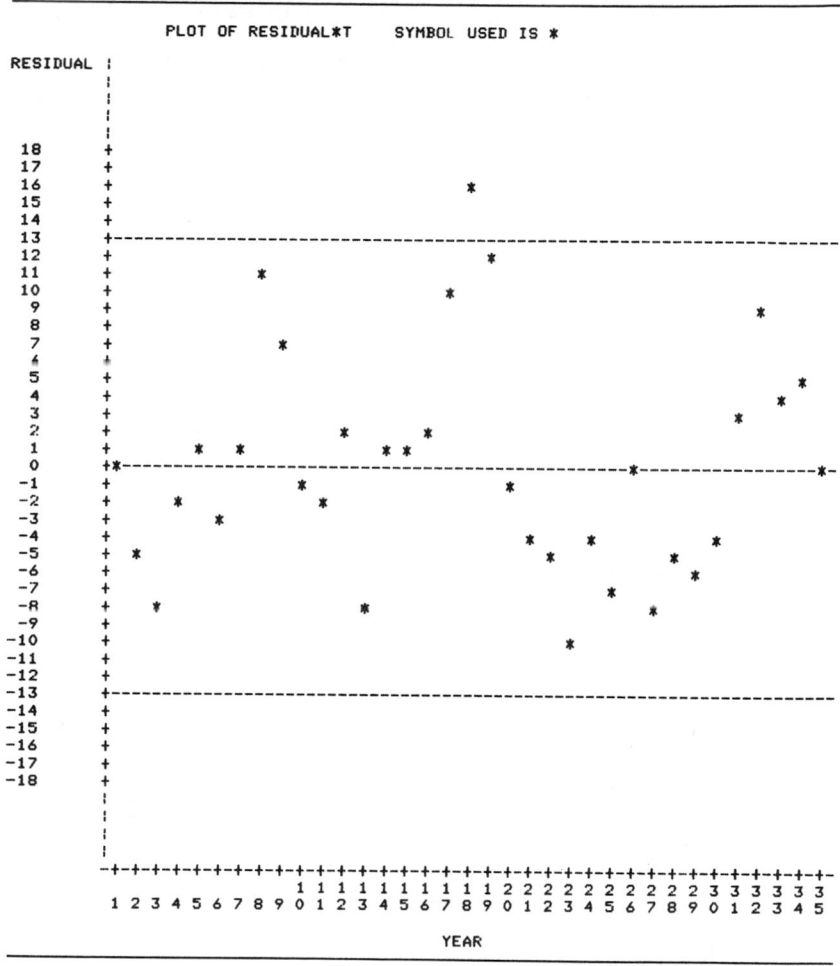

against the alternative

H_a: Positive residual correlation

since the hypothesis of positive residual correlation is consistent with economic trends and seasonal cycles.

The **Durbin–Watson d statistic** is used to test for the presence of residual correlation. This statistic is given by the formula

$$d = \frac{\sum_{t=2}^{n} (\hat{\varepsilon}_t - \hat{\varepsilon}_{t-1})^2}{\sum_{t=1}^{n} \hat{\varepsilon}_t^2}$$

where n is the number of observations and $(\hat{\varepsilon}_t - \hat{\varepsilon}_{t-1})$ represents the difference between a pair of successive residuals. By expanding the numerator of d, we can also write

$$d = \frac{\sum\limits_{t=2}^{n} \hat{\varepsilon}_t^2}{\sum\limits_{t=1}^{n} \hat{\varepsilon}_t^2} + \frac{\sum\limits_{t=2}^{n} \hat{\varepsilon}_{t-1}^2}{\sum\limits_{t=1}^{n} \hat{\varepsilon}_t^2} - \frac{2\sum\limits_{t=2}^{n} \hat{\varepsilon}_t\hat{\varepsilon}_{t-1}}{\sum\limits_{t=1}^{n} \hat{\varepsilon}_t^2} \approx 2 - \frac{2\sum\limits_{t=2}^{n} \hat{\varepsilon}_t\hat{\varepsilon}_{t-1}}{\sum\limits_{t=1}^{n} \hat{\varepsilon}_t^2}$$

If the residuals are uncorrelated,

$$\sum_{t=2}^{n} \hat{\varepsilon}_t\hat{\varepsilon}_{t-1} \approx 0$$

indicating no relationship between $\hat{\varepsilon}_t$ and $\hat{\varepsilon}_{t-1}$, the value of d will be close to 2. If the residuals are highly positively correlated,

$$\sum_{t=2}^{n} \hat{\varepsilon}_t\hat{\varepsilon}_{t-1} \approx \sum_{t=2}^{n} \hat{\varepsilon}_t^2$$

(since $\hat{\varepsilon}_t \approx \hat{\varepsilon}_{t-1}$), and the value of d will be near 0:

$$d \approx 2 - \frac{2\sum\limits_{t=2}^{n} \hat{\varepsilon}_t\hat{\varepsilon}_{t-1}}{\sum\limits_{t=1}^{n} \hat{\varepsilon}_t^2} \approx 2 - \frac{2\sum\limits_{t=2}^{n} \hat{\varepsilon}_t^2}{\sum\limits_{t=1}^{n} \hat{\varepsilon}_t^2} \approx 2 - 2 = 0$$

If the residuals are very negatively correlated, then $\hat{\varepsilon}_t \approx -\hat{\varepsilon}_{t-1}$, so that

$$\sum_{t=2}^{n} \hat{\varepsilon}_t\hat{\varepsilon}_{t-1} \approx -\sum_{t=2}^{n} \hat{\varepsilon}_t^2$$

and d will be approximately equal to 4. Thus, d ranges from 0 to 4, with interpretations as summarized in the box.

| Interpretation of Durbin–Watson d Statistic

DEFINITION

$$d = \frac{\sum\limits_{t=2}^{n} (\hat{\varepsilon}_t - \hat{\varepsilon}_{t-1})^2}{\sum\limits_{t=1}^{n} \hat{\varepsilon}_t^2}$$

Range of d: $0 \le d \le 4$

1. If residuals are uncorrelated, $d \approx 2$.
2. If residuals are positively correlated, $d < 2$, and if the correlation is very strong, $d \approx 0$.
3. If residuals are negatively correlated, $d > 2$, and if the correlation is very strong, $d \approx 4$.

Durbin and Watson (1951) have given tables for the lower-tail values of the *d* statistic, which we show in Table 8 ($\alpha = .05$) and Table 9 ($\alpha = .01$) of Appendix D. Part of Table 8 is reproduced in Table 7.10. For the sales example, we have $k = 1$ independent variable and $n = 35$ observations. Using $\alpha = .05$ for the one-tailed test for positive residual correlation, the table values are $d_L = 1.40$ and $d_U = 1.52$. The meaning of these values is illustrated in Figure 7.30. Because of the complexity of the sampling distribution of *d*, it is not possible to specify a single point that acts as a boundary between the rejection and nonrejection regions, as we did for the *z*, *t*, *F*, and other test statistics. Instead, an upper (d_U) and lower (d_L) bound are specified so that a *d* value less than d_L definitely *does* provide strong evidence of positive residual correlation at $\alpha = .05$ (recall that small *d* values indicate positive correlation), a *d* value greater than d_U *does not* provide evidence of positive correlation at $\alpha = .05$, but a value of *d* between d_L and d_U *might* be significant at the $\alpha = .05$ level. If $d_L < d < d_U$, more information is needed before we can reach any conclusion about the presence of residual correlation.

TABLE 7.10   **Reproduction of Part of Table 9 of Appendix D ($\alpha = .05$)**

n	$k = 1$		$k = 2$		$k = 3$		$k = 4$		$k = 5$	
	d_L	d_U	d_L	d_U	d_L	d_U	d_L	d_U	d_L	d_U
31	1.36	1.50	1.30	1.57	1.23	1.65	1.16	1.74	1.09	1.83
32	1.37	1.50	1.31	1.57	1.24	1.65	1.18	1.73	1.11	1.82
33	1.38	1.51	1.32	1.58	1.26	1.65	1.19	1.73	1.13	1.81
34	1.39	1.51	1.33	1.58	1.27	1.65	1.21	1.73	1.15	1.81
35	1.40	1.52	1.34	1.58	1.28	1.65	1.22	1.73	1.16	1.80
36	1.41	1.52	1.35	1.59	1.29	1.65	1.24	1.73	1.18	1.80
37	1.42	1.53	1.36	1.59	1.31	1.66	1.25	1.72	1.19	1.80
38	1.43	1.54	1.37	1.59	1.32	1.66	1.26	1.72	1.21	1.79
39	1.43	1.54	1.38	1.60	1.33	1.66	1.27	1.72	1.22	1.79
40	1.44	1.54	1.39	1.60	1.34	1.66	1.29	1.72	1.23	1.79

FIGURE 7.30

Rejection region for the Durbin–Watson *d* test: Sales example

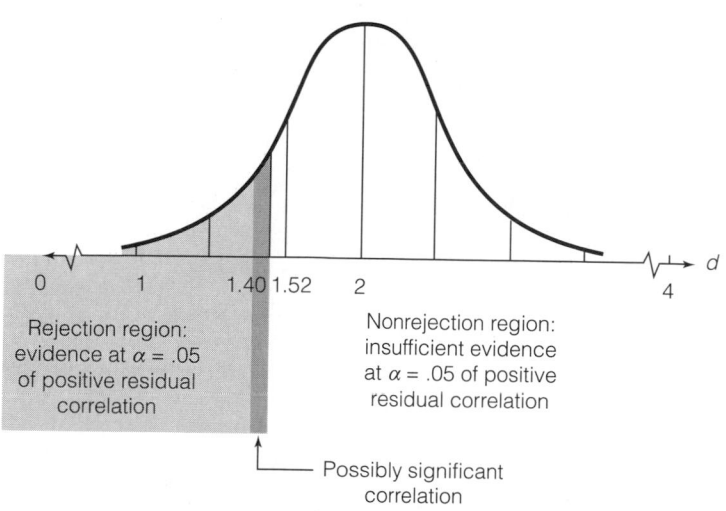

0   1   1.40 1.52   2   4 d

Rejection region: evidence at $\alpha = .05$ of positive residual correlation

Nonrejection region: insufficient evidence at $\alpha = .05$ of positive residual correlation

Possibly significant correlation

| Durbin–Watson d Test

ONE-TAILED TEST

H_0: No residual correlation
H_a: Positive residual correlation
      [or H_a: Negative residual
      correlation]

Test statistic:

$$d = \frac{\sum_{t=2}^{n} (\hat{\varepsilon}_t - \hat{\varepsilon}_{t-1})^2}{\sum_{t=1}^{n} \hat{\varepsilon}_t^2}$$

Rejection region: $d < d_{L,\alpha}$
   [or $(4 - d) < d_{L,\alpha}$ if
   H_a: Negative residual
   correlation]

Nonrejection region: $d > d_{U,\alpha}$
   [or $(4 - d) > d_{U,\alpha}$ if
   H_a: Negative residual
   correlation]

*Inconclusive ("possibly significant")
region:*
   $d_{L,\alpha} \le d \le d_{U,\alpha}$ (see Figure
   7.30) [or $d_{L,\alpha} \le (4 - d) \le$
   $d_{U,\alpha}$ if H_a: Negative residual
   correlation]

where $d_{L,\alpha}$ and $d_{U,\alpha}$ are the lower
and upper tabled values,
respectively, corresponding to k
independent variables and n
observations.

TWO-TAILED TEST

H_0: No residual correlation
H_a: Positive or negative residual
      correlation

Test statistic:

$$d = \frac{\sum_{t=2}^{n} (\hat{\varepsilon}_t - \hat{\varepsilon}_{t-1})^2}{\sum_{t=1}^{n} \hat{\varepsilon}_t^2}$$

Rejection region: $d < d_{L,\alpha/2}$
   or $(4 - d) < d_{L,\alpha/2}$

Nonrejection region: $d > d_{U,\alpha/2}$
   or $(4 - d) > d_{U,\alpha/2}$

*Inconclusive ("possibly significant")
region:*
   Any other result

where $d_{L,\alpha/2}$ and $d_{U,\alpha/2}$ are the
lower and upper tabled values,
respectively, corresponding to k
independent variables and n
observations.

Assumption: The residuals are normally distributed.

As indicated in the printout for the sales example (Figure 7.28), the computed value of d, .82, is less than the tabulated value of d_L, 1.40. Thus, we conclude that the residuals of the straight-line model for sales are positively correlated.

Tests for negative correlation and two-tailed tests can be conducted by making use of the symmetry of the sampling distribution of the d statistic about its mean, 2 (see Figure 7.30). That is, we compare $(4 - d)$ to d_L and d_U, and conclude that the residuals are negatively correlated if $(4 - d) < d_L$, that there is insufficient evidence to conclude that the residuals are negatively correlated if $(4 - d) >$

d_U, and that the test for negative residual correlation is *possibly* significant if $d_L < (4 - d) < d_U$.

Once strong evidence of residual correlation has been established, as in the case of the sales example, doubt is cast on the least squares results and any inferences drawn from them. In Chapter 9 we will present a time series model that accounts for the correlation of the random errors. The residual correlation can be taken into account in a time series model and thereby used to improve both the fit of the model and the reliability of model inferences.

EXERCISES

7.24 Find the values of d_L and d_U from Tables 9 and 10 of Appendix D for each of the following situations:
   a. $n = 30$, $k = 3$, $\alpha = .05$
   b. $n = 40$, $k = 1$, $\alpha = .01$
   c. $n = 35$, $k = 5$, $\alpha = .05$

7.25 Suppose you fit the time series model

$$E(y_t) = \beta_0 + \beta_1 t + \beta_2 t^2$$

to quarterly time series data collected over a 10-year period ($n = 40$ quarters), where $t =$ quarter ($t = 1, 2, 3, \ldots, 40$).
   a. Set up the test of hypothesis for positively correlated residuals. Specify H_0, H_a, the test statistic, and the rejection region.
   b. Suppose the Durbin–Watson d statistic is calculated to be 1.14. What is the appropriate conclusion?

7.26 Forecasts of automotive vehicle sales in the United States provide the basis for financial and strategic planning of large automotive corporations. Olson and Janakiraman developed a forecasting model for y, total monthly passenger car and light truck sales (in thousands):[*]

$$E(y) = \beta_0 + \beta_1 x_1 + \beta_2 x_2 + \beta_3 x_3 + \beta_4 x_4 + \beta_5 x_5$$

where

   $x_1 =$ Average monthly retail price of regular gasoline

   $x_2 =$ Annual percentage change in GNP per quarter

   $x_3 =$ Monthly consumer confidence index

   $x_4 =$ Total number of vehicles scrapped (millions) per month

   $x_5 =$ Vehicle seasonality

The model was fit to monthly data collected over a 12-year period (i.e., $n = 144$ months) with the following results:

$$\hat{y} = -676.42 - 1.93x_1 + 6.54x_2 + 2.02x_3 + .08x_4 + 9.82x_5$$
$$R^2 = .856 \qquad \text{Durbin–Watson } d = 1.01$$

[*]Olson, S. J. and Janakiraman, J. "Proposed U.S. passenger car and light truck sales forecast model." Paper presented at SAS User's Group International Conference, Reno, Nevada, 1985.

a. Is there sufficient evidence to indicate that the model contributes information for the prediction of y? Test using $\alpha = .05$.

b. Is there sufficient evidence to indicate that the regression errors are positively correlated? Test using $\alpha = .05$.

c. Comment on the validity of the inference concerning model adequacy in the light of the result of part **b**.

7.27 The consumer purchasing value of the dollar from 1970 to 1988 is illustrated by the data in the accompanying table, where the purchasing power of the dollar (compared to 1982) is listed for each year. The first-order model $E(y_t) = \beta_0 + \beta_1 t$ was fit to the data using SAS. The SAS printout is displayed here.

Analysis of Variance

Source	DF	Sum of Squares	Mean Square	F Value	Prob>F
Model	1	5.31474	5.31474	142.718	0.0001
Error	17	0.63307	0.03724		
C Total	18	5.94781			

Root MSE	0.19298	R-square	0.8936	
Dep Mean	1.47684	Adj R-sq	0.8873	
C.V.	13.06675			

Parameter Estimates

Variable	DF	Parameter Estimate	Standard Error	T for H0: Parameter=0	Prob > |T|
INTERCEP	1	192.571860	15.99601257	12.039	0.0001
T	1	-0.096561	0.00808285	-11.946	0.0001

Durbin-Watson D	0.173
(For Number of Obs.)	19
1st Order Autocorrelation	0.799

Obs	T	Dep Var Y	Predict Value	Residual
1	1970	2.5500	2.3459	0.2041
2	1971	2.4700	2.2493	0.2207
3	1972	2.3900	2.1528	0.2372
4	1973	2.1900	2.0562	0.1338
5	1974	1.9000	1.9596	-0.0596
6	1975	1.7200	1.8631	-0.1431
7	1976	1.6500	1.7665	-0.1165
8	1977	1.5500	1.6700	-0.1200
9	1978	1.4300	1.5734	-0.1434
10	1979	1.2900	1.4768	-0.1868
11	1980	1.1400	1.3803	-0.2403
12	1981	1.0400	1.2837	-0.2437
13	1982	1.0000	1.1872	-0.1872
14	1983	0.9800	1.0906	-0.1106
15	1984	0.9600	0.9940	-0.0340
16	1985	0.9500	0.8975	0.0525
17	1986	0.9700	0.8009	0.1691
18	1987	0.9500	0.7044	0.2456
19	1988	0.9300	0.6078	0.3222

a. Plot the regression residuals (shown on the printout) against t. Is there a tendency for the residuals to have long positive and negative runs? How do you account for this?

YEAR, t	VALUE, y	YEAR, t	VALUE, y
1970	$2.55	1980	$1.14
1971	2.47	1981	1.04
1972	2.39	1982	1.00
1973	2.19	1983	.98
1974	1.90	1984	.96
1975	1.72	1985	.95
1976	1.65	1986	.97
1977	1.55	1987	.95
1978	1.43	1988	.93
1979	1.29		

b. Locate the Durbin–Watson d statistic for this model on the printout. Test the hypothesis that the time series residuals are positively correlated. Test at $\alpha = .05$.

7.28 The table gives the factory sales (in thousands) of passenger cars in the United States for the years 1989 and 1990. The straight-line model $E(y_t) = \beta_0 + \beta_1 t$ is fit to the data using the method of least squares, with the following results:

$$\hat{y}_t = 607.17 - 5.71t \qquad \text{Durbin–Watson } d = 1.911$$

MONTH	TIME, t	1989 SALES, y	TIME, t	1990 SALES, y
January	1	616	13	335
February	2	606	14	488
March	3	654	15	617
April	4	651	16	509
May	5	672	17	625
June	6	612	18	627
July	7	275	19	346
August	8	540	20	530
September	9	576	21	542
October	10	618	22	625
November	11	453	23	438
December	12	535	24	369

Source: Survey of Current Business, U.S. Dept. of Commerce.

a. Calculate and plot the regression residuals against t. Is there a tendency for the residuals to have long positive and negative runs?

b. Is there evidence at $\alpha = .05$ level of significance that the residuals are autocorrelated?

7.29 B. N. Song compared annual consumption for two lower developed countries (LDCs)—Korea, a poor LDC, and Italy, a rich LDC (*Economic Development and Cultural Change*, Apr. 1981). Using data from the post–Korean War period, Song modeled annual consumption y_t as a function of total labor income x_{1t} and total property income x_{2t}, with the following results (assume data for $n = 40$ years were used in the analysis):

Korea: $\hat{y}_t = 7.81 + .91x_{1t} + .57x_{2t}$

$s = 1.29$

$d = 2.09$

Italy: $\hat{y}_t = 1{,}043.4 + .85x_{1t} + .40x_{2t}$

$s = 290.5$

$d = 1.07$

a. Is there evidence of positively correlated residuals in the consumption model for Korea? Test using $\alpha = .05$.

b. Is there evidence of positively correlated residuals in the consumption model for Italy? Test using $\alpha = .05$.

7.30   T. C. Chiang considered several time series forecasting models of future foreign exchange rates for U.S. currency (*The Journal of Financial Research*, Summer 1986). One popular theory among financial analysts is that the forward (90-day) exchange rate is a useful predictor of the future spot exchange rate. Using monthly data on exchange rates for the British pound for $n = 81$ months, Chiang fit the model

$$E(y_t) = \beta_0 + \beta_1 x_{t-1}$$

where

$y = \ln (\text{spot rate})$ in month t

$x_t = \ln (\text{forward rate})$ in month t

The method of least squares yielded the following results:

$\hat{y}_t = -.009 + .986x_{t-1}$ $(t = 47.9)$

$s = .0249$ $R^2 = .957$ Durbin–Watson $d = .962$

a. Is the model useful for predicting future spot exchange rates for the British pound? Test using $\alpha = .05$.

b. Interpret the values of s and R^2.

c. Is there evidence of positive autocorrelation among the residuals? Test using $\alpha = .05$.

d. Based on the results of parts **a–c**, would you recommend using the least squares model to forecast spot exchange rates?

Summary

An analysis of regression **residuals** can play an important role in the modeling process. Plots of the residuals against the independent variables can suggest modifications that will improve the model. These include the **addition of quadratic terms to allow for curvature** in the response surface, and **transforming the dependent variable to stabilize its variance**. **Histograms, stem-and-leaf displays**, and **normal probability plots** of residuals give visual clues as to whether the normality assumption is satisfied. **Plots of residuals** are also helpful for identification of **outliers**, which can then be traced to determine the cause of an unusually large or small observation. The influence the outlying observation has on the regression analysis can be measured using **leverage, Cook's distance D**, and **deleted residuals**.

When you use residual plots, you should always be aware that conclusions drawn from them are subjective. Therefore, they cannot substitute for formal tests to detect model inadequacies.

The F test (Section 1.9) can be used to detect differences in pairs of variances, and F tests for sets of parameters (Section 4.10) can be used to detect factor interactions and curvature in a response surface. The **Durbin–Watson d test** can be used to test for the presence of **residual correlation**. We will present methods for constructing time series models that allow for correlation of the random errors in Chapter 9.

SUPPLEMENTARY EXERCISES

7.31 Refer to Exercise 1.17. The first-order least squares model was found to be

$$\hat{y} = 95.75 - .3199x$$

where y is the average utility bill (dollars) for a standard-size home and x is the average monthly temperature (°F).

a. Use the data in the table that appears in Exercise 4.47 to calculate the regression residuals for this first-order model.

b. Display a plot of the residuals versus average monthly temperature, x.

c. Are there any observable trends in this plot? If so, what alternative model is suggested by the trend? Does this agree with your answer to part c in Exercise 4.47?

d. Fit the alternative model (part c), calculate the new regression residuals, and make a plot similar to that of part b. Comment on the result of this residual plot.

7.32 A naval base is considering modifying or adding to its fleet of 48 standard aircraft. The final decision regarding the type and number of aircraft to be added depends on a comparison of cost versus effectiveness of the modified fleet. Consequently, the naval base would like to model the projected percentage increase y in fleet effectiveness by the end of the decade as a function of the cost x of modifying the fleet. A first proposal is the quadratic model

$$E(y) = \beta_0 + \beta_1 x + \beta_2 x^2$$

The data provided in the accompanying table were collected on 10 naval bases of a similar size that recently expanded their fleets. The data were used to fit the model, and the SAS printout of the multiple regression analysis is reproduced on page 410.

PERCENTAGE IMPROVEMENT AT END OF DECADE	COST OF MODIFYING FLEET
y	x, millions of dollars
18	125
32	160
9	80
37	162
6	110
3	90
30	140
10	85
25	150
2	50

DEP VARIABLE: Y

ANALYSIS OF VARIANCE

SOURCE	DF	SUM OF SQUARES	MEAN SQUARE	F VALUE	PROB>F
MODEL	2	1368.77501	684.38750	33.079	0.0003
ERROR	7	144.82499	20.68928481		
C TOTAL	9	1513.60000			

ROOT MSE	4.548548	R-SQUARE	0.9043
DEP MEAN	17.2	ADJ R-SQ	0.8770
C.V.	26.44504		

PARAMETER ESTIMATES

VARIABLE	DF	PARAMETER ESTIMATE	STANDARD ERROR	T FOR H0: PARAMETER=0	PROB > :T:
INTERCEP	1	10.65903604	14.55009061	0.733	0.4876
X	1	-0.28160568	0.28087588	-1.003	0.3494
XX	1	0.002671936	0.001253832	2.131	0.0706

OBS	ACTUAL	PREDICT VALUE	STD ERR PREDICT	RESIDUAL	STD ERR RESIDUAL	STUDENT RESIDUAL	-2-1-0 1 2	COOK'S D
1	18.0000	17.2073	2.0627	0.7927	4.0539	0.1955	: : :	0.003
2	32.0000	34.0037	2.6525	-2.0037	3.6951	-0.5423	: *: :	0.051
3	9.0000	5.2310	2.0601	3.7690	4.0553	0.9294	: :* :	0.074
4	37.0000	35.1612	2.8397	1.8388	3.5532	0.5175	: :* :	0.057
5	6.0000	12.0128	2.2177	-6.0128	3.9713	-1.5141	: ***: :	0.238
6	3.0000	6.9572	2.0835	-3.9572	4.0433	-0.9787	: *: :	0.085
7	30.0000	23.6042	1.8468	6.3958	4.1568	1.5387	: :***	0.156
8	10.0000	6.0273	2.0492	3.9727	4.0608	0.9783	: :* :	0.081
9	25.0000	28.5367	2.0070	-3.5367	4.0818	-0.8665	: *: :	0.060
10	2.0000	3.2586	4.1920	-1.2586	1.7654	-0.7129	: *: :	0.955

SUM OF RESIDUALS	4.97380E-14
SUM OF SQUARED RESIDUALS	144.825
PREDICTED RESID SS (PRESS)	301.668

OBS	RESIDUAL	RSTUDENT	HAT DIAG H	COV RATIO	DFFITS	INTERCEP DFBETAS	X DFBETAS	XX DFBETAS
1	0.7927	0.1815	0.2057	1.9665	0.0924	-0.0594	0.0657	-0.0639
2	-2.0037	-0.5129	0.3401	2.1156	-0.3682	-0.1192	0.1505	-0.1871
3	3.7690	0.9190	0.2051	1.3457	0.4669	0.0286	0.0632	-0.1088
4	1.8388	0.4886	0.3898	2.3148	0.3904	0.1473	-0.1817	0.2199
5	-6.0128	-1.7093	0.2377	0.6336	-0.9545	0.5920	-0.7014	0.7211
6	-3.9572	-0.9753	0.2098	1.2924	-0.5025	0.1474	-0.2357	0.2724
7	6.3958	1.7511	0.1649	0.5511	0.7780	-0.3140	0.3143	-0.2585
8	3.9727	0.9748	0.2030	1.2818	0.4919	-0.0653	0.1582	-0.2007
9	-3.5367	-0.8490	0.1947	1.4030	-0.4174	0.0129	0.0093	-0.0503
10	-1.2586	-0.6854	0.8494	8.4082	-1.6276	-1.4594	1.2838	-1.1539

a. Calculate the regression residuals and construct a residual plot versus x. Do you detect any trends? Any outliers?

b. Interpret the influence diagnostics shown on the printout. Are there any observations that have large influence on the analysis?

7.33 In 1974, Congress adopted the Federal-Aid Highway Amendments that reduced the highway speed limit to 55 miles per hour (mph). Since that time, controversy over the social efficiency of the decision has grown. University of Colorado Professors T. H. Ferrester, R. F. McNown, and L. D. Singell conducted an analysis to estimate the effect of the 55-mph speed limit on traffic fatalities (*Southern Economic Journal*, Jan. 1984). Time series data for the United States from 1952 to 1979 ($n = 28$ years) were used to fit a regression model relating traffic fatalities y_t at time t to $k = 7$ independent variables:

x_{1t} = Real earned income

x_{2t} = Vehicle miles

x_{3t} = Ratio of number of youths to number of adults

x_{4t} = Percentage of all car purchases that are imported cars

x_{5t} = Average highway speed

x_{6t} = Percentage of cars traveling between 45 and 60 mph

$$x_{7t} = \begin{cases} 0 & \text{if 55-mph speed limit imposed} \\ 1 & \text{otherwise} \end{cases}$$

The results of the multiple regression are summarized as follows:

$$\hat{y}_t = -20,016.1 + 7,511.85x_{1t} - .01016x_{2t} - 36,750.0x_{3t} - 117.609x_{4t}$$
$$+ 1,325.22x_{5t} - 415.742x_{6t} + 9,678.08x_{7t}$$

$$R^2 = .987 \qquad F = 217.23 \qquad d = 1.97$$

a. Is there evidence that the model is useful for predicting annual traffic fatalities? Test using $\alpha = .05$.

b. Is there evidence that the regression residuals are positively correlated? Test using $\alpha = .05$.

7.34 A leading pharmaceutical company that produces a new hypertension pill would like to model annual revenue generated by this product. Company researchers utilized data collected over the past 15 years (1974–1988) to fit the model

$$E(y_t) = \beta_0 + \beta_1 x_t + \beta_2 t$$

where

y_t = Revenue in year t (in millions of dollars)

x_t = Cost per pill in year t

t = Year (1, 2, . . . , 15)

The SAS printout for the regression analysis is reproduced on page 412. A company statistician suspects that the assumption of independent errors may be violated and that, in fact, the regression residuals are positively correlated. Test this claim using $\alpha = .05$.

7.35 A 10-speed bicycle shop is located near a large southern university. The owner of the shop is having difficulty determining the quantity of bicycles to order each month from the manufacturer. To solve this problem, it is essential that the owner be able to predict the monthly demand for the bikes. The owner proposes the following model:

$$y = \beta_0 + \beta_1 x_1 + \beta_2 x_2 + \beta_3 x_3 + \beta_4 x_4 + \beta_5 x_5 + \varepsilon$$

where

y = Monthly demand for 10-speed bicycles

x_1 = Selling price of 10-speed bicycles

x_2 = Average price of lead-free gasoline

$$x_3 = \begin{cases} 1 & \text{if fall quarter (September–November)} \\ 0 & \text{if not} \end{cases}$$

SAS printout for Exercise 7.34

DEP VARIABLE: Y   ANALYSIS OF VARIANCE

	SOURCE	DF	SUM OF SQUARES	MEAN SQUARE	F VALUE	PROB>F
	MODEL	2	48.82325339	24.41162670	206.187	0.0001
	ERROR	12	1.42074661	0.11839555		
	C TOTAL	14	50.24400000			

ROOT MSE	0.3440865	R-SQUARE	0.9717
DEP MEAN	7.32	ADJ R-SQ	0.9670
C.V.	4.700636		

PARAMETER ESTIMATES

VARIABLE	DF	PARAMETER ESTIMATE	STANDARD ERROR	T FOR H0: PARAMETER=0	PROB > |T|
INTERCEP	1	3.26119109	1.87880228	1.736	0.1082
T	1	0.39158795	0.07045937	5.558	0.0001
X	1	1.58760907	4.12905034	0.384	0.7073

OBS	ID	ACTUAL	PREDICT VALUE	RESIDUAL
1	0.48	5.0000	4.4148	0.5852
2	0.45	4.9000	4.7588	0.1412
3	0.5	5.3000	5.2298	0.0702
4	0.5	5.7000	5.6213	0.0787
5	0.55	5.9000	6.0923	-0.1923
6	0.57	6.1000	6.5157	-0.4157
7	0.6	6.9000	6.9549	-0.0549
8	0.6	7.0000	7.3465	-0.3465
9	0.62	7.5000	7.7698	-0.2698
10	0.6	7.8000	8.1296	-0.3296
11	0.6	8.3000	8.5212	-0.2212
12	0.62	8.9000	8.9446	-0.0446
13	0.65	9.6000	9.3838	0.2162
14	0.7	10.5000	9.8547	0.6453
15	0.71	10.4000	10.2622	0.1378

SUM OF RESIDUALS   1.82077E-14
SUM OF SQUARED RESIDUALS   1.420747

DURBIN-WATSON D   0.776
(FOR NUMBER OF OBS.)   15
1ST ORDER AUTOCORRELATION 0.485

$$x_4 = \begin{cases} 1 & \text{if winter quarter (December–February)} \\ 0 & \text{if not} \end{cases}$$

$$x_5 = \begin{cases} 1 & \text{if spring quarter (March–May)} \\ 0 & \text{if not} \end{cases}$$

Data obtained from past records (given in the accompanying table) were used to fit the first-order model, and a portion of the SAS printout is also shown.

a. Construct and interpret plots of residuals against each of the independent variables, x_1 and x_2.

b. Calculate and plot the partial residuals for x_1. Interpret the plot.

c. Calculate and plot the partial residuals for x_2. Interpret the plot.

d. Check the plots in part **a** for residuals that lie more than 2 estimated standard deviations of ε away from the mean of 0. Can you classify any of the residuals as outliers?

e. Would you recommend that the owner use this model to predict monthly bicycle demand? If not, suggest an alternative model.

SAS printout for Exercise 7.35

Analysis of Variance

Source	DF	Sum of Squares	Mean Square	F Value	Prob>F
Model	5	4773.85553	954.77111	3.383	0.0538
Error	9	2539.74447	282.19383		
C Total	14	7313.60000			

Root MSE	16.79863	R-square	0.6527
Dep Mean	40.60000	Adj R-sq	0.4598
C.V.	41.37593		

Parameter Estimates

Variable	DF	Parameter Estimate	Standard Error	T for H0: Parameter=0	Prob > |T|
INTERCEP	1	-252.617103	308.22378892	-0.820	0.4336
X1	1	-0.789823	0.76446032	-1.033	0.3285
X2	1	285.969152	292.73703924	0.977	0.3542
X3	1	68.488937	23.28222353	2.942	0.0164
X4	1	31.432295	18.55433765	1.694	0.1245
X5	1	25.918087	15.17546604	1.708	0.1218

Durbin-Watson D   2.269
(For Number of Obs.)   15
1st Order Autocorrelation   -0.143

Obs	MONTH	Dep Var Y	Predict Value	Residual
1	MAY90	50.0000	48.7298	1.2702
2	JUN90	31.0000	18.3724	12.6276
3	JUL90	16.0000	26.2706	-10.2706
4	AUG90	22.0000	22.1309	-0.1309
5	SEP90	99.0	78.2823	20.7177
6	OCT90	80.0000	66.2443	13.7557
7	NOV90	37.0000	71.4735	-34.4735
8	DEC90	39.0000	39.0468	-0.0468
9	JAN91	51.0000	39.0468	11.9532
10	FEB91	30.0000	41.9065	-11.9065
11	MAR91	22.0000	31.3537	-9.3537
12	APR91	37.0000	33.1240	3.8760
13	MAY91	47.0000	42.7925	4.2075
14	JUN91	28.0000	23.6832	4.3168
15	JUL91	20.0000	26.5429	-6.5429

Sum of Residuals   1.090683E-12
Sum of Squared Residuals   2539.7445
Predicted Resid SS (Press)   6522.8552

YEAR	MONTH	DEMAND y	SELLING PRICE x_1, dollars	AVERAGE PRICE OF LEAD-FREE GASOLINE x_2, dollars	YEAR	MONTH	DEMAND y	SELLING PRICE x_1, dollars	AVERAGE PRICE OF LEAD-FREE GASOLINE x_2, dollars
1990	May	50	93	1.22	1991	January	51	105	1.20
	June	31	95	1.21		February	30	105	1.21
	July	16	85	1.21		March	22	115	1.22
	August	22	83	1.19		April	37	120	1.24
	September	99	95	1.18		May	47	115	1.26
	October	80	103	1.16		June	28	110	1.27
	November	37	100	1.17		July	20	110	1.28
	December	39	105	1.20					

7.36   The foreman of a printing shop is scheduling his work load for 1992, and he must estimate the number of employees available for work. He asks the company statistician to forecast the absentee rate for 1992. Since it is known that quarterly fluctuations exist, the following model is proposed:

$$y = \beta_0 + \beta_1 x_1 + \beta_2 x_2 + \beta_3 x_3 + \varepsilon$$

where

$$y = \text{Absentee rate} = \frac{\text{Total employees absent}}{\text{Total employees}}$$

$$x_1 = \begin{cases} 1 & \text{if quarter 1 (January–March)} \\ 0 & \text{if not} \end{cases}$$

$$x_2 = \begin{cases} 1 & \text{if quarter 2 (April–June)} \\ 0 & \text{if not} \end{cases}$$

$$x_3 = \begin{cases} 1 & \text{if quarter 3 (July–September)} \\ 0 & \text{if not} \end{cases}$$

YEAR	QUARTER 1	QUARTER 2	QUARTER 3	QUARTER 4
1987	.06	.13	.28	.07
1988	.12	.09	.19	.09
1989	.08	.18	.41	.07
1990	.05	.13	.23	.08
1991	.06	.07	.30	.05

a. Fit the model to the data given in the table.
b. Consider the nature of the response variable, y. Do you think that there may be possible violations of the usual assumptions about ε? Explain.
c. Suggest an alternative model that will approximately stabilize the variance of the error term ε.
d. Fit the alternative model. Check R^2 to determine whether model adequacy has improved.

7.37   Since the energy shortage, the price of foreign crude oil has skyrocketed. Consequently, crude oil imports into the United States have declined. The data in the accompanying table are the amounts of crude oil (millions of barrels) imported into the United States from the Organization of Petroleum Exporting Countries (OPEC) for the years 1973–1985.

YEAR	t	IMPORTS y_t	YEAR	t	IMPORTS y_t
1973	1	767	1980	8	1,414
1974	2	926	1981	9	1,067
1975	3	1,171	1982	10	633
1976	4	1,663	1983	11	540
1977	5	2,058	1984	12	553
1978	6	1,892	1985	13	479
1979	7	1,866			

Source: United States Bureau of the Census, *Statistical Abstracts of the United States, 1975–1987.*

a. Fit the model $E(y) = \beta_0 + \beta_1 t$.

b. Calculate the residuals for the model and plot the residuals against t. Do you detect any trends?

c. Test for correlated residuals using $\alpha = .01$.

7.38 The breeding ability of a thoroughbred horse is sometimes a more important consideration to prospective buyers than racing ability. Usually, the longer a horse lives, the greater its value for breeding purposes. Before marketing a group of horses, a breeder would like to be able to predict their lifelengths. The breeder believes that the gestation period of a thoroughbred horse may be an indicator of its lifelength. The information in the table was supplied to the breeder by various stables in the area. (Note that the horse has the greatest variation of gestation period of any species due to seasonal and feed factors.) The first-order (linear) model

$$y = \beta_0 + \beta_1 x + \varepsilon$$

was fit to the data. A portion of the SAS printout is shown here.

HORSE	GESTATION PERIOD x, days	LIFELENGTH y, years
1	403	30
2	279	22
3	307	7
4	416	31
5	265	21
6	356	27
7	298	25

ANALYSIS OF VARIANCE

SOURCE	DF	SUM OF SQUARES	MEAN SQUARE	F VALUE	PROB>F
MODEL	1	146.31556	146.31556	2.960	0.1459
ERROR	5	247.11301	49.42260285		
C TOTAL	6	393.42857			

ROOT MSE	7.030121	R-SQUARE	0.3719	
DEP MEAN	23.28571	ADJ R-SQ	0.2463	
C.V.	30.1907			

PARAMETER ESTIMATES

VARIABLE	DF	PARAMETER ESTIMATE	STANDARD ERROR	T FOR H0: PARAMETER=0	PROB > |T|
INTERCEP	1	-3.94341407	16.04679821	-0.246	0.8156
X	1	0.08201545	0.04766649	1.721	0.1459

OBS	ID	ACTUAL	PREDICT VALUE	RESIDUAL
1	403	30.0000	29.1088	0.8912
2	279	22.0000	18.9389	3.0611
3	307	7.0000	21.2353	-14.2353
4	416	31.0000	30.1750	0.8250
5	265	21.0000	17.7907	3.2093
6	356	27.0000	25.2541	1.7459
7	298	25.0000	20.4972	4.5028

SUM OF RESIDUALS 7.10543E-15
SUM OF SQUARED RESIDUALS 247.113

a. Check model adequacy by interpreting the F and R^2 statistics.

b. Construct a plot of the residuals versus x, gestation period.

c. Check for residuals that lie outside the interval $0 \pm 2s$ or $0 \pm 3s$.

d. The breeder has been informed that the short life span of horse number 3 (7 years) was due to a very rare disease. Omit the data for horse number 3 and refit the least squares line. Has the omission of this observation improved the model?

7.39   Taxes, a major source of income for the state of Florida, have grown steadily since 1970. The data in the table give the total state tax collections for the years 1971–1985. A tax economist fit the model

$$y = \beta_0 + \beta_1 x + \beta_2 x^2 + \varepsilon$$

to the data. Results from the SAS printout gave the least squares model

$$\hat{y} = 1{,}772.39 + 88.28x + 20.85x^2$$

YEAR	YEAR – 1970 x	TOTAL TAX COLLECTIONS y, million dollars
1971	1	1,587
1972	2	1,996
1973	3	2,488
1974	4	2,794
1975	5	2,791
1976	6	2,936
1977	7	3,275
1978	8	3,764
1979	9	4,291
1980	10	4,804
1981	11	5,314
1982	12	5,556
1983	13	6,225
1984	14	7,329
1985	15	7,883

Source: United States Bureau of the Census, *Statistical Abstracts of the United States 1972–1987.*

a. Calculate the regression residuals for this second-order (quadratic) model.

b. As a check on your calculations, compute the sum of the residuals. (The sum should be very close to 0.)

c. Graph the residuals against the independent variable x. What does the residual plot suggest about the validity of the usual assumptions about the error term ε?

7.40   The data in the table are the monthly market shares for a product over most of the past year. The least squares line relating market share to television advertising expenditure is found to be

$$\hat{y} = -1.56 + .687x$$

a. Calculate and plot the regression residuals in the manner outlined in this section.

b. The response variable y, market share, is recorded as a percentage. What does this lead you to believe about the least squares assumption of homoscedasticity? Does the residual plot substantiate this belief?

MONTH	MARKET SHARE y, %	TELEVISION ADVERTISING EXPENDITURE x, thousands of dollars
January	15	23
February	17	27
March	17	25
May	13	21
June	12	20
July	14	24
September	16	26
October	14	23
December	15	25

c. What variance-stabilizing transformation is suggested by the trend in the residual plot? If you have access to a computer package, refit the first-order model using the transformed responses. Calculate and plot these new regression residuals. Is there evidence that the transformation has been successful in stabilizing the variance of the error term, ε?

References

Barnett, V. and Lewis, T. *Outliers in Statistical Data*. New York: Wiley, 1978.

Belsley, D. A., Kuh, E., and Welsch, R. E. *Regression Diagnostics: Identifying Influential Data and Sources of Collinearity*. New York: Wiley, 1980.

Box, G. E. P. and Cox, D. R. "An analysis of transformations." *Journal of the Royal Statistical Society, Series B*, 1964, Vol. 26, pp. 211–243.

Cook, R. D. "Influential observations in linear regression." *Journal of the American Statistical Association*, 1979, Vol. 74, pp. 169–174.

Draper, N. and Smith, H. *Applied Regression Analysis*, 2nd ed. New York: Wiley, 1981.

Durbin, J. and Watson, G. S. "Testing for serial correlation in least squares regression, I." *Biometrika*, 1950, Vol. 37, pp. 409–428.

Durbin, J. and Watson, G. S. "Testing for serial correlation in least squares regression, II." *Biometrika*, 1951, Vol. 38, pp. 159–178.

Durbin, J. and Watson, G. S. "Testing for serial correlation in least squares regression, III." *Biometrika*, 1971, Vol. 58, pp. 1–19.

Granger, C. W. J. and Newbold, P. *Forecasting Economic Time Series*. New York: Academic Press, 1977.

Larsen, W. A. and McCleary, S. J. "The use of partial residual plots in regression analysis." *Technometrics*, Vol. 14, 1972, pp. 781–790.

Mansfield, E. R. and Conerly, M. D. "Diagnostic value of residual and partial residual plots." *The American Statistician*, Vol. 41, No. 2, May 1987, pp. 107–116.

Mendenhall, W. *Introduction to Linear Models and the Design and Analysis of Experiments*. Belmont, Ca.: Wadsworth, 1968.

Montgomery, D. C. and Peck, E. A. *Introduction to Linear Regression Analysis*. New York: Wiley, 1982.

Neter, J., Wasserman, W., and Kutner, M. H. *Applied Linear Statistical Models*, 3rd ed. Homewood, Ill.: Richard D. Irwin, 1990.

Stephens, M. A. "EDF statistics for goodness of fit and some comparisons." *Journal of the American Statistical Association*, 1974, Vol. 69, pp. 730–737.

Special Topics in Regression (Optional)

CONTENTS

8.1 Introduction

8.2 Piecewise Linear Regression

8.3 Inverse Prediction

8.4 Weighted Least Squares

8.5 Modeling Qualitative Dependent Variables

8.6 Logistic Regression

8.7 Ridge Regression

8.8 Robust Regression

8.9 Model Validation

OBJECTIVE

To introduce a number of special regression techniques for problems that require more advanced methods of analysis

8.1

Introduction

The procedures presented in Chapters 3–7 provide the tools basic to a regression analysis. An understanding of these techniques will enable you to successfully apply regression analysis to a variety of problems encountered in practice. For some studies, however, you may require more sophisticated techniques. In this chapter we introduce several special topics in regression for the advanced student.

8.2

Piecewise Linear Regression

Occasionally, the linear relationship between a dependent variable y and an independent variable x may differ for different intervals over the range of x. For example, it is known that the compressive strength y of concrete depends on the proportion x of water mixed with the cement. A certain type of concrete, when mixed in batches with varying water/cement ratios (measured as a percentage), may yield compressive strengths (measured in pounds per square inch) that follow the pattern shown in Figure 8.1. Note that the compressive strength decreases at a much faster rate for batches with water/cement ratios greater than 70%. That is, the slope of the relationship between compressive strength (y) and water/ cement ratio (x) changes when $x = 70$.

FIGURE 8.1

Relationship between compressive
strength (y) and water/cement ratio (x)

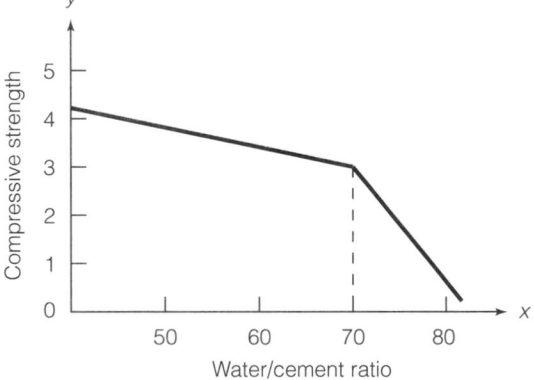

A model that proposes different straight-line relationships for different intervals over the range of x is called a **piecewise linear regression model**. As its name suggests, the linear regression model is fit in pieces. For the concrete example, the piecewise model would consist of two pieces, $x \leq 70$ and $x > 70$. The model can be expressed as follows:

$$y = \beta_0 + \beta_1 x_1 + \beta_2 (x_1 - 70) x_2 + \varepsilon$$

where

$x_1 =$ Water/cement ratio (x)

$$x_2 = \begin{cases} 1 & \text{if } x_1 > 70 \\ 0 & \text{if } x_1 \leq 70 \end{cases}$$

The value of the dummy variable x_2 controls the values of the slope and y-intercept for each piece. For example, when $x_1 \leq 70$, then $x_2 = 0$ and the equation is given by

$$y = \beta_0 + \beta_1 x_1 + \beta_2(x_1 - 70)(0) + \varepsilon$$
$$= \underbrace{\beta_0}_{y\text{-intercept}} + \underbrace{\beta_1 x_1}_{\text{Slope}} + \varepsilon$$

Conversely, if $x_1 > 70$, then $x_2 = 1$ and we have

$$y = \beta_0 + \beta_1 x_1 + \beta_2(x_1 - 70)(1) + \varepsilon$$
$$= \beta_0 + \beta_1 x_1 + \beta_2 x_1 - 70\beta_2 + \varepsilon$$

or

$$y = \underbrace{(\beta_0 - 70\beta_2)}_{y\text{-intercept}} + \underbrace{(\beta_1 + \beta_2)x_1}_{\text{Slope}} + \varepsilon$$

Thus, β_1 and $(\beta_1 + \beta_2)$ represent the slopes of the lines for the two intervals of x, $x \le 70$ and $x > 70$, respectively. Similarly, β_0 and $(\beta_0 - 70\beta_2)$ represent the respective y-intercepts. The slopes and y-intercepts of the two lines are illustrated graphically in Figure 8.2. [*Note:* The value at which the slope changes, 70 in this example, is often referred to as a **knot value**. Usually, the knot values of a piecewise regression are unknown and must be estimated from the sample data. This is often accomplished by visually inspecting the scattergram for the data and locating the points on the x-axis at which the slope appears to change.]

FIGURE 8.2

Slopes and y-intercepts for piecewise linear regression

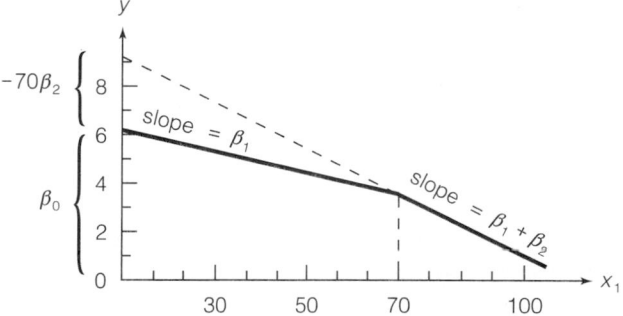

Piecewise regression models can be fit using the standard multiple regression routines of most computer packages by making the appropriate transformations on the independent variables. For example, consider the data on compressive strength (y) and water/cement ratio (x) for 18 batches of concrete recorded in Table 8.1 on page 422. (The water/cement ratio is computed by dividing the weight of water used in the mix by the weight of the cement.) To obtain the least squares fit of the piecewise linear regression model for the data of Table 8.1, we specify the model

$$E(y) = \beta_0 + \beta_1 x_1 + \beta_2 x_2^*$$

where

$$x_2^* = (x_1 - 70)x_2 \quad \text{and} \quad x_2 = \begin{cases} 1 & \text{if } x_1 > 70 \\ 0 & \text{if } x_1 \le 70 \end{cases}$$

The SAS printout for the piecewise linear regression is shown in Figure 8.3. From the printout we obtain the least squares prediction equation:

$$\hat{y} = 7.792 - .06633x_1 - .10119x_2^*$$

Note that the estimated mean change in compressive strength for a 1% increase in water/cement ratio is $\hat{\beta}_1 = -.06633$ for ratios less than or equal to 70% and $\hat{\beta}_1 + \hat{\beta}_2 = -.06633 + (-.10119) = -.16752$ for ratios greater than 70%.

TABLE 8.1   Data on Compressive Strength and Water/Cement Ratios for 18 Batches of Cement

BATCH	COMPRESSIVE STRENGTH y, pounds per square inch	WATER/CEMENT RATIO x, percent	BATCH	COMPRESSIVE STRENGTH y, pounds per square inch	WATER/CEMENT RATIO x, percent
1	4.67	47	10	2.21	73
2	3.54	68	11	4.10	60
3	2.25	75	12	1.13	85
4	3.82	65	13	1.67	80
5	4.50	50	14	1.59	75
6	4.07	55	15	3.91	63
7	.76	82	16	3.15	70
8	3.01	72	17	4.37	50
9	4.29	52	18	3.75	57

FIGURE 8.3

SAS printout for piecewise linear regression of data in Table 8.1

```
DEP VARIABLE: Y
                                          ANALYSIS OF VARIANCE

                                     SUM OF        MEAN
                   SOURCE      DF    SQUARES       SQUARE      F VALUE      PROB>F

                   MODEL        2  24.71775224  12.35887612   114.441      0.0001
                   ERROR       15   1.61989776   0.10799318
                   C TOTAL     17  26.33765000

                   ROOT MSE     0.3286232     R-SQUARE       0.9385
                   DEP MEAN        3.155      ADJ R-SQ       0.9303
                   C.V.        10.41595

                                        PARAMETER ESTIMATES

                                  PARAMETER     STANDARD    T FOR H0:
                   VARIABLE   DF   ESTIMATE       ERROR    PARAMETER=0    PROB > :T:

                   INTERCEP    1   7.79198302    0.67696058    11.510       0.0001
                   X1          1  -0.06633080    0.01123476    -5.904       0.0001
                   X2STAR      1  -0.10118610    0.02812449    -3.598       0.0026
```

Piecewise regression is not limited to two pieces, nor is it limited to straight lines. One or more of the pieces may require a quadratic or higher-order fit. Also, piecewise regression models can be proposed to allow for discontinuities or jumps in the regression function. Such models require additional dummy variables to be introduced. Several different piecewise linear regression models relating y to an independent variable x are shown in the box.

Piecewise Linear Regression Models Relating y to an Independent Variable x_1

TWO STRAIGHT LINES (CONTINUOUS)

$$E(y) = \beta_0 + \beta_1 x_1 + \beta_2(x_1 - k)x_2$$

where

$k = $ Knot value (i.e., the value of the independent variable x_1 at which the slope changes)

$$x_2 = \begin{cases} 1 & \text{if } x_1 > k \\ 0 & \text{if not} \end{cases}$$

	$x_1 \leq k$	$x_1 > k$
y-intercept	β_0	$\beta_0 - k\beta_2$
Slope	β_1	$\beta_1 + \beta_2$

THREE STRAIGHT LINES (CONTINUOUS)

$$E(y) = \beta_0 + \beta_1 x_1 + \beta_2(x_1 - k_1)x_2 + \beta_3(x_1 - k_2)x_3$$

where k_1 and k_2 are knot values of the independent variable x_1, $k_1 < k_2$, and

$$x_2 = \begin{cases} 1 & \text{if } x_1 > k_1 \\ 0 & \text{if not} \end{cases} \qquad x_3 = \begin{cases} 1 & \text{if } x_1 > k_2 \\ 0 & \text{if not} \end{cases}$$

	$x_1 \leq k_1$	$k_1 < x_1 \leq k_2$	$x_1 > k_2$
y-intercept	β_0	$\beta_0 - k_1\beta_2$	$\beta_0 - k_1\beta_2 - k_2\beta_3$
Slope	β_1	$\beta_1 + \beta_2$	$\beta_1 + \beta_2 + \beta_3$

TWO STRAIGHT LINES (DISCONTINUOUS)

$$E(y) = \beta_0 + \beta_1 x_1 + \beta_2(x_1 - k)x_2 + \beta_3 x_2$$

where

$k = $ Knot value (i.e., the value of the independent variable x_1 at which the slope changes—also the point of discontinuity)

$$x_2 = \begin{cases} 1 & \text{if } x_1 > k \\ 0 & \text{if not} \end{cases}$$

	$x_1 \leq k$	$x_1 > k$
y-intercept	β_0	$\beta_0 - k\beta_2 + \beta_3$
Slope	β_1	$\beta_1 + \beta_2$

Tests of model adequacy, tests and confidence intervals on individual β parameters, confidence intervals for $E(y)$, and prediction intervals for y for piecewise regression models are conducted in the usual manner.

EXERCISES

8.1 Consider a two-piece linear relationship between y and x with no discontinuity and a slope change at $x = 15$.
  a. Specify the appropriate piecewise linear regression model for y.
  b. In terms of the β coefficients, give the y-intercept and slope for observations with $x \leq 15$; for observations with $x > 15$.
  c. Explain how you could determine whether the two slopes proposed by the model are, in fact, different.

8.2 Consider a three-piece linear relationship between y and x with no discontinuity and slope changes at $x = 1.45$ and $x = 5.20$.
  a. Specify the appropriate piecewise linear regression model for y.
  b. In terms of the β coefficients, give the y-intercept and slope for each of the three intervals, $x \leq 1.45$, $1.45 < x \leq 5.20$, and $x > 5.20$.
  c. Explain how you could determine whether at least two of the three slopes proposed by the model are, in fact, different.

8.3 Consider a two-piece linear relationship between y and x with discontinuity and slope change at $x = 320$.
  a. Specify the appropriate piecewise linear regression model for y.
  b. In terms of the β coefficients, give the y-intercept and slope for observations with $x \leq 320$; for observations with $x > 320$.
  c. Explain how you could determine whether the two straight lines proposed by the model are, in fact, different.

8.4 The total amount y of corn produced in the United States is recorded (in millions of bushels) for the years 1950–1986 in the accompanying table.

YEAR	PRODUCTION	YEAR	PRODUCTION	YEAR	PRODUCTION
x	y	x	y	x	y
1950	3,075	1963	4,019	1976	6,289
1951	2,926	1964	3,484	1977	6,505
1952	3,292	1965	4,084	1978	7,268
1953	3,210	1966	4,117	1979	7,928
1954	3,058	1967	4,760	1980	6,639
1955	3,220	1968	4,450	1981	8,119
1956	3,445	1969	4,687	1982	8,235
1957	3,400	1970	4,152	1983	4,175
1958	3,725	1971	5,646	1984	7,674
1959	4,197	1972	5,580	1985	8,865
1960	4,314	1973	5,671	1986	8,253
1961	3,598	1974	4,701		
1962	3,606	1975	5,841		

Sources: *Agricultural Statistics, 1984*, U.S. Department of Agriculture (1968–1982); *Historical Statistics of the U.S., Colonial Times to 1970*, U.S. Department of Commerce, Bureau of the Census (1950–1967); *Commodity Year Book 1986*, Commodity Research Bureau (1983–1985); *Survey of Current Business*, U.S. Department of Commerce.

a. Plot the annual corn production y against year x. Note that corn production appears to increase at a much faster rate over the period 1971–1986.
b. Propose a piecewise linear model for annual corn production y with a knot at $x = 1970$.
c. If you have access to a computer program package, fit the model proposed in part b. Give the least squares prediction equation.
d. Is the model adequate for predicting annual corn production y? Test using $\alpha = .05$.
e. Give the estimates of the expected annual increases in corn production over the two periods, 1950–1970 and 1971–1986.
f. Is there sufficient evidence to indicate that annual corn production increases at a faster rate after 1970? Test using $\alpha = .05$.

8.5 The manager of a packaging plant wants to model the unit cost y of shipping lots of a semifragile product as a linear function of lot size x. Because of economies of scale, the manager believes that the cost per unit will decrease at a faster rate for lot sizes of more than 1,000. Data collected on unit cost and lot size for 15 recent shipments are given in the table.

SHIPPING COST y, \$ per unit	LOT SIZE x	SHIPPING COST y, \$ per unit	LOT SIZE x
1.29	1,150	2.90	520
2.20	840	2.63	670
2.26	900	.55	1,420
2.38	800	2.31	850
1.77	1,070	1.90	1,000
1.25	1,220	2.15	910
1.87	980	1.20	1,230
.71	1,300		

a. Specify the appropriate piecewise linear model for y.
b. If you have access to a computer program package, fit the model to the data. Give the least squares prediction equation.
c. Is the model adequate for predicting unit cost y? Test using $\alpha = .10$.
d. Give a 90% confidence interval for the mean increase in shipping cost per unit for every unit increase in lot size for lots with 1,000 or fewer units.

8.3

Inverse Prediction

Often, the goal of regression is to predict the value of one variable when another variable takes on a specified value. For most simple linear regression problems, we are interested in predicting y for a given x. We provided a formula for a prediction interval for y when $x = x_p$ in Section 3.9. In this section, we discuss **inverse prediction**—that is, predicting x for a given value of the dependent variable y.

Inverse prediction has many applications in the engineering and physical sciences, in medical research, and in business. For example, when calibrating a new instrument, scientists often search for approximate measurements y, which are easy and inexpensive to obtain, and which are related to the precise, but more expensive and time-consuming measurements x. If a regression analysis reveals that x and y are highly correlated, then the scientist could choose to use the quick and inexpensive approximate measurement value, say, $y = y_p$, to estimate the unknown precise measurement x. (In this context, the problem of inverse prediction is sometimes referred to as a **linear calibration** problem.) Physicians often use inverse prediction to determine the required level of dosage of a drug. Suppose a regression analysis conducted on patients with high blood pressure showed that a linear relationship exists between decrease in blood pressure y and dosage x of a new drug. Then a physician treating a new patient may want to determine what dosage x to administer to reduce the patient's blood pressure by an amount $y = y_p$. To illustrate inverse prediction in a business setting, consider a firm that sells a particular product. Suppose the firm's monthly market share y is linearly related to its monthly television advertising expenditure x. For a particular month, the firm may want to know how much it must spend on advertising x to attain a specified market share $y = y_p$.

The classical approach to inverse prediction is first to fit the familiar straight-line model

$$y = \beta_0 + \beta_1 x + \varepsilon$$

to a sample of n data points and obtain the least squares prediction equation

$$\hat{y} = \hat{\beta}_0 + \hat{\beta}_1 x$$

Solving the least squares prediction equation for x, we have

$$x = \frac{\hat{y} - \hat{\beta}_0}{\hat{\beta}_1}$$

Now let y_p be an observed value of y in the future with unknown x. Then a point estimate of x is given by

$$\hat{x} = \frac{y_p - \hat{\beta}_0}{\hat{\beta}_1}$$

Although no exact expression for the standard error of $\hat{x}$ (denoted $s_{\hat{x}}$) is known, we can algebraically manipulate the formula for a prediction interval for y given x (see Section 3.9) to form a prediction interval for x given y. It can be shown (proof omitted) that an approximate $(1 - \alpha)100\%$ prediction interval for x when $y = y_p$ is

$$\hat{x} \pm t_{\alpha/2} s_{\hat{x}} \approx \hat{x} \pm t_{\alpha/2} \left(\frac{s}{\hat{\beta}_1} \right) \sqrt{1 + \frac{1}{n} + \frac{(\hat{x} - \bar{x})^2}{SS_{xx}}}$$

where the distribution of t is based on $(n - 2)$ degrees of freedom, $s = \sqrt{MSE}$, and

$$SS_{xx} = \sum x^2 - n(\bar{x})^2$$

This approximation is appropriate as long as the quantity

$$D = \left(\frac{t_{\alpha/2} s}{\hat{\beta}_1} \right)^2 \cdot \frac{1}{SS_{xx}}$$

is small. The procedure for constructing an approximate confidence interval for x in inverse prediction is summarized in the box.

Inverse Prediction: Approximate $(1 - \alpha)100\%$ Prediction Interval for x when $y = y_p$ in Simple Linear Regression

$$\hat{x} \pm t_{\alpha/2}\left(\frac{s}{\hat{\beta}_1}\right)\sqrt{1 + \frac{1}{n} + \frac{(\hat{x} - \bar{x})^2}{SS_{xx}}}$$

where

$$\hat{x} = \frac{y_p - \hat{\beta}_0}{\hat{\beta}_1}$$

$\hat{\beta}_0$ and $\hat{\beta}_1$ are the y-intercept and slope, respectively, of the least squares line

$$n = \text{Sample size}$$
$$\bar{x} = \frac{\sum x}{n}$$
$$SS_{xx} = \sum x^2 - n(\bar{x})^2$$
$$s = \sqrt{MSE}$$

and the distribution of t is based on $(n - 2)$ degrees of freedom.

The approximation is appropriate when the quantity

$$D = \left(\frac{t_{\alpha/2}s}{\hat{\beta}_1}\right)^2 \cdot \frac{1}{SS_{xx}}$$

is small.*

EXAMPLE 8.1

A firm that sells copiers advertises regularly on television. One goal of the firm is to determine the amount it must spend on television advertising in a single month to gain a market share of 10%. For one year, the firm varied its monthly television advertising expenditures (x) and at the end of each month determined its market share (y). The data for the 12 months are recorded in Table 8.2 on page 428.

a. Fit the straight-line model $y = \beta_0 + \beta_1 x + \varepsilon$ to the data.
b. Is there evidence that television advertising expenditure x is linearly related to market share y? Test using $\alpha = .05$.
c. Use inverse prediction to estimate the amount that must be spent on television advertising in a particular month for the firm to gain a market share of $y = 10\%$. Construct an approximate 95% prediction interval for monthly television advertising expenditure x.

---

*Neter, Wasserman, and Kutner (1990) and others suggest using the approximation when D is less than .1.

TABLE 8.2 **A Firm's Market Share and Television Advertising Expenditure for 12 Months, Example 8.1**

MONTH	MARKET SHARE y, percent	TELEVISION ADVERTISING EXPENDITURE x, $ thousands
January	7.5	23
February	8.5	25
March	6.5	21
April	7.0	24
May	8.0	26
June	6.5	22
July	9.5	27
August	10.0	31
September	8.5	28
October	11.0	32
November	10.5	30
December	9.0	29

Solution

a. The SAS printout for the simple linear regression is shown in Figure 8.4. From the printout, we obtain the least squares line

$$\hat{y} = -1.975 + .397x$$

The least squares line is plotted along with 12 data points in Figure 8.5.

FIGURE 8.4

Least squares fit of straight-line model, Example 8.1

ANALYSIS OF VARIANCE

SOURCE	DF	SUM OF SQUARES	MEAN SQUARE	F VALUE	PROB>F
MODEL	1	22.52141608	22.52141608	83.174	0.0001
ERROR	10	2.70775058	0.27077506		
C TOTAL	11	25.22916667			

ROOT MSE	0.5203605	R-SQUARE	0.8927
DEP MEAN	8.541667	ADJ R-SQ	0.8819
C.V.	6.092025		

PARAMETER ESTIMATES

VARIABLE	DF	PARAMETER ESTIMATE	STANDARD ERROR	T FOR H0: PARAMETER=0	PROB > |T|
INTERCEP	1	-1.97494172	1.16288320	-1.698	0.1203
X	1	0.39685315	0.04351473	9.120	0.0001

b. To determine whether television advertising expenditure x is linearly related to market share y, we test the hypothesis

$$H_0: \quad \beta_1 = 0$$

against

$$H_a: \quad \beta_1 \neq 0$$

The value of the test statistic, shaded on the printout, is $t = 9.12$, and the associated p-value of the test is $p = .0001$ (also shaded). Thus, there is

FIGURE 8.5

Scattergram of data and least squares line, Example 8.1

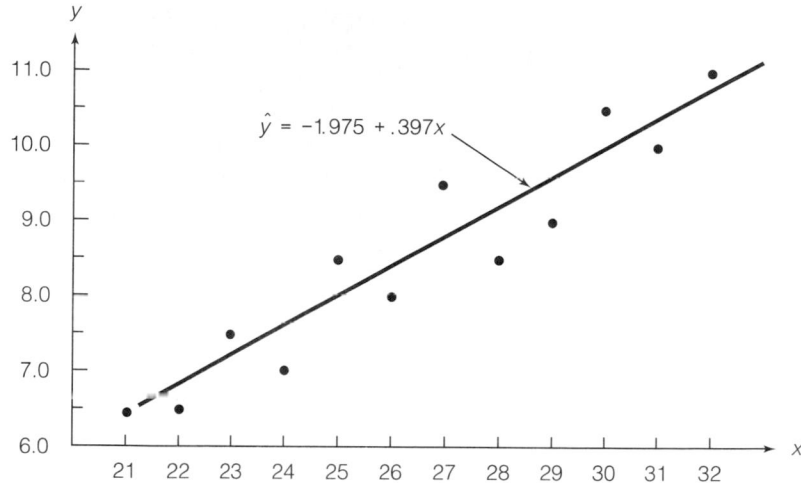

$$\hat{y} = -1.975 + .397x$$

sufficient evidence, at any value of α greater than .0001, to indicate that television advertising expenditure x and market share y are linearly related.

Caution: One should avoid using inverse prediction when there is insufficient evidence to reject the null hypothesis $H_0: \beta_1 = 0$. Inverse predictions made when x and y are *not* linearly related may lead to nonsensical results. Therefore, you should always conduct a test of model adequacy to be sure that x and y are linearly related before you carry out an inverse prediction.

c. Since the model is found to be adequate, we can use the model to predict x from y. For this example, we want to estimate the television advertising expenditure x that yields a market share of $y_p = 10\%$.

Substituting $y_p = 10$, $\hat{\beta}_0 = -1.975$, and $\hat{\beta}_1 = .397$ into the formula for $\hat{x}$ given in the box, we have

$$\hat{x} = \frac{y_p - \hat{\beta}_0}{\hat{\beta}_1}$$

$$= \frac{10 - (-1.975)}{.397} = 30.16$$

Thus, we estimate that the firm must spend $30,160 on television advertising in a particular month to gain a market share of 10%.

Before we construct an approximate 95% prediction interval for x, we check to determine whether the approximation is appropriate, that is, whether the quantity

$$D = \left(\frac{t_{\alpha/2}s}{\hat{\beta}_1}\right)^2 \cdot \frac{1}{SS_{xx}}$$

is small. For $\alpha = .05$, $t_{\alpha/2} = t_{.025} = 2.228$ for $n - 2 = 12 - 2 = 10$ degrees of freedom. From the printout (Figure 8.4), $s = $ **ROOT MSE** $= .5204$ and $\hat{\beta}_1 = .397$. The value of SS_{xx} is not shown on the printout and must be calculated:

$$SS_{xx} = \sum x^2 - n(\bar{x})^2$$

$$= 8{,}570 - 12\left(\frac{318}{12}\right)^2$$

$$= 8{,}570 - 8{,}427$$

$$= 143$$

Substituting these values into the formula for D, we have

$$D = \left(\frac{t_{\alpha/2}s}{\hat{\beta}_1}\right)^2 \cdot \frac{1}{SS_{xx}}$$

$$= \left[\frac{(2.228)(.5204)}{.397}\right]^2 \cdot \frac{1}{143}$$

$$= \frac{8.5295}{143}$$

$$= .0596$$

Since the value of D is small (i.e., less than .1), we may use the formula for the approximate 95% prediction interval given in the box:

$$\hat{x} \pm t_{\alpha/2}\left(\frac{s}{\hat{\beta}_1}\right)\sqrt{1 + \frac{1}{n} + \frac{(\hat{x} - \bar{x})^2}{SS_{xx}}}$$

$$30.16 \pm (2.228)\frac{(.5204)}{.397}\sqrt{1 + \frac{1}{12} + \frac{\left(30.16 - \frac{318}{12}\right)^2}{143}}$$

$$30.16 \pm (2.9205)(1.0849)$$

$$30.16 \pm 3.17$$

or (26.99, 33.33). Therefore, using the 95% prediction interval, we estimate that the amount of monthly television advertising expenditure required to gain a market share of 10% falls between \$26,999 and \$33,330.

Another approach to the inverse prediction problem is to regress x on y, i.e., fit the model (called the **inverse estimator model**)

$$x = \beta_0 + \beta_1 y + \varepsilon$$

and then use the standard formula for a prediction interval given in Section 3.9. However, in theory this method requires that x be a random variable. In many business applications, the value of x is set in advance (i.e., controlled) and therefore is *not* a random variable. (For example, the firm in Example 8.1 selected the amount x spent on advertising *prior* to each month.) Thus, the inverse model above may violate the standard least squares assumptions given in Chapter 4. Some researchers advocate the use of the inverse model despite this caution, whereas others have developed different estimators of x using a modification of the classical approach. Consult the references given at the end of this chapter for details on the various alternative methods of inverse prediction.

EXERCISES

8.6 The data for Exercise 3.22 are reproduced in the accompanying table. Use inverse prediction to estimate the subsidy rate x for a particular industry that laid off workers at a rate of $y = 6.50$ workers per 1,000. Construct an approximate 99% prediction interval for x.

INDUSTRY	SUBSIDY RATE x	LAYOFF RATE y
Apparel	57%	12.54
Chemicals	32	1.78
Construction	31	7.10
Electrical machinery	29	8.38
Fabricated metals	27	11.72
Food	36	5.10
Machinery	32	4.44
Miscellaneous manufacturing	61	9.82
Primary metals	23	7.34
Retail	27	1.98
Wholesale trade	33	1.86

Source: Tropel, R. H. "On layoffs and unemployment insurance." *American Economic Review*, 1983. Vol. 83, pp. 541–559.

8.7 The data for Exercise 3.24 are reproduced in the table.

STARTING SALARY y	GMAT x
$40,000	510
33,000	510
40,000	550
35,000	600
28,000	600
28,000	520
34,000	560
30,000	530
32,500	530
31,000	590

Source: Graduate College of Business Administration, University of Florida.

a. Use inverse prediction to estimate the GMAT score x of a former University of Florida MBA student with a starting salary of $y = \$35,000$. Construct an approximate 95% prediction interval for x.
b. Explain why the interval of part a is so wide.

8.8   The data in Table 3.8 are reproduced here. Use inverse prediction to estimate the distance from nearest fire station, x, for a residential fire that caused $18,200 in damages. Construct a 90% prediction interval for x.

DISTANCE FROM FIRE STATION x, miles	FIRE DAMAGE y, thousands of dollars
3.4	26.2
1.8	17.8
4.6	31.3
2.3	23.1
3.1	27.5
5.5	36.0
.7	14.1
3.0	22.3
2.6	19.6
4.3	31.3
2.1	24.0
1.1	17.3
6.1	43.2
4.8	36.4
3.8	26.1

8.9   A pharmaceutical company has developed a new drug designed to reduce a smoker's reliance on tobacco. Since certain dosages of the drug may reduce one's pulse rate to dangerously low levels, the product-testing division of the pharmaceutical company wants to model the relationship between decrease in pulse rate y (beats/minute) and dosage x (cubic centimeters). Different dosages of the drug were administered to eight randomly selected patients, and 30 minutes later the decrease in each patient's pulse rate was recorded, with the results given in the table.

PATIENT	DOSAGE x, cubic centimeters	DECREASE IN PULSE RATE y, beats/minute
1	2.0	12
2	4.0	20
3	1.5	6
4	1.0	3
5	3.0	16
6	3.5	20
7	2.5	13
8	3.0	18

a. Fit the straight-line model $E(y) = \beta_0 + \beta_1 x$ to the data.
b. Conduct a test for model adequacy. Use $\alpha = .05$.
c. Use inverse prediction to estimate the appropriate dosage x to administer to reduce a patient's pulse rate $y = 10$ beats per minute. Construct an approximate 95% prediction interval for x.

8.4

Weighted Least Squares

Consider the general linear model

$$y = \beta_0 + \beta_1 x_1 + \beta_2 x_2 + \cdots + \beta_k x_k + \varepsilon$$

To obtain the least squares estimates of the unknown β parameters, recall (from Section 4.3) that we minimize the quantity

$$SSE = \sum_{i=1}^{n} (y_i - \hat{y}_i)^2 = \sum_{i=1}^{n} [y_i - (\hat{\beta}_0 + \hat{\beta}_1 x_{1i} + \hat{\beta}_2 x_{2i} + \cdots + \hat{\beta}_k x_{ki})]^2$$

with respect to $\hat{\beta}_0, \hat{\beta}_1, \ldots, \hat{\beta}_k$.

The least squares criterion weighs each observation equally in determining the estimates of the β's. Sometimes we will want to weigh some observations more heavily than others. To do this we minimize

$$WSSE = \sum_{i=1}^{n} w_i (y_i - \hat{y}_i)^2$$

$$= \sum_{i=1}^{n} w_i [y_i - (\hat{\beta}_0 + \hat{\beta}_1 x_{1i} + \hat{\beta}_2 x_{2i} + \cdots + \hat{\beta}_k x_{ki})]^2$$

where w_i is the weight assigned to the ith observation. This procedure is known as **weighted least squares** and the resulting parameter estimates are called **weighted least squares estimates**. [Note that the ordinary least squares procedure assigns a weight of $w_i = 1$ to each observation.]

Weighted least squares has applications in the following two areas:

1. Stabilizing the variance of ε to satisfy the standard regression assumption of homoscedasticity
2. Dampening the influence of outlying observations on the regression analysis

Although the two applications are related, our discussion of weighted least squares in this section is geared toward the first application.

The regression routines of most statistical software packages have options for conducting a weighted least squares analysis. However, the weights w_i must be specified. When using weighted least squares as a variance-stabilizing technique, the weight for the ith observation should be the reciprocal of the variance of that observation's error term, σ_i^2, i.e.,

$$w_i = \frac{1}{\sigma_i^2}$$

In this manner, observations with larger error variances will receive less weight (and hence have less influence on the analysis) than observations with smaller error variances.

In practice, the actual variances σ_i^2 will usually be unknown. Fortunately, in many business applications, the error variance σ_i^2 is proportional to one or more of the levels of the independent variables. This fact will allow us to determine the appropriate weights to use. For example, in a simple linear regression problem, suppose we know that the error variance σ_i^2 increases proportionally with the value of the independent variable x_i, i.e.,

$$\sigma_i^2 = kx_i$$

where k is some unknown constant. Then the appropriate (albeit unknown) weight to use is

$$w_i = \frac{1}{kx_i}$$

Fortunately, it can be shown (proof omitted) that k can be ignored and the weights can be assigned as follows:

$$w_i = \frac{1}{x_i}$$

If the functional relationship between σ_i^2 and x_i is not known prior to conducting the analysis, the weights can be estimated based on the results of an ordinary (unweighted) least squares fit. For example, in simple linear regression, one approach is to divide the regression residuals into several groups of approximately equal size based on the value of the independent variable x and calculate the variance of the observed residuals in each group. An examination of the relationship between the residual variances and several different functions of x (such as x, x^2, and $\sqrt{x}$) may reveal the appropriate weights to use.

| EXAMPLE 8.2

A Department of Transportation (DOT) official is investigating the possibility of collusive bidding among the state's road construction contractors. One aspect of the investigation involves a comparison of the winning (lowest) bid price y on a job with the length x of new road construction, a measure of job size. The data listed in Table 8.3 were supplied by the DOT for a sample of 11 new road construction jobs with approximately the same number of bidders.

TABLE 8.3   Sample Data for New Road Construction Jobs, Example 8.2

JOB	LENGTH OF ROAD x, miles	WINNING BID PRICE y, $ thousands	JOB	LENGTH OF ROAD x, miles	WINNING BID PRICE y, $ thousands
1	2.0	10.1	7	7.0	71.1
2	2.4	11.4	8	11.5	132.7
3	3.1	24.2	9	10.9	108.0
4	3.5	26.5	10	12.2	126.2
5	6.4	66.8	11	12.6	140.7
6	6.1	53.8			

a. Use the method of least squares to fit the straight-line model

$$E(y) = \beta_0 + \beta_1 x$$

b. Calculate and plot the regression residuals against x. Do you detect any evidence of heteroscedasticity?

c. Use the method described in the preceding paragraph to find the approximate weights necessary to stabilize the error variances with weighted least squares.

d. Carry out the weighted least squares analysis using the weights determined in part c.

e. The **weighted least squares residuals** are defined as

$$\sqrt{w_i}(y_i - \hat{y}_i)$$

where $\hat{y}_i$ is the predicted value from the weighted least squares fit and w_i is the weight. Plot the weighted least squares residuals against x to determine whether the variances have stabilized.

Solution

a. The simple linear regression analysis was conducted using the SAS regression package. The SAS printout appears in Figure 8.6. The least squares line, obtained from the printout, is

$$\hat{y} = -15.11 + 12.07x$$

Note that the model is statistically useful (reject H_0: $\beta_1 = 0$) at $p = .0001$.

b. The regression residuals are calculated and reported in the bottom portion of the SAS printout (Figure 8.6). A plot of the residuals against the predictor

FIGURE 8.6

SAS printout for least squares fit of straight-line model, Example 8.2

DEP VARIABLE: Y

ANALYSIS OF VARIANCE

SOURCE	DF	SUM OF SQUARES	MEAN SQUARE	F VALUE	PROB>F
MODEL	1	24557.87884	24557.87884	850.451	0.0001
ERROR	9	259.88662	28.87629072		
C TOTAL	10	24817.76545			

ROOT MSE	5.373666	R-SQUARE	0.9895
DEP MEAN	70.13636	ADJ R-SQ	0.9884
C.V.	7.661741		

PARAMETER ESTIMATES

VARIABLE	DF	PARAMETER ESTIMATE	STANDARD ERROR	T FOR H0: PARAMETER=0	PROB > :T:
INTERCEP	1	-15.11237262	3.34221489	-4.522	0.0014
X	1	12.06867566	0.41384231	29.162	0.0001

OBS	ID	ACTUAL	PREDICT VALUE	RESIDUAL
1	2	10.1000	9.0250	1.0750
2	2.4	11.4000	13.8524	-2.4524
3	3.1	24.2000	22.3005	1.8995
4	3.5	26.5000	27.1280	-0.6280
5	6.4	66.8000	62.1272	4.6728
6	6.1	53.8000	58.5065	-4.7065
7	7	71.1000	69.3684	1.7316
8	11.5	132.7	123.7	9.0226
9	10.9	108.0	116.4	-8.4362
10	12.2	126.2	132.1	-5.9255
11	12.6	140.7	137.0	3.7471

SUM OF RESIDUALS   -2.86438E-14
SUM OF SQUARED RESIDUALS   259.8866

variable x is shown in Figure 8.7. The residual plot clearly shows that the residual variance increases as length of road x increases, strongly suggesting the presence of heteroscedasticity. A procedure such as weighted least squares is needed to stabilize the variances.

c. To apply weighted least squares, we must first determine the weights. Since it is not clear what function of x the error variance is proportional to, we will apply the procedure described previously to estimate the weights.

First, we must divide the data into several groups according to the value of the independent variable x. Ideally, we want to form one group of data points for each different value of x. However, unless each value of x is replicated, not all of the group residual variances can be calculated. Therefore, we resort to grouping the data according to "nearest neighbors." One choice would be to use three groups, $2 \leq x \leq 4$, $6 \leq x \leq 7$, and $10 \leq x \leq 13$. These groups have approximately the same numbers of observations (namely, 4, 3, and 4 observations, respectively).

Next, we calculate the sample variance s_i^2 of the residuals included in each group. The three residual variances are given in Table 8.4. These variances are compared to three different functions of $\bar{x}$ ($\bar{x}$, $\bar{x}^2$, and $\sqrt{x}$), as shown in Table 8.4, where $\bar{x}$ is the mean road length x for each group.

Note that the ratio $s_i^2/\bar{x}^2$ yields a value near .5 for each of the three groups. This result suggests that the residual variance of each group is proportional to $\bar{x}^2$, i.e.,

$$\sigma_i^2 = k\bar{x}_i^2, \quad i = 1, 2, 3$$

TABLE 8.4   **Comparison of Residual Variances to Three Functions of $\bar{x}$, Example 8.2**

GROUP	RANGE OF x	$\bar{x}_i$	s_i^2	$s_i^2/\bar{x}_i$	$s_i^2/\bar{x}_i^2$	$s_i^2/\sqrt{\bar{x}_i}$
1	$2 \leq x \leq 4$	2.75	3.722	1.353	.492	2.244
2	$6 \leq x \leq 7$	6.5	23.016	3.541	.545	9.028
3	$10 \leq x \leq 13$	11.8	67.031	5.681	.481	19.514

where k is approximately .5. Thus, a reasonable approximation to the weight for each group is

$$w_i = \frac{1}{\bar{x}_i^2}$$

With this weighting scheme, observations associated with large values of length of road x will have less influence on the regression residuals than observations associated with smaller values of x.

d. A weighted least squares analysis was conducted on the data in Table 8.3 using the weights

$$w_i = \frac{1}{\bar{x}_i^2}$$

FIGURE 8.7

Plot of least squares residuals versus x, Example 8.2

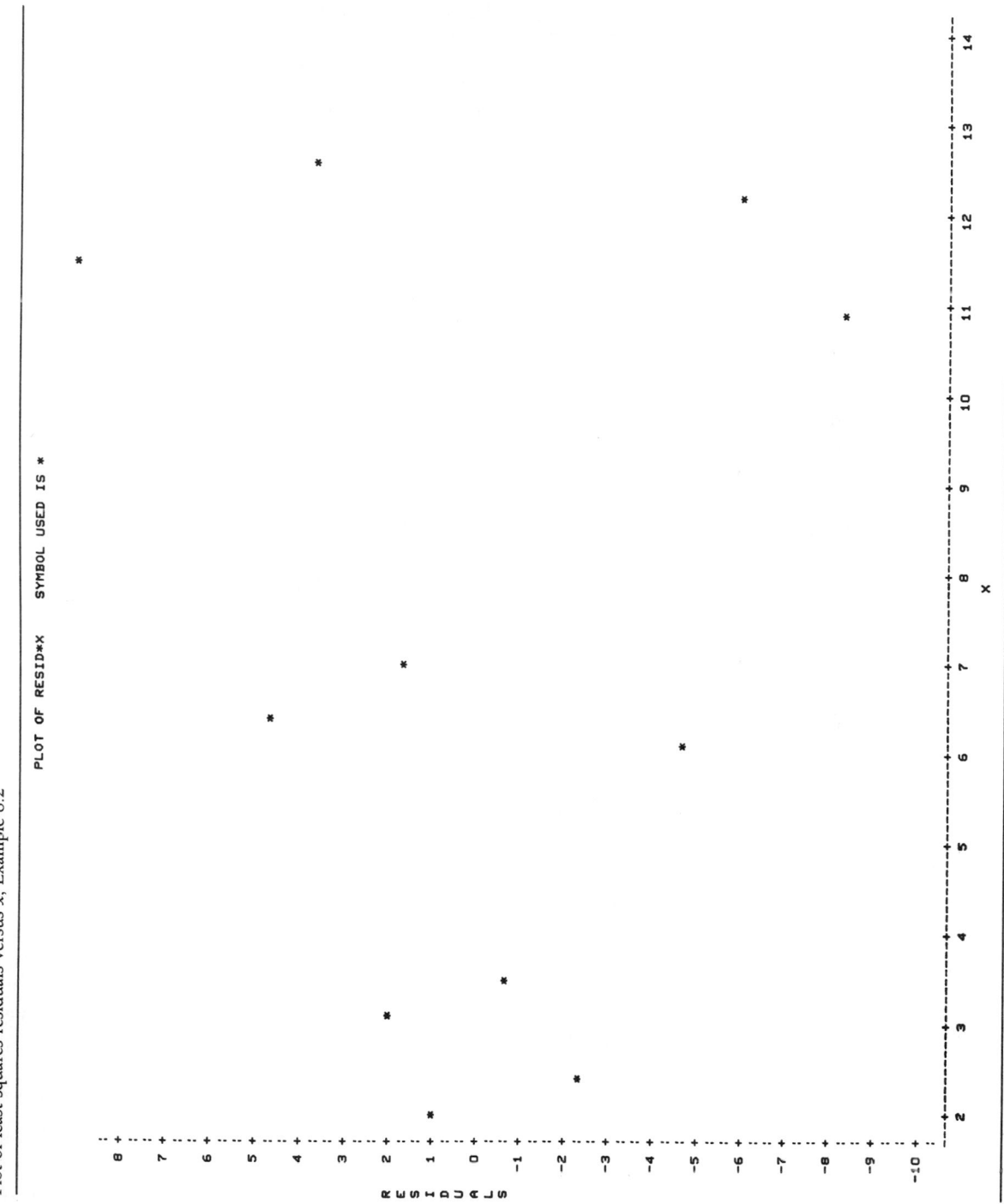

FIGURE 8.8

SAS printout of weighted least squares fit for straight-line model, Example 8.2

DEP VARIABLE: Y

ANALYSIS OF VARIANCE

SOURCE	DF	SUM OF SQUARES	MEAN SQUARE	F VALUE	PROB>F
MODEL	1	457.47787	457.47787	1021.773	0.0001
ERROR	9	4.02956477	0.44772942		
C TOTAL	10	461.50743			

ROOT MSE	0.6691259	R-SQUARE	0.9913	
DEP MEAN	28.20272	ADJ R-SQ	0.9903	
C.V.	2.372558			

PARAMETER ESTIMATES

VARIABLE	DF	PARAMETER ESTIMATE	STANDARD ERROR	T FOR H0: PARAMETER=0	PROB > :T:
INTERCEP	1	-15.27436091	1.60067934	-9.542	0.0001
X	1	12.12037303	0.37917418	31.965	0.0001

OBS	ID	ACTUAL	PREDICT VALUE	RESIDUAL
1	2	10.1000	8.9664	1.1336
2	2.4	11.4000	13.8145	-2.4145
3	3.1	24.2000	22.2988	1.9012
4	3.5	26.5000	27.1469	-0.6469
5	6.4	66.8000	62.2960	4.5040
6	6.1	53.8000	58.6599	-4.8599
7	7	71.1000	69.5683	1.5317
8	11.5	132.7	124.1	8.5901
9	10.9	108.0	116.8	-8.8377
10	12.2	126.2	132.6	-6.3942
11	12.6	140.7	137.4	3.2577

SUM OF RESIDUALS	-421.735
SUM OF SQUARED RESIDUALS	30358.76

The weighted least squares estimates are shown in the SAS printout reproduced in Figure 8.8. The prediction equation is

$$\hat{y} = -15.274 + 12.120x$$

Note that the test of model adequacy, H_0: $\beta_1 = 0$, is significant at $p = .0001$. Also, the standard error of the model, $s = \sqrt{\text{MSE}}$, is significantly smaller than the value of s for the unweighted least squares analysis (.669 compared to 5.37). This last result is expected because, in the presence of heteroscedasticity, the unweighted least squares estimates are subject to greater sampling error than the weighted least squares estimates.

e. A plot of the weighted least squares residuals against x is shown in Figure 8.9. The lack of a discernible pattern in the residual plot suggests that the weighted least squares procedure has corrected the problem of unequal variances.

Before concluding this section, we mention that the "nearest neighbor" technique, illustrated in Example 8.2, will not always be successful in finding the optimal or near-optimal weights in weighted least squares. First, it may not be easy to identify the appropriate groupings of data points, especially if more than

FIGURE 8.9

Plot of weighted residuals versus x, Example 8.2

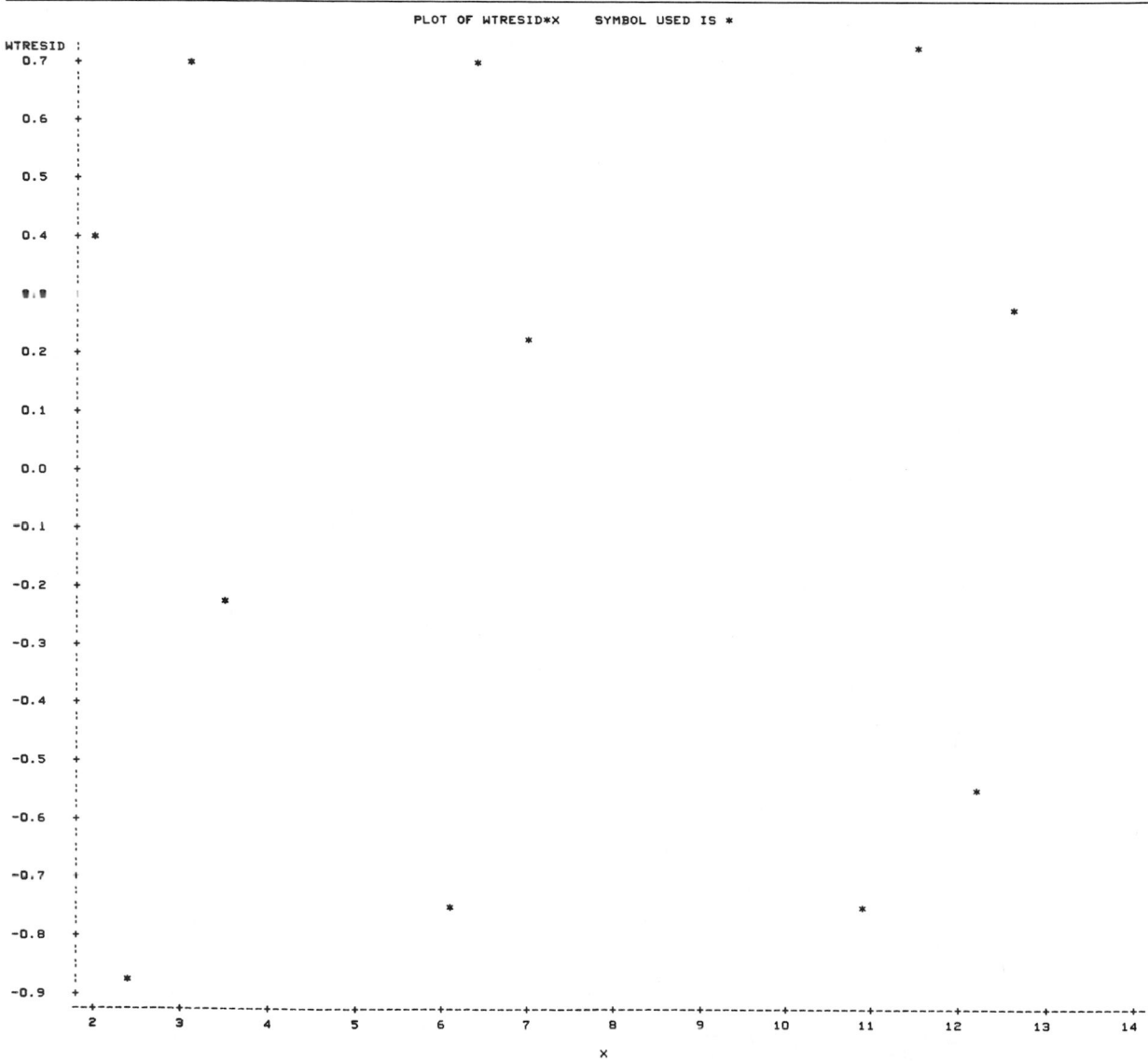

one independent variable is included in the regression. Second, the relationship between the residual variance and some preselected function of the independent variables may not reveal a consistent pattern over groups. In other words, unless the right function (or approximate function) of x is examined, the weights will be difficult to determine. More sophisticated techniques for choosing the weights in weighted least squares are available. Consult the references given at the end of this chapter for details on how to use these techniques.

EXERCISES

8.10   Consider the straight-line model $y_i = \beta_0 + \beta_1 x_i + \varepsilon_i$. Give the appropriate weights w_i to use in a weighted least squares regression if the variance of the random error ε_i, i.e., σ_i^2, is proportional to:

a. x_i^2 b. $\sqrt{x_i}$ c. x_i

d. $\dfrac{1}{n_i}$,   where n_i is the number of observations at level x_i e. $\dfrac{1}{x_i}$

8.11   A machine that mass produces rubber gaskets can be set at one of three different speeds: 100, 150, or 200 gaskets per minute. As part of a quality control study, the machine was monitored several different times at each of the three speeds and the number of defectives produced per hour was recorded. The data are provided in the accompanying table. Since the number of defectives (y) is thought to be linearly related to speed (x), the following straight-line model is proposed:

$$y = \beta_0 + \beta_1 x + \varepsilon$$

MACHINE SPEED x, gaskets per minute	NUMBER OF DEFECTIVES y	MACHINE SPEED x, gaskets per minute	NUMBER OF DEFECTIVES y
100	15	150	35
100	23	150	24
100	11	200	26
100	14	200	48
100	18	200	27
150	19	200	38
150	29	200	39
150	20		

a. Fit the model using the method of least squares. Is there evidence that the model is useful for predicting y? Test using $\alpha = .05$.
b. Plot the residuals from the least squares model against x. What does the plot reveal about the standard least squares assumption of homoscedasticity?
c. Estimate the appropriate weights to use in a weighted least squares regression. [*Hint*:   Calculate the variance of the least squares residuals at each level of x.]
d. Refit the model using weighted least squares. Compare the standard deviation of the weighted least squares slope to the standard deviation of the unweighted least squares slope.
e. Plot the weighted residuals against x to determine whether weighted least squares has corrected the problem of unequal variances.

8.12   Refer to the data on salary (y) and years of experience (x) for 50 executives, given in Table 7.7. (The data are reproduced here for convenience.) Recall that the least squares fit of the quadratic model $E(y) = \beta_0 + \beta_1 x + \beta_2 x^2$ yielded regression residuals with unequal variances (see Figure 7.13) Apply the method of weighted least squares to correct this problem. [*Hint*:   Estimate the weights using the "nearest neighbor" technique outlined in this section.]

YEARS OF EXPERIENCE	SALARY	YEARS OF EXPERIENCE	SALARY
x	y	x	y
7	$26,075	28	$64,785
28	79,370	26	61,581
23	65,726	27	70,678
18	41,983	20	51,301
19	62,309	18	39,346
15	41,154	1	24,833
24	53,610	26	65,929
13	33,697	20	41,721
2	22,444	26	82,641
8	32,562	28	99,139
20	43,078	23	52,624
21	56,000	17	50,594
18	58,667	25	53,272
7	22,210	26	65,343
2	20,521	19	46,216
18	49,727	16	54,288
11	33,233	3	20,844
21	43,628	12	32,586
4	16,105	23	71,235
24	65,644	20	36,530
20	63,022	19	52,745
20	47,780	27	67,282
15	38,853	25	80,931
25	66,537	12	32,303
25	67,447	11	38,371

8.5

Modeling Qualitative Dependent Variables

For all models discussed in the previous sections of this text, the response (dependent) variable y is a *quantitative* variable. In this section we consider models for which the response y is a **qualitative variable at two levels**, or, as it is sometimes called, a **binary variable**.

For example, an entrepreneur may want to relate the success or failure of a new business to the characteristics (such as age, years of experience, and years of education) of the owner. The value of the response of interest to the entrepreneur is either *yes*, the new business is a success, or *no*, the new business is a failure. (A success implies the business did not fail.) Similarly, a state attorney general investigating collusive practices among bidders for road construction contracts may want to determine what contract-related variables (such as number of bidders, bid amount, and cost of materials) are useful indicators of whether a bid is fixed (i.e., whether the bid price is intentionally set higher than the fair market value). Here, the value of the response variable is either *fixed* bid or *competitive* bid.

Just as with qualitative independent variables, we use **dummy** (i.e., coded 0–1) **variables** to represent the qualitative response variable. For example, the response of interest to the entrepreneur is recorded as

$$y = \begin{cases} 1 & \text{if new business a success} \\ 0 & \text{if new business a failure} \end{cases}$$

where the assignment of 0 and 1 to the two levels is arbitrary. The general linear model takes the usual form

$$y = \beta_0 + \beta_1 x_1 + \beta_2 x_2 + \cdots + \beta_k x_k + \varepsilon$$

However, when the response is binary, the expected response

$$E(y) = \beta_0 + \beta_1 x_1 + \beta_2 x_2 + \cdots + \beta_k x_k$$

has a special meaning. It can be shown* that $E(y) = \pi$, where π is the probability that $y = 1$ for given values of $x_1, x_2, \ldots, x_k$. Thus, for the entrepreneur, the mean response $E(y)$ represents the probability that a new business with certain owner-related characteristics will be a success.

When the ordinary least squares approach is used to fit models with a binary response, several well-known problems are encountered. A discussion of these problems and their solutions follows.

Problem 1

Nonnormal errors The standard least squares assumption of normal errors is violated since the response y and, hence, the random error ε can take on only two values. To see this, consider the simple model $y = \beta_0 + \beta_1 x + \varepsilon$. Then we can write

$$\varepsilon = y - (\beta_0 + \beta_1 x)$$

Thus, when $y = 1$, $\varepsilon = 1 - (\beta_0 + \beta_1 x)$ and when $y = 0$, $\varepsilon = -\beta_0 - \beta_1 x$.

When the sample size n is large, however, any inferences derived from the least squares prediction equation remain valid in most practical situations even though the errors are nonnormal.[†]

Problem 2

Unequal variances It can be shown[‡] that the variance σ^2 of the random error is a function of π, the probability that the response y equals 1. Specifically,

$$\sigma^2 = V(\varepsilon) = \pi(1 - \pi)$$

Since, for the general linear model,

$$\pi = E(y) = \beta_0 + \beta_1 x_1 + \beta_2 x_2 + \cdots + \beta_k x_k$$

this implies that σ^2 is not constant and, in fact, depends on the values of the independent variables; hence, the standard least squares assumption of equal

---

*The result is a straightforward application of the expectation theorem for a random variable. Let $\pi = P(y = 1)$ and $1 - \pi = P(y = 0)$, $0 \le \pi \le 1$. Then, by definition, $E(y) = \Sigma_y \, y_i \cdot p(y) = (1)P(y = 1) + (0)P(y = 0) = P(y = 1) = \pi$. Students familiar with discrete random variables will recognize y as the **Bernoulli random variable**, i.e., a binomial random variable with $n = 1$.

[†]This property is due to the asymptotic normality of the least squares estimates of the model parameters under very general conditions.

[‡]Using the properties of expected values with the Bernoulli random variable, we obtain $V(y) = E(y^2) - [E(y)]^2 = \Sigma y^2 \cdot p(y) - (\pi)^2 = (1)^2 P(y = 1) + (0)^2 P(y = 0) - \pi^2 = \pi - \pi^2 = \pi(1 - \pi)$. Since in regression, $V(\varepsilon) = V(y)$, the result follows.

variances is also violated. One solution to this problem is to use weighted least squares (see Section 8.4), where the weights are inversely proportional to σ^2, i.e.,

$$w_i = \frac{1}{\sigma_i^2}$$

$$= \frac{1}{\pi_i(1 - \pi_i)}$$

Unfortunately, the true proportion

$$\pi_i = E(y_i)$$
$$= \beta_0 + \beta_1 x_{1i} + \beta_2 x_{2i} + \cdots + \beta_k x_{ki}$$

is unknown since $\beta_0, \beta_1, \ldots, \beta_k$ are unknown population parameters. However, a technique called **two-stage least squares** can be applied to circumvent this difficulty. Two-stage least squares, as its name implies, involves conducting an analysis in two steps:

STAGE 1 Fit the regression model using the *ordinary least squares* procedure and obtain the predicted values $\hat{y}_i$, $i = 1, 2, \ldots, n$. Recall that $\hat{y}_i$ estimates π_i for the binary model.

STAGE 2 Refit the regression model using *weighted least squares*, where the estimated weights are calculated as follows:

$$w_i = \frac{1}{\hat{y}_i(1 - \hat{y}_i)}$$

Further iterations—revising the weights at each step—can be performed if desired. In most practical problems, however, the estimates of π_i obtained in stage 1 are adequate for use in weighted least squares.

Problem 3

Restricting the predicted response to be between 0 and 1   Since the predicted value $\hat{y}$ estimates $E(y) = \pi$, the probability that the response y equals 1, we would like $\hat{y}$ to have the property that $0 \leq \hat{y} \leq 1$. There is no guarantee, however, that the regression analysis will always yield predicted values in this range. Thus, the regression may lead to nonsensical results, i.e., negative estimated probabilities or predicted probabilities greater than 1. To avoid this problem, you may want to fit a model with a mean response function $E(y)$ that automatically falls between 0 and 1. (We consider one such model in the next section.)

In summary, the purpose of this section has been to identify some of the problems resulting from fitting a linear model with a binary response and to suggest ways in which to circumvent these problems. Another approach is to fit a model specially designed for a binary response, called a **logistic model**. Logistic models are the subject of Section 8.6.

EXERCISES

8.13   Discuss the problems associated with fitting a model where the response y is recorded as 0 or 1.

8.14   A retailer of home personal computers (PCs) conducted a study to relate PC ownership with annual income of heads of households. Data collected for a random sample of 20 households were used to fit the straight-line model $E(y) = \beta_0 + \beta_1 x$, where

$$y = \begin{cases} 1 & \text{if own PC} \\ 0 & \text{if not} \end{cases}$$

x = Annual income (in dollars)

The data are shown in the accompanying table. Fit the model using two-stage least squares. Is the model useful for predicting y? Test using $\alpha = .05$.

HOUSEHOLD	y	x	HOUSEHOLD	y	x
1	0	$16,300	11	1	$22,400
2	0	11,200	12	0	10,600
3	0	36,500	13	0	21,400
4	1	21,700	14	0	8,300
5	1	40,200	15	1	27,500
6	0	12,400	16	0	15,700
7	0	15,000	17	0	12,100
8	0	9,200	18	1	59,600
9	1	36,700	19	1	20,200
10	0	62,000	20	0	33,100

8.15   Suppose you are investigating allegations of sex discrimination in the hiring practices of a particular firm. An equal-rights group claims that females are less likely to be hired than males with the same background, experience, and other qualifications. Data (shown in the table) collected on 28 former applicants will be used to fit the model $E(y) = \beta_0 + \beta_1 x_1 + \beta_2 x_2 + \beta_3 x_3$, where

$$y = \begin{cases} 1 & \text{if hired} \\ 0 & \text{if not} \end{cases}$$

x_1 = Years of higher education   (4, 6, or 8)

x_2 = Years of experience

$$x_3 = \begin{cases} 1 & \text{if male applicant} \\ 0 & \text{if female applicant} \end{cases}$$

a. Interpret each of the β's in the multiple regression model.
b. If you have access to a statistical software package, fit the multiple regression model using two-stage least squares.
c. Conduct a test of model adequacy. Use $\alpha = .05$.
d. Is there sufficient evidence to indicate that sex is an important predictor of hiring status? Test using $\alpha = .05$.

HIRING STATUS	EDUCATION	EXPERIENCE	SEX	HIRING STATUS	EDUCATION	EXPERIENCE	SEX
y	x_1, years	x_2, years	x_3	y	x_1, years	x_2, years	x_3
0	6	2	0	1	4	5	1
0	4	0	1	0	6	4	0
1	6	6	1	0	8	0	1
1	6	3	1	1	6	1	1
0	4	1	0	0	4	7	0
1	8	3	0	0	4	1	1
0	4	2	1	0	4	5	0
0	4	4	0	0	6	0	1
0	6	1	0	1	8	5	1
1	8	10	0	0	4	9	0
0	4	2	1	0	8	1	0
0	8	5	0	0	6	1	1
0	4	2	0	1	4	10	1
0	6	7	0	1	6	12	0

e. Calculate a 95% confidence interval for the mean response $E(y)$ when $x_1 = 4$, $x_2 = 3$, and $x_3 = 0$. Interpret the interval.

8.6

Logistic Regression

Often, the relationship between a qualitative binary response y and a single predictor variable x is curvilinear. One particular curvilinear pattern frequently encountered in practice is the S-shaped curve shown in Figure 8.10. A model that accounts for this type of curvature is the **logistic** (or **logit**) **model**,

$$E(y) = \frac{\exp(\beta_0 + \beta_1 x)}{1 + \exp(\beta_0 + \beta_1 x)}$$

FIGURE 8.10

Graph of $E(y)$ for the logistic model

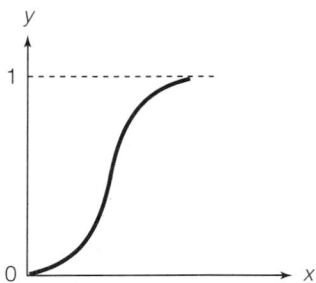

The logistic model was originally developed for use in **survival analysis**, where the response y is typically measured as 0 or 1, depending on whether or not the experimental unit (for example, a patient) "survives." Note that the curve shown in Figure 8.10 has asymptotes at 0 and 1—that is, the mean response $E(y)$ can never fall below 0 or above 1. Thus, the logistic model assures that the estimated response $\hat{y}$ (i.e., the estimated probability that $y = 1$) lies between 0 and 1.

In general, the logistic model can be written as shown in the box.

Logistic Regression Model for a Qualitative Dependent Variable

$$E(y) = \frac{\exp(\beta_0 + \beta_1 x_1 + \beta_2 x_2 + \cdots + \beta_k x_k)}{1 + \exp(\beta_0 + \beta_1 x_1 + \beta_2 x_2 + \cdots + \beta_k x_k)}$$

where

$$y = \begin{cases} 1 & \text{if category A occurs} \\ 0 & \text{if category B occurs} \end{cases}$$

$$E(y) = P(\text{category A occurs}) = \pi$$

$x_1, x_2, \ldots, x_k$ are quantitative or qualitative independent variables

---

In general, the logistic model can be written

$$E(y) = \frac{\exp(\beta_0 + \beta_1 x_1 + \beta_2 x_2 + \cdots + \beta_k x_k)}{1 + \exp(\beta_0 + \beta_1 x_1 + \beta_2 x_2 + \cdots + \beta_k x_k)}$$

Note that the general logistic model is not a linear function of the β parameters (see Section 4.1). Obtaining the parameter estimates of a **nonlinear regression model**, such as the logistic model, is a numerically tedious process and often requires sophisticated computer programs. In this section we briefly discuss two approaches to the problem, and give an example of a computer printout for the second.

1. *Least squares estimation using a transformation*   One method of fitting the model involves a transformation on the mean response $E(y)$. Recall (from Section 8.5) that for a binary response, $E(y) = \pi$, where π denotes the probability that $y = 1$. Then the logistic model

$$\pi = \frac{\exp(\beta_0 + \beta_1 x_1 + \cdots + \beta_k x_k)}{1 + \exp(\beta_0 + \beta_1 x_1 + \cdots + \beta_k x_k)}$$

implies (proof omitted) that

$$\ln\left(\frac{\pi}{1 - \pi}\right) = \beta_0 + \beta_1 x_1 + \cdots + \beta_k x_k$$

Set

$$\pi^* = \ln\left(\frac{\pi}{1 - \pi}\right)$$

The transformed logistic model

$$\pi^* = \beta_0 + \beta_1 x_1 + \cdots + \beta_k x_k$$

is now linear in the β's and the method of least squares can be applied.

Note: Since $\pi = P(y - 1)$, then $1 - \pi = P(y = 0)$. The ratio

$$\frac{\pi}{1 - \pi} = \frac{P(y = 1)}{P(y = 0)}$$

is known as the **odds** of the event, $y = 1$, occurring. (For example, if $\pi = .8$, then the odds of $y = 1$ occurring are $.8/.2 = 4$ to 1.) The transformed model, π^*, then, is a model for the log of the odds of $y = 1$ occurring and is often called the **log-odds model**.

Although the transformation succeeds in linearizing the response function, two other problems remain. First, since the true probability π is unknown, the values of the log-odds π^*, necessary for input into the regression, are also unknown. To carry out the least squares analysis, we must obtain estimates of π^* for each combination of the independent variables. A good choice is the estimator

$$\pi^* = \ln\left(\frac{\hat{\pi}}{1 - \hat{\pi}}\right)$$

where $\hat{\pi}$ is the sample proportion of 1's for the particular combination of x's. To obtain these estimates, however, *we must have replicated observations of the response y at each combination of the levels of the independent variables.* Thus, the least squares transformation approach is limited to replicated experiments, which occur infrequently in a practical business setting.

The second problem concerns the problem of unequal variances. The transformed logistic model yields error variances that are inversely proportional to $\pi(1 - \pi)$. Since π, or $E(y)$, is a function of the independent variables, the regression errors are heteroscedastic. To stabilize the variances, weighted least squares should be used. This technique also requires that replicated observations be available for each combination of the x's and, in addition, that the number of observations at each combination be relatively large. If the experiment is replicated, with n_j (large) observations at each combination of the levels of the independent variables, then the appropriate weights to use are

$$w_j = n_j \hat{\pi}_j (1 - \hat{\pi}_j)$$

where

$$\hat{\pi}_j = \frac{\text{Number of 1's for combination } j \text{ of the } x\text{'s}}{n_j}$$

2. *Maximum likelihood estimation* Estimates of the β parameters in the logistic model also can be obtained by applying a common statistical technique, called **maximum likelihood estimation**. Like the least squares estimators, the maximum likelihood estimators have certain desirable properties.[†] (In fact, when the errors of a linear regression model are normally distributed, the least

---

[†]For details on how to obtain maximum likelihood estimators and their distributional properties, consult the references given at the end of the chapter.

squares estimates and maximum likelihood estimates are equivalent.) Many of the available statistical computer program packages employ maximum likelihood estimation to fit logistic regression models. Therefore, one practical advantage of using the maximum likelihood method (rather than the transformation approach) to fit logistic regression models is that computer programs are readily available. Another advantage is that the data need not be replicated to apply maximum likelihood estimation.

The maximum likelihood estimates of the parameters of a logistic model have distributional properties that are different from the standard F and t distributions of least squares regression. Under certain conditions, the test statistics for testing individual parameters and overall model adequacy have approximate **chi-square (χ^2) distributions**. The χ^2 distribution is similar to the F distribution in that it depends on degrees of freedom and is nonnegative, as shown in Figure 8.11. (Critical values of the χ^2 distribution for various values of α and degrees of freedom are given in Table 10, Appendix D.) We illustrate the application of maximum likelihood estimation for logistic regression with an example.

FIGURE 8.11
Several chi-square probability distributions

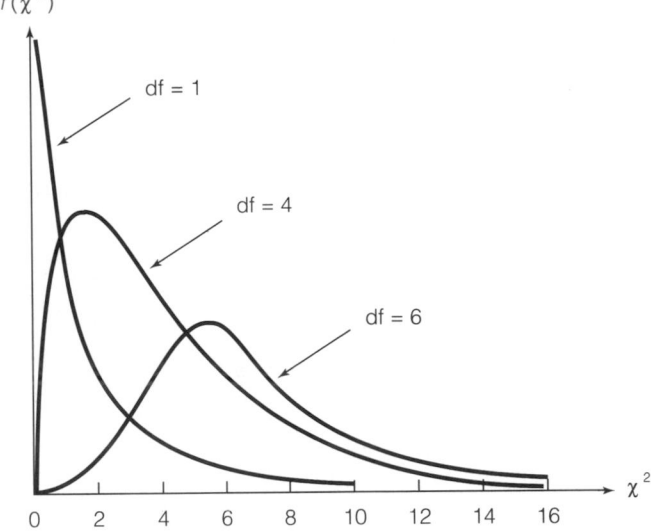

Consider the problem of collusive (i.e., noncompetitive) bidding among road construction contractors. Recall (from Section 8.5) that contractors sometimes scheme to set bid prices higher than the fair market (or competitive) price. Suppose an investigator has obtained information on the bid status (fixed or competitive) for a sample of 31 contracts. In addition, two variables thought to be related to bid status are also recorded for each contract: number of bidders x_1 and the difference between the winning (lowest) bid and the estimated competitive bid (called the engineer's estimate) x_2, measured as a percentage of the estimate. The data appear in Table 8.5, with the response y recorded as follows:

TABLE 8.5 **Data for a Sample of 31 Road Construction Bids**

CONTRACT	BID STATUS y	NUMBER OF BIDDERS x_1	DIFFERENCE BETWEEN WINNING BID AND ENGINEER'S ESTIMATE x_2, %	CONTRACT	BID STATUS y	NUMBER OF BIDDERS x_1	DIFFERENCE BETWEEN WINNING BID AND ENGINEER'S ESTIMATE x_2, %
1	1	4	19.2	17	0	10	6.6
2	1	2	24.1	18	1	5	−2.5
3	0	4	−7.1	19	0	13	24.2
4	1	3	3.9	20	0	7	2.3
5	0	9	4.5	21	1	3	36.9
6	0	6	10.6	22	0	4	−11.7
7	0	2	−3.0	23	1	2	22.1
8	0	11	16.2	24	1	3	10.4
9	1	6	72.8	25	0	2	9.1
10	0	7	28.7	26	0	5	2.0
11	1	3	11.5	27	0	6	12.6
12	1	2	56.3	28	1	5	18.0
13	0	5	−.5	29	0	3	1.5
14	0	3	−1.3	30	1	4	27.3
15	0	3	12.9	31	0	10	−8.4
16	0	8	34.1				

$$y = \begin{cases} 1 & \text{if fixed bid} \\ 0 & \text{if competitive bid} \end{cases}$$

An appropriate model for $E(y)$ is the logistic model

$$E(y) = \frac{\exp(\beta_0 + \beta_1 x_1 + \beta_2 x_2)}{1 + \exp(\beta_0 + \beta_1 x_1 + \beta_2 x_2)}$$

The model was fit to the data using the logistic regression option of SAS. The resulting printout is shown in Figure 8.12 on page 450.

The maximum likelihood estimates of β_0, β_1, and β_2 are given in Figure 8.12 under the column heading **Parameter Estimate**. These estimates (shaded in the printout) are $\hat{\beta}_0 = 1.4212$, $\hat{\beta}_1 = -.7553$, and $\hat{\beta}_2 = .1122$. Therefore, the prediction equation for the probability of a fixed bid (i.e., the probability that $y = 1$) is

$$\hat{y} = \frac{\exp(1.4212 - .7553x_1 + .1122x_2)}{1 + \exp(1.4212 - .7553x_1 + .1122x_2)}$$

In general, the coefficient $\hat{\beta}_i$ in the logistic model estimates the change in the log-odds when x_i is increased by one unit, holding all other x's in the model fixed. The antilog of the coefficient, $e^{\hat{\beta}_i}$, then estimates the odds-ratio

$$\frac{\pi_{x+1}/(1 - \pi_{x+1})}{\pi_x/(1 - \pi_x)}$$

FIGURE 8.12

SAS printout for logistic regression on bid status

Criteria for Assessing Model Fit

Criterion	Intercept Only	Intercept and Covariates	Chi-Square for Covariates
AIC	43.381	28.843	.
SC	44.815	33.145	.
-2 LOG L	41.381	22.843	18.538 with 2 DF (p=0.0001)
Score	.	.	13.466 with 2 DF (p=0.0012)

Analysis of Maximum Likelihood Estimates

Variable	Parameter Estimate	Standard Error	Wald Chi-Square	Pr > Chi-Square	Standardized Estimate
INTERCPT	1.4212	1.2867	1.2199	0.2694	.
NUMBIDS	-0.7553	0.3388	4.9708	0.0258	-1.231128
DOTEST	0.1122	0.0514	4.7670	0.0290	1.143067

Predicted Probabilities and 95% Confidence Limits

OBS	NUMBIDS	DOTEST	STATUS	PRED	CLLOWER	CLUPPER
1	4	19.2	1	0.63510	0.32984	0.86023
2	2	24.1	1	0.93180	0.53648	0.99384
3	4	-7.1	0	0.08342	0.01043	0.44010
4	3	3.9	1	0.39958	0.15868	0.70132
5	9	4.5	0	0.00760	0.00016	0.26825
6	6	10.6	0	0.12770	0.02582	0.44708
7	2	-3.0	0	0.39506	0.10273	0.78836
8	11	16.2	0	0.00624	0.00007	0.36810
9	6	72.8	1	0.99368	0.35201	0.99998
10	7	28.7	0	0.34391	0.06138	0.80776
11	3	11.5	1	0.60958	0.31579	0.84081
12	2	56.3	1	0.99803	0.69696	0.99999
13	5	-0.5	0	0.08229	0.01253	0.38782
14	3	-1.3	0	0.27078	0.07452	0.63131
15	3	12.9	0	0.64626	0.34076	0.86589
16	8	34.1	0	0.31103	0.03168	0.86168
17	10	6.6	0	0.00453	0.00006	0.26602
18	5	-2.5	1	0.06686	0.00852	0.37403
19	13	24.2	0	0.00339	0.00001	0.45712
20	7	2.3	0	0.02639	0.00166	0.30640
21	3	36.9	1	0.96428	0.54751	0.99834
22	4	-11.7	0	0.05152	0.00412	0.41602
23	2	22.1	1	0.91608	0.51886	0.99103
24	3	10.4	1	0.57984	0.29466	0.82011
25	2	9.1	0	0.71740	0.33904	0.92627
26	5	2.0	0	0.10611	0.02005	0.40784
27	6	12.6	0	0.15485	0.03485	0.48180
28	5	18.0	1	0.41683	0.17873	0.70127
29	3	1.5	0	0.33704	0.11480	0.66587
30	4	27.3	1	0.81200	0.40060	0.96541
31	10	-8.4	0	0.00085	0.00000	0.15843

where π_x is the value of $P(y = 1)$ for a fixed value x.* In other words, $e^{\hat{\beta}_i}$ is an estimate of the percentage increase (or decrease) in the odds $\pi = P(y = 1)/P(y = 0)$ for every 1-unit increase in x_i, holding the other x's fixed. This leads to the following interpretations on the β estimates:

---

*To see this, consider the model $\pi^* = \beta_0 + \beta_1 x$, where $x = 1$ or $x = 0$. When $x = 1$, we have $\pi_1^* = \beta_0 + \beta_1$; when $x = 0$, $\pi_0^* = \beta_0$. Now replace π_i^* with $\log[\pi_i/(1 + \pi_i)]$, and take the antilog of each side of the equation. Then we have $\pi_1/(1 - \pi_1) = e^{\beta_0}e^{\beta_1}$ and $\pi_0/(1 - \pi_0) = e^{\beta_0}$. Consequently, the odds-ratio is

$$\frac{\pi_1/(1 - \pi_1)}{\pi_0/(1 - \pi_0)} = e^{\beta_1}$$

$\hat{\beta}_1 = -.7553$ ($e^{\hat{\beta}_1} = .47$): For each additional bidder (x_1), we estimate the odds of a fixed contract to *decrease* by 53%, holding **DOTEST** (x_2) fixed.

$\hat{\beta}_2 = .1122$ ($e^{\hat{\beta}_2} = 1.12$): For every 1% increase in **DOTEST** (x_2), we estimate the odds of a fixed contract to *increase* by 12%, holding **NUMBIDS** (x_1) fixed.

The standard errors of the β estimates are given under the column **Standard Error**, and the (squared) ratios of the β estimates to their respective standard errors are given under the column **Wald Chi-Square**. As in regression with a linear model, this ratio provides a test statistic for testing the contribution of each variable to the model (i.e., for testing H_0: $\beta_i = 0$).* The observed significance levels of the tests (i.e., the *p*-values) are given under the column **Pr > Chi-Square**. Note that both independent variables, **NUMBIDS** (x_1) and **DOTEST** (x_2), have *p*-values less than .03 (implying that we would reject H_0: $\beta_1 = 0$ and H_0: $\beta_2 = 0$ for $\alpha = .03$).

The test statistic for testing the overall adequacy of the logistic model, i.e., for testing H_0: $\beta_1 = \beta_2 = 0$, is given in the upper portion of the printout (shaded) as $\chi^2 = 18.538$, with observed significance level (shaded) $p = .0001$.† Based on the *p*-value of the test, we can reject H_0 and conclude that at least one of the β coefficients is nonzero. Thus, the model is adequate for predicting bid status y.

Finally, the bottom portion of the printout gives predicted values and lower and upper 95% prediction limits for each observation used in the analysis in the columns titled **PRED**, **CLLOWER**, and **CLUPPER**, respectively. The 95% prediction interval for π for a contract with $x_1 = 3$ bidders and winning bid amount $x_2 = 11.5\%$ above the engineer's estimate is shaded on the printout. We estimate π, the probability of this particular contract being fixed, to fall between .3158 and .8408. Note that all the predicted values and limits lie between 0 and 1, a property of the logistic model.

In summary, we have presented two approaches to fitting logistic regression models. If the data are replicated, you may want to apply the transformation approach. The maximum likelihood estimation approach can be applied to any data set, but you need access to a statistical software package (such as SAS or SPSS) with logistic regression procedures.

This section should be viewed only as an overview of logistic regression. Many of the details of fitting logistic regression models using either technique have been omitted. Before conducting a logistic regression analysis, we strongly recommend that you consult the references given at the end of this chapter.

---

*In the logistic regression model, the ratio $(\hat{\beta}_i/s_{\hat{\beta}_i})^2$ has an approximate χ^2 distribution with 1 degree of freedom. Consult the references for more details of the χ^2 distribution and its use in logistic regression.

†The test statistic has an approximate χ^2 distribution with $k = 2$ degrees of freedom, where k is the number of β parameters in the model (excluding β_0).

EXERCISES

8.16   Refer to Exercise 8.14. The data for the random sample of 20 households were used to fit the logit model

$$E(y) = \frac{\exp(\beta_0 + \beta_1 x)}{1 + \exp(\beta_0 + \beta_1 x)}$$

An SPSS printout of the logistic regression is presented here. Interpret the results.

```
Dependent Variable..    Y

                        Chi-Square   df Significance
-2 Log Likelihood         22.969     18      .1918
Model Chi-Square           2.929      1      .0870
Improvement                2.929      1      .0870
Goodness of Fit           19.343     18      .3710

Classification Table for Y
                      Predicted
                   .00     1.00     Percent Correct
                    0   |   1
Observed         +-----+-----+
   .00     0     |  12 |  1  |      92.31%
                 +-----+-----+
  1.00     1     |   5 |  2  |      28.57%
                 +-----+-----+
                   Overall   70.00%

-------------------- Variables in the Equation ----------------------

Variable        B         S.E.      Wald      df      Sig       R     Exp(B)

X           5.47E-05   3.491E-05   2.4533     1     .1173    .1323   1.0001
Constant    -2.0188     1.0314     3.8314     1     .0503
-----------------------------------------------------------------------
```

8.17   Refer to Exercise 8.15. The data collected on the 28 former applicants was used to fit the logit model

$$E(y) = \frac{\exp(\beta_0 + \beta_1 x_1 + \beta_2 x_2 + \beta_3 x_3)}{1 + \exp(\beta_0 + \beta_1 x_1 + \beta_2 x_2 + \beta_3 x_3)}$$

where

$$y = \begin{cases} 1 & \text{if hired} \\ 0 & \text{if not} \end{cases}$$

x_1 = Years of higher education (4, 6, or 8)

x_2 = Years of experience

$$x_3 = \begin{cases} 1 & \text{if male applicant} \\ 0 & \text{if female applicant} \end{cases}$$

A SAS printout of the logistic regression is shown on page 453.

a. Conduct a test of model adequacy. Use $\alpha = .05$.

b. Is there sufficient evidence to indicate that sex is an important predictor of hiring status? Test using $\alpha = .05$.

c. Calculate a 95% confidence interval for the mean response $E(y)$ when $x_1 = 4$, $x_2 = 0$, and $x_3 = 1$. Interpret the interval.

```
                          Criteria for Assessing Model Fit

                                         Intercept
                          Intercept         and
        Criterion           Only        Covariates      Chi-Square for Covariates

        AIC                37.165         22.735              .
        SC                 38.497         28.064              .
        -2 LOG L           35.165         14.735         20.430 with 3 DF (p=0.0001)
        Score                .              .            15.032 with 3 DF (p=0.0018)

                   Analysis of Maximum Likelihood Estimates

                   Parameter      Standard        Wald          Pr >        Standardized
        Variable    Estimate        Error      Chi-Square    Chi-Square       Estimate

        INTERCPT    -14.2483        6.0805        5.4909        0.0191             .
        EDUC          1.1549        0.6023        3.6767        0.0552         1.001936
        EXP           0.9098        0.4293        4.1010        0.0341         1.890596
        SEX           5.6037        2.6028        4.6352        0.0313         1.569063

              Predicted Probabilities and 95% Confidence Limits

        OBS    EDUC     EXP     SEX     HIRED       PRED        CLL          CLU

          1      6       6       1        1        0.97688     0.42319      0.99959
          2      6       3       1        1        0.73385     0.26804      0.95405
          3      8       3       0        1        0.09282     0.00485      0.68232
          4      8      10       0        1        0.98352     0.27405      0.99989
          5      4       5       1        1        0.62813     0.11439      0.95669
          6      6       1       1        1        0.30886     0.07490      0.71155
          7      8       5       1        1        0.99420     0.50305      0.99997
          8      4      10       1        1        0.99378     0.29208      0.99998
          9      6      12       0        1        0.97338     0.19876      0.99981
         10      6       2       0        0        0.00407     0.00005      0.27086
         11      4       0       1        0        0.01755     0.00048      0.40027
         12      4       1       0        0        0.00016     0.00000      0.15324
         13      4       2       1        0        0.09927     0.00894      0.57370
         14      4       4       0        0        0.00250     0.00002      0.24465
         15      6       1       0        0        0.00164     0.00001      0.24049
         16      4       2       1        0        0.09927     0.00894      0.57370
         17      8       5       0        0        0.38699     0.04394      0.89661
         18      4       2       0        0        0.00041     0.00000      0.17559
         19      6       7       0        0        0.27888     0.04027      0.78091
         20      6       4       0        0        0.02461     0.00108      0.37067
         21      8       0       1        0        0.64439     0.11457      0.96209
         22      4       7       0        0        0.03698     0.00141      0.50996
         23      4       1       1        0        0.04248     0.00221      0.47030
         24      4       5       0        0        0.00618     0.00009      0.30151
         25      6       0       1        0        0.15248     0.02129      0.59808
         26      4       9       0        0        0.19153     0.01081      0.83708
         27      8       1       0        0        0.01631     0.00026      0.51595
         28      6       1       1        0        0.30886     0.07490      0.71155
```

8.7

Ridge Regression

When the sample data for regression exhibit multicollinearity, the least squares estimates of the β coefficients may be subject to extreme roundoff error as well as inflated standard errors (see Section 6.5). Since their magnitudes and signs may change considerably from sample to sample, the least squares estimates are said to be *unstable*. A technique developed for stabilizing the regression coefficients in the presence of multicollinearity is **ridge regression**.

Ridge regression is a modification of the method of least squares to allow *biased* estimators of the regression coefficients. At first glance, the idea of biased estimation may not seem very appealing. But consider the sampling distributions of two different estimators of a regression coefficient β, one unbiased and the other biased, shown in Figure 8.13 on page 454.

FIGURE 8.13

Sampling distributions of two estimators
of a regression coefficient β

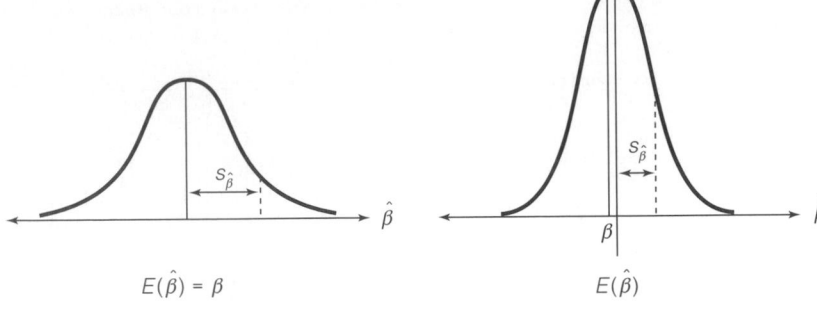

a. Unbiased estimator   **b.** Biased estimator

Figure 8.13a shows an unbiased estimator of β with a fairly large variance. In contrast, the estimator shown in Figure 8.13b has a slight bias but is much less variable. In this case, we would prefer the biased estimator over the unbiased estimator since it will lead to more precise estimates of the true β (i.e., narrower confidence intervals for β). One way to measure the "goodness" of an estimator of β is to calculate the **mean square error** of $\hat{\beta}$, denoted by MSE($\hat{\beta}$), where MSE($\hat{\beta}$) is defined as

$$\begin{aligned} \text{MSE}(\hat{\beta}) &= E[(\hat{\beta} - \beta)^2] \\ &= V(\hat{\beta}) + [E(\hat{\beta}) - \beta]^2 \end{aligned}$$

The difference $E(\hat{\beta}) - \beta$ is called the **bias** of $\hat{\beta}$. Therefore, MSE($\hat{\beta}$) is just the sum of the variance of $\hat{\beta}$ and the squared bias:

$$\text{MSE}(\hat{\beta}) = V(\hat{\beta}) + (\text{Bias in } \hat{\beta})^2$$

Let $\hat{\beta}_{\text{LS}}$ denote the least squares estimate of β. Then, since $E(\hat{\beta}_{\text{LS}}) = \beta$, the bias is 0 and

$$\text{MSE}(\hat{\beta}_{\text{LS}}) = V(\hat{\beta}_{\text{LS}})$$

We have previously stated that the variance of the least squares regression coefficient, and hence MSE($\hat{\beta}_{\text{LS}}$), will be quite large in the presence of multicollinearity. The idea behind ridge regression is to introduce a small amount of bias in the ridge estimator of β, denoted by $\hat{\beta}_{\text{R}}$, so that its mean square error is considerably smaller than the corresponding mean square error for least squares, i.e.,

$$\text{MSE}(\hat{\beta}_{\text{R}}) < \text{MSE}(\hat{\beta}_{\text{LS}})$$

In this manner, ridge regression will lead to narrower confidence intervals for the β coefficients, and hence, more stable estimates.

Although the mechanics of a ridge regression are beyond the scope of this text, we point out that some of the more sophisticated software packages (including SAS) are now capable of conducting this type of analysis. To obtain the ridge regression coefficients, the user must specify the value of a biasing constant c, where $c \geq 0$.* Researchers have shown that as the value of c increases, the bias in the ridge estimates increases while the variance decreases. The idea is to choose c so that the total mean square error for the ridge estimators is smaller than the total mean square error for the least squares estimates. Although such a c exists, the optimal value, unfortunately, is unknown.

Various methods for choosing the value of c have been proposed. One commonly used graphical technique employs a **ridge trace**. Values of the estimated ridge regression coefficients are calculated for different values of c ranging from 0 to 1 and are plotted. The plots for each of the independent variables in the model are overlaid to form the ridge trace. An example of ridge trace for a model with three independent variables is shown in Figure 8.14. Initially, the estimated coefficients may fluctuate dramatically as c is increased from 0 (especially if severe multicollinearity is present). Eventually, however, the ridge estimates will stabilize. After careful examination of the ridge trace, the analyst chooses the smallest value of c for which it appears that all the ridge estimates are stable. The choice of c, therefore, is subjective.

FIGURE 8.14

Ridge trace of β coefficients of a model with three independent variables

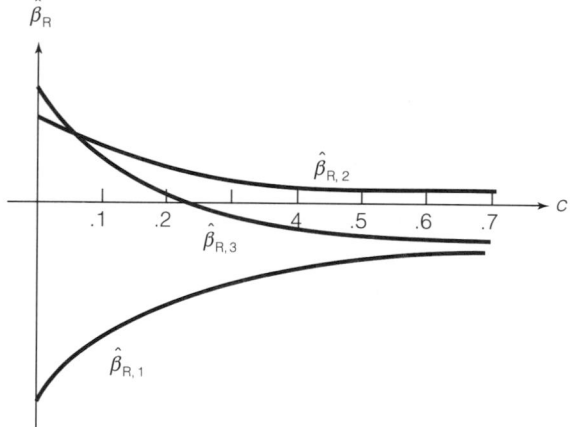

Once the value of c has been determined (using the ridge trace or some other analytical technique), the corresponding ridge estimates may be used in place of the least squares estimates. If the optimal (or near-optimal) value of c has been selected, the new estimates will have reduced variances (which lead to narrower confidence intervals for the β's). Also, some of the other problems associated with multicollinearity (e.g., incorrect signs on the β's) should have been corrected.

*In matrix notation, the ridge estimator $\hat{\beta}_R$ is calculated as follows:

$$\hat{\beta}_R = (\mathbf{X}'\mathbf{X}) + c\mathbf{I})^{-1}\mathbf{X}'\mathbf{Y}$$

When $c = 0$, the least squares estimator

$$\hat{\beta}_{LS} = (\mathbf{X}'\mathbf{X})^{-1}\mathbf{X}'\mathbf{Y}$$

is obtained. See Appendix A for details on the matrix mechanics of a regression analysis.

In conclusion, we caution that one should not assume that ridge regression is a panacea for multicollinearity or poor data. Although there are probably ridge regression estimates that are better than the least squares estimates when multicollinearity is present, the choice of the biasing constant c is crucial. Unfortunately, much of the controversy in ridge regression centers on how to find the optimal value of c. In addition, the exact distributional properties of the ridge estimators are unknown when c is estimated from the data. For these reasons, some statisticians recommend that ridge regression be used only as an exploratory data analysis tool for identifying unstable regression coefficients, and not for estimating parameters and testing hypotheses in a linear regression model.

8.8
Robust Regression

Consider the problem of fitting the linear regression model

$$y = \beta_0 + \beta_1 x_1 + \beta_2 x_2 + \cdots + \beta_k x_k + \varepsilon$$

by the method of least squares when the errors ε are nonnormal. In practice, moderate departures from the assumption of normality tend to have minimal effect on the validity of the least squares results (see Section 7.4). However, when the distribution of ε is **heavy-tailed** (longer-tailed) compared to the normal distribution, the method of least squares may not be appropriate. For example, the heavy-tailed error distribution shown in Figure 8.15 will most likely produce outliers with strong influence on the regression analysis. Furthermore, since they tend to "pull" the least squares fit too much in their direction, these outliers will have smaller than expected residuals and, consequently, are more difficult to detect.

FIGURE 8.15

Probability distribution of ε: Normal versus heavy-tailed

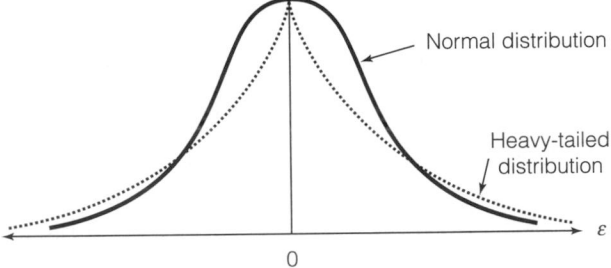

Robust regression procedures are available for errors that follow a nonnormal distribution. In the context of regression, the term *robust* describes a technique that yields estimates for the β's that are nearly as good as the least squares estimates when the assumption of normality is satisfied, and significantly better for a heavy-tailed distribution. Robust regression is designed to dampen the effect of outlying observations that otherwise would exhibit strong influence on the analysis. This has the effect of leaving the residuals of influential observations large so that they may be more easily identified.

A number of different robust regression procedures exist. They fall into one of three general classes: **M estimators**, **R estimators**, and **L estimators**. Of the three, robust techniques that produce M estimates of the β coefficients receive the most attention in the literature.

M estimates of the β coefficients are obtained by minimizing the quantity

$$\sum_{i=1}^{n} f(\hat{\varepsilon}_i)$$

where

$$\hat{\varepsilon}_i = y_i - (\hat{\beta}_0 + \hat{\beta}_1 x_{1i} + \hat{\beta}_2 x_{2i} + \cdots + \hat{\beta}_k x_{ki})$$

are the unobservable residuals and $f(\hat{\varepsilon}_i)$ is some function of the residuals. Note that the function $f(\hat{\varepsilon}_i) = \hat{\varepsilon}_i^2$ yields the ordinary least squares estimates since we are minimizing

$$\sum_{i=1}^{n} f(\hat{\varepsilon}_i) = \sum_{i=1}^{n} \hat{\varepsilon}_i^2$$
$$= \sum_{i=1}^{n} [y_i - (\hat{\beta}_0 + \hat{\beta}_1 x_{1i} + \hat{\beta}_2 x_{2i} + \cdots + \hat{\beta}_k x_{ki})]^2$$
$$= \text{SSE}$$

and, therefore, is appropriate when the errors are normal. For errors with heavier-tailed distributions, the analyst chooses some other function $f(\hat{\varepsilon}_i)$ that places less weight on the errors in the tails of the distribution. For example, the function $f(\hat{\varepsilon}_i) = |\hat{\varepsilon}_i|$ is appropriate when the errors follow the heavy-tailed distribution pictured in Figure 8.15. Since we are minimizing

$$\sum_{i=1}^{n} f(\hat{\varepsilon}_i) = \sum_{i=1}^{n} |\hat{\varepsilon}_i|$$
$$= \sum_{i=1}^{n} |y_i - (\hat{\beta}_0 + \hat{\beta}_1 x_{1i} + \hat{\beta}_2 x_{2i} + \cdots + \hat{\beta}_k x_{ki})|$$

the M estimators of robust regression yield the estimates obtained from the **method of absolute deviations** (see Section 7.5).

The other types of robust estimation, R estimation and L estimation, take a different approach. R estimators are obtained by minimizing the quantity

$$\sum_{i=1}^{n} [y_i - (\hat{\beta}_0 + \hat{\beta}_1 x_{1i} + \hat{\beta}_2 x_{2i} + \cdots + \hat{\beta}_k x_{ki})] R_i$$

where R_i is the rank of the ith residual when the residuals are placed in ascending order. L estimation is similar to R estimation because it involves ordering of the data, but it uses measures of location (such as the sample median) to estimate the regression coefficients.

The numerical techniques for obtaining robust estimates (M, R, or L estimates) are quite complex and require sophisticated computer programs. At present, statistical software packages for robust regression are not widely available. However, the growing demand for packaged robust regression programs, especially M estimation procedures, leads us to believe that these programs will be available in the near future.

Much of the current research in the area of robust regression is focused on the distributional properties of the robust estimators of the β coefficients. At present, there is little information available on robust confidence intervals, prediction intervals, and hypothesis testing procedures. For this reason, some researchers* recommend that robust regression be used in conjunction with and as a check on the method of least squares. If the results of the two procedures are substantially the same, use the least squares fit since confidence intervals and tests on the regression coefficients can be made. On the other hand, if the two analyses yield quite different results, use the robust fit to identify any influential observations. A careful examination of these data points may reveal the problem with the least squares fit.

8.9
Model Validation

Regression analysis is one of the most widely used statistical tools for estimation and prediction in the business world. All too frequently, however, a regression model deemed to be an adequate predictor of some response y performs poorly when applied in practice. For example, a model developed for forecasting new housing starts, although found to be statistically useful based on a test for overall model adequacy, may fail to take into account any extreme changes in future home mortgage rates because of new government policy. This points out an important problem. **There is no assurance that a model that fits the sample data well will be a successful predictor of y when applied to new data**. For this reason, it is important to assess the **validity** of the regression model in addition to its **adequacy** before using it in practice.

In the preceding chapters, we have presented several techniques for checking *model adequacy* (for example, tests of overall model adequacy, partial F tests, residual analysis, influence diagnostics). In short, checking model adequacy involves determining whether the regression model adequately fits the *sample data*. **Model validation**, however, involves an assessment of how the fitted regression model will perform in practice—that is, how successful it will be when applied to new or future data. A number of different model validation techniques have been proposed, several of which are briefly discussed in this section. You will need to consult the references for more details on how to apply these techniques.

1. *Examining the predicted values*   Sometimes, the predicted values $\hat{y}$ of the fitted regression model can help to identify an invalid model. Nonsensical or unreasonable predicted values may indicate that the form of the model is incorrect or that the β coefficients are poorly estimated. For example, a binary response model may yield predicted probabilities that are negative or greater than 1. In this case, the user may want to consider a model that produces predicted values between 0 and 1 (such as a logistic model) in practice. On the other hand, if the predicted values of the fitted model all seem reasonable, the user should refrain from using the model in practice until further checks of model validity are carried out.

---

*See Montgomery and Peck (1982).

2. *Examining the estimated model parameters*   Typically, the user of a regression model has some knowledge of the relative size and sign (positive or negative) of the model parameters. This information should be used as a check on the estimated β coefficients. Coefficients with signs opposite to what is expected or with unusually small or large values, or unstable coefficients (i.e., coefficients with large standard errors) forewarn that the final model may perform poorly when applied to new or different data.

3. *Collecting new data for prediction*   One of the most effective ways of validating a regression model is to use the model to predict y for a new sample. By directly comparing the predicted values to the observed values of the new data, we can determine the accuracy of the predictions and use this information to assess how well the model performs in practice.

    Several measures of model validity have been proposed for this purpose. One simple technique is to calculate the percentage of variability in the new data explained by the model, denoted $R^2_{\text{prediction}}$, and compare it to the coefficient of determination R^2 for the least squares fit. Let $y_1, y_2, \ldots, y_{n_1}$ represent the n_1 observations used to fit the final regression model and $y_{n_1+1}, y_{n_1+2}, \ldots, y_{n_1+n_2}$ represent the n_2 observations in the new data set. Then

$$R^2_{\text{prediction}} = 1 - \frac{\sum_{i=n_1+1}^{n_1+n_2} (y_i - \hat{y}_i)^2}{\sum_{i=n_1+1}^{n_1+n_2} (y_i - \bar{y})^2}$$

where $\hat{y}_i$ is the predicted value for the ith observation using the β estimates from the fitted model and $\bar{y}$ is the sample mean of the original data.* If $R^2_{\text{prediction}}$ compares favorably to R^2 from the least squares fit, we will have increased confidence in the usefulness of the model. However, if a significant drop in R^2 is observed, we should be cautious about using the model for prediction in practice.

    A similar type of comparison can be made between the mean square error, MSE, for the least squares fit and the mean squared prediction error

$$\text{MSE}_{\text{prediction}} = \frac{\sum_{i=n_1+1}^{n_1+n_2} (y_i - \hat{y}_i)^2}{n_2 - k}$$

where k is the number of β coefficients (excluding β_0) in the model. Whichever measure of model validity you decide to use, the number of observations in the new data set should be large enough to reliably assess the model's prediction performance. Montgomery and Peck (1982) recommend 15–20 new observations.

4. *Data-splitting (cross-validation)*   For those applications where it is impossible or impractical to collect new data, the original sample data can be split into two parts, with one part used to estimate the model parameters and the other

---

*Alternatively, the sample mean of the new data may be used.

part used to assess the fitted model's predictive ability. **Data-splitting** (or **cross-validation**, as it is sometimes known) can be accomplished in a variety of ways. A common technique is to randomly assign half the observations to the estimation data set and the other half to the prediction data set.* Measures of model validity, such as $R^2_{\text{prediction}}$ or $\text{MSE}_{\text{prediction}}$ can then be calculated. Of course, a sufficient number of observations must be available for data-splitting to be effective. For the estimation and prediction data sets of equal size, Snee (1977) recommends that the entire sample consist of at least $n = 2k + 25$ observations, where k is the number of β parameters in the model.

The appropriate model validation technique(s) will vary from application to application. Keep in mind that a favorable result is still no guarantee that the model will always perform successfully in practice. However, we have much greater confidence in a validated model than in one that simply fits the sample data well.

Summary

A number of special topics in regression have been introduced in this chapter. **Piecewise linear regression** can be employed when the theoretical relationship between the dependent variable y and a single independent variable x differs for different intervals over the range of x. **Inverse prediction** is a technique used for predicting a value of x for a given value of y. When the response y is **binary** (i.e., takes on only two values, 0 or 1), a **logistic regression model** may be more appropriate than a linear regression model.

Several methods are available for situations when the standard least squares assumptions about the random errors are violated. **Weighted least squares** is a variance-stabilizing technique that can also be used to dampen the influence of outlying observations. When the distribution of the errors is nonnormal, **robust regression** yields estimates of the β's with certain optimal properties. **Ridge regression** was developed to stabilize the estimated β coefficients in the presence of multicollinearity.

Finally, it is often important to assess a model's predictive performance before releasing it for use in the real world. Several **model validation techniques** are available for this purpose.

References

Agresti, A. *Categorical Data Analysis*. New York: Wiley, 1990.

Andrews, D. F. "A robust method for multiple linear regression." *Technometrics*, Vol. 16, 1974, pp. 523–531.

Cox, D. R. *The Analysis of Binary Data*. London: Methuen, 1970.

Draper, N. R. and Van Nostrand, R. C. "Ridge regression and James–Stein estimation: Review and comments." *Technometrics*. Vol. 21, 1979, p.451.

---

*Random splits are usually applied in cases where there is no logical basis for dividing the data. Consult the references for other, more formal, data-splitting techniques.

Geisser, S. "The predictive sample reuse method with applications." *Journal of the American Statistical Association*, Vol. 70, 1975, pp. 320–328.

Graybill, F. A. *Theory and Application of the Linear Model*. North Scituate, Mass.: Duxbury Press, 1976.

Halperin, M., Blackwelder, W. C., and Verter, J. I. "Estimation of the multivariate logistic risk function: A comparison of the discriminant function and maximum likelihood approaches." *Journal of Chronic Diseases*, Vol. 24, 1971, pp. 125–158.

Hauck, W. W. and Donner, A. "Wald's test as applied to hypotheses in logit analysis." *Journal of the American Statistical Association*, Vol. 72, 1977, pp. 851–853.

Hill, R. W. and Holland, P. W. "Two robust alternatives to least squares regression." *Journal of the American Statistical Association*, Vol. 72, 1977, pp. 828–833.

Hoerl, A. E. and Kennard, R. W. "Ridge regression: Biased estimation for nonorthogonal problems." *Technometrics*, Vol. 12, 1970, pp. 55–67.

Hoerl, A. E., Kennard, R. W., and Baldwin, K. F. "Ridge regression: Some simulations." *Communications in Statistics*, Vol. A5, 1976, pp. 77–88.

Hogg, R. V. "Statistical robustness: One view of its use in applications today." *The American Statistician*, Vol. 33, 1979, pp. 108–115.

Hosmer, D. W. and Lemeshow, S. *Applied Logistic Regression*. New York: Wiley, 1989.

Montgomery, D. C. and Peck, E. A. *Introduction to Linear Regression Analysis*. New York: Wiley, 1982.

Mosteller, F. and Tukey, J. W. *Data Analysis and Regression: A Second Course in Statistics*. Reading, Mass.: Addison-Wesley, 1977.

Neter, J., Wasserman, W., and Kutner, M. H. *Applied Linear Statistical Models*, 3rd ed. Homewood, Ill.: Richard D. Irwin, 1990.

Obenchain, R. L. "Classical F-tests and confidence intervals for ridge regression." *Technometrics*, Vol. 19, 1977, pp. 429–439.

Snee, R. D. "Validation of regression models: Methods and examples." *Technometrics*, Vol. 19, 1977, pp. 415–428.

Tsiatis, A. A. "A note on the goodness-of-fit test for the logistic regression model." *Biometrika*, Vol. 67, 1980, pp. 250–251.

Walker, S. H. and Duncan, D. B. "Estimation of the probability of an event as a function of several independent variables." *Biometrika*, Vol. 54, 1967, pp. 167–179.

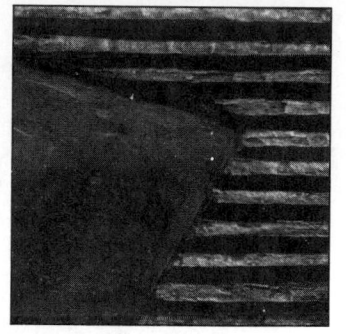

Time Series Modeling and Forecasting

CONTENTS

9.1   What Is a Time Series?

9.2   Time Series Components

9.3   Forecasting Using Smoothing Techniques (Optional)

9.4   Forecasting: The Regression Approach

9.5   Autocorrelation and Autoregressive Error Models

9.6   Other Models for Autocorrelated Errors (Optional)

9.7   Constructing Time Series Models

9.8   Fitting Time Series Models with Autoregressive Errors

9.9   Forecasting with Time Series Autoregressive Models

9.10 Seasonal Time Series Models: An Example

9.11 Forecasting Using Lagged Values of the Dependent Variable
     (Optional)

OBJECTIVE

To present models that allow for the correlation between observations taken sequentially over time; to show how these models can be used to forecast a future response

9.1
What Is a Time Series?

In many business and economic studies, the response variable y is measured sequentially in time. For example, we might record the number y of new housing starts for each month in a particular region. This collection of data is called a **time series**. Other examples of time series are data collected on the monthly production for a manufacturing company, the annual sales for a corporation, and the recorded month-end values of the prime interest rate.

Definition 9.1

A **time series** is a collection of data obtained by observing a response variable at periodic points in time.

Definition 9.2

If repeated observations on a variable produce a time series, the variable is called a **time series variable**. We use y_t to denote the value of the variable at time t.

If you were to develop a model relating the number of new housing starts to the prime interest rate over time, the model would be called a **time series model**, because both the dependent variable, new housing starts, and the independent variable, prime interest rate, are measured sequentially over time. Furthermore, time itself would probably play an important role in such a model, because the economic trends and seasonal cycles associated with different points in time would almost certainly affect both time series.

The construction of time series models is an important aspect of business and economic analyses, because many of the variables of most interest to business and economic researchers are time series. This chapter is an introduction to the very complex and voluminous body of material concerned with time series modeling and forecasting future values of a time series.

9.2
Time Series Components

Researchers often approach the problem of describing the nature of a time series y_t by identifying four kinds of change, or variation, in the time series values. These four components are commonly known as (1) secular trend, (2) cyclical effect, (3) seasonal variation, and (4) residual effect. The components of a time series are most easily identified and explained pictorially.

Figure 9.1a shows a **secular trend** in the time series values. The secular component describes the tendency of the value of the variable to increase or decrease over a long period of time. Thus, this type of change or variation is also known as the **long-term trend**. In Figure 9.1a, the long-term trend is of an increasing nature. However, this does not imply that the time series has always

FIGURE 9.1 Illustrating the components of a time series

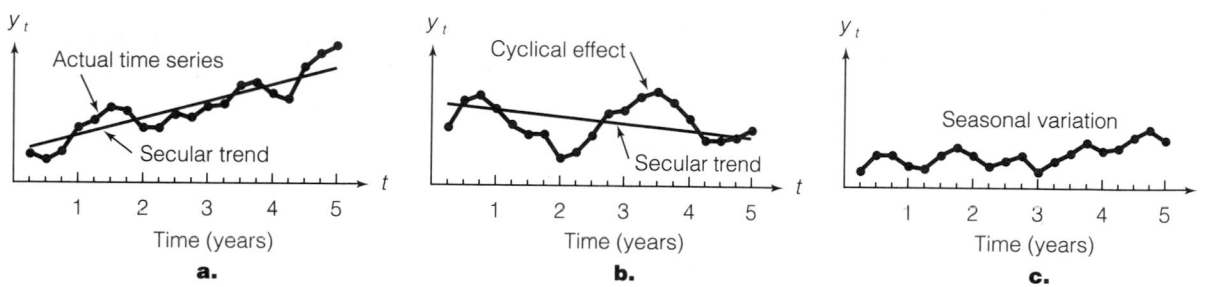

a.

b.

c.

moved upward from month to month and from year to year. You can see that the series fluctuates, but that the trend has been an increasing one over that period of time.

The **cyclical effect** in a time series, as shown in Figure 9.1b, generally describes the fluctuation about the secular trend that is attributable to business and economic conditions at the time. These fluctuations are sometimes called **business cycles**. During a period of general economic expansion, the business cycle lies above the secular trend, while during a recession, when business activity is likely to slump, the cycle lies below the secular trend. You can see that the cyclical variation does not follow any definite trend, but moves rather unpredictably.

The **seasonal variation** in a time series describes the fluctuations that recur during specific portions of each year (e.g., monthly or seasonally). In Figure 9.1c, you can see that the pattern of change in the time series within a year tends to be repeated from year to year, producing a wavelike or oscillating curve.

The final component, the **residual effect**, is what remains after the secular, cyclical, and seasonal components have been removed. This component is not systematic and may be attributed to unpredictable influences such as wars, hurricanes, presidential assassination, and randomness of human actions. Thus, the residual effect represents the random error component of a time series.

Definition 9.3

The **secular trend** (T_t) of a time series is the tendency of the series to increase or decrease over a long period of time. It is also known as the **long-term trend**.

Definition 9.4

The **cyclical fluctuation** (C_t) of a time series is the wavelike or oscillating pattern about the secular trend that is attributable to business and economic conditions at the time. It is also known as a **business cycle**.

> **Definition 9.5**
>
> The **seasonal variation** (S_t) of a time series describes the fluctuations that recur during specific portions of the year (e.g., monthly or seasonally).

> **Definition 9.6**
>
> The **residual effect** (R_t) of a time series is what remains after the secular, cyclical, and seasonal components have been removed.

In many practical applications of time series to business, the objective is to *forecast* (predict) some *future value or values* of the series. To obtain forecasts, some type of model that can be projected into the future must be used to describe the time series. One of the most widely used models is the **additive model***

$$y_t = T_t + C_t + S_t + R_t$$

where T_t, C_t, S_t, and R_t represent the secular trend, cyclical effect, seasonal variation, and residual effect, respectively, of the time series variable y_t. Various methods exist for estimating the components of the model and forecasting the time series. These range from simple **descriptive techniques**, which rely on smoothing the pattern of the time series, to complex **inferential models**, which combine regression analysis with specialized time series models. Several descriptive forecasting techniques are presented in optional Section 9.3, and forecasting using the general linear regression model of Chapter 4 is discussed in Section 9.4. The remainder of the chapter is devoted to the more complex and more powerful time series models.

9.3

Forecasting Using Smoothing Techniques (Optional)

Various descriptive methods are available for identifying and characterizing a time series. Generally, these methods attempt to remove the rapid fluctuations in a time series so that the secular trend can be seen. For this reason, they are sometimes called **smoothing techniques**. Once the secular trend is identified, forecasts for future values of the time series are easily obtained. In this section we present three of the more popular smoothing techniques.

Moving Average Method

A widely used smoothing technique is the **moving average method**. A moving average, M_t, at time t is formed by averaging the time series values over adjacent time periods. Moving averages aid in identifying the secular trend of a time series

---

*Another useful model is the **multiplicative model** $y_t = T_t C_t S_t R_t$. Recall (Section 4.9) that this model can be written in the form of an additive model by taking natural logarithms:

$$\ln y_t = \ln T_t + \ln C_t + \ln S_t + \ln R_t$$

TABLE 9.1 **Quarterly Power Loads, 1988–1991**

YEAR	QUARTER	TIME t	POWER LOAD y_t, megawatts
1988	I	1	103.5
	II	2	94.7
	III	3	118.6
	IV	4	109.3
1989	I	5	126.1
	II	6	116.0
	III	7	141.2
	IV	8	131.6
1990	I	9	144.5
	II	10	137.1
	III	11	159.0
	IV	12	149.5
1991	I	13	166.1
	II	14	152.5
	III	15	178.2
	IV	16	169.0

because the averaging modifies the effect of short-term (cyclical or seasonal) variation. That is, a plot of the moving averages yields a "smooth" time series curve that clearly depicts the long-term trend.

For example, consider the 1988–1991 quarterly power loads for a utility company located in a southern part of the United States, given in Table 9.1.

A graph of the quarterly time series, Figure 9.2, shows the pronounced seasonal variation, i.e., the fluctuation that recurs from year to year. The quarterly power loads are highest in the summer months (quarter III) with another smaller peak in the winter months (quarter I), and lowest during the spring and fall (quarters II and IV). To clearly identify the long-term trend of the series, we need to average, or "smooth out," these seasonal fluctuations. We will apply the moving average method for this purpose.

The first step in calculating a moving average for quarterly data is to sum the observed time values y_t—in this example, quarterly power loads—for the four quarters during the initial year 1988. Summing the values from Table 9.1, we have

$$y_1 + y_2 + y_3 + y_4 = 103.5 + 94.7 + 118.6 + 109.3$$
$$= 426.1$$

This sum is called a **4-point moving total**, which we denote by the symbol L_t. It is customary to use a subscript t to represent the time period at the midpoint of the four quarters in the total. Since for this sum, the midpoint is between $t = 2$ and $t = 3$, we will use the conventional procedure of "dropping it down one line" to $t = 3$. Thus, our first 4-point moving total is $L_3 = 426.1$.

We find the next moving total by eliminating the first quantity in the sum, $y_1 = 103.5$, and adding the next value in the time series sequence, $y_5 = 126.1$. This enables us to keep four quarters in the total of adjacent time periods. Thus, we have

$$L_4 = y_2 + y_3 + y_4 + y_5 = 94.7 + 118.6 + 109.3 + 126.1 = 448.7$$

FIGURE 9.2

Graph of quarterly power loads, Table 9.1

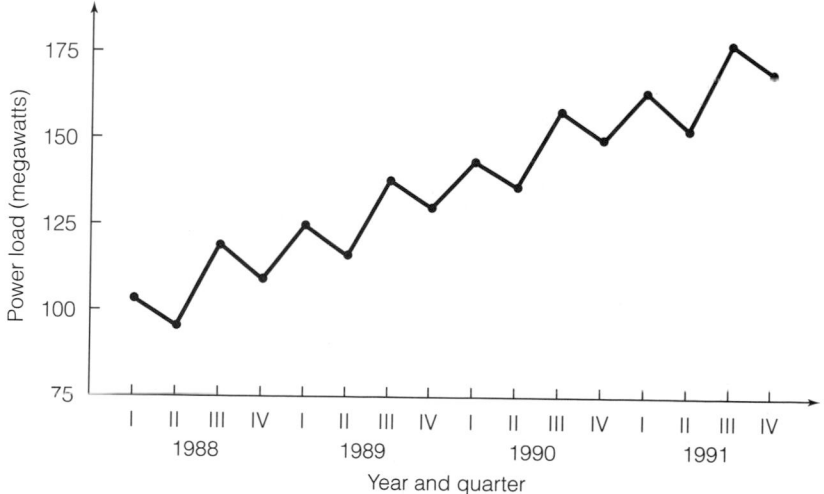

Continuing this process of "moving" the 4-point total over the time series until we have included the last value, we find

$$L_5 = y_3 + y_4 + y_5 + y_6 \quad = 118.6 + 109.3 + 126.1 + 116.0 = 470.0$$
$$L_6 = y_4 + y_5 + y_6 + y_7 \quad = 109.3 + 126.1 + 116.0 + 141.2 = 492.6$$
$$\vdots \qquad\qquad\qquad \vdots \qquad\qquad \vdots \qquad\qquad \vdots$$
$$L_{15} = y_{13} + y_{14} + y_{15} + y_{16} = 166.1 + 152.5 + 178.2 + 169.0 = 665.8$$

The complete set of 4-point moving totals is given in the appropriate column of Table 9.2. Notice that three data points will be "lost" in forming the moving totals.

T A B L E 9.2 **4-Point Moving Average for the Quarterly Power Load Data**

YEAR	QUARTER	TIME t	POWER LOAD y_t	4-POINT MOVING TOTAL L_t	4-POINT MOVING AVERAGE M_t	RATIO y_t/M_t
1988	I	1	103.5	—	—	—
	II	2	94.7	—	—	—
	III	3	118.6	426.1	106.5	1.113
	IV	4	109.3	448.7	112.2	.974
1989	I	5	126.1	470.0	117.5	1.073
	II	6	116.0	492.6	123.2	.942
	III	7	141.2	514.9	128.7	1.097
	IV	8	131.6	533.3	133.3	.987
1990	I	9	144.5	554.4	138.6	1.043
	II	10	137.1	572.2	143.1	.958
	III	11	159.0	590.1	147.5	1.078
	IV	12	149.5	611.7	152.9	.978
1991	I	13	166.1	627.1	156.8	1.059
	II	14	152.5	646.3	161.6	.944
	III	15	178.2	665.8	166.5	1.071
	IV	16	169.0	—	—	—

After the 4-point moving totals are calculated, the second step is to determine the **4-point moving average**, denoted by M_t, by dividing each of the moving totals by 4. For example, the first three values of the 4-point moving average for the quarterly power load data are:

$$M_3 = \frac{y_1 + y_2 + y_3 + y_4}{4} = \frac{L_3}{4} = \frac{426.1}{4} = 106.5$$

$$M_4 = \frac{y_2 + y_3 + y_4 + y_5}{4} = \frac{L_4}{4} = \frac{448.7}{4} = 112.2$$

$$M_5 = \frac{y_3 + y_4 + y_5 + y_6}{4} = \frac{L_5}{4} = \frac{470.0}{4} = 117.5$$

All of the 4-point moving averages are given in the appropriate column of Table 9.2.

Both the original power load time series and the 4-point moving average are graphed in Figure 9.3. Notice that the moving average has smoothed the time series, i.e., the averaging has modified the effects of the short-term or seasonal va.:ation. The plot of the 4-point moving average clearly depicts the secular (long-term) trend component of the time series.

F I G U R E 9.3   Quarterly power loads and 4-point moving average

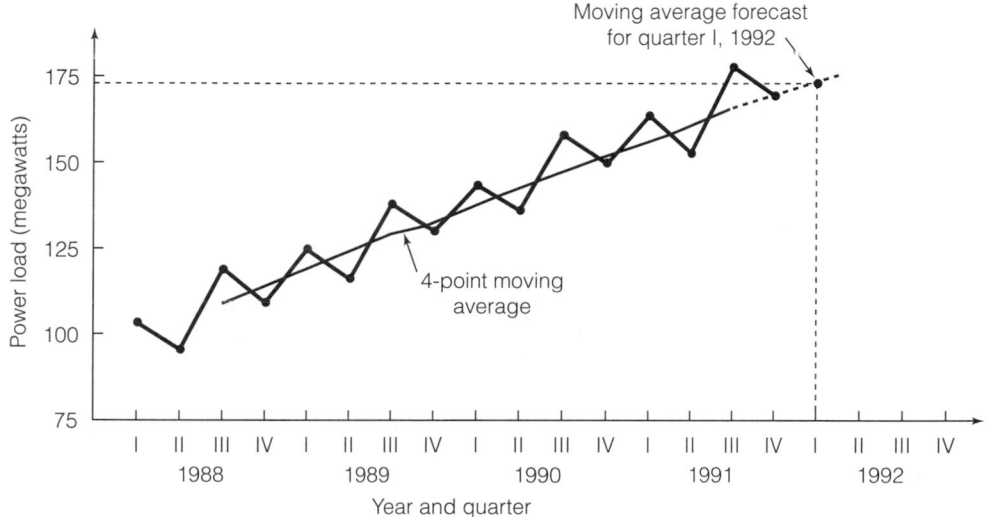

In addition to identifying a long-term trend, moving averages provide us with a measure of the seasonal effects in a time series. The ratio between the observed power load y_t and the 4-point moving average M_t for each quarter measures the seasonal effect (primarily attributable to temperature differences) for that quarter. The ratios y_t/M_t are shown in the last column of Table 9.2. Note that the ratio is always greater than 1 in quarters I and III, and always less than 1 in quarters II and IV. The average of the ratios for a particular quarter, multiplied by 100, can be used to form a **seasonal index** for that quarter. For example, the seasonal index for quarter I is

$$(100)\frac{1.073 + 1.043 + 1.059}{3} = 105.8$$

implying that the time series value in quarter I is, on the average, 105.8% of the moving average value for that time period.

To forecast a future value of the time series, simply extend the moving average M_t on the graph to the future time period. For example, a graphical extension of the moving average for the quarterly power loads to quarter I of 1992 ($t = 17$) yields a moving average of approximately $M_{17} = 172$ (see Figure 9.3). Thus, if there were no seasonal variation in the time series, we would expect the power load for quarter I of 1992 to be approximately 172 megawatts. To adjust the forecast for seasonal variation, multiply the future moving average value $M_{17} = 172$ by the seasonal index for quarter I, then divide by 100:

$$F_{17} = M_{17}\left(\frac{\text{Seasonal index for quarter I}}{100}\right)$$

$$= 172\left(\frac{105.8}{100}\right)$$

$$\approx 182$$

where F_{17} is the forecast of y_{17}. Therefore, the moving average forecast for the power load in quarter I of 1992 is approximately 182 megawatts.

Moving averages are not restricted to 4 points. For example, you may wish to calculate a 7-point moving average for daily data, a 12-point moving average for monthly data, or a 5-point moving average for yearly data. Although the choice of the number of points is arbitrary, you should search for the number N that yields a smooth series, but is not so large that many points at the end of the series are "lost." The method of forecasting with a general N-point moving average is outlined in the next box.

Exponential Smoothing

One problem with using a moving average to forecast future values of a time series is that values at the ends of the series are lost thereby requiring that we subjectively extend the graph of the moving average into the future. No exact calculation of a forecast is available since the moving average at a future time period t requires that we know one or more future values of the series. **Exponential smoothing** is a technique that leads to forecasts that can be explicitly calculated. Like the moving average method, exponential smoothing deemphasizes (or smooths) most of the residual effects. However, exponential smoothing averages only past and current values of the time series.

To obtain an exponentially smoothed time series, we first need to choose a weight w, between 0 and 1, called the **exponential smoothing constant**. The exponentially smoothed series, denoted E_t, is then calculated as follows:

$$E_1 = y_1$$
$$E_2 = wy_2 + (1 - w)E_1$$
$$E_3 = wy_3 + (1 - w)E_2$$
$$\vdots \quad \vdots$$
$$E_t = wy_t + (1 - w)E_{t-1}$$

You can see that the exponentially smoothed value at time t is simply a weighted average of the current time series value, y_t, and the exponentially smoothed value at the previous time period, E_{t-1}. Smaller values of w give less weight to the current value, y_t, whereas larger values give more weight to y_t.

> | Forecasting Using an N-Point Moving Average
>
> 1. Select N, the number of consecutive time series values $y_1, y_2, \ldots, y_n$ that will be averaged. (The time series values must be equally spaced.)
>
> 2. Calculate the N-point moving total, L_t, by summing the time series values over N adjacent time periods, where
>
> $$L_t = \begin{cases} y_{t-(N-1)/2} + \cdots + y_t + \cdots + y_{t+(N-1)/2} & \text{if } N \text{ is odd} \\ y_{t-N/2} + \cdots + y_t + \cdots + y_{t+N/2-1} & \text{if } N \text{ is even} \end{cases}$$
>
> 3. Compute the N-point moving average, M_t, by dividing the corresponding moving total by N:
>
> $$M_t = \frac{L_t}{N}$$
>
> 4. Graph the moving average M_t on the vertical axis with time t on the horizontal axis. (This plot should reveal a smooth curve that identifies the long-term trend of the time series.*) Extend the graph to a future time period to obtain the forecasted value of M_t.
>
> 5. For a future time period t, the forecast of y_t is
>
> $$F_t = \begin{cases} M_t & \text{if little or no seasonal variation exists in the time series} \\ M_t \cdot \dfrac{\text{Seasonal index}}{100} & \text{otherwise} \end{cases}$$
>
> where the seasonal index for a particular quarter (or month) is the average of past values of the ratios
>
> $$\frac{Y_t}{M_t}(100)$$
>
> for that quarter (or month).

For example, suppose we want to smooth the quarterly power loads given in Table 9.1 using an exponential smoothing constant of $w = .7$. Then we have

$$E_1 = y_1 = 103.5$$
$$E_2 = .7y_2 + (1 - .7)E_1$$
$$= .7(94.7) + .3(103.5) = 97.3$$
$$E_3 = .7y_3 + (1 - .7)E_2$$
$$= .7(118.6) + .3(97.3) = 112.2$$
$$\vdots$$

---

*When the number N of points is small, the plot may not yield a very smooth curve. However, the moving average will be smoother (or less variable) than the plot of the original time series values.

The exponentially smoothed values (using $w = .7$) for all the quarterly power loads are given in Table 9.3. Both the actual and the smoothed time series values are graphed in Figure 9.4.

Exponential smoothing forecasts are obtained by taking a weighted average of the most recent value of the time series, y_t, and the most recent exponentially smoothed value, E_t. If n is the last time period in which y_t is observed, then the forecast for a future time period t is given by

$$F_t = wy_n + (1 - w)E_n \quad \text{(see the box)}$$

TABLE 9.3   **Quarterly Power Load with Exponential Smoothing**

YEAR	QUARTER	TIME t	POWER LOAD y_t	EXPONENTIALLY SMOOTHED POWER LOAD E_t
1988	I	1	103.5	103.5
	II	2	94.7	97.3
	III	3	118.6	112.2
	IV	4	109.3	110.2
1989	I	5	126.1	121.3
	II	6	116.0	117.6
	III	7	141.2	134.1
	IV	8	131.6	132.4
1990	I	9	144.5	140.9
	II	10	137.1	138.2
	III	11	159.0	152.8
	IV	12	149.5	150.5
1991	I	13	166.1	161.4
	II	14	152.5	155.2
	III	15	178.2	171.3
	IV	16	169.0	169.7

FIGURE 9.4

Plot of exponentially smoothed power loads

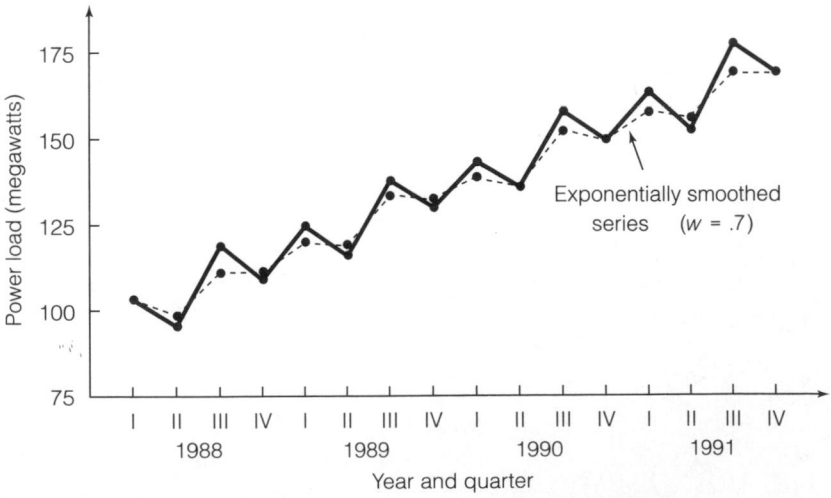

Note that the right-hand side of the forecast equation does not depend on t; hence, F_t is used to forecast *all* future values of y_t.

For example, the forecast for the power load in quarter I of 1992 ($t = 17$) is calculated as follows:

$$F_{17} = wy_{16} + (1 - w)E_{16}$$
$$= .7(169.0) + .3(169.7)$$
$$= 169.2$$

The forecasts for quarter II of 1992 ($t = 18$), quarter III of 1992 ($t = 19$), and all other future time periods will be the same:

$$F_{18} = 169.2$$
$$F_{19} = 169.2$$
$$F_{20} = 169.2$$
$$\vdots$$

This points out one disadvantage of the exponential smoothing forecasting technique. Since the exponentially smoothed forecast is constant for all future values, any changes in trend and/or seasonality are not taken into account. Therefore, exponentially smoothed forecasts are appropriate only when the trend and seasonal components of the time series are relatively insignificant.

Forecasting Using Exponential Smoothing

1. The data consist of n equally spaced time series values, $y_1, y_2, \ldots, y_n$.

2. Select a smoothing constant, w, between 0 and 1. (Smaller values of w give less weight to the current value of the series and yield a smoother series. Larger values of w give more weight to the current value of the series and yield a more variable series.)

3. Calculate the exponentially smoothed series, E_t, as follows:

$$E_1 = y_1$$
$$E_2 = wy_2 + (1 - w)E_1$$
$$E_3 = wy_3 + (1 - w)E_2$$
$$\vdots$$
$$E_n = wy_n + (1 - w)E_{n-1}$$

4. Calculate the forecast for any future time period t as follows:

$$F_t = wy_n + (1 - w)E_n, \quad t = n + 1, n + 2, \ldots$$

Holt–Winters Forecasting Model

One drawback to the exponential smoothing forecasting method is that the secular trend and seasonal components of a time series are not taken into account. The **Holt–Winters forecasting model** is an extension of the exponential smoothing method that explicitly recognizes the trend and seasonal variation in a time series.

Consider a time series with a trend component, but little or no seasonal variation. Then the Holt–Winters model for y_t is

$$E_t = wy_t + (1 - w)(E_{t-1} + T_{t-1})$$
$$T_t = v(E_t - E_{t-1}) + (1 - v)T_{t-1}$$

where E_t is the exponentially smoothed series, T_t is the trend component, and w and v are smoothing constants between 0 and 1. Note that the trend component T_t is a weighted average of the most recent change in the smoothed value (measured by the difference $E_t - E_{t-1}$) and the trend estimate of the previous time period (T_{t-1}). When seasonal variation is present in the time series, the Holt–Winters model takes the form

$$E_t = w(y_t/S_{t-P}) + (1 - w)(E_{t-1} + T_{t-1})$$
$$T_t = v(E_t - E_{t-1}) + (1 - v)T_{t-1}$$
$$S_t = u(y_t/E_t) + (1 - u)S_{t-P}$$

where S_t is the seasonal component, u is a constant between 0 and 1, and P is the number of time periods in a cycle (usually a year). The seasonal component S_t is a weighted average of the ratio y_t/E_t (i.e., the ratio of the actual time series value to the smoothed value) and the seasonal component for the previous cycle. For example, for the quarterly power loads, $P = 4$ (four quarters in a year) and the seasonal component for, say, quarter III of 1989 ($t = 7$) is a weighted average of the ratio y_7/E_7 and the seasonal component for quarter III of 1988 ($t = 3$). That is,

$$S_7 = u(y_7/E_7) + (1 - u)S_3$$

Forecasts for future time periods, $t = n + 1, n + 2, \ldots$, using the Holt–Winters models are obtained by summing the most recent exponentially smoothed component with an estimate of the expected increase (or decrease) attributable to trend. For seasonal models, the forecast is multiplied by the most recent estimate of the seasonal component (similar to the moving average method).

The Holt–Winters forecasting methodology is summarized in the accompanying box.

| EXAMPLE 9.1

Refer to the 1988–1991 quarterly power loads listed in Table 9.1. Use the Holt–Winters forecasting model with both trend and seasonal components to forecast the utility company's quarterly power loads in 1992. Use the smoothing constants $w = .7$, $v = .5$, and $u = .5$.

Solution

First note that $P = 4$ for the quarterly time series. Following the formulas for E_t, T_t, and S_t given in the box, we calculate

Forecasting Using the Holt–Winters Model

TREND COMPONENT ONLY

TREND AND SEASONAL COMPONENTS

1. The data consist of n equally spaced time series values, $y_1, y_2, \ldots, y_n$.

2. Select smoothing constants w and v, where $0 \leq w \leq 1$ and $0 \leq v \leq 1$.

2. Select smoothing constants w, v, and u, where $0 \leq w \leq 1$, $0 \leq v \leq 1$, and $0 \leq u \leq 1$.

3. Calculate the exponentially smoothed component, E_t, and the trend component, T_t, for $t = 2, 3, \ldots, n$ as follows:

$$E_t = \begin{cases} y_2, & t = 2 \\ wy_t + (1 - w)(E_{t-1} + T_{t-1}), & t > 2 \end{cases}$$

$$T_t = \begin{cases} y_2 - y_1, & t = 2 \\ v(E_t - E_{t-1}) + (1 - v)T_{t-1}, & t > 2 \end{cases}$$

[*Note:* E_1 and T_1 are not defined.]

3. Determine P, the number of time periods in a cycle. Usually, $P = 4$ for quarterly data and $P = 12$ for monthly data.

4. Calculate the exponentially smoothed component, E_t, the trend component, T_t, and the seasonal component, S_t, for $t = 2, 3, \ldots, n$ as follows:

$$E_t = \begin{cases} y_2, & t = 2 \\ wy_t + (1 - w)(E_{t-1} + T_{t-1}), \\ \quad t = 3, 4, \ldots, P + 2 \\ w(y_t/S_{t-P}) + (1 - w)(E_{t-1} + T_{t-1}), \\ \quad t > P + 2 \end{cases}$$

$$T_t = \begin{cases} y_2 - y_1, & t = 2 \\ v(E_t - E_{t-1}) + (1 - v)T_{t-1}, & t > 2 \end{cases}$$

$$S_t = \begin{cases} y_t/E_t, & t = 2, 3, \ldots, P + 2 \\ u(y_t/E_t) + (1 - u)S_{t-P}, & t > P + 2 \end{cases}$$

[*Note:* E_1, T_1, and S_1 are not defined.]

4. The forecast for a future time period t is given by

$$F_t = \begin{cases} E_n + T_n, & t = n + 1 \\ E_n + 2T_n, & t = n + 2 \\ \vdots \\ E_n + kT_n, & t = n + k \end{cases}$$

5. The forecast for a future time period t is given by

$$F_t = \begin{cases} (E_n + T_n)S_{n+1-P}, & t = n + 1 \\ (E_n + 2T_n)S_{n+2-P}, & t = n + 2 \\ \vdots \\ (E_n + kT_n)S_{n+k-P}, & t = n + k \end{cases}$$

$$E_2 = y_2 = 94.7$$
$$T_2 = y_2 - y_1 = 94.7 - 103.5 = -8.8$$
$$S_2 = y_2/E_2 = 94.7/94.7 = 1$$
$$E_3 = .7y_3 + (1 - .7)(E_2 + T_2)$$
$$\quad = .7(118.6) + .3(94.7 - 8.8) = 108.8$$
$$T_3 = .5(E_3 - E_2) + (1 - .5)T_2$$
$$\quad = .5(108.8 - 94.7) + .5(-8.8) = 2.6$$
$$S_3 = y_3/E_3 = 118.6/108.8 = 1.090$$

$$E_4 = .7y_4 + (1 - .7)(E_3 + T_3)$$
$$= .7(109.3) + .3(108.8 + 2.6) = 109.9$$
$$T_4 = .5(E_4 - E_3) + (1 - .5)T_3$$
$$= .5(109.9 - 108.8) + .5(2.6) = 1.9$$
$$S_4 = y_4/E_4 = 109.3/109.9 = .994$$

$$\vdots$$

(Remember that beginning with $t = P + 3 = 7$, the formulas for E_t and S_t, shown in the box, are slightly different.) All the values of E_t, T_t, and S_t are given in Table 9.4.

TABLE 9.4   **Holt–Winters Components for Quarterly Power Load Data**

YEAR	QUARTER	TIME t	POWER LOAD y_t	E_t $(w = .7)$	T_t $(v = .5)$	S_t $(u = .5)$
1988	I	1	103.5	—	—	—
	II	2	94.7	94.7	−8.8	1.000
	III	3	118.6	108.8	2.6	1.090
	IV	4	109.3	109.9	1.9	.994
1989	I	5	126.1	121.8	6.9	1.035
	II	6	116.0	119.8	2.5	.968
	III	7	141.2	127.4	5.1	1.100
	IV	8	131.6	132.3	5.0	.995
1990	I	9	144.5	138.9	5.8	1.038
	II	10	137.1	142.6	4.8	.965
	III	11	159.0	145.4	3.8	1.097
	IV	12	149.5	149.9	4.2	.996
1991	I	13	166.1	158.2	6.3	1.044
	II	14	152.5	160.0	4.1	.959
	III	15	178.2	162.9	3.5	1.095
	IV	16	169.0	168.7	4.7	.999

The forecast for quarter I of 1992 (i.e., y_{17}) is given by

$$F_{17} = (E_{16} + T_{16})S_{17-4}$$
$$= (E_{16} + T_{16})S_{13} = (168.7 + 4.7)(1.044)$$
$$= 181.0$$

Similarly, the forecasts for y_{18}, y_{19}, and y_{20} (quarters II, III, and IV, respectively) are

$$F_{18} = (E_{16} + 2T_{16})S_{18-4}$$
$$= (E_{16} + 2T_{16})S_{14} = [168.7 + 2(4.7)](.959)$$
$$= 170.8$$

$$F_{19} = (E_{16} + 3T_{16})S_{19-4}$$
$$= (E_{16} + 3T_{16})S_{15} = [168.7 + 3(4.7)](1.095)$$
$$= 200.2$$

$$F_{20} = (E_{16} + 4T_{16})S_{20-4}$$
$$= (E_{16} + 4T_{16})S_{16} = [168.7 + 4(4.7)](.999)$$
$$= 187.3$$

We conclude this section with a comment. A major disadvantage of forecasting with smoothing techniques (the moving average method, exponential smoothing, or the Holt–Winters models) is that no measure of the forecast error (or reliability) is known. Although forecast errors can be calculated *after* the future values of the time series have been observed, we prefer to have some measure of the accuracy of the forecast *before* the actual values are observed. For this reason, smoothing techniques are generally regarded as descriptive rather than as inferential procedures. On the other hand, forecasts with inferential models (such as regression models) are accompanied by measures of the *standard error of the forecast*, which allow us to construct prediction intervals for the future time series value. We discuss inferential time series forecasting models in the remaining sections of this chapter.

EXERCISES

9.1 The quarterly numbers of new privately owned housing starts (in thousands of dwellings) in the United States from Winter 1986 through Fall 1990 are recorded in the accompanying table.

YEAR	QUARTER	HOUSING STARTS	YEAR	QUARTER	HOUSING STARTS
1986	I	357.5	1989	I	290.6
	II	550.0		II	390.9
	III	165.5		III	346.2
	IV	384.6		IV	303.6
1987	I	336.3	1990	I	300.8
	II	463.2		II	320.8
	III	407.4		III	273.4
	IV	319.6		IV	200.9
1988	I	302.8			
	II	426.0			
	III	383.4			
	IV	336.6			

Source: Standard & Poor's Trade and Securities Statistics (Annual), New York, Standard & Poor's Corporation.

a. Plot the quarterly time series. Can you detect a long-term trend? Can you detect any seasonal variation?
b. Calculate the 4-point moving average for the quarterly housing starts.

    c. Graph the 4-point moving average on the same set of axes you used for the graph in part **a**. Is the long-term trend more evident? What effects has the moving average method removed or smoothed?

    d. Calculate the seasonal index for the number of housing starts in quarter I.

    e. Use the moving average method to forecast the number of housing starts in quarter I of 1991.

9.2   Refer to the quarterly housing starts data given in Exercise 9.1.

    a. Calculate the exponentially smoothed series for housing starts using a smoothing constant of $w = .2$.

    b. Use the exponentially smoothed series from part **a** to forecast the quarterly number of new housing starts in 1991.

    c. Use the Holt–Winters forecasting model with both trend and seasonal components to forecast the quarterly number of new housing starts in 1991. Use smoothing constants $w = .2$, $v = .5$, and $u = .7$.

9.3   Since the energy shortage, the price of foreign crude oil has skyrocketed. Consequently, crude oil imports into the United States have declined. The data in the table are the amounts of crude oil (millions of barrels) imported into the United States from the Organization of Petroleum Exporting Countries (OPEC) for the years 1973–1988.

YEAR	t	IMPORTS, y_t	YEAR	t	IMPORTS, y_t
1973	1	767	1981	9	1,067
1974	2	926	1982	10	633
1975	3	1,171	1983	11	540
1976	4	1,663	1984	12	553
1977	5	2,058	1985	13	479
1978	6	1,892	1986	14	771
1979	7	1,866	1987	15	876
1980	8	1,414	1988	16	987

Source: Statistical Abstract of the United States, U.S. Bureau of the Census, 1990.

    a. Plot the yearly time series. Can you detect a long-term trend?

    b. Calculate and plot a 3-point moving average for annual OPEC oil imports.

    c. Calculate and plot the exponentially smoothed series for annual OPEC oil imports using a smoothing constant of $w = .3$.

    d. Forecast OPEC oil imports in 1990 using the moving average method.

    e. Forecast OPEC oil imports in 1990 using exponential smoothing with $w = .3$.

    f. Forecast OPEC oil imports in 1990 using the Holt–Winters forecasting model with trend. Use smoothing constants $w = .3$ and $v = .8$.

9.4   The closing price of IBM common stock from January 1988 to December 1990 is shown in the table.

    a. Construct and plot a 12-point moving average for the stock price. Use the moving average method to forecast the price in January 1991.

    b. Construct and plot the exponentially smoothed series for the stock price using a smoothing constant of $w = .5$. Use the exponential smoothing technique to forecast the price in January 1991.

    c. Use the Holt–Winters forecasting model with trend and seasonal components to forecast the price in January 1991. Use smoothing constants $w = .5$, $v = .5$, and $u = .5$.

MONTH	IBM STOCK PRICE		
	1988	1989	1990
January	$112\frac{3}{8}$	$130\frac{5}{8}$	$98\frac{5}{8}$
February	$117\frac{1}{2}$	$121\frac{1}{2}$	$103\frac{7}{8}$
March	$107\frac{5}{8}$	$109\frac{1}{8}$	$106\frac{1}{8}$
April	$113\frac{3}{8}$	114	109
May	$112\frac{1}{2}$	$109\frac{5}{8}$	120
June	$127\frac{3}{8}$	$111\frac{7}{8}$	$117\frac{1}{2}$
July	$125\frac{3}{4}$	115	$111\frac{1}{2}$
August	$111\frac{1}{2}$	$117\frac{1}{8}$	$101\frac{7}{8}$
September	$115\frac{3}{8}$	$109\frac{1}{4}$	$106\frac{3}{8}$
October	$122\frac{5}{8}$	$100\frac{1}{4}$	$105\frac{3}{8}$
November	$118\frac{1}{2}$	$97\frac{5}{8}$	$113\frac{5}{8}$
December	$121\frac{7}{8}$	$94\frac{1}{8}$	113

Source: Security Owner's Stock Guide, Standard & Poor's Corporation.

d. The actual IBM stock price in January 1991 was $126\frac{3}{4}$. Calculate the forecast errors for the forecasts obtained in parts **a–c**. (The forecast error is measured by the difference, $y_t - F_t$.)

9.5 The Consumer Price Index (CPI) measures the increase (or decrease) in the prices of goods and services relative to a base year. The CPI for the years 1977–1990 (using 1967 as a base period) is shown in the table.

YEAR	CPI	YEAR	CPI
1977	181.5	1984	311.1
1978	195.4	1985	322.2
1979	217.4	1986	328.4
1980	246.8	1987	330.5
1981	272.4	1988	337.7
1982	289.1	1989	353.8
1983	298.4	1990	372.3

Source. Survey of Current Business, U.S. Department of Commerce, Bureau of Economic Analysis.

a. Graph the time series. Do you detect a long-term trend?

b. Calculate and plot a 5-point moving average for the CPI. Use the moving average to forecast the CPI in 1991.

c. Calculate and plot the exponentially smoothed series for the CPI using a smoothing constant of $w = .4$. Use the exponentially smoothed values to forecast the CPI in 1991.

d. Use the Holt–Winters forecasting model with trend to forecast the CPI in 1991. Use smoothing constants $w = .4$ and $v = .5$.

9.6 Standard & Poor's 500 Stock Composite Average (S&P 500) is a stock market index. Like the Dow Jones Industrial Average (DJA), it is an indicator of stock market activity. The table on page 480 contains end-of-quarter values of the S&P 500 for the years 1983–1990.

YEAR	QUARTER	S&P 500	YEAR	QUARTER	S&P 500
1983	I	151.07	1987	I	291.70
	II	165.11		II	304.00
	III	168.66		III	321.83
	IV	164.12		IV	247.08
1984	I	159.18	1988	I	258.89
	II	153.18		II	273.50
	III	166.10		III	271.91
	IV	167.24		IV	277.72
1985	I	180.66	1989	I	294.87
	II	191.85		II	317.98
	III	182.08		III	349.15
	IV	211.28		IV	353.40
1986	I	232.33	1990	I	339.94
	II	245.30		II	358.02
	III	238.27		III	306.05
	IV	248.61		IV	330.22

Source: Standard & Poor's Trade and Securities Statistics (Annual), New York, Standard & Poor's Corporation.

a. Calculate a 4-point moving average for the quarterly stock market index.

b. Plot the quarterly index and the 4-point moving average on the same graph. Can you identify the secular trend of the time series? Can you identify any seasonal variations about the secular trend?

c. Use the moving average method to forecast the quarterly S&P 500 for 1991.

d. Calculate and plot the exponentially smoothed series for the quarterly S&P 500 using a smoothing constant of $w = .3$.

e. Use the exponential smoothing technique with $w = .3$ to forecast the quarterly S&P 500 for 1991.

f. Use the Holt–Winters forecasting model with trend and seasonal components to forecast the quarterly S&P 500 for 1991. Use smoothing constants $w = .3$, $v = .8$, and $u = .5$.

9.7   Consider the gold price time series recorded in the table. (Gold prices are given in dollars per troy ounce.)

YEAR	PRICE OF GOLD	YEAR	PRICE OF GOLD	YEAR	PRICE OF GOLD
1971	41.25	1978	193.50	1985	317.30
1972	58.61	1979	307.80	1986	367.87
1973	97.81	1980	606.01	1987	408.91
1974	159.70	1981	450.63	1988	436.93
1975	161.40	1982	374.18	1989	381.21
1976	124.80	1983	449.03	1990	384.07
1977	148.30	1984	360.29		

Source: Survey of Current Business, United States Department of Commerce.

a. Calculate a 3-point moving average for the gold price time series. Plot the gold prices and the 3-point moving average on the same graph. Can you detect the long-term trend and any cyclical patterns in the time series?

b. Use the moving averages to forecast the price of gold in 1991.

c. Calculate and plot the exponentially smoothed gold price series using a smoothing constant of $w = .8$.

d. Use the exponentially smoothed series to forecast the price of gold in 1991.

e. Use the Holt–Winters forecasting model with trend to forecast the price of gold in 1991. Use smoothing constants $w = .8$ and $v = .4$.

9.4

Forecasting: The Regression Approach

Many firms use past sales to forecast future sales. Suppose a wholesale distributor of sporting goods is interested in forecasting its sales revenue for each of the next 5 years. Since an inaccurate forecast may have dire consequences to the distributor, some measure of the forecast's reliability is required. To make such forecasts and assess their reliability, an **inferential time series forecasting model** must be constructed. The familiar general linear regression model of Chapter 4 represents one type of inferential model since it allows us to calculate prediction intervals for the forecasts.

To illustrate the technique of forecasting with regression, consider the data in Table 9.5. The data are annual sales (in thousands of dollars) for a firm (say, the sporting goods distributor) in each of its 35 years of operation. A plot of the data (Figure 9.5 on page 482) reveals a linearly increasing trend, so the first-order (straight-line) model

$$E(y_t) = \beta_0 + \beta_1 t$$

seems plausible for describing the secular trend. The SAS printout for the model is shown in Figure 9.6 (page 483). Note that the model apparently provides an excellent fit to the data, with $R^2 = .98$, $F = 1,615.72$ (p-value $< .0001$), and $s = 6.39$. The least squares prediction equation, whose coefficients are shaded in Figure 9.6, is

$$\hat{y}_t = \hat{\beta}_0 + \hat{\beta}_1 t = .4015 + 4.2956t$$

TABLE 9.5 **A Firm's Yearly Sales Revenue (thousands of dollars)**

t	y_t	t	y_t	t	y_t
1	4.8	13	48.4	25	100.3
2	4.0	14	61.6	26	111.7
3	5.5	15	65.6	27	108.2
4	15.6	16	71.4	28	115.5
5	23.1	17	83.4	29	119.2
6	23.3	18	93.6	30	125.2
7	31.4	19	94.2	31	136.3
8	46.0	20	85.4	32	146.8
9	46.1	21	86.2	33	146.1
10	41.9	22	89.9	34	151.4
11	45.5	23	89.2	35	150.9
12	53.5	24	99.1		

FIGURE 9.5
Plot of sales data

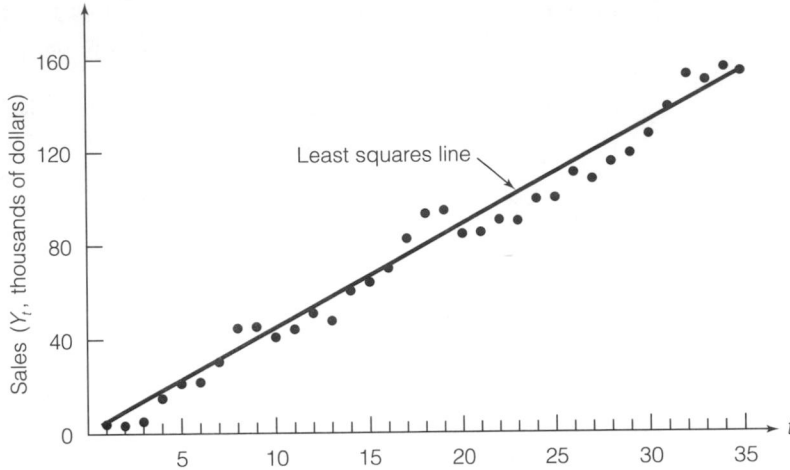

We can obtain sales forecasts and corresponding 95% prediction intervals for years 36–40 by employing the formulas given in Section 3.9. However, these values are given in the bottom portion of the SAS printout shown in Figure 9.6. For example, for $t = 36$, we have $\hat{y}_{36} = 155.0$ with the 95% prediction interval (141.3, 168.8). That is, we predict that sales revenue in year $t = 36$ will fall between \$141,300 and \$168,800 with 95% confidence.

Note that the prediction intervals for $t = 36, 37, \ldots, 40$ widen as we attempt to forecast farther into the future. Intuitively, we know that the farther into the future we forecast, the less certain we are of the accuracy of the forecast since some unexpected change in business and economic conditions may make the model inappropriate. Since we have less confidence in the forecast for, say, $t = 40$ than for $t = 36$, it follows that the prediction interval for $t = 40$ must be wider to attain a 95% level of confidence. For this reason, time series forecasting (regardless of the forecasting method) is generally confined to the short term.

Multiple regression models can also be used to forecast future values of a time series with seasonal variation. We illustrate with an example.

EXAMPLE 9.2

Refer to the 1988–1991 quarterly power loads listed in Table 9.1.

a. Propose a model for quarterly power load, y_t, that will account for both the secular trend and seasonal variation present in the series.
b. Fit the model to the data, and use the least squares prediction equation to forecast the utility company's quarterly power loads in 1992. Construct 95% prediction intervals for the forecasts.

Solution

a. A common way to describe seasonal differences in a time series is with dummy variables.* For quarterly data, a model that includes both trend and seasonal

---

*Another way to account for seasonal variation is with trigonometric (sine and cosine) terms. We discuss seasonal models with trigonometric terms in Section 9.7.

FIGURE 9.6

SAS printout for the straight-line model of yearly sales revenue

ANALYSIS OF VARIANCE

SOURCE	DF	SUM OF SQUARES	MEAN SQUARE	F VALUE	PROB>F
MODEL	1	65875.20817	65875.20817	1615.724	0.0001
ERROR	33	1345.45355	40.77131958		
C TOTAL	34	67220.66171			

ROOT MSE	6.385242	R-SQUARE	0.9800	
DEP MEAN	77.72286	ADJ R-SQ	0.9794	
C.V.	8.215398			

PARAMETER ESTIMATES

VARIABLE	DF	PARAMETER ESTIMATE	STANDARD ERROR	T FOR H0: PARAMETER=0	PROB > |T|
INTERCEP	1	0.40151261	2.20570829	0.182	0.8567
T	1	4.29563025	0.10686692	40.196	0.0001

OBS	ACTUAL	PREDICT VALUE	STD ERR PREDICT	LOWER95% PREDICT	UPPER95% PREDICT	RESIDUAL
1	4.8000	4.6971	2.1132	-8.9866	18.3809	0.1029
2	4.0000	8.9928	2.0220	-4.6338	22.6194	-4.9928
3	5.5000	13.2884	1.9325	-0.2843	26.8611	-7.7884
4	15.6000	17.5840	1.8448	4.0619	31.1061	-1.9840
5	23.1000	21.8797	1.7593	8.4048	35.3545	1.2203
6	23.3000	26.1753	1.6761	12.7444	39.6062	-2.8753
7	31.4000	30.4709	1.5959	17.0805	43.8613	0.9291
8	46.0000	34.7666	1.5189	21.4133	48.1198	11.2334
9	46.1000	39.0622	1.4457	25.7426	52.3818	7.0378
10	41.9000	43.3578	1.3769	30.0684	56.6472	-1.4578
11	45.5000	47.6534	1.3132	34.3908	60.9161	-2.1534
12	53.5000	51.9491	1.2554	38.7096	65.1886	1.5509
13	48.4000	56.2447	1.2043	43.0249	69.4645	-7.8447
14	61.6000	60.5403	1.1609	47.3366	73.7441	1.0597
15	65.6000	64.8360	1.1259	51.6448	78.0272	0.7640
16	71.4000	69.1316	1.1003	55.9494	82.3138	2.2684
17	83.4000	73.4272	1.0846	60.2504	86.6041	9.9728
18	93.6000	77.7229	1.0793	64.5478	90.8979	15.8771
19	94.2000	82.0185	1.0846	68.8416	95.1953	12.1815
20	85.4000	86.3141	1.1003	73.1319	99.4964	-0.9141
21	86.2000	90.6097	1.1259	77.4185	103.8	-4.4097
22	89.9000	94.9054	1.1609	81.7016	108.1	-5.0054
23	89.2000	99.2010	1.2043	85.9812	112.4	-10.0010
24	99.1000	103.5	1.2554	90.2571	116.7	-4.3966
25	100.3	107.8	1.3132	94.5296	121.1	-7.4923
26	111.7	112.1	1.3769	98.7985	125.4	-0.3879
27	108.2	116.4	1.4457	103.1	129.7	-8.1835
28	115.5	120.7	1.5189	107.3	134.0	-5.1792
29	119.2	125.0	1.5959	111.6	138.4	-5.7748
30	125.2	129.3	1.6761	115.8	142.7	-4.0704
31	136.3	133.6	1.7593	120.1	147.0	2.7339
32	146.8	137.9	1.8448	124.3	151.4	8.9383
33	146.1	142.2	1.9325	128.6	155.7	3.9427
34	151.4	146.5	2.0220	132.8	160.1	4.9471
35	150.9	150.7	2.1132	137.1	164.4	0.1514
36	.	155.0	2.2057	141.3	168.8	.
37	.	159.3	2.2995	145.5	173.1	.
38	.	163.6	2.3944	149.8	177.5	.
39	.	167.9	2.4903	154.0	181.9	.
40	.	172.2	2.5870	158.2	186.2	.

SUM OF RESIDUALS   -7.10543E-14
SUM OF SQUARED RESIDUALS   1345.454
PREDICTED RESID SS (PRESS)   1484.211

components is

$$E(y_t) = \beta_0 + \underbrace{\beta_1 t}_{\substack{\text{Secular} \\ \text{trend}}} + \underbrace{\beta_2 Q_1 + \beta_3 Q_2 + \beta_4 Q_3}_{\text{Seasonal component}}$$

where

t = Time period, ranging from $t = 1$ for quarter I of 1988 to $t = 16$ for quarter IV of 1991

y_t = Power load (megawatts) in time t

$Q_1 = \begin{cases} 1 & \text{if quarter I} \\ 0 & \text{if not} \end{cases}$ $Q_2 = \begin{cases} 1 & \text{if quarter II} \\ 0 & \text{if not} \end{cases}$

$Q_3 = \begin{cases} 1 & \text{if quarter III} \\ 0 & \text{if not} \end{cases}$ Base level = quarter IV

The β coefficients associated with the seasonal dummy variables determine the mean increase (or decrease) in power load for each quarter, relative to the base level quarter, quarter IV.

b. The model is fit to the data from Table 9.1 using the SAS multiple regression routine. The resulting SAS printout is shown in Figure 9.7. Note that the model appears to fit the data quite well: $R^2 = .997$, indicating that the model accounts for 99.7% of the sample variation in power loads over the 4-year period; $F = 968.96$ strongly supports the hypothesis that the model has predictive utility (p-value = .0001); and the standard deviation, **Root MSE** = 1.53, implies that the model predictions will usually be accurate to within approximately $\pm 2(1.53)$, or about ± 3.06 megawatts.

Forecasts and corresponding 95% prediction intervals for the 1992 power loads are reported in the bottom portion of the printout in Figure 9.7. For example, the forecast for power load in quarter I of 1992 is 184.7 megawatts with the 95% prediction interval (180.5, 188.9). Therefore, using a 95% prediction interval, we expect the power load in quarter I of 1992 to fall between 180.5 and 188.9 megawatts.

Many descriptive forecasting techniques have proved their merit by providing good forecasts for particular applications. Nevertheless, the advantage of forecasting using the regression approach is clear: Regression analysis provides us with a measure of reliability for each forecast through prediction intervals. However, there are two problems associated with forecasting time series using a multiple regression model.

Problem 1

We are using the least squares prediction equation to forecast values outside the region of observation of the independent variable, t. For example, in Example 9.2, we are forecasting for values of t between 17 and 20 (the four quarters of 1992), even though the observed power loads are for t values between 1 and 16.

FIGURE 9.7

SAS printout of least squares fit to quarterly power loads

Analysis of Variance

Source	DF	Sum of Squares	Mean Square	F Value	Prob>F
Model	4	9101.67800	2275.41950	968.962	0.0001
Error	11	25.83138	2.34831		
C Total	15	9127.50938			

Root MSE		1.53242	R-square	0.9972	
Dep Mean		137.30625	Adj R-sq	0.9961	
C.V.		1.11606			

Parameter Estimates

Variable	DF	Parameter Estimate	Standard Error	T for H0: Parameter=0	Prob > |T|
INTERCEP	1	90.906050	1.14331390	78.487	0.0001
T	1	4.964375	0.08566480	57.951	0.0001
Q1	1	10.093125	1.11364246	9.063	0.0001
Q2	1	-4.846250	1.09704478	-4.418	0.0010
Q3	1	14.364375	1.08696452	13.215	0.0001

Obs	ID	Dep Var Y	Predict Value	Std Err Predict	Lower95% Predict	Upper95% Predict	Residual
1	1988_1	103.5	105.3	0.923	101.3	109.2	-1.7637
2	1988_2	94.7	95.3	0.923	91.3518	99.2	-0.5887
3	1988_3	118.6	119.5	0.923	115.5	123.4	-0.8637
4	1988_4	109.3	110.1	0.923	106.1	114.0	-0.7637
5	1989_1	126.1	125.1	0.785	121.3	128.9	0.9788
6	1989_2	116.0	115.1	0.785	111.4	118.9	0.8538
7	1989_3	141.2	139.3	0.785	135.5	143.1	1.8788
8	1989_4	131.6	129.9	0.785	126.1	133.7	1.6788
9	1990_1	144.5	145.0	0.785	141.2	148.8	-0.4787
10	1990_2	137.1	135.0	0.785	131.2	138.8	2.0963
11	1990_3	159.0	159.2	0.785	155.4	163.0	-0.1787
12	1990_4	149.5	149.8	0.785	146.0	153.6	-0.2787
13	1991_1	166.1	164.8	0.923	160.9	168.8	1.2637
14	1991_2	152.5	154.9	0.923	150.9	158.8	-2.3612
15	1991_3	178.2	179.0	0.923	175.1	183.0	-0.8363
16	1991_4	169.0	169.6	0.923	165.7	173.6	-0.6362
17	1992_1	.	184.7	1.149	180.5	188.9	.
18	1992_2	.	174.7	1.149	170.5	178.9	.
19	1992_3	.	198.9	1.149	194.7	203.1	.
20	1992_4	.	189.5	1.149	185.3	193.7	.

Sum of Residuals   6.252776E-13
Sum of Squared Residuals   25.8314
Predicted Resid SS (Press)   55.6184

As noted in Chapter 6, it is risky to use a least squares regression model for prediction outside the range of the observed data because some unusual change, economic or political, may make the model inappropriate for predicting future events. Because forecasting always involves predictions about future values of a time series, this problem obviously cannot be avoided. However, it is important that the forecaster recognize the dangers of this type of prediction.

Problem 2

Recall the standard assumptions made about the random error component of a multiple regression model (Section 4.2). We assume that the errors have mean 0, constant variance, normal probability distributions, and are *independent*. The latter assumption is often violated in time series that exhibit short-term trends. As an illustration, refer to the plot of the sales revenue data shown in Figure 9.5.

Notice that the observed sales tend to deviate about the least squares line in positive and negative runs. That is, if the difference between the observed sales and predicted sales in year t is positive (or negative), the difference in year $t +$ 1 tends to be positive (or negative). Since the variation in the yearly sales is systematic, the implication is that the errors are correlated. (We gave a formal statistical test for correlated errors in Section 7.6.) Violation of this standard regression assumption could lead to unreliable forecasts.

Time series models have been developed specifically for the purpose of making forecasts when the errors are known to be correlated. These models include an **autoregressive term** for the correlated errors that result from cyclical, seasonal, or other short-term effects. Time series autoregressive models are the subject of Sections 9.5–9.11.

EXERCISES

9.8   The accompanying table records the volume of wheat (in thousands of bushels) harvested by members of a farmers' marketing cooperative for the period 1978–1991. The cooperative is interested in detecting the long-term linear trend of the wheat harvest.

YEAR	TIME	WHEAT HARVESTED	YEAR	TIME	WHEAT HARVESTED
1978	1	75	1985	8	91
1979	2	78	1986	9	92
1980	3	82	1987	10	92
1981	4	82	1988	11	93
1982	5	84	1989	12	96
1983	6	85	1990	13	101
1984	7	87	1991	14	102

a. Graph the wheat harvest time series.

b. Propose a model for the long-term linear trend of the time series.

c. Fit the model, using the method of least squares. Plot the least squares line on the graph of part a. Can you identify the long-term trend?

d. How well does the linear model describe the long-term trend? [*Hint:*   Check the value of R^2.]

e. Use the least squares model to forecast the volume of wheat harvested in 1992. Construct a 95% prediction interval for the forecast.

9.9   A realtor working in a large city wants to identify the secular trend in the weekly number of one-family houses sold by her firm. For the past 15 weeks she has collected data on her firm's home sales, as shown in the table.

a. Plot the time series. Is there visual evidence of a quadratic trend?

b. The realtor hypothesizes the model $E(y_t) = \beta_0 + \beta_1 t + \beta_2 t^2$ for the secular trend of the weekly time series. Fit the model to the data, using the method of least squares. (You will need access to a statistical software package.)

WEEK t	HOMES SOLD y_t	WEEK t	HOMES SOLD y_t	WEEK t	HOMES SOLD y_t
1	59	6	137	11	88
2	73	7	106	12	75
3	70	8	122	13	62
4	82	9	93	14	44
5	115	10	86	15	45

c. Plot the least squares model on the graph of part **a**. How well does the quadratic model describe the secular trend?

d. Use the model to forecast home sales in week 16 with a 95% prediction interval.

9.10 Refer to the quarterly S&P 500 values given in Exercise 9.6.
a. Hypothesize a time series model to account for trend and seasonal variation.
b. Fit the model in part **a** to the data.
c. Use the least squares model from part **b** to forecast the S&P 500 for all four quarters of 1991. Obtain 95% prediction intervals for the forecasts.

9.11 Information on intercity passenger traffic (excluding travel by private automobiles) since 1940 is given in the table. The data are recorded as percentages of total passenger-miles traveled.

YEAR	TIME	RAILROADS	BUSES	AIR CARRIERS
1940	1	67.1	26.5	2.8
1945	2	74.3	21.4	2.7
1950	3	46.3	37.7	14.3
1955	4	36.5	32.4	28.9
1960	5	28.6	25.7	42.1
1965	6	17.9	24.2	54.7
1970	7	7.3	16.9	73.1
1975	8	5.8	14.2	77.7
1980	9	4.7	11.4	83.9
1985	10	3.6	7.9	88.4

Source: *Statistical Abstract of the United States*, 1987. Interstate Commerce Commission, Civil Aeronautics Board.

a. Let y_t be the percentage of total passenger-miles at time t for a particular mode of transportation. Consider the linear model $E(y_t) = \beta_0 + \beta_1 t$. Which modes of transportation do you think have a secular trend adequately represented by this model?

b. Fit the model in part **a** to the data for each mode of transportation, using the method of least squares.

c. Plot the data and the least squares line for each mode of transportation. Which models adequately describe the secular trend of percentage of total passenger-miles traveled? Does this agree with your answer to part **a**?

d. Refer to your answer for part **c**. Use the least squares prediction equations to forecast the percentage of total passenger-miles to be traveled for the respective modes of transportation in 1990. If you have access to a statistical software package obtain 95% prediction intervals. What are the risks associated with this forecast procedure?

9.12   The annual price (in cents per pound) of galvanized steel from 1971 to 1989 is shown in the table.

YEAR	t	y_t	YEAR	t	y_t	YEAR	t	y_t
1971	1	9.61	1978	8	20.47	1985	15	30.30
1972	2	10.88	1979	9	22.32	1986	16	30.30
1973	3	10.59	1980	10	23.88	1987	17	30.49
1974	4	12.39	1981	11	26.88	1988	18	31.05
1975	5	14.80	1982	12	26.75	1989	19	31.05
1976	6	16.07	1983	13	28.43			
1977	7	18.10	1984	14	30.30			

Source: Standard & Poor's Trade and Securities Statistics (Annual), New York, Standard & Poor's Corporation.

a.   Plot the time series. Is there visual evidence of a linear trend?

b.   Fit the model $E(y_t) = \beta_0 + \beta_1 t$ to the data, using the method of least squares.

c.   Plot the least squares line on the graph of part a. How well does the model describe the time series?

d.   Use the fitted least squares model to forecast the price of galvanized steel for the years 1990–1994. Obtain 95% prediction intervals for the forecasts and verify that the width of the interval increases the farther you forecast into the future.

9.13   Refer to the monthly IBM common stock prices given in Exercise 9.4.

a.   Fit the model $E(y_t) = \beta_0 + \beta_1 t$ to the data.

b.   Plot the least squares line and the actual time series on the same graph. Is there visual evidence that the model provides a reasonable fit to the data?

c.   Test to determine whether the model is useful for predicting IBM monthly stock price. Use $\alpha = .05$.

d.   Use the fitted model to forecast the IBM stock price in January 1991 with a 95% prediction interval. Check to see that the actual value falls within the prediction interval. (Recall that the actual price in January 1991 was 126.75.).

9.14   Refer to Exercise 9.13.

a.   Propose a model for $E(y_t)$ that accounts for possible seasonal variation in the monthly series. [Hint: Consider a model with dummy variables for the 12 months, January, February, etc.]

b.   Fit the model of part a. Compare the regression results to the model with trend only, obtained in Exercise 9.13.

c.   Test the hypothesis that the monthly dummy variables are useful predictors of IBM stock price. [Hint: Conduct a partial F test.]

d.   Use the fitted least squares model from part b to forecast the IBM stock price in January 1991 with a 95% prediction interval. Compare the interval with the interval obtained in part d of Exercise 9.13.

9.15   The Employee Retirement Income Security Act (ERISA) of 1974 was originally established to enhance retirement security income. J. Ledolter (University of Iowa) and M. L. Power (Iowa State University) investigated the effects of ERISA on the growth in the number of private retirement plans (Journal of Risk and Insurance, Dec. 1983). Using quarterly data from 1956 through the third quarter of 1982 ($n = 107$ quarters), Ledolter and Power fit quarterly time series models for the number of pension qualifications and the number of profit-sharing plan qualifications. One of the various models investigated was the quadratic model $E(y_t) = \beta_0 + \beta_1 t + \beta_2 t^2$, where y_t is the logarithm of the dependent variable (number of pension or number of profit-sharing qualifications) in quarter t. The results (modified for the purpose of this exercise) are summarized below:

Pension plan qualifications:

$$\hat{y}_t = 6.19 + .039t - .00024t^2$$
$$t \ (\text{for } H_0: \beta_2 = 0) = -1.39$$

Profit-sharing plan qualifications:

$$\hat{y}_t = 6.22 + .035t - .00021t^2$$
$$t \ (\text{for } H_0: \beta_2 = 0) = -1.61$$

a. Is there evidence that the quarterly number of pension plan qualifications increases at a decreasing rate over time? Test using $\alpha = .05$. [*Hint:* Test $H_0: \beta_2 = 0$ against $H_a: \beta_2 < 0$.]
b. Forecast the number of pension plan qualifications for the fourth quarter of 1982 (i.e., $t = 108$). [*Hint:* Since y_t is the logarithm of the number of pension plan qualifications, to obtain the forecast you must take the antilogarithm of $\hat{y}_{108}$, i.e., $e^{\hat{y}_{108}}$.]
c. Is there evidence that the quarterly number of profit-sharing plan qualifications increases at a decreasing rate over time? Test using $\alpha = .05$. [*Hint:* Test $H_0: \beta_2 = 0$ against $H_a: \beta_2 < 0$.]
d. Forecast the number of profit-sharing plan qualifications for the fourth quarter of 1982 (i.e., $t = 108$). [*Hint:* Since y_t is the logarithm of the number of profit-sharing plan qualifications, to obtain the forecast you must take the antilogarithm of $\hat{y}_{108}$, i.e., $e^{\hat{y}_{108}}$.]

9.5
Autocorrelation and Autoregressive Error Models

In Chapter 7 we presented the Durbin–Watson test for detecting correlated residuals in a regression analysis. Correlated residuals are quite common when the response is a *time series* variable. Correlation of residuals for a regression model with a time series response is called **autocorrelation**, because the correlation is between residuals from the *same* time series model at different points in time.

A special case of autocorrelation that has many applications to business and economic phenomena is the case in which neighboring residuals one time period apart (say, at times t and $t + 1$) are correlated. This type of correlation is called **first-order autocorrelation**. In general, correlation between time series residuals m time periods apart is mth-order autocorrelation.

Definition 9.7
Autocorrelation is the correlation between time series residuals at different points in time. The special case in which neighboring residuals one time period apart (at times t and $t + 1$) are correlated is called **first-order autocorrelation**.

To see how autocorrelated residuals affect the regression model, we will assume a model similar to the general linear model of Chapter 4,

$$y_t = E(y_t) + R_t$$

where $E(y_t)$ is the regression model

$$E(y_t) = \beta_0 + \beta_1 x_1 + \cdots + \beta_k x_k$$

and R_t represents the random residual. We assume that the residual R_t has mean 0 and constant variance σ^2, but that it is autocorrelated. The effect of autocorrelation on the general linear model depends on the pattern of the autocorrelation. One of the most common patterns is that the autocorrelation between residuals at consecutive time points is positive. Thus, when the residual at time t, R_t, is indicating that the observed value y_t is more than the mean value $E(y_t)$, then the residual at time $(t + 1)$ will have a tendency (probability greater than .5) to be positive. This would occur, for example, if you were to model a monthly economic index (e.g., the Consumer Price Index) with a straight-line model. In times of recession, the observed values of the index will tend to be less than the predictions of a straight line for most or all of the months during the period. Similarly, in extremely inflationary periods, the residuals are likely to be positive because the observed value of the index will lie above the straight-line model. In either case, the fact that residuals at consecutive time points tend to have the same sign implies that they are **positively correlated**.

A second property commonly observed for autocorrelated residuals is that the size of the autocorrelation between values of the residual R at two different points in time diminishes rapidly as the distance between the time points increases. Thus, the autocorrelation between R_t and R_{t+m} becomes smaller (i.e., weaker) as the distance m between the time points becomes larger.

A residual model that possesses this property—positive autocorrelation diminishing rapidly as distance between time points increases—is the **first-order autoregressive error model**:

$$R_t = \phi R_{t-1} + \varepsilon_t, \quad -1 < \phi < 1$$

where ε_t, a residual called **white noise**, is uncorrelated with any and all other residual components. Thus, the value of the residual R_t is equal to a constant multiple, ϕ (Greek letter "phi"), of the previous residual, R_{t-1}, plus random error. In general, the constant ϕ is between -1 and $+1$, and the numerical value of ϕ determines the sign (positive or negative) and strength of the autocorrelation. In fact, it can be shown (proof omitted) that the autocorrelation between two residuals that are m time units apart, R_t and R_{t+m}, is

$$\text{Autocorrelation } (R_t, R_{t+m}) = \phi^m$$

Since the absolute value of ϕ will be less than 1, the autocorrelation between R_t and R_{t+m}, ϕ^m, will decrease as m increases. This means that neighboring values of R_t, i.e., $m = 1$, will have the highest correlation, and the correlation diminishes rapidly as the distance m between time points is increased. This points to an interesting property of the autoregressive time series model. The autocorrelation function depends only on the distance m between R values, and not on the time t. Time series models that possess this property are said to be **stationary**.

Definition 9.8

A **stationary time series model** for regression residuals is one that has mean 0, constant variance, and autocorrelations that depend only on the distance between time points.

The autocorrelation function of first-order autoregressive models is shown for several values of ϕ in Figure 9.8 on page 492. Note that positive values of ϕ yield positive autocorrelation for all residuals, whereas negative values of ϕ imply negative correlation for neighboring residuals, positive correlation between residuals two time points apart, negative correlation for residuals three time points apart, and so forth. The appropriate pattern will, of course, depend on the particular application, but the occurrence of a positive autocorrelation pattern is more common.

Although the first-order autoregressive error model provides a good representation for many autocorrelation patterns, more complex patterns can be described by higher-order autoregressive models. The general form of a pth-order autoregressive error model is

$$R_t = \phi_1 R_{t-1} + \phi_2 R_{t-2} + \cdots + \phi_p R_{t-p} + \varepsilon_t$$

The inclusion of p parameters, $\phi_1, \phi_2, \ldots, \phi_p$, permits more flexibility in the pattern of autocorrelations exhibited by a residual time series. When an autoregressive model is used to describe residual autocorrelations, the observed autocorrelations are used to estimate these parameters. Methods for estimating these parameters will be presented in Section 9.8.

EXERCISES

9.16 Suppose that the random component of a time series model follows the first-order autoregressive model $R_t = \phi R_{t-1} + \varepsilon_t$, where ε_t is a white-noise process. Consider four versions of this model: $\phi = .9$, $\phi = -.9$, $\phi = .2$, and $\phi = -.2$.

   a. Calculate the first 10 autocorrelations, Autocorrelation(R_t, R_{t+m}), $m = 1, 2, 3, \ldots, 10$, for each of the four models.

   b. Plot the autocorrelations against the distance in time separating the R values (m) for each case.

   c. Examine the rate at which the correlation diminishes in each plot. What does this imply?

9.17 When using time series to analyze quarterly data (data in which seasonal effects are present) it is highly possible that the random component of the model R_t, also exhibits the same seasonal variation as the dependent variable. In these cases, the following non–first-order autoregressive model is sometimes postulated for the correlated error term, R_t:

$$R_t = \phi R_{t-4} + \varepsilon_t$$

FIGURE 9.8
Autocorrelation functions for several first-order autoregressive error models: $R_t = \phi R_{t-1} + \varepsilon_t$

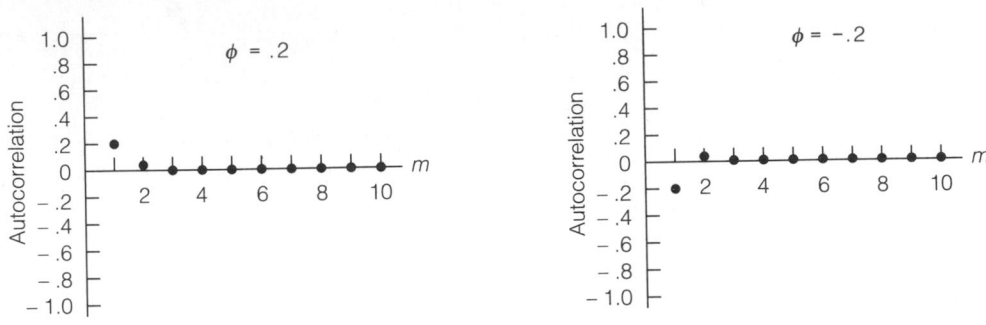

a. Weak autocorrelation

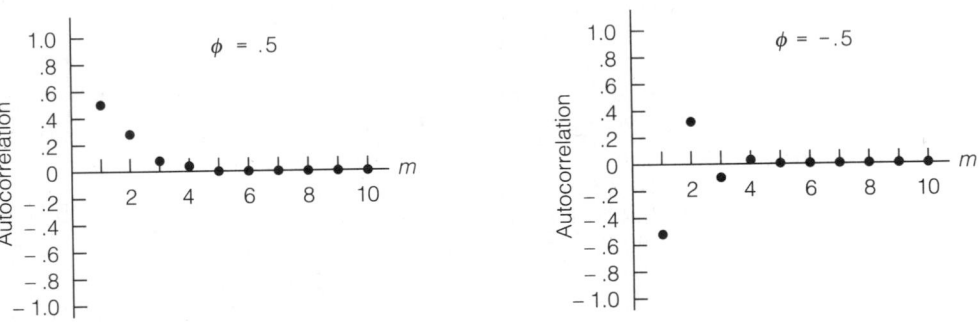

b. Moderate autocorrelation

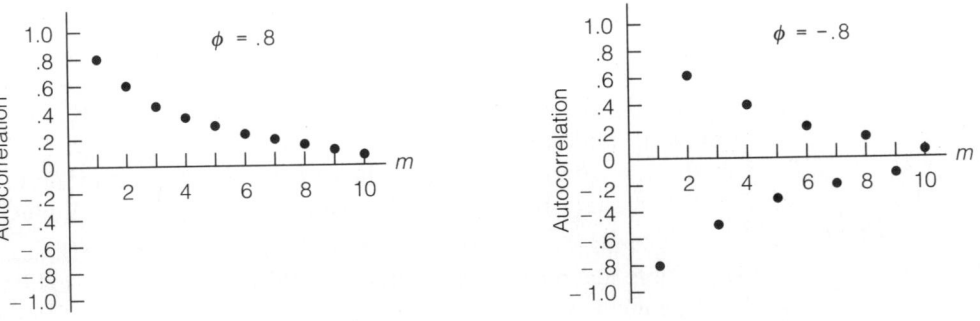

c. Strong autocorrelation

where $|\phi| < 1$ and ε_t is a white-noise process. The autocorrelation function for this model is given by

$$\text{Autocorrelation}(R_t, R_{t+m}) = \begin{cases} \phi^{m/4} & \text{if } m = 4, 8, 12, 16, 20, \ldots \\ 0 & \text{if otherwise} \end{cases}$$

a. Calculate the first 20 autocorrelations ($m = 1, 2, \ldots, 20$) for the model with constant coefficient $\phi = .5$.

b. Plot the autocorrelations against m, the distance in time separating the R values. Compare the rate at which the correlation diminishes with the first-order model $R_t = .5R_{t-1} + \varepsilon_t$.

9.18 Consider the autocorrelation pattern shown in the figure. Write a first-order autoregressive model that exhibits this pattern.

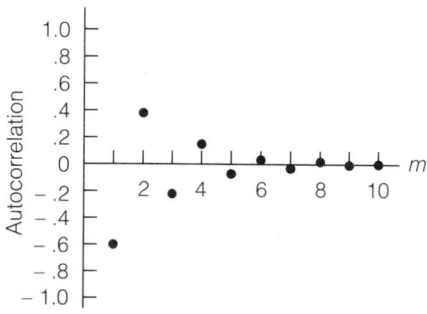

9.19 Write the general form for a fourth-order autoregressive model.

9.6
Other Models for Autocorrelated Errors (Optional)

There are many models for autocorrelated residuals in addition to the autoregressive model, but the autoregressive model provides a good approximation for the autocorrelation pattern in many applications. Recall that the autocorrelations for autoregressive models diminish rapidly as the time distance m between the residuals increases. Occasionally, residual autocorrelations appear to change abruptly from nonzero for small values of m to 0 for larger values of m. For example, neighboring residuals ($m = 1$) may be correlated, while residuals that are farther apart ($m > 1$) are uncorrelated. This pattern can be described by the **first-order moving average model**

$$R_t = \varepsilon_t + \theta\varepsilon_{t-1}$$

Note that the residual R_t is a linear combination of the current and previous *uncorrelated* (white-noise) residuals. It can be shown that the autocorrelations for this model are

$$\text{Autocorrelation}(R_t, R_{t+m}) = \begin{cases} \dfrac{\theta}{1 + \theta^2} & \text{if } m = 1 \\ 0 & \text{if } m > 1 \end{cases}$$

This pattern is shown in Figure 9.9 on page 494.

FIGURE 9.9

Autocorrelations for the first-order moving average model: $R_t = \varepsilon_t + \theta\varepsilon_{t-1}$

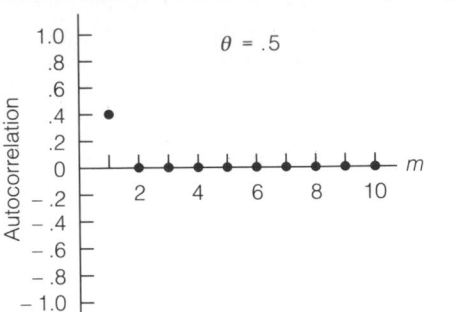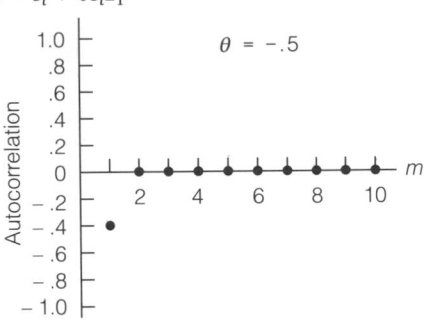

More generally, a qth-order moving average model is given by

$$R_t = \varepsilon_t + \theta\varepsilon_{t-1} + \theta_2\varepsilon_{t-2} + \cdots + \theta_q\varepsilon_{t-q}$$

Residuals within q time points are correlated, whereas those farther than q time points apart are uncorrelated. For example, a regression model for the quarterly earnings per share for a company may have residuals that are autocorrelated when within 1 year ($m = 4$ quarters) of one another, but uncorrelated when farther apart. An example of this pattern is shown in Figure 9.10.

FIGURE 9.10

Autocorrelations for a fourth-order moving average model

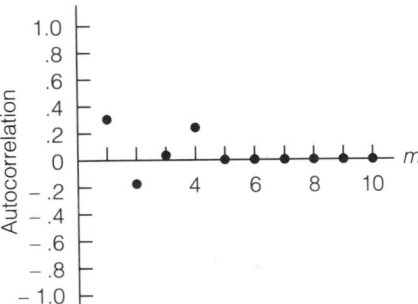

Some autocorrelation patterns require even more complex residual models. A more general model is a combination of the **autoregressive–moving average (ARMA) models**,

$$R_t = \phi_1 R_{t-1} + \cdots + \phi_p R_{t-p} + \varepsilon_t + \theta_1\varepsilon_{t-1} + \cdots + \theta_q\varepsilon_{t-q}$$

Like the autoregressive model, the ARMA model has autocorrelations that diminish as the distance m between residuals increases. However, the patterns that can be described by ARMA models are more general than those of either autoregressive or moving average models.

In Section 9.8, we present a method for estimating the parameters of an autoregressive residual model. The method for fitting time series models when the residual is either moving average or ARMA is more complicated, however. Consult the references at the end of the chapter for details of these methods.

9.7

Constructing Time Series Models

Recall that the general form of the times series model is

$$y_t = E(y_t) + R_t$$

We are assuming that the expected value of y_t is

$$E(y_t) = \beta_0 + \beta_1 x_1 + \beta_2 x_2 + \cdots + \beta_k x_k$$

where $x_1, x_2, \ldots, x_k$ are independent variables, which themselves may be time series, and the residual component, R_t, accounts for the pattern of autocorrelation in the residuals. Thus, a time series model consists of a pair of models: one model for the deterministic component $E(y_t)$ and one model for the autocorrelated residuals R_t.

Choosing the Deterministic Component

The deterministic portion of the model is chosen in exactly the same manner as the regression models of the preceding chapters except that some of the independent variables might be time series variables or might be trigonometric functions of time (such as $\sin t$ or $\cos t$). It is helpful to think of the deterministic component as consisting of the trend (T_t), cyclical (C_t), and seasonal (S_t) effects described in Section 9.2.

For example, we may want to model the number of new housing starts, y_t, as a function of the prime interest rate, x_t. Then, one model for the mean of y_t is

$$E(y_t) = \beta_0 + \beta_1 x_t$$

for which the mean number of new housing starts is a multiple β_1 of the prime interest rate, plus a constant β_0. Another possibility is a second-order relationship,

$$E(y_t) = \beta_0 + \beta_1 x_t + \beta_2 x_t^2$$

which permits the *rate* of increase in the mean number of housing starts to increase or decrease with the prime interest rate.

Yet another possibility is to model the mean number of new housing starts as a function of both the prime interest rate and the year, t. Thus, the model

$$E(y_t) = \beta_0 + \beta_1 x_t + \beta_2 t + \beta_3 x_t t$$

implies that the mean number of housing starts increases linearly in x_t, the prime interest rate, but the rate of increase depends on the year t. If we wanted to adjust for seasonal (cyclical) effects due to t, we might introduce time into the model using trigonometric functions of t. This topic will be explained in greater detail below.

Another important type of model for $E(y_t)$ is the **lagged independent variable model**. *Lagging* means that we are pairing observations on a dependent variable and independent variable at two different points in time, with the time corresponding to the independent variable lagging behind the time for the dependent variable. Suppose, for example, we believe that the monthly mean number of new housing starts is a function of the *previous* month's prime interest rate. Thus, we model y_t as a linear function of the lagged independent variable, prime interest rate, x_{t-1},

$$E(y_t) = \beta_0 + \beta_1 x_{t-1}$$

or, alternatively, as the second-order function,

$$E(y_t) = \beta_0 + \beta_1 x_{t-1} + \beta_2 x_{t-1}^2$$

For this example, the independent variable, prime interest rate x_t, is lagged 1 month behind the response y_t.

Many time series have distinct seasonal patterns. Retail sales are usually highest around Christmas, spring, and fall, with relative lulls in the winter and summer periods. Energy usage is highest in summer and winter, and lowest in spring and fall. Teenage unemployment rises in the summer months when schools are not in session, and falls near Christmas when many businesses hire part-time help.

When a time series' seasonality is exhibited in a relatively consistent pattern from year to year, we can model the pattern using trigonometric terms in the model for $E(y_t)$. For example, the model of a monthly series with mean $E(y_t)$ might be

$$E(y_t) = \beta_0 + \beta_1\left(\cos\frac{2\pi}{12}t\right) + \beta_2\left(\sin\frac{2\pi}{12}t\right)$$

This model would appear as shown in Figure 9.11. Note that the model is **cyclic**, with a **period** of 12 months. That is, the mean $E(y_t)$ completes a cycle every 12 months and then repeats the same cycle over the next 12 months. Thus, the **expected peaks and valleys** of the series remain the same from year to year. The coefficients β_1 and β_2 determine the **amplitude** and **phase shift** of the model. The amplitude is the magnitude of the seasonal effect, whereas the phase shift locates the peaks and valleys in time. For example, if we assume month 1 is January, the mean of the time series depicted in Figure 9.11 has a peak each April and a valley each October.

FIGURE 9.11

A seasonal time series model

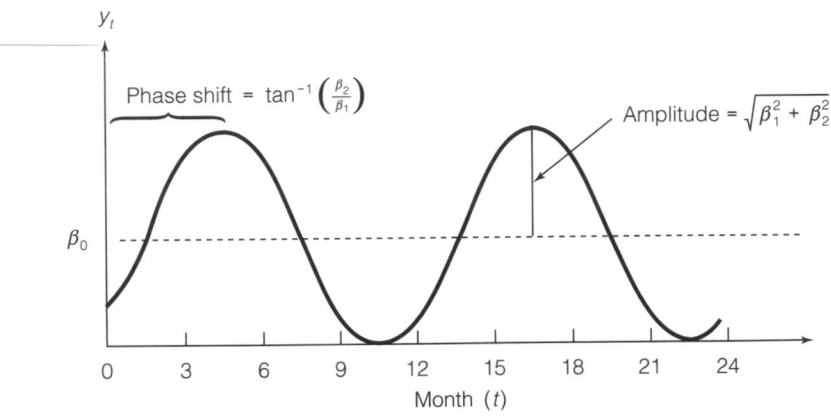

If the data are monthly or quarterly, we can treat the season as a qualitative independent variable (see Example 9.2), and write the model

$$E(y_t) = \beta_0 + \beta_1 S_1 + \beta_2 S_2 + \beta_3 S_3$$

where

$$S_1 = \begin{cases} 1 & \text{if season is spring (II)} \\ 0 & \text{otherwise} \end{cases} \qquad S_2 = \begin{cases} 1 & \text{if season is summer (III)} \\ 0 & \text{otherwise} \end{cases}$$

$$S_3 = \begin{cases} 1 & \text{if season is fall (IV)} \\ 0 & \text{otherwise} \end{cases}$$

Thus, S_1, S_2, and S_3 are dummy variables that describe the four levels of season, letting winter (I) be the base level. The β coefficients determine the mean value of y_t for each season, as shown in Figure 9.12. Note that for the dummy variable model and the trigonometric model, we assume the seasonal effects are approximately the same from year to year. If they tend to increase or decrease with time, an interaction of the seasonal effect with time may be necessary. (An example will be given in Section 9.10.)

FIGURE 9.12

Seasonal model for quarterly data using dummy variables

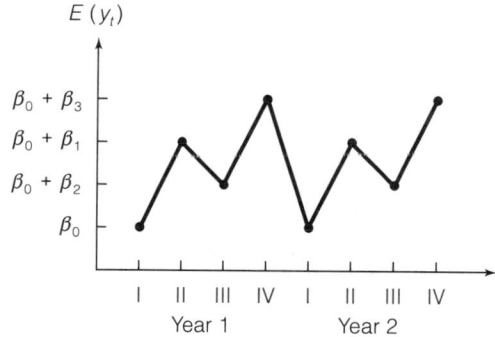

The appropriate form of the deterministic time series model will depend on both theory and data. Economic theory often provides several plausible models relating the mean response to one or more independent variables. The data can then be used to determine which, if any, of the models is best supported. The process is often an iterative one, beginning with preliminary models based on theoretical notions, using data to refine and modify these notions, collecting additional data to test the modified theories, and so forth.

Choosing the Residual Component

The appropriate form of the residual component R_t will depend on the pattern of autocorrelation in the residuals (see Sections 9.5 and 9.6). The autoregressive model of Section 9.5 is very useful for this aspect of time series modeling. The general form of an autoregressive model of order p is

$$R_t = \phi_1 R_{t-1} + \phi_2 R_{t-2} + \cdots + \phi_p R_{t-p} + \varepsilon_t$$

where ε_t is white noise (uncorrelated error). Recall that the name *autoregressive* comes from the fact that R_t is regressed on its own past values. As the order p is increased, more complex autocorrelation functions can be modeled. There are several other types of models that can be used for the random component, but the autoregressive model is very flexible and receives more application in business forecasting than the other models.

The simplest autoregressive error model is the **first-order autoregressive model**

$$R_t = \phi R_{t-1} + \varepsilon_t$$

Recall that the autocorrelation between residuals at two different points in time diminishes as the distance between the time points increases. Since many business and economic time series exhibit this property, the first-order autoregressive model is a popular choice for the residual component.

To summarize, we describe a general approach for constructing a time series:

1. Construct a regression model for the trend, seasonal, and cyclical components of $E(y_t)$. This model may be a polynomial in t for the trend (usually a straight-line or quadratic model) with trigonometric terms or dummy variables for the seasonal (cyclical) effects. The model may also include other time series variables as independent variables. For example, last year's rate of inflation may be used as a predictor of this year's GNP.

2. Next, construct a model for the random component (residual effect) of the model. A model that is widely used in practice is the first-order autoregressive error model

$$R_t = \phi R_{t-1} + \varepsilon_t$$

When the pattern of autocorrelation is more complex, use the general pth-order autoregressive model

$$R_t = \phi_1 R_{t-1} + \phi_2 R_{t-2} + \cdots + \phi_p R_{t-p} + \varepsilon_t$$

3. The two components are then combined so that the model can be used for forecasting:

$$y_t = E(y_t) + R_t$$

Prediction intervals are calculated to measure the reliability of the forecasts. In the following two sections we will demonstrate how time series models are fit to data and used for forecasting. In Section 9.10 we will present an example in which we fit a seasonal time series model to a set of data.

EXERCISES

9.20 Suppose you are interested in buying gold on the commodities market. Your broker has advised you that your best strategy is to sell back the gold at the first substantial jump in price. Hence, you are interested in a short-term investment. Before buying, you would like to model the closing price of gold, y_t, over time (in days), t.

a. Write a first-order model for the deterministic portion of the model, $E(y_t)$.

b. If a plot of the daily closing prices for the past month reveals a quadratic trend, write a plausible model for $E(y_t)$.

c. Since the closing price of gold on day $(t + 1)$ is very highly correlated with the closing price on day t, your broker suggests that the random error components of the model are not white noise. Given this information, postulate a model for the error term, R_t.

9.21 An economist wishes to model the GNP over time (in years) and also as a function of certain personal consumption expenditures. Let t = Time in years and let

y_t = GNP at time t

x_{1t} = Durable goods at time t

x_{2t} = Nondurable goods at time t

x_{3t} = Services at time t

a. The economist believes that y_t is linearly related to the independent variables x_{1t}, x_{2t}, x_{3t}, and t. Write the first-order model for $E(y_t)$.

b. Rewrite the model if interaction between the independent variables and time is present.

c. Postulate a model for the random error component, R_t. Explain why this model is appropriate.

9.22 Airlines sometimes overbook flights because of "no-show" passengers, i.e., passengers who have purchased a ticket but fail to board the flight. An airline supervisor wishes to be able to predict, for a Miami-to-New York flight, the monthly accumulation of no-show passengers during the upcoming year, using data from the past 3 years. Let y_t = Number of no-shows during month t.

a. Using dummy variables, propose a model for $E(y_t)$ that will take into account the seasonal (fall, winter, spring, summer) variation that may be present in the data.

b. Postulate a model for the error term R_t.

c. Write the full time series model for y_t (include random error terms).

d. Suppose the airline supervisor believes that the seasonal variation in the data is not constant from year to year; in other words, that there exists interaction between time and season. Rewrite the full model with the interaction terms added.

9.23 A farmer is interested in modeling the daily price of hogs at a livestock market. The farmer knows that the price varies over time (days) and also is reasonably confident that a seasonal effect is present.

a. Write a seasonal time series model with trigonometric terms for $E(y_t)$, where y_t = Selling price (in dollars) of hogs on day t.

b. Interpret the β parameters.

c. Include in the model an interaction between time and the trigonometric components. What does the presence of interaction signify?

d. Is it reasonable to assume that the random error component of the model, R_t, is white noise? Explain. Postulate a more appropriate model for R_t.

9.24 Numerous studies have been conducted to examine the relationship between seniority and productivity in business. A problem encountered in such studies is that individual output is often difficult to measure. G. A. Krohn developed a technique for estimating the experience–productivity relationship when such a measure is available (*Journal of Business & Economic Statistics*, Oct. 1983). Krohn modeled the batting average of a major league baseball player in year t (y_t) as a function of the player's age in year t (x_t) and an autoregressive error term (R_t).

a. Write a model for $E(y_t)$ that hypothesizes, as did Krohn, a curvilinear relationship with x_t.

b. Write a first-order autoregressive model for R_t.

c. Use the models from parts **a** and **b** to write the full time series autoregressive model for y_t.

9.8

Fitting Time Series Models with Autoregressive Errors

We have proposed a general form for a time series model:

$$y_t = E(y_t) + R_t$$

where

$$E(y_t) = \beta_0 + \beta_1 x_1 + \cdots + \beta_k x_k$$

and, using an autoregressive model for R_t,

$$R_t = \phi R_{t-1} + \phi_2 R_{t-2} + \cdots + \phi_p R_{t-p} + \varepsilon_t$$

We now want to develop estimators for the parameters β_0, β_1, ..., β_k of the regression model, and for the parameters ϕ_1, ϕ_2, ..., ϕ_p of the autoregressive model. The ultimate objective is to use the model to obtain forecasts (predictions) of future values of y_t, as well as to make inferences about the structure of the model itself.

We will introduce the techniques of fitting a time series model with a simple example. Refer to the data in Table 9.5, the annual sales for a firm in each of its 35 years of operation. Recall that the objective is to forecast future sales in years 36–40. In Section 9.4 we used a simple straight-line model for the mean sales

$$E(y_t) = \beta_0 + \beta_1 t$$

to make the forecasts.

The SAS printout showing the least squares estimates of β_0 and β_1 is reproduced in Figure 9.13. Although the model is useful for predicting annual sales (p-value for H_0: $\beta_1 = 0$ is less than .0001), the Durbin–Watson statistic is $d = .821$, which is less than the tabled value, $d_L = 1.40$ (Table 9 of Appendix D), for $\alpha = .05$, $n = 35$, and $k = 1$ independent variable. Thus, there is evidence that the residuals are positively correlated. The plot of the least squares residuals

FIGURE 9.13

SAS printout: Regression analysis for least squares sales data

DEP VARIABLE: Y

ANALYSIS OF VARIANCE

SOURCE	DF	SUM OF SQUARES	MEAN SQUARE	F VALUE	PROB>F
MODEL	1	65875.20817	65875.20817	1615.724	0.0001
ERROR	33	1345.45355	40.77131958		
C TOTAL	34	67220.66171			

ROOT MSE	6.385242	R-SQUARE	0.9800	
DEP MEAN	77.72286	ADJ R-SQ	0.9794	
C.V.	8.215398			

PARAMETER ESTIMATES

VARIABLE	DF	PARAMETER ESTIMATE	STANDARD ERROR	T FOR H0: PARAMETER=0	PROB > |T|
INTERCEP	1	0.40151261	2.20570829	0.182	0.8567
T	1	4.29563025	0.10686692	40.196	0.0001

DURBIN-WATSON D	0.821
(FOR NUMBER OF OBS.)	35
1ST ORDER AUTOCORRELATION	0.590

over time, in Figure 9.14, shows the pattern of positive autocorrelation. The residuals tend to cluster in positive and negative runs; if the residual at time t is positive, the residual at time $(t + 1)$ tends to be positive.

FIGURE 9.14

Annual sales time series example: Least squares residual plot

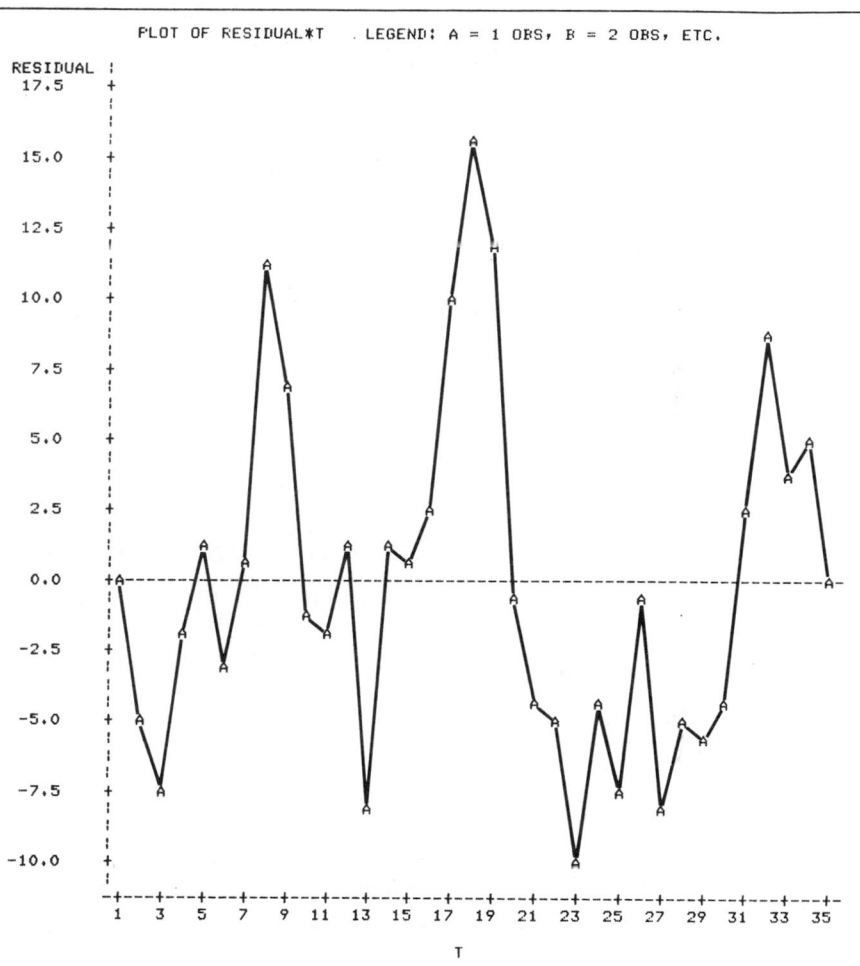

What are the consequences of fitting the least squares model when autocorrelated residuals are present? Although *the least squares estimators of β_0 and β_1 remain unbiased* even if the residuals are autocorrelated, i.e., $E(\hat{\beta}_0) = \beta_0$ and $E(\hat{\beta}_1) = \beta_1$, the *standard errors given by least squares theory are usually smaller than the true standard errors* when the residuals are positively autocorrelated. Consequently, t values computed by the methods of Chapter 4 (which apply when the errors are uncorrelated) will usually be inflated and will lead to a higher Type I error rate (α) than the value of α selected for a test. Thus, the application of standard least squares techniques to time series often produces statistical test results that are misleading and leads to overoptimistic evaluations of a model's predictive ability. There is a second reason for seeking methods that specifically take into account the autocorrelated residuals. If we can successfully model the residual autocorrelation, we should achieve a smaller MSE and correspondingly narrower prediction intervals than those given by the least squares model.

To account for the autocorrelated residual, we will postulate a first-order autoregressive model,

$$R_t = \phi R_{t-1} + \varepsilon_t$$

Thus, we use the pair of models

$$y_t = \beta_0 + \beta_1 t + R_t$$
$$R_t = \phi R_{t-1} + \varepsilon_t$$

to describe the yearly sales of the firm. To estimate the parameters of the time series model (β_0, β_1, and ϕ), a modification of the least squares method is required. To do this, we use a *transformation* that is much like the variance-stabilizing transformations discussed in Chapter 7.

First, we multiply the model

$$y_t = \beta_0 + \beta_1 t + R_t \tag{1}$$

by ϕ at time $(t - 1)$ to obtain

$$\phi y_{t-1} = \phi \beta_0 + \phi \beta_1 (t - 1) + \phi R_{t-1} \tag{2}$$

Taking the difference between equations (1) and (2), we have

$$y_t - \phi y_{t-1} = \beta_0(1 - \phi) + \beta_1[t - \phi(t - 1)] + (R_t - \phi R_{t-1})$$

or, since $R_t = \phi R_{t-1} + \varepsilon_t$, then

$$y_t^* = \beta_0^* + \beta_1 t^* + \varepsilon_t$$

where $y_t^* = y_t - \phi y_{t-1}$, $t^* = t - \phi(t - 1)$ and $\beta_0^* = \beta_0(1 - \phi)$. Thus, we can use the transformed dependent variable y_t^* and transformed independent variable t^* and obtain least squares estimates of β_0^* and β_1. The residual ε_t is uncorrelated, so that the assumptions necessary for the least squares estimators are all satisfied. The estimator of the original intercept, β_0, can be calculated by

$$\hat{\beta}_0 = \frac{\hat{\beta}_0^*}{1 - \phi}$$

This transformed model appears to solve the problem of first-order autoregressive residuals. However, making the transformation requires knowing the value of the parameter ϕ. Also, we lose the initial observation, since the values of y_t^* and t^* can be calculated only for $t \geq 2$. The methods for estimating ϕ and adjustments for the values at $t = 1$ will not be detailed here. Instead, we will present output from the SAS computer package, which both performs the transformation and estimates the model parameters, β_0, β_1, and ϕ.

The SAS printout of the straight-line, autoregressive time series model fit to the sales data is shown in Figure 9.15. Note that the format of the time series printout is different from that of the standard regression printout. The estimates

FIGURE 9.15

SAS printout for the combined straight-line autoregressive residual fit to the sales data

```
                    A U T O R E G   P R O C E D U R E



                        ORDINARY LEAST SQUARES ESTIMATES

                SSE           1345.454   DFE                   33
                MSE           40.77132   ROOT MSE        6.385242
                SBC           234.1562   AIC             231.0455
                REG RSQ         0.9800   TOTAL RSQ         0.9800
                DURBIN-WATSON   0.8207


        VARIABLE DF         B VALUE       STD ERROR    T RATIO APPROX PROB

        INTERCPT  1      0.40151261     2.20570829       0.182     0.8567
        T         1      4.29563025     0.10686692      40.196     0.0001

                        ESTIMATES OF AUTOCORRELATIONS

LAG   COVARIANCE   CORRELATION  -1 9 8 7 6 5 4 3 2 1 0 1 2 3 4 5 6 7 8 9 1
  0     38.4415     1.000000    |                   |*******************|
  1     22.6661     0.589624    |                   |***********        |

                        PRELIMINARY MSE=    25.07708

                ESTIMATES OF THE AUTOREGRESSIVE PARAMETERS
        LAG    COEFFICIENT      STD ERROR        T RATIO
          1    -0.58962415      0.14277861      -4.129639


                        YULE-WALKER ESTIMATES

                SSE           877.6854   DFE                   32
                MSE           27.42767   ROOT MSE        5.237143
                SBC           223.1868   AIC             218.5208
                REG RSQ         0.9412   TOTAL RSQ         0.9869



        VARIABLE DF         B VALUE       STD ERROR    T RATIO APPROX PROB

        INTERCPT  1      0.40575699     3.99697517       0.102     0.9198
        T         1      4.29593038     0.18983105      22.630     0.0001
```

of β_0 and β_1 in the deterministic component $E(y_t)$ appear at the bottom of the printout under the column heading **B VALUE**. The estimate of the first-order autoregressive parameter ϕ is given in the middle portion of the printout titled **ESTIMATES OF THE AUTOREGRESSIVE PARAMETERS** beneath the column heading **COEFFICIENT**.

The interpretations of two quantities shown on the SAS printout for the time series procedure differ from those described in the preceding sections. First, the quantity printed as **REG RSQ** (in the lower portion of the printout) is not the value of R^2 based on the values of y_t. Instead, it is based on the values of the transformed variable, y_t^*. When we refer to R^2 in this chapter, we will always mean the value R^2 based on the original time series variable y_t. This value, which will usually be larger than the **REG RSQ** value, is given on the printout as **TOTAL RSQ** (shaded). Second, the SAS time series model is defined so that ϕ takes the *opposite* sign from the value contained in our model. Consequently, you must multiply the estimate of ϕ shown in the printout by (-1) to obtain the estimate of ϕ for our model.

Therefore, the fitted models are

$$\hat{y}_t = .4058 + 4.2959t + \hat{R}_t$$
$$\hat{R}_t = .5896\hat{R}_{t-1}$$

with

$$MSE = 27.43$$

and

$$R^2 = .9869 \quad (\textbf{TOTAL RSQ} \text{ on the printout, Figure 9.15})$$

A comparison of the least squares (Figure 9.6) and autoregressive (Figure 9.15) computer printouts is given in Table 9.6. Note that the autoregressive model reduces MSE and increases R^2. The values of the estimators β_0 and β_1 change very little, but the estimated standard errors are considerably increased, thereby decreasing the t value for testing $H_0: \beta_1 = 0$. Note that the implication that the linear relationship between sales y_t and year t is of significant predictive value is the same using either method. However, you can see that the underestimation of standard errors by using least squares in the presence of residual autocorrelation could result in the inclusion of unimportant independent variables in the model, since the t values will usually be inflated.

T A B L E 9.6 **Comparison of Least Squares and Time Series Results**

	LEAST SQUARES	AUTOREGRESSIVE
R^2	.98	.99
MSE	40.77	27.43
$\hat{\beta}_0$	.4015	.4058
$\hat{\beta}_1$	4.2956	4.2959
Standard error ($\hat{\beta}_0$)	2.2057	3.9970
Standard error ($\hat{\beta}_1$)	.1069	.1898
t statistic for $H_0: \beta_1 = 0$	40.20	22.63
	($p < .0001$)	($p < .0001$)
$\hat{\phi}$	—	.5896
t statistic for $H_0: \phi = 0$	—	4.13

The estimated value of ϕ is .5896, and an approximate t test* of the hypothesis $H_0: \phi = 0$ yields a t value of 4.13. With 32 df, this value is significant at less than $\alpha = .01$. Thus, the result of the Durbin–Watson d test is confirmed: There is adequate evidence of positive residual autocorrelation.[†] Furthermore, the first-order autoregressive model appears to describe this residual correlation well.

---

*An explanation of this t test has been omitted. Consult the references at the end of the chapter for details of this test.

[†]This result is to be expected since it can be shown (proof omitted) that $\hat{\phi} \approx 1 - d/2$, where d is the value of the Durbin–Watson statistic.

The steps for fitting a time series model to a set of data are summarized in the box. Once the model is estimated, the model can be used to forecast future values of the time series y_t.

Steps for Fitting Time Series Models

1. Use the least squares approach to obtain initial estimates of the β parameters. Do *not* use the t or F tests to assess the importance of the parameters, since the estimates of their standard errors may be biased (often underestimated).

2. Analyze the residuals to determine whether they are autocorrelated. The Durbin–Watson test is one technique for making this determination.

3. If there is evidence of autocorrelation, construct a model for the residuals. The autoregressive model is one useful model. Consult the references at the end of the chapter for more types of residual models and for methods of identifying the most suitable model.

4. Reestimate the β parameters, taking the residual model into account. This involves a simple transformation if an autoregressive model is used; several statistical software packages have computer routines to accomplish this.

EXERCISES

9.25 The gross national product (GNP) is a measure of total U.S. output and is, therefore, an important indicator of the U.S. economy. The quarterly GNP values (in billions of dollars) from 1980 to 1989 are given in the table. Let y_t be the GNP in quarter t, $t = 1, 2, 3, \ldots, 40$.

YEAR	QUARTER			
	I	II	III	IV
1980	2,573	2,579	2,639	2,736
1981	2,867	2,913	3,005	3,032
1982	3,021	3,070	3,091	3,110
1983	3,172	3,272	3,362	3,432
1984	3,677	3,758	3,812	3,853
1985	3,909	3,965	4,031	4,088
1986	4,149	4,176	4,241	4,268
1987	4,389	4,476	4,567	4,648
1988	4,736	4,831	4,918	5,010
1989	5,101	5,174	5,239	5,289

Source: Standard & Poor's Trade and Securities Statistics (Annual), New York, Standard & Poor's Corporation.

a. Hypothesize a time series model for quarterly GNP that includes a straight-line long-term trend and autocorrelated residuals.

b. The SAS printout for the pair of models

$$y_t = \beta_0 + \beta_1 t + R_t$$
$$R_t = \phi R_{t-1} + \varepsilon_t$$

is shown here. Write the least squares prediction equation.

Autoreg Procedure

Dependent Variable = GNP

Ordinary Least Squares Estimates

SSE	214963.9	DFE	38
MSE	5656.944	Root MSE	75.21266
SBC	464.4667	AIC	461.0889
Reg Rsq	0.9920	Total Rsq	0.9920
Durbin-Watson	0.3128		

Variable	DF	B Value	Std Error	t Ratio	Approx Prob
Intercept	1	2405.03077	24.237	99.228	0.0001
T	1	70.70460	1.030	68.631	0.0001

Estimates of Autocorrelations

Lag	Covariance	Correlation	-1 9 8 7 6 5 4 3 2 1 0 1 2 3 4 5 6 7 8 9 1
0	5374.097	1.000000	| |********************|
1	4376.503	0.814370	| |**************** |

Preliminary MSE = 1810.004

Estimates of the Autoregressive Parameters

Lag	Coefficient	Std Error	t Ratio
1	-0.81436995	0.09540831	-8.535629

Yule-Walker Estimates

SSE	63669.95	DFE	37
MSE	1720.809	Root MSE	41.48264
SBC	420.5735	AIC	415.5069
Reg Rsq	0.9624	Total Rsq	0.9976

Variable	DF	B Value	Std Error	t Ratio	Approx Prob
Intercept	1	2428.47988	56.681	42.845	0.0001
T	1	70.23226	2.282	30.774	0.0001

---

c. Interpret the estimates of the model parameters, β_0, β_1, and ϕ.

d. Interpret the value of R^2.

9.26 Refer to Exercise 9.8.

a. Hypothesize a time series model for annual volume of wheat harvested, y_t, that takes into account the residual autocorrelation.

b. If you have access to a statistical software package that uses the modified least squares method, fit the autoregressive time series model. Interpret the estimates of the model parameters.

9.27 Refer to Exercises 9.4 and 9.13.

   a. Hypothesize a time series model for monthly price of IBM stock, y_t, that takes into account the residual autocorrelation.

   b. If you have access to a statistical software package that uses the modified least squares method, fit the autoregressive time series model. Interpret the estimates of the model parameters.

9.28 Refer to Exercise 9.15 and the study on the long-term effects of the Employment Retirement Income Security Act (ERISA). Ledolter and Power also fit quarterly time series models for the number of pension plan terminations and the number of profit-sharing plan terminations from the first quarter of 1956 through the third quarter of 1982 ($n = 107$ quarters). To account for residual correlation, they fit straight-line autoregressive models of the form

$$y_t = \beta_0 + \beta_1 t + \phi R_{t-1} + \varepsilon_t$$

The results were as follows:

Pension plan terminations: $\quad \hat{y}_t = 3.54 + .039t + .40\hat{R}_{t-1}$

Profit-sharing plan terminations: $\quad \hat{y}_t = 3.45 + .038t + .22\hat{R}_{t-1}$

   a. Interpret the estimates of the model parameters for pension plan terminations.

   b. Interpret the estimates of the model parameters for profit-sharing plan terminations.

9.29 The Dow Jones Industrial Average (DJA) is a widely followed stock market indicator. The values of the DJA from 1968 to 1987 are given in the accompanying table. Suppose we want to model the yearly DJA, y_t, as a function of t, where t is the number of years since 1967 (i.e., $t = 1$ for 1968, $t = 2$ for 1969, . . . , $t = 20$ for 1987). A time series model that includes a long-term trend and autocorrelated residuals is the regression–autoregression pair.

$$y_t = \beta_0 + \beta_1 t + R_t$$
$$R_t = \phi R_{t-1} + \varepsilon_t$$

The SAS printout for the time series model is also reproduced on page 508.

YEAR	DJA	YEAR	DJA
1968	944	1978	805
1969	800	1979	839
1970	839	1980	964
1971	885	1981	899
1972	951	1982	1,047
1973	924	1983	1,259
1974	759	1984	1,187
1975	802	1985	1,501
1976	975	1986	1,926
1977	835	1987	1,939

Source: Standard & Poor's Trade and Securities Statistics (Annual), New York, Standard & Poor's Corporation.

   a. Identify and interpret estimates of the model parameters.

   b. Interpret the value of R^2.

```
                           Autoreg Procedure

Dependent Variable = DJA


                    Ordinary Least Squares Estimates

            SSE            1073391    DFE               18
            MSE           59632.84    Root MSE    244.1984
            SBC            280.561    AIC         278.5696
            Reg Rsq         0.5388    Total Rsq     0.5388
            Durbin-Watson   0.4013


     Variable     DF      B Value    Std Error    t Ratio Approx Prob

     Intercept     1   598.078947       113.44      5.272      0.0001
     T             1    43.421053         9.47      4.585      0.0002


                    Estimates of Autocorrelations

Lag  Covariance  Correlation -1 9 8 7 6 5 4 3 2 1 0 1 2 3 4 5 6 7 8 9 1

  0   53669.56    1.000000  |                   |********************|
  1   35030.41    0.652705  |                   |*************        |


                    Preliminary MSE = 30805.02

            Estimates of the Autoregressive Parameters

      Lag    Coefficient      Std Error       t Ratio
        1    -0.65270535     0.18374787     -3.552179


                    Yule-Walker Estimates

            SSE           469529.7    DFE               17
            MSE           27619.39    Root MSE    166.1908
            SBC            267.575    AIC         264.5878
            Reg Rsq         0.3858    Total Rsq     0.7982


     Variable     DF      B Value    Std Error    t Ratio Approx Prob

     Intercept     1   621.870357       180.12      3.453      0.0030
     T             1    46.993781        14.38      3.267      0.0045
```

9.9

Forecasting with Time Series Autoregressive Models

The ultimate objective of fitting a time series model is often to forecast future values of the series. We will demonstrate the techniques for the simple model

$$y_t = \beta_0 + \beta_1 x_t + R_t$$

with the first-order autoregressive residual

$$R_t = \phi R_{t-1} + \varepsilon_t$$

Suppose we use the data $(y_1, x_1), (y_2, x_2), \ldots, (y_n, x_n)$ to obtain estimates of β_0, β_1, and ϕ, using the method presented in Section 9.8. We now want to forecast the value of y_{n+1}. From the model,

$$y_{n+1} = \beta_0 + \beta_1 x_{n+1} + R_{n+1}$$

where

$$R_{n+1} = \phi R_n + \varepsilon_{n+1}$$

Combining these, we obtain

$$y_{n+1} = \beta_0 + \beta_1 x_{n+1} + \phi R_n + \varepsilon_{n+1}$$

The forecast of y_{n+1} is obtained by estimating each of the unknown quantities in this equation:*

$$\hat{y}_{n+1} = \hat{\beta}_0 + \hat{\beta}_1 x_{n+1} + \hat{\phi}\hat{R}_n$$

where $\hat{\beta}_0$, $\hat{\beta}_1$, and $\hat{\phi}$ are the estimates based on the time series model-fitting approach presented in Section 9.8 and ε_{n+1} is estimated by its expected value 0. The estimate $\hat{R}_n$ of the residual R_n is obtained by noting that

$$R_n = y_n - (\beta_0 + \beta_1 x_n)$$

so that

$$\hat{R}_n = y_n - (\hat{\beta}_0 + \hat{\beta}_1 x_n)$$

The two-step-ahead forecast of y_{n+2} is similarly obtained. The true value of y_{n+2} is

$$\begin{aligned} y_{n+2} &= \beta_0 + \beta_1 x_{n+2} + R_{n+2} \\ &= \beta_0 + \beta_1 x_{n+2} + \phi R_{n+1} + \varepsilon_{n+2} \end{aligned}$$

and the forecast at $t = n + 2$ is

$$\hat{y}_{n+2} = \hat{\beta}_0 + \hat{\beta}_1 x_{n+2} + \hat{\phi}\hat{R}_{n+1}$$

The residual R_{n+1} (and all future residuals) can now be obtained from the recursive relation

$$R_{n+1} = \phi R_n + \varepsilon_{n+1}$$

so that

$$\hat{R}_{n+1} = \hat{\phi}\hat{R}_n$$

Thus, the forecasting of future y values is an iterative process, with each new forecast making use of the previous residual to obtain the estimated residual for the future time period. The general forecasting procedure using time series models with first-order autoregressive residuals is outlined in the box on page 510.

| EXAMPLE 9.3

Suppose we want to forecast the sales of the company for the data analyzed in Section 9.8. Recall that we fit the regression–autoregression pair of models

$$y_t = \beta_0 + \beta_1 t + R_t \qquad R_t = \phi R_{t-1} + \varepsilon_t$$

Using 35 years of sales data, we obtained the estimated models

$$\hat{y}_t = .4058 + 4.2959t + \hat{R}_t \qquad \hat{R}_t = .5896\hat{R}_{t-1}$$

---

*Note that the forecast requires the value of x_{n+1}. When x_t is itself a time series, the future value x_{n+1} will generally be unknown and must also be estimated. Often, $x_t = t$ (as in Example 9.3). In this case, the future time period (e.g., $t = n + 1$) is known and no estimate is required.

Forecasting Using Time Series Models with First-Order Autoregressive Residuals

$$y_t = \beta_0 + \beta_1 x_{1t} + \beta_2 x_{2t} + \cdots + \beta_k x_{kt} + R_t$$
$$R_t = \phi R_{t-1} + \varepsilon_t$$

STEP 1 Use a statistical software package to obtain the estimated model

$$\hat{y}_t = \hat{\beta}_0 + \hat{\beta}_1 x_{1t} + \hat{\beta}_2 x_{2t} + \cdots + \hat{\beta}_k x_{kt} + \hat{R}_t, \quad t = 1, 2, \ldots, n$$
$$\hat{R}_t = \hat{\phi} \hat{R}_{t-1}$$

STEP 2 Compute the estimated residual for the last time period in the data (i.e., $t = n$) as follows:

$$\hat{R}_n = y_n - \hat{y}_n$$
$$= y_n - (\hat{\beta}_0 + \hat{\beta}_1 x_{1n} + \hat{\beta}_2 x_{2n} + \cdots + \hat{\beta}_k x_{kn})$$

STEP 3 To forecast the value y_{n+1}, compute

$$\hat{R}_{n+1} = \hat{\phi} \hat{R}_n \quad \text{(where } \hat{R}_n \text{ is obtained from step 2)}$$
$$\hat{y}_{n+1} = \hat{\beta}_0 + \hat{\beta}_1 x_{1,n+1} + \hat{\beta}_2 x_{2,n+1} + \cdots + \hat{\beta}_k x_{k,n+1} + \hat{R}_{n+1}$$

STEP 4 To forecast the value y_{n+2}, compute

$$\hat{R}_{n+2} = \hat{\phi} \hat{R}_{n+1} \quad \text{(where } \hat{R}_{n+1} \text{ is obtained from step 3)}$$
$$\hat{y}_{n+2} = \hat{\beta}_0 + \hat{\beta}_1 x_{1,n+2} + \hat{\beta}_2 x_{2,n+2} + \cdots + \hat{\beta}_k x_{k,n+2} + \hat{R}_{n+2}$$

Future forecasts are obtained in a similar manner.

Combining these, we have

$$\hat{y}_t = .4058 + 4.2959t + .5896\hat{R}_{t-1}$$

a. Use the fitted model to forecast sales in years $t = 36$, 37, and 38.
b. Calculate approximate 95% prediction intervals for the forecasts.

Solution

a. The forecast for the 36th year requires an estimate of the residual R_{35},

$$\hat{R}_{35} = y_{35} - [\hat{\beta}_0 + \hat{\beta}_1(35)]$$
$$= 150.9 - [.4058 + 4.2959(35)]$$
$$= .1377$$

Then

$$\hat{R}_{36} = \hat{\phi} \hat{R}_{35} = (.5896)(.1377) = .0812$$

and

$$\hat{y}_{36} = \hat{\beta}_0 + \hat{\beta}_1(36) + \hat{R}_{36}$$
$$= .4058 + 4.2959(36) + .0812$$
$$= 155.14$$

To calculate the sales forecast for year 37, we first calculate

$$\hat{R}_{37} = \hat{\phi}\hat{R}_{36}$$
$$= (.5896)(.0812)$$
$$= .0479$$

The sales forecast is then

$$\hat{y}_{37} = \hat{\beta}_0 + \hat{\beta}_1(37) + \hat{R}_{37}$$
$$= .4058 + 4.2959(37) + .0479$$
$$= 159.40$$

Similarly, for $t = 38$,

$$\hat{R}_{38} = \hat{\phi}\hat{R}_{37}$$
$$= (.5896)(.0479)$$
$$= .0282$$

and

$$\hat{y}_{38} = \hat{\beta}_0 + \hat{\beta}_1(38) + \hat{R}_{38}$$
$$= .4058 + 4.2959(38) + .0282$$
$$= 163.68$$

We can proceed in this manner to generate sales forecasts as far into the future as desired. However, the potential for error increases as the distance into the future increases. Forecast errors are traceable to three primary causes:

1. The form of the model may change at some future time. This is an especially difficult source of error to quantify, since we will not usually know when or if the model changes, nor the extent of the change. The possibility of a change in the model structure is the primary reason we have consistently urged you to avoid predictions outside the observed range of the independent variables. However, time series forecasting leaves us little choice— by definition, the forecast will be a prediction at a future time.

2. A second source of forecast error is the uncorrelated residual, ε_t, with variance σ^2. For a first-order autoregressive residual, the forecast variance of the one-step-ahead prediction is σ^2, whereas that for the two-step-ahead prediction is $\sigma^2(1 + \phi^2)$, and, in general, for m steps ahead, the forecast variance* is $\sigma^2(1 + \phi^2 + \phi^4 + \cdots + \phi^{2(m-1)})$. Thus, the forecast variance increases as the distance is increased. These variances allow us to form approximate 95% prediction intervals for the forecasts (see the box on page 512).

3. A third source of variability is that attributable to the error of estimating the model parameters. This is generally of less consequence than the others, and is usually ignored in forming prediction intervals.

---

*See Fuller (1976).

Approximate 95% Forecasting Limits Using Time Series Models with First-Order Autoregressive Residuals

One-Step-Ahead Forecast

$$\hat{y}_{n+1} \pm 2\sqrt{\text{MSE}}$$

Two-Step-Ahead Forecast

$$\hat{y}_{n+2} \pm 2\sqrt{\text{MSE}(1 + \hat{\phi}^2)}$$

Three-Step-Ahead Forecast

$$\hat{y}_{n+3} \pm 2\sqrt{\text{MSE}(1 + \hat{\phi}^2 + \hat{\phi}^4)}$$

$$\vdots$$

m-Step-Ahead Forecast

$$\hat{y}_{n+m} \pm 2\sqrt{\text{MSE}(1 + \hat{\phi}^2 + \hat{\phi}^4 + \cdots + \hat{\phi}^{2(m-1)})}$$

[*Note:*   MSE estimates σ^2, the variance of the uncorrelated residual ε_t.]

---

b. To obtain a prediction interval, we first estimate σ^2 by the MSE, the mean square for error from the time series regression analysis. For the sales data, we form an approximate 95% prediction interval for the sales in year 36:

$$\hat{y}_{36} \pm z_{.025}\sqrt{\text{MSE}}$$
$$155.1 \pm 1.96\sqrt{27.42767}$$
$$155.1 \pm 10.1$$

or (144.8, 165.4). Thus, we forecast that the sales in year 36 will be between $145,000 and $165,000.

The approximate 95% prediction interval for year 37 is

$$\hat{y}_{37} \pm z_{.025}\sqrt{\text{MSE}(1 + \phi^2)}$$
$$159.4 \pm 196\sqrt{27.42767[1 + (.5896)^2]}$$
$$159.4 \pm 11.9$$

or (147.5, 171.3). Note that this interval is wider than that for the one-step-ahead forecast. The intervals will continue to widen as we attempt to forecast farther ahead.

The forecasts and prediction intervals for years 36–40 are shown in Figure 9.16. We again stress that the accuracy of these forecasts and intervals depends on the assumption that the model structure does not change during the forecasting period. If, for example, the company merges with another company during year 37, the structure of the sales model will almost surely change, and therefore, prediction intervals past year 37 are probably useless.

FIGURE 9.16

Forecasts and prediction intervals for
years 36–40: Straight-line model with
autoregressive residual

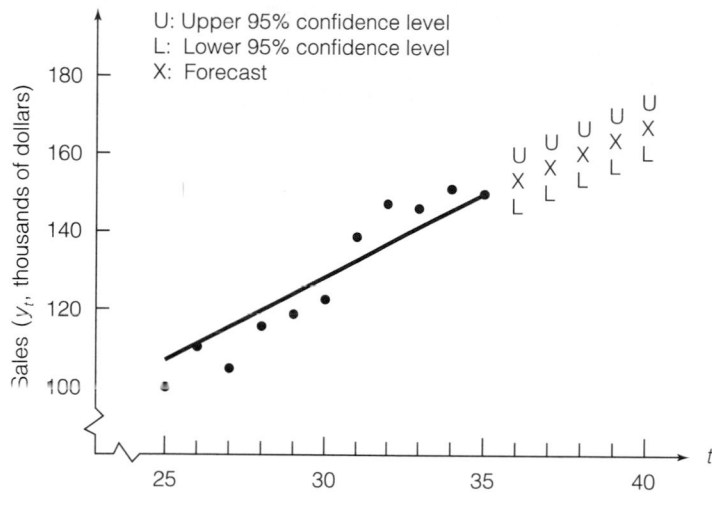

It is important to note that the forecasting procedure makes explicit use of
the residual autocorrelation. The result is a better forecast than would be obtained
using the standard least squares procedure of Chapter 4 (which ignores residual
correlation). Generally, this is reflected by narrower prediction intervals for the
time series forecasts than for the least squares prediction.* The end result, then,
of using a time series model when autocorrelation is present is that you obtain
more reliable estimates of the β coefficients, smaller residual variance, and more
accurate prediction intervals for future values of the time series.

EXERCISES

9.30 The annual time series model $y_t = \beta_0 + \beta_1 t + \phi R_{t-1} + \varepsilon_t$ was fit to data collected for $n = 30$ years
with the following results:

$$\hat{y}_t = 10 + 2.5t + .64\hat{R}_{t-1}$$
$$y_{30} = 82 \qquad \text{MSE} = 4.3$$

a. Calculate forecasts for y_t for $t = 31$, $t = 32$, and $t = 33$.
b. Construct approximate 95% prediction intervals for the forecasts obtained in part **a**.

9.31 The quarterly time series model $y_t = \beta_0 + \beta_1 t + \beta_2 t^2 + \phi R_{t-1} + \varepsilon_t$ was fit to data collected for $n = 48$ quarters, with the following results:

$$\hat{y}_t = 220 + 17t - .3t^2 + .82\hat{R}_{t-1}$$
$$y_{48} = 350 \qquad \text{MSE} = 10.5$$

*When n is large, approximate 95% prediction intervals obtained from the standard least squares
procedure reduce to $\hat{y}_t \pm 2\sqrt{\text{MSE}}$ for *all* future values of the time series. These intervals may actually
be narrower than the more accurate prediction intervals produced from the time series analysis.

a. Calculate forecasts for y_t for $t = 49$, $t = 50$, and $t = 51$.

b. Construct approximate 95% prediction intervals for the forecasts obtained in part **a**.

9.32   Use the fitted times series model of Exercise 9.25 to forecast GNP for the four quarters of 1990 and calculate approximate 95% forecast limits. Do these bounds contain the actual 1990 GNP values shown in the accompanying table?

QUARTER	1990 GNP
1	5,735
2	5,443
3	5,514
4	5,527

9.33   Use the fitted time series model of Exercise 9.26 to forecast annual volume of wheat harvest for 1992. Place approximate 95% confidence bounds on the forecast.

9.34   Use the fitted time series model of Exercise 9.27 to forecast the price of IBM common stock in February 1991. Place approximate 95% confidence bounds on the forecast.

9.35   Refer to Exercise 9.28. The values of MSE for the quarterly time series models of retirement plan terminations are as follows:

Pension plan termination: MSE = .0440

Profit-sharing plan termination: MSE = .0402

a. Forecast the number of pension plan terminations for the fourth quarter of 1982 (i.e., $t = 108$). Assume that $y_{107} = 7.5$. [*Hint:*   Remember that the forecasted number of pension plan terminations is $e^{\hat{y}_{108}}$.]

b. Place approximate 95% confidence bounds on the forecast obtained in part **a**. [*Hint:* First, calculate upper and lower confidence limits for y_{108}, then take antilogarithms.]

c. Repeat parts **a** and **b** for the number of profit-sharing plan terminations in the fourth quarter of 1982. Assume that $y_{107} = 7.6$.

9.36   Use the fitted time series model of Exercise 9.29 to forecast the DJA for the years 1988–1990 (i.e., $t = 21$, 22, and 23). Calculate approximate 95% prediction intervals for the forecasts. Do these intervals contain the actual 1988–1990 DJA values shown in the accompanying table? If not, give a plausible explanation.

YEAR	DJA
1988	2,169
1989	2,753
1990	2,761

9.37   Taxes, a major source of income for the state of Florida, have grown steadily since 1962. The data in the table give the total state tax collections for the years 1962–1985. A tax economist fit the model $y_t = \beta_0 + \beta_1 x + \beta_2 x^2 + \varepsilon$ to the data. Results from the SAS printout gave the least squares model

$$\hat{y} = 1{,}400.904 + 204.626t + 13.9949t^2$$

A plot of the residuals revealed that the usual least squares assumption of independent error terms may have been violated. Since the data were recorded over time (years), it seems reasonable to assume that the tax collections at time $(t + 1)$ are highly correlated with the tax collections at time t, and hence that the random errors in the model follow the same pattern.

YEAR	YEAR − 1970	TOTAL TAX COLLECTIONS (MILLION DOLLARS)	YEAR	YEAR − 1970	TOTAL TAX COLLECTIONS (MILLION DOLLARS)
	t	y		t	y
1962	−8	564	1974	4	2,794
1963	−7	592	1975	5	2,791
1964	−6	709	1976	6	2,936
1965	−5	762	1977	7	3,275
1966	−4	819	1978	8	3,764
1967	−3	877	1979	9	4,291
1968	−2	973	1980	10	4,804
1969	−1	1,269	1981	11	5,314
1970	0	1,421	1982	12	5,556
1971	1	1,587	1983	13	6,225
1972	2	1,996	1984	14	7,329
1973	3	2,488	1985	15	7,883

Source: Statistical Abstracts of the United States 1963–1987, United States Bureau of the Census.

a. Hypothesize a model for the correlated error term, R_t.
b. Rewrite the model fit by the tax economist, using the correlated error term.
c. Fit the time series model in part **b** to the data given in the table using an available statistical software package.
d. Use the fitted time series model from part **c** to forecast total tax collection dollars in 1992. Place an approximate 95% prediction interval around the forecast.

9.10

Seasonal Time Series Models: An Example

We have used a simple regression model to illustrate the methods of model estimation and forecasting when the residuals are autocorrelated. In this section we present a more realistic example that requires a seasonal model for $E(y_t)$, as well as an autoregressive model for the residual.

Critical water shortages have dire consequences for both business and private sectors of communities. Forecasting water usage for months in advance is essential to avoid such shortages. Suppose a community has monthly water usage records over the past 15 years. A plot of the last 6 years of the time series, y_t, is shown in Figure 9.17 on page 516. Note that both an increasing trend and a seasonal pattern appear prominent in the data. The water usage seems to peak during the summer months and decline during the winter months. Thus, we might propose the following model:

$$E(y_t) = \beta_0 + \beta_1 t + \beta_2\left(\cos\frac{2\pi}{12}t\right) + \beta_3\left(\sin\frac{2\pi}{12}t\right)$$

Since the amplitude of the seasonal effect (that is, the magnitude of the peaks and valleys) appears to increase with time, we include in the model an interaction between time and trigonometric components, to obtain

$$E(y_t) = \beta_0 + \beta_1 t + \beta_2\left(\cos\frac{2\pi}{12}t\right) + \beta_3\left(\sin\frac{2\pi}{12}t\right) + \beta_4 t\left(\cos\frac{2\pi}{12}t\right) + \beta_5 t\left(\sin\frac{2\pi}{12}t\right)$$

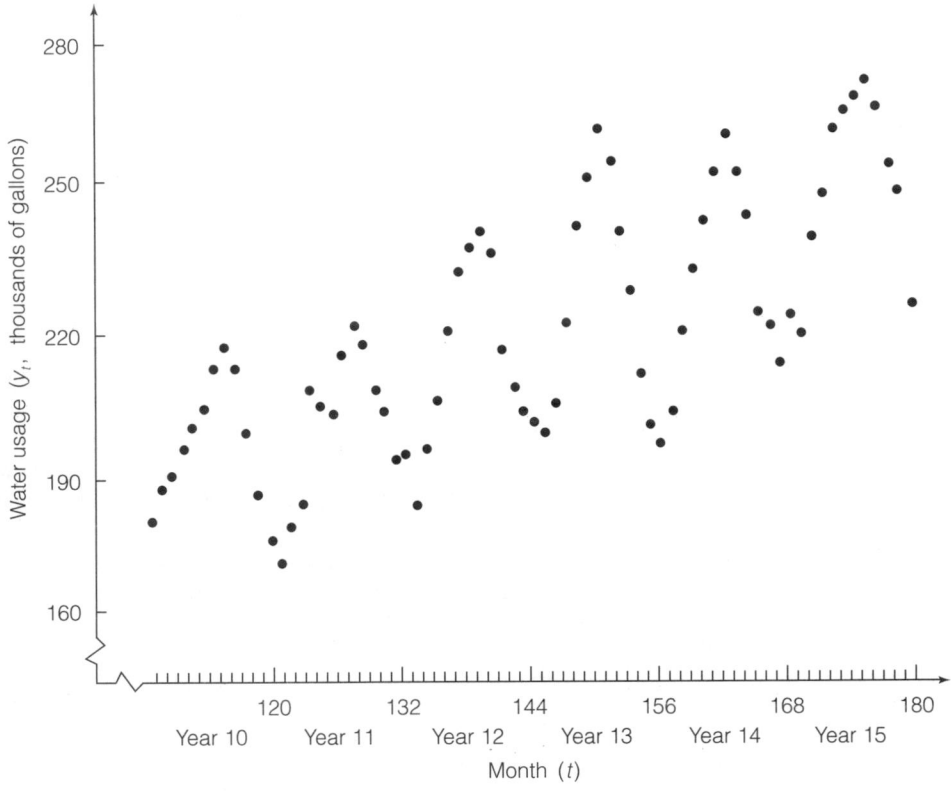

F I G U R E 9.17

Water usage time series

The model for the random component R_t must allow for short-term cyclic effects. For example, in an especially hot summer, if the water usage, y_t, exceeds the expected usage, $E(y_t)$, for July, we would expect the same thing to happen in August. Thus, we propose a first-order autoregressive model for the random component:*

$$R_t = \phi R_{t-1} + \varepsilon_t$$

---

*A more complex time series model may be more appropriate. We use the simple first-order autoregressive model so you can follow the modeling process more easily.

We now fit the models to the time series y_t, where y_t is expressed in thousands of gallons. The SAS printout is shown in Figure 9.18. The estimated models are given by

$$\hat{y}_t = 100.083 + .826t - 10.801\left(\cos\frac{2\pi}{12}t\right) - 7.086\left(\sin\frac{2\pi}{12}t\right) - .0556t\left(\cos\frac{2\pi}{12}t\right) - .0296t\left(\sin\frac{2\pi}{12}t\right) + \hat{R}_t$$

$$\hat{R}_t = .6617\hat{R}_{t-1}$$

FIGURE 9.18

SAS computer printout for water usage model

ESTIMATES OF THE AUTOREGRESSIVE PARAMETERS

LAG	COEFFICIENT	STD ERROR	T RATIO
1	-0.66167894	0.055886	-11.839831

YULE-WALKER ESTIMATES

SSE	4025.513	DFE	174
MSE	23.135	ROOT MSE	4.810
REG RSQ	0.9431	TOTAL RSQ	0.9900

VARIABLE	DF	B VALUE	STD ERROR	T RATIO	APPROX PROB
INTERCEP	1	100.083218977	2.07617706007	48.206	0.0001
T	1	0.826274293	0.01979498750	41.742	0.0001
CS	1	-10.801144	1.85586558083	-5.820	0.0001
SN	1	-7.0857642	1.89574083666	-3.738	0.0003
CST	1	-0.055634923	0.01771077652	-3.141	0.0020
SNT	1	-0.029630055	0.01820045673	-1.628	0.1053

with MSE = 23.135. The R^2 value of .99 (**TOTAL RSQ**) indicates that the model provides a good fit to the data.

We now use the models to forecast water usage for the next 12 months. The forecast for the first month is obtained as follows. The last residual value is $\hat{R}_{180} = -1.3247$, so that

$$\hat{R}_{181} = \hat{\phi}\hat{R}_{180} = (.6617)(-1.3247) = -.8766$$

Then,

$$\hat{y}_{181} = \hat{\beta}_0 + \hat{\beta}_1(181) + \hat{\beta}_2\left(\cos\frac{2\pi}{12}181\right) + \hat{\beta}_3\left(\sin\frac{2\pi}{12}181\right)$$

$$+ \hat{\beta}_4(181)\left(\cos\frac{2\pi}{12}181\right) + \hat{\beta}_5(181)\left(\sin\frac{2\pi}{12}181\right) + \hat{R}_{181} = 238.0$$

Approximate 95% prediction bounds on this forecast are given by $\pm 2\sqrt{\text{MSE}} = \pm 2\sqrt{23.135} = \pm 9.6.$* That is, we expect our forecast for 1 month ahead to be within 9,600 gallons of the actual water usage. This forecasting process is then repeated for the next 11 months. The forecasts and their bounds are shown in Figure 9.19 on page 518. Also shown are the actual values of water usage during year 16. Note that the forecast prediction intervals widen as we attempt to forecast farther into the future. This property of the prediction intervals makes long-term forecasts very unreliable.

---

*We are ignoring the errors in the parameter estimates in calculating the forecast reliability. These errors should be small for a series of this length.

FIGURE 9.19
Forecasts of water usage

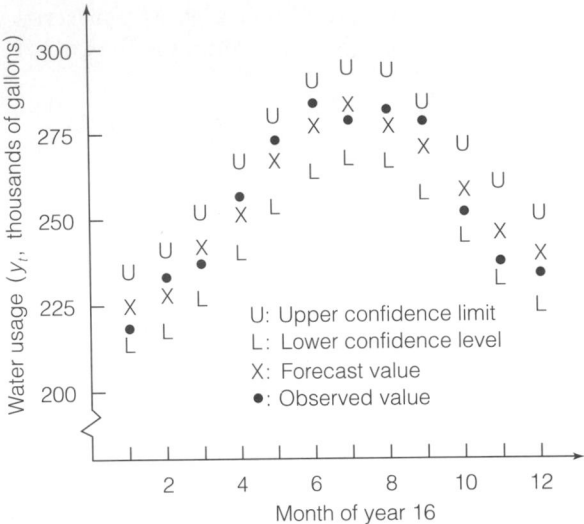

We have barely scratched the surface of time series modeling. The variety and complexity of available techniques are overwhelming. However, if we have convinced you that time series modeling is a useful and powerful tool for business forecasting, we have accomplished our purpose. The successful construction of time series models requires much experience, and, like regression modeling, entire texts are devoted to the subject (see the references at the end of the chapter).

We conclude with a warning: Many oversimplified forecasting methods have been proposed. They usually consist of graphical extensions of a trend or seasonal pattern to future time periods. Although such pictorial techniques are easy to understand and therefore are intuitively appealing, they should be avoided. There is no measure of reliability for these forecasts, and thus the risk associated with making decisions based on them is very high.

9.11

Forecasting Using Lagged Values of the Dependent Variable (Optional)

In Section 9.7, we discussed a variety of choices for the deterministic component, $E(y_t)$, of the time series models. All these models were functions of independent variables, such as t, x_t, x_{t-1}, and seasonal dummy variables. Often, the forecast of y_t can be improved by adding *lagged values of the dependent variable* to the model. For example, since the price y_t of a stock on day t is highly correlated with the price on the previous day, (i.e., on day $t - 1$), a useful model for $E(y_t)$ is

$$E(y_t) = \beta_0 + \beta_1 y_{t-1}$$

Models with lagged values of y_t tend to violate the standard regression assumptions outlined in Section 4.2; thus, they must be fit using specialized methods.

Box and Jenkins (1977) developed a method of analyzing time series models based on past values of y_t and past values of the random error ε_t. The general model, denoted **ARMA(p, q)**, takes the form:

$$y_t + \phi_1 y_{t-1} + \phi_2 y_{t-2} + \cdots + \phi_p y_{t-p} = \varepsilon_t + \theta_1 \varepsilon_{t-1} + \theta_2 \varepsilon_{t-2} + \cdots + \theta_q \varepsilon_{t-q}$$

Note that the left side of the equation is a **pth-order autoregressive model** for y_t (see Section 9.5), while the right side of the equation is a **qth-order moving average model** for the random error ε_t (see Section 9.6).

The analysis of an ARMA(p, q) model is divided into three stages: (1) identification, (2) estimation, and (3) forecasting. In the identification stage, the values of p and q are determined from the sample data. That is, the order of both the autoregressive portion and the moving average portion of the model are identified.* For example, the analyst may find the best fit to be an ARMA model with $p = 2$ and $q = 0$. Substituting $p = 2$ and $q = 0$ into the previous equation, we obtain the ARMA(2, 0) model

$$y_t + \phi_1 y_{t-1} + \phi_2 y_{t-2} = \varepsilon_t$$

Note that since $q = 0$, there is no moving average component to the model.

Once the model is identified, the second stage involves obtaining estimates of the model's parameters. In the case of the ARMA(2, 0) model, we require estimates of the autoregressive parameters ϕ_1 and ϕ_2. Tests for model adequacy are conducted, and, if the model is deemed adequate, the estimated model is used to forecast future values of y_t in the third stage.

Analysis of ARMA(p, q) models for y_t requires a level of expertise that is beyond the scope of this text. Even with this level of expertise, the analyst cannot hope to proceed without the aid of a sophisticated computer program. Procedures for identifying, estimating, and forecasting with ARMA(p, q) models are available in SAS, SPSS, and Minitab. Before attempting to run these procedures, however, you should consult the references provided at the end of this chapter.

Summary

Time series are often modeled as a combination of four components: **secular**, **seasonal**, **cyclical**, and **residual**. Both descriptive and inferential techniques are available for **estimating** the time series components and **forecasting** future values of the time series. The **moving average method** is a smoothing technique that uses estimates of the secular and seasonal components to forecast future values of a time series. However, the method requires you to extrapolate the moving average into the future to obtain the forecasts. Two alternative smoothing techniques that lead to explicit forecasts are **exponential smoothing** and the **Holt–Winters model**. Exponential smoothing is an adaptive forecasting method for time series with little or no secular or seasonal trends. The Holt–Winters model is an extension of the exponential smoothing technique that allows for trend and seasonal components.

One type of **inferential time series model** employs a combination of the **deterministic component** $E(y_t)$ of the typical multiple regression model with an autoregressive model for the **autocorrelated residual**. The deterministic portion

---

*This step involves a careful examination of a plot of the sample autocorrelations. Certain patterns in the plot (such as those shown in Figures 9.8–9.10) allow the analyst to identify p and q.

of the model accounts for the trend and seasonal components, and the autocorrelated residual deals with the problem of correlated errors.

The forecaster should be very careful to **distinguish between descriptive and inferential time series models**. If descriptive models (e.g., smoothing techniques) are used to predict future values of the series, no assessment of forecast reliability is possible. Only when a probabilistic model (e.g., a time series autoregressive model) is constructed can a prediction interval be used to evaluate the reliability of the forecast. Even then, if the structure of the model changes at some future time, forecasts beyond that point are probably useless. Careful application of time series modeling and forecasting will usually be rewarded with a better understanding of the phenomenon and with useful forecasts that assist in planning future strategy.

SUPPLEMENTARY EXERCISES

9.38 The level at which lending institutions set mortgage interest rates has a significant effect on the volume of buying, selling, and construction of residential and commercial real estate. The data in the table are the annual average mortgage interest rates on 48-month, fixed-rate loans on new automobiles for the period 1978–1988. Forecast the 1990 average new car loan interest rate using each of the methods listed below.

YEAR	INTEREST RATE (%)	YEAR	INTEREST RATE (%)
1978	11.02	1984	13.71
1979	12.02	1985	12.91
1980	14.30	1986	11.33
1981	16.54	1987	10.46
1982	16.83	1988	10.86
1983	13.92		

Source: Statistical Abstract of the United States, U.S. Bureau of the Census, 1990.

a. A 3-point moving average
b. The exponential smoothing technique ($w = .2$)
c. The Holt–Winters model with trend ($w = .2$ and $v = .5$)
d. Simple linear regression   (Obtain a 95% prediction interval.)
e. A straight-line, first-order autoregressive model   (Obtain an approximate 95% prediction interval.)

9.39 The table lists the total monthly retail sales (in $ billions) in the United States for the period January 1988 through December 1990.
a. Calculate and plot a 12-point moving average for the total retail sales time series. Can you detect the secular trend? Does there appear to be a seasonal pattern?
b. Use the moving average from part a to forecast retail sales in January 1991.
c. Calculate and plot the exponentially smoothed series using $w = .6$.
d. Obtain the forecast for January 1991 using the exponential smoothing technique.

	RETAIL SALES ($ BILLIONS)		
	1988	1989	1990
January	116.1	124.2	133.3
February	117.8	120.5	128.0
March	134.7	141.9	149.2
April	134.1	140.4	145.8
May	139.4	151.0	155.0
June	140.8	149.8	154.4
July	137.0	145.3	149.7
August	142.2	153.8	158.2
September	135.0	144.8	146.3
October	138.1	143.1	151.5
November	142.7	149.6	156.1
December	173.5	177.4	179.7

Source: *Standard & Poor's Trade and Securities Statistics* (Annual), New York, Standard & Poor's Corporation.

e. Obtain the forecast for January 1991 using the Holt–Winters model with trend and seasonal components and smoothing constants $w = .6$, $v = .7$, and $u = .5$.

f. Propose a time series model for total retail sales that accounts for secular trend, seasonal variation, and residual autocorrelation.

g. Fit the time series model specified in part f, using an available software package.

h. Use the time series model to forecast retail sales in January 1991. Obtain an approximate 95% prediction interval for the forecast.

9.40 A traditional pulse rate of the economic health of the accommodations (hotel–motel) industry is the trend in room occupancy. Average monthly occupancies for two recent years are given in the table for hotels and motels in the cities of Atlanta, Georgia, and Phoenix, Arizona.

MONTH	PERCENTAGE OF ROOMS OCCUPIED			
	YEAR 1		YEAR 2	
	Atlanta	Phoenix	Atlanta	Phoenix
January	59	67	64	72
February	63	85	69	91
March	68	83	73	87
April	70	69	67	75
May	63	63	68	70
June	59	52	71	61
July	68	49	67	46
August	64	49	71	44
September	62	56	65	63
October	73	69	72	73
November	62	63	63	71
December	47	48	47	51

Source: *Trends in the Hotel Industry.*

a. Graph the time series values for both cities on the same set of axes. Use two different colors so that you will be able to distinguish between the two time series.

b. Do you detect any differences in the seasonal and long-term trends of the two time series?

c. Postulate a time series model that will be useful in forecasting occupancy rates for Atlanta and Phoenix.

d. Fit the model in part c to each set of data using an available statistical computer program package.

e. Interpret the differences in the modified least squares parameter estimates for the two models. Does this support your answer to part b?

f. Would you recommend using the model to forecast monthly occupancy rates in year 3? Explain.

9.41 In May 1978, the first casino (Resorts International Hotel and Casino) opened in Atlantic City, New Jersey. In the first few years following the casino opening, employment in hotels and other lodging places accelerated along Atlantic City's boardwalk, as shown in the table.

YEAR	QUARTER	EMPLOYMENT IN ATLANTIC CITY HOTELS
1978	I	1,711
	II	4,065
	III	5,787
	IV	5,019
1979	I	5,459
	II	9,184
	III	12,168
	IV	11,842
1980	I	13,730
	II	14,964
	III	18,058
	IV	21,393

Source: Business Review, Jan.–Feb. 1982.

a. Use a smoothing technique to forecast employment in Atlantic City hotels in quarter I, 1981. Comment on the reliability of this forecast.

b. Propose a time series model for the quarterly series that will account for secular trend, seasonal variation, and residual autocorrelation. If you have access to a software package with a modified least squares routine, fit the model.

c. Use the fitted time series model of part b to forecast employment in quarter I, 1981. Place an approximate 95% prediction interval about the forecast.

d. Actual employment in quarter I of 1981 was 22,772. Check to see whether the forecasting technique of part c has captured this value.

e. Would you recommend using the fitted model in part b for forecasting quarterly employment in 1988? Explain.

9.42 The table lists NASA's yearly space shuttle system expenditures (in millions of dollars) for 1973 through 1988.

a. Plot the time series.

b. Calculate 3-point, 5-point, and 7-point moving averages. Plot each moving average series on the same graph. Which moving average best characterizes the long-term trend of space shuttle expenditures?

c. Graphically extend the moving average you chose in part b to forecast the outlay for the space shuttle in 1991. Comment on the reliability of this forecast. Is there a better forecasting method available? Explain.

YEAR	OUTLAY	YEAR	OUTLAY
1973	58	1981	2,724
1974	325	1982	2,932
1975	1,543	1983	3,014
1976	2,619	1984	3,127
1977	2,258	1985	2,636
1978	2,062	1986	2,606
1979	2,251	1987	4,165
1980	2,751	1988	2,378

Source. *Statistical Abstract of the United States*, U.S. Bureau of the Census, 1990.

9.43 The accompanying table shows U.S. beer production for the years 1973–1989. Suppose you are interested in forecasting U.S. beer production in 1990. Since a plot of the time series y_t reveals a linearly increasing trend, you hypothesize the model

$$E(y_t) = \beta_0 + \beta_1 t$$

for the secular trend.

YEAR	t	U.S. BEER PRODUCTION, y_t Millions of Barrels	YEAR	t	U.S. BEER PRODUCTION, y_t Millions of Barrels
1973	1	149	1982	10	196
1974	2	156	1983	11	196
1975	3	161	1984	12	193
1976	4	164	1985	13	194
1977	5	171	1986	14	197
1978	6	179	1987	15	195
1979	7	184	1988	16	197
1980	8	194	1989	17	199
1981	9	194			

Source: *Standard & Poor's Trade and Securities Statistics* (Annual), New York, Standard & Poor's Corporation.

a. Fit the model to the data using the method of least squares.
b. Plot the least squares model from part a and extend the line to forecast y_{18}, the U.S. beer production (in millions of barrels) in 1990. How reliable do you think this forecast is?
c. Calculate and plot the residuals for the model from part a. Is there visual evidence of residual autocorrelation?
d. How could you test to determine whether residual autocorrelation exists? If you have access to a software package, carry out the test. Use $\alpha = .05$.
e. Hypothesize a time series model that accounts for the residual autocorrelation. If you have access to a software package with a modified least squares routine, fit the model.
f. Compute a 95% prediction interval for y_{18}, the U.S. beer production in 1990. Why is this forecast preferred to that of part b?

9.44 Suppose a CPA firm wants to model its monthly income, y_t. The firm is growing at an increasing rate, so that the mean income will be modeled as a second-order function of t. In addition, the mean monthly income increases significantly each year from January through April because of processing tax returns.

a. Write a model for $E(y_t)$ to reflect both the second-order function of time, t, and the January–April jump in mean income.

b. Suppose the size of the January–April jump grows each year. How could this information be included in the model? Assume that 5 years of monthly data are available.

9.45 In 1974, Congress adopted the Federal-Aid Highway Amendments, which reduced the highway speed limit to 55 miles per hour (mph). Since that time, controversy over the social efficiency of the decision has grown. An analysis was conducted to estimate the effect of the 55-mph speed limit on traffic fatalities (*Southern Economic Journal*, Jan. 1984). Time series data for the United States from 1952 to 1979 ($n = 28$ years) were used to fit a regression model relating traffic fatalities y_t to $k = 7$ independent variables:

x_{1t} = Real earned income

x_{2t} = Vehicle miles

x_{3t} = Ratio of number of youths to number of adults

x_{4t} = Percentage of all car purchases that are imported cars

x_{5t} = Average highway speed

x_{6t} = Percentage of cars traveling between 45 and 60 mph

$x_{7t} = \begin{cases} 0 & \text{if 55-mph speed limit imposed} \\ 1 & \text{otherwise} \end{cases}$

The results of the multiple regression are summarized as follows:

$$\hat{y}_t = -20{,}016.4 + 7{,}544.85x_{1t} - .01046x_{2t} - 36{,}758.0x_{3t} - 117.609x_{4t}$$
$$+ 1{,}325.22x_{5t} - 415.742x_{6t} + 9{,}678.08x_{7t}$$

$$R^2 = .987 \qquad F = 217.23 \qquad d = 1.97$$

a. Is there evidence that the model is useful for predicting annual traffic fatalities? Test using $\alpha = .05$.

b. Is there evidence that the regression residuals are autocorrelated? Test using $\alpha = .05$. (Since Table 9 of Appendix A does not show critical values for $k = 7$ independent variables, use those based on $k = 5$.)

c. Based on your answer to part **b**, do you need to propose a time series model that accounts for autocorrelated residuals?

9.46 The data on annual OPEC oil imports, Exercise 9.3, are reproduced below.

YEAR	t	IMPORTS, y_t	YEAR	t	IMPORTS, y_t
1973	1	767	1981	9	1,067
1974	2	926	1982	10	633
1975	3	1,171	1983	11	540
1976	4	1,663	1984	12	553
1977	5	2,058	1985	13	479
1978	6	1,892	1986	14	771
1979	7	1,866	1987	15	876
1980	8	1,414	1988	16	987

Source: *Statistical Abstract of the United States*, U.S. Bureau of the Census, 1990.

a. Plot the time series.

b. Hypothesize a straight-line autoregressive time series model for annual amount of imported crude oil, y_t.

c. If you have access to a computer package, fit the proposed model to the data.

d. From the output, write the modified least squares prediction equation for y_t.

e. Forecast the amount of foreign crude oil imported into the United States from OPEC in 1990. Place approximate 95% prediction bounds on the forecast value.

References

Anderson, T. W. *The Statistical Analysis of Time Series*. New York: Wiley, 1971.

Box, G. E. P. and Jenkins, G. M. *Time Series Analysis: Forecasting and Control*, 2nd ed. San Francisco: Holden-Day, 1977.

Fuller, W. A. *Introduction to Statistical Time Series*. New York: Wiley, 1976.

Granger, C. W. J. *Spectral Analysis of Economic Time Series*. Princeton: Princeton University Press, 1964.

Granger, C. W. J. and Newbold, P. *Forecasting Economic Time Series*. New York: Academic Press, 1977

Makridakis, S. et al. *The Forecasting Accuracy of Major Time Series Methods*. New York: Wiley, 1984.

Nelson, C. R. *Applied Time Series Analysis for Managerial Forecasting*. San Francisco: Holden-Day, 1973.

Seitz, N. *Business Forecasting: Concepts and Microcomputer Applications*. Reston, Va.: Reston Publishing Company, 1984.

Principles of Experimental Design

CONTENTS

10.1 Introduction

10.2 Experimental Design Terminology

10.3 Controlling the Information in an Experiment

10.4 Noise-Reducing Designs

10.5 Volume-Increasing Designs

10.6 Selecting the Sample Size

10.7 The Importance of Randomization

OBJECTIVE

To present an overview of experiments designed to compare two or more population means; to explain the statistical principles of experimental design

10.1
Introduction

In Chapter 6 we learned that a regression analysis of observational data has some limitations. In particular, establishing a cause-and-effect relationship between an independent variable x and the response y is difficult since the values of other relevant independent variables—both those in the model and those omitted from the model—are not controlled. Recall that experimental data are data collected with the values of the x's set in advance of observing y (i.e., the values of the x's are controlled). With experimental data, we usually select the x's so that we can compare the mean responses, $E(y)$, for several different combinations of the x values.

The procedure for selecting sample data with the x's set in advance is called the **design of the experiment**. The statistical procedure for comparing the population means is called an **analysis of variance**. The objective of this chapter is to introduce some key aspects of experimental design. The analysis of the data from such experiments using an analysis of variance is the topic of Chapter 11.

10.2
Experimental Design Terminology

The study of experimental design originated in England and, in its early years, was associated solely with agricultural experimentation. The need for experimental design in agriculture was very clear: It takes a full year to obtain a single observation on the yield of a new variety of wheat. Consequently, the need to save time and money led to a study of ways to obtain more information using smaller samples. Similar motivations led to its subsequent acceptance and wide use in all fields of scientific experimentation. Despite this fact, the terminology associated with experimental design clearly indicates its early association with the biological sciences.

We will call the process of collecting sample data an **experiment** and the (dependent) variable to be measured, the **response** y. The planning of the sampling procedure is called the **design** of the experiment. The object upon which the response measurement y is taken is called an **experimental unit**.

Definition 10.1

The process of collecting sample data is called an **experiment**.

Definition 10.2

The plan for collecting the sample is called the **design** of the experiment.

Definition 10.3

The variable measured in the experiment is called the **response variable**.

> **Definition 10.4**
>
> The object upon which the response y is measured is called an **experimental unit**.

Independent variables that may be related to a response variable y are called **factors**. The value—that is, the intensity setting—assumed by a factor in an experiment is called a **level**. The combinations of levels of the factors for which the response will be observed are called **treatments**.

> **Definition 10.5**
>
> The independent variables, quantitative or qualitative, that are related to a response variable y are called **factors**.

> **Definition 10.6**
>
> The intensity setting of a factor (i.e., the value assumed by a factor in an experiment) is called a **level**.

> **Definition 10.7**
>
> A **treatment** is a particular combination of levels of the factors involved in an experiment.

EXAMPLE 10.1

A marketing study is conducted to investigate the effects of brand and shelf location on weekly coffee sales. Coffee sales are recorded for each of two brands (brand A and brand B) and three shelf locations (bottom, middle, and top) each week for a period of 20 weeks. For this experiment, identify

a. the experimental unit   b. the response
c. the factors   d. the factor levels
e. the treatments

Solution

a. Since the data will be collected each week for a period of 20 weeks, the experimental unit is 1 week.
b. The variable of interest, i.e., the response, is weekly coffee sales. Note that weekly coffee sales is a quantitative variable.
c. Since we are interested in investigating the effect of brand and shelf location on sales, brand and shelf location are the factors. Note that both factors are

qualitative variables, although, in general, they may be quantitative or qualitative.

d. For this experiment, brand is measured at two levels (A and B) and shelf location at three levels (bottom, middle, and top).

e. Since coffee sales are recorded for each of the six brand–shelf location combinations (brand A, bottom), (brand A, middle), (brand A, top), (brand B, bottom), (brand B, middle), and (brand B, top), the experiment involves six treatments. The term *treatments* is used to describe the factor level combinations to be included in an experiment because many experiments involve "treating" or doing something to alter the nature of the experimental unit. Thus, we might view the six brand–shelf location combinations as treatments on the experimental units in the marketing study involving coffee sales.

Now that you understand some of the terminology, it is helpful to think of the design of an experiment in four steps.

STEP 1 Select the factors to be included in the experiment, and identify the parameters that are the object of the study. Usually, the target parameters are the population means associated with the factor level combinations (i.e., treatments).

STEP 2 Choose the treatments (the factor level combinations to be included in the experiment).

STEP 3 Determine the number of observations (sample size) to be made for each treatment. [This will usually depend on the standard error(s) that you desire.]

STEP 4 Plan how the treatments will be assigned to the experimental units. That is, decide on which design to use.

By following these steps, you can control the quantity of information in an experiment. We shall explain how this is done in Section 10.3.

10.3
Controlling the Information in an Experiment

The problem of acquiring good experimental data is analogous to the problem faced by a communications engineer. The receipt of any signal, verbal or otherwise, depends on the volume of the signal and the amount of the background noise. The greater the volume of the signal, the greater will be the amount of information transmitted to the receiver. Conversely, the amount of information transmitted is reduced when the background noise is great. These intuitive thoughts about the factors that affect the information in an experiment are supported by the following fact: The standard errors of most estimators of the target parameters are proportional to σ (a measure of data variation or noise) and inversely proportional to the sample size (a measure of the volume of the signal). To illustrate, take the simple case where we wish to estimate a population mean μ by the sample mean $\bar{y}$. The standard error of the sampling distribution of $\bar{y}$ is

$$\sigma_{\bar{y}} = \frac{\sigma}{\sqrt{n}} \quad \text{(see Section 1.6)}$$

Note that, for a fixed sample size n, the smaller the value of σ, which measures the **variability (noise)** in the population of measurements, the smaller will be the standard error $\sigma_{\bar{y}}$. Similarly, by increasing the sample size n (**volume of the signal**) in a given experiment, you decrease $\sigma_{\bar{y}}$.

The first three steps in the design of an experiment (see Section 10.2), selecting the factors and treatments to be included in an experiment and specifying the sample sizes, determine the volume of the signal. You must select the treatments so that the observed values of y provide information on the parameters of interest. Then the larger the treatment sample sizes, the greater will be the quantity of information in the experiment. We present an example of a volume-increasing experiment in Section 10.5.

Is it possible to observe y and obtain no information on a parameter of interest? The answer is yes. To illustrate, suppose that you attempt to fit a first-order model

$$E(y) = \beta_0 + \beta_1 x$$

to a set of $n = 10$ data points, all of which were observed for a single value of x, say, $x = 5$. The data points might appear as shown in Figure 10.1. Clearly, there is no possibility of fitting a line to these data points. The only way to obtain information on β_0 and β_1 is to observe y for *different* values of x. Consequently, the $n = 10$ data points in this example contain absolutely no information on the parameters β_0 and β_1.

Step 4 in the design of an experiment provides an opportunity to reduce the noise (or experimental error) in an experiment. As we illustrate in Section 10.4, known sources of data variation can be reduced or eliminated by **blocking**—that is, observing all treatments within relatively homogeneous **blocks** of experimental material. When the treatments are compared within each block, any background noise produced by the block is canceled, or eliminated, allowing us to obtain better estimates of treatment differences.

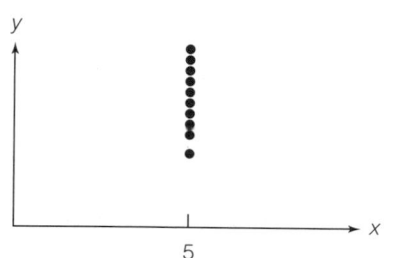

FIGURE 10.1

Data set with $n = 10$ responses, all at $x = 5$

Summary of Steps in Experimental Design	
Volume-increasing:	1. Select the factors.
	2. Choose the treatments (factor level combinations).
	3. Determine the sample size for each treatment.
Noise-reducing:	4. Assign the treatments to the experimental units.

In summary, it is useful to think of experimental designs as being either "noise reducers" or "volume increasers." We will learn, however, that most designs are multifunctional. That is, they tend to both reduce the noise and increase the volume of the signal at the same time. Nevertheless, we will find that specific designs lean heavily toward one or the other objective.

10.4
Noise-Reducing Designs

Noise reduction in an experimental design, i.e., the removal of extraneous experimental variation, can be accomplished by an appropriate assignment of treatments to the experimental units. The idea is to compare treatments within blocks of relatively homogeneous experimental units. The most common design of this type is called a **randomized block design**.

To illustrate, suppose we want to compare the mean length of time required to process a bank's daily receipts using three different computer programs, A, B, and C. Thus, we want to compare the three means μ_A, μ_B, and μ_C, where μ_i is the mean processing time for program i. One way to design the experiment is to select 15 days (where the days are the experimental units) and randomly assign one of the three programs (treatments) to process each day's receipts. A diagram of this design, called a **completely randomized design** (since the treatments are randomly assigned to the experimental units) is shown in Table 10.1.

TABLE 10.1 **Completely Randomized Design with $p = 3$ Treatments**

DAY	TREATMENT (PROGRAM) ASSIGNED
1	B
2	A
3	B
4	C
5	C
6	A
7	B
8	C
9	A
10	A
11	C
12	A
13	B
14	C
15	B

Definition 10.8

A **completely randomized design** to compare p treatments is one in which the treatments are randomly assigned to the experimental units.

This design has the obvious disadvantage that the processing times would vary greatly from day to day depending on the level of the day's business, the complexity of the transactions, etc. A better design—one that contains more information on the mean processing times—would be to utilize the receipts for only 5 days and process the data for each day using each of the three programs. This *randomized block* procedure acknowledges the fact that the length of time required to process a day's receipts varies substantially from day to day. By comparing the three processing times for each day, we eliminate day-to-day variation from the comparison.

The randomized block design that we have just described is shown diagrammatically in Figure 10.2. The figure shows that there are five days. Each day can be viewed as a **block** of three experimental units—runs on the computer—one corresponding to the use of each of the programs, A, B, and C. The blocks are said to be **randomized** because the treatments (computer programs) are randomly assigned to the experimental units within a block. For our example, the programs employed to process a day's receipts would be run in a random order to avoid bias introduced by other unknown and unmeasured variables that may affect the processing time.

In general, a randomized block design to compare p treatments will contain b relatively homogeneous blocks, with each block containing p experimental units. Each treatment appears once in every block with the p treatments randomly assigned to the experimental units within each block.

FIGURE 10.2

Diagram for a randomized block design containing $b = 5$ blocks and $p = 3$ treatments

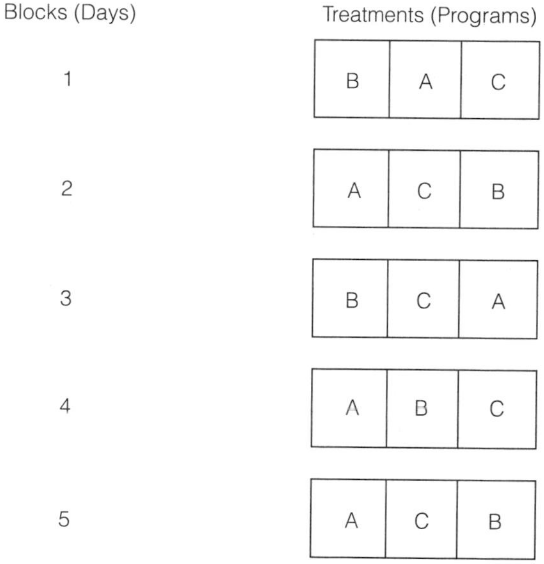

Blocks (Days)	Treatments (Programs)		
1	B	A	C
2	A	C	B
3	B	C	A
4	A	B	C
5	A	C	B

Definition 10.9

A **randomized block design** to compare p treatments involves b blocks, each containing p relatively homogeneous experimental units. The p treatments are randomly assigned to the experimental units within each block, with one experimental unit assigned per treatment.

EXAMPLE 10.2

Suppose you want to compare the abilities of four real estate appraisers, A, B, C, and D. One way to make the comparison would be to randomly allocate a number of pieces of real estate—say, 40—10 to each of the four appraisers. Each appraiser would appraise the property and you would record y, the difference between the appraised and selling prices expressed as a percentage of the selling price. Thus, y measures the appraisal error expressed as a percentage of selling price, and the treatment allocation to experimental units that we have described is a completely randomized design.

a. Discuss the problems with using a completely randomized design for this experiment.
b. Explain how you could employ a randomized block design.

Solution

a. The problem with using a completely randomized design for the appraisal experiment is that comparison of mean percentage errors will be influenced by the nature of the properties. Some properties will be easier to appraise than others, and the variation in percentage errors that can be attributed to this fact will make it more difficult to compare the treatment means.
b. To eliminate the effect of property-to-property variability in comparing appraiser means, you could select only 10 properties and require each appraiser to appraise the value of each of the 10 properties. Although in this case there is probably no need for randomization, it might be desirable to randomly assign the order (in time) of the appraisals. This randomized block design, consisting of $p = 4$ treatments and $b = 10$ blocks would appear as shown in Figure 10.3.

FIGURE 10.3

Diagram for a randomized block design: Example 10.2

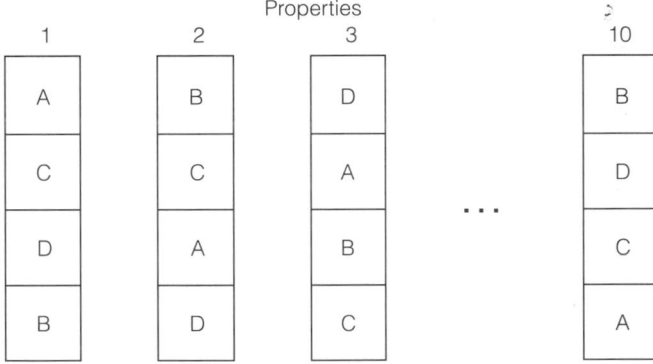

Each experimental design can be represented by a general linear model relating the response y to the factors (treatments, blocks, etc.) in the experiment. When the factors are qualitative in nature (as is often the case), the model includes dummy variables. For example, consider the completely randomized design portrayed in Table 10.1. Since the experiment involves three treatments (programs), we require two dummy variables. The model for this completely randomized design would appear as follows:

$$y = \beta_0 + \beta_1 x_1 + \beta_2 x_2 + \varepsilon$$

where

$$x_1 = \begin{cases} 1 & \text{if program A} \\ 0 & \text{if not} \end{cases}$$

$$x_2 = \begin{cases} 1 & \text{if program B} \\ 0 & \text{if not} \end{cases}$$

Note that we have arbitrarily selected program C as the base level. From our discussion of dummy-variable models in Chapter 5, we know that the mean responses for the three programs are

$$\mu_A = \beta_0 + \beta_1$$
$$\mu_B = \beta_0 + \beta_2$$
$$\mu_C = \beta_0$$

Recall that $\beta_1 = \mu_A - \mu_C$ and $\beta_2 = \mu_B - \mu_C$. Thus, to estimate the differences between the treatment means, we require estimates of β_1 and β_2.

Similarly, we can write the model for the randomized block design in Figure 10.2 as follows:

$$y = \beta_0 + \underbrace{\beta_1 x_1 + \beta_2 x_2}_{\text{Treatment effects}} + \underbrace{\beta_3 x_3 + \beta_4 x_4 + \beta_5 x_5 + \beta_6 x_6}_{\text{Block effects}} + \varepsilon$$

where

$$x_1 = \begin{cases} 1 & \text{if program A} \\ 0 & \text{if not} \end{cases} \quad x_2 = \begin{cases} 1 & \text{if program B} \\ 0 & \text{if not} \end{cases} \quad x_3 = \begin{cases} 1 & \text{if day 1} \\ 0 & \text{if not} \end{cases}$$

$$x_4 = \begin{cases} 1 & \text{if day 2} \\ 0 & \text{if not} \end{cases} \quad x_5 = \begin{cases} 1 & \text{if day 3} \\ 0 & \text{if not} \end{cases} \quad x_6 = \begin{cases} 1 & \text{if day 4} \\ 0 & \text{if not} \end{cases}$$

In addition to the treatment terms, the model includes four dummy variables representing the five blocks (days). Note that we have arbitrarily selected day 5 as the base level. Using this model, we can write each response y in the experiment of Figure 10.2 as a function of β's, as shown in Table 10.2.

TABLE 10.2 **The Response for the Randomized Block Design Shown in Figure 10.2**

BLOCKS (DAYS)	TREATMENTS (PROGRAMS)		
	A $(x_1 = 1, x_2 = 0)$	B $(x_1 = 0, x_2 = 1)$	C $(x_1 = 0, x_2 = 0)$
1 $(x_3 = 1, x_4 = x_5 = x_6 = 0)$	$y_{A1} = \beta_0 + \beta_1 + \beta_3 + \varepsilon_{A1}$	$y_{B1} = \beta_0 + \beta_2 + \beta_3 + \varepsilon_{B1}$	$y_{C1} = \beta_0 + \beta_3 + \varepsilon_{C1}$
2 $(x_4 = 1, x_3 = x_5 = x_6 = 0)$	$y_{A2} = \beta_0 + \beta_1 + \beta_4 + \varepsilon_{A2}$	$y_{B2} = \beta_0 + \beta_2 + \beta_4 + \varepsilon_{B2}$	$y_{C2} = \beta_0 + \beta_4 + \varepsilon_{C2}$
3 $(x_5 = 1, x_3 = x_4 = x_6 = 0)$	$y_{A3} = \beta_0 + \beta_1 + \beta_5 + \varepsilon_{A3}$	$y_{B3} = \beta_0 + \beta_2 + \beta_5 + \varepsilon_{B3}$	$y_{C3} = \beta_0 + \beta_5 + \varepsilon_{C3}$
4 $(x_6 = 1, x_3 = x_4 = x_5 = 0)$	$y_{A4} = \beta_0 + \beta_1 + \beta_6 + \varepsilon_{A4}$	$y_{B4} = \beta_0 + \beta_2 + \beta_6 + \varepsilon_{B4}$	$y_{C4} = \beta_0 + \beta_6 + \varepsilon_{C4}$
5 $(x_3 = x_4 = x_5 = x_6 = 0)$	$y_{A5} = \beta_0 + \beta_1 + \varepsilon_{A5}$	$y_{B5} = \beta_0 + \beta_2 + \varepsilon_{B5}$	$y_{C5} = \beta_0 + \varepsilon_{C5}$

For example, to obtain the model for the response y for treatment A in block 1 (denoted y_{A1}), we substitute $x_1 = 1$, $x_2 = 0$, $x_3 = 1$, $x_4 = 0$, $x_5 = 0$, and $x_6 = 0$ into the equation. The resulting model is

$$y_{A1} = \beta_0 + \beta_1 + \beta_3 + \varepsilon_{A1}$$

Now we will use Table 10.2 to illustrate how a randomized block design reduces experimental noise.

Since each treatment appears in each of the five blocks, there are five measured responses per treatment. Averaging the five responses for treatment A shown in Table 10.2, we obtain

$$\bar{y}_A = \frac{y_{A1} + y_{A2} + y_{A3} + y_{A4} + y_{A5}}{5}$$

$$= [(\beta_0 + \beta_1 + \beta_3 + \varepsilon_{A1}) + (\beta_0 + \beta_1 + \beta_4 + \varepsilon_{A2}) + (\beta_0 + \beta_1 + \beta_5 + \varepsilon_{A3})$$
$$+ (\beta_0 + \beta_1 + \beta_6 + \varepsilon_{A4}) + (\beta_0 + \beta_1 + \varepsilon_{A5})]/5$$

$$= \frac{5\beta_0 + 5\beta_1 + (\beta_3 + \beta_4 + \beta_5 + \beta_6) + (\varepsilon_{A1} + \varepsilon_{A2} + \varepsilon_{A3} + \varepsilon_{A4} + \varepsilon_{A5})}{5}$$

$$= \beta_0 + \beta_1 + \frac{(\beta_3 + \beta_4 + \beta_5 + \beta_6)}{5} + \bar{\varepsilon}_A$$

Similarly, the mean responses for treatments B and C are obtained:

$$\bar{y}_B = \frac{y_{B1} + y_{B2} + y_{B3} + y_{B4} + y_{B5}}{5}$$

$$= \beta_0 + \beta_2 + \frac{(\beta_3 + \beta_4 + \beta_5 + \beta_6)}{5} + \bar{\varepsilon}_B$$

$$\bar{y}_C = \frac{y_{C1} + y_{C2} + y_{C3} + y_{C4} + y_{C5}}{5}$$

$$= \beta_0 + \frac{(\beta_3 + \beta_4 + \beta_5 + \beta_6)}{5} + \bar{\varepsilon}_C$$

Since the objective is to compare treatment means, we are interested in the differences $\bar{y}_A - \bar{y}_B$, $\bar{y}_A - \bar{y}_C$, and $\bar{y}_B - \bar{y}_C$. These differences are calculated as follows:

$$\bar{y}_A - \bar{y}_B = [\beta_0 + \beta_1 + (\beta_3 + \beta_4 + \beta_5 + \beta_6)/5 + \bar{\varepsilon}_A]$$
$$- [\beta_0 + \beta_2 + (\beta_3 + \beta_4 + \beta_5 + \beta_6)/5 + \bar{\varepsilon}_B]$$
$$= (\beta_1 - \beta_2) + (\bar{\varepsilon}_A - \bar{\varepsilon}_B)$$

$$\bar{y}_A - \bar{y}_C = [\beta_0 + \beta_1 + (\beta_3 + \beta_4 + \beta_5 + \beta_6)/5 + \bar{\varepsilon}_A]$$
$$- [\beta_0 + (\beta_3 + \beta_4 + \beta_5 + \beta_6)/5 + \bar{\varepsilon}_C]$$
$$= \beta_1 + (\bar{\varepsilon}_A - \bar{\varepsilon}_C)$$

$$\bar{y}_B - \bar{y}_C = [\beta_0 + \beta_2 + (\beta_3 + \beta_4 + \beta_5 + \beta_6)/5 + \bar{\varepsilon}_B]$$
$$- [\beta_0 + (\beta_3 + \beta_4 + \beta_5 + \beta_6)/5 + \bar{\varepsilon}_C]$$
$$= \beta_2 + (\bar{\varepsilon}_B - \bar{\varepsilon}_C)$$

Note that for each pairwise comparison, the block β's (β_3, β_4, β_5, and β_6) cancel out, leaving only the treatment β's (β_1 and β_2). That is, the experimental noise resulting from differences between blocks is eliminated when treatment means are compared. The quantities $\bar{\varepsilon}_A - \bar{\varepsilon}_B$, $\bar{\varepsilon}_A - \bar{\varepsilon}_C$, and $\bar{\varepsilon}_B - \bar{\varepsilon}_C$ are the errors of estimation and represent the noise that tends to obscure the true differences between the treatment means.

What would occur if we employed the completely randomized design of Table 10.1 rather than the randomized block design? Since only one of the three computer programs is run each day, each treatment does not appear in each block. Consequently, when we compare the treatment means, the day-to-day variation (i.e., the block effects) will not cancel. For example, the difference between y_A and y_C would be

$$\bar{y}_A - \bar{y}_C = \beta_1 + \underbrace{\frac{(\text{Block } \beta\text{'s that do not cancel}) + (\bar{\varepsilon}_A - \bar{\varepsilon}_C)}{}}_{\text{Error of estimation}}$$

Thus, for the completely randomized design, the error of estimation will be increased by an amount involving the block effects (β_3, β_4, β_5, and β_6) that do not cancel. These effects, which inflate the error of estimation, cancel out for the randomized block design, thereby reducing the noise in the experiment.

EXAMPLE 10.3

Refer to Example 10.2 and the randomized block design employed to compare the mean percentage error rates for the four appraisers. The design is illustrated in Figure 10.3.

a. Write the model for the randomized block design.
b. Interpret the β parameters of the model, part **a**.
c. How can we use the model, part **a**, to test for differences among the mean percentage error rates of the four appraisers?

Solution

a. The experiment involves a qualitative factor, Appraisers, at four levels, which represent the treatments. The blocks for the experiment are the 10 properties. Therefore, the model is

$$E(y) = \beta_0 + \underbrace{\beta_1 x_1 + \beta_2 x_2 + \beta_3 x_3}_{\substack{\text{Treatments} \\ \text{(Appraisers)}}} + \underbrace{\beta_4 x_4 + \beta_5 x_5 + \cdots + \beta_{12} x_{12}}_{\substack{\text{Blocks} \\ \text{(Properties)}}}$$

where

$$x_1 = \begin{cases} 1 & \text{if appraiser A} \\ 0 & \text{if not} \end{cases} \quad x_2 = \begin{cases} 1 & \text{if appraiser B} \\ 0 & \text{if not} \end{cases} \quad x_3 = \begin{cases} 1 & \text{if appraiser C} \\ 0 & \text{if not} \end{cases}$$

$$x_4 = \begin{cases} 1 & \text{if property 1} \\ 0 & \text{if not} \end{cases} \quad x_5 = \begin{cases} 1 & \text{if property 2} \\ 0 & \text{if not} \end{cases}, \ldots, x_{12} = \begin{cases} 1 & \text{if property 9} \\ 0 & \text{if not} \end{cases}$$

b. Note that we have arbitrarily selected appraiser D and property 10 as the base levels. Following our discussion in Section 5.8, the interpretations of the β's are:

$$\beta_1 = \mu_A - \mu_D \quad \text{for a given property}$$
$$\beta_2 = \mu_B - \mu_D \quad \text{for a given property}$$
$$\beta_3 = \mu_C - \mu_D \quad \text{for a given property}$$
$$\beta_4 = \mu_1 - \mu_{10} \quad \text{for a given appraiser}$$
$$\beta_5 = \mu_2 - \mu_{10} \quad \text{for a given appraiser}$$
$$\vdots$$
$$\beta_{12} = \mu_9 - \mu_{10} \quad \text{for a given appraiser}$$

c. One way to determine whether the means for the four appraisers differ is to test the null hypothesis

$$H_0: \quad \mu_A = \mu_B = \mu_C = \mu_D$$

From our β interpretations in part **b**, this hypothesis is equivalent to testing

$$H_0: \quad \beta_1 = \beta_2 = \beta_3 = 0$$

To test this hypothesis, we drop the treatment β's (β_1, β_2, and β_3) from the complete model and fit the reduced model

$$E(y) = \beta_0 + \beta_4 x_4 + \beta_5 x_5 + \cdots + \beta_{12} x_{12}$$

Then we conduct the partial F test (see Section 4.10), where

$$F = \frac{(\text{SSE}_{\text{Reduced}} - \text{SSE}_{\text{Complete}})/3}{\text{MSE}_{\text{Complete}}}$$

The randomized block design represents one of the simplest types of noise-reducing designs. Other, more complex designs that employ the principle of blocking remove trends or variation in two or more directions. The **Latin square design** is useful when you want to eliminate two sources of variation, i.e., when you want to block in two directions. **Latin cube designs** allow you to block in three directions. A further variation in blocking occurs when the block contains fewer experimental units than the number of treatments. By properly assigning the treatments to a specified number of blocks, you can still obtain an estimate of the difference between a pair of treatments free of block effects. These are known as **incomplete block designs**. Consult the references for details on how to set up these more complex block designs.

EXERCISES

10.1   What two factors affect the quantity of information in an experiment?

10.2   How do block designs increase the quantity of information in an experiment?

10.3   Researchers recently conducted an experiment to compare the mean job satisfaction rating $E(y)$ of workers using three types of work scheduling: flextime (which allows workers to set their individual work schedules), staggered starting hours, and fixed hours.

a. Identify the treatments in the experiment.

b. Suppose 60 workers are available for the study. Explain how you would employ a completely randomized design for this experiment.

c. Write the linear model for the completely randomized design.

10.4 Retail store audits are periodic audits of a sample of retail sales to monitor inventory and purchases of a particular product. Such audits are often used by marketing researchers to estimate market share. A study was conducted to compare market shares of beer brands estimated by two different auditing methods.

a. Identify the treatments in the experiment.

b. Because of brand-to-brand variation in estimated market share, a randomized block design will be used. Explain how the treatments might be assigned to the experimental units if 10 beer brands are to be included in the study.

c. Write the linear model for the randomized block design.

10.5 Refer to the randomized block design of Examples 10.2 and 10.3.

a. Write the model for each observation of percentage appraisal error y for appraiser B. Sum the observations to obtain the average for appraiser B.

b. Repeat part **a** for appraiser D.

c. Show that $(\bar{y}_B - \bar{y}_D) = \beta_2 + (\bar{\varepsilon}_B - \bar{\varepsilon}_D)$. Note that the β's for blocks cancel when computing this difference.

10.5
Volume-Increasing Designs

In this section, we focus on how the proper choice of the treatments associated with two or more factors can increase the "volume" of information extracted from the experiment. The volume-increasing designs we will discuss are commonly known as **factorial designs** because they involve careful selection of the combinations of **factor levels** (i.e., treatments) in the experiment.

Consider a utility company that charges its customers a less expensive rate for using electricity during off-peak (less demanded) hours. The company is experimenting with several time-of-day pricing schedules. Two factors (i.e., independent variables) that the company can manipulate to form the schedule are price ratio, x_1, measured as the ratio of peak to off-peak prices, and peak period length, x_2, measured in hours. Suppose the utility company wants to investigate pricing ratio at two levels, 200% and 400%, and peak period length at two levels, 6 and 9 hours. The company will measure customer satisfaction, y, for several different schedules (i.e., combinations of x_1 and x_2) with the goal of comparing the mean satisfaction levels of the schedules. How should the company select the treatments for the experiment?

One method of selecting the price ratio–peak period length levels to be assigned to the experimental units (customers) would be to use the "one-at-a-time" approach. According to this procedure, one independent variable is varied while the remaining independent variables are held constant. This process is repeated for each of the independent variables in the experiment. This plan would *appear* to be extremely logical and consistent with the concept of blocking introduced in Section 10.4—that is, making comparisons within relatively homogeneous conditions—but this is not the case, as we will demonstrate.

The one-at-a-time approach applied to price ratio (x_1) and peak period length (x_2) is illustrated in Figure 10.4. When length is held constant at $x_2 = 6$ hours, we will observe the response y at a ratio of $x_1 = 200\%$ and $x_1 = 400\%$, thus yielding one pair of y values to estimate the average change in customer satisfaction as a result of changing the pricing ratio (x_1). Also, when pricing ratio is held constant at $x_1 = 200\%$, we observe the response y at a peak period length of $x_2 = 9$ hours. This observation, along with the one at (200%, 6 hours), allows us to estimate the average change in customer satisfaction as a result of a change in peak period length (x_2). The three treatments just described, (200%, 6 hours), (400%, 6 hours), and (200%, 9 hours), are indicated as points in Figure 10.4. Note that the figure shows two measurements (points) for each treatment. This is necessary to obtain an estimate of the standard deviation of the differences of interest.

FIGURE 10.4

"One-at-a-time" approach to selecting treatments

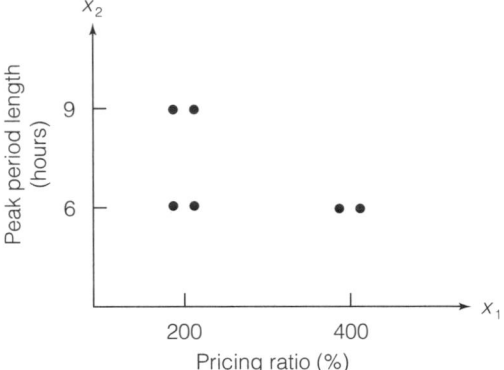

A second method of selecting the factor level combinations would be to choose the same three treatments as implied by the one-at-a-time approach and then to choose the fourth treatment at (400%, 9 hours) as shown in Figure 10.5. In other words, we have varied both variables, x_1 and x_2, at the same time.

FIGURE 10.5

Selecting all possible treatments

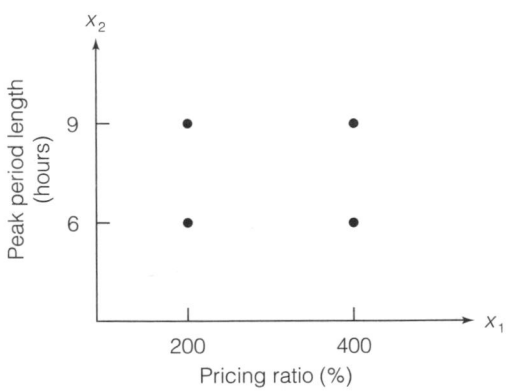

Which of the two designs yields the most information about the treatment differences? Surprisingly, the design of Figure 10.5, with only four observations, yields more accurate information than the one-at-a-time approach with its six observations. First, note that both designs yield two estimates of the difference between the mean response y at $x_1 = 200\%$ and $x_1 = 400\%$ when peak period length (x_2) is held constant, and both yield two estimates of the difference between the mean response y at $x_2 = 6$ hours and $x_2 = 9$ hours when pricing ratio (x_1) is held constant. But what if the difference between the mean response y at $x_1 = 200\%$ and at $x_1 = 400\%$ depends on which level of x_2 is held fixed, i.e., what if pricing ratio (x_1) and peak period length (x_2) *interact*? Then, we require estimates of the mean difference $(\mu_{200} - \mu_{400})$ when $x_2 = 6$ and the mean difference $(\mu_{200} - \mu_{400})$ when $x_2 = 9$. Estimates of both these differences are obtainable from the second design, Figure 10.5. However, since no estimate of the mean response for $x_1 = 400$ and $x_2 = 9$ is available from the one-at-a-time method, the interaction will go undetected for this design!

The importance of interaction between independent variables was emphasized in Section 4.9 and Chapter 5. If interaction is present, we cannot study the effect of one variable (or factor) on the response y independent of the other variable. Consequently, we require experimental designs that provide information on factor interaction.

Designs that accomplish this objective are called **factorial experiments**. A **complete factorial experiment** is one that includes all possible combinations of the levels of the factors as treatments. For the experiment on time-of-day pricing, we have two levels of pricing ratio (200% and 400%) and two levels of peak period length (6 and 9 hours). Hence, a complete factorial experiment will include $(2 \times 2) = 4$ treatments, as shown in Figure 10.5, and is called a **2 × 2 factorial design**.

| Definition 10.10

A **factorial design** is a method for selecting the treatments (that is, the factor level combinations) to be included in an experiment. A complete factorial experiment is one in which the treatments consist of all factor level combinations.

If we were to include a third factor, say, season, at four levels, then a complete factorial experiment would include all $2 \times 2 \times 4 = 16$ combinations of pricing ratio, peak period length, and season. The resulting collection of data would be called a **2 × 2 × 4 factorial design**.

EXAMPLE 10.4

Suppose you plan to conduct a marketing study of coffee sales at a supermarket to investigate the effect of three factors: shelf display (one of three types, A_1, A_2, and A_3), shelf location (bottom, B_1, middle, B_2, and top, B_3) and brand (C_1 and

C_2). Note that factor interactions could be very significant in this experiment. For example, brand C_1 might sell very well, regardless of the shelf display and shelf location, while the sales of C_2 may be greatly enhanced by using a particular display and locating it on a top shelf. Consequently, you will want to conduct a complete factorial experiment. Identify the 18 treatments for this $3 \times 3 \times 2$ factorial experiment.

Solution

The complete factorial experiment includes all possible combinations of shelf display, shelf location, and brand. We therefore would include the following treatments: $A_1B_1C_1$, $A_1B_1C_2$, $A_1B_2C_1$, $A_1B_2C_2$, $A_1B_3C_1$, $A_1B_3C_2$, $A_2B_1C_1$, $A_2B_1C_2$, $A_2B_2C_1$, $A_2B_2C_2$, $A_2B_3C_1$, $A_2B_3C_2$, $A_3B_1C_1$, $A_3B_1C_2$, $A_3B_2C_1$, $A_3B_2C_2$, $A_3B_3C_1$, $A_3B_3C_2$. These 18 treatments are shown diagrammatically in Figure 10.6.

FIGURE 10.6

The 18 treatments for the $3 \times 3 \times 2$ factorial of Example 10.4

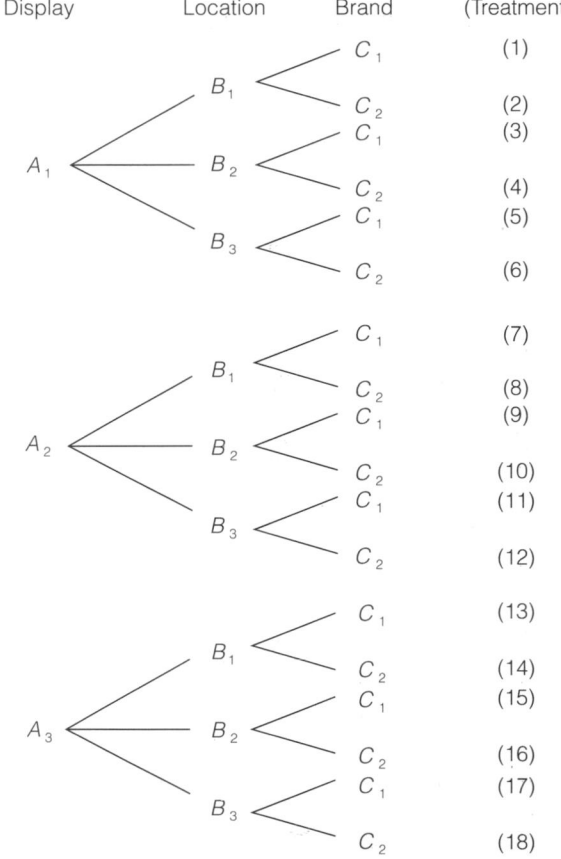

The linear model for a factorial design includes terms for each of the factors in the experiment—called **main effects**—and terms for factor interactions. For example, the model for the 2×2 factorial for the time-of-day pricing experiment includes a first-order term for the quantitative factor, pricing ratio (x_1), a first-order term for the quantitative factor, peak period length (x_2), and an interaction term:

$$y = \beta_0 + \underbrace{\beta_1 x_1 + \beta_2 x_2}_{\text{Main effects}} + \underbrace{\beta_3 x_1 x_2}_{\text{Interaction}} + \varepsilon$$

In general, the linear model for a complete factorial design for k factors contains terms for the following:

The main effects for each of the k factors

Two-way interaction terms for all pairs of factors

Three-way interaction terms for all combinations of three factors

$$\vdots$$

K-way interaction terms of all combinations of k factors.

If the factors are qualitative, then we set up dummy variables and proceed as in the next example.

EXAMPLE 10.5

Write the model for the $3 \times 3 \times 2$ factorial coffee sales experiment of Example 10.1.

Solution

Since the factors are qualitative, we set up dummy variables as follows:

$$x_1 = \begin{cases} 1 & \text{if display } A_1 \\ 0 & \text{if not} \end{cases} \qquad x_2 = \begin{cases} 1 & \text{if display } A_2 \\ 0 & \text{if not} \end{cases}$$

$$x_3 = \begin{cases} 1 & \text{if location } B_1 \\ 0 & \text{if not} \end{cases} \qquad x_4 = \begin{cases} 1 & \text{if location } B_2 \\ 0 & \text{if not} \end{cases}$$

$$x_5 = \begin{cases} 1 & \text{if brand } C_1 \\ 0 & \text{if brand } C_2 \end{cases}$$

Then the appropriate model is

$$y = \beta_0 + \underbrace{\beta_1 x_1 + \beta_2 x_2}_{\substack{\text{Display main} \\ \text{effects}}} + \underbrace{\beta_3 x_3 + \beta_4 x_4}_{\substack{\text{Location main} \\ \text{effects}}} + \underbrace{\beta_5 x_5}_{\substack{\text{Brand main} \\ \text{effect}}}$$

$$+ \underbrace{\beta_6 x_1 x_3 + \beta_7 x_1 x_4 + \beta_8 x_2 x_3 + \beta_9 x_2 x_4}_{\text{Display} \times \text{Location}} + \underbrace{\beta_{10} x_1 x_5 + \beta_{11} x_2 x_5}_{\text{Display} \times \text{Brand}}$$

$$+ \underbrace{\beta_{12} x_3 x_5 + \beta_{13} x_4 x_5}_{\text{Location} \times \text{Brand}}$$

$$+ \underbrace{\beta_{14} x_1 x_3 x_5 + \beta_{15} x_1 x_4 x_5 + \beta_{16} x_2 x_3 x_5 + \beta_{17} x_2 x_4 x_5}_{\text{Display} \times \text{Location} \times \text{Brand}}$$

Note that the number of parameters in the model for the $3 \times 3 \times 2$ factorial design of Example 10.5 is 18, which is equal to the number of treatments contained in the experiment. This is always the case for a complete factorial experiment. Consequently, if we fit the complete model to a single replication of the factorial treatments (i.e., one y observation measured per treatment), we will have no degrees of freedom available for estimating the error variance, σ^2. One way to solve this problem is to add additional data points to the sample. Researchers usually accomplish this by *replicating* the complete set of factorial treatments. That is, we collect two or more observed y values for each treatment in the experiment. This provides sufficient degrees of freedom for estimating σ^2.

One potential disadvantage of a complete factorial experiment is that it may require a large number of treatments. For example, an experiment involving 10 factors each at two levels would require $2^{10} = 1,024$ treatments! This might occur in an exploratory study where we are attempting to determine which of a large set of factors affect the response y. Several volume-increasing designs are available that employ only a fraction of the total number of treatments in a complete factorial experiment. For this reason, they are called **fractional factorial experiments**. Fractional factorials permit the estimation of the β parameters of lower-order terms (e.g., main effects and two-way interactions); however, β estimates of certain higher-order terms (e.g., three-way and four-way interactions) will be the same as some lower-order terms, thus confounding the results of the experiment. Consequently, a great deal of expertise is required to run and interpret fractional factorial experiments. Consult the references for details on fractional factorials and other more complex, volume-increasing designs.

EXERCISES

10.6   In what sense does a factorial experiment increase the quantity of information in an experiment?

10.7   Suppose you plan to investigate the effect of hourly pay rate and length of workday on some measure y of worker productivity. Both pay rate and length of workday will be set at three levels and y will be observed for all combinations of these factors.
   a. What type of experiment is this?
   b. Identify the factors and state whether they are quantitative or qualitative.
   c. Identify the treatments to be employed in the experiment.

10.8   A beverage company wants to investigate the effect of advertising expenditures and markets in two urban areas (M_1 and M_2) on the mean monthly sales of its best-selling diet cola. Advertising expenditure was set at \$15,000, \$18,000, \$21,000, \$24,000, and \$27,000, respectively, for the next 5 months in each of the two markets and the monthly sales were observed.
   a. List the factors involved in the experiment.
   b. For each factor, state whether it is quantitative or qualitative.
   c. How many treatments are involved in this experiment? List them.

10.9   Consider a factorial design with two factors, A and B, each at three levels. Suppose we select the following treatment (factor level) combinations to be included in the experiment: A_1B_1, A_2B_1, A_3B_1, A_1B_2, and A_1B_3.

a. Is this a complete factorial experiment? Explain.

b. Explain why it is impossible to investigate *AB* interaction in this experiment.

10.10   Write the complete factorial model for a 2 × 3 factorial experiment where both factors are qualitative.

10.11   Write the complete factorial model for a 2 × 3 × 3 factorial experiment where the factor at two levels is quantitative and the other two factors are qualitative.

10.12   Suppose you wish to investigate the effect of three factors on a response *y*. Explain why a factorial selection of treatments is better than varying each factor, one at a time, while holding the remaining two factors constant.

10.13   Why is the randomized block design a poor design to use to investigate the effect of two qualitative factors on a response *y*?

10.6

Selecting the Sample Size

We demonstrated how to select the sample size for estimating a single population mean or comparing two population means in Sections 1.7 and 1.9. We now show you how this problem can be solved for designed experiments.

As mentioned in Section 10.3, a measure of the quantity of information in an experiment that is pertinent to a particular population parameter is the standard error of the estimator of the parameter. A more practical measure is the half-width of the parameter's confidence interval, which will, of course, be a function of the standard error. For example, the half-width of a confidence interval for a population mean (given in Section 1.7) is

$$t_{\alpha/2}s_{\bar{y}} = t_{\alpha/2}\left(\frac{s}{\sqrt{n}}\right)$$

Similarly, the half-width of a confidence interval for the slope β_1 of a straight-line model relating *y* to *x* (given in Section 3.6) is

$$(t_{\alpha/2})s_{\hat{\beta}_1} = t_{\alpha/2}\left(\frac{s}{\sqrt{SS_{xx}}}\right) = t_{\alpha/2}\left(\sqrt{\frac{SSE}{n-2}}\right)\left(\frac{1}{\sqrt{SS_{xx}}}\right)$$

In both cases, the half-width is a function of the total number of data points in the experiment; each interval half-width gets smaller as the total number of data points *n* increases. The same is true for a confidence interval for a parameter β_i of a general linear model, for a confidence interval for $E(y)$, and for a prediction interval for *y*. Since each designed experiment can be represented by a linear model, this result can be used to select, approximately, the number of replications (i.e., the number of observations measured for each treatment) in the experiment.

For example, consider a designed experiment consisting of three treatments, A, B, and C. Suppose we want to estimate $(\mu_B - \mu_C)$, the difference between the treatment means for B and C. From our knowledge of linear models for designed experiments, we know this difference will be represented by one of the β parameters in the model, say, β_2. The confidence interval for β_2 for a single replication of the experiment is

$$\hat{\beta}_2 \pm (t_{\alpha/2})s_{\hat{\beta}_2}$$

If we repeat exactly the same experiment r times (we call this r **replications**), it can be shown (proof omitted) that the confidence interval for β_2 will be

$$\hat{\beta}_2 \pm \underbrace{t_{\alpha/2}\left(\frac{s_{\hat{\beta}_2}}{\sqrt{r}}\right)}_{B}$$

To find r, we first set the half-width of the interval to the largest value, B, we are willing to tolerate. Then we approximate $(t_{\alpha/2})$ and $s_{\hat{\beta}_2}$ and solve for the number of replications r.

EXAMPLE 10.6

Consider a 2×2 factorial experiment to investigate the effect of two factors on the light output y of flashbulbs used in cameras. The two factors (and their levels) are: x_1 = Amount of foil contained in the bulb (100 and 200 milligrams) and x_2 = Speed of sealing machine (1.2 and 1.3 revolutions per minute). The complete model for the 2×2 factorial experiment is

$$E(y) = \beta_0 + \beta_1 x_1 + \beta_2 x_2 + \beta_3 x_1 x_2$$

How many replicates of the 2×2 factorial are required to estimate β_3, the interaction β, to within .3 of its true value using a 95% confidence interval?

Solution

To solve for the number of replicates, r, we want to solve the equation

$$t_{\alpha/2}\left(\frac{s_{\hat{\beta}_3}}{\sqrt{r}}\right) = B$$

You can see that we need to have an estimate of $s_{\hat{\beta}_3}$, the standard error of $\hat{\beta}_3$ for a single replication. Suppose it is known from a previous experiment conducted by the manufacturer of the flashbulbs that $s_{\hat{\beta}_3} \approx .2$. For a 95% confidence interval, $\alpha = .05$ and $\alpha/2 = .025$. Since we want the half-width of the interval to be $B = .3$, we have

$$t_{.025}\left(\frac{.2}{\sqrt{r}}\right) = .3$$

The degrees of freedom for $t_{.025}$ will depend on the sample size $n = (2 \times 2)r = 4r$; consequently, we must approximate its value. In fact, since the model includes four parameters, the degrees of freedom for t will be df(Error) $= n - 4 = 4r - 4 = 4(r - 1)$. At minimum, we require two replicates; hence, we will have at least $4(2 - 1) = 4$ df. In Table 2 of Appendix D, we find $t_{.025} = 2.776$ for df $= 4$. We will use this conservative estimate of t in our calculations.

Substituting $t = 2.776$ into the equation, we have

$$\frac{2.776(.2)}{\sqrt{r}} = .3$$

$$\sqrt{r} = \frac{(2.776)(.2)}{.3} = 1.85$$

$$r = 3.42$$

Since we can run either three or four replications (but not 3.42), we should choose four replications to be reasonably certain that we will be able to estimate the interaction parameter, β_3, to within .3 of its true value. The 2×2 factorial with four replicates would be laid out as shown in Table 10.3.

T A B L E   10.3   **2 × 2 Factorial, with Four Replicates**

| | | AMOUNT OF FOIL, x_1 | |
		100	200
MACHINE SPEED,	1.2	4 observations on y	4 observations on y
x_2	1.3	4 observations on y	4 observations on y

EXERCISES

10.14   Why is replication important in a complete factorial experiment?

10.15   Consider a 2×2 factorial. How many replications are required to estimate the interaction β to within two units with a 90% confidence interval? Assume that the standard error of the estimate of the interaction β (based on a single replication) is approximately 3.

10.16   For a randomized block design with b blocks, the estimated standard error of the estimated difference between any two treatment means is $s\sqrt{2/b}$. Use this formula to determine the number of blocks required to estimate $(\mu_A - \mu_B)$, the difference between two treatment means, to within 10 units using a 95% confidence interval. Assume $s \approx 15$.

10.7
The Importance of Randomization

All the basic designs presented in this chapter involve randomization of some sort. In a completely randomized design and a basic factorial experiment, the treatments are randomly assigned to the experimental units. In a randomized block design, the blocks are randomly selected and the treatments within each block are assigned in random order. Why randomize? The answer is related to the assumptions we make about the random error ε in the linear model. Recall (Section 4.2) our assumption that ε follows a normal distribution with mean 0 and constant variance σ^2 for fixed settings of the independent variables (i.e., for each of the treatments). Further, we assume that the random errors associated with repeated observations are independent of each other in a probabilistic sense.

The experimenter rarely knows all of the important variables in a process and does not know the true functional form of the model. Hence, the functional form chosen to fit the true relation is only an approximation, and the variables included in the experiment form only a subset of the total. The random error, ε, is thus a composite error caused by the failure to include all of the important factors as well as the error in approximating the function.

Although many unmeasured and important independent variables affecting the response y do not vary in a completely random manner during the conduct of a designed experiment, we hope their behavior is such that their cumulative effect varies in a random manner and satisfies the assumptions upon which our inferential procedures are based. *The randomization in a designed experiment has the effect of randomly assigning these error effects to the treatments and assists in satisfying the assumptions on ε.*

Summary

Regression analysis based on observational data has at least one limitation: Even when the independent variables in a model are highly significant, we cannot infer a cause-and-effect relationship between the x's and y. The focus of this chapter was on data collected from a designed experiment in which the values of the independent variables are set in advance of observing y. By controlling the values of the x's, we hope to increase the amount of information extracted from the data.

Experimental design is a plan (or strategy) for collecting the experimental data. The goal is to increase the amount of information by controlling two factors:

1. **Volume** of the signal contained in the data
2. **Noise** or random variation in the data that is measured by σ^2

The first three steps in designing an experiment are as follows:

STEP 1 Select the **factors** (i.e., the independent variables) to be investigated.

STEP 2 Choose the factor level combinations (**treatments**).

STEP 3 Determine the number of observations for each treatment (i.e., the number of **replications** of the experiment).

These steps affect the volume of the signal contained in the data because they enable us to shift the information in the experiment so that it focuses on the parameter(s) of interest. An example of a volume-increasing design is a **factorial experiment**, in which all possible treatments (factor level combinations) are selected. With factorial designs, we shift the focus of the experiment to an investigation of factor interaction.

The fourth step in designing an experiment is

STEP 4 Decide how to assign the treatments to the experimental units.

Two basic methods of assigning treatments to the experimental units are the **completely randomized design** and the **randomized block design**. The latter is a noise-reducing design; by assigning treatments to relatively homogeneous blocks of experimental units, we can reduce the variation of treatment differences. The net effect of this action is to reduce experimental noise, measured by the variance of the random error ε that appears in the linear model.

The choice of design, noise-reducing or volume-increasing, will depend on your experimental objectives. In practice, researchers will attempt to employ both principles of design to increase the quantity of information in the experiment. For example, the treatments of a 2 × 2 factorial could be laid out in blocks to eliminate or reduce an unwanted source of variation. (An example of such a design is given in the next chapter.)

This chapter introduced the key principles of experimental design and presented some basic methods of collecting data in a designed experiment. Other, more complex designs, although beyond the scope of this text, may be more appropriate for your research problem. Consult the references listed at the end of the chapter to learn more about these designs. In Chapter 11, we demonstrate how to analyze experimental data using an **analysis of variance**.

SUPPLEMENTARY EXERCISES

10.17 How do you measure the quantity of information in a sample that is pertinent to a particular population parameter?

10.18 What steps in the design of an experiment affect the volume of the signal pertinent to a particular population parameter?

10.19 In what step in the design of an experiment can you possibly reduce the variation produced by extraneous and uncontrolled variables?

10.20 Explain the difference between a completely randomized design and a randomized block design. When is a randomized block design more advantageous?

10.21 Consider a two-factor factorial experiment where one factor is set at two levels and the other factor is set at four levels. How many treatments are included in the experiment? List them.

10.22 Write the complete factorial model for a 2 × 2 × 4 factorial experiment where both factors at two levels are quantitative and the third factor at four levels is qualitative. If you conduct one replication of this experiment, how many degrees of freedom will be available for estimating σ^2?

10.23 Refer to Exercise 10.22. Write the model for y assuming that you wish to enter main-effect terms for the factors, but no terms for factor interactions. How many degrees of freedom will be available for estimating σ^2?

References

Box, G. E. P., Hunter, W. G., & Hunter, J. S. *Statistics for Experimenters*. New York: Wiley, 1957.

Cochran, W. G. & Cox, G. M. *Experimental Designs*, 2nd ed. New York: Wiley, 1957.

Davies, O. L. *The Design and Analysis of Industrial Experiments*. New York: Hafner, 1956.

Kirk, R. E. *Experimental Design: Procedures for the Behavioral Sciences*. Pacific Grove, Calif.: Brooks/Cole, 1968.

Mendenhall, W. *Introduction to Linear Models and the Design and Analysis of Experiments*. Belmont, Calif.: Wadsworth, 1968.

Neter, J., Wasserman, W., & Kutner, M. H. *Applied Linear Statistical Models*, 3rd ed. Homewood, Ill.: Richard D. Irwin, 1988.

Winer, B. J. *Statistical Principles in Experimental Design*. New York: McGraw-Hill, 1962.

The Analysis of Variance
for Designed Experiments

CONTENTS

11.1 Introduction

11.2 The Logic Behind an Analysis of Variance

11.3 Completely Randomized Designs

11.4 Randomized Block Designs

11.5 Two-Factor Factorial Experiments

11.6 More Complex Factorial Designs (Optional)

11.7 Follow-Up Analysis: Tukey's Multiple Comparisons of Means

11.8 Other Multiple Comparisons Methods (Optional)

11.9 Checking ANOVA Assumptions

OBJECTIVE

To present a method for analyzing data collected from designed experiments for comparing two or more population means; to define the relationship of the analysis of variance to regression analysis and to identify their common features

11.1
Introduction

Once the data for a designed experiment have been collected, we will want to use the sample information to make inferences about the population means associated with the various treatments. The method used to compare the treatment means is traditionally known as **analysis of variance**, or **ANOVA**. The analysis of variance procedure provides a set of formulas that enable us to compute test statistics and confidence intervals required to make these inferences.

The formulas—one set for each experimental design—were developed in the early 1900s, well before the invention of computers. The formulas are easy to use, although the calculations can become quite tedious. However, you will recall from Chapter 10 that there is a linear model associated with each experimental design. Consequently, the same inferences derived from the ANOVA calculation formulas can be obtained by properly analyzing the model using a regression analysis and the computer.

In this chapter, the main focus is on the regression approach to analyzing data from a designed experiment. Several common experimental designs—some of which were presented in Chapter 10—are analyzed. We also provide the ANOVA calculation formulas for each design and show their relationship to regression. First, we provide the logic behind an analysis of variance and these formulas in Section 11.2.

11.2
The Logic Behind an Analysis of Variance

The concept behind an analysis of variance can be explained using the following simple example.

Consider an experiment with a single factor at two levels (that is, two treatments). Suppose we want to decide whether the two treatment means differ based on the means of two independent random samples, each containing $n_1 = n_2 = 5$ measurements, and that the y values appear as in Figure 11.1. Note that the five circles on the left are plots of the y values for sample 1 and the five solid dots on the right are plots of the y values for sample 2. Also, observe the horizontal lines that pass through the means for the two samples $\bar{y}_1$ and $\bar{y}_2$. Do you think the plots provide sufficient evidence to indicate a difference between the corresponding population means?

FIGURE 11.1

Plots of data for two samples

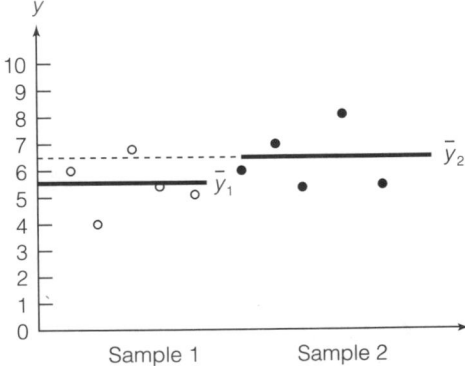

If you are uncertain whether the population means differ for the data in Figure 11.1, examine the situation for two different samples in Figure 11.2a. We think that you will agree that for these data, it appears that the population means differ. Examine a third case in Figure 11.2b. For these data, it appears that there is little or no difference between the population means.

FIGURE 11.2

Plots of data for two cases

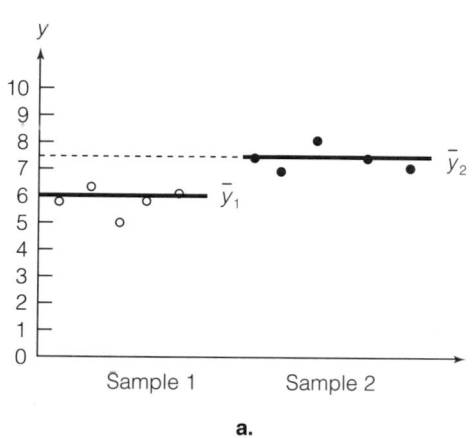 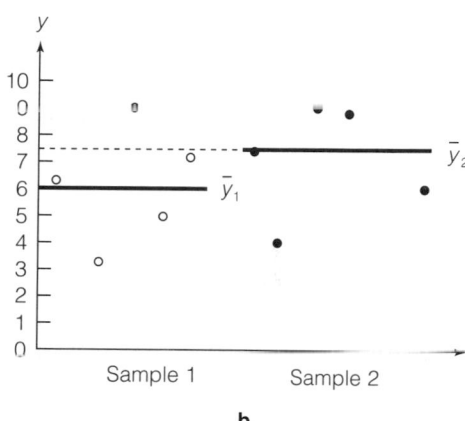

a.   b.

What elements of Figures 11.1 and 11.2 did we intuitively use to decide whether the data indicate a difference between the population means? The answer to the question is that we visually compared the distance (the variation) *between* the sample means to the variation *within* the y values for each of the two samples. Since the difference between the sample means in Figure 11.2a is large relative to the within-sample variation, we inferred that the population means differ. Conversely, in Figure 11.2b, the variation between the sample means is small relative to the within-sample variation, and therefore there is little evidence to infer that the means are significantly different.

The variation within samples is measured by the pooled s^2 that we computed for the independent random samples t test of Section 1.8, namely

$$\textit{Within-sample variation:} \quad s^2 = \frac{\sum_{i=1}^{n_1} (y_{i1} - \bar{y}_1)^2 + \sum_{i=1}^{n_2} (y_{i2} - \bar{y}_2)^2}{n_1 + n_2 - 2}$$

$$= \frac{\text{SSE}}{n_1 + n_2 - 2}$$

where y_{i1} is the ith observation in sample 1 and y_{i2} is the ith observation in sample 2. The quantity in the numerator of s^2 is often denoted **SSE**, the **sum of squared errors**. As with regression analysis, SSE measures unexplained variability. But in this case, it measures variability *unexplained* by the differences between the sample means.

A measure of the between-sample variation is given by the weighted sum of squares of deviations of the individual sample means about the mean for all 10 observations, $\bar{y}$, divided by the number of samples minus 1, i.e.,

$$\textit{Between-sample variation:}\quad \frac{n_1(\bar{y}_1 - \bar{y})^2 + n_2(\bar{y}_2 - \bar{y})^2}{2 - 1} = \frac{\text{SST}}{1}$$

The quantity in the numerator is often denoted **SST**, the **sum of squares for treatments**, since it measures the variability *explained* by the differences between the sample means of the two treatments.

For this experimental design, SSE and SST sum to a known total, namely

$$\text{SS(Total)} = \sum (y_i - \bar{y})^2$$

Also, the ratio

$$F = \frac{\text{Between-sample variation}}{\text{Within-sample variation}}$$

$$= \frac{\text{SST}/1}{\text{SSE}/(n_1 + n_2 - 2)}$$

has an F distribution with $\nu_1 = 1$ and $\nu_2 = n_1 + n_2 - 2$ degrees of freedom (df) and therefore can be used to test the null hypothesis of no difference between the treatment means. The additivity property of the sums of squares led early researchers to view this analysis as a **partitioning** of SS(Total) $= \Sigma(y_i - \bar{y})^2$ into sources corresponding to the factors included in the experiment and to SSE. The simple formulas for computing the sums of squares, the additivity property, and the form of the test statistic made it natural for this procedure to be called **analysis of variance**. We demonstrate the analysis of variance procedure and its relation to regression for several common experimental designs in Sections 11.3–11.6.

11.3

Completely Randomized Designs

Recall (Section 11.2) the first two steps in designing an experiment: (1) decide on the factors to be investigated and (2) select the factor level combinations (treatments) to be included in the experiment. For example, suppose you wish to compare the length of time to assemble a device in a manufacturing operation for workers who have completed one of three training programs, A, B, and C. Then this experiment involves a single factor, training program, at three levels, A, B, and C. Since training program is the only factor, these levels (A, B, and C) represent the treatments. Now we must decide the sample size for each treatment (step 3) and how to assign the treatments to the experimental units, namely the specific workers (step 4).

As we learned in Chapter 10, the most common assignment of treatments to experimental units is called a **completely randomized design**. To illustrate, suppose we wish to obtain equal amounts of information on the mean assembly times for the three training procedures, i.e., we decide to assign equal numbers of workers to each of the three training programs. Also, suppose we use the procedure of Section 1.7 (Example 1.11) to select the sample size, and we determine the number of workers in each of the three samples to be $n_1 = n_2 = n_3 = 10$. Then a completely randomized design is one in which the $n_1 + n_2 + n_3 = 30$ workers are **randomly assigned**, 10 to each of the three treatments. *A random assignment is one in which any one assignment is as probable as any other.* This eliminates the possibility of bias that might occur if the workers were assigned in some systematic manner. For example, a systematic assignment might accidentally assign most of the manually dexterous workers to training program A, thus underestimating the true mean assembly time corresponding to A.

Example 11.1 illustrates how a **random number table** can be used to assign the 30 workers to the three treatments.

EXAMPLE 11.1

Use the random number table, Table 7 in Appendix D, to assign $n = 30$ experimental units to three treatment groups.

Solution

The first step is to number the 30 workers from 1 to 30. We will then use Table 7 in Appendix D to select two-digit numbers, discarding those that are larger than 30 or are identical, until we have a total of 20 two-digit numbers. We will then have 20 of the integers between 1 and 30 arranged in random order. The workers who have been assigned the first 10 numbers in the sequence are assigned to training program A, the second group of ten workers are assigned to B, and the remaining workers are assigned to C.

To illustrate, suppose we start with the two-digit random number in row 5, column 6 of Table 7 and proceed down the column, selecting only two-digit numbers (the first two digits) less than or equal to 30 and deleting those that repeat. The first 20 are: 20, 18, 13, 16, 19, 04, 14, 06, 30, 25, 27, 17, 24, 21, 22, 02, 15, 05, 09, 08. The workers with the first 10 numbers are assigned to program A, the second 10 to B, and the remaining 10 to C. So the workers are assigned to the training program as follows:

A	B	C
20, 18, 13, 16, 19, 4, 14, 6, 30, 25	27, 17, 24, 21, 22, 2, 15, 5, 9, 8	1, 3, 7, 10, 11, 12, 23, 26, 28, 29

Example 11.2 will refresh your memory concerning the selection of sample sizes for comparing population means.

EXAMPLE 11.2

Refer to the experimental situation discussed in Example 11.1. Suppose you know from prior experience that the range of the assembly times for a given treatment will be roughly 8 minutes. Now, suppose you wish to estimate the difference in mean assembly times, corresponding to a pair of training programs, correct to within 1 minute. How many workers should be assigned to each program?

Solution

We know from Section 1.4 that one-fourth of the range of a set of measurements provides a rough approximation to their standard deviation. Using this approximation, it follows that the standard deviation of the assembly times for a given treatment is

$$\sigma \approx \frac{\text{Range}}{4} = \frac{8}{4} = 2$$

Since the bound on the error of estimation will be less than

$$2\sigma_{\bar{y}_1 - \bar{y}_2} = 2\sigma \sqrt{\frac{1}{n_1} + \frac{1}{n_2}}$$

with probability near .95, we will set

$$2\sigma \sqrt{\frac{1}{n_1} + \frac{1}{n_2}} = \text{Bound on the error} = 1$$

Thus, we will substitute $\sigma \approx 2$ into this equation and solve for $n_1 = n_2$. [*Note:* We agreed earlier to use equal treatment group sizes. Therefore $n_1 = n_2 = n_3$.] Then,

$$2(2) \sqrt{\frac{1}{n_1} + \frac{1}{n_1}} = 1 \text{ minute}$$
$$\sqrt{n_1} = 4\sqrt{2}$$
$$n_1 = 32$$

Therefore, to estimate the difference between a pair of mean assembly times to within 1 minute with 95% confidence, we must have a total of 96 workers, and we must randomly assign 32 to each of the three training programs.

EXAMPLE 11.3

Suppose a beverage bottler wished to compare the effect of three different advertising displays on the sales of a beverage in supermarkets. Identify the experimental units you would use for the experiment, and explain how you would employ a completely randomized design to collect the sales data.

Solution

Presumably, the bottler has a list of supermarkets in a number of different cities that market the beverage. If we decide to measure the sales increase (or decrease) as the monthly dollar increase in sales (over the previous month) for a given supermarket, then the experimental unit is a 1-month unit of time in a specific supermarket. Thus, we would randomly select a 1-month period of time for each of $n_1 + n_2 + n_3$ supermarkets and assign n_1 supermarkets to receive display D_1, n_2 to receive D_2, and n_3 to receive D_3.

In some experimental situations, we are unable to randomly assign the treatment to the experimental units because of the nature of the experimental units themselves. For example, suppose we want to compare the mean annual salaries of professors in two College of Business departments, accounting and economics.

Then the treatments—accounting and economics—cannot be "assigned" to the professors (experimental units). A professor is a member of either the accounting department or the economics (or some other) department and cannot be arbitrarily assigned one of the treatments. Rather, we view the treatments (departments) as populations from which we will select independent random samples of experimental units (professors). A completely randomized design involves a comparison of the means for a number, say, p, of treatments, based on independent random samples of $n_1, n_2, \ldots, n_p$ observations, drawn from populations associated with treatments $1, 2, \ldots, p$, respectively. We repeat our definition of a completely randomized design (given in Section 10.4) with this modification.

Definition 11.1

A **completely randomized design** to compare p treatment means is one in which the treatments are randomly assigned to the experimental units, or in which independent random samples are drawn from each of the p populations.

After collecting the data from a completely randomized design, we want to make inferences about p population means where μ_i is the mean of the population of measurements associated with treatment i, for $i = 1, 2, \ldots, p$. The null hypothesis to be tested is that the p treatment means are equal, i.e., $H_0: \mu_1 = \mu_2 = \cdots = \mu_p$, and the alternative hypothesis we wish to detect is that at least two of the treatment means differ. The appropriate linear model for the response y is

$$E(y) = \beta_0 + \beta_1 x_1 + \beta_2 x_2 + \cdots + \beta_{p-1} x_{p-1}$$

where

$$x_1 = \begin{cases} 1 & \text{if treatment 2} \\ 0 & \text{if not} \end{cases} \quad x_2 = \begin{cases} 1 & \text{if treatment 3} \\ 0 & \text{if not} \end{cases} \quad \cdots \quad x_{p-1} = \begin{cases} 1 & \text{if treatment } p \\ 0 & \text{if not} \end{cases}$$

and (arbitrarily) treatment 1 is the base level. Recall that this 0–1 system of coding implies that

$$\beta_0 = \mu_1$$
$$\beta_1 = \mu_2 - \mu_1$$
$$\beta_2 = \mu_3 - \mu_1$$
$$\vdots \qquad \vdots$$
$$\beta_{p-1} = \mu_p - \mu_1$$

The null hypothesis that the p population means are equal is equivalent to the null hypothesis that all the treatment differences equal 0, i.e.,

$$H_0: \quad \beta_1 = \beta_2 = \cdots = \beta_{p-1} = 0$$

To test this hypothesis using regression, we employ the technique of Section 4.10; that is, we compare the sum of squares for error, SSE_R, for the *reduced* model

$$E(y) = \beta_0$$

to the sum of squares for error, SSE_C, for the *complete* model

$$E(y) = \beta_0 + \beta_1 x_1 + \beta_2 x_2 + \cdots + \beta_{p-1} x_{p-1}$$

using the F statistic

$$F = \frac{(SSE_R - SSE_C)/\text{Number of } \beta \text{ parameters in } H_0}{SSE_C/[n - (\text{Number of } \beta \text{ parameters in the complete model})]}$$

$$= \frac{(SSE_R - SSE_C)/(p-1)}{SSE_C/(n-p)}$$

$$= \frac{(SSE_R - SSE_C)/(p-1)}{MSE_C}$$

where F is based on $\nu_1 = (p-1)$ and $\nu_2 = (n-p)$ df. If F exceeds the upper critical value, F_a, we reject H_0 and conclude that at least one of the treatment differences, $\beta_1, \beta_2, \ldots, \beta_{p-1}$, differs from zero, i.e., we conclude that at least two treatment means differ.

EXAMPLE 11.4

Show that the F statistic for testing the equality of treatment means in a completely randomized design is equivalent to a global F test of the complete model.

Solution

Since the reduced model contains only the β_0 term, the least squares estimate of β_0 is $\bar{y}$, and it follows that

$$SSE_R = \sum (y - \bar{y})^2 = SS_{yy}$$

We called this quantity the sum of squares for total in Chapter 4. The difference $(SSE_R - SSE_C)$ is simply $(SS_{yy} - SSE)$ for the complete model. Since in regression $(SS_{yy} - SSE) = SS(\text{Model})$, and the complete model has $(p-1)$ terms (excluding β_0),

$$F = \frac{(SSE_R - SSE_C)/(p-1)}{MSE_C} = \frac{SS(\text{Model})/(p-1)}{MSE} = \frac{MS(\text{Model})}{MSE}$$

Thus, it follows that the test statistic for testing the null hypothesis,

$$H_0: \quad \mu_1 = \mu_2 = \cdots = \mu_p$$

in a completely randomized design is the same as the F statistic for testing the global utility of the complete model for this design.

The regression approach to analyzing data from a completely randomized design is summarized in the box. Note that the test requires several assumptions about the distributions of the response y for the p treatments and that these assumptions are necessary regardless of the sizes of the samples. (We have more to say about these assumptions in Section 11.9.)

Model and F Test for a Completely Randomized Design with p Treatments

Complete model: $E(y) = \beta_0 + \beta_1 x_1 + \beta_2 x_2 + \cdots + \beta_{p-1} x_{p-1}$

where

$$x_1 = \begin{cases} 1 & \text{if treatment 2} \\ 0 & \text{if not} \end{cases} \qquad x_2 = \begin{cases} 1 & \text{if treatment 3} \\ 0 & \text{if not} \end{cases} , \ldots ,$$

$$x_{p-1} = \begin{cases} 1 & \text{if treatment } p \\ 0 & \text{if not} \end{cases}$$

H_0: $\beta_1 = \beta_2 = \cdots = \beta_{p-1} = 0$ (i.e., H_0: $\mu_1 = \mu_2 = \cdots = \mu_p$)

H_a: At least one of the β parameters listed in H_0 differs from 0
(i.e., H_a: At least two means differ).

Test statistic: $F = \dfrac{\text{MS(Model)}}{\text{MSE}}$

Rejection region: $F > F_\alpha$, where the distribution of F is based on $\nu_1 = p - 1$ and $\nu_2 = n - p$ degrees of freedom.

Assumptions: 1. All p population probability distributions corresponding to the p treatments are normal.
2. The population variances of the p treatments are equal.

EXAMPLE 11.5

Suppose a large chain of department stores wants to compare the mean dollar amounts owed by its delinquent credit card customers in three different annual income groups: under \$12,000 (group A), \$12,000–\$25,000 (group B), and over \$25,000 (group C). Samples of 10 customers with delinquent accounts are randomly selected from each group and the amount owed by each customer is recorded. The data are shown in Table 11.1 (page 560). Do the data provide sufficient evidence to indicate a difference in the mean amount owed for the three income groups? Test using $\alpha = .05$.

Solution

This experiment involves a single factor, income class, at three levels. Thus, we have a completely randomized design with $p = 3$ treatments. Let μ_A, μ_B, μ_C represent the mean amounts owed by customers with incomes under \$12,000, between \$12,000 and \$25,000, and over \$25,000, respectively. Then we want to test

H_0: $\mu_A = \mu_B = \mu_C$

against

H_a: At least two of the three treatment means differ.

TABLE 11.1   Income Class: Dollars Owed

	A	B	C
	Under $12,000	$12,000–$25,000	Over $25,000
	$148	$513	$335
	76	264	643
	393	433	216
	520	94	536
	236	535	128
	134	327	723
	55	214	258
	166	135	380
	415	280	594
	153	304	465
Total	$2,296	$3,099	$4,278

The appropriate linear model for $p = 3$ treatments is

Complete model: $E(y) = \beta_0 + \beta_1 x_1 + \beta_2 x_2$

where

$$x_1 = \begin{cases} 1 & \text{if income class B} \\ 0 & \text{if not} \end{cases} \quad \text{and} \quad x_2 = \begin{cases} 1 & \text{if income class C} \\ 0 & \text{if not} \end{cases}$$

Thus, we want to test H_0: $\beta_1 = \beta_2 = 0$.

The SAS regression analysis for the complete model is shown in Figure 11.3. The F statistic for testing the overall adequacy of the model (shaded on the printout) is $F = 3.48$, where the distribution of F is based on $\nu_1 = (p - 1) = 3 - 1 = 2$ and $\nu_2 = (n - p) = 30 - 3 = 27$ df. For $\alpha = .05$, the critical value (obtained from Table 4 of Appendix D) is $F_{.05} = 3.35$ (see Figure 11.4).

FIGURE 11.3

SAS printout for the completely randomized design, Example 11.5

```
Dependent Variable: AMOUNT

                        Analysis of Variance

                        Sum of        Mean
Source          DF      Squares       Square      F Value     Prob>F

Model            2  198772.46667   99386.23333     3.482      0.0452
Error           27  770670.90000   28543.36667
C Total         29  969443.36667

        Root MSE       168.94782     R-square     0.2050
        Dep Mean       322.43333     Adj R-sq     0.1462
        C.V.            52.39775

                        Parameter Estimates

                     Parameter      Standard     T for H0:
Variable   DF        Estimate         Error     Parameter=0    Prob > |T|

INTERCEP    1       427.800000     53.42599243      8.007        0.0001
X1          1      -198.200000     75.55576307     -2.623        0.0141
X2          1      -117.900000     75.55576307     -1.560        0.1303
```

FIGURE 11.4

Rejection region for Example 11.5; numerator df = 2, denominator df = 27, $\alpha = .05$

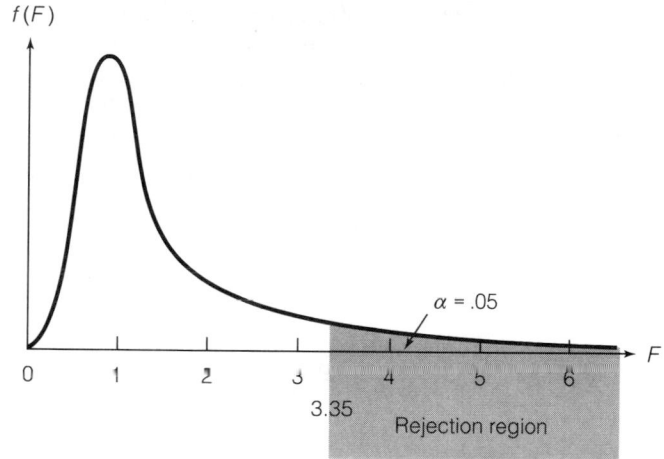

Since the computed value of F, 3.48, exceeds the critical value, $F_{.05} = 3.35$, we reject H_0 and conclude (at the $\alpha = .05$ level of significance) that the mean indebtedness of the delinquent credit card holders differs in at least two of the three income groups. We can arrive at the same conclusion by noting that $\alpha = .05$ is greater than the p-value (.0452) shaded on the printout.

The analysis of the data in Example 11.5 can also be accomplished using ANOVA computing formulas. In Section 11.2 we learned that an analysis of variance partitions SS(Total) $= \Sigma(y - \bar{y})^2$ into two components, SSE and SST (see Figure 11.5).

FIGURE 11.5

The partitioning of SS(Total) for a completely randomized design

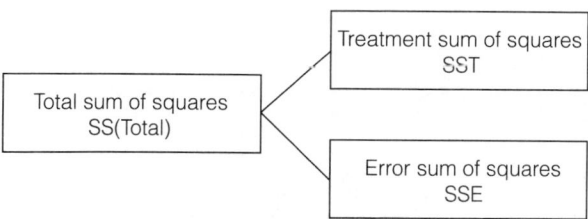

Recall that the quantity SST denotes the sum of squares for treatments and measures the variation explained by the differences between the treatment means. The sum of squares for error, SSE, is a measure of the unexplained variability, obtained by calculating a pooled measure of the variability *within* the p samples. If the treatment means truly differ, then SSE should be substantially smaller than SST. We compare the two sources of variability by forming an F statistic:

$$F = \frac{\text{SST}/(p - 1)}{\text{SSE}/(n - p)} = \frac{\text{MST}}{\text{MSE}}$$

where n is the total number of measurements. The numerator of the F statistic, $MST = SST/(p - 1)$, denotes **mean square for treatments** and is based on $(p - 1)$ degrees of freedom—one for each of the p treatments minus one for the estimation of the overall mean. The denominator of the F statistic, $MSE = SSE/(n - p)$, denotes **mean square for error** and is based on $(n - p)$ degrees of freedom—one for each of the n measurements minus one for each of the p treatment means being estimated. We have already demonstrated that this F statistic is identical to the global F value for the regression model specified earlier.

For completeness, we provide the computing formulas for an analysis of variance in the next box.

EXAMPLE 11.6

Refer to Example 11.5. Analyze the data of Table 11.1 using the ANOVA approach. Use $\alpha = .05$.

Solution

Rather than perform the tedious calculations by hand (we leave this for the student as an exercise), we resort to a statistical software package with an ANOVA routine. All three of the software packages discussed in this text (SAS, Minitab, and SPSS) have procedures that automatically compute the ANOVA sums of squares and the ANOVA F statistic.

The Minitab ANOVA printout is shown in Figure 11.6. The value of the test statistic (shaded on the printout) is $F = 3.48$. Note that this is identical to the F value obtained using the regression approach in Example 11.5. The p-value of the test (also shaded) is $p = .045$. (Likewise, this quantity is identical to that in Example 11.5.) Since $\alpha = .05$ exceeds this p-value, we have sufficient evidence to conclude that the treatments differ.

FIGURE 11.6

Minitab ANOVA printout for Example 11.6

```
ANALYSIS OF VARIANCE ON AMOUNT
SOURCE      DF        SS        MS         F         p
INCGROUP     2    198772     99386      3.48     0.045
ERROR       27    770671     28543
TOTAL       29    969443

                                  INDIVIDUAL 95 PCT CI'S FOR MEAN
                                  BASED ON POOLED STDEV
LEVEL        N      MEAN     STDEV  -+---------+---------+---------+-----
    1       10     229.6     158.2  (--------*--------)
    2       10     309.9     147.9            (--------*--------)
    3       10     427.8     196.8                      (--------*--------)
                                   -+---------+---------+---------+-----
POOLED STDEV =     168.9            120       240       360       480
```

The results of an analysis of variance are often summarized in tabular form. The general form of an ANOVA table for a completely randomized design is shown in Table 11.2. The column head **SOURCE** refers to the source of variation, and for each source, **DF** refers to the degrees of freedom, **SS** to the sum of squares, **MS** to the mean square, and **F** to the F statistic comparing the treatment mean square to the error mean square. Table 11.3 (page 564) is the ANOVA summary table corresponding to the analysis of variance data for Example 11.6, obtained from the Minitab printout.

Computing Formulas for the Analysis of Variance for a Completely Randomized Design

Sum of all n measurements $= \sum_{i=1}^{n} y_i$

Mean of all n measurements $= \bar{y}$

Sum of squares of all n measurements $= \sum_{i=1}^{n} y_i^2$

$\text{CM} = \text{Correction for mean}$

$$= \frac{(\text{Total of all observations})^2}{\text{Total number of observations}} = \frac{\left(\sum_{i=1}^{n} y_i \right)^2}{n}$$

$\text{SS(Total)} = \text{Total sum of squares}$

$\qquad = (\text{Sum of squares of all observations}) - \text{CM}$

$$= \sum_{i=1}^{n} y_i^2 - \text{CM}$$

$\text{SST} = \text{Sum of squares for treatments}$

$$= \left(\begin{array}{c} \text{Sum of squares of treatment totals with} \\ \text{each square divided by the number of} \\ \text{observations for that treatment} \end{array} \right) - \text{CM}$$

$$= \frac{T_1^2}{n_1} + \frac{T_2^2}{n_2} + \cdots + \frac{T_p^2}{n_p} - \text{CM}$$

$\text{SSE} = \text{Sum of squares for error}$

$\qquad = \text{SS(Total)} - \text{SST}$

$\text{MST} = \text{Mean square for treatments}$

$$= \frac{\text{SST}}{p - 1}$$

$\text{MSE} = \text{Mean square for error}$

$$= \frac{\text{SSE}}{n - p}$$

$$F = \frac{\text{MST}}{\text{MSE}}$$

T A B L E 11.2 ANOVA Summary Table for a Completely Randomized Design

SOURCE	df	SS	MS	F
Treatments	$p - 1$	SST	MST	MST/MSE
Error	$n - p$	SSE	MSE	
Total	$n - 1$	SS(Total)		

TABLE 11.3 ANOVA Summary Table for Example 11.6

SOURCE	df	SS	MS	F
Income group	2	198,772.467	99,386.233	3.48
Error	27	770,670.900	28,543.367	
Total	29	969,443.367		

Because the completely randomized design involves the selection of independent random samples, we can find a confidence interval for a single treatment mean using the method of Section 1.7 or for the difference between two treatment means using the methods of Section 1.9. The estimate of σ^2 will be based on the pooled sum of squares within all p samples; that is,

$$MSE = s^2 = \frac{SSE}{n - p}$$

This is the same quantity that is used as the denominator for the analysis of variance F test. The formulas for the confidence intervals of Sections 1.7 and 1.9 are reproduced in the box.

Confidence Intervals for Means: Completely Randomized Design

Single treatment mean (say, treatment i): $\bar{y}_i \pm t_{\alpha/2} \left(\frac{s}{\sqrt{n_i}} \right)$

Difference between two treatment means (say, treatments i and j):

$$(\bar{y}_i - \bar{y}_j) \pm t_{\alpha/2} s \sqrt{\frac{1}{n_i} + \frac{1}{n_j}}$$

where $\bar{y}_i$ is the sample mean response for population (treatment) i, $s = \sqrt{MSE}$, and $t_{\alpha/2}$ is the tabulated value of t (Table 2 of Appendix D) that locates $\alpha/2$ in the upper tail of the t distribution with $(n - p)$ df (the degrees of freedom associated with error in the ANOVA).

EXAMPLE 11.7

Refer to Example 11.5 and find a 95% confidence interval for μ_A, the mean indebtedness of customers with incomes less than $12,000 per year.

Solution

From Table 11.3, MSE = 28,543.367. Then

$$s = \sqrt{MSE} = \sqrt{28,543.367} = 168.9$$

The sample mean indebtedness for those with income levels under $12,000 is

$$\bar{y}_A = \frac{2,296}{10} = 229.6$$

where 2,296 is the total of the delinquent amounts for group A obtained from Table 11.1. The tabulated value of $t_{.025}$ for 27 df (the same as for MSE) is 2.052. Therefore, a 95% confidence interval for μ_A, the mean indebtedness of people with incomes less than \$12,000, is

$$\bar{y}_A \pm t_{\alpha/2}\left(\frac{s}{\sqrt{n_1}}\right) = 229.6 \pm 2.052\left(\frac{168.9}{\sqrt{10}}\right)$$

$$= 229.6 \pm 109.6$$

or (\$120.0, \$339.2).

Note that this confidence interval is quite wide—probably too wide to be of any practical value. The reason why the interval is so wide can be seen in the large amount of variation within each income class. For example, the indebtedness for customers with incomes under \$12,000 varies from \$55 to \$520. The more variable the data, the larger will be the value of s that appears in the confidence interval and the wider will be the confidence interval. Consequently, if you want to obtain a more accurate estimate of treatment means with a narrower confidence interval, you will have to select larger samples of customers from within each income class.

_____

Although we can use the formula given in the box to compare two treatment means in ANOVA, unless the two treatments are selected a priori (i.e., prior to conducting the ANOVA), it is more appropriate to apply one of the methods for comparing means presented in Sections 11.7 and 11.8.

EXERCISES

11.1 Refer to the completely randomized design of Exercise 10.3. Recall that the researchers want to compare the mean job satisfaction rating of workers using three types of work scheduling: flextime, staggered starting hours, and fixed hours. Use the random number table (Table 7 in Appendix D) to randomly assign the workers to the three work schedules.

11.2 A partially completed ANOVA table for a completely randomized design is shown here.

SOURCE	df	SS	MS	F
Treatments	4	24.7	_____	_____
Error	_____	_____	_____	
Total	34	62.4		

a. Complete the ANOVA table.
b. How many treatments are involved in the experiment?
c. Do the data provide sufficient evidence to indicate a difference among the treatment means? Test using $\alpha = .10$.

11.3   The data for a completely randomized design with two treatments are shown in the accompanying table.

TREATMENT 1	TREATMENT 2
10	12
7	8
8	13
11	10
10	10
9	11
9	

a. Give the linear model appropriate for analyzing the data using regression.

b. Fit the model, part **a**, to the data and conduct the analysis. [*Hint:*   You do not need a computer to fit the model. Use the formulas provided in Chapter 3.]

11.4   Refer to Exercise 11.3.

a. Calculate MST for the data using the ANOVA formulas. What type of variability is measured by this quantity?

b. Calculate MSE for the data using the ANOVA formulas. What type of variability is measured by this quantity?

c. How many degrees of freedom are associated with MST?

d. How many degrees of freedom are associated with MSE?

e. Compute the test statistic appropriate for testing $H_0: \mu_1 = \mu_2$ against the alternative that the two treatment means differ, using a significance level of $\alpha = .05$. (Compare the value to the test statistic obtained using regression in Exercise 11.3b.)

f. Summarize the results from parts **a–e** in an ANOVA table.

g. Specify the rejection region, using a significance level of $\alpha = .05$.

h. State the proper conclusion.

11.5   Exercises 11.3 and 11.4 involve a test of the null hypothesis $H_0: \mu_1 = \mu_2$ based on independent random sampling (recall the definition of a completely randomized design). This test was conducted in Section 1.9 using a Student's t statistic.

a. Use the Student's t test to test $H_0: \mu_1 = \mu_2$ against the alternative hypothesis $H_a: \mu_1 \neq \mu_2$. Test using $\alpha = .05$.

b. It can be shown (proof omitted) that an F statistic with $\nu_1 = 1$ numerator degree of freedom and ν_2 denominator degrees of freedom is equal to t^2, where t is a Student's t statistic based on ν_2 degrees of freedom. Square the value of t calculated in part **a**, and show that it is equal to the value of F calculated in Exercises 11.3b and 11.4e.

c. Is the analysis of variance F test for comparing two population means a one- or a two-tailed test of $H_0: \mu_1 = \mu_2$? [*Hint:*   Although the t test can be used to test for either $H_a: \mu_1 > \mu_2$ or $H_a: \mu_1 < \mu_2$, the alternative hypothesis for the F test is H_a: The two means are different.]

11.6   Each year *Business Week* reports the total cash compensations (salary plus bonus) for the top corporate executives. The data in the table (in thousands of dollars) were extracted from *Business Week*'s 1990 Executive Compensation Scoreboard. To compare the mean 1990 cash compensation, $E(y)$, of executives in the four groups, the following model was fit to the data:

$$E(y) = \beta_0 + \beta_1 x_1 + \beta_2 x_2 + \beta_3 x_3$$

where

$$x_1 = \begin{cases} 1 & \text{if consumer products} \\ 0 & \text{if not} \end{cases} \qquad x_2 = \begin{cases} 1 & \text{if utilities} \\ 0 & \text{if not} \end{cases}$$

$$x_3 = \begin{cases} 1 & \text{if industrial–high tech} \\ 0 & \text{if not} \end{cases}$$

Base level = financial services

The Minitab printout is presented here.

CONSUMER PRODUCTS	UTILITIES	INDUSTRIAL–HIGH TECH	FINANCIAL SERVICES
1,567	1,862	2,925	3,125
3,313	1,390	3,409	4,143
2,058	1,115	1,767	4,013
25,216	1,105	4,097	6,583
4,634	1,272	3,196	3,169
5,214	2,849	4,042	5,217
20,795	1,732	2,601	3,447
9,162	1,474	8,286	4,469

Source: "Executive compensation scoreboard." Business Week, May 3, 1990, pp. 65–108.

```
The regression equation is
Y = 4271 + 4724 X1 - 2671 X2 - 480 X3

Predictor       Coef       Stdev     t-ratio         p
Constant        4271        1651        2.59     0.015
X1              4724        2334        2.02     0.053
X2             -2671        2334       -1.14     0.262
X3              -480        2334       -0.21     0.838

s = 4669        R-sq = 27.6%     R-sq(adj) = 19.8%

Analysis of Variance

SOURCE         DF          SS          MS         F         p
Regression      3    232505648    77501880      3.56     0.027
Error          28    610272512    21795446
Total          31    842778176

SOURCE         DF      SEQ SS
X1              1    200071984
X2              1     31510622
X3              1       923041

Unusual Observations
Obs.      X1        Y      Fit Stdev.Fit  Residual   St.Resid
  4     1.00    25216     8995     1651     16221       3.71R
  7     1.00    20795     8995     1651     11800       2.70R

R denotes an obs. with a large st. resid.
```

a. Is there sufficient evidence to indicate that the model is useful for predicting cash compensation? Test using $\alpha = .01$.

b. What does the result from part a imply about the mean cash compensation for the four groups of executives?

c. Find a 99% confidence interval for the difference between the mean 1990 cash compensation of executives in the consumer products and financial services industries.

11.7   The display consoles of modern computer-based systems use many abbreviated words to accommodate the large volume of information to be displayed. Therefore, operators must learn to decode each abbreviation quickly and accurately. An experiment was conducted to determine the optimal method for abbreviating any specific set of words on the sonar consoles used at the Naval Submarine Medical Research Laboratory in Groton, Connecticut (*Human Factors*, Feb. 1984). Of the 20 Navy and civilian personnel who took part in the study, five were highly familiar with the sonar system. The 15 subjects unfamiliar with the system were randomly divided into three groups of five. Thus, the study consisted of a total of four groups (one experienced and three inexperienced groups), with five subjects per group. The experienced group and one inexperienced group (denoted TE and TI, respectively) were assigned to learn the simple method of abbreviation. One of the remaining inexperienced groups was assigned the conventional single abbreviation method (denoted CS), whereas the other was assigned the conventional multiple abbreviation method (denoted CM). Each subject was then given a list of 75 abbreviations to learn, one at a time, through the display console of a minicomputer. The number of trials until the subject accurately decoded at least 90% of the words on the list was recorded. Do the data provide sufficient evidence to indicate differences among the mean numbers of trials required for the four groups? Test using $\alpha = .05$. (Use the accompanying SAS printout to solve this problem.)

CM	CS	TE	TI
4	6	5	8
7	9	5	4
5	5	7	8
6	7	8	10
8	6	7	3

Source: Data are simulated values based on the group means reported in *Human Factors*, Feb. 1984. Copyright 1984 by the Human Factors Society, Inc. and reproduced by permission.

Analysis of Variance Procedure

Dependent Variable: TRIALS

Source	DF	Sum of Squares	Mean Square	F Value	Pr > F
Model	3	1.20000000	0.40000000	0.10	0.9566
Error	16	61.60000000	3.85000000		
Corrected Total	19	62.80000000			

	R-Square	C.V.	Root MSE	TRIALS Mean
	0.019108	30.658464	1.9621417	6.40000000

Source	DF	Anova SS	Mean Square	F Value	Pr > F
GROUP	3	1.20000000	0.40000000	0.10	0.9566

11.8   One of the selling points of golf balls is their durability. An independent testing laboratory is commissioned to compare the durability of four different brands of golf balls. Balls of each type will be put into a machine that hits the balls with the same force that a golfer does on the course. The number of hits required until the outer covering is cracked is recorded for each ball, with the results given in the table. Ten balls from each manufacturer are randomly selected for testing.

	BRAND		
A	B	C	D
310	261	233	275
235	219	289	290
279	263	301	265
306	247	264	284
237	288	273	239
284	197	208	257
259	207	245	232
273	221	271	251
219	244	298	266
301	228	276	287

a. Use regression to determine whether the mean durabilities of the four brands differ at $\alpha = .05$.

b. Estimate the difference between the mean durability of brands A and C using a 99% confidence interval.

11.9 How likely is a firm to hire a disabled, but rehabilitated, worker? Billions of dollars are spent annually in the United States on training disabled persons for the workplace, yet only 25% of all disabled served by vocational rehabilitation are gainfully employed. A study reported in the *Journal of Rehabilitation* (Apr./May/June 1986) investigated the low placement of disabled persons in the work force. Each employer in a sample of 124 companies located in central Kentucky was asked to rate the employability of an individual with a physical handicap in his or her company, using a 5-point scale (1 = cannot accommodate, 5 = can accommodate quite easily). One goal of the study was to determine whether company size (measured by the number of employees) affects the employability ratings of disabled people. To accomplish this, the companies were divided into three groups according to size—small (1–15 employees), medium (16–55 employees), and large (over 55 employees)—and the mean employability ratings compared.

a. Give the appropriate linear model for this study.

b. Give the appropriate null and alternative hypotheses for this study.

c. A partial ANOVA table for the data is shown here. Complete the table.

SOURCE	df	SS	MS	F
Company size	___	___	___	4.93
Error	___	190	___	
Total	123			

Source: Combs, I. H. and Omvig, C. P. "Accommodation of disabled people into employment: Perceptions of employers." *Journal of Rehabilitation*, Vol. 52, No. 2, Apr./May/June 1986, pp. 42–45.

d. Is there sufficient evidence to indicate that the mean employability ratings of physically handicapped persons differ among the three groups? Test using $\alpha = .05$.

11.10 A study was conducted to determine whether entrepreneurs, newly hired (transferred) managers, and newly promoted managers differ in their risk-taking propensities (*Academy of Management Journal*,

Sept. 1980). For the purposes of this study, entrepreneurs were defined as individuals who, within 3 months before the study, had ceased working for their employers to own and manage business ventures. Thirty-one individuals from each of the three groups were randomly selected to participate in the study. Each was administered a questionnaire that required the respondent to choose between a safe alternative and a more attractive but risky one. Test scores were designed to measure risk-taking propensity. (Lower scores are associated with greater conservatism in risk-taking situations.) The test scores for the three groups are summarized in the table. [*Note:* Although the individual scores are not provided, there is sufficient information in the table to conduct the analysis for this completely randomized design.]

GROUP	SAMPLE SIZE	SAMPLE MEAN	STANDARD DEVIATION	GROUP TOTALS
Entrepreneurs	31	71.00	11.94	2,201
Transferred managers	31	72.52	12.19	2,248
Promoted managers	31	66.97	10.84	2,076
	93			6,525

Source: Brockhaus, R. H. "Risk-taking propensity of entrepreneurs." *Academy of Management Journal*, Sept. 1980, Vol. 23, pp. 509–520.

a. Use the sum of the group totals and the total sample size to compute CM.

b. Use the individual group totals and sample sizes to compute SST.

c. Compute SSE using the pooled sum of squares formula:

$$SSE = \sum (y_{i1} - \bar{y}_1)^2 + \sum (y_{i2} - \bar{y}_2)^2 + \sum (y_{i3} - \bar{y}_3)^2$$
$$= (n_1 - 1)s_1^2 + (n_2 - 1)s_2^2 + (n_3 - 1)s_3^2$$

d. Find SS(Total).

e. Construct an ANOVA table for the data.

f. Do the data provide sufficient evidence to indicate differences in the mean risk-taking propensities among the three groups? Test using $\alpha = .05$.

11.11   Refer to Exercise 11.10

a. Give the linear model appropriate for analyzing the data using regression.

b. Use the information in the table, Exercise 11.10, to find the least squares prediction equation.

11.12   As oil drilling costs rise at unprecedented rates, the task of measuring drilling performance becomes essential to a successful oil company. One method of lowering drilling costs is to increase drilling speed. Researchers at Cities Service Co. have developed a drill bit, called the PD-1, which they believe penetrates rock at a faster rate than any other bit on the market. It is decided to compare the speed of the PD-1 with the two fastest drill bits known, the IADC 1-2-6 and the IADC 5-1-7, at 12 drilling locations in Texas. Four drilling sites were randomly assigned to each bit, and the rate of penetration (RoP) in feet per hour (fph) was recorded after drilling 3,000 feet at each site. The data are given in the table, followed by a Minitab ANOVA printout.

PD-1	IADC 1-2-6	IADC 5-1-7
35.2	25.8	14.7
30.1	29.7	28.9
37.6	26.6	23.3
34.3	30.1	16.2

```
ANALYSIS OF VARIANCE ON ROP
SOURCE      DF        SS        MS        F         p
DRILLBIT     2     366.6     183.3     9.50     0.006
ERROR        9     173.7      19.3
TOTAL       11     540.2
                                    INDIVIDUAL 95 PCT CI'S FOR MEAN
                                    BASED ON POOLED STDEV
LEVEL        N      MEAN      STDEV  --------+---------+---------+--------
  1          4     34.30ʊ     3.127                      (------*------)
  2          4     28.050     2.167              (------*------)
  3          4     20.775     6.589   (------*------)
                                    --------+---------+---------+--------
POOLED STDEV =      4.393            21.0      28.0      35.0
```

a. Based on this information, can Cities Service Co. conclude that the mean RoP differs for at least two of the three drill bits? Test at the $\alpha = .01$ level of significance.

b. Find a 95% confidence interval for μ_1, the mean rate of penetration for the new PD-1 drill bit. Interpret the interval.

c. Find a 95% confidence interval for $(\mu_1 - \mu_2)$, the difference between the mean rates of penetration for the PD-1 and the IADC 1-2-6 drill bits. Which of the two drill bits appears to have the faster mean rate? Explain.

11.13 When marketing its products in a foreign country, should a company use its own salespeople or salespeople from the target market country? To answer this question, a study was designed to investigate the effect of salesperson nationality on buyer attitudes (*Journal of Business Research*, Vol. 22, 1991). A sample of American MBA students was divided into two groups and shown a videotape of an advertisement for forklift trucks made in India. For group 1, an Indian sales representative made the presentation; for group 2, an American sales representative made the presentation. After viewing the tape, the subjects were asked whether the salesperson was trustworthy (measured on a 5-point scale). The mean scores were compared using an ANOVA.

a. The ANOVA resulted in an F value of 2.32, with an observed significance level of .13. Is there evidence of a difference between the mean trustworthiness scores of the two groups of MBA students? Use $\alpha = .10$.

b. The sample mean scores for the two groups are $\bar{y}_1 = 3.12$ and $\bar{y}_2 = 3.49$. Suppose you were to test H_0: $\mu_1 = \mu_2$ against H_a: $\mu_1 < \mu_2$ at $\alpha = .10$. Use the result, part a, to make the proper conclusion. [*Hint*: Use Exercise 4.5 and the fact that the p-value for a two-tailed t test is double the p-value for a one-tailed test.]

11.14 Because of growing concern about problems related to alcohol consumption, the Federal Trade Commission has considered banning alcohol advertising. It is unclear, however, whether alcohol advertising actually increases alcohol consumption. A study reported in the *Journal of Advertising Research* (Oct./Nov. 1985) examined the effect of price advertising on sales of beer in Lower Michigan. The state of Michigan was selected because it has prohibited retailers from advertising the price of beer products since 1975, except for a brief period (March 1982–May 1983) when the ban was temporarily lifted. The data in the table at the top of page 572 are the bimonthly total sales of brewed beverages (in thousands of 31-gallon barrels) over the period May 1981–April 1984. The data allow us to compare total sales of beer in three periods, before (period 1), during (period 2), and after (period 3) the lifting of the price advertising restrictions.

a. Treating this as a completely randomized design, identify the treatments for this experiment.

b. The data were subjected to an ANOVA using Minitab, which generated the printout on page 572. Identify the key elements on the printout.

PERIOD 1: PRICE ADVERTISING RESTRICTED (May/June 1981–Jan./Feb. 1982)	PERIOD 2: NO RESTRICTIONS (Mar./Apr. 1982–May/June 1983)	PERIOD 3: PRICE ADVERTISING RESTRICTED (July/Aug. 1983–Mar./Apr. 1984)
462	522	433
417	508	470
516	427	609
605	477	442
654	603	446
	692	
	584	
	496	

Source: Wilcox, G. B. "The effect of price advertising on alcoholic beverage sales." *Journal of Advertising Research*, Vol. 25, No. 5, Oct./Nov. 1985, pp. 33–37. Copyright © 1985 by the Advertising Research Foundation.

```
ANALYSIS OF VARIANCE ON SALES
SOURCE      DF        SS         MS        F         p
PERIOD       2      11358      5679     0.78     0.476
ERROR       15     109153      7277
TOTAL       17     120510
                                    INDIVIDUAL 95 PCT CI'S FOR MEAN
                                    BASED ON POOLED STDEV
LEVEL       N       MEAN      STDEV   ----+---------+---------+---------+--
  1         5     530.80      98.22              (------------*------------)
  2         8     538.63      83.68           (----------*---------)
  3         5     480.00      73.40   (-------------*------------)
                                    ----+---------+---------+---------+--
POOLED STDEV =     85.30            420       480       540       600
```

c. Is there sufficient evidence to indicate differences in the average total sales of beer in the three periods? Test using $\alpha = .10$.

d. Find a 95% confidence interval for the mean sales of period 2 on the printout and interpret the result.

11.15   Researchers have found that in U.S. businesses, setting specific and difficult goals improves individual output more than no goals, "do best" goals, or easy goals. A study was conducted to investigate whether this goal-setting theory can be generalized to other cultures, especially lesser developed countries (*Journal of Applied Psychology*, 1986). A sample of 92 rural women from a small Eastern Caribbean island took part in the experiment in which they worked at home smocking children's clothing for pay. Workers were free to complete any quantity and were paid on a piece-rate basis. Before beginning work, each subject was randomly assigned to one of three goal groups: Group 1 consisted of $n_1 = 30$ workers with a specific difficult goal—20% above the individual's previous high production; group 2 workers ($n_2 = 27$) were asked to "do their best"; and group 3 workers ($n_3 = 35$) received no goal instructions. The earnings per day were recorded for each worker, and the data

SOURCE	df	SS	MS	F
Groups	2	_____	7.3	_____
Error	89	_____	1.265	_____
Total	91	_____		

Source: Punnett, B. J. "Goal setting: An extension of the research." *Journal of Applied Psychology*, Vol. 71, No. 1, 1986, p. 172.

were subjected to an analysis of variance for a completely randomized design. A partial ANOVA summary table is reproduced here.

a. Fill in the missing entries in the ANOVA table.

b. Is there a significant difference among the mean earnings per day of the three groups? Use $\alpha = .01$.

11.4

Randomized Block Designs

A commonly used noise-reducing design is the randomized block design. Recall (Definition 10.9) that a randomized block design employs groups of homogeneous experimental units (matched as closely as possible) to compare the means of the populations associated with p treatments. The general layout of a randomized block design is shown in Figure 11.7. Note that there are b blocks of relatively homogeneous experimental units. Since each treatment must be represented in each block, the blocks each contain p experimental units. Although Figure 11.7 shows the p treatments in order within the blocks, in practice they would be assigned to the experimental units in random order (thus the name **randomized block design**).

FIGURE 11.7

General form of a randomized block design (treatment is denoted by T_p)

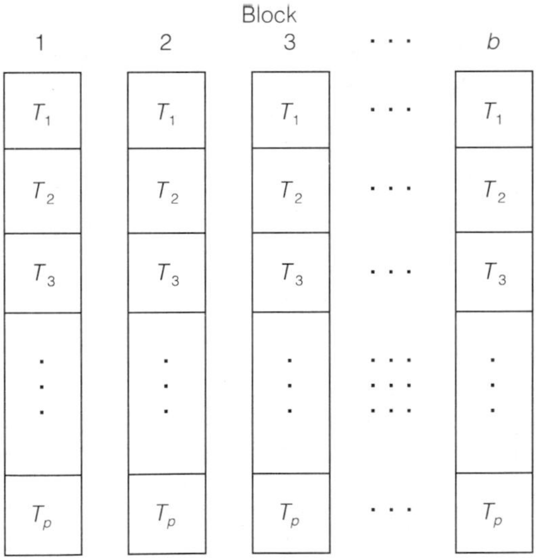

The complete model for a randomized block design contains $(p - 1)$ dummy variables for treatments and $(b - 1)$ dummy variables for blocks. Therefore, the total number of terms in the model, excluding β_0, is $(p - 1) + (b - 1) = p + b - 2$, as shown here.

Complete model:

$$E(y) = \beta_0 + \underbrace{\beta_1 x_1 + \beta_2 x_2 + \cdots + \beta_{p-1} x_{p-1}}_{\text{Treatment effects}} + \underbrace{\beta_p x_p + \cdots + \beta_{p+b-2} x_{p+b-2}}_{\text{Block effects}}$$

where

$$x_1 = \begin{cases} 1 & \text{if treatment 2} \\ 0 & \text{if not} \end{cases}, \quad x_2 = \begin{cases} 1 & \text{if treatment 3} \\ 0 & \text{if not} \end{cases}, \dots, \quad x_{p-1} = \begin{cases} 1 & \text{if treatment } p \\ 0 & \text{if not} \end{cases},$$

$$x_p = \begin{cases} 1 & \text{if block 2} \\ 0 & \text{if not} \end{cases}, \quad x_{p+1} = \begin{cases} 1 & \text{if block 3} \\ 0 & \text{if not} \end{cases}, \dots, \quad x_{p+b-2} = \begin{cases} 1 & \text{if block } b \\ 0 & \text{if not} \end{cases}$$

Note that the model does *not* include terms for treatment–block interaction. The reasons are twofold. First, the addition of these terms would leave 0 degrees of freedom for estimating σ^2. Second, the failure of the mean difference between a pair of treatments to remain the same from block to block is, by definition, experimental error. In other words, in a randomized block design, treatment–block interaction and experimental error are synonymous.

The primary objective of the analysis is to compare the p treatment means, $\mu_1, \mu_2, \dots, \mu_p$. That is, we want to test the null hypothesis

$$H_0: \quad \mu_1 = \mu_2 = \mu_3 = \cdots = \mu_p$$

Recall (Section 10.3) that this is equivalent to testing whether all the treatment parameters in the complete model are equal to 0, i.e.,

$$H_0: \quad \beta_1 = \beta_2 = \cdots = \beta_{p-1} = 0$$

To perform this test using regression, we drop the treatment terms and fit the reduced model:

Reduced model for testing treatments:

$$E(y) = \beta_0 + \underbrace{\beta_p x_p + \beta_{p+1} x_{p+1} + \cdots + \beta_{p+b-2} x_{p+b-2}}_{\text{Block effects}}$$

Then we compare the SSEs for the two models, SSE_R and SSE_C, using the "partial" F statistic:

$$F = \frac{(\text{SSE}_R - \text{SSE}_C)/\text{Number of } \beta\text{'s tested}}{\text{MSE}_C} = \frac{(\text{SSE}_R - \text{SSE}_C)/(p - 1)}{\text{MSE}_C}$$

A significant F value implies that the treatment means differ.

Occasionally, experimenters want to detemine whether blocking was effective in removing the extraneous source of variation, i.e., whether there is evidence of a difference among block means. In fact, if there are no differences among block means, the experimenter will lose information by blocking because blocking reduces the number of degrees of freedom associated with the estimated variance of the model, s^2. If blocking is *not* effective in reducing the variability, then the block parameters in the complete model will all equal 0 (i.e., there will be no differences among block means). Thus, we want to test

$$H_0: \quad \beta_p = \beta_{p+1} = \cdots = \beta_{p+b-2} = 0$$

Another reduced model, with the block β's dropped, is fit:

Reduced model for testing blocks:

$$E(y) = \beta_0 + \underbrace{\beta_1 x_1 + \beta_2 x_2 + \cdots + \beta_{p-1} x_{p-1}}_{\text{Treatment effects}}$$

The SSE for this second reduced model is compared to the SSE for the complete model in the usual fashion. A significant F test implies that blocking is effective in removing (or reducing) the targeted extraneous source of variation.

These two tests are summarized in the following boxes.

Models and ANOVA F Tests for a Randomized Block Design with p Treatments and b Blocks

Complete model:

$$E(y) = \beta_0 + \overbrace{\beta_1 x_1 + \cdots + \beta_{p-1} x_{p-1}}^{(p-1)\text{ treatment terms}} + \overbrace{\beta_p x_p + \cdots + \beta_{p+b-2} x_{p+b-2}}^{(b-1)\text{ block terms}}$$

where

$$x_1 = \begin{cases} 1 & \text{if treatment 2} \\ 0 & \text{if not} \end{cases} \cdots \quad x_{p-1} = \begin{cases} 1 & \text{if treatment } p \\ 0 & \text{if not} \end{cases}$$

$$x_p = \begin{cases} 1 & \text{if block 2} \\ 0 & \text{if not} \end{cases} \cdots x_{p+b-2} = \begin{cases} 1 & \text{if block } b \\ 0 & \text{if not} \end{cases}$$

TEST FOR COMPARING TREATMENT MEANS

H_0: $\beta_1 = \beta_2 = \cdots = \beta_{p-1} = 0$
(i.e., H_0: The p treatment means are equal)

H_a: At least one of the β parameters listed in H_0 differs from 0
(i.e., H_a: At least two treatment means differ)

Reduced model: $E(y) = \beta_0 + \beta_p x_p + \cdots + \beta_{p+b-2} x_{p+b-2}$

Test statistic: $F = \dfrac{(\text{SSE}_R - \text{SSE}_C)/(p-1)}{\text{SSE}_C/(n-p-b+1)}$

$$= \dfrac{(\text{SSE}_R - \text{SSE}_C)/(p-1)}{\text{MSE}_C}$$

where

$\text{SSE}_R = $ SSE for reduced model

$\text{SSE}_C = $ SSE for complete model

$\text{MSE}_C = $ MSE for complete model

Rejection region. $F > F_a$ where F is based on $\nu_1 = (p-1)$ and $\nu_2 = (n-p-b+1)$ degrees of freedom

---

TEST FOR COMPARING BLOCK MEANS

H_0: $\beta_p = \beta_{p+1} = \cdots = \beta_{p+b-2} = 0$
(i.e., H_0: The b block means are equal)

H_a: At least one of the β parameters listed in H_0 differs from 0
(i.e., H_a: At least two block means differ)

Reduced model: $E(y) = \beta_0 + \beta_1 x_1 + \beta_2 x_2 + \cdots + \beta_{p-1} x_{p-1}$

Test statistic: $F = \dfrac{(SSE_R - SSE_C)/(b-1)}{SSE_C/(n-p-b+1)}$

$= \dfrac{(SSE_R - SSE_C)/(b-1)}{MSE_C}$

where

SSE_R = SSE for reduced model

SSE_C = SSE for complete model

MSE_C = MSE for complete model

Rejection region: $F > F_a$ where F is based on $\nu_1 = (b-1)$ and
$\nu_2 = (n-p-b+1)$ degrees of freedom

ASSUMPTIONS:

1. The probability distribution of the difference between any pair of treatment observations within a block is approximately normal.

2. The variance of the difference is constant and the same for all pairs of observations.

EXAMPLE 11.8

Prior to submitting a bid for a construction job, cost engineers prepare a detailed analysis of the estimated labor and materials costs required to complete the job. This estimate will depend on the engineer who performs the analysis. An overly large estimate will reduce the chance of acceptance of a company's bid price, whereas an estimate that is too low will reduce the profit or even cause the company to lose money on the job. A company that employs three job cost engineers wanted to compare the mean level of the engineers' estimates. This was done by having each engineer estimate the cost of the same four jobs. The data (in hundreds of thousands of dollars) are shown in Table 11.4.

a. Perform an analysis of variance on the data, and test to determine whether there is sufficient evidence to indicate differences among treatment means. Test using $\alpha = .05$.

b. Test to determine whether blocking on jobs was successful in reducing the job-to-job variation in the estimates. Use $\alpha = .05$.

T A B L E 11.4 **Data for the Randomized Block Design of Example 11.8**

		JOB				TREATMENT MEANS
		1	2	3	4	
ENGINEER	1	4.6	6.2	5.0	6.6	5.60
	2	4.9	6.3	5.4	6.8	5.85
	3	4.4	5.9	5.4	6.3	5.50
BLOCK MEANS		4.63	6.13	5.27	6.57	

Solution

a. The data for this experiment were collected according to a randomized block design because we would expect estimates of the same job to be more nearly alike than estimates between jobs. Thus, the experiment involves three treatments (engineers) and four blocks (jobs).

The complete model for this design is

$$E(y) = \beta_0 + \underbrace{\beta_1 x_1 + \beta_2 x_2}_{\text{Treatments (engineers)}} + \underbrace{\beta_3 x_3 + \beta_4 x_4 + \beta_5 x_5}_{\text{Blocks (jobs)}}$$

where

$y = $ Cost estimate

$$x_1 = \begin{cases} 1 & \text{if engineer 2} \\ 0 & \text{if not} \end{cases} \qquad x_2 = \begin{cases} 1 & \text{if engineer 3} \\ 0 & \text{if not} \end{cases}$$

Base level = Engineer 1

$$x_3 = \begin{cases} 1 & \text{if block 2} \\ 0 & \text{if not} \end{cases} \qquad x_4 = \begin{cases} 1 & \text{if block 3} \\ 0 & \text{if not} \end{cases} \qquad x_5 = \begin{cases} 1 & \text{if block 4} \\ 0 & \text{if not} \end{cases}$$

Base level = Block 1

The SAS printout for the complete model is shown in Figure 11.8 (page 578). Note that $SSE_C = .18667$ and $MSE_C = .03111$ (shaded on the printout).

To test for differences among the treatment means, we will test

$H_0: \quad \mu_1 = \mu_2 = \mu_3$

where $\mu_i = $ mean cost estimate of engineer i. This is equivalent to testing

$H_0: \quad \beta_1 = \beta_2 = 0$

in the complete model. In order to proceed, we fit the reduced model

$$E(y) = \beta_0 + \underbrace{\beta_3 x_3 + \beta_4 x_4 + \beta_5 x_5}_{\text{Blocks (jobs)}}$$

The SAS printout for this reduced model is shown in Figure 11.9. Note that $SSE_R = .44667$ (shaded on the printout).

FIGURE 11.8

SAS printout for complete model of
Example 11.8

```
Dependent Variable: ESTCOST

                      Analysis of Variance

                              Sum of         Mean
     Source         DF       Squares        Square     F Value     Prob>F

     Model           5       7.02333       1.40467      45.150      0.0001
     Error           6       0.18667       0.03111
     C Total        11       7.21000

            Root MSE        0.17638     R-square     0.9741
            Dep Mean        5.65000     Adj R-sq     0.9525
            C.V.            3.12183

                      Parameter Estimates

                      Parameter      Standard     T for H0:
     Variable   DF     Estimate       Error      Parameter=0     Prob > |T|

     INTERCEP    1     4.583333     0.12472191      36.748         0.0001
     X1          1     0.250000     0.12472191       2.004         0.0919
     X2          1    -0.100000     0.12472191      -0.802         0.4533
     X3          1     1.500000     0.14401646      10.415         0.0001
     X4          1     0.633333     0.14401646       4.398         0.0046
     X5          1     1.933333     0.14401646      13.424         0.0001
```

FIGURE 11.9

SAS printout for reduced model for
testing treatments, Example 11.8

```
Dependent Variable: ESTCOST

                      Analysis of Variance

                              Sum of         Mean
     Source         DF       Squares        Square     F Value     Prob>F

     Model           3       6.76333       2.25444      40.378      0.0001
     Error           8       0.44667       0.05583
     C Total        11       7.21000

            Root MSE        0.23629     R-square     0.9380
            Dep Mean        5.65000     Adj R-sq     0.9148
            C.V.            4.18214

                      Parameter Estimates

                      Parameter      Standard     T for H0:
     Variable   DF     Estimate       Error      Parameter=0     Prob > |T|

     INTERCEP    1     4.633333     0.13642255      33.963         0.0001
     X3          1     1.500000     0.19293062       7.775         0.0001
     X4          1     0.633333     0.19293062       3.283         0.0111
     X5          1     1.933333     0.19293062      10.021         0.0001
```

The remaining elements of the test follow:

Test statistic:

$$F = \frac{(\text{SSE}_R - \text{SSE}_C)/(p-1)}{\text{MSE}_C} = \frac{(.44667 - .18667)/2}{.03111} = 4.18$$

Rejection region: $F > 5.14$, where $F_{.05} = 5.14$ (obtained from Table 4, Appendix D) is based on $\nu_1 = (p-1) = 2$ df and $\nu_2 = (n - p - b + 1) = 6$ df

Conclusion: Since $F = 4.18$ is less than the critical value, 5.14, there is insufficient evidence, at the $\alpha = .05$ level of significance, to indicate differences among the mean estimates for the three cost engineers.

b. To test for the effectiveness of blocking on jobs, we test

$$H_0: \quad \beta_3 = \beta_4 = \beta_5 = 0$$

in the complete model specified in part **a**. The reduced model is

$$E(y) = \beta_0 + \underbrace{\beta_1 x_1 + \beta_2 x_2}_{\text{Treatments (engineers)}}$$

The SAS printout for this second reduced model is shown in Figure 11.10. Note that $SSE_R = 6.95$ (shaded on the printout). The elements of the test follow.

FIGURE 11.10

SAS printout for reduced model for testing blocks, Example 11.8

```
Dependent Variable: ESTCOST

                         Analysis of Variance

                        Sum of          Mean
Source          DF      Squares         Square      F Value      Prob>F

Model           2       0.26000         0.13000      0.168       0.8477
Error           9       6.95000         0.77222
C Total        11       7.21000

        Root MSE        0.87876      R-square      0.0361
        Dep Mean        5.65000      Adj R-sq     -0.1781
        C.V.           15.55331

                        Parameter Estimates

                  Parameter      Standard     T for H0:
Variable   DF     Estimate       Error        Parameter=0     Prob > |T|

INTERCEP   1      5.600000       0.43938088      12.745        0.0001
X1         1      0.250000       0.62137840       0.402        0.6968
X2         1     -0.100000       0.62137840      -0.161        0.8757
```

Test statistic:

$$F = \frac{(SSE_R - SSE_C)/(b-1)}{MSE_C} = \frac{(6.95 - .18667)/3}{.03111} = 72.46$$

Rejection region: $F > 4.76$, where $F_{.05} = 4.76$ (from Table 4, Appendix D) is based on $\nu_1 = (b-1) = 3$ df and $\nu_2 = (n-p-b+1) = 6$ df.

Conclusion: Since $F = 72.46$ exceeds the critical value, 4.76, there is sufficient evidence (at $\alpha = .05$) to indicate differences among the block (job) means. It appears that blocking on jobs was effective in reducing the job-to-job variation in cost estimates.

Caution: The result of the test for the equality of block means must be interpreted with care, especially when the calculated value of the *F* test statistic does not fall in the rejection region. This does not necessarily imply that the block means are the same, i.e., that blocking is unimportant. Reaching this conclusion would be equivalent to accepting the null hypothesis, a practice we have carefully avoided because of the unknown probability of committing a Type II error (that is, of accepting H_0 when H_a is true). In other words, even when a test for block differences is inconclusive, we may still want to use the randomized block design in similar future experiments. If the experimenter believes that the experimental units are more homogeneous within blocks than among blocks, he or she should use the randomized block design regardless of whether the test comparing the block means shows them to be different.

The traditional analysis of variance approach to analyzing the data collected from a randomized block design is similar to the completely randomized design. The partitioning of SS(Total) for the randomized block design is most easily seen by examining Figure 11.11. Note that SS(Total) is now partitioned into *three* parts:

$$SS(Total) = SSB + SST + SSE$$

The formulas for calculating SST and SSB follow the same pattern as the formula for calculating SST for the completely randomized design.

FIGURE 11.11

Partitioning of the total sum of squares for the randomized block design

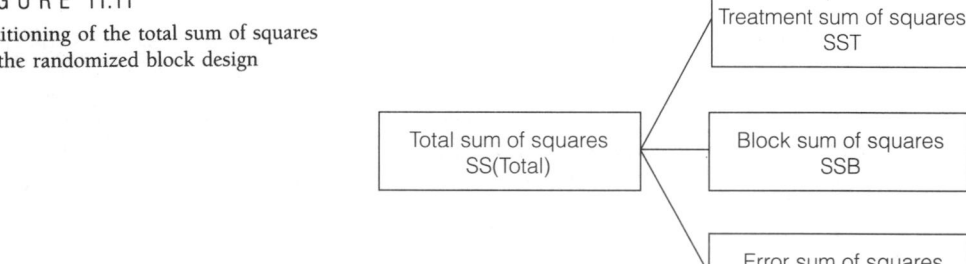

From these quantities, we obtain mean square for treatments, MST, mean square for blocks, MSB, and mean square for error, MSE, as shown in the box. The test statistics are

$$F = \frac{MST}{MSE} \quad \text{for testing treatments}$$

$$F = \frac{MSB}{MSE} \quad \text{for testing blocks}$$

These *F* values are equivalent to the "partial" *F* statistics of the regression approach.

EXAMPLE 11.9

Refer to Example 11.8. Perform an analysis of variance of the data in Table 11.4 using the ANOVA approach.

Solution

Rather than perform the calculations by hand (again, we leave this as an exercise for the student), we utilize a computer software package. The SPSS printout of the ANOVA is displayed in Figure 11.12. The F value for testing treatments, $F = 4.18$, and the F value for testing blocks, $F = 72.46$, are both shaded on the printout. Note that these values are identical to the F values computed using the regression approach, Example 11.8. The p-values of the tests (also shaded) lead to the same conclusions reached in Example 11.8. For example, the p-value for the test of treatment differences, $p = .073$, exceeds $\alpha = .05$; thus, there is insufficient evidence of differences among the treatment means.

FIGURE 11.12

SPSS printout for ANOVA of data in Table 11.4

Source of Variation	Sum of Squares	DF	Mean Square	F	Signif of F
Main Effects	7.023	5	1.405	45.150	.000
ENGINEER	.260	2	.130	4.179	.073
JOB	6.763	3	2.254	72.464	.000
Explained	7.023	5	1.405	45.150	.000
Residual	.187	6	.031		
Total	7.210	11	.655		

As with a completely randomized design, the sources of variation and their respective degrees of freedom, sums of squares, and mean squares for a randomized block design are summarized in an analysis of variance table. The format of an ANOVA table for a randomized block design is shown in Table 11.5; the ANOVA table for the data of Table 11.4 is shown in Table 11.6. (These quantities were obtained from the SPSS printout, Figure 11.12.) Note that the degrees of freedom for the three sources of variation, treatments, blocks, and error, sum to the degrees of freedom for SS(Total). Similarly, the sums of squares for the three sources will always sum to SS(Total).

TABLE 11.5 Analysis of Variance Table for a Randomized Block Design

SOURCE	df	SS	MS	F
Treatments	$p - 1$	SST	MST	MST/MSE
Blocks	$b - 1$	SSB	MSB	MSB/MSE
Error	$n - p - b + 1$	SSE	MSE	
Total	$n - 1$	SS(Total)		

TABLE 11.6 Analysis of Variance Table for Example 11.8

SOURCE	df	SS	MS	F
Treatments (Engineers)	2	.260	.130	4.18
Blocks (Jobs)	3	6.763	2.254	72.46
Error	6	.187	.031	
Total	11	7.210		

Computing Formulas for the Analysis of Variance for a Randomized Block Design

$$\sum_{i=1}^{n} y_i = \text{Sum of all } n \text{ measurements}$$

$$\sum_{i=1}^{n} y_i^2 = \text{Sum of squares of all } n \text{ measurements}$$

$$\text{CM} = \text{Correction for mean}$$

$$= \frac{(\text{Total of all measurements})^2}{\text{Total number of measurements}} = \frac{\left(\sum\limits_{i=1}^{n} y_i\right)^2}{n}$$

$$\text{SS(Total)} = \text{Total sum of squares}$$

$$= (\text{Sum of squares of all measurements}) - \text{CM}$$

$$= \sum_{i=1}^{n} y_i^2 - \text{CM}$$

$$\text{SST} = \text{Sum of squares for treatments}$$

$$= \left(\begin{array}{c}\text{Sum of squares of treatment totals with} \\ \text{each square divided by } b, \text{ the number of} \\ \text{measurements for that treatment}\end{array}\right) - \text{CM}$$

$$= \frac{T_1^2}{b} + \frac{T_2^2}{b} + \cdots + \frac{T_p^2}{b} - \text{CM}$$

$$\text{SSB} = \text{Sum of squares for blocks}$$

$$= \left(\begin{array}{c}\text{Sum of squares for block totals with} \\ \text{each square divided by } p, \text{ the number} \\ \text{of measurements in that block}\end{array}\right) - \text{CM}$$

$$= \frac{B_1^2}{p} + \frac{B_2^2}{p} + \cdots + \frac{B_b^2}{p} - \text{CM}$$

$$\text{SSE} = \text{Sum of squares for error} = \text{SS(Total)} - \text{SST} - \text{SSB}$$

$$\text{MST} = \text{Mean square for treatments} = \frac{\text{SST}}{p - 1}$$

$$\text{MSB} = \text{Mean square for blocks} = \frac{\text{SSB}}{b - 1}$$

$$\text{MSE} = \text{Mean square for error} = \frac{\text{SSE}}{n - p - b + 1}$$

$$F = \frac{\text{MST}}{\text{MSE}} \quad \text{for testing treatments}$$

$$F = \frac{\text{MSB}}{\text{MSE}} \quad \text{for testing blocks}$$

Confidence intervals for the difference between a pair of treatment means or block means for a randomized block design are shown in the accompanying box.

Confidence Intervals for the Difference ($\mu_i - \mu_j$) Between a Pair of Treatment Means or Block Means

Treatment means: $(\overline{T}_i - \overline{T}_j) \pm t_{\alpha/2}s\sqrt{\dfrac{2}{b}}$

Block means: $(\overline{B}_i - \overline{B}_j) \pm t_{\alpha/2}s\sqrt{\dfrac{2}{p}}$

where

b = Number of blocks

p = Number of treatments

$s = \sqrt{\text{MSE}}$

$\overline{T}_i$ = Sample mean for treatment i

$\overline{B}_i$ = Sample mean for block i

and $t_{\alpha/2}$ is based on $(n - b - p + 1)$ degrees of freedom

---

EXAMPLE 11.10

Refer to Example 11.8. Find a 90% confidence interval for the difference between the mean level of estimates for engineers 1 and 2.

Solution

From Example 11.8, we know that $b = 4$, $\overline{T}_1 = 5.60$, $\overline{T}_2 = 5.85$, and $s^2 = \text{MSE}_C = .03111$. The degrees of freedom associated with s^2 (and, therefore, with $t_{\alpha/2}$) is 6. Therefore, $s = \sqrt{s^2} = \sqrt{.03111} = .176$ and $t_{\alpha/2} = t_{.05} = 1.943$. Substituting these values into the formula for the confidence interval for $(\mu_1 - \mu_2)$, we obtain

$$(\overline{T}_1 - \overline{T}_2) \pm t_{\alpha/2}s\sqrt{\frac{2}{b}}$$

$$(5.60 - 5.85) \pm 1.943(.176)\sqrt{\frac{2}{4}}$$

$$-.25 \pm .24$$

or, $-.49$ to $-.01$. Since each unit represents \$100,000, we estimate the difference between the mean level of job estimates for estimators 1 and 2 to be enclosed by the interval, $-\$49,000$ to $-\$1,000$. [*Note:* At first glance, this result may appear to contradict the result of the F test for comparing treatment means. However, the observed significance level of the F test (.07) implies that significant differences exist between the means at $\alpha = .10$, which is consistent with the fact that 0 is not within the 90% confidence interval.]

There is one very important point to note when you block the treatments in an experiment. Recall from Section 10.3 that the block effects cancel. This fact enables us to calculate confidence intervals for the difference between treatment means using the formulas given in the box. But, if a sample treatment mean is used to estimate a *single treatment mean*, the block effects do not cancel. *Therefore, the only way that you can obtain an unbiased estimate of a single treatment mean (and corresponding confidence interval) in a blocked design is to randomly select the blocks from a large collection (population) of blocks and to treat the block effect as a second random component, in addition to random error.* Designs that contain two or more random components are called *nested designs* and are beyond the scope of this text. For more information on this topic, see the references at the end of this chapter.

EXERCISES

11.16 *Physical Therapy* (Aug. 1986) reported on a study to "determine whether the medial rotation that accompanies flexion of the shoulder took place during the performance of the flexion-abduction-lateral-rotation proprioceptive neuromuscular facilitation pattern (D_2F)." Ten college students, who exhibited no evidence of disease or limitation of movement in their shoulders, served as the subjects for the study. For each subject, the medial rotation was measured (in degrees) at each of three positions in the D_2F pattern: (1) beginning position, (2) point at which rotation changed directions, and (3) ending position. The goal of the analysis is to compare the mean medial rotation measurements of the three positions.

   a. Identify the treatments in this experiment.

   b. Identify the blocks in this experiment.

   c. Identify the response variable.

11.17 Refer to Exercise 11.16. Explain why a randomized block design is appropriate for this experiment.

11.18 The analysis of variance for a randomized block design produced the ANOVA table entries shown here.

SOURCE	df	SS	MS	F
Treatments	3	27.1	_____	_____
Blocks	5	_____	14.90	_____
Error	_____	33.4	_____	
Total	_____			

The sample means for the four treatments are as follows:

$$\bar{y}_A = 9.7 \qquad \bar{y}_B = 12.1 \qquad \bar{y}_C = 6.2 \qquad \bar{y}_D = 9.3$$

   a. Complete the ANOVA table.

   b. Do the data provide sufficient evidence to indicate a difference among the treatment means? Test using $\alpha = .01$.

   c. Do the data provide sufficient evidence to indicate that blocking was a useful design strategy to employ for this experiment? Explain.

d. Find a 95% confidence interval for $(\mu_A - \mu_B)$.

e. Find a 95% confidence interval for $(\mu_B - \mu_D)$.

11.19 A recent supermarket advertisement states: "Winn Dixie offers you the lowest total food bill! Here's the proof!" The "proof" (shown below) is a side-by-side listing of the prices of 60 grocery items purchased at Winn Dixie and two other supermarkets, Publix and Kash 'N Karry, on the same day.

ITEM	WINN-DIXIE	PUBLIX	KASH N' KARRY	ITEM	WINN-DIXIE	PUBLIX	KASH N' KARRY
Big Thirst Towel	1.21	1.49	1.59	Keb Graham Crust	.79	1.29	1.28
Camp Crm/Broccoli	.55	.67	.67	Spiffits Glass	1.98	2.19	2.59
Royal Oak Charcoal	2.99	3.59	3.39	Prog Lentil Soup	.79	1.13	1.12
Combo Chdr/Chz Snk	1.29	1.29	1.39	Lipton Tea Bags	2.07	2.17	2.17
3ure 3ak Trash Bag	1.29	1.79	1.89	Carnation Hot Coco	1.59	1.89	1.99
Dow Handi Wrap	1.59	2.39	2.29	Crystal Hot Sauce	.70	.87	.89
White Rain Shampoo	.96	.97	1.39	C/F/N/ Coffee Bag	1.17	1.15	1.55
Post Golden Crisp	2.78	2.99	3.35	Soup Start Bf Veg	1.39	2.03	1.94
Surf Detergent	2.29	1.89	1.89	Camp Pork & Beans	.44	.49	.58
Sacramento T/Juice	.79	.89	.99	Sunsweet Pit Prune	.98	1.33	1.10
SS Prune Juice	1.36	1.61	1.48	DM Vgcls Grdn Duet	1.07	1.13	1.29
V-8 Cocktail	1.18	1.29	1.28	Argo Corn Starch	.69	.89	.79
Rodd Kosher Dill	1.39	1.79	1.79	Sno Drop Bowl Clnr	.53	1.15	.99
Bisquick	2.09	2.19	2.09	Cadbury Milk Choc	.79	1.29	1.28
Kraft Italian Drs	.99	1.19	1.00	Andes Crm/De Ment	1.09	1.30	1.09
BC Hamburger Helper	1.46	1.75	1.75	Combat Ant & Roach	2.33	2.39	2.79
Comstock Chrry Pie	1.29	1.69	1.69	Joan/Arc Kid Bean	.45	.56	.38
Dawn Liquid King	2.59	2.29	2.58	La Vic Salsa Pican	1.22	1.75	1.49
DelMonte Ketchup	1.05	1.25	.59	Moist N Beef/Chz	2.39	3.19	2.99
Silver Floss Kraut	.77	.81	.69	Ortega Taco Shells	1.08	1.33	1.09
Trop Twist Beverag	1.74	2.15	2.25	Fresh Step Cat Lit	3.58	3.79	3.81
Purina Kitten Chow	1.09	1.05	1.29	Field Trial Dg/Fd	3.49	3.79	3.49
Niag Spray Starch	.89	.99	1.39	Tylenol Tablets	5.98	5.29	5.98
Soft Soap Country	.97	1.19	1.19	Rolaids Tablets	1.88	2.20	2.49
Northwood Syrup	1.13	1.37	1.37	Plax Rinse	2.88	3.14	2.53
Bumble Bee Tuna	.58	.65	.65	Correctol Laxative	3.44	3.98	3.59
Mueller Elbow/Mac	2.09	2.69	2.69	Tch Scnt Potpourri	1.50	1.89	1.89
Kell Nut Honey Crn	2.95	3.25	3.23	Chld Enema 2.250	.98	1.15	1.19
Cutter Spray	3.09	3.95	3.69	Gillette Atra Plus	5.00	5.24	5.59
Lawry Season Salt	2.28	2.97	2.85	Colgate Shave	.94	1.10	1.19

a. Suppose we want to use the data to compare the mean prices of grocery items at the three supermarkets. Identify the treatments and blocks for this randomized block design.

b. The data were subjected to an ANOVA using SAS. Use the information in the SAS printout at the top of page 586 to construct an ANOVA table for the data.

c. Test to determine whether the mean prices of grocery items differ among the three supermarkets. Use $\alpha = .01$.

d. Construct a 95% confidence interval for the difference between the mean prices per item at Winn-Dixie and Kash N' Karry.

e. Does the interval obtained in part **d** provide evidence of a significant difference in mean prices per item between Winn-Dixie and Kash N' Karry?

f. Test to determine whether blocking on grocery items was effective in reducing an extraneous source of variation.

SAS printout for Exercise 11.19

```
                      Analysis of Variance Procedure
Dependent Variable: PRICE
                                   Sum of          Mean
Source                DF          Squares        Square    F Value    Pr > F

Model                 61      218.2361989     3.5776426     106.27    0.0001
Error                118        3.9725322     0.0336655
Corrected Total      179      222.2087311

               R-Square            C.V.       Root MSE          PRICE Mean

               0.982123        9.989324       0.183482         1.83677778

Source                DF         Anova SS   Mean Square    F Value    Pr > F

SUPERMKT               2        2.6412678     1.3206339      39.23    0.0001
ITEM                  59      215.5949311     3.6541514     108.54    0.0001


               Level of        ------------PRICE------------
               SUPERMKT   N        Mean              SD

               KNKarry   60    1.92533333      1.14118503
               Publix    60    1.91950000      1.10339142
               WinnDix   60    1.66550000      1.09622376
```

11.20   Refer to Exercise 11.19.
   a. Give the complete model appropriate for analyzing the data.
   b. Give the reduced model for testing for differences among the mean prices at the three supermarkets.
   c. Fit the two models, parts **a** and **b**, using regression. Conduct the test for treatments. Compare the F value to the one shown on the SAS printout, Exercise 11.19.

11.21   Research into the human-engineering aspect of computing has grown tremendously because of the ever-increasing number of computer users. The end goal is to create "user-friendly" hardware and software that reduce as much as possible any form of human strain or stress resulting from computer usage. A recent topic of concern to computer terminal users is the color of the characters displayed on the video display screens. Early full-screen video display terminals presented the viewer with white characters on a black background. Initially, viewers found the high degree of contrast easy on the eyes. However, after an extended period of use, black and white displays were frequently found to cause temporary eye irritations. In the past few years, experimentation with other colors revealed that yellow/amber displays may be the easiest on the eyes. In one German study, video display

SUBJECT	GREEN/ BLACK	WHITE/ BLACK	YELLOW/ WHITE	ORANGE/ WHITE	YELLOW	YELLOW/ AMBER	YELLOW/ ORANGE
1	7	6	7	2	8	9	3
2	8	6	9	4	9	8	1
3	5	5	7	1	6	8	2
4	3	4	2	0	2	6	0
5	9	8	8	3	9	9	2
6	7	5	6	2	7	7	1
7	6	7	8	4	6	9	5
8	6	5	8	1	8	9	1
9	9	9	8	2	9	8	0
10	9	8	8	3	9	10	1

Source: Adapted from Solomon, L. and Burawa, A. "Maximize Your Computing Comfort and Efficiency." *Computers & Electronics,* Apr. 1983, pp. 35–40.

terminals were produced with white/black and six different symbol colors. Thirty test subjects were asked to specify which color combination they preferred by ranking each of the seven color combinations on a scale from 0 (no preference) to 10. Although the raw data were not revealed, the mean preference scores for each color were provided by the researchers. Based on these means, we have simulated the individual preference scores for 10 subjects in the accompanying table. The data were subjected to an ANOVA for a randomized block design using Minitab; the results are shown in the accompanying printout. Do the data provide sufficient evidence of a difference among the mean preference scores for the seven video display color combinations? Test using $\alpha = .05$.

```
ANALYSIS OF VARIANCE   SCORE

SOURCE        DF        SS        MS
COLOR          6      421.34     70.22
SUBJECT        9      114.30     12.70
ERROR         54       71.80      1.33
TOTAL         69      607.44
```

11.22   Refer to the study on the effect of price advertising on beer sales in Lower Michigan, described in Exercise 11.14. In that exercise you conducted the analysis as a completely randomized design. The fact that the experimental unit is a month (actually, a pair of successive months since we are measuring total bimonthly beer sales) could introduce an unwanted source of variation into the analysis—namely, the month-to-month variation in beer sales.

a. Explain how to set up a randomized block design to reduce this unwanted source of variation.

b. The data of Exercise 11.14 are reorganized in the accompanying table. Notice that the months in which the beer sales are recorded are now identified for each period. A randomized block ANOVA was conducted on only the data for those months that appear across all three periods. (Why is this necessary?) The Minitab printout is shown here. Construct an ANOVA summary table.

c. Is there sufficient evidence to indicate that the mean total bimonthly beer sales differ among the three periods? Test using $\alpha = .10$.

MONTHS	PERIOD 1	PERIOD 2	PERIOD 3
Jan./Feb.	654	692	442
Mar./Apr.	—	522, 584	446
May/June	462	508, 496	—
July/Aug.	417	427	433
Sept./Oct.	516	477	470
Nov./Dec.	605	603	609

Source: Wilcox, G. B. "The effect of price advertising on alcoholic beverage sales." *Journal of Advertising Research*, Vol. 25, No. 5, Oct./Nov. 1985, pp. 33–37.

```
ANALYSIS OF VARIANCE   SALES

SOURCE        DF        SS        MS
PERIOD         2       9726      4863
MONTH          3      68258     22753
ERROR          6      27947      4658
TOTAL         11     105932
```

    d. Calculate the value of *s* for this experiment and compare it to the value for the completely randomized design of Exercise 11.14. Does it appear that blocking on months was effective in reducing the month-to-month variation in beer sales?

    e. Conduct a test to determine whether blocking was effective in reducing month-to-month variation in beer sales. Use $\alpha = .10$. Does the result support your answer to part **d**?

11.23 The "in-tray" exercise aids in assessing the management potential of future administrators and executives. Developed 40 years ago as a training tool for officers in the U.S. Air Force, the in-tray or in-basket exercise simulates the typical contents of an executive's in-tray with a variety of everyday problems in a written form—letters, memoranda, notes, reports, and telephone messages—requiring decisions and action. Trainees are provided with instructions, information on the company, its organization, and the role to be played, and the in-tray contents, and are allotted a fixed amount of time to complete the tasks. After the tasks are completed, the trainees' performances are assessed by one or more expert raters, usually on a scale of 1 (high performance) to 6 (low performance). However, the reliability of the assessors' ratings should be determined before using the in-tray measure of managerial effectiveness. The phenomenon of rater reliability was investigated in the *Journal of Occupational Psychology*. Seven subjects, all candidates for a general management position in a manufacturing company in the British motor industry, were given the in-tray test. Overall in-tray performance of each candidate was assessed by three different raters. The results are given in the accompanying table. Note that each of the three raters judged the overall performance of all seven candidates. Thus, the candidates represent blocks, and the sampling design used is a randomized block design.

CANDIDATE	RATER 1	RATER 2	RATER 3
A	4.5	4.5	5.0
B	2.5	4.5	4.5
C	5.0	3.0	4.0
D	4.0	4.5	4.5
E	1.5	2.0	4.5
F	3.5	4.5	4.5
G	4.0	4.0	4.0

Source: Gill, R. W. T. "The in-tray (in-basket) exercise as a measure of management potential." *Journal of Occupational Psychology*, 1979, Vol. 52, pp. 185–195.

    a. Give the complete model appropriate for this design.

    b. Give the reduced model appropriate for testing for differences in the mean scores given by the three raters.

    c. Give the reduced model appropriate for determining whether blocking by candidates was effective in removing an unwanted source of variability.

11.24 Refer to Exercise 11.23. The models of parts **a**, **b**, and **c** were fit to data in the table using Minitab. The Minitab printouts are displayed on page 589.

    a. Construct an ANOVA summary table.

    b. Is there evidence of a difference among the mean performance scores assessed by the three raters? Use $\alpha = .05$.

    c. Is there evidence of a difference among the mean performance scores of the candidates? That is, is there evidence that blocking by candidates was effective in removing an unwanted source of variability? Use $\alpha = .05$.

Minitab printout for complete model,
Exercise 11.23a

```
The regression equation is
SCORE = 4.48 - 0.857 R1 - 0.571 R2
            + 0.667 C1 - 0.167 C2 + 0.000 C3 + 0.333 C4 - 1.33 C5 + 0.167 C6

Predictor        Coef        Stdev      t-ratio          p
Constant       4.4762       0.5401         8.29      0.000
R1            -0.8571       0.4410        -1.94      0.076
R2            -0.5714       0.4410        -1.30      0.219
C1             0.6667       0.6736         0.99      0.342
C2            -0.1667       0.6736        -0.25      0.809
C3             0.0000       0.6736         0.00      1.000
C4             0.3333       0.6736         0.49      0.630
C5            -1.3333       0.6736        -1.98      0.071
C6             0.1667       0.6736         0.25      0.809

s = 0.8250      R-sq = 54.5%      R-sq(adj) = 24.2%

Analysis of Variance

SOURCE          DF          SS           MS           F          p
Regression       8      9.7857       1.2232        1.80      0.174
Error           12      8.1667       0.6806
Total           20     17.9524
```

Minitab printout for reduced model,
Exercise 11.23b

```
The regression equation is
SCORE = 4.00 + 0.667 C1 - 0.167 C2 + 0.000 C3 + 0.333 C4 - 1.33 C5 + 0.167 C6

Predictor        Coef        Stdev      t-ratio          p
Constant       4.0000       0.5079         7.88      0.000
C1             0.6667       0.7182         0.93      0.369
C2            -0.1667       0.7182        -0.23      0.820
C3             0.0000       0.7182         0.00      1.000
C4             0.3333       0.7182         0.46      0.650
C5            -1.3333       0.7182        -1.86      0.085
C6             0.1667       0.7182         0.23      0.820

s = 0.8797      R-sq = 39.7%      R-sq(adj) = 13.8%

Analysis of Variance

SOURCE          DF          SS           MS           F          p
Regression       6      7.1190       1.1865        1.53      0.238
Error           14     10.8333       0.7738
Total           20     17.9524
```

Minitab printout for reduced model,
Exercise 11.23c

```
The regression equation is
SCORE = 4.43 - 0.857 R1 - 0.571 R2

Predictor        Coef        Stdev      t-ratio          p
Constant       4.4286       0.3483        12.71      0.000
R1            -0.8571       0.4926        -1.74      0.099
R2            -0.5714       0.4926        -1.16      0.261

s = 0.9215      R-sq = 14.9%      R-sq(adj) = 5.4%

Analysis of Variance

SOURCE          DF          SS           MS           F          p
Regression       2      2.6667       1.3333        1.57      0.235
Error           18     15.2857       0.8492
Total           20     17.9524
```

11.25   The traditional retail store audit is one of the most widely used marketing research tools among consumer package goods companies. It involves periodic audits of a sample of retail audits to monitor inventory and purchases of a particular product. V. K. Prasad, W. R. Casper, and R. J. Schieffer conducted a study to compare market data yielded by retail store audits with an alternative, less costly auditing procedure—weekend selldown audits and store purchases audits (*Journal of Marketing*, Winter 1984). The market shares of six major brands of beer distributed in eastern cities were estimated using each of the three store audit methods. The data are provided in the accompanying table, followed by a Minitab printout of the analysis of variance.

BRAND	TRADITIONAL STORE AUDIT	WEEKEND SELLDOWN AUDIT	STORE PURCHASES AUDIT
1	18.0	19.0	20.7
2	15.3	17.3	14.0
3	8.9	8.5	10.1
4	6.5	4.9	6.1
5	5.3	6.1	4.6
6	3.4	3.0	3.1

Source: Prasad, V. K., Casper, W. R., and Schieffer, R. J. "Alternatives to the traditional retail store audit: A field study." *Journal of Marketing*, Winter 1984, 48, pp. 54–61. Reprinted by permission of the American Marketing Association.

```
ANALYSIS OF VARIANCE   SHARE

SOURCE         DF        SS         MS
METHOD          2      0.19       0.10
BRAND           5    605.70     121.14
ERROR          10     13.05       1.30
TOTAL          17    618.94
```

a. Use the information in the Minitab printout to construct an ANOVA summary table for the data.
b. Verify the entries in the table, part a, using the ANOVA formulas.
c. Is there sufficient evidence to indicate a difference in the mean estimates of beer-brand market shares produced by the three auditing methods? Test using $\alpha = .05$.
d. Estimate the difference between the mean estimates of beer-brand market shares produced by the traditional store audit and the weekend selldown audit using a 95% confidence interval.

11.5

Two-Factor Factorial Experiments

In Section 10.4, we learned that factorial experiments are volume-increasing designs conducted to investigate the effect of two or more independent variables (factors) on the mean value of the response y. In this section, we focus on the analysis of two-factor factorial experiments.

For example, suppose we want to relate the mean number of defects on a finished item—say, a new desk top—to two factors, type of nozzle for the varnish spray gun and length of spraying time. Suppose further that we want to investigate the mean number of defects per desk for three types (three levels) of nozzles (N_1, N_2, and N_3) and for two lengths (two levels) of spraying time (S_1 and S_2). If we choose the treatments for the experiment to include all combinations of the three levels of nozzle type with the two levels of spraying time, i.e., we observe the number of defects for the factor level combinations $N_1S_1, N_1S_2, N_2S_1, N_2S_2, N_3S_1, N_3S_2$, our design is called a **complete 3×2 factorial experiment**. Note that the design will contain $3 \times 2 = 6$ treatments.

Factorial experiments, you will recall, are useful methods for selecting treatments because they permit us to make inferences about factor interactions. The complete model for the 3×2 factorial experiment contains $(3 - 1) = 2$ main effect terms for nozzles, $(2 - 1) = 1$ main effect term for spray time, and $(3 - 1)(2 - 1) = 2$ nozzle–spray time interaction terms:

$$E(y) = \beta_0 + \underbrace{\beta_1 x_1 + \beta_2 x_2}_{\substack{\text{Main effects} \\ \text{Nozzle}}} + \underbrace{\beta_3 x_3}_{\substack{\text{Main effect} \\ \text{Spray time}}} + \underbrace{\beta_4 x_1 x_2 + \beta_5 x_1 x_3}_{\substack{\text{Interaction} \\ \text{Nozzle} \times \text{Spray time}}}$$

The independent variables (factors) in the model can be either quantitative or qualitative. If they are quantitative, the main effects are represented by terms such as x, x^2, x^3, etc.; if qualitative, the main effects are represented by dummy variables. In our 3×2 factorial experiment, nozzle type is qualitative and spraying time is quantitative; hence, the x variables in the model are defined as follows:

$$x_1 = \begin{cases} 1 & \text{if nozzle } N_1 \\ 0 & \text{if not} \end{cases} \quad x_2 = \begin{cases} 1 & \text{if nozzle } N_2 \\ 0 & \text{if not} \end{cases} \quad \text{Base level} = N_3$$

$$x_3 = \text{Length of spraying time (in minutes)}$$

Note that the model for the 3×2 factorial contains a total of $3 \times 2 = 6$ β-parameters. If we observe only a single value of the response y for each of the $3 \times 2 = 6$ treatments, then $n = 6$ and df(Error) for the complete model is $(n - 6) = 0$. Consequently, for a factorial experiment, *the number r of observations per factor level combination (i.e., the number of replications of the factorial experiment) must always be 2 or more.* Otherwise, no degrees of freedom are available for estimating σ^2.

To test for factor interaction, we drop the interaction terms and fit the reduced model:

$$E(y) = \beta_0 + \underbrace{\beta_1 x_1 + \beta_2 x_2}_{\substack{\text{Main effect} \\ \text{Nozzle}}} + \underbrace{\beta_3 x_3}_{\substack{\text{Main effect} \\ \text{Spray time}}}$$

The null hypothesis of no interaction, $H_0: \beta_4 = \beta_5 = 0$, is tested by comparing the SSEs for the two models in a partial F statistic. This test for interaction is summarized, in general, in the box on pages 592–593.

Tests for factor main effects are conducted in a similar manner. The main effect terms of interest are dropped from the complete model and the reduced model is fit. The SSEs for the two models are compared in the usual fashion.

Before we work through a numerical example of an analysis of variance for a factorial experiment, we need to understand the practical significance of the tests for factor interaction and factor main effects. We illustrate these concepts in Example 11.11.

Models and ANOVA F Test for Interaction in a Two-Factor Factorial Experiment with Factor A at a Levels and Factor B at b Levels

Complete model:

$$E(y) = \overbrace{\beta_0 + \beta_1 x_1 + \cdots + \beta_{a-1} x_{a-1}}^{\text{Main effect } A \text{ terms}} + \overbrace{\beta_a x_a + \cdots + \beta_{a+b-2} x_{a+b-2}}^{\text{Main effect } B \text{ terms}}$$

$$+ \overbrace{\beta_{a+b-1} x_1 x_a + \beta_{a+b} x_1 x_{a+1} + \cdots + \beta_{ab-1} x_{a-1} x_{a+b-2}}^{AB \text{ interaction terms}}$$

where*

$$x_1 = \begin{cases} 1 & \text{if level 2 of factor } A \\ 0 & \text{if not} \end{cases} \cdots$$

$$x_{a-1} = \begin{cases} 1 & \text{if level } a \text{ of factor } A \\ 0 & \text{if not} \end{cases}$$

$$x_a = \begin{cases} 1 & \text{if level 2 of factor } B \\ 0 & \text{if not} \end{cases} \cdots$$

$$x_{a+b-2} = \begin{cases} 1 & \text{if level } b \text{ of factor } B \\ 0 & \text{if not} \end{cases}$$

H_0: $\beta_{a+b-1} = \beta_{a+b} = \cdots = \beta_{ab-1} = 0$
(i.e., H_0: No interaction between factors A and B)

H_a: At least one of the β parameters listed in H_0 differs from 0
(i.e., H_a: Factors A and B interact)

Reduced model:

$$E(y) = \overbrace{\beta_0 + \beta_1 x_1 + \cdots + \beta_{a-1} x_{a-1}}^{\text{Main effect } A \text{ terms}} + \overbrace{\beta_a x_a + \cdots + \beta_{a+b-2} x_{a+b-2}}^{\text{Main effect } B \text{ terms}}$$

Test statistic: $F = \dfrac{(SSE_R - SSE_C)/[(a-1)(b-1)]}{SSE_C/[ab(r-1)]}$

$$= \dfrac{(SSE_R - SSE_C)/[(a-1)(b-1)]}{MSE_C}$$

where

SSE_R = SSE for reduced model

SSE_C = SSE for complete model

MSE_C = MSE for complete model

r = Number of replications (i.e., number of y measurements per cell of the $a \times b$ factorial)

(continued)

Rejection region: $F > F_a$, where F is based on $\nu_1 = (a-1)(b-1)$ and $\nu_2 = ab(r-1)$ df

Assumptions:
1. The population probability distribution of the observations for any factor level combination is approximately normal.
2. The variance of the probability distribution is constant and the same for all factor level combinations.

*Note: The independent variables, $x_1, x_2, \ldots, x_{a+b-2}$, are defined for an experiment in which both factors represent *qualitative* variables. When a factor is *quantitative*, you may choose to represent the main effects with quantitative terms such as x, x^2, x^3, and so forth.

EXAMPLE 11.11

A company that stamps gaskets out of sheets of rubber, plastic, and other materials, wants to compare the mean number of gaskets produced per hour for two different types of stamping machines. Practically, the manufacturer wants to determine whether one machine is more productive than the other and, even more important, whether one machine is more productive in making rubber gaskets while the other is more productive in making plastic gaskets. To answer these questions, the manufacturer decides to conduct a 2×3 factorial experiment using three types of gasket material, B_1, B_2, and B_3, with each of the two types of stamping machines, A_1 and A_2. Each machine is operated for three 1-hour time periods for each of the gasket materials, with the eighteen 1-hour time periods assigned to the six machine–material combinations in random order. (The purpose of the randomization is to eliminate the possibility that uncontrolled environmental factors might bias the results.) Suppose we have calculated and plotted the six treatment means. Two hypothetical plots of the six means are shown in Figures 11.13a and 11.13b (page 594). The three means for stamping machine A_1 are connected by solid line segments and the corresponding three means for machine A_2 by dashed line segments. What do these plots imply about the productivity of the two stamping machines?

Solution

Figure 11.13a suggests that machine A_1 produces a larger number of gaskets per hour, regardless of the gasket material, and is therefore superior to machine A_2. On the average, machine A_1 stamps more cork (B_1) gaskets per hour than rubber or plastic, but the *difference* in the mean numbers of gaskets produced by the two machines remains approximately the same, regardless of the gasket material. Thus, the difference in the mean number of gaskets produced by the two machines is *independent* of the gasket material used in the stamping process.

In contrast to Figure 11.13a, Figure 11.13b shows the productivity for machine A_1 to be greater than that for machine A_2, when the gasket material is cork (B_1) or plastic (B_3). But the means are reversed for rubber (B_2) gasket material. For this material, machine A_2 produces, on the average, more gaskets per hour than

FIGURE 11.13

Hypothetical plot of the means for the six machine–material combinations

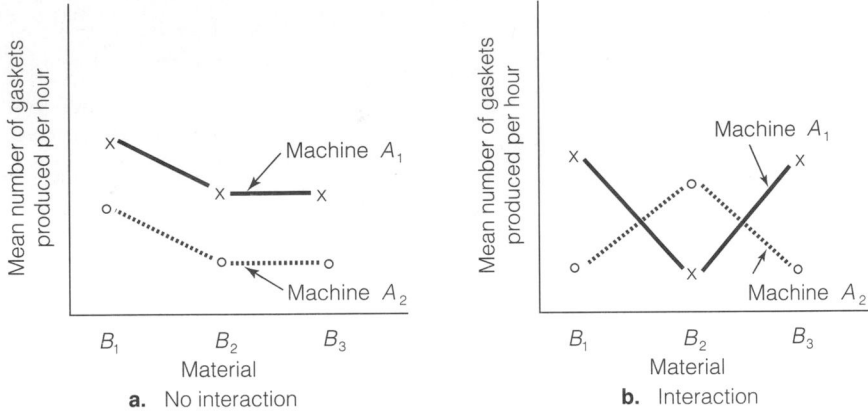

a. No interaction

b. Interaction

machine A_1. Thus, Figure 11.13b illustrates a situation where the mean value of the response variable *depends* on the combination of the factor levels. When this situation occurs, we say that the factors *interact*. Thus, one of the most important objectives of a factorial experiment is to detect factor interaction if it exists.

Definition 11.2

In a factorial experiment, when the difference in the mean levels of factor A depends on the different levels of factor B, we say that the factors A and B **interact**. If the difference is independent of the levels of B, then there is **no interaction** between A and B.

Tests for main effects are relevant only when no interaction exists between factors. Generally, the test for interaction is performed first. *If there is evidence of factor interaction, then we will not perform the tests on the main effects.* Rather, we will want to focus attention on the individual cell (treatment) means, perhaps locating one that is the largest or the smallest.

EXAMPLE 11.12

A manufacturer, whose daily supply of raw materials is variable and limited, can use the material to produce two different products in various proportions. The profit per unit of raw material obtained by producing each of the two products depends on the length of a product's manufacturing run and, hence, on the amount of raw material assigned to it. Other factors, such as worker productivity and machine breakdown, affect the profit per unit as well, but their net effect on profit is random and uncontrollable. The manufacturer has conducted an experiment to investigate the effect of the level of supply of raw materials, S, and the ratio of its assignment, R, to the two product manufacturing lines on the profit

y per unit of raw material. The ultimate goal would be to be able to choose the best ratio R to match each day's supply of raw materials, S. The levels of supply of the raw material chosen for the experiment were 15, 18, and 21 tons; the levels of the ratio of allocation to the two product lines were $\frac{1}{2}$, 1, and 2. The response was the profit (in dollars) per unit of raw material supply obtained from a single day's production. Three replications of a complete 3×3 factorial experiment were conducted in a random sequence (i.e., a completely randomized design). The data for the 27 days are shown in Table 11.7.

TABLE 11.7 **Data for Example 11.12**

| | | RAW MATERIAL SUPPLY, TONS (S) | | |
		15	18	21
RATIO OF	$\frac{1}{2}$	23, 20, 21	22, 19, 20	19, 18, 21
RAW MATERIAL	1	22, 20, 19	24, 25, 22	20, 19, 22
ALLOCATION (R)	2	18, 18, 16	21, 23, 20	20, 22, 24

a. Write the complete model for the experiment.
b. Do the data present sufficient evidence to indicate an interaction between supply S and ratio R? Use $\alpha = .05$.
c. Based on the result, part b, should we perform tests for main effects?

Solution

a. Both factors, supply and ratio, are quantitative. Accordingly, when the factors in a factorial experiment are quantitative, the main effects can be represented by terms such as x, x^2, x^3, and so forth. Since each factor has three levels, we require two main effects, x and x^2, for each factor. (In general, the number of main effect terms will be one less than the number of levels for a factor.) Consequently, the complete factorial model for this 3×3 factorial experiment is

$$y = \beta_0 + \underbrace{\beta_1 x_1 + \beta_2 x_1^2}_{\text{Supply main effects}} + \underbrace{\beta_3 x_2 + \beta_4 x_2^2}_{\text{Ratio main effects}}$$
$$+ \underbrace{\beta_5 x_1 x_2 + \beta_6 x_1 x_2^2 + \beta_7 x_1^2 x_2 + \beta_8 x_1^2 x_2^2}_{\text{Supply–Ratio interaction}} + \varepsilon$$

where

x_1 = Supply of raw material (in tons)

x_2 = Ratio of allocation

Note that the interaction terms for the model are constructed by taking the products of the various main effect terms, one from each factor. For example, we included terms involving the products of x_1 with x_2 and x_2^2. The remaining interaction terms were formed by multiplying x_1^2 by x_2 and by x_2^2.

b. To test the null hypothesis that supply and ratio do not interact, we must test the null hypothesis that the interaction terms are not needed in the linear model of part **a**:

$$H_0: \quad \beta_5 = \beta_6 = \beta_7 = \beta_8 = 0$$

This requires that we fit the reduced model

$$y = \beta_0 + \beta_1 x_1 + \beta_2 x_1^2 + \beta_3 x_2 + \beta_4 x_2^2$$

and perform the partial F test outlined in Section 4.10. The test statistic is

$$F = \frac{(SSE_R - SSE_C)/4}{MSE_C}$$

where

$$SSE_R = \text{SSE for reduced model}$$
$$SSE_C = \text{SSE for complete model}$$
$$MSE_C = \text{MSE for complete model}$$

The complete model of part **a** and the reduced model presented here were fit to the data in Table 11.7 using SAS. The SAS printouts are displayed in Figures 11.14a and 11.14b. The pertinent quantities, shaded on the printout, are

$$SSE_C = 43.33333 \quad \text{(see Figure 11.14a)}$$
$$MSE_C = 2.40741 \quad \text{(see Figure 11.14a)}$$
$$SSE_R = 89.55556 \quad \text{(see Figure 11.14b)}$$

Substituting these values into the formula for the test statistic, we obtain

$$F = \frac{(SSE_R - SSE_C)/4}{MSE_C} = \frac{(89.55556 - 43.33333)/4}{2.40741} = 4.80$$

FIGURE 11.14a

SAS printout for complete factorial model

Analysis of Variance

Source	DF	Sum of Squares	Mean Square	F Value	Prob>F
Model	8	74.66667	9.33333	3.877	0.0081
Error	18	43.33333	2.40741		
C Total	26	118.00000			

Root MSE	1.55158	R-square	0.6328	
Dep Mean	20.66667	Adj R-sq	0.4696	
C.V.	7.50766			

Parameter Estimates

Variable	DF	Parameter Estimate	Standard Error	T for H0: Parameter=0	Prob > |T|
INTERCEP	1	245.333333	130.49665074	1.880	0.0764
X1	1	-25.074074	14.71842356	-1.704	0.1057
X1SQ	1	0.679012	0.40837272	1.663	0.1137
X2	1	-534.333333	252.45534989	-2.117	0.0485
X2SQ	1	192.666667	97.17010948	1.983	0.0629
X1X2	1	60.555556	28.47387077	2.127	0.0475
X1X2SQ	1	-22.148148	10.95959797	-2.021	0.0584
X1SQX2	1	-1.666667	0.79002700	-2.110	0.0492
X1SQX2SQ	1	0.617284	0.30408153	2.030	0.0574

FIGURE 11.14b

SAS printout for reduced factorial model

```
                          Analysis of Variance

                             Sum of        Mean
        Source      DF       Squares       Square    F Value    Prob>F

        Model        4       28.44444      7.11111    1.747     0.1757
        Error       22       89.55556      4.07071
        C Total     26      118.00000


             Root MSE        2.01760    R-square      0.2411
             Dep Mean       20.66667    Adj R-sq      0.1031
             C.V.            9.76258


                          Parameter Estimates

                        Parameter     Standard    T for H0:
        Variable  DF     Estimate       Error     Parameter=0    Prob > |T|

        INTERCEP   1    -43.481481    29.32960686    -1.483        0.1524
        X1         1      6.814815     3.29853705     2.066        0.0508
        X1SQ       1     -0.185185     0.09152016    -2.023        0.0553
        X2         1      5.666667     4.35851270     1.300        0.2070
        X2SQ       1     -2.296296     1.67759232    -1.369        0.1849
```

The rejection region for the test is $F > F_{.05}$, where $\nu_1 = 4$ (the number of β's tested in H_0), $\nu_2 = 18$ (the degrees of freedom for error for the complete model), and $F_{.05} = 2.93$ (obtained from Table 4, Appendix D). Since the computed test statistic, $F = 4.80$, exceeds $F_{.05}$, we reject H_0 and conclude that supply and ratio interact.

c. The presence of interaction tells you that the mean profit depends on the particular combination of levels of supply S and ratio R. Consequently, there is little point in checking to see whether the means differ for the three levels of supply or whether they differ for the three levels of ratio (i.e., we will not perform the tests for main effects). For example, the supply level that gave the highest mean profit (over all levels of R) might not be the same supply–ratio level combination that produces the largest mean profit per unit of raw material.

The traditional analysis of variance approach to analyzing a complete two-factor factorial with factor A at a levels and factor B at b levels utilizes the fact that the total sum of squares, SS(Total), can be partitioned into four parts, SS(A), SS(B), SS(AB), and SSE (see Figure 11.15 on page 598). The first two sums of squares, SS(A) and SS(B), are called **main effect sums of squares** to distinguish them from the **interaction sum of squares**, SS(AB).

Since the sums of squares and the degrees of freedom for the analysis of variance are additive, the analysis of variance table appears as shown in Table 11.8.

TABLE 11.8 **ANOVA Table for an $a \times b$ Factorial Design with r Observations per Cell** (*Note:* $n = abr$)

SOURCE	df	SS	MS	F
Main effects A	$(a - 1)$	SS(A)	MS(A) = SS(A)/$(a - 1)$	MS(A)/MSE
Main effects B	$(b - 1)$	SS(B)	MS(B) = SS(B)/$(b - 1)$	MS(B)/MSE
AB interaction	$(a - 1)(b - 1)$	SS(AB)	MS(AB) = SS(AB)/$[(a - 1)(b - 1)]$	MS(AB)/MSE
Error	$ab(r - 1)$	SSE	MSE = SSE/$[ab(r - 1)]$	
Total	$abr - 1$	SS(Total)		

FIGURE 11.15

Partitioning of the total sum of squares
for a complete two-factor factorial
experiment

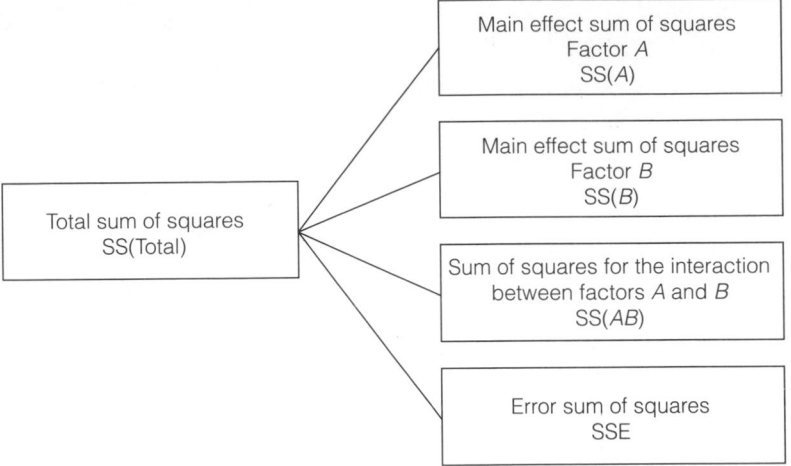

Note that the F statistics for testing factor main effects and factor interaction
are obtained by dividing the appropriate mean square by MSE. The numerator
df for the test of interest will equal the df of the source of variation being tested;
the denominator df will always equal df(Error). These F tests are equivalent to
the F tests obtained by fitting complete and reduced models in regression.*

For completeness, the formulas for calculating the ANOVA sums of squares
for a complete two-factor factorial experiment are given in the box on pages 599–
600.

EXAMPLE 11.13

Refer to Example 11.12.

a. Construct an ANOVA summary table for the analysis.
b. Conduct the test for supply $\times$ ratio interaction using the traditional analysis
of variance approach.

Solution

a. Although the formulas given in the previous box are straightforward, they
can become quite tedious to use. Therefore, we resort to a statistical software
package to conduct the ANOVA. A SAS printout of the ANOVA is displayed
in Figure 11.16 (page 600). The value of SS(Total), given in the SAS printout
under **Sum of Squares** in the **Corrected Total** row, is SS(Total) = 118. The
sums of squares, mean squares, and F values for the factors S, R, and $S \times R$

---

*The ANOVA F tests for main effects shown in Table 11.8 are equivalent to those of the regression
approach only when the reduced model includes interaction terms. Since we usually test for main
effects only after determining that interaction is nonsignificant, some statisticians favor dropping the
interaction terms from both the complete and reduced models prior to conducting the main effect
tests. For example, to test for main effect A, the complete model includes terms for main effects A
and B, whereas the reduced model includes terms for main effect B only. To obtain the equivalent
result using the ANOVA approach, the sums of squares for AB interaction and error are "pooled"
and a new MSE computed, where

$$MSE = \frac{SS(AB) + SSE}{n - a - b + 1}$$

Computing Formulas for the Analysis of Variance for a Two-Factor Factorial Experiment

CM = Correction for the mean

$$= \frac{(\text{Total of all } n \text{ measurements})^2}{n}$$

$$= \frac{\left(\sum_{i=1}^{n} y_i\right)^2}{n}$$

$SS(\text{Total})$ = Total sum of squares

= Sum of squares of all n measurements − CM

$$= \sum_{i=1}^{n} y_i^2 - CM$$

$SS(A)$ = Sum of squares for main effects, independent variable 1

$$= \left(\begin{array}{c} \text{Sum of squares of the totals } A_1, A_2, \ldots, A_a \\ \text{divided by the number of measurements} \\ \text{in a single total, namely } br \end{array} \right) - CM$$

$$= \frac{\sum_{i=1}^{a} A_i^2}{br} - CM$$

$SS(B)$ = Sum of squares for main effects, independent variable 2

$$= \left(\begin{array}{c} \text{Sum of squares of the totals } B_1, B_2, \ldots, B_b \\ \text{divided by the number of measurements} \\ \text{in a single total, namely } ar \end{array} \right) - CM$$

$$= \frac{\sum_{i=1}^{b} B_i^2}{ar} - CM$$

$SS(AB)$ = Sum of squares for AB interaction

$$= \left(\begin{array}{c} \text{Sum of squares of the cell totals} \\ AB_{11}, AB_{12}, \ldots, AB_{ab} \text{ divided by} \\ \text{the number of measurements} \\ \text{in a single total, namely } r \end{array} \right) - SS(A) - SS(B) - CM$$

$$= \frac{\sum_{i=1}^{b} \sum_{i=1}^{a} AB_{ij}^2}{r} - SS(A) - SS(B) - CM$$

(continued)

where

a = Number of levels of independent variable 1

b = Number of levels of independent variable 2

r = Number of measurements for each pair of levels of independent variables 1 and 2

n = Total number of measurements
  $= a \times b \times r$

A_i = Total of all measurements of independent variable 1 at level i $(i = 1, 2, \ldots, a)$

B_j = Total of all measurements of independent variable 2 at level j $(j = 1, 2, \ldots, b)$

AB_{ij} = Total of all measurements at the ith level of independent variable 1 and at the jth level of independent variable 2 $(i = 1, 2, \ldots, a;$ $j = 1, 2, \ldots, b)$

interaction are given under the **Anova SS**, **Mean Square**, and **F value** columns, respectively, in the bottom portion of the printout. These values are shown in the ANOVA table given in Table 11.9.

FIGURE 11.16

SAS printout for ANOVA of data in Example 11.13

Analysis of Variance Procedure

Dependent Variable: PROFIT

Source	DF	Sum of Squares	Mean Square	F Value	Pr > F
Model	8	74.66666667	9.33333333	3.88	0.0081
Error	18	43.33333333	2.40740741		
Corrected Total	26	118.00000000			

	R-Square	C.V.	Root MSE	PROFIT Mean
	0.632768	7.507656	1.551582	20.6666667

Source	DF	Anova SS	Mean Square	F Value	Pr > F
SUPPLY	2	20.22222222	10.11111111	4.20	0.0318
RATIO	2	8.22222222	4.11111111	1.71	0.2094
SUPPLY*RATIO	4	46.22222222	11.55555556	4.80	0.0082

TABLE 11.9   **ANOVA Table for Example 11.13**

SOURCE	df	SS	MS
Supply	2	20.22	10.11
Ratio	2	8.22	4.11
Supply–Ratio interaction	4	46.22	11.56
Error	18	43.33	2.41
Total	26	118.00	

b. To test the hypothesis that supply and ratio interact, we use the test statistic

$$F = \frac{MS(SR)}{MSE} = \frac{11.56}{2.41} = 4.80 \quad \text{(shaded on the SAS printout)}$$

Note that this value is identical to the test statistic obtained in Example 11.12 using regression. The p-value of the test (also shaded on the SAS printout) is .0082. Since this value is less than the selected value of $\alpha = .05$, we conclude that supply and ratio interact.

Confidence intervals for a single treatment mean and for the difference between two treatment means in a factorial experiment are provided in the following boxes.

$100(1 - \alpha)\%$ Confidence Interval for the Mean of a Single Treatment: Factorial Experiment

$$\bar{y}_{ij} \pm t_{\alpha/2}\left(\frac{s}{\sqrt{r}}\right)$$

where

$\bar{y}_{ij}$ = Sample mean for the treatment identified by level i of the first factor and level j of the second factor

r = Number of measurements per treatment

$s = \sqrt{MSE}$

and $t_{\alpha/2}$ is based on $ab(r - 1)$df.

$100(1 - \alpha)\%$ Confidence Interval for the Difference Between a Pair of Treatment Means: Factorial Experiment

Let

$\bar{y}_1$ = Sample mean of the r measurements for the first treatment

$\bar{y}_2$ = Sample mean of the r measurements for the second treatment

Then, the $100(1 - \alpha)\%$ confidence interval for the difference between the treatment means is

$$(\bar{y}_1 - \bar{y}_2) \pm t_{\alpha/2}s\sqrt{\frac{2}{r}}$$

where $s = \sqrt{MSE}$ and $t_{\alpha/2}$ is based on $ab(r - 1)$df.

EXAMPLE 11.14

Refer to Examples 11.12 and 11.13.

a. Find a 95% confidence interval to estimate the mean profit per unit of raw materials when $S = 18$ tons and the ratio of production is $R = 1$.
b. Find a 95% confidence interval to estimate the difference in mean profit per unit of raw materials when $\left(S = 18, R = \frac{1}{2}\right)$ and $(S = 18, R = 1)$.

Solution

a. A 95% confidence interval for the mean $E(y)$ when supply $S = 18$ and $R = 1$ is

$$\bar{y}_{18,1} \pm t_{.025}\left(\frac{s}{\sqrt{r}}\right)$$

where $\bar{y}_{18,1}$ is the mean of the $r = 3$ values of y for $S = 18$, $R = 1$ (obtained from Table 11.7), and $t_{.025} = 2.101$ is based on 18 df. Substituting, we obtain

$$\frac{71}{3} \pm 2.101\left(\frac{1.55}{\sqrt{3}}\right)$$

$$23.67 \pm 1.88$$

Therefore, our interval estimate for the mean profit per unit of raw material where $S = 18$ and $R = 1$ is \$21.79 to \$25.55.

b. A 95% confidence interval for the difference in mean profit per unit of raw material for two different combinations of levels of S and R is

$$(\bar{y}_1 - \bar{y}_2) \pm t_{.025}s\sqrt{\frac{2}{r}}$$

where $\bar{y}_1$ and $\bar{y}_2$ represent the means of the $r = 3$ replications for the factor level combinations $\left(S = 18, R = \frac{1}{2}\right)$ and $(S = 18, R = 1)$, respectively. From Table 11.7, the sums of the three measurements for these two treatments are 61 and 71. Substituting, we obtain

$$\left(\frac{61}{3} - \frac{71}{3}\right) \pm (2.101)(1.55)\sqrt{\frac{2}{3}}$$

$$-3.33 \pm 2.66$$

Therefore, the interval estimate for the difference in mean profit per unit of raw material for the two factor level combinations is $(-\$5.99, -\$.67)$. The negative values indicate that we estimate the mean for $\left(S = 18, R = \frac{1}{2}\right)$ to be less than the mean for $(S = 18, R = 1)$ by between \$.67 and \$5.99.

Throughout this chapter we have presented two methods for analyzing data from a designed experiment: the regression approach and the traditional ANOVA approach. In a factorial experiment, the two methods yield identical results when both factors are qualitative; however, regression will provide more information when at least one of the factors is quantitative. For example, the analysis of variance in Example 11.13 enables us to estimate the mean profit per unit of supply for *only* the nine combinations of supply–ratio levels. It will not permit us to estimate the mean response for some other combination of levels of the independent variables not included among the nine used in the factorial experiment. Alternatively, the prediction equation obtained from the regression analysis in Example 11.12 enables us to estimate the mean profit per unit of supply when $(S = 17, R = 1)$. We could not obtain this estimate from the analysis of variance in Example 11.12.

The prediction equation found by regression analysis also contributes other information not provided by traditional analysis of variance. For example, we might wish to estimate the rate of change in the mean profit, $E(y)$, for unit changes in S, R, or both for specific values of S and R. Or, we might want to determine whether the third- and fourth-order terms in the complete model of Example 11.12 really contribute additional information for the prediction of profit, y.

We illustrate some of these applications in the final two examples of this section.

EXAMPLE 11.15

Do the data provide sufficient information to indicate that third- and fourth-order terms in the complete factorial model given in Example 11.12 contribute information for the prediction of y? Use $\alpha = .05$.

Solution

If the response to the question is yes, then at least one of the parameters, β_6, β_7, or β_8, of the complete factorial model differs from 0 (i.e., they are needed in the model). Consequently, the null hypothesis is

$$H_0: \quad \beta_6 = \beta_7 = \beta_8 = 0$$

and the alternative hypothesis is

$$H_a: \quad \text{At least one of the three } \beta\text{'s is nonzero.}$$

To test this hypothesis, we compute the drop in SSE between the appropriate reduced and complete model.

For this application the complete model is the complete factorial model of Example 11.12:

Complete model: $E(y) = \beta_0 + \beta_1 x_1 + \beta_2 x_1^2 + \beta_3 x_2 + \beta_4 x_2^2 + \beta_5 x_1 x_2$
$$+ \beta_6 x_1 x_2^2 + \beta_7 x_1^2 x_2 + \beta_8 x_1^2 x_2^2$$

The reduced model is the complete model above, minus the third- and fourth-order terms, i.e., the reduced model is the second-order model:

Reduced model: $E(y) = \beta_0 + \beta_1 x_1 + \beta_2 x_1^2 + \beta_3 x_2 + \beta_4 x_2^2 + \beta_5 x_1 x_2$

Recall (from Figure 11.14a) that the SSE and MSE for the complete model are $\text{SSE}_C = 43.3333$ and $\text{MSE}_C = 2.4704$. A SAS printout of the regression analysis of the reduced model is shown in Figure 11.17 (page 604). The SSE for the reduced model (shaded) is $\text{SSE}_R = 54.49206$.

FIGURE 11.17

SAS printout for the reduced (second-order) model

```
Dependent Variable: PROFIT

                       Analysis of Variance

                         Sum of          Mean
     Source        DF    Squares        Square    F Value    Prob>F

     Model          5    63.50794      12.70159     4.895    0.0040
     Error         21    54.49206       2.59486
     C Total       26   118.00000

          Root MSE       1.61086    R-square     0.5382
          Dep Mean      20.66667    Adj R-sq     0.4283
          C.V.           7.79447

                       Parameter Estimates

                    Parameter      Standard    T for H0:
     Variable   DF   Estimate         Error   Parameter=0   Prob > |T|

     INTERCEP    1  -27.814815   23.80152168      -1.169       0.2557
     S           1    5.944444    2.64418353       2.248       0.0354
     R           1   -7.761905    5.04522969      -1.538       0.1389
     SR          1    0.746032    0.20294890       3.676       0.0014
     SS          1   -0.185185    0.07306996      -2.534       0.0193
     RR          1   -2.296296    1.33939441      -1.714       0.1012
```

Consequently, the test statistic required to conduct the test is

Test statistic:

$$F = \frac{(SSE_R - SSE_C)/(\text{number of } \beta\text{'s tested})}{MSE_C} = \frac{(54.49206 - 43.3333)/3}{2.4704}$$

$$= 1.54$$

The F statistic is based on $\nu_1 = 3$ numerator df, $\nu_2 = 18$ denominator df, and the critical value (obtained from Table 4 of Appendix D) is $F_{.05} = 3.16$. Thus, the rejection region is:

Rejection region: $F > 3.16$

Conclusion: Since the computed value of F (1.54) is less than the critical value, $F_{.05} = 3.16$, we cannot reject the null hypothesis that $\beta_6 = \beta_7 = \beta_8 = 0$. That is, there is insufficient evidence (at $\alpha = .05$) to indicate that the third- and fourth-order terms associated with β_6, β_7, and β_8 contribute information for the prediction of y. Since the complete factorial model contributes no more information about y than the reduced (second-order) model, we recommend using the second-order model in practice.

EXAMPLE 11.16

Use the second-order model of Example 11.15 and find a 95% confidence interval for the mean profit per unit supply of raw material when $S = 17$ and $R = 1$.

Solution

The portion of the SAS printout for the second-order model with 95% confidence intervals for $E(y)$ is shown in Figure 11.18.

FIGURE 11.18

SAS printout of confidence intervals for mean profit

Obs	S	R	Dep Var PROFIT	Predict Value	Std Err Predict	Lower95% Mean	Upper95% Mean	Residual
1	15	0.5	23.0000	20.8254	0.803	19.1549	22.4959	2.1746
2	15	0.5	20.0000	20.8254	0.803	19.1549	22.4959	-0.8254
3	15	0.5	21.0000	20.8254	0.803	19.1549	22.4959	0.1746
4	18	0.5	22.0000	21.4444	0.693	20.0029	22.8860	0.5556
5	18	0.5	19.0000	21.4444	0.693	20.0029	22.8860	-2.4444
6	18	0.5	20.0000	21.4444	0.693	20.0029	22.8860	-1.4444
7	21	0.5	19.0000	18.7302	0.803	17.0596	20.4007	0.2698
8	21	0.5	18.0000	18.7302	0.803	17.0596	20.4007	-0.7302
9	21	0.5	21.0000	18.7302	0.803	17.0596	20.4007	2.2698
10	15	1	22.0000	20.8175	0.701	19.3605	22.2744	1.1825
11	15	1	20.0000	20.8175	0.701	19.3605	22.2744	-0.8175
12	15	1	19.0000	20.8175	0.701	19.3605	22.2744	-1.8175
13	18	1	24.0000	22.5556	0.693	21.1140	23.9971	1.4444
14	18	1	25.0000	22.5556	0.693	21.1140	23.9971	2.4444
15	18	1	22.0000	22.5556	0.693	21.1140	23.9971	-0.5556
16	21	1	20.0000	20.9603	0.701	19.5034	22.4173	-0.9603
17	21	1	19.0000	20.9603	0.701	19.5034	22.4173	-1.9603
18	21	1	22.0000	20.9603	0.701	19.5034	22.4173	1.0397
19	15	2	18.0000	17.3571	0.859	15.5707	19.1436	0.6429
20	15	2	18.0000	17.3571	0.859	15.5707	19.1436	0.6429
21	15	2	16.0000	17.3571	0.859	15.5707	19.1436	-1.3571
22	18	2	21.0000	21.3333	0.693	19.8918	22.7749	-0.3333
23	18	2	23.0000	21.3333	0.693	19.8918	22.7749	1.6667
24	18	2	20.0000	21.3333	0.693	19.8918	22.7749	-1.3333
25	21	2	20.0000	21.9762	0.859	20.1897	23.7627	-1.9762
26	21	2	22.0000	21.9762	0.859	20.1897	23.7627	0.0238
27	21	2	24.0000	21.9762	0.859	20.1897	23.7627	2.0238
28	17	1	.	22.3466	0.663	20.9687	23.7244	.

The confidence interval for $E(y)$ when $S = 17$ and $R = 1$ is given in the last row of the printout. You can see that the interval is (20.97, 23.72). Thus, we estimate (with confidence coefficient equal to .95) that the mean profit per unit of supply will lie between $20.97 and $23.72 when $S = 17$ tons and $R = 1$. Beyond this immediate result, you will note that this example illustrates the power and versatility of a regression analysis. In particular, there is no way to obtain this estimate from the analysis of variance in Example 11.13. However, a computerized regression package can be easily programmed to include the confidence interval automatically.

EXERCISES

11.26   The analysis of variance for a 3×2 factorial experiment, with four observations per treatment, produced the ANOVA table entries shown here.

SOURCE	df	SS	MS	F
A	——	100	——	——
B	1	——	——	——
AB	2	——	2.5	——
Error	——	——	2.0	
Total	——	700		

a. Complete the ANOVA table.
b. Test for interaction between factor A and factor B. Use $\alpha = .05$.
c. Test for differences in main effect means for factor A. Use $\alpha = .05$.
d. Test for differences in main effect means for factor B. Use $\alpha = .05$.

11.27   Refer to Example 11.11 and the factorial experiment designed to measure the effect of two factors, gasket material and stamping machine, on productivity of a manufacturing process. The data for the 2×3 factorial experiment, number of gaskets produced per hour (in thousands), are shown in the accompanying table.

		GASKET MATERIAL			TOTAL
		Cork, B_1	Rubber, B_2	Plastic, B_3	
STAMPING MACHINE	A_1	4.31 4.27 4.40	3.36 3.42 3.48	4.01 3.94 3.89	35.08
	A_2	3.94 3.81 3.99	3.91 3.80 3.85	3.48 3.53 3.42	33.73
TOTAL		24.72	21.82	22.27	68.81

a. Construct an ANOVA summary table using the ANOVA calculation formulas.

b. A SAS printout of the analysis is shown here. Check to see that the entries in the table, part **a**, agree with the values shown on the printout.

```
                        Analysis of Variance Procedure

Dependent Variable: NUMBER
                                 Sum of            Mean
Source                 DF        Squares          Square    F Value    Pr > F

Model                   5      1.68122778      0.33624556     76.52    0.0001
Error                  12      0.05273333      0.00439444
Corrected Total        17      1.73396111

                R-Square            C.V.        Root MSE         NUMBER Mean

                0.969588         1.734095       0.066291          3.82277778


Source                 DF       Anova SS     Mean Square    F Value    Pr > F

MACHINE                 1      0.10125000      0.10125000     23.04    0.0004
MATERIAL                2      0.81194444      0.40597222     92.38    0.0001
MACHINE*MATERIAL        2      0.76803333      0.38401667     87.39    0.0001
```

c. Is there evidence of interaction between gasket material and stamping machine? Test using $\alpha = .01$.

d. Explain the practical significance of the result obtained in part **b**.

e. Based on parts **c** and **d**, would you recommend that tests for main effects be conducted? Explain.

f. Find a 95% confidence interval for the difference in the mean number of gaskets produced by machines A_1 and A_2, when stamping cork (B_1) gaskets. Interpret the interval.

11.28   Video games have revolutionized children's leisure time activities. However, many parents, including the U.S. surgeon general, believe that video games are a bad influence on their children. A study was conducted to examine the effect of playing video games on fifth-graders' free play (*Journal of Applied Social Psychology*, Vol. 16, 1986). Eighty-four fifth-graders were paired randomly, and then each pair was randomly assigned to one of three types of games, an aggressive video game (Missile Command), a nonaggressive video game (Pac-Man), or a pen-and-paper maze-solving game (control) in equal numbers. One member of each pair was then randomly chosen to play the designated game (player) for 8 minutes, while the other member watched (observer). Thus, 14 fifth-graders were assigned to each of the 3 × 2 = 6 experimental conditions. After video play was concluded, the children were sent to a toy room for free play. The goal of the experiment was to investigate the effect of type of game (Missile Command, Pac-Man, and control) and position (player or observer) on degree of aggressive play in the toy room.

a. Identify the factors in this factorial experiment.

b. Identify the levels of the factors.

c. What are the treatments?

d. Give the sources of variation and their respective degrees of freedom in an ANOVA table for this experiment.

e. Give the complete model appropriate for analyzing the data for this experiment.

f. A significant interaction was found between type of game and position (p-value) < .01. Interpret this result in the words of the problem.

11.29 The *Accounting Review* (Jan. 1991) reported on a study of the effect of two factors, confirmation of accounts receivable and verification of sales transactions, on account misstatement risk by auditors. Both factors were held at the same two levels: completed or not completed. Thus, the experimental design is a 2 × 2 factorial design.

a. Identify the factors, factor levels, and treatments for this experiment.

b. Explain what factor interaction means for this experiment.

c. A graph of the hypothetical mean misstatement risks for each of the 2 × 2 = 4 treatments is shown here. In this hypothetical case, does it appear that interaction exists?

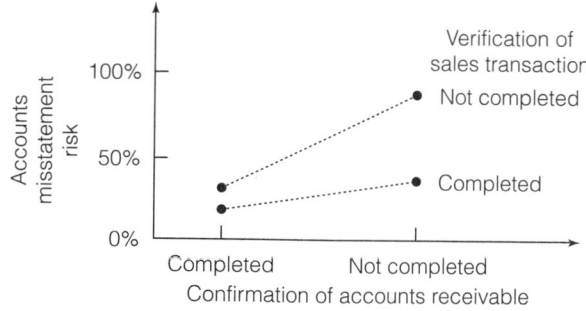

Source: Brown, C. E. and Solomon, I. "Configural information processing in auditing: The role of domain-specific knowledge." *The Accounting Review,* Vol. 66, No. 1, Jan. 1991, p. 105 (Figure 1).

11.30 *Time-of-day pricing* is a plan by which electric customers are charged a less expensive rate for using electricity during off-peak (less demanded) hours. One experiment (reported in *Journal of Consumer Research,* June 1982) was conducted to measure customer satisfaction with several time-of-day pricing schemes. The experiment consisted of two factors, price ratio (the ratio of peak to off-peak prices) and peak period length, each at three levels. The 3 × 3 = 9 combinations of price ratio and peak period length represent the nine time-of-day pricing schemes. For each pricing scheme, customers were randomly selected and asked to rate satisfaction with the plan using an index from 10 to 38, with 38 indicating extreme satisfaction. Suppose four customers were sampled for each pricing scheme. The table at the top of page 608 gives the satisfaction scores for these customers. [*Note:* The data are based on mean scores provided in the *Journal of Consumer Research* article.]

		PRICING RATIO					
		2:1		4:1		8:1	
PEAK PERIOD LENGTH	6 hours	25 26	28 27	31 26	29 27	24 25	28 26
	9 hours	26 29	27 30	25 30	24 26	33 25	28 27
	12 hours	22 25	20 21	33 25	27 27	30 26	31 27

a. The data were subjected to an analysis of variance using Minitab. Use the accompanying Minitab printout to construct an ANOVA table for the data.

```
ANALYSIS OF VARIANCE   SATSCORE

SOURCE        DF        SS        MS
PERIOD         2     10.67      5.33
RATIO          2     32.00     16.00
INTERACTION    4     95.33     23.83
ERROR         27    161.00      5.96
TOTAL         35    299.00
```

b. Compute the nine customer satisfaction index means.
c. Plot the nine means from part b on a graph similar to Figure 11.13. Does it appear that the two factors, price ratio and peak period length, interact? Explain.
d. Do the data provide sufficient evidence of interaction between price ratio and peak period length? Test using $\alpha = .05$.
e. Do the data provide sufficient evidence that mean customer satisfaction differs for the three peak period lengths? Test using $\alpha = .05$.
f. When is the test of part e appropriate?
g. Find a 90% confidence interval for the mean customer satisfaction rating of a pricing scheme with a peak period length of 9 hours and pricing ratio of 2:1.
h. Find a 95% confidence interval for the difference between the mean customer satisfaction ratings of pricing schemes 9 hours, 8:1 ratio and 6 hours, 8:1 ratio. Interpret the interval.

11.31 Many television and radio commercials utilize high intensity stimulation (e.g., rapid changes in visual imagery, quick movements, bright lights, and loud sounds) to increase the general arousal of the viewer or listener. In theory, the more aroused the viewer, the more likely he or she is to remember the advertised product. *Psychology & Marketing* (Summer 1986) reported on a study designed to examine the effect of high-intensity advertising on the audience. The study was conducted at a small midwestern liberal arts college. Two groups of psychology students (10 introverts and 10 extroverts) took part in the experiment. (Scores on the Eysenck Personality Questionnaire were used to classify the students into one of the two groups.) Five subjects from each group were then randomly assigned to one of two experimental conditions: high volume commercial or normal volume commercial. Thus, the experiment consists of two factors, personality type and commercial volume, each at two levels. All students listened to a 5-minute tape-recorded "radio program" that included a commercial for a fictitious brand of chewing gum played at the assigned volume, and each subject's attitude toward the product was measured on a 14-point scale (1 = strongly dislike, 14 = strongly like).

Since five responses were recorded for each of the $2 \times 2 = 4$ factor level combinations, the factorial experiment includes five replicates of each treatment.

a. A partial ANOVA table for the experiment is shown here. Calculate as many of the missing entries as possible.

SOURCE	df	SS	MS	F
Personality (P)	1	___	___	___
Volume (V)	___	___	___	4.61
VP	___	___	___	11.39
Error	___	___	15	
Total	___	___		

Source: Cetola, H. and Prinkley, K. "Introversion-extraversion and loud commercials." *Psychology & Marketing*, Vol. 3, No. 2, Summer 1986, pp. 123–132.

b. The means of the four treatments are given in the table. Plot the means in a graph similar to Figure 11.13. Does it appear that the two factors, personality type and commercial volume, interact?

	High Volume	Normal Volume
Extroverts	10.6	7.0
Introverts	7.2	6.0

c. Conduct a test to determine whether personality type interacts with commercial volume. Use $\alpha = .05$. Interpret the result.

11.32 Computer-based management information systems (MIS) is one of the fastest growing industries in the United States, yet little is known about the ethical decision-making processes of persons involved in creating and maintaining these systems. An empirical investigation was conducted to determine whether MIS majors, on average, exhibit ethical decision-making processes that differ from other business students (*Journal of Business Ethics*, Vol. 10, 1991). A large sample of business students were divided into two groups based on their major (MIS or non-MIS). Within each group, half of the students were administered the regular form of the Defining Issues Test (DIT) and the other half a form of the DIT modified to incorporate MIS. Thus, a 2×2 factorial experimental design was used, with the two factors, major and form. The dependent (response) variable measured was the "principled morality," or P score, expressed as a percentage. (High scores indicate morally conscious decisions.)

SOURCE	df	SS	MS	F	p-VALUE
Form	1	.13	.13	.00	.98
Major	1	1,290.10	1,290.10	6.25	.01
Form × Major	1	56.38	56.38	.27	.60
Error	233	48,120.15	206.52		
Total	236	49,466.76			

Source: Paradice, D. B. and Dejoie, R. M. "The ethical decision-making processes of information systems workers." *Journal of Business Ethics*, Vol. 10, 1991, p. 9 (Table V).

a. Write the complete model for this 2×2 factorial design.

b. Interpret the results of the ANOVA summarized in the table.

11.33 Refer to Exercise 11.21. Another important concern of designers and users of computer display monitors is readability of the text on the screen. Two factors thought to affect readability of scrolling texts is window size (i.e., number of characters displayed per line) and jump length (i.e., number of characters advanced per jump). To investigate this phenomenon, a 2×3 factorial experiment was conducted, with window size at two levels (20 and 40 characters) and jump length at three levels (1, 5, and 9 characters). The response variable of interest was reading rate (measured as number of words per minute) of Chinese college students who participated in the study (*Human Factors*, June 1988).

a. The ANOVA F test for interaction between the two factors resulted in a p-value exceeding .10. Interpret this result.

b. The ANOVA F test for the main effect of window size resulted in a p-value exceeding .10. Interpret this result.

c. The ANOVA F test for the main effect of jump length resulted in a p-value less than .05. Interpret this result.

11.34 A trade-off study regarding the inspection and test of transformer parts was conducted by the quality department of a major defense contractor. The investigation was structured to examine the effects of varying inspection levels and incoming test times to detect early part failure or fatigue. The levels of inspection selected were full military inspection (*A*), reduced military specification level (*B*), and commercial grade (*C*). Operational burn-in test times chosen for this study were at 1-hour increments from 1 hour to 9 hours. The response was failures per thousand pieces obtained from samples taken from lot sizes inspected to a specified level and burned-in over a prescribed time length. Three replications were randomly sequenced under each condition making this a complete 3×9 factorial experiment (a total of 81 observations). The accompanying table contains the data collected for the study. Conduct a complete analysis of the data. One of the main objectives is to determine the most optimal burn-in hours, given a specified inspection level.

INSPECTION LEVELS

Burn-in, hours	Full Military Specification, A			Reduced Military Specification, B			Commercial, C		
1	7.60	7.50	7.67	7.70	7.10	7.20	6.16	6.13	6.21
2	6.54	7.46	6.84	5.85	6.15	6.15	6.21	5.50	5.64
3	6.53	5.85	6.38	5.30	5.60	5.80	5.41	5.45	5.35
4	5.66	5.98	5.37	5.38	5.27	5.29	5.68	5.47	5.84
5	5.00	5.27	5.39	4.85	4.99	4.98	5.65	6.00	6.15
6	4.20	3.60	4.20	4.50	4.56	4.50	6.70	6.72	6.54
7	3.66	3.92	4.22	3.97	3.90	3.84	7.90	7.47	7.70
8	3.76	3.68	3.80	4.37	3.86	4.46	8.40	8.60	7.90
9	3.46	3.55	3.45	5.25	5.63	5.25	8.82	9.76	9.52

Source: Danny La Nuez, College of Business Administration, graduate student, University of South Florida, 1989–1990.

11.35 A field experiment was conducted at a not-for-profit research and development organization to examine the expectations, attitudes, and decisions of employees with regard to training programs (*Academy of Management Journal*, Sept. 1987). In particular, the study was aimed at determining how prior information managers received about a 2-day workshop on performance reviews and how much freedom they had in choosing to enter the workshop affected their evaluation of the training program.

These two factors, prior information and degree of choice, were each varied at two levels. The prior information managers received about the training program was either a realistic preview of the program and its benefits or a traditional announcement that tended to exaggerate the workshop's benefits. Degree of choice was either low (mandatory attendance) or high (little pressure from supervisors to attend). Twenty-one managers were randomly assigned to each of the 2 × 2 = 4 experimental conditions; thus, a 2 × 2 factorial design was employed. At the end of the training program, each manager was asked to rate his or her satisfaction with the workshop on a 7-point scale (1 = no satisfaction, 7 = extremely satisfied). The ratings were subjected to an analysis of variance, with the results partially summarized in the accompanying table.

SOURCE	df	SS	MS	F
Prior information (P)	1	‾‾‾	1.55	‾‾‾
Degree of choice (D)	1	‾‾‾	22.26	‾‾‾
PD interaction	1	‾‾‾	.61	‾‾‾
Error	80	‾‾‾	1.43	
Total	83	‾‾‾		

Source: Hicks, W. D. and Klimoski, R. J. "Entry into training programs and its effects on training outcomes: A field experiment." *Academy of Management Journal*, Vol. 30, No. 3, Sept. 1987, p. 548.

a. Complete the ANOVA summary table.
b. Conduct the appropriate ANOVA F tests (use $\alpha = .05$). Interpret the results.
c. The sample mean ratings for the 42 managers in each of the two degree-of-choice conditions were 6.14 (high) and 5.13 (low). Construct a 95% confidence interval for the difference between the mean satisfaction ratings of managers in these two groups. Interpret the result.

11.36 How does a worker's gender or rank in a company affect other people's evaluations of the worker's performance or qualifications? *Personnel Psychology* (Spring 1983) reported on a study to determine the effects of a writer's gender and organizational position on evaluations of business memos. In one portion of the study, each person in a sample of approximately 100 subjects (all members of the Wisconsin Personnel and Industrial Relations Association) was asked to rate a poorly written memo on a 7-point scale (1 = very poor, 7 = very good). The attributed gender (male or female) of the supposed author of the memo and the attributed memo author's organizational position (executive or assistant) were varied from memo to memo. Thus, the experiment consists of two factors, author gender and author position, with each at two levels. Assume that 25 subjects are sampled for each of the four memo types. The accompanying table gives the totals of the rating scores for each memo type.

		AUTHOR GENDER		TOTALS
		Male	Female	
AUTHOR POSITION	Executive	76	94	170
	Assistant	89	82	171
TOTALS		165	176	341

Source: Morrow, W. R. and Lowenberg, G. "Evaluation of business memos: Effects of writer sex on organizational position, memo quality, and rater sex." *Personnel Psychology*, Spring 1983, *36*, pp. 73–85.

a. Compute the main effect and interaction sums of squares for the two factors, gender and position.

b. Assume that the value of s^2 obtained from the analysis of variance is 2.10. Calculate SSE and SS(Total).

c. Construct an ANOVA table for the data.

d. What do we mean when we say that the two factors, gender and position, interact? Illustrate with a graph.

e. Is there sufficient evidence to indicate an interaction between gender and position? Test using $\alpha = .01$.

11.37   A recent trend in advertising and marketing is the use of the chief executive officer (CEO) as the spokesperson for the company (e.g., Lee Iacocca for Chrysler and Dave Thomas for Wendy's). A study was conducted to determine the perceived effectiveness of CEOs as marketing spokespersons (*Journal of Advertising*, Jan. 1986). Eleven magazine advertisements in which CEOs acted as spokespersons were selected for analysis. Each ad was rated by a sample of 58 MBA students on each of eight items designed to measure credibility. Thus, the experiment consists of two factors, CEO and credibility, with CEO at 11 levels and credibility at 8 levels (well known, believable, likeable, integrity, knowledgeable, expert, influential, and powerful). The 58 ratings for each of the $8 \times 11 = 88$ treatments (a total of 5,104 observations) were subjected to an analysis of variance, with the results shown in the accompanying ANOVA table. Interpret these results.

SOURCE	df	SS	MS	F
CEO	10	2,609.5	260.95	140.34
Credibility	7	1,012.5	144.64	77.79
CEO–Credibility	70	636.7	9.10	4.89
Error	5,016	9,326.2	1.86	
Total	5,103	13,584.9		

Source: Reidenbach, R. E. and Pitts, R. E. "Not all CEOs are created equal as advertising spokespersons: Evaluating the effective CEO spokesperson." *Journal of Advertising*, Vol. 15, No. 1, January 1986, pp. 30–36.

11.6

More Complex Factorial Designs (Optional)

In this optional section we present some useful factorial designs that are more complex than the basic two-factor factorial of Section 11.5. These designs fall under the general category of a **k-way classification of data**. A k-way classification of data arises when we run all combinations of the levels of k independent variables. These independent variables can be factors or blocks.

For example, consider a replicated $2 \times 3 \times 3$ factorial experiment in which the $2 \times 3 \times 3 = 18$ treatments are assigned to the experimental units according to a completely randomized design. Since every combination of the three factors (a total of 18) is examined, the design is often called a three-way classification of data, Similarly, a k-way classification of data would result if we randomly assign the treatments of a $(k - 1)$-factor factorial experiment to the experimental units of a randomized block design. For example, if we assigned the $2 \times 3 = 6$ treatments of a complete 2×3 factorial experiment to blocks containing six experimental units each, the data would be arranged in a three-way classification, i.e., according to the two factors and the blocks.

The formulas required for calculating the sums of squares for main effects and interactions for an analysis of variance for a k-way classification of data are complicated and, therefore, are not given here. If you are interested in the computational formulas, see the references. As with the designs in the previous three sections, we provide the appropriate linear model for these more complex designs and use either regression or the standard ANOVA output of a statistical software package to analyze the data.

EXAMPLE 11.17

Consider a $2 \times 3 \times 3$ factorial experiment with qualitative factors and $r = 3$ experimental units randomly assigned to each treatment.

a. Write the appropriate linear model for the design.
b. Indicate the sources of variation and their associated degrees of freedom in a partial ANOVA table.

Solution

a. Denote the three qualitative factors as A, B, and C, with A at two levels, and B and C at three levels. Then the linear model for the experiment will contain one parameter corresponding to main effects for A, two each for B and C, $(1)(2) = 2$ each for the AB and AC interactions, $(2)(2) = 4$ for the BC interaction, and $(1)(2)(2) = 4$ for the three-way ABC interaction. Three-way interaction terms measure the failure of two-way interaction effects to remain the same from one level to another level of the third factor.

$$E(y) = \beta_0 + \underbrace{\beta_1 x_1}_{\substack{\text{Main effect}\\A}} + \underbrace{\beta_2 x_2 + \beta_3 x_3}_{\substack{\text{Main effects}\\B}} + \underbrace{\beta_4 x_4 + \beta_5 x_5}_{\substack{\text{Main effects}\\C}}$$

$$+ \underbrace{\beta_6 x_1 x_2 + \beta_7 x_1 x_3}_{A \times B \text{ interaction}} + \underbrace{\beta_8 x_1 x_4 + \beta_9 x_1 x_5}_{A \times C \text{ interaction}}$$

$$+ \underbrace{\beta_{10} x_2 x_4 + \beta_{11} x_2 x_5 + \beta_{12} x_3 x_4 + \beta_{13} x_3 x_5}_{B \times C \text{ interaction}}$$

$$+ \underbrace{\beta_{14} x_1 x_2 x_4 + \beta_{15} x_1 x_3 x_4 + \beta_{16} x_1 x_2 x_5 + \beta_{17} x_1 x_3 x_5}_{A \times B \times C \text{ interaction}}$$

where

$$x_1 = \begin{cases} 1 & \text{if level 1 of } A \\ 0 & \text{if level 2 of } A \end{cases} \qquad x_2 = \begin{cases} 1 & \text{if level 1 of } B \\ 0 & \text{if not} \end{cases}$$

$$x_3 = \begin{cases} 1 & \text{if level 2 of } B \\ 0 & \text{if not} \end{cases} \qquad x_4 = \begin{cases} 1 & \text{if level 1 of } C \\ 0 & \text{if not} \end{cases}$$

$$x_5 = \begin{cases} 1 & \text{if level 2 of } C \\ 0 & \text{if not} \end{cases}$$

b. The sources of variation and the respective degrees of freedom corresponding to these sets of parameters are shown in Table 11.10.

T A B L E 11.10 Analysis of Variance Table for Example 11.17

SOURCE	df
Main effect A	1
Main effect B	2
Main effect C	2
AB interaction	2
AC interaction	2
BC interaction	4
ABC interaction	4
Error	36
Total	53

The degrees of freedom for SS(Total) will always equal $(n - 1)$—that is, n minus 1 degree of freedom for β_0. Since the degrees of freedom for all sources must sum to the degrees of freedom for SS(Total), it follows that the degrees of freedom for error will equal the degrees of freedom for SS(Total), minus the sum of the degrees of freedom for main effects and interactions, i.e., $(n - 1) - 17$. Our experiment will contain three observations for each of the $2 \times 3 \times 3 = 18$ treatments; therefore, $n = (18)(3) = 54$, and the degrees of freedom for error will equal $53 - 17 = 36$.

If data for this experiment were analyzed on a computer, the computer printout would show the analysis of variance table that we have constructed and would include the associated mean squares, values of the F test statistics, and their observed significance levels. Each F statistic would represent the ratio of the source mean square to MSE $= s^2$.

EXAMPLE 11.18

A transistor manufacturer conducted an experiment to investigate the effects of three factors on productivity (measured in thousands of dollars of items produced) per 40-hour week. The factors were as follows:

1. Length of work week (two levels): five consecutive 8-hour days or four consecutive 10-hour days
2. Shift (two levels): day or evening shift
3. Number of coffee breaks (three levels): 0, 1, or 2

The experiment was conducted over a 24-week period with the $2 \times 2 \times 3 = 12$ treatments assigned in a random manner to the 24 weeks. The data for this completely randomized design are shown in Table 11.11. Perform an analysis of variance for the data.

T A B L E 11.11   **Data for Example 11.18**

		DAY SHIFT			NIGHT SHIFT		
		COFFEE BREAKS			COFFEE BREAKS		
		0	1	2	0	1	2
LENGTH OF	4 days	94	105	96	90	102	103
		97	106	91	89	97	98
WORK WEEK	5 days	96	100	82	81	90	94
		92	103	88	84	92	96

Solution

The data were subjected to an analysis of variance. The SAS printout is shown in Figure 11.19.

Pertinent sections of the SAS printout are boxed and numbered, as follows:

1. The value of SS(Total), shown in the **Corrected Total** row of box 1, is 1,091.833333. The degrees of freedom associated with this quantity is equal to $(n - 1) = (24 - 1) = 23$. Box 1 gives the partitioning (the analysis of variance) of this quantity into two sources of variation. The first source, **Model,**

FIGURE 11.19

SAS printout for ANOVA of data in Table 11.11

```
                        Analysis of Variance Procedure
Dependent Variable: DOLLARS
                                  Sum of            Mean
Source                  DF        Squares          Square    F Value    Pr > F

Model                   11     1009.833333       91.803030     13.43     0.0001
Error                   12       82.000000        6.833333       ①
Corrected Total         23     1091.833333

                R-Square            C.V.         Root MSE          DOLLARS Mean
          ④   0.924897          2.768647    ③   2.614065          94.4166667


Source                  DF        Anova SS      Mean Square    F Value    Pr > F

SHIFT                    1      48.1666667      48.1666667       7.05      0.0210
DAYS                     1     204.1666667     204.1666667      29.88   ②  0.0001
SHIFT*DAYS               1       8.1666667       8.1666667       1.20      0.2958
BREAKS                   2     334.0833333     167.0416667      24.45      0.0001
SHIFT*BREAKS             2     385.5833333     192.7916667      28.21      0.0001
DAYS*BREAKS              2       8.0833333       4.0416667       0.59      0.5689
SHIFT*DAYS*BREAKS        2      21.5833333      10.7916667       1.58      0.2461
```

corresponds to the 11 parameters (all except β_0) in the model. The second source is **Error**. The degrees of freedom, sums of squares, and mean squares for these quantities are shown in their respective columns. For example, MSE = 6.833333. The F statistic for testing

$$H_0: \quad \beta_1 = \beta_2 = \cdots = \beta_{11} = 0$$

is based on $\nu_1 = 11$ and $\nu_2 = 12$ degrees of freedom and is shown on the printout as $F = 13.43$. The observed significance level, shown under **Pr > F**, is .0001. This small observed significance level presents ample evidence to indicate that at least one of the three independent variables—shifts, number of days in a working week, or number of coffee breaks per day—contributes information for the prediction of mean productivity.

2. To determine which sets of parameters are actually contributing information for the prediction of y, we examine the breakdown (box 2) of SS(Model) into components corresponding to the sets of parameters for main effects **SHIFTS**, **DAYS**, and **BREAKS**, and parameters for two-way interactions, **SHIFT*DAYS**, **SHIFT*BREAKS**, and **DAYS*BREAKS**. The last **Model** source of variation corresponds to the set of all three-way **SHIFT*DAYS*BREAKS** parameters. Note that the degrees of freedom for these sources sum to 11, the number of degrees of freedom for **Model**. Similarly, the sum of the component sums of squares is equal to SS(Model). Box 2 does not give the mean squares associated with the sources, but it does give the F values associated with testing hypotheses concerning the set of parameters associated with each source. Box 2 also gives the observed significance levels of these tests. You can see that there is ample evidence to indicate the presence of a **SHIFT*BREAKS** interaction. The F tests associated with all three main effect parameter sets are also statistically significant at the $\alpha = .05$ level of significance. The practical implication of these results is that there is evidence to indicate that all three independent variables, shift, number of work days per week, and number of coffee breaks

per day, contribute information for the prediction of productivity. The presence of a **SHIFT*BREAKS** interaction means that the effect of the number of breaks on productivity is not the same from shift to shift. Thus, the specific number of coffee breaks that might achieve maximum productivity on one shift might be different from the number of breaks that will achieve maximum productivity on the other shift.

3. Box 3 gives the value of $s = \sqrt{\text{MSE}} = 2.614065$. This value would be used to construct a confidence interval to compare the difference between two of the 12 treatment means. The confidence interval for the difference between a pair of means, $(\mu_i - \mu_j)$, would be

$$(\bar{y}_i - \bar{y}_j) \pm t_{\alpha/2} s \sqrt{\frac{2}{r}}$$

where r is the number of replications of the factorial experiment within a completely randomized design. There were $r = 2$ observations for each of the 12 treatments (factor level combinations) in this example.

4. Box 4 gives the value of R^2, a measure of how well the model fits the experimental data. It is of value primarily when the number of degrees of freedom for error is large—say, at least 5 or 6. The larger the number of degrees of freedom for error, the greater will be its practical importance. The value of R^2 for this analysis, .924897, indicates that the model provides a fairly good fit to the data. It also suggests that the model could be improved by adding new predictor variables or, possibly, by including higher-order terms in the variables originally included in the model.

EXAMPLE 11.19

In a manufacturing process, a plastic rod is produced by heating a granular plastic to a molten state and then extruding it under pressure through a nozzle. An experiment was conducted to investigate the effect of two factors, extrusion temperature (°F) and pressure (pounds per square inch), on the rate of extrusion (inches per second) of the molded rod. A complete 2×2 factorial experiment (that is, with each factor at two levels) was conducted. Three batches of granular plastic were used for the experiment, with each batch (viewed as a block) divided into four equal parts. The four portions of granular plastic for a given batch were randomly assigned to the four treatments; this was repeated for each of the three batches, resulting in a 2×2 factorial experiment laid out in three blocks. The data are shown in Table 11.12. Perform an analysis of variance for these data.

TABLE 11.12   **Data for Example 11.19**

		BATCH (BLOCK)					
		1		**2**		**3**	
		PRESSURE		PRESSURE		PRESSURE	
		40	60	40	60	40	60
TEMPERATURE	200°	1.35	1.74	1.31	1.67	1.40	1.86
	300°	2.48	3.63	2.29	3.30	2.14	3.27

Solution

This experiment consists of a three-way classification of the data corresponding to batches (blocks), pressure, and temperature. The analysis of variance for this 2×2 factorial experiment (four treatments) laid out in a randomized block design (three blocks) yields the sources and degrees of freedom shown in Table 11.13.

The linear model for the experiment is

$$E(y) = \beta_0 + \overbrace{\beta_1 x_1}^{\substack{\text{Main}\\\text{effect}\\P}} + \overbrace{\beta_2 x_2}^{\substack{\text{Main}\\\text{effect}\\T}} + \overbrace{\beta_3 x_1 x_2}^{\substack{PT\\\text{inter-}\\\text{action}}} + \overbrace{\beta_4 x_3 + \beta_5 x_4}^{\substack{\text{Block}\\\text{terms}}}$$

where

$x_1 = $ Pressure $x_2 = $ Temperature

$x_3 = \begin{cases} 1 & \text{if block 2} \\ 0 & \text{otherwise} \end{cases}$ $x_4 = \begin{cases} 1 & \text{if block 3} \\ 0 & \text{otherwise} \end{cases}$

The SAS printout for the analysis of variance is shown in Figure 11.20. You can see from the printout that the F test for the model was highly significant (observed significance level is .0001). Thus, there is ample evidence to indicate differences among the block means, or the treatment means, or both. Proceeding to the breakdown of the model sources, you can see that the values of the F statistics for pressure, temperature, and the temperature–pressure interaction are all highly significant (that is, their observed significance levels are very small). Therefore, all of the terms ($\beta_1 x_1$, $\beta_2 x_2$, and $\beta_3 x_1 x_2$) contribute information for the prediction of y.

TABLE 11.13 Table of Sources and Degrees of Freedom for Example 11.19

SOURCE	df
Pressure (P)	1
Temperature (T)	1
Blocks	2
Pressure–temperature interaction	1
Error	6
Total	11

FIGURE 11.20

SAS printout for ANOVA of data in Table 11.12

```
                    Analysis of Variance Procedure
Dependent Variable: RATE
                                Sum of          Mean
Source              DF         Squares         Square    F Value    Pr > F
Model                5       7.14938333     1.42987667     83.23    0.0001
Error                6       0.10308333     0.01718056
Corrected Total     11       7.25246667

              R-Square         C.V.         Root MSE        RATE Mean
              0.985786      5.948924        0.131075        2.20333333

Source              DF       Anova SS     Mean Square    F Value    Pr > F
PRESSURE             1     1.68750000     1.68750000      98.22    0.0001
TEMP                 1     5.04403333     5.04403333     293.59    0.0001
PRESSURE*TEMP        1     0.36053333     0.36053333      20.98    0.0038
BATCH                2     0.05731667     0.02865833       1.67    0.2654
```

The treatments in the experiment were assigned according to a randomized block design. Thus, we expected the extrusion of the plastic to vary from batch to batch. Because the F test for testing differences among block means was not statistically significant (the observed significance level, **Pr > F**, was as large as .2654), there is insufficient evidence to indicate a difference in the mean extrusion of the plastic from batch to batch. Blocking does not appear to have increased the amount of information in the experiment.

Many other complex designs, such as fractional factorials, Latin square designs, and incomplete blocks designs, fall under the general k-way classification of data. Consult the references for the layout of these designs and the linear models appropriate for analyzing them.

EXERCISES

11.38 In increasingly severe oil well environments, oil producers are interested in high-strength nickel alloys that are corrosion-resistant. Since nickel alloys are especially susceptible to hydrogen embrittlement, an experiment was conducted to compare the yield strengths of nickel alloy tensile specimens cathodically charged in a 4% sulfuric acid solution saturated with carbon disulfide, a hydrogen recombination poison. Two alloys were combined: inconel alloy (75% nickel composition) and incoloy (30% nickel composition). The alloys were tested under two material conditions (cold rolled and cold drawn), each at three different charging times (0, 25, and 50 days). Thus, a $2 \times 2 \times 3$ factorial experiment was conducted, with alloy type at two levels, material condition at two levels, and charging time at three levels. Two hydrogen-charged tensile specimens were prepared for each of the $2 \times 2 \times 3 = 12$ factor level combinations. Their yield strengths (kilograms per square inch) are recorded in the table. The SAS analysis of variance printout for the data is also shown here.

		ALLOY TYPE							
		INCONEL				INCOLOY			
		Cold rolled		Cold drawn		Cold rolled		Cold drawn	
CHARGING TIME	0 days	53.4	52.6	47.1	49.3	50.6	49.9	30.9	31.4
	25 days	55.2	55.7	50.8	51.4	51.6	53.2	31.7	33.3
	50 days	51.0	50.5	45.2	44.0	50.5	50.2	29.7	28.1

Dependent Variable: YIELD

Source	DF	Sum of Squares	Mean Square	F Value	Pr > F
Model	11	1931.734583	175.612235	258.73	0.0001
Error	12	8.145000	0.678750		
Corrected Total	23	1939.879583			

R-Square	C.V.	Root MSE	YIELD Mean
0.995801	1.801942	0.823863	45.7208333

Source	DF	Anova SS	Mean Square	F Value	Pr > F
ALLOY	1	552.0004167	552.0004167	813.26	0.0001
MATCOND	1	956.3437500	956.3437500	1408.98	0.0001
ALLOY*MATCOND	1	339.7537500	339.7537500	500.56	0.0001
TIME	2	71.0408333	35.5204167	52.33	0.0001
ALLOY*TIME	2	7.9858333	3.9929167	5.88	0.0166
MATCOND*TIME	2	4.1725000	2.0862500	3.07	0.0836
ALLOY*MATCOND*TIME	2	0.4375000	0.2187500	0.32	0.7306

a. Is there evidence of any interactions among the three factors? Test using $\alpha = .05$. [*Note:* This means that you must test all the interaction parameters. The drop in SSE appropriate for the test would be the sum of all interaction sums of squares.]

b. Now examine the F tests shown on the printout for the individual interactions. Which, if any, of the interactions are statistically significant at the .05 level of significance?

11.39 Refer to Exercise 11.38. Since charging time is a quantitative factor, we could plot the strength y versus charging time x_1 for each of the four combinations of alloy type and material condition. This suggests that a prediction equation relating mean strength $E(y)$ to charging time x_1 may be useful. Consider the model

$$E(y) = \beta_0 + \beta_1 x_1 + \beta_2 x_1^2 + \beta_3 x_2 + \beta_4 x_3 + \beta_5 x_2 x_3$$
$$+ \beta_6 x_1 x_2 + \beta_7 x_1 x_3 + \beta_8 x_1 x_2 x_3$$
$$+ \beta_9 x_1^2 x_2 + \beta_{10} x_1^2 x_3 + \beta_{11} x_1^2 x_2 x_3$$

where

$x_1 = $ Charging time

$x_2 = \begin{cases} 1 & \text{if inconel alloy} \\ 0 & \text{if incoloy alloy} \end{cases}$ $x_3 = \begin{cases} 1 & \text{if cold rolled} \\ 0 & \text{if cold drawn} \end{cases}$

a. Using the model shown here, give the relationship between mean strength $E(y)$ and charging time x_1 for cold-drawn incoloy alloy.

b. Using the model shown here, give the relationship between mean strength $E(y)$ and charging time x_1 for cold-drawn inconel alloy.

c. Using the model shown here, give the relationship between mean strength $E(y)$ and charging time x_1 for cold-rolled inconel alloy.

d. The SAS multiple regression analysis for fitting the model to the data is shown at the top of page 620. Find the prediction equation.

e. Refer to part d. Find the prediction equations for each of the four combinations of alloy type and material condition.

f. Refer to part d. Plot the data points for each of the four combinations of alloy type and material condition. Graph the respective prediction equations.

11.40 Refer to Exercises 11.38–11.39. If the relationship between mean strength $E(y)$ and charging time x_1 is the same for all four combinations of alloy type and material condition, the appropriate model for $E(y)$ is

$$E(y) = \beta_0 + \beta_1 x_1 + \beta_2 x_1^2$$

Use the SAS printout on page 620 for fitting this model to the data, together with the information in the printout of Exercise 11.39, to decide whether the data provide sufficient evidence to indicate differences among the second-order models relating $E(y)$ to x_1 for the four categories of alloy type and material condition. Test using $\alpha = .05$.

SAS printout for Exercise 11.39

Dependent Variable: YIELD

Analysis of Variance

Source	DF	Sum of Squares	Mean Square	F Value	Prob>F
Model	11	1931.73458	175.61223	258.729	0.0001
Error	12	8.14500	0.67875		
C Total	23	1939.87958			

Root MSE	0.82386	R-square	0.9958	
Dep Mean	45.72083	Adj R-sq	0.9920	
C.V.	1.80194			

Parameter Estimates

Variable	DF	Parameter Estimate	Standard Error	T for H0: Parameter=0	Prob > |T|
INTERCEP	1	31.150000	0.58255901	53.471	0.0001
X1	1	0.153000	0.05940960	2.575	0.0243
X1SQ	1	-0.003960	0.00114158	-3.469	0.0046
X2	1	17.050000	0.82386285	20.695	0.0001
X3	1	19.100000	0.82386285	23.183	0.0001
X2X3	1	-14.300000	1.16511802	-12.273	0.0001
X1X2	1	0.151000	0.08401786	1.797	0.0975
X1X3	1	0.017000	0.08401786	0.202	0.8430
X1X2X3	1	-0.080000	0.11881919	-0.673	0.5135
X1SQX2	1	-0.003560	0.00161443	-2.205	0.0477
X1SQX3	1	0.000600	0.00161443	0.372	0.7166
X1SQX2X3	1	0.001200	0.00228316	0.526	0.6087

SAS printout for Exercise 11.40

Dependent Variable: YIELD

Analysis of Variance

Source	DF	Sum of Squares	Mean Square	F Value	Prob>F
Model	2	71.04083	35.52042	0.399	0.6759
Error	21	1868.83875	88.99232		
C Total	23	1939.87958			

Root MSE	9.43357	R-square	0.0366	
Dep Mean	45.72083	Adj R-sq	-0.0551	
C.V.	20.63299			

Parameter Estimates

Variable	DF	Parameter Estimate	Standard Error	T for H0: Parameter=0	Prob > |T|
INTERCEP	1	45.650000	3.33527213	13.687	0.0001
X1	1	0.217000	0.34013235	0.638	0.5304
X1SQ	1	-0.005140	0.00653577	-0.786	0.4404

11.41   A study was conducted to evaluate the use of computer-assisted instruction (CAI) in teaching an introductory FORTRAN programming course (GE 102) in the College of Engineering at Oregon State University (*Engineering Education*, Feb. 1986). One of the objectives was to investigate the effect of four factors on a student's final exam score (y) in the course. The factors and their respective levels are as follows:

> *Group* (3 levels):
> Control group (student receives no CAI)
> Guided CAI (student receives CAI, but proceeds at same pace as normal class)
> Self-paced CAI (student receives CAI and proceeds at his or her own pace)

Math background (4 levels):
High school algebra
Trigonometry
Differential calculus
Integral calculus

Computer background (3 levels):
None (little or no exposure)
Some (one programming language)
Extensive (more than one programming language)

Grade in prerequisite course, GE 101 (3 levels):
A, B, or C

a. How many treatments are associated with this four-way factorial experiment?

b. Write the complete factorial model for this experiment. [*Hint:* Use dummy variables to represent the factors.]

c. What hypothesis would you test to determine whether interaction exists among the four factors?

11.42 A 2×2 factorial experiment was conducted for each of 3 weeks to determine the effect of two factors, temperature and pressure, on the yield of a chemical. Temperature was set at 300° and 500°. The pressure maintained in the reactor was set at 100 and 200 pounds per square inch. Four days were randomly selected within each week and the four factor level combinations were randomly assigned to them.

a. What type of design was used for this experiment?

b. Construct an analysis of variance table showing all sources and their respective degrees of freedom.

11.7

Follow-Up Analysis: Tukey's Multiple Comparisons of Means

Many practical experiments are conducted to determine the largest (or the smallest) mean in a set. For example, suppose a drugstore is considering five floor displays for a new product. The drugstore would want to determine which display yields the greatest weekly sales of the product. Similarly, a production engineer might want to determine which among six machines or which among three foremen achieves the highest mean productivity per hour. A stockbroker might want to choose one stock, from among four, that yields the highest mean return, and so on.

Once differences among, say, five treatment means have been detected in an ANOVA, choosing the treatment with the largest mean might appear to be a simple matter. We could, for example, obtain the sample means $\bar{y}_1, \bar{y}_2, \ldots, \bar{y}_5$, and compare them by constructing a $(1 - \alpha)100\%$ confidence interval for the difference between each pair of treatment means. However, there is a problem associated with this procedure: **A confidence interval for $\mu_i - \mu_j$, with its corresponding value of α, is valid only when the two treatments (i and j) to be compared are selected prior to experimentation.** After you have looked at the data, you cannot use a confidence interval to compare the treatments for the largest and smallest sample means because they will always be farther apart, on the average, than any pair of treatments selected at random. Furthermore, if you construct a series of confidence intervals, each with a chance α of indicating a difference between a pair of means if no difference exists, then the risk of making *at least one* Type I error in the series of inferences will be larger than the value of α specified for a single interval.

There are a number of procedures for comparing and ranking a group of treatment means as part of a **follow-up analysis** to the ANOVA. The one that we present in this section, known as **Tukey's method for multiple comparisons**, utilizes the Studentized range

$$q = \frac{\bar{y}_{\max} - \bar{y}_{\min}}{s/\sqrt{n}}$$

(where $\bar{y}_{\max}$ and $\bar{y}_{\min}$ are the largest and smallest sample means, respectively) to determine whether the difference in any pair of sample means implies a difference in the corresponding treatment means. The logic behind this **multiple comparisons procedure** is that if we determine a critical value for the difference between the largest and smallest sample means, $|\bar{y}_{\max} - \bar{y}_{\min}|$, one that implies a difference in their respective treatment means, then any other pair of sample means that differ by as much as or more than this critical value would also imply a difference in the corresponding treatment means. Tukey's (1949) procedure selects this critical distance, ω, so that the probability of making one or more Type I errors (concluding that a difference exists between a pair of treatment means if, in fact, they are identical) is α. Therefore, the risk of making a Type I error applies to the whole procedure, i.e., to the comparisons of all pairs of means in the experiment, rather than to a single comparison. Consequently, the value of α selected by the researchers is called an **experimentwise error rate** (in contrast to a **comparisonwise error rate**).

Tukey's procedure relies on the assumption that the p sample means are based on independent random samples, *each containing an equal number n_t of observations*. Then if $s = \sqrt{\text{MSE}}$ is the computed standard deviation for the analysis, the distance ω is

$$\omega = q(p, \nu)\frac{s}{\sqrt{n_t}}$$

The tabulated statistic $q_\alpha(p, \nu)$ is the critical value of the Studentized range, the value that locates α in the upper tail of the q distribution. This critical value depends on α, the number of treatment means involved in the comparison, and ν, the number of degrees of freedom associated with MSE, as shown in the box. Values of $q(p, \nu)$ for $\alpha = .05$ and $\alpha = .01$ are given in Tables 11 and 12, respectively, of Appendix D.

| E X A M P L E 11.20

Refer to the ANOVA for the completely randomized design, Examples 11.5 and 11.6. Recall that we rejected the null hypothesis of no differences among the mean amounts owed for the three income groups. Use Tukey's method to compare the three treatment means.

Solution

STEP 1 For this follow-up analysis, we will select an experimentwise error rate of $\alpha = .05$.

| Tukey's Multiple Comparisons Procedure: Equal Sample Sizes

1. Select the desired experimentwise error rate, α.

2. Calculate

$$\omega = q_\alpha(p, \nu)\frac{s}{\sqrt{n_t}}$$

where

$\quad p$ = Number of sample means

$\quad s = \sqrt{MSE}$

$\quad \nu$ = Number of degrees of freedom associated with MSE

$\quad n_t$ = Number of observations in each of the p samples

$\quad q_\alpha(p, \nu)$ = Critical value of the Studentized range (Tables 11 and 12 of Appendix D)

3. Calculate and rank the p sample means.

4. Place a bar over those pairs of treatment means that differ by less than ω. A pair of treatments not connected by an overbar (i.e., differing by more than ω) implies a difference in the corresponding population means.

Note: The confidence level associated with all inferences drawn from the analysis is $(1 - \alpha)$.

STEP 2 From previous examples, we have $(p - 3)$ treatments, $\nu = 27$ df for error, $s = \sqrt{MSE} = 168.95$, and $n_t = 10$ observations per treatment. The critical value of the Studentized range (obtained from Table 11, Appendix D) is $q_{.05}(3, 27) \approx 3.5$. Substituting these values into the formula for ω, we obtain

$$\omega = q_{.05}(3, 27)\left(\frac{s}{\sqrt{n_t}}\right) = 3.5\left(\frac{168.95}{\sqrt{10}}\right) = 187.0$$

STEP 3 The sample means for the three income groups (obtained from Table 11.1) are

$$\bar{y}_A = 229.6 \qquad \bar{y}_B = 309.8 \qquad \bar{y}_C = 427.8$$

STEP 4 Based on the critical difference $\omega = 187$, the three treatment means are ranked as follows:

Sample means	229.6	309.9	427.8
Treatments	Under 12,000	12,000–25,000	Over 25,000
	(A)	(B)	(C)

This same information can be obtained using a statistical software package. The SAS printout of the Tukey analysis is shown in Figure 11.21. Tukey's critical

FIGURE 11.21
SAS printout of Tukey analysis,
Example 11.20

```
                    Analysis of Variance Procedure

        Tukey's Studentized Range (HSD) Test for variable: AMOUNT

   NOTE: This test controls the type I experimentwise error rate, but
         generally has a higher type II error rate than REGWQ.

                Alpha= 0.05  df= 27  MSE= 28543.37
             Critical Value of Studentized Range= 3.506
                Minimum Significant Difference= 187.33

   Means with the same letter are not significantly different.

        Tukey Grouping              Mean     N  INCCLASS

                        A         427.80     10  GROUP_C
                        A
                   B    A         309.90     10  GROUP_B
                   B
                   B              229.60     10  GROUP_A
```

difference, $\omega = 187.33$, is shaded on the printout. (This value differs slightly from our calculated value because of rounding.) Note that SAS lists the treatment means vertically in descending order. Treatment means connected by the same letter (A, B, C, etc.) are *not* significantly different.

From this information we infer that the mean amount owed for income group C is significantly larger than the mean amount owed for income group A, since $\bar{y}_C$ exceeds $\bar{y}_A$ by more than the critical value. However, the treatment pairs (A, B) and (B, C) are connected by a bar (or the same letter) since neither $(\bar{y}_B - \bar{y}_A)$ nor $(\bar{y}_C - \bar{y}_B)$ exceeds ω. This indicates that the sample means for these pairs of treatments are not significantly different. Practically, these results imply that income group C has the highest mean amount owed and income group A has the lowest. The mean for income group B, however, is not significantly different from either of the other two means. These inferences are made with an overall confidence level of $(1 - \alpha) = .95$.

EXAMPLE 11.21

A transistor manufacturer conducted an experiment to investigate the effects of two factors on productivity (measured in thousands of dollars of items produced) per 40-hour week. The factors were:

Length of work week (two levels):   five consecutive 8-hour days or
                                      four consecutive 10-hours days

Number of coffee breaks (three levels):   0, 1, or 2

The experiment was conducted over a 12-week period with the $2 \times 3 = 6$ treatments assigned in a random manner to the 12 weeks. The data for this two-factor factorial experiment are shown in Table 11.14. (Note that this experiment is a simpler version of the design used in Example 11.18.)

a. Perform an analysis of variance for the data.

b. Compare the six population means using Tukey's multiple comparisons procedure. Use $\alpha = .05$.

TABLE 11.14 Data for Example 11.21

| | | COFFEE BREAKS | | |
		0	1	2
LENGTH OF WORK WEEK	4 days	101 102	104 107	95 92
	5 days	95 93	109 110	83 87

Solution

a. The SAS printout of the ANOVA for the 2 × 3 factorial is displayed in Figure 11.22. Note that the test for interaction between the two factors, length (L) and breaks (B), is significant at $\alpha = .01$. (The p-value, .0051, is shaded on the printout.) Since interaction implies that the level of length (L) that yields the highest mean productivity may differ across different levels of breaks (B), we ignore the tests for main effects and focus our investigation on the individual treatment means.

FIGURE 11.22

SAS ANOVA printout for Example 11.21

Analysis of Variance Procedure

Dependent Variable: PRODUCT

Source	DF	Sum of Squares	Mean Square	F Value	Pr > F
Model	5	811.6666667	162.3333333	48.70	0.0001
Error	6	20.0000000	3.3333333		
Corrected Total	11	831.6666667			

R-Square	C.V.	Root MSE	PRODUCT Mean
0.975952	1.859839	1.825742	98.1666667

Source	DF	Anova SS	Mean Square	F Value	Pr > F
LENGTH	1	48.0000000	48.0000000	14.40	0.0090
BREAKS	2	667.1666667	333.5833333	100.07	0.0001
LENGTH*BREAKS	2	96.5000000	48.2500000	14.48	0.0051

b. The sample means for the six factor level combinations are shown in Table 11.15. Since the sample means in the table represent measures of productivity in the manufacture of transistors, we would want to find the length of work week and number of coffee breaks that yield the highest mean productivity.

TABLE 11.15 Sample Means for the $p = 6$ Treatments of Example 11.21

| | | COFFEE BREAKS, B | | |
		0	1	2
LENGTH OF WORK WEEK, L	4 days	101.5	105.5	93.5
	5 days	94.0	109.5	85.0

The first step in the ranking procedure is to calculate ω for $p = 6$ (we are ranking six treatment means), $n_t = 2$ (two observations per treatment), $\alpha = .05$, and $s = \sqrt{MSE} = \sqrt{3.33} = 1.83$ (where MSE is given in Figure 11.22). Since MSE is based on $\nu = 6$ degrees of freedom, we have

$$q_{.05}(6, 6) = 5.63$$

and

$$\omega = q_{.05}(6, 6)\left(\frac{s}{\sqrt{n_t}}\right)$$

$$= (5.63)\left(\frac{1.83}{\sqrt{2}}\right)$$

$$= 7.27$$

Therefore, population means corresponding to pairs of sample means that differ by more than $\omega = 7.27$ will be judged to be different. The six sample means are ranked as follows:

Sample means	85.0	93.5	94.0	101.5	105.5	109.5
Treatments	(5, 2)	(4, 2)	(5, 0)	(4, 0)	(4, 1)	(5, 1)
(Length, Breaks)						

Using $\omega = 7.27$ as a yardstick to determine differences between pairs of treatments, we have placed connecting bars over those means that *do not* significantly differ. The following conclusions can be drawn:

1. We see that there is evidence of a difference between the population mean of the treatment corresponding to a 5-day work week with two coffee breaks (with the smallest sample mean of 85.0) and every other treatment mean. Therefore, we can conclude that the 5-day, 2-break work week yields the lowest mean productivity among all length–break combinations.
2. The population mean of the treatment corresponding to a 5-day, 1-break work week (with the largest sample mean of 109.5) is significantly larger than the treatments corresponding to the four smallest sample means. However, there is no evidence of a difference between the 5-day, 1-break treatment mean and the 4-day, 1-break treatment mean (with a sample mean of 105.5).
3. There is no evidence of a difference between the 4-day, 1-break treatment mean (with a sample mean of 105.5) and the 4-day, 0-break treatment mean (with a sample mean of 101.5). Both these treatments, though, have significantly larger means than the treatments corresponding to the three smallest sample means.
4. There is no evidence of a difference among the treatments corresponding to the sample means 93.5 and 94.0. Further experimentation would be required to determine whether the observed differences in these means really imply a difference in the corresponding sample means.

In summary, the treatment means appear to fall into four groups, as shown below:

	TREATMENTS (LENGTH, BREAKS)
Group 1 (lowest mean productivity)	(5, 2)
Group 2	(4, 2) and (5, 0)
Group 3	(4, 0) and (4, 1)
Group 4 (highest mean productivity)	(4, 1) and (5, 1)

Notice that it is unclear where we should place the treatment corresponding to a 4-day, 1-break work week because of the overlapping bars above its sample mean, 105.5. That is, although there is sufficient evidence to indicate that treatments (4, 0) and (5, 1) differ, neither has been shown to differ significantly from treatment (4, 1). Tukey's method guarantees that the probability of making one or more Type I errors in these pairwise comparisons is only $\alpha = .05$.

Remember that Tukey's multiple comparisons procedure requires the sample sizes associated with the treatments to be equal. This, of course, will be satisfied for the randomized block designs and factorial experiments described in Sections 11.4 and 11.5, respectively. The sample sizes, however, may not be equal in a completely randomized design (Section 11.3). In this case a modification of Tukey's method (sometimes called the Tukey–Kramer method) is necessary, as described in the box. The technique requires that the critical difference ω_{ij} be calculated for each pair of treatments (i, j) in the experiment and pairwise comparisons made based on the appropriate value of ω_{ij}. However, when Tukey's method is used with unequal sample sizes, the value of α selected a priori by the researcher only approximates the true experimentwise error rate. In fact, when applied to unequal sample sizes, the procedure has been found to be more conservative, i.e., less likely to detect differences between pairs of treatment means when they exist, than in the case of equal sample sizes. For this reason, researchers sometimes look to alternative methods of multiple comparisons when the sample sizes are unequal. Two of these methods are presented in optional Section 11.8.

In general, multiple comparisons of treatment means should be performed only as a follow-up analysis to the ANOVA, i.e., only after we have conducted the appropriate analysis of variance F test(s) and determined that sufficient evidence exists of differences among the treatment means. Be wary of conducting multiple comparisons when the ANOVA F test indicates no evidence of a difference among a small number of treatment means—this may lead to confusing and contradictory results.*

---

*When a large number of treatments are to be compared, a borderline, nonsignificant F value (e.g., $.05 < p$-value $< .10$) may mask differences between some of the means. In this situation, it is better to ignore the F test and proceed directly to a multiple comparisons procedure.

Tukey's Approximate Multiple Comparisons Procedure for Unequal Sample Sizes

1. Calculate for each treatment pair (i, j)

$$\omega_{ij} = q_\alpha(p, \nu) \frac{s}{\sqrt{2}} \sqrt{\frac{1}{n_i} + \frac{1}{n_j}}$$

where

p = Number of sample means

$s = \sqrt{MSE}$

ν = Number of degrees of freedom associated with MSE

n_i = Number of observations in sample for treatment i

n_j = Number of observations in sample for treatment j

$q_\alpha(p, \nu)$ = Critical value of the Studentized range (Tables 11 and 12 of Appendix D)

2. Rank the p sample means and place a bar over any treatment pair (i, j) that differs by less than ω_{ij}. Any pair of sample means not connected by an overbar (i.e., differing by more than ω) implies a difference in the corresponding population means.

Note: This procedure is approximate, i.e., the value of α selected by the researcher approximates the true probability of making at least one Type I error.

Warning

In practice, it is advisable to avoid conducting multiple comparisons of a small number of treatment means when the corresponding ANOVA F test is nonsignificant; otherwise, confusing and contradictory results may occur.

EXERCISES

11.43   In business, the prevailing theory is that companies can be categorized into one of the following four types based on their strategic profile: reactors (are marginal competitors, unstable, victims of industry forces); defenders (specialize in established products, lower costs while maintaining quality); prospectors (develop new/improved products); and analyzers (operate in two product areas—one stable, one dynamic). The *American Business Review* (Jan. 1990) reported on a study that proposes a fifth organization type, balancers, who operate in three product spheres—one stable and two dynamic.

Each firm in a sample of 78 glassware firms was categorized into one of these five types, and the level of performance (process research and development ratio) of each was measured.

a. A completely randomized design ANOVA of the data resulted in a significant (at $\alpha = .05$) F value for treatments (organization types). Interpret this result.

b. Multiple comparisons of the five mean performance levels (using a procedure similar to Tukey's, at $\alpha = .05$) are summarized in the following table. Interpret the results.

Mean	.138	.235	.820	.826	.911
Type	Reactor	Prospector	Defender	Analyzer	Balancer

Source: Wright, P., et al. "Business performance and conduct of organization types: A study of select special-purpose and laboratory glassware firms." *American Business Review*, Jan. 1990, p. 95 (Table 4).

11.44 Refer to Exercise 11.12. Use Tukey's multiple comparisons procedure to compare the mean rates of penetration for the three types of drill bits. Identify the means that appear to differ. Use $\alpha = .05$.

11.45 Refer to the *Human Factors* (Apr. 1990) study of the performance of a computerized speech recognizer, Exercise 4.32. Accuracy was measured at three levels (90%, 95%, and 99%) and vocabulary size at three levels (75%, 87.5%, and 100%). The data on task completion times (minutes) were subjected to an analysis of variance for a 3×3 factorial design. The F test for accuracy–vocabulary interaction resulted in a p-value less than .0003.

a. Interpret the result of the test for interaction.

b. As a follow-up to the test for interaction, the mean task completion times for the three levels of accuracy were compared under each level of vocabulary. Do you agree with this method of analysis? Explain.

c. Refer to part **b.** Tukey's multiple comparisons method was used to compare the three accuracy means within each level of vocabulary at an experimentwise error rate of $\alpha = .05$. The results are summarized here. Interpret these results.

VOCABULARY SIZE	ACCURACY LEVEL		
	99%	95%	90%
75%	15.49	19.29	22.19
87.5%	12.77	14.31	16.48
100%	8.67	9.68	11.88

Source: Casali, S. P., Williges, B. H., and Dryden, R. D. "Effects of recognition accuracy and vocabulary size of a speech recognition system on task performance and user acceptance." *Human Factors*, Vol. 32, No. 2, Apr. 1990, p. 190 (Figure 2).

11.46 Refer to the *Journal of Applied Psychology* study introduced in Exercise 11.15. The sample mean earnings per day for each of the three groups of workers are provided here. Use Tukey's method to rank the treatment means. Use $\alpha = .09$.

Group 1 (*Specific difficult*) $1.91
Group 2 (*Do best*) $1.30
Group 3 (*No goal*) $ .96

11.47 Refer to Exercise 11.19. Use Tukey's multiple comparisons procedure to compare the mean prices of grocery items for the three supermarkets. Identify the means that appear to differ. Use $\alpha = .01$.

11.48   Refer to Exercise 11.23. Use Tukey's multiple comparisons procedure to compare the mean in-tray performance scores assessed by the three raters. Identify the means that appear to differ. Use $\alpha = .05$.

11.49   Refer to Exercise 11.27. Use Tukey's multiple comparisons procedure to compare the productivity means for the $2 \times 3 = 6$ machine–material combinations. Identify the means that appear to differ. Use $\alpha = .05$.

11.50   Refer to Exercise 8.30. Use Tukey's multiple comparisons procedure to compare the mean satisfaction scores for the three peak period lengths under each of the three pricing ratios. Identify the means that appear to differ under each pricing ratio. Use $\alpha = .01$.

11.51   Refer to the *Academy of Management Journal* study of Exercise 11.35. The sample mean satisfaction ratings of managers for the four combinations of prior information and degree of choice are shown in the accompanying table. Use Tukey's method to rank the four means. Use $\alpha = .05$.

		PRIOR INFORMATION	
		Realistic Preview	Traditional Announcement
DEGREE OF CHOICE	High	6.20	6.06
	Low	5.33	4.82

Source: Hicks, W. D. and Klimoski, R. J. "Entry into training programs and its effects on training outcomes: A field experiment." *Academy of Management Journal*, Vol. 30, No. 3, Sept. 1987, p. 548.

11.52   Refer to the 8×11 factorial experiment in Exercise 11.37. The mean ratings for the 11 CEOs are given in the table for the credibility item "well-known person." Compare the means using Tukey's multiple comparisons procedure. Use $\alpha = .05$. [*Note:* Higher ratings indicate a greater degree of credibility.]

CEO (COMPANY)	MEAN RATING
1. David Mahoney (Avis)	3.74
2. Franco Bolla (Bolla Wines)	4.57
3. Lee Iacocca (Chrysler)	6.47
4. L. Stanley Crane (Conrail)	2.21
5. Constantine Demmas (Demmas Financial Services)	2.16
6. W. H. Bricker (Diamond Shamrock)	2.67
7. Frank Borman (Eastern)	6.16
8. Herb Imhoff (General Employment Enterprises)	2.07
9. Guy Milner (Norrell Temporary Services)	2.88
10. Frank B. Hall (F. B. Hall & Company)	2.31
11. L. S. Shoen (U-Haul)	3.66

11.8
Other Multiple Comparisons Methods (Optional)

In this optional section we present two alternatives to Tukey's method of multiple comparisons of treatment means. The choice of methods will depend on the type of experimental design used and the particular error rate that the researcher wants to control.

Scheffé Method

Recall that Tukey's method of multiple comparisons is designed to control the experimentwise error rate, i.e., the probability of making at least one Type I error in the comparison of *all pairs* of treatment means in the experiment. Therefore, Tukey's method should be applied when you are interested in pairwise comparisons only.

Scheffé (1953) developed a more general procedure for comparing all possible linear combinations of the treatment means, called **contrasts**.

| Definition 11.3

A **contrast** L is a linear combination of the p treatment means in a designed experiment, i.e.,

$$L = \sum_{i=1}^{p} c_i \mu_i$$

where the constants $c_1, c_2, \ldots, c_p$ sum to 0, i.e., $\sum_{i=1}^{p} c_i = 0$.

For example, in an experiment with four treatments (A, B, C, D), you might want to compare the following contrasts, where μ_i represents the population mean for treatment i:

$$L_1 = \frac{\mu_A + \mu_B}{2} - \frac{\mu_C + \mu_D}{2}$$

$$L_2 = \mu_A - \mu_D$$

$$L_3 = \frac{\mu_B + \mu_C + \mu_D}{3} - \mu_A$$

The contrast L_2 involves a comparison of a pair of treatment means, whereas L_1 and L_3 are more complex comparisons of the treatments. Thus, pairwise comparisons are special cases of general contrasts.

As in Tukey's method, the value of α selected by the researcher using Scheffé's method applies to the procedure as a whole, i.e., to the comparisons of all possible contrasts (not just those considered by the researcher). Unlike Tukey's method, however, the probability of at least one Type I error, α, is exact regardless of whether the sample sizes are equal. For this reason, some researchers prefer Scheffé's method to Tukey's method in the case of unequal samples, even if only pairwise comparisons of treatment means are made.

The Scheffé method for general contrasts is outlined in the box on page 632.

In the special case of all pairwise comparisons in an experiment with four treatments, the relevant contrasts reduce to $L_1 = \mu_A - \mu_B$, $L_2 = \mu_A - \mu_C$, $L_3 = \mu_A - \mu_D$, and so forth. Notice that for each of these contrasts $\sum c_i^2/n_i$ reduces to $(1/n_i + 1/n_j)$, where n_i and n_j are the sizes of treatments i and j, respectively. [For example, for contrast L_1, $c_1 = 1$, $c_2 = -1$, $c_3 = c_4 = 0$, and $\sum c_i^2/n_i = (1/n_1 + 1/n_2)$.] Consequently, the formula for S in the box (page 633) can be simplified and pairwise comparisons made using the technique of Section 8.8.

Scheffé's Multiple Comparisons Procedure for General Contrasts

1. For each contrast $L = \sum_{i=1}^{p} c_i \mu_i$, calculate

$$\hat{L} = \sum_{i=1}^{p} c_i \bar{y}_i$$

and

$$S = \sqrt{(p - 1)F_\alpha(p - 1, \nu)\text{MSE} \sum_{i=1}^{p} \left(\frac{c_i^2}{n_i} \right)}$$

where

$p = $ Number of sample (treatment) means

$\text{MSE} = $ Mean squared error

$n_i = $ Number of observations in sample for treatment i

$\bar{y}_i = $ Sample mean for treatment i

$F_\alpha(p - 1, \nu) = $ Critical value of F distribution with $\nu_1 = p - 1$ numerator df and $\nu_2 = \nu$ denominator df (Tables 3, 4, 5, and 6 of Appendix D)

$\nu = $ Number of degrees of freedom associated with MSE

2. Calculate the confidence interval $\hat{L} \pm S$ for each contrast. The confidence coefficient, $1 - \alpha$, applies to the procedure as a whole, i.e., to the entire set of confidence intervals for all possible contrasts.

EXAMPLE 11.22

Refer to the 2×3 factorial experiment to investigate the effects of length of work week and number of coffee breaks on mean productivity in Example 11.21. Compare the population means corresponding to the six length–break combinations (treatments) using the Scheffé method of multiple comparisons. Use $\alpha = .05$.

Solution

From Figure 11.22 in Example 11.21, we have MSE $= 3.33$, $p = 6$, and $\nu = $ df(Error) $= 6$. Also, there are two observations per treatment, $n_i = n_j = 2$, for all treatment pairs. Consequently, the critical difference S_{ij} will be the same for all treatment pairs (i, j). From Table 4 in Appendix D, we have $F_{.05} = 4.39$ (based on $p - 1 = 5$ numerator df and $\nu = 6$ denominator df). Then, Scheffé's critical difference S_{ij} is

$$S_{ij} = \sqrt{(p - 1)F_{.05}\text{MSE}\left(\frac{1}{n_i} + \frac{1}{n_j} \right)}$$

$$= \sqrt{(5)(4.39)(3.33)\left(\frac{1}{2} + \frac{1}{2} \right)} = 8.55$$

Scheffé's Multiple Comparisons Procedure for Pairwise Comparisons of Treatment Means

1. Calculate Scheffé's critical difference for each pair of treatments (i, j):

$$S_{ij} = \sqrt{(p - 1)F_\alpha(p - 1, \nu)\text{MSE}\left(\frac{1}{n_i} + \frac{1}{n_j}\right)}$$

where

p = Number of sample (treatment) means

MSE = Mean squared error

n_i = Number of observations in sample for treatment i

n_j = Number of observations in sample for treatment j

$F_\alpha(p - 1, \nu)$ = Critical value of F distribution with $\nu_1 = p - 1$ numerator df and $\nu_2 = \nu$ denominator df (Tables 3, 4, 5, and 6 of Appendix D)

ν = Number of degrees of freedom associated with MSE

2. Rank the p sample means and place a bar over any treatment pair (i, j) that differs by less than S_{ij}. Any pair of sample means not connected by an overbar implies a difference in the corresponding population means.

Treatment means differing by more than $S = 8.55$ will imply a significant difference between the corresponding population means. The rankings of the treatment means, with overbars indicating "no evidence of a difference," are shown here:

Sample means	85.0	93.5	94.0	101.5	105.5	109.5
Treatments (Length, Breaks)	(5, 2)	(4, 2)	(5, 0)	(4, 0)	(4, 1)	(5, 1)

Recall that the manufacturer's goal is to determine the treatment(s) yielding the highest mean productivity. Based on this result, the six means can be grouped as follows:

	TREATMENTS (LENGTH, BREAKS)
Group 1 (lowest mean productivity)	(5, 2) and (4, 2)
Group 2	(4, 2), (5, 0), and (4, 0)
Group 3 (highest mean productivity)	(4, 0), (4, 1), and (5, 1)

You can see that the treatments corresponding to 5-day, 1-break and 4-day, 1-break work weeks produce the highest mean productivity. The Scheffé method did not detect a significant difference between these treatments and treatment (4, 0), or a significant difference between treatment (4, 0) and either treatment (4, 2) or (5, 0). Thus, we are not certain where to place treatment (4, 0)—in the group with the highest mean productivity (group 3) or in the middle group (group 2).

Note that in Example 11.22, the Scheffé method produced a critical difference of $S = 8.55$—a value larger than Tukey's critical difference of $\omega = 7.27$ (Example 11.21). This implies that Tukey's method produces narrower confidence intervals than Scheffé's method for differences in pairs of treatment means. Therefore, if only pairwise comparisons of treatments are to be made, Tukey's is the preferred method as long as the sample sizes are equal. The Scheffé method, on the other hand, yields narrower confidence intervals (i.e., smaller critical differences) for situations in which the goal of the researchers is to make comparisons of general contrasts.

Bonferroni Approach

As noted earlier, Tukey's multiple comparisons procedure in the case of unequal sample sizes is approximate. That is, the value of α selected a priori only approximates the true probability of making at least one Type I error. The Bonferroni approach is an exact method that is applicable in either the equal or the unequal sample size case. Further, Bonferroni's procedure covers all possible comparisons of treatments, including pairwise comparisons, general contrasts, or combinations of pairwise comparisons and more complex contrasts.

The Bonferroni approach is based on the following result (proof omitted): If g comparisons are to be made, each with confidence coefficient $1 - \alpha/g$, then the overall probability of making one or more Type I errors (i.e., the experimentwise error rate) is at most α. That is, the set of intervals constructed using the Bonferroni method yields an overall confidence level of at least $1 - \alpha$. For example, if you want to construct $g = 2$ confidence intervals with an experimentwise error rate of at most $\alpha = .05$, then each individual interval must be constructed using a confidence level of $1 - .05/2 = .975$.

When applied to only pairwise comparisons of treatments, the Bonferroni approach can be carried out as shown in the box on page 636.

EXAMPLE 11.23

Refer to the two-factor factorial experiment in Example 11.21. Use Bonferroni's method to perform pairwise comparisons of the six treatment means. Use $\alpha = .05$.

Solution

From Examples 11.21 and 11.22, we have $p = 6$, $s = \sqrt{MSE} = 1.83$, $\nu = 6$, and $n_i = n_j = 2$ for all treatment pairs (i, j). For $p = 6$ means, the number of pairwise comparisons to be made is

$$g = \frac{p(p - 1)}{2} = \frac{6(5)}{2} = 15$$

> ## Bonferroni Multiple Comparisons Procedure for General Contrasts
>
> 1. For each contrast $L = \sum_{i=1}^{p} c_i \mu_i$, calculate
>
> $$\hat{L} = \sum_{i=1}^{p} c_i \bar{y}_i$$
>
> and
>
> $$B = t_{\alpha/(2g)} s \sqrt{\sum_{i=1}^{p} \left(\frac{c_i^2}{n_i} \right)}$$
>
> where
>
> p = Number of sample (treatment) means
>
> g = Number of contrasts
>
> $s = \sqrt{\text{MSE}}$
>
> ν = Number of degrees of freedom associated with MSE
>
> n_i = Number of observations in sample for treatment i
>
> $\bar{y}_i$ = Sample mean for treatment i
>
> $t_{\alpha/(2g)}$ = Critical value of t distribution with ν df and tail area $\alpha/(2g)$ (Table 2 in Appendix D)
>
> 2. Calculate the confidence interval $\hat{L} \pm B$ for each contrast. The confidence coefficient for the procedure as a whole, i.e., for the entire set of confidence intervals, is *at least* $1 - \alpha$.

Thus, we need to find the critical value, $t_{\alpha/(2g)} = t_{.05/[2(15)]} = t_{.0017}$, for the t distribution based on $\nu = 6$ df. This value, although not shown in Table 2 in Appendix D, is approximately 4.7.* Substituting $t_{.0017} \approx 4.7$ into the equation for Bonferroni's critical difference B_{ij}, we have

$$B_{ij} \approx t_{.0017} s \sqrt{\frac{1}{n_i} + \frac{1}{n_j}} = (4.7)(1.83) \sqrt{\frac{1}{2} + \frac{1}{2}} = 8.60$$

for any treatment pair (i, j).

Using the value $B = 8.60$ to detect significant differences between treatment means, we obtain the following results:

Sample means	85.0	93.5	94.0	101.5	105.5	109.5
Treatments (Length, Breaks)	(5, 2)	(4, 2)	(5, 0)	(4, 0)	(4, 1)	(5, 1)

---

*We obtained the value using the SAS probability generating function for a Student's t distribution.

Bonferroni Multiple Comparisons Procedure for Pairwise Comparisons of Treatment Means

1. Calculate for each treatment pair (i, j)

$$B_{ij} = t_{\alpha/(2g)}s \sqrt{\frac{1}{n_i} + \frac{1}{n_j}}$$

where

p = Number of sample (treatment) means in the experiment

g = Number of pairwise comparisons
[*Note:*   If all pairwise comparisons are to be made, then $g = p(p - 1)/2$]

$s = \sqrt{\text{MSE}}$

ν = Number of degrees of freedom associated with MSE

n_i = Number of observations in sample for treatment i

n_j = Number of observations in sample for treatment j

$t_{\alpha/(2g)}$ = Critical value of t distribution with ν df and tail area $\alpha/(2g)$ (Table 2 in Appendix D)

2. Rank the sample means and place a bar over any treatment pair (i, j) whose sample means differ by less than B_{ij}. Any pair of means not connected by an overbar implies a difference in the corresponding population means.

Note: The level of confidence associated with all inferences drawn from the analysis is at least $(1 - \alpha)$.

You can see that the group of treatments with the highest mean productivity includes treatments (4, 0), (4, 1), and (5, 1). The bar over these three means, however, indicates that we are unable to detect differences between any pair of these treatments. All inferences derived from this analysis can be made at an overall confidence level of at least $1 - \alpha = .95$.

When applied to pairwise comparisons of treatments, the Bonferroni method, like the Scheffé procedure, produces wider confidence intervals (reflected by the magnitude of the critical difference) than Tukey's method. (In Example 11.23, Bonferroni's critical difference is $B \approx 8.60$ compared to Tukey's $\omega = 7.27$.) Therefore, if only pairwise comparisons are of interest, Tukey's procedure is again superior. However, if the sample sizes are unequal or more complex contrasts are to be compared, the Bonferroni technique may be preferred. Unlike the Tukey and Scheffé methods, however, you must know in advance how many contrasts are to be compared to properly use Bonferroni's procedure. Also, the value needed to calculate the critical difference, B, $t_{\alpha/(2g)}$, may not be available in the t tables provided in most texts, and you will have to estimate its value.

In this section we have presented two alternatives to Tukey's multiple comparisons procedure. The technique you select will depend on several factors, including the sample sizes and the type of comparisons to be made. Keep in mind, however, that many other methods of making multiple comparisons are available, and one or more of these techniques may be more appropriate to use in your particular application. Consult the references given at the end of this chapter for details on other techniques.

EXERCISES

11.53   Refer to Exercises 11.12 and 11.44.
   a. Use the Scheffé method to perform all pairwise comparisons of the mean rates of penetration for the three types of drill bits. Use $\alpha = .05$.
   b. Use the Bonferroni approach to perform all pairwise comparisons of the mean rates of penetration for the three types of drill bits. Use $\alpha = .05$.
   c. Compare the results in parts **a** and **b** to the results of Tukey's method in Exercise 11.44.

11.54   Refer to Exercises 11.27 and 11.49.
   a. Use the Scheffé method to perform all pairwise comparisons of the productivity means for the $2 \times 3 = 6$ machine–materials combinations. Use $\alpha = .05$.
   b. Use the Bonferroni approach to perform all pairwise comparisons of the productivity means for the $2 \times 3 = 6$ machine–materials combinations. Use $\alpha = .05$.
   c. Compare the results in parts **a** and **b** to the results of Tukey's method in Exercise 11.49.

11.55   Refer to Exercises 11.35 and 11.51.
   a. Use the Scheffé method to perform all pairwise comparisons of the four treatment means. Use $\alpha = .05$.
   b. Use the Bonferroni approach to perform all pairwise comparisons of the four treatment means. Use $\alpha = .05$.
   c. Compare the results in parts **a** and **b** to the results of Tukey's method in Exercise 11.51.

11.9
Checking ANOVA Assumptions

For each of the experiments and designs discussed in this chapter, we listed in the relevant boxes the assumptions underlying the analysis in the terminology of ANOVA. For example, in the box on page 559, the assumptions for a completely randomized design are (1) the p probability distributions of the response y corresponding to the p treatments are normal and (2) the population variances of the p treatments are equal. Similarly, for randomized block designs and factorial designs, the data for the treatments must come from normal probability distributions with equal variances.

These assumptions are equivalent to those required for a regression analysis (see Section 4.2). The reason, of course, is that the probabilistic model for the response y that underlies each design is the familiar general linear regression model of Chapter 4. Consequently, checks on the ANOVA assumptions can be performed by examining the regression residuals, as described in Chapter 7. A brief overview of these techniques follows.

Detecting Nonnormal Populations

1. For each treatment, construct a histogram or stem-and-leaf display of the residuals. Look for highly skewed distributions. (Remember that ANOVA, like regression, is robust with respect to the normality assumption. That is, slight departures from normality will have little impact on the validity of the inferences derived from the analysis.) [*Note*:   If the sample size for each treatment is small, then these graphs will probably be of limited use.]
2. Formal statistical tests of normality (such as the Shapiro–Wilk test) are also available. The null hypothesis is that the probability distribution of the response is normal. These tests, however, are sensitive to slight departures from normality. Since in most scientific applications the normality assumption will not be satisfied exactly, these tests will likely result in a rejection of the null hypothesis and, consequently, are of limited use in practice. Consult the references for more information on these formal tests.
3. If the distribution of the residuals departs greatly from normality, a *normalizing transformation* may be necessary. For example, for highly skewed distributions, transformations on the response y such as $\log(y)$ or $\sqrt{y}$ tend to "normalize" the data since these functions "pull" the observations in the tail of the distribution back toward the mean.

Detecting Unequal Variances

1. For each treatment, construct a residual frequency plot and look for differences in the spread (variability) of the residuals shown in the plots. Residual frequency plots for the three income groups (treatments) in the completely randomized design ANOVA of Example 11.5 are shown in Figure 11.23. Note that the variability of the residuals in each plot is about the same; thus, the assumption of equal variances appears to be satisfied.

F I G U R E 11.23
Residual frequency plot for ANOVA of Example 11.5

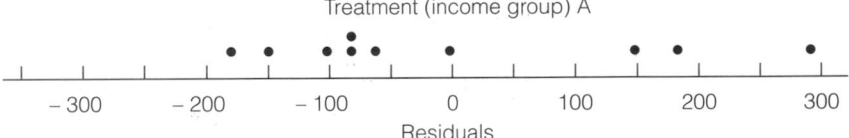

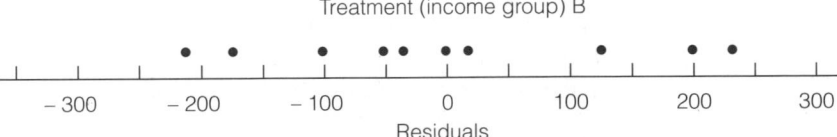

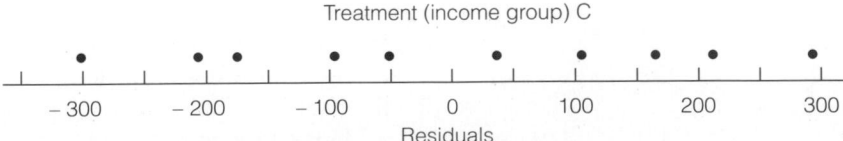

2. When the sample sizes are small for each treatment, only a few points are plotted on the residual frequency plots, making it difficult to detect differences in variation. In this situation, you may want to use one of several formal statistical tests of homogeneity of variances that are available. For p treatments, the null hypothesis is $H_0: \sigma_1^2 = \sigma_2^2 = \cdots = \sigma_p^2$, where σ_i^2 is the population variance of the response y corresponding to the ith treatment. If all p populations are approximately normal, **Bartlett's test for homogeneity of variances** can be applied. The elements of the test are shown in the box (page 640). Note that the test statistic depends on whether the sample sizes are equal or unequal.

We apply Bartlett's test to the ANOVA of Example 11.5. The null hypothesis is

$$H_0: \quad \sigma_1^2 = \sigma_2^2 = \sigma_3^2$$

where σ_1^2, σ_2^2 and σ_3^2 are the population debt variances for the three income groups, A, B, and C, respectively. Since the sample sizes are the same ($n_1 = n_2 = n_3 = 10$), the formula for the test statistic is

$$B = \frac{(n-1)[p \ln \bar{s}^2 - \Sigma \ln s_i^2]}{1 + \dfrac{(p+1)}{3p(n-1)}}$$

To obtain B, we first compute the sample variances, s_1^2, s_2^2, s_3^2, and their average, $\bar{s}^2$. The sample variances, shown (shaded) on the SAS printout displayed in Figure 11.24, are $s_1^2 = 25,026$, $s_2^2 = 21,867$, and $s_3^2 = 38,737$. Then, the average sample variance is $\bar{s}^2 = (25,026 + 21,867 + 38,737)/3 = 28,543$. Substituting these values into the formula, we obtain

$$B = \frac{(10-1)\{3 \ln(28543) - [\ln(25026) + \ln(21867) + \ln(38737)]\}}{1 + \dfrac{(3+1)}{3(3)(10-1)}} = .79$$

FIGURE 11.24

SAS descriptive statistics for residuals of Example 11.5

```
                      Analysis Variable : RESID

---------------------------------- GROUP=A ----------------------------------

     N Obs           Mean        Variance        Std Dev
     ---------------------------------------------------------
      10     -2.27374E-14       25026.04      158.1962213
     ---------------------------------------------------------


---------------------------------- GROUP=B ----------------------------------

     N Obs           Mean        Variance        Std Dev
     ---------------------------------------------------------
      10     -3.41061E-14       21866.77      147.8741582
     ---------------------------------------------------------


---------------------------------- GROUP=C ----------------------------------

     N Obs           Mean        Variance        Std Dev
     ---------------------------------------------------------
      10     -1.13687E-14       38737.29      196.8179079
     ---------------------------------------------------------
```

Since this value is less than the critical value, $\chi^2_{.05} = 5.99147$ (based on $p - 1 = 2$ degrees of freedom), we fail to reject the null hypothesis of equal variances; there is insufficient evidence (at $\alpha = .05$) of differences among the variances in debt for the three treatments. This result supports our conclusion we derived from the residual frequency plots of Figure 11.23.

Bartlett's test works well when the data come from normal (or near normal) distributions. The results, however, can be misleading for nonnormal data.

Bartlett's Test of Homogeneity of Variance

H_0: $\sigma^2_1 = \sigma^2_2 = \cdots = \sigma^2_p$

H_a: At least two variances differ.

Test statistic (equal sample sizes):

$$B = \frac{(n - 1)[p \ln \bar{s}^2 - \Sigma \ln s^2_i]}{1 + \dfrac{(p + 1)}{3p(n - 1)}}$$

where

$n = n_1 = n_2 = \cdots = n_p$

s^2_i = Sample variance for sample i

$\bar{s}^2$ = Average of the p sample variances = $(\Sigma s^2_i)/p$

$\ln x$ = Natural logarithm (i.e., log to the base e) of the quantity x

Test statistic (unequal sample sizes):

$$B = \frac{[\Sigma(n_i - 1)]\ln \bar{s}^2 - \Sigma(n_i - 1)\ln s^2_i}{1 + \dfrac{1}{3(p - 1)}\left\{\Sigma\dfrac{1}{(n_i - 1)} - \dfrac{1}{\Sigma(n_i - 1)}\right\}}$$

where

n_i = Sample size for sample i

s^2_i = Sample variance for sample i

$\bar{s}^2$ = Weighted average of the p sample variances = $\dfrac{\Sigma(n_i - 1) s^2_i}{\Sigma(n_i - 1)}$

$\ln x$ = Natural logarithm (i.e., log to the base e) of the quantity x

Rejection region: $B > \chi^2_\alpha$,
where χ^2_α locates an area α in the upper tail of a χ^2 distribution with $(p - 1)$ degrees of freedom

Assumptions: 1. Independent random samples are selected from the p populations.

2. All p populations are normally distributed.

In this case, we can apply a test that is much less sensitive to nonnormality in the data. See the references at the end of this chapter for more information on alternative tests.

3. When unequal variances are detected, use one of the **variance-stabilizing transformations** of the response y discussed in Section 5.3.

   In most business applications, the assumptions will not be satisfied exactly. These analysis of variance procedures are flexible, however, in the sense that slight departures from the assumptions will not significantly affect the analysis or the validity of the resulting inferences. On the other hand, gross violations of the assumptions (e.g., a nonconstant variance) will cast doubt on the validity of the inferences. Therefore, you should make it standard practice to conduct an analysis of the residuals from the ANOVA using the techniques of Chapter 7 to verify that the assumptions are (approximately) satisfied.

EXERCISES

[*Note:* The exercises require the use of a computer to calculate and plot residuals.]

11.56   Check the assumptions for the completely randomized design ANOVA of Exercise 11.8.

11.57   Check the assumptions for the completely randomized design ANOVA of Exercise 11.14.

11.58   Check the assumptions for the completely randomized design ANOVA of Exercises 11.19 and 11.20.

11.59   Check the assumptions for the factorial design ANOVA of Exercise 11.30.

11.60   Check the assumptions for the factorial design ANOVA of Exercise 11.34.

Summary

In this chapter we demonstrated how to analyze data collected from a designed experiment using either regression analysis or **analysis of variance (ANOVA)**. An analysis of variance partitions the total sum of squares, SS(Total), into SSE and, depending on the design, sums of squares for treatments (SST), blocks (SSB), and factor main effects and interaction. Tests for treatment means, block means, factor interactions, and so forth are obtained by calculating the ratio of the appropriate mean square to MSE and conducting an F test. These F tests can also be conducted by fitting the appropriate linear models to the data using regression. The models differ depending on the specific design employed.

Throughout this chapter we showed that an ANOVA and a regression analysis yield equivalent results. But an analysis of variance possesses both advantages and disadvantages compared with a regression analysis. One major advantage of an analysis of variance is that it is easy to perform on a pocket or desk calculator. Second, an analysis of variance is essential when y is affected by more than one source of random variation, and then the experimental design permits us to separate these sources and to estimate the variances of their respective random components. (A discussion of these models is not included in this text.) The disadvantages are its restrictions and limitations, as follows:

1. **The set of analysis of variance formulas appropriate for a particular experimental design applies only to that design.** If the data collected are observational (i.e., the independent variables are uncontrolled), an analysis of variance is inappropriate. No deviations from the design are permitted. Consequently, the method is of value only for special types of designed experiments.

2. **In contrast to a regression analysis, the analysis of variance formulas change from one design to another.** (There is a pattern, as indicated in Section 11.3, but the pattern is usually not apparent to a beginner.)

3. **The ANOVA formulas for a factorial experiment and many other experimental designs can be used only when the sample sizes are equal.** However, the regression approach applies to both equal and unequal sample sizes for the various factor level combinations.

4. **An analysis of variance does not give you a prediction equation.** This is a great handicap when one (or more) of the independent variables is quantitative.

5. **Although a linear model is always implied in an analysis of variance, it is rarely presented or discussed when analyses of data have been performed.** Consequently, the thrust of an analysis of variance is often counter (although it need not be) to the notion of modeling, which is the modern quantitative way of analyzing business phenomena.

In conclusion, perhaps the most important point for you to note in this chapter is the following: If your data can be modeled using a linear model that contains a single random error component (which is the model used throughout this text), then a regression analysis can do everything that an analysis of variance can do and it can do more! But you will probably need a computer to do it. Regression analysis is simple, uses the same formulas for all analyses, is programmed for both mainframe and personal computers, and can be used to analyze data obtained from both designed and undesigned experiments. Consequently, a beginner may be advised to stick to regression analyses for analyzing the relationship between a set of independent variables and a response y if a computer is available to perform the computations.

SUPPLEMENTARY EXERCISES

[*Note:*   Most of these exercises require you to compute the entries in an ANOVA table. We recommend that you conduct the analysis using an available statistical software package. Exercises marked with an asterisk(*) are from the optional sections in this chapter.]

11.61   A fast-food chain that specializes in Mexican food (tacos, burritos, etc.) is opening a new franchise in a university town. An important consideration in determining where the franchise will be located is traffic density. Five possible locations (each near a major intersection) are under consideration by the chain. To compare the average density of traffic at the possible sites, company employees are placed at each of the five locations to count the number of cars passing each location daily for a period of 10 randomly selected days. (At location IV, the counter assigned to record traffic density became ill and could obtain data for only 8 of the days.) The results are listed in the accompanying table. Assuming the samples of days were independently selected, conduct a complete analysis of the data.

| | | LOCATION | | |
I	II	III	IV	V
344	412	237	518	367
382	441	390	501	445
353	607	365	577	480
395	531	355	642	323
207	486	217	489	366
312	508	268	475	325
407	337	117	532	316
421	419	273	540	381
366	499	288		407
222	387	351		339

11.62   A nuclear power plant, which uses water from the surrounding bay for cooling its condensers, is required by the Environmental Protection Agency (EPA) to determine whether discharging its heated water into the bay has a detrimental effect on the plant life in the water. The EPA requests that the power plant investigate three strategically chosen locations, called *stations*. Stations 1 and 2 are located near the plant's discharge tubes, while station 3 is located farther out in the bay. During one randomly selected day in each of six months, a diver descends to each of the stations, randomly samples a square meter area of the bottom, and counts the number of blades of the different types of grasses present. The results for one important grass type are listed in the table.

| MONTH | STATION | | |
	1	2	3
April	32	40	30
May	28	31	53
June	25	22	61
July	37	30	56
August	20	26	48
September	18	21	30

Is there sufficient evidence to indicate that the mean number of blades found per square meter per month differs for at least two of the three stations? Use $\alpha = .05$.

11.63   In the United States, two basic types of management attitudes prevail: Theory X bosses believe that workers are basically lazy and untrustworthy, and Theory Y managers hold that employees are hard-working, dependable individuals. Japanese firms take a third approach: Theory Z companies emphasize long-range planning, consensus decision making, and strong, mutual worker–employer loyalty. Suppose we want to compare the mean hourly wage rates of workers at Theory X-, Y-, and Z-style corporations. Independent random samples of six engineering firms of each managerial philosophy were selected, and the starting hourly wage rates for laborers at each were recorded, as shown in the table at the top of page 644. Analyze the data.

MANAGERIAL ATTITUDE		
Theory X	Theory Y	Theory Z
5.20	6.25	5.50
5.20	6.80	5.75
6.10	6.87	4.60
6.00	7.10	5.36
5.75	6.30	5.85
5.60	6.35	5.90

11.64   A company conducted an experiment to determine the effects of three types of incentive pay plans on worker productivity for both union and non-union workers. The company used plants in adjacent towns; one was unionized and the other was not. One-third of the production workers in each plant were assigned to each incentive plan. Then six workers were randomly selected from each group, and their productivity (in number of items produced) was measured for a 1-week period. The six productivity measures for the 2×3 factor combinations are listed in the accompanying table. Conduct the analysis for the company.

		INCENTIVE PLAN					
		A		B		C	
	Union	337	328	346	373	317	341
		362	319	351	338	335	329
UNION		305	344	355	365	310	315
AFFILIATION		359	346	371	377	350	336
	Nonunion	345	396	352	401	349	351
		381	373	399	378	374	340

11.65   In hopes of attracting more riders, a city transit company plans to have express bus service from a suburban terminal to the downtown business district. These buses will travel along a major city street where there are numerous traffic lights that will affect travel time. The city decides to study the effect of four different plans on the travel times for buses: (1) a special bus express lane; (2) traffic signal progression; (3) express lane plus traffic signal progression; and (4) control—no special travel arrangements. Travel times (in minutes) are measured for randomly selected weekdays during a morning rush-hour trip while each plan is in effect. The results are recorded in the table. Carry out a complete analysis of the data.

PLAN			
1	2	3	4
27	25	24	30
25	28	19	33
29	30	22	31
26	27	21	
	24	26	

11.66   The chemical element antimony is sometimes added to tin—lead solder to replace the more expensive tin and to reduce the cost of soldering. A factorial experiment was conducted to determine how antimony affects the strength of the tin—lead solder joint (*Journal of Materials Science*, May 1986).

Tin–lead solder specimens were prepared using one of four possible cooling methods (water-quenched, WQ; oil-quenched, OQ; air-blown, AB; and furnace-cooled, FC) and with one of four possible amounts of antimony (0%, 3%, 5%, and 10%) added to the composition. Three solder joints were randomly assigned to each of the $4 \times 4 = 16$ treatments and the shear strength of each measured. The experiment results are shown in the accompanying table. Conduct a complete analysis of the data.

AMOUNT OF ANTIMONY % weight	COOLING METHOD	SHEAR STRENGTH MPa
0	WQ	17.6, 19.5, 18.3
0	OQ	20.0, 24.3, 21.9
0	AB	18.3, 19.8, 22.9
0	FC	19.4, 19.8, 20.3
3	WQ	18.6, 19.5, 19.0
3	OQ	20.0, 20.9, 20.4
3	AB	21.7, 22.9, 22.1
3	FC	19.0, 20.9, 19.9
5	WQ	22.3, 19.5, 20.5
5	OQ	20.9, 22.9, 20.6
5	AB	22.9, 19.7, 21.6
5	FC	19.6, 16.4, 20.5
10	WQ	15.2, 17.1, 16.6
10	OQ	16.4, 19.0, 18.1
10	AB	15.8, 17.3, 17.1
10	FC	16.4, 17.6, 17.6

Source: Tomlinson, W. J. and Cooper, G. A. "Fracture Mechanism of Brass/Sn-Pb-Sb Solder Joints and the Effect of Production Variables on the Joint Strength." *Journal of Materials Science*, Vol. 21, No. 5, May 1986, p. 1731 (Table II). Copyright 1986 Chapman and Hall.

11.67 Methods of displaying goods can affect their sales. The manager of a supermarket would like to try three different display types for a certain diet beverage. The manager chooses the locations for the three displays in such a way that each display type will be equally accessible to the customers. Each display will be set up for eight 1-week periods. Between each of these experimental periods, a standard display will be used for 2 weeks. For each of the eight experimental weeks, the sales (in dollars) of diet beverage from each display is determined, with the results shown in the table. Analyze the data to determine which of the three display types should be used.

PERIOD	DISPLAY A	B	C
1	$225	$253	$208
2	237	235	213
3	210	222	205
4	219	233	212
5	241	244	236
6	207	199	231
7	252	260	244
8	214	216	201

11.68   The percentage of water removed from paper as it passes through a dryer depends on the temperature of the dryer and the speed of the paper passing through it. A laboratory experiment was conducted to investigate the relationship between dryer temperature T at three levels (100°, 120°, and 140°F) and exposure time E (which is related to speed) also at three levels (10, 20, and 30 seconds). Four paper specimens were prepared for each of the $3 \times 3 = 9$ conditions. The data (percentages of water removed) are shown in the accompanying table. Carry out a complete analysis of the data.

		TEMPERATURE (T)					
		100		120		140	
EXPOSURE TIME (E)	10	24	26	33	33	45	49
		21	25	36	32	44	45
	20	39	34	51	50	67	64
		37	40	47	52	68	65
	30	58	55	75	71	89	87
		56	53	70	73	86	83

11.69   Researchers at the University of Iowa conducted a study to evaluate the effectiveness of performance appraisal training in an organizational setting (*Personnel Psychology*, Autumn 1984). Each member of a sample of 345 middle-level managers was randomly assigned to one of three training conditions: no training ($n_1 = 122$), computer-assisted instruction ($n_2 = 135$), and computer-assisted training plus a behavior modeling workshop ($n_3 = 88$). After formal training, the managers were administered a 25-question multiple choice test of managerial knowledge, and the number of correct answers was recorded for each. The data are summarized in the table. Analyze the data for the researchers. [*Hint:* Use the technique outlined in Exercise 11.10.]

TRAINING GROUP	SAMPLE SIZE	SAMPLE MEAN	STANDARD DEVIATION
No training	122	16.75	1.37
Computer assisted training	135	18.35	1.33
Computer training plus workshop	88	18.88	1.20

Source: Davis, B. L. and Mount, M. K. "Effectiveness of performance appraisal training using computer-assisted instruction and modeling behavior." *Personnel Psychology*, Autumn 1984, Vol. 37, pp. 439–451.

11.70   A real estate broker wanted to investigate the relationship between a response y, the per-square-foot increase in sales value per year of residential property, and two qualitative independent variables: region in the broker's community and salesperson who negotiated the sale. Properties included in the study were those that had been recently sold after being held for approximately 2 years. The broker divided the area of the community into four regions and then classified the properties according to salesperson (one of three) and region (one of four). Three properties were randomly selected from within each of the 12 salesperson–region categories, and the annual increase in value per square foot per year was calculated for each. The data are shown in the table. Analyze the data for the real estate broker.

		REGION				TOTALS
		R_1	R_2	R_3	R_4	
SALESPERSON	S_1	4.22, 3.90, 4.36	2.96, 3.31, 3.13	3.25, 3.73, 3.56	1.81, 2.22, 2.21	38.66
	S_2	4.86, 4.42, 4.53	2.60, 2.84, 2.49	3.01, 3.61, 3.45	1.96, 1.96, 2.10	37.83
	S_3	3.69, 4.46, 4.08	2.50, 2.51, 2.06	3.55, 3.10, 3.34	2.49, 2.23, 2.78	36.79
TOTALS		38.52	24.40	30.60	19.76	113.28

11.71 From time to time, one branch office of a company must make shipments to a certain branch office in another state. There are three package delivery services between the two cities where the branch offices are located. Since the price structures for the three delivery services are quite similar, the company wants to compare the delivery times. The company plans to make several different types of shipments to its branch office. To compare the carriers, each shipment will be sent in triplicate, one with each carrier. The results listed in the table are the delivery times in hours. Compare the mean delivery times of the three carriers.

SHIPMENT	CARRIER		
	I	II	III
1	15.2	16.9	17.1
2	14.3	16.4	16.1
3	14.7	15.9	15.7
4	15.1	16.7	17.0
5	14.0	15.6	15.5

11.72 To be able to provide its clients with comparative information on two large suburban residential communities, a realtor wants to know the average home value in each community. Eight homes are selected at random within each community and are appraised by the realtor. The appraisals (in thousands of dollars) are given in the accompanying table. Can you conclude that the average home value is different in the two communities? You have three ways of analyzing this problem, as described in parts a–c.

COMMUNITY	
A	B
43.5	73.5
49.5	62.0
38.0	47.5
66.5	36.5
57.5	44.5
32.0	56.0
67.5	68.0
71.5	63.5

a. Use the two-sample t statistic (Section 1.9) to test $H_0: \mu_A = \mu_B$.

b. Consider the regression model

$$y = \beta_0 + \beta_1 x + \varepsilon$$

where

$$x = \begin{cases} 1 & \text{if community B} \\ 0 & \text{if community A} \end{cases}$$

y = Appraised price

Since $\beta_1 = \mu_B - \mu_A$, testing $H_0: \beta_1 = 0$ is equivalent to testing $H_0: \mu_A = \mu_B$. Use the partial reproduction of the SAS printout shown on page 648 to test $H_0: \beta_1 = 0$. Use $\alpha = .05$.

ANALYSIS OF VARIANCE

SOURCE	DF	SUM OF SQUARES	MEAN SQUARE	F VALUE	PROB>F
MODEL	1	40.64062500	40.64062500	0.215	0.6501
ERROR	14	2648.71875	189.19420		
C TOTAL	15	2689.35938			

ROOT MSE	13.75479	R-SQUARE	0.0151
DEP MEAN	54.84375	ADJ R-SQ	-0.0552
C.V.	25.07996		

PARAMETER ESTIMATES

VARIABLE	DF	PARAMETER ESTIMATE	STANDARD ERROR	T FOR H0: PARAMETER=0	PROB > |T|
INTERCEP	1	53.25000000	4.86305198	10.950	0.0001
X	1	3.18750000	6.87739406	0.463	0.6501

c. Use the ANOVA method to test $H_0: \mu_A = \mu_B$. Use $\alpha = .05$.

d. Using the results of these three tests, verify that the tests are the equivalent (for this special case, $p = 2$) of the completely randomized design in terms of the test statistic value and rejection region. For the three methods used, what are the advantages and disadvantages (limitations) of using each in analyzing results for this type of experimental design?

11.73   In purchasing negotiations, buyers and sellers tend to have conflicting bargaining goals. A buyer wants to obtain the best value at minimum cost, whereas the seller wants to maximize profit. In addition, the seller's commission often is directly contingent on his or her selling effectiveness. In contrast, buyers are rarely compensated directly for their purchasing performance in any one transaction. What effect do these different compensation systems have on the motivation of salespeople and professional buyers? Researchers conducted a study to determine what effects contingent and noncontingent pay systems might have on bargaining behavior and outcomes. They employed a 2×2 factorial experiment with two factors, buyer condition and seller condition, each at two levels (contingent and noncontingent reward). A sample of 160 industrial purchasers was randomly divided into two groups of buyers and sellers, with 80 subjects in each group. Each buyer was matched with a seller to form one of 80 different bargaining "dyads." The dyads were then divided equally among the $2 \times 2 = 4$ buyer–seller reward conditions. Thus, 20 bargaining dyads were assigned to each of the four "treatments" of the experiment.

All subjects participated in a game in which the sellers were told they represented an individual who wanted to sell an office building (for no less than $100,000) and buyers were informed they represented an individual interested in purchasing the building (for no more than $150,000). Sellers in the contingent reward group (SC) were told that they would be rewarded in cash in proportion to the final negotiated price, whereas sellers in the noncontingent reward group (SNC) were told they would be paid a fixed fee. Similarly, buyers in the contingent reward group (BC) were led to believe that they would be compensated according to the final agreed price, while buyers in the noncontingent reward group (BNC) received a fixed sum regardless of the outcome. One of the key bargaining variables recorded in the experiment was seller's initial offer. The accompanying table gives the mean of the seller's initial offer (in thousands of dollars) for each of the four buyer–seller reward conditions.

a. Construct an ANOVA table for the 2×2 factorial experiment. (For this experiment, $s = \sqrt{MSE} \approx 21.00$.)

		BUYER CONDITION	
		Contingent (BC)	Noncontingent (BNC)
SELLER	Contingent (SC)	160.3	173.05
CONDITION	Noncontingent (SNC)	155.5	149.50

Source: McFillen, J. M., Reck, R. R., and Benton, W. C. "An experiment in purchasing negotiations." *Journal of Purchasing and Materials Management*, Summer 1983, Vol. 19, No. 1, pp. 2–8.

b. Plot the four cell means on the same graph. Does it appear that the two factors, buyer condition and seller condition, interact?

c. Test for interaction between the two factors, buyer condition and seller condition. Use $\alpha = .10$.

d. Give the complete and reduced models appropriate for testing for interaction using the regression approach.

e. Assuming that contingently rewarded sellers open with higher initial offers than noncontingently rewarded sellers, the researchers hypothesized that this difference is due to reward manipulation and therefore that the means for BC/SC and BNC/SC will not differ significantly. Construct a 90% confidence interval for the difference between mean seller's initial offer for the BC/SC and BNC/SC groups. Does the result refute or support the researchers' hypothesis?

11.74 A production manager who supervises an assembly operation wants to investigate the effect of the incoming rate (parts per minute) x_1 of components and room temperature x_2 on the productivity (number of items produced per minute) y. The component parts approach the worker on a belt and return to the worker if not selected on the first trip past the assembly point. It is thought that an increase in the arrival rate of components has a positive effect on the assembly rate, up to a point, after which increases may annoy the assembler and reduce productivity. Similarly, it is suspected that lowering the room temperature is beneficial to a point, after which reductions may reduce productivity. The experimenter used the same assembly position for each worker. Thirty-two workers were used for the experiment, two each assigned to the 16 factor level combinations of a 4 × 4 factorial experiment. The data, in parts per minute averaged over a 5-minute period, are shown in the table.

		RATE OF INCOMING COMPONENTS, PARTS PER MINUTE (x_1)			
		40	50	60	70
	65	24.0, 23.8	25.6, 25.4	29.2, 29.4	28.4, 27.6
ROOM	70	25.0, 26.0	28.8, 28.8	31.6, 32.0	30.2, 30.0
TEMPERATURE, °F (x_2)	75	25.6, 25.0	27.6, 28.0	29.8, 28.6	28.0, 27.0
	80	24.0, 24.6	27.6, 26.2	27.6, 28.6	26.0, 24.4

a. Perform an analysis of variance for the data. Display the computed quantities in an ANOVA table.

b. Write the linear model implied by the analysis of variance. [*Hint:* For a quantitative variable recorded at four levels, main effects include terms for x, x^2, and x^3.]

c. Do the data provide sufficient evidence to indicate differences among the mean responses for the 16 treatments of the 4 × 4 factorial experiment? Test using $\alpha = .05$.

d. Do the data provide sufficient evidence to indicate an interaction between arrival rate x_1 and room temperature x_2 on worker productivity? Test using $\alpha = .05$.

e. Find the value of R^2 that would be obtained if you were to fit the linear model in part **b** to the data.

f. Explain why a regression analysis would be a useful addition to the inferential methods used in parts **a–e**.

11.75 A second-order model would be a reasonable choice to model the data of Exercise 11.74. To simplify the analysis, we will code the arrival rate and temperature values as follows:

$$x_1 = \frac{\text{Arrival rate} - 55}{5} \qquad x_2 = \frac{\text{Temperature} - 72.5}{2.5}$$

A printout of the SAS regression analysis is shown here.

ANALYSIS OF VARIANCE

SOURCE	DF	SUM OF SQUARES	MEAN SQUARE	F VALUE	PROB>F
MODEL	5	130.80680	26.16136000	27.726	0.0001
ERROR	26	24.53320000	0.94358462		
C TOTAL	31	155.34000			

ROOT MSE	0.9713828	R-SQUARE	0.8421	
DEP MEAN	27.325	ADJ R-SQ	0.8117	
C.V.	3.554923			

PARAMETER ESTIMATES

VARIABLE	DF	PARAMETER ESTIMATE	STANDARD ERROR	T FOR H0: PARAMETER=0	PROB > |T|
INTERCEP	1	29.85625000	0.34876060	85.607	0.0001
X1	1	0.56000000	0.07679456	7.292	0.0001
X2	1	-0.16250000	0.07679456	-2.116	0.0441
X1X2	1	-0.11350000	0.03434357	-3.305	0.0028
X1X1	1	-0.27500000	0.04292946	-6.406	0.0001
X2X2	1	-0.23125000	0.04292946	-5.387	0.0001

a. Write the second-order linear model for the response. Note the difference between this model and the ANOVA model in Exercise 11.74, part **a**.

b. Give the prediction relating the response y to the coded independent variables x_1 and x_2.

c. Why does the SSE given in the computer printout differ from the SSE obtained in Exercise 11.74?

d. Find the value of R^2 appropriate for your second-order model and interpret its value.

e. Do the data provide sufficient evidence to indicate that the complete factorial model provides more information for predicting y than a second-order model?

*11.76 Suppose you want to investigate the effect of two factors—arrival rate of product components, A, and temperature of the room, T—on the length of time, y, required by individual workers to perform a product assembly operation. Each factor will be held at two levels: arrival rate at .5 and 1.0 component per second, and temperature at 70° and 80°F. Thus, a 2×2 factorial experiment will be employed. To block out worker-to-worker variability, each of 10 randomly selected workers will be required to assemble the product under all four experimental conditions. Therefore, the four treatments (working conditions) will be assigned to the experimental units (workers) using a randomized block design, where the workers represent the blocks.

The appropriate complete model for the randomized block design is

$$\text{Complete model:} \quad E(y) = \beta_0 + \overbrace{\beta_1 x_1 + \beta_2 x_2 + \beta_3 x_1 x_2}^{\substack{\text{Treatment effects (main} \\ \text{effects and interaction terms} \\ \text{for arrival rate and temperature)}}}$$

$$+ \overbrace{\beta_4 x_3 + \beta_5 x_4 + \cdots + \beta_{12} x_{11}}^{\text{Block (worker) effects}}$$

where

$x_1 = $ Arrival rate $\qquad x_2 = $ Temperature

$$x_3 = \begin{cases} 1 & \text{if worker 1} \\ 0 & \text{if not} \end{cases}$$

$$x_4 = \begin{cases} 1 & \text{if worker 2} \\ 0 & \text{if not} \end{cases} \qquad \cdots$$

$$x_{11} = \begin{cases} 1 & \text{if worker 9} \\ 0 & \text{if not} \end{cases}$$

[Note that $x_3 = x_4 = \cdots = x_{11} = 0$ if worker (block) 10 is the assembler.]

The assembly time data for the 2×2 factorial experiment with a randomized block design are given in the table. The SAS printouts for the complete model and the reduced model

$$\text{Reduced model:} \quad E(y) = \beta_0 + \overbrace{\beta_4 x_3 + \beta_5 x_4 + \cdots + \beta_{12} x_{11}}^{\text{Workers}}$$

are also provided on page 652.

			WORKER									
			1	2	3	4	5	6	7	8	9	10
ROOM TEMPERATURE	70°F	ARRIVAL RATE, COMPONENT PER SECOND	.5	1.7	1.3	1.7	2.0	2.0	2.3	2.0	2.8	1.5	1.6
			1.0	.8	.8	1.5	1.2	1.2	1.7	1.1	1.5	.5	1.0
	80°F		.5	1.3	1.5	2.3	1.6	2.2	2.1	1.8	2.4	1.3	1.8
			1.0	1.8	1.5	2.3	2.0	2.7	2.2	2.3	2.6	1.3	1.8

a. Do the data provide sufficient evidence to indicate a difference among the four treatment means?

b. Does the effect of a change in arrival rate on assembly time depend on temperature (i.e., do arrival rate and temperature interact)?

c. Estimate the mean loss (or gain) in assembly time as arrival rate is increased from .5 to 1.0 component per second and temperature is held at 70°F. What inference can you make based on this estimate?

SAS printout: Complete model for Exercise 11.76

ANALYSIS OF VARIANCE

SOURCE	DF	SUM OF SQUARES	MEAN SQUARE	F VALUE	PROB>F
MODEL	12	10.17400000	0.84783333	21.176	0.0001
ERROR	27	1.08100000	0.04003704		
C TOTAL	39	11.25500000			

ROOT MSE	0.2000926	R-SQUARE	0.9040
DEP MEAN	1.725	ADJ R-SQ	0.8613
C.V.	11.59957		

PARAMETER ESTIMATES

VARIABLE	DF	PARAMETER ESTIMATE	STANDARD ERROR	T FOR H0: PARAMETER=0	PROB > |T|
INTERCEP	1	9.75500000	1.50701723	6.473	0.0001
X1	1	-15.24000000	1.90245845	-8.011	0.0001
X2	1	-0.10400000	0.02000926	-5.198	0.0001
X1X2	1	0.19600000	0.02530993	7.744	0.0001
X3	1	-0.15000000	0.14148681	-1.060	0.2985
X4	1	-0.27500000	0.14148681	-1.944	0.0624
X5	1	0.40000000	0.14148681	2.827	0.0087
X6	1	0.15000000	0.14148681	1.060	0.2985
X7	1	0.47500000	0.14148681	3.357	0.0024
X8	1	0.52500000	0.14148681	3.711	0.0010
X9	1	0.25000000	0.14148681	1.767	0.0885
X10	1	0.77500000	0.14148681	5.478	0.0001
X11	1	-0.40000000	0.14148681	-2.827	0.0087

SAS printout: Reduced model for Exercise 11.76

ANALYSIS OF VARIANCE

SOURCE	DF	SUM OF SQUARES	MEAN SQUARE	F VALUE	PROB>F
MODEL	9	5.19500000	0.57722222	2.858	0.0147
ERROR	30	6.06000000	0.20200000		
C TOTAL	39	11.25500000			

ROOT MSE	0.4494441	R-SQUARE	0.4616
DEP MEAN	1.725	ADJ R-SQ	0.3000
C.V.	26.05473		

PARAMETER ESTIMATES

VARIABLE	DF	PARAMETER ESTIMATE	STANDARD ERROR	T FOR H0: PARAMETER=0	PROB > |T|
INTERCEP	1	1.55000000	0.22472205	6.897	0.0001
X3	1	-0.15000000	0.31780497	-0.472	0.6404
X4	1	-0.27500000	0.31780497	-0.865	0.3937
X5	1	0.40000000	0.31780497	1.259	0.2179
X6	1	0.15000000	0.31780497	0.472	0.6404
X7	1	0.47500000	0.31780497	1.495	0.1455
X8	1	0.52500000	0.31780497	1.652	0.1090
X9	1	0.25000000	0.31780497	0.787	0.4377
X10	1	0.77500000	0.31780497	2.439	0.0209
X11	1	-0.40000000	0.31780497	-1.259	0.2179

11.77   An experiment was conducted to compare the abilities of four real estate appraisers. To eliminate the effect of property-to-property variation in comparing appraiser means, 10 properties were randomly selected, and each appraiser was required to appraise each property. Thus, a randomized block experiment was conducted. The appraised values (in thousands of dollars) obtained from the four appraisers for each of the 10 properties are shown in the table.

a. Write the complete model for this randomized block design.

b. Modify the model for y in part a by deleting the treatment (appraiser) parameters.

						PROPERTY					
		1	2	3	4	5	6	7	8	9	10
APPRAISER	A	46.3	63.0	43.1	84.6	51.0	49.2	60.0	31.5	47.2	55.0
	B	50.9	66.2	46.7	88.4	50.0	54.0	54.8	32.0	50.5	60.3
	C	48.1	64.7	46.0	83.0	48.8	51.4	62.1	30.5	46.0	57.1
	D	49.2	65.0	48.3	87.8	52.7	52.9	66.0	33.4	49.0	58.0

c. Modify the model in part **a** by deleting the block (property) parameters.

d. The values of SSE associated with the models in parts **a**, **b**, and **c** are shown in the accompanying table. Do the data provide sufficient evidence to indicate differences in the mean appraisal levels for the four appraisers? Test using $\alpha = .05$.

	MODEL		
	a	b	c
SSE	105.28175	167.61250	7,397.88700

e. Intuitively, we know that blocking was necessary in this experiment because of the large amount of variation in the property values. As an exercise, use the information in part **d** to verify the contention that blocking increased the information in the experiment. Test the null hypothesis of "no difference in block means" using $\alpha = .05$.

11.78 Suppose you plan to investigate the effect of hourly pay rate and length of workday on some measure y of worker productivity. Both pay rate and length of workday will be set at three levels, and y will be observed for all combinations of these factors. Thus, a 3×3 factorial experiment will be employed.

a. Identify the factors and state whether they are quantitative or qualitative.

b. Identify the treatments to be employed in the experiment.

c. Write a complete factorial model for the experiment. [*Hint:* When the factors are quantitative, main effect terms include x and x^2 terms for each factor.]

d. What is the order of the model specified in part **c**?

e. Suppose you want to fit a second-order model to the data. Give the appropriate model.

f. If you have only one observation for each combination and you fit a complete factorial model to the data, how many degrees of freedom will be available for estimating σ^2?

g. Refer to part **f**. If you fit a second-order model to the data, how many degrees of freedom will be available for estimating σ^2?

h. Suppose you replicated the experiment and hence obtained two observations for each treatment. How many degrees of freedom would be available for estimating σ^2 if you fit: (1) the complete factorial model? (2) a second-order model?

i. Which model—complete factorial or second-order—do you think would be more appropriate for this experiment?

j. Explain how to conduct a test for interaction between pay rate and length of workday using the complete factorial model.

k. Explain how to conduct a test for interaction between pay rate and length of workday using the second-order model.

11.79   The data for the 3 × 3 factorial experiment described in Exercise 11.78 are shown in the accompanying table, as well as the SAS computer printout of the regression analysis for the second-order model.

| | | HOURLY PAY RATE, DOLLARS (x_1) | | |
		6.50	7.00	7.50
	8	350	375	402
		377	390	411
LENGTH OF WORKDAY, HOURS (x_2)	9	398	423	434
		386	424	429
	10	345	377	394
		351	381	389

ANALYSIS OF VARIANCE

SOURCE	DF	SUM OF SQUARES	MEAN SQUARE	F VALUE	PROB>F
MODEL	5	11355.01389	2271.00278	33.863	0.0001
ERROR	12	804.76389	67.06365741		
C TOTAL	17	12159.77778			

ROOT MSE	8.18924	R-SQUARE	0.9338	
DEP MEAN	390.8889	ADJ R-SQ	0.9062	
C.V.	2.09503			

PARAMETER ESTIMATES

VARIABLE	DF	PARAMETER ESTIMATE	STANDARD ERROR	T FOR H0: PARAMETER=0	PROB > |T|
INTERCEP	1	-4026.63889	939.43751	-4.286	0.0011
X1	1	385.08333	235.19426	1.637	0.1275
X1X1	1	-24.66666667	16.37848069	-1.506	0.1579
X2	1	661.58333	84.14751627	7.862	0.0001
X2X2	1	-37.16666667	4.09462017	-9.077	0.0001
X1X2	1	0.25000000	5.79066738	0.043	0.9663

a. Do the data provide sufficient evidence to indicate that the model contributes information for the prediction of y?

b. Find R^2 and interpret its meaning.

c. Does a large value for R^2 imply that the second-order model is a good predictor of productivity?

d. If the value for R^2 were small, would this imply that the model is inadequate, i.e., that we need to change the form of the model and/or add other independent variables to the model that may be related to worker productivity?

e. Is there evidence of interaction between the two factors, pay rate and length of workday? Test using $\alpha = .05$.

f. Extract the parameter estimates from the computer printout and give the prediction equation.

g. Graph the predicted productivity $\hat{y}$ as a function of pay rate x_1 for $x_2 = 8$ hours. Then obtain the predicted productivity curves for $x_2 = 9$ and $x_2 = 10$ hours.

*11.80   A $2 \times 2 \times 2 \times 2 = 2^4$ factorial experiment was conducted to investigate the effect of four factors on the light output, y, of flashbulbs. Two observations were taken for each of the factorial treatments. The factors are: amount of foil contained in a bulb (100 and 120 milligrams); speed of sealing machine (1.2 and 1.3 revolutions per minute); shift (day or night); machine operator (A or B). The data for the two replications of the 2^4 factorial experiment are shown here, and the SAS computer printout for the regression analysis follows.

		AMOUNT OF FOIL			
		100 milligrams		120 milligrams	
		SPEED OF MACHINE			
		1.2 rpm	1.3 rpm	1.2 rpm	1.3 rpm
DAY SHIFT	Operator B	6; 5	5; 4	16; 14	13; 14
	Operator A	7; 5	6; 5	16; 17	16; 15
NIGHT SHIFT	Operator B	8; 6	7; 5	15; 14	17; 14
	Operator A	5; 4	4; 3	15; 13	13; 14

ANALYSIS OF VARIANCE

SOURCE	DF	SUM OF SQUARES	MEAN SQUARE	F VALUE	PROB>F
MODEL	15	745.46875	49.69791667	40.778	0.0001
ERROR	16	19.50000000	1.21875000		
C TOTAL	31	764.96875			

ROOT MSE	1.10397	R-SQUARE	0.9745	
DEP MEAN	10.03125	ADJ R-SQ	0.9506	
C.V.	11.00531			

PARAMETER ESTIMATES

VARIABLE	DF	PARAMETER ESTIMATE	STANDARD ERROR	T FOR H0: PARAMETER=0	PROB > |T|
INTERCEP	1	10.03125000	0.19515619	51.401	0.0001
X1	1	4.71875000	0.19515619	24.179	0.0001
X2	1	-0.34375000	0.19515619	-1.761	0.0973
X3	1	0.21875000	0.19515619	1.121	0.2789
X4	1	-0.15625000	0.19515619	-0.801	0.4351
X1X2	1	0.09375000	0.19515619	0.480	0.6375
X1X3	1	0.15625000	0.19515619	0.801	0.4351
X1X4	1	0.28125000	0.19515619	1.441	0.1688
X2X3	1	-0.15625000	0.19515619	-0.801	0.4351
X2X4	1	-0.03125000	0.19515619	-0.160	0.8748
X3X4	1	0.78125000	0.19515619	4.003	0.0010
X1X2X3	1	-0.21875000	0.19515619	-1.121	0.2789
X1X2X4	1	-0.09375000	0.19515619	-0.480	0.6375
X1X3X4	1	-0.03125000	0.19515619	-0.160	0.8748
X2X3X4	1	0.15625000	0.19515619	0.801	0.4351
X1X2X3X4	1	0.09375000	0.19515619	0.480	0.6375

To simplify computations, we let

$$x_1 = \frac{\text{Amount of foil} - 110}{10} \qquad x_2 = \frac{\text{Speed of machine} - 1.25}{.05}$$

so that x_1 and x_2 will take values -1 and $+1$. Also,

$$x_3 = \begin{cases} -1 & \text{if night shift} \\ 1 & \text{if day shift} \end{cases} \qquad x_4 = \begin{cases} -1 & \text{if machine operator } B \\ 1 & \text{if machine operator } A \end{cases}$$

a. Do the data provide sufficient evidence to indicate that any of the factors contribute information for the prediction of y? Give the results of a statistical test to support your answer.
b. Identify the factors that appear to affect the amount of light y in the flashbulbs.
c. Give the complete factorial model for y.
d. How many degrees of freedom will be available for estimating σ^2?

ON YOUR OWN...

As a result of ever-increasing food costs, consumers are becoming more discerning in their choice of supermarkets. It usually is more convenient to shop at only one market, as opposed to buying different items at different markets. Thus, it would be useful to compare the mean food expenditure for a market basket of food items from store to store. Since there is a great deal of variability in the prices of products sold at any supermarket, we will consider an experiment that blocks on products.

Choose three (or more) supermarkets in your area that you want to compare; then choose approximately ten (or more) food products you typically purchase. For each food item, record the price each store charges in the following manner:

FOOD ITEM 1	FOOD ITEM 2 . . . FOOD ITEM 10
Price store 1	Price store 1 . . . Price store 1
Price store 2	Price store 2 . . . Price store 2
Price store 3	Price store 3 . . . Price store 3

Use the data you obtain to test

H_0:   Mean expenditures at the stores are the same.

H_a: Mean expenditures for at least two of the stores are different.

Also, test to determine whether blocking on food items is advisable in this kind of experiment. Fully interpret the results of your analysis.

References

Box, G. E. P., Hunter, W. G., and Hunter, J. S. *Statistics for Experimenters*. New York: Wiley, 1978.

Cochran, W. G. and Cox, G. M. *Experimental Designs*, 2nd ed. New York: Wiley, 1957.

Davies, O. L. *The Design and Analysis of Industrial Experiments*. New York: Hafner, 1956.

Kirk, R. E. *Experimental Design: Procedures for the Behavioral Sciences*. Monterey, Calif.: Brooks/ Cole, 1968.

Kramer, C. Y. "Extension of multiple range tests to group means with unequal number of replications." *Biometrics*, Vol. 12, 1956, pp. 307–310.

Mendenhall, W. *Introduction to Linear Models and the Design and Analysis of Experiments*. Belmont, Calif.: Wadsworth, 1968.

Miller, R. G. *Simultaneous Statistical Inference*, 2nd ed. New York: Springer-Verlag, 1981.

Neter, J., Wasserman, W., and Kutner, M. H. *Applied Linear Statistical Models*, 3rd ed. Homewood, Ill.: Richard D. Irwin, 1990.

Scheffé, H. "A method for judging all contrasts in the analysis of variance." *Biometrika*, Vol. 40, 1953, pp. 87–104.

Scheffé, H. *The Analysis of Variance*. New York: Wiley, 1959.

Tukey, J. W. "Comparing individual means in the analysis of variance." *Biometrics*, Vol. 5, 1949, pp. 99–114.

Winer, B. J. *Statistical Principles in Experimental Design*. New York: McGraw-Hill, 1962.

Modeling the Sale Prices of Residential Properties in Four Neighborhoods

CONTENTS

12.1 The Problem
12.2 The Data
12.3 The Models
12.4 Model Comparisons
12.5 Interpreting the Prediction Equation
12.6 Predicting the Sale Price of a Property
12.7 Conclusions

OBJECTIVE

To demonstrate how regression analysis can be used to model and compare the relationship between the sale prices and assessed values of residential properties for four different city neighborhoods

12.1
The Problem

This case study concerns a problem of interest to real estate appraisers, tax assessors, real estate investors, and home buyers—namely, the relationship between the appraised value of a property and its sale price. The sale price for any given property will vary depending on the price set by the seller, the strength of appeal of the property to a specific buyer, and the state of the money and real estate markets. Therefore, we can think of the sale price of a specific property as possessing a relative frequency distribution. The mean of this distribution might be regarded as a measure of the fair value of the property. Presumably, this is the value that a property appraiser or a tax assessor would like to attach to a given property.

The purpose of this case study is to examine the relationship between the mean sale price $E(y)$ of a property and the following independent variables:

1. Appraised land value of the property
2. Appraised value of the improvements on the property
3. Neighborhood in which the property is listed

The objectives of the study are twofold:

1. To determine whether the data indicate that appraised values of land and improvements are related to sale prices. That is, do the data supply sufficient evidence to indicate that these variables contribute information for the prediction of sale price?
2. To acquire the prediction equation relating appraised value of land and improvements to sale price and to determine whether this relationship is the same for a variety of neighborhoods. In other words, do the appraisers use the same appraisal criteria for various types of neighborhoods?

12.2
The Data

The data for the study were supplied by the property appraiser's office of Hillsborough County, Florida, and consist of the appraised land and improvement values and sale prices for residential properties sold in the city of Tampa, Florida, during 1990. Four neighborhoods, each relatively homogeneous but differing sociologically and in property types and values, were identified within the city and surrounding area. The subset of sales and appraisal data pertinent to these four neighborhoods was used to develop a prediction equation relating sale prices to appraised land and improvement values. The data (recorded in thousands of dollars) are given in Appendix E.

12.3
The Models

If the mean sale price $E(y)$ of a property were, in fact, equal to its appraised value x, the relationship between $E(y)$ and x would be a straight line with slope equal to 1, as shown in Figure 12.1. But it is unlikely that this ideal situation will exist. The property appraiser's data could be several years old and consequently may represent (because of inflation) only a percentage of the actual mean sale price. In fact, it is common for realtors and real estate appraisers to model the natural logarithm of sales price, $\log(y)$, as a function of appraised value, x.

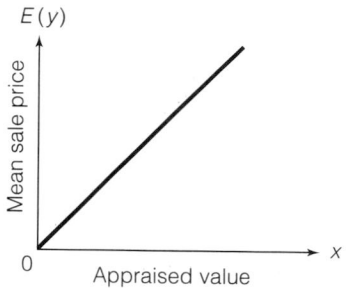

FIGURE 12.1

The theoretical relationship between mean sale price and appraised value x

Model 1

MODEL 1

First-order log model, identical for all neighborhoods

Model 2

MODEL 2

First-order log model, constant differences between neighborhoods

We know (Section 4.9) that the antilogarithm of the slope of the log model represents a percentage change in sale price y for every 1-unit increase in appraised value x. Also, experience has shown that the relationship between $E(y)$ and x is usually not a straight line, but curvilinear. One reason is that appraisers have a tendency to overappraise or underappraise properties in specific price ranges, say, very low-priced or very high-priced properties. Recall (Section 6.7) that modeling $\log(y)$ as a linear function of x introduces a curvilinear relationship between y and x. Consequently, we will use $y^* = \log(y)$ as the dependent variable.

Recall that we want to relate y^* to three independent variables: the qualitative factor, neighborhood (four levels), and the two quantitative factors, appraised land value and appraised improvement value. We consider the following four models as candidates for this relationship:

Model 1 is a first-order log model that will trace a response plane for mean log of sale price, $E(y^*)$, as a function of x_1 = appraised land value and x_2 = appraised improvement value. This model will assume that the response planes are identical for all four neighborhoods, i.e., that a first-order log model is appropriate for relating y^* to x_1 and x_2 and that the relationship between the sale price and the appraised value of a property is the same for all neighborhoods. This model is

$$E(y^*) = \beta_0 + \overbrace{\beta_1 x_1}^{\substack{\text{Appraised land} \\ \text{value}}} + \overbrace{\beta_2 x_2}^{\substack{\text{Appraised} \\ \text{improvement value}}}$$

Note that in Model 1, we are assuming that the percentage change in sale price y for every \$1,000 (1-unit) increase in appraised land value x_1 (represented by $e^{\beta_1} - 1$) is constant for a fixed appraised improvements value, x_2. Likewise, the percentage change in y for every \$1,000 increase in x_2 (represented by $e^{\beta_2} - 1$) is constant for fixed x_1.

Model 2 will assume that the relationship between $E(y^*)$ and x_1 and x_2 is first-order (a planar response surface), but that the planes' y^*-intercepts differ depending on the neighborhood. This model would be appropriate if the appraiser's procedure for establishing appraised values produced a relationship between mean sale price and x_1 and x_2 that differed in at least two neighborhoods, but the differences remained constant for different values of x_1 and x_2. Model 2 is

$$E(y^*) = \beta_0 + \overbrace{\beta_1 x_1}^{\substack{\text{Appraised} \\ \text{land value}}} + \overbrace{\beta_2 x_2}^{\substack{\text{Appraised} \\ \text{improvement value}}} + \overbrace{\beta_3 x_3 + \beta_4 x_4 + \beta_5 x_5}^{\substack{\text{Main effect terms} \\ \text{for neighborhoods}}}$$

where

x_1 = Appraised land value

x_2 = Appraised improvement value

$x_3 = \begin{cases} 1 & \text{if neighborhood A} \\ 0 & \text{if not} \end{cases} \qquad x_4 = \begin{cases} 1 & \text{if neighborhood B} \\ 0 & \text{if not} \end{cases}$

$x_5 = \begin{cases} 1 & \text{if neighborhood C} \\ 0 & \text{if not} \end{cases}$

The fourth neighborhood, neighborhood D, was chosen as the base level. Consequently, the model will predict $E(y^*)$ for neighborhood D when $x_3 = x_4 = x_5 = 0$.

Although it allows for neighborhood differences, Model 2 assumes that percentage change in sale price y for every \$1,000 increase in either x_1 or x_2 does not depend on neighborhood.

Model 3

Model 3 is similar to Model 2 except that we will add interaction terms between the neighborhood dummy variables and x_1 and between the neighborhood dummy variables and x_2. These interaction terms allow the percentage change in y for increases in x_1 or x_2 to vary depending on the neighborhood. The equation of Model 3 is

M O D E L 3
First-order log model, no restrictions on neighborhood differences

$$E(y) = \beta_0 + \overbrace{\beta_1 x_1}^{\substack{\text{Appraised land} \\ \text{value}}} + \overbrace{\beta_2 x_2}^{\substack{\text{Appraised} \\ \text{improvement value}}}$$

$$+ \overbrace{\beta_3 x_3 + \beta_4 x_4 + \beta_5 x_5}^{\text{Main effect terms for neighborhoods}}$$

$$+ \overbrace{\beta_6 x_1 x_3 + \beta_7 x_1 x_4 + \beta_8 x_1 x_5}^{\text{Interaction, appraised land by neighborhood}}$$

$$+ \overbrace{\beta_9 x_2 x_3 + \beta_{10} x_2 x_4 + \beta_{11} x_2 x_5}^{\text{Interaction, appraised improvement by neighborhood}}$$

Note that for Model 3, the percentage change in sale price y for every \$1,000 increase in appraised land value x_1 (holding x_2 fixed) is $(e^{\beta_1} - 1)$ in neighborhood D and $(e^{\beta_1 + \beta_6} - 1)$ in neighborhood A.

Model 4

Model 4 differs from the previous three models by the addition of terms for x_1, x_2-interaction. Thus, Model 4 is a second-order (interaction) model that will trace (geometrically) a second-order response surface, one for each neighborhood. The interaction model follows:

M O D E L 4
Interaction model in x_1 and x_2 that differs from one neighborhood to another

$$E(y^*) = \overbrace{\beta_0 + \beta_1 x_1 + \beta_2 x_2 + \beta_3 x_1 x_2}^{\text{Interaction model in } x_1 \text{ and } x_2} + \overbrace{\beta_4 x_3 + \beta_5 x_4 + \beta_6 x_5}^{\text{Main effect terms for neighborhoods}}$$

$$\left. \begin{array}{l} + \beta_7 x_1 x_3 + \beta_8 x_1 x_4 + \beta_9 x_1 x_5 + \beta_{10} x_2 x_3 \\ + \beta_{11} x_2 x_4 + \beta_{12} x_2 x_5 + \beta_{13} x_1 x_2 x_3 \\ + \beta_{14} x_1 x_2 x_4 + \beta_{15} x_1 x_2 x_5 \end{array} \right\} \begin{array}{l} \text{Interaction terms: } x_1, \\ x_2, \text{ and } x_1 x_2 \text{ terms by} \\ \text{neighborhood} \end{array}$$

Unlike Models 1–3, Model 4 allows the percentage change in y for increases in x_1 to depend on x_2, and vice versa. For example, the percentage change in sale price for a \$1,000 increase in appraised land value in neighborhood D is $(e^{\beta_1 + \beta_3 x_2} - 1)$.

We will fit Models 1–4 to the data. Then, we will compare the models using the partial F test outlined in Section 4.10.

12.4

Model Comparisons

The SAS printouts for Models 1–4 are shown in Figures 12.2, 12.3, 12.4, and 12.5, respectively. Checking these printouts, you will note that the sum of squares for error, the mean square for error, R^2, s, and degrees of freedom for error for each of the four models are as listed in Table 12.1 (page 662).

FIGURE 12.2

Model 1: First-order log model in land and improvements appraised values, identical for all neighborhoods

```
Model: MODEL1
Dependent Variable: LNSALES

                        Analysis of Variance

                              Sum of         Mean
        Source       DF      Squares        Square     F Value     Prob>F

        Model         2     343.63712     171.81856    810.043     0.0001
        Error      1151     244.13903       0.21211
        C Total    1153     587.77615

             Root MSE      0.46055      R-square      0.5846
             Dep Mean      3.84928      Adj R-sq      0.5839
             C.V.         11.96469

                        Parameter Estimates

                      Parameter      Standard     T for H0:
        Variable  DF   Estimate         Error    Parameter=0    Prob > |T|

        INTERCEP   1   3.177222      0.02184722     145.429      0.0000
        LAND       1   0.013508      0.00156276       8.644      0.0001
        IMP        1   0.013566      0.00067701      20.038      0.0001
```

FIGURE 12.3

Model 2: First-order log model, constant differences between neighborhoods

```
Model: MODEL2
Dependent Variable: LNSALES

                        Analysis of Variance

                              Sum of         Mean
        Source       DF      Squares        Square     F Value     Prob>F

        Model         5     371.11471      74.22294    393.277     0.0001
        Error      1148     216.66144       0.18873
        C Total    1153     587.77615

             Root MSE      0.43443      R-square      0.6314
             Dep Mean      3.84928      Adj R-sq      0.6298
             C.V.         11.28601

                        Parameter Estimates

                      Parameter      Standard     T for H0:
        Variable  DF   Estimate         Error    Parameter=0    Prob > |T|

        INTERCEP   1   3.077202      0.02455712     125.308      0.0000
        LAND       1   0.009902      0.00156114       6.343      0.0001
        IMP        1   0.012338      0.00074255      16.616      0.0001
        NA         1   0.219871      0.09838735       2.235      0.0256
        NB         1   0.403035      0.04368094       9.227      0.0001
        NC         1   0.309930      0.03089784      10.031      0.0001

Test: M1VSM2     Numerator:      9.1592  DF:     3   F value:  48.5308
                 Denominator:   0.188729 DF: 1148   Prob>F:    0.0001
```

FIGURE 12.4

Model 3: First-order log model, no
restriction on differences between
neighborhoods

Model: MODEL3
Dependent Variable: LNSALES

Analysis of Variance

Source	DF	Sum of Squares	Mean Square	F Value	Prob>F
Model	11	398.77922	36.25266	219.054	0.0001
Error	1142	188.99693	0.16550		
C Total	1153	587.77615			

Root MSE	0.40681	R-square	0.6785	
Dep Mean	3.84928	Adj R-sq	0.6754	
C.V.	10.56854			

Parameter Estimates

Variable	DF	Parameter Estimate	Standard Error	T for H0: Parameter=0	Prob > |T|
INTERCEP	1	2.603613	0.05135028	50.703	0.0001
LAND	1	0.034691	0.00459212	7.554	0.0001
IMP	1	0.025225	0.00189898	13.283	0.0001
NA	1	1.883621	0.18814600	10.011	0.0001
NB	1	0.904007	0.07613408	11.874	0.0001
NC	1	0.502904	0.09398386	5.351	0.0001
LAND_NA	1	-0.029413	0.00769242	-3.824	0.0001
LAND_NB	1	-0.026067	0.00496368	-5.251	0.0001
LAND_NC	1	-0.015645	0.00572188	-2.734	0.0063
IMP_NA	1	-0.021167	0.00313130	-6.760	0.0001
IMP_NB	1	-0.012820	0.00210641	-6.086	0.0001
IMP_NC	1	-0.006992	0.00288657	-2.422	0.0156

Test: M2VSM3	Numerator:	4.6108	DF:	6	F value:	27.8601
	Denominator:	0.165496	DF:	1142	Prob>F:	0.0001

T A B L E 12.1   **Summary of Regressions of the Models**

MODEL	SSE	MSE	R^2	s	df(ERROR)
1	244.14	.212	.585	.461	1151
2	216.66	.189	.631	.434	1148
3	189.00	.166	.679	.407	1142
4	184.54	.162	.686	.403	1138

T E S T # 1
Model 1 versus model 2

To test the hypothesis that a single first-order log model is appropriate for all neighborhoods, we wish to test the null hypothesis that the neighborhood parameters in Model 2 are all equal to 0, i.e.,

$$H_0: \quad \beta_3 = \beta_4 = \beta_5 = 0$$

That is, we want to compare the complete model, Model 2, to the reduced model, Model 1. The test statistic is

$$F = \frac{(\text{SSE}_R - \text{SSE}_C)/\text{Number of } \beta \text{ parameters in } H_0}{\text{MSE}_C}$$

$$= \frac{(\text{SSE}_1 - \text{SSE}_2)/3}{\text{MSE}_2}$$

FIGURE 12.5

Model 4: Log model with
(land × improvements) interaction,
different for all neighborhoods

```
Model: MODEL4
Dependent Variable: LNSALES

                         Analysis of Variance

                             Sum of        Mean
      Source        DF       Squares       Square      F Value     Prob>F

      Model         15      403.23145     26.88210     165.769     0.0001
      Error       1138      184.54470      0.16217
      C Total     1153      587.77615

           Root MSE        0.40270     R-square      0.6860
           Dep Mean        3.84928     Adj R-sq      0.6819
           C.V.           10.46165

                         Parameter Estimates

                      Parameter      Standard      T for H0:
      Variable   DF    Estimate        Error      Parameter=0     Prob > |T|

      INTERCEP    1    2.360251      0.09395936      25.120         0.0001
      LAND        1    0.064617      0.01072794       6.023         0.0001
      IMP         1    0.035114      0.00372098       9.437         0.0001
      LAND_IMP    1   -0.001170      0.00037992      -3.080         0.0021
      NA          1    1.574260      0.52798183       2.982         0.0029
      NB          1    0.928262      0.13105999       7.083         0.0001
      NC          1    0.430979      0.16755155       2.572         0.0102
      LAND_NA     1   -0.051477      0.01416073      -3.635         0.0003
      LAND_NB     1   -0.048085      0.01119860      -4.294         0.0001
      LAND_NC     1   -0.024762      0.01356050      -1.826         0.0681
      IMP_NA      1   -0.025959      0.00633615      -4.097         0.0001
      IMP_NB      1   -0.019035      0.00401708      -4.739         0.0001
      IMP_NC      1   -0.003832      0.00640650      -0.598         0.5498
      L_I_NA      1    0.001104      0.00038439       2.872         0.0042
      L_I_NB      1    0.001057      0.00038174       2.770         0.0057
      L_I_NC      1    0.000298      0.00049513       0.602         0.5476


Test: M3VSM4   Numerator:      1.1131   DF:    4   F value:  6.8637
               Denominator:   0.162166  DF: 1138   Prob>F:   0.0001
```

where the mean square for the numerator is based on $\nu_1 = 3$ df, and MSE_2, the estimate of σ^2 using Model 2, is based on $\nu_2 = 1,148$ df. Substituting SSE_1, SSE_2, and MSE_2 into the formula for F, we obtain

$$F = \frac{(SSE_1 - SSE_2)/3}{MSE_2} = \frac{(244.14 - 216.66)/3}{.189} = 48.53$$

The tabulated critical value of F for $\alpha = .01$ and $\nu_1 = 3$ and $\nu_2 = 1,148$ df is not shown in Table 6 of Appendix D, but you can see that it is approximately equal to 3.78. Since the computed value of F, 48.53, exceeds this value, we have evidence to indicate that the addition of the neighborhood dummy variables in Model 2 contributes significantly to the prediction of y^*. (Note that the p-value of the test, .0001, is shown at the bottom of the SAS printout, Figure 12.3.) The practical implication of this result is that the appraiser is not assigning appraised values to properties in such a way that the first-order relationship between log(sales), y^*, and appraised values x_1 and x_2 is the same for all neighborhoods.

TEST #2

Model 2 versus Model 3

Can the prediction equation be improved by the addition of interactions between neighborhood and x_1 and between neighborhood and x_2? That is, do the data provide sufficient evidence to indicate that Model 3 is a better predictor of sale price than Model 2? To answer this question, we will test the null hypothesis that the parameters associated with all neighborhood interaction terms in Model 3 equal 0. Thus, Model 2 is now the reduced model and Model 3 is the complete model.

Checking the equation of Model 3, you will see that there are six neighborhood interaction terms and that the parameters included in H_0 will be

$$H_0: \quad \beta_6 = \beta_7 = \beta_8 = \beta_9 = \beta_{10} = \beta_{11} = 0$$

To test H_0, we form the test statistic

$$F = \frac{(SSE_R - SSE_C)/\text{Number of } \beta \text{ parameters in } H_0}{MSE_C} = \frac{(SSE_2 - SSE_3)/6}{MSE_3}$$

where MSE_3 is the estimate of σ^2 obtained from the complete second-order model (Model 3). Substituting the values of SSE_2, SSE_3, and MSE_3 into the formula for F, we obtain

$$F = \frac{(216.66 - 189.00)/6}{.166} = 27.86$$

The degrees of freedom for F are $\nu_1 = 6$ and $\nu_2 = 1,142$ (the degrees of freedom for MSE_3 are given on the computer printout, Figure 12.4) and the tabulated F value for $\alpha = 0.1$ (Table 6 of Appendix D) is approximately 2.80. Since the computed value of F, 27.86, exceeds the critical value, $F_{.01} = 2.80$, there is sufficient evidence to indicate that the neighborhood interaction terms of Model 3 contribute information for the prediction of y^*. (The p-value for this test, .0001, is also given at the bottom of the SAS printout, Figure 12.4.) Practically, this test implies that the rate of change (measured as a percentage) of sale price y with either appraised value, x_1 or x_2, differs for each of the four neighborhoods.

TEST # 3
Model 3 versus Model 4

We have already shown that the first-order prediction equations vary among neighborhoods. To determine whether the (second-order) interaction terms involving the appraised values, x_1 and x_2, contribute significantly to the prediction of y^*, we test the hypothesis that the four parameters involving $x_1 x_2$ in Model 4 all equal 0. The null hypothesis is

$$H_0: \quad \beta_3 = \beta_{13} = \beta_{14} = \beta_{15} = 0$$

and the alternative hypothesis is that at least one of these parameters does not equal 0. Using Model 4 as the complete model and Model 3 as the reduced model, the F statistic is

$$F = \frac{(SSE_R - SSE_C)/\text{Number of } \beta \text{ parameters in } H_0}{MSE_C} = \frac{(SSE_3 - SSE_4)/4}{MSE_4}$$

$$= \frac{(189.00 - 184.54)/4}{.162} = 6.86$$

Again, the tabulated value of $F_{.01}$ for $\nu_1 = 4$ and $\nu_2 = 1,138$ df is not given in the $F_{.01}$ table, but you can see that it is approximately equal to 3.32. The computed

F value, 6.86, exceeds this critical value, supporting the alternative hypothesis that the $x_1 x_2$ interaction terms of Model 4 contribute significantly to the prediction of y^*. [*Note:* The p-value of the test, .0001, is shown at the bottom of the SAS printout, Figure 12.5.]

The results of the preceding tests suggest that Model 4 is the best of the four models for modeling sale price y. The global F value for testing H_0: $\beta_1 = \beta_2 = \cdots = \beta_{15} = 0$ is highly significant ($F = 165.769$, p-value $< .0001$) and the R^2 value indicates that the model is explaining almost 70% of the variability in the log of sale prices. Although a few of the t tests involving the individual β parameters in Model 4 are nonsignificant, this does not imply that these terms should be dropped from the model.

Whenever a model includes a large number of interactions (as in Model 4) and/or squared terms, several t tests will often be nonsignificant even if the global F test is highly significant. This result is due partly to the unavoidable intercorrelations among the main effects for a variable, its interactions, and its squared terms (see the discussion on multicollinearity in Section 6.3). We warned in Chapter 4 of the dangers of conducting a series of t tests to determine model adequacy. For a model with a large number of β's, such as Model 4, you should avoid conducting any t tests at all and rely on the global F test and partial F tests to determine the important terms for predicting y.

12.5
Interpreting the Prediction Equation

Substituting the estimates of the Model 4 parameters (Figure 12.5) into the prediction equation, we have

$$\hat{y}^* = 2.3603 + .0646x_1 + .0351x_2 - .00117x_1 x_2 + 1.5743x_3 + .9283x_4$$
$$+ .4310x_5 - .0515x_1 x_3 - .0481x_1 x_4 - .0248x_1 x_5 - .0260x_2 x_3$$
$$- .0190x_2 x_4 - .0038x_2 x_5 + .0011x_1 x_2 x_3 + .00106x_1 x_2 x_4 + .0003x_1 x_2 x_5$$

We have noted that the model yields four response surfaces, one for each neighborhood. One way to interpret the prediction equation is to first find the equation of the response surface for each neighborhood. Substituting the appropriate values of the neighborhood dummy variables, x_3, x_4, and x_5, into the equation and combining like terms, we obtain the following:

Neighborhood A: $(x_3 = 1, x_4 = x_5 = 0)$

$$\hat{y}^* = (2.3603 + 1.5743) + (.0646 - .0515)x_1 + (.0351 - .0260)x_2$$
$$+ (-.00117 + .00110)x_1 x_2$$
$$= \mathbf{3.9346 + .0131x_1 + .0091x_2 - .00007x_1 x_2}$$

Neighborhood B: $(x_3 = 0, x_4 = 1, x_5 = 0)$

$$\hat{y}^* = (2.3603 + .9283) + (.0646 - .0481)x_1 + (.0351 - .0190)x_2$$
$$+ (-.00117 + .00106)x_1 x_2$$
$$= \mathbf{3.2889 + .0165x_1 + .0161x_2 - .00011x_1 x_2}$$

Neighborhood C: $(x_3 = x_4 = 0, x_5 = 1)$

$$\hat{y}^* = (2.3603 + .4310) + (.0646 - .0248)x_1 + (.0351 - .0038)x_2$$
$$+ (-.00117 + .00030)x_1x_2$$
$$= 2.7913 + .0398x_1 + .0313x_2 - .00087x_1x_2$$

Neighborhood D: $(x_3 = x_4 = x_5 = 0)$

$$\hat{y}^* = 2.3603 + .0646x_1 + .0351x_2 - .00117x_1x_2$$

Note that each equation is in the form of an interaction model involving appraised land value x_1 and appraised improvements x_2. To interpret the β estimates of each interaction equation, we hold one independent variable fixed, say, x_1, and focus on the slope of the line relating y^* to x_2. For example, holding appraised land value constant at \$10,000 ($x_1 = 10$), the slope of the y^*–x_2 line for neighborhood D is

$$\hat{\beta}_2 + \hat{\beta}_3x_1 = .0351 - .00117(10) = .0234$$

Since the dependent variable, y^*, is a log of sales price, the antilog of the slope minus 1 represents the percentage change in sale price y. Since $e^{.0234} - 1 = .0237$, our interpretation follows: For residential properties in neighborhood D with appraised land value of \$10,000, the sale price will increase 2.37% for every \$1,000 increase in appraised improvements.

Similar interpretations can be made for the slopes for other combinations of neighborhoods and appraised land value x_1. The estimated slopes for several of these combinations are computed and shown in Table 12.2.

T A B L E 12.2   **Estimated Percentage Increase in Sale Price for \$1,000 Increase in Appraised Improvements**

		NEIGHBORHOOD			
		A	B	C	D
APPRAISED LAND VALUE	\$10,000	.84	1.51	2.28	2.37
	\$15,000	.81	1.46	1.84	1.77
	\$20,000	.77	1.40	1.40	1.18

Because of the interaction terms in the model, the percentage increases in sale price for a \$1,000 increase in appraised improvements, x_2, differ for each neighborhood and for different levels of land value, x_1.

Some trends are evident from Table 12.2. For fixed appraised land value, x_1, the percentage increase in sales price for every \$1,000 increase in appraised improvements is smallest for neighborhood A. For a given neighborhood, the percentage increase decreases as appraised land value increases.

Another way to describe the prediction equation would be to graph predicted sale price for each neighborhood as a function of appraised improvements value

x_2 for different levels of appraised land value x_1. The curves (one curve for each neighborhood) are shown in Figure 12.6 for appraised land value $x_1 = \$10,000$. Similar sets of curves are shown in Figures 12.7 and 12.8 (pages 668–669) for $x_1 = \$25,000$ and $\$50,000$, respectively.* Note that not all figures have curves corresponding to all four neighborhoods. This is because the appraised land values for sales in some neighborhoods might not have been as low as $\$10,000$ or as high as $\$50,000$. Some curves are shortened for similar reasons, i.e., the appraised improvements values for some neighborhoods might not span the range shown on the x_2-axis.

FIGURE 12.6

Graph of predicted sale price versus appraised improvements (x_2) when appraised land value (x_1) is $10,000

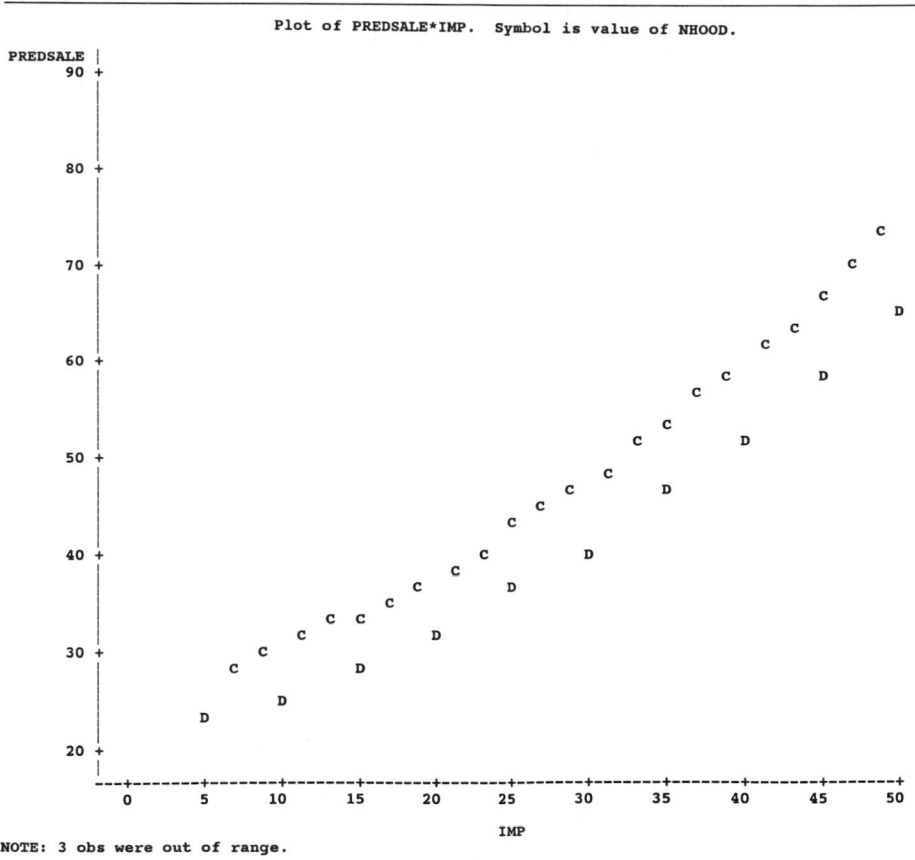

NOTE: 3 obs were out of range.

---

*The curves for predicted sale price, $\hat{y}$, were obtained as follows: First, the predicted log of sale price, $\hat{y}^*$, was obtained from the fitted regression equation for each combination of x_1, x_2, and neighborhood. Then, antilogs of the predicted log sale price were computed to obtain $\hat{y}$, i.e., $\hat{y} = e^{\hat{y}^*}$. The predicted value, $\hat{y}$, was then plotted against x_2 for the different levels of x_1 using neighborhood as a plotting symbol.

FIGURE 12.7

Graph of predicted sale price versus appraised value of improvements when appraised land value is $25,000

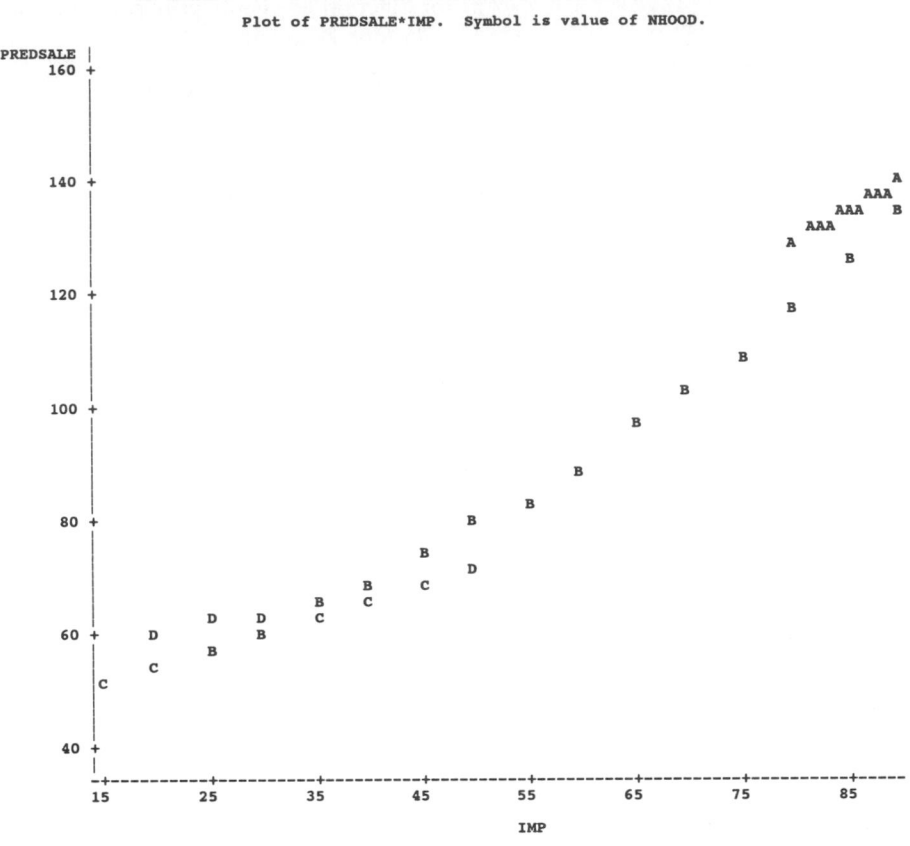

Plot of PREDSALE*IMP.   Symbol is value of NHOOD.

NOTE: 5 obs hidden.   1 obs were out of range.

Before we examine the curves in Figures 12.6–12.8, you will want some information concerning the nature of the neighborhoods. Figure 12.9 (page 670) gives the means, standard deviations, and other descriptive statistics for the sale price y and the appraised values x_1 and x_2 for each of the four neighborhoods. The mean sale prices confirm what we (the authors) know to be true, i.e., that neighborhoods A and B are two of the relatively expensive residential areas in the city. Most of the inhabitants are older, established professional or business people. In contrast, neighborhood C is a less expensive residential area inhabited primarily by young married couples who are starting their careers. Neighborhood D is considered to be one of the least expensive residential areas in the city and is inhabited by mostly low-income families.

In Figures 12.6–12.8, you will note that the estimated mean sale price $\hat{y}$ increases as the appraised improvements value x_2 increases, but the increase is not always linear. The relationship appears to be nearly a straight line for some neighborhoods (for example, neighborhood A) but is curvilinear for other neighborhoods.

There are some other interesting features to note. For example, in Figure 12.7 observe the different predicted sale prices for an appraised improvement value

FIGURE 12.8 Graph of predicted sale price versus appraised value of improvements when appraised land value is $50,000

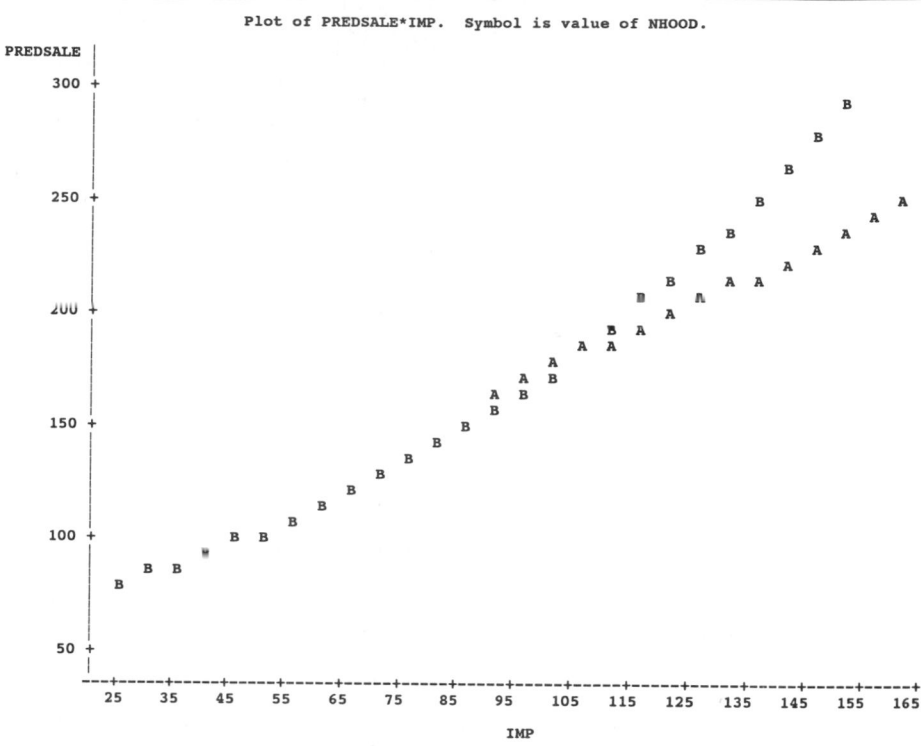

Plot of PREDSALE*IMP. Symbol is value of NHOOD.

NOTE: 1 obs hidden. 1 obs were out of range.

of $25,000. You can see that properties in neighborhood D (a lower-priced neighborhood) sell at a higher predicted price than those in the more expensive neighborhood B when the appraised land value is low, i.e., $x_1 = \$25,000$. This would suggest that the appraised values of properties in this price range in neighborhood B are too high compared with similar properties in neighborhood D. One reason for this could be that a low appraised property value in neighborhood B might correspond to a very small lot; this might have a strong depressive effect on sale prices. In contrast, you can see in Figure 12.7 that properties in neighborhood B bring higher prices for an appraised home value of $50,000 than for similarly appraised homes in neighborhood D. Thus, neighborhood B homes are appraised at a lower value than homes in neighborhood D that sell for the same price.

The three major points to be derived from an analysis of the sale price–appraised value curves are as follows:

1. The percentage rate at which sale price increases with appraised value differs for different neighborhoods. This percentage increase is smaller for the more expensive neighborhoods (A and B) than for the less expensive neighborhoods (C and D).

FIGURE 12.9

Real estate analyses: 1990 sales in four neighborhoods

Descriptive Statistics for Four Neighborhoods

-------------------------------- NHOOD=A --------------------------------

N Obs	Variable	N	Minimum	Maximum	Mean	Std Dev
42	SALES	42	93.9000000	422.0000000	193.7380952	65.5110788
	LNSALES	42	4.5422304	6.0450053	5.2196121	0.3005363
	LAND	42	24.6370000	108.9960000	49.4794762	18.8797839
	IMP	42	50.7650000	311.1210000	116.1116667	46.7971693

-------------------------------- NHOOD=B --------------------------------

N Obs	Variable	N	Minimum	Maximum	Mean	Std Dev
258	SALES	258	11.0000000	355.0000000	100.9616279	58.3184038
	LNSALES	258	2.3978953	5.8721178	4.4600507	0.5770448
	LAND	258	3.2100000	120.0000000	24.5626744	15.6917178
	IMP	258	2.0000000	189.9310000	59.7009612	32.4382254

-------------------------------- NHOOD=C --------------------------------

N Obs	Variable	N	Minimum	Maximum	Mean	Std Dev
372	SALES	372	6.0000000	253.5000000	50.3372231	19.7725356
	LNSALES	372	1.7917595	5.5353638	3.8424021	0.4242907
	LAND	372	5.5200000	68.2500000	14.2968629	6.1978100
	IMP	372	5.0000000	58.4180000	25.4255618	9.7317395

-------------------------------- NHOOD=D --------------------------------

N Obs	Variable	N	Minimum	Maximum	Mean	Std Dev
482	SALES	482	5.5000000	152.0000000	34.9618734	17.8140348
	LNSALES	482	1.7047481	5.0238805	3.4082539	0.5818234
	LAND	482	1.8000000	51.1920000	8.8472324	4.1717595
	IMP	482	2.5000000	76.3490000	19.7313320	10.0881887

2. The curves for neighborhoods A and B frequently lie above those for neighborhoods C and D, indicating that properties in the higher-priced neighborhoods are being underappraised relative to sale price compared with properties in the lower-priced neighborhoods.

3. For most neighborhoods, the graphs curve upward as the appraised improvements value increases, indicating that more expensive properties within most neighborhoods are underappraised compared with the appraised values of less expensive properties.

12.6
Predicting the Sale Price of a Property

How well do appraised land value x_1 and appraised improvements value x_2 predict residential property sale price? Recall that from Model 4 (Figure 12.5), we obtained $R^2 = .686$, indicating that the model accounts for approximately 70% of the sample variability in the log sale price values, y^*. This seems to indicate that the model provides a reasonably good fit to the data, but note that $s = .4027$ (see Figure 12.5). Our interpretation is that approximately 95% of the predicted sale price values will fall within $(e^{2s} - 1)100\% = 123\%$ of their actual values. This large standard deviation may lead to large errors of prediction for some residential properties if the model is used in practice.

Figure 12.10 is a portion of a SAS printout showing 95% prediction intervals for the log sale price of several residential properties from each neighborhood. By taking antilogs of the endpoints of each interval, we obtain 95% prediction intervals for sale price, y. For example, the last row of the printout gives the 95% prediction interval for $\log(y)$ for a property in neighborhood D with an appraised land value of $10,000 and an appraised improvements value of $20,000. The interval is (2.6836, 4.2658). Taking antilogs of the endpoints, we obtain

$$e^{2.6836} = 14.638$$
$$e^{4.2658} = 71.222$$

FIGURE 12.10

95% prediction intervals for four residential properties

Obs	NHOOD	LAND	IMP	Dep Var LNSALES	Predict Value	Std Err Predict	Lower95% Predict	Upper95% Predict	Residual
1	A	50	150	.	5.4684	0.140	4.6320	6.3049	.
2	B	25	100	.	5.0282	0.049	4.2323	5.8241	.
3	C	15	30	.	3.9350	0.023	3.1436	4.7265	.
4	D	10	20	.	3.4747	0.020	2.6836	4.2658	.

Because y is recorded in thousands of dollars, a 95% prediction interval for the sale price of this particular residential property is ($14,638, $71,222).

Such a wide prediction interval casts doubt on whether the prediction equation could be of practical value in predicting property sale prices. We feel certain that a much more accurate predictor of sale price could be developed by relating y to the variables that describe the property (such as location, square footage, and number of bedrooms) and those that describe the market (mortgage interest rates, availability of money, and so forth).

12.7 Conclusions

The results of the regression analyses described in Section 12.4 indicate that the relationships between property sale prices and appraised values are not consistent from one neighborhood to another. Further, the widths of the prediction intervals in Section 12.6 are sizable, thus indicating that there is room for improvement in the methods used to determine appraised property values.

EXERCISES

12.1 Explain why the tests of model adequacy conducted in Section 12.4 give no assurances that Model 4 will be a successful predictor of sale price in the future.

12.2 If you have access to a statistical software package, use the data-splitting technique of Section 8.9 to assess the validity of Model 4.

ON YOUR OWN...

Recall that the data for this case study are given in Appendix E.* The full data set contains sale price information for the four neighborhoods (A, B, C, and D) compared in this case study, as well as sale price information for three additional neighbor-hoods (E, F, and G). As a class project, use the full data set to build a regression model for sale price of a residential property. Part of your analysis will involve a comparison of the seven neighborhoods.

---

*This data set, as well as all appendix data sets, can be obtained on a $3\frac{1}{2}''$ or $5\frac{1}{4}''$ diskette by contacting the publisher.

An Analysis of Bidding Competition

CONTENTS

13.1 The Problem

13.2 The Data

13.3 Models for Stable and Unstable Market Conditions

13.4 The Regression Analyses: Testing to Detect Unstable Market Conditions

13.5 A Residual Analysis of the Data

13.6 Summary

OBJECTIVE

To present an example of regression modeling that can be used to detect the possibility of collusive bidding

13.1
The Problem

Many products and services are purchased by governments and businesses on the basis of competitive bids, and frequently contracts are awarded to the lowest bidders. This process works extremely well in competitive markets, but it has the potential to increase the cost of purchasing if the markets are noncompetitive or if collusive practices are present.

Numerous methods exist for detecting the possibility of collusive practices among bidders. In general, these procedures involve the detection of significant departures from normal market conditions such as (1) systematic rotation of the winning bid, (2) stable market shares over time, (3) geographic market divisions, (4) lack of relationship between delivery costs and bid levels, (5) high degree of uniformity and stability in bid levels over time, and (6) presence of a baseline point pricing scheme (Rothrock and McClave, 1979). This chapter demonstrates two applications of regression analysis to detect collusive bidding. In particular, we will show you how to use regression analysis to detect statistically significant changes in the annual pattern of mean low-bid prices for different markets. Then we will examine the residuals, the deviations between the predicted and the actual low-bid prices, to detect unusually large positive residuals.

13.2
The Data

Each county in the state of Florida negotiates an annual contract for bread to supply the county's public schools. Sealed bids are submitted by the vendors and the lowest bid (price per pound of bread) is selected as the bid winner. The data of interest in this study are the annual low-bid prices for each of the three major bread purchases—white bread, hamburger buns, and hot dog buns—for most of the 67 county school boards in the state of Florida for the years 1970–1975. To reduce the size of this analysis, we will confine our attention to only one of these products—white bread.

The state was divided into eight geographic market areas. Each region contained counties that were adjacent to one another, possessed similar economic characteristics, and were served by the same group of vendors. There were 13 vendors that baked and supplied the bread, but only three of these possessed multiplant operations and sold bread throughout most of the state. The other vendors confined their selling efforts to a much smaller number of the eight geographic market areas. The total number of winning bids included in the analysis is 303. The following data were recorded for each of the 303 low bids and are provided in Appendix F.

1. LBPRICE: Low-bid price (y)
2. MKT: Geographic region containing a group of buyers with essentially the same set of vendors
3. YEAR: Year in which contract was let

13.3
Models for Stable and Unstable Market Conditions

The low-bid price to a particular school system should be proportional to the cost of producing and delivering a pound of bread. This cost will vary from year to year and from market to market. The annual variation in the cost per pound of bread would generally be inflationary because of increasing costs of fuel, labor, and raw materials over the period 1970–1975, and would rise in unequal jumps from year to year. The year-to-year effect of cost increases on low-bid price for a single market might appear as shown in Figure 13.1.

FIGURE 13.1

Hypothetical relationship between low-bid price and year for a single market

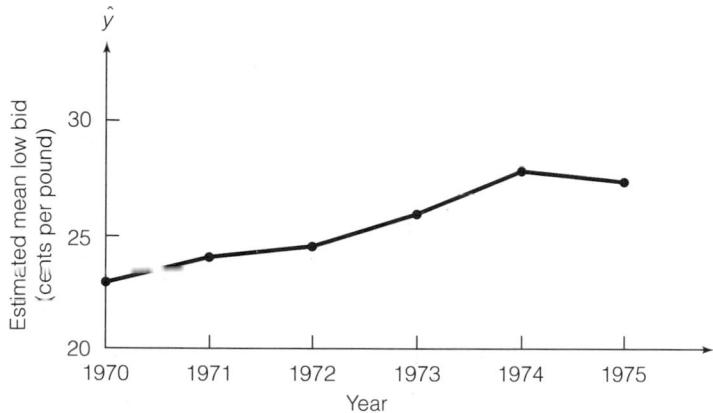

A second source of variation in cost per pound of bread (and hence bid prices) would be the different characteristics of the eight markets. The distances required to transport the bread would vary from market to market as would the costs of labor, rent, and other forms of overhead. The net effect of these costs would be to raise or lower the intercept of the low-bid price–year curve in Figure 13.1, depending on the particular market.

Annual changes in the cost of fuel, labor, and other purchases would not produce identical changes in the cost per pound of bread in all eight markets, but we would expect the overall changes in cost to be roughly of the same magnitude. Consequently, assuming that no unusual disturbance affects the cost in any of the markets (we will call this a **stable market condition**), the cost–year curves for the eight markets would be identical except for a vertical shift and might appear as shown in Figure 13.2 (page 676). Thus, Figure 13.2 shows the effect of the two sources of variation, the variation between markets and the inflationary variation as a result of time, on low-bid price in a stable market. The fact that the pattern of low-bid prices from market to market remains the same from year to year implies that the two factors, market and time, do not interact. The linear model that characterizes this situation is

MODEL 1

A stable market condition

$$E(y) = \beta_0 + \overbrace{\beta_1 x_1 + \beta_2 x_2 + \cdots + \beta_7 x_7}^{\text{Main effects, market}}$$

$$+ \overbrace{\beta_8 x_8 + \beta_9 x_9 + \cdots + \beta_{12} x_{12}}^{\text{Main effects, time}}$$

$$x_1 = \begin{cases} 1 & \text{if market 1} \\ 0 & \text{if not} \end{cases} \quad x_2 = \begin{cases} 1 & \text{if market 2} \\ 0 & \text{if not} \end{cases} \cdots \quad x_7 = \begin{cases} 1 & \text{if market 7} \\ 0 & \text{if not} \end{cases}$$

$$x_8 = \begin{cases} 1 & \text{if 1970} \\ 0 & \text{if not} \end{cases} \quad x_9 = \begin{cases} 1 & \text{if 1971} \\ 0 & \text{if not} \end{cases} \cdots \quad x_{12} = \begin{cases} 1 & \text{if 1974} \\ 0 & \text{if not} \end{cases}$$

FIGURE 13.2

Hypothetical pattern of annual low-bid prices for a stable market condition

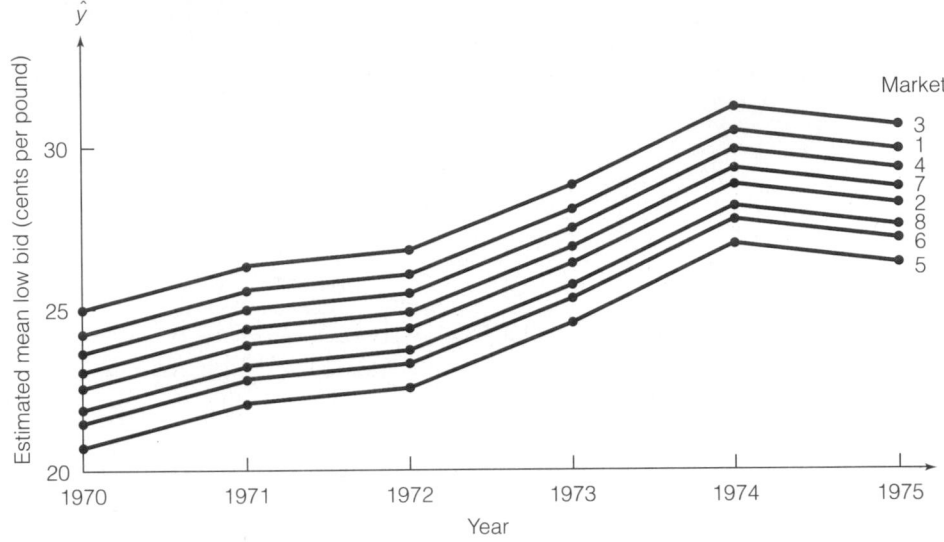

You will notice that time, a quantitative factor, was entered into the model as if it were a qualitative factor. The reason for this is that we expected an irregular pattern of annual inflationary increases in the mean low-bid prices. If we expected $E(y)$ to increase over time in a smooth and regular pattern, we would have entered time into the model with $x_8 = \text{Year} - c$ (where c is a constant, say, 1972) and with terms involving x_8, x_8^2, and so forth.

Suppose the markets are not stable over time—that is, suppose the low-bid price in a particular market increases substantially in a given year in relation to the low-bid prices for the other markets. This situation, which represents a market–time interaction, signals a very unusual increase in costs in that market, or it may be a warning of the possibility of collusive bidding. A graph of this situation might appear as shown in Figure 13.3.

An interactive model that would characterize the situation described in Figure 13.3 is

MODEL 2

Unstable market condition

$$
\begin{aligned}
E(y) = \beta_0 + &\overbrace{\beta_1 x_1 + \beta_2 x_2 + \cdots + \beta_7 x_7}^{\text{Main effects, market}} \\
&\overbrace{+ \beta_8 x_8 + \beta_9 x_9 + \cdots + \beta_{12} x_{12}}^{\text{Main effects, year}} \\
&\left.\begin{aligned}
&+ \beta_{13} x_1 x_8 + \beta_{14} x_2 x_8 + \cdots + \beta_{19} x_7 x_8 \\
&+ \beta_{20} x_1 x_9 + \beta_{21} x_2 x_9 + \cdots + \beta_{26} x_7 x_9 \\
&+ \beta_{27} x_1 x_{10} + \beta_{28} x_2 x_{10} + \cdots + \beta_{33} x_7 x_{10} \\
&+ \beta_{34} x_1 x_{11} + \beta_{35} x_2 x_{11} + \cdots + \beta_{40} x_7 x_{11} \\
&+ \beta_{41} x_1 x_{12} + \beta_{42} x_2 x_{12} + \cdots + \beta_{47} x_7 x_{12}
\end{aligned}\right\} \begin{aligned}&\text{Year–market} \\ &\text{interaction}\end{aligned}
\end{aligned}
$$

FIGURE 13.3

Hypothetical pattern of annual low-bid prices for an unstable market condition

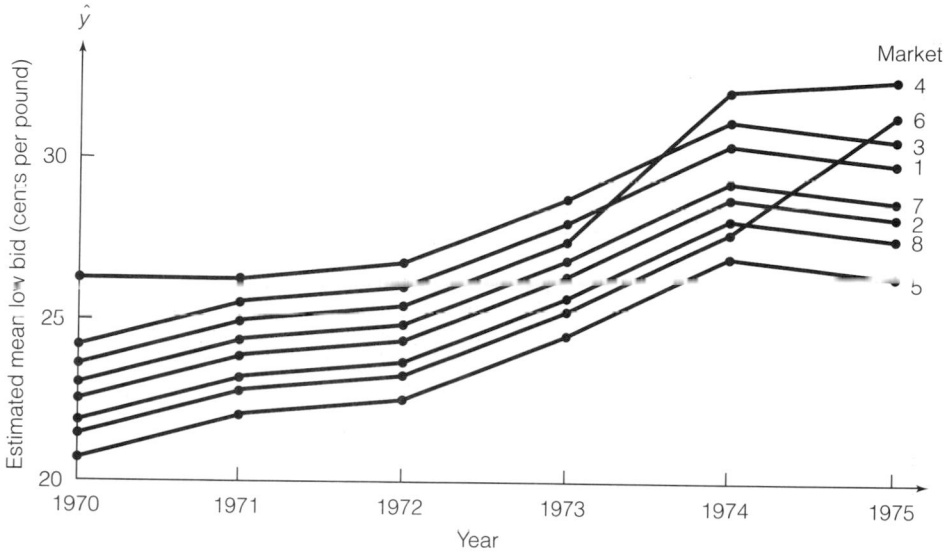

13.4

The Regression Analyses: Testing to Detect Unstable Market Conditions

We have presented two models in Section 13.3. Model 1 represents a stable market condition, the situation where the differences in mean low-bid prices between any pair of markets remain the same from year to year. Model 2 represents an unstable market condition, a situation where differences in the mean low-bid prices between any pair of markets is allowed to change from year to year. Model 2 contains all the terms contained in Model 1 and, in addition, contains 35 year–market interaction terms. Testing the null hypothesis that "the market is stable" is equivalent to testing the null hypothesis that the 35 interaction parameters of Model 2 equal 0, i.e.,

$$H_0: \quad \beta_{13} = \beta_{14} = \beta_{15} = \cdots = \beta_{47} = 0$$

Support of the alternative hypothesis means that *at least one* of the parameters specified in H_0 differs from 0. The practical implication of this situation is that during at least one of the years, some external factors (such as economic factors or collusive bidding) produced unusually large (or small) mean low-bid prices for some markets.

The regression analyses for Models 1 and 2 were conducted using the SAS GLM procedure (as opposed to SAS REG), and their printouts are shown in Figures 13.4 and 13.5 (pages 678–679), respectively. Note that the GLM printout looks very similar to the usual REG printouts presented in earlier chapters, but includes some additional information under the ANOVA table which we will discuss in the following paragraphs.* A summary of the values of SSE, MSE, and R^2 for Models 1 and 2 is shown in Table 13.1 (page 680).

---

*The appropriate SAS GLM and SAS REG commands are given in Appendix C.

FIGURE 13.4

Model 1: Low-bid price versus market and year

DEPENDENT VARIABLE: LBPRICE

SOURCE	DF	SUM OF SQUARES	MEAN SQUARE	F VALUE	PR > F	R-SQUARE	C.V.
MODEL	12	0.59109360	0.04925780	66.09	0.0001	0.732232	11.2455
ERROR	290	0.21615516	0.00074536			ROOT MSE	LBPRICE MEAN
CORRECTED TOTAL	302	0.80724876				0.02730133	0.24277662

SOURCE	DF	TYPE I SS	F VALUE	PR > F	DF	TYPE III SS	F VALUE	PR > F
MKT	7	0.04055373	7.77	0.0001	7	0.03899047	7.47	0.0001
YEAR	5	0.55053987	147.72	0.0001	5	0.55053987	147.72	0.0001

PARAMETER		ESTIMATE	T FOR H0: PARAMETER=0	PR > |T|	STD ERROR OF ESTIMATE
INTERCEPT		0.28273797 B	38.07	0.0001	0.00742624
MKT	1	0.01646959 B	2.16	0.0317	0.00762948
	2	0.01487183 B	1.88	0.0611	0.00790900
	3	-0.01322826 B	-1.82	0.0702	0.00728017
	4	0.00831731 B	1.07	0.2874	0.00780301
	5	-0.00114983 B	-0.13	0.8945	0.00865911
	6	-0.00537925 B	-0.64	0.5224	0.00839998
	7	-0.00198967 B	-0.21	0.8347	0.00952599
	8	0.00000000 B	.	.	.
YEAR	70	-0.09001053 B	-16.05	0.0001	0.00560670
	71	-0.07795952 B	-14.43	0.0001	0.00540136
	72	-0.07607686 B	-14.40	0.0001	0.00528409
	73	-0.05303058 B	-9.99	0.0001	0.00531023
	74	0.02274620 B	4.37	0.0001	0.00520704
	75	0.00000000 B	.	.	.

FIGURE 13.5

Model 2: Low-bid price versus market and year

SOURCE	DF	SUM OF SQUARES	MEAN SQUARE	F VALUE	PR > F	R-SQUARE	C.V.
MODEL	47	0.62971154	0.01339812	19.24	0.0001	0.780071	10.8685
ERROR	255	0.17753722	0.00069622				
CORRECTED TOTAL	302	0.80724876				ROOT MSE	LBPRICE MEAN
						0.02638606	0.24277662

SOURCE	DF	TYPE I SS	F VALUE	PR > F	DF	TYPE III SS	F VALUE	PR > F
MKT	7	0.04055373	8.32	0.0001	7	0.03951572	8.11	0.0001
YEAR	5	0.55053987	158.15	0.0001	5	0.38679965	111.11	0.0001
MKT*YEAR	35	0.03861794	1.58	0.0243	35	0.03861794	1.58	0.0243

PARAMETER		ESTIMATE	T FOR H0: PARAMETER=0	PR > |T|	STD ERROR OF ESTIMATE
INTERCEPT		0.28377037 B	18.63	0.0001	0.01523400
MKT	1	0.05439012 B	3.09	0.0022	0.01759071
	2	0.01807041 B	0.97	0.3354	0.01865776
	3	-0.00991037 B	-0.59	0.5551	0.01666801
	4	-0.01054815 B	-0.60	0.5493	0.01759071
	5	-0.02950370 B	-1.53	0.1270	0.01926966
	6	-0.01657037 B	-0.86	0.3906	0.01926966
	7	-0.01710370 B	-0.79	0.4280	0.02154413
	8	0.00000000 B			
YEAR	70	-0.09780741 B	-4.54	0.0001	0.02154413
	71	-0.07643703 B	-3.17	0.0017	0.02408707
	72	-0.07080741 B	-3.29	0.0012	0.02154413
	73	-0.04896296 B	-2.27	0.0239	0.02154413
	74	0.01434074 B	0.67	0.5062	0.02154413
	75	0.00000000 B			
MKT*YEAR	1 70	-0.03593945 B	-1.43	0.1529	0.02507062
	1 71	-0.04763596 B	-1.75	0.0821	0.02728687
	1 72	-0.05305258 B	-2.13	0.0339	0.02487702
	1 73	-0.04908642 B	-1.97	0.0496	0.02487702
	1 74	-0.04398754 B	-1.77	0.0782	0.02487702
	1 75	0.00000000 B			
	2 70	-0.00539259 B	-0.20	0.8382	0.02638606
	2 71	-0.00980265 B	-0.35	0.7285	0.02820789
	2 72	-0.01738942 B	-0.67	0.5054	0.02607005
	2 73	-0.00289939 B	-0.11	0.9117	0.02607005
	2 74	0.01511957 B	0.58	0.5625	0.02607005
	2 75	0.00000000 B			
	3 70	0.00063630 B	0.03	0.9787	0.02384498
	3 71	-0.00658963 B	-0.25	0.8002	0.02607639
	3 72	-0.01213712 B	-0.51	0.6086	0.02375055
	3 73	-0.00630519 B	-0.27	0.7909	0.02375122
	3 74	0.00307354 B	0.13	0.8968	0.02367055
	3 75	0.00000000 B			
	4 70	0.04467407 B	1.71	0.0881	0.02609124
	4 71	0.02844444 B	0.82	0.4122	0.02781335
	4 72	0.02548994 B	1.01	0.3150	0.02513136
	4 73	0.00990741 B	0.40	0.6930	0.02557062
	4 74	0.02251111 B	0.90	0.3664	0.02487702
	4 75	0.00000000 B			
	5 70	0.05234843 B	1.81	0.0713	0.02890448
	5 71	0.03833704 B	1.24	0.2151	0.03084650
	5 72	0.02101991 B	0.75	0.4516	0.02788280
	5 73	0.01136296 B	0.41	0.6840	0.02788280
	5 74	0.05552592 B	2.04	0.0426	0.02725141
	5 75	0.00000000 B			
	6 70	0.02827407 B	1.01	0.3115	0.02788280
	6 71	0.02523703 B	0.86	0.3899	0.02930319
	6 72	0.01960741 B	0.72	0.4725	0.02725141
	6 73	0.02976296 B	1.07	0.2868	0.02788280
	6 74	0.03045926 B	1.12	0.2647	0.02725141
	6 75	0.00000000 B			
	7 70	0.02314074 B	0.62	0.5357	0.03731553
	7 71	0.00754814 B	0.23	0.8155	0.03231620
	7 72	0.02591851 B	0.79	0.4332	0.03046800
	7 73	0.01040740 B	0.34	0.7329	0.03046800
	7 74	0.02565925 B	0.84	0.4005	0.03046800
	7 75	0.00000000 B			
	8 70	0.00000000 B			
	8 71	0.00000000 B			
	8 72	0.00000000 B			
	8 73	0.00000000 B			
	8 74	0.00000000 B			
	8 75	0.00000000 B			

T A B L E 13.1   **A Summary of the Values (Rounded) of SSE, MSE, and R^2 for Models 1 and 2**

MODEL	SSE	MSE	df	R^2
1	.216155	.000745	290	.732
2	.177537	.000696	255	.780

To test the null hypothesis that stable market conditions exist, i.e.,

$$H_0: \quad \beta_{13} = \beta_{14} = \cdots = \beta_{47} = 0$$

we obtain the drop in SSE from the reduced model (Model 1) to the complete model (Model 2), and calculate the value of the F statistic:

$$F = \frac{\dfrac{(\text{SSE}_1 - \text{SSE}_2)}{\text{Number of } \beta \text{ parameters in } H_0}}{\text{MSE}_2} = \frac{\dfrac{.216155 - .177537}{35}}{.000696}$$

$$= 1.58$$

The critical value of F for $\alpha = .05$, and $\nu_1 = 35$ and $\nu_2 = 255$ df is approximately equal to 1.50 (see Table 4 of Appendix D). Since the observed value of F, 1.58, exceeds this tabulated value, we reject the null hypothesis and conclude that there is evidence of a year–market interaction.

In conducting this test for factor interaction, we employed the standard procedure of fitting complete and reduced models (Chapter 4) so that our test would not be tied to a single computer program package. However, as explained in Chapter 4, some computer programs (e.g., the SAS GLM procedure) automatically compute the drop in SSE associated with sets of parameters in the complete model. In particular, the SAS computer printout in Figure 13.5 gives these quantities under the column titled **TYPE III SS**. Reading in that column in the **MKT∗YEAR** row, you obtain the drop in SSE associated with the 35 interaction terms as .03861794.* The next column gives the computed value of the F statistic based on this drop in SSE, namely, 1.58. The last column gives the significance level for the F statistic, .0243. In other words, the probability of observing a value of F as large as or larger than 1.58 is only .0243, assuming the null hypothesis is true. Consequently, there is ample evidence to indicate that something is causing the mean low-bid price in some markets to move in an unstable pattern from one year to another.

Now that we have evidence to indicate a year–market interaction, we will examine computer plots of the curves tracing estimated mean low-bid price versus year for the eight markets (see Figure 13.6). The 1's trace the estimated mean

---

*The **TYPE III SS** and **TYPE I SS** for the source of variation listed last (e.g., **MKT∗YEAR**) will always be equal. However, the two **SS** values will differ for the other sources. For **TYPE III SS**, the complete model is always the model that is fit, whereas the reduced model for the row (source) in question drops the terms in that row. (For this reason, the **TYPE III SS** is often called "partial" sums of squares.) Alternatively, the complete model for the **TYPE I SS** column is built sequentially; the model for any row includes the terms specified in that row and all preceding rows. (For this reason, **TYPE I SS** is often called "sequential" sums of squares.)

FIGURE 13.6

Plot of predicted values of price versus market for each year: Numbers identify the market associated with a plotted predicted value

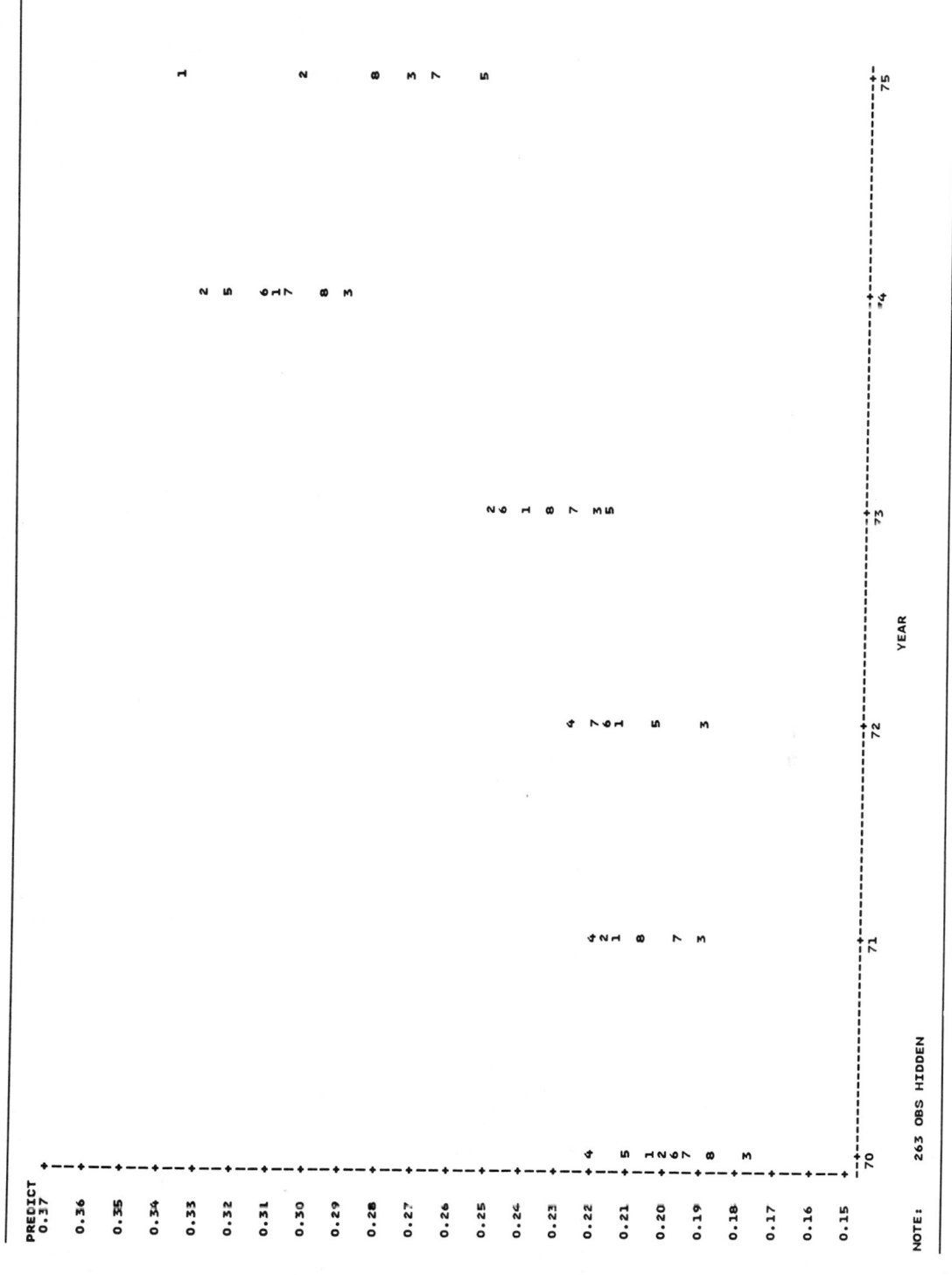

low-bid prices for market 1 over the years 1970–1975. Similarly, the 2's correspond to the estimated mean low-bid prices for market 2, the 3's correspond to market 3, . . . , and the 8's correspond to market 8. The first thing to notice about the estimated mean low-bid plots is the grouping of estimated low bids for any given year. The range of the estimated mean low-bid prices for the years 1970–1974, approximately 4¢ per pound, may seem small, but in this high-volume business, a 1¢-per-pound increase in contract price can produce a substantial increase in income. Observe the changes in the relative position of the estimated mean low-bid prices for the various markets as you move from year to year. We have already shown that some of these changes cannot be explained solely by random error. Rather, they represent a combination of random variation and year–market interaction.

The second interesting aspect of Figure 13.6 is the inflationary year-to-year shift of the groups—relatively gradual through 1973, rising sharply in 1974, and then dropping substantially during the latter part of the 1974–1975 recession.

When you attempt to locate specific deviations from stable market conditions, you are led immediately to market 1. Notice the large positive deviation of the estimated mean low-bid price for market 1 in 1975. In fact, the estimated mean low-bid price for market 1 is only slightly larger than the mean for all markets for each of the years 1970–1974, but it is much larger than the mean of all markets for 1975. In fact, as the estimated mean low-bid price decreases substantially from 1974 to 1975 for all other markets, the estimated mean low-bid price for market 1 increases. If you examine the computer printout for Figure 13.5, you will find that our speculation of a market 1 by year interaction is not a figment of our imagination. Most of the individual t tests associated with the market 1 by year interaction parameters are statistically significant.

The high estimated mean low-bid price for market 1 in 1975 may not be the only deviation from stable market conditions that is important, but it is certainly the most obvious. What does it mean? The answer to this question would have to come from some economists and accountants after a thorough review of the suppliers' financial accounts and an examination of the various economic factors at play in the markets. Perhaps the seemingly excessive estimated mean low-bid price in market 1 in 1974 can be explained by unusual economic factors that affected only market 1 in 1975. Or, perhaps the large deviation in bid price from the market means for 1975 is a sign of collusive bidding.

13.5
A Residual Analysis of the Data

An analysis of residuals provides a second method for detecting possible collusive bidding by examining the pattern of low bids *within* given markets. Low bids in a particular year and within a given market should lie, with high probability, within 2 estimated standard deviations of the estimated market mean for that year, i.e., within $2s$ of $\hat{y}$. Large positive deviations exceeding this value ($\hat{y} \pm 2s$) would be suspect and should be examined for source of cause. Notice the difference between this technique and the technique of Section 13.4, where we employed a regression analysis to detect unusual year-to-year changes in the relative values of the *mean* low-bid prices for a set of markets. In contrast, a residual analysis examines the behavior of low bids about the means within market–years.

What type of residual pattern within market–years might provide a clue to the possibility of collusive bidding? This is a question for a specialist to answer, someone who has studied bidding patterns within market–years for proven cases of collusive bidding. Variation in low bids within a market–year would probably be due to varying unit costs of transportation or differing cost-capacity considerations within each firm. Consequently, we would question unusually large deviations from the estimated mean for a predicted market–year, in other words, unusually large residuals.

A computer plot of the relative frequency distribution of the 303 Model 2 regression analysis residuals is shown in Figure 13.7, and a computer printout of the complete set of residuals is shown in Figure 13.8 (pages 684–688).

FIGURE 13.7

The relative frequency distribution of residuals for Model 2

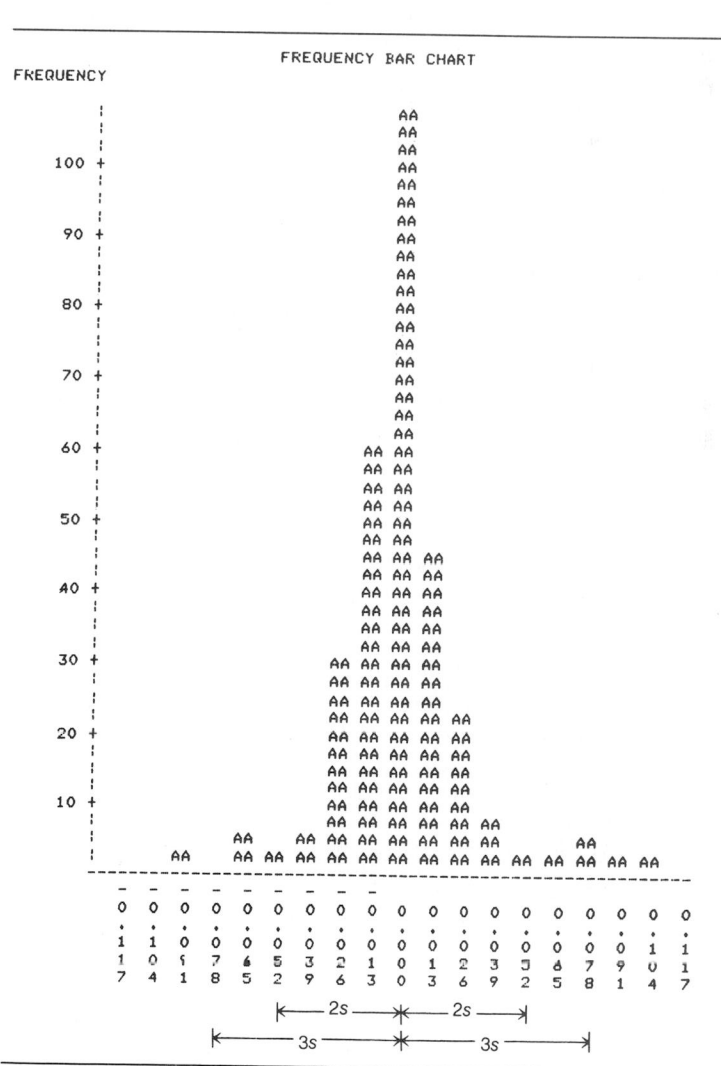

FIGURE 13.8

SAS computer printout of the residuals for Model 2

OBS	ID	ACTUAL	PREDICT VALUE	STD ERR PREDICT	RESIDUAL	STD ERR RESIDUAL	STUDENT RESIDUAL	COOK'S D
1	70_3	0.1503	0.1767	0.007617	-0.0264	0.0253	-1.0459	0.002
2	71_3	0.1667	0.1908	0.007052	-0.0242	0.0254	-0.9505	0.001
3	72_3	0.1833	0.1909	0.007052	-0.007582	0.0254	-0.2982	0.000
4	73_3	0.2133	0.2186	0.0073182	-.005261	0.0254	-0.2075	0.000
5	74_3	0.3067	0.2913	0.007052	0.0154	0.0254	0.6054	0.000
6	75_3	0.2800	0.2739	0.0068129	0.00614	0.0255	0.2409	0.000
7	70_3	0.2000	0.1767	0.007617	0.0233	0.0253	0.9227	0.001
8	71_3	0.2100	0.1908	0.007052	0.0192	0.0254	0.7538	0.002
9	73_3	0.1799	0.1909	0.007052	-0.0110	0.0254	-0.4345	0.001
10	74_3	0.1533	0.2186	0.007052	-0.0653	0.0254	-2.5743	0.012
11	70_1	0.2000	0.2913	0.007052	-0.0913	0.0254	-3.5898	0.021
12	71_1	0.2068	0.2044	0.0068129	0.0023864	0.0255	0.0967	0.000
13	72_1	0.2209	0.2141	0.0093289	0.0068125	0.0247	0.2760	0.000
14	73_1	0.2209	0.2143	0.0093289	0.0065995	0.0247	0.2653	0.000
15	75_1	0.2550	0.2401	0.0087954	0.0149	0.0249	0.5985	0.003
16	75_3	0.3640	0.3382	0.0087954	0.0258	0.0249	1.0387	0.000
17	70_8	0.2800	0.2739	0.0068129	0.00614	0.0255	0.2409	0.000
18	71_8	0.2480	0.2543	0.0118	-0.006267	0.0236	-0.2655	0.012
19	72_8	0.2090	0.1860	0.0152	0.0230	0.0215	1.0693	0.026
20	73_8	0.2280	0.2073	0.0187	0.0207	0.0187	1.1077	0.007
21	74_8	0.2300	0.2130	0.0152	0.0170	0.0215	0.7908	0.010
22	75_8	0.2307	0.2348	0.0152	-0.004074	0.0215	-0.1891	0.038
23	70_1	0.2767	0.2981	0.0152	-0.0214	0.0215	-0.9954	0.000
24	71_1	0.2425	0.2838	0.0152	-0.0412	0.0215	-1.9141	0.001
25	72_1	0.2029	0.2044	0.0093289	-0.001505	0.0247	-0.0610	0.000
26	73_1	0.2280	0.2141	0.0093289	0.0139	0.0247	0.5637	0.000
27	74_1	0.2255	0.2143	0.0087954	0.0112	0.0249	0.4484	0.003
28	75_1	0.2400	0.2401	0.0087954	-1.1E-04	0.0249	-0.004466	0.000
29	71_7	0.3345	0.3382	0.0087954	-0.0260	0.0249	-1.0464	0.029
30	73_7	0.3400	0.3085	0.0152	0.0018395	0.0215	0.0739	0.017
31	74_7	0.2033	0.1978	0.0152	0.0055556	0.0215	0.2579	0.064
32	75_7	0.2560	0.2198	0.0152	0.0362	0.0215	1.6813	0.000
33	73_7	0.2560	0.2281	0.0152	0.0279	0.0215	1.2945	0.001
34	74_7	0.3600	0.3067	0.0152	0.0533	0.0215	2.4755	0.001
35	75_7	0.2667	0.2667	0.0152	6.0E-16	0.0215	2.8E-14	0.000
36	70_3	0.1913	0.1767	0.007617	0.0146	0.0253	0.5797	0.001
37	71_3	0.2067	0.1908	0.007052	0.0158	0.0254	0.6227	0.001
38	72_3	0.2133	0.1909	0.007052	0.0224	0.0254	0.8817	0.001
39	73_3	0.2267	0.2186	0.0073182	0.0080728	0.0254	0.3184	0.000
40	74_3	0.3040	0.2913	0.007052	0.0127	0.0254	0.5005	0.000
41	75_3	0.2667	0.2739	0.0068129	-0.007193	0.0255	-0.2822	0.000
42	70_3	0.1900	0.1767	0.007617	0.0133	0.0253	0.5269	0.001
43	71_3	0.2150	0.1908	0.007052	0.0241	0.0254	0.9505	0.001
44	72_3	0.2200	0.1909	0.007052	0.0291	0.0254	1.1439	0.002
45	74_3	0.3534	0.2913	0.007052	0.0622	0.0254	2.4449	0.010
46	75_3	0.2600	0.2739	0.0068129	-0.0139	0.0255	-0.5437	0.000
47	71_7	0.2000	0.1978	0.0152	0.0022222	0.0215	0.1031	0.001
48	74_7	0.2133	0.2281	0.0152	-0.0044444	0.0215	-0.2991	0.000
49	75_7	0.2333	0.2198	0.0152	0.0052222	0.0215	0.2424	0.001
50	73_7	0.2700	0.3067	0.0152	-0.0367	0.0215	-1.7019	0.030
51	75_7	0.2667	0.2667	0.0152	6.0E-16	0.0215	2.8E-14	0.001
52	70_3	0.1600	0.1767	0.007617	-0.0167	0.0253	-0.6606	0.000
53	71_3	0.1733	0.1908	0.007052	-0.0175	0.0254	-0.6883	0.001
54	72_3	0.1867	0.1909	0.007052	-0.004249	0.0254	-0.1671	0.001
55	73_3	0.2407	0.2186	0.0073182	0.0221	0.0254	0.8707	0.001
56	74_3	0.2977	0.2913	0.007052	0.063924	0.0254	0.2514	0.000
57	75_3	0.2760	0.2739	0.0068129	0.0021	0.0255	0.0840	0.000
58	70_8	0.1956	0.1860	0.0152	0.0095926	0.0215	0.4453	0.002
59	72_8	0.2222	0.2130	0.0152	0.0092593	0.0215	0.4298	0.002
60	73_8	0.2356	0.2348	0.0152	7.5E-04	0.0215	0.0347	0.000
61	74_8	0.3200	0.2981	0.0152	0.0219	0.0215	1.0160	0.000
62	75_8	0.3111	0.2838	0.0152	0.0273	0.0215	1.2691	0.017
63	73_4	0.2300	0.2342	0.0093289	-.004167	0.0247	-0.1688	0.000
64	74_4	0.3133	0.3101	0.0087954	0.0032593	0.0249	-0.1310	0.000
65	75_4	0.2667	0.2732	0.0087954	-.006556	0.0249	-0.2635	0.000

(A residual plot with axis labels -2 -1 0 1 2 appears between the STD ERR RESIDUAL and STUDENT RESIDUAL columns.)

FIGURE 13.8
(continued)

67	70_3	0.1757	0.1767	0.007617	-.001022	0.0253	-0.0405			0.000
68	71_3	0.2167	0.1908	0.007052	0.0258	0.0254	1.0160		**	0.002
69	72_3	0.1700	0.1909	0.007052	-0.0209	0.0254	-0.8226	*		0.001
70	73_3	0.1533	0.2186	.0073182	-0.0653	0.0254	-2.5743	*****		0.012
71	74_3	0.3133	0.2913	0.007052	0.0221	0.0254	0.8676		*	0.001
72	75_3	0.2660	0.2739	.0068129	-0.00786	0.0255	-0.3083			0.000
73	70_1	0.2000	0.2044	.0093289	-.004414	0.0247	-0.1788			0.000
74	71_1	0.2000	0.2141	.0093289	-0.0141	0.0247	-0.5708	*		0.001
75	72_1	0.2000	0.2143	.0087954	-0.0143	0.0249	-0.5748	*		0.001
76	73_1	0.2160	0.2401	.0087954	-0.0241	0.0249	-0.9692	*		0.002
77	74_1	0.3289	0.3085	.0087954	0.0204	0.0249	0.8190		*	0.002
78	75_1	0.3289	0.3382	.0087954	-.009272	0.0249	-0.3727			0.000
79	72_3	0.2050	0.1909	0.007052	0.0141	0.0254	0.5539		*	0.000
80	73_3	0.2979	0.2186	.0073182	0.0793	0.0254	3.1291		******	0.017
81	74_3	0.3133	0.2913	0.007052	0.0221	0.0254	0.8676		*	0.001
82	75_3	0.2533	0.2739	.0068129	-0.0205	0.0255	-0.8052	*		0.001
83	71_2	0.2088	0.2155	0.009973	-.006738	0.0244	-0.2758			0.000
84	72_2	0.2088	0.2136	0.009973	-.004781	0.0244	-0.1957			0.000
85	73_2	0.2550	0.2499	0.009973	.0050786	0.0244	0.2079			0.000
86	74_2	0.3200	0.3312	0.009973	-0.0112	0.0244	-0.4600			0.001
87	75_2	0.2560	0.3018	0.0108	-0.0458	0.0241	-1.9005	***		0.015
88	70_2	0.1848	0.1986	0.0108	-0.0138	0.0241	-0.5720	*		0.001
89	71_2	0.2068	0.2155	0.009973	-.008738	0.0244	-0.3577			0.000
90	72_2	0.2068	0.2136	0.009973	-.006781	0.0244	-0.2776			0.000
91	73_2	0.2550	0.2499	0.009973	.0050786	0.0244	0.2079			0.000
92	74_2	0.3300	0.3312	0.009973	-.001238	0.0244	-0.0507			0.000
93	75_2	0.3200	0.3018	0.0108	0.0182	0.0241	0.7565		*	0.002
94	70_1	0.2200	0.2044	.0093289	0.0156	0.0247	0.6315		*	0.001
95	71_1	0.2115	0.2141	.0093289	-.002587	0.0247	-0.1048			0.000
96	72_1	0.2115	0.2143	.0087954	-.002801	0.0249	-0.1126			0.000
97	74_1	0.3000	0.3085	.0087954	-.008514	0.0249	-0.3422			0.000
98	75_1	0.3267	0.3382	.0087954	-0.0115	0.0249	-0.4620			0.001
99	70_3	0.1733	0.1767	0.007617	-.003356	0.0253	-0.1328			0.000
100	71_3	0.1850	0.1908	0.007052	-.005833	0.0254	-0.2294			0.000
101	72_3	0.1850	0.1909	0.007052	-.005915	0.0254	-0.2327			0.000
102	73_3	0.2250	0.2186	.0073182	.0064062	0.0254	0.2527			0.000
103	74_3	0.2897	0.2913	0.007052	-.001524	0.0254	-0.0599			0.000
104	75_3	0.2910	0.2739	.0068129	0.0171	0.0255	0.6724		*	0.001
105	70_4	0.2160	0.2201	0.0118	-.004089	0.0236	-0.1733			0.000
106	71_4	0.1900	0.2196	0.0108	-0.0296	0.0241	-1.2301	**		0.006
107	72_4	0.2200	0.2279	0.009973	-.007905	0.0244	-0.3236			0.000
108	73_4	0.2347	0.2342	.0093289	5.0E-04	0.0247	0.0203			0.000
109	74_4	0.3133	0.3101	.0087954	.0032593	0.0249	0.1310			0.000
110	75_4	0.2657	0.2732	.0087954	-.007556	0.0249	-0.3037			0.000
111	70_7	0.1920	0.1920	0.0264	1.4E-17	0	.			.
112	70_4	0.2067	0.2201	0.0118	-0.0134	0.0236	-0.5687	*		0.002
113	71_4	0.2133	0.2196	0.0108	-.006296	0.0241	-0.2614			0.000
114	72_4	0.2133	0.2279	0.009973	-0.0146	0.0244	-0.5965	*		0.001
115	73_4	0.1933	0.2342	.0093289	-0.0408	0.0247	-1.6544	***		0.008
116	74_4	0.3067	0.3101	.0087954	-.003407	0.0249	-0.1370			0.000
117	75_4	0.2667	0.2732	.0087954	-.006556	0.0249	-0.2635			0.000
118	72_4	0.2280	0.2279	0.009973	9.5E-05	0.0244	.0038986			0.000
119	73_4	0.2280	0.2342	.0093289	-.006167	0.0247	-0.2498			0.000
120	74_4	0.2867	0.3101	.0087954	-0.0234	0.0249	-0.9409	*		0.002
121	75_4	0.2600	0.2732	.0087954	-0.0132	0.0249	-0.5315	*		0.001
122	70_1	0.2134	0.2044	.0093289	.0089864	0.0247	0.3641			0.000
123	71_1	0.2279	0.2141	.0093289	0.0139	0.0247	0.5616		*	0.001
124	72_1	0.2232	0.2143	.0087954	.0089495	0.0249	0.3597			0.000
125	73_1	0.2550	0.2401	.0087954	0.0149	0.0249	0.5985		*	0.001
126	74_1	0.3067	0.3085	.0087954	-.001847	0.0249	-0.0742			0.000
127	75_1	0.3400	0.3382	.0087954	.0018395	0.0249	0.0739			0.000
128	70_6	0.1667	0.1977	0.0132	-0.0310	0.0229	-1.3566	**		0.013
129	71_6	0.2080	0.2160	0.0118	-0.008	0.0236	-0.3390			0.001
130	72_6	0.2240	0.2160	0.0118	0.008	0.0236	0.3390			0.001
131	73_6	0.2480	0.2480	0.0132	2.8E-16	0.0229	1.2E-14			0.000
132	74_6	0.3120	0.3120	0.0118	3.1E-16	0.0236	1.3E-14			0.000

(continued)

FIGURE 13.8
(continued)

OBS	ID	ACTUAL	PREDICT VALUE	STD ERR PREDICT	RESIDUAL	STD ERR RESIDUAL	STUDENT RESIDUAL	-2-1-0 1 2	COOK'S D
133	75_6	0.2640	0.2672	0.0118	-0.0032	0.0236	-0.1356		0.000
134	70_1	0.1922	0.2044	.0093289	-0.0122	0.0247	-0.4948		0.001
135	71_1	0.2185	0.2141	.0093289	.0044625	0.0247	0.1808		0.000
136	72_1	0.2185	0.2143	.0087954	.0042495	0.0249	0.1708		0.000
137	73_1	0.2401	0.2401	.0087954	-1.1E-04	0.0249	-.004466		0.000
138	74_1	0.3067	0.3085	.0087954	-.001847	0.0249	-0.0742		0.000
139	70_2	0.2052	0.1986	0.0108	.0066222	0.0241	0.2749		0.000
140	71_2	0.2200	0.2155	0.0108	.0044619	0.0241	0.1826		0.000
141	72_2	0.2200	0.2136	.009973	.006419	0.0244	0.2628		0.000
142	73_2	0.2600	0.2499	.009973	0.0101	0.0244	0.4126		0.001
143	74_2	0.2640	0.3312	.009973	-0.0672	0.0244	-2.7524	******	0.026
144	70_3	0.3067	0.3018	0.0108	.0048889	0.0241	0.2030		0.000
145	71_3	0.1450	0.1767	.007617	-0.0317	0.0253	-1.2544	***	0.003
146	72_3	0.1650	0.1908	.007052	-0.0258	0.0254	-1.0160	**	0.002
147	73_3	0.1750	0.1909	.007052	-0.0159	0.0254	-0.6259	*	0.001
148	70_5	0.1969	0.2088	0.0152	-0.0119	0.0215	-0.5516	*	0.003
149	72_5	0.2000	0.2162	0.0152	-0.0162	0.0215	-0.7504	*	0.006
150	73_5	0.2400	0.2045	0.0132	-.004479	0.0229	1.0211	**	0.000
151	74_5	0.3000	0.2167	0.0132	-0.0241	0.0229	-1.0226	**	0.007
152	71_5	0.2667	0.3241	0.0118	-0.0124	0.0236	-0.0233	*	0.005
153	71_7	0.1900	0.2543	0.0118	-.007778	0.0236	-0.5254	**	0.001
154	72_7	0.1950	0.1978	0.0152	-0.0298	0.0215	-0.3610	**	0.020
155	74_7	0.2900	0.2198	0.0152	-0.0331	0.0215	-1.3822	****	0.025
156	75_7	0.2667	0.2281	0.0152	-0.0167	0.0215	-1.5369	****	0.006
157	75_7	0.1848	0.3067	0.0152	6.0E-16	0.0215	-0.7736	*	0.000
158	75_2	0.2068	0.1986	0.0108	-0.0138	0.0215	2.8E-14		0.001
159	70_2	0.2524	0.2136	.009973	-.004038	0.0244	-0.5720		0.000
160	71_2	0.3300	0.2499	.009973	-.006781	0.0244	-0.2776		0.000
161	72_2	0.1867	0.3312	.009973	.0025286	0.0244	-0.1035		0.000
162	73_2	0.2000	0.1767	.007617	-.001238	0.0253	-0.0507		0.000
163	74_2	0.2000	0.2499	.009973	.0099778	0.0244	0.3950		0.000
164	70_3	0.2407	0.1909	.007052	.0091667	0.0254	0.3605		0.000
165	71_3	0.2977	0.1909	.007052	.0090845	0.0254	0.3573		0.000
166	72_3	0.2760	0.2186	.007052	0.0221	0.0254	0.8707	*	0.001
167	74_3	0.1867	0.2739	.007182	.0063924	0.0254	0.2514		0.000
168	74_3	0.2067	0.2913	.0068129	-0.0119	0.0255	0.0840		0.001
169	75_3	0.2040	0.1986	0.0108	-.008871	0.0241	-0.4945		0.000
170	70_2	0.3067	0.2155	.009973	-.006914	0.0244	-0.3632		0.000
171	71_2	0.2560	0.2136	.009973	-0.0459	0.0244	-0.2830		0.012
172	73_2	0.2000	0.2499	.009973	-0.0246	0.0244	-1.8798	***	0.004
173	75_2	0.2067	0.3312	.009973	-0.0458	0.0244	-1.0058	**	0.015
174	71_4	0.2267	0.2201	0.0108	-0.0201	0.0236	-1.9005	***	0.004
175	70_4	0.3000	0.3018	0.0108	-.0279	0.0241	-0.8512	**	0.001
176	73_4	0.2667	0.2196	0.0118	-0.0130	0.0244	-0.5382	**	0.005
177	74_4	0.1842	0.2279	.009973	-0.0075	0.0244	-1.1423	***	0.001
178	75_4	0.1859	0.2342	.0093289	-.0101	0.0247	-0.3039		0.000
179	71_3	0.2208	0.3101	.0087954	-.006556	0.0249	-0.4050		0.000
180	73_3	0.3040	0.2732	.0087954	-.006667	0.0249	-0.2635		0.000
181	73_3	0.2944	0.1908	.007052	-.005049	0.0254	-0.2622		0.000
182	75_3	0.2080	0.1909	.007052	.0022062	0.0254	-0.1986		0.000
183	70_6	0.2160	0.2186	.007182	0.0127	0.0254	0.5005	*	0.001
184	72_6	0.2240	0.2913	.007182	0.0205	0.0254	0.0870		0.000
185	70_6	0.2480	0.2739	.0068129	0.0103	0.0255	0.8058	*	0.001
186	72_6	0.3120	0.1977	0.0132	1.4E-16	0.0229	0.4522		0.000
187	74_6	0.2800	0.2160	0.0118	-0.008	0.0236	5.9E-15		0.001
188	75_6	0.1553	0.2460	0.0132	2.8E-16	0.0236	0.3390		0.001
189	74_6	0.1867	0.3120	0.0118	3.1E-16	0.0236	1.2E-14		0.000
190	75_6	0.1867	0.2672	0.0118	-0.0326	0.0236	1.3E-14		0.000
191	70_8	0.2381	0.1860	0.0152	-0.0207	0.0215	0.5424	*	0.002
192	72_8	0.2977	0.2073	0.0187	-.0263	0.0187	-1.1077	****	0.024
193	73_8	0.2977	0.2130	0.0152	.003259	0.0215	-1.2206	***	0.016
194	74_8		0.2348	0.0152	-4.4E-04	0.0215	-1.1544	***	0.000
195	72_8		0.2981	0.0152		0.0215	-0.0206		0.000
196	73_8		0.2838	0.0152	0.0139	0.0215	0.6450	*	0.004

FIGURE 13.8
(continued)

199	71_3	0.1725	0.1908	0.007052	-0.0183	0.0254	-0.7210	*	0.001
200	72_3	0.1887	0.1909	0.007052	-0.002165	0.0254	-0.0852		0.000
201	73_3	0.2100	0.2186	0.0073182	-0.008594	0.0254	-0.3390		0.000
202	74_3	0.2900	0.2913	0.007052	-0.001274	0.0254	-0.0501		0.000
203	75_3	0.2700	0.2739	0.0068129	-0.00386	0.0255	-0.1514		0.000
204	70_1	0.2000	0.2044	0.0093289	-0.004414	0.0247	-0.1788		0.000
205	71_1	0.2058	0.2141	0.0087954	-0.008287	0.0247	-0.3358		0.000
206	72_1	0.2058	0.2143	0.0087954	-0.008501	0.0249	-0.3417		0.004
207	73_1	0.2350	0.2401	0.0087954	-0.005111	0.0249	-0.2055		0.003
208	74_1	0.2793	0.3085	0.0087954	-0.0292	0.0249	-1.1743		0.001
209	75_1	0.3111	0.3382	0.0087954	-0.0270	0.0249	-1.1743	**	0.001
210	70_6	0.2080	0.1977	0.0132	0.0103	0.0229	0.4522		0.000
211	71_6	0.2240	0.2160	0.0118	-0.008	0.0236	-0.3390		0.000
212	72_6	0.2080	0.2160	0.0118	-0.008	0.0236	-0.3390		0.000
213	73_6	0.2480	0.2480	0.0132	2.8E-16	0.0229	1.2E-14		0.000
214	74_6	0.3120	0.3120	0.0118	-0.0032	0.0236	-0.1356		0.000
215	75_6	0.2640	0.2672	0.0118	3.1E-16	0.0229	1.3E-14		0.000
216	72_5	0.2067	0.2045	0.0132	0.0021875	0.0229	0.7294	***	0.004
217	73_5	0.2333	0.2167	0.0132	-0.0167	0.0236	-0.8531		0.004
218	74_5	0.3040	0.3241	0.0118	-0.0201	0.0236	-0.1808		0.029
219	75_5	0.2500	0.2543	0.0118	-0.004267	0.0236	1.6799	***	0.034
220	70_5	0.2450	0.2088	0.0152	0.0362	0.0215	1.8025	***	0.007
221	71_5	0.2550	0.2162	0.0152	0.0388	0.0215	0.2416	****	0.069
222	72_5	0.2100	0.2045	0.0132	0.0055208	0.0229	1.0211	****	0.000
223	73_5	0.2400	0.2167	0.0132	0.0233	0.0236	3.6384	****	0.000
224	74_5	0.4100	0.3241	0.0118	0.0859	0.0236	5.9E-15	****	0.039
225	71_5	0.2120	0.2160	0.0118	1.4E-16	0.0236	-0.1695		0.044
226	72_6	0.3120	0.3120	0.0118	-0.004	0.0236	1.3E-14		0.034
227	74_6	0.2640	0.2672	0.0118	3.1E-16	0.0236	-0.1356		0.038
228	75_6	0.2844	0.2201	0.0118	0.0644	0.0236	2.7269	******	0.023
229	70_4	0.2978	0.2196	0.0108	0.0781	0.0241	3.2444	******	0.000
230	71_4	0.3040	0.2279	0.009973	0.0698	0.0244	3.1150	******	0.002
231	72_4	0.3040	0.2342	0.0093289	0.0953	0.0247	2.8293	******	0.001
232	73_4	0.4053	0.3101	0.0087954	0.0734	0.0249	3.8292	******	0.003
233	74_4	0.3467	0.2732	0.0087954	-0.006074	0.0249	2.9523	******	0.000
234	75_4	0.3040	0.3101	0.0087954	-0.0167	0.0249	-0.2442	******	0.000
235	74_4	0.2533	0.2732	0.0087954	-0.0199	0.0249	-0.7995		0.001
236	75_4	0.3040	0.1767	0.007617	-0.0208	0.0253	-0.6606		0.001
237	70_3	0.1600	0.1908	0.007052	-0.0109	0.0254	-0.8194	***	0.002
238	71_3	0.1700	0.1909	0.007052	-0.0347	0.0254	-0.4293	***	0.001
239	72_3	0.1800	0.2186	0.0073182	-0.007624	0.0254	1.3703	***	0.003
240	73_3	0.2533	0.2913	0.007052	6.4E-04	0.0254	-0.2999		0.000
241	74_3	0.2836	0.2739	0.0068129	-0.004414	0.0255	-0.0251		0.000
242	75_3	0.2745	0.2044	0.0093289	-0.0141	0.0247	-0.1788		0.000
243	70_1	0.2000	0.2141	0.0087954	-0.0143	0.0249	-0.5708		0.001
244	71_1	0.2000	0.2143	0.0087954	-1.1E-04	0.0249	-0.5748	**	0.001
245	72_1	0.2143	0.2401	0.0087954	0.0204	0.0249	-0.004466	**	0.001
246	73_1	0.2400	0.3085	0.0087954	-0.009272	0.0249	0.8190		0.002
247	74_1	0.2401	0.2201	0.0118	-0.0268	0.0236	-0.3727	**	0.007
248	75_1	0.3382	0.2279	0.0108	-0.002963	0.0241	-1.1337	**	0.001
249	70_4	0.3289	0.2342	0.009973	-0.0146	0.0244	-0.1230	*	0.005
250	71_4	0.1933	0.3101	0.0093289	-0.0075	0.0247	0.5965	*	0.000
251	72_4	0.2167	0.3241	0.0087954	-0.0221	0.0249	-0.3039		0.001
252	73_4	0.2133	0.2543	0.0118	-0.006556	0.0236	-0.8873	*	0.001
253	74_4	0.2267	0.1908	0.0118	0.0118	0.0249	-0.2635	*	0.002
254	75_4	0.2880	0.2186	0.007052	-9.3E-04	0.0236	-0.0395	*	0.000
255	74_5	0.2667	0.2913	0.007052	0.0313	0.0253	1.2394	**	0.004
256	75_5	0.3033	0.2739	0.0073182	0.0158	0.0254	0.6227	**	0.003
257	70_3	0.2553	0.1977	0.0073182	0.0347	0.0254	1.3703	**	0.003
258	71_3	0.2080	0.2186	0.0068129	-0.0063924	0.0254	-0.2514		0.000
259	72_3	0.2067	0.2913	0.0068129	-0.00386	0.0255	-0.1514		0.001
260	73_3	0.2977	0.2739	0.0132	0.0103	0.0229	-0.4522		0.003
261	75_3	0.2700	0.1977	0.0132	1.4E-16	0.0254	5.9E-15		0.000
262	70_6	0.2080	0.2160	0.0118	-0.004	0.0236	-0.1695		0.001
263	75_3	0.2160	0.2160	0.0118					0.000
264	72_6	0.2120	0.2160	0.0118					

(continued)

FIGURE 13.8
(continued)

OBS	ID	ACTUAL	PREDICT VALUE	STD ERR PREDICT	RESIDUAL	STD ERR RESIDUAL	STUDENT RESIDUAL	-2-1-0 1 2	COOK'S D
265	73_6	0.2480	0.2480	0.0132	2.8E-16	0.0229	1.2E-14		0.000
266	74_6	0.3120	0.3120	0.0118	3.1E-16	0.0236	1.3E-14		0.000
267	75_6	0.2640	0.2672	0.0118	-0.0032	0.0236	-0.1356		0.000
268	71_4	0.1933	0.2196	0.0108	-0.0263	0.0241	-1.0917		0.005
269	72_4	0.2167	0.2279	0.009973	-0.0112	0.0244	-0.4600		0.001
270	73_4	0.2300	0.2342	0.0093289	-.004167	0.0247	-0.1688		0.000
271	74_4	0.2733	0.3101	.0087954	-0.0367	0.0249	-1.4769		0.006
272	75_4	0.2667	0.2732	.0087954	-.006556	0.0249	-0.2635		0.000
273	70_2	0.2320	0.1986	0.0108	0.0334	0.0241	1.3876		0.008
274	71_2	0.2480	0.2155	0.009973	0.0325	0.0244	1.3288		0.006
275	72_2	0.2480	0.2136	0.009973	0.0344	0.0244	1.4090		0.007
276	73_2	0.2680	0.2499	0.009973	0.0181	0.0244	0.7401		0.002
277	74_2	0.3280	0.3312	0.009973	-.003238	0.0244	-0.1326		0.000
278	75_2	0.3520	0.3018	0.0108	0.0502	0.0241	2.0850		0.018
279	70_3	0.1800	0.1767	0.007617	.0033111	0.0253	0.1311		0.000
280	71_3	0.2000	0.1908	0.007052	.0091667	0.0254	0.3605		0.000
281	72_3	0.2000	0.1909	0.007052	.0090845	0.0254	0.3573		0.000
282	73_3	0.1533	0.2186	.0073182	-0.0653	0.0254	-2.5743		0.012
283	74_3	0.2267	0.2913	0.007052	-0.0646	0.0254	-2.5410		0.010
284	75_3	0.2700	0.2739	.0068129	-0.00386	0.0255	-0.1514		0.000
285	70_5	0.1845	0.2088	0.0152	-0.0243	0.0215	-1.1283		0.013
286	71_5	0.1935	0.2162	0.0152	-0.0227	0.0215	-1.0521		0.012
287	72_5	0.2012	0.2045	0.0132	-.003229	0.0229	-0.1413		0.000
288	73_5	0.1533	0.2167	0.0118	-0.0633	0.0236	-2.7716		0.053
289	74_5	0.3033	0.3241	0.0118	-0.0208	0.0236	-0.8813		0.004
290	75_5	0.2533	0.2543	0.0108	-9.3E-04	0.0241	-0.0395		0.000
291	70_2	0.1980	0.1986	0.009973	-5.8E-04	0.0241	-0.0240		0.000
292	71_2	0.2070	0.2155	0.009973	-.008538	0.0244	-0.3495		0.001
293	72_2	0.1980	0.2136	0.009973	-0.0156	0.0244	-0.6378		0.001
294	73_2	0.2550	0.2499	0.009973	.0050786	0.0244	0.2079		0.000
295	74_2	0.4400	0.3312	0.009973	0.1088	0.0244	4.4522		0.069
296	75_2	0.3200	0.3018	0.0108	0.0182	0.0241	0.7565		0.002
297	73_1	0.2400	0.2401	.0087954	-1.1E-04	0.0249	-.004466		0.000
298	74_1	0.2850	0.3085	.0087954	-0.0235	0.0249	-0.9452		0.002
299	75_1	0.3289	0.3382	.0087954	-.009272	0.0249	-0.3727		0.000
300	72_1	0.2232	0.2143	.0087954	.0089495	0.0249	0.3597		0.000
301	73_1	0.2400	0.2401	.0087954	-1.1E-04	0.0249	-.004466		0.000
302	74_1	0.3067	0.3085	.0087954	-.001847	0.0249	-0.0742		0.000
303	75_1	0.3750	0.3382	.0087954	0.0368	0.0249	1.4809		0.006

SUM OF RESIDUALS   5.10980E-13
SUM OF SQUARED RESIDUALS   0.1775372
PREDICTED RESID SS (PRESS)   0.2549934

The value of s given in the Model 2 regression analysis (Figure 13.5) is approximately .026. We have used this value to construct intervals of $\pm 2s$ and $\pm 3s$ about the mean of the residuals, which is 0, in the relative frequency distribution in Figure 13.7. You can see that a few of the low bids lie more than 3 standard deviations below the mean 0, and several more than 3 standard deviations above the mean. A person checking for clues to collusive bidding can identify the large residuals by locating them on the computer printout in Figure 13.8.

The residual patterns for years and markets can be seen in Figures 13.9 and 13.10 (pages 690–691). Both these figures are plots of the 303 residuals against the predicted estimated low-bid price $\hat{y}$. The points locating the individual residual values are identified by numbers. The number that identifies a residual on Figure 13.9 is the last digit of the year in which the low bid occurred; the number that identifies a residual in Figure 13.10 indicates the market in which the low bid occurred. Since both Figures 13.9 and 13.10 are plots of the residuals versus $\hat{y}$, they are identical (except for the identifying numbers). Consequently, you can determine the market–year associated with any residual by locating its position on both Figures 13.9 and 13.10 and reading the year and market identification numbers. For example, the large positive residual located in the upper right-hand corner of the distribution is identified by a 4 in Figure 13.9 and a 2 in Figure 13.10. Consequently, this is the residual for a low bid obtained in 1974 in market 2.

Three horizontal lines are shown on Figures 13.9 and 13.10. The middle line locates the mean of the residuals, which is 0. The upper and lower lines locate the mean of the residuals $\pm 2s$, respectively. You can tell whether individual residuals are outliers by observing their location relative to the $\pm 2s$ band.

The residual patterns shown in Figures 13.9 and 13.10 are interesting, but they may or may not be relevant. For example, note that most of the residuals lie well within the $\pm 2s$ band. Notice the tight vertical line of 3's (corresponding to market 3) plotted above $\hat{y} = .27$ in Figure 13.10. Checking Figure 13.9, you can see that these residuals correspond to low bids made in 1975. Notice how closely the low bids group about the predicted value of the mean for that market–year. An investigation of the supplier associated with these low bids might show that most of the low bids were submitted by the same supplier, or it might indicate a very small variation in the low bids among a group of different suppliers. Unusual regularities or patterns of this type indicate unusual market conditions that may warrant further investigation.

13.6 Summary

We have demonstrated two applications of regression analysis for the detection of possible collusive bidding. The first procedure seeks to detect an overly large deviation of the mean low bid for a particular market from the mean low bid for all markets for a given year. The second procedure investigates the deviation of individual low bids *within* a given market–year.

The first procedure is based on a model for a stable market. This model, which assumes that inflationary year-to-year increases in costs should apply uniformly

FIGURE 13.9

A plot of residuals versus predicted value $\hat{y}$: Numbers identify the year associated with the residual

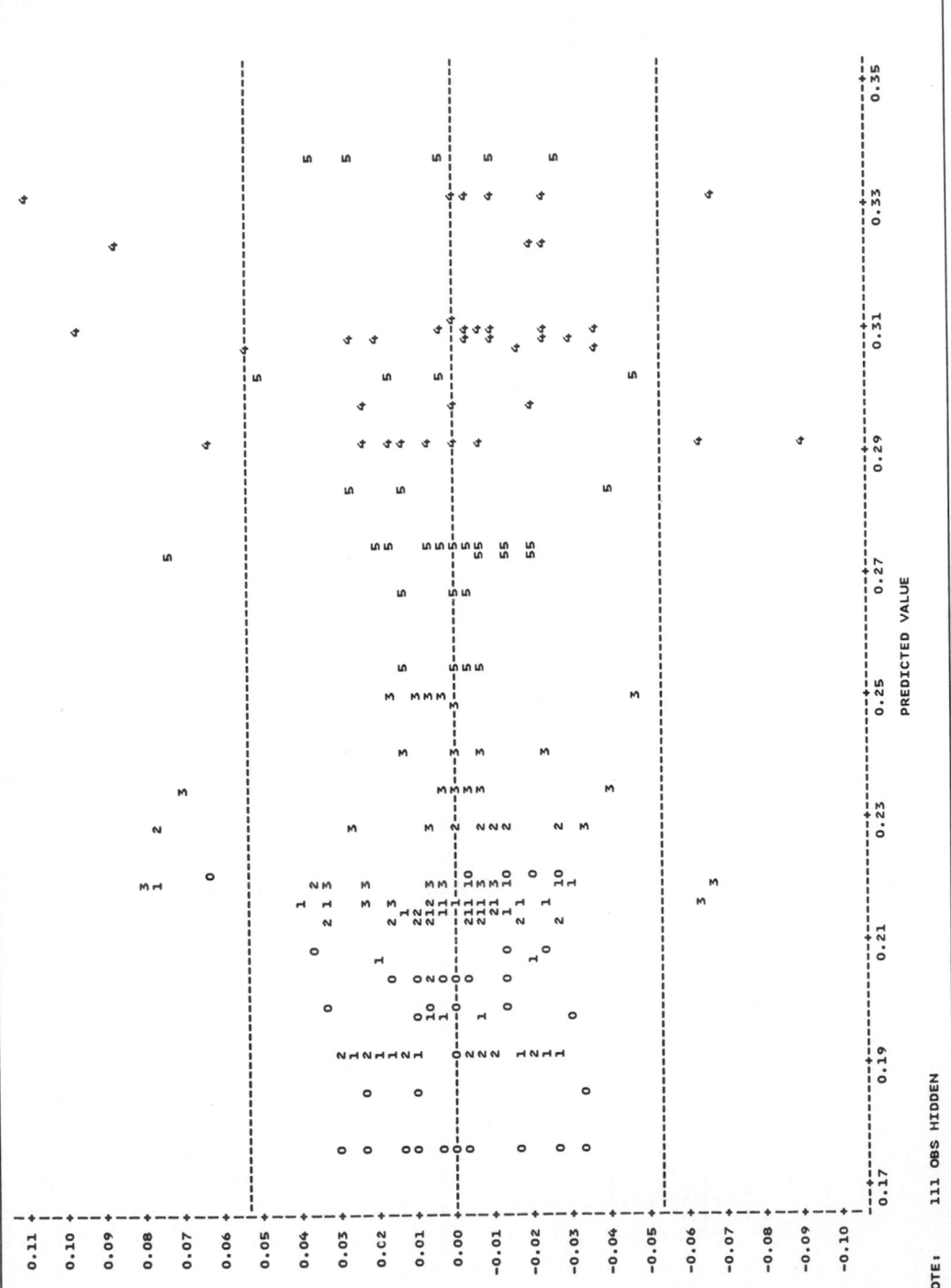

FIGURE 13.10

A plot of residuals versus predicted value ŷ: Numbers identify the market associated with the residual

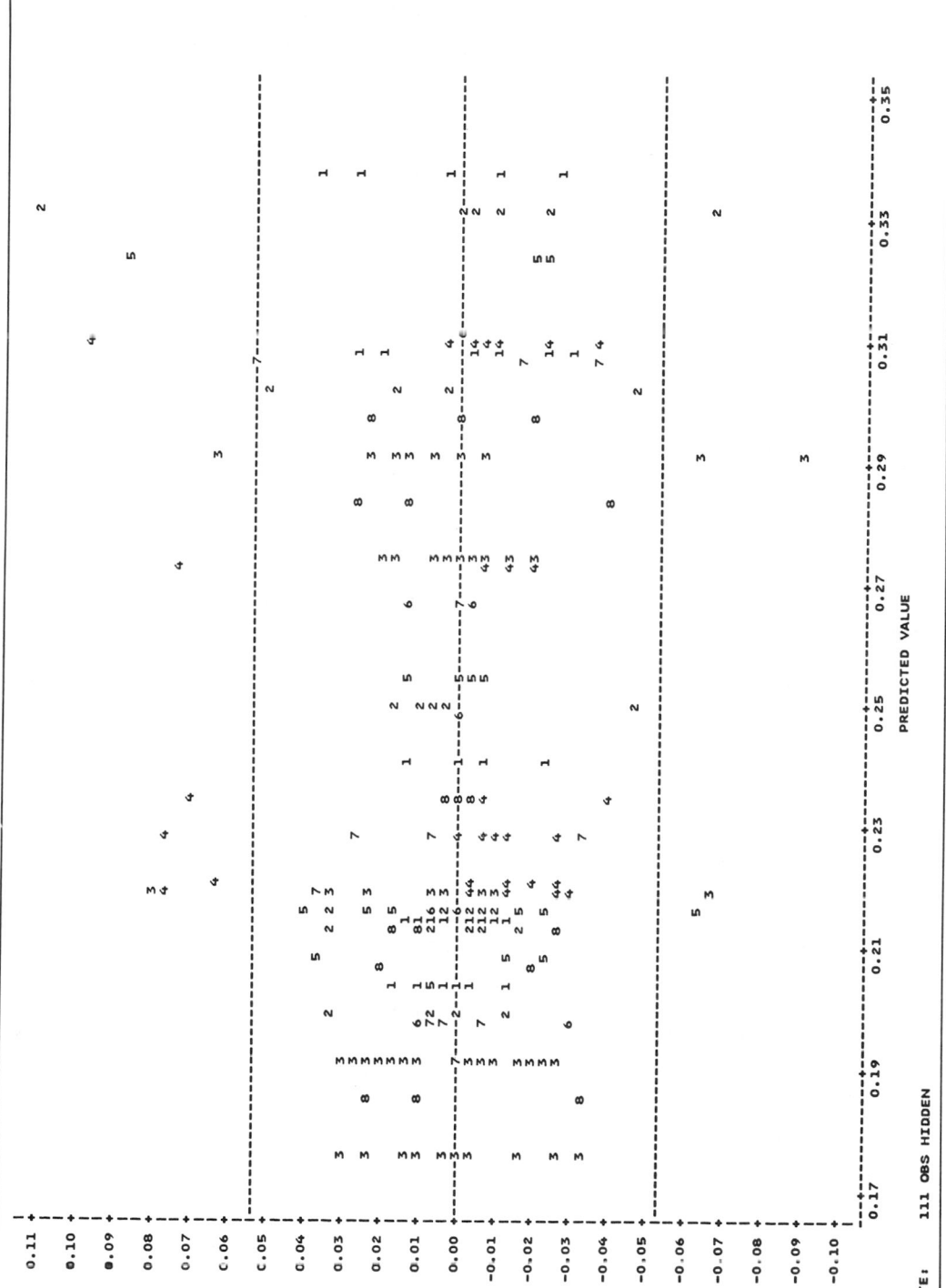

(more or less) to all markets and affect the mean low-bid prices for all markets in a similar way, contains only main effect terms for market and year. An unstable market model—one that allows for market–year interactions—models the situation where the low-bid price for one or more markets deviates substantially from the mean of the low-bid prices for all markets in one or more years. It is our contention that forces are operating in one or more market–years, and that these forces might be explained by unusual market conditions or, in the case of large positive deviations, by collusive bidding. Note that we do not mean to imply that collusive bidding could not occur when market conditions appear to be stable. We are saying that unstable market conditions may be a clue to the presence of collusive bidding, and the procedure that we have described will detect this condition if it exists.

The second procedure that we employed required an examination of the regression residuals, the deviations of the observed low bids about the estimated mean low bid for a particular market in a particular year. We would expect these residuals to vary in a random manner and would expect most of them to lie within $2s$ or $3s$ of their mean (0). Unusually large positive residuals or unusual patterns of residuals might provide a clue to the presence of collusive bidding.

EXERCISES

13.1   In addition to the predicted values and residuals, Figure 13.8 lists the value of Cook's distance, D_i, for each of the 303 observations in the analysis. Recall that D_i measures the overall influence for observation i on the analysis (see Section 6.5). Check the printout in Figure 13.8 and identify those observations with the largest Cook's D values. These observations might be the result of collusive bidding in the corresponding market and year.

13.2   Refer to observations selected in Exercise 13.1. Do any of these observations have *unusual* influence on the regression results? Use the fact that the 50th percentile of an F distribution with $\nu_1 = 48$ and $\nu_2 = 255$ degrees of freedom is .9887.

Reference

Rothrock, T. P. and McClave, J. T. "An analysis of bidding competition in the Florida school bidding competition using a statistical model."   Paper presented at the TIMS/ORSA Joint National Meeting, Chicago, 1979.

Reluctance to Transmit Bad News: The MUM Effect

CONTENTS

14.1 The Problem
14.2 The Design
14.3 Analysis of Variance Models and Results
14.4 Follow-up Analysis
14.5 Conclusions

OBJECTIVE

To present a designed experiment to investigate the effects of two manipulated factors on the reluctance of human subjects to transmit bad news to others

14.1
The Problem

In a 1970 experiment, psychologists S. Rosen and A. Tesser found that people were reluctant to transmit bad news to peers in a nonprofessional setting. Rosen and Tesser termed this phenomenon the "MUM effect."* Since that time, numerous studies have investigated the impact of the MUM effect in a professional setting, e.g., on doctor–patient relationships, organizational functioning, and group psychotherapy. The consensus: The reluctance to transmit bad news continues to be a major professional concern.

Why do people keep mum when given an opportunity to transmit bad news to others? Two theories have emerged from this research. The first maintains that the MUM effect is an *aversion to private discomfort*. To avoid discomforts such as empathy with the victim's distress or guilt feelings for their own good fortune, would-be communicators of bad news keep mum. The second theory is that the MUM effect is a *public display*. People experience little or no discomfort when transmitting bad news, but keep mum to avoid an unfavorable impression or to pay homage to a social norm.

The subject of this case study is a recent article by C. F. Bond and E. L. Anderson (*Journal of Experimental Social Psychology*, Vol. 23, 1987). Bond and Anderson conducted a controlled experiment to determine which of the two explanations for the MUM effect is more plausible. "If the MUM effect is an aversion to private discomfort," they state, "subjects should show the effect whether or not they are visible [to the victim]. If the effect is a public display, it should be stronger if the subject is visible than if the subject cannot be seen."

14.2
The Design

Forty undergraduates (25 males and 15 females) at Duke University participated in the experiment to fulfill an introductory psychology course requirement. Each subject was asked to administer an IQ test to another student and then provide the test taker with his or her percentile score. Unknown to the subject, the test taker was a confederate student working with the researchers.

The experiment manipulated two factors, *subject visibility* and *confederate success*, each at two levels. Subject visibility was manipulated by written instructions that told some subjects that they were *visible* to the test taker through a glass plate and the others that they were *not visible* through a one-way mirror. Confederate success was manipulated by supplying the subject with one of two bogus answer keys. With one answer key, the confederate would always seem to succeed at the test, placing him or her in the top 20% of all Duke undergraduates; when the other answer key was used, the confederate would always seem to fail, ranking in the bottom 20%.

Ten subjects were randomly assigned to each of the $2 \times 2 = 4$ experimental conditions; thus, a 2×2 factorial design with 10 replications was employed. The design is diagrammed in Table 14.1.

---

*Rosen, S. and Tesser, A. "On reluctance to communicate undesirable information: The MUM effect." *Journal of Communication*, Vol. 22, 1970, pp. 124–141.

TABLE 14.1 2 × 2 Factorial Design

| | | CONFEDERATE SUCCESS | |
		Success	Failure
SUBJECT VISIBILITY	Visible	Subject 1 2 . . . 10	Subject 21 22 . . . 30
	Not Visible	Subject 11 12 . . 20	Subject 31 32 . . 40

One of several behavioral variables that were measured during the experiment was *latency to feedback*, defined as time (in seconds) between the end of the test and delivery of feedback (i.e., the percentile score) from the subject to the test taker. This case focuses on an analysis of variance of the dependent variable, latency to feedback. The longer it takes the subject to deliver the score, presumably the greater the MUM effect. With this analysis, the researchers hope to determine whether either one of the two factors, subject visibility or confederate success, has an impact on the MUM effect, and, if so, whether the factors are independent.

14.3
Analysis of Variance Models and Results

Since both factors, subject visibility and confederate success, are qualitative, the complete model for this 2 × 2 factorial experiment is written as follows:

Complete model: $E(y) = \beta_0 + \underbrace{\beta_1 x_1}_{\substack{\text{Visibility} \\ \text{main} \\ \text{effect}}} + \underbrace{\beta_2 x_2}_{\substack{\text{Success} \\ \text{main} \\ \text{effect}}} + \underbrace{\beta_3 x_1 x_2}_{\substack{\text{Visibility} \times \text{success} \\ \text{interaction}}}$

where

y = Latency to feedback

$x_1 = \begin{cases} 1 & \text{if subject visible} \\ 0 & \text{if not} \end{cases}$ $x_2 = \begin{cases} 1 & \text{if confederate success} \\ 0 & \text{if confederate failure} \end{cases}$

To test for factor interaction, we can compare the complete model to the reduced model

Reduced model: $E(y) = \beta_0 + \beta_1 x_1 + \beta_2 x_2$

using the partial F test, or, equivalently, we can conduct a t test on the interaction parameter, β_3. Either way, the null hypothesis to be tested is

H_0: $\beta_3 = 0$

Although the raw data for the experiment were not provided in the journal article, a summary of the results was given in the form of an ANOVA table. The table is reproduced in Table 14.2.

T A B L E 14.2   **ANOVA Table for the 2 × 2 Factorial Experiment**

SOURCE	df	SS	MS	F
Subject visibility	1	1,380.24	1,380.24	4.26
Confederate success	1	1,325.16	1,325.16	4.09
Visibility × success	1	3,385.80	3,385.80	10.45
Error	36	11,664.00	324.00	
Total	39	17,755.20		

The F statistic for testing the visibility–success interaction reported in the table is $F = 10.45$. Since the p-value of the test is not provided, we are required to find the rejection region for a given value of α. The critical F value depends on $\nu_1 = 1$ numerator degree of freedom (i.e., df for visibility × success) and $\nu_2 = 36$ denominator degrees of freedom (i.e., df for error). Using $\alpha = .05$, the critical F value (found in Table 4, Appendix D) is $F_{.05} \approx 4.12$. Thus, the rejection region is

Rejection region: $F > 4.12$

Since the calculated F (10.45) falls into the rejection region, we reject H_0 at $\alpha = .05$ and conclude that the two factors, subject visibility and confederate success, interact.

Practically, this result implies that the effect of confederate success on mean latency to feedback, $E(y)$, depends on whether the subject is visible. Similarly, the effect of subject visibility on $E(y)$ depends on the success or failure of the confederate student. In other words, we cannot examine the effect of one factor on latency to feedback without knowing the level of the second factor. Consequently, we ignore the F test for factor main effects and focus on the nature of the differences among the means of the $2 \times 2 = 4$ experimental conditions.

14.4
Follow-up Analysis

The sample latency to feedback means (in seconds) for each of the four experimental conditions are provided in Table 14.3. These four means are plotted in Figure 14.1.

T A B L E 14.3   **Sample Means for the Four Experimental Conditions**

		CONFEDERATE SUCCESS	
		Success	Failure
SUBJECT	Visible	73.1	147.2
	Not visible	89.6	72.5

We will conduct a follow-up analysis of the ANOVA by comparing the two confederate success means within each level of subject visibility. Since a balanced design is employed ($n = 10$ subjects per treatment), Tukey's method of multiple comparisons of means will be used.

FIGURE 14.1

Plot of sample means for 2×2 factorial

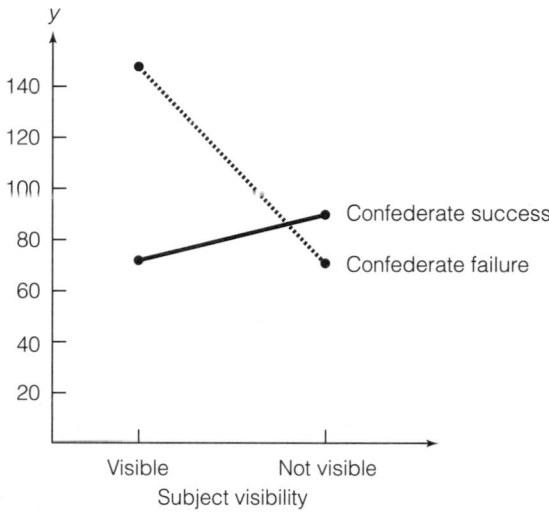

Recall (Section 11.7) that Tukey's method requires that we find the critical value of the Studentized range, $q_\alpha(p, \nu)$, where p is the number of means to be compared and ν is the degrees of freedom associated with MSE (i.e., df for error). Since we want to compare two means at each level of visibility, $p = 2$; also, from Table 14.2, $\nu = 36$. Using $\alpha = .05$, the critical value of q (obtained from Table 11, Appendix D) is $q_{.05}(2, 36) \approx 2.87$.

Tukey's critical difference is

$$\omega = q_\alpha(p, \nu)\frac{s}{\sqrt{n_t}}$$

where $s = \sqrt{\text{MSE}}$ and n_t is the number of observations per treatment. Substituting the appropriate values into the formula for ω, we have

$$\omega = q_{.05}(2, 36)\sqrt{\frac{3{,}385.80}{10}} = (2.87)\sqrt{338.58} = 52.81$$

Consequently, the difference between the two sample means must exceed 52.81 for the corresponding population means to be considered different.

In Table 14.3, we observe that the two confederate success means for nonvisible subjects, $\bar{y}_{\text{success}} = 89.6$ and $\bar{y}_{\text{failure}} = 72.5$, are not significantly different (at $\alpha = .05$) since their difference is less than 52.81 in absolute value. Alternatively, the two confederate success means for visible subjects, $\bar{y}_{\text{success}} = 73.1$ and $\bar{y}_{\text{failure}} = 147.2$ are significantly different at $\alpha = .05$. Thus, only when the subject is visible can we conclude that confederate success has an effect on the mean latency to feedback. Furthermore, since $\bar{y}_{\text{failure}} = 147.2$ is over twice as large as $\bar{y}_{\text{success}} = 73.1$ at this level of subject visibility, the researchers conclude that "subjects appear reluctant to transmit bad news—but only when they are visible to the news recipient."

14.5
Conclusions

In their discussion of these results, the researchers conclude:

In this experiment, subjects were required to give a test taker either success or failure feedback. While doing so, they presumed themselves to be visible to the test taker or visible to no one. Subjects who were visible took twice as long to deliver failure feedback as success feedback; those who were not visible delivered failure and success feedback with equal speed.

These results are not consistent with the discomfort explanation as originally conceived. We had imagined that subjects might empathize with another's failure, that mere observation of the failure would be sufficient to arouse vicarious distress. We found no behavioral evidence of such discomfort. . . . We also imagined that subjects would be reluctant to induce discomfort by announcing a poor intelligence performance, and that they would defer the announcement while checking the IQ test score. We found evidence of this deferral—but only when the subject could be seen. In private, subjects seemed blithe to others' misfortune—as quick to relay bad as good news. As the latency results suggest, there is no inherent discomfort in the transmission of bad news.

EXERCISES

14.1   Use Table 14.2 to determine SSEs for the complete and reduced ANOVA models. Then use these values to obtain the F statistic for testing interaction.

14.2   The journal article made no mention of an analysis of the ANOVA residuals. Discuss the potential problems of an ANOVA when no residual analysis is conducted.

14.3   A second dependent variable measured in the study was *gaze*, defined as the proportion of time the subject was looking toward the confederate test taker on the other side of the glass plate. Gaze was measured at four points in time using videotape segments: early in the test, late in the test, during the wait for feedback, and after the feedback. Construct a complete model for analyzing gaze as a function of subject visibility, confederate success, and videotape segment. Identify the important tests to conduct.

Reference

Bond, C. F. and Anderson, E. L. "The reluctance to transmit bad news: Private discomfort or public display?" *Journal of Experimental Social Psychology*, Vol. 23, 1987, pp. 176–187.

An Investigation of Factors Affecting the Sale Price of Condominium Units Sold at Public Auction

CONTENTS

15.1 The Problem

15.2 The Data

15.3 The Models

15.4 The Regression Analyses

15.5 An Analysis of the Residuals from Model 3

15.6 What the Model 3 Regression Analysis Tells Us

15.7 Comparing the Mean Sale Price for Two Types of Units (Optional)

15.8 Conclusions

OBJECTIVE

To show how regression analysis can be used to develop a model relating sale price of condominium units to a set of independent variables and, particularly, to show how the model can be used to reveal some interesting relationships among these variables

15.1

The Problem

This chapter contains a partial investigation of the factors that affect the sale price of oceanside condominium units. It represents an extension of an analysis of the same data by Herman Kelting (1979). Because there are many different theories (and models) that might be proposed in this type of study, we present the complete data set for further analysis in Appendix G.

The sales data were obtained for a new oceanside condominium complex consisting of two adjacent and connected eight-floor buildings. The complex contains 200 units of equal size (approximately 500 square foot each). The locations of the buildings relative to the ocean, the swimming pool, the parking lot, etc., are shown in Figure 15.1. There are several features of the complex that you should note. The units facing south, called *ocean-view*, face the beach and ocean. In addition, units in building 1 have a good view of the pool. Units to the rear of the building, called *bay-view*, face the parking lot and an area of land that ultimately borders a bay. The view from the upper floors of these units is primarily of wooded, sandy terrain. The bay is very distant and barely visible.

The only elevator in the complex is located at the east end of building 1, as are the office and the game room. People moving to or from the higher floor units in building 2 would likely use the elevator and move through the passages to their units. Thus, units on the higher floors and at a greater distance from the elevator would be less convenient; they would require greater effort in moving

FIGURE 15.1   Layout of condominium complex

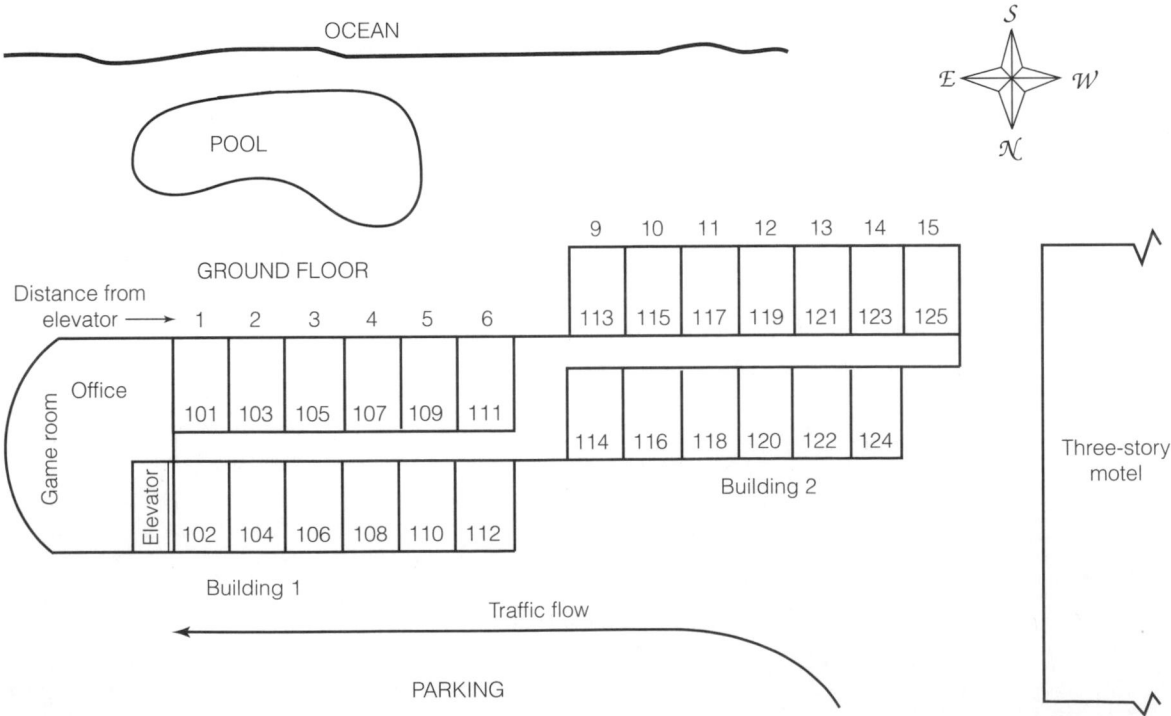

baggage, groceries, etc., and would be farther away from the game room, the office, and the swimming pool. These units also possess an advantage: There would be the least amount of traffic through the hallways in the area, and hence they are the most private.

Lower-floor oceanside units are most suited to active people; they open onto the beach, ocean, and pool. They are within easy reach of the game room and they are easily reached from the parking area.

Checking Figure 15.1, you will see that some of the units in the center of the complex, units numbered __11 and __14, have part of their view blocked. We would expect this to be a disadvantage. We will show you later that this expectation is true for the ocean-view units and that these units sold at a lower price than adjacent ocean-view units.

The condominium complex was completed at the time of the 1975 recession; sales were slow and the developer was forced to sell most of the units at auction approximately 18 months after opening. Many unsold units were furnished by the developer and rented prior to the auction.

This condominium complex was particularly suited to our study. The single elevator located at one end of the complex produces a remarkably high level of both inconvenience and privacy for the people occupying units on the top floors in building 2. Consequently, the data provide a good opportunity to investigate the relationship between sale price, height of the unit (floor number), distance of the unit from the elevator, and presence or absence of an ocean view. The presence or absence of furniture in each of the units also enables us to investigate the effect of the availability of furniture on sale price. Finally, the auction data are completely buyer-specified and hence consumer-oriented in contrast to most other real estate sales data, which are, to a high degree, seller- and broker-specified.

15.2
The Data

In addition to the sale price, the following data were recorded for each of the 106 units sold at the auction:

1. *Floor height*   The floor location of the unit; this variable, x_1, could take values 1, 2, . . . , 8.

2. *Distance from elevator*   This distance, measured along the length of the complex, was expressed in number of condominium units. An additional two units of distance was added to the units in building 2 to account for the walking distance in the connecting area between the two buildings. Thus, the distance of unit 105 from the elevator would be 3, and the distance between unit 113 and the elevator would be 9. This variable, x_2, could take values 1, 2, . . . , 15.

3. *View of ocean*   The presence or absence of an ocean view was recorded for each unit and entered into the model with a dummy variable, x_3, where $x_3 = 1$ if the unit possessed an ocean view and $x_3 = 0$ if not. Note that units not possessing an ocean view would face the parking lot.

4. *End unit*   We expected the partial reduction of view of end units on the ocean side (numbers ending in 11) to reduce their sale price. The ocean view of these end units is partially blocked by building 2. This qualitative variable was entered into the model with a dummy variable, x_4, where $x_4 = 1$ if the unit has a unit number ending in 11 and $x_4 = 0$ if not.

5. *Furniture*   The presence or absence of furniture was recorded for each unit. This qualitative variable was entered into the model using a single dummy variable, x_5, where $x_5 = 1$ if the unit was furnished and $x_5 = 0$ if not.

The raw data used in this analysis are presented in Appendix G.

15.3
The Models

We recorded data on five independent variables, two quantitative (floor height x_1 and distance from elevator x_2) and three qualitative (view of ocean, end unit, and furniture). We postulated four models relating mean sale price to these five factors. The models, numbered 1–4, are developed in sequence, Model 1 being the simplest and Model 4, the most complex. Each of Models 2 and 3 contains all the terms of the preceding models along with new terms that we think will improve their predictive ability. Thus, Model 2 contains all the terms contained in Model 1 plus some new terms, and hence it should predict mean sale price as well as or better than Model 1. Similarly, Model 3 should predict as well as or better than Model 2. Model 4 does not contain all the terms contained in Model 3, but that is only because we have entered floor height into Model 4 as a qualitative independent variable. Consequently, Model 4 contains all the predictive ability of Model 3 and it could be an improvement over Model 3 if our theory is correct. The logic employed in this sequential model building procedure will be explained in the following discussion.

The simplest theory that we might postulate is that the five factors affect the price in an independent manner and that the effect of the two quantitative factors on sale price is linear. Thus, we envision a set of planes, each identical except for their y-intercepts. We would expect sale price planes for ocean-view units to be higher than those with a bay view, those corresponding to end units (__11) would be lower than for non-end units, and those with furniture would be higher than those without.

Model 1

$$E(y) = \beta_0 + \beta_1 x_1 + \beta_2 x_2 + \beta_3 x_3 + \beta_4 x_4 + \beta_5 x_5$$

where

$x_1 =$ Floor height $(x_1 = 1, 2, \ldots, 8)$

$x_2 =$ Distance from elevator $(x_2 = 1, 2, \ldots, 15)$

$$x_3 = \begin{cases} 1 & \text{if an ocean view} \\ 0 & \text{if not} \end{cases} \qquad x_4 = \begin{cases} 1 & \text{if an end unit} \\ 0 & \text{if not} \end{cases} \qquad x_5 = \begin{cases} 1 & \text{if furnished} \\ 0 & \text{if not} \end{cases}$$

The second theory that we considered was that the effects on sale price of floor height and distance from elevator might not be linear. Consequently, we constructed Model 2, which is similar to Model 1 except that second-order terms are included for x_1 and x_2. This model envisions a single second-order response surface for $E(y)$ in x_1 and x_2 that possesses identically the same shape, regardless of the view, whether the unit is an end unit, and whether the unit is furnished. Expressed in other terminology, Model 2 assumes that there is no interaction between any of the qualitative factors (view of ocean, end unit, and furniture) and the quantitative factors (floor height and distance from elevator).

Model 2

$$E(y) = \beta_0 + \overbrace{\beta_1 x_1 + \beta_2 x_2 + \beta_3 x_1 x_2 + \beta_4 x_1^2 + \beta_5 x_2^2}^{\text{Second-order model in } x_1 \text{ and } x_2}$$

$$+ \quad \overbrace{\beta_6 x_3}^{\text{View of ocean}} \quad + \quad \overbrace{\beta_7 x_4}^{\text{End unit}} \quad + \quad \overbrace{\beta_8 x_5}^{\text{Furniture}}$$

Model 2 may possess a serious shortcoming. It assumes that the shape of the second-order response surface, relating mean sale price $E(y)$ to x_1 and x_2 is identical for ocean-view and bay-view units. Since we think that there is a strong possibility that completely different preference patterns may govern the purchase of these two groups of units, we will construct a model that provides for two completely different second-order response surfaces—one for ocean-view units and one for bay-view units. Further, we will assume that the effects of the two qualitative factors, end unit and furniture, are additive; i.e., their presence or absence will simply shift the mean sale price response surface up or down by a fixed amount. Thus, Model 3 is given as follows:

Model 3

$$E(y) = \beta_0 + \overbrace{\beta_1 x_1 + \beta_2 x_2 + \beta_3 x_1 x_2 + \beta_4 x_1^2 + \beta_5 x_2^2}^{\text{Second-order model in } x_1 \text{ and } x_2}$$

$$+ \quad \overbrace{\beta_6 x_3}^{\text{View of ocean}} \quad + \quad \overbrace{\beta_7 x_4}^{\text{End unit}} \quad + \quad \overbrace{\beta_8 x_5}^{\text{Furniture}}$$

$$+ \overbrace{\beta_9 x_1 x_3 + \beta_{10} x_2 x_3 + \beta_{11} x_1 x_2 x_3 + \beta_{12} x_1^2 x_3 + \beta_{13} x_2^2 x_3}^{\substack{\text{Interaction of the second-order model} \\ \text{with view of ocean}}}$$

As a fourth possibility, we constructed a model similar to Model 3 but entered floor height as a qualitative factor at eight levels. The reasons for doing this are:

1. Higher-floor units have better views but less accessibility to the outdoors. This latter characteristic could be a particularly undesirable feature for these units.
2. The views of some lower-floor bayside units were blocked by a nearby three-story motel.

If our supposition is correct and if these features would have a depressive effect on the sale price of these units, then the relationship between floor height and mean sale price would not be second-order (a smooth curvilinear relationship). Entering floor height as a qualitative factor would permit a better fit to this irregular relationship and improve the prediction equation. Thus, Model 4 is identical to Model 3 except that Model 4 contains seven main effect terms for floor height (in contrast to two for Model 3), and it also contains the corresponding interactions of these variables with the other variables included in Model 3. We will subsequently show that there was no evidence to indicate that Model 4 contributes more information for the prediction of y than Model 3. For this reason and also because Model 4 contains so many terms (49), we will not give the formula for this model.*

15.4
The Regression Analyses

This section gives the regression analyses for the four models described in Section 15.3. You will see that our approach is to build the model in a sequential manner. In each case we use an F test to see whether a particular model contributes more information for the prediction of sale price than its predecessor.

This procedure is more conservative than if we were to employ a step-down procedure, i.e., assume Model 4 to be the appropriate model and then test and possibly delete terms. Deleting terms can be particularly risky because, in doing so, you are tacitly accepting the null hypothesis. Thus, you risk deleting important terms from the model and do so with an unknown probability of committing a Type II error.

Do not be unduly influenced by the individual t tests associated with an analysis. As you will see, it is possible for a set of terms to contribute information for the prediction of y when none of their respective t values are statistically significant. This is because the t test tests the contribution of a single term, given that all the other terms are retained in the model. Therefore, if a set of terms contributes overlapping information, it is possible that none of the terms individually would be statistically significant when the set, as a whole, contributes information for the prediction of y.

The SAS regression analysis computer printouts for fitting Models 1, 2, 3, and 4 to the data are shown in Figures 15.2, 15.3, 15.4, and 15.5, respectively. A summary containing SSE values for these models, their respective degrees of freedom, and R^2 values is provided in Table 15.1.

Examining the computer printout for the first-order model (Model 1) in Figure 15.2, you can see that the value of the F statistic for testing the null hypothesis

$$H_0: \quad \beta_1 = \beta_2 = \cdots = \beta_5 = 0$$

is 48.42. This is statistically significant at a level of $\alpha = .0001$. Consequently, there is ample evidence to indicate that the model contributes information for the prediction of y. At least one of the five factors contributes information for the prediction of sale price.

---

*Some of these terms were not estimable because sales were not consummated for some combinations of the independent variables. This is why the computer printout in Figure 15.5 shows only 41 df for the model.

FIGURE 15.2

SAS regression analysis: Model 1

ANALYSIS OF VARIANCE

SOURCE	DF	SUM OF SQUARES	MEAN SQUARE	F VALUE	PROB>F
MODEL	5	236620761	47324152.11	48.419	0.0001
ERROR	100	97737975.28	977379.75		
C TOTAL	105	334358736			

ROOT MSE	988.6252	R-SQUARE	0.7077	
DEP MEAN	19176.04	ADJ R-SQ	0.6931	
C.V.	5.155524			

PARAMETER ESTIMATES

VARIABLE	DF	PARAMETER ESTIMATE	STANDARD ERROR	T FOR H0: PARAMETER=0	PROB > |T|
INTERCEP	1	17770.00813	416.07044	42.709	0.0001
X1	1	-73.60311471	52.97036540	-1.391	0.1674
X2	1	-86.41176445	24.44923524	-3.534	0.0006
X3	1	3136.59235	222.70788	14.084	0.0001
X4	1	-1781.05526	397.45816	-4.481	0.0001
X5	1	986.99033	204.77016	4.820	0.0001

FIGURE 15.3

SAS regression analysis: Model 2

ANALYSIS OF VARIANCE

SOURCE	DF	SUM OF SQUARES	MEAN SQUARE	F VALUE	PROB>F
MODEL	8	244325938	30540742.24	32.904	0.0001
ERROR	97	90032797.92	928173.17		
C TOTAL	105	334358736			

ROOT MSE	963.4174	R-SQUARE	0.7307	
DEP MEAN	19176.04	ADJ R-SQ	0.7085	
C.V.	5.024069			

PARAMETER ESTIMATES

VARIABLE	DF	PARAMETER ESTIMATE	STANDARD ERROR	T FOR H0: PARAMETER=0	PROB > |T|
INTERCEP	1	19461.81838	764.78661	25.447	0.0001
X1	1	-683.94788	245.08261	-2.791	0.0063
X2	1	-264.54977	122.58866	-2.158	0.0334
X1X2	1	4.81613443	13.54099144	0.356	0.7229
X1X1	1	57.93750913	23.74605653	2.440	0.0165
X2X2	1	11.33982388	7.70152793	1.472	0.1441
X3	1	3051.55998	219.25992	13.918	0.0001
X4	1	-1680.93326	409.72999	-4.103	0.0001
X5	1	1114.78991	205.04676	5.437	0.0001

TABLE 15.1 A Summary of the Values (Rounded) of SSE, MSE, and R^2 for Models 1, 2, 3, and 4

MODEL	SSE	df(Error)	MSE	R^2
1	97,737,975	100	977,380	.7077
2	90,032,798	97	928,173	.7307
3	78,428,545	92	852,484	.7654
4	51,647,128	64	806,986	.8455

If you examine the t tests for the individual parameters, you will see that they are all statistically significant except the test for β_1, the parameter associated

FIGURE 15.4

SAS regression analysis: Model 3

ANALYSIS OF VARIANCE

SOURCE	DF	SUM OF SQUARES	MEAN SQUARE	F VALUE	PROB>F
MODEL	13	255930191	19686937.74	23.094	0.0001
ERROR	92	78428545.24	852484.19		
C TOTAL	105	334358736			

ROOT MSE	923.3007	R-SQUARE	0.7654	
DEP MEAN	19176.04	ADJ R-SQ	0.7323	
C.V.	4.814867			

PARAMETER ESTIMATES

VARIABLE	DF	PARAMETER ESTIMATE	STANDARD ERROR	T FOR H0: PARAMETER=0	PROB > |T|
INTERCEP	1	14412.04197	2795.75178	5.155	0.0001
X1	1	819.81446	941.47799	0.871	0.3861
X2	1	-413.28219	303.20570	-1.363	0.1762
X1X2	1	33.53465808	34.14413805	0.982	0.3286
X1X1	1	-54.41406576	80.65633152	-0.675	0.5016
X2X2	1	11.87166612	16.59695367	0.715	0.4762
X3	1	8247.71451	2885.53387	2.858	0.0053
X4	1	-1632.24166	400.11602	-4.079	0.0001
X5	1	1242.52585	204.36346	6.080	0.0001
X1X3	1	-1350.92966	974.75335	-1.386	0.1691
X2X3	1	50.74158830	334.14998	0.152	0.8796
X1X2X3	1	-29.98215571	37.25553940	-0.805	0.4230
X1X1X3	1	91.37334655	84.49650785	1.081	0.2824
X2X2X3	1	5.55440578	19.06625192	0.291	0.7715

FIGURE 15.5

SAS regression analysis: Model 4

ANALYSIS OF VARIANCE

SOURCE	DF	SUM OF SQUARES	MEAN SQUARE	F VALUE	PROB>F
MODEL	41	282711608	6895405.07	8.545	0.0001
ERROR	64	51647128.00	806986.37		
C TOTAL	105	334358736			

ROOT MSE	898.3242	R-SQUARE	0.8455	
DEP MEAN	19176.04	ADJ R-SQ	0.7466	
C.V.	4.684618			

with floor height x_1 (p-value $= .1674$). The failure of floor height x_1 to reveal itself as an important information contributor goes against our intuition, and it demonstrates the pitfalls that can occur when you attempt to interpret the results of t tests in a regression analysis. Intuitively, we would expect floor height to be an important factor. You might argue that units on the higher floors possess a better view and hence should command a higher mean sale price. Or, you might argue that units on the lower floors have greater accessibility to the pool and ocean and, consequently, should be in greater demand. Why, then, is the t test for floor height not statistically significant? The answer is that both of the preceding arguments are correct, one for the oceanside and one for the bayside. Thus, you will subsequently see that there is an interaction between floor height and view of ocean. Ocean-view units on the lower floors sell at higher prices than ocean-view units on the higher floors. In contrast, bay-view units on the higher floors command higher prices than bay-view units on the lower floors. These two contrasting effects tend to cancel (because we have not included interaction terms in the model) and thereby give the false impression that floor height is not an important variable for predicting mean sale price.

But, of course, we are looking ahead. Our next step is to determine whether Model 2 is better than Model 1.

Are floor height x_1 and distance from elevator x_2 related to sale price in a curvilinear manner, i.e., should we be using a second-order response surface instead of a first-order surface to relate $E(y)$ to x_1 and x_2? To answer this question, we will examine the drop in SSE from Model 1 to Model 2. The null hypothesis "Model 2 contributes no more information for the prediction of y than Model 1" is equivalent to testing

$$H_0: \quad \beta_3 = \beta_4 = \beta_5 = 0$$

where β_3, β_4, and β_5 appear in Model 2. The F statistic for the test, based on 3 and 97 df, is

$$F = \frac{\dfrac{(\text{SSE}_1 - \text{SSE}_2)}{\text{Number of } \beta \text{ parameters in } H_0}}{\text{MSE}_2} = \frac{\dfrac{97{,}737{,}975 - 90{,}032{,}798}{3}}{928{,}173.17}$$

$$= 2.77$$

An approximate value for $F_{.05}$ based on 3 and 97 df, obtained from Table 4 in Appendix D, is 2.70. Since the computed F value exceeds the tabulated value, $F_{.05} = 2.70$, we reject H_0. There is evidence to indicate that Model 2 contributes more information for the prediction of y than Model 1. This tells us that there is evidence of curvature in the response surfaces relating mean sale price, $E(y)$, to floor height x_1 and distance from elevator x_2.

You will recall that the difference between Models 2 and 3 is that Model 3 allows for two differently shaped second-order surfaces, one for ocean-view units and another for bay-view units; Model 2 employs a single surface to represent both types of units. Consequently, we wish to test the null hypothesis that "a single second-order surface adaquately characterizes the relationship between $E(y)$, floor height x_1, and distance from elevator x_2 for both ocean-view and bay-view units" [i.e., Model 2 adequately models $E(y)$] against the alternative hypothesis that you need two different second-order surfaces, i.e, you need Model 3. Thus,

$$H_0: \quad \beta_9 = \beta_{10} = \cdots = \beta_{13} = 0$$

where β_9, β_{10}, ..., β_{13} are parameters in Model 3. The F statistic for this test, based on 5 and 92 df, is

$$F = \frac{\dfrac{(\text{SSE}_2 - \text{SSE}_3)}{\text{Number of } \beta \text{ parameters in } H_0}}{\text{MSE}_3}$$

$$= \frac{\dfrac{90{,}032{,}798 - 78{,}428{,}545}{5}}{852{,}484.19}$$

$$= 2.72$$

An approximate value for $F_{.05}$ based on 5 and 92 df, obtained from Table 4 in Appendix D, is 2.33. Since the computed F value exceeds this tabulated value, we reject H_0 and conclude that there is evidence to indicate that we need two different second-order surfaces to relate $E(y)$ to x_1 and x_2, one each for ocean-view and bay-view units.

Finally, we question whether Model 4 will provide an improvement over Model 3, i.e., will we gain information for predicting y by entering floor height into the model as a qualitative factor at eight levels? The F statistic to test the null hypothesis "Model 4 contributes no more information for predicting y than does Model 3" compares the drop in SSE from Model 3 to Model 4 with s_4^2. This F statistic, based on 28 df (the difference in the numbers of parameters in Models 4 and 3) and 64 df, is

$$F = \frac{\dfrac{(\text{SSE}_3 - \text{SSE}_4)}{\text{Number of } \beta \text{ parameters in } H_0}}{\text{MSE}_4}$$

$$= \frac{\dfrac{78{,}428{,}545 - 51{,}647{,}128}{28}}{806{,}986}$$

$$= 1.19$$

Checking Table 4 in Appendix D, you will find that the value for $F_{.05}$, based on 28 and 64 df, is approximately 1.65. Since the computed F is less than this value, there is not sufficient evidence to indicate that Model 4 is a significant improvement over Model 3.

Having checked the four theories of Section 15.3, we have evidence to indicate that Model 2 is significantly better than Model 1 and that Model 3 is better than Model 2. We will examine the prediction equation for Model 3 and see what it tells us about the relationship between the mean sale price $E(y)$ and the five factors employed in our study; but first, it is important that we examine the residuals for Model 3 to determine whether the standard least squares assumptions about the random error term are satisfied.

15.5

An Analysis of the Residuals from Model 3

The four standard least squares assumptions about the random error term ε are (from Chapter 4) the following:

1. The mean is 0.
2. The variance (σ^2) is constant for all settings of the independent variables.
3. The errors follow a normal distribution.
4. The errors are independent.

If one or more of these assumptions are violated, any inferences derived from the Model 3 regression analysis are suspect. It is unlikely that assumption 1 (0 mean) is violated because the method of least squares guarantees that the mean of the residuals is 0. The same can be said for assumption 4 (independent errors) since the sale price data are not a time series. However, verifying assumptions 2 and 3 requires a thorough examination of the residuals from Model 3.

Recall that we can check for heteroscedastic errors (i.e., errors with unequal variances) by plotting the residuals against the predicted values. This residual plot is shown in Figure 15.6 (page 710). If the variances were not constant, we would expect to see a cone-shaped pattern (since the response is sale price) in Figure 15.6, with the spread of the residuals increasing as $\hat{y}$ increases. Note, however, that the residuals appear to be randomly scattered around 0. Therefore, assumption 2 (constant variance) appears to be satisfied.

To check the normality assumption (assumption 3), we have generated a histogram of the residuals in Figure 15.7 (page 711). It is very evident that the distribution of residuals is not normal, but skewed to the right. At this point, we could opt to use a transformation on the response (similar to the variance-stabilizing transformations discussed in Section 6.3) to normalize the residuals. However, a nonnormal error distribution is often due to the presence of a single outlier. If this outlier is eliminated (or corrected) the normality assumption may then be satisfied.

We can use the residual plot in Figure 15.6 to detect outliers in the analysis. In Section 6.4, we defined outliers to be residuals that exceed $3s$ (in absolute value) and suspect outliers as residuals that fall between $2s$ and $3s$ away from 0. The $\pm 2s$ and $\pm 3s$ lines are marked on Figure 15.6 (where $s = 923$ from Figure 15.4). Note that there is one outlier and one suspect outlier, both with large *positive* residuals (approximately 5,500 and 2,500, respectively). Should we automatically eliminate these two observations from the analysis and refit Model 3? Although many analysts adopt such an approach, we should carefully examine the observations before deciding to eliminate them. We may discover a correctable recording (or coding) error, or we may find that the outliers are very influential and are due to an inadequate model (in which case, it is the model that needs fixing, not the data).

An examination of the SAS printout of the Model 3 residuals (not shown) reveals that the two observations in question are identified by observation numbers 35 and 49 (where the observations are numbered from 1 to 106). The sale prices, floor heights, and so forth, for these two data points were found to be recorded and coded correctly. To determine how influential these outliers are on the analysis, influence diagnostics (i.e., Cook's D and leverage) were generated using SAS. The results are summarized in Table 15.2.

T A B L E 15.2 **Influence Diagnostics for Two Outliers in Model 3**

OBSERVATION	RESPONSE y	PREDICTED VALUE $\hat{y}$	RESIDUAL $y - \hat{y}$	LEVERAGE h	COOK'S DISTANCE D
35	26,500	21,070.4	5,429.6	.0605	.169
49	21,000	18,414.3	2,585.7	.1607	.128

Based on the "rules of thumb" given in Section 6.5, neither observation has strong influence on the analysis. Both leverage (h) values fall below $2(k + 1)/n = 2(14)/106 = .264$, indicating that the observations are not influential with respect to their x values; and both Cook's D values fall below .96 [the 50th

FIGURE 15.6
Plot of residuals versus $\hat{y}$ for Model 3

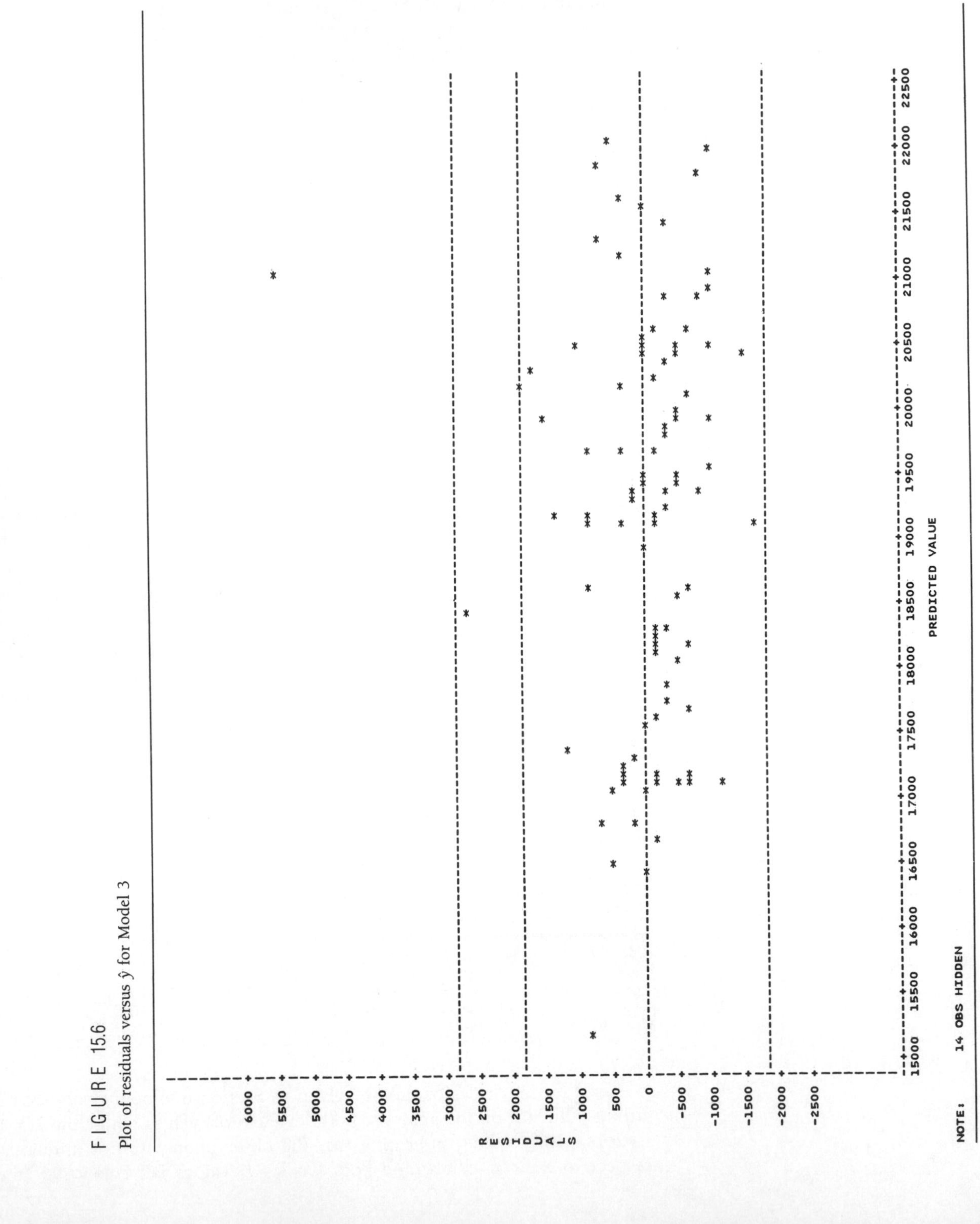

FIGURE 15.7

Histogram of residuals from Model 3

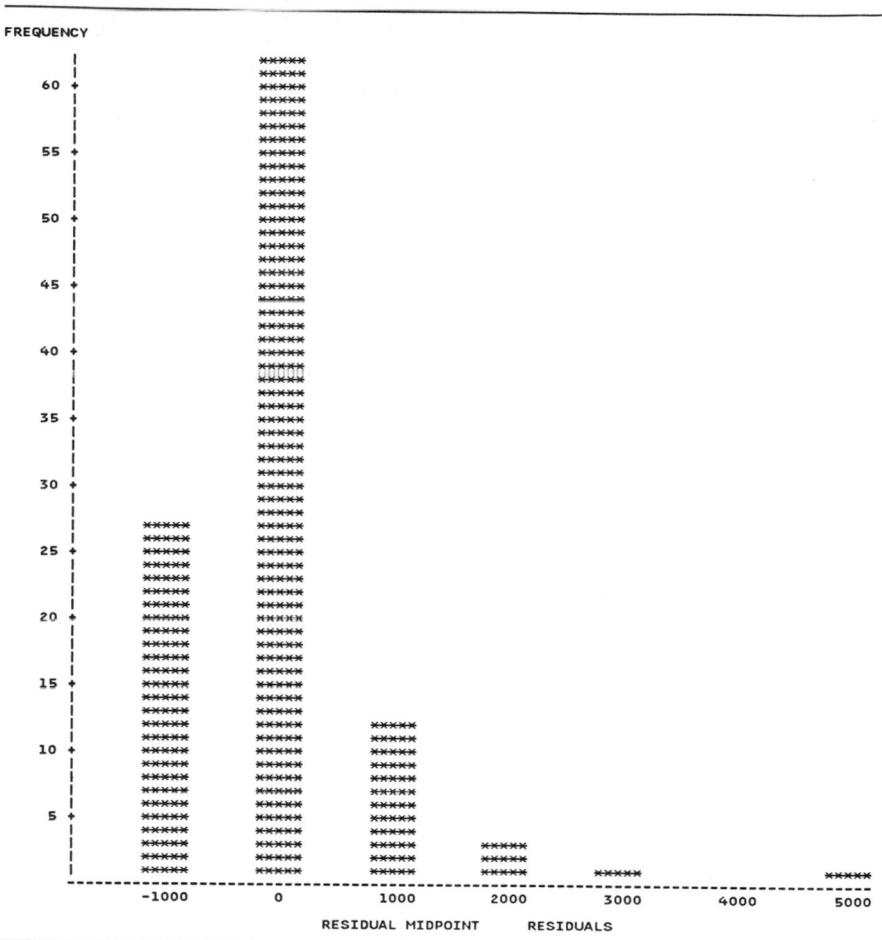

percentile of an F distribution with $\nu_1 = k + 1 = 14$ and $\nu_2 = n - (k + 1) =$ $106 - 14 = 92$ degrees of freedom], implying that they do not exhibit strong overall influence on the regression results (e.g., the β estimates). Consequently, if we remove these outliers from the data and refit Model 3, the least squares prediction equation will not be greatly affected and the normality assumption will probably be more nearly satisfied.

The SAS printout for the refitted model is shown in Figure 15.8 on page 712. Note that df(error) is reduced from 92 to 90 (since we eliminated the two outlying observations), and the β estimates remain relatively unchanged. The model standard deviation, however, is decreased significantly from 923 to 658, implying that the refitted model will yield more accurate predictions of sale price. A residual plot for the refitted model is shown in Figure 15.9 (page 713) and a histogram of the residuals in Figure 15.10 (page 714). The residual plot (Figure 15.9) reveals no evidence of outliers, and the histogram of the residuals (Figure 15.10) is now approximately normal.

FIGURE 15.8

SAS printout for Model 3 with outliers deleted

ANALYSIS OF VARIANCE

SOURCE	DF	SUM OF SQUARES	MEAN SQUARE	F VALUE	PROB>F
MODEL	13	237649572	18280736.29	42.254	0.0001
ERROR	90	38937243.63	432636.04		
C TOTAL	103	276586815			

ROOT MSE	657.7507	R-SQUARE	0.8592	
DEP MEAN	19088.08	ADJ R-SQ	0.8389	
C.V.	3.445872			

PARAMETER ESTIMATES

VARIABLE	DF	PARAMETER ESTIMATE	STANDARD ERROR	T FOR H0: PARAMETER=0	PROB > |T|
INTERCEP	1	15390.19258	2002.11309	7.687	0.0001
X1	1	310.30678	682.48668	0.455	0.6504
X2	1	-203.35236	222.11810	-0.916	0.3624
X1X2	1	26.08674126	24.39118628	1.070	0.2877
X1X1	1	-10.85790288	58.55677607	-0.185	0.8533
X2X2	1	1.09206281	12.11658757	0.090	0.9284
X3	1	7537.23222	2065.74801	3.649	0.0004
X4	1	-1500.60982	285.47441	-5.257	0.0001
X5	1	1075.38112	146.64993	7.333	0.0001
X1X3	1	-1024.61067	706.31066	-1.451	0.1504
X2X3	1	-167.24636	243.44976	-0.687	0.4939
X1X2X3	1	-25.43791062	26.60757746	-0.956	0.3416
X1X1X3	1	67.67866117	61.32668383	1.104	0.2727
X2X2X3	1	19.21642228	13.82949599	1.390	0.1681

15.6

What the Model 3 Regression Analysis Tells Us

We have settled on Model 3 (with two observations deleted) as our choice to relate mean sale price $E(y)$ to five factors: the two quantitative factors, floor height x_1 and distance from elevator x_2; and the three qualitative factors, view of ocean, end unit, and furniture. This model postulates two different second-order surfaces relating mean sale price $E(y)$ to x_1 and x_2, one for ocean-view units and one for bay-view units. The effect of each of the two qualitative factors, end unit (numbered __11) and furniture, is to produce a change in mean sale price that is identical for all combinations of values of x_1 and x_2. In other words, assigning a value of 1 to one of the dummy variables increases (or decreases) the estimated mean sale price by a fixed amount. The net effect is to push the second-order surface upward or downward, with the direction depending on the level of the specific qualitative factor. The estimated increase (or decrease) in mean sale price because of a given qualitative factor is given by the estimated value of the β parameter associated with its dummy variable.

For example, the prediction equation (with rounded values given for the parameter estimates) obtained from Figure 15.8 is

$$\hat{y} = 15{,}390.2 + 310.3x_1 - 203.4x_2$$
$$+ 26.1x_1x_2 - 10.9x_1^2 + 1.1x_2^2$$
$$+ 7{,}537.2x_3 - 1{,}500.6x_4 + 1{,}075.4x_5$$
$$- 1{,}024.6x_1x_3 - 167.2x_2x_3 - 25.4x_1x_2x_3$$
$$+ 67.7x_1^2x_3 + 19.2x_2^2x_3$$

FIGURE 15.9
Plot of residuals versus ŷ for Model 3
with outliers deleted

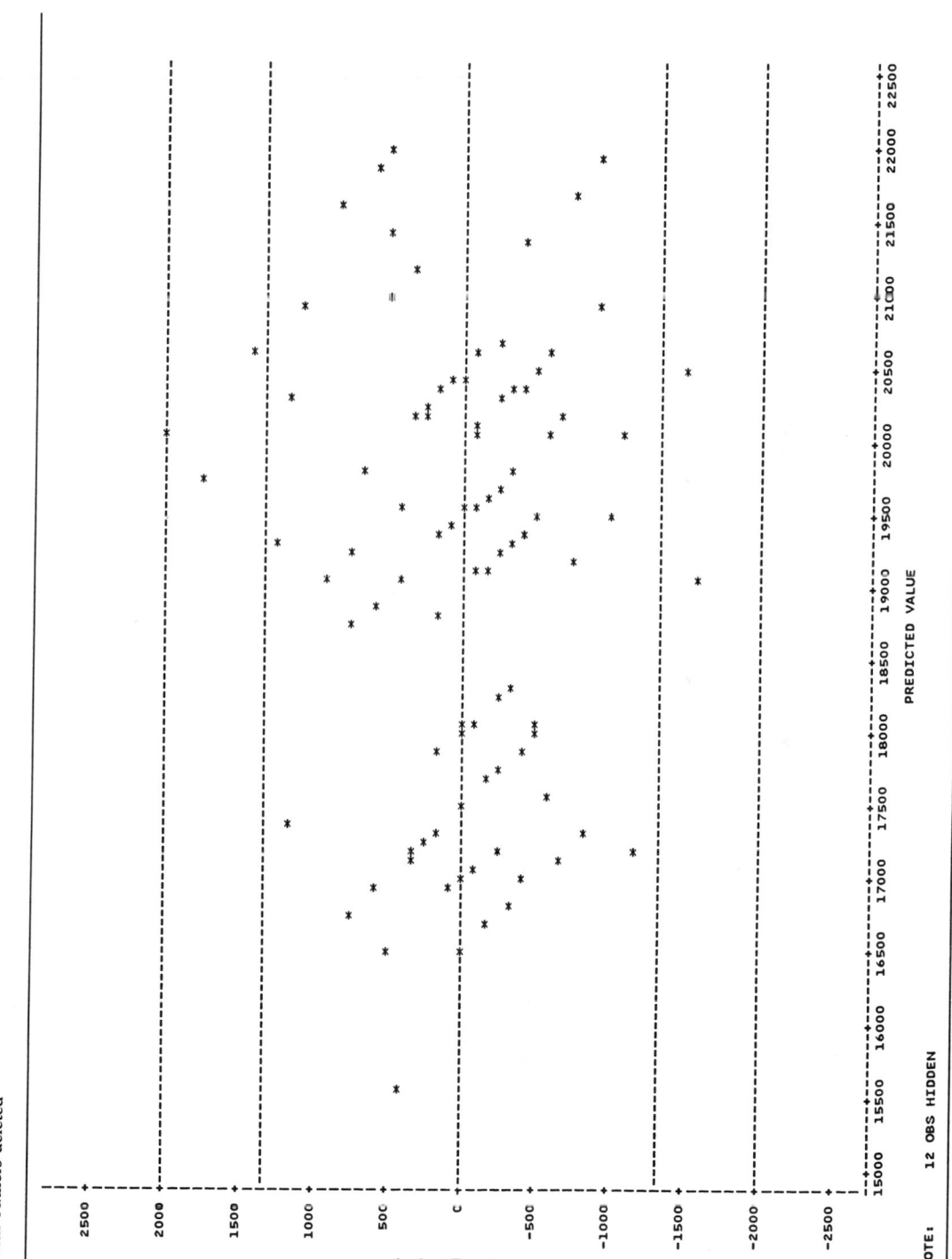

FIGURE 15.10   Histogram of residuals from Model 3 with outliers deleted

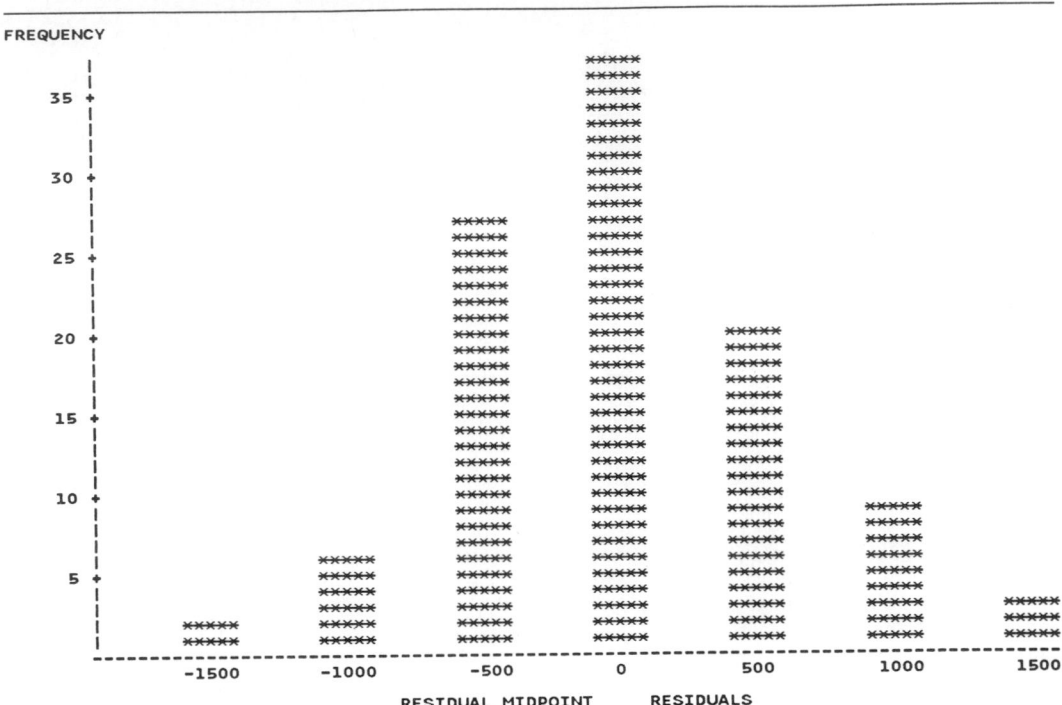

Since the dummy variables for end unit and furniture are, respectively, x_4 and x_5, the estimated changes in mean sale price for these qualitative factors are:

End unit $(x_4 = 1)$: $\hat{\beta}_7 = -\$1,500.6$
Furnished $(x_5 = 1)$: $\hat{\beta}_8 = +\$1,075.4$

Thus, if you substitute $x_4 = 1$ into the prediction equation, the estimated mean decrease in sale price for an end unit is \$1,500.6, regardless of the view, floor, and whether it is furnished.

The effect of floor height x_1 and distance from elevator x_2 can be determined by plotting $\hat{y}$ as a function of one of the variables for given values of the other. For example, suppose we wish to examine the relationship between $\hat{y}$, x_1, and x_2 for bay-view $(x_3 = 0)$, non-end units $(x_4 = 0)$ with no furniture $(x_5 = 0)$. The prediction curve relating $\hat{y}$ to distance from elevator x_2 can be graphed for each floor by first setting $x_1 = 1$, then $x_1 = 2, \ldots, x_1 = 8$. The graphs of these curves are shown in Figure 15.11; the 1's indicate the curve for the first floor, the 2's the curve for the second floor, and so forth. In Figure 15.11, we can see some interesting patterns in the estimated mean sale prices:

FIGURE 15.11   Plot of predicted price versus distance from elevator: Bay-view units

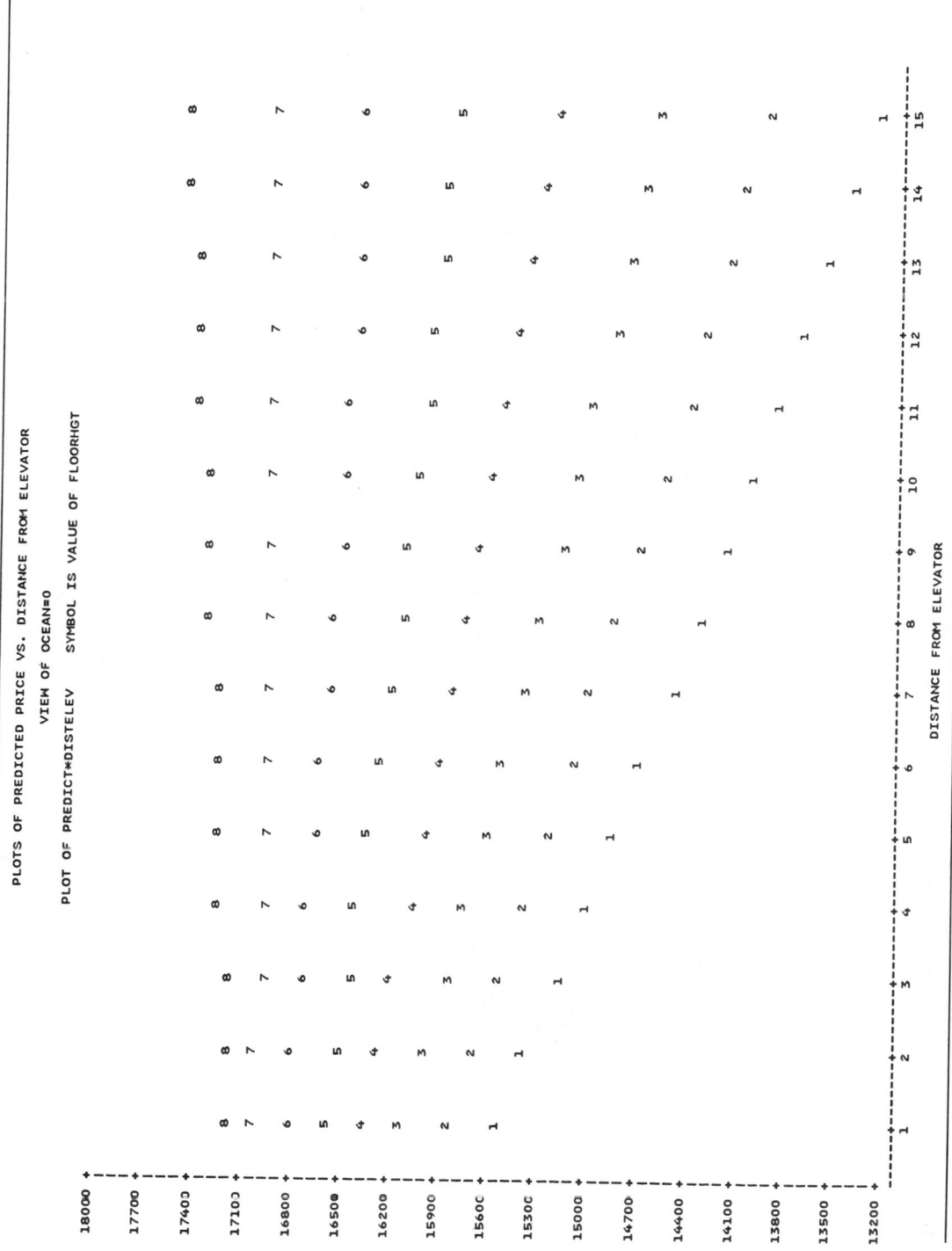

1. The higher the floor of a bay-view unit, the higher will be the mean sale price. Low floors look out onto the parking lot and, all other variables held constant, are least preferred.
2. The relationship is curvilinear and is not the same for each floor.
3. Units on the first floor near the office have a higher estimated mean sale price than second- or third-floor units located in the west end of the complex. Perhaps the reason for this is that these units are close to the pool and the game room, and these advantages outweigh the disadvantage of a poor view.
4. The mean sale price decreases as the distance from the elevator and center of activity increases for the lower floors, but the decrease is less as you move upward, floor to floor. Finally, note that the estimated mean sale price increases substantially for units on the highest floor that are farthest away from the elevator. These units are subjected to the least human traffic and are, therefore, the most private. Consequently, a possible explanation for their high price is that buyers place a higher value on the privacy provided by the units than the negative value that they assign to their inconvenience. One additional explanation for the generally higher estimated sale price for units at the ends of the complex may be that they possess more windows.

A similar set of curves is shown in Figure 15.12 for ocean-view units ($x_3 = 1$). You will note some amazing differences between these curves and those for the bay-view units in Figure 15.11 (these differences explain why we needed two separate second-order surfaces to describe these two sets of units). The preference for floors is completely reversed on the ocean side of the complex: the lower the floor, the higher the estimated mean sale price. Apparently, people selecting the ocean-view units are primarily concerned with accessibility to the ocean, pool, beach, and game room. Note that the estimated mean sale price is highest near the elevator. It drops and then rises as you reach the units farthest from the elevator. An explanation for this phenomenon is similar to the one that we used for the bayside units. Units near the elevator are more accessible and nearer to the recreational facilities. Those farthest from the elevators afford the greatest privacy. Units near the center of the complex offer reduced amounts of both accessibility *and* privacy. Notice that units adjacent to the elevator command a higher estimated mean sale price than those near the west end of the complex, thus suggesting that accessibility has a greater influence on price than privacy.

Rather than examine the graphs of $\hat{y}$ as a function of distance from elevator x_2, you may want to see how $\hat{y}$ behaves as a function of floor height x_1 for units located at various distances from the elevator. These estimated mean sale price curves are shown for bay-view units in Figure 15.13 (page 718) and for ocean-view units in Figure 15.14 (page 719). To avoid congestion in the graphs, we have shown only the curves for distances $x_2 = 1, 3, 5, \ldots, 13, 15$. A curve traced by a sequence of 1's indicates that it applies to units located a distance of 1 from the elevator. Similarly, a curve traced by a sequence of 5's applies to units located a distance of 5 from the elevator. We leave it to you and to the real estate experts to deduce the practical implications of these curves.

FIGURE 15.12 Plot of predicted price versus distance from elevator: Ocean-view units

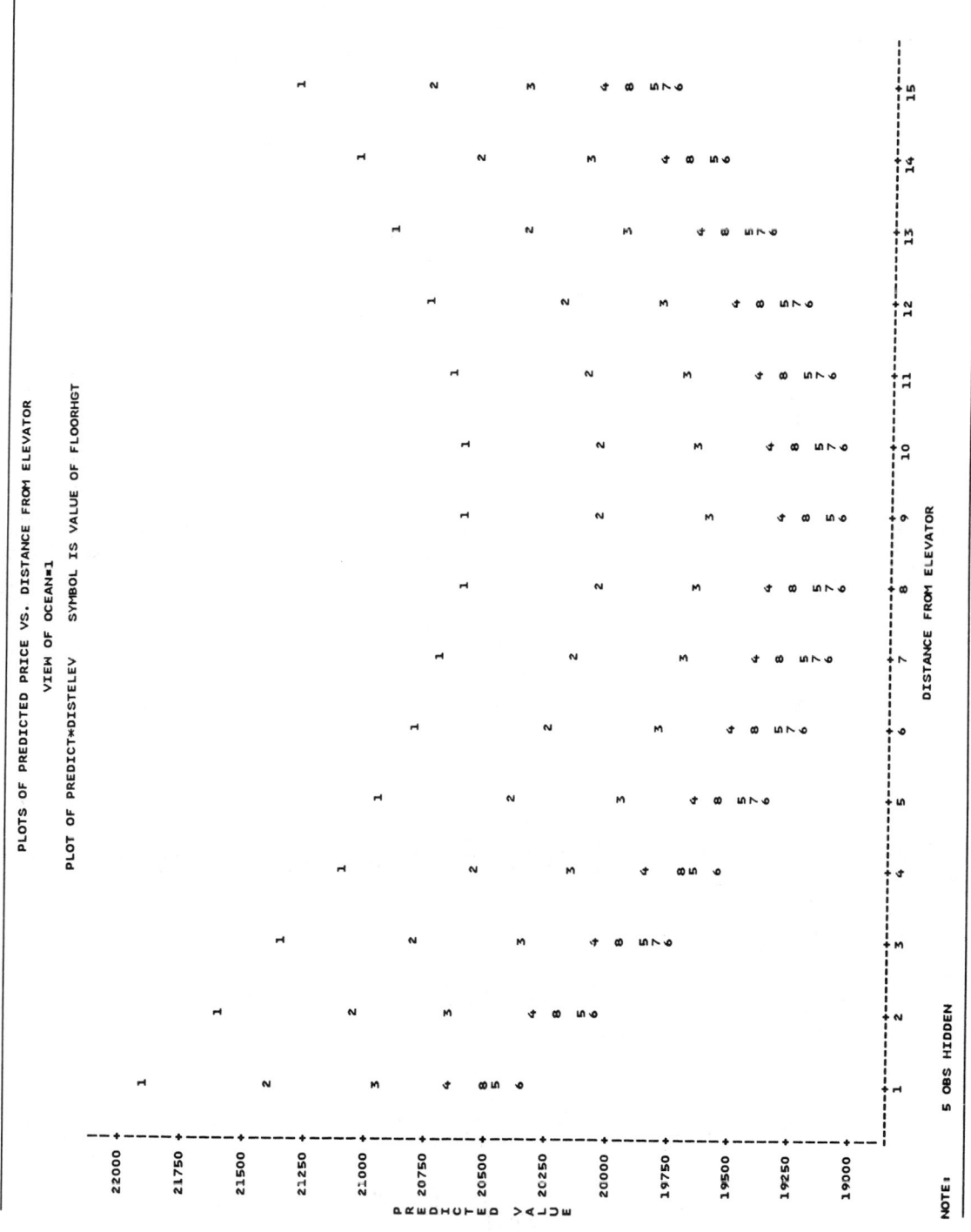

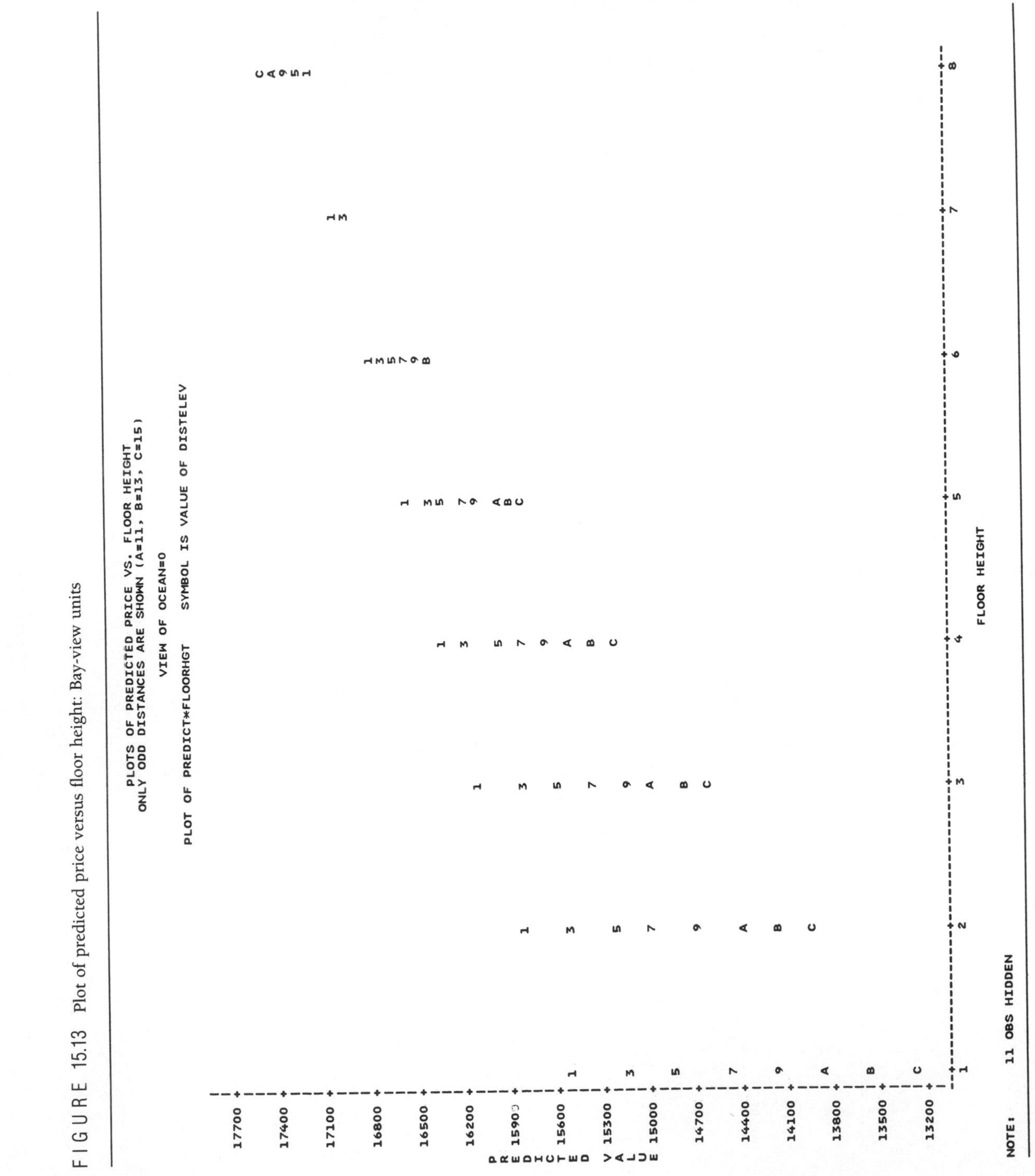

FIGURE 15.13 Plot of predicted price versus floor height: Bay-view units

FIGURE 15.14 Plot of predicted price versus floor height: Ocean-view units

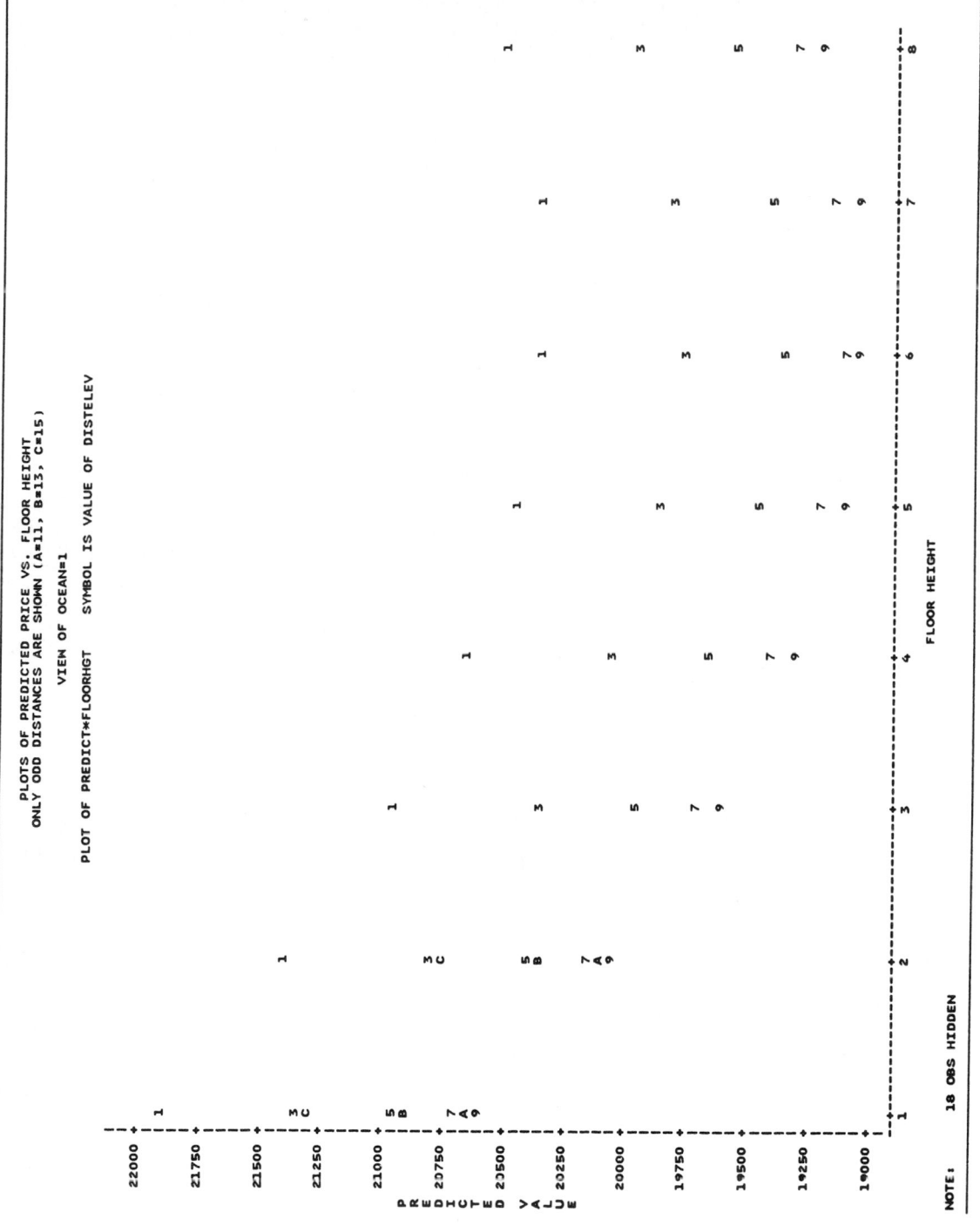

15.7

Comparing the Mean Sale Price for Two Types of Units (Optional)

[*Note:* This section requires an understanding of the mechanics of a multiple regression analysis presented in Appendix A.]

Comparing the mean sale price for two types of units might seem a useless endeavor, considering that all the units have been sold and that we will never be able to sell the units in the same economic environment that existed at the time the data were collected. Nevertheless, this information might be useful to a real estate appraiser or to a developer who is pricing units in a new and similar condominium complex. We will assume that a comparison is useful and show you how it can be accomplished.

Suppose you want to estimate the difference in mean sale price for units in two different locations and with or without furniture. For example, suppose you wish to estimate the difference in mean sale price between the first-floor ocean-view and bay-view units located at the east end of building 1 (units 101 and 102 in Figure 15.1). Both these units afford a maximum of accessibility, but they possess different views. We assume that both are furnished. The estimate of the mean sale price $E(y)$ for the first-floor, bay-view unit will be the value of $\hat{y}$ when $x_1 = 1$, $x_2 = 1$, $x_3 = 0$, $x_4 = 0$, $x_5 = 1$. Similarly, the estimated value of $E(y)$ for the first-floor, ocean-view unit is obtained by substituting $x_1 = 1$, $x_2 = 1$, $x_3 = 1$, $x_4 = 0$, and $x_5 = 1$ into the prediction equation.

We will let $\hat{y}_o$ and $\hat{y}_b$ represent the estimated mean sale prices for the first-floor ocean-view and bay-view units, respectively. Then the estimator of the difference in mean sale prices for the two units is

$$\ell = \hat{y}_o - \hat{y}_b$$

We represent this estimator by the symbol ℓ, because it is a linear function of the parameter estimators $\hat{\beta}_0, \hat{\beta}_1, \ldots, \hat{\beta}_{13}$; i.e.,

$$\begin{aligned}
\hat{y}_o = {} & \hat{\beta}_0 + \hat{\beta}_1(1) + \hat{\beta}_2(1) + \hat{\beta}_3(1)(1) + \hat{\beta}_4(1)^2 + \hat{\beta}_5(1)^2 \\
& + \hat{\beta}_6(1) + \hat{\beta}_7(0) + \hat{\beta}_8(1) + \hat{\beta}_9(1)(1) + \hat{\beta}_{10}(1)(1) \\
& + \hat{\beta}_{11}(1)(1)(1) + \hat{\beta}_{12}(1)^2(1) + \hat{\beta}_{13}(1)^2(1) \\
\hat{y}_b = {} & \hat{\beta}_0 + \hat{\beta}_1(1) + \hat{\beta}_2(1) + \hat{\beta}_3(1)(1) + \hat{\beta}_4(1)^2 + \hat{\beta}_5(1)^2 \\
& + \hat{\beta}_6(0) + \hat{\beta}_7(0) + \hat{\beta}_8(1) + \hat{\beta}_9(1)(0) + \hat{\beta}_{10}(1)(0) \\
& + \hat{\beta}_{11}(1)(1)(0) + \hat{\beta}_{12}(1)^2(0) + \hat{\beta}_{13}(1)^2(0)
\end{aligned}$$

then

$$\ell = \hat{y}_o - \hat{y}_b = \hat{\beta}_6 + \hat{\beta}_9 + \hat{\beta}_{10} + \hat{\beta}_{11} + \hat{\beta}_{12} + \hat{\beta}_{13}$$

A 95% confidence interval for the mean value of a linear function of the estimators $\hat{\beta}_0, \hat{\beta}_1, \ldots, \hat{\beta}_k$, given in Section A.7 of Appendix A, is

$$\ell \pm t_{.025} s \sqrt{\mathbf{a}'(\mathbf{X}'\mathbf{X})^{-1}\mathbf{a}}$$

where in our case, $\ell = \hat{y}_o - \hat{y}_b$ is the estimate of the difference in mean values for the two units, $E(y_o) - E(y_b)$; s is the least squares estimate of the standard deviation from the regression analysis of Model 3 (Figure 15.8); and $(\mathbf{X}'\mathbf{X})^{-1}$, the inverse matrix for the Model 3 regression analysis, is shown in Figure 15.15

(page 722). The **a** matrix is a column matrix containing elements $a_0, a_1, a_2,$ $\ldots, a_{13}$, where $a_0, a_1, a_2, \ldots, a_{13}$ are the coefficients of $\hat{\beta}_0, \hat{\beta}_1, \ldots, \hat{\beta}_{13}$ in the linear function ℓ, i.e.,

$$\ell = a_0\hat{\beta}_0 + a_1\hat{\beta}_1 + \cdots + a_{13}\hat{\beta}_{13}$$

Since our linear function is

$$\ell = \hat{\beta}_6 + \hat{\beta}_9 + \hat{\beta}_{10} + \hat{\beta}_{11} + \hat{\beta}_{12} + \hat{\beta}_{13}$$

it follows that $a_6 = a_9 = a_{10} = a_{11} = a_{12} = a_{13} = 1$ and $a_0 = a_1 = a_2 = a_3 = a_4 = a_5 = a_7 = a_8 = 0$.

Substituting the values of $\hat{\beta}_6, \hat{\beta}_9, \hat{\beta}_{10}, \ldots, \hat{\beta}_{13}$ (given in Figure 15.8) into ℓ, we have

$$\begin{aligned}
\ell = \hat{y}_o - \hat{y}_b &= \hat{\beta}_6 + \hat{\beta}_9 + \hat{\beta}_{10} + \hat{\beta}_{11} + \hat{\beta}_{12} + \hat{\beta}_{13} \\
&= 7{,}537.23 - 1{,}024.61 - 167.25 - 25.44 + 67.68 + 19.22 \\
&= \$6{,}406.83
\end{aligned}$$

The value of $t_{.025}$ needed for the confidence interval is approximately equal to 1.96 (because of the large number of degrees of freedom), and the value of s, given in Figure 15.8, is $s = 657.75$. Finally, the matrix product $\mathbf{a'(X'X)^{-1}a}$ can be obtained by multiplying the **a** matrix (described in the preceding paragraph) and the $\mathbf{(X'X)^{-1}}$ matrix given in Figure 15.15. Substituting these values into the formula for the confidence interval, we obtain (using a computer):

$$\overbrace{\hat{y}_o - \hat{y}_b}^{\ell} \pm t_{.025}s\sqrt{\mathbf{a'(X'X)^{-1}a}}$$
$$\$6{,}406.83 \pm \$2{,}733.46$$

Therefore, we estimate the difference in the mean sale prices of first-floor ocean-view and bay-view units (units 101 and 102) to lie within the interval \$3,673.37 to \$9,140.29.

You can use the technique described above to compare the mean sale prices for any pair of units.

15.8
Conclusions

You may be able to propose a better model for mean sale price than Model 3, but we think that Model 3 provides a good fit to the data. Further, it reveals some interesting information on the preferences of buyers of oceanside condominium units.

Lower floors are preferred on the ocean side; the closer they lie to the elevator and pool, the higher the estimated price. Some preference is given to the privacy of units located in the upper floor west-end.

Higher floors are preferred on the bay-view side (the side facing away from the ocean) with maximum preference given to units near the elevator (convenient and close to activities) and, to a lesser degree, to the privacy afforded by the west-end units.

FIGURE 15.15 SAS printout of the $(X'X)^{-1}$ matrix for Model 3

INVERSE	INTERCEP	X1	X2	X1X2	X1X1	X2X2	X3	X4
INTERCEP	9.265194	-2.99228	-0.25651	0.01708831	0.2368571	0.01335886	-9.26105	-0.00350909
X1	-2.99228	1.076628	-0.00585467	0.002321133	-0.090368	-0.00156845	2.995569	-0.00109312
X2	-0.25651	-0.00585467	0.1140368	-0.00896821	-0.004048291	-0.00445162	0.2560235	0.000412168
X1X2	0.01708831	0.002321133	-0.00896821	0.001375128	-0.000755996	.0000311222	-0.0170757	-0.000107253
X1X1	0.2368571	-0.090368	-0.004048291	-0.000755996	0.0001270988	-0.000393423	-0.23705	-0.000019801
X2X2	0.01335886	-0.00156845	-0.00445162	.0000311222	-0.000393423	0.001688068	-0.0133355	0.00020436
X3	-9.26105	2.995569	0.2560235	-0.0170757	-0.23705	-0.0133355	9.863521	-0.00346016
X4	-0.00350909	-0.00109312	0.000412168	-0.000107253	-0.000019801	0.00020436	-0.00346016	0.18837
X5	-0.0362161	-0.0112817	0.00425385	0.001688068	-0.00020436	0.001577825	0.03053109	0.004816517
X1X3	2.99394	-1.07611	0.005659513	-0.00231605	0.09029058	-0.0004476544	-3.16263	0.01236905
X2X3	0.2609271	-0.00225471	-0.114556	0.008981714	-0.00425418	-0.000129609	-0.325309	-0.020075
X1X1X3	-0.237302	0.09022947	-0.00399605	-0.00137448	-0.00790486	-0.000341465	0.2213208	-0.00318277
X2X2X3	-0.0137351	0.00145126	0.004495806	-0.000032272	-0.000109564	-0.000341465	0.2497183	0.001543755
PRICPAID	15390.19	310.3068	-203.352	26.08674	-10.8579	1.092063	7537.232	-1500.61

INVERSE	X5	X1X3	X2X3	X1X1X3	X2X2X3	PRICPAID
INTERCEP	-0.0362161	2.99394	0.2609271	-0.237302	-0.0137351	15390.19
X1	-0.0112817	-1.07611	-0.00225471	0.09022947	0.00145126	310.3068
X2	0.00425385	0.005659513	-0.114556	-0.00399605	0.004495806	-203.352
X1X2	0.001688068	-0.00231605	0.008981714	-0.00137448	-0.000032272	26.08674
X1X1	-0.00020436	0.09029058	-0.00425418	-0.00790486	-0.000109564	-10.8579
X2X2	-0.00020436	0.001577825	-0.0004476544	-0.000129609	-0.000341465	1.092063
X3	0.03053109	-3.16263	-0.325309	0.2213208	0.2497183	7537.232
X4	0.004816517	0.01236905	-0.020075	-0.00318277	0.001543755	-1500.61
X5	0.004970968	0.009001128	-0.000181965	-0.000107756	-0.0007207221	1075.381
X1X3	0.009001128	1.153105	-0.00425868	-0.0973915	0.0008693132	-1024.61
X2X3	-0.000181965	-0.00425868	0.1369923	-0.00972204	0.004412018	-167.246
X1X1X3	-0.000107756	-0.0973915	-0.00972204	0.008693132	0.0001387393	-25.4379
X2X2X3	-0.0007207221	0.0008693132	0.004412018	0.0001387393	0.0004420689	19.21642
PRICPAID	1075.381	-1024.61	-167.246	-25.4379	19.21642	38937244

EXERCISE

15.1 The data used in this study are presented in Appendix G.* If you have access to a computer, fit Models 1, 2, and 3 to the data set. Note that a computer program package other than the one used in the study may produce slightly different answers because of rounding.

ON YOUR OWN...

Postulate some models that you think might be an improvement over Model 3. Fit these models to the data set in Appendix G. Test to see whether they do, in fact, contribute more information for predicting sale price than Model 3

Reference

Kelting, H. "Investigation of condominium sale prices in three market scenarios: Utility of stepwise, interactive, multiple regression analysis and implications for design and appraisal methodology." Unpublished paper, University of Florida, Gainesville, 1979.

---

*Of the 200 units in the condominium complex, 106 were sold at auction and the remainder were sold (some more than once) at the developer's fixed price. The 106 units analyzed in this study are identified by observation (OBS) numbers 77–182 in Appendix G.

Modeling Daily Peak Electricity Demands

CONTENTS

16.1 The Problem

16.2 The Data

16.3 The Models

16.4 The Regression and Autoregression Analyses

16.5 Forecasting Daily Peak Electricity Demand

16.6 Summary

OBJECTIVE

To present a time series approach to modeling daily peak electricity demands on a power corporation and to show how to use the time series model for short-term forecasting

16.1

The Problem

To operate effectively, power companies must be able to predict daily peak demand for electricity. *Demand* (or *load*) is defined as the rate (measured in megawatts) at which electric energy is delivered to customers. Since demand is normally recorded on an hourly basis, daily peak demand refers to the maximum hourly demand in a 24-hour period. Power companies are continually developing and refining statistical models of daily peak demand.

Models of daily peak demand serve a twofold purpose. First, the models provide short-term *forecasts* that will assist in the economic planning and dispatching of electric energy. Second, models that relate peak demand to one or more weather variables provide estimates of historical peak demands under a set of alternative weather conditions. That is, since changing weather conditions represent the primary source of variation in peak demand, the model can be used to answer the often-asked question, "What would the peak daily demand have been had normal weather prevailed?" This second application, commonly referred to as *weather normalization*, is mainly an exercise in *backcasting* (i.e., adjusting historical data) rather than forecasting (Jacob, 1985).

Since peak demand is recorded over time (days), the dependent variable is a time series and one approach to modeling daily peak demand is to use a time series model. This chapter presents key results of a study designed to compare several alternative methods of modeling 1983 daily peak demand for the Florida Power Corporation (FPC). For this case study, we focus on two time series models and a multiple regression model proposed in the original FPC study. Then we demonstrate how to forecast daily peak demand using one of the time series models. (We leave the problem of backcasting as an exercise.)

16.2

The Data

The data for the study consist of daily observations on peak demand recorded by the FPC for the period beginning November 1, 1982, and ending October 31, 1983, and several factors that are known to influence demand. It is typically assumed that demand consists of two components, a non–weather-sensitive "base" demand that is not influenced by temperature changes and a weather-sensitive demand component that is highly responsive to changes in temperature.

The principal factor that affects the usage of non–weather-sensitive appliances (such as refrigerators, generators, lights, and computers) is the *day of the week*. Typically, Saturdays have lower peak demands than weekdays due to decreased commercial and industrial activity, whereas Sundays and holidays exhibit even lower peak demand levels as commercial and industrial activity declines even further.

The single most important factor affecting the usage of weather-sensitive appliances (such as heaters and air conditioners) is *temperature*. During the winter months, as temperatures drop below comfortable levels, customers begin to operate their electric heating units, thereby increasing the level of demand placed on the system. Similarly, during the summer months, as temperatures climb above comfortable levels, the use of air conditioning drives demand upward. Since the FPC serves 32 counties along west-central and northern Florida, it was necessary to capture temperature conditions from multiple weather stations. This was accomplished by identifying three primary weather stations within the FPC service area and recording the temperature value at the hour of peak demand each day at each station. A weighted average of these three daily temperatures was used to represent coincident temperature (i.e., temperature at the hour of peak demand) for the entire FPC service area, where the weights were proportional to the percentage of total electricity sales attributable to the weather zones surrounding each of the three weather stations.

To summarize, the independent variable (y_t) and the independent variables recorded for each of the 365 days of the November 1982–October 1983 year were

Dependent Variable:

y_t = Peak demand (in megawatts) observed on day t

Independent Variables:

Day of the week: Weekday, Saturday, or Sunday/holiday

Temperature: Coincident temperature (in degrees), i.e., the temperature recorded at the hour of the peak demand on day t, calculated as a weighted average of three daily temperatures.

16.3
The Models

In any modeling procedure, it is often helpful to plot the data in a scattergram. Figure 16.1 (page 728) shows a graph of the daily peak demand (y_t) from November 1982 through October 1983. The effects of seasonal weather on peak demand are readily apparent from the figure. One way to account for this seasonal variation is to include dummy variables for months or trigonometric terms in the model (refer to Section 9.6). However, since temperature is such a strong indicator of the weather, the FPC chose a simpler model with temperature as the sole seasonal weather variable.

Figure 16.2 (page 729) presents a scatterplot of daily peak demands versus coincident temperature. Note the nonlinear relationship that exists between the two variables. During the cool winter months, peak demand is inversely related to temperature; lower temperatures cause increased usage of heating equipment, which in turn cause higher peak demands. In contrast, the summer months reveal a positive relationship between peak demand and temperature; higher temperatures yield higher peak demands because of greater usage of air conditioners. You might think that a second-order (quadratic) model would be a good choice to account for the U-shaped distribution of peak demands shown in Figure 16.2. The FPC, however, rejected such a model for two reasons:

1. A quadratic model yields a symmetrical shape (i.e., a parabola) and would, therefore, not allow independent estimates of the winter and summer peak demand–temperature relationship.

2. In theory, there exists a mild temperature range where peak demand is assumed to consist solely of the non–weather-sensitive base demand component. For this range, a temperature change will not spur any additional heating or cooling and, consequently, has no impact on demand. The lack of linearity in the bottom portion of the U-shaped parabola fit by the quadratic model would yield overestimates of peak demand at the extremes of the mild temperature range and underestimates for temperatures in the middle of this range (see Figure 16.3 on page 729).

FIGURE 16.1 Daily peak megawatt demands, November 1982–October 1983

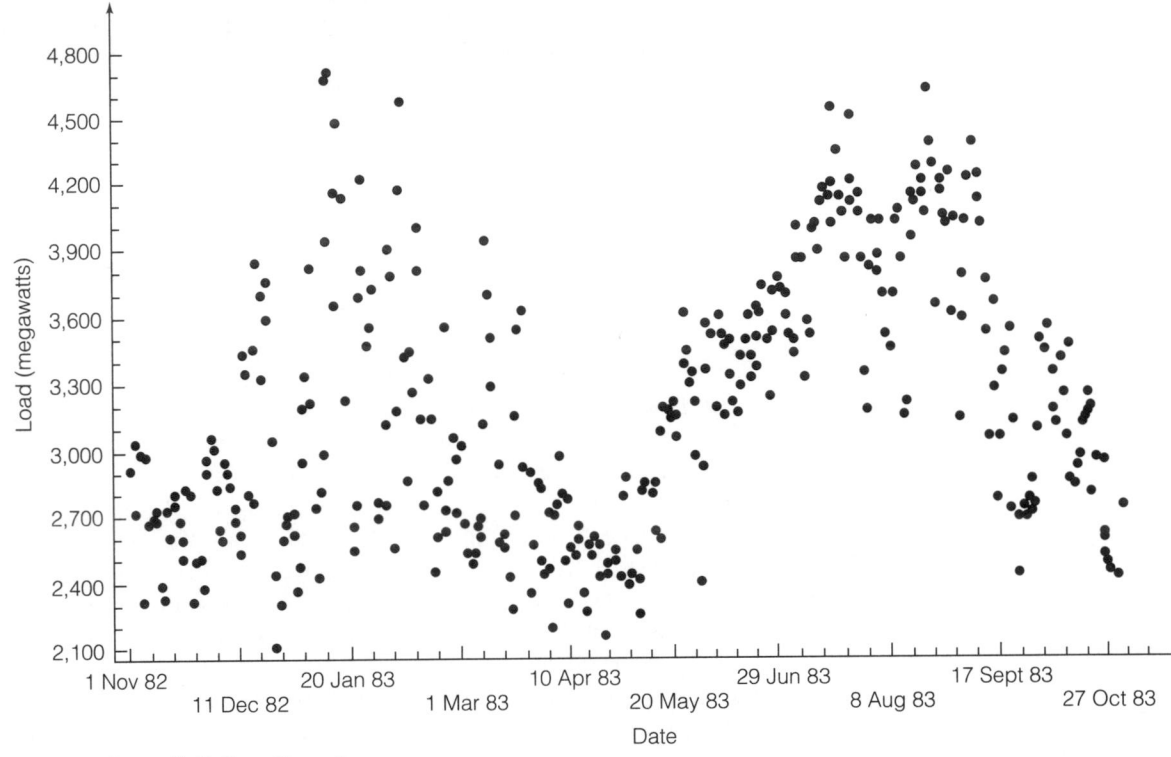

Source: Florida Power Corporation.

The solution was to model daily peak demand with a piecewise linear regression model (see Section 8.2). This approach has the advantage of allowing the peak demand–temperature relationship to vary between some prespecified temperature ranges, as well as providing a mechanism for joining the separate pieces.

Using the piecewise linear specification as the basic model structure, the following model of daily peak demand was proposed:

Model 1

$$y_t = \beta_0 + \underbrace{\beta_1(x_{1t} - 59)x_{2t} + \beta_2(x_{1t} - 78)x_{3t}}_{\text{Temperature}} + \underbrace{\beta_3 x_{4t} + \beta_4 x_{5t}}_{\text{Day of the week}} + \varepsilon_t$$

where

x_{1t} = Coincident temperature on day t

$x_{2t} = \begin{cases} 1 & \text{if } x_{1t} < 59 \\ 0 & \text{if not} \end{cases}$ $x_{3t} = \begin{cases} 1 & \text{if } x_{1t} > 78 \\ 0 & \text{if not} \end{cases}$

$x_{4t} = \begin{cases} 1 & \text{if Saturday} \\ 0 & \text{if not} \end{cases}$ $x_{5t} = \begin{cases} 1 & \text{if Sunday or holiday} \\ 0 & \text{if not} \end{cases}$ (Base level = Weekday)

ε_t = Uncorrelated error term

FIGURE 16.2   Daily peak demand versus temperature, November 1982–October 1983

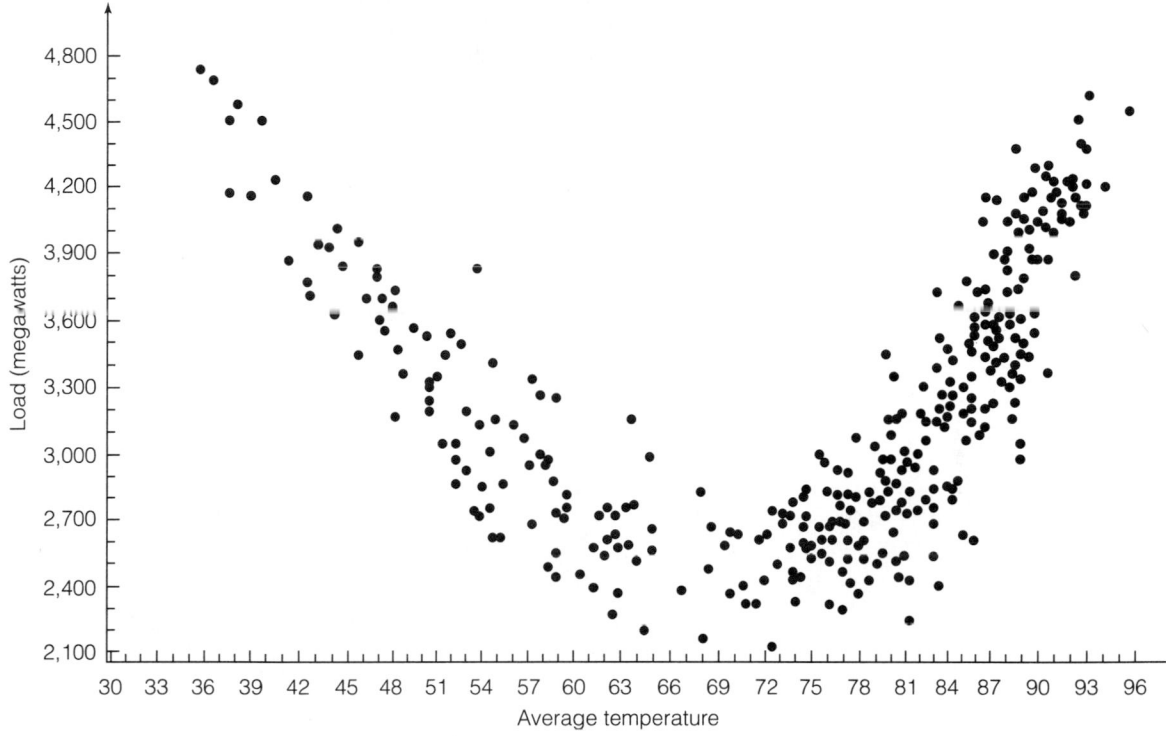

Source: Florida Power Corporation.

FIGURE 16.3

Theoretical relationship between daily peak demand and temperature

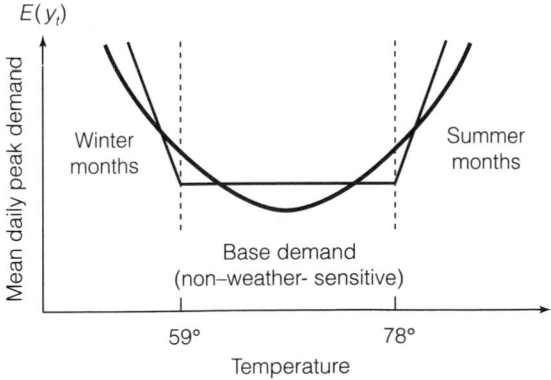

Model 1 proposes three different straight-line relationships between peak demand (y_t) and coincident temperature x_{1t}, one for each of the three temperature ranges corresponding to winter months (less than 59°), non–weather-sensitive

months (between 59° and 78°), and summer months (greater than 78°).* The model also allows for variations in demand because of day of the week (Saturday, Sunday/holiday, or weekday). Since interaction between temperature and day of the week is omitted, the model is assuming that the differences between mean peak demand for weekdays and weekends/holidays is constant for the winter-sensitive, summer-sensitive, and non–weather-sensitive months.

We will illustrate the mechanics of the piecewise linear terms by finding the equations of the three demand–temperature lines for weekdays (i.e., $x_{4t} = x_{5t} = 0$). Substituting $x_{4t} = 0$ and $x_{5t} = 0$ into the model, we have

Winter-sensitive months ($x_{1t} < 59°$, $x_{2t} = 1$, $x_{3t} = 0$):

$$E(y_t) = \beta_0 + \beta_1(x_{1t} - 59)(0) + \beta_2(x_{1t} - 78)(1) + \beta_3(0) + \beta_4(0)$$
$$= \beta_0 + \beta_1(x_{1t} - 59)$$
$$= (\beta_0 - 59\beta_1) + \beta_1 x_{1t}$$

Summer-sensitive months ($x_{1t} > 78°$, $x_{2t} = 0$, $x_{3t} = 1$):

$$E(y_t) = \beta_0 + \beta_1(x_{1t} - 59)(0) + \beta_2(x_{1t} - 78)(1) + \beta_3(0) + \beta_4(0)$$
$$= \beta_0 + \beta_2(x_{1t} - 78)$$
$$= (\beta_0 - 78\beta_2) + \beta_2 x_{1t}$$

Non–weather-sensitive months ($59° \leq x_{1t} \leq 78°$, $x_{2t} = x_{3t} = 0$):

$$E(y_t) = \beta_0 + \beta_1(x_{1t} - 59)(0) + \beta_2(x_{1t} - 78)(0) + \beta_3(0) + \beta_4(0)$$
$$= \beta_0$$

Note that the slope of the demand–temperature line for winter-sensitive months (when $x_{1t} < 59$) is β_1 (which we expect to be negative), whereas the slope for summer-sensitive months (when $x_{1t} > 78$) is β_2 (which we expect to be positive). The intercept term β_0 represents the mean daily peak demand observed in the non–weather-sensitive period (when $59 \leq x_{1t} \leq 78$). Notice also that the peak demand during non–weather-sensitive days does not depend on temperature (x_{1t}).

Model 1 is a multiple regression model that relies on the standard regression assumptions of independent errors (ε_t uncorrelated). This may be a serious shortcoming in view of the fact that the data are in the form of a time series. To account for possible autocorrelated residuals, two time series models were proposed:

Model 2

$$y_t = \beta_0 + \beta_1(x_{1t} - 59)x_{2t} + \beta_2(x_{1t} - 78)x_{3t} + \beta_3 x_{4t} + \beta_4 x_{5t} + R_t$$
$$R_t = \phi_1 R_{t-1} + \varepsilon_t$$

---

*The temperature values, 59° and 78°, identify where the winter- and summer-sensitive portions of demand join the base demand component. These "knot values" were determined from visual inspection of the graph in Figure 16.2.

Model 2 proposes a regression–autoregression pair of models for daily peak demand (y_t). The deterministic component, $E(y_t)$, is identical to the deterministic component of Model 1; however, a first-order autoregressive model is chosen for the random error component. Recall (from Section 9.5) that a first-order autoregressive model is appropriate when the correlation between residuals diminishes as the distance between time periods (in this case, days) increases.

Model 3

$$y_t = \beta_0 + \beta_1(x_{1t} - 59)x_{2t} + \beta_2(x_{1t} - 78)x_{3t} + \beta_3 x_{4t} + \beta_4 x_{5t} + R_t$$
$$R_t = \phi_1 R_{t-1} + \phi_2 R_{t-2} + \phi_5 R_{t-5} + \phi_7 R_{t-7} + \varepsilon_t$$

Model 3 extends the first-order autoregressive error model of Model 2 to a seventh-order autoregressive model with lags at 1, 2, 5, and 7. In theory, the peak demand on day t will be highly correlated with the peak demand on day $t + 1$. However, there also may be significant correlation between demand 2 days, 5 days, and/or 1 week (7 days) apart. This more general error model is proposed to account for any residual correlation that may occur as a result of the week-to-week variation in peak demand, in addition to the day-to-day variation.

16.4

The Regression and Autoregression Analyses

The multiple regression computer printout for Model 1 is shown in Figure 16.4, and a plot of the least squares fit is shown in Figure 16.5 (page 732). The model appears to provide a good fit to the data, with $R^2 = .8307$ and $F = 441.729$ (significant at $p = .0001$). The value of **ROOT MSE**, $s = 245.585$, implies that we can expect to predict daily peak demand accurate to within $2s \approx 491$ megawatts of its true value. However, we must be careful not to conclude at this point that the model is useful for predicting peak demand. Recall that in the presence of autocorrelated residuals, the standard errors of the regression coefficients are

FIGURE 16.4

SAS printout for multiple regression model of daily peak demand, Model 1

```
DEP VARIABLE: LOAD

                   SUM OF          MEAN
SOURCE      DF     SQUARES         SQUARE       F VALUE      PROB > F
MODEL        4     106565982       26641496     441.729      0.0001
ERROR      360      21712247       60311.797
C TOTAL    364     128278229

       ROOT MSE      245.585       R-SQUARE
       DEP MEAN     3191.863       ADJ R-SQ              0.8289
       C.V.            7.694083

                   PARAMETER       STANDARD     T FOR H0:
VARIABLE DF        ESTIMATE        ERROR        PARAMETER = 0    PROB > !T!
INTERCEP  1         2670.171       21.251829     125.644         0.0001
AVTW      1          -82.039853     2.941928     -27.886         0.0001
AVTS      1          114.443        3.050468      37.516         0.0001
SAT       1         -164.932       37.990216      -4.341         0.0001
SUN       1         -285.114       35.328293      -8.070         0.0001

DURBIN-WATSON D                     0.705
(NUMBER OF OBS)                       365
1ST ORDER AUTOCORRELATION           0.648
```

FIGURE 16.5 Daily peak demand versus temperature: actual versus fitted piecewise linear model

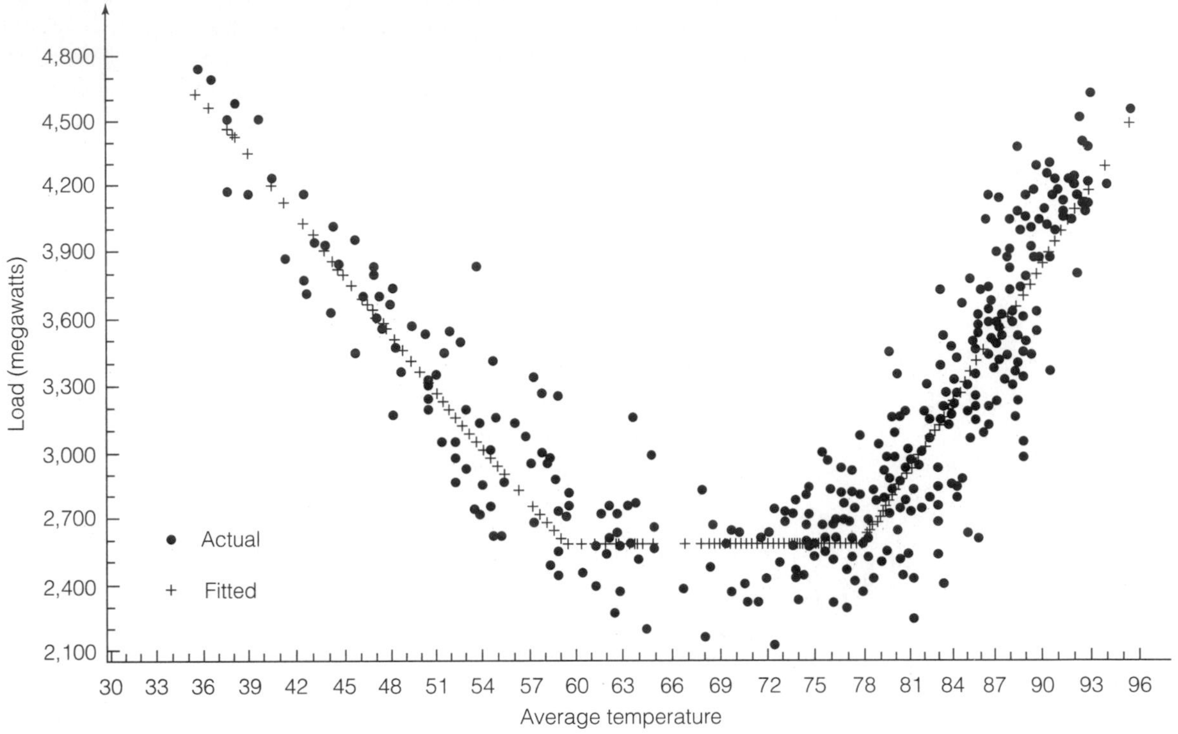

Source: Florida Power Corporation.

underestimated, thereby inflating the corresponding t statistics for testing H_0: β_i = 0. At worst, this could lead to the false conclusion that a β parameter is significantly different from 0; at best, the results, although significant, give an overoptimistic view of the predictive ability of the model.

To determine whether the residuals of the multiple regression model are positively autocorrelated, we conduct the Durbin–Watson test:

H_0: Uncorrelated residuals

H_a: Positive residual correlation

Recall that the Durbin–Watson test is designed specifically for detecting first-order autocorrelation in the residuals, R_t. Thus, we can write the null and alternative hypotheses as

H_0: $\phi_1 = 0$

H_a: $\phi_1 > 0$

where $R_t = \phi_1 R_{t-1} + \varepsilon_t$, and ε_t = uncorrelated error (white noise).

The test statistic, given at the bottom of the printout in Figure 16.4, is $d = .705$. Recall that small values of d lead us to reject H_0: $\phi_1 = 0$ in favor of the alternative H_a: $\phi_1 > 0$. For $\alpha = .01$, $n = 365$, and $k = 4$ (the number of β parameters in the model, excluding β_0), the lower bound on the critical value (obtained from Table 9 of Appendix D) is approximately $d_L = 1.46$. Since the value of the test statistic, $d = .705$, falls well below the lower bound, there is strong evidence at $\alpha = .01$ of positive (first-order) autocorrelated residuals. Thus, we need to incorporate terms for residual autocorrelation into the model.

The time series printouts for Models 2 and 3 are shown in Figures 16.6 and 16.7 (page 734), respectively. A summary of the results for all three models is given in Table 16.1.

TABLE 16.1 **Summary of Results for Models 1, 2, and 3**

	MODEL 1	MODEL 2	MODEL 3
R^2	.8307	.9225	.9351
MSE	60,311.797	27,687.44	23,398.43
s	245.585	166.394	152.966

FIGURE 16.6 SAS printout for first-order autoregressive time series model of daily peak demand, Model 2

```
                    DEPENDENT VARIABLE = LOAD

                 ORDINARY LEAST SQUARES ESTIMATES

                    VARIABLE  DF      B VALUE
                    INTERCPT  1      2670.171
                    AVTW      1       -82.0399
                    AVTS      1       114.4427
                    SAT       1      -164.932
                    SUN       1      -285.114


                 ESTIMATES OF AUTOCORRELATIONS

  LAG    COVARIANCE    CORRELATION   -1 9 8 7 6 5 4 3 2 1 0 1 2 3 4 5 6 7 8 9 1
   0      59485.6       1.000000      :                    |********************|
   1      38519.4       0.647541      :                    |************        :


                 PRELIMINARY MSE =      34542.75

              ESTIMATES OF THE AUTOREGRESSIVE PARAMETERS

            LAG      COEFFICIENT        STD ERROR      T RATIO
             1      -0.64754083          0.039887    -16.234581
                 YULE-WALKER ESTIMATES
          SSE          9939789              DFE            359
          MSE          27687.44             ROOT MSE   166.3943
          REG RSQ       0.7626              TOTAL RSQ    0.9225


  VARIABLE DF        B VALUE          STD ERROR      T RATIO    APPROX PROB
  INTERCPT  1    2812.96710162     29.8790879359      94.145      0.0001
  AVTW      1     -65.337453        2.6639248330     -24.527      0.0001
  AVTS      1      83.45523858      3.8531993234      21.659      0.0001
  SAT       1    -130.82831        22.4136023517      -5.837      0.0001
  SUN       1    -275.55071        21.3736779578     -12.892      0.0001
```

FIGURE 16.7

SAS printout for seventh-order autoregressive time series model of daily peak demand, Model 3

DEPENDENT VARIABLE = LOAD

ORDINARY LEAST SQUARES ESTIMATES

VARIABLE	DF	B VALUE
INTERCPT	1	2670.171
AVTW	1	-82.0399
AVTS	1	114.4427
SAT	1	-164.932
SUN	1	-285.114

ESTIMATES OF AUTOCORRELATIONS

LAG	COVARIANCE	CORRELATION	-1 9 8 7 6 5 4 3 2 1 0 1 2 3 4 5 6 7 8 9 1
0	59485.6	1.000000	⏐*******************⏐
1	38519.4	0.647541	⏐************ ⏐
2	35741	0.600834	⏐*********** ⏐
3	32868.2	0.552540	⏐********** ⏐
4	29917.9	0.502943	⏐********* ⏐
5	31340.9	0.526865	⏐********** ⏐
6	30061.9	0.505364	⏐********* ⏐
7	31508	0.529674	⏐********** ⏐

PRELIMINARY MSE =   28841.4

ESTIMATES OF THE AUTOREGRESSIVE PARAMETERS

LAG	COEFFICIENT	STD ERROR	T RATIO
1	-0.36793644	0.049902	-7.373164
2	-0.20702784	0.051722	-4.002705
3	0.00000000	0.000000	
4	0.00000000	0.000000	
5	-0.13526445	0.049072	-2.756459
6	0.00000000	0.000000	
7	-0.15338478	0.048430	-3.167144

EXPECTED AUTOCORRELATIONS

LAG	AUTOCORR
0	0.9668
1	0.6117
2	0.5640
3	0.4794
4	0.4494
5	0.4819
6	0.4469
7	0.4888

YULE-WALKER ESTIMATES

SSE	8329842	DFE	356	
MSE	23398.43	ROOT MSE	152.9655	
REG RSQ	0.8112	TOTAL RSQ	0.9351	

VARIABLE	DF	B VALUE	STD ERROR	T RATIO	APPROX PROB
INTERCPT	1	2809.95021301	58.2345577940	48.252	0.0001
AVTW	1	-71.28248	2.2620998226	-31.512	0.0001
AVTS	1	79.12014515	4.1805721109	18.926	0.0001
SAT	1	-150.52375	23.4727721121	-6.413	0.0001
SUN	1	-262.27342	21.6832940758	-12.096	0.0001

The addition of the first-order autoregressive error term in Model 2 yielded a drastic improvement to the fit of the model. The value of R^2 (i.e., **TOTAL RSQ**) increased from .83 for Model 1 to .92, and the standard deviation s (i.e., **ROOT MSE**) decreased from 245.6 to 166.4. These results support the conclusion reached by the Durbin–Watson test—namely, that the first-order autoregressive lag parameter ϕ_1 is significantly different from 0.

Does the more general autoregressive error model (Model 3) provide a better approximation to the pattern of correlation in the residuals than the first-order autoregressive model (Model 2)? To test this hypothesis, we would need to test H_0: $\phi_2 = \phi_5 = \phi_7 = 0$. Although we omit discussion of tests on autoregressive parameters in this text,* we can arrive at a decision from a pragmatic point of view by again comparing the values of R^2 and s for the two models. The more complex autoregressive model proposed by Model 3 yields a slight increase in R^2 (.935 compared to .923 for Model 2) and a slight decrease in the value of s (153.00 compared to 166.4 for Model 2). The additional lag parameters, although they may be statistically significant, may not be practically significant. The practical analyst may decide that the first-order autoregressive process proposed by Model 2 is the more desirable option since it is easier to use to forecast peak daily demand (and therefore more explainable to managers) while yielding approximate prediction errors (as measured by $2s$) that are only slightly larger than those for Model 3.

For the purposes of illustration, we use Model 2 to forecast daily peak demand in the following section.

16.5

Forecasting Daily Peak Electricity Demand

Suppose the FPC decided to use Model 2 to forecast daily peak demand for the first seven days of November 1983. The estimated model,[†] obtained from Figure 16.6, is given by

$$\hat{y}_t = 2{,}812.967 - 65.337(x_{1t} - 59)x_{2t} + 83.455(x_{1t} - 78)x_{3t}$$
$$- 130.828x_{4t} - 275.551x_{5t} + \hat{R}_t$$

$$\hat{R}_t = .6475\hat{R}_{t-1}$$

The forecast for November 1, 1983 ($t = 366$), requires an estimate of the residual R_{365}, where $\hat{R}_{365} = y_{365} - \hat{y}_{365}$. The last day of the November 1982–October 1983 time period ($t = 365$) was October 31, 1983, a Monday. On this day the peak demand was recorded as $y_{365} = 2{,}752$ megawatts and the coincident temperature as $x_{1,365} = 77°$. Substituting the appropriate values of the dummy variables into the equation for $\hat{y}_t$ (i.e., $x_{2t} = 0$, $x_{3t} = 0$, $x_{4t} = 0$, and $x_{5t} = 0$),

---

*For details of tests on autoregressive parameters, see Fuller (1976).

†Remember that the estimate of ϕ_1 is obtained by multiplying the value reported on the SAS printout by (-1).

we have

$$
\begin{aligned}
\hat{R}_{365} &= y_{365} - \hat{y}_{365} \\
&= 2{,}752 - [2{,}812.967 - 65.337(77 - 59)(0) + 83.455(77 - 78)(0) \\
&\qquad\qquad\qquad - 130.828(0) - 275.551(0)] \\
&= 2{,}752 - 2{,}812.967 = -60.967
\end{aligned}
$$

Then the formula for calculating the forecast for Tuesday, November 1, 1983, is

$$
\begin{aligned}
\hat{y}_{366} &= 2{,}812.967 - 65.337(x_{1,366} - 59)x_{2,366} + 83.455(x_{1,366} - 78)x_{3,366} \\
&\qquad - 130.828x_{4,366} - 275.551x_{5,366} + \hat{R}_{366}
\end{aligned}
$$

where

$$
\hat{R}_{366} = \hat{\phi}_1 \hat{R}_{365} = (.6475)(-60.967) = -39.476
$$

Note that the forecast requires an estimate of coincident temperature on that day, $\hat{x}_{1,366}$. If the FPC wants to forecast demand under normal weather conditions, then this estimate can be obtained from historical data for that day. Or, the FPC may choose to rely on a meteorologist's weather forecast for that day. For this example, assume that $\hat{x}_{1,366} = 76°$ (the actual temperature recorded by the FPC). Then $x_{2,366} = x_{3,366} = 0$ (since $59 \leq \hat{x}_{1,366} \leq 78$) and $x_{4,366} = x_{5,366} = 0$ (since the target day is a Tuesday). Substituting these values and the value of $\hat{R}_{366}$ into the equation, we have

$$
\begin{aligned}
\hat{y}_{366} &= 2{,}812.967 - 65.337(76 - 59)(0) + 83.455(76 - 78)(0) \\
&\qquad\qquad - 130.828(0) - 275.551(0) - 39.476 \\
&= 2{,}773.49
\end{aligned}
$$

Similarly, a forecast for Wednesday, November 2, 1983 (i.e., $t = 367$), can be obtained:

$$
\begin{aligned}
\hat{y}_{367} &= 2{,}812.967 - 65.337(x_{1,367} - 59)x_{2,367} + 83.455(x_{1,367} - 78)x_{3,367} \\
&\qquad - 130.828x_{3,367} - 275.551x_{4,367} + \hat{R}_{367}
\end{aligned}
$$

where $\hat{R}_{367} = \hat{\phi}_1 \hat{R}_{366} = (.6475)(-39.476) = -25.561$, and $x_{3,367} = x_{4,367} = 0$. For an estimated coincident temperature of $\hat{x}_{1,367} = 77°$ (again, this is the actual temperature recorded on that day), we have $x_{2,367} = 0$ and $x_{3,367} = 0$. Substituting these values into the prediction equation, we obtain

$$
\begin{aligned}
\hat{y}_{367} &= 2{,}812.967 - 65.337(77 - 59)(0) + 83.455(77 - 78)(0) \\
&\qquad\qquad - 130.828(0) - 275.551(0) - 25.561 \\
&= 2{,}812.967 - 25.561 \\
&= 2{,}787.41
\end{aligned}
$$

Approximate 95% prediction intervals for the two forecasts are calculated as follows:

Tuesday, Nov. 1, 1983:

$$\hat{y}_{366} \pm 1.96 \sqrt{MSE}$$

$$= 2{,}773.49 \pm 1.96\sqrt{27{,}687.44}$$

$$= 2{,}773.49 \pm 326.14 \quad \text{or} \quad (2{,}447.35, 3{,}099.63)$$

Wednesday, Nov. 2, 1983:

$$\hat{y}_{367} \pm 1.96 \sqrt{MSE(1 + \hat{\phi}_1^2)}$$

$$= 2{,}787.41 \pm 1.96\sqrt{(27{,}687.44)[1 + (.6475)^2]}$$

$$= 2{,}787.41 \pm 388.53 \quad \text{or} \quad (2{,}398.88, 3{,}175.94)$$

The forecasts, approximate 95% prediction intervals, and actual daily peak demands (recorded by the FPC) for the first seven days of November 1983 are given in Table 16.2. Note that actual peak demand y_t falls within the corresponding prediction interval for all seven days. Thus, the model appears to be useful for making short-term forecasts of daily peak demand. Of course, if the prediction intervals were extremely wide, this result would be of no practical value. For example, the forecast error $y_t - \hat{y}_t$, measured as a percentage of the actual value y_t, may be large even when y_t falls within the prediction interval. Various techniques, such as the percent forecast error, are available for evaluating the accuracy of forecasts. Consult the references given at the end of Chapter 9 for details on these techniques.

T A B L E 16.2   **Forecasts and Actual Peak Demands for the First Seven Days of November 1983**

DATE	t	FORECAST $\hat{y}_t$	APPROXIMATE 95% PREDICTION INTERVAL	ACTUAL DEMAND y_t	ACTUAL TEMPERATURE x_{1t}
Tues., Nov. 1	366	2,773.49	(2,447.35, 3,099.63)	2,799	76
Wed., Nov. 2	367	2,787.41	(2,398.88, 3,175.94)	2,784	77
Thurs., Nov. 3	368	2,796.42	(2,384.53, 3,208.31)	2,845	77
Fri., Nov. 4	369	2,802.25	(2,380.92, 3,223.58)	2,701	76
Sat., Nov. 5	370	2,675.20	(2,249.97, 3,100.43)	2,512	72
Sun., Nov. 6	371	2,532.92	(2,106.07, 2,959.77)	2,419	71
Mon., Nov. 7	372	2,810.06	(2,382.59, 3,237.53)	2,749	68

16.6 Summary

This case study presents a time series approach to modeling and forecasting daily peak demand observed at Florida Power Corporation. A graphical analysis of the data provided the means of identifying and formulating a piecewise linear regression model relating peak demand to temperature and day of the week. The multiple regression model, although providing a good fit to the data, exhibited strong signs of positive residual autocorrelation.

Two autoregressive time series models were proposed to account for the auto-correlated errors. Both models were shown to provide a drastic improvement in model adequacy. Either could be used to provide reliable short-term forecasts of daily peak demand or for weather normalization (i.e., estimating the peak demand if normal weather conditions had prevailed).

EXERCISES

16.1   All three models discussed in this case study make the underlying assumption that the peak demand–temperature relationship is independent of day of the week. Write a model that includes interaction between temperature and day of the week. Show the effect the interaction has on the straight-line relationships between peak demand and temperature. Explain how you could test the significance of the interaction terms.

16.2   Consider the problem of using Model 2 for weather normalization. Suppose the temperature on Saturday, March 5, 1983 (i.e., $t = 125$), was $x_{1,125} = 25°$, unusually cold for that day. Normally, temperatures range from 40° to 50° on March 5 in the FPC service area. Substitute $x_{1,125} = 45°$ into the prediction equation to obtain an estimate of the peak demand expected if normal weather conditions had prevailed on March 5, 1983. Calculate an approximate 95% prediction interval for the estimate. [*Hint*:   Use $\hat{y}_{125} \pm 1.96\sqrt{\text{MSE}}$.]

References

Fuller, W. A. *Introduction to Statistical Time Series*. New York: Wiley, 1976.

Jacob, M. F. "A time series approach to modeling daily peak electricity demands." Paper presented at the SAS Users Group International Annual Conference, Reno, Nevada, 1985.

The Mechanics of a Multiple Regression Analysis

CONTENTS

A.1 Introduction

A.2 Matrices and Matrix Multiplication

A.3 Identity Matrices and Matrix Inversion

A.4 Solving Systems of Simultaneous Linear Equations

A.5 The Least Squares Equations and Their Solutions

A.6 Calculating SSE and s^2

A.7 Standard Errors of Estimators, Test Statistics, and Confidence Intervals for $\beta_0, \beta_1, \ldots, \beta_k$

A.8 A Confidence Interval for a Linear Function of the β Parameters; A Confidence Interval for $E(y)$

A.9 A Prediction Interval for Some Value of y to Be Observed in the Future

A.1

Introduction

The rationale behind a multiple regression analysis and the types of inferences it permits you to make were the subjects of Chapter 4. We noted that the method of least squares most often leads to a very difficult computational problem, namely, the solution of a set of $(k + 1)$ simultaneous linear equations in the unknown values of the estimates $\hat{\beta}_0, \hat{\beta}_1, \ldots, \hat{\beta}_k$, and that the formulas for the estimated standard errors $s_{\hat{\beta}_0}, s_{\hat{\beta}_1}, \ldots, s_{\hat{\beta}_k}$ were too complicated to express as ordinary algebraic formulas. We circumvented both these problems in a very easy way— we relied on the least squares estimates, confidence intervals, tests, etc., provided by a standard regression analysis software package. Thus, Chapter 4 provides a basic working knowledge of the types of inferences you might wish to make from a multiple regression analysis and explains how to interpret the results. If we can do this, why would we wish to know the actual process performed by the computer?

There are several answers to this question:

1. Some multiple regression statistical software packages do not print all the information you may want. As one illustration, we noted in Chapter 4 that very often the objective of a regression analysis is to develop a prediction equation that can be used to estimate the mean value of y (say, mean profit or mean yield) for given values of the predictor variables $x_1, x_2, \ldots, x_k$. Some software packages do not give the confidence interval for $E(y)$ or a prediction interval for y. Thus, you might need to know how to find the necessary quantities from the analysis and perform the computations yourself.

2. A multiple regression software package may possess the capability of computing some specific quantity that you desire, but you may find the instructions on how to "call" for this special calculation difficult to understand. It may be easier to identify the components required for your computation and do it yourself.

3. For some designed experiments, finding the least squares equations and solving them is a trivial operation. Understanding the process by which the least squares equations are generated and understanding how they are solved will help you understand how experimental design affects the results of a regression analysis. Thus, a knowledge of the computations involved in performing a regression analysis will help you to better understand the contents of Chapters 10 and 11.

To summarize, "knowing how it is done" is not essential for performing an ordinary regression analysis or interpreting its results. But "knowing how" helps, and it is essential for an understanding of many of the finer points associated with a multiple regression analysis. This appendix explains "how it is done" without getting into the unpleasant task of performing the computations for solving the least squares equations. This mechanical and tedious procedure can be left to a computer (the solutions are verifiable). We illustrate the procedure in Appendix B.

A.2
Matrices and Matrix Multiplication

Although it is very difficult to give the formulas for the multiple regression least squares estimators and for their estimated standard errors in ordinary algebra, it is easy to do so using **matrix algebra**. Thus, by arranging the data in particular rectangular patterns called **matrices** and by performing various operations with them, we can obtain the least squares estimates and their estimated standard errors. In this section and Sections A.3 and A.4, we will define what we mean by a matrix and explain various operations that can be performed with matrices. We will explain how to use this information to conduct a regression analysis in Section A.5.

Three matrices, **A**, **B**, and **C**, are shown below. Note that each matrix is a rectangular arrangement of numbers with one number in every row–column position.

$$A = \begin{bmatrix} 2 & 3 \\ 0 & 1 \\ -1 & 6 \end{bmatrix} \quad B = \begin{bmatrix} 3 & 0 & 1 \\ -1 & 0 & 1 \\ 4 & 2 & 0 \end{bmatrix} \quad C = \begin{bmatrix} 1 \\ 2 \\ 1 \end{bmatrix}$$

> ### Definition A.1
>
> A **matrix** is a rectangular array of numbers.*

The numbers that appear in a matrix are called **elements** of the matrix. If a matrix contains r rows and c columns, there will be an element in each of the row–column positions of the matrix, and the matrix will have $r \times c$ elements. For example, the matrix **A** shown above contains $r = 3$ rows, $c = 2$ columns, and $rc = (3)(2) = 6$ elements, one in each of the six row–column positions.

> ### Definition A.2
>
> A number in a particular row–column position is called an **element** of the matrix.

Notice that the matrices **A**, **B**, and **C** contain different numbers of rows and columns. The numbers of rows and columns give the **dimensions** of a matrix.

When we give a formula in matrix notation, the elements of a matrix will be represented by symbols. For example, if we have a matrix

$$\mathbf{A} = \begin{bmatrix} a_{11} & a_{12} & a_{13} \\ a_{21} & a_{22} & a_{23} \end{bmatrix}$$

the symbol a_{ij} will denote the element in the ith row and jth column of the matrix. The first subscript always identifies the row and the second identifies the column in which the element is located. For example, the element a_{12} is in the first row and second column of the matrix **A**. The rows are always numbered from top to bottom, and the columns are always numbered from left to right.

> ### Definition A.3
>
> A matrix containing r rows and c columns is said to be an $r \times c$ **matrix** where r and c are the **dimensions** of the matrix.

> ### Definition A.4
>
> If $r - c$, a matrix is said to be a **square matrix**.

---

*For our purpose, we assume that the numbers are real.

Matrices are usually identified by capital letters, such as **A**, **B**, **C**, corresponding to the letters of the alphabet employed in ordinary algebra. The difference is that in ordinary algebra, a letter is used to denote a single real number, whereas in matrix algebra, *a letter denotes a rectangular array of numbers.* The operations of matrix algebra are very similar to those of ordinary algebra—you can add matrices, subtract them, multiply them, and so on. But since we are concerned only with the applications of matrix algebra to the solution of the least squares equations, we will define only the operations and types of matrices that are pertinent to that subject.

The most important operation for us is matrix multiplication, which requires **row–column multiplication**. To illustrate this process, suppose we wish to find the product **AB**, where

$$A = \begin{bmatrix} 2 & 1 \\ 4 & -1 \end{bmatrix} \qquad B = \begin{bmatrix} 2 & 0 & 3 \\ -1 & 4 & 0 \end{bmatrix}$$

We will always multiply the rows of **A** (the matrix on the left) by the columns of **B** (the matrix on the right). The product formed by the first row of **A** times the first column of **B** is obtained by multiplying the elements in corresponding positions and summing these products. Thus, the first row, first column product, shown diagrammatically here, is

$$(2)(2) + (1)(-1) = 4 - 1 = 3$$

$$AB = \begin{bmatrix} 2 & 1 \\ 4 & -1 \end{bmatrix} \begin{bmatrix} 2 & 0 & 3 \\ -1 & 4 & 0 \end{bmatrix} = \begin{bmatrix} 3 & \\ & \end{bmatrix}$$

Similarly, the first row, second column product is

$$(2)(0) + (1)(4) = 0 + 4 = 4$$

So far we have

$$AB = \begin{bmatrix} 3 & 4 & \\ & & \end{bmatrix}$$

To find the complete matrix product **AB**, all we need to do is find each element in the **AB** matrix. Thus, we will define an element in the ith row, jth column of **AB** as the product of the ith row of **A** and the jth column of **B**. We complete the process in Example A.1.

EXAMPLE A.1

Find the product **AB**, where

$$A = \begin{bmatrix} 2 & 1 \\ 4 & -1 \end{bmatrix} \qquad B = \begin{bmatrix} 2 & 0 & 3 \\ -1 & 4 & 0 \end{bmatrix}$$

Solution

If we represent the product **AB** as

$$C = \begin{bmatrix} c_{11} & c_{12} & c_{13} \\ c_{21} & c_{22} & c_{23} \end{bmatrix}$$

we have already found $c_{11} = 3$ and $c_{12} = 4$. Similarly, the element c_{21}, the element in the second row, first column of **AB**, is the product of the second row of **A** and the first column of **B**:

$$(4)(2) + (-1)(-1) = 8 + 1 = 9$$

Proceeding in a similar manner to find the remaining elements of **AB**, we have

$$\mathbf{AB} = \begin{bmatrix} 2 & 1 \\ 4 & -1 \end{bmatrix} \begin{bmatrix} 2 & 0 & 3 \\ -1 & 4 & 0 \end{bmatrix} = \begin{bmatrix} 3 & 4 & 6 \\ 9 & -4 & 12 \end{bmatrix}$$

Now, try to find the product **BA**, using matrices **A** and **B** from Example A.1. You will observe two very important differences between multiplication in matrix algebra and multiplication in ordinary algebra:

1. You cannot find the product **BA** because you cannot perform row–column multiplication. You can see that the dimensions do not match by placing the matrices side-by-side.

$$\underset{2 \times 3 \quad 2 \times 2}{\overset{\mathbf{BA}}{\diagup \diagdown}} \qquad \text{does not exist}$$

The number of elements (3) in a row of **B** (the matrix on the left) does not match the number of elements (2) in a column of **A** (the matrix on the right). Therefore, you cannot perform row–column multiplication, and the matrix product **BA** does not exist. The point is, not all matrices can be multiplied. You can find products for matrices **A** and **B** only when **A** is $r \times d$ and **B** is $d \times c$. That is:

Requirement for Multiplication

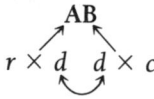

The two inner dimension numbers must be equal. The dimensions of the product will always be given by the outer dimension numbers:

Dimensions of **AB** Are $r \times c$

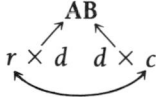

2. The second difference between ordinary and matrix multiplication is that in ordinary algebra, $ab = ba$. In matrix algebra, AB usually does not equal BA. In fact, as noted in item 1 above, it may not even exist.

Definition A.5

The product AB of an $r \times d$ matrix A and a $d \times c$ matrix B is an $r \times c$ matrix C, where the element c_{ij} ($i = 1, 2, \ldots, r; j = 1, 2, \ldots, c$) of C is the product of the ith row of A and the jth column of B.

EXAMPLE A.2

Given the matrices below, find IA and IB.

$$A = \begin{bmatrix} 2 \\ 1 \\ 3 \end{bmatrix} \quad B = \begin{bmatrix} 3 & 0 \\ 1 & 2 \\ 4 & -1 \end{bmatrix} \quad I = \begin{bmatrix} 1 & 0 & 0 \\ 0 & 1 & 0 \\ 0 & 0 & 1 \end{bmatrix}$$

Solution

Notice that the product

$$\underset{3 \times 3 \quad 3 \times 1}{IA}$$

exists and that it will be of dimensions 3×1. Performing the row–column multiplication yields

$$IA = \begin{bmatrix} 1 & 0 & 0 \\ 0 & 1 & 0 \\ 0 & 0 & 1 \end{bmatrix} \begin{bmatrix} 2 \\ 1 \\ 3 \end{bmatrix} = \begin{bmatrix} 2 \\ 1 \\ 3 \end{bmatrix}$$

Similarly,

$$\underset{3 \times 3 \quad 3 \times 2}{IB}$$

exists and is of dimensions 3×2. Performing the row–column multiplications, we obtain

$$IB = \begin{bmatrix} 1 & 0 & 0 \\ 0 & 1 & 0 \\ 0 & 0 & 1 \end{bmatrix} \begin{bmatrix} 3 & 0 \\ 1 & 2 \\ 4 & -1 \end{bmatrix} = \begin{bmatrix} 3 & 0 \\ 1 & 2 \\ 4 & -1 \end{bmatrix}$$

Notice that the I matrix possesses a special property. We have $IA = A$ and $IB = B$. We will comment further on this property in Section A.3.

EXERCISES

A.1   Given the matrices **A**, **B**, and **C**:

$$A = \begin{bmatrix} 3 & 0 \\ -1 & 4 \end{bmatrix} \quad B = \begin{bmatrix} 2 & 1 \\ 0 & -1 \end{bmatrix} \quad C = \begin{bmatrix} 1 & 0 & 3 \\ -2 & 1 & 2 \end{bmatrix}$$

a. Find **AB**.   b. Find **AC**.   c. Find **BA**.

A.2   Given the matrices **A**, **B**, and **C**:

$$A = \begin{bmatrix} 3 & 1 & 3 \\ 2 & 0 & 4 \\ -4 & 1 & 2 \end{bmatrix} \quad B = [1 \quad 0 \quad 2] \quad C = \begin{bmatrix} 3 \\ 0 \\ 2 \end{bmatrix}$$

a. Find **AC**.   b. Find **BC**.
c. Is it possible to find **AB**? Explain.

A.3   If **A** is a 3 × 2 matrix and **B** is a 2 × 4 matrix:
a. What are the dimensions of **AB**?
b. Is it possible to find the product **BA**? Explain.

A.4   If matrices **B** and **C** are of dimensions 1 × 3 and 3 × 1, respectively:
a. What are the dimensions of the product **BC**?
b. What are the dimensions of **CB**?
c. If **B** and **C** are the matrices shown in Exercise A.2, find **CB**.

A.5   Given the matrices **A**, **B**, and **C**:

$$A = \begin{bmatrix} 1 & 0 & 0 \\ 0 & 3 & 0 \\ 0 & 0 & 2 \end{bmatrix} \quad B = \begin{bmatrix} 2 & 3 \\ -3 & 0 \\ 4 & -1 \end{bmatrix} \quad C = [3 \quad 0 \quad 2]$$

a. Find **AB**.   b. Find **CA**.   c. Find **CB**.

A.6   Given the matrices:

$$A = [3 \quad 0 \quad -1 \quad 2] \quad B = \begin{bmatrix} 2 \\ -1 \\ 0 \\ 3 \end{bmatrix}$$

a. Find **AB**.   b. Find **BA**.

A.3

Identity Matrices and Matrix Inversion

In ordinary algebra, the number 1 is the identity element for the multiplication operation. That is, 1 is the element such that any other number, say, c, multiplied by the identity element is equal to c. Thus, $4(1) = 4$, $(-5)(1) = -5$, and so forth.

The corresponding identity element for multiplication in matrix algebra, identified by the symbol **I**, is a matrix such that

$$AI = IA = A \quad \text{for any matrix } A$$

The difference between identity elements in ordinary algebra and matrix algebra is that in ordinary algebra there is only one identity element, the number 1. In matrix algebra, the identity matrix must possess the correct dimensions for the product **IA** to exist. Consequently, there is an infinitely large number of identity matrices—all square and possessing the same pattern. The 1×1, 2×2, and 3×3 identity matrices are

$$\underset{1 \times 1}{\mathbf{I}} = [1] \qquad \underset{2 \times 2}{\mathbf{I}} = \begin{bmatrix} 1 & 0 \\ 0 & 1 \end{bmatrix} \qquad \underset{3 \times 3}{\mathbf{I}} = \begin{bmatrix} 1 & 0 & 0 \\ 0 & 1 & 0 \\ 0 & 0 & 1 \end{bmatrix}$$

In Example A.2, we demonstrated the fact that this matrix satisfies the property

$$\mathbf{IA} = \mathbf{A}$$

Definition A.6

If **A** is any matrix, then a matrix **I** is defined to be an **identity matrix** if **AI** = **IA** = **A**. The matrices that satisfy this definition possess the pattern

$$\mathbf{I} = \begin{bmatrix} 1 & 0 & 0 & \ldots & 0 \\ 0 & 1 & 0 & \ldots & 0 \\ 0 & 0 & 1 & \ldots & 0 \\ \cdot & \cdot & \cdot & \cdots & \cdot \\ \cdot & \cdot & \cdot & \cdots & \cdot \\ \cdot & \cdot & \cdot & \cdots & \cdot \\ 0 & 0 & 0 & \ldots & 1 \end{bmatrix}$$

EXAMPLE A.3

If **A** is the matrix shown below, find **IA** and **AI**.

$$A = \begin{bmatrix} 3 & 4 & -1 \\ 1 & 0 & 2 \end{bmatrix}$$

Solution

$$\underset{2 \times 2 \quad 2 \times 3}{\mathbf{IA}} = \begin{bmatrix} 1 & 0 \\ 0 & 1 \end{bmatrix}\begin{bmatrix} 3 & 4 & -1 \\ 1 & 0 & 2 \end{bmatrix} = \begin{bmatrix} 3 & 4 & -1 \\ 1 & 0 & 2 \end{bmatrix} = \mathbf{A}$$

$$\underset{2 \times 3 \quad 3 \times 3}{\mathbf{AI}} = \begin{bmatrix} 3 & 4 & -1 \\ 1 & 0 & 2 \end{bmatrix}\begin{bmatrix} 1 & 0 & 0 \\ 0 & 1 & 0 \\ 0 & 0 & 1 \end{bmatrix} = \begin{bmatrix} 3 & 4 & -1 \\ 1 & 0 & 2 \end{bmatrix} = \mathbf{A}$$

Notice that the identity matrices used to find the products **IA** and **AI** were of different dimensions. This was necessary for the products to exist.

The identity element assumes importance when we consider the process of division and its role in the solution of equations. In ordinary algebra, division is essentially multiplication using the reciprocals of elements. For example, the equation

$$2X = 6$$

can be solved by dividing both sides of the equation by 2, *or* it can be solved by *multiplying* both sides of the equation by $\frac{1}{2}$, which is the reciprocal of 2. Thus,

$$\left(\frac{1}{2}\right)2X = \frac{1}{2}(6)$$
$$X = 3$$

What is the reciprocal of an element? It is the element such that the reciprocal times the element is equal to the identity element. Thus, the reciprocal of 3 is $\frac{1}{3}$ because

$$3\left(\frac{1}{3}\right) = 1$$

The identity matrix plays the same role in matrix algebra. Thus, the reciprocal of a matrix A, called the **inverse of A** and denoted by the symbol A^{-1}, is a matrix such that $AA^{-1} = A^{-1}A = I$.

Inverses are defined only for square matrices, but not all square matrices possess inverses. Those that do have inverses play an important role in solving the least squares equations and in other aspects of a regression analysis. We will show you one important application of the inverse matrix in Section A.4. The procedure for finding the inverse of a matrix is demonstrated in Appendix B.

Definition A.7

The square matrix A^{-1} is said to be the **inverse** of the square matrix A if

$$A^{-1}A = AA^{-1} = I.$$

The procedure for finding an inverse matrix is computationally quite tedious and is performed most often using a computer. There are several exceptions. For example, finding the inverse of one type of matrix, called a **diagonal matrix**, is easy. A diagonal matrix is one that has nonzero elements down the **main diagonal** (running from top left of the matrix to bottom right) and 0 elements elsewhere. For example, the identity matrix is a diagonal matrix (with 1's along the main diagonal), as are the following matrices:

$$A = \begin{bmatrix} 3 & 0 & 0 \\ 0 & 1 & 0 \\ 0 & 0 & 2 \end{bmatrix} \qquad B = \begin{bmatrix} 5 & 0 & 0 & 0 \\ 0 & 2 & 0 & 0 \\ 0 & 0 & 1 & 0 \\ 0 & 0 & 0 & 5 \end{bmatrix}$$

Definition A.8

A **diagonal matrix** is one that contains nonzero elements on the main diagonal and 0 elements elsewhere.

You can verify that the inverse of

$$\mathbf{A} = \begin{bmatrix} 3 & 0 & 0 \\ 0 & 1 & 0 \\ 0 & 0 & 2 \end{bmatrix} \quad \text{is} \quad \mathbf{A}^{-1} = \begin{bmatrix} \frac{1}{3} & 0 & 0 \\ 0 & 1 & 0 \\ 0 & 0 & \frac{1}{2} \end{bmatrix}$$

i.e., $\mathbf{AA}^{-1} = \mathbf{I}$. In general, the inverse of a diagonal matrix is given by the following theorem, which is stated without proof:

Theorem A.1

The **inverse of a diagonal matrix**

$$\mathbf{D} = \begin{bmatrix} d_{11} & 0 & 0 & \cdots & 0 \\ 0 & d_{22} & 0 & \cdots & 0 \\ 0 & 0 & d_{33} & \cdots & 0 \\ \cdot & \cdot & \cdot & \cdots & \cdot \\ \cdot & \cdot & \cdot & \cdots & \cdot \\ \cdot & \cdot & \cdot & \cdots & \cdot \\ 0 & 0 & 0 & \cdots & d_{nn} \end{bmatrix} \quad \text{is} \quad \mathbf{D}^{-1} = \begin{bmatrix} 1/d_{11} & 0 & 0 & \cdots & 0 \\ 0 & 1/d_{22} & 0 & \cdots & 0 \\ 0 & 0 & 1/d_{33} & \cdots & 0 \\ \cdot & \cdot & \cdot & \cdots & \cdot \\ \cdot & \cdot & \cdot & \cdots & \cdot \\ \cdot & \cdot & \cdot & \cdots & \cdot \\ 0 & 0 & 0 & \cdots & 1/d_{nn} \end{bmatrix}$$

A second type of matrix that is easy to invert is a 2×2 matrix. The following theorem shows how to find the inverse of this type of matrix.

Theorem A.2

The **inverse of a 2 × 2 matrix**

$$\mathbf{A} = \begin{bmatrix} a & b \\ c & d \end{bmatrix} \quad \text{is} \quad \mathbf{A}^{-1} = \begin{bmatrix} \dfrac{d}{ad - bc} & \dfrac{-b}{ad - bc} \\ \dfrac{-c}{ad - bc} & \dfrac{a}{ad - bc} \end{bmatrix}$$

You can verify that the inverse of

$$\mathbf{A} = \begin{bmatrix} 1 & -2 \\ -2 & 6 \end{bmatrix} \quad \text{is} \quad \mathbf{A}^{-1} = \begin{bmatrix} 3 & 1 \\ 1 & \frac{1}{2} \end{bmatrix}$$

We demonstrate another technique for finding $\mathbf{A}^{-1}$ in Appendix B.

EXERCISES

A.7 Let $A = \begin{bmatrix} 3 & 0 & 2 \\ -1 & 1 & 4 \end{bmatrix}$.

    a. Give the identity matrix that will be used to obtain the product IA.
    b. Show that $IA = A$.
    c. Give the identity matrix that will be used to find the product AI.
    d. Show that $AI = A$.

A.8 Given the following matrices A and B, show that $AB = I$, that $BA = I$, and consequently, verify that $B = A^{-1}$

$$A = \begin{bmatrix} 1 & 0 & 0 \\ 0 & 2 & 0 \\ 0 & 0 & 3 \end{bmatrix} \qquad B = \begin{bmatrix} 1 & 0 & 0 \\ 0 & \frac{1}{2} & 0 \\ 0 & 0 & \frac{1}{3} \end{bmatrix}$$

A.9 If

$$A = \begin{bmatrix} 12 & 0 & 0 & 8 \\ 0 & 12 & 0 & 0 \\ 0 & 0 & 8 & 0 \\ 8 & 0 & 0 & 8 \end{bmatrix}$$

verify that

$$A^{-1} = \begin{bmatrix} \frac{1}{4} & 0 & 0 & -\frac{1}{4} \\ 0 & \frac{1}{12} & 0 & 0 \\ 0 & 0 & \frac{1}{8} & 0 \\ -\frac{1}{4} & 0 & 0 & \frac{3}{8} \end{bmatrix}$$

A.10 If

$$A = \begin{bmatrix} 3 & 0 & 0 \\ 0 & 5 & 0 \\ 0 & 0 & 7 \end{bmatrix}$$

show that

$$A^{-1} = \begin{bmatrix} \frac{1}{3} & 0 & 0 \\ 0 & \frac{1}{5} & 0 \\ 0 & 0 & \frac{1}{7} \end{bmatrix}$$

A.11 Verify Theorem A.1.

A.12 Verify Theorem A.2.

A.13 Find the inverse of

$$A = \begin{bmatrix} 2 & -1 \\ 2 & 3 \end{bmatrix}$$

A.4

Solving Systems of Simultaneous Linear Equations

Consider the following set of simultaneous linear equations in two unknowns:

$$2v_1 + v_2 = 7$$

$$v_1 - v_2 = 2$$

Note that the solution for these equations is $v_1 = 3$, $v_2 = 1$.

Now define the matrices

$$A = \begin{bmatrix} 2 & 1 \\ 1 & -1 \end{bmatrix} \qquad V = \begin{bmatrix} v_1 \\ v_2 \end{bmatrix} \qquad G = \begin{bmatrix} 7 \\ 2 \end{bmatrix}$$

Thus, A is the matrix of coefficients of v_1 and v_2, V is a column matrix containing the unknowns (written in order, top to bottom), and G is a column matrix containing the numbers on the right-hand side of the equal signs.

Now, the given system of simultaneous equations can be rewritten as a **matrix equation**:

$$AV = G$$

By a matrix equation, we mean that the product matrix AV, is equal to the matrix G. *Equality of matrices means that corresponding elements are equal.* You can see that this is true for the expression $AV = G$, since

$$\underset{2 \times 2 \quad 2 \times 1}{\overset{AV}{\overbrace{\qquad}}} = \begin{bmatrix} 2 & 1 \\ 1 & -1 \end{bmatrix} \begin{bmatrix} v_1 \\ v_2 \end{bmatrix} = \begin{bmatrix} (2v_1 + v_2) \\ (v_1 - v_2) \end{bmatrix} = \underset{2 \times 1}{G}$$

The matrix procedure for expressing a system of two simultaneous linear equations in two unknowns can be extended to express a set of k simultaneous equations in k unknowns. If the equations are written in the orderly pattern

$$a_{11}v_1 + a_{12}v_2 + \cdots + a_{1k}v_k = g_1$$
$$a_{21}v_1 + a_{22}v_2 + \cdots + a_{2k}v_k = g_2$$
$$\vdots \qquad \vdots \qquad \qquad \vdots \qquad \vdots$$
$$a_{k1}v_1 + a_{k2}v_2 + \cdots + a_{kk}v_k = g_k$$

then the set of simultaneous linear equations can be expressed as the matrix equation $AV = G$, where

$$A = \begin{bmatrix} a_{11} & a_{12} & \cdots & a_{1k} \\ a_{21} & & \cdots & a_{2k} \\ \vdots & & & \vdots \\ a_{k1} & & \cdots & a_{kk} \end{bmatrix} \qquad V = \begin{bmatrix} v_1 \\ v_2 \\ \vdots \\ v_k \end{bmatrix} \qquad G = \begin{bmatrix} g_1 \\ g_2 \\ \vdots \\ g_k \end{bmatrix}$$

Now let us solve this system of simultaneous equations. (If they are uniquely solvable, it can be shown that A^{-1} exists.) Multiplying both sides of the matrix equation by A^{-1}, we have

$$(A^{-1})AV = (A^{-1})G$$

But since $A^{-1}A = I$, we have

$$(I)V = A^{-1}G$$
$$V = A^{-1}G$$

In other words, if we know A^{-1}, we can find the solution to the set of simultaneous linear equations by obtaining the product $A^{-1}G$.

Matrix Solution to a Set of Simultaneous Linear Equations, $AV = G$

Solution: $V = A^{-1}G$

EXAMPLE A.4

Apply the result from the box to find the solution to the set of simultaneous linear equations

$$2v_1 + v_2 = 7$$
$$v_1 - v_2 = 2$$

Solution

The first step is to obtain the inverse of the coefficient matrix,

$$A = \begin{bmatrix} 2 & 1 \\ 1 & -1 \end{bmatrix}$$

namely,

$$A^{-1} = \begin{bmatrix} \frac{1}{3} & \frac{1}{3} \\ \frac{1}{3} & -\frac{2}{3} \end{bmatrix}$$

(This matrix can be found using a packaged computer program for matrix inversion or, for this simple case, you could use the procedure explained in Appendix B.) As a check, note that

$$A^{-1}A = \begin{bmatrix} \frac{1}{3} & \frac{1}{3} \\ \frac{1}{3} & -\frac{2}{3} \end{bmatrix} \begin{bmatrix} 2 & 1 \\ 1 & -1 \end{bmatrix} = \begin{bmatrix} 1 & 0 \\ 0 & 1 \end{bmatrix} = I$$

The second step is to obtain the product $A^{-1}G$. Thus,

$$V = A^{-1}G = \begin{bmatrix} \frac{1}{3} & \frac{1}{3} \\ \frac{1}{3} & -\frac{2}{3} \end{bmatrix} \begin{bmatrix} 7 \\ 2 \end{bmatrix} = \begin{bmatrix} 3 \\ 1 \end{bmatrix}$$

Since

$$V = \begin{bmatrix} v_1 \\ v_2 \end{bmatrix} = \begin{bmatrix} 3 \\ 1 \end{bmatrix}$$

it follows that $v_1 = 3$ and $v_2 = 1$. You can see that these values of v_1 and v_2 satisfy the simultaneous linear equations and are the values that we specified as a solution at the beginning of this section.

EXERCISES

A.14 Suppose the simultaneous linear equations

$$3v_1 + v_2 = 5$$
$$v_1 - v_2 = 3$$

are expressed as a matrix equation,

$$AV = G$$

a. Find the matrices A, V, and G.

b. Verify that

$$A^{-1} = \begin{bmatrix} \frac{1}{4} & \frac{1}{4} \\ \frac{1}{4} & -\frac{3}{4} \end{bmatrix}$$

[*Note*: A procedure for finding A^{-1} is given in Appendix B.]

c. Solve the equations by finding $V = A^{-1}G$.

A.15 For the simultaneous linear equations

$$10v_1 + 20v_3 - 60 = 0$$
$$20v_2 - 60 = 0$$
$$20v_1 + 68v_3 - 176 = 0$$

a. Find the matrices A, V, and G.

b. Verify that

$$A^{-1} = \begin{bmatrix} \frac{17}{70} & 0 & -\frac{1}{14} \\ 0 & \frac{1}{20} & 0 \\ -\frac{1}{14} & 0 & \frac{1}{28} \end{bmatrix}$$

c. Solve the equations by finding $V = A^{-1}G$.

A.5

The Least Squares Equations and Their Solutions

To apply matrix algebra to a regression analysis, we must place the data in matrices in a particular pattern. We will suppose that the linear model is

$$y = \beta_0 + \beta_1 x_1 + \beta_2 x_2 + \cdots + \beta_k x_k + \varepsilon$$

where (from Chapter 4) $x_1, x_2, \ldots, x_k$ could actually represent the squares, cubes, cross products, or other functions of predictor variables, and ε is a random error. We will assume that we have collected n data points, i.e., n values of y and corresponding values of $x_1, x_2, \ldots, x_k$, and that these are denoted as shown in the table:

DATA POINT	y Value	x_1	x_2	$\cdots$	x_k
1	y_1	x_{11}	x_{21}		x_{k1}
2	y_2	x_{12}	x_{22}		x_{k2}
.	.	.	.		.
.	.	.	.		.
n	y_n	x_{1n}	x_{2n}		x_{kn}

Then the two data matrices $\mathbf{Y}$ and $\mathbf{X}$ are as shown in the next box.

The Data Matrices $\mathbf{Y}$ and $\mathbf{X}$ and the $\hat{\boldsymbol{\beta}}$ Matrix

$$\mathbf{Y} = \begin{bmatrix} y_1 \\ y_2 \\ y_3 \\ \cdot \\ \cdot \\ \cdot \\ y_n \end{bmatrix} \qquad \mathbf{X} = \begin{bmatrix} 1 & x_{11} & x_{21} & \cdots & x_{k1} \\ 1 & x_{12} & x_{22} & \cdots & x_{k2} \\ 1 & x_{13} & x_{23} & \cdots & x_{k3} \\ \cdot & \cdot & \cdot & & \cdot \\ \cdot & \cdot & \cdot & & \cdot \\ \cdot & \cdot & \cdot & & \cdot \\ 1 & x_{1n} & x_{2n} & \cdots & x_{kn} \end{bmatrix} \qquad \hat{\boldsymbol{\beta}} = \begin{bmatrix} \hat{\beta}_0 \\ \hat{\beta}_1 \\ \hat{\beta}_2 \\ \cdot \\ \cdot \\ \cdot \\ \hat{\beta}_k \end{bmatrix}$$

Notice that the first column in the $\mathbf{X}$ matrix is a column of 1's. Thus, we are inserting a value of x, namely, x_0, as the coefficient of β_0, where x_0 is a variable always equal to 1. Therefore, there is one column in the $\mathbf{X}$ matrix for each β parameter. Also, remember that a particular data point is identified by specific rows of the $\mathbf{Y}$ and $\mathbf{X}$ matrices. For example, the y value y_3 for data point 3 is in the third row of the $\mathbf{Y}$ matrix, and the corresponding values of $x_1, x_2, \ldots, x_k$ appear in the third row of the $\mathbf{X}$ matrix.

The $\hat{\boldsymbol{\beta}}$ matrix shown in the box contains the least squares estimates (which we are attempting to obtain) of the coefficients $\beta_0, \beta_1, \ldots, \beta_k$ of the linear model

$$y = \beta_0 + \beta_1 x_1 + \beta_2 x_2 + \cdots + \beta_k x_k + \varepsilon$$

To write the least squares equation, we need to define what we mean by the **transpose of a matrix**. If

$$\mathbf{Y} = \begin{bmatrix} 5 \\ 1 \\ 0 \\ 4 \\ 2 \end{bmatrix} \qquad \mathbf{X} = \begin{bmatrix} 1 & 0 \\ 1 & 1 \\ 1 & 4 \\ 1 & 2 \\ 1 & 6 \end{bmatrix}$$

then the transpose matrices of the **Y** and **X** matrices, denoted as **Y'** and **X'**, respectively, are

$$\mathbf{Y'} = [5 \quad 1 \quad 0 \quad 4 \quad 2] \qquad \mathbf{X'} = \begin{bmatrix} 1 & 1 & 1 & 1 & 1 \\ 0 & 1 & 4 & 2 & 6 \end{bmatrix}$$

| **Definition A.9**

The **transpose of a matrix A**, denoted as **A'**, is obtained by interchanging corresponding rows and columns of the **A** matrix. That is, the *i*th row of the **A** matrix becomes the *i*th column of the **A'** matrix.

Using the **Y** and **X** data matrices, their transposes, and the $\hat{\boldsymbol{\beta}}$ matrix, we can write the least squares equations (proof omitted) as:

| **Least Squares Matrix Equation**

$$(\mathbf{X'X})\hat{\boldsymbol{\beta}} = \mathbf{X'Y}$$

Thus, $(\mathbf{X'X})$ is the coefficient matrix of the least squares estimates $\hat{\beta}_0$, $\hat{\beta}_1, \ldots, \hat{\beta}_k$, and $\mathbf{X'Y}$ gives the matrix of constants that appear on the right-hand side of the equality signs. In the notation of Section A.4,

$$\mathbf{A} = \mathbf{X'X} \qquad \mathbf{V} = \hat{\boldsymbol{\beta}} \qquad \mathbf{G} = \mathbf{X'Y}$$

The solution, which follows from Section A.4, is:

| **Least Squares Solution**

$$\hat{\boldsymbol{\beta}} = (\mathbf{X'X})^{-1}\mathbf{X'Y}$$

Thus, to solve the least squares matrix equation, the computer calculates $(\mathbf{X'X})$, $(\mathbf{X'X})^{-1}$, $\mathbf{X'Y}$, and, finally, the product $(\mathbf{X'X})^{-1}\mathbf{X'Y}$. We will illustrate this process using the data for the advertising example from Section 3.3.

| EXAMPLE A.5   Find the least squares line for the data given in Table A.1.

TABLE A.1

MONTH	ADVERTISING EXPENDITURE x, hundreds of dollars	SALES REVENUE y, thousands of dollars
1	1	1
2	2	1
3	3	2
4	4	2
5	5	4

Solution

The model is

$$y = \beta_0 + \beta_1 x_1 + \varepsilon$$

and the $\mathbf{Y}$, $\mathbf{X}$, and $\hat{\boldsymbol{\beta}}$ matrices are

$$\mathbf{Y} = \begin{bmatrix} 1 \\ 1 \\ 2 \\ 2 \\ 4 \end{bmatrix} \qquad \mathbf{X} = \begin{matrix} \quad x_0 \quad x_1 \\ \begin{bmatrix} 1 & 1 \\ 1 & 2 \\ 1 & 3 \\ 1 & 4 \\ 1 & 5 \end{bmatrix} \end{matrix} \qquad \hat{\boldsymbol{\beta}} = \begin{bmatrix} \hat{\beta}_0 \\ \hat{\beta}_1 \end{bmatrix}$$

Then,

$$\mathbf{X}'\mathbf{X} = \begin{bmatrix} 1 & 1 & 1 & 1 & 1 \\ 1 & 2 & 3 & 4 & 5 \end{bmatrix} \begin{bmatrix} 1 & 1 \\ 1 & 2 \\ 1 & 3 \\ 1 & 4 \\ 1 & 5 \end{bmatrix} = \begin{bmatrix} 5 & 15 \\ 15 & 55 \end{bmatrix}$$

$$\mathbf{X}'\mathbf{Y} = \begin{bmatrix} 1 & 1 & 1 & 1 & 1 \\ 1 & 2 & 3 & 4 & 5 \end{bmatrix} \begin{bmatrix} 1 \\ 1 \\ 2 \\ 2 \\ 4 \end{bmatrix} = \begin{bmatrix} 10 \\ 37 \end{bmatrix}$$

The last matrix that we need is $(\mathbf{X}'\mathbf{X})^{-1}$. This matrix, which can be found by using Theorem A.2 (or by using the method of Appendix B), is

$$(\mathbf{X}'\mathbf{X})^{-1} = \begin{bmatrix} 1.1 & -.3 \\ -.3 & .1 \end{bmatrix}$$

Then the solution to the least squares equation is

$$\hat{\boldsymbol{\beta}} = (\mathbf{X}'\mathbf{X})^{-1}\mathbf{X}'\mathbf{Y} = \begin{bmatrix} 1.1 & -.3 \\ -.3 & .1 \end{bmatrix} \begin{bmatrix} 10 \\ 37 \end{bmatrix} = \begin{bmatrix} -.1 \\ .7 \end{bmatrix}$$

Thus, $\hat{\beta}_0 = -.1$, $\hat{\beta}_1 = .7$, and the prediction equation is

$$\hat{y} = -.1 + .7x$$

You can verify that this is the same answer as obtained in Section 3.3.

EXAMPLE A.6

Find the least squares solution for fitting the monthly kilowatt-hour usage y to size of home x for the model

$$y = \beta_0 + \beta_1 x + \beta_2 x^2 + \varepsilon$$

(The computer printout for this regression analysis was presented and discussed in Section 4.3.) The data are shown in Table A.2.

T A B L E A.2 **Data for Power Usage Study**

SIZE OF HOME x, square feet	MONTHLY USAGE y, kilowatt-hours	SIZE OF HOME x, square feet	MONTHLY USAGE y, kilowatt-hours
1,290	1,182	1,840	1,711
1,350	1,172	1,980	1,804
1,470	1,264	2,230	1,840
1,600	1,493	2,400	1,956
1,710	1,571	2,930	1,954

Solution

The **Y**, **X**, and $\hat{\boldsymbol{\beta}}$ matrices are shown below:

$$\mathbf{Y} = \begin{bmatrix} 1,182 \\ 1,172 \\ 1,264 \\ 1,493 \\ 1,571 \\ 1,711 \\ 1,804 \\ 1,840 \\ 1,956 \\ 1,954 \end{bmatrix} \quad \mathbf{X} = \begin{bmatrix} x_0 & x & x^2 \\ 1 & 1,290 & 1,664,100 \\ 1 & 1,350 & 1,822,500 \\ 1 & 1,470 & 2,160,900 \\ 1 & 1,600 & 2,560,000 \\ 1 & 1,710 & 2,924,100 \\ 1 & 1,840 & 3,385,600 \\ 1 & 1,980 & 3,920,400 \\ 1 & 2,230 & 4,972,900 \\ 1 & 2,400 & 5,760,000 \\ 1 & 2,930 & 8,584,900 \end{bmatrix}$$

Then:

$$\mathbf{X}'\mathbf{X} = \begin{bmatrix} 10 & 18,800 & 37,755,400 \\ 18,800 & 37,755,400 & 8,093.9 \times 10^7 \\ 37,755,400 & 8,093.9 \times 10^7 & 1.843 \times 10^{14} \end{bmatrix}$$

$$\mathbf{X}'\mathbf{Y} = \begin{bmatrix} 15,947 \\ 31,283,250 \\ 6.53069 \times 10^{10} \end{bmatrix}$$

And (obtained using a statistical software package):

$$(X'X)^{-1} = \begin{bmatrix} 26.9156 & -.027027 & 6.3554 \times 10^{-6} \\ -.027027 & 2.75914 \times 10^{-5} & -6.5804 \times 10^{-9} \\ 6.3554 \times 10^{-6} & -6.5804 \times 10^{-9} & 1.5934 \times 10^{-12} \end{bmatrix}$$

Finally, performing the multiplication, we obtain

$$\hat{\boldsymbol{\beta}} = (X'X)^{-1}X'Y$$

$$= \begin{bmatrix} 26.9156 & -.027027 & 6.3554 \times 10^{-6} \\ -.027027 & 2.75914 \times 10^{-5} & -6.5804 \times 10^{-9} \\ 6.3554 \times 10^{-6} & -6.5804 \times 10^{-9} & 1.5934 \times 10^{-12} \end{bmatrix} \begin{bmatrix} 15,947 \\ 31,283,250 \\ 6.53069 \times 10^{10} \end{bmatrix}$$

$$= \begin{bmatrix} -1,216.14389 \\ 2.39893 \\ -.00045 \end{bmatrix}$$

Thus, $\hat{\beta}_0 = -1,216.14389$, $\hat{\beta}_1 = 2.39893$, $\hat{\beta}_2 = -.00045$, and the prediction equation is

$$\hat{y} = -1,216.14389 + 2.39893x - .00045x^2$$

These are the values shown in the SAS printout, Figure 4.2.

EXERCISES

A.16   Use the method of least squares to fit a straight line to the five data points:

x	−2	−1	0	1	2
y	4	3	3	1	−1

  a. Construct Y and X matrices for the data.
  b. Find $X'X$ and $X'Y$.
  c. Find the least squares estimates $\hat{\boldsymbol{\beta}} = (X'X)^{-1}X'Y$. [*Note:* See Theorem A.1 for information on finding $(X'X)^{-1}$.]
  d. Give the prediction equation.

  Note that the matrix procedure gives the same solution as obtained in Exercise 3.7.

A.17   Use the method of least squares to fit the model $E(y) = \beta_0 + \beta_1 x$ to the six data points:

x	1	2	3	4	5	6
y	1	2	2	3	5	5

a. Construct **Y** and **X** matrices for the data.
b. Find **X'X** and **X'Y**.
c. Verify that

$$(\mathbf{X'X})^{-1} = \begin{bmatrix} \frac{13}{15} & -\frac{7}{35} \\ -\frac{7}{35} & \frac{2}{35} \end{bmatrix}$$

d. Find the $\hat{\boldsymbol{\beta}}$ matrix.
e. Give the prediction equation.

Note that the matrix procedure gives the same solution as obtained in Exercise 3.6.

A.18   An experiment was conducted in which two y observations were collected for each of five values of x:

x	\multicolumn{2}{c}{-2}	\multicolumn{2}{c}{-1}	\multicolumn{2}{c}{0}	\multicolumn{2}{c}{1}	\multicolumn{2}{c}{2}
y	1.1	1.3	2.0	2.1	2.7	2.8	3.4	3.6	4.1	4.0

Use the method of least squares to fit the second-order model, $E(y) = \beta_0 + \beta_1 x + \beta_2 x^2$, to the 10 data points.
a. Give the dimensions of the **Y** and **X** matrices.
b. Verify that

$$(\mathbf{X'X})^{-1} = \begin{bmatrix} \frac{17}{70} & 0 & -\frac{1}{14} \\ 0 & \frac{1}{20} & 0 \\ -\frac{1}{14} & 0 & \frac{1}{28} \end{bmatrix}$$

c. Both **X'X** and $(\mathbf{X'X})^{-1}$ are symmetric matrices. What is a symmetric matrix?
d. Find the $\hat{\boldsymbol{\beta}}$ matrix and the least squares prediction equation.
e. Plot the data points and graph the prediction equation.

A.6
Calculating SSE and s^2

You will recall that the variances of the estimators of all the β parameters and of $\hat{y}$ will depend on the value of σ^2, the variance of the random error ε that appears in the linear model. Since σ^2 will rarely be known in advance, we must use the sample data to estimate its value.

Formulas for SSE and s^2

$$\text{SSE} = \mathbf{Y'Y} - \hat{\boldsymbol{\beta}}'\mathbf{X'Y}$$

$$s^2 = \frac{\text{SSE}}{n - \text{Number of } \beta \text{ parameters in model}}$$

We will demonstrate the use of these formulas with the advertising–sales data of Example A.5.

EXAMPLE A.7

Solution

Find the SSE for the advertising–sales data of Example A.5.

From Example A.5,

$$\hat{\beta} = \begin{bmatrix} -.1 \\ .7 \end{bmatrix} \quad \text{and} \quad X'Y = \begin{bmatrix} 10 \\ 37 \end{bmatrix}$$

Then,

$$Y'Y = \begin{bmatrix} 1 & 1 & 2 & 2 & 4 \end{bmatrix} \begin{bmatrix} 1 \\ 1 \\ 2 \\ 2 \\ 4 \end{bmatrix} = 26$$

and

$$\hat{\beta}'X'Y = \begin{bmatrix} -.1 & .7 \end{bmatrix} \begin{bmatrix} 10 \\ 37 \end{bmatrix} = 24.9$$

So

$$\text{SSE} = Y'Y - \hat{\beta}'X'Y = 26 - 24.9 = 1.1$$

(Note that this is the same answer as was obtained in Section 3.3.)
Finally,

$$s^2 = \frac{\text{SSE}}{n - \text{Number of } \beta \text{ parameters in model}} = \frac{1.1}{5 - 2} = .367$$

This estimate is needed to construct a confidence interval for β_1, to test a hypothesis concerning its value, or to construct a confidence interval for the mean sales for a given advertising expenditure.

A.7

Standard Errors of Estimators, Test Statistics, and Confidence Intervals for $\beta_0, \beta_1, \ldots, \beta_k$

The importance of this Appendix is evidenced by the fact that all the relevant information pertaining to the standard errors of the sampling distributions of $\hat{\beta}_0, \hat{\beta}_1, \ldots, \hat{\beta}_k$ (and hence of $\hat{Y}$) is contained in $(X'X)^{-1}$. Thus, if we denote the $(X'X)^{-1}$ matrix as

$$(X'X)^{-1} = \begin{bmatrix} c_{00} & c_{01} & \cdots & c_{0k} \\ c_{10} & c_{11} & \cdots & c_{1k} \\ c_{20} & c_{21} & \cdots & c_{2k} \\ . & . & . & . \\ . & . & . & . \\ . & . & . & . \\ c_{k0} & c_{k1} & \cdots & c_{kk} \end{bmatrix}$$

then it can be shown (proof omitted) that the standard errors of the sampling distributions of $\hat{\beta}_0, \hat{\beta}_1, \ldots, \hat{\beta}_k$ are

$$\sigma_{\hat{\beta}_0} = \sigma\sqrt{c_{00}}$$
$$\sigma_{\hat{\beta}_1} = \sigma\sqrt{c_{11}}$$
$$\sigma_{\hat{\beta}_2} = \sigma\sqrt{c_{22}}$$
$$\vdots$$
$$\sigma_{\hat{\beta}_k} = \sigma\sqrt{c_{kk}}$$

where σ is the standard deviation of the random error ε. In other words, the diagonal elements of $(\mathbf{X}'\mathbf{X})^{-1}$ give the values of $c_{00}, c_{11}, \ldots, c_{kk}$ that are required for finding the standard errors of the estimators $\hat{\beta}_0, \hat{\beta}_1, \ldots, \hat{\beta}_k$. The estimated values of the standard errors are obtained by replacing σ by s in the formulas for the standard errors. Thus, for example, the estimated standard error of $\hat{\beta}_1$ is $s_{\hat{\beta}_1} = s\sqrt{c_{11}}$.

The confidence interval for a single β parameter, β_i, is given in the box.

Confidence Interval for β_i

$$\hat{\beta}_i \pm t_{\alpha/2}(\text{Estimated standard error of } \hat{\beta}_i)$$

or

$$\hat{\beta}_i \pm t_{\alpha/2}s\sqrt{c_{ii}}$$

where $t_{\alpha/2}$ is based on the number of degrees of freedom associated with s.

Similarly, the test statistic for testing the null hypothesis H_0: $\beta_i = 0$ is as shown in the next box.

Test Statistic for H_0: $\beta_i = 0$

$$t = \frac{\hat{\beta}_i}{s\sqrt{c_{ii}}}$$

EXAMPLE A.8

Refer to Example A.5 and find the estimated standard error for the sampling distribution of $\hat{\beta}_1$, the estimator of the slope of the line β_1. Then give a 95% confidence interval for β_1.

Solution

The $(\mathbf{X}'\mathbf{X})^{-1}$ matrix for the least squares solution of Example A.5 was

$$(\mathbf{X}'\mathbf{X})^{-1} = \begin{bmatrix} 1.1 & -.3 \\ -.3 & .1 \end{bmatrix}$$

Therefore, $c_{00} = 1.1$, $c_{11} = .1$, and the estimated standard error for $\hat{\beta}_1$ is

$$s_{\hat{\beta}_1} = s\sqrt{c_{11}} = \sqrt{.367}(\sqrt{.1}) = .192$$

The value for s, $\sqrt{.367}$, was obtained from Example A.7.

A 95% confidence interval for β_1 is

$$\hat{\beta}_1 \pm t_{\alpha/2}s\sqrt{c_{11}}$$

$$.7 \pm (3.182)(.192) = (.09, 1.31)$$

The t value, $t_{.025}$, is based on $(n - 2) = 3$ df. Observe that this is the same confidence interval as the one obtained in Section 3.6.

EXAMPLE A.9

Refer to Example A.6 and the least squares solution for fitting power usage y to the size of a home x using the model

$$y = \beta_0 + \beta_1 x + \beta_2 x^2 + \varepsilon$$

a. Compute the estimated standard error for $\hat{\beta}_1$, and compare this result with the regression analysis printout shown in Figure A.1.
b. Compute the value of the test statistic for testing $H_0: \beta_2 = 0$. Compare this with the value given in the computer printout shown in Figure A.1.

FIGURE A.1   SAS printout for the power usage data, Examples A.6 and A.9

ANALYSIS OF VARIANCE

SOURCE	DF	SUM OF SQUARES	MEAN SQUARE	F VALUE	PROB>F
MODEL	2	831069.55	415534.77	189.710	0.0001
ERROR	7	15332.55363	2190.36480		
C TOTAL	9	846402.10			

ROOT MSE	46.80133	R-SQUARE	0.9819	
DEP MEAN	1594.7	ADJ R-SQ	0.9767	
C.V.	2.934805			

PARAMETER ESTIMATES

VARIABLE	DF	PARAMETER ESTIMATE	STANDARD ERROR	T FOR H0: PARAMETER=0	PROB > |T|
INTERCEP	1	-1216.14389	242.80637	-5.009	0.0016
X	1	2.39893018	0.24583560	9.758	0.0001
XX	1	-0.000450040	0.000059077	-7.618	0.0001

Solution

The fitted model is

$$\hat{y} = -1{,}216.14389 + 2.39893x - .00045x^2$$

The $(X'X)^{-1}$ matrix, obtained in Example A.6, is

$$(X'X)^{-1} = \begin{bmatrix} 26.9156 & -.027027 & 6.3554 \times 10^{-6} \\ -.027027 & 2.75914 \times 10^{-5} & -6.5804 \times 10^{-9} \\ 6.3554 \times 10^{-6} & -6.5804 \times 10^{-9} & 1.5934 \times 10^{-12} \end{bmatrix}$$

Note from $(X'X)^{-1}$ that

$$c_{00} = 26.9156$$
$$c_{11} = 2.75914 \times 10^{-5}$$
$$c_{22} = 1.5934 \times 10^{-12}$$

and that from the printout, $s = 46.801$.

a. The estimated standard error of $\hat{\beta}_1$ is

$$s_{\hat{\beta}_1} = s\sqrt{c_{11}}$$
$$= (46.801)\sqrt{2.75914 \times 10^{-1}} = .24583$$

Notice that this agrees with the value of $s_{\hat{\beta}_1}$ shown in the computer printout (Figure A.1).

b. The value of the test statistic for testing $H_0: \beta_2 = 0$ is

$$t = \frac{\hat{\beta}_2}{s\sqrt{c_{22}}} = \frac{-.00045}{(46.801)\sqrt{1.5934 \times 10^{-12}}} = -7.62$$

Notice that this value of the t statistic agrees with the value given in the column headed **T FOR H0: PARAMETER = 0** shown in the printout (Figure A.1).

EXERCISES

A.19 Do the data given in Exercise A.16 provide sufficient evidence to indicate that x contributes information for the prediction of y? Test $H_0: \beta_1 = 0$ against $H_a: \beta_1 \neq 0$ using $\alpha = .05$.

A.20 Find a 90% confidence interval for the slope of the line in Exercise A.19.

A.21 The term in the second-order model $E(y) = \beta_0 + \beta_1 x + \beta_2 x^2$ that controls the curvature in its graph is $\beta_2 x^2$. If $\beta_2 = 0$, $E(y)$ graphs as a straight line. Do the data given in Exercise A.18 provide sufficient evidence to indicate curvature in the model for $E(y)$? Test $H_0: \beta_2 = 0$ against $H_a: \beta_2 \neq 0$ using $\alpha = .10$.

A.8

A Confidence Interval for a Linear Function of the β Parameters; a Confidence Interval for $E(y)$

Suppose we were to postulate that the mean value of the productivity, y, of a company is related to the size of the company, x, and that the relationship could be modeled by the expression

$$E(y) = \beta_0 + \beta_1 x + \beta_2 x^2$$

A graph of $E(y)$ might appear as shown in Figure A.2.

FIGURE A.2

Graph of mean productivity $E(y)$

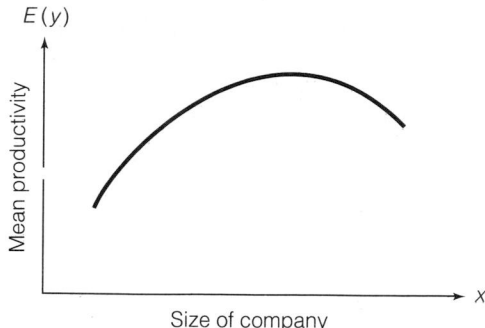

We might have several reasons for collecting data on the productivity and size of a set of n companies and finding the least squares prediction equation,

$$\hat{y} = \hat{\beta}_0 + \hat{\beta}_1 x + \hat{\beta}_2 x^2$$

For example, we might wish to estimate the mean productivity for a company of a given size (say, $x = 2$). That is, we might wish to estimate

$$E(y) = \beta_0 + \beta_1 x + \beta_2 x^2$$
$$= \beta_0 + 2\beta_1 + 4\beta_2 \qquad \text{where} \quad x = 2$$

Or we might wish to estimate the marginal increase in productivity, the slope of a tangent to the curve, when $x = 2$ (see Figure A.3). The marginal productivity for y when $x = 2$ is the rate of change of $E(y)$ with respect to x, evaluated at $x = 2$.* The marginal productivity for a value of x, denoted by the symbol $dE(y)/dx$, can be shown (proof omitted) to be

$$\frac{dE(y)}{dx} = \beta_1 + 2\beta_2 x$$

FIGURE A.3

Marginal productivity

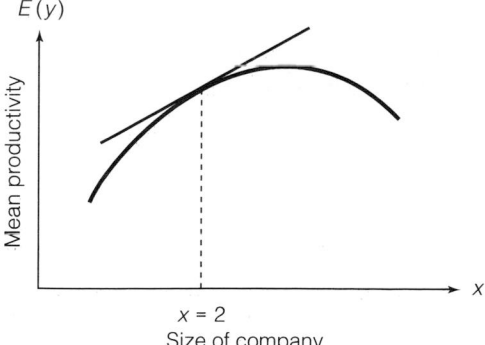

Therefore, the marginal productivity at $x = 2$ is

$$\frac{dE(y)}{dx} = \beta_1 + 2\beta_2(2) = \beta_1 + 4\beta_2$$

---

*If you have had calculus, note that the marginal productivity for y given x is the first derivative of $E(y) = \beta_0 + \beta_1 x + \beta_2 x^2$ with respect to x.

Note that for $x = 2$, both $E(y)$ and the marginal productivity are *linear* functions of the unknown parameters β_0, β_1, β_2 in the model. The problem we pose in this section is that of finding confidence intervals for linear functions of β parameters or testing hypotheses concerning their values. The information necessary to solve this problem is rarely given in a standard multiple regression analysis computer printout, but we can find these confidence intervals or values of the appropriate test statistics from knowledge of $(\mathbf{X'X})^{-1}$.

We will suppose that we have a model,

$$y = \beta_0 + \beta_1 x_1 + \cdots + \beta_k x_k + \varepsilon$$

and that we are interested in making an inference about a linear function of the β parameters, say,

$$a_0\beta_0 + a_1\beta_1 + \cdots + a_k\beta_k$$

where $a_0, a_1, \ldots, a_k$ are known constants. Further, we will use the corresponding linear function of least squares estimates,

$$\ell = a_0\hat{\beta}_0 + a_1\hat{\beta}_1 + \cdots + a_k\hat{\beta}_k$$

as our best estimate of $a_0\beta_0 + a_1\beta_1 + \cdots + a_k\beta_k$.

Then, for the assumptions on the random error ε (stated in Section 4.2), the sampling distribution for the estimator ℓ will be normal, with mean and standard error as given in the next box.

Mean and Standard Error of ℓ

$$E(\ell) = a_0\beta_0 + a_1\beta_1 + \cdots + a_k\beta_k$$
$$\sigma_\ell = \sqrt{\sigma^2 \mathbf{a'}(\mathbf{X'X})^{-1}\mathbf{a}}$$

where σ^2 is the variance of ε, $(\mathbf{X'X})^{-1}$ is the inverse matrix obtained in fitting the least squares model to the set of data, and

$$\mathbf{a} = \begin{bmatrix} a_0 \\ a_1 \\ a_2 \\ . \\ . \\ . \\ a_k \end{bmatrix}$$

This indicates that ℓ is an unbiased estimator of

$$E(\ell) = a_0\beta_0 + a_1\beta_1 + \cdots + a_k\beta_k$$

and that its sampling distribution would appear as shown in Figure A.4.

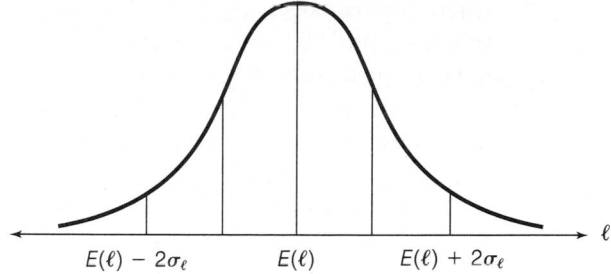

$$E(\ell) - 2\sigma_\ell \qquad E(\ell) \qquad E(\ell) + 2\sigma_\ell \qquad \ell$$

It can be shown that a $100(1 - \alpha)\%$ confidence interval for $E(\ell)$ is:

A $100(1 - \alpha)\%$ Confidence Interval for $E(\ell)$

$$\ell \pm t_{\alpha/2}\sqrt{s^2 \mathbf{a}'(\mathbf{X}'\mathbf{X})^{-1}\mathbf{a}}$$

where

$$E(\ell) = a_0\beta_0 + a_1\beta_1 + \cdots + a_k\beta_k$$

$$\ell = a_0\hat{\beta}_0 + a_1\hat{\beta}_1 + \cdots + a_k\hat{\beta}_k \qquad \mathbf{a} = \begin{bmatrix} a_0 \\ a_1 \\ a_2 \\ \cdot \\ \cdot \\ \cdot \\ a_k \end{bmatrix}$$

s^2 and $(\mathbf{X}'\mathbf{X})^{-1}$ are obtained from the least squares procedure, and $t_{\alpha/2}$ is based on the number of degrees of freedom associated with s^2.

The linear function of the β parameters that is most often the focus of our attention is

$$E(y) = \beta_0 + \beta_1 x_1 + \cdots + \beta_k x_k$$

That is, we want to find a confidence interval for $E(y)$ for specific values of x_1, $x_2, \ldots, x_k$. For this special case,

$$\ell = \hat{y}$$

and the $\mathbf{a}$ matrix is

$$\mathbf{a} = \begin{bmatrix} 1 \\ x_1 \\ x_2 \\ \cdot \\ \cdot \\ \cdot \\ x_k \end{bmatrix}$$

where the symbols $x_1, x_2, \ldots, x_k$ in the **a** matrix indicate the specific numerical values assumed by these variables. Thus, the procedure for forming a confidence interval for $E(y)$ is as shown in the box.

A $100(1 - \alpha)\%$ Confidence Interval for $E(y)$

$$\ell \pm t_{\alpha/2}\sqrt{s^2 \mathbf{a}'(\mathbf{X}'\mathbf{X})^{-1}\mathbf{a}}$$

where

$$E(y) = \beta_0 + \beta_1 x_1 + \beta_2 x_2 + \cdots + \beta_k x_k$$

$$\ell = \hat{y} = \hat{\beta}_0 + \hat{\beta}_1 x_1 + \cdots + \hat{\beta}_k x_k \qquad \mathbf{a} = \begin{bmatrix} 1 \\ x_1 \\ x_2 \\ \cdot \\ \cdot \\ \cdot \\ x_k \end{bmatrix}$$

s^2 and $(\mathbf{X}'\mathbf{X})^{-1}$ are obtained from the least squares analysis, and $t_{\alpha/2}$ is based on the number of degrees of freedom associated with s^2, namely, $n - (k + 1)$.

EXAMPLE A.10

Refer to the data of Example A.5 for sales revenue y and advertising expenditure x. Find a 95% confidence interval for the mean sales revenue $E(y)$ when advertising expenditure is $x = 4$.

Solution

The confidence interval for $E(y)$ for a given value of x is

$$\hat{y} \pm t_{\alpha/2}\sqrt{s^2 \mathbf{a}'(\mathbf{X}'\mathbf{X})^{-1}\mathbf{a}}$$

Consequently, we need to find and substitute the values of $\mathbf{a}'(\mathbf{X}'\mathbf{X})^{-1}\mathbf{a}$, $t_{\alpha/2}$, and $\hat{y}$ into this formula. Since we wish to estimate

$$\begin{aligned} E(y) &= \beta_0 + \beta_1 x \\ &= \beta_0 + \beta_1(4) \qquad \text{when} \quad x = 4 \\ &= \beta_0 + 4\beta_1 \end{aligned}$$

it follows that the coefficients of β_0 and β_1 are $a_0 = 1$ and $a_1 = 4$, and thus,

$$\mathbf{a} = \begin{bmatrix} 1 \\ 4 \end{bmatrix}$$

From Examples A.5 and A.7, $\hat{y} = -.1 + .7x$,

$$(\mathbf{X}'\mathbf{X})^{-1} = \begin{bmatrix} 1.1 & -.3 \\ -.3 & .1 \end{bmatrix}$$

and $s^2 = .367$. Then,

$$\mathbf{a}'(\mathbf{X}'\mathbf{X})^{-1}\mathbf{a} = \begin{bmatrix} 1 & 4 \end{bmatrix} \begin{bmatrix} 1.1 & -.3 \\ -.3 & .1 \end{bmatrix} \begin{bmatrix} 1 \\ 4 \end{bmatrix}$$

We first calculate

$$\mathbf{a}'(\mathbf{X}'\mathbf{X})^{-1} = \begin{bmatrix} 1 & 4 \end{bmatrix} \begin{bmatrix} 1.1 & -.3 \\ -.3 & .1 \end{bmatrix} = \begin{bmatrix} -.1 & .1 \end{bmatrix}$$

Then,

$$\mathbf{a}'(\mathbf{X}'\mathbf{X})^{-1}\mathbf{a} = \begin{bmatrix} -.1 & .1 \end{bmatrix} \begin{bmatrix} 1 \\ 4 \end{bmatrix} = .3$$

The t value, $t_{.025}$, based on 3 df is 3.182. So, a 95% confidence interval for the mean sales revenue with an advertising expenditure of 4 is

$$\hat{y} \pm t_{\alpha/2}\sqrt{s^2\mathbf{a}'(\mathbf{X}'\mathbf{X})^{-1}\mathbf{a}}$$

Since $\hat{y} = -.1 + .7x = -.1 + (.7)(4) = 2.7$, the 95% confidence interval for $E(y)$ when $x = 4$ is

$$2.7 \pm (3.182)\sqrt{(.367)(.3)}$$

$$2.7 \pm 1.1$$

Notice that this is exactly the same result as obtained in Example 3.4.

EXAMPLE A.11

An economist recorded a measure of productivity y and the size x for each of 100 companies producing cement. A regression model,

$$y = \beta_0 + \beta_1 x + \beta_2 x^2 + \varepsilon$$

fit to the $n = 100$ data points produced the following results:

$$\hat{y} = 2.6 + .7x - .2x^2$$

where x is coded to take values in the interval $-2 < x < 2$,* and

$$(\mathbf{X}'\mathbf{X})^{-1} = \begin{bmatrix} .0025 & .0005 & -.0070 \\ .0005 & .0055 & 0 \\ -.0070 & 0 & .0050 \end{bmatrix} \qquad s = .14$$

Find a 95% confidence interval for the marginal increase in productivity given that the coded size of a plant is $x = 1.5$.

---

*We give a formula for *coding* observational data in Section 5.6.

Solution

The mean value of y for a given value of x is

$$E(y) = \beta_0 + \beta_1 x + \beta_2 x^2$$

Therefore, the marginal increase in y for $x = 1.5$ is

$$\frac{dE(y)}{dx} = \beta_1 + 2\beta_2 x$$
$$= \beta_1 + 2(1.5)\beta_2$$

Or,

$$E(\ell) = \beta_1 + 3\beta_2 \qquad \text{when} \quad x = 1.5$$

Note from the prediction equation, $\hat{y} = 2.6 + .7x - .2x^2$, that $\hat{\beta}_1 = .7$ and $\hat{\beta}_2 = -.2$. Therefore,

$$\ell = \hat{\beta}_1 + 3\hat{\beta}_2 = .7 + 3(-.2) = .1$$

and

$$\mathbf{a} = \begin{bmatrix} a_0 \\ a_1 \\ a_2 \end{bmatrix} = \begin{bmatrix} 0 \\ 1 \\ 3 \end{bmatrix}$$

We next calculate

$$\mathbf{a}'(\mathbf{X}'\mathbf{X})^{-1}\mathbf{a} = \begin{bmatrix} 0 & 1 & 3 \end{bmatrix} \begin{bmatrix} .0025 & .0005 & -.0070 \\ .0005 & .0055 & 0 \\ -.0070 & 0 & .0050 \end{bmatrix} \begin{bmatrix} 0 \\ 1 \\ 3 \end{bmatrix} = .0505$$

Then, since s is based on $n - (k + 1) = 100 - 3 = 97$ df, $t_{.025} \approx 1.96$, and a 95% confidence interval for the marginal increase in productivity when $x = 1.5$ is

$$\ell \pm t_{.025}\sqrt{(s^2)\mathbf{a}'(\mathbf{X}'\mathbf{X})^{-1}\mathbf{a}}$$

or

$$.1 \pm (1.96)\sqrt{(.14)^2(.0505)}$$
$$.1 \pm .062$$

Thus, the marginal increase in productivity, the slope of the tangent to the curve

$$E(y) = \beta_0 + \beta_1 x + \beta_2 x^2$$

is estimated to lie in the interval $.1 \pm .062$ at $x = 1.5$. A graph of $\hat{y} = 2.6 + .7x - .2x^2$ is shown in Figure A.5.

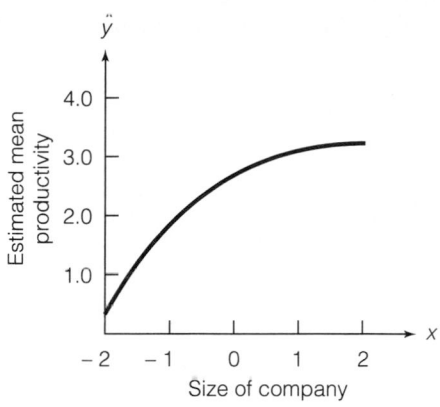

FIGURE A.5
A graph of $\hat{y} = 2.6 + .7x - .2x^2$

A.9

A Prediction Interval for Some Value of y to Be Observed in the Future

We have indicated in Sections 3.9 and 4.7 that two of the most important applications of the least squares predictor $\hat{y}$ are estimating the mean value of y (the topic of the preceding section) and predicting a new value of y, yet unobserved, for specific values of $x_1, x_2, \ldots, x_k$. The difference between these two inferential problems (when each would be pertinent) was explained in Chapters 3 and 4, but we will give another example to make certain that the distinction is clear.

Suppose you are the manager of a manufacturing plant and that y, the daily profit is a function of various process variables $x_1, x_2, \ldots, x_k$. Suppose you want to know how much money you would make *in the long run* if the x's are set at specific values. For this case, you would be interested in finding a confidence interval for the mean profit per day, $E(y)$. In contrast, suppose you planned to operate the plant for only one more day! Then you would be interested in predicting the value of y, the profit associated with tomorrow's production.

We have indicated that the error of prediction is always larger than the error of estimating $E(y)$. You can see this by comparing the formula for the prediction interval (shown in the box on page 770) with the formula for the confidence interval for $E(y)$ that was given in Section A.8.

EXAMPLE A.12

Refer to the sales–advertising expenditure example (Example A.10). Find a 95% prediction interval for the sales revenue next month, if it is known that next month's advertising expenditure will be $x = 4$.

Solution

The 95% prediction interval for sales revenue y is

$$\hat{y} \pm t_{\alpha/2}\sqrt{s^2(1 + \mathbf{a}'(\mathbf{X}'\mathbf{X})^{-1}\mathbf{a})}$$

From Example A.10, when $x = 4$, $\hat{y} = -.1 + .7x = -.1 + (.7)(4) = 2.7$, $s^2 = .367$, $t_{.025} = 3.182$, and $\mathbf{a}'(\mathbf{X}'\mathbf{X})^{-1}\mathbf{a} = .3$. Then the 95% prediction interval for y is

$$2.7 \pm (3.182)\sqrt{(.367)(1 + .3)}$$
$$2.7 \pm 2.2$$

You will find that this is the same solution as obtained in Example 3.4.

A $100(1 - \alpha)\%$ Prediction Interval for y

$$\hat{y} \pm t_{\alpha/2}\sqrt{s^2 + s^2 a'(X'X)^{-1}a} = \hat{y} \pm t_{\alpha/2}\sqrt{s^2(1 + a'(X'X)^{-1}a)}$$

where

$$\hat{y} = \hat{\beta}_0 + \hat{\beta}_1 x_1 + \cdots + \hat{\beta}_k x_k$$

s^2 and $(X'X)^{-1}$ are obtained from the least squares analysis,

$$a = \begin{bmatrix} 1 \\ x_1 \\ x_2 \\ \cdot \\ \cdot \\ \cdot \\ x_k \end{bmatrix}$$

contains the numerical values of $x_1, x_2, \ldots, x_k$, and $t_{\alpha/2}$ is based on the number of degrees of freedom associated with s^2, namely, $n - (k + 1)$.

EXERCISES

A.22 Refer to Exercise A.16. Find a 90% confidence interval for $E(y)$ when $x = 1$. Interpret the interval.

A.23 Refer to Exercise A.16. Suppose you plan to observe y for $x = 1$. Find a 90% prediction interval for that value of y. Interpret the interval.

A.24 Refer to Exercise A.17. Find a 90% confidence interval for $E(y)$ when $x = 2$. Interpret the interval.

A.25 Refer to Exercise A.17. Find a 90% prediction interval for a value of y to be observed in the future when $x = 2$. Interpret the interval.

A.26 Refer to Exercise A.18. Find a 90% confidence interval for the mean value of y when $x = 1$. Interpret the interval.

A.27 Refer to Exercise A.18. Find a 90% prediction interval for a value of y to be observed in the future when $x = 1$.

A.28 The productivity (items produced per hour) per worker on a manufacturing assembly line is expected to increase as piecework pay rate (in dollars) increases; it is expected to stabilize after a certain pay rate has been reached. The productivity of five different workers was recorded for each of five piecework pay rates, $.80, $.90, $1.00, $1.10, $1.20, thus giving $n = 25$ data points. A multiple regression analysis using a second-order model,

$$E(y) = \beta_0 + \beta_1 x + \beta_2 x^2$$

gave

$$\hat{y} = 2.08 + 8.42x - 1.65x^2$$

$$\text{SSE} = 26.62, \text{SS}_{yy} = 784.11, \text{ and}$$

$$(X'X)^{-1} = \begin{bmatrix} .020 & -.010 & .015 \\ -.010 & .040 & -.006 \\ .015 & -.006 & .028 \end{bmatrix}$$

a. Find s^2.

b. Find a 95% confidence interval for the mean productivity when the pay rate is $1.10. Interpret this interval.

c. Find a 95% prediction interval for the production of an individual worker who is paid at a rate of $1.10 per piece. Interpret the interval.

d. Find R^2 and interpret the value.

Summary

Except for the tedious process of inverting a matrix (given in Appendix B), we have covered the major steps performed by a computer in fitting a linear statistical model to a set of data using the method of least squares. We have also explained how to find the confidence intervals, prediction intervals, and values of test statistics that would be pertinent in a regression analysis.

In addition to providing a better understanding of a multiple regression analysis, the most important contributions of this Appendix are contained in Sections A.8 and A.9. If you want to make a specific inference concerning the mean value of y or any linear function of the β parameters and if you are unable to obtain the results from the computer package you are using, you will find the contents of Sections A.8 and A.9 very useful. Since you will almost always be able to find a computer program package to find $(X'X)^{-1}$, you will be able to calculate the desired confidence interval(s) and so forth on your own.

SUPPLEMENTARY EXERCISES

A.29 Use the method of least squares to fit a straight line to the six data points:

x	−5	−3	−1	1	3	5
y	1.1	1.9	3.0	3.8	5.1	6.0

a. Construct **Y** and **X** matrices for the data.

b. Find $\mathbf{X'X}$ and $\mathbf{X'Y}$.

c. Find the least squares estimates,

$$\hat{\beta} = (\mathbf{X'X})^{-1}\mathbf{X'Y}$$

[*Note*:   See Theorem A.1 for information on finding $(\mathbf{X'X})^{-1}$.]

d. Give the prediction equation.

e. Find SSE and s^2.

f. Does the model contribute information for the prediction of y? Test H_0: $\beta_1 = 0$. Use $\alpha = .05$.

g. Find r^2 and interpret its value.

h. Find a 90% confidence interval for $E(y)$ when $x = .5$. Interpret the interval.

A.30   An experiment was conducted to investigate the effect of extrusion pressure P and temperature at extrusion T on the strength y of a new type of plastic. Two plastic specimens were prepared for each of five combinations of pressure and temperature. The specimens were then tested in random order, and the breaking strength for each specimen was recorded. The independent variables were coded to simplify computations, i.e.,

$$x_1 = \frac{P - 200}{10} \qquad x_2 = \frac{T - 400}{25}$$

The $n = 10$ data points are listed in the table.

y	x_1	x_2
5.2; 5.0	-2	2
.3; $-.1$	-1	-1
-1.2; -1.1	0	-2
2.2; 2.0	1	-1
6.2; 6.1	2	2

a. Give the $\mathbf{Y}$ and $\mathbf{X}$ matrices needed to fit the model $y = \beta_0 + \beta_1 x_1 + \beta_2 x_2 + \varepsilon$.

b. Find the least squares prediction equation.

c. Find SSE and s^2.

d. Does the model contribute information for the prediction of y? Test using $\alpha = .05$.

e. Find R^2 and interpret its value.

f. Test the null hypothesis that $\beta_1 = 0$. Use $\alpha = .05$. What is the practical implication of the test?

g. Find a 90% confidence interval for the mean strength of the plastic for $x_1 = -2$ and $x_2 = 2$.

h. Suppose a single specimen of the plastic is to be installed in the engine mount of a Douglas DC-10 aircraft. Find a 90% prediction interval for the strength of this specimen if $x_1 = -2$ and $x_2 = 2$.

A.31   Suppose we obtained two replications of the experiment described in Exercise A.17, i.e., two values of y were observed for each of the six values of x. The data are shown here:

x	1		2		3		4		5		6	
y	1.1	.5	1.8	2.0	2.0	2.9	3.8	3.4	4.1	5.0	5.0	5.8

a. Suppose (as in Exercise A.17) you wish to fit the model $E(y) = \beta_0 + \beta_1 x$. Construct $\mathbf{Y}$ and $\mathbf{X}$ matrices for the data. [*Hint*:   Remember, the $\mathbf{Y}$ matrix must be of dimension 12×1.]

b. Find $X'X$ and $X'Y$.

c. Compare the $X'X$ matrix for two replications of the experiment with the $X'X$ matrix obtained for a single replication (part b of Exercise A.17). What is the relationship between the elements in the two matrices?

d. Observe the $(X'X)^{-1}$ matrix for a single replication (see part c of Exercise A.17). Verify that the $(X'X)^{-1}$ matrix for two replications contains elements that are equal to $\frac{1}{2}$ of the values of the corresponding elements in the $(X'X)^{-1}$ matrix for a single replication of the experiment. [*Hint:* Show that the product of the $(X'X)^{-1}$ matrix (for two replications) and the $X'X$ matrix from part c equals the identity matrix I.]

e. Find the prediction equation.

f. Find SSE and s^2.

g. Do the data provide sufficient information to indicate that x contributes information for the prediction of y? Test using $\alpha = .05$.

h. Find r^2 and interpret its value.

A.32 Refer to Exercise A.31.

a. Find a 90% confidence interval for $E(y)$ when $x = 4.5$. Interpret the interval.

b. Suppose we wish to predict the value of y if, in the future, $x = 4.5$. Find a 90% prediction interval for y and interpret the interval.

A.33 Refer to Exercise A.31. Suppose you replicated the experiment described in Exercise A.17 three times, i.e., you collected three observations on y for each value of x. Then $n = 18$.

a. What would be the dimensions of the Y matrix?

b. Write the X matrix for three replications. Compare with the X matrices for one and for two replications. Note the pattern.

c. Examine the $X'X$ matrices obtained for one and two replications of the experiment (obtained in Exercises A.17 and A.31, respectively). Deduce the values of the elements of the $X'X$ matrix for three replications.

d. Look at your answer to Exercise A.31, part d. Deduce the values of the elements in the $(X'X)^{-1}$ matrix for three replications.

e. Suppose you wanted to find a 90% confidence interval for $E(y)$ when $x = 4.5$ based on three replications of the experiment. Find the value of $a'(X'X)^{-1}a$ that appears in the confidence interval and compare with the value of $a'(X'X)^{-1}a$ that would be obtained for a single replication of the experiment.

f. Approximately how much of a reduction in the width of the confidence interval is obtained by using three versus two replications? [*Note:* The values of s computed from the two sets of data will almost certainly be different.]

References

Draper, N. and Smith, H. *Applied Regression Analysis*. New York: Wiley, 1966.

Graybill, F. A. *Theory and Application of the Linear Model*. North Scituate, Mass.: Duxbury, 1976.

Kleinbaum, D. and Kupper, L. *Applied Regression Analysis and Other Multivariable Methods*. North Scituate, Mass.: Duxbury, 1978.

Mendenhall, W. *Introduction to Linear Models and the Design and Analysis of Experiments*. Belmont, Calif.: Wadsworth, 1968.

Neter, J., Wasserman, W., and Kutner, M. H. *Applied Linear Statistical Models*, 3rd ed. Homewood, Ill.: Richard D. Irwin, 1989.

Younger, M. S. *A First Course in Linear Regression*, 2nd ed. Boston: Duxbury, 1985.

A Procedure for Inverting a Matrix

There are several different methods for inverting matrices. All are tedious and time-consuming. Consequently, in practice, you will invert almost all matrices using an electronic computer. The purpose of this section is to present one method so that you will be able to invert small (2×2 or 3×3) matrices manually and so that you will appreciate the enormous computing problem involved in inverting large matrices (and, consequently, in fitting linear models containing many terms to a set of data). In particular, you will be able to understand why rounding errors creep into the inversion process and, consequently, why two different computer programs might invert the same matrix and produce inverse matrices with slightly different corresponding elements.

The procedure we will demonstrate to invert a matrix $\mathbf{A}$ requires us to perform a series of operations on the rows of the $\mathbf{A}$ matrix. For example, suppose

$$\mathbf{A} = \begin{bmatrix} 1 & -2 \\ -2 & 6 \end{bmatrix}$$

We will identify two different ways to operate on a row of a matrix:*

1. We can multiply every element in one particular row by a constant, c. For example, we could operate on the first row of the $\mathbf{A}$ matrix by multiplying every element in the row by a constant, say, 2. Then the resulting row would be $[2 \quad -4]$.

2. We can operate on a row by multiplying another row of the matrix by a constant and then adding (or subtracting) the elements of that row to elements in corresponding positions in the row operated upon. For example, we could operate on the first row of the $\mathbf{A}$ matrix by multiplying the second row by a constant, say, 2:

$$2[-2 \quad 6] = [-4 \quad 12]$$

Then we add this row to row 1:

$$[(1 - 4) \quad (-2 + 12)] = [-3 \quad 10]$$

---

*We omit a third row operation, because it would add little and could be confusing.

Note one important point. We operated on the *first* row of the A matrix. Although we used the second row of the matrix to perform the operation, *the second row would remain unchanged*. Therefore, the row operation on the A matrix that we have just described would produce the new matrix,

$$\begin{bmatrix} -3 & 10 \\ -2 & 6 \end{bmatrix}$$

Matrix inversion using row operations is based on an elementary result from matrix algebra. It can be shown (proof omitted) that performing a series of row operations on a matrix A is equivalent to multiplying A by a matrix B, i.e., row operations produce a new matrix, BA. This result is used as follows: Place the A matrix and an identity matrix I of the same dimensions side by side. Then perform the same series of row operations on both A and I until the A matrix has been changed into the identity matrix I. This means that you have multiplied both A and I by some matrix B such that:

$$A = \begin{bmatrix} \end{bmatrix} \qquad I = \begin{bmatrix} 1 & 0 & 0 & \cdots & 0 \\ 0 & 1 & 0 & \cdots & 0 \\ 0 & 0 & 1 & \cdots & 0 \\ \vdots & \vdots & \vdots & & \vdots \\ 0 & 0 & 0 & \cdots & 1 \end{bmatrix}$$

$$\downarrow \qquad \leftarrow \text{Row operations change A to I} \rightarrow \qquad \downarrow$$

$$I = \begin{bmatrix} \end{bmatrix} \qquad B = \begin{bmatrix} \end{bmatrix}$$

$$BA = I \quad \text{and} \quad BI = B$$

Since $BA = I$, it follows that $B = A^{-1}$. Therefore, as the A matrix is transformed by row operations into the identity matrix I, the identity matrix I is transformed into A^{-1}, i.e.,

$$BI = B = A^{-1}$$

We will show you how this procedure works with two examples.

EXAMPLE B.1

Find the inverse of the matrix

$$A = \begin{bmatrix} 1 & -2 \\ -2 & 6 \end{bmatrix}$$

Solution

Place the A matrix and a 2×2 identity matrix side by side and then perform the following series of row operations (we will indicate by arrow the row operated upon in each operation):

$$A = \begin{bmatrix} 1 & -2 \\ -2 & 6 \end{bmatrix} \qquad I = \begin{bmatrix} 1 & 0 \\ 0 & 1 \end{bmatrix}$$

OPERATION 1 Multiply the first row by 2 and add to the second row:

$$\rightarrow \begin{bmatrix} 1 & -2 \\ 0 & 2 \end{bmatrix} \qquad \begin{bmatrix} 1 & 0 \\ 2 & 1 \end{bmatrix}$$

OPERATION 2 Multiply the second row by $\frac{1}{2}$:

$$\rightarrow \begin{bmatrix} 1 & -2 \\ 0 & 1 \end{bmatrix} \qquad \begin{bmatrix} 1 & 0 \\ 1 & \frac{1}{2} \end{bmatrix}$$

OPERATION 3 Multiply the second row by 2 and add it to the first row:

$$\rightarrow \begin{bmatrix} 1 & 0 \\ 0 & 1 \end{bmatrix} \qquad \begin{bmatrix} 3 & 1 \\ 1 & \frac{1}{2} \end{bmatrix}$$

Thus,

$$A^{-1} = \begin{bmatrix} 3 & 1 \\ 1 & \frac{1}{2} \end{bmatrix}$$

(Note that our solution matches the one obtained using Theorem A.2.)

The final step in finding an inverse is to check your solution by finding the product $A^{-1}A$ to see if it equals the identity matrix I. To check:

$$A^{-1}A = \begin{bmatrix} 3 & 1 \\ 1 & \frac{1}{2} \end{bmatrix}\begin{bmatrix} 1 & -2 \\ -2 & 6 \end{bmatrix}$$

$$= \begin{bmatrix} 1 & 0 \\ 0 & 1 \end{bmatrix}$$

Since this product is equal to the identity matrix, it follows that our solution for A^{-1} is correct.

EXAMPLE B.2

Find the inverse of the matrix

$$A = \begin{bmatrix} 2 & 0 & 3 \\ 0 & 4 & 1 \\ 3 & 1 & 2 \end{bmatrix}$$

Solution

Place an identity matrix alongside the **A** matrix and perform the row operations:

OPERATION 1 Multiply row 1 by $\frac{1}{2}$:

$$\rightarrow \begin{bmatrix} 1 & 0 & \frac{3}{2} \\ 0 & 4 & 1 \\ 3 & 1 & 2 \end{bmatrix} \qquad \begin{bmatrix} \frac{1}{2} & 0 & 0 \\ 0 & 1 & 0 \\ 0 & 0 & 1 \end{bmatrix}$$

OPERATION 2 Multiply row 1 by 3 and subtract from row 3:

$$\begin{bmatrix} 1 & 0 & \frac{3}{2} \\ 0 & 4 & 1 \\ \rightarrow 0 & 1 & -\frac{5}{2} \end{bmatrix} \qquad \begin{bmatrix} \frac{1}{2} & 0 & 0 \\ 0 & 1 & 0 \\ -\frac{3}{2} & 0 & 1 \end{bmatrix}$$

OPERATION 3 Multiply row 2 by $\frac{1}{4}$:

$$\rightarrow \begin{bmatrix} 1 & 0 & \frac{3}{2} \\ 0 & 1 & \frac{1}{4} \\ 0 & 1 & -\frac{5}{2} \end{bmatrix} \qquad \begin{bmatrix} \frac{1}{2} & 0 & 0 \\ 0 & \frac{1}{4} & 0 \\ -\frac{3}{2} & 0 & 1 \end{bmatrix}$$

OPERATION 4 Subtract row 2 from row 3:

$$\begin{bmatrix} 1 & 0 & \frac{3}{2} \\ 0 & 1 & \frac{1}{4} \\ \rightarrow 0 & 0 & -\frac{11}{4} \end{bmatrix} \qquad \begin{bmatrix} \frac{1}{2} & 0 & 0 \\ 0 & \frac{1}{4} & 0 \\ -\frac{3}{2} & -\frac{1}{4} & 1 \end{bmatrix}$$

OPERATION 5 Multiply row 3 by $-\frac{4}{11}$:

$$\begin{bmatrix} 1 & 0 & \frac{3}{2} \\ 0 & 1 & \frac{1}{4} \\ \rightarrow 0 & 0 & 1 \end{bmatrix} \qquad \begin{bmatrix} \frac{1}{2} & 0 & 0 \\ 0 & \frac{1}{4} & 0 \\ \frac{12}{22} & \frac{1}{11} & -\frac{4}{11} \end{bmatrix}$$

OPERATION 6 Operate on row 2 by subtracting $\frac{1}{4}$ of row 3:

$$\rightarrow \begin{bmatrix} 1 & 0 & \frac{3}{2} \\ 0 & 1 & 0 \\ 0 & 0 & 1 \end{bmatrix} \qquad \begin{bmatrix} \frac{1}{2} & 0 & 0 \\ -\frac{3}{22} & \frac{5}{22} & \frac{1}{11} \\ \frac{12}{22} & \frac{1}{11} & -\frac{4}{11} \end{bmatrix}$$

OPERATION 7 Operate on row 1 by subtracting $\frac{3}{2}$ of row 3:

$$\rightarrow \begin{bmatrix} 1 & 0 & 0 \\ 0 & 1 & 0 \\ 0 & 0 & 1 \end{bmatrix} \quad \begin{bmatrix} -\frac{7}{22} & -\frac{3}{22} & \frac{6}{11} \\ -\frac{3}{22} & \frac{5}{22} & \frac{1}{11} \\ \frac{6}{11} & \frac{1}{11} & -\frac{4}{11} \end{bmatrix} = \mathbf{A}^{-1}$$

To check the solution, we find the product:

$$\mathbf{A}^{-1}\mathbf{A} = \begin{bmatrix} -\frac{7}{22} & -\frac{3}{22} & \frac{6}{11} \\ -\frac{3}{22} & \frac{5}{22} & \frac{1}{11} \\ \frac{6}{11} & \frac{1}{11} & -\frac{4}{11} \end{bmatrix}\begin{bmatrix} 2 & 0 & 3 \\ 0 & 4 & 1 \\ 3 & 1 & 2 \end{bmatrix}$$

$$= \begin{bmatrix} 1 & 0 & 0 \\ 0 & 1 & 0 \\ 0 & 0 & 1 \end{bmatrix}$$

Since the product $\mathbf{A}^{-1}\mathbf{A}$ is equal to the identity matrix, it follows that our solution for $\mathbf{A}^{-1}$ is correct.

Examples B.1 and B.2 indicate the strategy employed when performing row operations on the $\mathbf{A}$ matrix to change it into an identity matrix. Multiply the first row by a constant to change the element in the top left row into a 1. Then perform operations to change all elements in the first column into 0's. Then operate on the second row and change the second diagonal element into a 1. Then operate to change all elements in the second column beneath row 2 into 0's. Then operate on the diagonal element in row 3, etc. When all elements on the main diagonal are 1's and all below the main diagonal are 0's, perform row operations to change the last column to 0; then the next-to-last, etc., until you get back to the first column. The procedure for changing the off-diagonal elements to 0's is indicated diagrammatically as shown:

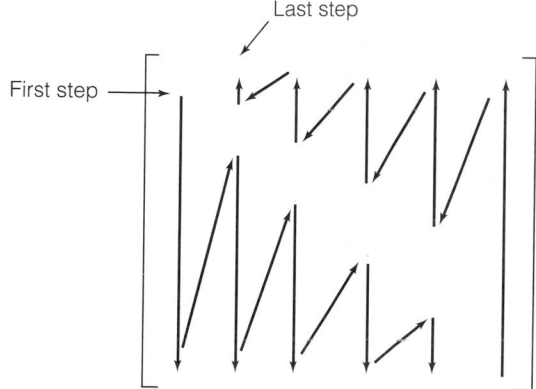

The preceding instructions on how to invert a matrix using row operations suggest that the inversion of a large matrix would involve many multiplications, subtractions, and additions and, consequently, could produce large rounding errors in the calculations unless you carry a large number of significant figures in the calculations. This explains why two different multiple regression analysis computer programs may produce different estimates of the same β parameters, and it emphasizes the importance of carrying a large number of significant figures in all computations when inverting a matrix.

Other methods for inverting matrices are discussed in the references. All work—exactly—in theory. It is only in the actual process of performing the calculations that the rounding errors occur.

EXERCISE

B.1   Invert the following matrices and check your answers to make certain that $A^{-1}A = AA^{-1} = I$:

a. $A = \begin{bmatrix} 3 & 2 \\ 4 & 5 \end{bmatrix}$ b. $A = \begin{bmatrix} 3 & 0 & -2 \\ 1 & 4 & 2 \\ 5 & 1 & 1 \end{bmatrix}$

c. $A = \begin{bmatrix} 1 & 0 & 1 \\ 0 & 2 & 1 \\ 1 & 1 & 3 \end{bmatrix}$ d. $A = \begin{bmatrix} 4 & 0 & 10 \\ 0 & 10 & 0 \\ 10 & 0 & 5 \end{bmatrix}$

[*Note*:   No answers are given to these exercises. You will know if your answer is correct if $A^{-1}A = I$.]

Performing Regression Analysis Using the Computer: SAS, SPSS, and Minitab Commands

CONTENTS

C.1 Introduction

C.2 Data Entry: SAS

C.3 Data Entry: SPSS

C.4 Data Entry: Minitab

C.5 Relative Frequency Distributions, Descriptive Statistics, Correlations, and Plots

C.6 Multiple Regression

C.7 Stepwise Regression

C.8 Residual Analysis and Regression Diagnostics

C.9 Analysis of Variance

C.1

Introduction

This appendix explains how to perform a regression analysis using each of three popular statistical software packages: SAS, SPSS, and Minitab. At least one of these programs is available at most college and university computing centers. The packages are also compatible with the computers used by many private corporations, foundations, and government agencies; moreover, all three packages have software versions available for use on an IBM-compatible personal computer (PC).

A statistical software package provides "canned" programs for a variety of statistical methods, including regression analysis. SAS, SPSS, and Minitab are similar in that they require the user to enter a list of instructions in a specific form and a specific order. Each utilizes the following three basic types of instructions:*

---

*When using these packages on a mainframe computer, additional instructions are required. The basic function of these statements, called **job control language (JCL) instructions**, is to inform the computer which statistical program package to execute. The appropriate JCL may vary from one computing center to another. Your instructor will inform you of the appropriate JCL commands to use at your institution.

1. *Data entry commands:* instructions on how the data will be entered
2. *Input data values:* the numeric values of the variables in the data set
3. *Statistical analysis commands:* instructions on the type of analysis to be performed on the data

The data entry instructions of each of the three packages, SAS, SPSS, and Minitab, are provided in Sections C.2, C.3, and C.4, respectively. The appropriate statistical analysis commands are given in the remaining sections of this appendix.

C.2
Data Entry: SAS

All SAS programs consist of at least two steps: (1) DATA steps—used to create data sets (data entry instructions), and (2) PROC (procedure) steps—used to analyze these data sets (statistical analysis instructions).*

Let us now create a SAS data set consisting of the sale prices for a sample of five residential properties listed in Table C.1.

TABLE C.1 **Sale Prices of Five Residential Properties**

$81,300
54,100
96,900
72,900
51,200

Each of the nine statements in Program C.1 is entered as a line of type on the computer screen, beginning in any column.

PROGRAM C.1

SAS statements for creating a data set consisting of five sale prices

Command line		
1	DATA SALES;	⎫
2	INPUT SALEPRIC;	⎬ Data entry instructions
3	CARDS;	⎭
4	81300	⎫
5	54100	⎪
6	96900	⎬ Input data values (1 observation per line)
7	72900	⎪
8	51200	⎭
9	PROC PRINT;	} Print instruction

---

*When SAS is used on a mainframe computer, the necessary JCL statements precede all the SAS data entry and statistical analysis instructions. For example, the JCL instruction

    // EXEC SAS

commands the computer to execute the SAS program package. Your instructor will give you the JCL instructions required at your computing center.

COMMAND 1   The first command in any SAS program is usually the DATA statement. In Program C.1, the DATA statement asks SAS to create a data set named SALES. In a DATA statement, the word DATA must be followed by a blank and the name of the data set that you want to create. The name SALES was chosen arbitrarily for our example. SAS will accept any data set or variable name with a maximum of eight characters (letters or numbers), as long as the characters are *not* separated by blanks and the name does *not* begin with a number. For example, MYHOUSE is a legitimate SAS name, but MY HOUSE, 7ELEVEN, and APPRAISERS are not. The name MY HOUSE includes a blank, 7ELEVEN begins with a number, and APPRAISERS includes more than eight characters.

COMMAND 2   The INPUT statement follows the DATA statement and describes the variables that were measured to form the data set. The variable names must follow the word INPUT and must be separated by blanks. In this example, the data set SALES contains measurements on the single variable called SALEPRIC, which again was arbitrarily named.

COMMAND 3   The CARDS statement signals SAS that the input data values follow immediately. The computer reads the numeric value of the variable SALEPRIC on each of the five data lines, thereby creating a data set with a total of five observations.

COMMANDS 4–8   The data lines contain the actual raw data—in this case, the five sale prices listed in Table C.1. Each data line includes only a single observation—that is, a single value of SALEPRIC.

COMMAND 9   The PRINT procedure (PROC) will produce a listing of the entire data set.

Note that all SAS statements, with the exception of the data values, end with a semicolon. In general, each SAS statement *must* end in a semicolon. The *only* exceptions to this rule are the input data lines (in this example, lines 4–8).

EXAMPLE C.1

a. Write the appropriate SAS statements to create a data set that contains information on the location (neighborhood), sale price, appraised land value, and appraised value of improvements for the three residential properties described in Table C.2.

TABLE C.2

PROPERTY	NBRHOOD	SALEPRIC	LANDVAL	IMPROVAL
1	B	72,500	12,000	56,000
2	B	97,000	15,000	79,500
3	A	121,300	22,100	85,100

b. Write the appropriate SAS statements that will compute the sum of the appraised land and improvement values, add this new variable to the data set, and print the results.

Solution

a. The six statements shown in Program C.2 will create the desired data set. The data set, which we have called SALES, has three observations (properties), with four variables—NBRHOOD, SALEPRIC, LANDVAL, and IMPROVAL— per observation. Notice that a dollar sign ($) follows the variable named NBRHOOD on the INPUT statement (line 2). This is because NBRHOOD is a **character** variable—that is, a qualitative variable—whose values (A or B) are characters, not numbers. In contrast, SALEPRIC, LANDVAL, and IMPROVAL are read as **numeric** variables, since their values are quantitative. Whenever you want to include a character (nonnumeric) variable in a SAS data set, a dollar sign must follow the variable name in the INPUT statement. This informs the computer that the values of this variable to be read on the data lines are observations on a qualitative variable.

PROGRAM C.2   SAS statements for creating the property data set

Command
line

```
1     DATA SALES;                                        Data entry
2     INPUT NBRHOOD   $  SALEPRIC LANDVAL IMPROVAL;      instructions
3     CARDS;
4     B    72500   12000   56000      Input data values
5     B    97000   15000   79500      (1 observation per line)
6     A   121300   22100   85100
```

Note that the order in which the names of the variables appear in the INPUT statement must be the same as the order in which the values of the variables appear on the data lines (lines 4–6). Thus, the value of NBRHOOD appears first on the data lines, followed by the values of SALEPRIC, LANDVAL, and IMPROVAL, respectively. As long as these values are separated by at least one blank and appear in the proper order, they may begin in any of the 80 columns of the screen. For example, lines 4–6 could have been typed as follows:

Command   Column 1
line   ↓

```
4          B   72500        12000        56000
5      B                97000        15000        79500
6                         A          121300 22100 85100
```

b. The SAS statements shown in Program C.3 expand the data set of part **a** by creating an additional variable, named TOTVAL, which represents the sum of the appraised land and improvement values. The additional statement needed is that shown on line 3:

```
TOTVAL = LANDVAL + IMPROVAL;
```

SAS will now automatically compute TOTVAL for each of the three observations (residential properties) and include this variable on the data set.

P R O G R A M C.3 SAS statements for creating and printing the property sales data set with total appraised value added

Command
line

```
1     DATA SALES;
2     INPUT NBRHOOD  $  SALEPRIC LANDVAL IMPROVAL;   ⎫ Data entry
3     TOTVAL = LANDVAL + IMPROVAL;                   ⎬ instructions
4     CARDS;                                         ⎭
5     B    72500   12000   56000  ⎫
6     B    97000   15000   79500  ⎬ Input data values
7     A   121300   22100   85100  ⎭
8     PROC PRINT; }  Print instruction
```

The SAS data set, which we have named SALES, is stored in the computer as shown in Table C.3.*

T A B L E C.3 **SAS Data Set SALES Stored in the Computer**

OBSERVATION	NBRHOOD	SALEPRIC	LANDVAL	IMPROVAL	TOTVAL
1	B	72500	12000	56000	68000
2	B	97000	15000	79500	94500
3	A	121300	22100	85100	107200

Any SAS statements that are used to create variables in addition to those on the INPUT statement should follow the INPUT statement, but precede the CARDS statement. When creating or transforming variables in SAS or in any of the other computer packages we discuss in this appendix, you should use the standard arithmetic operation symbols, $+$, $-$, $*$, and $/$, for addition, subtraction, multiplication, and division, respectively. Two other important transformations in regression are the natural log and square root transformations. The SAS functions for the natural log and square root transformations follow:

```
LOGPRIC  = LOG(SALEPRIC);
SQRTPRIC = SQRT(SALEPRIC);
```

---

*The storage is temporary, not permanent. That is, SAS will store the data set on the computer temporarily until the program run is complete. Once the program is finished, the data set is no longer available for use and must be recreated in future programs. Consult the SAS references if you want to store a permanent SAS data set.

Once a data set has been created, you may begin analyzing it using PROC statements (e.g., PROC PRINT). These statistical analysis instructions will follow the input data values in the SAS program.

We end this section with a word of caution: Unlike some computer programming languages, a SAS statement may begin in *any* of the 80 columns of the computer terminal screen. Also, it is not necessary to begin each SAS statement on a new line. SAS statements can run continuously on a line, as long as each of the separate commands ends in a semicolon. However, we strongly recommend that beginning users of SAS follow the approach of Programs C.1–C.3: (1) Start typing in column 1, and (2) Begin each new SAS command on a new line. Our experience has indicated that those who follow this simple approach will make fewer programming errors and have an easier time "debugging" their programs.

C.3
Data Entry: SPSS

Commands or control statements for the **Statistical Package for the Social Sciences (SPSS)** may be classified into one of two groups: (1) data creation commands and (2) statistical procedure commands. Data creation commands are data entry instructions used to create a data set, and statistical procedure commands define the statistical procedure to be used to analyze the data set.*

Consider the SPSS statements of Program C.4, which create a data set containing the neighborhood, sale price, appraised land value, appraised improvements, and total appraised value of the three properties of Example C.1. Each SPSS command begins with a **command keyword** (which may contain more than one word). Although a few commands (such as BEGIN DATA and END DATA) are complete in themselves, most require **specifications**. Specifications are made up of variable names, subcommands, numbers, and other keywords required to complete an SPSS command.

P R O G R A M C.4   SPSS statements for creating the property sales data set

Command line

```
1    DATA LIST   FREE/SALEPRIC LANDVAL IMPROVAL NBRHOOD (A1),    ⎫  Data entry
2    COMPUTE      TOTVAL = LANDVAL + IMPROVAL                     ⎬  instructions
3    BEGIN DATA,                                                 ⎭
4     72500   12000    56000    B  ⎫  Input data values
5     97000   15000    79500    B  ⎬  (1 observation per line)
6    121300   22100    85100    A  ⎭
7    END DATA,
8    LIST,      } Print instruction
9    FINISH,
```

*The appropriate JCL instructions for a mainframe computer must precede all the SPSS control statements. For example, the JCL statement

```
// EXEC SPSS
```

commands the computer to begin executing the SPSS program package. See your instructor for the JCL instructions required at your institution.

COMMAND 1   The first SPSS statement, the DATA LIST command, identifies the variables to be read from the input data lines. Specifications for this command include formats and names for the variables in the data set.

SPSS offers various options for arranging your data on the data lines. We have entered the SPSS keyword FREE after the DATA LIST command. The FREE option enables you to enter the data anywhere on the data line, as long as the variable values are separated by at least one blank and the order in which the values appear is consistent with the order given in the variable list (described in the following paragraph). Since the freefield-format mode of data entry is the simplest to describe and use, we will illustrate this method of entry in all our SPSS examples. If you want to use another method of data entry, you will need to consult the SPSS references given at the end of this appendix.

The names of the variables in the data set are listed following the slash (/) in the DATA LIST statement. These variable names (each up to eight characters long) must be separated by blanks. As in Example C.1, we have named the variables NBRHOOD, SALEPRIC, LANDVAL, and IMPROVAL. An alphanumeric format of the form (An) must be specified, in parentheses, after the name of any nonnumeric variable. In Program C.4, the format A1 implies that values of the nonnumeric variable NBRHOOD will occupy a single column on the input data lines.*

COMMAND 2   The COMPUTE command is used to create a new variable, arbitrarily called TOTVAL, by summing the values of LANDVAL and IMPROVAL. The algebraic expression specified in the COMPUTE statement may begin in any column following the COMPUTE command.†

SPSS COMPUTE statements use the standard arithmetic operation symbols, $+$, $-$, $*$, and $/$, for addition, subtraction, multiplication, and division, respectively. The natural log and square root transformations in SPSS are shown here:

```
COMPUTE    LOGPRIC    =    LN(SALEPRIC)
COMPUTE    SQRTPRIC   =    SQRT(SALEPRIC)
```

COMMAND 3   After compiling the necessary SPSS data entry instructions, the computer is ready to begin reading the input data. The BEGIN DATA command instructs SPSS to start this process and signals the computer that the input data values (lines 4–6 of Program C.4) follow immediately.

COMMANDS 4–6   The input data values follow the BEGIN DATA command. Each data line represents a single observation (or **case**, as it is called in SPSS). The only variables whose values must appear on the data lines are those named on

---

*In the mainframe SPSS environment, an asterisk (*) separates the numeric variables from the alphanumeric variables, with the numeric variables listed first.

†The COMPUTE command is an optional SPSS statement. It is necessary only when you want to create additional variables or transform existing variables.

the DATA LIST command. In our example, we do *not* enter the value of TOT-VAL—the sum of the appraised values—on the data lines. The COMPUTE statement guarantees that SPSS will automatically calculate TOTVAL for each observation and include it in the data set.

Since the freefield format was specified in command 1, the values of the variables can appear anywhere on the data lines, as long as they are separated by at least one blank and the order of appearance is consistent with the order specified in the DATA LIST command. Thus, the first value on the data line is the value of SALEPRIC, the second is the value of LANDVAL, and so forth.

COMMAND 7   To alert SPSS that all the input data values have been read, insert an END DATA statement after the last data line.

COMMAND 8   The LIST command will produce a listing of the data for all the variables in the data set, including variables created using COMPUTE commands.

COMMAND 9   ALL SPSS programs terminate with a FINISH command.

The (temporary) SPSS data file that results from Program C.4 is stored in the computer as indicated in Table C.4.* Now that we have created the SPSS data file, we are ready to begin analyzing it using SPSS statistical procedure commands. These analysis instructions will be inserted *after* the input data values.

T A B L E C.4   **SPSS Data File Stored in the Computer**

CASE	NBRHOOD	SALEPRIC	LANDVAL	IMPROVAL	TOTVAL
1	B	72500	12000	56000	68000
2	B	97000	15000	79500	94500
3	A	121300	22100	85100	107200

Note: In the PC environment, all SPSS commands must end with a command terminator (usually a period); the only exceptions to this rule are input data values. Omit the periods when using mainframe SPSS.

C.4
Data Entry: Minitab

Minitab is an easy-to-use statistical software package that has been designed especially for those who have no previous experience with computers. Minitab commands can be given in English; that is, the commands can be expressed in nearly the same language you would use to instruct someone to do the computations, tasks, or analyses by hand. Also, Minitab commands may begin and end in any column of the command line. We illustrate with Example C.2.

*Consult the SPSS references for instructions on how to save a permanent SPSS data file.

EXAMPLE C.2

Refer to Example C.1. Write the Minitab commands that will create the property sales data set, which is reproduced in Table C.5. Include the total of the appraised land and improvements values on the data set.

TABLE C.5

PROPERTY	NBRHOOD	SALEPRIC	LANDVAL	IMPROVAL
1	B	72500	12000	56000
2	B	97000	15000	79500
3	A	121300	22100	85100

Solution

The appropriate Minitab statements are given in Program C.5.

PROGRAM C.5   Minitab statements for creating property sales data set

Command
line

```
1    READ  C1 C2 C3  C4 }  Data entry instruction
2    1      72500    12000    56000 ⎤
3    1      97000    15000    79500 ⎬  Input data values
4    0     121300    22100    85100 ⎦  (1 observation per line)
5    ADD   C3 C4 PUT INTO C5 }  Data entry instruction
6    NAME  C1='NBRDUMMY' C2='SALEPRIC' C3='LANDVAL' ⎤
7    NAME  C4='IMPROVAL' C5='TOTVAL'                ⎦  Name variables
8    PRINT C1-C5 }  Print instruction
9    STOP
```

COMMAND 1   Minitab stores data in a "worksheet" that it maintains inside the computer. The Minitab READ statement instructs the computer to read the data on lines 2–4 and to put the values of the variables into specific "columns" of the worksheet. (In contrast to the columns of a terminal screen, the columns of a Minitab worksheet may contain values with more than one digit, for example, 90, 128, 10772.) The columns are designated by the letter C and a corresponding number. Instead of referring to a variable by its name, we refer to it in Minitab by the column into which it is placed (for example, C1, C2, . . .). In this example, the computer reads the value of the first variable (neighborhood) and places it into column 1 (C1) of the worksheet. Likewise, the value of the second variable (sale price) is read and placed into column 2 (C2); the value of the third variable (land value) goes into column 3 (C3); and the value of the fourth variable (improvements value) goes into column 4 (C4). In Minitab, the input data values always follow the READ command.

COMMANDS 2–4   The input data lines contain the actual values of the variables read in the worksheet columns. These values must be separated by at least one blank. Notice, however, that the values of the qualitative variable neighborhood in C1 are entered as numbers (0 or 1) rather than as letters (A or B). Minitab requires that all data used in statistical analyses be numerical. Thus, if we want to use the qualitative variable neighborhood in a statistical procedure, we need to convert its possible values to numbers. In this example, we arbitrarily let 1 represent neighborhood B and 0 represent neighborhood A.

COMMAND 5   The Minitab command ADD instructs the computer to add the values of the variables stored in columns 3 and 4 and to place the sum in column 5 (a new column in the worksheet that represents the total appraised value).

Minitab commands for the other usual arithmetic operations (subtraction, multiplication, and division) and for natural log and square root transformations are illustrated here.

```
SUBTRACT C2 FROM C3 PUT INTO C6
MULTIPLY C2 BY 2 PUT INTO C7
DIVIDE C4 BY 10 PUT INTO C8
LOGE OF C2 PUT INTO C9
SQRT OF C2 PUT INTO C10
```

(Note the words FROM, BY, OF, PUT, and INTO may be omitted from the Minitab commands for arithmetic operations.)

COMMANDS 6–7   The NAME command is used to assign descriptive names to the columns of the Minitab worksheet for labeling printouts. In future Minitab statements, you may refer to the columns by these names (e.g., NBRDUMMY) or by the column numbers (e.g., C1).

COMMAND 8   The PRINT command will produce a listing of the data in the Minitab worksheet for the specified variables (columns).

COMMAND 9   All Minitab programs terminate with the STOP command.

The Minitab worksheet, stored (temporarily) in the computer as shown in Table C.6,* is now ready for data analysis.

T A B L E C.6   **Minitab Worksheet Stored in the Computer**

OBSERVATION	NBRDUMMY C1	SALEPRIC C2	LANDVAL C3	IMPROVAL C4	TOTVAL C5
1	1	72500	12000	56000	68000
2	1	97000	15000	79500	94500
3	0	121300	22100	85100	107200

[*Note:* The commands of Program C.5 were written using the minimal amount of coded text required by Minitab. However, you may insert key words

*Consult the Minitab references for instructions on how to save a permanent Minitab data file.

within each command to help you follow the logic of the program. For example, line 1 of Program C.5 could be written as follows:

```
READ NBRDUMMY IN C1, SALE PRICE IN C2, APPRAISED VALUES IN C3 AND C4
```

Minitab ignores the extraneous words (NBRDUMMY, SALE PRICE, and so forth), and reads only the command name (READ) and its arguments (C1, C2, C3, C4).]

C.5

Relative Frequency Distributions, Descriptive Statistics, Correlations, and Plots

Prior to actually running the regression analysis, it is often helpful to conduct a preliminary analysis of the data. This may include examining relative frequency distributions and descriptive statistics (such as the mean, median, and standard deviation) for both the dependent and independent variables, correlations between pairs of variables, and plots of the dependent variable against each of the independent variables.

Suppose we randomly select 50 condominium units from Appendix G and record

Dependent variable:

y = Sale price of the unit in thousands of dollars (PRICPAID)

Independent variables:

x_1 = Floor height (FLOORHGT)

x_2 = Distance from elevator (DISTELEV)

$$x_3 = \begin{cases} 1 & \text{if ocean-view (VIEWOCN)} \\ 0 & \text{if not} \end{cases}$$

The necessary statements for conducting a preliminary analysis on these data are given for each of the three software packages, SAS, SPSS, and Minitab, in Programs C.6–C.8 (pages 792–793), respectively. Each program will produce the following:

1. A relative frequency distribution for the sale price (y) of a unit
2. Descriptive (univariate) statistics for the quantitative variables sale price (y), floor height (x_1), and distance from elevator (x_2)
3. A table of frequencies for the values of the qualitative variable ocean-view (x_3)
4. Pairwise correlations between sale price (y), floor height (x_1), distance from elevator (x_2), and ocean-view (x_3)
5. Plots of sale price (y) against floor height (x_1) and distance from elevator (x_2)

Some of the statements shown in Programs C.6–C.8 are optional statements included solely for the purpose of labeling the printouts, and thus may be omitted. Consult the references at the end of this appendix for details on optional commands and their purpose.

P R O G R A M C.6   SAS statements for conducting a preliminary analysis on the sample data from Appendix G

Command
line

```
1     DATA CONDO;                                    ⎫
2     INPUT PRICPAID FLOORHGT DISTELEV VIEWOCN @@;   ⎬ Data entry instructions
3     CARDS;                                         ⎭
4     199   1   2   0   204   1   13   0   ⎫
.       ·   ·   ·   ·    ·    ·    ·   ·   ⎪  Input data values
.       ·   ·   ·   ·    ·    ·    ·   ·   ⎬  (2 observations per line)
.       ·   ·   ·   ·    ·    ·    ·   ·   ⎪
28    169   3   1   0   306   3   15   1   ⎭
29    PROC CHART;                          ⎫ Relative frequency distribution
30    VBAR PRICPAID/TYPE=PERCENT;          ⎭ for sale price
31    PROC FREQ;                           ⎫
32    TABLES VIEWOCN;                       ⎭ Frequency table for ocean-view
33    PROC UNIVARIATE;                     ⎫ Descriptive (univariate) statistics
34    VAR PRICPAID FLOORHGT DISTELEV;      ⎭
35    PROC CORR;                           ⎫ Matrix of pairwise corre-
36    VAR PRICPAID FLOORHGT DISTELEV VIEWOCN;  ⎭ lations between variables
37    PROC PLOT;                           ⎫ Plots of sale price against
38    PLOT PRICPAID*(FLOORHGT DISTELEV)=VIEWOCN;  ⎭ independent variables
```

COMMAND 2   The @@ symbols at the end of the INPUT statement permit multiple observations on the input data lines.

COMMAND 30   TYPE=PERCENT produces a percentage bar chart (that is, a relative frequency distribution). If TYPE is omitted, a frequency distribution is constructed.

COMMAND 38   The option =VIEWOCN produces scatterplots with the values of ocean-view (0 or 1) as plotting symbols.

P R O G R A M C.7   SPSS statements for conducting a preliminary analysis on the sample data from Appendix G

Command
line

```
1     DATA LIST FREE/PRICPAID FLOORHGT DISTELEV VIEWOCN. } Data entry instructions
2     BEGIN DATA.
3     199   1   2   0   204   1   13   0   ⎫
.       ·   ·   ·   ·    ·    ·    ·   ·   ⎪  Input data vlues
.       ·   ·   ·   ·    ·    ·    ·   ·   ⎬  (2 observations per line)
.       ·   ·   ·   ·    ·    ·    ·   ·   ⎪
27    169   3   1   0   306   3   15   1   ⎭
28    END DATA.
29    FREQUENCIES   VARIABLES = PRICPAID/    ⎫ Relative frequency distribution
30                  HISTOGRAM = PERCENT.     ⎭ for sale price
31    FREQUENCIES   VARIABLES = VIEWOCN/     ⎫ Frequency bar chart for ocean-view
32                  BARCHART.                ⎭
33    FREQUENCIES   VARIABLES = PRICPAID FLOORHGT DISTELEV/  ⎫ Descriptive statistics
34                  STATISTICS = ALL.        ⎭
35    CORRELATION   VARIABLES = PRICPAID FLOORHGT DISTELEV VIEWOCN.  ⎫ Pairwise corre-
                                                                      ⎭ lation matrix
                                                             ⎫ Plots of sale
36    PLOT   /PLOT PRICPAID WITH FLOORHGT DISTELEV BY VIEWOCN.  ⎬ price against inde-
                                                             ⎭ pendent variables
37    FINISH.
```

COMMAND 1   The FREE subcommand permits multiple observations (cases) on the input data lines.

COMMAND 30   HISTOGRAM=PERCENT produces a percentage histogram (i.e., relative frequency distribution). If PERCENT is omitted, a frequency distribution is ̲constructed.

COMMAND 36   The option BY VIEWOCN produces scatterplots with the values of ocean-view (0 or 1) as plotting symbols.

P R O G R A M C.8   Minitab statements for conducting a preliminary analysis on the sample data from Appendix G

**Command
line**

```
 1    READ C1-C4 }  Data entry instructions
 2    199  1   2   0  ⎫
 3    204  1  13   0  ⎪
 .     ∙   ∙   ∙   ∙  ⎬  Input data values
 .     ∙   ∙   ∙   ∙  ⎪  (1 observation per line)
 .     ∙   ∙   ∙   ∙  ⎭
51    306  3  15   1
52    NAME C1='PRICPAID' C2='FLOORHGT' C3='DISTELEV' C4='VIEWOCN' }  Name variables
53    HISTOGRAM C1       }  Frequency distribution for sale price
54    HISTOGRAM C4       }  Frequency bar chart of ocean-view
55    DESCRIBE C1-C3     }  Descriptive (univariate) statistics
56    CORRELATION C1-C4  }  Pairwise correlation matrix
57    PLOT C1 VS. C2  ⎫
58    PLOT C1 VS. C3  ⎬  Plots of sale price against independent variables
59    PLOT C1 VS. C4  ⎭
60    STOP
```

C.6
Multiple Regression

For all three of the statistical software packages discussed in this appendix, simple and multiple regression analyses are performed using similar commands.

Suppose we want to fit the multiple regression model

$$y = \beta_0 + \beta_1 x_1 + \beta_2 x_2 + \beta_3 x_1 x_2 + \beta_4 x_1^2 + \beta_5 x_1^2 x_2 + \varepsilon$$

to data collected for a sample of residential properties, where

y = Sale price of a residential property

x_1 = Total appraised value of the property

$x_2 = \begin{cases} 1 & \text{if property located in neighborhood B} \\ 0 & \text{if property located in neighborhood A} \end{cases}$

The SAS, SPSS, and Minitab commands for conducting the multiple regression analysis are given in Programs C.9–C.11 (pages 794–795), respectively. Note that in all three programs, the values of the qualitative variable, neighborhood, are specified as 0 and 1 on the input data lines, and the higher-order terms in the model ($x_1 x_2$, x_1^2, and $x_1^2 x_2$) have been created through data transformation statements. The SPSS and Minitab regression procedures require the user to create the necessary qualitative dummy variables and higher-order terms for the independent variable. SAS is capable of creating dummy variables and higher-order terms internally, but this requires the use of a more general model-fitting procedure.* For the purposes of this appendix, we give only the basic SAS regression commands.

P R O G R A M C.9   SAS statements for fitting the multiple regression model $y = \beta_0 + \beta_1 x_1 + \beta_2 x_2 + \beta_3 x_1 x_2 + \beta_4 x_1^2 + \beta_5 x_1^2 x_2 + \varepsilon$

Command line

```
1    DATA SALES;                    ⎫
2    INPUT Y X1 X2 @@;              ⎪
3    X1X2=X1*X2;                    ⎬  Data entry instructions
4    X1SQ=X1*X1;                    ⎪
5    X1SQX2=X1*X1*X2;               ⎭
6    CARDS;

     [Input data values]

7    PROC REG;                                      ⎫  Regression analysis
8    MODEL Y = X1 X2 X1X2 X1SQ X1SQX2               ⎬  instructions
9              /CLI;                                ⎭
```

COMMAND 9   The CLI option prints predicted values, residuals, and corresponding lower and upper 95% prediction limits. Specify CLI to get 95% confidence intervals for $E(y)$.

P R O G R A M C.10   SPSS statements for fitting the multiple regression model $y = \beta_0 + \beta_1 x_1 + \beta_2 x_2 + \beta_3 x_1 x_2 + \beta_4 x_1^2 + \beta_5 x_1^2 x_2 + \varepsilon$

Command line

```
1    DATA LIST    FREE/ Y X1 X2,        ⎫
2    COMPUTE      X1X2=X1*X2,           ⎪
3    COMPUTE      X1SQ=X1*X1,           ⎬  Data entry instructions
4    COMPUTE      X1SQX2=X1*X1*X2,      ⎪
5    BEGIN DATA,                        ⎭

     [Input data values]

6    END DATA,
7    REGRESSION   VARIABLES = Y, X1, X2, X1X2, X1SQ, X1SQX2/  ⎫  Regression
8                 CRITERIA=TOLERANCE(.00001)/                 ⎬  analysis
9                 DEPENDENT=Y/METHOD=ENTER/                   ⎪  instructions
10                CASEWISE=ALL,                               ⎭
11   FINISH,
```

COMMAND 8   A low tolerance level (in this case, .00001) is specified to guarantee that all the independent variables will be entered into the regression equation. (SPSS will omit an independent variable from the regression if its tolerance is less than the level specified.)

---

*The GLM procedure in SAS will create dummy variables and higher-order terms internally. Consult the SAS references for details on how to use GLM in regression.

COMMAND 9 Specifying ENTER on the METHOD subcommand forces all independent variables into the model.

COMMAND 10 The CASEWISE=ALL option prints predicted values and residuals for all observations (cases) in the analysis.

P R O G R A M C.11 Minitab statements for fitting the multiple regression model $y = \beta_0 + \beta_1 x_1 + \beta_2 x_2 + \beta_3 x_1 x_2 + \beta_4 x_1^2 + \beta_5 x_1^2 x_2 + \varepsilon$

Command line

1	`READ Y IN C1, PREDICTORS IN C2-C3` ⎫
2	`MULTIPLY C2 BY C3 PUT IN C4` ⎪
3	`MULTIPLY C2 BY C2 PUT IN C5` ⎬ Data entry instructions
4	`MULTIPLY C3 BY C5 PUT IN C6` ⎭
5	`NAME C1='Y', C2='X1', C3='X2'` ⎫
6	`NAME C4='X1X2', C5='X1SQ', C6='X2SQ'` ⎬ Name variables

[Input data values]

7	`REGRESS C1 ON 5 PREDICTORS, C2-C6` } Regression analysis instructions

C.7
Stepwise Regression

Stepwise regression routines are available in each of the three software packages discussed in this text. Although their methodologies and outputs may differ slightly, each provides an objective statistical screening procedure for eliminating unimportant variables from a large set of potential predictor variables.

The commands necessary to call the stepwise regression routines of SAS, SPSS, and Minitab are given in Programs C.12–C.14 (pages 796–797), respectively. In these examples, stepwise regression is applied to data collected for a sample of executives of a management consultant firm to determine which of the following 10 independent variables should be included in the construction of the final model for executive salary (y):

x_1 = Years of experience

x_2 = Years of education

$x_3 = \begin{cases} 1 & \text{if male} \\ 0 & \text{if female} \end{cases}$

x_4 = Number of employees supervised

x_5 = Corporate assets (millions of dollars)

$x_6 = \begin{cases} 1 & \text{if board member} \\ 0 & \text{if not} \end{cases}$

x_7 = Years of age

x_8 = Company profits of past 12 months (millions of dollars)

$x_9 = \begin{cases} 1 & \text{if international responsibility} \\ 0 & \text{if not} \end{cases}$

x_{10} = Company total sales of past 12 months (millions of dollars)

All three programs utilize the conventional "stepwise" variable selection technique. That is, after a variable has been added to the model, the program will recheck the previously entered variables and remove those that are not statistically significant at some prespecified α.

PROGRAM C.12   SAS stepwise regression commands for the executive salary data

Command line

```
1    DATA EXECS;          ⎫
2    INPUT Y X1-X10;      ⎬ Data entry instructions
3    CARDS;               ⎭

     [Input data values]

4    PROC STEPWISE;                                ⎫ Stepwise regression
5    MODEL Y=X1-X10/STEPWISE SLE=.15  SLS=.15;    ⎬ instructions
                                                   ⎭
```

COMMAND 4   The printed output includes R^2, the C_p statistic, and the F statistic for testing adequacy of the model at each step.

COMMAND 5   The SLE option specifies the significance level α required for a variable to enter into the model. (The default SLE is .15.) The SLS option specifies the significance level α required for a variable to stay in the model. (The default SLS is .15.) As an option, variable selection techniques other than STEPWISE are available. These include FORWARD (forward selection), BACKWARD (backward elimination), MAXR, and MINR. Consult the SAS references for more details on the use of these techniques. (The default selection technique is STEPWISE.)

PROGRAM C.13   SPSS stepwise regression commands for the executive salary data

Command line

```
1    DATA LIST    FREE/Y, X1 TO X10. } Data entry instructions
2    BEGIN DATA.

     [Input data values]

3    END DATA.
4    REGRESSION    VARIABLES=Y, X1 TO X10/          ⎫
5                  CRITERIA=PIN(.05) POUT(.10)/     ⎬ Stepwise regression
6                  DEPENDENT=Y/METHOD=STEPWISE.     ⎭ instructions
7    FINISH.
```

COMMAND 5   The PIN option of the CRITERIA subcommand specifies the significance level α required for a variable to enter into the model. (The default PIN is .05.) The POUT option specifies the significance level α required for a variable to stay in the equation. (The default POUT is .10.)

COMMAND 6   As an option, variable selection techniques other than STEPWISE are available. These include FORWARD (forward selection), BACKWARD (backward elimination), and TEST. Consult the SPSS references for more details on the use of these techniques.

PROGRAM C.14   Minitab stepwise regression commands for the executive salary data

Command line		
1	`READ Y IN C1, INDEPENDENTS IN C2-C11 }`	**Data entry instructions**
	[Input data values]	
2	`STEPWISE ON C1, PREDICTORS IN C2-C11;` ⎫	Stepwise regression
3	`FENTER=4  FREMOVE=4,` ⎭	instructions

COMMAND 3   The FENTER subcommand specifies a value of the F statistic required for a variable to enter into the model. If $F >$ FENTER, the variable is entered. (The default FENTER is 4.) The FREMOVE subcommand specifies a value of the F statistic required for a variable to stay in the model. If $F <$ FREMOVE, the variable is removed. (The default FREMOVE is 4.) Forward selection and backward elimination techniques are also available. Specify FREMOVE=0 for forward selection and FREMOVE=100000 ENTER=C2–C11 for backward elimination.

C.8
Residual Analysis and Regression Diagnostics

The regression routines of the three software packages have options for conducting a basic residual analysis—residual plots, histograms of residuals, and normal probability plots. In addition, SAS, SPSS, and Minitab are capable of producing variance inflation and influence diagnostics, including deleted residuals, leverage, and Cook's distance.

Programs C.15–C.17 (pages 798–799), contain the SAS, SPSS, and Minitab commands, respectively, required to produce regression diagnostics and a residual analysis of the interaction model

$$y = \beta_0 + \beta_1 x_1 + \beta_2 x_2 + \beta_3 x_1 x_2 + \varepsilon$$

where

$y =$ Sale price of a residential property

$x_1 =$ Total appraised value of the property

$x_2 = \begin{cases} 1 & \text{if property located in neighborhood B} \\ 0 & \text{if property located in neighborhood A} \end{cases}$

P R O G R A M C.15 SAS statements for producing diagnostics and a residual analysis of the model $y = \beta_0 + \beta_1 x_1 + \beta_2 x_2 + \beta_3 x_1 x_2 + \varepsilon$

Command
line

```
1    DATA SALES;            ⎫
2    INPUT Y X1 X2 @@;      ⎬  Data entry instructions
3    X1X2=X1*X2;            ⎪
4    CARDS;                 ⎭

     [Input data values]

5    PROC REG;                            ⎫  Regression analysis, listing of
6    MODEL Y = X1 X2 X1X2                 ⎬  residuals, influence diagnostics,
7          /R INFLUENCE VIF;             ⎪  variance inflation factors
8    OUTPUT OUT=RESIDS P=YHAT R=RESID;   ⎭
9    PROC PLOT;                 ⎫  Residual plots
10   PLOT RESID*(YHAT X1);     ⎭
11   PROC CHART;                         ⎫  Histogram of residuals
12   VBAR RESID/TYPE=PERCENT;           ⎭
13   PROC UNIVARIATE PLOT NORMAL;   ⎫  Descriptive statistics for residuals, stem-
14   VAR RESID;                      ⎭  and-leaf plot, and normal probability plot
```

COMMAND 7   The R option produces a list of residuals, predicted values, Studentized residuals, and Cook's D statistic for all observations in the analysis. The INFLUENCE option requests a detailed analysis of the influence of each observation on the β estimates. This includes leverage values and Studentized deleted residuals. The VIF option produces variance inflation factors for the independent variables.

COMMAND 13   The PLOT option produces a stem-and-leaf plot for the variable specified (for example, RESID). The NORMAL option produces a normal probability plot for the variable.

P R O G R A M C.16 SPSS statements for prodocing diagnostics and a residual analysis of the model $y = \beta_0 + \beta_1 x_1 + \beta_2 x_2 + \beta_3 x_1 x_2 + \varepsilon$

Command
line

```
1    DATA LIST    FREE/Y X1 X2.   ⎫  Data entry instructions
2    COMPUTE      X1X2=X1*X2.     ⎬
3    BEGIN DATA.                  ⎭

     [Input data values]

4    END DATA.
5    REGRESSION   VARIABLES=Y, X1, X2, X1X2/       ⎫
6                 CRITERIA=TOLERANCE(.00001)/       ⎬  Regression analysis
7                 DEPENDENT=Y/METHOD=ENTER/         ⎪
8                 RESIDUALS/                        ⎭
9                 CASEWISE=DEFAULT ALL SRESID SDRESID COOK LEVER/  ⎫  Residual
10                SCATTERPLOT=(*RESID,*PRED)(*RESID,X1).           ⎬  analysis,
                                                                   ⎭  diagnostics
11   FINISH.
```

COMMAND 8   The RESIDUALS subcommand produces a histogram and normal probability plot of the standardized residuals, and a list of the 10 "worst" outliers.

COMMAND 9   The CASEWISE subcommand produces a list of the residuals and predicted values (DEFAULT), Studentized residuals (SRESID), Studentized deleted residuals (SDRESID), Cook's distances (COOK), and leverage values (LEVER) for all observations (ALL) used in the analysis.

COMMAND 10   The SCATTERPLOT subcommand produces two scatterplots—the residuals against the predicted values and the residuals against x_1.

PROGRAM C.17   Minitab statements for conducting a residual analysis of the model $y = \beta_0 + \beta_1 x_1 + \beta_2 x_2 + \beta_3 x_1 x_2 + \varepsilon$

Command line	
1	`READ Y IN C1, PREDICTORS IN C2 C3 }` Data entry instruction
	[Input data values]
2	`MULTIPLY C2 BY C3, PUT IN C4 }` Data entry instruction
3	`REGRESS Y IN C1, USING 3 PREDICTORS IN C2-C4;` ⎫ Regression analysis
4	`    RESIDUALS IN C5, PREDICTED VALUES IN C6;` ⎬ (storing residuals and
5	`    HI IN C7, TRESIDUALS IN C8, COOKD IN C9;` ⎭ predicted values)
6	`    VIF.`
7	`PLOT C5 VS. C6 }`
8	`PLOT C5 VS. C2 }` Residual analysis (plots and histograms)
9	`HISTOGRAM OF C5 }`
10	`PRINT C7 C8 C9. }` List of influence diagnostics
11	`STOP.`

COMMAND 4   Residuals and predicted values are stored in columns C5 and C6, respectively, for the purposes of plotting.

COMMAND 5   Leverage (HI), Studentized deleted residuals (TRESIDUALS), and Cook's distance values (COOKD) are stored in columns C7–C9.

COMMAND 6   The VIF subcommand produces variance inflation factors on the regression output.

C.9

Analysis of Variance

All three computer software packages described in this text contain analysis of variance routines for designed experiments, ranging from simple one-way classifications of data (completely randomized designs) to the more sophisticated k-way classifications (k-factor factor experiments).

As noted in Chapter 11, each experimental design possesses an underlying probabilistic regression model for the response y. These models can become quite complex, especially for large factorial experiments. Fortunately, the three software packages we discuss do not require the exact model to be specified. Rather, only the sources of variation for the analysis of variance need to be given in the appropriate command lines. For example, for the 3×3 factorial of Example 11.12, we need to specify three sources of variation: (1) ratio of the raw material allocation (main effect); (2) supply of the raw material (main effect); and (3) ratio–supply interaction. The software package will automatically fit the complete factorial model that corresponds to these sources of variation and produce an ANOVA summary table.

The SAS, SPSS, and Minitab commands for conducting the analysis of variance for the factorial experiment of Example 11.12 are given in Programs C.18–C.20, respectively. The ANOVA commands for the completely randomized and randomized block designs are similar in nature, except where noted.

P R O G R A M C.18   SAS ANOVA commands for a 3×3 factorial experiment

Command line

```
 1    DATA MANUFACT;                                ⎫ Data entry commands
 2    INPUT RATIO SUPPLY PROFIT @@;                 ⎭
 3    CARDS;
 4    .5   15   23   .5   15   20   .5   15   21    ⎫
 .     .    .    .    .    .    .    .    .    .     ⎪ Input data values
 .     .    .    .    .    .    .    .    .    .     ⎬ (3 observations per line)
 .     .    .    .    .    .    .    .    .    .     ⎪
12    2    21   20   2    21   22   2    21   24     ⎭
13    PROC ANOVA;                                   ⎫
14    CLASSES RATIO SUPPLY;                         ⎬ ANOVA instructions
15    MODEL PROFIT=RATIO SUPPLY RATIO*SUPPLY;       ⎭
16    MEANS RATIO SUPPLY/TUKEY ALPHA =.01 LINES;    ⎫ Multiple comparison
                                                    ⎭ of means
```

COMMAND 14   The CLASSES statement identifies the independent variables (factors) for the experiment. In SAS, the factors can be either quantitative or qualitative variables.

COMMAND 15   The sources of variation are specified to the right of the equals sign (=) in the MODEL statement, the dependent (response) variable to the left. Interactions are specified by placing an asterisk (*) between the factors (e.g., RATIO*SUPPLY).

COMMAND 16   The MEANS command produces a multiple comparisons analysis of the means of the specified sources. The TUKEY option selects the Tukey multiple comparisons procedure. (Specify BON or SCHEFFE to use the Bonferroni or Scheffé method.) The ALPHA option specifies the experimentwise error rate to be used. (The default is $\alpha = .05$.) The LINES option requests that the results be presented by showing line segments connecting means that are not significantly different. (If LINES is omitted, confidence intervals on the differences between all pairs of means are displayed.)

P R O G R A M C.19 SPSS ANOVA commands for a 3 × 3 factorial experiment

Command
line

```
 1    DATA LIST FREE/RATIO SUPPLY PROFIT. }  Data entry instruction
 2    BEGIN DATA.
 3    0  0  23  0  0  20  0  0  21
 4    0  1  22  0  1  19  0  1  20
 .    .  .   .  .  .   .  .  .   .      Input data values
 .    .  .   .  .  .   .  .  .   .      (3 observations per line)
 .    .  .   .  .  .   .  .  .   .
11    2  2  20  2  2  22  2  2  24
12    END DATA.
13    ANOVA   PROFIT BY RATIO(0,2) SUPPLY (0,2). }  ANOVA instruction
14    FINISH.
```

COMMAND 13 The dependent (response) variable is listed to the left of BY in the ANOVA command, whereas the main effects (factors) are listed to the right. The range of the coded values of the factors on the data lines must be specified in parentheses after each factor. (Note that the three values of RATIO and SUPPLY are coded as 0, 1, and 2 on the input data lines.*) The SPSS default is to fit a full factorial model, which includes sources for main effects and all interactions. If you want to omit the factor interactions, use the subcommand /OPTIONS=3. (This option is useful in a randomized block design, where you want to omit treatment–block interaction.)

Note: Multiple comparisons of means are not available with the general SPSS ANOVA procedure. However, you can perform this type of analysis using the ONEWAY procedure. For example, the statement

```
ONEWAY PROFIT BY RATIO(0,2)/RANGES TUKEY.
```

will perform a one-way analysis of variance for the dependent variable PROFIT and a single factor, RATIO, at three levels. The RANGES subcommand requests that a multiple comparison of means be performed using Tukey's method. Alternatively, specify MODLSD for the Bonferroni method or SCHEFFE for Scheffé's method.

P R O G R A M C.20 Minitab commands for a 3 × 3 factorial experiment

Command
line

```
 1    READ RATIO IN C1, SUPPLY IN C2, PROFIT IN C3 }  Data entry instruction
 2    .5  15  23
 .     .   .   .      Input data lines
 .     .   .   .      (1 observation per line)
 .     .   .   .
29    2   21  24
30    TWOWAY AOV ON PROFIT IN C3, FACTORS IN C1, C2 }  ANOVA instructions
```

---

*In mainframe SPSS, the sources of variation must also be specified in a DESIGN subcommand (e.g., DESIGN=RATIO, SUPPLY, RATIO BY SUPPLY).

COMMAND 30   The TWOWAY procedure in Minitab automatically includes interaction between the factors as a source of variation. When conducting an ANOVA for a randomized block design, you will need to pool the interaction sum of squares with the error sum of squares. This can be accomplished by including the subcommand ADDITIVE. [*Note:*   For a completely randomized design, use the Minitab procedure ONEWAY.]

GENERAL   Multiple comparisons of means are not available in Minitab.

References

Minitab Reference Manual, Release 8, Minitab, Inc., 1989.

Norusis, M. J. *SPSS/PC+ 4.0 Base Manual*, SPSS, Inc., 1990.

Norusis, M. J. *SPSS/PC+ Statistics 4.0*, SPSS, Inc., 1990.

Ryan, B. F., Joiner, B. L., and Ryan, T. A. *Minitab Handbook*, 2nd ed. Boston: PWS-Kent, 1990 (Revised printing).

SAS Procedures Guide for Personal Computers, Version 6 ed., 1986. SAS Institute, Inc.

SAS User's Guide: Basics, Version 6 ed., 1986. SAS Institute, Inc.

SAS User's Guide: Statistics, Version 6 ed., 1986. SAS Institute, Inc.

Useful Statistical Tables

CONTENTS

Table 1   Normal Curve Areas
Table 2   Critical Values for Student's t
Table 3   Critical Values for the F Statistic: $F_{.10}$
Table 4   Critical Values for the F Statistic: $F_{.05}$
Table 5   Critical Values for the F Statistic: $F_{.025}$
Table 6   Critical Values for the F Statistic: $F_{.01}$
Table 7   Random Numbers
Table 8   Critical Values for the Durbin–Watson d Statistic ($\alpha = .05$)
Table 9   Critical Values for the Durbin–Watson d Statistic ($\alpha = .01$)
Table 10 Critical Values for the χ^2 Statistic
Table 11 Percentage Points of the Studentized Range, $q(p, \nu)$, Upper 5%
Table 12 Percentage Points of the Studentized Range, $q(p, \nu)$, Upper 1%

TABLE 1   Normal Curve Areas

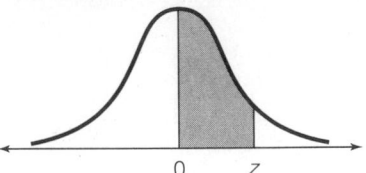

z	.00	.01	.02	.03	.04	.05	.06	.07	.08	.09
.0	.0000	.0040	.0080	.0120	.0160	.0199	.0239	.0279	.0319	.0359
.1	.0398	.0438	.0478	.0517	.0557	.0596	.0636	.0675	.0714	.0753
.2	.0793	.0832	.0871	.0910	.0948	.0987	.1026	.1064	.1103	.1141
.3	.1179	.1217	.1255	.1293	.1331	.1368	.1406	.1443	.1480	.1517
.4	.1554	.1591	.1628	.1664	.1700	.1736	.1772	.1808	.1844	.1879
.5	.1915	.1950	.1985	.2019	.2054	.2088	.2123	.2157	.2190	.2224
.6	.2257	.2291	.2324	.2357	.2389	.2422	.2454	.2486	.2517	.2549
.7	.2580	.2611	.2642	.2673	.2704	.2734	.2764	.2794	.2823	.2852
.8	.2881	.2910	.2939	.2967	.2995	.3023	.3051	.3078	.3106	.3133
.9	.3159	.3186	.3212	.3238	.3264	.3289	.3315	.3340	.3365	.3389
1.0	.3413	.3438	.3461	.3485	.3508	.3531	.3554	.3577	.3599	.3621
1.1	.3643	.3665	.3686	.3708	.3729	.3749	.3770	.3790	.3810	.3830
1.2	.3849	.3869	.3888	.3907	.3925	.3944	.3962	.3980	.3997	.4015
1.3	.4032	.4049	.4066	.4082	.4099	.4115	.4131	.4147	.4162	.4177
1.4	.4192	.4207	.4222	.4236	.4251	.4265	.4279	.4292	.4306	.4319
1.5	.4332	.4345	.4357	.4370	.4382	.4394	.4406	.4418	.4429	.4441
1.6	.4452	.4463	.4474	.4484	.4495	.4505	.4515	.4525	.4535	.4545
1.7	.4554	.4564	.4573	.4582	.4591	.4599	.4608	.4616	.4625	.4633
1.8	.4641	.4649	.4656	.4664	.4671	.4678	.4686	.4693	.4699	.4706
1.9	.4713	.4719	.4726	.4732	.4738	.4744	.4750	.4756	.4761	.4767
2.0	.4772	.4778	.4783	.4788	.4793	.4798	.4803	.4808	.4812	.4817
2.1	.4821	.4826	.4830	.4834	.4838	.4842	.4846	.4850	.4854	.4857
2.2	.4861	.4864	.4868	.4871	.4875	.4878	.4881	.4884	.4887	.4890
2.3	.4893	.4896	.4898	.4901	.4904	.4906	.4909	.4911	.4913	.4916
2.4	.4918	.4920	.4922	.4925	.4927	.4929	.4931	.4932	.4934	.4936
2.5	.4938	.4940	.4941	.4943	.4945	.4946	.4948	.4949	.4951	.4952
2.6	.4953	.4955	.4956	.4957	.4959	.4960	.4961	.4962	.4963	.4964
2.7	.4965	.4966	.4967	.4968	.4969	.4970	.4971	.4972	.4973	.4974
2.8	.4974	.4975	.4976	.4977	.4977	.4978	.4979	.4979	.4980	.4981
2.9	.4981	.4982	.4982	.4983	.4984	.4984	.4985	.4985	.4986	.4986
3.0	.4987	.4987	.4987	.4988	.4988	.4989	.4989	.4989	.4990	.4990

Source: Abridged from Table I of A. Hald, *Statistical Tables and Formulas* (New York: John Wiley & Sons, Inc.), 1952. Reproduced by permission of the publisher.

TABLE 2 Critical Values for Student's t

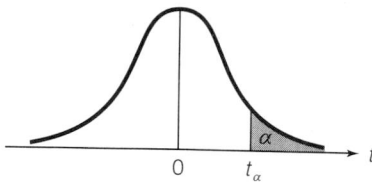

ν	$t_{.100}$	$t_{.050}$	$t_{.025}$	$t_{.010}$	$t_{.005}$	$t_{.001}$	$t_{.0005}$
1	3.070	6.314	12.706	31.821	63.657	318.31	636.62
2	1.886	2.920	4.303	6.965	9.925	22.326	31.598
3	1.638	2.353	3.182	4.541	5.841	10.213	12.924
4	1.533	2.132	2.776	3.747	4.604	7.173	8.610
5	1.476	2.015	2.571	3.365	4.032	5.893	6.869
6	1.440	1.943	2.447	3.143	3.707	5.208	5.959
7	1.415	1.895	2.365	2.998	3.499	4.785	5.408
8	1.397	1.860	2.306	2.896	3.355	4.501	5.041
9	1.383	1.833	2.262	2.821	3.250	4.297	4.781
10	1.372	1.812	2.228	2.764	3.169	4.144	4.587
11	1.363	1.796	2.201	2.718	3.106	4.025	4.437
12	1.356	1.782	2.179	2.681	3.055	3.930	4.318
13	1.350	1.771	2.160	2.650	3.012	3.852	4.221
14	1.345	1.761	2.145	2.624	2.977	3.787	4.140
15	1.341	1.753	2.131	2.602	2.947	3.733	4.073
16	1.337	1.746	2.120	2.583	2.921	3.686	4.015
17	1.333	1.740	2.110	2.567	2.898	3.646	3.965
18	1.330	1.734	2.101	2.552	2.878	3.610	3.922
19	1.328	1.729	2.093	2.539	2.861	3.579	3.883
20	1.325	1.725	2.086	2.528	2.845	3.552	3.850
21	1.323	1.721	2.080	2.518	2.831	3.527	3.819
22	1.321	1.717	2.074	2.508	2.819	3.505	3.792
23	1.319	1.714	2.069	2.500	2.807	3.485	3.767
24	1.318	1.711	2.064	2.492	2.797	3.467	3.745
25	1.316	1.708	2.060	2.485	2.787	3.450	3.725
26	1.315	1.706	2.056	2.479	2.779	3.435	3.707
27	1.314	1.703	2.052	2.473	2.771	3.421	3.690
28	1.313	1.701	2.048	2.467	2.763	3.408	3.674
29	1.311	1.699	2.045	2.462	2.756	3.396	3.659
30	1.310	1.697	2.042	2.457	2.750	3.385	3.646
40	1.303	1.684	2.021	2.423	2.704	3.307	3.551
60	1.296	1.671	2.000	2.390	2.660	3.232	3.460
120	1.289	1.658	1.980	2.358	2.617	3.160	3.373
∞	1.282	1.645	1.960	2.326	2.576	3.090	3.291

Source: This table is reproduced with the kind permission of the Trustees of Biometrika from E. S. Pearson and H. O. Hartley (eds.), *The Biometrika Tables for Statisticians*, Vol. 1, 3rd ed., *Biometrika*, 1966.

TABLE 3   Critical Values for the F Statistic: $F_{.10}$

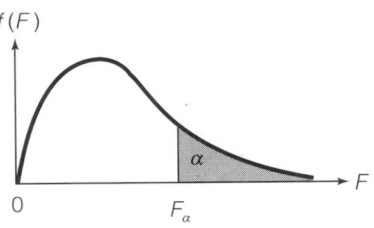

ν_2 \ ν_1	NUMERATOR DEGREES OF FREEDOM								
	1	2	3	4	5	6	7	8	9
1	39.86	49.50	53.59	55.83	57.24	58.20	58.91	59.44	59.86
2	8.53	9.00	9.16	9.24	9.29	9.33	9.35	9.37	9.38
3	5.54	5.46	5.39	5.34	5.31	5.28	5.27	5.25	5.24
4	4.54	4.32	4.19	4.11	4.05	4.01	3.98	3.95	3.94
5	4.06	3.78	3.62	3.52	3.45	3.40	3.37	3.34	3.32
6	3.78	3.46	3.29	3.18	3.11	3.05	3.01	2.98	2.96
7	3.59	3.26	3.07	2.96	2.88	2.83	2.78	2.75	2.72
8	3.46	3.11	2.92	2.81	2.73	2.67	2.62	2.59	2.56
9	3.36	3.01	2.81	2.69	2.61	2.55	2.51	2.47	2.44
10	3.29	2.92	2.73	2.61	2.52	2.46	2.41	2.38	2.35
11	3.23	2.86	2.66	2.54	2.45	2.39	2.34	2.30	2.27
12	3.18	2.81	2.61	2.48	2.39	2.33	2.28	2.24	2.21
13	3.14	2.76	2.56	2.43	2.35	2.28	2.23	2.20	2.16
14	3.10	2.73	2.52	2.39	2.31	2.24	2.19	2.15	2.12
15	3.07	2.70	2.49	2.36	2.27	2.21	2.16	2.12	2.09
16	3.05	2.67	2.46	2.33	2.24	2.18	2.13	2.09	2.06
17	3.03	2.64	2.44	2.31	2.22	2.15	2.10	2.06	2.03
18	3.01	2.62	2.42	2.29	2.20	2.13	2.08	2.04	2.00
19	2.99	2.61	2.40	2.27	2.18	2.11	2.06	2.02	1.98
20	2.97	2.59	2.38	2.25	2.16	2.09	2.04	2.00	1.96
21	2.96	2.57	2.36	2.23	2.14	2.08	2.02	1.98	1.95
22	2.95	2.56	2.35	2.22	2.13	2.06	2.01	1.97	1.93
23	2.94	2.55	2.34	2.21	2.11	2.05	1.99	1.95	1.92
24	2.93	2.54	2.33	2.19	2.10	2.04	1.98	1.94	1.91
25	2.92	2.53	2.32	2.18	2.09	2.02	1.97	1.93	1.89
26	2.91	2.52	2.31	2.17	2.08	2.01	1.96	1.92	1.88
27	2.90	2.51	2.30	2.17	2.07	2.00	1.95	1.91	1.87
28	2.89	2.50	2.29	2.16	2.06	2.00	1.94	1.90	1.87
29	2.89	2.50	2.28	2.15	2.06	1.99	1.93	1.89	1.86
30	2.88	2.49	2.28	2.14	2.05	1.98	1.93	1.88	1.85
40	2.84	2.44	2.23	2.09	2.00	1.93	1.87	1.83	1.79
60	2.79	2.39	2.18	2.04	1.95	1.87	1.82	1.77	1.74
120	2.75	2.35	2.13	1.99	1.90	1.82	1.77	1.72	1.68
∞	2.71	2.30	2.08	1.94	1.85	1.77	1.72	1.67	1.63

DENOMINATOR DEGREES OF FREEDOM

Source: From M. Merrington and C. M. Thompson, "Tables of percentage points of the inverted beta (F)-distribution," *Biometrika*, 1943, 33, 73–88. Reproduced by permission of the *Biometrika* Trustees.

ν_1		NUMERATOR DEGREES OF FREEDOM								
ν_2	10	12	15	20	24	30	40	60	120	∞
1	60.19	60.71	61.22	61.74	62.00	62.26	62.53	62.79	63.06	63.33
2	9.39	9.41	9.42	9.44	9.45	9.46	9.47	9.47	9.48	9.49
3	5.23	5.22	5.20	5.18	5.18	5.17	5.16	5.15	5.14	5.13
4	3.92	3.90	3.87	3.84	3.83	3.82	3.80	3.79	3.78	3.76
5	3.30	3.27	3.24	3.21	3.19	3.17	3.16	3.14	3.12	3.10
6	2.94	2.90	2.87	2.84	2.82	2.80	2.78	2.76	2.74	2.72
7	2.70	2.67	2.63	2.59	2.58	2.56	2.54	2.51	2.49	2.47
8	2.54	2.50	2.46	2.42	2.40	2.38	2.36	2.34	2.32	2.29
9	2.42	2.38	2.34	2.30	2.28	2.25	2.23	2.21	2.18	2.16
10	2.32	2.28	2.24	2.20	2.18	2.16	2.13	2.11	2.08	2.06
11	2.25	2.21	2.17	2.12	2.10	2.08	2.05	2.03	2.00	1.97
12	2.19	2.15	2.10	2.06	2.04	2.01	1.99	1.96	1.93	1.90
13	2.14	2.10	2.05	2.01	1.98	1.96	1.93	1.90	1.88	1.85
14	2.10	2.05	2.01	1.96	1.94	1.91	1.89	1.86	1.83	1.80
15	2.06	2.02	1.97	1.92	1.90	1.87	1.85	1.82	1.79	1.76
16	2.03	1.99	1.94	1.89	1.87	1.84	1.81	1.78	1.75	1.72
17	2.00	1.96	1.91	1.86	1.84	1.81	1.78	1.75	1.72	1.69
18	1.98	1.93	1.89	1.84	1.81	1.78	1.75	1.72	1.69	1.66
19	1.96	1.91	1.86	1.81	1.79	1.76	1.73	1.70	1.67	1.63
20	1.94	1.89	1.84	1.79	1.77	1.74	1.71	1.68	1.64	1.61
21	1.92	1.87	1.83	1.78	1.75	1.72	1.69	1.66	1.62	1.59
22	1.90	1.86	1.81	1.76	1.73	1.70	1.67	1.64	1.60	1.57
23	1.89	1.84	1.80	1.74	1.72	1.69	1.66	1.62	1.59	1.55
24	1.88	1.83	1.78	1.73	1.70	1.67	1.64	1.61	1.57	1.53
25	1.87	1.82	1.77	1.72	1.69	1.66	1.63	1.59	1.56	1.52
26	1.86	1.81	1.76	1.71	1.68	1.65	1.61	1.58	1.54	1.50
27	1.85	1.80	1.75	1.70	1.67	1.64	1.60	1.57	1.53	1.49
28	1.84	1.79	1.74	1.69	1.66	1.63	1.59	1.56	1.52	1.48
29	1.83	1.78	1.73	1.68	1.65	1.62	1.58	1.55	1.51	1.47
30	1.82	1.77	1.72	1.67	1.64	1.61	1.57	1.54	1.50	1.46
40	1.76	1.71	1.66	1.61	1.57	1.54	1.51	1.47	1.42	1.38
60	1.71	1.66	1.60	1.54	1.51	1.48	1.44	1.40	1.35	1.29
120	1.65	1.60	1.55	1.48	1.45	1.41	1.37	1.32	1.26	1.19
∞	1.60	1.55	1.49	1.42	1.38	1.34	1.30	1.24	1.17	1.00

DENOMINATOR DEGREES OF FREEDOM

TABLE 4   Critical Values for the F Statistic: $F_{.05}$

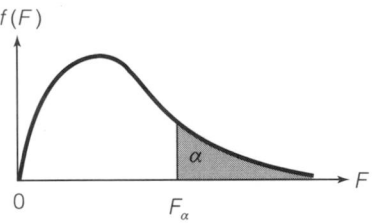

ν_1	NUMERATOR DEGREES OF FREEDOM								
ν_2	1	2	3	4	5	6	7	8	9
1	161.4	199.5	215.7	224.6	230.2	234.0	236.8	238.9	240.5
2	18.51	19.00	19.16	19.25	19.30	19.33	19.35	19.37	19.38
3	10.13	9.55	9.28	9.12	9.01	8.94	8.89	8.85	8.81
4	7.71	6.94	6.59	6.39	6.26	6.16	6.09	6.04	6.00
5	6.61	5.79	5.41	5.19	5.05	4.95	4.88	4.82	4.77
6	5.99	5.14	4.76	4.53	4.39	4.28	4.21	4.15	4.10
7	5.59	4.74	4.35	4.12	3.97	3.87	3.79	3.73	3.68
8	5.32	4.46	4.07	3.84	3.69	3.58	3.50	3.44	3.39
9	5.12	4.26	3.86	3.63	3.48	3.37	3.29	3.23	3.18
10	4.96	4.10	3.71	3.48	3.33	3.22	3.14	3.07	3.02
11	4.84	3.98	3.59	3.36	3.20	3.09	3.01	2.95	2.90
12	4.75	3.89	3.49	3.26	3.11	3.00	2.91	2.85	2.80
13	4.67	3.81	3.41	3.18	3.03	2.92	2.83	2.77	2.71
14	4.60	3.74	3.34	3.11	2.96	2.85	2.76	2.70	2.65
15	4.54	3.68	3.29	3.06	2.90	2.79	2.71	2.64	2.59
16	4.49	3.63	3.24	3.01	2.85	2.74	2.66	2.59	2.54
17	4.45	3.59	3.20	2.96	2.81	2.70	2.61	2.55	2.49
18	4.41	3.55	3.16	2.93	2.77	2.66	2.58	2.51	2.46
19	4.38	3.52	3.13	2.90	2.74	2.63	2.54	2.48	2.42
20	4.35	3.49	3.10	2.87	2.71	2.60	2.51	2.45	2.39
21	4.32	3.47	3.07	2.84	2.68	2.57	2.49	2.42	2.37
22	4.30	3.44	3.05	2.82	2.66	2.55	2.46	2.40	2.34
23	4.28	3.42	3.03	2.80	2.64	2.53	2.44	2.37	2.32
24	4.26	3.40	3.01	2.78	2.62	2.51	2.42	2.36	2.30
25	4.24	3.39	2.99	2.76	2.60	2.49	2.40	2.34	2.28
26	4.23	3.37	2.98	2.74	2.59	2.47	2.39	2.32	2.27
27	4.21	3.35	2.96	2.73	2.57	2.46	2.37	2.31	2.25
28	4.20	3.34	2.95	2.71	2.56	2.45	2.36	2.29	2.24
29	4.18	3.33	2.93	2.70	2.55	2.43	2.35	2.28	2.22
30	4.17	3.32	2.92	2.69	2.53	2.42	2.33	2.27	2.21
40	4.08	3.23	2.84	2.61	2.45	2.34	2.25	2.18	2.12
60	4.00	3.15	2.76	2.53	2.37	2.25	2.17	2.10	2.04
120	3.92	3.07	2.68	2.45	2.29	2.17	2.09	2.02	1.96
∞	3.84	3.00	2.60	2.37	2.21	2.10	2.01	1.94	1.88

DENOMINATOR DEGREES OF FREEDOM

Source: From M. Merrington and C. M. Thompson, "Tables of percentage points of the inverted beta (F)-distribution," *Biometrika*, 1943, *33*, 73–88. Reproduced by permission of the *Biometrika* Trustees.

ν_1	NUMERATOR DEGREES OF FREEDOM									
ν_2	10	12	15	20	24	30	40	60	120	∞
1	241.9	243.9	245.9	248.0	249.1	250.1	251.1	252.2	253.3	254.3
2	19.40	19.41	19.43	19.45	19.45	19.46	19.47	19.48	19.49	19.50
3	8.79	8.74	8.70	8.66	8.64	8.62	8.59	8.57	8.55	8.53
4	5.96	5.91	5.86	5.80	5.77	5.75	5.72	5.69	5.66	5.63
5	4.74	4.68	4.62	4.56	4.53	4.50	4.46	4.43	4.40	4.36
6	4.06	4.00	3.94	3.87	3.84	3.81	3.77	3.74	3.70	3.67
7	3.64	3.57	3.51	3.44	3.41	3.38	3.34	3.30	3.27	3.23
8	3.35	3.28	3.22	3.15	3.12	3.08	3.04	3.01	2.97	2.93
9	3.14	3.07	3.01	2.94	2.90	2.86	2.83	2.79	2.75	2.71
10	2.98	2.91	2.85	2.77	2.74	2.70	2.66	2.62	2.58	2.54
11	2.85	2.79	2.72	2.65	2.61	2.57	2.53	2.49	2.45	2.40
12	2.75	2.69	2.62	2.54	2.51	2.47	2.43	2.38	2.34	2.30
13	2.67	2.60	2.53	2.46	2.42	2.38	2.34	2.30	2.25	2.21
14	2.60	2.53	2.46	2.39	2.35	2.31	2.27	2.22	2.18	2.13
15	2.54	2.48	2.40	2.33	2.29	2.25	2.20	2.16	2.11	2.07
16	2.49	2.42	2.35	2.28	2.24	2.19	2.15	2.11	2.06	2.01
17	2.45	2.38	2.31	2.23	2.19	2.15	2.10	2.06	2.01	1.96
18	2.41	2.34	2.27	2.19	2.15	2.11	2.06	2.02	1.97	1.92
19	2.38	2.31	2.23	2.16	2.11	2.07	2.03	1.98	1.93	1.88
20	2.35	2.28	2.20	2.12	2.08	2.04	1.99	1.95	1.90	1.84
21	2.32	2.25	2.18	2.10	2.05	2.01	1.96	1.92	1.87	1.81
22	2.30	2.23	2.15	2.07	2.03	1.98	1.94	1.89	1.84	1.78
23	2.27	2.20	2.13	2.05	2.01	1.96	1.91	1.86	1.81	1.76
24	2.25	2.18	2.11	2.03	1.98	1.94	1.89	1.84	1.79	1.73
25	2.24	2.16	2.09	2.01	1.96	1.92	1.87	1.82	1.77	1.71
26	2.22	2.15	2.07	1.99	1.95	1.90	1.85	1.80	1.75	1.69
27	2.20	2.13	2.06	1.97	1.93	1.88	1.84	1.79	1.73	1.67
28	2.19	2.12	2.04	1.96	1.91	1.87	1.82	1.77	1.71	1.65
29	2.18	2.10	2.03	1.94	1.90	1.85	1.81	1.75	1.70	1.64
30	2.16	2.09	2.01	1.93	1.89	1.84	1.79	1.74	1.68	1.62
40	2.08	2.00	1.92	1.84	1.79	1.74	1.69	1.64	1.58	1.51
60	1.99	1.92	1.84	1.75	1.70	1.65	1.59	1.53	1.47	1.39
120	1.91	1.83	1.75	1.66	1.61	1.55	1.50	1.43	1.35	1.25
∞	1.83	1.75	1.67	1.57	1.52	1.46	1.39	1.32	1.22	1.00

DENOMINATOR DEGREES OF FREEDOM

TABLE 5   Critical Values for the F Statistic: $F_{.025}$

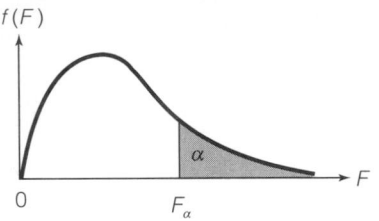

ν_1	NUMERATOR DEGREES OF FREEDOM								
ν_2	1	2	3	4	5	6	7	8	9
1	647.8	799.5	864.2	899.6	921.8	937.1	948.2	956.7	963.3
2	38.51	39.00	39.17	39.25	39.30	39.33	39.36	39.37	39.39
3	17.44	16.04	15.44	15.10	14.88	14.73	14.62	14.54	14.47
4	12.22	10.65	9.98	9.60	9.36	9.20	9.07	8.98	8.90
5	10.01	8.43	7.76	7.39	7.15	6.98	6.85	6.76	6.68
6	8.81	7.26	6.60	6.23	5.99	5.82	5.70	5.60	5.52
7	8.07	6.54	5.89	5.52	5.29	5.12	4.99	4.90	4.82
8	7.57	6.06	5.42	5.05	4.82	4.65	4.53	4.43	4.36
9	7.21	5.71	5.08	4.72	4.48	4.32	4.20	4.10	4.03
10	6.94	5.46	4.83	4.47	4.24	4.07	3.95	3.85	3.78
11	6.72	5.26	4.63	4.28	4.04	3.88	3.76	3.66	3.59
12	6.55	5.10	4.47	4.12	3.89	3.73	3.61	3.51	3.44
13	6.41	4.97	4.35	4.00	3.77	3.60	3.48	3.39	3.31
14	6.30	4.86	4.24	3.89	3.66	3.50	3.38	3.29	3.21
15	6.20	4.77	4.15	3.80	3.58	3.41	3.29	3.20	3.12
16	6.12	4.69	4.08	3.73	3.50	3.34	3.22	3.12	3.05
17	6.04	4.62	4.01	3.66	3.44	3.28	3.16	3.06	2.98
18	5.98	4.56	3.95	3.61	3.38	3.22	3.10	3.01	2.93
19	5.92	4.51	3.90	3.56	3.33	3.17	3.05	2.96	2.88
20	5.87	4.46	3.86	3.51	3.29	3.13	3.01	2.91	2.84
21	5.83	4.42	3.82	3.48	3.25	3.09	2.97	2.87	2.80
22	5.79	4.38	3.78	3.44	3.22	3.05	2.93	2.84	2.76
23	5.75	4.35	3.75	3.41	3.18	3.02	2.90	2.81	2.73
24	5.72	4.32	3.72	3.38	3.15	2.99	2.87	2.78	2.70
25	5.69	4.29	3.69	3.35	3.13	2.97	2.85	2.75	2.68
26	5.66	4.27	3.67	3.33	3.10	2.94	2.82	2.73	2.65
27	5.63	4.24	3.65	3.31	3.08	2.92	2.80	2.71	2.63
28	5.61	4.22	3.63	3.29	3.06	2.90	2.78	2.69	2.61
29	5.59	4.20	3.61	3.27	3.04	2.88	2.76	2.67	2.59
30	5.57	4.18	3.59	3.25	3.03	2.87	2.75	2.65	2.57
40	5.42	4.05	3.46	3.13	2.90	2.74	2.62	2.53	2.45
60	5.29	3.93	3.34	3.01	2.79	2.63	2.51	2.41	2.33
120	5.15	3.80	3.23	2.89	2.67	2.52	2.39	2.30	2.22
∞	5.02	3.69	3.12	2.79	2.57	2.41	2.29	2.19	2.11

DENOMINATOR DEGREES OF FREEDOM

ν_1	NUMERATOR DEGREES OF FREEDOM									
ν_2	10	12	15	20	24	30	40	60	120	∞
1	968.6	976.7	984.9	993.1	997.2	1001	1006	1010	1014	1018
2	39.40	39.41	39.43	39.45	39.46	39.46	39.47	39.48	39.49	39.50
3	14.42	14.34	14.25	14.17	14.12	14.08	14.04	13.99	13.95	13.90
4	8.84	8.75	8.66	8.56	8.51	8.46	8.41	8.36	8.31	8.26
5	6.62	6.52	6.43	6.33	6.28	6.23	6.18	6.12	6.07	6.02
6	5.46	5.37	5.27	5.17	5.12	5.07	5.01	4.96	4.90	4.85
7	4.76	4.67	4.57	4.47	4.42	4.36	4.31	4.25	4.20	4.14
8	4.30	4.20	4.10	4.00	3.95	3.89	3.84	3.78	3.73	3.67
9	3.96	3.87	3.77	3.67	3.61	3.56	3.51	3.45	3.39	3.33
10	3.72	3.62	3.52	3.42	3.37	3.31	3.26	3.20	3.14	3.08
11	3.53	3.43	3.33	3.23	3.17	3.12	3.06	3.00	2.94	2.88
12	3.37	3.28	3.18	3.07	3.02	2.96	2.91	2.85	2.79	2.72
13	3.25	3.15	3.05	2.95	2.89	2.84	2.78	2.72	2.66	2.60
14	3.15	3.05	2.95	2.84	2.79	2.73	2.67	2.61	2.55	2.49
15	3.06	2.96	2.86	2.76	2.70	2.64	2.59	2.52	2.46	2.40
16	2.99	2.89	2.79	2.68	2.63	2.57	2.51	2.45	2.38	2.32
17	2.92	2.82	2.72	2.62	2.56	2.50	2.44	2.38	2.32	2.25
18	2.87	2.77	2.67	2.56	2.50	2.44	2.38	2.32	2.26	2.19
19	2.82	2.72	2.62	2.51	2.45	2.39	2.33	2.27	2.20	2.13
20	2.77	2.68	2.57	2.46	2.41	2.35	2.29	2.22	2.16	2.09
21	2.73	2.64	2.53	2.42	2.37	2.31	2.25	2.18	2.11	2.04
22	2.70	2.60	2.50	2.39	2.33	2.27	2.21	2.14	2.08	2.00
23	2.67	2.57	2.47	2.36	2.30	2.24	2.18	2.11	2.04	1.97
24	2.64	2.54	2.44	2.33	2.27	2.21	2.15	2.08	2.01	1.94
25	2.61	2.51	2.41	2.30	2.24	2.18	2.12	2.05	1.98	1.91
26	2.59	2.49	2.39	2.28	2.22	2.16	2.09	2.03	1.95	1.88
27	2.57	2.47	2.36	2.25	2.19	2.13	2.07	2.00	1.93	1.85
28	2.55	2.45	2.34	2.23	2.17	2.11	2.05	1.98	1.91	1.83
29	2.53	2.43	2.32	2.21	2.15	2.09	2.03	1.96	1.89	1.81
30	2.51	2.41	2.31	2.20	2.14	2.07	2.01	1.94	1.87	1.79
40	2.39	2.29	2.18	2.07	2.01	1.94	1.88	1.80	1.72	1.64
60	2.27	2.17	2.06	1.94	1.88	1.82	1.74	1.67	1.58	1.48
120	2.16	2.05	1.94	1.82	1.76	1.69	1.61	1.53	1.43	1.31
∞	2.05	1.94	1.83	1.71	1.64	1.57	1.48	1.39	1.27	1.00

DENOMINATOR DEGREES OF FREEDOM

TABLE 6   Critical Values for the F Statistic: $F_{.01}$

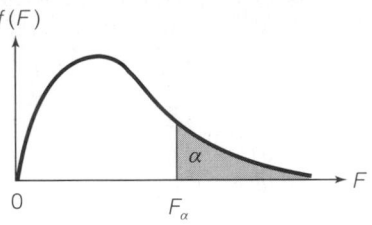

	ν_1	NUMERATOR DEGREES OF FREEDOM								
	ν_2	1	2	3	4	5	6	7	8	9
	1	4,052	4,999.5	5,403	5,625	5,764	5,859	5,928	5,982	6,022
	2	98.50	99.00	99.17	99.25	99.30	99.33	99.36	99.37	99.39
	3	34.12	30.82	29.46	28.71	28.24	27.91	27.67	27.49	27.35
	4	21.20	18.00	16.69	15.98	15.52	15.21	14.98	14.80	14.66
	5	16.26	13.27	12.06	11.39	10.97	10.67	10.46	10.29	10.16
	6	13.75	10.92	9.78	9.15	8.75	8.47	8.26	8.10	7.98
	7	12.25	9.55	8.45	7.85	7.46	7.19	6.99	6.84	6.72
	8	11.26	8.65	7.59	7.01	6.63	6.37	6.18	6.03	5.91
	9	10.56	8.02	6.99	6.42	6.06	5.80	5.61	5.47	5.35
	10	10.04	7.56	6.55	5.99	5.64	5.39	5.20	5.06	4.94
	11	9.65	7.21	6.22	5.67	5.32	5.07	4.89	4.74	4.63
	12	9.33	6.93	5.95	5.41	5.06	4.82	4.64	4.50	4.39
	13	9.07	6.70	5.74	5.21	4.86	4.62	4.44	4.30	4.19
	14	8.86	6.51	5.56	5.04	4.69	4.46	4.28	4.14	4.03
	15	8.68	6.36	5.42	4.89	4.56	4.32	4.14	4.00	3.89
	16	8.53	6.23	5.29	4.77	4.44	4.20	4.03	3.89	3.78
	17	8.40	6.11	5.18	4.67	4.34	4.10	3.93	3.79	3.68
	18	8.29	6.01	5.09	4.58	4.25	4.01	3.84	3.71	3.60
DENOMINATOR DEGREES OF FREEDOM	19	8.18	5.93	5.01	4.50	4.17	3.94	3.77	3.63	3.52
	20	8.10	5.85	4.94	4.43	4.10	3.87	3.70	3.56	3.46
	21	8.02	5.78	4.87	4.37	4.04	3.81	3.64	3.51	3.40
	22	7.95	5.72	4.82	4.31	3.99	3.76	3.59	3.45	3.35
	23	7.88	5.66	4.76	4.26	3.94	3.71	3.54	3.41	3.30
	24	7.82	5.61	4.72	4.22	3.90	3.67	3.50	3.36	3.26
	25	7.77	5.57	4.68	4.18	3.85	3.63	3.46	3.32	3.22
	26	7.72	5.53	4.64	4.14	3.82	3.59	3.42	3.29	3.18
	27	7.68	5.49	4.60	4.11	3.78	3.56	3.39	3.26	3.15
	28	7.64	5.45	4.57	4.07	3.75	3.53	3.36	3.23	3.12
	29	7.60	5.42	4.54	4.04	3.73	3.50	3.33	3.20	3.09
	30	7.56	5.39	4.51	4.02	3.70	3.47	3.30	3.17	3.07
	40	7.31	5.18	4.31	3.83	3.51	3.29	3.12	2.99	2.89
	60	7.08	4.98	4.13	3.65	3.34	3.12	2.95	2.82	2.72
	120	6.85	4.79	3.95	3.48	3.17	2.96	2.79	2.66	2.56
	∞	6.63	4.61	3.78	3.32	3.02	2.80	2.64	2.51	2.41

Source: From M. Merrington and C. M. Thompson, "Tables of percentage points of the inverted beta (F)-distribution," *Biometrika*, 1943, *33*, 73–88. Reproduced by permission of the *Biometrika* Trustees.

ν_1	NUMERATOR DEGREES OF FREEDOM									
ν_2	10	12	15	20	24	30	40	60	120	∞
1	6,056	6,106	6,157	6,209	6,235	6,261	6,287	6,313	6,339	6,366
2	99.40	99.42	99.43	99.45	99.46	99.47	99.47	99.48	99.49	99.50
3	27.23	27.05	26.87	26.69	26.60	26.50	26.41	26.32	26.22	26.13
4	14.55	14.37	14.20	14.02	13.93	13.84	13.75	13.65	13.56	13.46
5	10.05	9.89	9.72	9.55	9.47	9.38	9.29	9.20	9.11	9.02
6	7.87	7.72	7.56	7.40	7.31	7.23	7.14	7.06	6.97	6.88
7	6.62	6.47	6.31	6.16	6.07	5.99	5.91	5.82	5.74	5.65
8	5.81	5.67	5.52	5.36	5.28	5.20	5.12	5.03	4.95	4.86
9	5.26	5.11	4.96	4.81	4.73	4.65	4.57	4.48	4.40	4.31
10	4.85	4.71	4.56	4.41	4.33	4.25	4.17	4.08	4.00	3.91
11	4.54	4.40	4.25	4.10	4.02	3.94	3.86	3.78	3.69	3.60
12	4.30	4.16	4.01	3.86	3.78	3.70	3.62	3.54	3.45	3.36
13	4.10	3.96	3.82	3.66	3.59	3.51	3.43	3.34	3.25	3.17
14	3.94	3.80	3.66	3.51	3.43	3.35	3.27	3.18	3.09	3.00
15	3.80	3.67	3.52	3.37	3.29	3.21	3.13	3.05	2.96	2.87
16	3.69	3.55	3.41	3.26	3.18	3.10	3.02	2.93	2.84	2.75
17	3.59	3.46	3.31	3.16	3.08	3.00	2.92	2.83	2.75	2.65
18	3.51	3.37	3.23	3.08	3.00	2.92	2.84	2.75	2.66	2.57
19	3.43	3.30	3.15	3.00	2.92	2.84	2.76	2.67	2.58	2.49
20	3.37	3.23	3.09	2.94	2.86	2.78	2.69	2.61	2.52	2.42
21	3.31	3.17	3.03	2.88	2.80	2.72	2.64	2.55	2.46	2.36
22	3.26	3.12	2.98	2.83	2.75	2.67	2.58	2.50	2.40	2.31
23	3.21	3.07	2.93	2.78	2.70	2.62	2.54	2.45	2.35	2.26
24	3.17	3.03	2.89	2.74	2.66	2.58	2.49	2.40	2.31	2.21
25	3.13	2.99	2.85	2.70	2.62	2.54	2.45	2.36	2.27	2.17
26	3.09	2.96	2.81	2.66	2.58	2.50	2.42	2.33	2.23	2.13
27	3.06	2.93	2.78	2.63	2.55	2.47	2.38	2.29	2.20	2.10
28	3.03	2.90	2.75	2.60	2.52	2.44	2.35	2.26	2.17	2.06
29	3.00	2.87	2.73	2.57	2.49	2.41	2.33	2.23	2.14	2.03
30	2.98	2.84	2.70	2.55	2.47	2.39	2.30	2.21	2.11	2.01
40	2.80	2.66	2.52	2.37	2.29	2.20	2.11	2.02	1.92	1.80
60	2.63	2.50	2.35	2.20	2.12	2.03	1.94	1.84	1.73	1.60
120	2.47	2.34	2.19	2.03	1.95	1.86	1.76	1.66	1.53	1.38
∞	2.32	2.18	2.04	1.88	1.79	1.70	1.59	1.47	1.32	1.00

DENOMINATOR DEGREES OF FREEDOM

TABLE 7　**Random Numbers**

ROW \ COLUMN	1	2	3	4	5	6	7	8	9	10	11	12	13	14
1	10480	15011	01536	02011	81647	91646	69179	14194	62590	36207	20969	99570	91291	90700
2	22368	46573	25595	85393	30995	89198	27982	53402	93965	34095	52666	19174	39615	99505
3	24130	48360	22527	97265	76393	64809	15179	24830	49340	32081	30680	19655	63348	58629
4	42167	93093	06243	61680	07856	16376	39440	53537	71341	57004	00849	74917	97758	16379
5	37570	39975	81837	16656	06121	91782	60468	81305	49684	60672	14110	06927	01263	54613
6	77921	06907	11008	42751	27756	53498	18602	70659	90655	15053	21916	81825	44394	42880
7	99562	72905	56420	69994	98872	31016	71194	18738	44013	48840	63213	21069	10634	12952
8	96301	91977	05463	07972	18876	20922	94595	56869	69014	60045	18425	84903	42508	32307
9	89579	14342	63661	10281	17453	18103	57740	84378	25331	12566	58678	44947	05585	56941
10	85475	36857	53342	53988	53060	59533	38867	62300	08158	17983	16439	11458	18593	64952
11	28918	69578	88231	33276	70997	79936	56865	05859	90106	31595	01547	85590	91610	78188
12	63553	40961	48235	03427	49626	69445	18663	72695	52180	20847	12234	90511	33703	90322
13	09429	93969	52636	92737	88974	33488	36320	17617	30015	08272	84115	27156	30613	74952
14	10365	61129	87529	85689	48237	52267	67689	93394	01511	26358	85104	20285	29975	89868
15	07119	97336	71048	08178	77233	13916	47564	81056	97735	85977	29372	74461	28551	90707
16	51085	12765	51821	51259	77452	16308	60756	92144	49442	53900	70960	63990	75601	40719
17	02368	21382	52404	60268	89368	19885	55322	44819	01188	65255	64835	44919	05944	55157
18	01011	54092	33362	94904	31273	04146	18594	29852	71585	85030	51132	01915	92747	64951
19	52162	53916	46369	58586	23216	14513	83149	98736	23495	64350	94738	17752	35156	35749
20	07056	97628	33787	09998	42698	06691	76988	13602	51851	46104	88916	19509	25625	58104
21	48663	91245	85828	14346	09172	30168	90229	04734	59193	22178	30421	61666	99904	32812
22	54164	58492	22421	74103	47070	25306	76468	26384	58151	06646	21524	15227	96909	44592
23	32639	32363	05597	24200	13363	38005	94342	28728	35806	06912	17012	64161	18296	22851
24	29334	27001	87637	87308	58731	00256	45834	15398	46557	41135	10367	07684	36188	18510
25	02488	33062	28834	07351	19731	92420	60952	61280	50001	67658	32586	86679	50720	94953
26	81525	72295	04839	96423	24878	82651	66566	14778	76797	14780	13300	87074	79666	95725
27	29676	20591	68086	26432	46901	20849	89768	81536	86645	12659	92259	57102	80428	25280
28	00742	57392	39064	66432	84673	40027	32832	61362	98947	96067	64760	64584	96096	98253
29	05366	04213	25669	26422	44407	44048	37937	63904	45766	66134	75470	66520	34693	90449
30	91921	26418	64117	94305	26766	25940	39972	22209	71500	64568	91402	42416	07844	69618
31	00582	04711	87917	77341	42206	35126	74087	99547	81817	42607	43808	76655	62028	76630
32	00725	69884	62797	56170	86324	88072	76222	36086	84637	93161	76038	65855	77919	88006
33	69011	65795	95876	55293	18988	27354	26575	08625	40801	59920	29841	80150	12777	48501
34	25976	57948	29888	88604	67917	48708	18912	82271	65424	69774	33611	54262	85963	03547
35	09763	83473	73577	12908	30883	18317	28290	35797	05998	41688	34952	37888	38917	88050

TABLE 7 Continued

COLUMN ROW	1	2	3	4	5	6	7	8	9	10	11	12	13	14
36	91576	42595	27958	30134	04024	86385	29880	99730	55536	84855	29080	09250	79656	73211
37	17955	56349	90999	49127	20044	59931	06115	20542	18059	02008	73708	83517	36103	42791
38	46503	18584	18845	49618	02304	51038	20655	58727	28168	15475	56942	53389	20562	87338
39	92157	89634	94824	78171	84610	82834	09922	25417	44137	48413	25555	21246	35509	20468
40	14577	62765	35605	81263	39667	47358	56873	56307	61607	49518	89656	20103	77490	18062
41	98427	07523	33362	64270	01638	92477	66969	98420	04880	45585	46565	04102	46880	45709
42	34914	63976	88720	82765	34476	17032	87589	40836	32427	70002	70663	88863	77775	69348
43	70060	28277	39475	46473	23219	53416	94970	25832	69975	94884	19661	72828	00102	66794
44	53976	54914	06990	67245	68350	82948	11398	42878	80287	88267	47363	46634	06541	97809
45	76072	29515	40980	07391	58745	25774	22987	80059	39911	96189	41151	14222	60697	59583
46	90725	52210	83974	29992	65831	38857	50490	83765	55657	14361	31720	57375	56228	41546
47	64364	67412	33339	31926	14883	24413	59744	92351	97473	89286	35931	04110	23726	51900
48	08962	00358	31662	25388	61642	34072	81249	35648	56891	69352	48373	45578	78547	81788
49	95012	68379	93526	70765	10592	04542	76463	54328	02349	17247	28865	14777	62730	92277
50	15664	10493	20492	38391	91132	21999	59516	81652	27195	48223	46751	22923	32261	85653
51	16408	81899	04153	53381	79401	21438	83035	92350	36693	31238	59649	91754	72772	02338
52	18629	81953	05520	91962	04739	13092	97662	24822	94730	06496	35090	04822	86774	98289
53	73115	35101	47498	87637	99016	71060	88824	71013	18735	20286	23153	72924	35165	43040
54	57491	16703	23167	49323	45021	33132	12544	41035	80780	45393	44812	12515	98931	91202
55	30405	83946	23792	14422	15059	45799	22716	19792	09983	74353	68668	30429	70735	25499
56	16631	35006	85900	98275	32388	52390	16815	69298	82732	38480	73817	32523	41961	44437
57	96773	20206	42559	78985	05300	22164	24369	54224	35083	19687	11052	91491	60383	19746
58	38935	64202	14349	82674	66523	44133	00697	35552	35970	19124	63318	29686	03387	59846
59	31624	76384	17403	53363	44167	64486	64758	75366	76554	31601	12614	33072	60332	92325
60	78919	19474	23632	27889	47914	02584	37680	20801	72152	39339	34806	08930	85001	87820
61	03931	33309	57047	74211	63445	17361	62825	39908	05607	91284	68833	25570	38818	46920
62	74426	33278	43972	10119	89917	15665	52872	73823	73144	88662	88970	74492	51805	99378
63	09066	00903	20795	95452	92648	45454	09552	88815	16553	51125	79375	97596	16296	66092
64	42238	12426	87025	14267	20979	04508	64535	31355	86064	29472	47689	05974	52468	16834
65	16153	08002	26504	41744	81959	65642	74240	56302	00033	67107	77510	70625	28725	34191
66	21457	40742	29820	96783	29400	21840	15035	34537	33310	06116	95240	15957	16572	06004
67	21581	57802	02050	89728	17937	37621	47075	42080	97403	48626	68995	43805	33386	21597
68	55612	78095	83197	33732	05810	24813	86902	60397	16489	03264	88525	42786	05269	92532
69	44657	66999	99324	51281	84463	60563	79312	93454	68876	25471	93911	25650	12682	73572

(continued)

TABLE 7   Continued

ROW \ COLUMN	1	2	3	4	5	6	7	8	9	10	11	12	13	14
70	91340	84979	46949	81973	37949	61023	43997	15263	80644	43942	89203	71795	99533	50501
71	91227	21199	31935	27022	84067	05462	35216	14486	29891	68607	41867	14951	91696	85065
72	50001	38140	66321	19924	72163	09538	12151	06878	91903	18749	34405	56087	82790	70925
73	65390	05224	72958	28609	81406	39147	25549	48542	42627	45233	57202	94617	23772	07896
74	27504	96131	83944	41575	10573	08619	64482	73923	36152	05184	94142	25299	84387	34925
75	37169	94851	39117	89632	00959	16487	65536	49071	39782	17095	02330	74301	00275	48280
76	11508	70225	51111	38351	19444	66499	71945	05422	13442	78675	84081	66938	93654	59894
77	37449	30362	06694	54690	04052	53115	62757	95348	78662	11163	81651	50245	34971	52924
78	46515	70331	85922	38329	57015	15765	97161	17869	45349	61796	66345	81073	49106	79860
79	30986	81223	42416	58353	21532	30502	32305	86482	05174	07901	54339	58861	74818	46942
80	63798	64995	46583	09785	44160	78128	83991	42865	92520	83531	80377	35909	81250	54238
81	82486	84846	99254	67632	43218	50076	21361	64816	51202	88124	41870	52689	51275	83556
82	21885	32906	92431	09060	64297	51674	64126	62570	26123	05155	59194	52799	28225	85762
83	60336	98782	07408	53458	13564	59089	26445	29789	85205	41001	12535	12133	14645	23541
84	43937	46891	24010	25560	86355	33941	25786	54990	71899	15475	95434	98227	21824	19585
85	97656	63175	89303	16275	07100	92063	21942	18611	47348	20203	18534	03862	78095	50136
86	03299	01221	05418	38982	55758	92237	26759	86367	21216	98442	08303	56613	91511	75928
87	79626	06486	03574	17668	07785	76020	79924	25651	83325	88428	85076	72811	22717	50585
88	85636	68335	47539	03129	65651	11977	02510	26113	99447	68645	34327	15152	55230	93448
89	18039	14367	61337	06177	12143	46609	32989	74014	64708	00533	35398	58408	13261	47908
90	08362	15656	60627	36478	65648	16764	53412	09013	07832	41574	17639	82163	60859	75567
91	79556	29068	04142	16268	15387	12856	66227	38358	22478	73373	88732	09443	82558	05250
92	92608	82674	27072	32534	17075	27698	98204	63863	11951	34648	88022	56148	34925	57031
93	23982	25835	40055	67006	12293	02753	14827	23235	35071	99704	37543	11601	35503	85171
94	09915	96306	05908	97901	28395	14186	00821	80703	70426	75647	76310	88717	37890	40129
95	59037	33300	26695	62247	69927	76123	50842	43834	86654	70959	79725	93872	28117	19233
96	42488	78077	69882	61657	34136	79180	97526	43092	04098	73571	80799	76536	71255	64239
97	46764	86273	63003	93017	31204	36692	40202	35275	57306	55543	53203	18098	47625	88684
98	03237	45430	55417	63282	90816	17349	88298	90183	36600	78406	06216	95787	42579	90730
99	86591	81482	52667	61582	14972	90053	89534	76036	49199	43716	97548	04379	46370	28672
100	38534	01715	94964	87288	65680	43772	39560	12918	86537	62738	19636	51132	25739	56947

Source:　Abridged from W. H. Beyer (ed.). CRC Standard Mathematical Tables, 24th edition. (Cleveland: The Chemical Rubber Company), 1976.

TABLE 8 Critical Values for the Durbin–Watson d Statistic ($\alpha = .05$)

n	$k = 1$		$k = 2$		$k = 3$		$k = 4$		$k = 5$	
	d_L	d_U	d_L	d_U	d_L	d_U	d_L	d_U	d_L	d_U
15	1.08	1.36	.95	1.54	.82	1.75	.69	1.97	.56	2.21
16	1.10	1.37	.98	1.54	.86	1.73	.74	1.93	.62	2.15
17	1.13	1.38	1.02	1.54	.90	1.71	.78	1.90	.67	2.10
18	1.16	1.39	1.05	1.53	.93	1.69	.82	1.87	.71	2.06
19	1.18	1.40	1.08	1.53	.97	1.68	.86	1.85	.75	2.02
20	1.20	1.41	1.10	1.54	1.00	1.68	.90	1.83	.79	1.99
21	1.22	1.42	1.13	1.54	1.03	1.67	.93	1.81	.83	1.96
22	1.24	1.43	1.15	1.54	1.05	1.66	.96	1.80	.86	1.94
23	1.26	1.44	1.17	1.54	1.08	1.66	.99	1.79	.90	1.92
24	1.27	1.45	1.19	1.55	1.10	1.66	1.01	1.78	.93	1.90
25	1.29	1.45	1.21	1.55	1.12	1.66	1.04	1.77	.95	1.89
26	1.30	1.46	1.22	1.55	1.14	1.65	1.06	1.76	.98	1.88
27	1.32	1.47	1.24	1.56	1.16	1.65	1.08	1.76	1.01	1.86
28	1.33	1.48	1.26	1.56	1.18	1.65	1.10	1.75	1.03	1.85
29	1.34	1.48	1.27	1.56	1.20	1.65	1.12	1.74	1.05	1.84
30	1.35	1.49	1.28	1.57	1.21	1.65	1.14	1.74	1.07	1.83
31	1.36	1.50	1.30	1.57	1.23	1.65	1.16	1.74	1.09	1.83
32	1.37	1.50	1.31	1.57	1.24	1.65	1.18	1.73	1.11	1.82
33	1.38	1.51	1.32	1.58	1.26	1.65	1.19	1.73	1.13	1.81
34	1.39	1.51	1.33	1.58	1.27	1.65	1.21	1.73	1.15	1.81
35	1.40	1.52	1.34	1.58	1.28	1.65	1.22	1.73	1.16	1.80
36	1.41	1.52	1.35	1.59	1.29	1.65	1.24	1.73	1.18	1.80
37	1.42	1.53	1.36	1.59	1.31	1.66	1.25	1.72	1.19	1.80
38	1.43	1.54	1.37	1.59	1.32	1.66	1.26	1.72	1.21	1.79
39	1.43	1.54	1.38	1.60	1.33	1.66	1.27	1.72	1.22	1.79
40	1.44	1.54	1.39	1.60	1.34	1.66	1.29	1.72	1.23	1.79
45	1.48	1.57	1.43	1.62	1.38	1.67	1.34	1.72	1.29	1.78
50	1.50	1.59	1.46	1.63	1.42	1.67	1.38	1.72	1.34	1.77
55	1.53	1.60	1.49	1.64	1.45	1.68	1.41	1.72	1.38	1.77
60	1.55	1.62	1.51	1.65	1.48	1.69	1.44	1.73	1.41	1.77
65	1.57	1.63	1.54	1.66	1.50	1.70	1.47	1.73	1.44	1.77
70	1.58	1.64	1.55	1.67	1.52	1.70	1.49	1.74	1.46	1.77
75	1.60	1.65	1.57	1.68	1.54	1.71	1.51	1.74	1.49	1.77
80	1.61	1.66	1.59	1.69	1.56	1.72	1.53	1.74	1.51	1.77
85	1.62	1.67	1.60	1.70	1.57	1.72	1.55	1.75	1.52	1.77
90	1.63	1.68	1.61	1.70	1.59	1.73	1.57	1.75	1.54	1.78
95	1.64	1.69	1.62	1.71	1.60	1.73	1.58	1.75	1.56	1.78
100	1.65	1.69	1.63	1.72	1.61	1.74	1.59	1.76	1.57	1.78

Source: From J. Durbin and G. S. Watson, "Testing for serial correlation in least squares regression, II," *Biometrika*, 1951, *30*, 159–178. Reproduced by permission of the *Biometrika* Trustees.

T A B L E 9   Critical Values for the Durbin–Watson d Statistic ($\alpha = .01$)

n	$k = 1$		$k = 2$		$k = 3$		$k = 4$		$k = 5$	
	d_L	d_U	d_L	d_U	d_L	d_U	d_L	d_U	d_L	d_U
15	.81	1.07	.70	1.25	.59	1.46	.49	1.70	.39	1.96
16	.84	1.09	.74	1.25	.63	1.44	.53	1.66	.44	1.90
17	.87	1.10	.77	1.25	.67	1.43	.57	1.63	.48	1.85
18	.90	1.12	.80	1.26	.71	1.42	.61	1.60	.52	1.80
19	.93	1.13	.83	1.26	.74	1.41	.65	1.58	.56	1.77
20	.95	1.15	.86	1.27	.77	1.41	.68	1.57	.60	1.74
21	.97	1.16	.89	1.27	.80	1.41	.72	1.55	.63	1.71
22	1.00	1.17	.91	1.28	.83	1.40	.75	1.54	.66	1.69
23	1.02	1.19	.94	1.29	.86	1.40	.77	1.53	.70	1.67
24	1.04	1.20	.96	1.30	.88	1.41	.80	1.53	.72	1.66
25	1.05	1.21	.98	1.30	.90	1.41	.83	1.52	.75	1.65
26	1.07	1.22	1.00	1.31	.93	1.41	.85	1.52	.78	1.64
27	1.09	1.23	1.02	1.32	.95	1.41	.88	1.51	.81	1.63
28	1.10	1.24	1.04	1.32	.97	1.41	.90	1.51	.83	1.62
29	1.12	1.25	1.05	1.33	.99	1.42	.92	1.51	.85	1.61
30	1.13	1.26	1.07	1.34	1.01	1.42	.94	1.51	.88	1.61
31	1.15	1.27	1.08	1.34	1.02	1.42	.96	1.51	.90	1.60
32	1.16	1.28	1.10	1.35	1.04	1.43	.98	1.51	.92	1.60
33	1.17	1.29	1.11	1.36	1.05	1.43	1.00	1.51	.94	1.59
34	1.18	1.30	1.13	1.36	1.07	1.43	1.01	1.51	.95	1.59
35	1.19	1.31	1.14	1.37	1.08	1.44	1.03	1.51	.97	1.59
36	1.21	1.32	1.15	1.38	1.10	1.44	1.04	1.51	.99	1.59
37	1.22	1.32	1.16	1.38	1.11	1.45	1.06	1.51	1.00	1.59
38	1.23	1.33	1.18	1.39	1.12	1.45	1.07	1.52	1.02	1.58
39	1.24	1.34	1.19	1.39	1.14	1.45	1.09	1.52	1.03	1.58
40	1.25	1.34	1.20	1.40	1.15	1.46	1.10	1.52	1.05	1.58
45	1.29	1.38	1.24	1.42	1.20	1.48	1.16	1.53	1.11	1.58
50	1.32	1.40	1.28	1.45	1.24	1.49	1.20	1.54	1.16	1.59
55	1.36	1.43	1.32	1.47	1.28	1.51	1.25	1.55	1.21	1.59
60	1.38	1.45	1.35	1.48	1.32	1.52	1.28	1.56	1.25	1.60
65	1.41	1.47	1.38	1.50	1.35	1.53	1.31	1.57	1.28	1.61
70	1.43	1.49	1.40	1.52	1.37	1.55	1.34	1.58	1.31	1.61
75	1.45	1.50	1.42	1.53	1.39	1.56	1.37	1.59	1.34	1.62
80	1.47	1.52	1.44	1.54	1.42	1.57	1.39	1.60	1.36	1.62
85	1.48	1.53	1.46	1.55	1.43	1.58	1.41	1.60	1.39	1.63
90	1.50	1.54	1.47	1.56	1.45	1.59	1.43	1.61	1.41	1.64
95	1.51	1.55	1.49	1.57	1.47	1.60	1.45	1.62	1.42	1.64
100	1.52	1.56	1.50	1.58	1.48	1.60	1.46	1.63	1.44	1.65

Source: From J. Durbin and G. S. Watson, "Testing for serial correlation in least squares regression, II," *Biometrika*, 1951, *30*, 159–178. Reproduced by permission of the *Biometrika* Trustees.

TABLE 10 Critical Values for the χ^2 Statistic

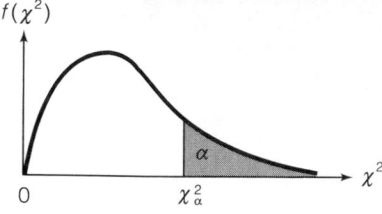

DEGREES OF FREEDOM	$\chi^2_{.995}$	$\chi^2_{.990}$	$\chi^2_{.975}$	$\chi^2_{.950}$	$\chi^2_{.900}$
1	.0000393	.0001571	.0009821	.0039321	.0157908
2	.0100251	.0201007	.0506356	.102587	.210720
3	.0717212	.114832	.215795	.351846	.584375
4	.206990	.297110	.484419	.710721	1.063623
5	.411740	.554300	.831211	1.145476	1.61031
6	.675727	.872085	1.237347	1.63539	2.20413
7	.989265	1.239043	1.68987	2.16735	2.83311
8	1.344419	1.646482	2.17973	2.73264	3.48954
9	1.734926	2.087912	2.70039	3.32511	4.16816
10	2.15585	2.55821	3.24697	3.94030	4.86518
11	2.60321	3.05347	3.81575	4.57481	5.57779
12	3.07382	3.57056	4.40379	5.22603	6.30380
13	3.56503	4.10691	5.00874	5.89186	7.04150
14	4.07468	4.66043	5.62872	6.57063	7.78953
15	4.60094	5.22935	6.26214	7.26094	8.54675
16	5.14224	5.81221	6.90766	7.96164	9.31223
17	5.69724	6.40776	7.56418	8.67176	10.0852
18	6.26481	7.01491	8.23075	9.39046	10.8649
19	6.84398	7.63273	8.90655	10.1170	11.6509
20	7.43386	8.26040	9.59083	10.8508	12.4426
21	8.03366	8.89720	10.28293	11.5913	13.2396
22	8.64272	9.54249	10.9823	12.3380	14.0415
23	9.26042	10.19567	11.6885	13.0905	14.8479
24	9.88623	10.8564	12.4011	13.8484	15.6587
25	10.5197	11.5240	13.1197	14.6114	16.4734
26	11.1603	12.1981	13.8439	15.3791	17.2919
27	11.8076	12.8786	14.5733	16.1513	18.1138
28	12.4613	13.5648	15.3079	16.9279	18.9392
29	13.1211	14.2565	16.0471	17.7083	19.7677
30	13.7867	14.9535	16.7908	18.4926	20.5992
40	20.7065	22.1643	24.4331	26.5093	29.0505
50	27.9907	29.7067	32.3574	34.7642	37.6886
60	35.5346	37.4848	40.4817	43.1879	46.4589
70	43.2752	45.4418	48.7576	51.7393	55.3290
80	51.1720	53.5400	57.1532	60.3915	64.2778
90	59.1963	61.7541	65.6466	69.1260	73.2912
100	67.3276	70.0648	74.2219	77.9295	82.3581
150	109.142	112.668	117.985	122.692	128.275
200	152.241	156.432	162.728	168.279	174.835
300	240.663	245.972	253.912	260.878	269.068
400	330.903	337.155	346.482	354.641	364.207
500	422.303	429.388	439.936	449.147	459.926

(*continued*)

T A B L E 10   **Continued**

DEGREES OF FREEDOM	$\chi^2_{.100}$	$\chi^2_{.050}$	$\chi^2_{.025}$	$\chi^2_{.010}$	$\chi^2_{.005}$
1	2.70554	3.84146	5.02389	6.63490	7.87944
2	4.60517	5.99147	7.37776	9.21034	10.5966
3	6.25139	7.81473	9.34840	11.3449	12.8381
4	7.77944	9.48773	11.1433	13.2767	14.8602
5	9.23635	11.0705	12.8325	15.0863	16.7496
6	10.6446	12.5916	14.4494	16.8119	18.5476
7	12.0170	14.0671	16.0128	18.4753	20.2777
8	13.3616	15.5073	17.5346	20.0902	21.9550
9	14.6837	16.9190	19.0228	21.6660	23.5893
10	15.9871	18.3070	20.4831	23.2093	25.1882
11	17.2750	19.6751	21.9200	24.7250	26.7569
12	18.5494	21.0261	23.3367	26.2170	28.2995
13	19.8119	22.3621	24.7356	27.6883	29.8194
14	21.0642	23.6848	26.1190	29.1413	31.3193
15	22.3072	24.9958	27.4884	30.5779	32.8013
16	23.5418	26.2962	28.8454	31.9999	34.2672
17	24.7690	27.5871	30.1910	33.4087	35.7185
18	25.9894	28.8693	31.5264	34.8053	37.1564
19	27.2036	30.1435	32.8523	36.1908	38.5822
20	28.4120	31.4104	34.1696	37.5662	39.9968
21	29.6151	32.6705	35.4789	38.9321	41.4010
22	30.8133	33.9244	36.7807	40.2894	42.7956
23	32.0069	35.1725	38.0757	41.6384	44.1813
24	33.1963	36.4151	39.3641	42.9798	45.5585
25	34.3816	37.6525	40.6465	44.3141	46.9278
26	36.5631	38.8852	41.9232	45.6417	48.2899
27	36.7412	40.1133	43.1944	46.9630	49.6449
28	37.9159	41.3372	44.4607	48.2782	50.9933
29	39.0875	42.5569	45.7222	49.5879	52.3356
30	40.2560	43.7729	46.9792	50.8922	53.6720
40	51.8050	55.7585	59.3417	63.6907	66.7659
50	63.1671	67.5048	71.4202	76.1539	79.4900
60	74.3970	79.0819	83.2976	88.3794	91.9517
70	85.5271	90.5312	95.0231	100.425	104.215
80	96.5782	101.879	106.629	112.329	116.321
90	107.565	113.145	118.136	124.116	128.299
100	118.498	124.342	129.561	135.807	140.169
150	172.581	179.581	185.800	193.208	198.360
200	226.021	233.994	241.058	249.445	255.264
300	331.789	341.395	349.874	359.906	366.844
400	436.649	447.632	457.305	468.724	476.606
500	540.930	553.127	563.852	576.493	585.207

Source: From C. M. Thompson, "Tables of the Percentage Points of the χ^2-Distribution." *Biometrika*, 1941, Vol. 32, pp. 188–189. Reproduced by permission of the *Biometrika* trustees.

TABLE 11 Percentage Points of the Studentized Range $q(p, \nu)$, Upper 5%

ν \ p	2	3	4	5	6	7	8	9	10	11
1	17.97	26.98	32.82	37.08	40.41	43.12	45.40	47.36	49.07	50.59
2	6.08	8.33	9.80	10.88	11.74	12.44	13.03	13.54	13.99	14.39
3	4.50	5.91	6.82	7.50	8.04	8.48	8.85	9.18	9.46	9.72
4	3.93	5.04	5.76	6.29	6.71	7.05	7.35	7.60	7.83	8.03
5	3.64	4.60	5.22	5.67	6.03	6.33	6.58	6.80	6.99	7.17
6	3.46	4.34	4.90	5.30	5.63	5.90	6.12	6.32	6.49	6.65
7	3.34	4.16	4.68	5.06	5.36	5.61	5.82	6.00	6.16	6.30
8	3.26	4.04	4.53	4.89	5.17	5.40	5.60	5.77	5.92	6.05
9	3.20	3.95	4.41	4.76	5.02	5.24	5.43	5.59	5.74	5.87
10	3.15	3.88	4.33	4.65	4.91	5.12	5.30	5.46	5.60	5.72
11	3.11	3.82	4.26	4.57	4.82	5.03	5.20	5.35	5.49	5.61
12	3.08	3.77	4.20	4.51	4.75	4.95	5.12	5.27	5.39	5.51
13	3.06	3.73	4.15	4.45	4.69	4.88	5.05	5.19	5.32	5.43
14	3.03	3.70	4.11	4.41	4.64	4.83	4.99	5.13	5.25	5.36
15	3.01	3.67	4.08	4.37	4.60	4.78	4.94	5.08	5.20	5.31
16	3.00	3.65	4.05	4.33	4.56	4.74	4.90	5.03	5.15	5.26
17	2.98	3.63	4.02	4.30	4.52	4.70	4.86	4.99	5.11	5.21
18	2.97	3.61	4.00	4.28	4.49	4.67	4.82	4.96	5.07	5.17
19	2.96	3.59	3.98	4.25	4.47	4.65	4.79	4.92	5.04	5.14
20	2.95	3.58	3.96	4.23	4.45	4.62	4.77	4.90	5.01	5.11
24	2.92	3.53	3.90	4.17	4.37	4.54	4.68	4.81	4.92	5.01
30	2.89	3.49	3.85	4.10	4.30	4.46	4.60	4.72	4.82	4.92
40	2.86	3.44	3.79	4.04	4.23	4.39	4.52	4.63	4.73	4.82
60	2.83	3.40	3.74	3.98	4.16	4.31	4.44	4.55	4.65	4.73
120	2.80	3.36	3.68	3.92	4.10	4.24	4.36	4.47	4.56	4.64
∞	2.77	3.31	3.63	3.86	4.03	4.17	4.29	4.39	4.47	4.55

(*continued*)

T A B L E 11   **Continued**

p / ν	12	13	14	15	16	17	18	19	20
1	51.96	53.20	54.33	55.36	56.32	57.22	58.04	58.83	59.56
2	14.75	15.08	15.38	15.65	15.91	16.14	16.37	16.57	16.77
3	9.95	10.15	10.35	10.52	10.69	10.84	10.98	11.11	11.24
4	8.21	8.37	8.52	8.66	8.79	8.91	9.03	9.13	9.23
5	7.32	7.47	7.60	7.72	7.83	7.93	8.03	8.12	8.21
6	6.79	6.92	7.03	7.14	7.24	7.34	7.43	7.51	7.59
7	6.43	6.55	6.66	6.76	6.85	6.94	7.02	7.10	7.17
8	6.18	6.29	6.39	6.48	6.57	6.65	6.73	6.80	6.87
9	5.98	6.09	6.19	6.28	6.36	6.44	6.51	6.58	6.64
10	5.83	5.93	6.03	6.11	6.19	6.27	6.34	6.40	6.47
11	5.71	5.81	5.90	5.98	6.06	6.13	6.20	6.27	6.33
12	5.61	5.71	5.80	5.88	5.95	6.02	6.09	6.15	6.21
13	5.53	5.63	5.71	5.79	5.86	5.93	5.99	6.05	6.11
14	5.46	5.55	5.64	5.71	5.79	5.85	5.91	5.97	6.03
15	5.40	5.49	5.57	5.65	5.72	5.78	5.85	5.90	5.96
16	5.35	5.44	5.52	5.59	5.66	5.73	5.79	5.84	5.90
17	5.31	5.39	5.47	5.54	5.61	5.67	5.73	5.79	5.84
18	5.27	5.35	5.43	5.50	5.57	5.63	5.69	5.74	5.79
19	5.23	5.31	5.39	5.46	5.53	5.59	5.65	5.70	5.75
20	5.20	5.28	5.36	5.43	5.49	5.55	5.61	5.66	5.71
24	5.10	5.18	5.25	5.32	5.38	5.44	5.49	5.55	5.59
30	5.00	5.08	5.15	5.21	5.27	5.33	5.38	5.43	5.47
40	4.90	4.98	5.04	5.11	5.16	5.22	5.27	5.31	5.36
60	4.81	4.88	4.94	5.00	5.06	5.11	5.15	5.20	5.24
120	4.71	4.78	4.84	4.90	4.95	5.00	5.04	5.09	5.13
∞	4.62	4.68	4.74	4.80	4.85	4.89	4.93	4.97	5.01

Source: *Biometrika Tables for Statisticians*, Vol. I, 3rd ed., edited by E. S. Pearson and H. O. Hartley (Cambridge University Press, 1966). Reproduced by permission of Professor E. S. Pearson and the *Biometrika* Trustees.

TABLE 12 Percentage Points of the Studentized Range $q(p, \nu)$, Upper 1%

ν \ p	2	3	4	5	6	7	8	9	10	11
1	90.03	135.0	164.3	185.6	202.2	215.8	227.2	237.0	245.6	253.2
2	14.04	19.02	22.29	24.72	26.63	28.20	29.53	30.68	31.69	32.59
3	8.26	10.62	12.17	13.33	14.24	15.00	15.64	16.20	16.69	17.13
4	6.51	8.12	9.17	9.96	10.58	11.10	11.55	11.93	12.27	12.57
5	5.70	6.98	7.80	8.42	8.91	9.32	9.67	9.97	10.24	10.48
6	5.24	6.33	7.03	7.56	7.97	8.32	8.61	8.87	9.10	9.30
7	4.95	5.92	6.54	7.01	7.37	7.68	7.94	8.17	8.37	8.55
8	4.75	5.64	6.20	6.62	6.96	7.24	7.47	7.68	7.86	8.03
9	4.60	5.43	5.96	6.35	6.66	6.91	7.13	7.33	7.49	7.65
10	4.48	5.27	5.77	6.14	6.43	6.67	6.87	7.05	7.21	7.36
11	4.39	5.15	5.62	5.97	6.25	6.48	6.67	6.84	6.99	7.13
12	4.32	5.05	5.50	5.84	6.10	6.32	6.51	6.67	6.81	6.94
13	4.26	4.96	5.40	5.73	5.98	6.19	6.37	6.53	6.67	6.79
14	4.21	4.89	5.32	5.63	5.88	6.08	6.26	6.41	6.54	6.66
15	4.17	4.84	5.25	5.56	5.80	5.99	6.16	6.31	6.44	6.55
16	4.13	4.79	5.19	5.49	5.72	5.92	6.08	6.22	6.35	6.46
17	4.10	4.74	5.14	5.43	5.66	5.85	6.01	6.15	6.27	6.38
18	4.07	4.70	5.09	5.38	5.60	5.79	5.94	6.08	6.20	6.31
19	4.05	4.67	5.05	5.33	5.55	5.73	5.89	6.02	6.14	6.25
20	4.02	4.64	5.02	5.29	5.51	5.69	5.84	5.97	6.09	6.19
24	3.96	4.55	4.91	5.17	5.37	5.54	5.69	5.81	5.92	6.02
30	3.89	4.45	4.80	5.05	5.24	5.40	5.54	5.65	5.76	5.85
40	3.82	4.37	4.70	4.93	5.11	5.26	5.39	5.50	5.60	5.69
60	3.76	4.28	4.59	4.82	4.99	5.13	5.25	5.36	5.45	5.53
120	3.70	4.20	4.50	4.71	4.87	5.01	5.12	5.21	5.30	5.37
∞	3.64	4.12	4.40	4.60	4.76	4.88	4.99	5.08	5.16	5.23

(*continued*)

TABLE 12 **Continued**

ν \ p	12	13	14	15	16	17	18	19	20
1	260.0	266.2	271.8	277.0	281.8	286.3	290.0	294.3	298.0
2	33.40	34.13	34.81	35.43	36.00	36.53	37.03	37.50	37.95
3	17.53	17.89	18.22	18.52	18.81	19.07	19.32	19.55	19.77
4	12.84	13.09	13.32	13.53	13.73	13.91	14.08	14.24	14.40
5	10.70	10.89	11.08	11.24	11.40	11.55	11.68	11.81	11.93
6	9.48	9.65	9.81	9.95	10.08	10.21	10.32	10.43	10.54
7	8.71	8.86	9.00	9.12	9.24	9.35	9.46	9.55	9.65
8	8.18	8.31	8.44	8.55	8.66	8.76	8.85	8.94	9.03
9	7.78	7.91	8.03	8.13	8.23	8.33	8.41	8.49	8.57
10	7.49	7.60	7.71	7.81	7.91	7.99	8.08	8.15	8.23
11	7.25	7.36	7.46	7.56	7.65	7.73	7.81	7.88	7.95
12	7.06	7.17	7.26	7.36	7.44	7.52	7.59	7.66	7.73
13	6.90	7.01	7.10	7.19	7.27	7.35	7.42	7.48	7.55
14	6.77	6.87	6.96	7.05	7.13	7.20	7.27	7.33	7.39
15	6.66	6.76	6.84	6.93	7.00	7.07	7.14	7.20	7.26
16	6.56	6.66	6.74	6.82	6.90	6.97	7.03	7.09	7.15
17	6.48	6.57	6.66	6.73	6.81	6.87	6.94	7.00	7.05
18	6.41	6.50	6.58	6.65	6.72	6.79	6.85	6.91	6.97
19	6.34	6.43	6.51	6.58	6.65	6.72	6.78	6.84	6.89
20	6.28	6.37	6.45	6.52	6.59	6.65	6.71	6.77	6.82
24	6.11	6.19	6.26	6.33	6.39	6.45	6.51	6.56	6.61
30	5.93	6.01	6.08	6.14	6.20	6.26	6.31	6.36	6.41
40	5.76	5.83	5.90	5.96	6.02	6.07	6.12	6.16	6.21
60	5.60	5.67	5.73	5.78	5.84	5.89	5.93	5.97	6.01
120	5.44	5.50	5.56	5.61	5.66	5.71	5.75	5.79	5.83
∞	5.29	5.35	5.40	5.45	5.49	5.54	5.57	5.61	5.65

Source: Biometrika Tables for Statisticians, Vol. I, 3rd ed., edited by E. S. Pearson and H. O. Hartley (Cambridge University Press, 1966). Reproduced by permission of Professor E. S. Pearson and the *Biometrika* Trustees.

Data Set: Real Estate Appraisals and Sales Data for Seven Neighborhoods in Tampa, Florida (See Case Study 12)

Note: The following pages present information for the first 100 property sales in each neighborhood. The complete data set is available on $5\frac{1}{4}''$ floppy or $3\frac{1}{2}''$ micro diskette from the publisher.

1990 RESIDENTIAL PROPERTY SALES FOR 7 NEIGHBORHOODS

NBRHOOD A

OBS	LANDVAL	IMPROVAL	SALEPRIC	OBS	LANDVAL	IMPROVAL	SALEPRIC
1	$30,000	$64,831	$118,500	2	$30,000	$50,765	$93,900
3	$46,651	$118,573	$191,500	4	$45,990	$91,402	$184,000
5	$42,394	$98,181	$168,000	6	$47,751	$103,351	$169,000
7	$63,596	$102,182	$208,500	8	$56,658	$153,806	$255,000
9	$51,428	$172,451	$264,000	10	$93,200	$311,121	$422,000
11	$76,125	$178,172	$290,000	12	$54,360	$161,934	$237,000
13	$65,376	$134,458	$286,500	14	$42,400	$115,046	$202,500
15	$40,800	$92,606	$168,000	16	$108,996	$205,864	$305,900
17	$102,170	$222,786	$375,000	18	$24,637	$90,598	$169,900
19	$30,600	$80,858	$135,000	20	$44,730	$99,047	$176,000
21	$50,563	$92,615	$176,500	22	$40,128	$100,713	$179,500
23	$48,112	$113,269	$199,500	24	$66,429	$117,771	$205,000
25	$54,077	$121,848	$205,000	26	$43,494	$113,757	$189,000
27	$55,792	$121,736	$180,800	28	$48,000	$103,289	$175,000
29	$62,291	$102,857	$173,000	30	$53,878	$104,768	$174,900
31	$50,275	$112,207	$202,500	32	$49,306	$101,420	$171,100
33	$42,742	$82,397	$139,000	34	$36,454	$92,442	$145,000
35	$30,720	$85,244	$138,000	36	$30,602	$82,498	$128,500
37	$30,720	$87,738	$145,000	38	$31,270	$79,233	$125,000
39	$31,270	$85,479	$162,000	40	$48,979	$105,946	$190,000
41	$49,974	$141,493	$182,000	42	$25,200	$79,938	$130,500

1990 RESIDENTIAL PROPERTY SALES FOR 7 NEIGHBORHOODS

NBRHOOD B

OBS	LANDVAL	IMPROVAL	SALEPRIC	OBS	LANDVAL	IMPROVAL	SALEPRIC
43	$22,922	$89,618	$180,000	44	$11,891	$76,590	$75,600
45	$11,550	$32,897	$35,300	46	$29,441	$70,548	$108,000
47	$33,072	$63,587	$119,000	48	$43,987	$13,316	$57,500
49	$24,840	$47,068	$60,000	50	$21,563	$55,489	$90,000
51	$15,435	$70,309	$105,000	52	$40,250	$52,022	$109,000
53	$16,433	$83,880	$120,000	54	$17,976	$77,126	$120,000
55	$15,340	$85,552	$118,000	56	$15,115	$79,174	$113,000
57	$15,529	$97,910	$131,000	58	$15,444	$90,852	$131,000
59	$15,115	$89,224	$117,500	60	$15,400	$94,057	$115,000
61	$14,366	$78,432	$115,000	62	$13,989	$91,427	$123,900
63	$17,560	$98,445	$140,000	64	$15,744	$78,001	$107,000
65	$14,563	$118,788	$130,000	66	$9,639	$52,815	$86,500
67	$8,526	$47,166	$69,000	68	$8,036	$7,685	$16,300
69	$10,049	$48,042	$80,100	70	$10,049	$39,235	$73,000
71	$10,049	$39,684	$73,000	72	$12,222	$52,904	$87,500
73	$8,568	$51,758	$79,000	74	$8,568	$43,920	$75,000
75	$8,568	$18,824	$25,000	76	$20,617	$104,797	$167,900
77	$15,932	$38,641	$85,000	78	$3,210	$14,940	$15,000
79	$11,093	$43,812	$85,000	80	$8,027	$31,802	$59,000
81	$17,399	$61,850	$60,000	82	$11,588	$54,358	$83,500
83	$14,100	$53,629	$75,000	84	$13,110	$37,017	$50,000
85	$13,110	$41,305	$71,700	86	$13,110	$41,305	$72,700
87	$13,110	$40,104	$59,900	88	$13,110	$41,088	$74,600
89	$13,110	$41,305	$59,000	90	$18,066	$41,305	$68,000
91	$13,358	$20,215	$51,900	92	$9,108	$19,031	$43,000
93	$9,715	$17,344	$25,700	94	$9,715	$17,164	$13,000
95	$9,108	$20,944	$50,000	96	$9,412	$20,413	$38,200
97	$9,306	$23,672	$48,000	98	$9,108	$23,295	$51,900
99	$9,108	$36,110	$61,900	100	$10,626	$21,500	$41,500
101	$9,108	$21,195	$31,100	102	$9,108	$26,892	$20,000
103	$9,412	$26,706	$42,600	104	$10,488	$25,980	$48,500
105	$10,488	$32,121	$33,600	106	$10,488	$35,069	$56,900
107	$10,488	$27,706	$45,900	108	$13,433	$43,484	$66,000
109	$10,488	$20,189	$52,900	110	$10,389	$29,016	$48,000
111	$10,948	$33,672	$63,500	112	$12,043	$34,651	$60,000
113	$12,784	$39,271	$73,400	114	$14,744	$42,222	$78,500
115	$11,186	$53,061	$71,200	116	$11,186	$51,104	$62,800
117	$11,543	$50,381	$71,600	118	$12,799	$50,607	$70,100
119	$14,300	$41,136	$62,500	120	$59,660	$2,000	$38,500
121	$17,940	$59,417	$73,500	122	$21,420	$71,848	$90,000
123	$19,354	$78,948	$117,000	124	$21,218	$80,688	$100,000
125	$21,218	$89,179	$142,000	126	$21,218	$52,156	$86,000
127	$21,726	$70,897	$107,000	128	$20,800	$49,959	$105,000
129	$18,216	$63,915	$100,000	130	$18,216	$87,754	$132,500
131	$25,200	$93,291	$113,000	132	$15,840	$57,291	$87,800
133	$15,840	$76,668	$103,700	134	$15,520	$52,186	$81,500
135	$12,288	$66,301	$94,000	136	$14,400	$54,180	$78,000
137	$14,356	$61,877	$87,300	138	$14,208	$65,575	$99,000
139	$16,694	$66,357	$106,000	140	$19,002	$58,738	$101,000
141	$16,000	$77,063	$103,000	142	$15,840	$76,668	$96,900

1990 RESIDENTIAL PROPERTY SALES FOR 7 NEIGHBORHOODS

NBRHOOD C

OBS	LANDVAL	IMPROVAL	SALEPRIC	OBS	LANDVAL	IMPROVAL	SALEPRIC
301	$15,593	$14,551	$30,200	302	$15,593	$19,011	$43,000
303	$15,840	$15,658	$54,000	304	$16,583	$15,657	$57,000
305	$15,593	$16,532	$50,000	306	$15,593	$12,453	$24,900
307	$15,593	$13,309	$35,500	308	$15,593	$15,818	$33,000
309	$15,593	$17,459	$49,300	310	$15,593	$13,320	$42,900
311	$15,593	$11,180	$49,900	312	$15,840	$18,955	$46,900
313	$16,830	$15,981	$44,500	314	$15,593	$16,203	$41,000
315	$18,563	$24,364	$63,700	316	$15,840	$15,773	$49,000
317	$18,563	$19,477	$62,000	318	$16,932	$25,113	$65,000
319	$16,932	$16,936	$39,000	320	$16,544	$13,993	$35,900
321	$20,831	$20,225	$62,000	322	$16,110	$26,309	$65,000
323	$19,905	$27,900	$67,000	324	$17,061	$26,120	$65,000
325	$18,609	$21,234	$59,000	326	$21,175	$26,351	$69,000
327	$16,665	$16,442	$49,000	328	$17,776	$15,931	$34,900
329	$20,098	$23,579	$65,700	330	$14,950	$31,120	$72,000
331	$11,500	$27,505	$57,500	332	$17,250	$25,886	$53,000
333	$19,013	$29,770	$64,500	334	$16,422	$25,815	$39,900
335	$24,150	$36,131	$73,000	336	$15,939	$26,449	$50,000
337	$15,939	$24,145	$47,000	338	$15,939	$26,112	$49,000
339	$15,939	$20,335	$40,000	340	$12,075	$22,209	$50,000
341	$14,438	$30,267	$73,700	342	$10,971	$31,049	$58,000
343	$10,971	$22,069	$44,900	344	$11,702	$21,617	$51,100
345	$12,408	$31,630	$58,000	346	$13,200	$27,093	$46,900
347	$10,824	$36,413	$12,000	348	$12,144	$33,749	$73,500
349	$12,672	$23,872	$49,900	350	$13,464	$23,092	$55,000
351	$17,063	$30,163	$38,300	352	$14,326	$30,035	$51,000
353	$17,250	$30,324	$55,900	354	$17,250	$42,943	$64,500
355	$17,250	$27,011	$54,500	356	$15,520	$25,342	$52,100
357	$12,144	$17,205	$32,000	358	$12,549	$19,938	$48,000
359	$13,156	$23,735	$58,000	360	$12,549	$25,855	$52,500
361	$14,625	$25,835	$42,900	362	$17,480	$25,958	$49,800
363	$14,168	$20,650	$53,500	364	$14,168	$18,639	$57,000
365	$11,440	$19,055	$43,500	366	$11,440	$20,688	$19,000
367	$11,330	$19,805	$55,000	368	$15,180	$51,322	$71,900
369	$25,520	$37,811	$59,900	370	$11,440	$21,871	$57,000
371	$14,300	$34,466	$60,000	372	$22,880	$48,847	$72,000
373	$17,335	$42,061	$65,000	374	$11,440	$27,335	$35,500
375	$11,440	$28,535	$60,000	376	$17,160	$14,687	$40,900
377	$11,440	$19,478	$31,000	378	$11,440	$13,980	$33,000
379	$11,440	$16,194	$39,900	380	$22,880	$17,256	$20,500
381	$11,440	$36,806	$63,100	382	$13,613	$29,045	$51,700
383	$13,750	$12,276	$35,000	384	$14,080	$13,240	$37,000
385	$14,080	$12,794	$46,000	386	$13,750	$13,096	$38,500
387	$13,613	$13,766	$25,200	388	$13,776	$14,989	$41,800
389	$13,640	$14,799	$52,000	390	$12,954	$22,435	$35,000
391	$12,144	$22,550	$46,000	392	$13,728	$9,014	$29,200
393	$15,182	$8,892	$44,000	394	$14,956	$13,587	$42,000
395	$13,608	$16,564	$80,000	396	$17,940	$24,013	$55,500
397	$17,940	$23,632	$61,500	398	$17,940	$26,291	$51,900
399	$17,940	$22,556	$59,100	400	$17,940	$21,523	$45,000

1990 RESIDENTIAL PROPERTY SALES FOR 7 NEIGHBORHOODS

NBRHOOD D

OBS	LANDVAL	IMPROVAL	SALEPRIC	OBS	LANDVAL	IMPROVAL	SALEPRIC
673	$6,670	$8,920	$29,900	674	$6,670	$7,997	$23,300
675	$6,670	$12,672	$13,000	676	$6,670	$9,879	$17,800
677	$6,670	$36,186	$42,500	678	$7,468	$19,198	$68,700
679	$6,670	$25,282	$49,600	680	$6,670	$14,605	$28,000
681	$13,340	$11,711	$25,000	682	$6,670	$18,140	$36,900
683	$6,525	$13,862	$44,300	684	$13,340	$15,247	$20,000
685	$6,525	$10,554	$14,500	686	$6,525	$9,501	$29,500
687	$6,525	$7,640	$14,000	688	$6,786	$38,392	$48,000
689	$6,670	$11,685	$12,000	690	$6,670	$19,731	$42,000
691	$7,178	$9,468	$27,000	692	$6,670	$11,210	$23,400
693	$6,670	$16,118	$40,200	694	$4,314	$6,288	$8,000
695	$6,786	$15,047	$10,000	696	$6,670	$9,513	$19,000
697	$6,670	$11,959	$27,000	698	$6,670	$8,751	$21,000
699	$7,468	$10,371	$16,500	700	$6,525	$15,179	$26,000
701	$13,340	$14,139	$39,900	702	$6,670	$16,601	$54,000
703	$10,556	$24,242	$40,909	704	$8,934	$17,130	$35,000
705	$9,504	$18,804	$49,000	706	$10,692	$21,700	$51,500
707	$9,720	$21,877	$51,900	708	$9,720	$33,396	$55,000
709	$9,504	$21,469	$36,300	710	$9,504	$24,897	$43,500
711	$26,624	$48,364	$90,000	712	$10,080	$39,115	$40,000
713	$9,360	$27,101	$50,000	714	$9,360	$19,239	$41,000
715	$8,363	$17,708	$29,500	716	$13,459	$37,393	$52,500
717	$17,298	$26,162	$50,000	718	$17,945	$31,952	$61,500
719	$7,518	$14,369	$22,000	720	$10,928	$25,835	$34,600
721	$9,920	$41,092	$59,400	722	$7,750	$23,378	$35,600
723	$8,119	$8,873	$27,000	724	$11,860	$19,223	$40,000
725	$15,982	$20,688	$53,500	726	$7,518	$30,122	$46,500
727	$7,518	$35,966	$48,000	728	$8,089	$15,358	$35,000
729	$6,345	$10,995	$23,000	730	$15,525	$11,084	$40,000
731	$6,028	$16,691	$39,000	732	$6,345	$10,085	$40,000
733	$6,345	$20,275	$32,500	734	$6,345	$11,049	$14,000
735	$6,345	$6,727	$15,000	736	$6,345	$11,270	$26,000
737	$6,345	$9,354	$28,000	738	$4,600	$14,439	$28,000
739	$6,210	$6,444	$9,900	740	$9,453	$21,544	$38,400
741	$8,538	$16,663	$40,000	742	$6,670	$11,334	$10,500
743	$6,670	$15,017	$15,300	744	$11,310	$21,360	$30,000
745	$7,487	$16,744	$30,500	746	$6,894	$12,781	$14,955
747	$17,215	$26,752	$13,000	748	$6,815	$10,785	$37,100
749	$11,722	$35,821	$31,300	750	$6,815	$14,679	$10,000
751	$14,570	$20,698	$46,500	752	$6,075	$13,066	$37,400
753	$8,662	$10,987	$22,000	754	$6,210	$18,231	$10,000
755	$6,210	$24,736	$38,300	756	$6,210	$9,706	$5,500
757	$6,210	$9,438	$15,000	758	$6,804	$21,587	$15,000
759	$10,125	$10,841	$25,400	760	$6,210	$10,516	$31,700
761	$6,210	$23,822	$41,900	762	$9,794	$23,600	$51,000
763	$8,239	$23,791	$42,000	764	$6,790	$9,790	$42,000
765	$17,000	$27,062	$42,900	766	$9,400	$54,096	$74,900
767	$12,739	$56,809	$79,900	768	$8,908	$25,328	$35,500
769	$8,908	$19,644	$51,300	770	$7,762	$22,808	$43,500
771	$8,940	$26,033	$55,000	772	$8,940	$20,115	$42,000

1990 RESIDENTIAL PROPERTY SALES FOR 7 NEIGHBORHOODS

NBRHOOD E

OBS	LANDVAL	IMPROVAL	SALEPRIC	OBS	LANDVAL	IMPROVAL	SALEPRIC
1155	$18,832	$65,345	$88,000	1156	$18,260	$67,165	$107,000
1157	$22,869	$49,222	$80,000	1158	$21,259	$55,764	$84,000
1159	$19,781	$68,694	$107,000	1160	$16,080	$44,092	$74,900
1161	$16,320	$45,599	$33,500	1162	$19,920	$45,727	$79,900
1163	$22,320	$65,622	$71,000	1164	$21,600	$68,463	$110,800
1165	$20,544	$49,404	$77,100	1166	$21,463	$57,257	$84,000
1167	$24,970	$65,599	$69,600	1168	$16,490	$42,226	$70,000
1169	$16,733	$45,953	$82,600	1170	$17,000	$45,486	$79,900
1171	$17,400	$68,375	$95,800	1172	$19,200	$58,089	$86,200
1173	$16,800	$55,763	$83,500	1174	$19,200	$73,845	$97,000
1175	$19,500	$67,560	$98,000	1176	$17,472	$54,828	$80,600
1177	$17,472	$75,902	$120,000	1178	$17,520	$57,601	$88,500
1179	$19,200	$76,415	$109,900	1180	$16,080	$56,603	$70,000
1181	$18,120	$68,793	$99,000	1182	$20,400	$77,702	$112,900
1183	$16,320	$47,113	$79,900	1184	$17,945	$76,647	$90,000
1185	$18,000	$70,407	$98,000	1186	$16,080	$72,072	$91,500
1187	$17,945	$72,967	$93,000	1188	$17,363	$54,637	$66,900
1189	$17,387	$71,084	$108,000	1190	$18,816	$72,580	$109,700
1191	$18,309	$76,591	$115,000	1192	$18,309	$48,449	$80,000
1193	$40,590	$93,149	$120,000	1194	$18,864	$72,482	$95,000
1195	$16,800	$48,846	$82,500	1196	$20,568	$77,276	$107,500
1197	$16,800	$61,379	$78,000	1198	$16,800	$43,104	$69,000
1199	$16,800	$60,129	$91,700	1200	$17,325	$71,738	$87,000
1201	$19,824	$47,595	$78,000	1202	$16,800	$81,739	$106,000
1203	$17,136	$80,518	$105,000	1204	$16,920	$51,933	$85,000
1205	$14,420	$36,506	$65,000	1206	$14,800	$46,595	$74,800
1207	$14,800	$58,141	$87,500	1208	$13,580	$61,097	$89,000
1209	$13,598	$57,723	$82,000	1210	$12,432	$36,671	$62,000
1211	$13,640	$41,120	$71,900	1212	$12,950	$41,504	$69,900
1213	$17,325	$49,198	$70,000	1214	$18,021	$48,588	$75,000
1215	$18,750	$45,820	$76,000	1216	$15,792	$50,945	$87,000
1217	$16,589	$55,249	$80,000	1218	$16,589	$65,940	$89,500
1219	$17,832	$49,683	$85,800	1220	$16,128	$53,120	$64,700
1221	$33,286	$52,629	$110,000	1222	$31,120	$78,118	$159,500
1223	$26,640	$68,859	$150,000	1224	$29,160	$58,769	$93,900
1225	$32,960	$51,458	$95,000	1226	$28,824	$59,921	$125,000
1227	$27,820	$46,396	$90,000	1228	$14,922	$39,368	$68,000
1229	$14,162	$41,366	$67,500	1230	$7,650	$29,579	$47,800
1231	$7,470	$29,424	$57,900	1232	$11,720	$29,424	$45,564
1233	$7,138	$33,004	$39,000	1234	$17,424	$61,884	$88,500
1235	$16,051	$56,203	$75,000	1236	$14,362	$52,380	$77,500
1237	$14,362	$57,146	$87,900	1238	$22,810	$62,079	$78,000
1239	$17,600	$47,262	$75,000	1240	$22,876	$69,052	$120,500
1241	$23,074	$49,097	$83,700	1242	$17,600	$69,579	$100,000
1243	$20,020	$55,676	$76,900	1244	$18,040	$44,565	$84,500
1245	$18,513	$52,569	$84,400	1246	$18,700	$49,508	$50,500
1247	$18,700	$65,454	$79,500	1248	$16,280	$53,599	$18,500
1249	$17,776	$48,065	$79,000	1250	$15,151	$35,737	$57,700
1251	$15,578	$41,705	$76,500	1252	$17,054	$53,967	$88,200
1253	$17,503	$52,982	$91,000	1254	$18,828	$54,107	$91,400

1990 RESIDENTIAL PROPERTY SALES FOR 7 NEIGHBORHOODS

NBRHOOD F

OBS	LANDVAL	IMPROVAL	SALEPRIC	OBS	LANDVAL	IMPROVAL	SALEPRIC
1655	$14,861	$59,258	$74,900	1656	$14,976	$48,957	$57,300
1657	$13,440	$41,177	$61,000	1658	$45,765	$181,227	$440,000
1659	$15,244	$55,169	$87,500	1660	$18,260	$59,267	$82,000
1661	$16,680	$55,525	$78,000	1662	$16,680	$72,867	$99,500
1663	$17,020	$61,935	$93,000	1664	$53,421	$119,792	$175,000
1665	$31,417	$99,413	$185,000	1666	$32,311	$75,343	$123,000
1667	$26,817	$78,726	$108,000	1668	$25,751	$82,259	$110,000
1669	$25,751	$64,568	$100,500	1670	$24,564	$66,533	$108,000
1671	$24,564	$71,149	$112,900	1672	$27,640	$85,347	$106,000
1673	$29,656	$78,968	$147,500	1674	$37,720	$63,785	$96,000
1675	$27,554	$77,837	$132,000	1676	$29,900	$85,451	$114,000
1677	$57,446	$121,702	$145,000	1678	$46,589	$99,862	$155,000
1679	$46,863	$116,059	$93,000	1680	$27,636	$105,596	$140,000
1681	$68,270	$104,022	$196,000	1682	$48,325	$180,974	$198,000
1683	$43,165	$89,063	$134,800	1684	$40,656	$104,673	$172,000
1685	$32,017	$81,717	$98,500	1686	$19,911	$91,286	$110,000
1687	$19,256	$92,299	$112,000	1688	$19,475	$93,133	$114,000
1689	$19,670	$67,559	$99,000	1690	$54,506	$125,062	$164,500
1691	$13,825	$60,319	$88,000	1692	$12,960	$56,778	$79,000
1693	$19,289	$83,953	$118,500	1694	$14,171	$78,234	$92,000
1695	$14,868	$72,621	$105,000	1696	$15,000	$99,392	$112,900
1697	$14,420	$65,163	$87,900	1698	$19,920	$56,605	$75,000
1699	$14,103	$49,616	$77,500	1700	$14,623	$49,616	$82,700
1701	$14,171	$55,354	$80,500	1702	$16,286	$59,846	$85,500
1703	$16,288	$58,046	$96,000	1704	$14,111	$51,100	$84,400
1705	$18,039	$74,920	$114,000	1706	$14,700	$71,989	$113,000
1707	$17,500	$72,090	$115,000	1708	$22,656	$91,701	$164,000
1709	$16,975	$68,936	$128,000	1710	$17,325	$69,265	$135,000
1711	$17,500	$79,493	$133,000	1712	$18,191	$92,708	$134,900
1713	$30,388	$88,946	$145,000	1714	$17,325	$97,991	$139,000
1715	$17,807	$106,152	$147,200	1716	$18,431	$84,099	$143,300
1717	$24,522	$98,733	$134,300	1718	$20,098	$96,149	$139,000
1719	$20,701	$85,513	$113,000	1720	$22,770	$85,812	$112,800
1721	$23,460	$83,276	$122,000	1722	$20,098	$85,446	$120,000
1723	$20,852	$86,394	$122,500	1724	$22,425	$96,828	$127,500
1725	$20,852	$85,701	$137,000	1726	$22,194	$87,696	$133,000
1727	$22,194	$86,457	$137,000	1728	$22,194	$96,214	$134,900
1729	$22,194	$111,917	$150,000	1730	$22,194	$95,211	$135,000
1731	$26,611	$109,695	$170,000	1732	$28,270	$105,166	$210,000
1733	$28,077	$116,925	$167,000	1734	$32,041	$119,129	$190,000
1735	$29,553	$89,361	$151,000	1736	$30,455	$88,315	$155,000
1737	$28,274	$91,797	$145,000	1738	$30,138	$106,410	$161,900
1739	$50,000	$113,504	$185,000	1740	$52,479	$126,092	$197,000
1741	$42,263	$106,334	$170,000	1742	$40,068	$86,171	$144,000
1743	$40,241	$109,908	$161,000	1744	$31,918	$71,157	$96,900
1745	$30,608	$88,669	$127,500	1746	$30,513	$79,839	$94,000
1747	$31,310	$96,018	$132,000	1748	$27,636	$110,295	$163,900
1749	$25,166	$130,412	$162,500	1750	$26,614	$64,815	$97,000
1751	$26,614	$81,469	$105,000	1752	$36,173	$92,463	$111,000
1753	$28,839	$95,959	$139,000	1754	$33,922	$86,714	$120,000

1990 RESIDENTIAL PROPERTY SALES FOR 7 NEIGHBORHOODS

NBRHOOD G

OBS	LANDVAL	IMPROVAL	SALEPRIC	OBS	LANDVAL	IMPROVAL	SALEPRIC
1974	$18,883	$68,293	$100,000	1975	$19,080	$71,466	$87,000
1976	$19,558	$58,925	$104,400	1977	$19,423	$66,323	$100,000
1978	$23,118	$61,062	$92,800	1979	$18,883	$71,240	$88,000
1980	$23,885	$72,163	$95,000	1981	$22,930	$66,363	$86,000
1982	$19,473	$58,852	$84,900	1983	$24,629	$58,741	$84,300
1984	$18,555	$56,675	$84,000	1985	$19,419	$64,484	$97,900
1986	$15,147	$55,528	$88,000	1987	$17,672	$53,380	$58,300
1988	$20,808	$40,023	$66,500	1989	$18,080	$70,350	$102,500
1990	$25,312	$53,865	$83,500	1991	$19,829	$67,731	$87,300
1992	$27,770	$69,544	$99,000	1993	$22,032	$61,216	$84,900
1994	$19,584	$57,721	$81,000	1995	$16,891	$47,326	$74,500
1996	$19,486	$68,417	$90,600	1997	$19,641	$75,968	$98,000
1998	$19,312	$60,601	$78,500	1999	$23,011	$62,618	$89,900
2000	$18,360	$52,645	$68,500	2001	$28,432	$66,110	$82,500
2002	$19,541	$80,604	$95,000	2003	$23,409	$52,497	$74,000
2004	$19,691	$67,785	$98,500	2005	$20,035	$53,532	$84,900
2006	$19,584	$66,138	$90,818	2007	$19,584	$58,377	$89,900
2008	$30,887	$76,002	$113,500	2009	$25,728	$90,319	$118,500
2010	$16,519	$59,691	$92,500	2011	$16,101	$66,431	$85,000
2012	$19,300	$55,085	$81,000	2013	$19,320	$78,567	$86,000
2014	$22,409	$79,581	$105,000	2015	$19,339	$92,928	$112,300
2016	$19,799	$57,624	$82,500	2017	$18,850	$63,401	$82,000
2018	$21,978	$91,140	$115,000	2019	$21,770	$57,933	$87,000
2020	$33,048	$66,432	$97,500	2021	$28,416	$86,777	$110,500
2022	$31,450	$97,357	$171,500	2023	$29,896	$90,721	$140,000
2024	$27,592	$60,747	$99,000	2025	$55,932	$140,924	$243,800
2026	$35,501	$107,068	$177,800	2027	$30,192	$166,919	$239,900
2028	$39,397	$153,645	$216,500	2029	$30,153	$75,980	$120,000
2030	$24,887	$65,681	$91,000	2031	$25,097	$89,630	$118,000
2032	$21,321	$92,020	$125,500	2033	$21,883	$76,287	$114,900
2034	$14,550	$53,431	$79,900	2035	$22,176	$58,656	$80,000
2036	$21,483	$53,489	$63,000	2037	$21,455	$62,617	$95,000
2038	$20,984	$59,435	$74,500	2039	$21,380	$59,714	$89,500
2040	$20,560	$53,907	$74,900	2041	$20,268	$60,455	$81,000
2042	$24,918	$58,981	$76,100	2043	$20,375	$54,221	$85,000
2044	$29,896	$62,297	$109,000	2045	$42,636	$75,215	$115,000
2046	$35,788	$86,089	$142,000	2047	$34,644	$131,430	$189,900
2048	$28,861	$73,059	$107,900	2049	$29,896	$77,303	$125,000
2050	$24,276	$74,913	$115,500	2051	$22,877	$84,293	$128,000
2052	$22,725	$126,422	$165,000	2053	$23,361	$93,968	$124,600
2054	$11,303	$47,262	$59,000	2055	$11,283	$40,767	$63,500
2056	$11,967	$38,882	$60,000	2057	$11,220	$32,955	$60,500
2058	$13,980	$36,494	$46,909	2059	$11,893	$39,085	$62,900
2060	$11,730	$32,393	$56,900	2061	$13,207	$32,655	$55,000
2062	$17,735	$36,494	$59,000	2063	$19,719	$87,151	$117,700
2064	$20,571	$64,956	$84,000	2065	$17,533	$82,195	$102,500
2066	$16,560	$75,399	$99,900	2067	$18,584	$79,027	$120,000
2068	$21,883	$74,377	$118,500	2069	$20,045	$68,444	$105,000
2070	$17,443	$61,207	$85,000	2071	$17,002	$85,444	$110,500
2072	$17,002	$55,224	$80,400	2073	$15,456	$73,764	$107,500

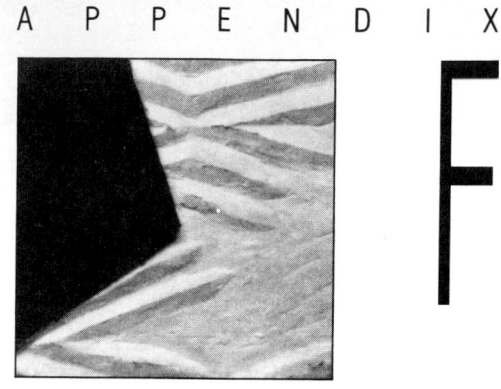

Data Set: 1970–1975 Low Bid Prices for Bread Supplied to Florida Public Schools (See Case Study 13)

**1970-1975 LOW BID PRICES (DOLLARS) FOR BREAD
SUPPLIED TO FLORIDA PUBLIC SCHOOLS**

OBS	MARKET	YEAR	LBPRICE
1	3	70	.150267
2	3	71	.166667
3	3	72	.183333
4	3	73	.213333
5	3	74	.306667
6	3	75	.280000
7	3	70	.200000
8	3	71	.210000
9	3	72	.179867
10	3	73	.153333
11	3	74	.200000
12	3	75	.280000
13	1	70	.206800
14	1	71	.220900
15	1	72	.220900
16	1	73	.255000
17	1	75	.364000
18	3	75	.280000
19	5	75	.248000
20	8	70	.209000
21	8	71	.228000
22	8	72	.230000
23	8	73	.230733
24	8	74	.276667
25	8	75	.242533
26	1	70	.202909
27	1	71	.228000
28	1	72	.225455
29	1	73	.240000
30	1	74	.334545
31	1	75	.340000
32	7	71	.203333
33	7	72	.256000
34	7	73	.256000
35	7	74	.360000
36	7	75	.266667
37	3	70	.191333
38	3	71	.206667
39	3	72	.213333
40	3	73	.226667
41	3	74	.304000
42	3	75	.266667
43	3	70	.190000
44	3	71	.215000
45	3	72	.220000
46	3	74	.353440
47	3	75	.260000
48	7	71	.200000
49	7	72	.213333
50	7	73	.233333
51	7	74	.270000
52	7	75	.266667
53	3	70	.160000
54	3	71	.173333
55	3	72	.186667
56	3	73	.240667
57	3	74	.297667
58	3	75	.276000
59	8	70	.195556
60	8	72	.222222
61	8	73	.235556
62	8	74	.320000
63	8	75	.311111
64	4	73	.230000
65	4	74	.313333
66	4	75	.266667
67	3	70	.175667
68	3	71	.216667
69	3	72	.170000
70	3	73	.153333
71	3	74	.313333
72	3	75	.266000

1970-1975 LOW BID PRICES (DOLLARS) FOR BREAD
SUPPLIED TO FLORIDA PUBLIC SCHOOLS

OBS	MARKET	YEAR	LBPRICE
73	1	70	.200000
74	1	71	.200000
75	1	72	.200000
76	1	73	.216000
77	1	74	.328889
78	1	75	.328889
79	3	72	.205000
80	3	73	.297920
81	3	74	.313333
82	3	75	.253333
83	2	71	.208800
84	2	72	.208800
85	2	73	.255000
86	2	74	.320000
87	2	75	.256000
88	2	70	.184800
89	2	71	.206800
90	2	72	.206800
91	2	73	.255000
92	2	74	.330000
93	2	75	.320000
94	1	70	.220000
95	1	71	.211500
96	1	72	.211500
97	1	74	.300000
98	1	75	.326667
99	3	70	.173333
100	3	71	.185000
101	3	72	.185000
102	3	73	.225000
103	3	74	.289750
104	3	75	.291000
105	4	70	.216000
106	4	71	.190000
107	4	72	.220000
108	4	73	.234667
109	4	74	.313333
110	4	75	.265667
111	7	70	.192000
112	4	70	.206667
113	4	71	.213333
114	4	72	.213333
115	4	73	.193333
116	4	74	.306667
117	4	75	.266667
118	4	72	.228000
119	4	73	.228000
120	4	74	.286667
121	4	75	.260000
122	1	70	.213400
123	1	71	.227950
124	1	72	.223250
125	1	73	.255000
126	1	74	.306667
127	1	75	.340000
128	6	70	.166667
129	6	71	.208000
130	6	72	.224000
131	6	73	.248000
132	6	74	.312000
133	6	75	.264000
134	1	70	.192200
135	1	71	.218550
136	1	72	.218550
137	1	73	.240000
138	1	74	.306667
139	2	70	.205200
140	2	71	.220000
141	2	72	.220000
142	2	73	.260000
143	2	74	.264000
144	2	75	.306667

(continued)

**1970-1975 LOW BID PRICES (DOLLARS) FOR BREAD
SUPPLIED TO FLORIDA PUBLIC SCHOOLS**

OBS	MARKET	YEAR	LBPRICE
145	3	70	.145000
146	3	71	.165000
147	3	72	.175000
148	5	70	.196923
149	5	71	.200000
150	5	72	.200000
151	5	73	.240000
152	5	74	.300000
153	5	75	.266667
154	7	71	.190000
155	7	72	.190000
156	7	73	.195000
157	7	74	.290000
158	7	75	.266667
159	2	70	.184800
160	2	71	.211500
161	2	72	.206800
162	2	73	.252450
163	2	74	.330000
164	3	70	.186667
165	3	71	.200000
166	3	72	.200000
167	3	73	.240667
168	3	74	.297667
169	3	75	.276000
170	2	70	.186667
171	2	71	.206667
172	2	72	.206667
173	2	73	.204000
174	2	74	.306667
175	2	75	.256000
176	4	70	.200000
177	4	71	.206667
178	4	72	.200000
179	4	73	.226667
180	4	74	.300000
181	4	75	.266667
182	3	71	.184167
183	3	72	.185867
184	3	73	.220800
185	3	74	.304000
186	3	75	.294400
187	6	70	.208000
188	6	71	.216000
189	6	72	.224000
190	6	73	.248000
191	6	74	.312000
192	6	75	.280000
193	8	70	.153333
194	8	71	.186667
195	8	72	.186667
196	8	73	.238133
197	8	74	.297667
198	8	75	.297667
199	3	71	.172500
200	3	72	.188750
201	3	73	.210000
202	3	74	.290000
203	3	75	.270000
204	1	70	.200000
205	1	71	.205800
206	1	72	.205800
207	1	73	.235000
208	1	74	.279300
209	1	75	.311111
210	6	70	.208000
211	6	71	.224000
212	6	72	.208000
213	6	73	.248000
214	6	74	.312000
215	6	75	.264000
216	5	72	.206667

**1970-1975 LOW BID PRICES (DOLLARS) FOR BREAD
SUPPLIED TO FLORIDA PUBLIC SCHOOLS**

OBS	MARKET	YEAR	LBPRICE
217	5	73	.233333
218	5	74	.304000
219	5	75	.250000
220	5	70	.245000
221	5	71	.255000
222	5	72	.210000
223	5	73	.240000
224	5	74	.410000
225	6	71	.216000
226	6	72	.212000
227	6	74	.312000
228	6	75	.264000
229	4	70	.284444
230	4	71	.297778
231	4	72	.304000
232	4	73	.304000
233	4	74	.405333
234	4	75	.346667
235	4	74	.304000
236	4	75	.253333
237	3	70	.160000
238	3	71	.170000
239	3	72	.180000
240	3	73	.253333
241	3	74	.283650
242	3	75	.274500
243	1	70	.200000
244	1	71	.200000
245	1	72	.200000
246	1	73	.240000
247	1	74	.328889
248	1	75	.328889
249	4	70	.193333
250	4	71	.216667
251	4	72	.213333
252	4	73	.226667
253	4	74	.288000
254	4	75	.266667
255	5	74	.303333
256	5	75	.253333
257	3	70	.208000
258	3	71	.206667
259	3	73	.253333
260	3	74	.297667
261	3	75	.270000
262	6	70	.208000
263	6	71	.216000
264	6	72	.212000
265	6	73	.248000
266	6	74	.312000
267	6	75	.264000
268	4	71	.193333
269	4	72	.216667
270	4	73	.230000
271	4	74	.273333
272	4	75	.266667
273	2	70	.232000
274	2	71	.248000
275	2	72	.248000
276	2	73	.268000
277	2	74	.328000
278	2	75	.352000
279	3	70	.180000
280	3	71	.200000
281	3	72	.200000
282	3	73	.153333
283	3	74	.226667
284	3	75	.270000
285	5	70	.184500
286	5	71	.193500
287	5	72	.201250
288	5	73	.153333

(continued)

**1970-1975 LOW BID PRICES (DOLLARS) FOR BREAD
SUPPLIED TO FLORIDA PUBLIC SCHOOLS**

OBS	MARKET	YEAR	LBPRICE
289	5	74	.303333
290	5	75	.253333
291	2	70	.198000
292	2	71	.207000
293	2	72	.198000
294	2	73	.255000
295	2	74	.440000
296	2	75	.320000
297	1	73	.240000
298	1	74	.285000
299	1	75	.328889
300	1	72	.223250
301	1	73	.240000
302	1	74	.306667
303	1	75	.375000

Data Set: Condominium Sales Data (See Case Study 15)

OBS	PRICPAID	FLOORHGT	DISTELEV	VIEWOCEN	ENDUNT	FURN
1	$19,900	1	2	0	0	0
2	$20,400	1	13	0	0	0
3	$27,500	1	15	1	0	0
4	$20,900	2	14	0	0	0
5	$25,900	1	2	1	0	0
6	$26,400	1	9	1	0	0
7	$26,400	1	12	1	0	0
8	$26,400	1	13	1	0	0
9	$25,400	3	9	1	0	0
10	$26,400	3	15	1	0	0
11	$27,900	8	15	1	0	0
12	$20,400	1	12	0	0	0
13	$19,900	2	13	0	0	0
14	$24,900	2	15	1	0	0
15	$26,400	3	11	1	0	0
16	$20,900	3	14	0	0	0
17	$25,900	4	15	1	0	0
18	$20,400	1	14	0	0	0
19	$27,400	7	15	1	0	0
20	$19,900	1	3	0	0	0
21	$19,900	1	11	0	0	0
22	$27,400	2	14	1	0	0
23	$18,000	3	11	0	0	0
24	$25,400	3	12	1	0	0
25	$28,400	4	14	1	0	0
26	$26,900	3	5	1	0	0
27	$15,900	2	3	0	0	0
28	$19,900	1	6	0	0	0
29	$20,400	1	9	0	0	0
30	$19,000	2	11	0	0	0
31	$26,100	2	12	1	0	0
32	$15,900	2	12	0	0	0
33	$26,100	2	13	1	0	0
34	$16,900	3	6	0	0	0
35	$19,900	4	14	0	0	0
36	$18,900	5	14	0	0	0
37	$29,400	6	15	1	0	0
38	$19,900	6	14	0	0	0
39	$20,400	2	6	0	0	0
40	$27,900	3	14	1	0	0
41	$15,900	6	9	0	0	0
42	$15,100	2	4	0	0	0
43	$15,100	2	5	0	0	0
44	$15,900	2	10	0	0	0
45	$19,000	8	12	0	0	0
46	$19,000	8	13	0	0	0
47	$17,500	1	4	0	0	0
48	$17,500	1	5	0	0	0
49	$17,500	1	10	0	0	0
50	$15,900	2	2	0	0	0
51	$16,300	3	13	0	0	0
52	$21,900	8	14	0	0	0
53	$16,100	3	5	0	0	0
54	$16,900	3	3	0	0	0
55	$16,900	3	4	0	0	0
56	$16,500	3	10	0	0	0
57	$16,900	3	12	0	0	0
58	$26,900	3	13	1	0	0
59	$17,600	4	11	0	0	0
60	$18,600	5	11	0	0	0
61	$19,500	8	11	0	0	0
62	$25,900	1	3	1	0	0
63	$23,500	1	4	1	0	0
64	$24,000	1	11	1	0	0
65	$15,900	2	1	0	0	0
66	$16,900	3	1	0	0	0
67	$26,000	4	1	1	0	0

OBS	PRICPAID	FLOORHGT	DISTELEV	VIEWOCEN	ENDUNT	FURN
68	$20,900	4	1	0	0	0
69	$17,600	4	6	0	0	0
70	$17,900	4	9	0	0	0
71	$17,900	4	12	0	0	0
72	$27,400	4	13	1	0	0
73	$17,900	5	13	0	0	0
74	$18,000	6	2	0	0	0
75	$19,900	6	6	0	0	0
76	$19,000	6	12	0	0	0
77	$21,000	1	1	1	0	0
78	$22,500	1	5	1	0	1
79	$20,000	1	6	1	1	1
80	$22,000	1	10	1	0	0
81	$21,000	2	1	1	0	0
82	$22,500	2	3	1	0	1
83	$22,500	2	4	1	0	1
84	$22,000	2	5	1	0	1
85	$19,500	2	6	1	1	0
86	$19,000	2	9	1	0	0
87	$19,500	2	10	1	0	0
88	$20,000	3	1	1	0	0
89	$21,000	3	2	1	0	1
90	$17,000	3	2	0	0	1
91	$20,000	3	3	1	0	0
92	$18,000	3	6	1	1	0
93	$20,500	3	10	1	0	1
94	$21,500	4	2	1	0	0
95	$17,500	4	2	0	0	1
96	$20,000	4	3	1	0	0
97	$17,500	4	3	0	0	1
98	$19,500	4	4	1	0	0
99	$16,500	4	4	0	0	1
100	$19,500	4	5	1	0	0
101	$17,500	4	6	1	1	0
102	$20,000	4	9	1	0	1
103	$19,000	4	10	1	0	0
104	$16,000	4	10	0	0	0
105	$20,500	4	11	1	0	1
106	$19,000	4	12	1	0	0
107	$20,500	5	1	1	0	0
108	$17,500	5	1	0	0	1
109	$21,500	5	2	1	0	1
110	$22,000	5	3	1	0	1
111	$26,500	5	4	1	0	1
112	$17,500	5	4	0	0	1
113	$20,000	5	5	1	0	1
114	$19,500	5	6	1	1	1
115	$18,500	5	6	0	0	1
116	$19,500	5	9	1	0	1
117	$16,000	5	9	0	0	1
118	$20,500	5	10	1	0	1
119	$17,500	5	10	0	0	1
120	$20,500	5	11	1	0	1
121	$20,500	5	12	1	0	1
122	$16,600	5	12	0	0	1
123	$19,500	5	13	1	0	0
124	$20,500	5	15	1	0	0
125	$21,000	6	1	0	0	1
126	$16,500	6	1	0	0	0
127	$22,000	6	2	1	0	0
128	$18,000	6	2	0	0	1
129	$21,500	6	3	1	0	0
130	$17,500	6	3	0	0	0
131	$20,000	6	4	1	0	1
132	$16,500	6	4	0	0	0
133	$17,500	6	5	0	0	1
134	$18,500	6	6	1	0	0
135	$20,000	6	9	1	0	1
136	$16,500	6	9	0	0	0
137	$17,500	6	10	1	0	0
138	$17,000	6	10	0	0	1

(continued)

ORS	PRICPAID	FLOORHGT	DISTELEV	VIEWOCEN	ENDUNT	FURN
139	$19,000	6	11	1	0	0
140	$17,000	6	11	0	0	0
141	$20,000	6	12	1	0	1
142	$19,500	6	13	1	0	0
143	$19,500	6	14	1	0	0
144	$20,500	7	1	1	0	0
145	$18,000	7	1	0	0	1
146	$19,500	7	2	1	0	0
147	$17,000	7	2	0	0	0
148	$19,500	7	3	1	0	0
149	$18,060	7	3	0	0	1
150	$20,000	7	4	1	0	0
151	$17,500	7	4	0	0	1
152	$19,000	7	5	1	0	0
153	$19,000	7	6	1	1	1
154	$19,500	7	9	1	0	0
155	$17,000	7	9	0	0	0
156	$20,000	7	10	1	0	0
157	$18,000	7	10	0	0	1
158	$19,000	7	11	1	0	0
159	$18,000	7	11	0	0	1
160	$19,000	7	12	1	0	0
161	$17,500	7	12	0	0	0
162	$19,500	7	14	1	0	0
163	$19,000	8	1	1	0	0
164	$18,000	8	1	0	0	1
165	$20,500	8	2	1	0	0
166	$17,500	8	2	0	0	0
167	$21,500	8	3	1	0	1
168	$17,500	8	3	0	0	0
169	$19,500	8	4	1	0	0
170	$17,000	8	4	0	0	0
171	$18,500	8	5	1	0	0
172	$17,500	8	5	0	0	0
173	$17,500	8	6	1	1	0
174	$17,500	8	6	0	0	0
175	$20,500	8	9	1	0	1
176	$16,500	8	9	0	0	0
177	$20,000	8	10	1	0	0
178	$17,500	8	10	0	0	0
179	$20,500	8	11	1	0	0
180	$19,000	8	12	1	0	0
181	$20,500	8	13	1	0	1
182	$19,500	8	14	1	0	0
183	$15,000	3	13	0	0	1
184	$14,000	5	1	1	0	1
185	$13,000	5	2	0	0	1
186	$18,200	6	5	1	0	1
187	$17,500	5	1	0	0	1
188	$16,400	5	14	1	0	1
189	$24,900	4	1	1	0	1
190	$21,500	7	13	1	0	1
191	$19,900	2	12	0	0	1
192	$19,000	8	5	0	0	1
193	$29,500	1	13	1	0	1
194	$21,700	2	13	0	0	1
195	$26,500	3	10	1	0	1
196	$20,200	4	1	0	0	1
197	$17,700	4	4	0	0	1
198	$18,900	8	11	0	0	1
199	$19,700	3	12	0	0	1
200	$20,500	3	1	0	0	1
201	$20,600	2	13	0	0	1
202	$23,100	7	4	1	0	1
203	$25,500	4	10	1	0	1
204	$25,000	7	14	1	0	1
205	$23,000	1	14	0	0	1
206	$27,900	4	12	1	0	1
207	$21,500	6	1	0	0	1
208	$30,600	3	15	1	0	1
209	$19,000	3	6	0	0	1

Answers to Odd-Numbered Exercises

Note: Your answers may vary slightly from those shown here due to rounding.

Chapter 1

1.1a. Audit status (audited or not) of all 1987 income tax returns   **b.** Tax returns examined in the RIA study   **c.** An income tax return
d. Estimate the true percentage of returns audited in a particular city
1.3 (1) Quantitative (2) Qualitative (3) Quantitative (4) Qualitative (5) Quantitative (6) Qualitative (7) Qualitative (8) Quantitative
(9) Quantitative (10) Quantitative
1.5a. Job status (quit or not) of all state lottery winners   **b.** Job status of 576 lottery winners who returned questionnaire
c. 11% of all state lottery winners quit their job during first year after winning

1.7a.	1	24, 96	**c.** $7/20 = .35$
	2	05, 61, 66	
	3	29, 32, 46, 98, 99	
	4	09, 23, 33	
	5	23, 33, 41, 41, 89	
	6	47	
	7	32	

1.9a. 9   **b.** 5   **1.11a.** Sample   **1.13a.** 10; 91.5; 9.566   **b.** 52.0; 3,336; 57.76   **c.** -2; 1.6; 1.265   **d.** .333; .0587; .242   **1.15a.** (3.95, 12.03)
b. No   **1.17** $R = 900 - 50 = 850$; $s \approx 850/4 = 212.5$ **1.19a.** 14.682   **b.** 199.974; 14.141   **c.** $\approx .95$ **d.** $60/66 = .909$
e. Decrease; decrease   **f.** 13.74; 11.96   **g.** $59/65 = .908$ **1.21a.** .6826   **b.** .9500   **c.** .9000   **d.** .9974   **1.23a.** -3 **b.** 1.96   **c.** 1.645
d. 1.0   **e.** $-.15$ **1.25a.** .95   **b.** .90   **c.** .9974   **d.** .9759   **e.** .1574   **f.** .9319   **1.27a.** .9406   **b.** .0068   **1.29a.** .1423   **b.** .0526
c. -50.50 **1.31b.** 4.68   **c.** 7.931   **1.33a.** Approximately normal; $\mu_{\bar{y}} = 24.7$; $\sigma_{\bar{y}} = 1.93$ **b.** .0055
c. Waiting time distribution under the new operating procedure has changed (either the mean has decreased or the standard deviation has increased).
1.35 In repeated sampling, 95% of all such confidence intervals contain μ.   **1.37a.** $1.94 \pm .13$ **b.** Increase n or decrease $(1 - \alpha)$
1.41a. $53,000 \pm 35,760.16$ **b.** Relative frequency distribution of the sampled population is approximately normal
c. Increase n or decrease confidence coefficient; increase n **1.43** $.0817 \pm .0071$ **1.45a.** Rejecting H_0 when H_0 is true
b. Accepting H_0 when H_0 is false   **c.** Probability of a Type I error   **d.** Probability of a Type II error   **1.47** No
1.51 Reject H_0; $z = 8.82$; $p \approx 0$ **1.53** $z = 3.54$; yes   **1.55** Do not reject H_0; $t = -1.39$ **1.57** $t = -12.60$, reject H_0 **1.59a.** .1
b. $.1 \pm .27$ **d.** Increase n's   **1.61** $t = 1.30$; no   **1.63a.** H_0: $\mu_1 - \mu_2 = 0$, H_a: $\mu_1 - \mu_2 > 0$ **b.** $z > 1.645$ **c.** Reject H_0
1.65 No, $p = .50$; both investment/quad populations are approximately normal with equal variances   **1.67a.** $z = 1.29$; no   **b.** 2.92 ± 3.715
1.69 The two sampled populations are normally distributed; the samples are randomly and independently selected from their respective
populations.   **1.71** $F = 3.41$; reject H_0 **1.73** No (at $\alpha = .05$), $F = 2.43$ **1.75a.** 5; 4.637; 21.5   **b.** 16.75; 6.021; 36.25
c. 4.857; 5.460; 29.81   **d.** 4; 0; 0   **1.77** Brand name   **1.79a.** 4 below the mean   **b.** .5 above the mean   **c.** 0   **d.** 6 above
1.81a. 0   **b.** .05   **c.** .44   **d.** -1.09 **1.83a.** .095   **b.** .9987   **c.** .3085   **1.85a.** .2033   **b.** No   **c.** ≈ 0 **d.** No   **1.87a.** 41.135 ± 6.173
c. Relative frequency distribution of foreign revenues is approximately normal   **1.89a.** $\bar{y} = 12,869$, $s = 12,027$
b. Approximately normal with $\mu_{\bar{y}} = \mu$ and $\sigma_{\bar{y}} = \sigma/\sqrt{n}$ **c.** .95   **1.91** Small   **1.93** $F = 4.06$; reject H_0 **1.95** Yes; $z = 6.15$
1.97 Yes; $z = 5.68$

Chapter 3

3.3a. $\beta_0 = 2$; $\beta_1 = 2$ **b.** $\beta_0 = 4$; $\beta_1 = 1$ **c.** $\beta_0 = -2$; $\beta_1 = 4$ **d.** $\beta_0 = -4$; $\beta_1 = -1$ **3.5a.** $\beta_1 = 2$; $\beta_0 = 3$ **b.** $\beta_1 = 1$; $\beta_0 = 1$
c. $\beta_1 = 3$; $\beta_0 = -2$ **d.** $\beta_1 = 5$; $\beta_0 = 0$ **e.** $\beta_1 = -2$; $\beta_0 = 4$ **3.7a.** $\hat{\beta}_0 = 2$; $\hat{\beta}_1 = -1.2$ **3.9b.** $\hat{y} = 4.79 + .014x$ **3.11a.** Yes
b. $\hat{\beta}_0 = 1.192$, $\hat{\beta}_1 = .987$ **3.13a.** 1.143; .286; .535 **b.** 1.60; .533; .730 **c.** 740.72; 92.59; 9.622 **d.** 30.768; 2.051; 1.432
e. 77.414; 4.301; 2.074 **f.** .2509; .0502; .2240 **3.15b.** $\hat{y} = .3537 + .000004426x$ **d.** .3892 **e.** SSE = .10218, s^2 = .01022 **f.** .10108
3.17a. $t = 6.708$; yes **b.** $t = -5.196$; yes **c.** $t = -1.33$; no **d.** $t = 5.05$; yes **e.** $t = 4.48$; yes **f.** $t = 6.91$; yes **3.19** $t = .256$; no
3.21a. $t = -.44$; do not reject H_0 **b.** $-.01748 \pm .06536$ **3.23** $\hat{\beta}_1 = 7.70$, $t = 2.556$; yes ($\alpha = .05$) **3.25** Yes; $t = 10.59$
3.27 Yes; $t = 3.88$ **3.29a.** Yes; $t = -6.64$ **b.** Yes; $t = -2.85$ **c.** .8464 **d.** .5041 **3.31a.** Yes; $t = 7.31$ **b.** .139
3.33b. Probably; however, we cannot conduct a test since n is not given **3.35** 9.106 ± 3.2 **3.39** $.3891 \pm .2423$
3.41a. $\hat{y} = 14{,}192 - 148.978x$ **b.** $t = -4.36$; reject H_0 **c.** $8{,}977.9 \pm 1{,}388.6$ **d.** $x = 35$ is outside the range of the sample data
3.43a. $\hat{y} = -.535 + 15.526x$ **b.** SSE = 6.974, s^2 = .8718, s = .9337, r^2 = .9760 **c.** $t = 18.05$ **d.** .0001 **e.** (3.357, 7.994)
3.45a. $\hat{y} = 30.117 + .316x$ **b.** SSE = 281.487, s^2 = 15.638, s = 3.954, r^2 = .1716 **c.** $t = 1.93$ **d.** .0694 **e.** (34.0568, 51.4619)
3.47a. $\hat{y} = -9.2667x$ **b.** SSE = 12.8667; s^2 = 3.2167; s = 1.7935 **c.** Yes; $t = -28.30$ **d.** $-9.2667 \pm .9090$ **e.** $-9.2667 \pm .9090$
f. -9.2667 ± 5.0611 **3.49a.** $\hat{y} = 5.364x$ **b.** Yes; $t = 25.28$ **c.** 18.774 ± 6.300 **3.51a.** $\hat{y} = 51.18x$ **b.** Yes; $t = 154.56$
c. $\hat{y} = 1{,}855.35 + 47.07x$; yes, $t = 93.36$ **d.** $y = \beta_0 + \beta_1 x + \varepsilon$ (reject H_0: $\beta_0 = 0$, $t = 8.37$) **3.53a.** $t = 24.0$; yes **b.** $t = 9.33$; yes
c. .941, .774 **3.55a.** Yes; $t = -6.69$ **b.** .04 **3.57a.** $\hat{y} = 14{,}012 - 1{,}783x$; no (at $\alpha = .05$), $t = -.98$
b. $\hat{y} = 19{,}680 - 3{,}887x$; yes (at $\alpha = .05$), $t = -2.29$ **c.** Predicting outside range of x (1.52–4.11)
3.59a. Aggressive: $\hat{y} = 1.4508 + .0163x$; defensive: $\hat{y} = .4594 - .0046x$; neutral: $\hat{y} = .9112 + .00873x$
b. Aggressive: reject H_0, $t = 4.38$; defensive: reject H_0, $t = -5.49$; neutral: reject H_0, $t = 5.67$
c. Aggressive: $.0163 \pm .0088$; defensive: $-.0046 \pm .0020$; neutral: $.00873 \pm .00364$; aggressive and neutral stocks increase linearly; defensive
stocks decrease linearly **3.61a.** Yes **b.** Yes **d.** 8.29 **e.** 21.15 **3.63** Yes

Chapter 4

4.1 df = n − (Number of independent variables + 1) **4.3a.** Yes; positive **b.** Yes; positive **c.** $t = 24.56$; yes **d.** $1.278 \pm .505$
4.5b. No; $t = -.15$ **4.7a.** -75.51 ± 26.17 **b.** Do not reject H_0; $t = 1.38$ **4.9a.** $F = 1.056$; do not reject H_0 **b.** .05
4.11a. $F = 9.6893$; reject H_0: $\beta_1 = \beta_2 = \beta_3 = 0$ **b.** $t = 1.444$; reject H_0: $\beta_1 = 0$ in favor of H_a: $\beta_1 > 0$ **4.13a.** $F = 17.8$; yes
b. $t = -3.50$; reject H_0: $\beta_1 = 0$ **c.** -6.38 ± 4.723 **4.15a.** (7.4528, 26.8903) **c.** No; $x_2 = 1.10$ is outside the range of the sample data
4.17 $E(y) = \beta_0 + \beta_1 x_1 + \beta_2 x_2 + \beta_3 x_3 + \beta_4 x_4$
4.19 $E(y) = \beta_0 + \beta_1 x_1 + \beta_2 x_2 + \beta_3 x_3 + \beta_4 x_1 x_2 + \beta_5 x_1 x_3 + \beta_6 x_2 x_3 + \beta_7 x_1^2 + \beta_8 x_2^2 + \beta_9 x_3^2$
4.21 $E(y) = \beta_0 + \beta_1 x_1 + \beta_2 x_2 + \beta_3 x_3$, where $x_1 = \begin{cases} 1 \text{ if A} \\ 0 \text{ if not} \end{cases}$, $x_2 = \begin{cases} 1 \text{ if B} \\ 0 \text{ if not} \end{cases}$, $x_3 = \begin{cases} 1 \text{ if C} \\ 0 \text{ if not} \end{cases}$
$\beta_0 = \mu_D$, $\beta_1 = \mu_A - \mu_D$, $\beta_2 = \mu_B - \mu_D$, $\beta_3 = \mu_C - \mu_D$
4.23c. Parallel lines **4.25b.** Second-order **c.** Different shapes **d.** Yes **e.** Shift curves along the x_1-axis **4.27a.** $E(y) = \beta_0 + \beta_1 x_1 + \beta_2 x_2$
b. $E(y) = \beta_0 + \beta_1 x_1 + \beta_2 x_2 + \beta_3 x_1 x_2$ **c.** $E(y) = \beta_0 + \beta_1 x_1 + \beta_2 x_2 + \beta_3 x_1 x_2 + \beta_4 x_1^2 + \beta_5 x_2^2$ **d.** .6004
e. No; $F = 1.80$ (p-value = .2465) **f.** (.1354, .5774); inadequate model
4.29a. $\hat{y} = 22.019 - .181x_1 - .25x_2 - 4.691x_3 + 3.674x_4 + 22.52x_5$ **b.** .599 **c.** 8.657 **d.** Yes; $F = 87.45$ **e.** Yes; $t = -4.64$
f. $3.674 \pm .789$ **g.** 22.52 ± 7.048 **4.31a.** $F = 3{,}909.25$; reject H_0 **c.** H_0: $\beta_4 = 0$; H_a: $\beta_4 < 0$ **e.** No, since β_4 is positive
4.33b. Number of bedrooms; complex size; covered parking; traffic congestion **c.** Inflated Type I error rate **d.** Yes; $F = 21.65$
4.35a. H_0: $\beta_5 = \beta_6 = \beta_7 = \beta_8 = 0$; $F > 2.37$ **b.** Yes **c.** Reject H_0 for both tests **4.37a.** $F = 2.94$; yes **b.** $F = 1.50$; no
c. $F = 5.73$; reject H_0 **4.39a.** $\beta_3, \beta_4, \ldots, \beta_{11}$ **b.** H_0: $\beta_3 = \beta_4 = \cdots = \beta_{11} = 0$ **c.** H_0: $\beta_2 = \beta_9 = \beta_{10} = \beta_{11} = 0$
4.41a. Use all but size (acres) and farm land (yes–no), testing at $\alpha = .01$
b. $E(y) = \beta_0 + \beta_1 x_1 + \beta_2 x_2 + \beta_3 x_3 + \beta_4 x_4 + \beta_5 x_1 x_2 + \beta_6 x_1 x_3 + \beta_7 x_1 x_4 +$
$\underbrace{\beta_8 x_2 x_3 + \beta_9 x_2 x_4 + \beta_{10} x_3 x_4 + \beta_{11} x_1^2 + \beta_{12} x_2^2 + \beta_{13} x_3^2 + \beta_{14} x_4^2}$
quantitative terms
$+ \beta_{15} x_5 + \beta_{16} x_6 + \beta_{17} x_5 x_6 \}$ qualitative terms
$+$ (quantitative × qualitative) interaction terms
where x_1 = seedlings and saplings, x_2 = percent ponds,
x_3 = distance to state park, x_4 = site index,
$x_5 = \begin{cases} 1 \text{ if residential land} \\ 0 \text{ if not} \end{cases}$, $x_6 = \begin{cases} 1 \text{ if branches or springs} \\ 0 \text{ if not} \end{cases}$

4.43a. (i) 4; (ii) 6; (iii) 6; (iv) 1   **b.** (i) max $R^2 = .2130$, min MSE $= 193.8$, min $C_p = 2.54$, min PRESS $= 10,507.4$; (ii) max $R^2 = .2473$, min MSE $= 189.1$, min $C_p = 2.34$, min PRESS $= 10,461.0$; (iii) max $R^2 = .2663$, min MSE $= 188.2$, min $C_p = 3.12$, min PRESS $= 10,489.3$; (iv) max $R^2 = .2682$, min MSE $= 191.7$, min $C_p = 5.00$, min PRESS $= 10,710.4$ **4.45a.** $F = 106.48$; yes   **b.** -725 ± 198.7
d. $t = 1.72$; reject H_0: $\beta_6 = 0$ **4.47b.** $\hat{y} = 95.75 - .3199x$ **c.** Quadratic model   **4.49** $t = -1.104$; do not reject H_0
4.51a. $\hat{y} = 2.14 - .15x_1 + .03x_2 + 2.54x_3 - .34x_4 - .26x_5 - .72x_6$ **b.** Yes; $F = 25.27$ **c.** Yes; one-tailed p-value $< .05/2 = .025$
d. No; p-value $> .05$ **e.** H_0: $\beta_4 = \beta_5 = \beta_6 = 0$ **4.53b.** $t = 5$; yes   **c.** 825   **d.** Support
e. The estimate of the difference in mean attendance between weekends and weekdays is $15x_3 - 700$ **4.55a.** Yes; $F = 69.17$
b. Yes; $F = 91.39$ **c.** Yes; $F = 205.54$ **d.** Yes; $F = 129.89$ **4.57a.** Test H_0: $\beta_2 = \beta_3 = 0$ **c.** $t = -3.333$; yes

Chapter 5

5.1 a, b, g quantitative; others qualitative   **5.3a.** Quantitative   **b.** Quantitative   **c.** Qualitative   **d** Qualitative   **e.** Qualitative
f. Quantitative   **g.** Qualitative   **h.** Qualitative   **i.** Qualitative   **j.** Quantitative   **k.** Qualitative   **5.5** $E(y) = \beta_0 + \beta_1 x + \beta_2 x^2$, where $\beta_2 > 0$
5.7a. Both are quantitative   **b.** $E(y) = \beta_0 + \beta_1 x_1 + \beta_2 x_2$ **c.** $E(y) = \beta_0 + \beta_1 x_1 + \beta_2 x_2 + \beta_3 x_1 x_2$
d. $E(y) = \beta_0 + \beta_1 x_1 + \beta_2 x_2 + \beta_3 x_1 x_2 + \beta_4 x_1^2 + \beta_5 x_2^2$
5.9a. $E(y) = \beta_0 + \beta_1 x_1 + \beta_2 x_2 + \beta_3 x_3 + \beta_4 x_1 x_2 + \beta_5 x_1 x_3 + \beta_6 x_2 x_3$
b. $E(y) = \beta_0 + \beta_1 x_1 + \beta_2 x_2 + \beta_3 x_3 + \beta_4 x_1 x_2 + \beta_5 x_1 x_3 + \beta_6 x_2 x_3 + \beta_7 x_1^2 + \beta_8 x_2^2 + \beta_9 x_3^2$
5.11a. $E(y) = \beta_0 + \beta_1 x_1 + \beta_2 x_2 + \beta_3 x_1 x_2 + \beta_4 x_1^2 + \beta_5 x_2^2$ **b.** $E(y) = \beta_0 + \beta_1 x_1 + \beta_2 x_2$ **c.** $E(y) = \beta_0 + \beta_1 x_1 + \beta_2 x_2 + \beta_3 x_1 x_2$
d. $\beta_1 + \beta_3 x_2$ **e.** $\beta_2 + \beta_3 x_1$ **5.13a.** $u = (x - 33)/2.16$ **b.** $-1.389, -.926, -.463, 0, .463, .926, 1.389$ **c.** $.99966$ **d.** 0
e. $\hat{y} = 37.5714 - .4629u - 5.3333u^2$ **5.15a.** $\mu_{\text{(wife employed)}} < \mu_{\text{(wife unemployed)}}$
b. 2% of variation in job satisfaction is explained by the model
5.17a. $E(y) = \beta_0 + \beta_1 x_1 + \beta_2 x_2 + \beta_3 x_3$,

where $x_1 = \begin{cases} 1 & \text{if moderate concentration} \\ 0 & \text{if not} \end{cases}$, $x_2 = \begin{cases} 1 & \text{if high concentration} \\ 0 & \text{if not} \end{cases}$, $x_3 = \begin{cases} 1 & \text{if consumer type} \\ 0 & \text{if not} \end{cases}$

c. $E(y) = \beta_0 + \beta_1 x_1 + \beta_2 x_2 + \beta_3 x_3 + \beta_4 x_1 x_3 + \beta_5 x_2 x_3$ **e.** Test H_0: $\beta_4 = \beta_5 = 0$ using a partial F test   **5.19a.** 5   **b.** -1 **c.** -1 versus 1
5.21a. $F = 14.42$, reject H_0: $\beta_1 = \beta_2 = \beta_3 = 0$; $R^2 = .812$; $s = 8.057$ **b.** $F = 8.79$; reject H_0: $\beta_4 = \beta_5 = 0$
c. DF-2: 2.14; blended: 4.865; adv. timing: 7.815   **5.23a.** $E(y) = \beta_0 + \beta_1 x_1 + \beta_2 x_2 + \beta_3 x_3$
b. Test H_0: $\beta_3 = 0$ versus H_a: $\beta_3 < 0$ using a t test
c. $E(y) = \beta_0 + \beta_1 x_1 + \beta_2 x_2 + \beta_3 x_1 x_2 + \beta_4 x_1^2 + \beta_5 x_2^2 + \beta_6 x_3 + \beta_7 x_1 x_3 + \beta_8 x_2 x_3 + \beta_9 x_1 x_2 x_3 + \beta_{10} x_1^2 x_3 + \beta_{11} x_2^2 x_3$
d. Test H_0: $\beta_1 = \beta_3 = \beta_4 = \beta_7 = \beta_9 = \beta_{10} = 0$ using a partial F test   **5.25a.** H_0: $\beta_3 = \beta_5 = 0$ **b.** H_0: $\beta_1 = \beta_3 = \beta_5 = 0$
5.27a. 1, quantitative, 0–4 years; 2, qualitative, two levels; 3, qualitative, two levels; 4, quantitative, 6–24 inches; 5, qualitative, three levels
b. $E(y) = \beta_0 + \beta_1 x_1 + \beta_2 x_2 + \beta_3 x_3 + \beta_4 x_4 + \beta_5 x_5 + \beta_6 x_6$ **c.** No interaction terms; include $\beta_7 x_4 x_5 + \beta_8 x_4 x_6$ in model
5.29a. Qualitative
b. $E(y) = \beta_0 + \beta_1 x_1 + \beta_2 x_2$, where

$x_1 = \begin{cases} 1 & \text{if brand B}_2 \\ 0 & \text{otherwise} \end{cases}$, $x_2 = \begin{cases} 1 & \text{if brand B}_3 \\ 0 & \text{if otherwise} \end{cases}$

d. $E(y) = \beta_0 + \beta_2$

5.31a. $E(y) = \beta_0 + \beta_1 x_1 + \beta_2 x_2$, where $x_1 = $ years of education, $x_2 = \begin{cases} 1 & \text{if certified} \\ 0 & \text{if not} \end{cases}$

b. $E(y) = \beta_0 + \beta_1 x_1 + \beta_2 x_2 + \beta_3 x_1 x_2$
c. $E(y) = \beta_0 + \beta_1 x_1 + \beta_2 x_1^2 + \beta_3 x_2 + \beta_4 x_1 x_2 + \beta_5 x_1^2 x_2$ **5.33b.** H_0: $\beta_2 = \beta_5 = 0$ **c.** H_0: $\beta_3 = \beta_4 = \beta_5 = 0$
5.35a. $E(y) = \beta_0 + \beta_1 x_1 + \beta_2 x_2 + \beta_3 x_3 + \beta_4 x_4$; no interaction   **b.** Include $\beta_5 x_1 x_2 + \beta_6 x_1 x_3 + \beta_7 x_1 x_4$ in model
5.37a. $E(y) = \beta_0 + \beta_1 x_1 + \beta_2 x_1^2 + \beta_3 x_2 + \beta_4 x_3 + \beta_5 x_1 x_2 + \beta_6 x_1 x_3 + \beta_7 x_1^2 x_2 + \beta_8 x_1^2 x_3$
b. $\hat{y} = 3.803 + .5563u_1 - .5816u_1^2 - .6657x_2 - 1.2771x_3 - .1464u_1 x_2 - .01464u_1 x_3 - .09184u_1^2 x_2 + .16837u_1^2 x_3$
c. Yes; reject H_0: $\beta_3 = \beta_4 = \beta_5 = \beta_6 = \beta_7 = \beta_8 = 0$, $F = 24.66$

Chapter 6

6.7 Unable to test model adequacy since there are no degrees of freedom available for estimating σ^2 (i.e., df $= n - 3 = 0$)
6.9a. $\hat{y} = 5.7106 + .62597x$ **b.** 193.501 ± 26.767
c. The prediction interval should be used cautiously, or not at all, since $x = 300$ lies outside the range of the sample data.
6.11 The least squares prediction equation should be used cautiously, or not at all, since the data point $(x_1 = 95.18, x_2 = 1.19)$ lies outside the experimental region.   **6.13** 3; n greater than 3   **6.15** 3 levels each; n greater than 6   **6.17a.** Curvilinear   **b.** Straight-line
c. $\widehat{\log(y)} = 10.6364 - 2.16985 \log(x)$; yes, $t = -13.44$ **d.** 25.95   **6.19a.** $\hat{y} = .01907 + .00589x_1$; reject H_0: $\beta_1 = 0$, $t = 2.73$
b. $\hat{y} = -.40202 + .000309x_1 + .003905x_2 + .18476x_3$; reject H_0: $\beta_1 = \beta_2 = \beta_3 = 0$; $F = 4.53$
c. Do not reject H_0: $\beta_1 = 0$, $t = .098$; yes; possible multicollinearity   **d.** VIF $= 2.295$; $R^2 = .5642$

Chapter 7

7.1a. $-.406, -.206, -.047, .053, .112, .212, .271, .471, .330, .230, -.411, -.611$ **b.** Yes; needs curvature **7.3a.** $\hat{y} = 40.35 - .207x$
b. $-4.64, -3.94, -1.83, .57, 2.58, 1.28, 4.69, 4.09, 4.39, 2.79, .50, 1.10, -6.09, -5.49$ **c.** Yes; model needs curvature
d. $\hat{y} = -1,051.108 + 66.186x - 1.006x^2$; yes, $t = -11.80$ **7.5a.** $.796, -.219, -1.004, .077, -1.071, 1.116, .868, .827, .223, -1.612$
d. $43.83, 20.36, 36.42, 31.88, 26.99, 40.41, 45.77, 25.15, 35.77, 45.16$ **e.** $7.19, 2.34, 10.50, 3.91, 9.15, 6.23, 9.81, 8.49, 9.17, .94$
7.7 Yes; needs curvature **7.9a.** Yes; assumption of equal variances violated **b.** Use variance-stabilizing transformation $y^* = \sqrt{y}$
7.11a. $\hat{y} = .94 - .214x$ **b.** $0, .02, -.026, .034, .088, -.112, -.058, .002, .036, .016$ **c.** Football shape; unequal variances
d. Use the transformation $y^* = \sin^{-1} \sqrt{y}$ and fit the model $y^* = \beta_0 + \beta_1 x + \varepsilon$ **e.** $\hat{y}^* = 1.307 - .2496x$; possibly
7.13 Residuals are approximately normal **7.15** Residuals are approximately normal **7.17** No outliers **7.19** No outliers **7.21** No outliers
7.23a. No **b.** Observation (child) #24 is influential
7.25a. H_0: No residual correlation; H_a: Positive residual correlation; test statistic: d; rejection region: $d < 1.39$ **b.** Reject H_0
7.27a. Yes **b.** $d = .173$; reject H_0 ($d_L = 1.18$) **7.29a.** No, do not reject H_0; $d_U = 1.60$ **b.** Yes; reject H_0; $d_L = 1.39$
7.31a. Residuals: $15.41, 9.65, -2.07, -16.51, -18.67, -7.79, 11.12, 27.72, -5.47, -19.27, -4.47, 10.37$
c. Plot suggests second-order model
d. $\hat{y} = 322.738 - 8.051x + .061x^2$, $R^2 = .920$; $F = 51.66$ (significant at $\alpha = .01$); residuals: $-6.15, 6.67, 2.87, -1.60, -3.49, -3.96, 1.88,$
$4.22, -3.51, -2.06, 7.62, -2.49$ **7.33a.** Yes; $F_{.05} = 2.51$ **b.** No; $d_L = 1.03$, $d_U = 1.85$
7.35a. Residuals: $1.27, 12.63, -10.27, -.13, 20.72, 13.76, -34.47, -.05, 11.95, -11.91, -9.35, 3.88, 4.21, 4.32, -6.54$
b. $-72.18, -62.41, -77.41, -65.69, -54.32, -67.60, -113.46, -82.98, -70.98, -94.84, -100.18, -90.90, -86.62, -82.56, -93.42$
c. $350.2, 358.7, 335.8, 340.2, 358.2, 345.5, 300.1, 343.1, 355.1, 334.1, 339.5, 358.5, 364.5, 367.5, 359.5$ **7.37a.** $\hat{y} = 1,604.81 - 64.1044t$
b. Residuals: $-773.7, -550.6, -241.5, 314.6, 773.7, 671.8, 709.9, 322.0, 39.1, -330.8, -359.7, -282.6, -292.5$; yes, runs of positive and
negative residuals **c.** $d = .337$, reject H_0
7.39a. $-294.52, -36.35, 263.12, 334.89, 55.96, -116.67, -137.00, -49.03, 35.24, 63.81, 47.68, -278.15, -218.68, 234.09, 95.16$ **b.** $-.45$
c. No trends; assumptions appear to be satisfied

Chapter 8

8.1a. $E(y) = \beta_0 + \beta_1 x_1 + \beta_2(x_1 - 15)x_2$, where $x_1 = x$ and $x_2 = \begin{cases} 1 & \text{if } x_1 > 15 \\ 0 & \text{if not} \end{cases}$

b.

	y-INTERCEPT	SLOPE
$x \le 15$	β_0	β_1
$x > 15$	$\beta_0 - 15\beta_2$	$\beta_1 + \beta_2$

c. Test H_0: $\beta_2 = 0$

8.3a. $E(y) = \beta_0 + \beta_1 x_1 + \beta_2(x_1 - 320)x_2 + \beta_3 x_2$, where $x_1 = x$, $x_2 = \begin{cases} 1 & \text{if } x_1 > 320 \\ 0 & \text{if not} \end{cases}$

b.

	y-INTERCEPT	SLOPE
$x \le 320$	β_0	β_1
$x > 320$	$\beta_0 - 320\beta_2 + \beta_3$	$\beta_1 + \beta_2$

c. Test H_0: $\beta_2 = \beta_3 = 0$

8.5a. $E(y) = \beta_0 + \beta_1 x_1 + \beta_2(x_1 - 1,000)x_2$, where $x_1 = x$ and $x_2 = \begin{cases} 1 & \text{if } x_1 > 1,000 \\ 0 & \text{if not} \end{cases}$ **b.** $\hat{y} = 4.024 - .0021x_1 - .00139(x_1 - 1,000)x_2$
c. Yes; $F = 363.44$ (p-value $= .0001$) **d.** $-.0021 \pm .000366$ **8.7a.** 473.3 ± 540.2 **b.** Model is inadequate
8.9a. $\hat{y} = -2.03 + 6.06x$ **b.** Reject H_0: $t = 10.35$ **c.** $1.985 \pm .687$
8.11a. $\hat{y} = -1.2667 + .176x$; yes, $t = 4.09$ (p-value $= .0013$)
b. Residuals: $-1.33, 6.67, -5.33, -2.33, 1.67, -6.13, 3.87, -5.13, 9.87, -1.13, -7.93, 14.07, -6.93, 4.07, -3.93$; appears to be violated

c.

x	VARIANCE OF RESIDUALS; $w_i = 1/x_i^2$
100	20.7
150	44.3
200	85.2

d. $\hat{y} = -1.5143 + .1777x$

8.15b. $\hat{y} = -.5279 + .0750x_1 + .0747x_2 + .3912x_3$ **c.** Reject H_0; $F = 21.79$ **d.** Yes; $t = 4.01$ **e.** $(-.2122, .2047)$
8.17a. Reject H_0; $X^2 = 20.43$ **b.** Yes; $X^2 = 4.63$ **c.** $(.00048, .40027)$

Chapter 9

9.1b.

	I	II	III	IV
1986	—	—	439.4	434.1
1987	412.4	397.9	381.6	373.3
1988	364.0	358.0	362.2	359.2
1989	350.4	341.1	332.8	335.4
1990	317.9	299.7	274.0	—

d. 85.57   **e.** 200 (using $M_{21} \approx 234$)

9.3b.

1973	—	1981	1,038.0
1974	954.7	1982	746.7
1975	1,253.3	1983	575.3
1976	1,630.7	1984	524.0
1977	1,871.0	1985	601.0
1978	1,938.7	1986	708.7
1979	1,724.0	1987	878.0
1980	1,449.0	1988	—

c.

1973	767.0	1981	1,426.5
1974	814.7	1982	1,188.5
1975	921.6	1983	993.9
1976	1,144.0	1984	861.7
1977	1,418.2	1985	746.9
1978	1,560.4	1986	754.1
1979	1,652.0	1987	790.7
1980	1,580.6	1988	849.6

d. 1,228   **e.** 890.8   **f.** 1,219

9.5b.

1977	—	1984	309.8
1978	—	1985	318.1
1979	222.7	1986	326.0
1980	244.2	1987	334.5
1981	264.8	1988	344.5
1982	283.6	1989	—
1983	298.6	1990	—
		1991	forecast = 374

c.

1977	181.5	1984	289.5
1978	187.1	1985	302.6
1979	199.2	1986	312.9
1980	218.2	1987	319.9
1981	239.9	1988	327.0
1982	259.6	1989	337.7
1983	275.1	1990	351.6
		1991	forecast = 359.9

d. 371.7

9.7a.

1971	1972	1973	1974	1975	1976	1977	1978	1979	1980
—	65.89	105.37	139.64	148.63	144.83	155.53	216.53	369.10	454.81

1981	1982	1983	1984	1985	1986	1987	1988	1989	1990
476.94	424.61	394.50	375.54	348.49	364.69	404.57	409.02	400.74	—

b. Approximately 400

c.

1971	1972	1973	1974	1975	1976	1977	1978	1979	1980
41.25	55.14	89.28	145.62	158.24	131.49	144.94	183.79	283.00	541.41

1981	1982	1983	1984	1985	1986	1987	1988	1989	1990
468.79	393.10	437.84	375.80	329.00	360.10	399.15	429.37	390.84	385.43

d. 384.3 **e.** 381.4

9.9b. $\hat{y}_t = 39.4879 + 19.13032t - 1.31529t^2$ **d.** $(-31.25, 48.97)$

9.11b. Railroads: $\hat{y}_t = 74.527 - 8.239t$; buses: $\hat{y}_t = 35.260 - 2.442t$; air carriers: $\hat{y}_t = -13.173 + 10.915t$

d. Railroads: -16.11, $(-40.82, 8.60)$; buses: 8.40, $(-8.69, 25.49)$; air carriers: 106.90, $(90.78, 123.00)$ **9.13a.** $\hat{y}_t = 118.68 - .344t$ **b.** Yes

c. $t = -2.79$; reject H_0 **d.** $(89.5, 122.4)$ **9.15a.** No; $t = -1.39$ $(-t_{.05} = -1.645)$ **b.** 2,003.48 **c.** No; $t = -1.61$ $(-t_{.05} = -1.645)$

b. 1,901.81 **9.17a.** 0, 0, 0, .5, 0, 0, 0, .25, 0, 0, 0, .125, 0, 0, 0, .0625, 0, 0, 0, .03125

b. Autocorrelations for first-order model: .5, .25, .125, .0625, .03125, .0156, .0078, .0039, .0019, .0010, .0005, .0002, .0001, 0, 0, 0, 0, 0, 0, 0

9.19. $R_t = \phi_1 R_{t-1} + \phi_2 R_{t-2} + \phi_3 R_{t-3} + \phi_4 R_{t-4} + \varepsilon_t$ **9.21a.** $E(y_t) = \beta_0 + \beta_1 x_{1t} + \beta_2 x_{2t} + \beta_3 x_{3t} + \beta_4 t$

b. $E(y_t) = \beta_0 + \beta_1 x_{1t} + \beta_2 x_{2t} + \beta_3 x_{3t} + \beta_4 t + \beta_5 x_{1t} t + \beta_6 x_{2t} t + \beta_7 x_{3t} t$ **c.** $R_t = \phi R_{t-1} + \varepsilon_t$

9.23a. $E(y_t) = \beta_0 + \beta_1\left[\cos\left(\dfrac{2\pi}{365}\right)t\right] + \beta_2\left[\sin\left(\dfrac{2\pi}{365}\right)t\right]$

c. $E(y_t) = \beta_0 + \beta_1\left[\cos\left(\dfrac{2\pi}{365}\right)t\right] + \beta_2\left[\sin\left(\dfrac{2\pi}{365}\right)t\right] + \beta_3 t + \beta_4 t\left[\cos\left(\dfrac{2\pi}{365}\right)t\right] + \beta_5 t\left[\sin\left(\dfrac{2\pi}{365}\right)t\right]$ **d.** No; $R_t = \phi R_{t-1} + \varepsilon_t$

9.25a. $y_t = \beta_0 + \beta_1 t + \phi R_{t-1} + \varepsilon_t$ **b.** $\hat{y} = 2,428.48 + 70.23t + .814\hat{R}_{t-1}$ **d.** $R^2 = .9976$, $s = 41.48$ **9.27a.** $y_t = \beta_0 + \beta_1 t + \phi R_{t-1} + \varepsilon_t$

b. $\hat{y} = 117.41 - .274t + .574\hat{R}_{t-1}$ **9.29a.** $\hat{\beta}_0 = 621.87$; $\hat{\beta}_1 = 46.99$; $\hat{\phi} = .653$ **b.** .7982 **9.31a.** $\hat{y}_{49} = 336.91$; $\hat{y}_{50} = 323.41$; $\hat{y}_{51} = 309.46$

b. y_{49}: 336.91 ± 6.48; y_{50}: 323.41 ± 8.34; y_{51}: 309.46 ± 9.36 **9.33** 103.29 ± 2.73 **9.35a.** 2,136.2 **b.** $(1,404.3, 3,249.7)$

c. 1,944.0; $(1,301.8, 2,902.9)$ **9.37a.** $R_t = \phi R_{t-1} + \varepsilon_t$ **b.** $y_t = \beta_0 + \beta_1 t + \beta_2 t^2 + \phi R_{t-1} + \varepsilon_t$

c. $\hat{y}_t = 1,381.79 + 205.69t + 14.32t^2 + .504\hat{R}_{t-1}$ **d.** $12,839 \pm 375$

9.39a.

YEAR	JAN.	FEB.	MAR.	APR.	MAY	JUNE
1988	—	—	—	—	—	—
1989	141.4	142.1	143.0	143.8	144.3	144.8
1990	148.3	148.7	149.0	149.2	149.9	150.4

YEAR	JULY	AUG.	SEPT.	OCT.	NOV.	DEC.
1988	137.6	138.3	138.5	139.1	139.6	140.6
1989	145.2	145.9	146.5	147.1	147.6	147.9
1990	150.6	—	—	—	—	—

b. 137.7 (using $M_{37} = 155$)

YEAR	JAN.	FEB.	MAR.	APR.	MAY	JUNE
1988	116.1	117.1	127.7	131.5	136.3	139.0
1989	138.7	127.8	136.2	138.7	146.1	148.3
1990	146.2	135.3	143.6	144.9	151.0	153.0

YEAR	JULY	AUG.	SEPT.	OCT.	NOV.	DEC.
1988	137.8	140.4	137.2	137.7	140.7	160.4
1989	146.5	150.9	147.2	144.8	147.7	165.5
1990	151.0	155.3	149.9	150.9	154.0	169.4

d. 175.6 **e.** 135.8

f. $y_t = \beta_0 + \beta_1 t + \beta_2 M_1 + \beta_3 M_2 + \cdots + \beta_{12} M_{11} + \phi R_{t-1} + \varepsilon_t$, where $M_1, M_2, \ldots, M_{11}$ are dummy variables for months
g. $\hat{y}_t = 163.86 + .540t - 46.43M_1 - 49.31M_2 - 30.03M_3 - 32.40M_4 - 24.58M_5 - 25.25M_6 - 30.13M_7 - 23.27M_8 - 33.17M_9 - 31.51M_{10} - 26.82M_{11} - .066\hat{R}_{t-1}$ **h.** 137.17 ± 4.02
9.41b. $y_t = \beta_0 + \beta_1 t + \beta_2 Q_1 + \beta_3 Q_2 + \beta_4 Q_3 + \phi R_{t-1} + \varepsilon_t$, where $Q_1, Q_2,$ and Q_3 are dummy variables for quarters; $\hat{y}_t = -173.933 + 1,623.684t - 859.249Q_1 - 141.867Q_2 + 826.733Q_3 + .1702\hat{R}_{t-1}$ **c.** $20,429.19; 20,429.19 \pm 2,915.69$ **9.43a.** $\hat{y}_t = 156.65 + 2.98t$ **b.** 210
c. Residuals: $-10.63, -6.61, -4.59, -4.57, -.55, 4.47, 6.49, 13.51, 10.53, 9.55, 6.57, .59, -1.39, -1.37, -6.35, -7.33, -8.31$
d. Durbin–Watson test; $d = .234$, reject H_0 **e.** $y_t = \beta_0 + \beta_1 t + \phi R_{t-1} + \varepsilon_t$; $\hat{y}_t = 153.08 + 3.07t + .777\hat{R}_{t-1}$ **f.** 203.5 ± 7.93
9.45a. Yes, $F = 217.23$ **b.** No, $d = 1.97$
c. An AR(1) error model is not required; however, the pattern of residual correlation may be greater than first-order.

Chapter 10

10.1 Noise (variability) and volume (n) **10.3a.** Flextime, staggered, and fixed work schedules
b. Randomly assign 20 workers to each work schedule
c. $E(y) = \beta_0 + \beta_1 x_1 + \beta_2 x_2$, where $x_1 = \begin{cases} 1 & \text{if flextime} \\ 0 & \text{if not} \end{cases}$, $x_2 = \begin{cases} 1 & \text{if staggered} \\ 0 & \text{if not} \end{cases}$
10.5a. $y_{B1} = \beta_0 + \beta_2 + \beta_4 + \varepsilon_{B1}; y_{B2} = \beta_0 + \beta_2 + \beta_5 + \varepsilon_{B2}; \ldots, y_{B,10} = \beta_0 + \beta_2 + \varepsilon_{B,10};$
$\bar{y}_B = \beta_0 + \beta_2 + (\beta_4 + \beta_5 + \cdots + \beta_{12})/10 + \bar{\varepsilon}_B$ **b.** $y_{D1} = \beta_0 + \beta_4 + \varepsilon_{D1}; y_{D2} = \beta_0 + \beta_5 + \varepsilon_{D2}; \ldots, y_{D,10} = \beta_0 + \varepsilon_{D,10};$
$\bar{y}_D = \beta_0 + (\beta_4 + \beta_5 + \cdots + \beta_{12})/10 + \bar{\varepsilon}_D$ **10.7a.** 3×3 factorial   **b.** Factors: pay rate (quantitative) and length of workday (quantitative)
c. Treatments: $P_1L_1, P_1L_2, P_1L_3, P_2L_1, P_2L_2, P_2L_3, P_3L_1, P_3L_2, P_3L_3$ **10.9a.** No
10.11 $E(y) = \beta_0 + \beta_1 x_1 + \beta_2 x_2 + \beta_3 x_3 + \beta_4 x_4 + \beta_5 x_5 + \beta_6 x_1 x_2 + \beta_7 x_1 x_3 + \beta_8 x_1 x_4 + \beta_9 x_1 x_5 + \beta_{10} x_2 x_4 + \beta_{11} x_2 x_5 + \beta_{12} x_3 x_4 + \beta_{13} x_3 x_5 + \beta_{14} x_1 x_2 x_4 + \beta_{15} x_1 x_2 x_5 + \beta_{16} x_1 x_3 x_4 + \beta_{17} x_1 x_3 x_5$, where
$x_1 =$ quantitative factor A; x_2, x_3 are dummy variables for qualitative factor B; x_4, x_5 are dummy variables for qualitative factor C
10.13 Cannot investigate factor interaction   **10.15** 7   **10.21** 8 treatments: $A_1B_1, A_1B_2, A_1B_3, A_1B_4, A_2B_1, A_2B_2, A_2B_3, A_2B_4$
10.23 $E(y) = \beta_0 + \beta_1 x_1 + \beta_2 x_2 + \beta_3 x_3 + \beta_4 x_4 + \beta_5 x_5$; 10

Chapter 11

11.3a. $E(y) = \beta_0 + \beta_1 x$, where $x = \begin{cases} 1 & \text{if treatment 1} \\ 0 & \text{if treatment 2} \end{cases}$ **b.** $\hat{y} = 10.667 - 1.524x; t = -1.775$, do not reject H_0
11.5a. $t = -1.78$; do not reject H_0 **c.** Two-tailed   **11.7** No; do not reject H_0, $F = .10$ $(p = .9566)$
11.9a. $E(y) = \beta_0 + \beta_1 x_1 + \beta_2 x_2$, where $x_1 = \begin{cases} 1 & \text{if small company,} \\ 0 & \text{if not} \end{cases}$, $x_2 = \begin{cases} 1 & \text{if medium company} \\ 0 & \text{if not} \end{cases}$ **b.** $H_0: \beta_1 = \beta_2 = 0$

c.

SOURCE	df	SS	MS	F
Company size	2	15.4822	7.7411	4.93
Error	121	190	1.5702	
Total	123	205.4822		

d. Yes, $F = 4.93$
11.11a. $E(y) = \beta_0 + \beta_1 x_1 + \beta_2 x_2$, where
$x_1 = \begin{cases} 1 & \text{if entrepreneur} \\ 0 & \text{if not} \end{cases}$, $x_2 = \begin{cases} 1 & \text{if transferred} \\ 0 & \text{if not} \end{cases}$ **b.** $\hat{y} = 66.97 + 4.03x_1 + 5.55x_2$
11.13a. No, do not reject H_0 **b.** Reject H_0, p-value $= .065$

11.15a.

SOURCE	df	SS	MS	F
Groups	2	14.6	7.3	5.77
Error	89	112.585	1.265	
Total	91	127.185		

b. Yes, reject H_0

11.17 Attempt to remove extraneous source of variation due to differences in medial rotation of college students
11.19a. Treatments: Winn-Dixie, Publix, and Kash 'N Karry; blocks: 60 grocery items

b.

SOURCE	df	SS	MS	F
Supermarkets	2	2.641	1.321	39.23
Items	59	215.595	3.654	108.54
Error	118	3.97	.034	
Total	179	222.206		

c. Reject H_0, $F = 39.23$ **d.** $-.26 \pm .066$ **e.** Yes   **f.** Reject H_0, $F = 108.54$

11.21 Yes; $F = 52.8$

11.23a. $E(y) = \beta_0 + \beta_1 x_1 + \beta_2 x_2 + \beta_3 x_3 + \cdots + \beta_8 x_8$, where
$x_1 = \begin{cases} 1 \text{ if Rater 1} \\ 0 \text{ if not} \end{cases}$, $x_2 = \begin{cases} 1 \text{ if Rater 2} \\ 0 \text{ if not} \end{cases}$, $x_3 - x_8$ are dummy variables for candidates
b. $E(y) = \beta_0 + \beta_3 x_3 + \cdots + \beta_8 x_8$ **c.** $E(y) = \beta_0 + \beta_1 x_1 + \beta_2 x_2$

11.25a.

SOURCE	df	SS	MS	F
Methods	2	.19	.10	.08
Brands	5	605.70	121.14	93.18
Error	10	13.05	1.30	
Total	17	618.94		

c. No, do not reject H_0; $F = .08$ **d.** $-.233 \pm 1.469$

11.27a.

SOURCE	df	SS	MS	F
Machines (A)	1	.101	.101	23.04
Materials (B)	2	.812	.406	92.38
AB	2	.768	.384	87.39
Error	12	.053	.0044	
Total	17	1.734		

c. Yes, reject H_0; $F = 87.39$ **e.** No; interaction is present   **f.** $.414 \pm .118$

11.29a. Factors (levels): accounts receivable (completed, not completed); verification (completed, not completed); treatments: CC, CN, NC, NN   **c.** Yes

11.31a. $df(V) = 1$; $df(VP) = 1$; $df(\text{Error}) = 16$; $df(\text{Total}) = 19$; $SSE = 240$; $SS(V) = 69.15$; $SS(VP) = 170.85$; $MS(V) = 69.15$; $MS(VP) = 170.85$; remaining missing entries cannot be determined from the table   **b.** Yes   **c.** $F = 11.39$; reject H_0

11.33a. No evidence of interaction   **b.** No evidence of window size main effect   **c.** Evidence of jump length main effect

11.35a.

SOURCE	df	SS	MS	F
P	1	1.55	1.55	1.08
D	1	22.26	22.26	15.57
$P \times D$	1	.61	.61	.43
Error	80	114.40	1.43	
Total	83	138.82		

b. No evidence of $P \times D$ interaction, $F = .43$; no evidence of P main effect, $F = 1.08$; evidence of D main effect, $F = 15.57$ **c.** $1.01 \pm .52$

11.37 Evidence of factor interaction; $F = 4.89$, reject H_0 ($F_{.05} \approx 1.32$)

11.39a. $E(y) = \beta_0 + \beta_1 x_1 + \beta_2 x_1^2$ **b.** $E(y) = (\beta_0 + \beta_3) + (\beta_1 + \beta_6)x_1 + (\beta_2 + \beta_9)x_1^2$
c. $E(y) = (\beta_0 + \beta_3 + \beta_4 + \beta_5) + (\beta_1 + \beta_6 + \beta_7 + \beta_8)x_1 + (\beta_2 + \beta_9 + \beta_{10} + \beta_{11})x_1^2$
d. $\hat{y} = 31.15 + .153x_1 - .00396x_1^2 + 17.05x_2 + 1.91x_3 - 14.3x_2x_3 + .151x_1x_2 + .017x_1x_3 - .08x_1x_2x_3 - .00356x_1^2x_2 + .0006x_1^2x_3 + .0012x_1^2x_2x_3$
e. Rolled/inconel: $\hat{y} = 53 + .241x_1 - .00572x_1^2$; Rolled/incoloy: $\hat{y} = 50.25 + .17x_1 + .00336x_1^2$; Drawn/inconel: $\hat{y} = 48.2 + .304x_1 - .00752x_1^2$; Drawn/incoloy: $\hat{y} = 31.15 + .153x_1 - .00396x_1^2$

11.41a. $3 \times 4 \times 3 \times 3 = 108$
b. The complete model has 108 terms, including β_0, 9 main effect terms, 30 two-way interactions, 44 three-way interactions, and 24 four-way interactions   **c.** H_0: $\beta_{10} = \beta_{11} = \cdots = \beta_{107} = 0$ **11.43a.** Reject H_0: $\mu_R = \mu_P = \mu_D = \mu_A = \mu_B$ **b.** $(\mu_R, \mu_P) < (\mu_D, \mu_A, \mu_B)$; $\mu_D < \mu_B$
11.45a. Evidence of accuracy $\times$ vocabulary interaction   **b.** Yes
c. 75%: means for all three accuracy levels are significantly different; 87.5%: means for all three accuracy levels are significantly different;

100%: 99% and 95% accuracy level means are not significantly different, whereas 90% accuracy level mean is significantly larger than the others
11.47 $\omega = .099$; the means of Publix and Kash 'N Karry are not significantly different, whereas the mean for Winn-Dixie is significantly lower than the other two means
11.49 $\omega = .182$; means for the following treatment pairs are significantly different: (A_1B_1, A_1B_3), (A_1B_1, A_2B_1), (A_1B_1, A_2B_2), (A_1B_1, A_2B_3), (A_1B_1, A_1B_2), (A_1B_3, A_2B_3), (A_1B_3, A_1B_2), (A_2B_1, A_2B_3), (A_2B_1, A_1B_2), (A_2B_2, A_2B_3), (A_2B_2, A_1B_2)
11.51 $\omega = .97$; the following treatment pairs are significantly different: (high/realistic, low/traditional) and (high/traditional, low/traditional)
11.53a. $S = 9.07$; the PD-1 and IADC 5-1-7 means are significantly different
b. $B = 9.22$; the PD-1 and IADC 5-1-7 means are significantly different
11.55a. $S = 1.05$; the following treatment pairs are significantly different: (high/realistic, low/traditional) and (high/traditional, low/traditional)
b. $B \approx .97$; the following treatment pairs are significantly different: (high/realistic, low/traditional) and (high/traditional, low/traditional)
c. Results of Tukey, Scheffé, and Bonferroni are identical   **11.57** Residuals appear to be normal; variances appear to be equal ($B = .314$)
11.59 Small samples make it difficult to check normality of residuals; variances appear to be equal ($B = 4.13$)   **11.61** Reject H_0; $F = 17.66$
11.63a. Yes; $F = 13.00$ **b.** $(\mu_X - \mu_Y)$: $-.97 \pm .702$; $(\mu_X - \mu_Z)$: $.148 \pm .702$; $(\mu_Y - \mu_Z)$: $1.118 \pm .702$ **c.** Theory Y

11.65

SOURCE	df	SS	MS	F
Plans	3	154.112	51.371	10.21
Error	13	65.417	5.032	
Total	16	219.529		

Reject H_0: $\mu_1 = \mu_2 = \mu_3 = \mu_4$ at $\alpha = .10$; Bonferroni comparisons ($\alpha = .06$): $\mu_3 < (\mu_2, \mu_4)$

11.67

SOURCE	df	SS	MS	F
Displays	2	784.083	392.042	2.99
Periods	7	4,602.292	657.470	5.02
Error	14	1,832.583	130.899	
Total	23	7,218.958		

Do not reject H_0: $\mu_A = \mu_B = \mu_C$ at $\alpha = .05$

11.69

SOURCE	df	SS	MS	F
Groups	2	273.068	136.534	79.22
Error	342	589.418	1.723	
Total	344	862.486		

Reject H_0: $\mu_{NT} = \mu_{CAT} = \mu_{CTW}$ at $\alpha = .05$; Bonferroni comparisons ($\alpha = .06$): All three means are significantly different

11.71

SOURCE	df	SS	MS	F
Carriers	2	8.8573	4.4287	83.82
Shipments	4	3.9773	.9943	18.82
Error	8	.4227	.0528	
Total	14	13.2573		

Reject H_0: $\mu_I = \mu_{II} = \mu_{III}$ at $\alpha = .05$; Tukey comparisons: $\mu_I < (\mu_{II}, \mu_{III})$

11.73a

SOURCE	df	SS	MS	F
Buyers (B)	1	227.8125	227.8126	.52
Sellers (S)	1	4,018.6125	4,018.6126	9.11
(BS)	1	1,757.8125	1,757.8122	3.99
Error	76	33,516.0000	441.0000	
Total	79	39,520.2375		

c. Reject H_0: $F = 3.99$ **d.** Complete model: $E(y) = \beta_0 + \beta_1x_1 + \beta_2x_2 + \beta_3x_1x_2$, where   **e.** -12.75 ± 10.92

$$x_1 = \begin{cases} 1 \text{ if BC} \\ 0 \text{ if BNC} \end{cases}, \quad x_2 = \begin{cases} 1 \text{ if SC} \\ 0 \text{ if SNC} \end{cases}; \quad \text{Reduced model: } E(y) = \beta_0 + \beta_1x_1 + \beta_2x_2$$

11.75a. $E(y) = \beta_0 + \beta_1 x_1 + \beta_2 x_2 + \beta_3 x_1 x_2 + \beta_4 x_1^2 + \beta_5 x_2^2$ **b.** $\hat{y} = 29.86 + .56x_1 - .1625x_2 - .1135x_1 x_2 - .275x_1^2 - .23125x_2^2$
c. The two models are different   **d.** $R^2 = .842$ **e.** Yes; $F = 5.67$
11.77a. $E(y) = \beta_0 + \beta_1 x_1 + \beta_2 x_2 + \beta_3 x_3 + \beta_4 x_4 + \beta_5 x_5 + \beta_6 x_6 + \beta_7 x_7 + \beta_8 x_8 + \beta_9 x_9 + \beta_{10} x_{10} + \beta_{11} x_{11} + \beta_{12} x_{12}$
b. $E(y) = \beta_0 + \beta_4 x_4 + \beta_5 x_5 + \beta_6 x_6 + \beta_7 x_7 + \beta_8 x_8 + \beta_9 x_9 + \beta_{10} x_{10} + \beta_{11} x_{11} + \beta_{12} x_{12}$ **c.** $E(y) = \beta_0 + \beta_1 x_1 + \beta_2 x_2 + \beta_3 x_3$
d. Yes; $F = 5.33$; reject H_0 **e.** $F = 207.80$; reject H_0 **11.79a.** Yes; $F = 33.86$ is significant at $\alpha = .01$
b. $R^2 = .9338$; 93.38% of total variability is explained by the model   **c.** Yes   **d.** Yes   **e.** No; $t = .04$
f. $\hat{y} = -4,026.64 + 385.08x_1 - 24.67x_1^2 + 661.58x_2 - 37.17x_2^2 + .25x_1 x_2$

Appendix A

A.1a. $\begin{bmatrix} 6 & 3 \\ -2 & -5 \end{bmatrix}$ **b.** $\begin{bmatrix} 3 & 0 & 9 \\ -9 & 4 & 5 \end{bmatrix}$ **c.** $\begin{bmatrix} 5 & 4 \\ 1 & -4 \end{bmatrix}$

A.3a. 3×4 **b.** No; the number of elements in a row of B does not match the number of elements in a column of A

A.5a. $\begin{bmatrix} 2 & 3 \\ -9 & 0 \\ 8 & 2 \end{bmatrix}$ **b.** $\begin{bmatrix} 3 & 0 & 4 \end{bmatrix}$ **c.** $\begin{bmatrix} 14 & 7 \end{bmatrix}$ **A.7a.** $\begin{bmatrix} 1 & 0 \\ 0 & 1 \end{bmatrix}$ **c.** $\begin{bmatrix} 1 & 0 & 0 \\ 0 & 1 & 0 \\ 0 & 0 & 1 \end{bmatrix}$ **A.13** $\begin{bmatrix} 3/8 & 1/8 \\ -1/4 & 1/4 \end{bmatrix}$

A.15a. $A = \begin{bmatrix} 10 & 0 & 20 \\ 0 & 20 & 0 \\ 20 & 0 & 68 \end{bmatrix}$; $V = \begin{bmatrix} v_1 \\ v_2 \\ v_3 \end{bmatrix}$; $G = \begin{bmatrix} 60 \\ 60 \\ 176 \end{bmatrix}$ **c.** $\begin{bmatrix} 2 \\ 3 \\ 2 \end{bmatrix}$

A.17a. $Y = \begin{bmatrix} 1 \\ 2 \\ 2 \\ 3 \\ 5 \\ 5 \end{bmatrix}$; $X = \begin{bmatrix} 1 & 1 \\ 1 & 2 \\ 1 & 3 \\ 1 & 4 \\ 1 & 5 \\ 1 & 6 \end{bmatrix}$ **b.** $X'X = \begin{bmatrix} 6 & 21 \\ 21 & 91 \end{bmatrix}$; $X'Y = \begin{bmatrix} 18 \\ 78 \end{bmatrix}$ **d.** $\hat{\beta} = \begin{bmatrix} 0 \\ .8571 \end{bmatrix}$ **e.** $\hat{y} = .8571x$

A.19 $t = -5.196$; reject H_0 **A.21** $t = -2.222$; reject H_0 **A.23** $(-1.1593, 2.7593)$ **A.25** $(.4153, 3.0131)$ **A.27** $(3.2719, 3.6581)$

A.29a. $X = \begin{bmatrix} 1 & -5 \\ 1 & -3 \\ 1 & -1 \\ 1 & 1 \\ 1 & 3 \\ 1 & 5 \end{bmatrix}$; $Y = \begin{bmatrix} 1.1 \\ 1.9 \\ 3.0 \\ 3.8 \\ 5.1 \\ 6.0 \end{bmatrix}$ **b.** $X'X = \begin{bmatrix} 6 & 0 \\ 0 & 70 \end{bmatrix}$; $X'Y = \begin{bmatrix} 20.9 \\ 34.9 \end{bmatrix}$ **c.** $\hat{\beta} = \begin{bmatrix} 3.4833 \\ .4986 \end{bmatrix}$ **d.** $\hat{y} = 3.4833 + .4986x$

e. SSE $= .0682$; $s^2 = .0170$ **f.** $t = 31.95$; yes   **g.** $r^2 = .9961$ **h.** $(3.62, 3.85)$

A.31a. $X = \begin{bmatrix} 1 & 1 \\ 1 & 1 \\ 1 & 2 \\ 1 & 2 \\ 1 & 3 \\ 1 & 3 \\ 1 & 4 \\ 1 & 4 \\ 1 & 5 \\ 1 & 5 \\ 1 & 6 \\ 1 & 6 \end{bmatrix}$; $Y = \begin{bmatrix} 1.1 \\ .5 \\ 1.8 \\ 2.0 \\ 2.0 \\ 2.9 \\ 3.8 \\ 3.4 \\ 4.1 \\ 5.0 \\ 5.0 \\ 5.8 \end{bmatrix}$ **b.** $X'X = \begin{bmatrix} 12 & 42 \\ 42 & 182 \end{bmatrix}$; $X'Y = \begin{bmatrix} 37.4 \\ 163.0 \end{bmatrix}$

c. The elements of $X'X$ are increased by a factor of 2.

d. $(X'X)^{-1} = \begin{bmatrix} .4333 & -.1 \\ -.1 & .02857 \end{bmatrix}$ **e.** $\hat{\beta} = \begin{bmatrix} -.09333 \\ .91714 \end{bmatrix}$; $\hat{y} = -.09333 + .91714x$ **f.** SSE $= 1.5564$; $s^2 = .15564$ **g.** $t = 13.75$; yes

h. $r^2 = .9498$

A.33a. 18×1 **b.**
$$X = \begin{bmatrix} 1 & 1 \\ 1 & 1 \\ 1 & 1 \\ 1 & 2 \\ 1 & 2 \\ 1 & 2 \\ 1 & 3 \\ 1 & 3 \\ 1 & 3 \\ 1 & 4 \\ 1 & 4 \\ 1 & 4 \\ 1 & 5 \\ 1 & 5 \\ 1 & 5 \\ 1 & 6 \\ 1 & 6 \\ 1 & 6 \end{bmatrix}$$

c. $\begin{bmatrix} 18 & 63 \\ 63 & 273 \end{bmatrix}$ **d.** $(X'X)^{-1} = \begin{bmatrix} .28889 & -.06667 \\ -.06667 & .019048 \end{bmatrix}$ **e.** $a'(X'X)^{-1}a = .0746$

f. Width is reduced by approximately 21%.

Index

Absolute deviations, method of, 395, 457
Additive model, 367, 467
Adjusted R^2 criterion, 232
All-possible-regressions selection procedure, 232
Alternative hypothesis, 43
Amplitude, 496
Analysis of variance, 552, 554
  assumptions, 637
  compared to regression analysis, 554
  completely randomized design, 554, 563
  k-way classification, 612
  Minitab printout, 562
  overall utility of a multiple regression model, 184
  randomized block design, 573, 582
  SAS printout, 560, 578–579, 596–597, 604, 615, 617, 624–625
  SPSS printout, 581
  two-factor factorial experiments, 590
Autocorrelation, 489
Autoregressive error model, 489, 493, 498, 500, 508
Autoregressive moving average (ARMA) model, 493–494, 519

β risk index, 312
β value, 312
Backward elimination, 229
Bartlett's test for homogeneity of variances, 639–640
Base level, 202
Bernoulli random variable, 442
Bias, 454
Biased estimator, 34, 453
Binary variable, 441
Binomial experiment, 367
Binomial probability distribution, 368

Block, 533
Bonferroni multiple comparisons procedure, 316, 634
  general contrasts, 635
  pairwise comparisons, 636
Break-even analysis, 29
Business cycles, 465

C_p statistic, 233
Case studies
  bidding competition, 673
  daily peak electricity demands, 725
  reluctance to transmit bad news, 693
  sale prices and assessed values of properties in four neighborhoods, 657
  sale prices of condominium units, 699
Categorical variable, 201
Central limit theorem, 31
Chi-square distribution, 448
Class, 10
Coding, 275
  orthogonal, 279
Coefficient of correlation (*see* Correlation, coefficient of)
Coefficient of determination, 121–122
  adjusted multiple, 181, 232
  multiple, 179, 232
Coefficient matrix, 276
Competitive strategy, 55
Complete factorial experiment, 591
Complete model, 218
Completely randomized design, 532, 554
  assumptions, 559
  calculations for, 563
  confidence intervals for, 545, 564
  definition, 557
  regression approach, 559
  test of hypothesis for, 559

Confidence coefficient, 34
Confidence interval, 34
  difference between means, completely randomized design, 564
  difference between means, randomized block design, 547, 583
  difference between two population means, large independent samples, 52
  difference between two population means, small independent samples, 55
  difference between two treatment means, factorial experiment, 545, 601
  mean of a dependent variable in a multiple regression model, 191–192
  mean of a dependent variable in a regression line through the origin, 143
  mean of a dependent variable in a straight-line model, 128
  mean of a single treatment, factorial experiment, 601
  population mean, large sample, 35
  population mean, small sample, 38
  single coefficient in a multiple regression model, 171
  single mean, completely randomized design, 564
  slope of regression line through origin, 143
  slope of a straight-line model, 112
Contour lines, 267
Contrast, 631
Cook's distance, 392
Coordinative strategy, 56
Correlated errors, 325, 399
Correlation, coefficient of, 115
  Pearson product moment, 116
  population, 118
Cross-validation, 459
Cyclical effect, 465

Cyclical fluctuation, 465

d-statistic, 401–402
Data
  experimental, 90
  observational, 89
  qualitative, 3
  quantitative, 3
  time series, 325, 399, 464
Data-splitting, 459
Data transformations, 337, 369
Degrees of freedom, 36
Deleted residual, 234, 391
  Studentized, 394
Dependent variable, 82
Descriptive techniques, 466
Design of an experiment, 528, 531
  completely randomized design, 532, 554,
    557
  factorial experiment, 539, 590, 612
  randomized block design, 532–533, 573
  selecting sample size for, 545
Designed experiment, 322
Determination, coefficient of, 121–122
Deterministic model, 85
Deviations, 98
Distribution, probability (*see* Probability
    distribution)
Dummy variables, 201, 282, 441
Durbin–Watson test, 399, 404

Error
  nonnormal, 442
  probability of Type I, 46
  probability of Type II, 46
  random, 82–83
Estimability, 325
Estimator
  biased, 34, 453
  of the difference between two population
      means, 51–52
  maximum likelihood, 447
  of the mean of dependent variable, sam-
      pling error for, 128
  of a population mean, 34
  of the random error variance in multiple
      regression, 167–168
  of the random error variance in simple lin-
      ear regression, 107
  of regression coefficients, 99–100
  ridge, 455
  unbiased, 34
Expected value, 16, 82
Experiment, 528

Experimental data, 90
Experimental design, 528, 531
  completely randomized, 532
  factorial, 539
  randomized block, 532–533
Experimental unit, 2, 529
Exponential relative frequency distribution,
    31
Exponential smoothing, 470, 473
  constant, 470
Extrapolation, 102, 335

F distribution, 64
F statistic, 64, 182, 219
Factor, 529
Factor level, 539
Factorial design, 539, 541
  three-way ($2 \times 2 \times 4$), 541
  two-way (2×2), 541
Factorial experiment, 541, 590, 612
First-order model, 95, 196, 260, 265
Forecasting, 466
  exponential smoothing, 470, 473
  Holt–Winters, 474–475
  lagged dependent variable values, 518
  moving average method, 466, 471
  regression approach, 481
  time series model, 481, 508, 512
Forecasting errors, 477, 511
Forward selection, 229
Fractional factorial experiment, 544

General linear model, 162
Geometric mean, 374
Goodness of fit, 256

Hat matrix, 389
Heavy-tailed distribution, 456
Heteroscedasticity, 325, 366
Histogram, 8
  constructing, 11
Holt–Winters model, 474–475
Homoscedasticity, 366
Hypothesis testing (*see also* Tests of hypothe-
    ses), 43–44

Incomplete block designs, 538
Independent variables, 84
  lagged, 495
  levels, 258
  qualitative, 201, 257
  quantitative, 195, 257
Indicator variables, 282
Inference (*see* Confidence interval; Tests of
    hypotheses)

Inferential models, 466, 481
Influence diagnostic, 392
Influential observations, 384
Interaction, 185, 197, 268, 287, 541, 543,
    594, 597
  terms, 288
Interval estimate, 34
Inverse estimator model, 430
Inverse prediction, 424, 427

Jackknife, 390–391

k-way classification, 612
Knot value, 421

L-estimators, 456
Lack of fit, 352
  test for, 363
Lagged dependent variable values, 518
Lagged independent variable model, 495
Latin cube design, 538
Latin square design, 538
Least squares estimates, 99–100
Least squares line, 99
Least squares method, 94, 97, 165
  formulas for straight-line model, 100
  two-stage, 443
  weighted, 433
Least squares prediction equation, 99
Level of an independent variable, 201, 258,
    529
Level of significance, 48
Leverage, 235, 388
Linear calibration, 426
Linear statistical models, 162
  first-order, 196
  second-order, 197–198
Logarithmic transformation, 371
Logistic model, 443, 445
Logistic regression, 445–446
Logit model, 445
Log-odds model, 447
Long-term trend, 464–465

M-estimators, 456
Main effect, 284, 286, 542, 597
Matrices,
  coefficient, 276
  hat, 389
Maximum likelihood estimation, 447
Mean
  arithmetic (*see also* Confidence interval *and*
      Tests of hypotheses), 16
  geometric, 374
Mean square for treatments (MST), 562

Mean squared error (MSE), 168, 454, 562
  total (TMSE), 233
Minimum variance unbiased estimator
  (MVUE), 34
Model, 82
  additive, 367, 467
  autoregressive, 486, 489, 508
  autoregressive moving average (ARMA),
    519
  complete, 218
  deterministic, 85
  first-order, 95, 196, 260, 265–266
  inferential, 466
  interaction, 185, 197, 270, 287
  inverse estimator, 430
  lagged independent variable, 495
  linear statistical, 162
  logistic, 445–446
  logit, 445
  log-odds, 447
  main effects, 286
  moving average, 493, 519
  multiple regression, 162
  multiplicative, 205, 368, 466
  nonlinear, 446
  piecewise linear, 420, 423
  polynomial, 260, 327
  probabilistic, 83
  quadratic, 165, 260
  reduced, 218
  regression, 85, 87
  seasonal time series, 496, 515
  second-order, 165, 197–198, 219, 260, 268
  simple linear, 94
  stationary, 491
  straight-line, 94–95, 219
  time series, 464, 486, 501, 508
  unbiased, 233
  white noise, 490
  with qualitative dependent variable, 441
  with qualitative independent variables, 203,
    281, 284, 295
  with qualitative and quantitative indepen-
    dent variables, 299
  with quantitative independent variables,
    196, 259, 265, 268
Model building, 256
Model validation, 458
Moving average
  method, 466
  model, 493
Multicollinearity, 254, 276, 329
  detection of, 331
Multiple coefficient of determination, 179,
  232

Multiple comparisons, 622
  Bonferroni approach, 634
  Scheffé method, 631
  Tukey procedure, 622
Multiple regression (see Regression, multiple)
Multiplicative model, 205, 368, 466

N-point moving average, 471
Noise, 531
Noise-reducing designs, 532
Nonlinear regression model, 446
Nonnormal error, 442
Normal probability distribution, 24
Normal probability plot, 301
Null hypothesis, 43

Observational data, 89, 322
  coding, 277
Observed significance level, 48
Odds, 447
One-tailed statistical test, 44
Origin, regression through, 141
Orthogonal coding, 279
Outlier, 325, 384

p-value, 48
Parameters, 20
  estimability of, 325
Partial regression residuals, 358
Pearson product moment coefficient of corre-
  lation, 116
Phase shift, 496
Piecewise linear regression, 420
Poisson distribution, 367
Polynomial model, 260, 327
  orthogonal, 279
Population, 4
Population correlation coefficient, 118
Prediction equation, least squares, 99
Prediction interval
  for a forecast of time series, 484, 512
  for an individual value of the dependent
    variable in a multiple regression
    model, 192–193
  for an individual value of the dependent
    variable in a straight-line model, 128,
    143
  for an individual value of the dependent
    variable in a straight-line model
    (inverse prediction), 427
Prediction sum of squares (PRESS) criterion,
  233
Probabilistic model, 83
Probability distribution, 7
  binomial, 368

Probability distribution (continued)
  chi-square, 448
  exponential, 31
  F, 64
  normal, 24
  Poisson, 367
  standard normal, 25
  t, 36

Quadratic model, 165, 260
Qualitative data, 3
Qualitative dependent variable, 441
Qualitative independent variable, 201, 257
Quantitative data, 3
Quantitative independent variable, 195, 257

R-estimators, 456
Random error, 82–83
  additive model, 367
  assumptions, 105, 164
  multiplicative model, 368
Random number table, 555
Random sample, 5
Random variable, 7
  Bernoulli, 442
Randomized block design, 532, 573
  assumptions, 576
  calculations for, 582
  confidence intervals for, 545, 583
  regression approach, 575–576
  test of hypothesis for, 575
Range, 16
Reduced model, 218
Regression, linear, 94
  assumptions, 105
  estimation of error variance, 107
  least squares estimates, 99–100
  least squares method, 94
  line, 99
  model, 85
  residual, 99
  SAS printout, 135, 139–141, 151
  test for linear correlation, 119
  test of model utility, 109
  through the origin, 141
  using model for estimation and prediction,
    127
Regression, multiple, 162
  assumptions, 164
  detecting lack of fit, 352
  estimation of error variance, 167
  inferences about parameters, 169
  least squares method, 165
  Minitab printout, 177, 185–186, 204, 214
  model, 162

Regression, multiple (*continued*)
  residuals, 352
  SAS printout, 167, 171, 173, 180, 183,
    189, 192–195, 200, 207, 211–212,
    220, 253
  SPSS^x printout, 176
  test of model utility, 187
  using model for estimation and prediction,
    191
Regression, stepwise, 224
  backward elimination, 229
  forward selection, 229
  SAS printout, 227–228, 241
Regression analysis, 82, 85
  all-possible-regressions selection, 232
  logistic, 445
  multicollinearity, 329
  multiple, 94
  piecewise linear, 420
  residual, 99, 352–353
  ridge, 335, 453
  robust, 456
  simple linear, 94
  stepwise, 224
  time series, 464
  variable selection techniques, 224, 232
Regression modeling, 87
Rejection region, 44
Relative frequency distribution, 7
  exponential, 31
Reliability of inference, 46
Replications, 546
Research hypothesis, 43
Residual, 99, 352
  deleted, 391
  first-order autocorrelation of, 489
  first-order autoregressive, 508, 510, 512
  partial, 358
  positively correlated, 490
  Studentized deleted, 394
  weighted least squares, 435
Residual analysis, 352
  to check normality assumption, 379
  to detect lack of fit, 352
  to detect outliers, 384
  to detect residual correlation, 399
  to detect unequal variances, 366
  to identify influential observations, 384
Residual correlation, 399
Residual effect, 465
Response surface, 85
Response variable, 82, 528
Ridge estimator, 455
Ridge regression, 335, 453
Ridge trace, 455
Robust regression, 456

Sample, 5
  random, 5
Sample size, determination of
  for designed experiments, 545
  for estimating the difference between two
    population means, 58
  for estimating the population mean, 39
Sample statistic, 20
Sampling distribution, 30
  of the difference between independent sam-
    ple means, 51
  of the estimator of the mean of the depen-
    dent variable, 128
  of the predictor of an individual value of
    the dependent variable, 128
  of the sample mean, 30–31
Scattergram, 97
Scheffé multiple comparisons procedure, 631
  general contrasts, 632
  pairwise comparisons, 633
Seasonal index, 469
Seasonal time series model, 496, 515
Seasonal variation, 465
Second-order model, 197–198, 260, 268
Secular trend, 464–465
Significance level, 48
Simple linear regression (*see* Regression,
  linear)
Smoothing techniques, 466
Square-root data transformation, 369
Stabilizing transformations, 369
Stable market condition, 675
Standard deviation, 18
Standard error, 30–31
  of the forecast, 477
Standard normal distribution, 25
Standardized regression coefficients, 327
Stationary model, 491
Statistical inference (*see* Confidence interval;
  Tests of hypotheses)
Statistics, 2, 20
Stem-and-leaf-plot, 8
  constructing, 10
Stepwise regression, 224
Student's *t* (*see* *t* statistic)
Studentized deleted residual, 394
Studentized range, 622–623
Sum of squared errors (SSE), 98, 553
Sum of squares for treatments (SST), 554
Survival analysis, 445

t distribution, 36
t statistic, 46, 54, 111, 170
Tchebysheff's theorem, 18
Test statistic, 43

Tests of hypotheses, 43–44
  all coefficients in a multiple regression
    model, 184
  correlation coefficient, 119
  difference between two population means,
    large independent samples, 52
  difference between two population means,
    small independent samples, 55
  Durbin–Watson *d* test, 404
  factorial experiment, 592
  global utility of multiple regression model,
    183
  heteroscedasticity, 374
  lack of fit, 363
  linear correlation, 119
  model utility, 111, 184
  p population means, completely random-
    ized design, 559
  p treatment means and *b* blocks, random-
    ized block design, 575
  population mean, large sample, 45
  population mean, small sample, 46
  residual correlation, 404
  set of coefficients in a multiple regression
    model, 219
  single coefficient in a multiple regression
    model, 172
  slope of a regression line through origin,
    143
  two population variances, 67
Time series
  analysis, 464
  components, 464
  constructing models for, 495
  data, 325, 399, 464
  definition, 464
  forecasting, 481, 508
  models, 481, 486, 491, 500, 512, 515
  smoothing techniques, 466
  variable, 464
Tolerance, 331
Total mean squared error (TMSE), 233
Transformations of data, 337, 368–369
Treatment, 529
Tukey's multiple comparisons procedure,
  621–622
  equal sample sizes, 623
  unequal sample sizes, 628
Two-factor factorial experiments, 590
  assumptions, 593
  calculations for, 599–600
  confidence intervals for, 601
  regression approach, 592–593
  tests of hypothesis for, 592
Two-stage least squares, 443

Two-tailed statistical test, 44
Type I error, 46
Type II error, 46

Unbiased estimator, 34
Unbiased model, 233

Variability, 531
Variable, 2
    Bernoulli random, 442
    binary, 441
    categorical, 201
    dependent, 82
    dummy 201, 282, 441
    independent, 84
    indicator, 282
    lagged independent, 495
    qualitative, 201, 257
    quantitative, 195, 257
    random, 7
    response, 82, 528
    time series, 464
Variable selection techniques
    all-possible-regressions, 232
    stepwise regression, 224
Variance, 17
Variance inflation factors, 330
Variance-stabilizing transformations, 368, 641
Volume, 531
Volume-increasing designs, 539

Weighted least squares, 433
    estimates, 433
    residuals, 435
White noise, 490

z score, 25
z test statistic, 44–45, 52